Symbol	Meaning
N_A	Density of acceptor atoms
N_D	Density of donor atoms
n_i	Intrinsic carrier concentration
nsec	Nanosecond (10^{-9} second)
P	Percentage tilt
	Power
p	Hole carrier concentration
P_D	Power dissipation
pF	Picofarad (10^{-12} Farad)
P.M.	Phase margin
Q	Quality factor of a circuit or inductor
	Stored charge
	Transistor
q	Charge of an electron
q/m	Charge to mass ratio of an electron
Q_s	Stored charge
R	Resistance
	Resistor
	Static resistance
r	Dynamic resistance
	Radius
R.F.	Ripple factor
R_i	Input resistance
R_o	Output resistance
s	Complex variable
S_β, S_E, S_I, S_V	Stability factors
T	Loop gain
	Period of a wave
	Specific time interval

Symbol	Meaning
	Temperature
	Vacuum tube
t	Time
T_j	Junction temperature
T_m	Melting point
V, V	Volt
	Voltage
v	Velocity
	Voltage
V_B	Potential barrier voltage
v_d	Drift velocity
V_δ	Disturbance signal
V_g	Generator ac voltage signal
V_i	Input ac voltage signal
V_o	Output ac voltage signal
V_P	Pinchoff voltage (FET)
V_ψ	Correction signal
V_T	Turn-on voltage (MOSFET)
V_t	Test generator signal
W, W	Watts
	Width
X	Distance
	Generalized signal
	Reactance
	Transformer
x	Distance
	Independent variable
y	Dependent variable
y_f, y_i, y_o, y_r	Admittance parameters
Z	Impedance

Symbol	Meaning
Z_o	Coaxial cable characteristic impedance
α	Common-base short-circuit current gain
β	Common-emitter short-circuit current gain
γ	Emitter efficiency
δ	Collector multiplication ratio
ϵ	Electric field
ϵ_0	Dielectric constant for free space
η	Conversion efficiency, dc to ac
θ	Angle
	Thermal resistance
λ	Wavelength
μ	Micron
	Mobility
	Vacuum tube voltage gain parameter
μH	Microhenry (10^{-6} Henry)
μ_n	Electron mobility
μ_p	Hole mobility
μsec	Microsecond (10^{-6} second)
ξ	Doping ratio
ρ	Resistivity
σ	Conductivity
τ	Carrier lifetime
	Time constant
τ_B	Base transient time
τ_r	Risetime
ϕ_m	Maximum phase shift
ω	Radian frequency

Applied Electronics

Applied Electronics

J. F. Pierce
T. J. Paulus

The University of Tennessee

Charles E. Merrill Publishing Company
A Bell & Howell Company
Columbus, Ohio

To Our Families

EDITH	SUE
Virginia	Mike
George	Cathy
	Rick

Published by
Charles E. Merrill Publishing Co.
A Bell & Howell Company
Columbus, Ohio 43216

International Standard Book Number: 0-675-09139-x

Library of Congress Catalog Card Number: 77-184058

1 2 3 4 5 6 7–77 76 75 74 73 72
Printed in the United States of America

Preface

As the title implies, this book was written for readers who need to use the knowledge that they acquire. Therefore, the primary objective of this introductory text on linear electronic circuits is to provide the reader with the necessary theoretical and applied material so that after completing the book, he will be able to analyze and design useful, multistage, linear amplifiers. This objective is accomplished by using a balanced combination of theory, analysis, and design. Real engineering applications are introduced as soon as possible to stimulate the interest of the reader and to stress many important aspects of the theory. Numerous worked-out examples with corresponding experimental results stress the degree of correlation that can be expected between ideal textbook circuits and actual experimental circuits.

The wide range of theoretical and applied material in this text is developed at three levels. First, the theory is presented. Second, the theory is explained in detail and reduced to practical applications. Third, more than 130 examples and more than 400 problems are included to illustrate major theoretical points and practical applications. One may choose to emphasize any of these levels without jeopardizing continuity. With a background in dc and ac circuits, together with complex algebra, the reader may gain a good working knowledge of linear active circuit analysis and design by relying heavily on the detailed discussions and examples. The reader with a background in calculus and circuit theory will find the theoretical developments and more difficult problems somewhat challenging. In both cases, the large number of practical examples with experimental results add a dimension in understanding that is rarely found in a textbook.

While the emphasis in this text is on electronic circuits, Chap. 1 provides background material on solid state physics that is essential to the understanding of solid state devices. Starting with the structure of matter and the related energy bands, the concepts and equations for both diffusion and drift currents are developed. The conductivity characteristics of different types of materials are discussed, with particular emphasis on semiconductors. For those who have had a course in solid state physics, this chapter may be omitted; or, alternately, it may be used as review material.

Building on the material in Chap. 1, the characteristics and operation of junction diodes are discussed in Chap. 2. The potential barrier and its effect on the flow of current carriers across a junction for various bias condictions is introduced first. From this discussion, the large-signal, non-linear and rectifying characteristics of diodes are developed. In an effort to introduce practical applications as early as possible, simple dc power supplies are included at this point. Small-signal parameters and diode models are then discussed to lay the foundation for future work in transistor parameters. The graphical analysis of circuits containing both linear and non-linear elements is introduced for circuits that are neither large-signal nor small-signal cases. Diode logic circuits are also discussed. Voltage-reference diodes and metal-semiconductor diodes are included to illustrate special cases and applications of diodes. At the end of Chap. 2, the reader should be familiar with the theory and application of diodes. He should be able to design power supplies that are reasonably well regulated with voltage-reference diodes and also some types of logic circuits.

Parallel treatments of field-effect transistors (JFET and MOSFET) and bipolar transistors are contained in Chaps. 3 and 4. The physical operation and static characteristics are discussed first. Bias circuits and large-signal analysis are then discussed to emphasize the necessary compatibility of the dc and ac circuits. Device models and small-signal equivalent circuits are developed, with emphasis on the three most often used configurations. With this background, the small-signal analysis and design of single and multistage amplifiers in the mid-band frequency region are discussed in detail. Special emphasis is placed on differential amplifiers because of the many practical applications of this configuration. As design aids, tables that summarize the characteristics of various types of amplifier configurations are included. At the end of these chapters, the reader should be able to design practical multistage amplifiers in the mid-band region.

Integrated circuits are introduced in Chap. 5 as logical extensions of the previous discussions of solid-state technology and amplifier design. The fabrication and layout of monolithic passive and active devices are discussed in detail. Monolithic circuit design and layout are also discussed. For completeness, this chapter includes treatments of MSI, LSI, hybrid, and compatible devices and circuits. At the end of this chapter, the reader should be able to design and lay out multistage amplifiers in integrated circuit form.

Conduction and control of current in a vacuum are still important areas in electronics. A discussion of vacuum diodes, triodes, tetrodes, and pentodes is included in Chap. 6. This treatment closely parallels those in Chaps. 3 and 4 for transistors. Cathode ray tubes with both magnetic and electrostatic focusing are included for completeness.

At this point, the reader should be rather proficient in analyzing and designing single- and multistage amplifiers in the mid-band range using any of the three types of active devices. In Chap. 7 the study is extended to the frequency and transient response of single-stage amplifiers of the three important configurations. In this discussion, all three types of active devices are considered. The study is then extended to multistage amplifiers in Chap. 8. This chapter contains several design examples to illustrate the process of reducing a problem to specifications and of choosing combinations of circuits to meet these specifications. At the end of Chap. 8, the reader should be able to design multistage amplifiers to meet gain, input impedance, output impedance, and response specifications.

An important class of amplifiers that were not included in the first eight chapters is the tuned amplifier. Tuned amplifiers using various types of devices are discussed in Chap. 9. A brief treatment of the fundamentals of tuned circuits is followed by single- and multistage tuned amplifiers. Since tuned amplifiers have a tendency to be unstable, the causes of instability and methods of neutralizing an amplifier are also included in this chapter.

Because of the inherent nonlinearities in all active devices, negative feedback is used in almost all high-quality circuits. Consequently, heavy emphasis is placed on the analysis and design of negative-feedback amplifiers in Chaps. 10 and 11. This treatment, which assumes no previous background in feedback circuits, is introduced by comparing the effects of parameter variations on circuit performance with and without feedback. After demonstrating the beneficial effects of negative feedback, the basic concepts and equations for negative-feedback amplifiers are developed and then simplified with appropriate approximations. Methods for calculating the loop gain, input impedance and output impedance are discussed in detail and illustrated with examples. The two most often used types of negative-feedback amplifiers, voltage sample-voltage sum and voltage sample-current sum, are given special attention.

Only the mid-band case is treated in Chap. 10. Using this material as a base, the response, stability and compensation of negative-feedback amplifiers are discussed in Chap. 11. Starting with the case of dominant high and low corner frequencies, the discussion is extended to the multipole cases and the resulting stability problems. Various methods of compensating feedback amplifiers together with a discussion of their advantages and disadvantages are also included. Since operational amplifiers are so widely used, a brief discussion of their characteristics and applications is a logical continuation of the feedback amplifier study. At the end of Chap. 11, the reader should be able to design several types of multistage feedback amplifiers and compensate them for the desired response.

The very popular subject of audio power amplifiers and their associated preamplifiers are introduced in Chap. 12 to provide the reader with some background in systems. After a discussion of a general sound reproduction system, the ac power handling requirements for power amplifiers are established. Using these requirements, the necessary ac current and voltage swings are determined. Various classes of amplifiers that can produce significant amounts of output power are discussed and compared for distortion and efficiency. With this background, practical power amplifiers are discussed. These amplifiers include the push-pull configurations and also the more popular complementary configurations. Since dc regulated power supplies are simply special cases of negative-feedback power amplifiers, they are also included in this chapter.

Every effort has been made to produce a readable, interesting, and useful text for the students. Literally hundreds of written, student critiques on each section have been analyzed, and corresponding changes have been made to clarify the material. In addition, suggestions by numerous teachers have been incorporated into the text. The authors believe that the resulting text is one that will please both the teachers and the students.

ACKNOWLEDGMENTS

In extensively class testing the manuscript for this text, the authors incurred many obligations. We are especially indebted to the hundreds of students whose interest and constructive criticisms helped to develop the text to its present form. We are also indebted to Professors W. P. Albritton, Jr., James Barrett, T. V. Blalock, J. A. Connelly, E. J. Kennedy, Buford Smith, J. W. Waller, and also to Messrs. R. J. Roberts and R. D. McKnight for their helpful suggestions and for using this material in their classes. We are grateful to Mrs. Mary Rose Bearden for typing the final manuscript as well as the countless revisions, and to Mrs. Lynne Wakefield for her patience and help in converting the manuscript into book form.

April, 1972

J. F. Pierce
T. J. Paulus

Contents

Appendixes

1

Solid State Physics

Transistors are made from semiconducting materials. As the name implies, semiconductors are neither good conductors nor good insulators. In this chapter the first goal is to develop an intuitive understanding of the conduction processes in solids with emphasis on semiconducting solids. In addition, temperature effects, doping levels, and other factors that affect the conduction process will be treated in some detail.

An in-depth treatment of semiconductor physics is not possible in one short chapter. Consequently, many approximations are made to keep the first-order effects in focus. Reference books on this subject are listed at the end of this chapter for those who wish a more rigorous and detailed treatment of this subject.

1.1 The Structure of Matter

The world around us is composed of *matter*. Matter is anything that occupies space and has mass. Aluminum, water, and air are common examples and illustrate the three states of matter: solid, liquid, and gas. All matter is made up of various elements, of which more than one hundred are known. An element is a substance which can neither be formed nor decomposed by an ordinary chemical reaction.

Elements can combine chemically to make *compounds* whose properties are radically different from the properties of the individual elements. For example, hydrogen and oxygen are gases, but they combine to form water. Two extremely active and dangerous elements, sodium and chlorine, combine to form common table salt. Elements can also be found as *mixtures*. For example, air contains many gases including oxygen, nitrogen, carbon dioxide, and the inert gases.

The gases in air are not chemically bound together and can be separated by a distillation process to obtain such things as LOX (liquid oxygen) for rocket fuel and neon for signs. Finally, many useful elements can be found in a pure form, typical examples being iron, copper, and silver.

The smallest part of an element that retains the specific properties of that element is an *atom*. However, all atoms may be broken down into several different subatomic particles, the three most important being *electrons*, *protons*, and *neutrons*. Under similar conditions, all electrons are identical, all protons are identical, and all neutrons are identical. Therefore, the inherent characteristics of a particular atom are determined to a large extent by the number of each of these particles and other less important particles contained in the atom.

Both an electron and a proton carry an electrical charge. These charges are equal in magnitude but are opposite in sign; i.e., the charge on an electron is negative while that on a proton is positive. However, the proton has a mass that is 1836 times greater than that of an electron. A neutron has no net charge (or is neutral) and has approximately the same mass as that of a proton-electron combination.

Many models have been developed in attempts to explain the construction and dynamics of individual atoms. One of the first and probably the most easily understood model was introduced by Bohr in the early part of this century. A Bohr model for lithium is shown in Fig. 1–1. In this model, an atom is a miniature solar system with the nucleus located at the center. The nucleus is composed of protons, neutrons, and several minor particles. Electrons are visualized as traveling in various orbits around this nucleus in somewhat the same way that planets travel around the sun. The Bohr atom has been refined by the theories of modern physics.* Some of these refinements will be introduced later as they contribute to an understanding of the properties of electron devices.

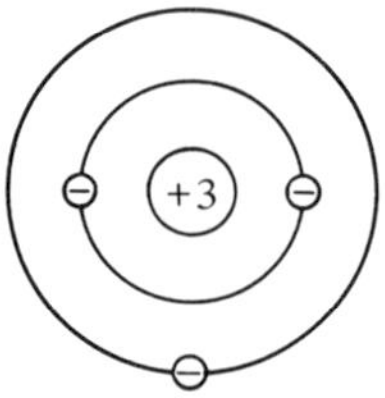

Lithium

FIGURE 1–1 *The Bohr model of an atom of lithium.*

1.2 Energy Levels in the Isolated Atom

A simple classical model of the atom is one which has negatively charged electrons orbiting around a positively charged nucleus. Using this model, the hydrogen atom is easiest to describe since its nucleus has a single positive charge (a proton) neutralizing the charge of a single electron in orbit. Thus, our first important generalization about an isolated atom is that it has a neutral charge.

A Coulomb force (or a force of attraction) exists between the nucleus and the orbiting electron which tends to pull the electron into the nucleus. The Coulomb

---

* Interesting and readable treatments of the principles of quantum mechanics can be found in Adler, Martin, Rusk, and Shive in the references at the end of the chapter.

force, F_c, is equal to Kq_eq_n/r^2 where q_e and q_n are the charges on the electron and on the nucleus, respectively, r is the distance between the two charges, and K is a proportionality constant whose value depends upon the system of units used. The Coulomb force is a radial force since the only component of force is that which acts on a line between the nucleus and the electron.

According to *Newton's Second Law*, F_c is ma where m is the mass of the electron in our case, and a is the acceleration of the electron toward the nucleus. But since the radial component of acceleration is (v^2/r), and no other component exists because we have a radial force, we may equate F_c to $m(v^2/r)$.* From this relationship, one might assume that an orbit of any radius could be maintained simply by an adjustment of the velocity of the electron to make the two forces equal. However, experimental evidence indicates that this assumption is not correct.

Electromagnetic radiation has two aspects: a wave aspect and a particle or quantum aspect. In 1924, Louis deBroglie conjectured that because of the symmetry of nature, a material particle with mass might well exhibit wave properties. Subsequent experiments have shown that material particles, specifically electrons, exhibit wave characteristics similar to electromagnetic waves. However, loss of energy by radiation does not normally occur unless an atom is excited with external energy. Consequently, the orbit of an electron is normally stable.

With wave properties attributed to the electron and assuming a circular orbit, the only possible orbits are those in which the wavelength associated with the electron fits into the orbit an integral (or whole) number of times. This statement is equivalent to saying that the momentum (mv) of an electron in an orbit is inversely proportioned to a deBroglie wavelength** (λ) which fits a whole number of times (n) into the perimeter of the orbit, or

$$n\lambda = \frac{nh}{mv} = 2\pi r \tag{1-1}$$

where n is an integer (1, 2, 3...), h is Planck's constant (4.13×10^{-15} eV-sec) and $v = nh/(2\pi mr)$. The spaces between allowable orbits are called forbidden regions because electrons cannot remain in these regions. Allowable orbits and the associated forbidden regions are shown in Fig. 1–2 for a hydrogen atom.

The many allowable orbitals in which an electron can stay for a period of time may be found from Schrödinger's wave equation. One or more orbitals can occur at the same discrete energy level with that level being thought of as an energy shell. As shown in Fig. 1–2, the K-shell is nearest the nucleus. The next shells are L, M, N, O, etc., in that order. Our second generalization is that the isolated atom has many possible shells in which electrons may orbit, and these shells exist only at discrete energy levels.

In hydrogen, the normal orbital for the single electron is in the K-shell. Energy is required to move an electron from the K-shell (quantum number $n = 1$) to

---

* We have assumed in arriving at this expression that the nucleus, which is our force center, is stationary in an inertial frame.

** Electron diffraction experiments indicate that the wavelengths, λ, and the momentum, mv, are related by $\lambda = h/(mv)$ where λ is called the deBroglie wavelength. In 1926, Schrödinger postulated a wave equation for the electron, and in doing so he united the classical theory of waves and the deBroglie principle.

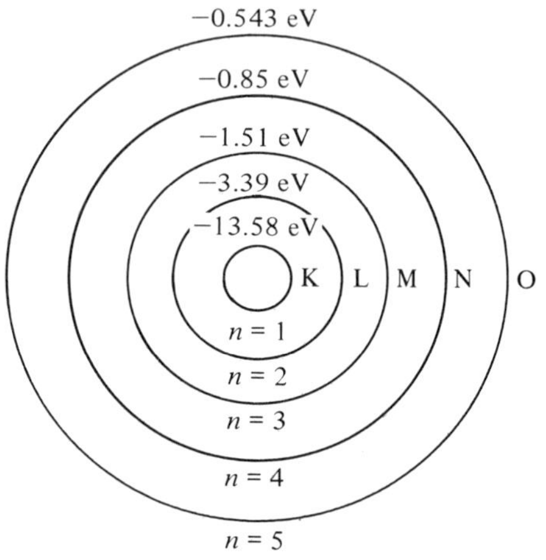

FIGURE 1–2 *The energy shell diagram for the hydrogen atom.*

the L-shell (quantum number $n = 2$) or to any higher shell. This energy is often expressed in terms of electron volts where one electron volt (eV) is the energy acquired by an electron in falling through a potential difference of one volt, or

$$1 \text{ eV} = (1.602 \times 10^{-19} \text{ coulomb})(1 \text{ volt}) = 1.602 \times 10^{-19} \text{ joules}$$

As indicated in Fig. 1–2, the K-shell of the hydrogen atom is associated with an energy level of −13.58 eV where the zero-energy reference level is located at infinity. This value means that an electron in the K-shell must acquire an energy of at least 13.58 eV in order to move to an orbit with an infinite radius or to escape completely from the attraction of the nucleus. Conversely, an electron starting at infinity and falling into the K-shell of the isolated hydrogen atom would lose an energy of 13.58 eV. Therefore, the minus signs in Fig. 1–2 indicate a loss of energy as the electron moves closer to the nucleus.

Energy lost by an electron when moving from a given shell to a lower shell appears in the form of electromagnetic radiation. The frequency of this radiation is related to the deBroglie wavelength and is given by

$$f = E/h = c/\lambda \tag{1–2}$$

where E is the energy either absorbed or lost (emitted) in electron volts, and c is the speed of light (3×10^{10} cm/sec). This relationship is illustrated in the following example.

EXAMPLE 1–1 An electron is initially in the K-shell of an isolated hydrogen atom. The electron becomes energized so that it moves to the L-shell for a short period of time. Find (a) the energy required to move the electron to the L-shell, and (b) the wavelength of the x-ray produced when the electron returns to the K-shell.

Solution: (a) The difference between the L- and K-shell energy levels is

$$E_L - E_K = -3.39 - (-13.58) = 10.19 \text{ eV}$$

The electron must obtain an energy of 10.19 eV from some external source to make the transition. When the electron drops from the L-shell to the K-shell, this same amount of energy is released in the form of an x-ray. (b) The wavelength of the x-ray produced is

$$\lambda = \frac{c}{f} = \frac{ch}{E} = \frac{(3 \times 10^{10})(4.13 \times 10^{-15})}{10.19} \simeq 1.22 \times 10^{-5} \text{ cm}$$

The Bohr model of an atom has been refined by identifying each state uniquely with four quantum numbers. The primary quantum number, n, denotes the shell; $n = 1$ for the first or K-shell; $n = 2$ for the second or L-shell, etc. Each shell contains one or more orbitals that are identified by orbital quantum numbers, ℓ, where values of ℓ range from zero to $(n - 1)$. The orbitals in a shell have slightly different energy levels, and each orbital has two energy states. For the first shell, $\ell = 0$, and the orbital path is spherical and concentric with the nucleus. Orbitals with this geometry are designated as s orbitals. Each shell has one s-type orbital. For $\ell = 1$, three orbitals are present with one orbital path symmetrical with each of three mutually perpendicular axes (the x, y, and z axes). These orbitals are called p-type orbitals. In the second shell ($n = 2$), ℓ has values of 0 and 1 so that this shell contains one s-type orbital and three p-type orbitals which may be designated as p_x, p_y, and p_z. The third shell contains one s-type ($\ell = 0$), three p-type ($\ell = 1$), and five d-type ($\ell = 2$) orbitals.

The third quantum number, m_ℓ, specifies the orientation of the magnetic field that is generated by an orbiting electron. In addition to the magnetic field produced by an electron moving in an orbit, the spinning motion of the electron itself also produces a magnetic field. Since the spin may be either clockwise or counterclockwise, the spin is designated by a fourth quantum number, m_s, which may have an arbitrary value of either $+\frac{1}{2}$ or $-\frac{1}{2}$. The significance of m_s is that two electrons may share one orbital only if their spins are opposite, since each orbital has two states as indicated above.

Electrons can occupy any of these orbitals under favorable conditions. However, no more than two electrons can occupy an orbital simultaneously, or no more than one in each state, as indicated by Pauli's exclusion principle. The data in Table 1–1 indicate how the shells fill up as the atoms become more complex. Hydrogen has one electron, and it normally occupies the s orbital in the K-shell ($n = 1$). Helium has two electrons which fill the single orbital in the K-shell. Lithium has three electrons, two filling the K-shell, and one occupying the s orbital in the L-shell (or the $2s$ orbital). When the $2s$ orbital is filled, the next electron goes into a $2p$ orbital, the p orbitals having a slightly higher energy level than the s orbital. Each of the three p orbitals in a given shell accumulates one electron before any of them receive two. When the L-shell contains eight electrons, one in each of its states, the next electron must establish an orbit in the M-shell ($n = 3$) which has 18 states or 9 orbitals (one s-type, three p-type, and five d-type).

When eight electrons are in the M-shell, the next electron must go to the N-shell, although 10 of the 18 states in the M-shell are not occupied. This "octet theory" of eight electrons in the outer shell is illustrated by all of the inert gases. An important point in semiconductor theory is that when eight electrons are established in an outer shell, the next electron must go to the next higher energy shell. Table 1–1 further indicates that after one or two electrons are in the next higher shell, the lower shell may accept more electrons until all its states are filled.

The electrons in the outermost normally occupied shell of an atom are of particular importance because they determine most of the chemical and physical properties of an element. Since these electrons are responsible for the chemical valence properties, they are referred to as the *valence electrons*.

TABLE 1–1 *Partial Table of Elements*

	(2) K	(8) L	(18) M	(32) N		(2) K	(8) L	(18) M	(32) N	(50) O	(72) P	(98) Q
1 Hydrogen	1				30 Zinc	2	8	18	2			
2 Helium	2				31 Gallium	2	8	18	3			
3 Lithium	2	1			32 Germanium	2	8	18	4			
4 Beryllium	2	2			33 Arsenic	2	8	18	5			
5 Boron	2	3			34 Selenium	2	8	18	6			
6 Carbon	2	4			35 Bromine	2	8	18	7			
7 Nitrogen	2	5			36 Krypton	2	8	18	8			
8 Oxygen	2	6			37 Rubidium	2	8	18	8	1		
9 Fluorine	2	7			38 Strontium	2	8	18	8	2		
10 Neon	2	8			39 Yttrium	2	8	18	9	2		
11 Sodium	2	8	1		40 Zirconium	2	8	18	10	2		
12 Magnesium	2	8	2		41 Niobium	2	8	18	12	1		
13 Aluminum	2	8	3		42 Molybdenum	2	8	18	13	1		
14 Silicon	2	8	4		43 Technetium	2	8	18	13	2		
15 Phosphorus	2	8	5		44 Ruthenium	2	8	18	15	2		
16 Sulfur	2	8	6		45 Rhodium	2	8	18	16	1		
17 Chlorine	2	8	7		46 Palladium	2	8	18	18			
18 Argon	2	8	8		47 Silver	2	8	18	18	1		
19 Potassium	2	8	8	1	48 Cadmium	2	8	18	18	2		
20 Calcium	2	8	8	2	49 Indium	2	8	18	18	3		
21 Scandium	2	8	9	2	50 Tin	2	8	18	18	4		
22 Titanium	2	8	10	2	51 Antimony	2	8	18	18	5		
23 Vanadium	2	8	11	2	52 Tellurium	2	8	18	18	6		
24 Chromium	2	8	13	1	53 Iodine	2	8	18	18	7		
25 Manganese	2	8	13	2	54 Xenon	2	8	18	18	8		
26 Iron	2	8	14	2	55 Cesium	2	8	18	18	8	1	
27 Cobalt	2	8	15	2	56 Barium	2	8	18	18	8	2	
28 Nickel	2	8	16	2	57 Lanthanum	2	8	18	18	9	2	
29 Copper	2	8	18	1	58 Cerium	2	8	18	20	8	2	

1.3 Energy Levels in Crystals

Crystals are solids in which the atoms are arranged in an orderly manner. Three simple examples of ordered arrangements of crystal structures are shown in Fig. 1–3. The atoms in crystal structures can be bonded together in at least three different ways: ionic, covalent, and metallic.

An example of *ionic binding* is found in sodium chloride or table salt. Chlorine has seven electrons in its outermost occupied (or valence) shell, whereas sodium has only one (see Table 1–1). A chlorine atom may take the valence electron* from a sodium atom to form an "octet" of electrons in its outer orbit, thereby creating a very stable condition for both atoms. In this process, however, the sodium atom is left with a net positive charge and the chlorine atom with a net negative charge. The two atoms are attracted to each other by an electrostatic

* The electron is probably not lost completely by the sodium atom. The close proximity of these two atoms deforms the orbits so that the electron spends most of its time with the chlorine atom.

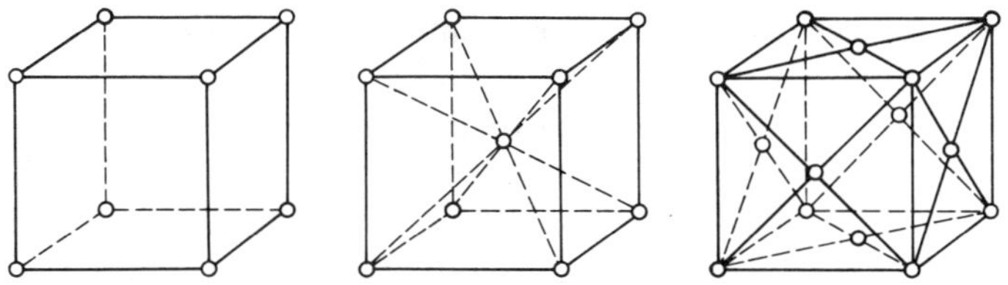

FIGURE 1–3 *Three types of simple crystal structure. (Adapted from Thomas L. Martin, Jr.,* Physical Basis for Electrical Engineering, *Prentice-Hall, Inc., 1957.)*

force. If the two atoms moved too close to each other, their filled orbitals would overlap. However, from Pauli's exclusion principle, we know that two filled orbitals cannot overlap. As the two atoms (ions) move closer to each other, a separation distance is reached wherein the quantum mechanical repulsion force becomes equal to the force due to the electrostatic attraction. At this point a stable separation distance exists, and the total energy of the two atoms (ions) is a minimum.

A *covalent bond* is formed when each of two atoms contributes one electron which spins in a direction opposite to that of the other. As a result, the two electrons, one from each atom, share a region between the two atoms. The simplest case is that of the H_2 molecule where each hydrogen atom contributes one electron to the bond, and the two electrons have opposite spins. Covalent bonding is very important in semiconductor theory. Atoms having four valence electrons (tetravalent) tend to form crystal structures in which one atom shares an electron with each of its four neighbors. This sharing effectively gives each atom eight electrons in its valence band and results in a closed covalent "octet." Electrons in this "octet" are hundreds of times more difficult to dislodge than the same electrons would be in an isolated atom. Examples of tetravalent materials that form covalent bonds are carbon (diamond), germanium, and silicon. The latter two materials are frequently used in semiconductor diodes and transistors. In covalent bonding, two adjacent atoms (ions) are held together by their mutual attraction for two localized electrons.

In *metallic bonding* the atoms are more closely packed together than can be accounted for by simple covalent bonding. Instead of the valence electrons being concentrated in pairs between adjacent atoms as in covalent bonding, they are spread out or de-localized in metallic bonding so as to be shared by many atoms. Metallic bonding usually occurs when the outer shell vacancies outnumber the outer shell electrons. In this way, maximum use is made of the available orbitals causing the binding force produced by the shared electrons to be maximized. This binding force makes the system very stable. In metallic bonding, a symmetrical group of atoms (ions) is held together by their mutual attraction for the same de-localized electrons.

Current molecular-orbital theory indicates that the interaction of an orbital of one atom with that of another changes the molecular orbitals. For large numbers of atoms the interaction of N atomic orbitals will produce N changed molecular orbitals with the total number remaining constant. The molecular orbitals produced by atomic orbital interactions constitute energy bands that exist throughout the crystal with the width of these bands being determined by the type of interacting atomic orbitals. An overly simplified example of the formation of bands with 15 atoms is shown in Fig. 1–4. On this basis, many combinations can be expected, including wide bands, narrow bands, and bands separated by forbidden gaps. The resulting valence band (outermost band with at least some occupied orbitals) may be either completely or incompletely filled. Metals usually have incompletely filled outermost occupied bands. At absolute zero the lower molecular orbitals of a partially filled band are occupied. This part is called the valence band. The upper part containing the unoccupied orbitals is called the conduction band.* At room temperature, many electrons are excited into the conduction band because adjacent orbitals are separated by a very small energy difference. These electrons can then migrate under a potential difference; i.e., conduction can occur easily.

If the highest occupied band is completely filled with valence electrons and if this band is separated from the next higher (conduction) band by a large energy gap, conduction cannot occur because free electrons are not available to migrate. A material having this type of band structure is a good insulator.

If the forbidden gap (or band gap) is not very large, then at room temperature, a few electrons will gain sufficient energy to jump the gap and occupy orbitals in the conduction band. These electrons can be used as current carriers. Materials having this type of band structure are called semiconductors. These relationships will be discussed in more detail in the next section.

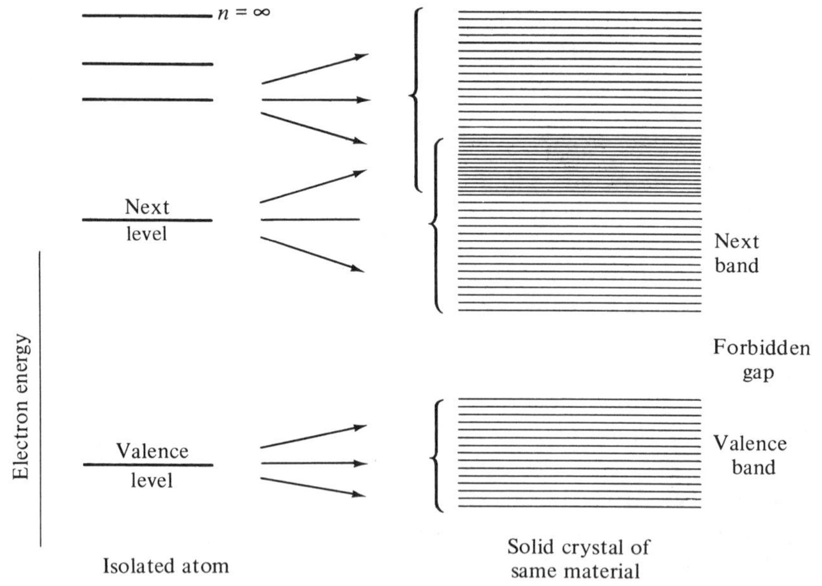

FIGURE 1–4 *The discrete energy levels of the isolated atoms (left) merge into bands of energy (right) when a crystal is formed. The arrangement of the energy bands is a characteristic of the crystal. (Adapted from J. N. Shive,* Physics of Solid State Electronics, *Charles E. Merrill Publishing Company, 1966.)*

---

* The conduction band is a band or part of a band that lies above the outermost occupied band at absolute zero.

How the orbitals in silicon change as several atoms are moved together to form a crystal is an example of how energy bands and band gaps are formed in a semiconductor. Table 1–1 shows that a silicon atom has both the K- and L-shells filled and has four electrons in the M-shell. The 9 orbitals of the M-shell appear in three levels: one in the 3*s*-level, three in the 3*p*-level, and five in the 3*d*-level. Two valence electrons fill the 3*s*-level, and the remaining two are in the 3*p*-level.

If the silicon atoms are many atomic diameters apart, they act as isolated atoms. This condition is shown for one atom at the right side of Fig. 1–5. As an atom moves in the direction of the reference atoms (from right to left in Fig. 1–5), the M-band orbitals begin to interact at a separation distance of two or three atomic diameters. This interaction causes the unfilled states in the 3*p* and 3*d* subshells to spread and move up in energy while the four filled states (two in the 3*s* and two in the 3*p* orbitals) move down* (decrease) in energy. Thus the interaction splits the filled states from the unfilled states, and band splitting results. At normal atomic spacing in a silicon crystal, the filled states are in a covalent bond containing 8 shared electrons for each atom, 4 of its own and one shared from each of its four neighbors. This arrangement produces a stable octet. As shown in Fig. 1–5, at normal atomic spacing, the nearest unfilled states (which are in the conduction band) are separated from the filled states in the

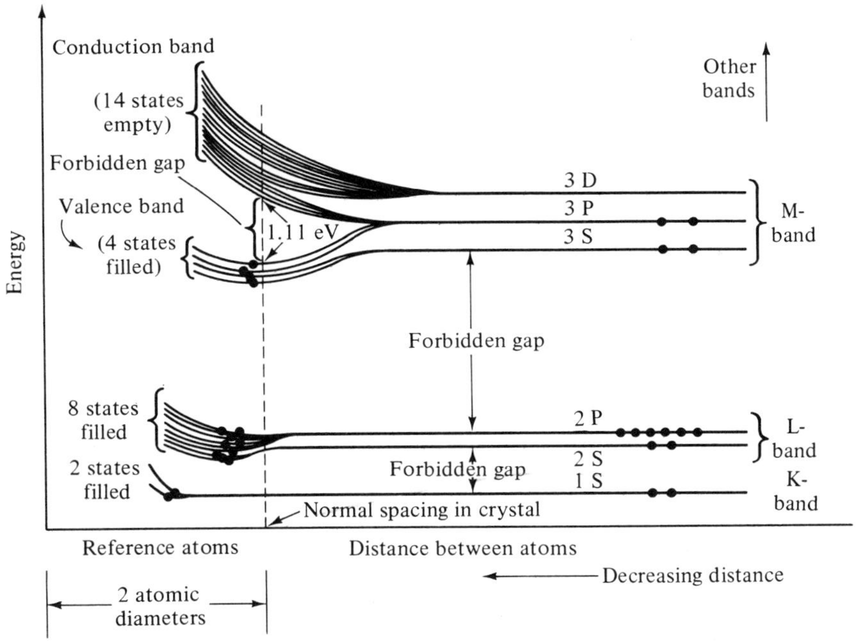

FIGURE 1–5 *Band spreading and splitting as a function of atomic spacing for silicon. This figure shows how the orbitals of one atom are affected by the interaction with the orbitals of a second atom as the first atom is moved from right to left. The second atom is stationary at the reference.*

* This change indicates that the 4 electrons have entered into covalent bonds with a subsequent decrease in the energy and an increase in the stability of the system.

valence band by a forbidden gap (band gap) having a width of 1.11 eV. Energy band conditions at normal atomic spacings are illustrated in Fig. 1–6.

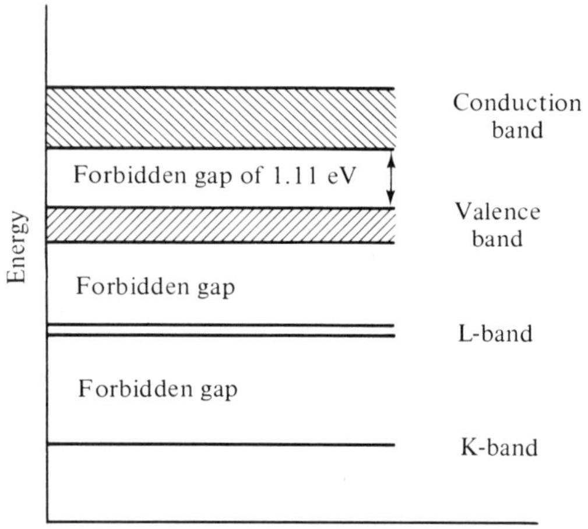

FIGURE 1–6 *Energy bands in a silicon crystal. At absolute zero, all energy states in the valence band are filled, and all states in the conduction band are empty.*

1.4 Conductors, Semiconductors and Insulators

As indicated in the preceding section, the ability of a solid to conduct electric current is determined to a large extent by the relationship between the valence and conduction energy bands. Thinking of a current as a net movement of electrons in a given direction, the valence electrons must be able to change states if a net movement in a given direction is to occur. Movement of an electron around an orbital path does not constitute a net flow of current. If the valence electrons can change states easily, the material is a conductor. If only a small percentage of the valence electrons can change states at room temperature, then the material is a semiconductor. If almost none of the valence electrons can change states, the material is an insulator. Let us now explore each of these cases.

In a good conductor, the valence and conduction bands are not separated by a forbidden energy gap. When copper atoms form a solid, band formation occurs such that each atom contributes more states than electrons to the highest energy band that contains electrons. Since the number of available states in the band exceeds the number of electrons available to fill all of the states, electrons can change states with very little change in energy because the states are so close together. At a temperature of absolute zero, the valence electrons occupy the lowest orbitals in this band (two electrons to an orbital) with the fully occupied part of the band being the valence band. The conduction band containing all of the unoccupied orbitals in the same band at absolute zero is adjacent to and immediately above the valence band as shown in Fig. 1–7(a). However, at room temperature, thermal energy is sufficient to cause the orbitals to be occupied somewhat randomly over the entire band.

In a semiconductor, electrons cannot move from the valence band into the conduction band so easily because the two bands are separated by an appreciable forbidden gap as shown in Fig. 1–7(b). For example, when silicon atoms form a

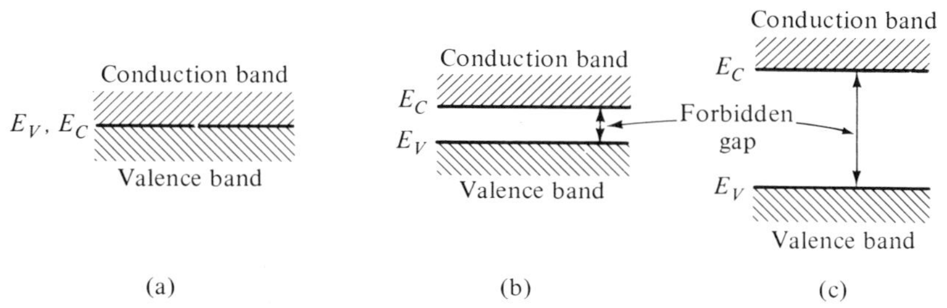

FIGURE 1–7 *A comparison of the band structure for three types of materials: (a) a conductor, (b) a semiconductor, and (c) an insulator. The conduction band is empty, and the valence band is filled at absolute zero.*

crystal, the band splitting effect as shown in Fig. 1–5 produces a forbidden band gap between the valence and conduction bands of 1.11 eV. At a temperature of absolute zero, all of the electrons are in the valence band, filling all of the orbitals in this band. Consequently, at this temperature silicon is a good insulator because current carriers are not available in the conduction band. At room temperature, a small percentage of the electrons acquire sufficient thermal energy to jump the 1.11 eV energy gap and occupy orbitals in the conduction band for a short period of time. While in the conduction band, electrons are available as current carriers. After a few lattice collisions, electrons in the conduction band lose their acquired thermal energy and drop back into the valence band. Other electrons take their place so that at a given temperature, the average number of electrons in the conduction band per unit volume is relatively constant for a given material. The effect of the 1.11 eV band gap in reducing the number of carriers available at room temperature may be appreciated by noting that a cubic centimeter of copper has approximately 4×10^{12} times more electrons in the conduction band than silicon.

In an insulator, the energy band gap between the valence and conduction bands is several electron volts. Consequently, at room temperature almost no valence electrons can obtain sufficient thermal energy to jump across the band gap to the conduction band. With all of the orbitals in the valence band filled and no electrons in the conduction band, electrons cannot change states; therefore, current cannot flow.

Since the number of electrons that can obtain sufficient thermal energy to jump across a forbidden gap increases with temperature, we can expect semiconductors to have more current carriers available at elevated temperatures. In fact, materials that make good insulators at room temperature may conduct easily at highly elevated temperatures.

The ability of a material to carry current is specified in terms of conductivity; i.e., a good conductor has a high value of conductivity. A comparison of the conductivity values for several materials is shown in Fig. 1–8.

No fine line exists between insulators and semiconductors or between conductors and semiconductors. Research is constantly being carried on to develop semiconducting materials that have more desirable properties.

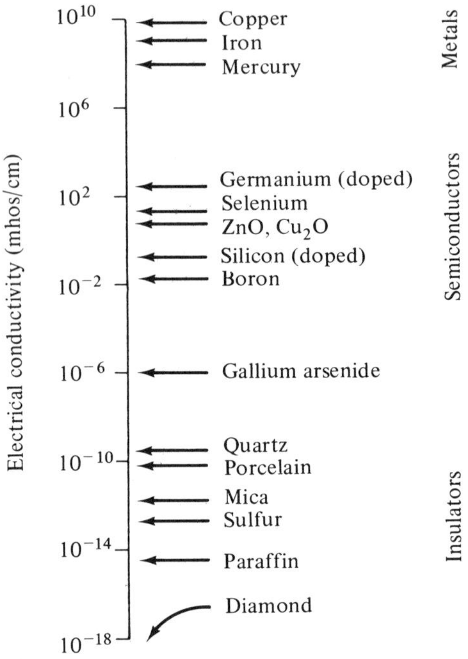

FIGURE 1–8 *A comparison of the conductivity of several materials. (Adapted from J. N. Shive,* Physics of Solid State Electronics, *Charles E. Merrill Publishing Company, 1966.)*

1.5 Intrinsic Semiconductors

A covalent semiconductor crystal that is free of lattice imperfections and impurities is referred to as an *intrinsic* semiconductor. While perfect crystals are difficult to obtain, semiconductors of reasonable purity are required as basic materials in the fabrication of solid state diodes, transistors, and integrated circuits. An understanding of intrinsic semiconductors is necessary to effectively use these devices. Two important objectives of this section are to find the number of available carriers in intrinsic material and to find the average energy level of a crystal. To accomplish these objectives, we will study the density of states in a band and also the probability that some of these states are occupied at a given temperature. Starting with equations for these two conditions, we will derive an equation for the average number of electrons in the conduction band at a given temperature (the number of electrons available as current carriers). However, we must first recognize the existence of another type of current carrier, the *hole.*

Consider a crystal of intrinsic silicon at 0°K. All of the electrons are in the lowest available orbitals. The valence band is completely filled, and the conduction band is completely empty. When the temperature rises to room temperature, atoms in the crystal lattice vibrate significantly. A few electrons in the valence band will gain the 1.11 eV of energy required to break the covalent bond and will move from the valence band into the conduction band. These electrons are now relatively free to move about in the silicon crystal. An individual electron will stay in the conduction band for only a short period of time, ranging from less than one microsecond to more than one millisecond before it drops back into a covalent bond. The average period of time that electrons stay in the conduction band is called the carrier *lifetime.*

When an electron moves to the conduction band, it leaves a vacancy in a covalent bond. This vacancy can easily be occupied by an electron from a neighboring covalent bond since little energy is required for the transition. In this way the vacancy, or *hole*, can move from one atom to the next in the crystal. The possible paths of a hole and electron in a crystal are shown in Fig. 1–9.

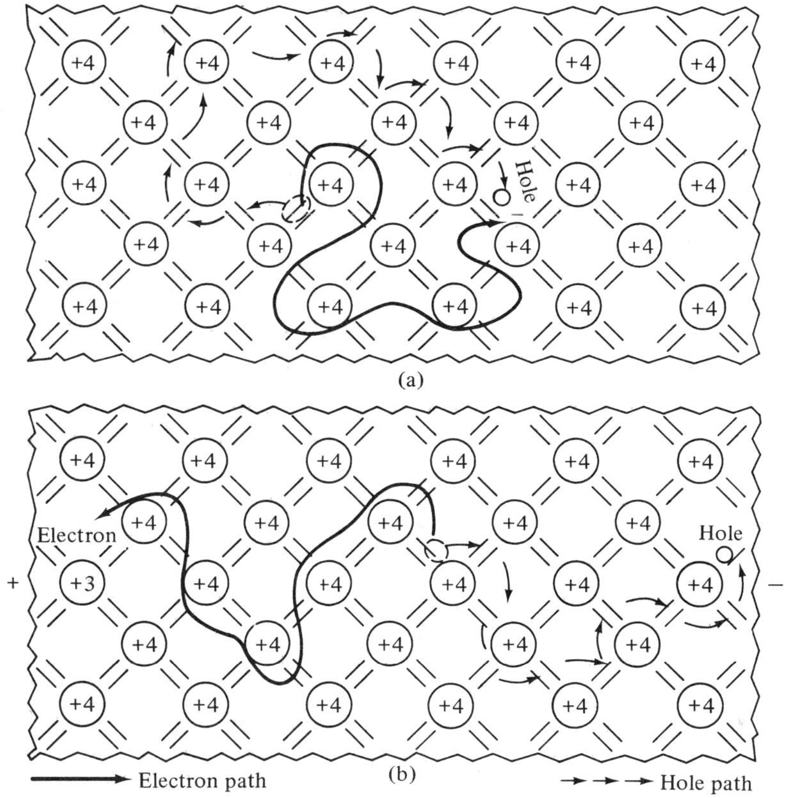

FIGURE 1–9 *Hole and electron movement in a crystal: (a) random motion and (b) motion under the influence of an electric field. Note the two covalent electrons between an atom and each of its four neighboring atoms. In a crystal, the atoms are arranged in a tetrahedral configuration.*

The two types of motion are different. Electron motion in the conduction band of silicon is similar to that in copper; i.e., the electron can move many atomic diameters before falling back into the valence band. In contrast, each time a hole moves, a covalent bond must be broken, and another one remade. Many electrons are involved in the motion of a single hole. When searching for a parking space for our automobile, we examine the vacancies rather than all the automobiles filling the spaces. In a similar manner, we consider the behavior of a single hole rather than the many electrons involved in its motion.

The properties of a hole can be summarized as follows. (1) It behaves as a particle with a charge equal in magnitude but opposite in polarity to that of an electron. (2) In an electric field, a hole moves in a direction opposite to that of an electron or in the same direction as conventional current. Holes exist only in the valence band.

The activity discussed above depends upon the formation and recombination of *electron-hole pairs*. As the temperature increases, more electrons gain sufficient energy to break free from a covalent bond and move to the conduction band. An equal number of holes are left behind or are "formed" in the valence band. Simultaneously, electrons in the conduction band are losing energy through collisions and dropping back into the valence band where they recombine with holes. An equilibrium condition is reached when the electron-hole pair generation and recombination rates are equal. This dynamic process causes some average number of holes and electrons to exist in a semiconductor at any given temperature, with the average number being different for different materials.

Let us now find the number of free electrons and holes available under conditions of thermal equilibrium. The distribution of states within the various energy bands of the semiconductor is shown symbolically in Fig. 1–10(a). Only a small number of states are shown in this example for clarity. In an actual crystal, the energy bands contain millions of states per CC which form an almost

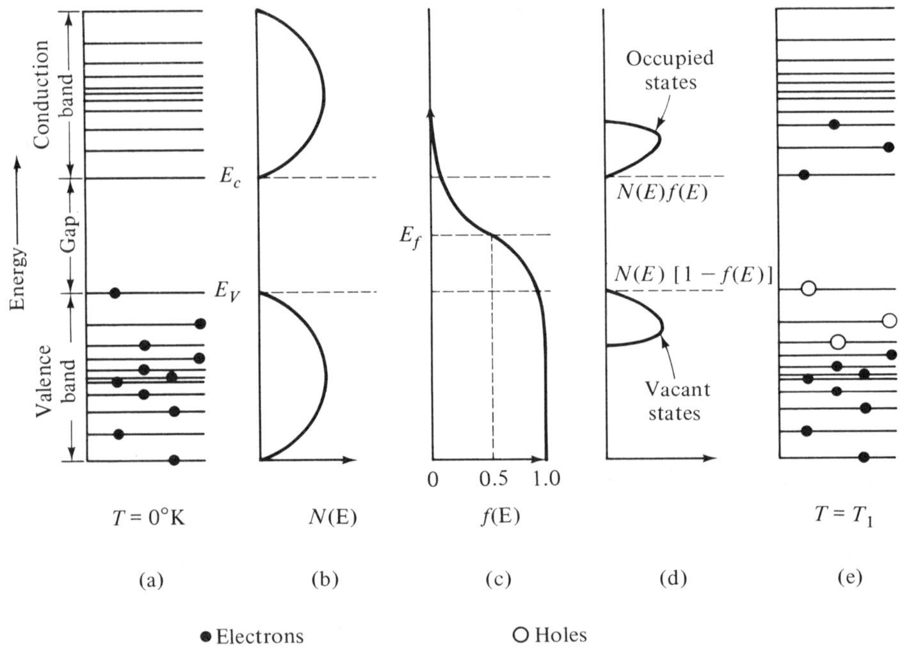

FIGURE 1–10 *A simplified energy band model of an intrinsic crystal. (a) At absolute zero, the valence band states are filled; the conduction band states are empty. (b) The density of available state,* N(E), *as a function of energy. (c)* f(E) *is the probability of occupancy. (d) The product* N(E)f(E) *is the probable density, at a finite temperature, of holes in the valence band and electrons in the conduction band. (e) Electrons in the conduction band and holes in the valence band at a higher temperature,* T, *than 0°K as shown in (a).*

continuous distribution across a band as a function of energy. Note that the maximum density occurs at the center of a band. The distribution function, $N(E)$, is defined as the density of available states (number per unit volume) per unit energy. To find the number of states within an energy band whose upper limit is E_2 and whose lower limit is E_1, $N(E)$ is integrated over this energy range, or

$$\text{Number of states between } E_1 \text{ and } E_2 = \int_{E_1}^{E_2} N(E)\, dE \qquad \textbf{(1–3)}$$

If this integration is performed over the valence band of a real semiconductor, the total number of valence-band energy states per unit volume for the particular crystal is obtained. Curves of the density of states for two energy bands are shown in Fig. 1–10(b). The smooth curves in Fig. 1–10(b) are obtained as the number of states becomes large (real semiconductor). From quantum theory,* the function $N(E)$ for the conduction band in a semiconductor is

$$N(E) = \frac{8\sqrt{2}\,\pi m_e^{3/2}}{h^3}\sqrt{(E - E_c)} \qquad \textbf{(1–4)}$$

where m_e is the mass of an electron; h is Planck's constant, 4.13×10^{-15} eV-sec, and E_c is the energy level at the bottom edge of the conduction band. This equation can be used to find the number of states per cubic centimeter at any energy level (E) in the conduction band.

Having defined the distribution of available states, the next step in finding the total number of carriers is to determine the probability that these states will be occupied. This probability of occupation is determined by the *Fermi-Dirac distribution function* which gives the percentage of these states that are occupied under conditions of thermal equilibrium at any energy level E. For example, if there are 10 available states at a given energy level and the Fermi-Dirac function has a value of 0.8 at this level, then an average of 8 of the 10 states are occupied. At some given instant in time, however, only 5 states may be occupied, or possibly all 10. But when a large number of available states is considered and when the period is taken over a time that is long compared to the average lifetime of an electron in the conduction band, the Fermi-Dirac function**gives accurate results. The Fermi-Dirac function is plotted in Fig. 1–10(c), and is given by***

$$f(E) = \frac{1}{\exp{(E - E_F)/kT} + 1} = \frac{1}{\exp{(V - V_F)q/kT} + 1} \qquad \textbf{(1–5)}$$

where $f(E)$ is the probability of finding an electron with energy E; E_F is the

---

* For a derivation of this equation, see William Shockley, *Electrons and Holes in Semiconductors* (New York: D. Van Nostrand Company, Inc., 1950), pp. 220–23. Equations (1–4), (1–5), and (1–6) are included for background information. The important equations in this section are Eqs. (1–7) through (1–10). A reader without a background in density functions and statistics should go directly to Eq. (1–7).

** Spin $\frac{1}{2}$ particles (electron, protons) require a Fermi-Dirac distribution while spin 1 particles (protons) require a Bose-Einstein distribution.

*** *Op. cit.*, pp. 461–64.

statistical reference called the *Fermi-level* (to be discussed later); V is the voltage; q is the charge on an electron (1.6×10^{-19} coulomb); E is qV; k is Boltzman's constant (1.38×10^{-23} joule/°K), and T is the temperature in degrees Kelvin.

At room temperature, the Fermi-Dirac function is shown as a solid line in Fig. 1–10(c). Most of the change in this function occurs in the region of the band gap, although electrons with energies up to infinity are possible. However, significant probabilities occur only a few multiples of kT/q on either side of E_F. The density of electrons, dn, in some small energy range, dE, at a particular temperature, is the product of the density of available states and the probability that these states will be occupied, or

$$dn = N(E)f(E)\,dE \qquad \textbf{(1–6)}$$

The number of electrons in the conduction band of a semiconductor can be calculated by integrating Eq. (1–6) from the bottom of the conduction band to the top of the conduction band. This operation is equivalent to integrating from E_C to approximately $+\infty$, since electrons are not likely to be found in the higher bands. Then, using Eq. (1–4) for the density of available states and Eq. (1–5) for the probability of occupation, the average number of electrons in the conduction band of a unit volume of a crystal at a given temperature is

$$n_i = \frac{4\sqrt{2}}{h^3}(\pi m_e kT)^{3/2} \exp\,[(E_F - E_C)/kT] \qquad \textbf{(1–7)}$$

This equation can be used to find the number of electrons per cubic centimeter that are in the conduction band at a given temperature. These electrons are available as current carriers.

The hole concentration in the valence band (holes per cm³) may be calculated by using a technique similar to that by which the electron density was found. The equation for the density of available states in the valence band will be the same as the corresponding equation for the conduction band [Eq. (1–4)] except that the mass of the electron is replaced by the effective mass of the hole, m_h. The probability that a state in the valence band is vacant, which means a hole is present, is equal to one minus the probability that it is occupied. Finding the product of these two functions as before, and integrating over the valence band (E_V to $-\infty$) yields

$$p_i = \frac{4\sqrt{2}}{h^3}(\pi m_h kT)^{3/2} \exp\,[(E_V - E_F)/kT] \qquad \textbf{(1–8)}$$

In this equation, p_i is the number of holes that is available as current carriers in a cubic centimeter of intrinsic material.

The number of holes in the valence band is equal to the number of electrons in the conduction band in an intrinsic crystal. Consequently, Eqs. (1–7) and (1–8) can be equated to find the Fermi level. If the mass of a hole and the mass of an electron are assumed to be approximately equal, then the Fermi level in intrinsic material is

$$E_{Fi} = \frac{E_C + E_V}{2} \qquad \textbf{(1–9)}$$

This equation indicates that the Fermi level is located at the middle of the band gap in an intrinsic semiconductor. In Eq. (1–7), note that $E_F - E_C$ is $(-E_g/2)$ where E_g is $E_C - E_V$. Therefore, n_i or p_i in intrinsic material is

$$n_i = p_i = AT^{3/2} \exp(-E_g/2kT) = AT^{3/2} \exp[(-qV_g)/2kT] \quad \textbf{(1–10)}$$

where $E_g = qV_g$ and A is a constant. Equation (1–10) is easier to apply than either Eq. (1–7) or (1–8), since the width of the band gap is usually known.

EXAMPLE 1–2 Find the intrinsic carrier concentration for a silicon crystal at 300°K (room temperature). The constant A in Eq. (1–10) is approximately 6×10^{15} for silicon.

Solution: The band gap voltage (V) for silicon is 1.1 V, and (kT/q) is 25.8 mV at 300°K. Substituting these values into Eq. (1–10) gives

$$\begin{aligned} n_i &= (6 \times 10^{15})(300)^{3/2} \exp[-(1.1)/(2 \times 0.0258)] \\ &= (6 \times 10^{15})(5.18 \times 10^3)/\exp(21.4) \\ &\simeq 1.6 \times 10^{10} \text{ electrons/cm}^3 \end{aligned}$$

The Fermi level is a measure of the electrochemical potential of a composite system, or a crystal in our case. For a given temperature, and under equilibrium conditions, the Fermi level will have a constant value throughout an intrinsic crystal. Proof of this statement depends on thermodynamic considerations which are well beyond the scope of this text. For our purposes, the following simplified version of the Fermi level is helpful. Let $(E_{e1} - E_F)$ be the energy difference between the Fermi level and an electron in the conduction band, and let $(E_{e2} - E_F)$ be the energy difference for another, etc. Then sum all of the energy differences for all of the electrons in the conduction band to obtain a total excess energy represented by the electrons in the conduction band. Now let $(E_F - E_{h1})$ be the energy difference between the Fermi level and a hole in the valence band, and let $(E_F - E_{h2})$ be the difference for another, etc. Consider the sum of all of these energies as a total energy deficiency caused by holes in the valence band. Then the Fermi level lies at an energy level such that the total excess energy is equal in magnitude to the total energy deficiency. In this sense the Fermi level is an average energy reference.

As indicated above, the number of electrons in the conduction band is equal to the number of holes in the valence band for intrinsic material. Since the distribution of holes relative to the top of the valence band is somewhat the same as that for electrons relative to the bottom of the conduction band, we would expect to find the Fermi level in the center of the band gap for intrinsic materials. However, if electrons could be injected into the conduction band without adding holes in the valence band, then the total excess energy would be increased without increasing the total deficiency. Consequently, the Fermi level would move up closer to the conduction band so that the excess and deficiency relative to the Fermi level would balance. On the other hand, if holes could be added to the valence band without adding electrons in the conduction band, the Fermi level would move closer to the valence band. Either of these conditions can be achieved by adding impurities to an intrinsic semiconductor crystal.

1.6 Doped Semiconductors

Most semiconductors have various impurities added to them which radically change their carrier concentrations. The addition of a small amount of known impurities is called *doping*. Impurity doping levels range from one doping atom for each 10^7 to one doping atom for each 10^9 semiconductor atoms. This section deals with the effects of adding these minute amounts of impurities.*

Consider a silicon crystal in which an antimony atom has replaced one of the silicon atoms in the lattice structure. This condition is shown in a two dimensional array in Fig. 1–11. Antimony is a pentavalent element (it has five electrons in its valence band). Only four of these electrons can take part in covalent bonds with electrons from the neighboring silicon atoms to form an "octet." As a result, the fifth electron is only slightly attracted to the antimony atom and only a small amount of additional energy is required to move this electron into the conduction band. The energy band diagram in Fig. 1–12 shows the result of the presence of the antimony atom in the crystal lattice. The fifth valence electron occupies an energy level in the forbidden region, typically less than 0.05 eV below the lower edge of the conduction band. At room temperature almost all of these electrons are excited into the conduction band. Since each pentavalent impurity atom contributes one electron to the number of available carriers in the conduction band, pentavalent atoms are called *donors*, and their density is designated as N_D.

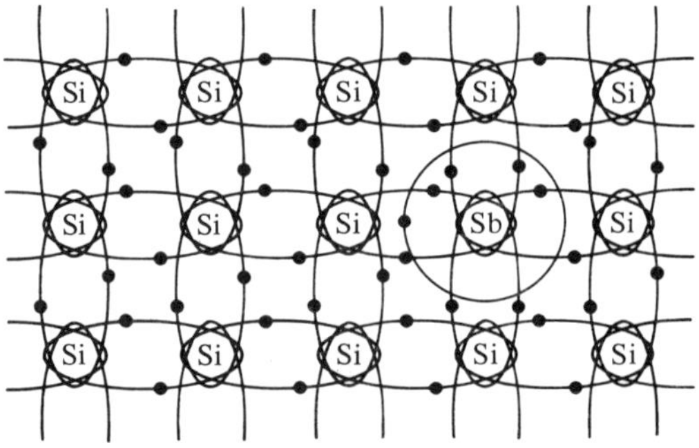

FIGURE 1–11 *An antimony atom in a silicon crystal supports its fifth valence electron in a non-binding orbit. (Adapted from J. N. Shive,* Physics of Solid State Electronics, *Charles E. Merrill Publishing Co., 1966.)*

Assume that a silicon crystal is doped with one antimony atom for each 10^7 silicon atoms. Since silicon has approximately 5×10^{22} atoms/cc, this doping corresponds to a doping density, N_D, of approximately 5×10^{15} atoms/cc. Thus, the relatively few donor atoms increased the electron concentration in the conduction band from about 10^{10} for intrinsic silicon to more than 10^{15} electrons/cc for doped silicon. This change drastically increases the conductivity of the crystal. Since each antimony atom has lost one electron, the atom becomes a positive ion which is locked into the crystal lattice by the four covalent bonds. Other pentavalent elements, such as arsenic and phosphorus, can also act as donor atoms and produce similar results.

* The methods by which these impurities are added to the semiconductor crystal are treated in the sections on diode, transistor, and integrated circuit construction.

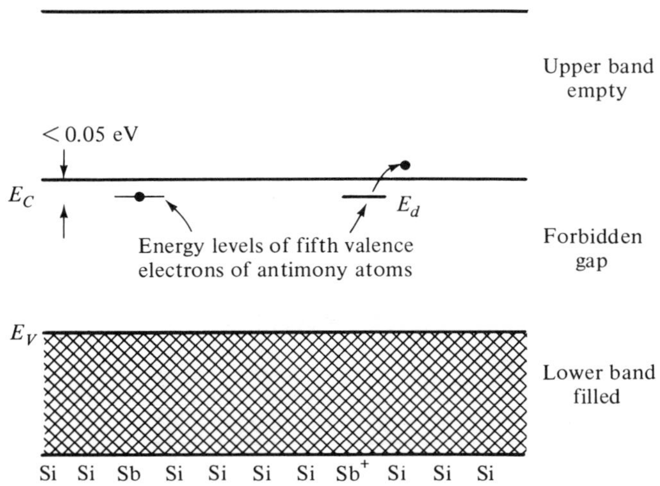

FIGURE 1–12 *The fifth valence electron of an antimony impurity atom in silicon occupies an energy level slightly below the bottom of the conduction band. It can be transferred easily into the conduction band, where it becomes available as a current carrier. The antimony atom then becomes a positive ion. (Adapted from J. N. Shive,* Physics of Solid State Electronics, *Charles E. Merrill Publishing Co., 1966.)*

The carrier concentration and conductivity of a semiconductor can also be changed by adding trivalent atoms such as indium, boron, aluminum, or gallium to the lattice structure. These atoms have only three valence electrons. An indium atom locked into a silicon crystal lattice is shown in Fig. 1–13. Since only three valence electrons are available from the indium atom, one covalent bond is incomplete, thus producing a hole. This hole or vacancy can be filled by an electron from a neighboring bond with the addition of very little energy. In fact, the vacancy corresponds to an energy level just above the edge of the valence band as shown in Fig. 1–14. When an electron moves from a silicon covalent bond into the empty indium bond, it leaves behind a positive hole and makes the indium atom negative; i.e., the atom has one electron that is not neutralized by a proton in the nucleus. Trivalent atoms are called *acceptor* atoms, and their concentration is denoted by N_A. The addition of acceptor atoms to a semicon-

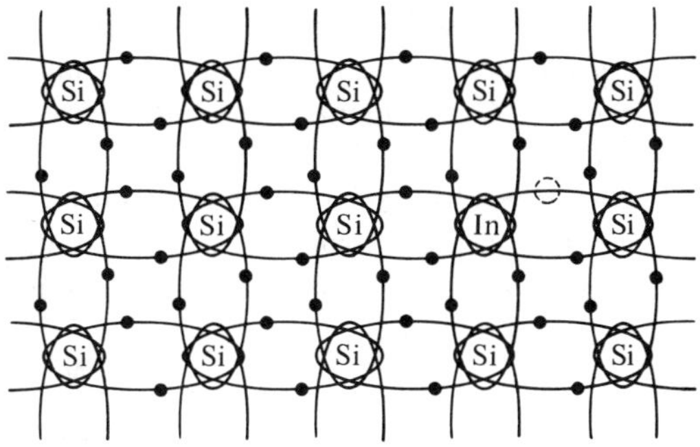

FIGURE 1–13 *An indium atom locked into a silicon crystal lattice gives rise to a bonding-electron deficit. (Adapted from J. N. Shive,* Physics of Solid State Electronics, *Charles E. Merrill Publishing Co., 1966.)*

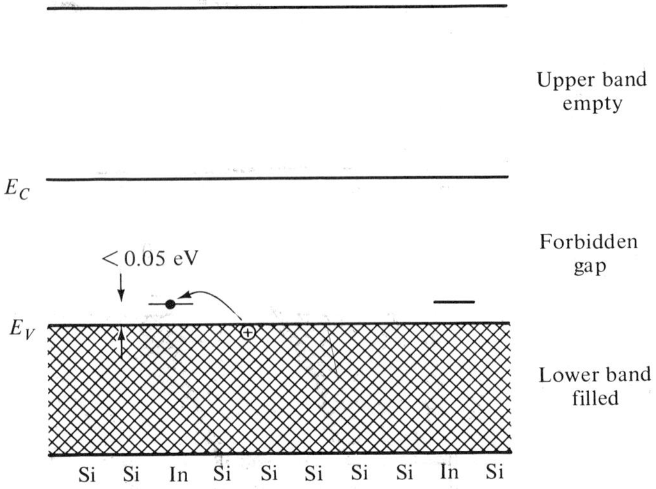

FIGURE 1–14 *The bonding valence electron deficit of an indium atom dissolved in a silicon crystal may be thought of as a local energy level a short distance above the top of the valence band. An electron from the surrounding silicon matrix can be activated easily into this level, producing a free positive hole. The indium atom becomes a negative ion. (Adapted from J. N. Shive,* Physics of Solid State Electronics, *Charles E. Merrill Publishing Co., 1966.)*

ductor crystal increases the free hole concentration in the valence band and hence the conductivity.

Adding donor or acceptor atoms to a pure semiconductor material produces a drastic increase in the concentration of free carriers. Doping with donor atoms increases the density of negative charge carriers (electrons) in the conduction band and produces an N-type semiconductor. Doping with acceptor atoms increases the positive charge carriers (holes) in the valence band and produces a P-type semiconductor. Germanium and silicon can be made either N-type or P-type, depending on the impurity added.

The dominant type of carrier in a semiconductor is referred to as a majority carrier, and the other type a minority carrier. Holes are majority carriers in P-type material, and electrons are minority carriers. This relationship is reversed in N-type crystals.

Both the semiconductor atoms and also the impurity atoms which make up a crystal of doped semiconductor initially had a neutral charge before the doping process. Consequently, the entire crystal is charged neutrally after the doping process; i.e., the total number of electrons and protons in the crystal are equal. Negative charge is associated with the free electrons, n, and also with the ionized acceptor atoms, N_A. Positive charge is associated with the free holes, p, and also with the ionized donor atoms, N_D. The charge neutrality relationship becomes

$$n + N_A = p + N_D \tag{1–11}$$

For a moderately doped N-type crystal, N_A is zero, and the hole concentration is small. Consequently, the electron concentration is

$$n = p + N_D \simeq N_D \tag{1–12}$$

The hole concentration in P-type material is

$$p = n + N_A \simeq N_A \tag{1–13}$$

Note that the subscript "i" has been dropped from the carrier concentrations for doped semiconductors.

Assuming that all of the impurity atoms are ionized at room temperature, the number of electrons added in the conduction band by doping with pentavalent material is equal to the number of impurities added. With respect to the Fermi level, these are electrons with excess energy. Since the lifetime of an electron in the conduction band is relatively constant up to reasonably high doping levels, electrons are falling back into the valence band at a higher rate than for intrinsic material. The electrons falling into the valence band quickly recombine with holes. Consequently, the hole density is reduced below the intrinsic value since the rate of electron-hole formations is relatively constant at a given temperature. Since the number of electrons in the conduction band increases while the number of holes decreases, the Fermi level moves closer to the conduction band in order to equalize the total excess and total deficit energies. The actual location of the Fermi level depends on the doping density and temperature. An increase in donor doping density causes the Fermi level to move closer to the conduction band while an increase in temperature causes it to move closer to the center of the band gap.

In P-type material, the Fermi level lies between the top of the valence band and the center of the band gap. An expression for the Fermi level in doped material may be obtained as follows. Equations (1–7) and (1–8) are valid for doped material as well as for intrinsic material. Collecting coefficients in these equations,

$$n = K \exp\left[(E_F - E_C)/kT\right] \tag{1–14}$$

and

$$p = K \exp\left[(E_V - E_F)/kT\right] \tag{1–15}$$

Taking the natural logarithm of n/p and simplifying the results by using Eq. (1–9), the Fermi level, in general, is

$$E_F = E_{Fi} + \frac{kT}{2} \ln \frac{n}{p} \tag{1–16}$$

where E_{Fi} is located at the center of the band gap. N-type material contains more electrons than holes. Consequently, the second term in Eq. (1–16) is positive, thereby making E_F for N-type material larger than that for intrinsic material. Conversely, the second term is negative for P-type. The location of the Fermi level in the band gap may be summarized as follows:

(a) The Fermi level is located at the center of the band gap for intrinsic materials
(b) The Fermi level lies between the center of the band gap and the bottom of the conduction band in N-type materials
(c) The Fermi levels lie between the center of the band gap and the top of the valence band in P-type material

These results are shown in Fig. 1–15.

A very useful result occurs if the product of the hole concentration, p, and the electron concentration, n, is formed in doped material. From Eqs. (1–10), (1–14), and (1–15),

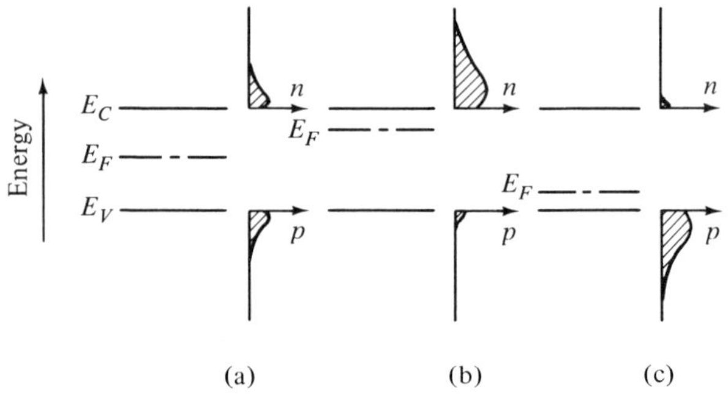

FIGURE 1–15 *Carrier concentrations and energy band diagrams for intrinsic and doped material: (a) intrinsic, (b) N-type and (c) P-type semiconductors.*

$$np = K_n K_p \exp\left[-(E_C - E_V)/kT\right] = n_i^2 \tag{1–17}$$

Thus, the product of holes and electrons is always a constant which is equal to the intrinsic carrier concentration squared. This product is approximately 5.7×10^{26} for germanium and 2.3×10^{20} for silicon at 300°K.

EXAMPLE 1–3 Find the average number of holes and electrons in a cubic centimeter of silicon at 300°K if the sample is doped with one donor impurity atom for each 10^7 intrinsic atoms.

Solution: The number of donor atoms added per cc, N_D, is

$$N_D = \frac{5 \times 10^{22}\ \text{atoms/cm}^3}{10^7\ \text{intrinsic atoms/donor}} = 5 \times 10^{15}\ \text{donor atoms added/cm}^3$$

$$n = n_i + N_D = 1.5 \times 10^{10} + 5 \times 10^{15} \simeq 5 \times 10^{15}\ \text{electrons/cm}^3$$

$$p = \frac{n_i^2}{n} = \frac{2.3 \times 10^{20}}{5 \times 10^{15}} \simeq 4.6 \times 10^4\ \text{holes/cm}^3$$

1.7 Current Flow in a Crystal

As indicated in Sec. 1.4, an electric current is a net flow of charge carriers in a given direction. This net flow can occur only if charge carriers can change states easily. Assuming that this condition exists, then two types of current can flow, *drift current* and *diffusion current*. Drift current flows under the influence of an electric field, whereas diffusion current flows because the charge is more concentrated in one region than in an adjacent region. Both types are important in semiconductor studies. Drift current will be considered first.

A *crystal* connected between two electrodes is shown in Fig. 1–16. Assume that some average number of electrons has gained sufficient energy from thermal

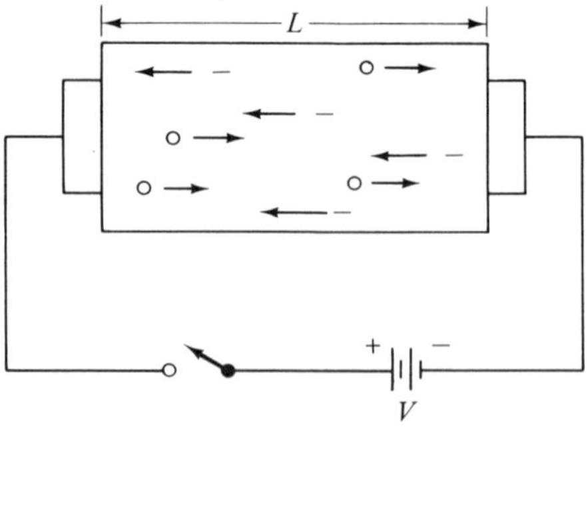

FIGURE 1–16 *A circuit in which drift current flows.*

sources to be in the conduction band of the solid. The conduction band is nearly empty and has many states available for the electrons to occupy. With the switch open, the motion of each electron is dependent upon thermal energy and consequently is random as shown in Fig. 1–17(a). No net charge movement occurs, and hence no net current flows.

When the switch is closed, voltage is applied across the crystal. The electric field produced by this voltage causes the available charge carriers to move or drift in a net direction. Holes in the valence band* drift to the negative end of the semiconductor or in the same direction as conventional current flow. Electrons in the conduction band drift to the positive end. Each of these two drift motions constitutes a flow of current. However, each component of current is in the same direction** so that the total drift current through the crystal is the sum of the two components.

The electric field causes the free electrons to accelerate; however, an electron can not travel very far before it has a collision with the crystal lattice and loses some of its energy. The average distance between collisions is called the *mean free path* and is typically 10^{-7} cm for an electron in silicon. The loss of energy causes the electron to drop to a lower energy level [see Fig. 1–17(b)]. The motion of an electron in an applied field is a series of start-stop motions which can be approximated by an average drift velocity, v_d. This drift velocity is proportional to the electric field, ε, where the constant of proportionality for electrons is the mobility factor, μ_n. Thus,

$$v_d(\text{cm/sec}) = \mu_n[(\text{cm/sec})/(\text{volts/cm})]\varepsilon(\text{volts/cm}) = \mu_n V/L \qquad \textbf{(1–18)}$$

where V is the voltage drop across the sample, and L is the length of the sample ($\varepsilon = V/L$). The total current flow depends on the number of electrons per cm³

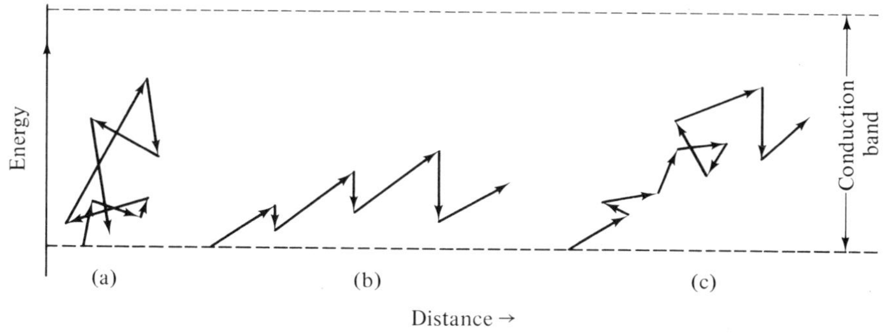

FIGURE 1–17 *Motion of electrons in a conduction band: (a) random motion due to thermal energy, (b) unidirectional motion due to electric field, and (c) combined motion. Numerous collisions cause changes in the direction of motion.*

* Electrons are extracted at the positive end of the crystal. Some of these electrons were in the valence band and leave a hole behind. This process is called *hole injection.* Other electrons were in the conduction band. Holes that reach the negative end combine with an electron from the conductor so that the total current flowing in the external metal conductor is composed entirely of electrons.

** Current flow was initially conceived as a flow of positive charge with the direction being from the positive terminal to the negative terminal. Since holes are positive, the flow of holes is in the same direction as conventional current flow. However, electrons are negative, and a flow of electrons in one direction constitutes a current in the opposite direction.

(n) that are acting as carriers, the charge on the carrier (q in coulombs), the cross sectional area [A(cm^2)], and the drift velocity [v_d(cm/sec)]. The current component due to electron drift is

$$I_e = nqAv_d = nqA\mu_n\mathcal{E} = nqA\mu_n(V/L) \text{ amperes} \qquad \textbf{(1–19)}$$

The resistance of a uniform bar of material in terms of its physical parameters is

$$R = \rho(\text{ohm-cm})\frac{L(\text{cm})}{A(\text{cm}^2)} = \frac{L}{\sigma A} \text{ ohms} \qquad \textbf{(1–20)}$$

where ρ is the resistivity, and its reciprocal, σ, is the conductivity of the material. Using Ohm's law ($V = IR$) together with Eqs. (1–19) and (1–20), the resistance of the sample in terms of the physical parameters is

$$R = \frac{L}{nqA\mu_n} \qquad \textbf{(1–21)}$$

The conductivity of a unit sample is

$$\sigma = nq\mu_n \qquad \textbf{(1–22)}$$

These equations are valid when only electron flow is considered.

If holes are present but electrons are not, the conductivity is

$$\sigma = pq\mu_p \qquad \textbf{(1–23)}$$

In a sample containing both holes and electrons, both current components are present. Since both components are independent of each other, the conductivity components can be added. Therefore, the conductivity of a unit sample is

$$\sigma = q(n\mu_n + p\mu_p) \qquad \textbf{(1–24)}$$

where σ is drift conductance in mhos per centimeter; q is the charge on an electron (1.6×10^{-19} coul); n is the concentration of electrons available for conduction (electrons per cubic centimeter); p is the concentration of holes (holes per cubic centimeter); μ_n is the mobility factor for electrons (centimeters per second for a voltage gradient of one volt per centimeter or cm^2/volt-sec), and μ_p is the mobility factor for holes (centimeter per second for voltage gradient of one volt per centimeter). The mobility constant indicates the relative ease with which a carrier can move through the lattice and is a function of the arrangement of the atoms in the crystal, the dielectric constant, and the amount of vibration exhibited by the atoms.

EXAMPLE 1–4 Find the number of impurity atoms that must be added to one cm^3 of intrinsic silicon in order to convert it to (a) 10 ohm-cm, P-type silicon, and (b) 10 ohm-cm, N-type silicon. (See Table 1–3 for parameter values.)

Solution: (a) The hole density, p, is approximately equal to the number of acceptor atoms added per cm^3 or

$$p \simeq N_A$$

From Eq. (1–23),

$$\rho = \frac{1}{\sigma} = \frac{1}{qp\mu_p} = \frac{1}{qN_A\mu_p}$$

Solving for N_A and substituting appropriate values,

$$N_A = \frac{1}{q\rho\mu_p} = \frac{1}{(1.6 \times 10^{-19} \text{ coulombs})(10 \text{ ohm-cm})(480 \text{ cm}^2/\text{volt-sec})}$$
$$= 1.5 \times 10^{15} \text{ atoms/cc}$$

(b) For N-type material, the electron density, n, is approximately equal to the number of donor type atoms added per cc, N_D.

$$N_D = \frac{1}{q\rho\mu_n} = \frac{1}{(1.6 \times 10^{-19})(10)(1350)} = 4.63 \times 10^{14} \text{ atoms/cc}$$

EXAMPLE 1–5 The mobility constant for a sample of copper at 25°C is approximately 35 cm²/volt-sec, and the number of electrons available for conduction is approximately $8.2 \times 10^{22}/\text{cm}^3$. Find the conductivity and resistivity.

Solution: Since copper has no hole-type carriers, the conductivity is

$$\sigma = qn\mu_n = (1.6 \times 10^{-19})(8.2 \times 10^{22})(35) \simeq 4.6 \times 10^5 \text{ mho/cm}$$

Resistivity is the reciprocal of the conductivity; therefore,

$$\rho = 1/(4.6 \times 10^5) \simeq 2.16 \times 10^{-6} \text{ ohm-cm}$$

The second type of current, diffusion current, results from the familiar diffusion effect. A simple example is the diffusion of smoke in air. This mixing occurs even if no wind is present and is nature's way of equalizing a disturbance. Diffusion is a property of particles in general and is independent of the amount and type of charge on a particle.

A quantity of charge has been dumped or injected into one end of a crystal in Fig. 1–18. The process by which this is accomplished will be discussed in the next two chapters. The probability that a current carrier near the plane *a-b* will move to the left is equal to the probability that it will move to the right. However, there are more carriers on the left than on the right. Therefore, on the average, more carriers cross the plane from left to right than from right to left. The result is a net movement of charge (or flow of current) across plane *a-b*. Eventually, the current carriers will be uniformly distributed as shown in Fig. 1–18(c). Until this condition exists, diffusion current will flow.

Mathematically, the current at any cross section, *a-b*, as shown in Fig. 1–18, is

$$i = (qDA)(-dn/dx) = \frac{\text{coulombs}}{\text{electron}}\left(\frac{\text{cm}^2}{\text{sec}}\right)(\text{cm}^2)\left(\frac{\text{electrons}}{\text{cm}^4}\right) \quad \textbf{(1–25)}$$

where q is the charge on an electron or hole; $D = (\mu kT/q)$ and is the diffusion constant for carriers (cm²/sec); A is the cross-sectional area (cm²); and dn/dx is the slope of the density curve.* In Fig. 1–18, a given quantity of charge has been injected into the end of the bar at $t = 0$, so that charge distributions are shown for three different instants; t_1 occurs a relatively short time after t_0, and t_2 a very

---

* In order to simplify the process, the assumption is made that no recombinations occur.

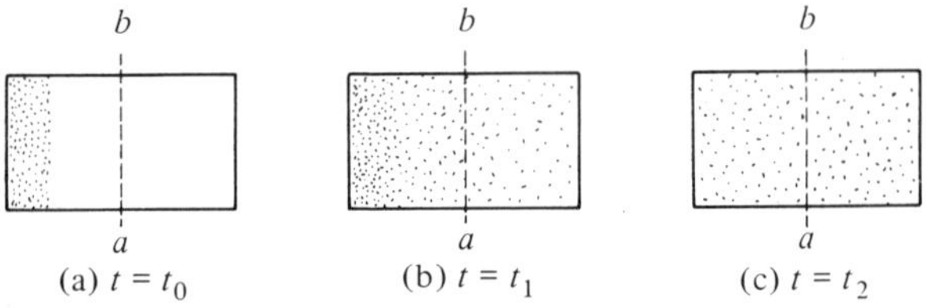

FIGURE 1–18 *Current flow by diffusion of carriers from an area of high concentration to an area of low concentration: (a) initial distribution; (b) distribution a short time later; (c) eventual distribution.*

long time later. The current decays to zero as a function of time on an exponential curve.*

Even more important in our future work is the case in which we continue to inject carriers into the left end and remove them from the right end. For this case, if we assume that there are no recombinations of holes and electrons in the material, the diffusion current must equal the external current.

In order to maintain this diffusion current, a nonuniform distribution of charge must be established along the bar. This situation is similar to a metal bar that is heated with a torch on one end and attached to a heat sink on the other end. The torch end will be much hotter than the sink end. Consequently, a temperature gradient will exist along the bar. Equilibrium conditions are established when the quantity of heat accepted by the torch end of the bar per unit time is equal to the quantity delivered to the sink per unit time, assuming no radiation from the bar. The temperature gradient along the bar will increase until this condition exists. In the same way, charge can flow through a crystal by diffusion at a given rate, provided the charge distribution can increase the probability of net charge movement to a sufficient magnitude.

The important points here are:

1. Current can flow in a field-free region.
2. In a crystal this is called diffusion current.
3. Before diffusion current can flow, a nonuniform distribution of charge must be established.
4. A charge distribution with a linear slope will produce a constant current (assuming no recombination).
5. Diffusion current cannot be increased any faster than the necessary charge distribution can be established.

EXAMPLE 1–6 Assuming that the magnitude of the slope of the charge distribution curve, (dn/dx), at the plane *a-b* in Fig. 1–18 is 10^{14} electrons/cm^4 and that the cross-sectional area is 1.0 cm^2, find the instantaneous current flow across the plane *a-b*. The block is germanium and has a diffusion constant for electrons of 100 cm^2/sec.

Solution: From Eq. (1–25),

$$i = (1.6 \times 10^{-19})(100)(1.0)(-10^{14}) = -1.6 \text{ mA}$$

or a current of 1.6 mA to the left. The minus sign on dn/dx means that the density decreases as distance increases.

---

* If the electron carrier density is uniform, then dn/dx is zero, and the diffusion current is zero. If the density is greater in one place than in another, then the change in density, dn, as a function of distance, dx, determines the diffusion current for a particular material and carrier type.

The total current in a semiconductor may be composed of two components, a hole current and an electron current. In addition, each component can be made up of two types of current, drift and diffusion. A general expression for the total current is

$$\begin{aligned} I_T &= (I_n) + (I_p) = (I_{n\text{drift}} + I_{n\text{diffusion}}) + (I_{p\text{drift}} + I_{p\text{diffusion}}) \\ &= \left(nqA\mu_n\mathcal{E} + qAD_n\frac{dn}{dx}\right) + \left(pqA\mu_p\mathcal{E} + qAD_p\frac{dp}{dx}\right) \qquad \textbf{(1–26)} \end{aligned}$$

1.8 Summary

A brief review of matter and atoms in general provided the basis for a study of crystals. Crystals are characterized by a regular lattice structure. The valence and conduction bands in crystals are particularly important because their relationship to each other determines the electrical properties of the crystal.

Semiconductors have electrical conductivity values that lie between those of metals and insulators. The current carriers in an intrinsic semiconductor are holes and electrons that are thermally generated. At room temperature, relatively few carriers are available, and the conductivity has a low value (high resistance). The conductivity can be increased (resistance decreased) by doping the semiconductor with either trivalent (acceptor) or pentavalent (donor) atoms. Acceptor atoms increase the hole density in the valence band and form negative ion centers which are locked into the crystal lattice. Donor atoms increase the electron density in the conduction band and form positive ions which are locked into the crystal lattice.

The charge neutrality relationship, Eq. (1–11), is used to calculate the majority carrier concentration in a semiconductor. Since the product of the free holes and electrons is a constant regardless of doping, Eq. (1–17) is used to calculate the minority carrier density. These densities are important in the study of diodes and transistors.

In intrinsic material, the Fermi level is located in the center of the band gap. In N-type material, this level lies between the middle of the band gap and the bottom of the conduction band with heavier doping levels placing it near the latter level. In P-type material the Fermi level lies between the middle of the band gap and the top of the valence band. For a particular crystal at a given temperature, the Fermi level is constant throughout the crystal. This relationship is most helpful in understanding the physical operation of semiconductor devices.

The two types of current carriers in a semiconductor, holes and electrons, can have a net movement due to drift in an electric field or due to diffusion. Consequently, any combination of four components of current can exist simultaneously in a semiconductor: (1) electron drift, (2) electron diffusion, (3) hole drift, and (4) hole diffusion. These current components can be calculated using Eq. (1–26). All four currents can be important in the operation of diodes and transistors.

The purpose of this first chapter was to develop an intuitive understanding of semiconductor physics. Some useful constants are summarized in Table 1–2 and some approximate semiconductor parameters are summarized in Table 1–3. No attempt was made to develop an in-depth mathematical treatment of the subject. Entire areas of semiconductor physics were not discussed. The treatment given, however, is adequate for the study of semiconductor devices and their application in circuits. The next step in this study is to discuss the construction, properties, and applications of a semiconductor diode. This device is constructed by doping one part of a semiconductor crystal with an N-type material and another part of the same crystal with P-type material.

TABLE 1–2 *Useful Constants and Values*

Name	Symbol	Value	Units
Charge on an electron	q	1.602×10^{-19}	coulomb
Rest mass of an electron	m	9.106×10^{-31}	kilogram (kg)
Charge to mass ratio	q/m	1.759×10^{11}	coulombs/kg
Velocity to light	c	3.0×10^{8}	meters/second
Planck's constant	h	4.13×10^{-15}	eV-sec
Boltzmann's constant	k	1.38×10^{-23}	joule/°K
Angstrom unit	A	10^{-10}	meter
Space permittivity	ϵ_0	8.85×10^{-12}	farads/meter
Electron volt	eV	1.602×10^{-19}	joule

TABLE 1–3 *Approximate Semiconductor Parameters at 300°C*

	Symbol	Germanium	Silicon	Units
Energy gap	E_g	0.69	1.11	eV
Electron mobility	μ_n	3900	1350	cm^2/V-sec.
Hole mobility	μ_p	1900	480	cm^2/V-sec.
Electron diffusion constant	D_n	100	35	cm^2/sec.
Hole diffusion constant	D_p	49	12	cm^2/sec.
Melting point	T_m	936	1420	°C
Dielectric constant	K	15.8	11.7	—
Intrinsic carrier concentration	n_i	1.5×10^{10}	2.4×10^{13}	electrons/cm^3
Atoms/cm^3		4.4×10^{22}	5×10^{22}	

Problems

Section 1.1

1–1 A silicon atom has 14 neutrons in its nucleus. (a) Determine the approximate mass of a silicon atom in terms of the mass of an electron. (b) Repeat for a germanium atom which has 40 neutrons in its nucleus.

Section 1.2

1–2 What valence shell characteristic is shared by the atoms of carbon, germanium, silicon, and tin?

1–3 A hydrogen atom radiates a wave having a frequency of 3.16×10^{15} Hz. Describe the origin of this radiation.

1–4 How many unoccupied energy states are available in the valence shell of (a) a silicon atom? (b) a germanium atom?

Section 1.3

1–5 (a) Give one reason why tin is a poor semiconductor even though it has four electrons in its valence shells. (b) Repeat for carbon. (*Hint:* Examine the band gap and primary quantum number of the valence shells of silicon and germanium.)

1–6 (a) One hundred silicon atoms are frozen into a solid. How many energy states are present in the valence band? (b) Repeat for germanium.

1–7 The ionization potential of a neon atom is approximately 21.5 eV. Find the minimum velocity of an electron that will ionize this gas.

1–8 When an electron returns to its normal orbit in neon, an energy of 21.5 eV will be released. Find the wavelength of the x-ray produced.

Section 1.4

1–9 Explain why an electron must be able to change energy states in a semiconductor before it can contribute to conduction.

1–10 At temperatures much lower than room temperature, a semiconductor becomes a good insulator. Why does a metal not become a good semiconductor at low temperature?

Section 1.5

1–11 Derive the expression for E_{Fi} for the case when the mass of a hole and the mass of an electron are not assumed equal.

1–12 Find the value of the constant A in Eq. (1–10) for (a) silicon and (b) germanium at 300°K. The band gap for silicon is 1.1 eV and for germanium is 0.7 eV.

Section 1.6

1–13 Verify Eq. (1–16).

1–14 Give an intuitive explanation of why the electron density (n) in a P-type semiconductor is less than that for intrinsic material (less than n_i).

1–15 Verify Eq. (1–17).

1–16 What density and type of impurity must be added to a silicon crystal to produce an electron density of 10^6 electrons/cc at 300°K?

1–17 What density and type of impurity must be added to a silicon crystal to produce a hole density of 10^5 holes/cc at 300°K?

Section 1.7

1–18 Find the conductivity and resistivity of intrinsic silicon at 300°K.

1–19 Find the conductivity and resistivity of a sample of N-type silicon doped with one N-type impurity atom for each 10^7 intrinsic atoms. Assume $T = 300°K$; $\mu_n = 1135$, and $\mu_p = 410$. Compare this answer with the answer to Prob. 1–18 and the value given in Fig. 1–8. (Note that μ_n and μ_p change with doping levels.)

1–20 Find the current flow in a sample cube measuring one centimeter on each side when 0.1 volt is applied to opposite faces of the cube. The material of the cube is given in Prob. 1–19. What percentage of the current is carried by holes?

1–21 Find the instantaneous current in Ex. 1–6 if the carriers are holes. Assume that D_p is approximately 49 cm²/sec for germanium.

1–22 A steady diffusion current of 1 mA flows in a block of silicon which has a cross-sectional area of 0.1 cm^2. Find the charge gradient that is required to produce this flow if (a) the carriers are electrons, and (b) the carriers are holes.

1–23 An electron gun is used to inject electrons into the face of a cube of intrinsic germanium one centimeter on each side. The opposite face is grounded. Find the slope of the charge distribution when a steady diffusion of 100 μA flows.

References

Adler, R. B., A. C. Smith, and R. L. Longini, *Introduction to Semiconductor Physics*, Vol. 1, Semiconductor Electronics Education Committee. New York: John Wiley & Sons, Inc., 1964.

Lindmayer, Joseph and Charles Y. Wrigley, *Fundamentals of Semiconductor Devices*. New York: D. Van Nostrand Company, Inc., 1965.

Martin, Thomas L., Jr., *Physical Basis for Electrical Engineering*. Englewood Cliffs, N. J.: Prentice-Hall, Inc., 1957.

Pierce, J. F., *Transistor Circuit Theory and Design*. Columbus, Ohio: Charles E. Merrill Publishing Co., 1963.

Rusk, Rogers D., *Introduction to College Physics*. New York: Appleton-Century-Crofts, Inc., 1960.

Sanderson, R. T., *Inorganic Chemistry*. New York: Reinhold Publishing Corporation, 1967.

Shive, J. N., *Physics of Solid State Electronics*. Columbus, Ohio: Charles E. Merrill Publishing Co., 1966.

2

Junction Diodes

As indicated in Chapter 1, a junction diode is formed by joining a piece of P-type crystal to a piece of N-type crystal in such a way that the lattice is continuous across the junction. The electrical characteristics of a diode are radically different from those of either component material. For example, either P-type or N-type material will conduct equally well in either direction. However, a diode will conduct well in only one direction. This difference is due to the nature of the region in the neighborhood of the junction.

In this chapter we will first investigate the junction and its characteristics. Then equivalent circuits for diodes will be developed so that the diode's performance in electrical circuits can be understood easily. Several typical applications will illustrate the techniques used in practice to analyze circuits containing diodes.

The junction diode is an important circuit element. Equally important is the fact that junction diodes are basic building blocks for almost all semiconductor devices. A good understanding of the P-N junctions is necessary in all semiconductor device studies.

2.1 The Potential Barrier

The continuity of the crystal lattice across a P-N junction may be established in many different ways. Three of the most frequently used processes are:

1. A growing process in which the impurity concentration is varied while a crystal is being "pulled" or grown
2. An alloying process in which the impurity is melted into a chip of semiconductor
3. A diffusion process in which the impurity gas is diffused into a piece of semiconductor material which is held at a high temperature

The process employed determines to some extent the characteristics of a diode. For example, an alloy junction is rather abrupt while a grown junction is not. The immediate problem is to develop the theory of the P-N junction. In this development, an abrupt junction will be assumed; consequently, deviations from this theory may be expected for the gradual junction case.

Assume that a block of N-type semiconductor is joined to a block of P-type material (as shown in Fig. 2–1) to form a single-crystal P-N junction. Prior to the formation of the junction, each block was neutrally charged since each block contained an equal number of electrons and protons. The N-type material contained a relatively large number of free electrons as majority carriers and an almost equal number of immobile, ionized atoms, each with one plus charge. This block also contained a few holes as minority carriers. The P-type block contained a large number of available holes as majority carriers, a few free electrons as minority carriers, and negative immobile impurity ions.

When the junction is formed, electrons in the N-type material and holes in the P-type material diffuse across the junction in an attempt to equalize both the free-electron density and the free-hole density throughout the crystal. However, both the N-type and the P-type materials were initially neutral. When an electron diffuses across the junction into the P-type block, the N-side has a deficiency of one electron, or a net positive charge, and the P-side has an excess of one electron, or a net negative charge. In a similar manner a hole moving from the P-side into the N-side adds a positive charge to the N-side and leaves a negative charge on the P-side. The net result of this *initial diffusion current* or forward current, I_F, is to produce a large number of immobile ions with positive charges (uncovered impurity atoms) close to the junction on the N-side and an equal number of negative charges (also impurity atoms) on the P-side (as shown in Fig. 2–1). An electric field is established between these two groups of charged immobile ions that reduces the diffusion of majority carriers across the junction. The voltage associated with this field is called the *potential barrier*, V_B.

Minority carriers on either side of the junction enter the picture as soon as the electric field is established. Minority carriers (holes on the N-side or electrons on the P-side) that wander into the field set up by the charged immobile ions are immediately swept to the other side of the junction. Thus, electrons are forced to move into the N-side and holes into the P-side. This action reduces the net charge on each side and reduces the magnitude of the potential barrier. The flow of minority carriers across the junction constitutes drift current since it is caused by an electric field. This component of current is often called the *reverse saturation current*, I_S.

The potential barrier increases in magnitude as the initial diffusion current leaves more ions (uncovered impurity atoms) on each side of the junction. The region containing these uncovered (charged) atoms is located adjacent to and on each side of the junction and is called the *depletion region*. This region contains very few current carriers, due to the presence of the electric field set up by positive ions on the N-side and negative ions on the P-side.

Note that the diffusion current, I_F, is composed of two parts, holes diffusing from the P-side to the N-side, and electrons diffusing from the N-side to the P-side. Carriers that cross the junction must obtain sufficient thermal energy to surmount the barrier or to overcome the repelling force of the field at the junc-

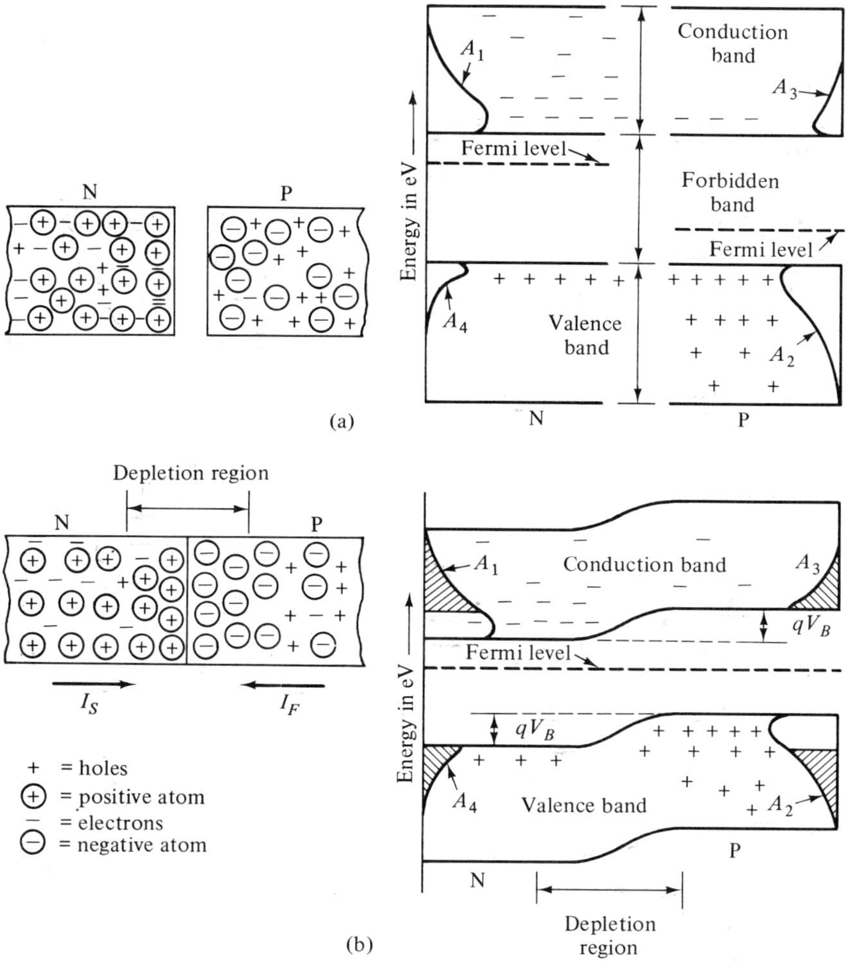

FIGURE 2–1 *Energy diagrams for N- and P-type materials (a) before creating a junction, and (b) after the junction is formed. Note that in (a) the Fermi levels occur at different energy levels for the two materials, and in (b) the Fermi levels are at the same energy level. The areas* A_1 *and* A_3 *are proportional to electron densities, and* A_2 *and* A_4 *are proportional to hole densities at the various energy levels.*

tion. The reverse current, I_S, also contains two components, holes drifting from the N-side to the P-side, and electrons drifting from the P-side to the N-side. These carriers are produced by electron-hole pair formations in the intrinsic material in or near the junction. As the barrier voltage increases, due to majority carrier diffusion, a point is reached where the diffusion or forward current, I_F, is equal in magnitude to the drift or reverse current, I_S. When this condition occurs, equilibrium is established with zero net current in the (unbiased) diode; i.e., the electron components in each direction are equal, and the hole components in each direction are equal. The electron currents are equal when the shaded portion of A_1 in Fig. 2–1(b) is equal to the total area of A_3. The hole currents are equal when the shaded area of A_2 is equal to the total area of A_4.

As indicated in Sec. 1.6, the Fermi level of N-type material is higher than that of P-type material. Under zero-bias conditions (no external voltage applied), these two levels must equalize after the junction is formed. The initial diffusion of electrons across the junction into the P-side, and also that of holes across the junction into the N-side, is essentially a process in which the Fermi levels of the two sides are equalized. The final result is shown in Fig. 2–1(b). Electrons diffusing into (and holes out of) P-type material raise the energy level of the P-side relative to the N-side. Therefore, the energy level of the conduction band on the P-side is raised, and the energy level on the N-side is lowered. The total relative displacement is equal to the magnitude of qV_B. This value is also the difference (in electron-volts) between the two Fermi levels before the junction was formed.

The value of the potential barrier, V_B, can be calculated using the fact that the total electron (or hole) current is zero in thermal equilibrium. From Eq. (1–26),

$$I_n = qA\mu_n \mathcal{E} n + qAD_n\left(-\frac{dn}{dx}\right) = 0 \tag{2–1}$$

The electric field can be found by equating the drift component and the diffusion component of current yielding

$$\mathcal{E} = \frac{D_n}{\mu_n}\frac{1}{n}\left(\frac{dn}{dx}\right) \tag{2–2}$$

The voltage associated with this electric field* is

$$V_B = \frac{D_n}{\mu_n} \ln \frac{n_{NO}}{n_{PO}} \tag{2–3}$$

The mobility and diffusion constants are related by the Einstein relation,

$$\frac{D_n}{\mu_n} = \frac{D_p}{\mu_p} = \frac{kT}{q} \tag{2–4}$$

where the factor kT/q is often referred to as the thermal voltage and has a value of approximately 26 millivolts at room temperature (300°K). The barrier potential can now be written as

$$V_B = \frac{kT}{q} \ln \frac{n_{NO}}{n_{PO}} = \frac{kT}{q} \ln \frac{p_{PO}}{p_{NO}} \tag{2–5}$$

where n_{NO} and n_{PO} are equilibrium densities of free electrons in the N- and P-type materials, and P_{PO} and P_{NO} are equilibrium hole densities in P- and N-type materials.

Very briefly, the equilibrium conditions at the junction of an unbiased diode may be summarized as follows:

1. Charged immobile impurity atoms are arrayed on each side of the junction with positive charges on the N-side and negative charges on the P-side.
2. A field exists between these unlike charges.

---

* From Eq. (2–2) and the definition of a voltage in terms of a field, the voltage associated with an electric field is

$$V = -\int \mathcal{E}\, dx = -\int_{n_{NO}}^{n_{PO}} \frac{D_n}{\mu_n}\left(\frac{dn}{n}\right)$$

where n_{NO} is the equilibrium electron density on the N-side, and n_{PO} is the equilibrium electron density on the P-side.

3. A voltage called the barrier potential is associated with this field. The magnitude of this voltage, under normal conditions, is approximately 0.3 volt for germanium and 0.7 volt for silicon.
4. Majority carriers (holes on the P-side and electrons on the N-side) can cross this barrier only if they have energies equal to or greater than approximately 0.3 eV for germanium and 0.7 eV for silicon. Many majority carriers diffuse across the junction at normal temperatures.
5. An equal number of minority carriers formed in or near the depletion region are drifting, under the influence of the electric field of the barrier, in the opposite direction.
6. When an electron (a majority carrier in the N-material) leaves the N-region and moves into the P-region, it becomes a minority carrier. An electron in P-type material does not travel far before it combines with a hole. In a similar manner, a hole moving from the P-side into the N-side quickly combines with an electron. The distance that minority carriers travel before the total number which crosses the junction is reduced to $1/\epsilon$ (or 1/2.71) of the original value is called the *diffusion distance*, L.
7. Regions beyond several diffusion distances from the junction are essentially undisturbed and may be treated simply as N-type or P-type materials.

EXAMPLE 2–1 Find the magnitude of the potential barrier for a silicon P-N junction in which the P-side is doped with one impurity atom for each 10^6 silicon atoms, and the N-side is doped with one impurity atom for each 5×10^7 silicon atoms. Assume that $T = 300°\text{K}$.

Solution: Assume that silicon has 4.96×10^{22} atoms/cm³ and that each impurity atom contributes one free carrier at 300°K. Then neglecting intrinsic densities, the equilibrium densities are

$$p_{PO} = \frac{4.96 \times 10^{22} \text{ silicon atoms/cm}^3}{10^6 \text{ silicon atoms/hole}} = 4.96 \times 10^{16} \text{ holes/cm}^3$$

$$n_{NO} = \frac{4.96 \times 10^{22} \text{ silicon atoms/cm}^3}{5 \times 10^7 \text{ silicon atoms/electron}} = 9.94 \times 10^{14} \text{ electrons/cm}^3$$

$$p_{NO} = \frac{n_i^2}{n_{NO}} = \frac{2.3 \times 10^{20}}{9.94 \times 10^{14}} = 2.32 \times 10^5 \text{ holes/cm}^3$$

and the magnitude of V_B is

$$V_B = \frac{kT}{q} \ln \frac{p_{PO}}{p_{NO}} = 0.026 \ln(2.14 \times 10^{11}) = 0.68 \text{ V}$$

As indicated above, the potential barrier in the unbiased diode will adjust the forward or diffusion current so that it is equal to the reverse or drift current. If the drift current changes for any reason, the magnitude of the barrier will change. For example, an increase in temperature will increase the number of minority carriers generated. Therefore, the reverse current will increase. The forward current component will also increase because more majority carriers have sufficient energy to cross the potential barrier. However, the reverse current increases* more than the forward current. To equalize the two currents, the potential barrier decreases approximately two millivolts per centigrade degree rise.

---

* The reverse current in a typical diode increases by a factor of two (doubles) for each 10°C increase in temperature.

2.2 Current Flow in a Biased Diode

As indicated in Sec. 2.1, the net current flowing across an unbiased P-N junction is zero because the barrier potential, V_B, equates the forward (diffusion) current and the reverse (drift) current. An external voltage may be applied to the diode to increase the barrier potential as shown in Fig. 2–2, or to decrease the barrier potential as shown in Fig. 2–3.

A reverse-bias condition exists when the externally applied voltage V_R increases the junction voltage to a value of $(V_B + V_R)$. For this condition, the forward current is reduced while the reverse current increases only slightly, if at all. The resulting changes in the current components may be explained in terms of the energy diagram shown in Fig. 2–2. At ambient temperature, energies of the free electrons (those in the conduction band) are distributed over an appreciable energy range as shown by the energy distribution function in the conduction band on the N-side of the junction. *Only those free electrons on the N-side with an energy level equal to or greater than that at the bottom of the conduction band on the P-side can participate in the flow of the electron-component of the forward current.** The relative number having the necessary energy to diffuse past the junction is indicated by the shaded portion of the distribution function, A_1. Of those electrons that have sufficient energy, only a small percentage actually cross the junction, those that are both near the junction and moving in the proper direction. Holes in the valence band behave in similar fashion except that *only holes on the P-side having energies as low as the top of the valence band on the N-side* (those in the shaded portion of A_2) *may cross the junction.* Raising the Fermi level on the P-side relative to that on the N-side by an amount qV_R with an external reverse bias reduces the number of electrons on the N-side and holes on the P-side that are available for forward current conduction and thereby reduces the I_F component. A reverse voltage of only a few tenths of a volt is usually sufficient to reduce I_F to nearly zero. The total current is then the reverse saturation current, I_S.

If the barrier voltage, V_B, is decreased with an external voltage, V_F, to a value of $(V_B - V_F)$, the junction is forward biased. As indicated by the shaded area of A_1 in Fig. 2–3, many more electrons in the conduction band can move from the N- to the P-side than from the P- to the N-side. A similar unbalance applies to holes moving from the P- to the N-side. Therefore, the forward current increases while the reverse current remains relatively constant. The net result is a forward current that increases with an increase in forward-bias voltage, V_F.

The net current $(I_F - I_S)$ is not a linear function of the forward-bias voltage, V_F. This nonlinear relationship may be understood by comparing Figs. 2–2 and 2–3. In both cases, the forward current is proportional to the shaded areas of A_1 and A_2, while the reverse current is proportional to the shaded areas (or total areas) of A_3 and A_4. Since almost all of the external voltage measured across the

---

* An electron remains at a given energy level unless it gains or loses energy. All of the electrons in the conduction band on the P-side have sufficient energy to move to the N-side regardless of the bias. However, only those electrons in the conduction band on the N-side that have energies as high as the shaded portion of A_1 can move to the P-side. Consequently, the electron component of I_R is theoretically independent of the bias while the electron component of I_F can be varied by changing the area of A_1 that is above the dashed line. Compare the shaded parts of A_1 in Figs. 2–2 and 2–3 for reverse and forward bias.

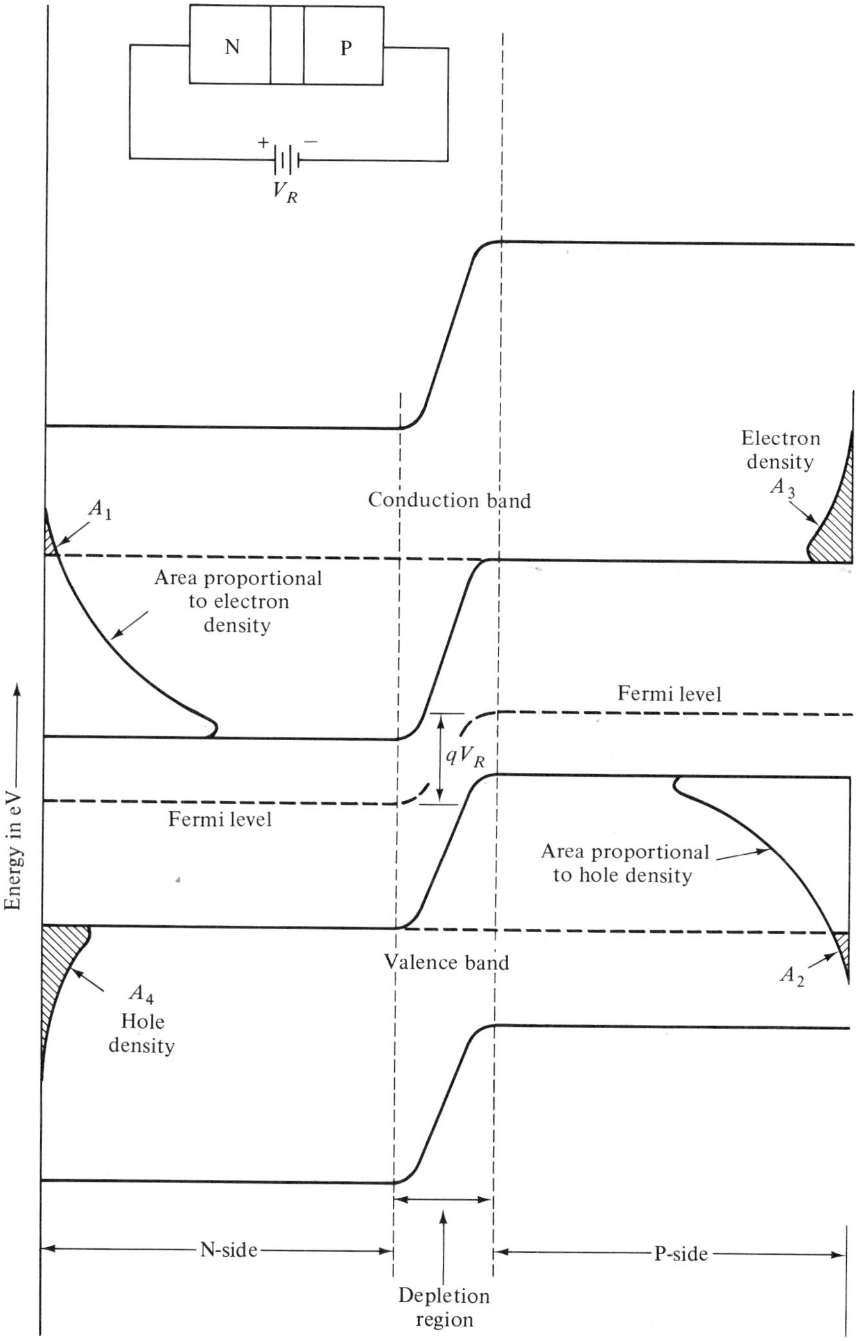

FIGURE 2–2 *A reverse-biased diode and the associated energy diagram. Forward current carriers are obtained at energy levels indicated by shaded areas* A_1 *(electrons) and* A_2 *(holes). Reverse current carriers are obtained at energy levels indicated by* A_3 *(electrons) and* A_4 *(holes).*

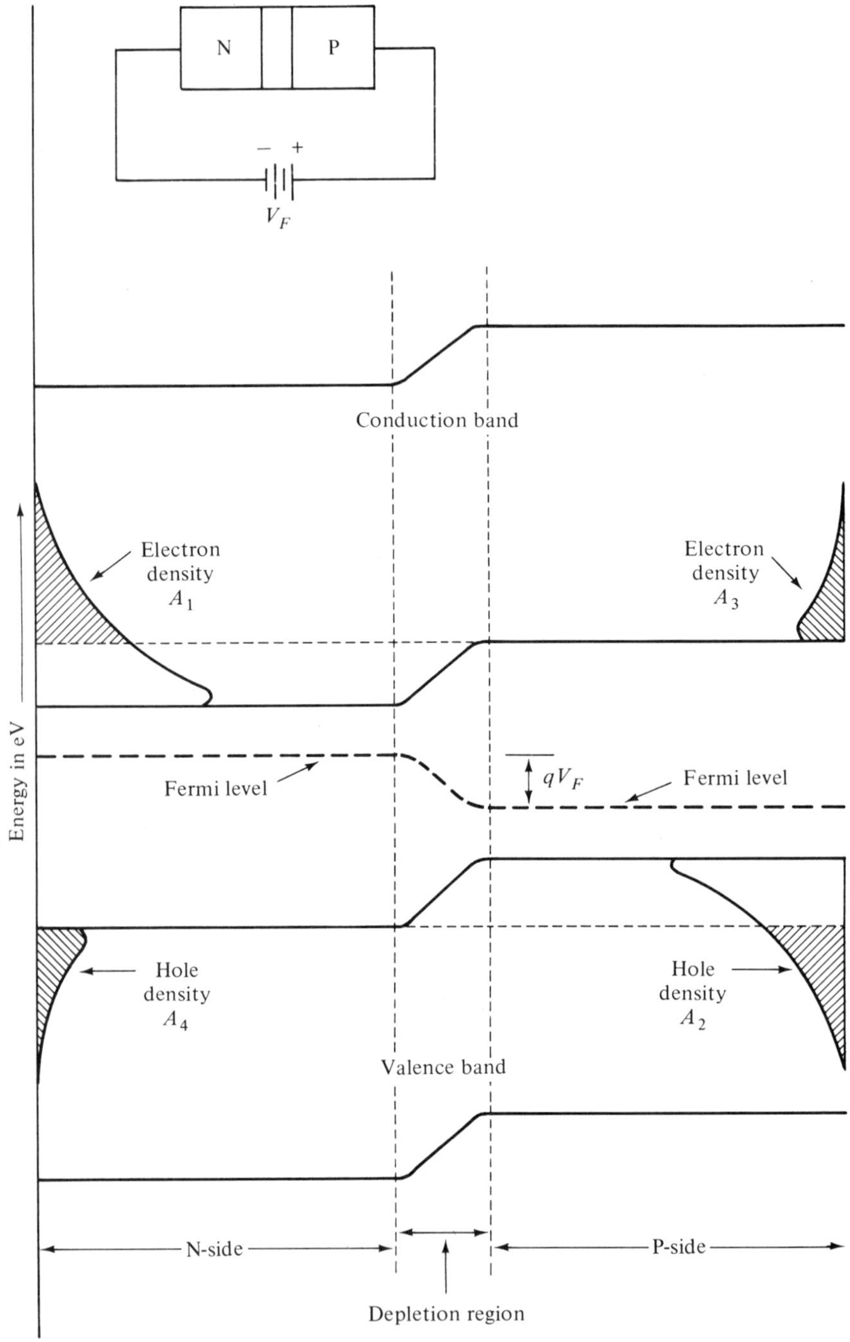

FIGURE 2–3 *A forward biased diode and the associated energy diagram. Forward current carriers are obtained at energy levels indicated by shaded areas* A_1 *(electrons) and* A_2 *(holes). Reverse current carriers are obtained at energy levels indicated by* A_3 *(electrons) and* A_4 *(holes).*

diode is developed across the junction (adding to V_B for reverse bias and subtracting from V_B for forward bias), the net voltage across the barrier is a linear function of the external bias. Therefore, a linear increase in the forward-bias voltage lowers the potential barrier linearly, and causes the forward current to increase exponentially.* Attempts to reduce the net barrier voltage to zero with a forward bias would result in very large forward currents which could easily produce sufficient heat (I^2R loss) in the diode to melt the material and destroy the junction.

An equation relating the external voltage across a diode to the net current flowing through the diode may be obtained by evaluating charge densities on each side of the depletion region and determining the magnitudes of the diffusion currents from the slopes of these densities as discussed in Sec. 1.7. The equilibrium densities of electrons (and holes) on each side of the junction are related by the barrier voltage, V_B, in Eq. (2–5). Solving for the electron density in the conduction band of the P-side (n_{PO}),

$$n_{PO} = n_{NO} \exp(-qV_B/kT) \qquad \textbf{(2–6)}$$

For the zero-bias condition, this equation is valid up to the edge of the depletion region on each side of the junction but is not valid inside the depletion region. The electron density in the conduction band on the N-side at the edge of the depletion region is the same as that at several diffusion lengths into the N-material. This equilibrium condition exists up to the edge of the depletion region because the number of electrons flowing across the junction in one direction is equal to the number flowing across the junction in the other direction during a given period. The same reasoning may be applied to the hole density on each side of the junction. These densities are shown in Fig. 2–4(a) for the zero-bias condition.

For any other bias condition, the number of electrons flowing in one direction is not equal to the number flowing in the other direction during the same period of time. As a result of this inequality, both electron and hole densities near the edge of the depletion region deviate from the equilibrium values. Using the junction as a reference, and letting W_P be the distance that the depletion region extends into the P-side, the electron density at the edge of the depletion region, n_{WP}, may be obtained by including the external-bias voltage in Eq. (2–6).**This density is

$$\begin{aligned} n_{WP} &= n_{NO} \exp[q(V - V_B)/kT] \\ &= n_{NO} \exp(-qV_B/kT) \exp(qV/kT) \\ &= n_{PO} \exp(qV/kT) \qquad \textbf{(2–7)} \end{aligned}$$

* Note that a rectangular distribution would be required to obtain a linear change in forward current with a linear change in junction voltage. Since the energy states in a band are not uniformly distributed (the density in the center of the band is greater than at the edges) and also since the carriers are not uniformly distributed (electron density is greatest near the bottom of the conduction band and holes near the top of the valence band), the carriers are distributed as shown in A_1, A_2, A_3, and A_4 in Figs. 2–2 and 2–3.

** This approach is valid only when the N and P regions are much wider than several diffusion lengths. For short N or P regions, the boundary conditions are different, causing the equations to be different.

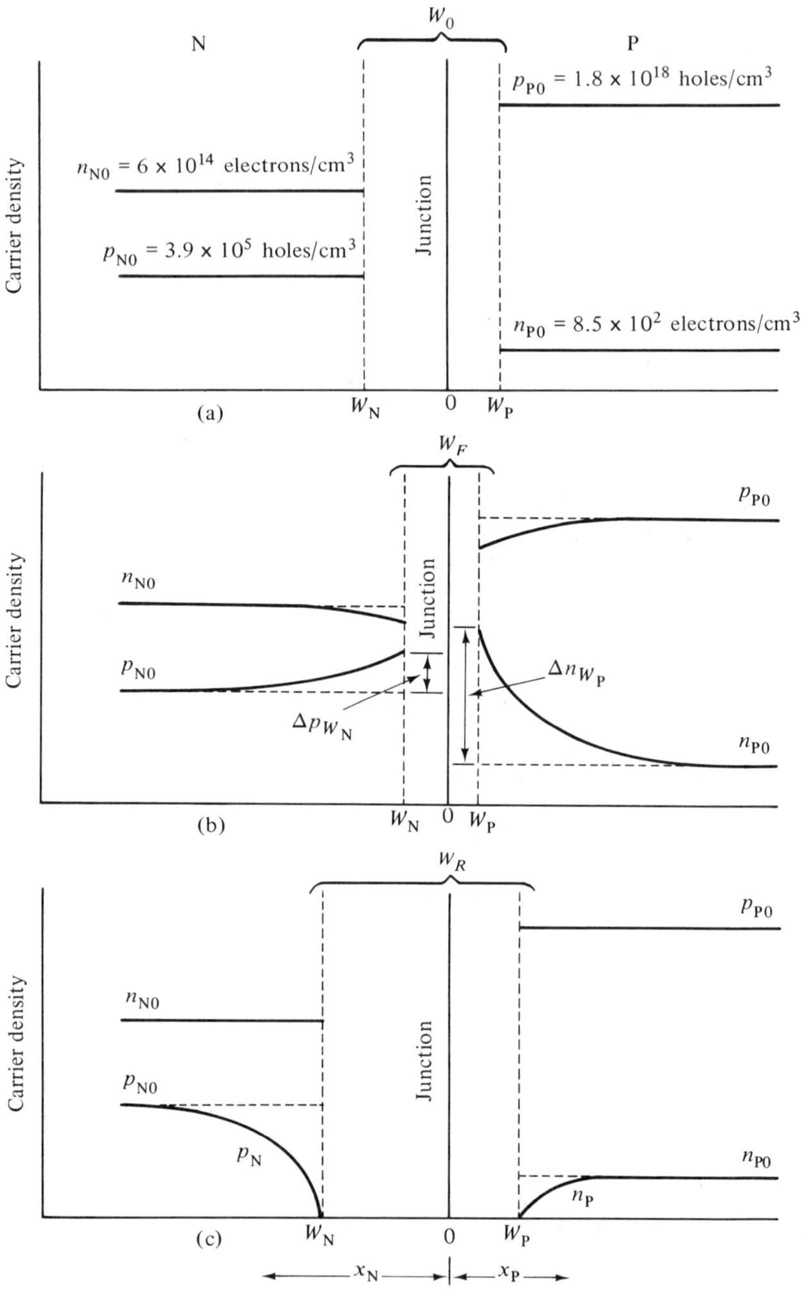

FIGURE 2–4 *Carrier densities in a silicon diode made with 0.05 ohm-cm, P-type material and 7.5 ohm-cm, N-type material. Densities near the junction are shown for three different bias conditions: (a) zero-bias, (b) forward-bias, and (c) reverse-bias. In all cases, equilibrium densities exist several diffusion lengths from the junction. Note how the width of the depletion region,* W, *changes for different bias conditions.*

where V is the external voltage across the diode and is positive for forward bias and negative for reverse bias. The electron density at the edge of the depletion region in excess of the equilibrium density, n_{PO}, is

$$\Delta n_{WP} = n_{WP} - n_{PO} = n_{PO}\,[\exp(qV/kT) - 1] \qquad \textbf{(2–8)}$$

A similar equation for the excess hole density on the N-side at the edge of the depletion region (a distance W_N from the junction) is

$$\Delta p_{WN} = p_{NO}\,[\exp(qV/kT) - 1] \qquad \textbf{(2–9)}$$

These excess densities are shown in Fig. 2–4(b) for a forward-bias voltage.

The excess electrons on the P-side diffuse into the P-material and combine with holes. The rate of recombination is proportional to the number of electrons remaining. This relationship means that the electron distribution decays on an exponential curve as a function of distance. Representing the diffusion length of electrons in P-type material by L_n, the excess electron density at any distance,* x_P, in the P-material is

$$\Delta n_{xP} = n_{PO}\,[\exp(qV/kT) - 1]\,\exp\,[(W_P - x_P)/L_n] \qquad \textbf{(2–10)}$$

and the excess hole density at any distance x_N in the N-material is**

$$\Delta p_{xN} = p_{NO}\,[\exp(qV/kT) - 1]\,\exp\,[W_N - x_N)/L_p] \qquad \textbf{(2–11)}$$

A nonuniform distribution of charge produces a diffusion current that is proportional to the slope of the charge density function. Consequently, the hole and electron densities near the junction of a forward-biased diode, Fig. 2–4(b), will produce diffusion currents. To determine the total diffusion current, note that this current is the sum of the hole current and the electron current at any point in the diode as shown in Fig. 2–5. Also, as indicated in this figure, the electron and hole components do not change across the depletion region. To determine the total diffusion current, we will find the electron component at the edge of the depletion region on the P-side and the hole component at the edge of the depletion region on the N-side. Since the two components do not change across the depletion region, I_T is the sum of these two components. Each component will be evaluated by finding the slope of the appropriate density at the edge of the depletion region and using this value in the diffusion equation, Eq. (1–25).

The initial slope of any decaying exponential having the form $A\exp(-x/B)$ is $-A/B$. In our case, the coefficient, A, is the excess carrier concentration (Δn_{WP} or Δp_{WN}), and the exponent constant, B, is the diffusion length, L. From Fig. 2–4(b), the slope of the hole concentration at the edge of the depletion region on the N-side is

$$(dp/dx)_{WN} = \Delta p_{WN}/L_p \qquad \textbf{(2–12)}$$

The magnitude of the slope of the electron concentration at the edge of the depletion region on the P-side is

$$|(dn/dx)_{WP}| = |\Delta n_{WP}/L_n| \qquad \textbf{(2–13)}$$

---

* The distances x_P and W_P are measured from the junction. Since the carrier density is very low in the depletion region, values of x_P less than W_P should not be considered.

** Note that both x_N and x_P are taken as positive when measured away from the junction.

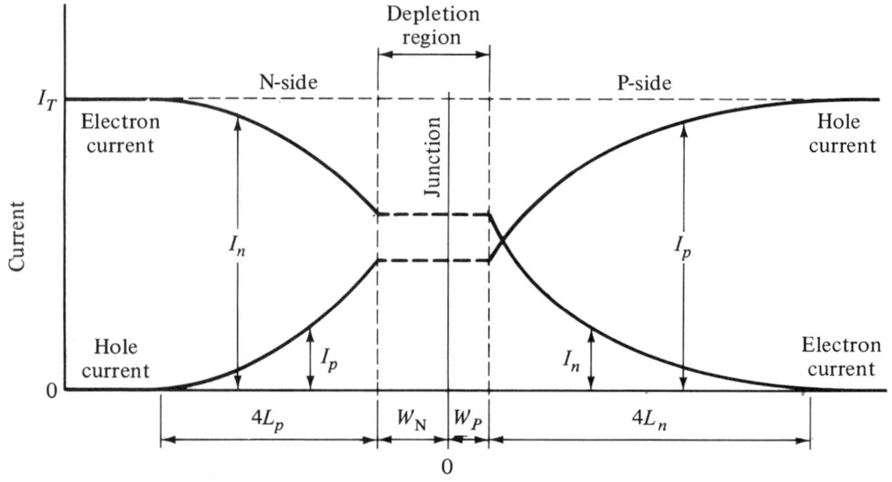

FIGURE 2–5 *Forward current components in a diode. Note that the total current,* I_T, *is constant and is equal to the sum of the hole current,* I_p, *and the electron current,* I_n, *at any point.*

A general expression for electron and hole diffusion currents is, from Eq. (1–25),

$$I_n = -qD_nA(dn/dx) \quad \text{and} \quad I_p = qD_pA(dp/dx) \qquad \textbf{(2–14)}$$

Since very few recombinations occur in the depletion region, the total current is the sum of the electron diffusion current on the P-side at the edge of the depletion region and the hole diffusion current on the N-side at the edge of the depletion region, as indicated above. Substituting Eqs. (2–12) and (2–13) into (2–14), the total current becomes

$$\begin{aligned} I_T &= I_{pWN} + I_{nWP} \\ &= \frac{qAD_pp_{NO}}{L_p}[\exp(qV/kT) - 1] + \frac{qAD_nn_{PO}}{L_n}[\exp(qV/kT) - 1] \\ &= qA\left[\frac{D_pp_{NO}}{L_p} + \frac{D_nn_{PO}}{L_n}\right][\exp(qV/kT) - 1] \end{aligned} \qquad \textbf{(2–15)}$$

The current components in Eq. (2–15) are shown in Fig. 2–5. Current continuity is established on both sides of the junction by electron-hole recombinations. Starting on the left side of this figure, current is carried almost entirely by electrons in the N-type material up to within a few hole-diffusion lengths, L_p, of the edge of the depletion region. From here to the edge of the depletion region, electron-hole recombinations reduce the number of free electrons, thereby causing the electron current to decrease exponentially. The hole component (holes from the P-side) increases by the same amount and keeps the total current, I_T, constant. In the P-side, due to more recombinations, the electron current component continues to decrease and the hole component continues to increase until the electron component decreases to a negligible value at a few* diffusion lengths, L_n, from the edge of the depletion region in the P-side. From

* The value of the electron component is less than 1 percent of its value at the edge of the depletion region on the P-side at a point $5L_n$ from the edge of the depletion region.

this point to the outside edge, current is carried almost entirely by holes. At the right-hand end of the diode, holes combine with electrons from the metallic contact so that only electrons flow in the external metal circuit.

For a reverse-bias condition, almost all of the holes on the N-side and electrons on the P-side (minority carriers produced by electron-hole pair formations in the semiconductor material) that are close to the junction diffuse into the depletion region and are swept across to the other side by the net barrier voltage. Thus, holes moving to the P-side add to the equilibrium value, p_{PO}, at the edge of the depletion region. A similar condition applies to electrons moving into the N-side. An order of magnitude comparison indicates that, although the minority density is reduced to almost zero at the edge of the depletion region, the effect on the majority carrier density on the other side is scarcely detectable because I_S is so small. Consequently, the majority carrier density is essentially unchanged from the equilibrium value as shown in Fig. 2–4(c). However, the reverse external bias causes the width of the depletion region to increase.

The reverse saturation current, I_S, may be obtained by adding electron and hole current components at the edge of the depletion region. However, Eq. (2–15) is valid for both forward- and reverse-bias conditions. Therefore, I_S may be obtained by making V a large negative number in this equation. The resulting equation for I_S is

$$I_S = qA\left[\frac{D_p p_{NO}}{L_p} + \frac{D_n n_{PO}}{L_n}\right] \qquad \textbf{(2–16)}$$

By combining Eqs. (2–15) and (2–16), the following well-known theoretical diode equation is obtained.

$$I = I_S\,[\exp{(qV/kT)} - 1] \qquad \textbf{(2–17)}$$

The curve generated by Eq. (2–17) is shown in Fig. 2–6. In practice, the forward characteristic of a diode is different from the predicted curve due to the voltage drop in the bulk material. While this difference is small at low-current

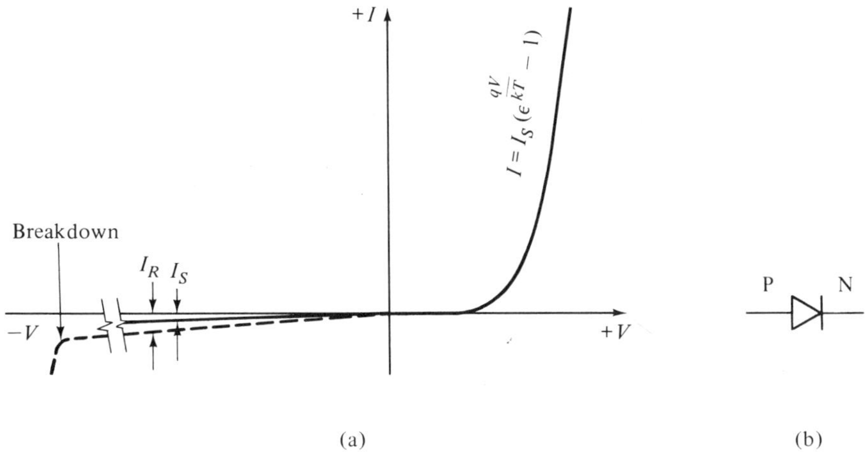

FIGURE 2–6 *(a) Volt-ampere characteristic curve for a P-N junction diode. The actual reverse curve is different from the predicted curve because of leakage current and breakdown effects. (b) The diode symbol. The arrow points in the direction of easy current flow.*

levels, the difference can become significant for high-current levels. The value of I_S is in the order of microamperes (10^{-6} A) for a germanium diode and nanoamperes (10^{-9} A) for a silicon diode. The circuit symbol for a diode is also shown in Fig. 2–6. The arrow points in the direction of forward current flow.

For reverse-bias conditions, three factors can cause the actual reverse current (I_R) to deviate from the theoretical value (I_S) as V_R is increased. All three of these effects combine to produce a slight slope on the reverse part of the volt-ampere curve as shown in Fig. 2–6. First, leakage current* at the surface of the junction increases as V_R increases. This effect has been reduced tremendously by improved manufacturing techniques (surface passivation). The second factor is the slight increase in the rate at which electron-hole pairs are thermally generated. This factor is associated with the increase in the power dissipated at the junction as V_R increases. The third factor is an increase in the rate of formation of electron-hole pairs that occurs in the depletion region as V_R is increased. While this last effect is small for values of V_R that are less than one volt, the field intensity in the depletion region increases to very large values for reasonable values of V_R because this region is very narrow. A high field accelerates the carriers as they move across the depletion region. Part of the resulting increase in energy ($\frac{1}{2}$ mv^2) is transferred to atoms in the depletion region when collisions occur. This additional energy may assist the thermal energy in producing an electron-hole pair. Field intensities on the order of 10^5 to 10^6 volts/cm can cause avalanche multiplication or breakdown. An attempt to increase V_R beyond the breakdown voltage will cause a very large increase in the reverse current for a very small increase in V_R. Breakdown effects are treated in Sec. 2.7.

EXAMPLE 2–2 Find the current flowing through a germanium diode when a forward bias of 0.2 V is applied across the diode. Repeat for a reverse bias of 0.2 V. Assume that $I_S = 5\ \mu$A.

Solution: The value of q/kT is approximately 39 at room temperature. For the forward-bias condition:

$$I_F = I_S[\exp(39\ V) - 1] = 5 \times 10^{-6}[\exp(7.8) - 1] = 12.2 \text{ mA}$$

For the reverse-bias condition:

$$I_R = 5 \times 10^{-6}[\exp(-0.2 \times 39) - 1] = -5\ \mu\text{A}$$

By calculating the ratio I_F/I_R, we find that approximately 2440 times more current flows for a forward bias of 0.2 V than for a reverse bias of 0.2 V.

2.3 Diode Rectifier Circuits

The ideal volt-ampere characteristic of a junction diode was developed in the preceding section and is shown in Fig. 2–6. The actual characteristics of a germanium and also a silicon diode are shown in Fig. 2–7 for comparison. From

* Leakage currents flow around the junction rather than through the junction.

Fig. 2–7 we observe that the reverse saturation current of a typical diode is very small and that the forward voltage drop is on the order of one volt, even at high currents. If the circuit voltages and currents are much larger than these values, the diode may be treated as a simple switch that is open if the diode is reverse biased and is closed if the diode is forward biased. One very important application of semiconductor diodes that falls into this category is that of a rectifier in a dc power supply. In this application the primary function of the diode is to convert ac into dc by allowing current to flow in only one direction.

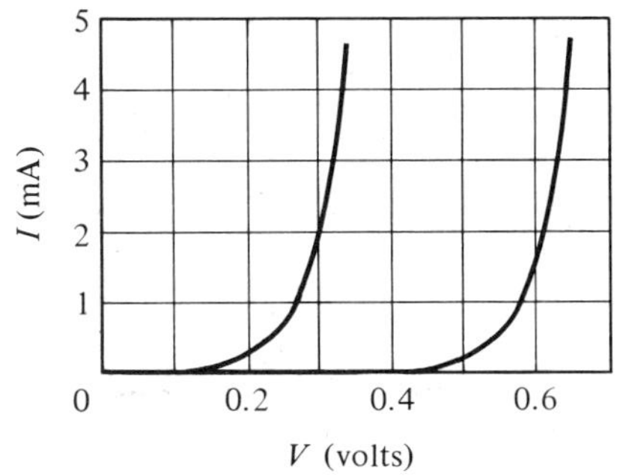

FIGURE 2–7 *The volt-ampere characteristic of a germanium and a silicon diode. Typical values of cut-in voltage are 0.1 V for germanium and 0.5 V for silicon. The voltage drop approaches 1 V at high current levels.*

Three principal rectifier circuits are useful in converting single-phase ac to dc: (1) the half-wave rectifier, (2) the full-wave rectifier, and (3) the bridge rectifier. Polyphase ac inputs are used in some industrial applications which require large amounts of dc current.

One of the simplest forms of a rectifier circuit is a half-wave rectifier as shown in Fig. 2–8(a). When the input signal is positive, the diode is approximately a

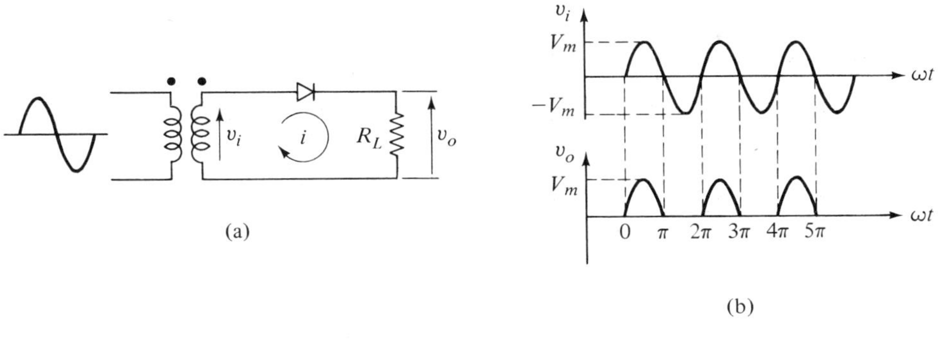

FIGURE 2–8 *A half-wave rectifier with resistive load: (a) circuit diagram and (b) input and output voltage waveforms.*

short circuit, and all of the input voltage appears across R_L. On the negative swing, the diode is approximately an open circuit, and the circuit current is approximately zero. During this part of the cycle, the input voltage is dropped across the diode.

The performance characteristics of this circuit may be obtained as follows. Assuming an ideal diode and a sine wave input, v_i, with a maximum value of V_m, the instantaneous circuit current in Fig. 2–8(a) is given by

$$i = \begin{cases} \dfrac{V_m}{R_L} \sin \omega t, & 0 \leq \omega t \leq \pi \\ 0, & \pi \leq \omega t \leq 2\pi \end{cases} \tag{2–18}$$

This equation indicates that current flows for the positive half of the input sine wave and has a maximum value of V_m/R_L. During the negative portion of the input wave, the current is zero. The dc, or average, component of the current is the area under the current waveform divided by the period of the input wave, or

$$I_{dc} = \frac{1}{2\pi} \int_{\omega t=0}^{\omega t=2\pi} i \, d\omega t = \frac{V_m}{R_L} \times \frac{1}{\pi} = \frac{I_m}{\pi} \tag{2–19}$$

The dc component of the output voltage is

$$V_{dc} = I_{dc} R_L = \frac{V_m}{\pi} \tag{2–20}$$

and the ac or rms* component of the output current is

$$I_{rms} = \sqrt{\frac{1}{2\pi} \int_{\omega t=0}^{\omega t=2\pi} i^2 \, d\omega t} = \frac{I_m}{2} \tag{2–21}$$

One important rating on a diode is the maximum reverse voltage that it can withstand without breaking down. The *peak-inverse-voltage* (PIV) applied across a diode in a half-wave rectifier occurs when the input is at its maximum negative value, $-V_m$. Consequently, the breakdown voltage rating for the diode must be greater than the peak inverse voltage in the circuit.

The ideal dc power supply has no ac component in the output; only the dc component is present in the load. In this respect, the circuit in Fig. 2–8 is far from ideal. A measure of the purity of the rectifier dc output is the *ripple factor* (R.F.), which is defined as

$$\text{R.F.} = \frac{\text{rms value of the ac harmonic content}}{\text{average or dc value}} = \frac{\sqrt{I^2_{rms} - I^2_{dc}}}{I_{dc}} = \frac{\sqrt{V^2_{rms} - V^2_{dc}}}{V_{dc}} \tag{2–22}$$

The ripple factor of the half-wave rectifier may be found by substituting Eqs. (2–19) and (2–21) into (2–22).** This value is 1.21, or the ac ripple is 1.21 times larger than the dc component.

---

* RMS means root mean square. To obtain the rms value, square the wave, find the mean value, and then find the square root of the mean value.

** Many ac meters are not true rms meters. Most ac meters measure some property of the waveform, such as its peak value, and are calibrated in terms of the rms value of a sine wave. This approach gives an accurate reading only when the wave is sinusoidal. Consequently, with the exception of the true rms meter, the reading of an ac meter may not give a correct rms value. This complicates the experimental determination of the ripple factor. However, an order of magnitude comparison can be made between the experimental and theoretical values of a ripple factor. The experimental values are obtained by measuring the ac and dc components of the load voltage (or current) with a standard laboratory meter and finding the ratio of V_{ac}/V_{dc}. This procedure is illustrated in Ex. 2–4.

The ac ripple factor in the output of a half-wave rectifier circuit is too large for most applications. However, the magnitude of the ripple may be reduced considerably by adding a filter circuit. One simple filter circuit that is often used in practice is the shunt-capacitor filter shown in Fig. 2–9 (a). The capacitor accumulates charge during the period that the diode conducts and discharges through R while the diode is nonconducting. The charging period occurs only when the input voltage is more positive than the capacitor voltage. When the diode is nonconducting, the voltage across C decays on an exponential curve that is determined by the RC product. An analysis* of this circuit yields a curve of V_{dc}/V_m as a function of ωRC. These relationships are shown in Fig. 2–10. The effect of C on the ripple factor is shown in Fig. 2–11. The effect of a shunt capacitance filter is shown in the following example.

EXAMPLE 2–3 The output of the rectifier circuit in Fig. 2–9(a) without a capacitance filter is a 60-Hz, half-rectified sine wave, with a peak value of 4 V, as shown in Fig. 2–9(b). If R_L is 270 ohms, calculate the dc component and the ripple factor of the output voltage when $C = 82\ \mu\text{F}$ and also when $C = 460\ \mu\text{F}$.

Solution: The ratio V_{dc}/V_m is available from Fig. 2–10, and the ripple factor is available from Fig. 2–11. The value for ω is $2\pi f$ or 377 for a 60-Hz wave. These calculations indicate that an increase in capacitance increases the dc component and decreases the ripple in the output.

(a) For $C = 82\ \mu\text{F}$,

$$\omega R_L C = (377)(270)(82 \times 10^{-6}) \simeq 8.4$$

For this value on the half-wave curve,

$$V_{dc}/V_m \simeq 0.78$$
$$V_{dc} \simeq (0.78)(4) = 3.1\ \text{V}$$
$$\text{R.F.} \simeq 0.2$$

(b) For $C = 460\ \mu\text{F}$,

$$\omega R_L C = (377)(270)(460 \times 10^{-6}) \simeq 47$$
$$V_{dc} \simeq (0.95)(4) = 3.8\ \text{V}$$
$$\text{R.F.} \simeq 0.04$$

One simple method of reducing the ripple factor in the output waveform of a rectifier is to provide an output during both the positive and negative portions of the input signal. The full-wave rectifier accomplishes this using a center-tapped transformer as shown in Fig. 2–12. When the input signal is positive, D_1 conducts, and D_2 is reverse biased. During the negative portion of the input signal, the polarity of v_i is reversed. Therefore, D_2 conducts, and D_1 is reverse biased. The direction of the current through R_L is the same in both cases.

The performance characteristics of the full-wave rectifier without a filter in the output can be determined in a manner similar to that used for the half-wave rectifier. These relationships are:

$$\left.\begin{aligned} I_{rms} &= I_m/\sqrt{2} \qquad & \text{R.F.} &= 0.48 \\ I_{dc} &= 2I_m/\pi \qquad & \text{PIV} &= 2V_m \end{aligned}\right\} \tag{2–23}$$

---

* For derivation see Ryder, pp. 118–121, in the references at the end of the chapter. In this case, V_m is the maximum voltage at the input to the filter.

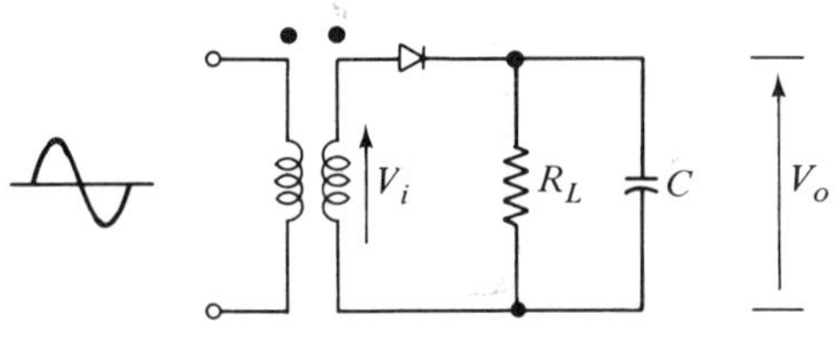

(a)

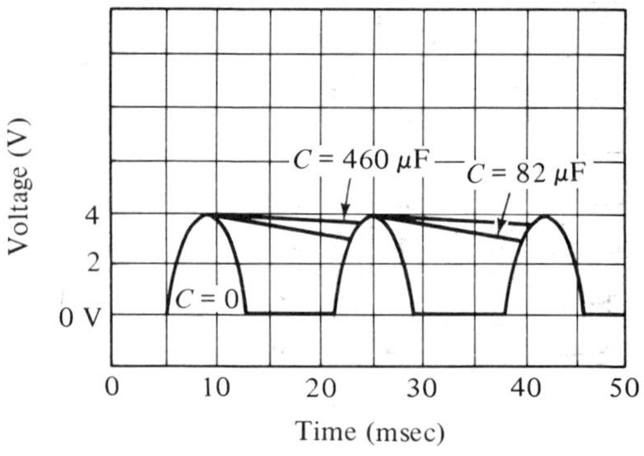

(b)

FIGURE 2–9 *A half-wave rectifier with a shunt capacitance filter from Ex. 2–3: (a) circuit diagram and (b) waveforms.*

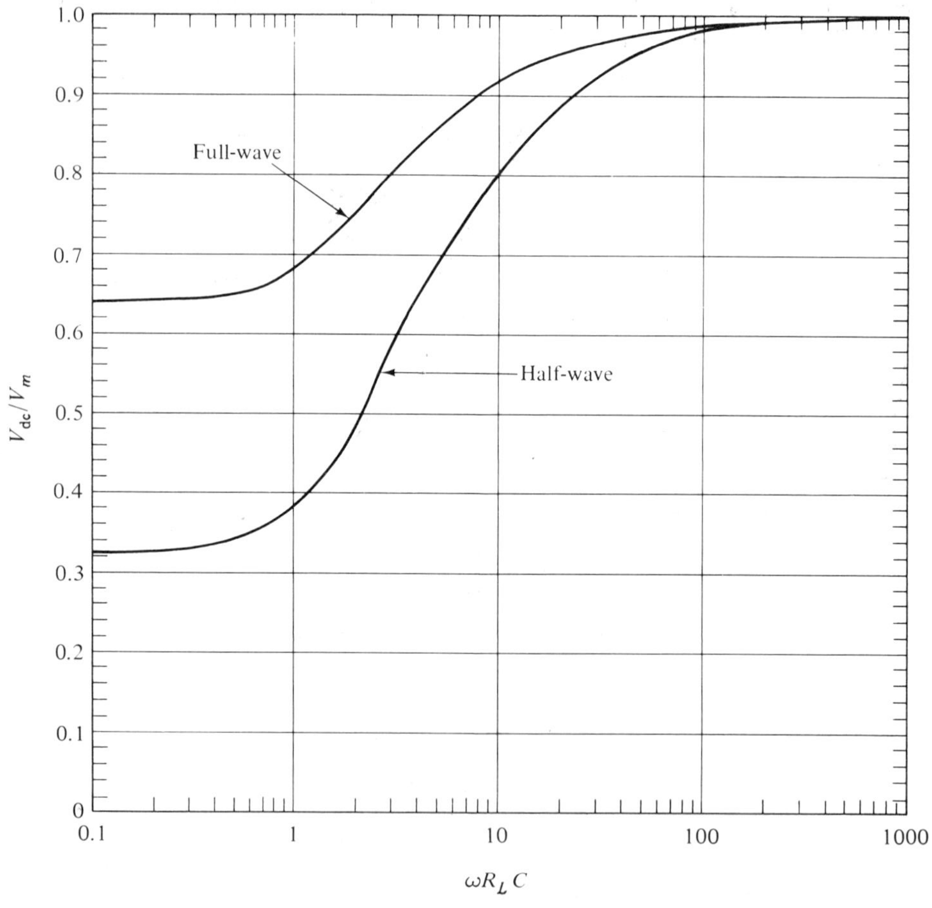

FIGURE 2–10 *The ratio* V_{dc}/V_m *as a function of* $\omega R_L C$ *for a half-wave and a full-wave rectifier with a shunt capacitor filter. The resistance is in ohms, and the capacitance is in farads.*

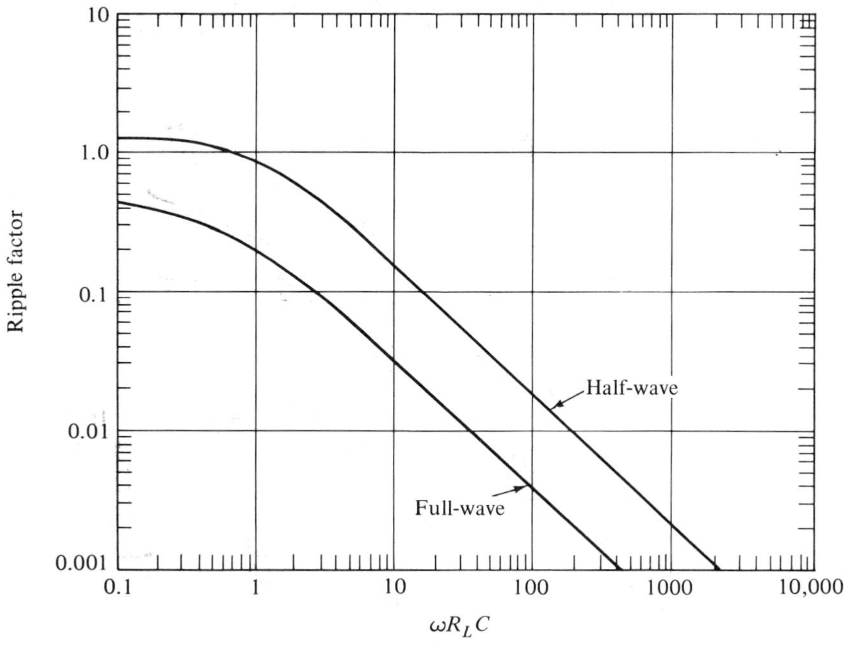

FIGURE 2–11 *Ripple factor as a function of* $\omega R_L C$ *for a half-wave and a full-wave rectifier with a shunt capacitor filter.*

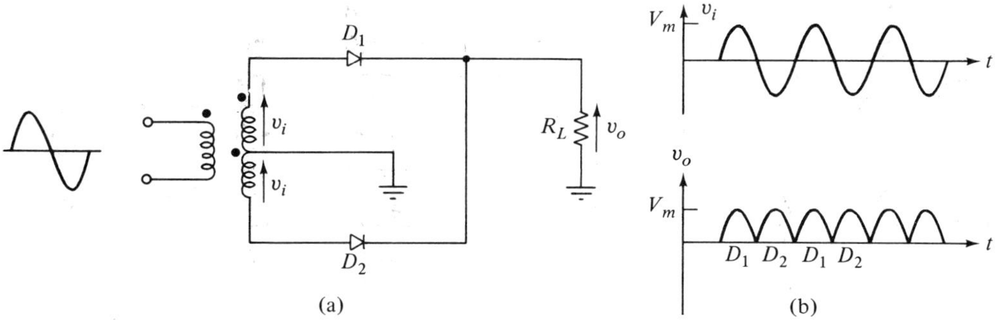

FIGURE 2–12 *Full-wave rectifier circuit: (a) circuit diagram and (b) waveforms.*

A shunt capacitance filter is added by placing a capacitor in parallel with R_L. The effect of the filter on the performance is shown in Figs. 2–10 and 2–11. Note that for a given value of $\omega R_L C$, the full-wave rectifier has less ripple in the output than the half-wave rectifier.

A second type of full-wave rectifier is the bridge type shown in Fig. 2–13. This circuit eliminates the need for a center-tapped transformer and is the most common type of rectifier circuit.

The operation of the bridge rectifier at specific points on the ac cycle is shown in Fig. 2–14.* Let v_i be positive and equal to V_m. At this point on the input

* For a bridge rectifier circuit, avoid grounding the transformer secondary and one side of the load resistor simultaneously because this will short circuit one of the diodes.

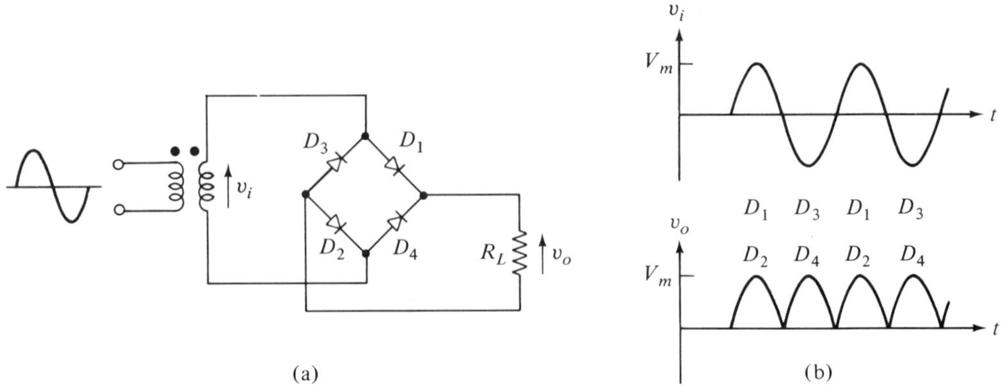

FIGURE 2–13 *A bridge rectifier circuit* (a) *circuit diagram and* (b) *the output waveform.*

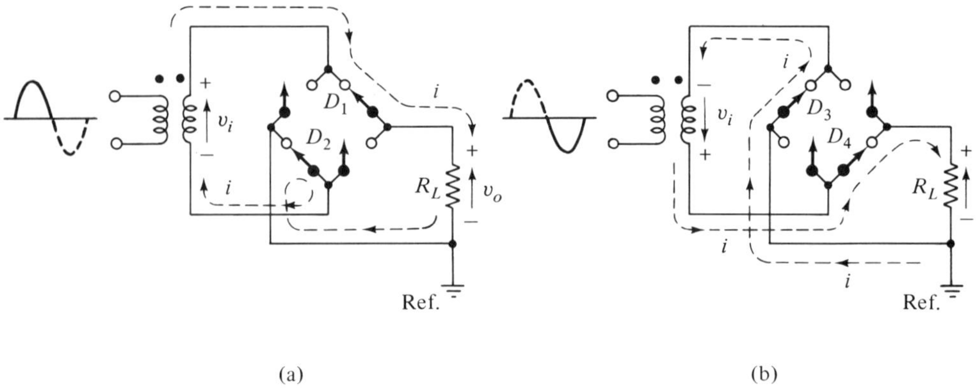

FIGURE 2–14 *Circuit voltages and currents for a bridge rectifier when* (a) v_i *equals* $+V_m$ (b) v_i *equals* $-V_m$.

waveform, as shown in Fig. 2–14(a), the output voltage equals V_m (assuming ideal diodes), and D_1 and D_2 are forward biased. The circuit current flows through the transformer, D_1, R_L, D_2, and back through the transformer. Both D_3 and D_4 are reverse biased by a voltage equal to V_m. When v_i is negative and is equal to $-V_m$, D_3 and D_4 conduct as shown in Fig. 2–14(b).

The values of I_{dc}, I_{rms}, and the ripple factor for the bridge rectifier are the same as for the full-wave rectifier [see Eq. (2–23)]. The value of the PIV, however, is only V_m.

EXAMPLE 2–4 A bridge rectifier circuit is shown in Fig. 2–15(a). The primary-to-secondary voltage rating of the transformer is 110 V/26.5 V, and the secondary current rating is 0.6 A. Calculate the dc output voltage and the ripple factor.

Solution: The dc voltage can be found from Fig. 2–10, and the ripple factor can be found from Fig. 2–11. Since the transformer is connected directly to a power line, the source resistance, R_s, is negligible. The forward voltage drop across the diodes, usually less than one volt, will be considered for greater accuracy.

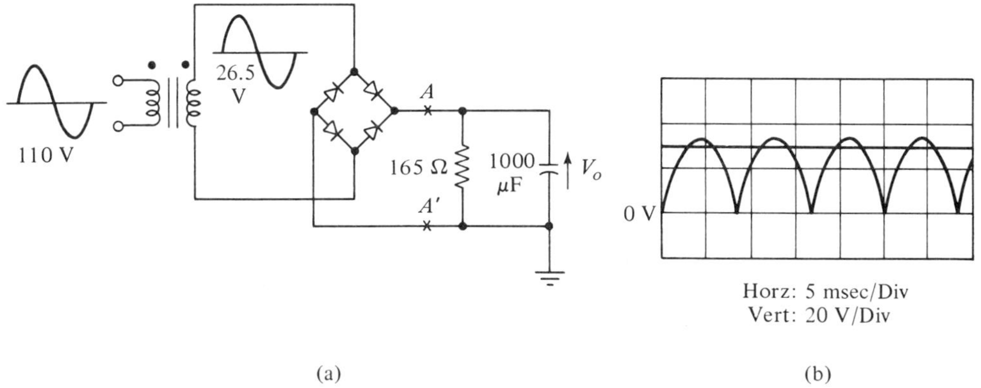

FIGURE 2–15 *A bridge rectifier circuit with a shunt capacitance filter for Ex. 2–4: (a) actual circuit and (b) output waveforms.*

For the filter shown in Fig. 2–15(a),

$$\omega R_L C = (377)(165)(1000 \times 10^{-6}) \simeq 62$$

$$V_{dc}/V_m \simeq 0.98 \qquad \text{R.F.} \simeq 0.006$$

The maximum value of the transformer secondary voltage is

$$V_m \simeq (26.5)(\sqrt{2}) \simeq 37.5 \text{ V}$$

Assuming a 1-V drop across each conducting diode, the maximum value of voltage across R_L is approximately 35.5 V. Consequently,

$$V_{dc} \simeq (0.98)(35.5) \simeq 35 \text{ V}$$

The oscilloscope waveform of the output voltage of the bridge rectifier with and without the filter capacitor is shown in Fig. 2–15(b). With the filter capacitor connected, the dc meter reading was 35.2 V, and an ac meter reading was 0.165 V. The ac meter was not a true rms meter. Consequently, the ripple factor can only be approximated as

$$\text{R.F.} \simeq \frac{0.165}{35.2} \simeq 0.0047$$

which is reasonably close to the calculated value.

A few of the many filter circuits other than the shunt capacitor filter are shown in Fig. 2–16. The *L*- and π- section filters are used when a very low ripple factor

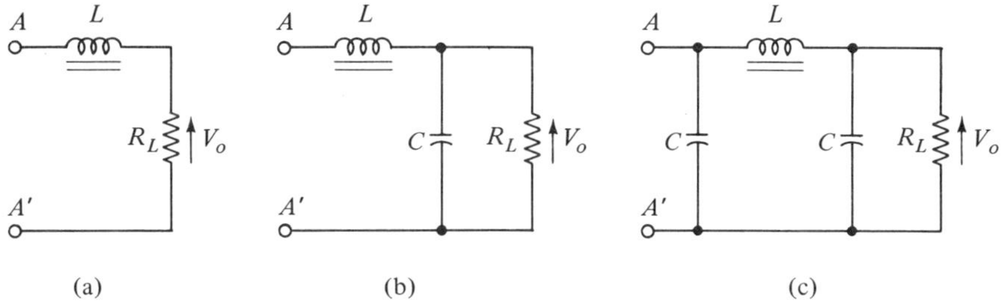

FIGURE 2–16 *Various rectifier filter circuits: (a) series inductor, (b)* L-*section, and (c) π-section.*

is required. Often the inductor is replaced by a power resistor having a low value, usually less than 10 ohms.

2.4 Junction Diode Parameters

In the preceding section, the diode was treated as a switch because the circuit currents and voltages were large. In that case, the deviations of the diode characteristics from those of an ideal switch could be neglected without serious error. However, in many applications, the currents and voltages are small, making the characteristics of a diode very important. The purpose of this section is to discuss the parameters of a diode in great detail so that it can be treated analytically in small-signal and high-frequency circuits. These parameters include the cut-in voltage, forward and reverse resistance, junction capacitance, and diffusion capacitance. Not all of these parameters are useful in any one problem. However, the circuit designer must have the flexibility of being able to select those that are pertinent.

One important parameter of a diode is the *cut-in voltage*, V_γ. This parameter is the forward-bias voltage that causes a significant forward current to flow in the diode (microampere range). Values of V_γ range from 0.05 to 0.15 V for a germanium diode and from 0.4 to 0.55 V for a silicon diode. Typical values of cut-in voltage are 0.1 V for germanium and 0.5 V for silicon. Also of importance is the typical value of forward bias which produces a diode forward current in the milliampere range. This value is called V_A, the active region voltage. In future calculations, the value of V_A will be assumed as 0.2 V for germanium and 0.6 V for silicon. These values are also very useful in bias calculations for bipolar transistors. However, the exact value of forward bias required for a given forward current depends on the particular diode. The value of cut-in voltage is also temperature dependent, a typical change being -2 mV/C°. To maintain a given current, the bias voltage must be reduced approximately 2 mV for each 1°C rise in temperature.

When used as circuit elements in the low-frequency range, diodes are nonlinear resistors. In circuit work, four resistance parameters may be useful: (1) a forward static resistance, (2) a forward dynamic resistance, (3) a reverse static resistance, and (4) a reverse dynamic resistance. The static values are useful in dc problems, and the dynamic values are useful in circuits with low-level ac signals. These resistance parameters are discussed below in the order listed.

The forward static or dc resistance may be determined graphically by finding the inverse slope (V_1/I_1) of a line drawn from the origin to a point on the curve corresponding to the dc current through the diode as shown in Fig. 2–17. The forward dynamic or ac resistance may be determined graphically by finding the inverse slope of a line drawn tangent to the curve at the operating point, Q. Both the ac and the dc forward resistances are rather high for low values of current, and decrease in value as the dc current increases.

If the effect of the bulk resistance of a diode is neglected, both the ac and dc resistance values at a given current may be obtained analytically. As indicated above, the forward dynamic resistance may be found from the inverse slope of the characteristic curve for the diode and is

$$r_f = dV/dI \qquad \textbf{(2–24)}$$

The dynamic resistance is obtained by taking the derivative of Eq. (2–17) and inverting the results. This derivative is

$$\frac{dI}{dV} = \frac{I_S q}{kT}[\exp(qV/kT)] = \frac{q(I + I_S)}{kT} \quad \textbf{(2–25)}$$

Therefore, the forward dynamic resistance is

$$r_f = \frac{kT}{q(I + I_S)} \simeq \frac{kT}{qI} \quad \textbf{(2–26)}$$

Since k/q is a constant, kT/q is a constant at a given temperature. At room temperature (300°K), kT/q is approximately 26 mV. Assuming room temperature and assuming that I is 1 mA of dc current, the dynamic resistance is approximately 26 ohms. If the dc current increases to 2 mA, the ac resistance decreases to 13 ohms. This simple relationship is very useful, not only in working with diodes but also with bipolar transistors.

In practice, the dynamic resistance is slightly higher than the value predicted by Eq. (2–26) because the bulk resistance of the P- and N-materials must be added to r_f. This deviation is usually so slight in low-current diodes that it can be neglected, especially in the case of a diode made from low-resistivity or heavily doped materials.

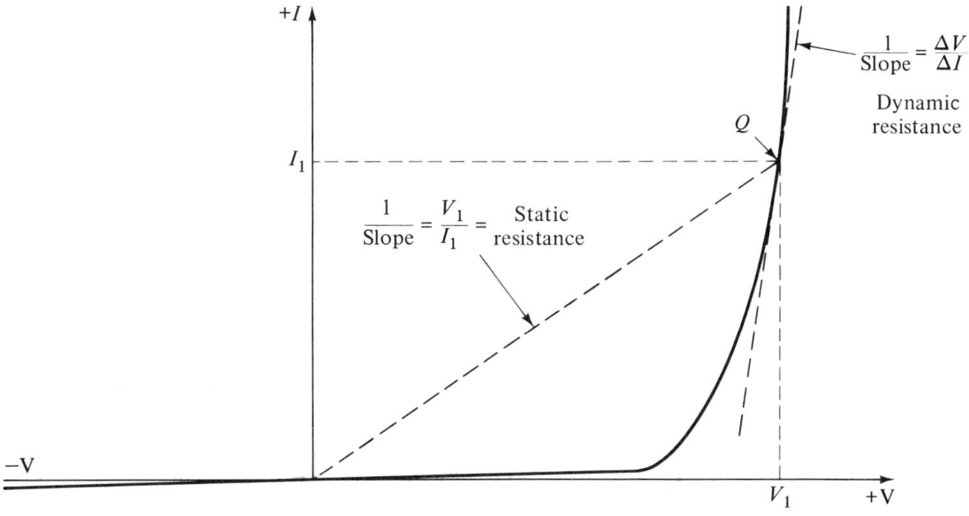

FIGURE 2–17 *Graphical determination of the static and dynamic forward resistance of a diode at a given operating point,* Q.

EXAMPLE 2–5 Find the static and dynamic forward resistances of the diode in Ex. 2–2.

Solution: A dc current of 12.2 mA flowed in the diode with 0.2 V forward bias. The forward static resistance, R_F, is the ratio of the total voltage to the total current. In this case,

$$R_F = \frac{200 \times 10^{-3}}{12.2 \times 10^{-3}} = 16.4\,\Omega$$

The dynamic resistance, as found from Eq. (2–26), is

$$r_f = \frac{26 \text{ mV}}{12.2 \text{ mA}} \simeq 2.1 \ \Omega$$

Diodes are frequently used to isolate circuits from voltage waveforms having a given polarity or exceeding a given magnitude. Under these conditions, the reverse resistance determines the degree of isolation that can be obtained. Since the reverse-biased diode is essentially a constant-current source, the dc reverse resistance, R_R, will be a function of the voltage across the diode, or

$$R_R = V/I_R \qquad \textbf{(2–27)}$$

The ac resistance of a reverse-biased diode, r_r, is rather high and can be determined graphically by finding the inverse slope of the characteristics curve at the operating voltage.

EXAMPLE 2–6 Find the values of the ac and dc reverse resistances for a reverse bias of 5 and 10 V respectively. The reverse characteristic curve is shown in Fig. 2–18.

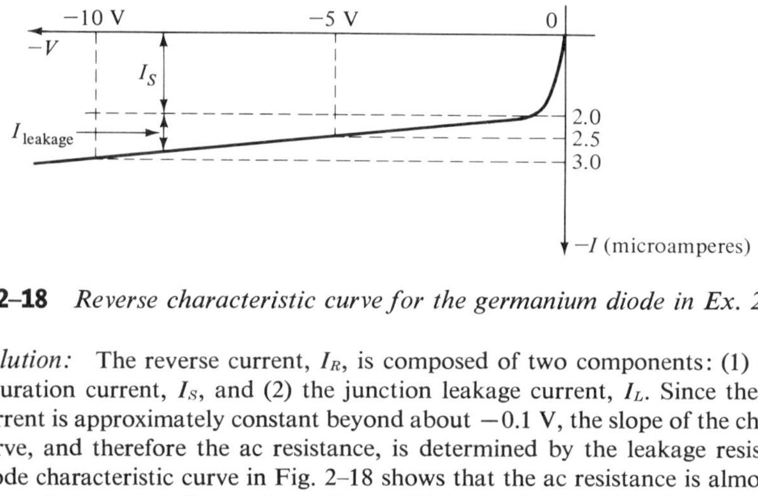

FIGURE 2–18 *Reverse characteristic curve for the germanium diode in Ex. 2–6.*

Solution: The reverse current, I_R, is composed of two components: (1) the reverse saturation current, I_S, and (2) the junction leakage current, I_L. Since the saturation current is approximately constant beyond about −0.1 V, the slope of the characteristic curve, and therefore the ac resistance, is determined by the leakage resistance. The diode characteristic curve in Fig. 2–18 shows that the ac resistance is almost constant over a large range of operating voltages. The ac resistance at each operating point is the inverse slope of the curve at that point. Since the slope is approximately constant, the ac reverse resistance is

$$r_r \simeq \frac{\Delta V}{\Delta I} = \frac{5}{0.5 \times 10^{-6}} = 10 \times 10^6 \ \Omega$$

The dc resistance for a bias of −5 V is

$$R_R = 5/(2.5 \times 10^{-6}) = 2 \times 10^6 \ \Omega$$

and at −10 V the dc resistance is

$$R_R = 10/(3 \times 10^{-6}) = 3.33 \times 10^6 \ \Omega$$

Now that the low-frequency parameters have been developed, several diode equivalent circuits can be structured that can be used to replace the diode in an equivalent circuit. The choice of a diode equivalent circuit depends upon the particular application of the diode being replaced. Three common low-frequency equivalent circuits are shown in Fig. 2–19. An ideal diode would be a perfect switch with a zero cut-in voltage, a zero forward resistance, and an infinite reverse resistance. This approximation [shown in Fig. 2–19(a)] is often useful in initial or rough calculations for a diode circuit. As indicated in the preceding section, the ideal switch approximation is also useful in rectifier circuits which have large voltage swings.

The characteristics of an actual diode deviate from those of an ideal diode in several respects. The reverse resistance ranges from less than 50 kilohms for a poor diode to more than 100 megohms for a good silicon diode. The forward resistance varies from several hundred ohms for very low-current operation to a fraction of an ohm for high-current operation. The dynamic resistance is included in the equivalent circuit shown in Fig. 2–19(b). This form of the equivalent circuit is useful in a *small-signal analysis*. This type of analysis is used extensively in circuits where the signals of importance are low-level ac signals.

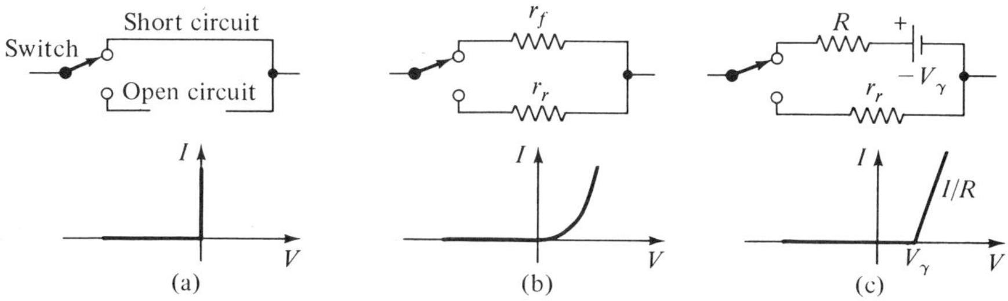

FIGURE 2–19 *Simple low-frequency equivalent circuits for a diode and the associated characteristics: (a) ideal diode, (b) actual diode, and (c) piece-wise linear model.*

An alternate equivalent circuit that has a *V-I* characteristic which approximates an actual diode is shown in Fig. 2–19(c). This circuit is a piece-wise linear model in which the forward bias conditions are represented by the cut-in voltage and a fixed resistor *R* having a value equal to the inverse of the average slope of the forward characteristic. The reverse characteristic is approximated by either the reverse dynamic resistance or an open circuit. Since both dc and ac parameters are incorporated into this model, it is useful when the input is either a dc level or an ac voltage which swings over a wide range.

EXAMPLE 2–7 For the circuit shown in Fig. 2–20(a) find (a) the circuit current

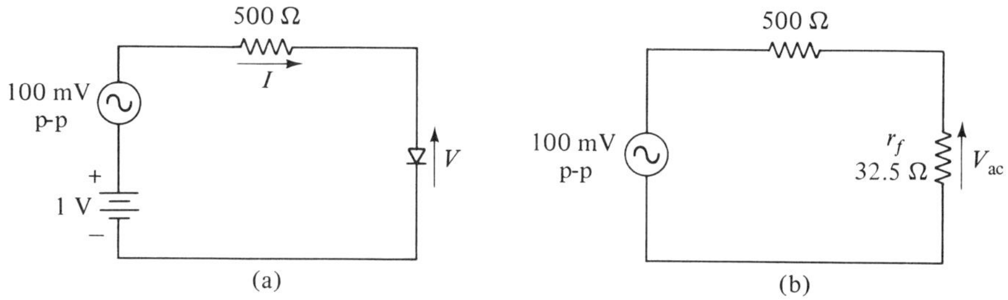

FIGURE 2–20 *The simple diode circuit discussed in Ex. 2–7: (a) actual circuit and (b) small-signal equivalent circuit.*

and (b) the peak-to-peak ac voltage across the silicon diode.

Solution: Divide the analysis into two parts. First, find the dc or bias conditions in the circuit using the bias voltage. Second, replace the diode by its ac equivalent circuit (find r_f at the dc current), and then solve for the diode ac voltage drop. This division of the analysis into a bias part and an ac part is common practice and forms the basis of small-signal analysis. This approach is used extensively in transistor circuits.

(a) The dc current for a 1-V bias is

$$I \simeq \frac{V_{\text{dc}} - V_A}{R} \simeq \frac{1 - 0.6}{500} \simeq 0.8 \text{ mA}$$

(b) To find the ac voltage across the diode, first evaluate r_f at 0.8 mA dc, and then divide the input voltage across r_f and the 500-Ω resistor. At $I = 0.8$ mA, r_f is

$$r_f \simeq \frac{26 \text{ mV}}{0.8 \text{ mA}} \simeq 32.5 \ \Omega$$

Using voltage division and the circuit in Fig. 2–20(b),

$$V_{\text{ac}} \simeq (100 \text{ mV})\left(\frac{32.5}{500 + 32.5}\right) \simeq 6.1 \text{ mV peak-to-peak}$$

The voltage across the diode, as measured using an oscilloscope, was approximately 8 mV peak-to-peak.

The characteristics of a junction diode are modified in the high-frequency range by the presence of two capacitors: (1) the junction or transition capacitance, C_T, and (2) the storage or diffusion capacitance, C_D. The low-frequency equivalent circuits shown in Fig. 2–19 may be converted to high-frequency equivalent circuits by the addition of these capacitors. Junction capacitance will be discussed first.

The depletion region associated with a P-N junction consists of a positively charged region in the N-material adjacent to a negatively charged region in the P-material as shown in Fig. 2–21. This condition is similar to that of two oppositely charged parallel plates since the semiconductor material acts like a dielectric. However, the charge associated with a given voltage across the junction is not on parallel plates. This charge consists of negative ions (acceptor atoms) in the depletion region on the P-side and positive ions (donor atoms) in the depletion region on the N-side. If the junction voltage increases, the width of the depletion region increases to uncover more ions on each side of the junction. In this way, the charge is a function of the applied voltage.

The depth of the depletion region is a function of both the junction voltage and the doping level on each side of the junction. As shown in Fig. 2–21(a), the diode is doped more heavily on the N-side than on the P-side. Consequently, the depletion region extends farther into the P-side in order to uncover the same number of charges on each side. Since the effective thickness of the dielectric increases with an increase in voltage, the capacitance, C_T, is a function of the applied voltage; i.e., if V doubles, we should not expect q to double as it does in a parallel-plate capacitor. To derive an expression* for C_T, we must find the change

---

* The junction capacitance is calculated as follows. The total charge within the depletion region on the N-side is equal to that on the P-side. Therefore,

in charge, dq, for a change in voltage, dv.

For a given diode having an abrupt junction,

$$C_T = K_1(V_T)^{-1/2} \tag{2-28}$$

where K_1 is a constant for a given diode, and V_T is the algebraic sum of the unbiased barrier voltage, V_B, and externally applied voltage, V. This equation indicates that the junction capacitance is a nonlinear function of the voltage developed across the junction. A similar approach for a graded junction (diffusion type)** indicates that

$$C_T = K_2(V_T)^{-1/3} \tag{2-29}$$

For both types the junction capacitance is smallest with a large reverse bias and increases to a relatively large value for a forward bias. However, for forward-bias conditions, C_T is shunted by the low forward resistance (r_f) of the diode.

As indicated above, the junction capacitance is a function of the distribution of oppositely charged immobile atoms (ions) that are bound into the crystal lattice inside the depletion region. The diffusion capacitance, C_D, is a function

---

$$|qN_AW_PA| = |qN_DW_nA|$$

where A is the junction area, and WA is the volume. The voltage across the junction is found in terms of these charge densities by a double integration of Poisson's equation, which relates charge and voltage in a dielectric.

$$\frac{d^2V_n}{dx^2} = -\frac{qN_D}{K\varepsilon_o} \quad \text{and} \quad \frac{d^2V_P}{dx^2} = \frac{qN_A}{K\varepsilon_o}$$

where K is the relative dielectric constant of the material, and ε_o is the dielectric constant for free space. Applying the appropriate boundary conditions, these integrations yield

$$V_N = \frac{qN_DW_N^2}{2K\varepsilon_o} \quad \text{and} \quad V_P = \frac{qN_AW_P^2}{2K\varepsilon_o}$$

The width of the depletion region is related to the total voltage by

$$W_P = \left(\frac{2K\varepsilon_oV_T}{qN_A\left(1 + \frac{N_A}{N_D}\right)}\right)^{1/2} \quad \text{and} \quad W_N\left(\frac{2K\varepsilon_oV_T}{qN_D\left(1 + \frac{N_D}{N_A}\right)}\right)^{1/2}$$

and the total width is

$$W = W_P + W_N = \left(\frac{2K\varepsilon_oV_T}{q}\right)^{1/2} \times \left(\frac{N_A + N_D}{N_AN_D}\right)^{1/2}$$

On the P-side, the change in width, dW_P, for a small change in voltage, dV, is

$$\frac{dW_P}{dV} = \left(\frac{K\varepsilon_oN_D}{2qV_TN_A(N_D + N_A)}\right)^{1/2}$$

and the change in charge associated with a small change in width is

$$dQ_p = qAN_AdW$$

Similar results occur on the N-side. The final result is

$$C_T = \frac{dQ}{dV} = \frac{dQ}{dW}\frac{dW}{dV} = A\left(\frac{qK\varepsilon_o}{2V_T} \times \frac{N_AN_D}{N_A + N_D}\right)^{1/2} = K_1(V_T)^{-1/2}$$

** Junctions formed by the diffusion method differ from other types of junctions in several respects. One principal difference is that the doping level is not uniform or homogeneous throughout the N-type or P-type material, but varies in a complicated manner (usually according to an error function, *erf*, or to a complementary error function, *cerf*). The use of such impurity profiles generally enhances the performance of diodes and transistors. However, we will assume homogeneous doping in our discussions to keep the required mathematics as simple as possible.

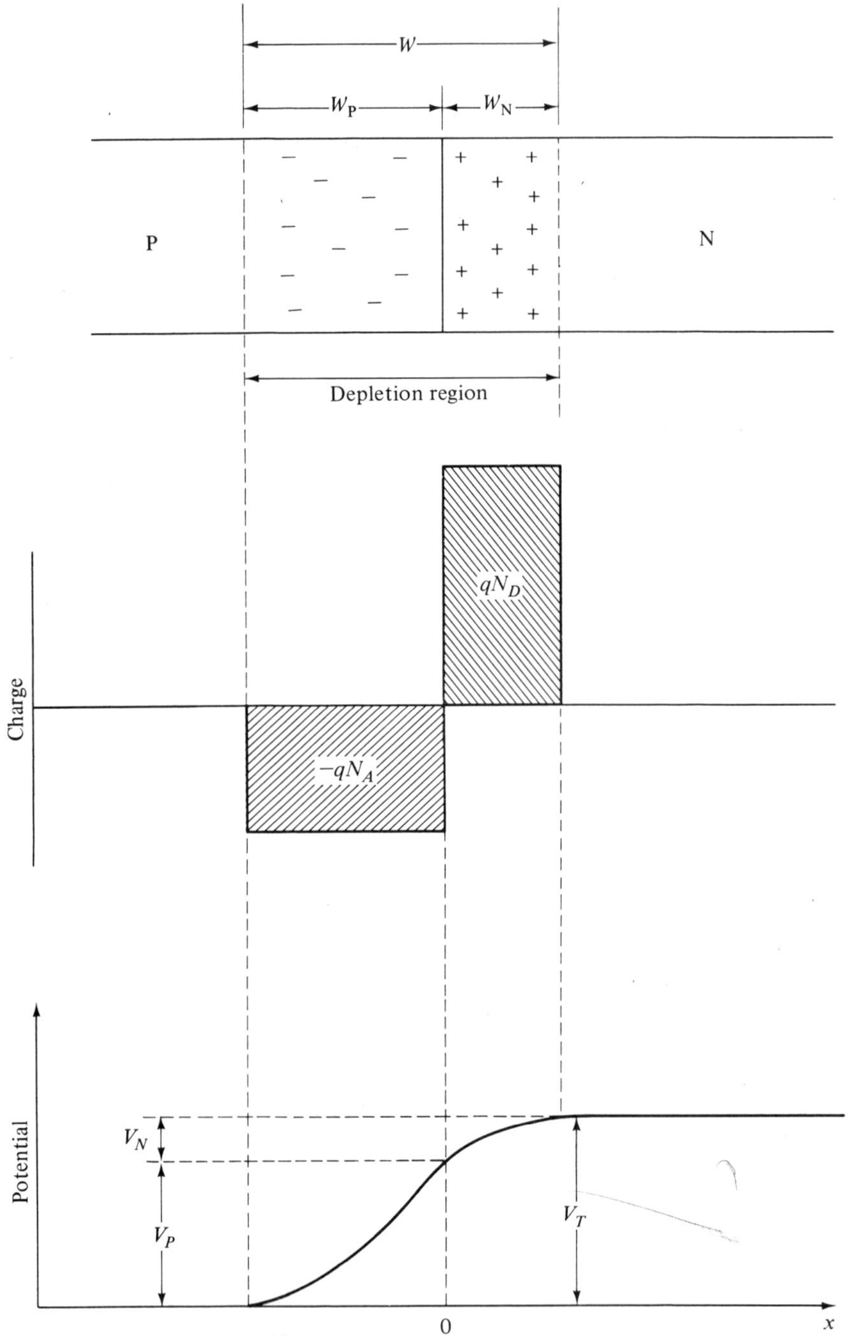

FIGURE 2–21 *Charge density and potential distribution for an abrupt P-N junction, in which the N-side is more heavily doped than the P-side.*

of the distribution of current carriers near, but not inside, the depletion region. The distribution involved is that of the *excess minority charges* shown in Fig. 2–4(b) for a forward-bias condition. If the forward voltage is suddenly increased as shown in Fig. 2–22(a), the forward current increases, but does not have the same rate of rise as the voltage. Before the current can increase, both the depletion region and the minority charge distribution must undergo changes. The width of the depletion region must first decrease to accommodate the decrease in the net barrier potential ($V_B - V_F$). Some delay occurs before the current increases, because this change in width does not occur instantaneously and also because of the time required for carriers to diffuse across the junction and combine with the opposite carrier type. Before the full change in current can occur, the excess charge density must increase as shown by the shaded area in Fig. 2–22(c). Only the change in electron density on the P-side is shown. A similar, but not necessarily equal, change takes place in the excess hole density on the N-side. Since diffusion is an exponential process, the current increases on an exponential curve. The change in stored charge, dq, that is associated with a change in the forward-bias voltage, dv, gives rise to the effect known as the *storage capacitance* or the *diffusion capacitance*, C_D.

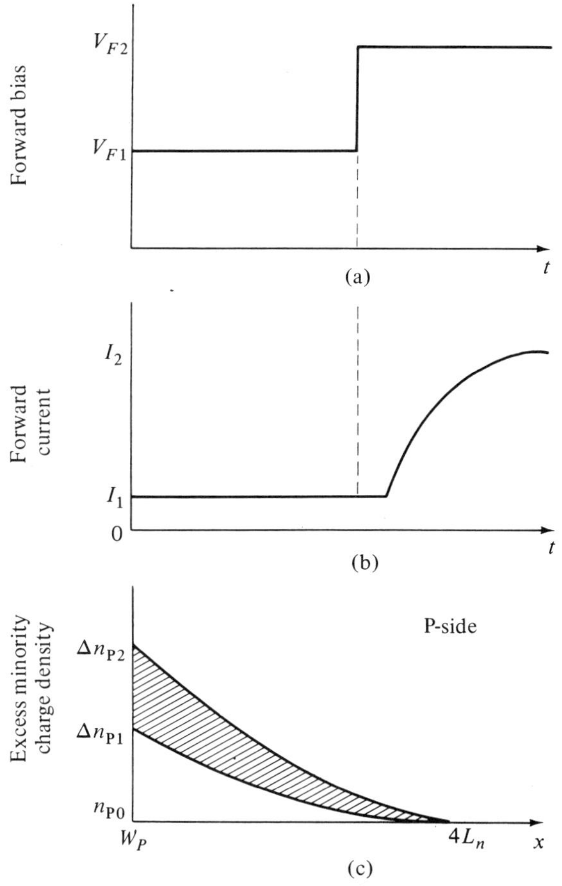

FIGURE 2–22 *Change in excess electron density on the P-side of the P-N junction necessary to support the increase in current produced by an increase in forward bias: (a) forward-bias change, (b) resulting current change, and (c) change in stored charge as indicated by shaded area.*

An expression for C_D may be derived from the excess charge [Eqs. (2–10) and (2–11)]. The analysis* yields

$$C_D = \frac{q}{kT}(I_{pWN}\,\tau_p + I_{nWP}\,\tau_n) \qquad \textbf{(2–30)}$$

where I_{pWN} and I_{nWP} are given in Eq. (2–15); τ_p is the hole lifetime in the N-type material, and τ_n is the electron lifetime in the P-type material. One component of current frequently will dominate at the junction, particularly when one side of the diode is more heavily doped (has much lower resistivity) than the other side, with the heavily doped side furnishing the larger current component. If the N-side is heavily doped compared to the P-side, the expression for C_D becomes

$$C_D \simeq \left(\frac{qI_{nWP}}{kT}\right)\tau_n \simeq \frac{\tau_n}{r_f} \qquad \textbf{(2–31)}$$

Two small-signal, high-frequency equivalent circuits are shown in Fig. 2–23, one for forward bias and one for reverse bias. For the forward-bias condition, C_D is usually much larger than C_T at current levels larger than 0.5 mA. Under reverse-bias conditions, C_D becomes negligible, and only C_T appears in the equivalent circuit. The bulk resistance values, R_P and R_N, although small, may be significant in the high-frequency range.

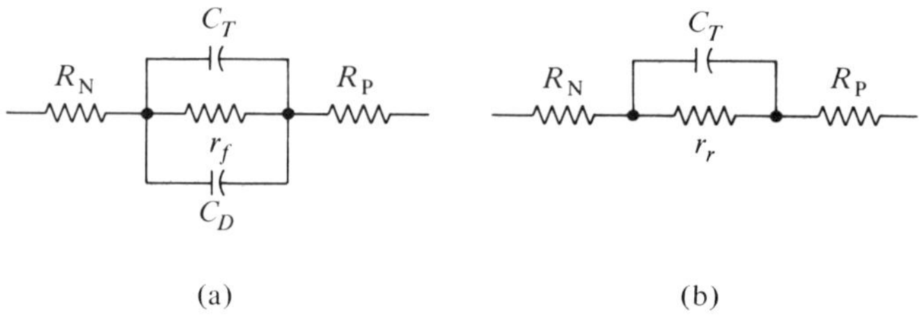

FIGURE 2–23 *Small-signal, high-frequency equivalent circuits for (a) a forward-biased diode, and (b) a reverse-biased diode.*

EXAMPLE 2–8 Sketch the waveform of the ac voltage across the diode in Fig. 2–24(a). Find the time required for the output to reach 98 percent of its final value. The diode is a silicon unit with equal doping levels and approximately equal lifetimes of 10 nsec. The value of C_T is approximately 50 pF.

---

* An expression for C_D may be derived by integrating separately Eqs. (2–10) and (2–11) to find the total excess charge, and then finding the change in stored charge for a change in applied voltage. The limits of integration are (1) the edge of the depletion region, and (2) infinity. For a junction having area A, the integration yields

$$Q_T = Q_N + Q_P = qA(L_p p_{NO} + L_n n_{PO}) \exp(qV/KT)$$

Then

$$C_D = \frac{dQ_T}{dV} = \frac{qA}{kT}(qL_p p_{NO} + qL_n n_{PO}) \exp(qV/KT)$$

Eq. (2–30) is obtained by substituting $(D\tau)^{1/2}$ for L and combining the results with Eq. (2–15).

Solution: Using a small-signal approach, the parameters are evaluated first. Then the time constant for the circuit can be estimated. The risetime is a direct function of this time constant. The circuit current is

$$I \simeq \frac{5 - 0.6}{4.7\text{ K}} \simeq 0.94 \text{ mA}$$

For this current, the forward dynamic resistance is

$$r_f \simeq \frac{26 \text{ mV}}{0.94 \text{ mA}} \simeq 28\ \Omega$$

Since each side of the diode is equally doped, both the hole and the electron components of the current are significant. From Eq. (2–30),

$$C_D = \frac{q}{kT}(I_{PWN}\,\tau_p + I_{NWP}\,\tau_n) \simeq \frac{qI}{kT}\tau = \frac{\tau}{r_f} = \frac{10 \times 10^{-9}}{28} = 360 \text{ pF}$$

The ac equivalent circuit is shown in Fig. 2–24(b). The final amplitude of the voltage step across the diode is approximately

$$V_o \simeq (2\text{ V})\left(\frac{28}{4.7\text{ K} + 28}\right) \simeq 12 \text{ mV}$$

The circuit time constant is given by*

$$\tau_{CKT} = (R \,||\, r_f)(C_D + C_T) \simeq (28)(410 \times 10^{-12}) \simeq 11.4 \text{ nsec}$$

The output will reach approximately 98 percent of its final value in four time constants, or 46 nsec.

The 10 to 90 percent response time is a useful figure-of-merit in step response analyses and is called the *risetime*. The risetime is directly related to the circuit time constant and high frequency response. Experimental determinations of these responses are relatively easy to make and are treated in Chap. 7.

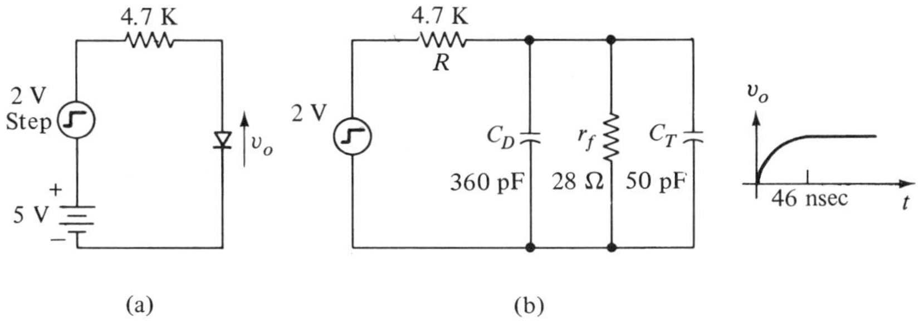

FIGURE 2–24 *A simple diode circuit with a voltage step input for Ex. 2–8: (a) actual circuit, and (b) ac equivalent circuit.*

* The parallel lines, ||, indicate the parallel value of the two resistors.

2.5 Graphical Analysis

In Sec. 2.3, diodes were used with input voltage levels that were large compared with the voltage drop across the diodes. In these circuits V_A was neglected, and the diode was considered as a switch. Circuits having input voltages of 10 V or higher might fall into this class. In Sec. 2.4, the diode was characterized for small-signal applications of approximately 0.1 V or less. In this range the excursions are so small that the diode parameters are considered constants.

The range of input voltages between approximately 0.1 V and 10 V does not fall into either of the above classes. The voltage drop across a diode is not negligible compared with the input signal. In addition, the parameters of a diode vary widely. The diode nonlinearities, however, can be incorporated into a graphical analysis as discussed below.

Consider the diode circuit shown in Fig. 2–25(a). The diode bias conditions are found by setting the ac input signal at zero and solving for the circuit current. Summing the dc voltages around the loop gives

$$V_{DC} = IR + V_D \tag{2–32}$$

where V_{DC} is the dc source voltage, and V_D is the dc drop across the diode. The relationship between the diode voltage and current [see Eq. (2–17)] is nonlinear as shown in Fig. 2–25. A graphical approach will be used to find the circuit current. From Eq. (2–32), the current in the series circuit is

$$I = \frac{V_{DC}}{R} - \frac{V_D}{R} \tag{2–33}$$

This expression is in the form of an equation for a straight line with a slope of $-1/R$, an I-axis intercept* of V_{DC}/R, and a V-axis intercept of V_{DC}. When plotted on the same graph with the diode characteristic, this straight line is called the *load line*. The intersection of this load line with the diode characteristic curve gives the values of the circuit current and the voltage drop across the diode. This point of intersection is called the *quiescent conditions*, or simply the Q-point.

When an ac voltage is superimposed on V_{DC}, the instantaneous source voltage, V_i, is the algebraic sum of V_{DC} and the instantaneous value of the ac wave as shown in Fig. 2–25(b). The load line must terminate on the V-axis at the instantaneous source voltage; i.e., the sum of the voltages around the circuit must be zero at each instant of time. Therefore, since R is constant, the load line swings on each side of and parallel to the dc load line as the ac signal swings around V_{DC}. In Fig. 2–25(b), the V-axis intercept swings from 2 V at the ac positive peak to 0 V at the ac negative peak. The point of intersection of the instantaneous load line with the diode curve gives the instantaneous circuit current, i_i. In this way, a cycle of ac voltage traces out a cycle of ac current. As indicated in Fig. 2–25(b), the resulting current waveform does not have the same shape as the voltage waveform because the diode curve is nonlinear. For the conditions shown, the circuit current is zero over part of the cycle (partial rectification) and is not sinusoidal over the remainder of the cycle.

---

* At the I-axis intercept, the voltage drop across the diode is zero. At the V-axis intercept, the circuit current is zero, and all of the voltage is dropped across the diode. These are theoretical limits for a given R and V_{DC}.

The graphical analysis is introduced here to give the reader a better appreciation of the distortion that the nonlinear characteristics of a diode can produce with intermediate signal levels. In addition, the concept of a load line has been introduced. Both points will be important in future studies. The following example illustrates the graphical approach.

EXAMPLE 2–9 Find the waveform of the voltage across the 500-Ω resistor in the circuit shown in Fig. 2–25(a).

Solution: The dc load line is drawn from 1 V (dc source voltage) on the voltage axis to [1 V/(500 ohms)] on the current axis. The dc current with no ac signal occurs at the intersection of this dc load line and the diode curve. The ac signal has a maximum value of +1 V and a minimum value of −1 V. These extreme values can be used to construct new load lines which are parallel to the dc load line (the load resistance is constant at 500 ohms) and set the limits on the diode current. These two extreme load lines are shown as dashed lines in Fig. 2–25(b). The maximum value of current occurs when $V_{ac} = 1$ V, and is approximately 2.3 mA. The current is approximately zero when $V_{ac} \leqq -0.5$ V. The voltage across the 500-ohm resistor has the same waveform as the current, with a maximum value of

$$V_{Rm} = i_{max} R \simeq (2.3 \text{ mA})(0.5 \text{ K}) = 1.15 \text{ V}$$

and a minimum value of zero. The measured value of the maximum voltage across the resistor was 1.2 V.

No fine line of distinction exists between small- and large-signal analyses. Strictly speaking, a small-signal approach should be used only when the param-

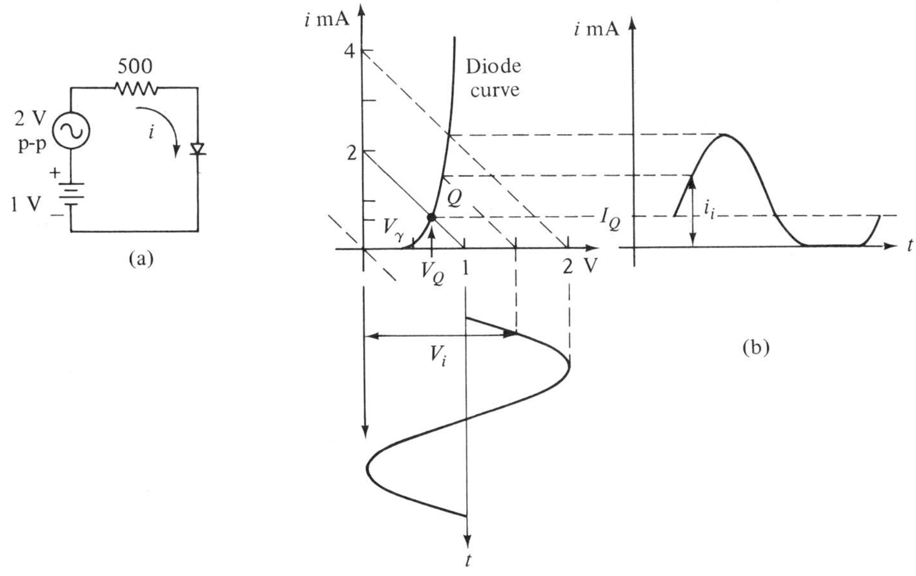

FIGURE 2–25 *A simple diode circuit for use in Ex. 2–9: (a) actual circuit and (b) diode characteristic and waveforms. The diode Q-point can be found using the diode characteristic and the dc load line, the latter being a plot of the voltage equation for the circuit.*

eters of importance, such as the forward or reverse dynamic resistance, change very little as the input signal changes. However, small-signal analyses are often used under moderately large-signal conditions to obtain an approximate solution. The graphical approach has the disadvantage of requiring an accurate *V-I* characteristic curve for the diode under study. Since the characteristics of most semiconductor devices vary from unit to unit, even within a particular device type, the characteristic curve should be determined experimentally for each individual diode. Consequently, equivalent circuits are widely used even when the results are only approximate.

2.6 Diode Logic Circuits

Another important area in which diodes are used extensively is in logic or digital circuits. In this area, diodes are called *gates*. Before considering the operation of diode gates, a short discussion of the binary number system and diode logic is appropriate.

A digital system usually operates in a binary, or base 2, number system, which has only two digits: 0 and 1. Any number can be represented in the base 2 system as well as in the familiar base 10 system. Since the binary system has only two digits, a binary number can be represented by devices which can be biased in either of two possible states. These two states can be designated in a number of ways: true-false, high-low, on-off, 1-0, etc. The 1-0 notation is used to represent the two binary states in this discussion.

Diode logic can be either *pulse logic* or *level logic*. In pulse logic, the information is conveyed by the presence or absence of a pulse. In level logic, the information is conveyed as one of two possible voltage levels. Logic circuits are further divided into *positive logic* and *negative logic*. In positive logic, the 1 level, $V(1)$, is the more positive voltage and the 0 level, $V(0)$, is the less positive or more negative voltage. Negative logic is just the opposite; the voltage associated with $V(0)$ is more positive than that associated with $V(1)$. Note that the actual value of the two voltages has no significance. Positive and negative logic are illustrated in Fig. 2–26. For simplicity, only positive pulse logic will be used here.

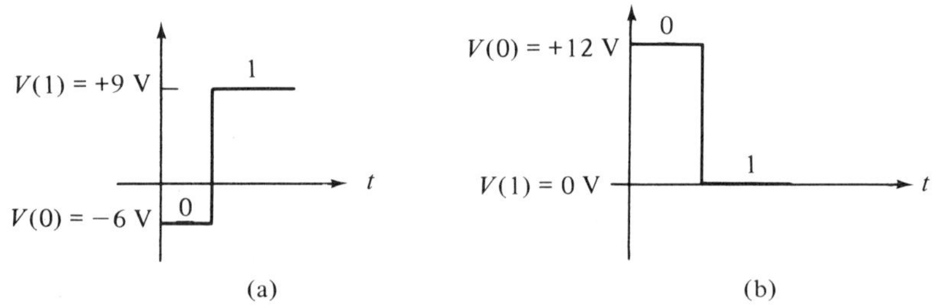

FIGURE 2–26 *An illustration of two types of logic: (a) positive and (b) negative logic.*

Logic gates usually have multiple input terminals and a single output terminal. The function of the gate is to generate an output signal only when certain predetermined input signals are present. Two gates often encountered in logic circuits are the OR gate and the AND gate. Other important logic circuits require transistors.

The output of an OR gate assumes the 1 state when *one or more* inputs assume the 1 state. A two-input, positive-logic OR gate* is shown in Fig. 2–27. If all inputs are in the $V(0)$ state, and if $V_R = V(0)$ or is at ground potential, the diodes have zero bias and do not conduct. Since the current through R is zero, the output voltage is zero or

$$V_o = V(0) \tag{2–34}$$

Now let the V_1 input change to the $V(1)$ state. This change will forward bias D_1, allowing current to flow in R. Using voltage division between R_S and R, the output is

$$V_o = V_R + \left[\frac{V(1) - V_R - V_A}{R + R_S}\right] R \tag{2–35}$$

When R is much greater than R_S, the output is approximately

$$V_o \simeq V(1) - V_A \tag{2–36}$$

If both inputs are present, the output will be slightly different because the current in each diode must flow through R. However, the output will still be given approximately by Eq. (2–36).

EXAMPLE 2–10 Sketch the output waveform of the OR gate in Fig. 2–27 if $V(0) = V_R = 0$ V, $V(1) = 4.4$ V, and the diodes are silicon.

Solution: An output is present when either input is present. From Eq. (2–36),

$$V_o \simeq V(1) - V_A = 4.4 \text{ V} - 0.6 \text{ V} = 3.8 \text{ V}$$

The measured value of V_o was 3.5 V with a single input in the $V(1)$ state, and 3.6 V with both inputs in the $V(1)$ state.

The second principal type of diode gate is the AND gate which can also have multiple inputs and a single output. The output will assume the $V(1)$ state only when *all* inputs are in the $V(1)$ state. Otherwise the output is in the $V(0)$ state. The AND gate is also called a *coincidence* gate since all inputs must be present to obtain an output signal. A two-input positive-logic AND gate is shown in Fig. 2–28. When all inputs are in the $V(0)$ state, both diodes are forward biased,

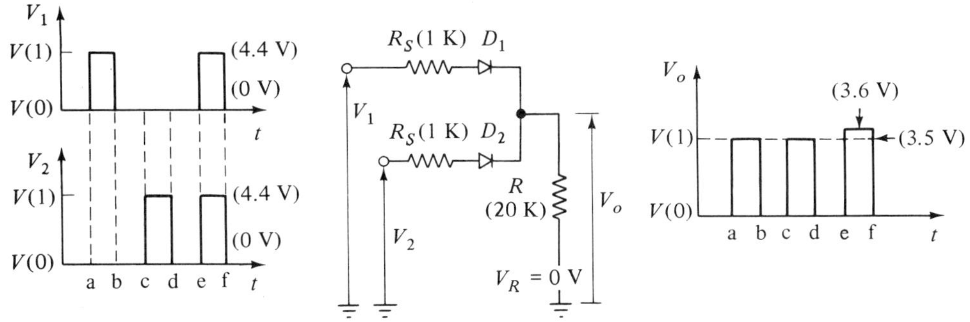

FIGURE 2–27 *A positive logic OR gate. The values in parentheses apply to Ex. 2–10.*

---

* A negative logic OR gate can be made by connecting all diodes in Fig. 2–27 in the other direction and reversing all voltage polarities.

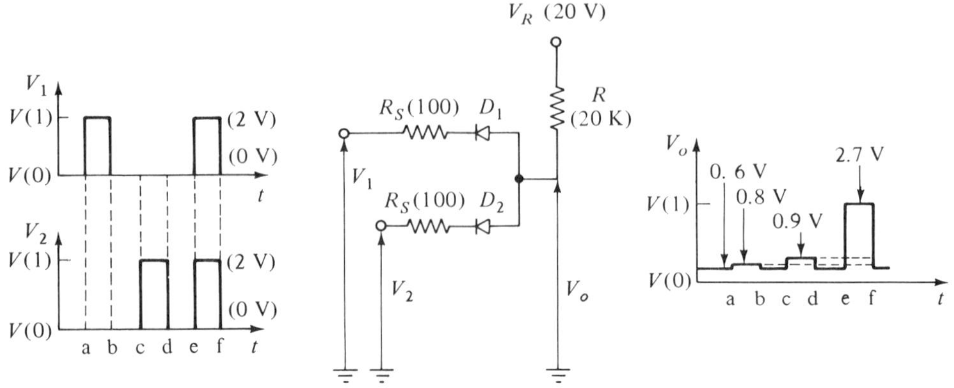

FIGURE 2–28 *A positive logic AND gate. The values in parentheses apply to Ex. 2–11.*

and the output is

$$V_o = V_R - \left[\frac{V_R - V(0) - V_A}{R + R_S/2}\right] R \tag{2–37}$$

When R is much greater than R_S, the output is approximately

$$V_o \simeq V(0) + V_A \tag{2–38}$$

If one input changes to the $V(1)$ state, the corresponding diode is no longer forward biased and does not conduct. The remaining diode, however, is still forward biased and will hold the output to approximately $V(0)$. With both inputs in the $V(1)$ state and with V_R greater than $V(1) + V_A$, the output is available from Eq. (2–37) by replacing $V(0)$ with $V(1)$ giving

$$V_o \simeq V(1) + V_A \tag{2–39}$$

EXAMPLE 2–11 Sketch the output waveform of the AND gate in Fig. 2–28 if $V_R = 20$ V, $V(0) = 0$ V, $V(1) = 2$ V, and the diodes are silicon.

Solution: An output will be present only when both inputs are present. When only one input is present, the output is given by Eq. (2–38) as

$$V_o \simeq V(0) + V_A = 0\text{ V} + 0.6\text{ V} = 0.6\text{ V}$$

With both inputs in the $V(1)$ state, the output is given by Eq. (2–39) as

$$V_o \simeq V(1) + V_A = 2\text{ V} + 0.6\text{ V} = 2.6\text{ V}$$

The measured value of V_o was 0.9 V with one input in the $V(1)$ state, and 2.7 V with both inputs in the $V(1)$ state. With both inputs in the $V(0)$ state, the output was 0.8 V. This lower value of V_o results because both diodes are forward biased when both inputs are in the $V(0)$ state, and only one diode is forward biased when only one input is in the $V(0)$ state. With only one diode conducting, all the current in R flows through the diode resulting in a higher value of V_A.

The experimental results of Exs. 2–10 and 2–11 show that the output varies slightly depending on the number of diodes conducting. This output variation is called *logic noise*. This noise determines the smallest signal amplitude that can be used in logic circuits.

2.7 Voltage-Reference Diodes

Until breakdown occurs, the reverse current in a diode remains relatively constant as the reverse voltage is increased (see Sec. 2.2). At that point on the characteristic curve, the current increases drastically for a small increase in the reverse voltage (shown in Fig. 2–29). The region beyond breakdown can make the diode useful as a voltage-reference diode because the voltage remains relatively constant for a wide range of current values. Two physical phenomena are responsible for diode breakdown: (1) avalanche effects, and (2) tunneling or Zener effects. Avalanche breakdown occurs in diodes having breakdown voltages in excess of approximately five volts. Below five volts, tunneling effects dominate the breakdown characteristics. Both phenomena can be significant in the three to eight volt range.

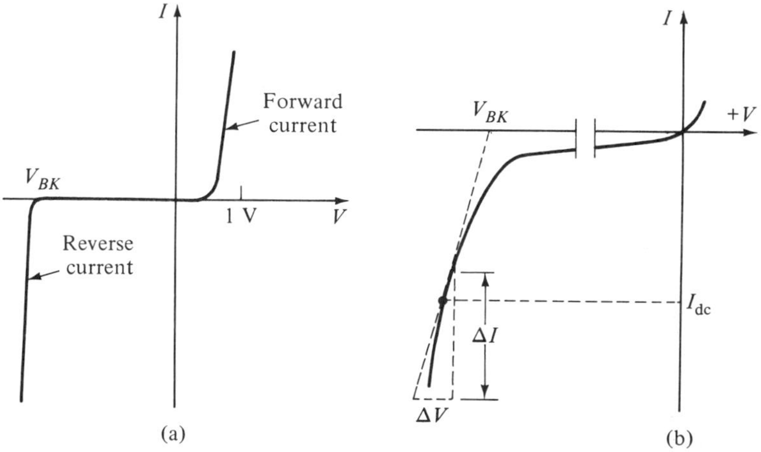

FIGURE 2–29 *The voltage-current characteristics of a voltage-reference diode showing the breakdown characteristics at* V_{BK}: *(a) the forward and reverse characteristics, and (b) an expanded view of the knee of the reverse characteristic curve.*

A voltage-reference diode* is usually fabricated as an abrupt junction with one side much more heavily doped than the other. For example, a 10-V silicon reference diode might be doped with 10^{19} donor impurity atoms/cm^3 on the N-side and 10^{17} acceptor impurity atoms/cm^3 on the P-side. Reducing the doping level on the P-side to 5×10^{15} acceptor impurity atoms/cm^3 increases the breakdown voltage to approximately 50 V. In both cases, the depletion region is very narrow and lies almost entirely in the side with the lighter doping. The field intensity required to produce breakdown in the depletion region is approximately 100 kV/cm or greater for semiconductors.

If the doping is very heavy on both sides, the depletion region is very narrow, and the barrier voltage alone is sufficient to produce breakdown conditions. An increase in the doping beyond this level produces *tunnel diodes* which are in a breakdown condition even when a small forward bias is applied.

---

* Also known as voltage regulator diodes and Zener diodes.

With the range of breakdown conditions in perspective, the two mechanisms responsible for breakdown will now be considered. Since avalanche breakdown is probably more important, it will be discussed first. The reverse current, I_S, is a drift current and flows across the depletion region. Current carriers making up I_S have numerous collisions with the lattice atoms as they move through this region. The path length between collisions is somewhat random; however, a mean free path length exists in a given material for a particular field intensity and temperature. If a carrier can obtain sufficient energy from the electric field between collisions, its impact with an atom may free an electron. When this event occurs, the number of carriers is increased by an electron and a hole. Both of these carriers, as well as the incident carrier, are accelerated and may generate additional electron-hole pairs. Some multiplication occurs below the breakdown voltage, V_{BK}, for occasional longer path lengths. As the reverse voltage increases, the multiplication process becomes more pronounced. At V_{BK} the mean free path and the electric field intensity enable most of the carriers to obtain sufficient energy between collisions to produce electron-hole pairs. A chain reaction or an avalanche condition occurs at reverse-voltage levels slightly above V_{BK} in which the normal reverse saturation current can produce current levels many times larger than I_S. The effect of this multiplication is shown in Fig. 2–29.

Zener breakdown occurs in diodes having heavy doping on both sides. Diodes with doping levels of 10^{18} atoms/cm^3 or higher have depletion regions only a few hundred angstroms wide. A small external voltage combined with the barrier voltage is sufficient to produce a very high field intensity. When the field intensity reaches a level of approximately 10^5 V/cm, the covalent bonds are disrupted; i.e., electrons that do not have sufficient thermal energy are assisted by the high field to escape from the covalent bond. These electrons move from the valence band on the P-side directly to the conduction band on the N-side. Avalanche effects are not pronounced because the width of the depletion region is less than a mean free path length.

The temperature coefficient of the reference voltage depends on the dominant breakdown effect. Avalanche breakdown has a positive temperature coefficient because the mean free path decreases as temperature increases. Zener breakdown has a negative temperature coefficient because the width of the forbidden gap decreases with temperature. The two effects tend to cancel in diodes with reference voltages in the four to six volt range as shown in Fig. 2–30. However, note that a zero temperature coefficient range occurs over a very narrow current range.

When used as a voltage-reference diode, the *V-I* characteristic curve should break sharply downward and have a very steep slope. The sharpness of breakdown is a function of the dominant breakdown effect. The reverse characteristics of both a 3.6 V and a 7.5 V breakdown diode, as displayed on a curve tracer, are shown in Fig. 2–31. Zener breakdown is characterized by a soft knee, and avalanche breakdown is characterized by a sharp knee.

An equivalent circuit for a voltage-reference diode in the breakdown region is a simple series circuit containing a resistor and a battery as shown in Fig. 2–32. The resistor, r_z, represents the dynamic resistance at the operating current and should be as low as possible. The battery voltage is the voltage intercept of a line tan-

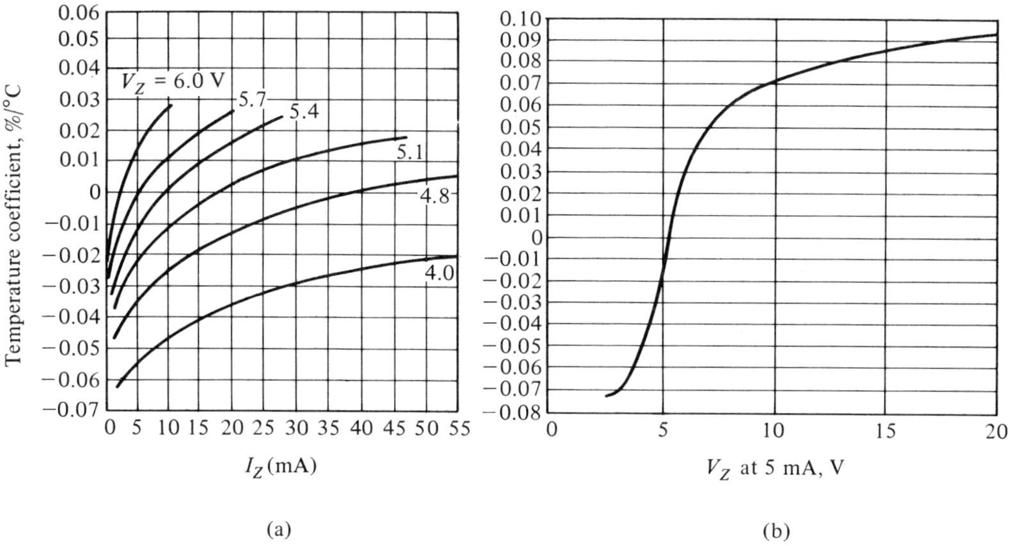

FIGURE 2–30 *Temperature coefficients for a number of reference diodes having different operating voltages (a) as a function of operating current and (b) as a function of operating voltage. The voltage* V_Z *is measured at* $I_Z = 5$ *mA (from 25 to 100°C).*

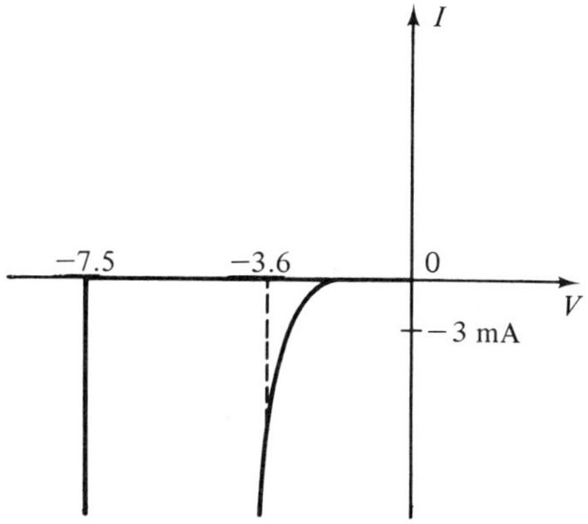

FIGURE 2–31 *Reverse characteristic of a 3.6 V and 7.5 V reference diode.*

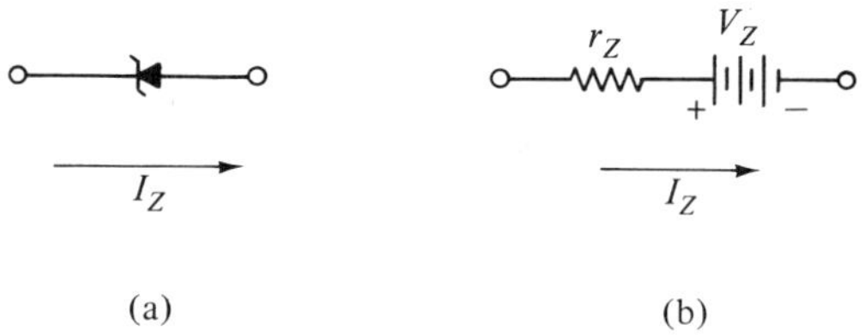

FIGURE 2–32 *(a) A symbol for voltage-reference diodes, and (b) an equivalent circuit under breakdown conditions.*

gent to the curve at the operating point. Graphical methods for determining both r_z and V_z (or V_{BK}) are shown in Fig. 2–29. Values of V_z quoted for a particular type number are nominal and may vary as much as a volt or more from the quoted value.

EXAMPLE 2–12 The input voltage to the circuit in Fig. 2–33(a) is a full-wave rectified sine wave as shown in Fig. 2–33(b). Sketch the output waveform.

Solution: Assuming that the value of r_z is much less than 165 ohms, the reference diode will hold the load voltage at approximately 12 V when the input voltage exceeds $V_{in(min)}$ where

$$V_{in(min)} = V_o + I_s R_s = 12 + (12/820)165 = 14.4 \text{ V}$$

When the input voltage is less than $V_{in(min)}$, the reference diode is approximately an open circuit and can be neglected. The output voltage will then be proportional to the input voltage and is given by

$$V_0 = V_{in}\left(\frac{820}{820 + 165}\right) \simeq 0.83 V_{in}, \qquad V_{in} < V_{in(min)}$$

The output waveform is shown in Fig. 2–33(c).

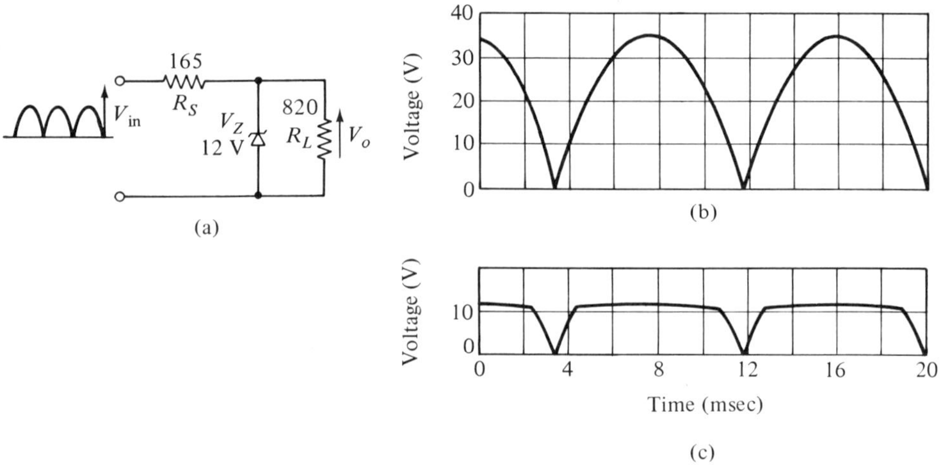

FIGURE 2–33 *Reference diode circuit for use in Ex. 2–12: (a) actual circuit, (b) input waveform, and (c) output waveform.*

Voltage-reference diodes are found in many applications, one of the most important being regulated low-current dc power supplies. When used as a shunt element across the load, a voltage-reference diode will hold the load voltage relatively constant against input voltage changes as well as load variations. In addition, a reference diode will also help reduce the ac ripple on the dc voltage. The following example illustrates these functions.

EXAMPLE 2–13 The bridge rectifier circuit of Ex. 2–4 [see Fig. 2–15(a)] has been modified to include a voltage-reference diode as shown in Fig. 2–34. For this regulated supply, find the output ripple and the minimum values of $V_{in(dc)}$ and R_L required to maintain regulation of the output voltage. The dynamic resistance of the Zener diode is approximately 2 ohms for diode currents between 5 and 150 mA.

Solution: The effective resistance in parallel with the capacitor is

$$R \simeq 165 + (r_z \parallel 820) \simeq 167\ \Omega$$

Since this value of resistance is approximately the same as that used in Ex. 2–4, the results of Ex. 2–4 apply. The dc voltage across the capacitor, V_C, is 35.5 V with a ripple factor of 0.006. This ripple factor indicates an ac voltage or ripple component of

$$V_{ac} \simeq (0.006)(35.5) \simeq 212\text{ mV}$$

The output ripple voltage divides across R_S and the parallel combination of r_z and R_L or 2 ohms. Therefore, the output ripple voltage is

$$V_{o(ac)} = V_{C(ac)} \frac{r_z \parallel R_L}{r_z \parallel R_L + R_S} \simeq (212)\left(\frac{2}{2 + 165}\right) \simeq 2.5\text{ mV}$$

The measured ac voltage across the capacitor (using a peak reading ac meter) was 165 mV. The measured ac output voltage, using the same ac meter, was 1.5 mV. The discrepancy between the calculated and measured ripple voltages is due primarily to the use of a peak reading meter rather than a true rms meter. The measured dc voltage was 12.1 V.

The dc current through R_S is

$$I_S = \frac{(35.2 - 12.1)\text{ V}}{165} = 140\text{ mA}$$

The Zener diode current is

$$I_Z = 0.140 - \frac{12\text{ V}}{820\ \Omega} \simeq 125\text{ mA}$$

The power supply output is regulated against changes in input voltage and load resistance. Assuming $r_z = 2$ ohms, the output dc voltage will remain constant provided that the dc current does not fall below 5 mA. This condition means that the dc voltage across the capacitor must be equal to or greater than $V_{C(\min)}$ where

$$V_{C(\min)} = R_S(I_{Z(\min)} + I_L) + V_{0(dc)}$$
$$\simeq 165(5 + 15)\text{ mA} + 12\text{ V} \simeq 15.3\text{ V}$$

Assuming a constant value of V_C, output regulation is maintained provided the load current is less than

$$I_{L(\max)} = I_S - I_{Z(\min)} = 140 - 5 = 135\text{ mA}$$

The minimum value of R_L is

$$R_{L(\min)} = \frac{V_{0(dc)}}{I_{L(\max)}} = \frac{12\text{ V}}{135\text{ mA}} \simeq 39\ \Omega$$

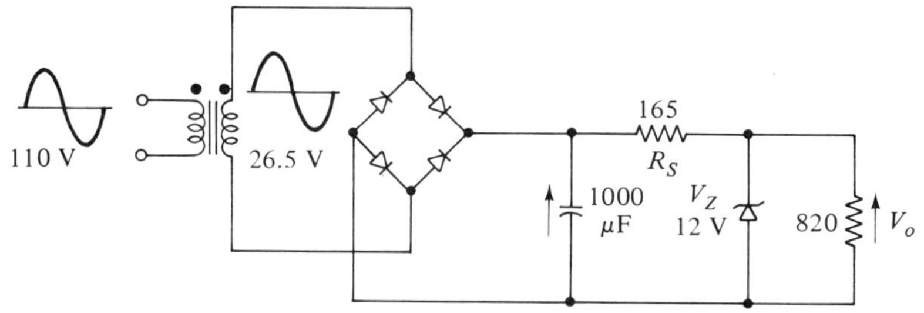

FIGURE 2–34 *A regulated dc power supply from Ex. 2–13.*

2.8 Metal-Semiconductor Junctions

The junctions discussed up to this point have been those formed when one part of a crystal is doped with P-type material, and an adjacent part is doped with N-type material. For normal doping levels (below 10^{18} impurity atoms/cc) and reasonable purity of the base material, these junctions always conduct much better in one direction than the other, and are therefore called rectifying junctions. Another important type of junction can be formed between metal and either P-type or N-type semiconductor material. Depending upon the type of metal used and the doping density in the semiconductor, this type of junction can be either rectifying or ohmic. Both types will be discussed briefly with the rectifying type being considered first.

Consider the case of a junction between a metal and N-type silicon. The energy band diagrams on each side of the junction are shown in Fig. 2–35 for several different conditions three of which are: (a) thermal equilibrium, (b) forward bias, and (c) reverse bias. With no external voltage applied across the junction, the Fermi levels of the two materials must be equal as shown in (a). As with a P-N junction, this condition is associated with band bending at the junction and a depletion region, W, in the semiconductor at the junction. With a doping density less than 10^{17}/cc, W is too wide to allow a significant amount of tunneling. Consequently, current can flow only if electrons on the semiconductor side have energies greater than qV_B above the bottom of the conduction band for forward current flow or if electrons on the metal side have energies greater than qV_M for reverse current flow. Both components can be made very small by the proper selection of the metal and doping level in the semiconductor. Since these two components must be equal under thermal equilibrium, this selection also determines the amount of band bending shown in (a).

When a forward bias voltage, V_F, is applied to the junction,* the energy bands on the semiconductor side are shifted up (in N-type material) by an amount of qV_F. Electrons in the conduction band need very little additional energy above E_c to move into the metal. At the top of the barrier, these electrons have energies of qV_M or greater and correspond to very high energy electrons in the metal. The density of electrons with energies this large or larger is very low unless the metal is very hot. Consequently, these diodes are called *hot carrier* diodes. However, Schottky first modeled metal-semiconductor barriers approximately seventy years before the P-N junction was discovered. Therefore, these diodes are also called *Schottky barriers* or *Schottky diodes*. As in the case of a P-N junction, the amount of forward current that flows is a logarithmic function of the forward bias.

When a reverse bias voltage, V_R, is applied to the junction, the energy bands are bent more than in thermal equilibrium, and the forward current components become essentially zero. Since very few electrons in the metal have energies greater than qV_M above the Fermi level, the reverse current is rather small and almost independent of the bias until reverse breakdown occurs.

The gross current-voltage characteristics of a Schottky diode are similar to those of a P-N junction. However, a metal-silicon diode will usually have a

* For forward bias, the metal is positive for N-type and negative for P-type semiconductors. This bias causes less band bending than zero bias.

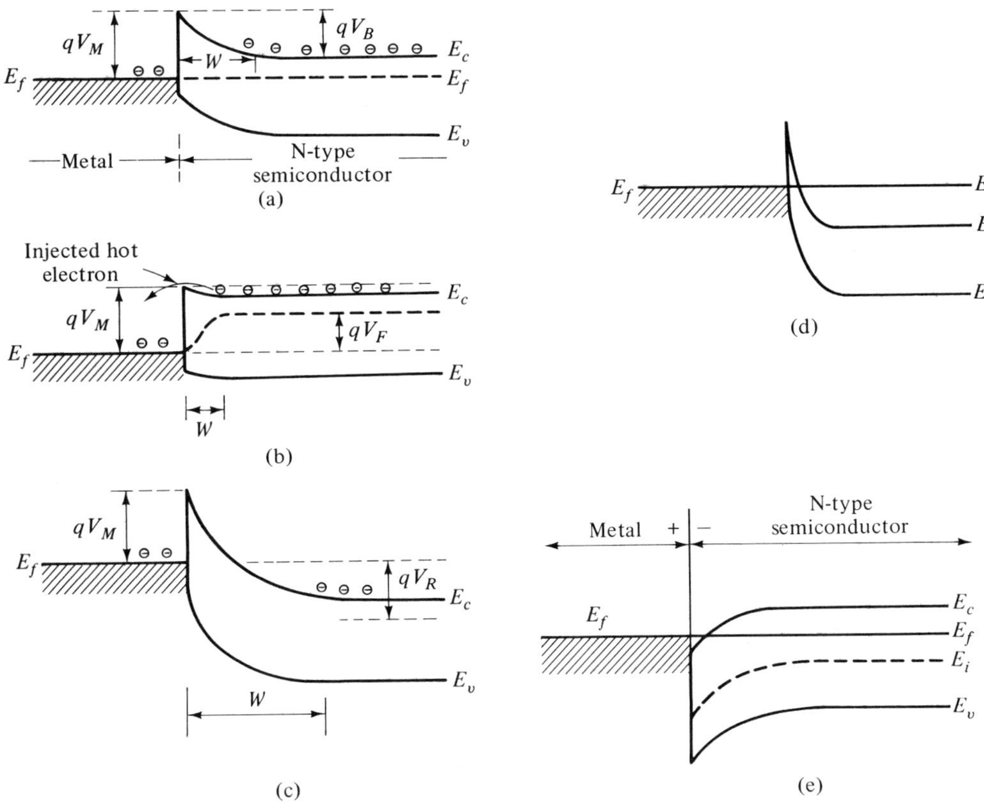

FIGURE 2–35 *Energy band diagrams for metal to semiconductor (N-type) contacts: (a) Schottky barrier in thermal equilibrium, (b) Schottky barrier with a forward bias for easy current flow, (c) Schottky barrier with a reverse bias, (d) ohmic type contact resulting from very heavy doping of the semiconductor (in thermal equilibrium), and (e) ohmic type contact resulting from the Fermi level of the metal being higher than that of the semiconductor before the junction was formed.*

higher forward current than the P-N junction for the same forward bias voltage when both diodes have equal areas. The primary advantage of a Schottky diode over a P-N diode is the speed with which it can be switched.

The ability to switch rapidly is a very important characteristic of diodes that are used in logic circuits. In a P-N junction diode, the flow of forward current is associated with minority carrier storage on each side of the junction [see Fig. 2–4(b)]. A change in current requires changes in these minority charge distributions as shown in Fig. 2–22. The time required to effect these changes limits the speed with which a diode can switch from an ON to an OFF state or from an OFF to an ON state. As a result of minority charge storage, fast P-N junction diodes have switching times that are limited to a few nanoseconds or longer. In the Schottky diode, this limitation is minimized because the current is carried by majority carriers. For example, electrons are majority carriers in both the metal and N-type silicon. The resulting charge storage is negligible. Consequently, switching times as short as a few picoseconds have been achieved with Schottky diodes.

The general equations for Schottky diodes have somewhat the same form as those for P-N junction diodes. The current component due to majority carriers is

$$I = I_o[\exp(qV_F/kT) - 1] \qquad \textbf{(2–40)}$$

where

$$I_o = [4\pi m^* qk^2/h^3][AT^2] \exp(-V_B/kT) \qquad \textbf{(2–41)}$$

where A is the effective diode area, and m^* is the effective mass of the majority carrier. Minority carriers also flow in a Schottky diode. However, these current components are considerably less (typically six orders of magnitude less) than the majority components and are usually neglected.

As indicated in Fig. 2–35, the depletion (space charge) region lies in the semiconductor side of the junction. The width of this region is

$$W = (2K\varepsilon_o V_T qN)^{1/2} \qquad \textbf{(2–42)}$$

where N is the doping density.

Some metal-semiconductor junctions are ohmic rather than rectifying types. For example, if the N-type semiconductor in Fig. 2–35 is doped heavily (greater than 10^{19} impurity atoms/cc), the width of the depletion region is very narrow. In addition, the Fermi level in the semiconductor lies above E_c as shown in Fig. 2–35(d). For these conditions, electrons can tunnel directly through the depletion region in either direction. The current is a rather linear function of voltage in either direction over a wide current range. Consequently, the metal-semiconductor junction acts like a low value of resistance.

Another type of ohmic junction results if the Fermi level of the metal is higher than that of the doped N-type semiconductor as shown in Fig. 2–35(e). Electrons can move from the metal to the conduction band of the semiconductor or from the conduction band to the metal with little energy. Consequently, current flows almost equally well in either direction.

Other types of ohmic contacts can be fabricated. However, they are not well understood. One method that apparently produces reasonably good ohmic contacts is to damage or strain the surface of a semiconductor prior to plating or bonding the metal on the surface. Another method is to alloy a metal to the surface. This process apparently introduces many energy states into the band gap so that an effective rectifying junction is not formed. Probably one of the easiest methods used to attach external leads to semiconductor devices is to first dope the contact region heavily (degenerate doping) and then to either plate or thermally bond leads to the degenerate material. This method results in an almost metal-to-metal junction. The depletion region, if significant, is very narrow so that tunneling occurs (as explained above). This type of contact is not truly ohmic, however, especially at high current densities. True ohmic contacts are rather difficult to fabricate between an external metal lead and a semiconductor material.

2.9 Summary

In Chapter 1, the basic properties of both N-type and P-type semiconductor materials were discussed. In this chapter, the two materials were arranged in the same crystalline structure to form a P-N junction. The concept of a potential barrier was introduced, and the effect of this potential barrier on the flow of current was discussed. A forward bias lowers the potential barrier, and current flows easily through the diode. A reverse bias increases the potential barrier and allows very little current to flow through the diode.

The diode has nonlinear characteristics, thereby making the exact analysis of a circuit containing a diode very difficult. Approximate analyses make use of the following important parameters: the cut-in voltage ($V\gamma$), the forward and reverse static resistances (R_F and R_R), the forward and reverse dynamic resistances (r_f and r_r), the diffusion capacitance (C_D), and the transition capacitance (C_T). Of these parameters, r_f is probably the one most frequently used. It has a value of (26 mV)/I_f. Combinations of these parameters are used to replace diodes in circuits to facilitate an approximate analysis.

The best approach in analyzing circuits containing diodes is determined by the signal level. For large signals, the diode may be treated as a voltage-operated switch. For medium signal levels, the cut-in voltage may be considered. For low-level signals, equivalent circuits may be used. The diffusion and transition capacitance must be considered when the signal frequencies lie in the high-frequency range. A graphical analysis is advantageous for the medium signal case if the characteristic (V-I) curve for the diode is available.

Two important applications have been considered to illustrate some of these techniques; (1) rectifiers and (2) logic circuits. In power rectifier circuits, the forward voltage drop across the diode is usually so small (approximately one volt) that it can be neglected. Consequently, the diode is treated as a switch. In logic circuits, the voltage drops across the diodes are usually an appreciable part of the signal voltage and must be considered.

The reverse breakdown voltage of a diode is a function of the doping levels on each side of the junction. Breakdown diodes are a class of diodes that can dissipate the heat generated in the diodes under reverse breakdown conditions. These diodes are widely used as voltage regulators because their dynamic resistance is low in the breakdown region.

The material in this chapter forms a base for our future study. The explanation of the physical operation of the field-effect transistor (Chapter 3) requires a knowledge of the properties of a reverse biased P-N junction. The bipolar transistor (Chapter 4) is best explained in terms of a forward biased P-N junction and a reverse biased P-N junction formed in close proximity. Integrated circuits (Chapter 5) contain numerous P-N junctions formed in a single crystal. The study of the theory of operation and application of vacuum tubes (Chapter 6) is aided by comparing them to semiconductor devices. Most special purpose semiconductor devices are dependent on some property of P-N junction. Consequently, the material in this chapter will be referred to often in this study of applied electronics.

Problems

Section 2.1

2–1 A germanium diode is made of 10 ohm-cm, P-type material and 1 ohm-cm, N-type material. What is the magnitude of the potential barrier at room temperature if μ_n is 3900, and μ_p is 1900 cm²/volt-sec? Assume that the resistivity is controlled by a majority carrier.

2–2 Find the magnitude of the potential barrier for a silicon diode which has an impurity concentration of 10^{17} on the N-side and 10^{18} on the P-side.

2–3 Derive Eq. (2–5) using the fact that the total hole current is zero in thermal equilibrium.

2–4 How will the magnitude of the potential barrier change as the doping level is increased on (a) the N-side, (b) the P-side, and (c) both sides of the diode?

Section 2.2

2–5 Design a diode which will have a forward current of 1 mA when the applied forward bias voltage is 0.6 V. Assume that the cross-sectional area is 0.01 cm^2; L_n is 10^{-2} cm; L_p is 5×10^{-2} cm; μ_n is 1300 cm^2/volt-sec; and μ_p is 400 cm^2/volt-sec. Assume a reasonable value for the doping level on one side. The design is complete when the doping level of each side is specified. (*Hint:* First find I_S.)

2–6 A germanium diode is made with 10 ohm-cm, P-type material and 1 ohm-cm, N-type material. Assume that D_n is 90, and D_p is 60 cm^2/sec; L_n and L_p are 10^{-2}; the cross-sectional area is 0.01 cm^2; μ_n is 3900; and μ_p is 1900 cm^2/volt-cm. Find the current that flows when a forward bias of 0.15 V is applied to the diode.

2–7 Find the maximum and minimum current that flows in a germanium diode when a 0.05 volt-peak sine wave is superimposed on a +0.1 dc voltage across the diode terminals. Assume that $I_S = 10\ \mu$A.

Section 2.3

2–8 Verify that the ripple factor of a half-wave rectifier is 1.21.

2–9 Design a half-wave rectifier which can supply 100 mA at 20 V with a ripple factor less than 0.1. Use components which are commercially available. Start with a 110-V, 60-Hz source, and specify the transformer turns ratio, component values, and power ratings.

2–10 Verify that the ripple factor of a full-wave rectifier is 0.48.

2–11 Design a bridge rectifier which can provide 1 A at 25 V with a ripple factor less than 0.005. Use components which are commercially available. Assume a 110-V, 60-Hz source, and specify the transformer turns ratio, component values, and power ratings.

2–12 Why does a series inductor reduce the ripple factor?

Section 2.4

2–13 For the silicon diode characteristic shown in Fig. 2–7, find the forward static and dynamic resistance with a 0.6-V forward bias.

2–14 Repeat Prob. 2–13 for the germanium diode with a 0.3-V forward bias.

2–15 A semiconductor diode is characterized by $I_S = 10^{-9}$ amp. Find the dynamic resistance for forward-bias voltages of (a) 0 V, (b) 0.3 V, (c) 0.6 V, and (d) 1.0 V.

2–16 Find the forward dynamic resistance of (a) a silicon diode and (b) a germanium diode for forward currents of 1 mA, and 10 mA.

2–17 A 22.5-volt battery is connected in series with a diode and a 1-M resistor. The voltage across the resistor is to be less than 1 mV. (a) Should the diode be germanium or silicon? (b) Sketch the circuit.

2–18 A diode is connected in series with a 1-K resistor and a 6-V battery. The voltage drop across the diode is 0.6 V. The diode connections are then reversed. (a) What voltage will appear across the resistor? (b) Estimate the diode reverse static resistance. Neglect leakage currents.

2–19 Draw the total *V-I* characteristic for the circuit shown in Fig. 2–36(a) assuming

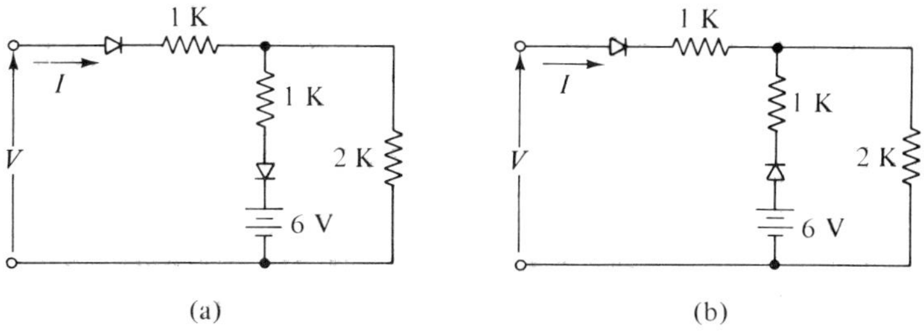

FIGURE 2–36 *Circuit for Probs. 2–19 and 2–20.*

silicon diodes. Label all important points and slopes on the curve as the input voltage, V, varies from 0 to +20 V.

2–20 Repeat Prob. 2–19 for the circuit in Fig. 2–36(b).

2–21 Repeat Ex. 2–7 for a germanium diode.

2–22 (a) Draw the *V-I* characteristic of a germanium diode with $I_S = 10^{-6}$ amp by calculating the current with forward-bias voltages of 0.1, 0.2, 0.3, 0.4, and 0.5 V. (b) Repeat part (a) assuming the material has bulk resistance values of $R_N = R_P = 100$ ohms.

2–23 The transition capacitance of a silicon diode is 10 pF at a reverse bias of 10 V. What is the value of this capacitance at a reverse bias of 20 V?

2–24 The diffusion capacitance of a germanium diode is 50 pF when the diode current is 2 mA. What is the value of this capacitance when the current is 3 mA?

2–25 A germanium diode is made of 10 ohm-cm, N-type material and 1 ohm-cmr P-type material. The carrier lifetime is 10^{-6} sec for holes and 0.5×10^{-6} sec fo, electrons. Assume that the diffusion constant is 46.8 for holes and 98.8 for electrons. What is the diffusion capacitance at a current of 1 mA if the junction area is 0.01 cm^2?

2–26 A silicon diode made of P-type material doped to one part in 10^5 and N-type material doped to one part in 10^7 has a carrier lifetime of $\tau_p = 10^{-8}$ sec. What is the diffusion capacitance if the current is 5 mA?

Section 2.5

2–27 Repeat Ex. 2–9 using a germanium diode whose characteristic is given in Fig. 2–7. What maximum voltage appears across the resistor?

2–28 Repeat Ex. 2–9 with $V_{dc} = 0$ V.

2–29 Repeat Ex. 2–9 with the ac voltage 1.0 V peak-to-peak.

Section 2.6

2–30 (a) Sketch a two-input negative logic OR gate. (b) Repeat for an AND gate. (c) Compare the negative logic gates to the positive logic gates shown in Figs. 2–27 and 2–28.

2–31 The binary levels for the input to the AND gate in Fig. 2–37 are $V(0) = 0$

V and $V(1) = 20$ V. Assuming ideal diodes, find the value of V_o when: (a) $V_1 = V_2 = V(0)$; (b) $V_1 = V_2 = V(1)$; and (c) $V_1 = V(0)$, and $V_2 = V(1)$.

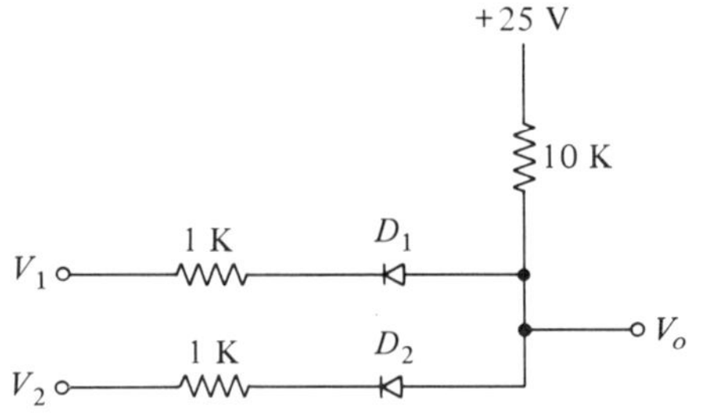

FIGURE 2–37 *Circuit for Prob. 2–31.*

2–32 Often diode gates are combined to form various combinations of AND and OR gates. Such a combination is shown in Fig. 2–38. Describe the "logic" performed by this circuit. What conditions are necessary for the output to be in the $V(1)$ state? Assume positive logic with $V(1) = 20$ V, and $V(0) = 0$ V.

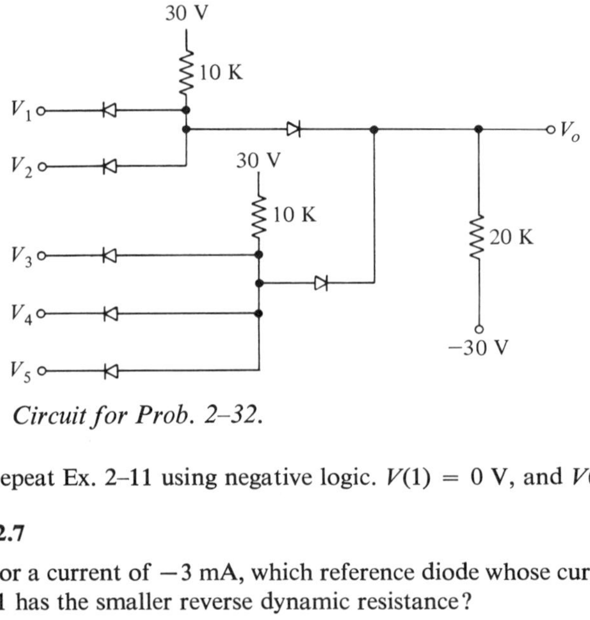

FIGURE 2–38 *Circuit for Prob. 2–32.*

2–33 Repeat Ex. 2–11 using negative logic. $V(1) = 0$ V, and $V(0) = 2$ V.

Section 2.7

2–34 For a current of −3 mA, which reference diode whose curves are shown in Fig. 2–31 has the smaller reverse dynamic resistance?

2–35 The circuit in Fig. 2–33(a) is supplied with a dc input V_{in}. (a) What minimum value of V_{in} is required in order for the output to be regulated if the reference diode requires at least 1 mA for good regulation? (b) Repeat for $R_S = 50$ ohms.

2–36 The circuit in Fig. 2–33(a) is supplied with a dc input of 24 V. Calculate the allowable variation in R_L before the value of V_o begins to change appreciably. The reference diode requires a minimum of 1 mA for good regulation.

2–37 The minimum current through the reference diode in Fig. 2–39 is to be 2 mA. (a) Find the value of R_S which will allow good voltage regulation across the load if the current demand ranges from 1 to 8 mA. Assume V_{dc} is constant. (b) What current will flow if the output is accidentally short circuited ($R_L = 0$)?

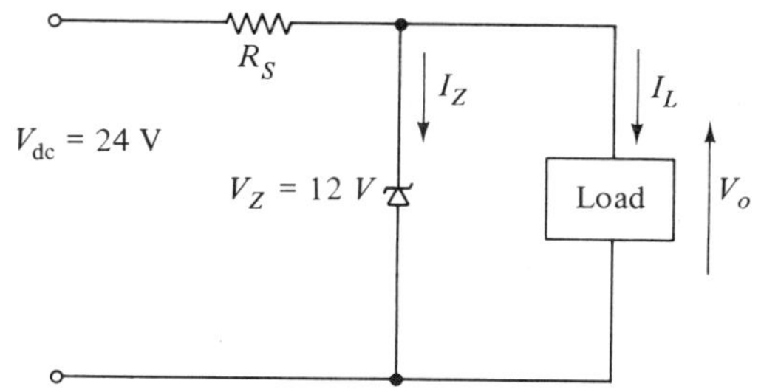

FIGURE 2–39 *Circuit for Prob. 2–37.*

2–38 The circuit shown in Fig. 2–40 is a Zener regulated ±12-V power supply. The input transformer is a 110/26.5 V CT (Center Tapped so that half of the secondary voltage appears between each end terminal and the center terminal). The dynamic resistance of the Zener diode is approximately 1 ohm at 100 mA. (a) Redraw the circuit to show that it is made up of two full-wave rectifier circuits each similar to the circuit in Fig. 2–12(a). (b) Estimate the peak-to-peak value of the ac component of the output voltage. (c) What should the power rating of the 60-ohm resistors be to allow for short-circuit protection? (Experimental results: the magnitude of the voltage across each capacitor was approximately 18.2 V dc with a 0.5 V peak-to-peak ac ripple. The output voltage was 12.5 V dc with a 5 mV peak-to-peak ac ripple.)

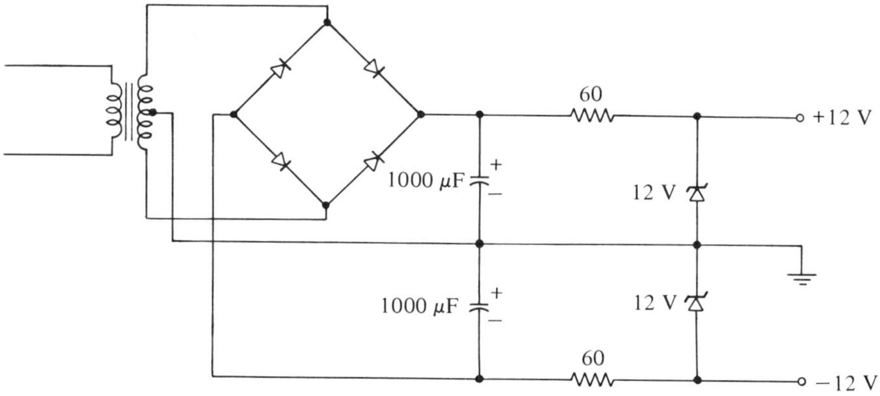

FIGURE 2–40 *Circuit for Prob. 2–38.*

References

Conwell, E. M., "Properties of Silicon and Germanium," *Proceedings of the IRE*, 40 (1952), 1327.

Gray, P. E., D. DeWitt, A. R. Boothroyd, and J. F. Gibbons, *Physical Electronics and Circuit Models for Transistors*. SEEC Series, vol. 2. New York: John Wiley & Sons, Inc., 1964.

Hunter, L. P., *Handbook of Semiconductor Electronics* (2nd ed.). New York: McGraw-Hill Book Company, 1962. (This book has a very extensive bibliography.)

Lindmayer, Joseph and C. Y. Wrigley, *Fundamentals of Semiconductor Devices*. New York: D. Van Nostrand Company, Inc., 1965.

Pierce, J. F., *Semiconductor Junction Devices*. Columbus, Ohio: Charles E. Merrill Publishing Company, 1967.

Pierce, J. F., *Transistor Circuit Theory and Design*. Columbus, Ohio: Charles E. Merrill Publishing Company, 1963.

Ryder, J. D., *Electronic Fundamentals and Applications* (3rd ed.). Englewood Cliffs, N. J.: Prentice-Hall, Inc., 1964.

Shive, J. N., *Physics of Solid State Electronics*. Columbus, Ohio: Charles E. Merrill Publishing Company, 1967.

Shive, J. H., *Semiconductor Devices*. New York: D. Van Nostrand Company, Inc., 1959.

Shockley, W., *Electrons and Holes in Semiconductors*. New York: D. Van Nostrand Company, Inc., 1950.

Yu, Albert Y. C., "The Metal-Semiconductor Contact: An Old Device with a New Future," *IEEE Spectrum*. (March, 1970), pp. 83–89.

Zettler, Robert A. and A. Michael Cowley, "Hybrid Hot Carrier Diodes," *Hewlett-Packard Journal*, 20, No. 6 (February, 1969), pp. 13–19.

3

Field-Effect Transistors

A field-effect transistor (FET) is a device consisting of a semiconductor current path whose resistance is a function of the intensity of a transverse electric field. Because of its many desirable characteristics, the FET is one of the most promising active (amplifying) semiconductor devices that has been developed in recent years.

The first attempts to construct active devices based on the field-effect principle were disclosed in three patents issued to J. E. Lillienfeld in 1930–35. Further attempts were made by a group led by W. Shockley in the latter part of the 1940s. Although some degree of success was achieved, the devices produced at that time did not perform well enough to be practical. However, the point-contact transistor, the PNP and NPN junction transistors, and many other solid-state devices were apparently an outgrowth of this work.

In 1952, Shockley proposed a P-N junction structure for the FET that significantly improved its characteristics. Because of the better understanding of surface states and improved fabrication techniques, his proposed device made possible the excellent field-effect devices that are now commercially available.

The operation of this single-junction device is based on a process called *conductance modulation.* In this process, the depletion region associated with a reverse-biased P-N junction causes the conductance of a small semiconductor bar to vary as the depth of the depletion region is varied. Since the depth of the depletion region is a function of the reverse-bias voltage, the conductance of the current path can be made to vary as a function of the applied reverse voltage.

In 1959, techniques for growing thin films of silicon dioxide (SiO_2) on the surface of semiconductors were introduced. This development paved the way for the insulated-gate field-effect transistor (IGFET), now called the metal-oxide-semiconductor field-effect transistor (MOSFET). Unlike the junction FET (abbreviated JFET), the control element in the MOSFET transistor is electrically insulated from the semiconductor channel.

JFET and the MOSFET units are discussed separately in this chapter. In each case, we will begin with the physical operation and the resulting static characteristics of the device. The bias problem will then be discussed, ac equivalent circuits will be developed, and finally, some simple applications will be presented.

3.1 The Junction Field-Effect Transistor (JFET)

The junction field-effect transistor (JFET) is basically a reverse-biased P-N junction which has one side of the junction lightly doped. The complete structure of one form of a JFET is shown in Fig. 3–1. Unlike the current flow in a junction diode, the principal current flow is restricted to the lighter doped material through the center of the device; i.e., the major part of the current does not cross a junction. One way to construct a JFET is to start with a block of lightly doped semiconductor (5 to 10 ohm-cm) and form a P-N junction by heavily doping the two regions on each side of the block. Metal-to-semiconductor ohmic contacts are then attached to each end of the block and are designated the *source* and *drain* terminals. The lightly doped material between the source and drain terminals is called the *channel*. The heavily doped regions are called *gates* and have ohmic contacts to accommodate external connections.

In a two-gate JFET, the gate leads are normally connected together. Since the gate regions are more heavily doped than the channel, the depletion region of the P-N junction will lie mainly in the channel. For an N-type channel and a given bias voltage, the depletion region extends into the channel a distance* of W_N where

$$W_N \simeq \sqrt{\frac{2K\varepsilon_o V}{qN_D}} \tag{3–1}$$

where the voltage V is the total junction voltage and is the algebraic sum of three voltages: (1) the barrier voltage, V_B, (2) the bias voltage, V_{GS}, and (3) any voltage drop produced by current flowing in the channel.

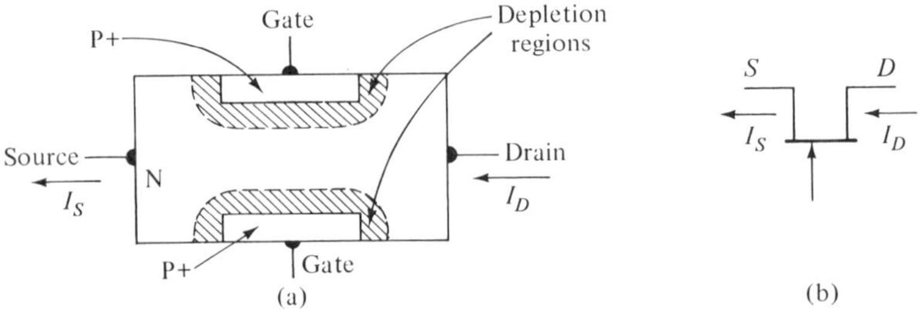

FIGURE 3–1 *An N-channel JFET: (a) cross-sectional view showing the channel and the depletion regions, and (b) symbol for an N-channel JFET. Note that the dc current enters the drain terminal for an N-channel unit.*

In Fig. 3–2(a), the switch in the drain lead is open, and no appreciable channel current flows. The width of the depeletion region increases as the value of V_{GS} increases. Note that V_{GS} has a polarity that reverse biases the gate-channel P-N junction. The depletion region will extend completely across the channel [Fig.

* The width of the depletion region is given in a footnote in Sec. 2.4.

3–2(b)] if V_{GS} is made sufficiently large. The condition of a completely depleted channel is called *pinch-off*. The minimum value of V_{GS} which will cause pinch-off is called the *pinch-off voltage*, V_P. The channel resistance then approaches that of an equivalent bar of intrinsic material with practical values of resistivity ranging from 5,000 to 10,000 ohm-cm as compared with 5 to 10 ohm-cm for a non-depleted channel.

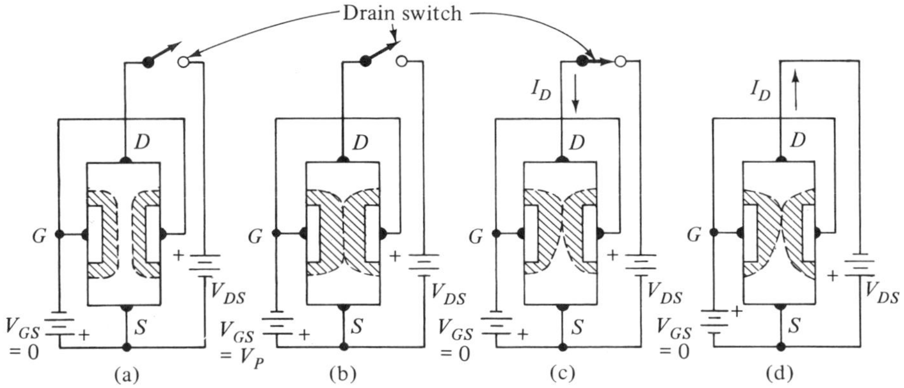

FIGURE 3–2 *Junction field-effect transistors for different bias conditions: (a) depletion region for an N-channel unit for* $V_{GS} = 0$ *and no channel current flowing, (b)* $V_{GS} = V_P$ *and no channel current flowing, (c)* $V_{GS} = 0$ *and channel current flowing, and (d) the depletion region for a P-channel device with* $V_{GS} = 0$ *and channel current flowing.*

With the switch closed in the drain circuit as shown in Fig. 3–2(c), channel current, I_D, flows. This current produces a voltage drop along the channel resistance. If the drain-to-source voltage, V_{DS}, increases, an increase in the drain or channel current, I_D, occurs. An increase in I_D increases the *IR* drop along the channel which increases the reverse bias of the gate-source junction and, in turn, increases the depth of the depletion region into the channel. This additional penetration increases the channel resistance and tends to reduce I_D. Consequently, I_D is a function of two voltages, V_{DS} and V_{GS}. The major theoretical problem associated with a JFET is to determine the relationship between I_D and these two voltages. Some appreciation of this problem may be obtained by relating the terminal characteristics of a JFET with the physical operation discussed above.*

One curve of a set of characteristic curves is shown in Fig. 3–3. This curve shows how I_D increases as the magnitude of V_{DS} increases for a P-channel unit when V_{GS} is zero or when the gate is connected directly to the source. Physically, the $V_{GS} = 0$ curve is generated in the following manner. Let V_{DS} increase from zero to some small value, ΔV_{DS}. This change causes I_D to increase** an amount

---

* At this point we can attach some meaning to the designation of *source* and *drain*. The channel majority carriers flow from the source terminal to the drain terminal. For an N-channel device [Fig. 3–2(c)], the majority carriers are electrons which move from the source to the drain, corresponding to the conventional current flow, I_D, shown in Fig. 3–2(c). For a P-channel device [Fig. 3–2(d)], the majority carriers are holes which also enter at the source and move in the same direction as conventional current. A study of Fig. 3–2(c) and (d) will show that pinch-off always occurs in that part of the channel nearest to the drain.

** I_D increases almost linearly with V_{DS} for small values of V_{DS} as shown in Fig. 3–3. However, this linear relationship does not hold for values of V_{DS} in the knee of the curve.

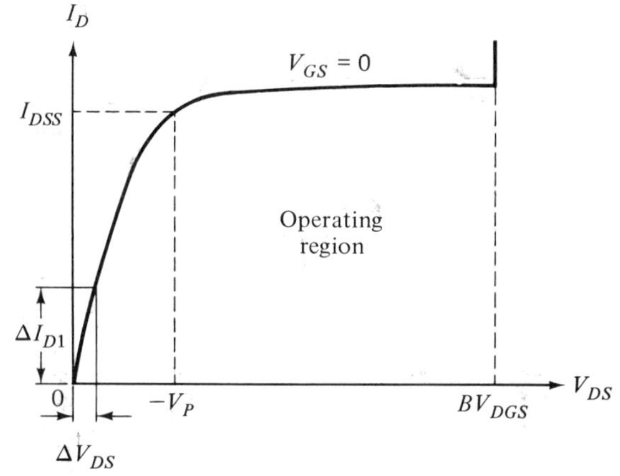

FIGURE 3–3 *The static characteristic curve for* $V_{GS} = 0$ *showing symbols used in equations.*

equal to ΔI_{D1}. However, as I_D increases, the incremental voltage drop increases along the channel. This voltage drop provides a reverse bias on the gate-to-channel junction that increases from zero at the gate end to a magnitude of ΔV_{DS} near the drain end of the channel. Since this increase in reverse-bias increases the width of the depletion region progressively along the channel [Fig. 3–2(c)], the channel width is reduced progressively along the channel causing an increase in channel resistance. Now let V_{DS} increase by an additional amount ΔV_{DS}. The drain current, I_D, also increases, but the increase will be less than ΔI_{D1} since the channel resistance has increased. Consequently, as V_{DS} increases in steps of ΔV_{DS}, I_D increases less and less (due to the continuing increase in channel resistance) until it is essentially independent of increases in V_{DS}, as shown in Fig. 3–3. This independence occurs beyond pinch-off, or when V_{DS} becomes greater* than $|V_P|$ and causes the depletion region to extend completely across the channel at the drain end. For a particular transistor, the pinch-off voltage, V_P, may be estimated from the knee of the $V_{GS} = 0$ curve as shown in Fig. 3–3. The value of I_D with $V_{DS} = -V_P$ and $V_{GS} = 0$ is designated I_{DSS} as shown in Fig. 3–3.

Increasing V_{DS} to relatively high values causes a reverse-voltage avalanche breakdown in the gate-to-channel junction. This phenomenon is discussed in Sec. 2.7. On the curve for which V_{GS} is zero, the breakdown voltage is symbolized as BV_{DGS} and refers to the breakdown voltage between drain and gate with the source disconnected. As shown in later sections, values of V_P, I_{DSS}, and BV_{DGS} are sufficient to characterize the JFET for small-signal amplifier analysis and design.**

Two sets of static drain characteristic curves are shown in Fig. 3–4, one for a P-channel JFET and one for an N-channel unit. In each case the top curve is the one discussed above, or the $V_{GS} = 0$ curve. Additional curves are shown for constant values of V_{GS}. In addition, a pinch-off curve is shown for each set. As

---

* The sign of V_P is the same as that of V_{GS} and hence opposite in sign to V_{DS}.

** Experimental determination of V_P and I_{DSS} can be difficult if the $V_{GS} = 0$ curve has a rounded knee. A more exact method, which involves cross-checking device parameters, is outlined in Sec. 3.5.

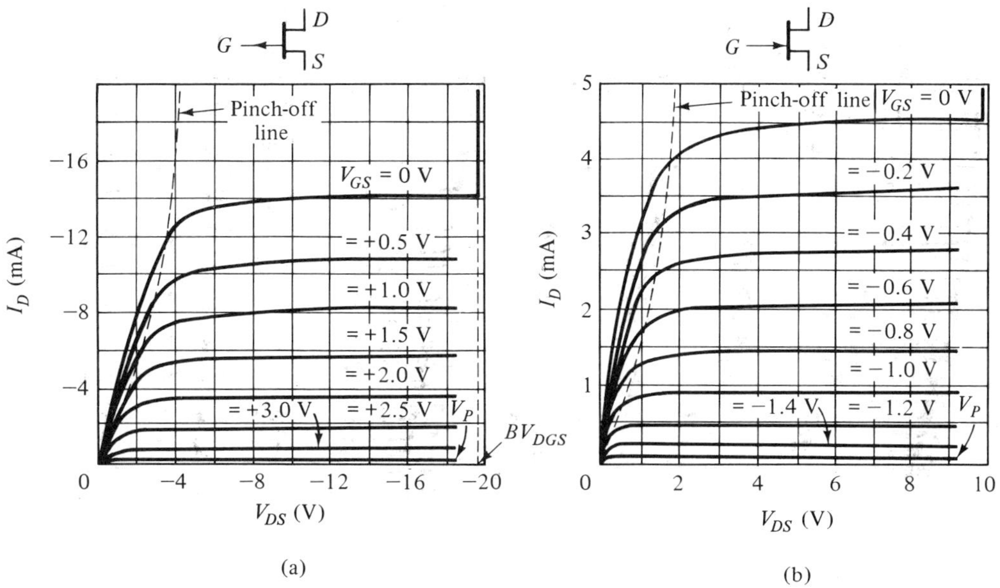

FIGURE 3–4 *The static drain characteristic curves and symbols for (a) a P-channel JFET and (b) an N-channel JFET. Note that the pinch-off line is the locus of the* $V_{DS} + V_{GS}$ *values that extend the depletion region completely across the channel at the drain end.*

indicated above, the pinch-off voltage, V_P, occurs at the knee of the $V_{GS} = 0$ curve. However, if a reverse bias voltage, V_{GS}, is applied to the junction, a smaller value of V_{DS} is required to deplete the channel at the drain end. When the depletion region extends across the channel [see Fig. 3–2(c) and (d)], I_D becomes relatively independent of V_{DS}, and the slope of the curve approaches zero. Note that when the reverse bias voltage, V_{GS}, is sufficiently large, the depletion regions extend completely across the channel. This value of V_{GS} is equal in magnitude to V_P.

EXAMPLE 3–1 Estimate the values of V_P, I_{DSS}, and BV_{DGS} for the device whose characteristics are given in Fig. 3–4(b).

Solution: The value of V_P can be estimated by either finding the knee of the $V_{GS} = 0$ curve or by finding the value of V_{GS} which reduces I_D to approximately zero. Using the latter method, V_P is approximately -1.6 V. From the $V_{GS} = 0$ curve with $V_{DS} = -V_P$, I_{DSS} is approximately 4 mA, and BV_{DGS} is 10 V.

In Fig. 3–4, the family of drain characteristic curves is generated by assigning values to V_{GS} ranging from 0 to V_P and then varying V_{DS} from zero to values less than BV_{DGS} while measuring I_D. The pinch-off voltage depends on both V_{GS} and V_{DS} and will occur when

$$|V_{GS}| + |V_{DS}| = |V_P| \quad \textbf{(3–2)}$$

This relationship is shown as the pinch-off line in Fig. 3–4. The region to the left of this line is called the *ohmic* region. In this area, the JFET can be used as a voltage-controlled resistor, and the operation is true conductance modulation. The region to the right of the pinch-off line is called the *saturated* or *pinch-off* region.

Equations relating I_D to V_{GS} in the saturated region can be developed if the details of the device construction and geometry are known. Most modern JFETs have a planar structure and are fabricated by a diffusion process. The details of this manufacturing process for a P-channel unit are shown in Fig. 3–5. Fabrication begins with a heavily doped N-type (N+) substrate [Fig. 3–5(a)]. A layer of silicon dioxide (SiO_2) is then grown on the substrate surface* [Fig. 3–5(b)]. A portion of this layer is then removed by a photolithographic etching process, thereby creating an opening through which P-type impurities in the gaseous state can diffuse into the N+ substrate [Fig. 3–5(c)]. The remaining oxide layer forms a mask which confines the impurities to those regions of the substrate under the etched opening. A second oxide layer is then grown [Fig. 3–5(d)], and a second, smaller window etched [Fig. 3–5(e)]. N-type impurities are then diffused into the P-type region [Fig. 3–5(e)]. A third oxide layer [Fig. 3–5(f)] is grown, and openings etched to provide access to the isolated N+ regions and each end of the P region [Fig. 3–5(g)]. Ohmic contacts are formed (often by the vacuum deposition of aluminum) to provide source, gate, and drain terminals [Fig. 3–5(g)]. The depletion regions of the completed P-channel JFET are shown in Fig. 3–5(h).

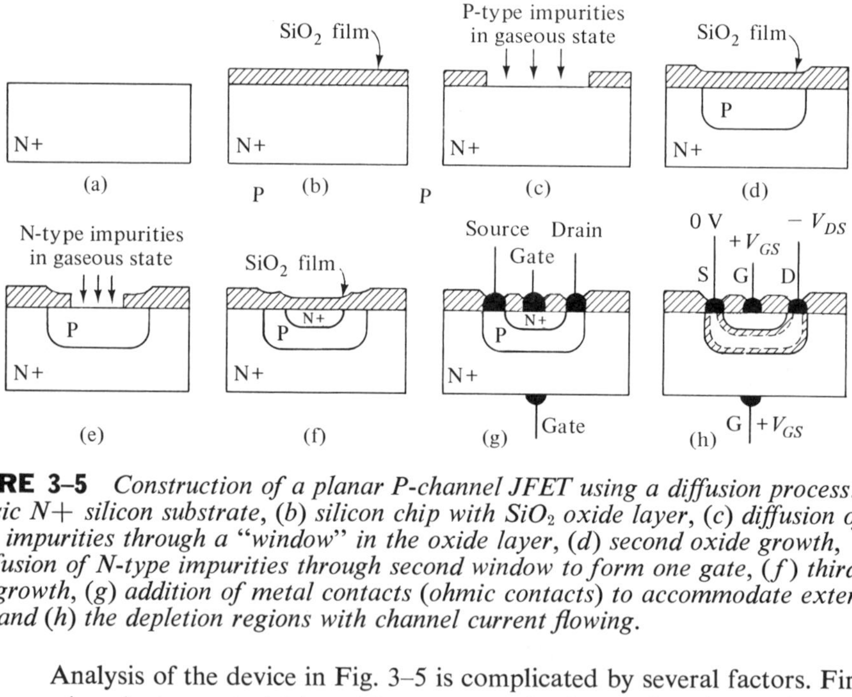

FIGURE 3–5 *Construction of a planar P-channel JFET using a diffusion process: (a) basic N+ silicon substrate, (b) silicon chip with SiO_2 oxide layer, (c) diffusion of P-type impurities through a "window" in the oxide layer, (d) second oxide growth, (e) diffusion of N-type impurities through second window to form one gate, (f) third oxide growth, (g) addition of metal contacts (ohmic contacts) to accommodate external leads, and (h) the depletion regions with channel current flowing.*

Analysis of the device in Fig. 3–5 is complicated by several factors. First, the semiconductor material is not homogeneously doped since it was formed by a diffusion process. Second, the mobility is not constant due to the high electric fields present in the channel. An analysis which includes these effects, and which

* Growing this oxide layer is referred to as "surface passivation."

is beyond the scope of this text, indicates that the drain current in the saturated region for planar JFETs is given by

$$I_D = I_{DSS}\left(1 - \frac{V_{GS}}{V_P}\right)^2 \tag{3-3}$$

Beyond pinch off, the curves of I_D rise slowly with increasing V_{DS}. Since the physical action in the channel beyond pinch off is not well understood, no attempt will be made here to explain this increase in current.

3.2 The Metal-Oxide-Semiconductor Field-Effect Transistor (MOSFET)

As indicated in Sec. 3.1, JFETs are constructed so that the depletion region associated with a reverse biased P-N junction can change the channel conductivity when the reverse voltage is changed. A similar effect can be achieved by using a gate that is separated from the channel with a very thin layer of insulating material. The channel conductivity is then varied by attracting or repelling carriers into or out of the channel by an opposite charge on the gate. As shown in Fig. 3–6, this device is constructed with a metal gate, an insulator, and a semiconductor channel, and is called a *metal-oxide-semiconductor field-effect transistor* (MOSFET). The gate and channel constitute two plates of a capacitor which are separated by a layer of oxide having a thickness in the order of 1000 angstroms (10^{-5} cm). Placing a charge on the gate attracts an equal but opposite charge into the semiconductor channel. If the channel is N-type, and a positive charge is placed on the gate, more electrons are attracted into the channel. With more carriers in the channel, the resistance of the channel decreases and allows more current to flow for a given value of drain-to-source voltage, V_{DS}. However, if a negative charge is placed on the gate, some of the electrons are repelled out of the channel, and the resistance of the channel increases. With fewer carriers available, less channel current can flow for a given value of V_{DS}. Since the gate is electrically insulated from the channel,* the gate voltage may be either positive or negative.

The insulated gate device allows the designer an added degree of freedom. The channel may be doped opposite that of the lightly-doped substrate as shown

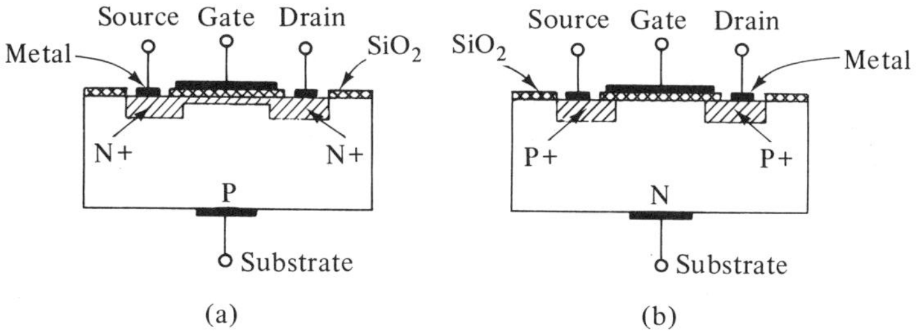

FIGURE 3–6 *Construction of MOSFET devices: (a) N-channel depletion-mode device, and (b) P-channel enhancement-mode device. Note that an enhancement-mode MOSFET has no channel in the unbiased state.*

* The gate of a JFET is not electrically insulated from the channel. Consequently, the gate-channel junction must be reverse biased.

in Fig. 3–6(a), or the substrate may extend all the way to the oxide as shown in Fig. 3–6(b). The first case is called a *depletion-mode device*. In this device, drain current flows with zero gate voltage as shown in Fig. 3–7(a) for an N-channel unit. Note that for this unit, drain current can be controlled with both plus and minus values of gate voltage. If the lightly-doped substrate extends to the insulator, little or no drain current flows for a positive gate voltage and an N-type substrate* [as shown in Fig. 3–7(b)] because no channel exists. However, negative gate voltages attract holes into the region under the insulator and create a P-type channel connecting the P^+-type source and drain materials. These holes allow drain current to flow when V_{DS} is applied. This type of device is called an *enhancement-mode MOS transistor*. Note that in Fig. 3–7(b), −6 V must be applied to the gate of this unit to support a small amount of drain current. This value is called the *turn-on voltage*, V_T, and is a design parameter of the device. Some enhancement-mode devices are fabricated with a combination silicon dioxide (SiO_2) and silicon nitride (Si_3N_4) insulator and a doped silicon gate. These changes tend to reduce the magnitude of the turn-on voltage and, in general, improve device performance.

As in the JFET, the drain current in MOSFET devices is a function of both V_{GS} and V_{DS}. In addition, the flow of drain current produces a voltage drop along the channel that reduces the cross section of the conducting part of the channel. Thus, for a given value of V_{GS}, some value of channel current will produce pinch-off. Consequently, for a fixed value of V_{GS}, I_D increases with V_{DS} up to pinch-off. Beyond pinch-off, I_D remains relatively constant as V_{DS} increases (see Fig. 3–7).

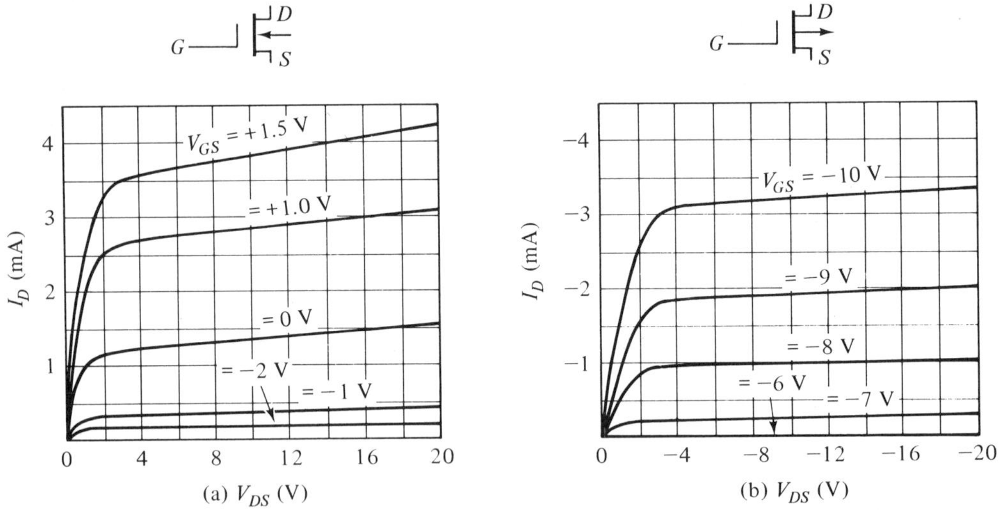

FIGURE 3–7 *Static characteristics and symbols for MOSFET devices:* (*a*) *N-channel depletion-mode unit, and* (*b*) *P-channel enhancement-mode unit.*

* Current cannot flow in a MOSFET unless a channel connects the source and drain inside the device. In addition, the channel must be the same type as the source and drain (P-type channel for a P^+ source and drain). For a positive voltage and an N-type substrate the P-N junction extends to the insulator and blocks the flow of current.

For the reader who is interested in band theory, some insight into the physical operation of the MOSFET can be obtained by examining the energy levels at the surface of a semiconductor and also at the semiconductor-oxide interface. Inside the crystal, each atom contributes four electrons to the covalent bonds. On the surface, an atom is missing on the outside; therefore, each surface atom has only 7 covalent electrons instead of 8. The empty space is sometimes called a "dangling" bond. Since a cubic centimeter of silicon contains approximately 10^{22} atoms, each face of a centimeter cube could have a maximum* of 10^{14} to 10^{15} dangling bonds. These dangling bonds trap electrons and produce a sheet of negative charge on the surface. These electrons, in turn, attract holes to the region just under the surface. The resulting energy-band diagram near the surface of N-type material is shown in Fig. 3–8(a). Note that the energy levels near the surface indicate that this region is P-type.** This condition is called an *inversion layer* and is produced by the neutralization of the large number of negative charges on the surface by holes just below the surface.

When an oxide is grown on the surface of a crystal, many of the surface atoms of the crystal that would normally have dangling bonds are used in the formation of the SiO_2. A "clean" interface is produced with an equivalent surface charge of approximately 10^{11} charged particles per square centimeter. In growing the oxide at high temperatures, the region under the oxide becomes more N-type than the substrate. This condition is shown in Fig 3–8(b) for a P-type substrate.

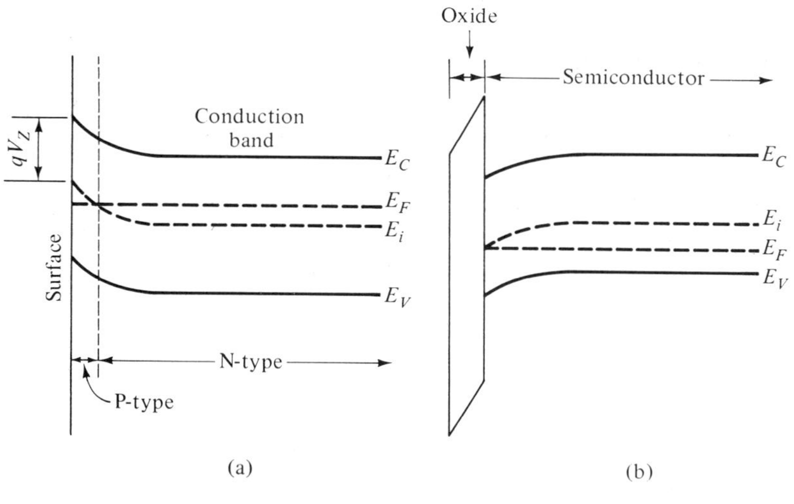

FIGURE 3–8 *Energy-band diagrams near a discontinuity: (a) an inversion layer near a surface for an N-type substrate, and (b) changes near an oxide-semiconductor interface for a P-type substrate. Note that various degrees of band bending can be produced by modifying the doping level.*

* This number might be expected on an ideal surface. In practice, the surface attracts such impurities as water vapor, gases, etc. These impurities contaminate the surface so that the number of dangling bonds on an exposed surface is not easily predictable.

** From Chapter 1, we noted that the Fermi level is below the band center (or intrinsic Fermi level) for P-type and above for N-type.

Since E_i and E_F intersect at the interface, the region just under the oxide is intrinsic. This equilibrium condition would be desired for an enhancement-mode MOS transistor with an off-set voltage of approximately zero.

A depletion-mode unit with a P-type substrate requires that the region under the oxide layer be N-type when the gate voltage is zero as shown in Fig. 3–9(a).

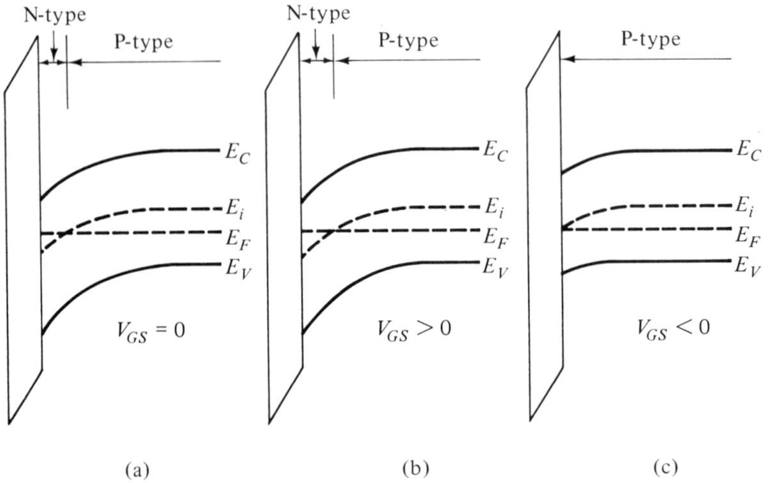

FIGURE 3–9 *Energy-band diagrams near an oxide-semiconductor interface for a built-in N-type channel: (a) equilibrium conditions, (b) positive gate voltage, and (c) negative gate voltage equal to the cutoff value,* V_P. *Note how the gate voltage changes the point of intersection of* E_i *and* E_F.

Increasing the gate voltage above zero causes the bands to bend more [as shown in Fig. 3–9(b)] thereby increasing the depth of the channel. A negative gate voltage causes the bands to become less curved and decreases the depth of the channel as shown in Fig. 3–9(c). The value of negative gate voltage that produces the condition shown in Fig. 3–9(c) is the cutoff voltage, V_P, since this voltage just makes the region under the oxide intrinsic.

Theoretical analyses and experimental data indicate that the depletion-mode MOSFET obeys the same square-law relationship as the JFET. For a depletion-mode MOSFET, the drain current is

$$I_D = I_{DSS}\left(1 - \frac{V_{GS}}{V_P}\right)^2 \tag{3–4}$$

where the values of V_P and I_{DSS} can be determined from the $V_{GS} = 0$ curve in the same manner as for the JFET. An important difference between the JFET and the depletion-mode unit is that the latter can have both positive and negative values of V_{GS}.

Since the enhancement-mode MOSFET does not conduct for values of V_{GS} less than the off-set voltage, values of current for V_{GS} equal to zero have no meaning. The parameter I_{DSS} was defined as the drain current when $V_{GS} = 0$, and $V_{DS} = -V_P$ and, consequently, does not apply to the enhancement-mode MOSFET. However, I_D is still related to V_{GS} by a square-law relationship, namely,

$$I_D = I_{DO}\left(1 - \frac{V_{GS}}{V_T}\right)^2 \tag{3-5}$$

where V_T is the turn-on voltage, and I_{DO} is the saturated value of drain current for $V_{DS} = V_{GS} - V_T$. Determination of the value of I_{DO} and the use of Eq. (3–5) is demonstrated in the following example.

EXAMPLE 3–2 Estimate the value of I_{DO} for the enhancement-mode MOSFET whose characteristics are given in Fig. 3–7(b). Using this value of I_{DO}, predict the saturated value of I_D for $V_{GS} = -8$ V.

Solution: From Fig. 3–7(b), the value of turn-on voltage is approximately

$$V_T \simeq -6 \text{ V}$$

The value of I_{DO} can be estimated using any V_{GS} curve. For instance, set V_{GS} equal to -10 V. Pinch-off will occur when

$$V_{DS} = V_{GS} - V_T = -10 - (-6) = -4 \text{ V}$$

The corresponding value of I_D at the knee of the -10-V curve is

$$I_D \simeq -3.2 \text{ mA}$$

Substituting these values into Eq. (3–5) and solving for I_{DO} yields

$$I_{DO} = \frac{I_D}{\left(1 - \frac{V_{GS}}{V_T}\right)^2} = \frac{-3.2 \text{ mA}}{\left(1 - \frac{-10}{-6}\right)^2} = \frac{3.2}{(-0.67)^2} \simeq -7.2 \text{ mA}$$

This value of I_{DO} can now be used to estimate other values of I_D in the saturation region. At

$$V_{GS} = -8 \text{ V}$$

$$I_D = I_{DO}\left(1 - \frac{V_{GS}}{V_T}\right)^2 = -7.2\left(1 - \frac{8}{6}\right)^2 \simeq -0.8 \text{ mA}$$

which compares with approximately -0.9 mA obtained from the curves in Fig. 3–7(b) for V_{DS} equal to $(V_{GS} - V_T)$ or -2 V. As with the JFET, the MOSFET curves of I_D beyond pinch-off rise slowly as V_{DS} increases.

In this and the preceding section, the physical operation, characteristic curves, and important mathematical relationships of two field-effect devices have been studied. Let us compare these devices on the basis of their potential application. The MOSFET, both depletion and enhancement units, has much lower gate current values than the JFET since the channel and gate are separated by an insulator rather than a P-N junction. Consequently, the input impedance is considerably higher. However, fabrication difficulties have been encountered with the MOSFET. Often their characteristics are functions of both time and use, making it difficult to predict their performance with accuracy. The changes in device characteristics are produced primarily by ion drift in the oxide film. The oxide layer of the MOSFET can also be punctured easily by the static charge accumulated by an individual.* In contrast, the JFET has stable characteristics as a function of both time and use and has no oxide layer to be punctured. However, the MOSFET possesses certain economic advantages. These units can be fabricated with a single diffusion process followed by the formation of multiple oxide layers and then an etching process, the latter being easy to achieve with reliability. This sequence can be compared with the multiple diffusion processes that are required to fabricate a JFET. The depletion-mode MOSFET has greater freedom in biasing since it can accept both positive and negative values of V_{GS}. The circuit simplicity, small size, and theoretically high manufacturing yield of the enhancement-mode MOSFET make it attractive for use in logic or switching circuit applications, particularly in large-scale integration (LSI) of logic functions. These reasons have prompted extensive research aimed at improving the characteristics and reliability of MOSFET devices. However, emphasis in this chapter will be placed on the JFET because of its ease of handling and commercial availability. In most applications, a MOSFET could be substituted for a JFET with little change in circuitry.

3.3 Bias Circuits for Field-Effect Transistors

All active (amplifying) devices must be biased properly before they will perform a given function. However, the desired bias condition is determined by the particular application. In this chapter we will be concerned primarily with linear applications.

In linear amplifiers, the active device is biased so that a desired amount of dc current flows through the device. For an FET, V_{GS} and V_{DS} are adjusted to cause a desired amount of channel current (or I_D) to flow. One circuit that can be used for this purpose is shown in Fig. 3–10(a). Once the FET is properly biased, an ac voltage can be applied to the input terminals and cause the channel current to vary. The bias conditions selected should allow the variation in the channel current to exactly follow the input voltage changes; i.e., if the input voltage is a sine wave, then the variation in I_D should also be a sine wave. The value of I_D at the bias point**is selected with this linear relationship in mind. If the input signal is extremely small, then the selection of I_D and V_{GS} is not critical. However, if the input signal is reasonably large (one volt or more), I_D and V_{GS} must be selected carefully to avoid nonlinear effects. While the choice of a bias point is generally not unique, the factors that influence a logical choice will be discussed for many conditions in the next few chapters.

Bias requirements can be conveniently divided into two catagories: (1) those imposed by the device, and (2) those imposed by the ac operation. From the static characteristic curves of an FET (as shown in Fig. 3–10), we observe that the values of the dc bias parameters, V_{DS}, V_{GS}, and I_D are restricted to certain

---

* If a person touches the gate lead with either the source or drain terminal grounded, sufficient static voltage may be applied to the oxide film to cause dielectric breakdown. Several precautions that can be taken are: (1) twist all leads together when the device is not in use, (2) connect all leads together with an alligator clip when putting the device into or removing it from a circuit, and (3) have the operator ground himself at the hand or wrist when handling a MOSFET. A simple method of grounding is to rest one hand on a grounded part of the equipment while handling a MOSFET.

** The bias point is also called the operating point and the Q point. We will use all three terms interchangeably.

ranges if the device is to function as a linear, active element. For example, the drain of an N-channel JFET must be positive with respect to both the source and gate terminals while the drain of a P-channel JFET must be negative with respect to these two terminals. In addition, all FET devices require a dc connection to the gate terminal to provide a path for any leakage current.

Like other semiconductor devices, the FET has two important maximum ratings: (1) the drain-to-gate breakdown voltage, and (2) the maximum device power dissipation. The first generally requires that the dc supply voltage, V_{DD}, be less than the breakdown voltage, BV_{DGS}, since V_{DG} is approximately equal to V_{DD} when I_D is reduced to zero or very small values. The second maximum rating imposes restrictions on the product of V_{DS} and I_D. For example, if $P_{D\max}$ is 300 mW, and V_{DS} is 20 V, then I_D should not exceed 15 mA.

Ac considerations add further restrictions on the selection of V_{DS} and V_{GS}. If the ac input voltage is applied to the gate, the peak values of the ac signal plus V_{GS} should not cause the drain current to go to zero at one extreme or force the gate diode of a JFET into forward conduction at the other extreme. In addition, the Q point (or operating point) should be selected to give the desired ac gain. In general, linear amplifiers are not biased in the triode region because the nonlinearities in this region produce excessive signal distortion. These problems will be discussed later in this chapter. To summarize for a JFET: (1) values of I_D may range from zero to I_{DSS}, (2) values of V_{GS} may range from approximately zero to V_P, (3) values of V_{DS} may range from $|V_P|$ to BV_{DGS}, and (4) the device dissipation may range from zero to $P_{D\max}$. With these and possibly other factors involved in the selection of the Q-point, the choice is usually not unique.

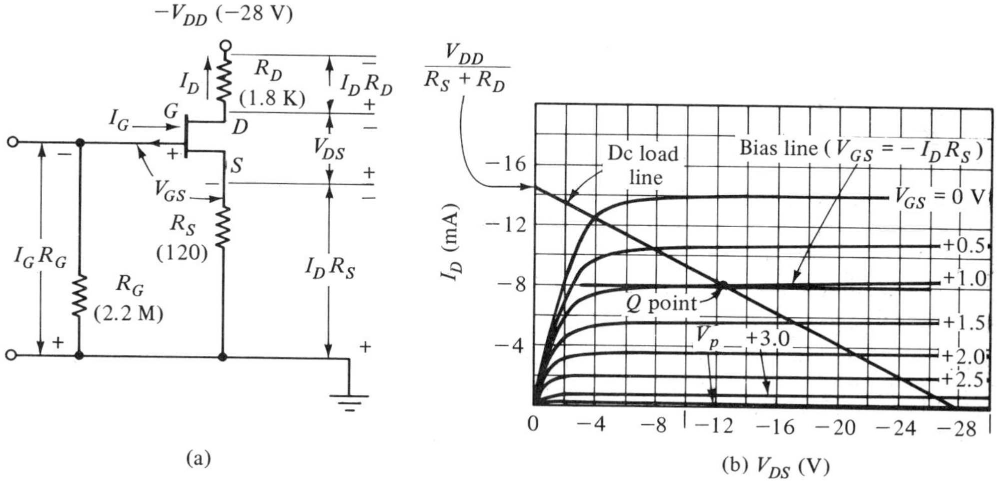

FIGURE 3–10 *(a) A dc bias circuit for a P-channel JFET, and (b) the graphical construction necessary for finding the Q-point values.*

Considering these restrictions on the bias point, let us study the entire dc circuit for the P-channel JFET shown in Fig. 3–10(a). One important simplification that can be made is that the gate current, I_G, is very small (10^{-13} to 10^{-8} A for a JFET and as low as 10^{-18} A for a MOSFET) and usually can be neglected. This approximation implies that the drain and source currents are equal and also that the voltage drop $I_G R_G$ is approximately zero. With this simplification, the following two dc circuit equations can be written:

$$V_{DD} = I_D R_D + V_{DS} + I_D R_S \tag{3–6}$$

$$V_{GS} = -I_D R_S \tag{3–7}$$

where the values of I_D, V_{DS}, and V_{GS} are those of the operating point. If any one of the Q-point values is known along with the circuit parameters, the other two can be found using Eqs. (3–6) and (3–7). If only the circuit is given, all voltages and currents can be found provided that either the static curves for the device or the values of V_P and I_{DSS} are known.

A graphical solution is necessary if only the static curves and circuit values are available. The necessary graphical construction is shown in Fig. 3–10(b). Eq. (3–6) can be rewritten as

$$I_D = \frac{V_{DD} - V_{DS}}{R_D + R_S} \tag{3–8}$$

which is an equation for a straight line with a slope of $-1/(R_D + R_S)$. Equation (3–8) is plotted in Fig. 3–10(b) and is called the *dc load line*. Equation (3–7) is also plotted in Fig. 3–10(b) by assuming values of I_D and solving for V_{GS}. This curve is called the *bias line*. The Q-point is located at the point of intersection of these two lines.* Construction of the bias line is often unnecessary since V_{GS} is usually chosen on the basis of ac performance.

EXAMPLE 3–3 In the circuit in Fig. 3–10(a), $R_D = 1.8$ K, $R_S = 120\ \Omega$, and $V_{DD} = -28$ V. The device characteristics are given in Fig. 3–10(b). Find the Q point values.

Solution: The necessary graphical construction is shown in Fig. 3–10(b). When I_D is zero, the full dc source voltage is developed across the transistor, or

$$V_{DS} = V_{DD} = -28 \text{ V}$$

Consequently, one point on the straight line occurs at $I_D = 0$ and $V_{DS} = -28$ V. Another point could occur when $V_{DS} = 0$. This condition implies that the voltage drop across $(R_D + R_S)$ is V_{DD}, or

$$I_D = \frac{V_{DD}}{R_D + R_S} = -14.4 \text{ mA}$$

The dc load line is drawn between these two points. Two points on the bias line can be obtained by assuming two values for I_D and finding corresponding values for V_{GS} from Eq. (3–8). Thus, for $I_D = -8$ mA,

$$V_{GS} = -(-8 \times 10^{-3} \times 120) = 0.96 \text{ V}$$

and for $I_D = -8.5$ mA,

$$V_{GS} = -(-8.5 \times 10^{-3} \times 120) = 1.02 \text{ V}$$

From Fig. 3–10, the load line and bias line intersect at approximately

$$V_{DS} = -12.5 \text{ V}, \qquad I_D = -8 \text{ mA}, \qquad V_{GS} = +1 \text{ V}$$

The bias values are usually measured with respect to the circuit ground because one

---

* Essentially, this technique amounts to a graphical solution of these simultaneous equations: the two circuit equations and the device characteristics. This approach is similar to the procedure used in Sec. 2.5 for diodes.

side of the measuring instrument is usually grounded. In this circuit, the measured voltages from source to ground, V_S, and from drain to ground, V_D, were

$$V_S = -1.0 \text{ V}, \qquad V_D = -12.5 \text{ V}$$

The bias values can be obtained from these two measured values as follows:

$$V_{DS} = V_D - V_S = -12.5 + 1.0 = -11.5 \text{ V}$$
$$I_D = V_S/R_S = -1 \text{ V}/120 = -8.3 \text{ mA}$$
$$V_{GS} = -V_S = +1 \text{ V}$$

The circuit shown in Fig. 3–10(a) is called a *self-bias* circuit because the voltage necessary to reverse bias the input P-N junction is developed by I_D flowing through R_S. This type of circuit is often encountered in practice and can be designed easily, as illustrated in the following example.

EXAMPLE 3–4 An N-channel JFET has values of $V_P = -3$ V, $I_{DSS} = 6$ mA, and $BV_{DGS} = 30$ V. Design a circuit so that the device will be biased at approximately $V_{DS} = 10$ V and $I_D = 4$ mA.

Solution: The bias circuit can be the same as shown in Fig. 3–10(a) except that the power supply, V_{DD}, is of opposite polarity. The value of V_{DD} is chosen to be less than BV_{DGS}, say 24 V. The required value of V_{GS} is found by substituting the known values into Eq. (3–4):

$$I_D = I_{DSS}\left(1 - \frac{V_{GS}}{V_P}\right)^2$$

$$4 \text{ mA} = 6 \text{ mA}\left(1 - \frac{V_{GS}}{3}\right)^2$$

or

$$V_{GS} = 0.54 \text{ V}$$

Solving for R_S and R_D,

$$R_S = \frac{V_{GS}}{I_D} \simeq 135\ \Omega \qquad R_D = \frac{V_{DD} - (V_{DS} + I_D R_S)}{I_D} \simeq 3.4 \text{ K}$$

The nearest values of standard 5 percent carbon resistors are $R_S = 120$ ohms, and $R_D = 3.3$ K.

Several FET bias circuit arrangements other than that shown in Fig. 3–10(a), are shown in Fig. 3–11. All devices shown are N-channel. For P-channel units, the polarity of all supply voltages should be changed. The circuit in Fig. 3–11(a) requires two power supplies to provide the necessary reverse bias for the gate-to-channel junction. The input resistance of the circuit is $R_1||R_2$ if the JFET input resistance is extremely large.

A "universal bias circuit" is shown in Fig. 3–11(b). The current I_D flowing through R_S establishes the source voltage V_S with respect to ground. Since the gate current is negligible, R_1 and R_2 may be selected to place the gate at any point from zero to V_{DD}. Consequently, V_{GS} may be positive, zero or negative. The purpose of R_G is to increase the ac input impedance and has very little effect on the dc bias circuit. The circuit in Fig. 3–11(c) also has a high input resistance.*

---

* Proof of this statement requires some knowledge of negative feedback. This topic is treated extensively in Chapter 10.

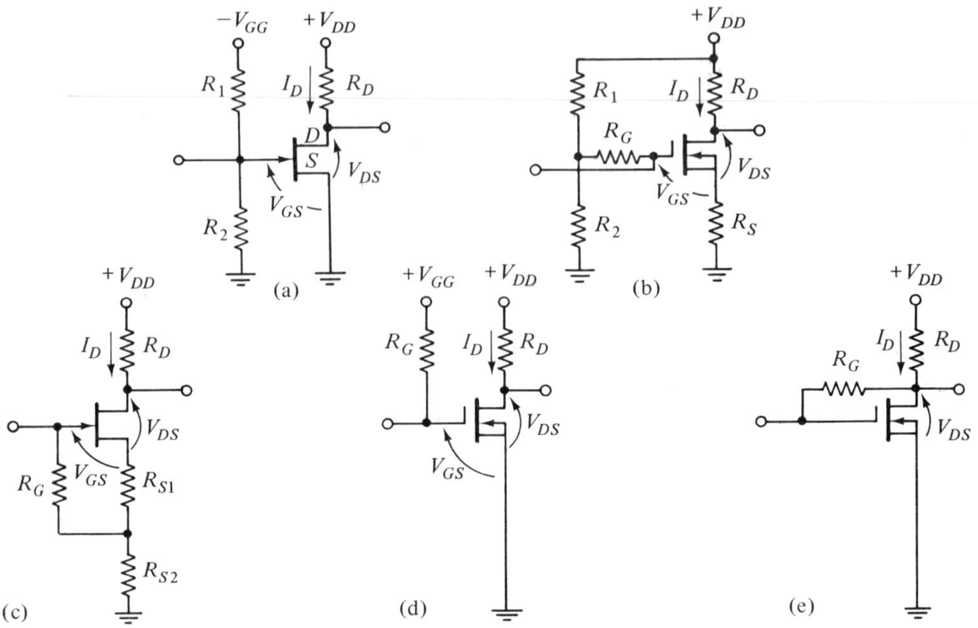

FIGURE 3–11 *Additional bias circuits for N-channel devices. See text for descriptions.*

The bias circuits in Fig. 3–11(d) and (e) are well suited for both enhancement mode and depletion mode units that require positive values of gate voltage. In Fig. 3–11(e), V_{GS} is equal to V_{DS}, a satisfactory condition for some applications. The resistor and power supply values for the circuits in Fig. 3–11 are specified in a manner similar to that for the bias circuit in Fig. 3–10(a).

All semiconductor devices exhibit some degree of temperature sensitivity. In a JFET, the barrier voltage and channel conductivity are both functions of temperature. The MOSFET drain and gate currents are temperature dependent. Unless the operating point, or Q point, is stabilized, it has a tendency to move up along the load line as the temperature of the FET increases. Circuits with self bias [Figs. 3–10(a), 3–11(b), and 3–11(c)] provide some degree of current stabilization for I_D through the $I_D R_S$ voltage drop. If I_D increases, $I_D R_S$ (and therefore V_{GS}) increases. An increase in V_{GS} reduces I_D. Thus, these circuits reduce the variations in the drain current, I_D, whether due to temperature changes or differences between units.

The two temperature effects noted above for the JFET produce opposite effects and, under certain bias conditions, can cancel each other. Then the drain current is independent of temperature changes: that is, it has a zero temperature coefficient. A plot of the drain current variations as a function of temperature for a particular N-channel JFET is shown in Fig. 3–12. The specific value of drain current in the saturated region which has a zero temperature coefficient is given by

$$I_D = I_{DSS}(0.59/V_P)^2 \tag{3–9}$$

for a P-channel silicon planar JFET, and by

$$I_D = I_{DSS}(0.63/V_P)^2 \tag{3–10}$$

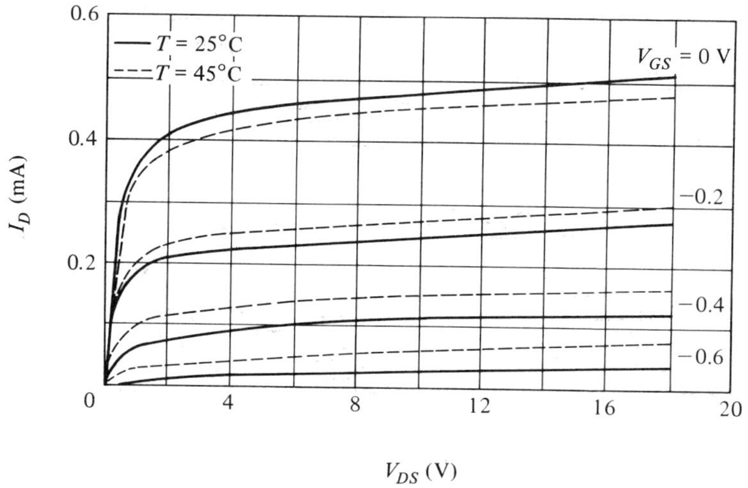

FIGURE 3–12 *A plot of drain current variations with temperature for a particular N-channel JFET.*

for an N-channel silicon planar JFET. Equations (3–9) and (3–10) are useful in designing circuits where the temperature effects are critical.

3.4 Large-Signal Analysis

Having biased the transistor for linear operation, the next problem is to determine the ac characteristics of the circuit. The analysis of FET amplifier circuits can be divided into two catagories: large-signal analysis and small-signal analysis. No fine line of distinction exists between the two types. A large-signal analysis is required when the input ac signal causes the FET current, I_D, to vary over a wide range and will be treated in this section. In this approach, a graphical analysis is used. If I_D varies over a small range, a small-signal analysis using equivalent circuits is a better approach and will be treated in the remainder of the chapter. In both types of analyses only linear signals are considered although amplitude distortion can be present under large signal conditions.

A three terminal device such as an FET is connected in an amplifier circuit with one terminal common to both the input and the output ac signal. This condition allows six possible configurations, three of which have practical value. The three practical FET configurations are the *common-source* (CS), the *common-gate* (CG), and the *common-drain* (CD) amplifiers with the designation determined by the common terminal. All three configurations will be discussed in this chapter. However, only the CS amplifier will be treated in this section.

A complete single-stage CS amplifier is shown in Fig. 3–13(a). An ac generator with an internal impedance R_g and an open circuit voltage V_g drives the gate of the JFET through C_1. The combination of R_g and V_g may represent either a signal generator or a Thévenin equivalent of the preceding circuit. Resistor R_G is the gate leakage resistor and provides a dc path to ground for any gate current. Capacitors C_1 and C_2 are coupling capacitors which allow different dc levels to exist on each side of each capacitor. C_S is a by-pass capacitor for the self-biasing source resistor R_S. These three capacitors act as short circuits to the ac signal in the mid-frequency range.* Resistor R_D determines the value of V_{DS} and is part of the ac load, R_{Leq}, on the amplifier. The other part of the ac load is R_L which, in general, is the input impedance to the next stage or to the measuring device. The dc power supply is an ac short circuit to ground. From this discussion, an ac mid-band equivalent circuit can be constructed as shown in Fig. 3–14(a) where the circuit capacitors have been replaced by short circuits and

$$R_{Leq} = R_D || R_L \qquad \textbf{(3–11)}$$

The objective of this section is to calculate the ac output voltage, V_o, when V_g and the circuit values are known.

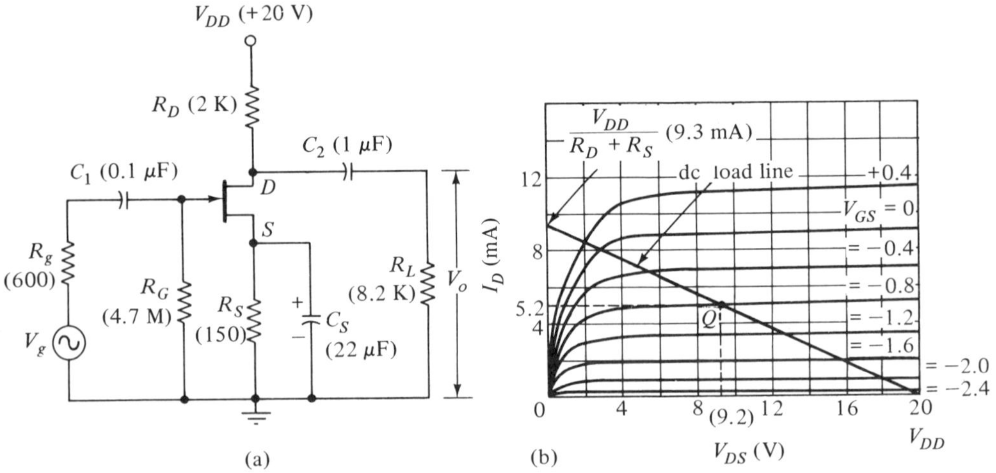

FIGURE 3–13 *A Common-Source (CS) Amplifier circuit using an N-channel JFET as the active device: (a) complete circuit, and (b) the device characteristics and dc load line. Values in parentheses apply to Ex. 3–5.*

The amplifying action of the circuit in Fig. 3–13 can be developed as follows. The JFET is dc biased at some value of V_{GS}, V_{DS}, and I_D. The input ac signal causes the instantaneous value of V_{GS} to change, which in turn causes I_D to change. The ac component of I_D develops an ac voltage across R_{Leq}. The waveform at the drain of the JFET will consist of two components: (1) a dc component equal to the bias voltage $V_D = V_{DD} - I_D R_D$, and (2) an ac component which is V_o. The voltage developed across R_L will not have a dc component (due to the blocking action of C_2). When V_g goes positive, it reduces the instantaneous value of V_{GS} causing I_D to increase. An increase in I_D increases the voltage drop across R_D, thereby causing the voltage at the drain terminal (and at the output

* Two types of capacitors are important in amplifier circuits: (1) the coupling and by-pass capacitors, and (2) the parasitic capacitance associated with the active device (which is similar to C_D and C_T for a diode and will be studied in Sec. 3.5). The first type is generally in the μF (10^{-6} F) range, and the second type in the pF (10^{-12} F) range. The magnitude of the impedance of a capacitor is given by $|Z| = |1/(2\pi fC)|$. Due to the order-of-magnitude difference between the two types of capacitors, there is some frequency range in which the coupling and by-pass capacitors are approximately ac short-circuits, and the device parasitic capacitors are approximately ac open circuits. This is the mid-frequency range. The extent of the mid-frequency range and the complete amplifier response as a function of frequency is developed in Chap. 7. Until then, we will assume all input signals are in the mid-frequency range.

terminal) to decrease. Consequently, the CS amplifier inverts the signals; i.e., a positive going input signal produces a negative going output signal.

The steps in a graphical analysis of the CS amplifier in Fig. 3–13(a) are as follows. First the dc load line is constructed, as shown in Fig. 3–13(b). Then the Q point is determined. Since the ac load is different from the dc load, a separate ac load line must be constructed. The ac load line passes through the Q point (corresponding to a zero input signal) and has a slope of $-1/R_{Leq}$. A convenient method for establishing a second point on the ac load line is to assume a change in current, ΔI_D, from the Q point and find the corresponding change in voltage, $\Delta V_{DS} = \Delta I_D R_{Leq}$. The ac load line is shown in Fig. 3–14(b).

The ac load line is the locus of all ac or instantaneous operating points. That is, when V_{GS} is changed by superimposing on it an instantaneous voltage V_g, the corresponding instantaneous change in either I_D or V_{DS} may be determined by projecting from the intersection of the $(V_{GS} + V_g)$ curve and the ac load line to the current or voltage axis. These points are illustrated in the following example.

EXAMPLE 3–5 Find the output voltage V_o for the circuit in Fig. 3–13(a) when $V_{GS} = -0.8$ V, $V_g = 2$ V peak-to-peak, and $V_{DD} = +20$ V. The JFET characteristics are shown in Fig. 3–13(b).

Solution: (a) The dc load line: Construct the dc load line starting at $V_{DD} = 20$ V on the voltage axis and terminating on the current axis at

$$I_D = \frac{V_{DD}}{R_S + R_D} = \frac{20\text{ V}}{2\text{ K} + 150} \simeq 9.3\text{ mA}$$

The Q point is located at the intersection of the dc load line and the V_{GS} bias curve (−0.8 V given in this case). The remaining bias values, obtained from the coordinates of the Q point, are

$$I_D \simeq 5.2\text{ mA} \qquad V_{DS} \simeq 9.2\text{ V}$$

(b) The ac load line: The ac load is

$$R_{Leq} = R_D \,||\, R_L = 2\text{ K} \,||\, 8.2\text{ K} \simeq 1.6\text{ K}$$

As indicated, the ac load passes through the Q point. To find another point on the ac load line, assume a convenient value for a change in current from the Q point, and find the corresponding change in voltage. A decrease in current of 5.2 mA from the Q point places I_D on the voltage axis since I_D at the Q point is 5.2 mA. A decrease in current causes a decrease in the voltage drop across R_{Leq} and increases the drop across the transistor. The increase in voltage across the transistor for a $\Delta I_D = 5.2$ mA is

$$\Delta V_{DS} = \Delta I_D R_{Leq} = (5.2\text{ mA})(1.6\text{ K}) = 8.3\text{ V}$$

Therefore, the decrease in current, ΔI_D, increases V_{DS} from 9.2 V to (9.2 + 8.3) or 17.5 V at $I_D = 0$. The ac load line is drawn through these two points as shown in Fig. 3–14(b).

(c) The output ac voltage: Since the JFET input impedance is very high (10 to 1000 M), the load on the generator is approximately equal to R_G or 4.7 M. Almost all of V_g is available across the gate-to-source terminals to swing V_{GS} around its Q point. The ac component of the gate-voltage swings V_{GS} to (−0.8 − 1.0) or −1.8 V on one peak of the ac input wave and to (−0.8 + 1.0) or +0.2 V on the other peak.* When

---

* The value of V_{GS} can go slightly positive provided the GS junction does not conduct a large forward current. However, driving V_{GS} positive for a JFET is not good practice.

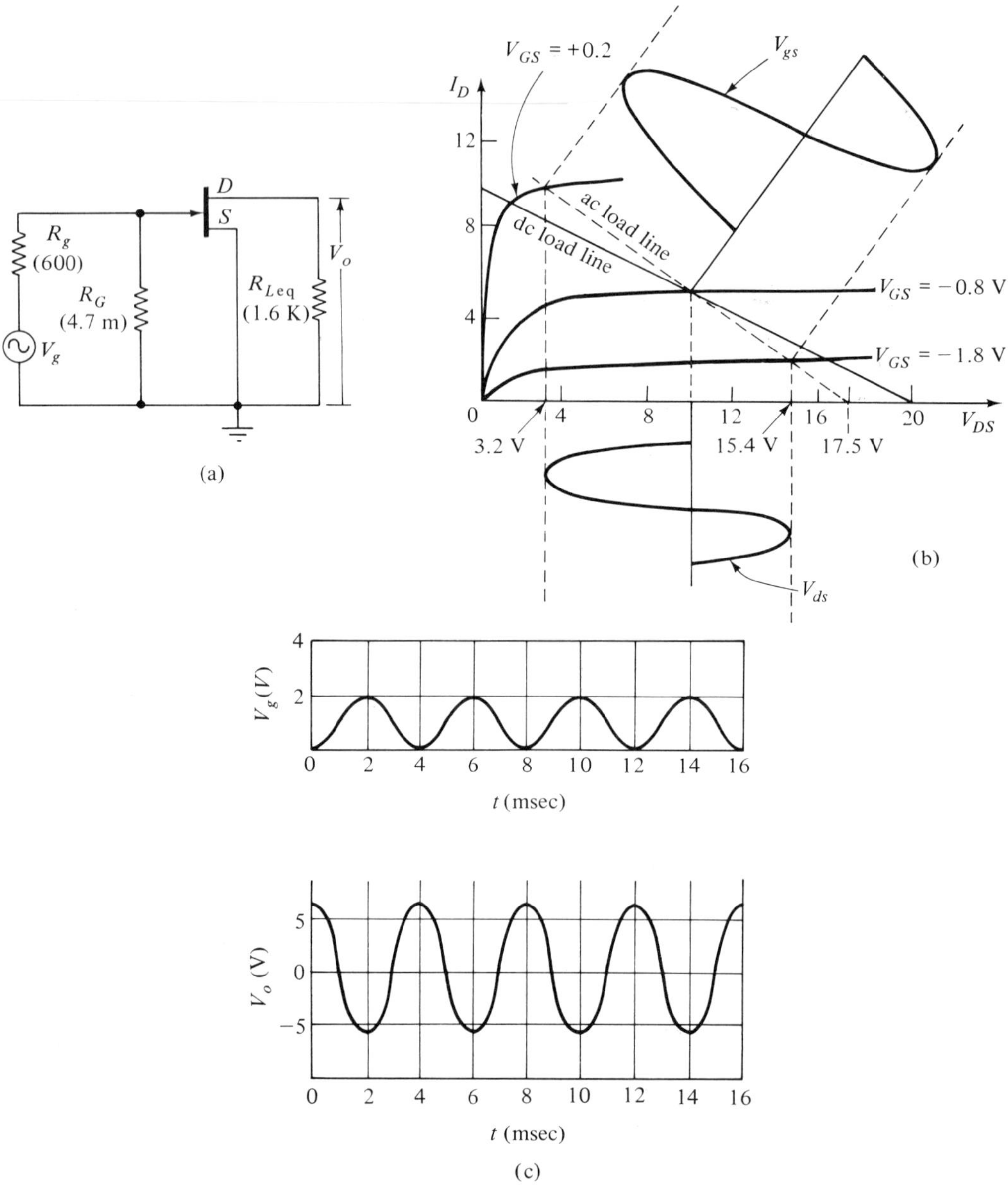

FIGURE 3–14 *The CS amplifier from Ex. 3–5: (a) mid-band equivalent circuit, (b) graphical analysis, and (c) input and output waveforms.*

$V_{GS} = +0.2$ V, the corresponding drain-to-ground* voltage is found by projecting the point of intersection of the $V_{GS} = +0.2$ V curve and the ac load line to the voltage axis as shown in Fig. 3–14(b). This voltage is 3.2 V. The corresponding voltage on the negative peak of the input signal is 15.4 V. The output voltage swings between these two values as shown in Fig. 3–14(b). This swing is not symmetrical around the Q point and is best described by

---

* C_S is a short circuit to the ac current; therefore, the instantaneous drain-to-ground voltage is also drain-to-source voltage.

$$V_o \text{ positive} = 15.4 - 9.2 = 6.2 \text{ V}$$
$$V_o \text{ negative} = 9.2 - 3.2 = 6 \text{ V}$$

Note that signal inversion has occurred. The gain of the amplifier is

$$A_{vo} = V_o/V_g \simeq -(6 + 6.2)/2 = -6.1$$

(d) Experimental results: The measured dc voltages with respect to ground were

$$V_G = 0 \text{ V} \qquad V_S = 0.84 \text{ V} \qquad V_D = 9.2 \text{ V}$$

The corresponding dc values at the Q point were

$$V_{GS} = -V_S = -0.84 \text{ V} \qquad I_D = V_{GS}/R_S = 5.6 \text{ mA}$$
$$V_{DS} = V_D - V_S = 9.2 - 0.84 \simeq 8.4 \text{ V}$$

These values are slightly different from the calculated values due to component tolerance (± 10 percent) and the tolerance of the JFET curves. The input and output waveforms are shown in Fig. 3–14(c). The reader should note that (1) the signal is inverted, (2) the output signal varies around zero volts dc, and (3) distortion is apparent on the negative peak (corresponding to a V_{GS} value slightly above zero). The measured gain for this signal is approximately 12.2/2 or 6.1.

3.5 FET Equivalent Circuits and Small-Signal Analysis

The output voltage in Ex. 3–5 was not symmetrical, thereby indicating that the nonlinearities of the device produced some amplitude distortion. This distortion is produced by the nonlinear characteristics of the active device. The relative distortion may be reduced, but not entirely eliminated, by reducing the amplitude of the input voltage swing. However, if the input signal is reduced to a small value, the output voltage values obtained graphically are difficult to read accurately.

Improved accuracy can be obtained by using a small-signal analysis in which the device is represented by a linear model valid over some limited operating region. In this approach, we assume that the device is properly biased for amplifier operation. In addition, only the variations from the Q points are considered. Consequently, the treatment is divided into two parts: (1) a dc or bias part in which the Q-point values are determined, and (2) an ac part in which the variations from the Q-point values are determined. A small-signal analysis treats only the ac problem.

Each active device can be represented by many different circuit models. The basic requirement on each model, however, is the same. Given an input ac voltage and current, the output ac voltage and current for the model must be the same as for the actual device. A model may be structured from a mathematical base or a physical base. In this section we will develop the equivalent circuit of an FET based on its physical operation. The resulting circuit can be compared to the general two-port treated in Appendix A.* Since both types of FETs have similar equivalent circuits, both will be treated in this section with the JFET treated first.

* Readers familiar with the two-port network representation should find it easy to correlate the techniques of small signal analysis to the more general two-port analysis. Readers not familiar with this branch of circuit theory may find it advantageous to refer to Appendix A either prior to Sec. 3.5 or at the end of Chap. 3 if the more general analysis is of interest. Although small-signal analyses are based on two-port network theory, a detailed knowledge of this theory is not necessary for successful circuit analysis and design.

An ac equivalent circuit for the JFET that is based on its physical operation is shown in Fig. 3–15(a). This circuit is obtained by approximating various sections of the JFET with equivalent lumped components. The physical model of the JFET in Fig. 3–15(b) identifies the various points in the transistor. Starting at the gate, the ac gate current, I_g, flows to the channel through the transition capacitance and resistance of a reverse-biased diode. These parameters are distributed along the channel; however, for simplicity* they are lumped into two branches (G to D' and G to S') as shown in Fig. 3–15(a). These RC branches are simply the equivalent circuits of a reverse-biased diode as shown in Fig. 2–23(b). Resistors $r_{ss'}$ and $r_{dd'}$ represent the resistance of the bulk material between the end of the channel and the respective ohmic contact. These resistance values are usually small and are neglected. With this approximation, the primes are dropped since S and S' become the same point.

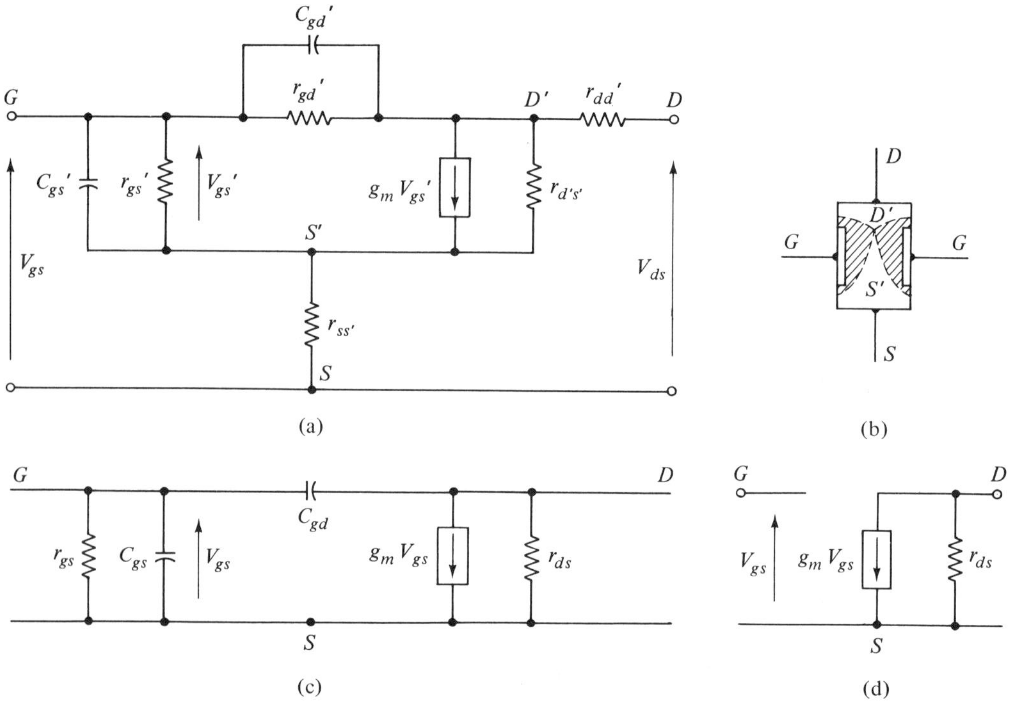

FIGURE 3–15 *Circuit models of a JFET: (a) The complete ac equivalent of the JFET based on its physical operation, (b) the physical model of the JFET, (c) a simplified high-frequency model, and (d) an approximate mid- and low-frequency model.*

The equivalent circuit component for the region between D' and S' are obtained from the device curves as shown in Fig. 3–16. The channel resistance beyond pinch-off is represented by $r_{d's'}$ (or r_{ds} if both $r_{dd'}$ and $r_{ss'} \simeq 0$) and is determined by the inverse slope of the constant V_{GS} curve which passes through the Q point. The ac component of I_D is proportional to the ac component of

---

* To obtain a practical or usable equivalent circuit, many approximations are made. The final circuit is accurate only if the approximations produce negligible errors.

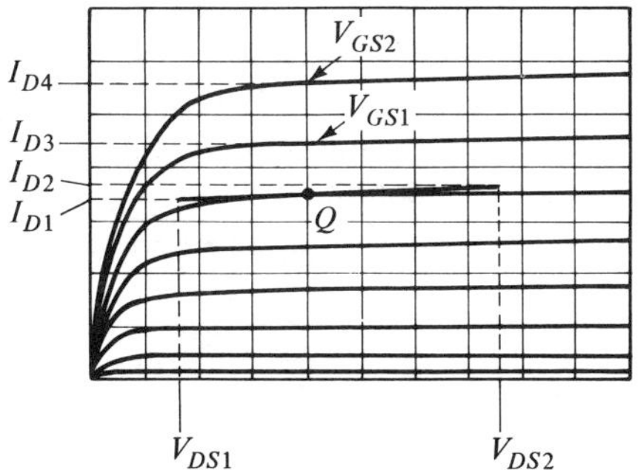

$$r_{ds} = \frac{\Delta V_{DS}}{\Delta I_D} = \frac{V_{DS2} - V_{DS1}}{I_{D2} - I_{D1}}$$

$$g_m = \frac{\Delta I_D}{\Delta V_{GS}} = \frac{I_{D4} - I_{D3}}{V_{GS2} - V_{GS1}}$$

FIGURE 3–16 *A method of determining the values of* r_{ds} *and* g_m *for an FET using the device curves. Both quantities are evaluated at the dc operating point.*

V_{GS} with the transconductance, g_m, being the constant of proportionality; i.e., $\Delta I_D = g_m \Delta V_{GS}$ or $I_d = g_m V_{gs}$. The dependent current generator, $g_m V_{gs}$, represents the effect of the ac voltage in changing the channel current.

The simplified equivalent circuit shown in Fig. 3–15(c) is obtained by making two assumptions. First, as indicated above, $r_{ss'}$ and $r_{dd'}$ usually have small values, 100 ohms or less for planar JFETs, and are neglected as zero.* Second, $r_{gd'}$ is the resistance of a reverse biased diode and is usually greater than fifty megohms. For reasonable values of R_S and R_L, the effect of $r_{gd'}$ is negligible if its value exceeds a few megohms. The resulting equivalent circuit is shown in Fig. 3–15(c) with the primes omitted from the symbols. This circuit is useful in high-frequency analyses and will be used frequently in Chap. 7.

For problems in the mid- and low-frequency range, C_{gd} and C_{gs} have high values of reactance and can be neglected as open circuits in comparison with the usual circuit values of R_S and R_L. In addition, r_{gs} is the resistance of a reverse biased diode and is usually much larger than the circuit bias resistor, R_G. Consequently, r_{gs} can also be neglected. The resulting small-signal, mid-frequency equivalent circuit is shown in Fig. 3–15(d).**

The circuit in Fig. 3–15(d) requires only two transistor parameters, g_m and r_{ds}. While the value of g_m can be determined from the characteristic curves as $\Delta I_D/\Delta V_{GS}$ (see Fig. 3–16), it can also be determined by finding the change in I_D due to a change in V_{GS} using Eq. (3–4).*** With this approach,

$$g_m = \frac{2}{V_P}\sqrt{I_D I_{DSS}} \tag{3–12}$$

The maximum value of g_m occurs when I_D equals I_{DSS} and is designated g_{mo}. From Eq. (3–12),

$$g_{mo} = 2I_{DDS}/V_P \tag{3–13}$$

* This assumption does not result in a good approximation unless (1) $r_{dd'}$ is much less than R_L, and (2) $g_m r_{ss'}$ is much less than unity.

** This circuit is similar to the general y-parameter equivalent circuit in Appendix A if we assume an infinite input impedance ($y_i = 0$) and use only the real part of y_f, which is g_m.

*** This approach corresponds to taking the partial derivative of Eq. (3–4) with respect to V_{GS}, or

$$g_m = \left|\frac{\partial I_D}{\partial V_{GS}}\right| = \frac{2I_{DSS}}{V_P}\left(1 - \frac{V_{GS}}{V_P}\right) = \frac{2}{V_P}\sqrt{I_D I_{DSS}}$$

The value of r_{ds} must be determined graphically as shown in Fig. 3–16. However, the value of r_{ds} is usually greater than 50 K and can often be neglected for values of R_{Leq} less than about 5 K since r_{ds} will appear in parallel with R_{Leq}. The values of g_m and r_{ds} have been determined in a manner which is consistent with the general two-port methods treated in Appendix A.

The two principal equations needed for the simplified small-signal analysis of a JFET are

$$I_D = I_{DSS}\left(1 - \frac{V_{GS}}{V_P}\right)^2 \qquad g_m = \frac{2}{V_P}\sqrt{I_D I_{DSS}} \tag{3–14}$$

The first equation is used to calculate the bias or dc conditions, and the second is used in ac gain calculations. Since JFETs and both types of MOSFETs obey the same square-law relationship for drain current, Eq. (3–14) applies to all three FET types* provided they are biased for linear operation.

In Eq. (3–14), the FET is described by two constants, V_P and I_{DSS}. However, accurate values of these two constants are difficult to obtain. Actual FETs exhibit wide variations in the values of these two quantities from unit-to-unit, even for devices having the same designation. For example, a typical specification sheet may have values of I_{DSS} ranging from 0.5 to 15 mA and V_P less than 8 V. A given device described by this specification sheet will have particular values of V_P and I_{DSS} somewhere within these broad limits. Assuming that the curves for a particular device are available or can be displayed on a curve tracer, an experimental determination of V_P and I_{DSS} from the $V_{GS} = 0$ curve is difficult for an FET with a rounded knee. Reasonably accurate results can be obtained, however, by checking the value of g_{mo} using the following procedure. First, estimate the value of V_P from either the value of V_{GS} which causes the drain current to go to zero or from the knee of the $V_{GS} = 0$ curve. Second, find the corresponding value of I_{DSS} from the $V_{GS} = 0$ curve. Third, estimate the value of g_{mo} from the curves in the region near the $V_{GS} = 0$ curve when I_D is approximately I_{DSS} using the method shown in Fig. 3–16. Fourth, compare this value of g_{mo} with the value given by Eq. (3–13). Accurate values of V_P and I_{DSS} are obtained when the two calculated values of g_{mo} are in close agreement; i.e., when

$$\frac{\Delta I_D}{\Delta V_{GS}} \simeq \frac{2I_{DSS}}{V_P} \tag{3–15}$$

This procedure is illustrated in the following example.

EXAMPLE 3–6 Determine the values of V_P and I_{DSS} for the N-channel JFET whose characteristics are given in Fig. 3–13(b). Verify these values by calculating g_{mo} by two methods.

Solution: From Fig. 3–13(b), the value of V_{GS} which causes I_D to be zero is approximately -2.4 V. At $V_{GS} = 0$ and $V_{DS} = -V_P = 2.4$ V, $I_D = I_{DSS} = 7.6$ mA. The experimental value of g_{mo} can be estimated by assuming a small increment in the neighborhood of $V_{GS} = 0$ V (say 0 V to -0.4 V) and finding the corresponding change in I_D. Then

$$g_{mo} = \frac{\Delta I_D}{\Delta V_{GS}} = \frac{(9 - 7)\text{ mA}}{0 - (-0.4)\text{ V}} \simeq 5 \text{ millimhos}$$

---

* For the enhancement MOSFET, V_T is substituted for V_P, and I_{DO} for I_{DSS}.

From Eq. (3–13),

$$g_{mo} = \frac{2I_{DSS}}{V_P} = \frac{2(7.6)}{2.4} \simeq 6.3 \text{ millimhos}$$

A 26 percent difference exists between the two values of g_{mo}.

A second trial of $V_P = -3.2$ V and $I_{DSS} = 8.3$ mA yields

$$g_{mo} = \frac{2I_{DSS}}{V_P} = \frac{2(8.3)}{3.2} \simeq 5.2 \text{ millimhos}$$

This value is within 5 percent of the experimental value of g_{mo} using the delta method. Hence the values of V_P and I_{DSS} are within the tolerance of most circuit components.

The complete ac equivalent circuit of the JFET was developed from the physical operation of the device. Even though the construction of the MOSFET is different from that of the JFET, the ac equivalent circuits are very similar since their characteristic curves are similar, and the value of I_D depends primarily on the value of V_{GS} for both devices. A complete small-signal equivalent circuit for the MOSFET, biased in the linear region, is shown in Fig. 3–17(a). In this circuit the parallel RC network ($r_{gs'}C_{gs'}$, and $r_{gd'}C_{gd'}$) represent the leakage resistance and capacitance of the oxide film. The diodes D_1 and D_2 represent

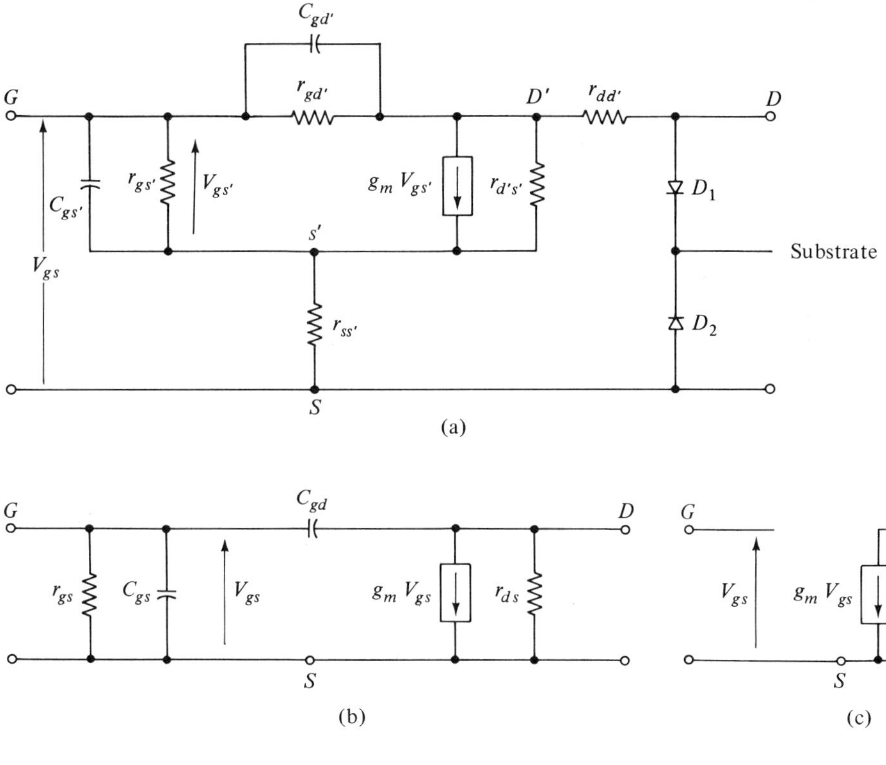

FIGURE 3–17 *Equivalent circuits of the MOSFET biased for linear operation: (a) a complete circuit, (b) a simplified high-frequency model, and (c) an approximate mid- and low-frequency model.*

P-N junctions formed by the substrate* and source and the substrate and drain respectively. The other components in Fig. 3–17(a) are analogous to those of the JFET. Approximations similar to those for the JFET can be made to obtain the simplified high-frequency model shown in Fig. 3–17(b). As shown in Fig. 3–17(c), the equivalent circuit for the MOSFET in the mid- and low-frequency ranges is identical to that of the JFET shown in Fig. 3–15(c).

In the mid-frequency region, both forms of the FET are governed by Eq. (3–14) and can be represented by the same approximate equivalent circuit, Fig. 3–15(d) or 3–17(c). The use of linear equivalent circuits is restricted to devices biased for linear operation and to small variations around the Q point where the device curves can be considered to be approximately linear. The application of equivalent circuits to the analysis of small-signal linear amplifiers is treated in the remainder of this chapter.

3.6 The Common-Source (CS) Amplifier

The small-signal equivalent circuit of an FET was developed in the last section. In this section, we will use the equivalent circuit to determine the small-signal behavior of a CS amplifier. The procedure is as follows. First the bias conditions are determined. Next the mid-band equivalent circuit is structured by replacing all coupling and by-pass capacitors with short circuits and replacing the FET with its mid-band equivalent circuit. The mid-band performance of the amplifier can now be determined by circuit analysis techniques. The objective of this section is to develop the expressions for the voltage gain, input impedance, and output impedance of the CS amplifier in the mid-band frequency range.

A complete CS amplifier is shown in Fig. 3–18(a);** and the mid-band circuit of the amplifier is shown in Fig. 3–18(b). A complete mid-band equivalent circuit is obtained by substituting the FET equivalent circuit [Fig. 3–15(d)] for the transistor as shown in Fig. 3–18(c).

The mid-band response of an amplifier may be characterized by the following: (1) the generator-to-output voltage gain, A_{vo}, (2) the input impedance of the amplifier, R_i, and (3) the output impedance of the amplifier, R_o. The generator-to-output gain, A_{vo}, is the ratio of V_o to V_g.

In general, the impedance looking into a pair of terminals is the ratio of the voltage between the terminals to the current entering one of the terminals. For example, from Fig. 3–18(a), R_i is the ratio of V_i to the ac current flowing in C_1. The impedances R_{ig} and R_{od} pertain only to the transistor while the impedances R_i and R_o include the effects of any bias resistors present in the circuit. The performance terms of interest are calculated below. In addition to these general terms, three other performance terms are usually desirable: (1) the CS terminal-to-terminal voltage gain, A_{vt} (which is the ratio of V_o to V_i), (2) the CS input impedance, R_{ig}, and (3) the CS output impedance, R_{od}.

Referring to Fig. 3–18(c), the output voltage is given by

$$V_o = -I_o R_{Leq} \tag{3–16}$$

* The substrate is normally reverse biased relative to both the source and drain. The diodes become important only if the substrate becomes forward biased relative to either the source or drain.

** This is the same circuit as shown in Fig. 3–14(a), and all components perform the same function.

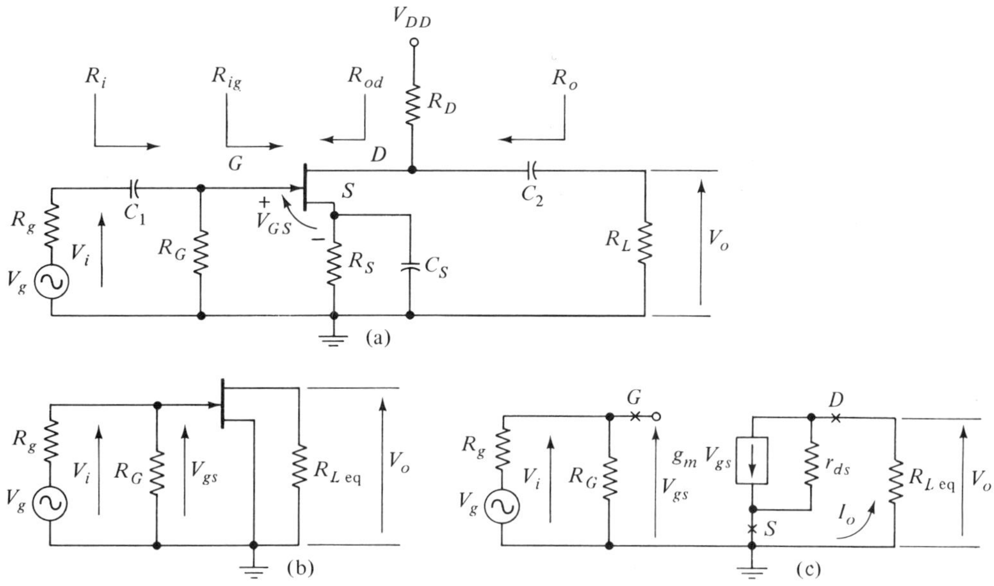

FIGURE 3–18 *A common-source FET amplifier: (a) a complete CS amplifier, (b) a mid-band equivalent circuit for the amplifier, and (c) a complete mid-band equivalent circuit.*

where I_o is the current flowing in R_{Leq} due to the current generator $g_m V_{gs}$. Applying the current splitting theorem* to r_{ds} and R_{Leq} gives I_o as

$$I_o = (g_m V_{gs}) \left(\frac{r_{ds}}{r_{ds} + R_{Leq}} \right) \tag{3–17}$$

The voltage V_{gs} is the ac gate-to-source voltage and is the component of the generator voltage, V_g, that is dropped across R_G. Using the voltage division theorem,**

$$V_{gs} = V_g \left(\frac{R_G}{R_G + R_g} \right) \tag{3–18}$$

Substituting Eqs. (3–17) and (3–18) into (3–16), the output voltage becomes

$$V_o = -V_g \left(\frac{R_G}{R_G + R_g} \right) (g_m) \left(\frac{r_{ds} R_{Leq}}{r_{ds} + R_{Leq}} \right) \tag{3–19}$$

---

* The current splitting theorem for two parallel resistors is

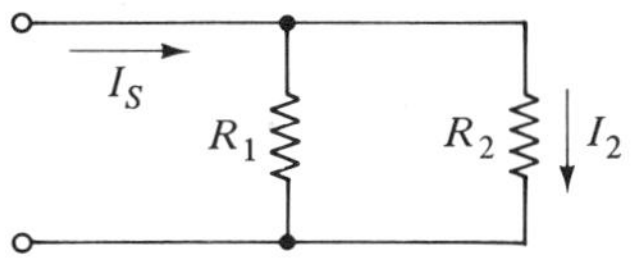

$$I_2 = I_S \frac{R_1}{R_1 + R_2} = I_S \frac{\text{(opposite } R)}{\text{(sum of } R\text{'s)}}$$

** The voltage division theorem for two resistors is

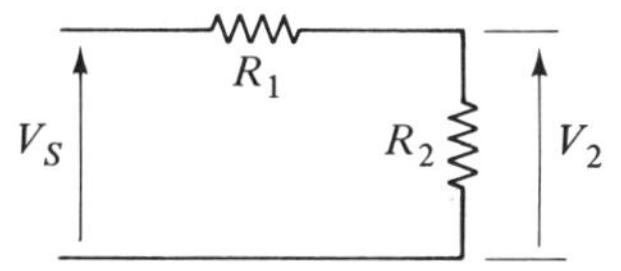

$$V_2 = V_S \left(\frac{R_2}{R_1 + R_2} \right)$$

Consequently, the generator-to-output voltage gain is

$$A_{vo} = \frac{V_o}{V_g} = -\left(\frac{R_G}{R_G + R_g}\right)(g_m R_{L\text{tot}}) \tag{3–20}$$

where $R_{L\text{tot}}$ is $R_{L\text{eq}}||r_{ds}$ and represents the total equivalent resistance between the drain terminal and ground. The negative sign in Eq. (3–20) indicates that the output signal of a CS amplifier is 180° out of phase (or inverted) relative to V_g.

The CS input impedance, R_{ig}, can be calculated using Fig. 3–19(a). In this

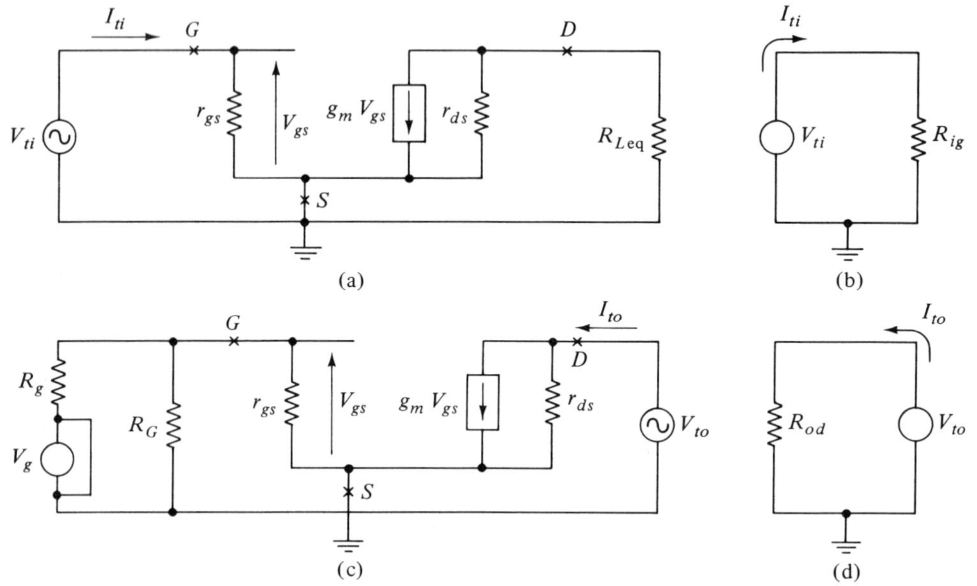

FIGURE 3–19 *The input and output impedance of a CS stage: (a) the input circuit for finding* R_{ig}, *(b) the equivalent input circuit, (c) the output circuit for finding* R_{od}, *and (d) the equivalent output circuit.*

figure, the circuit to the left of the gate terminal as shown in Fig. 3–18(b), including R_G, R_g, and V_g, has been replaced by a test voltage generator, V_{ti}. In addition, the equivalent resistor, r_{gs}, has been included in the equivalent circuit of the JFET as shown in Fig. 3–15(c). The current I_{ti} flowing as a result of V_{ti} is

$$I_{ti} = V_{ti}/r_{gs} \tag{3–21}$$

Solving for R_{ig},

$$R_{ig} = V_{ti}/I_{ti} = r_{gs} \tag{3–22}$$

With R_{ig} known, the input impedance to the amplifier can be determined from Fig. 3–18(a) as

$$R_i = R_G \,||\, R_{ig} = R_G \,||\, r_{gs} \simeq R_G \tag{3–23}$$

The value of r_{gs} is in the high megohm range, and can usually be neglected in calculating R_i.

The CS output impedance, R_{od}, is calculated using Fig. 3–19(c). In this figure, the circuit to the right of the drain terminal as shown in Fig. 3–18(b), including R_D and R_L, has been replaced by a test voltage generator, V_{to}. Also, the input generator, V_g, is short circuited, and r_{gs} is included in the JFET equivalent circuit. The current flowing in the test generator is

$$I_{to} = V_{to}/r_{ds} \tag{3–24}$$

Thus, R_{od} is equal to r_{ds}. With R_{od} known, the output impedance of the amplifier can be determined from Fig. 3–18(a).

$$R_o = R_D \,||\, R_{od} = R_D \,||\, r_{ds} \simeq R_D \tag{3–25}$$

Neglecting r_{ds} in the calculation of R_o is generally valid only when R_D is less than about 10 K since r_{ds} has values typically in the 50 K to 200 K range when the FET is biased in the linear region.

The expression for A_{vo} can now be simplified. The value of A_{vo}, as given by Eq. (3–20), can be divided into two terms: (1) the voltage gain from the generator to the transistor input terminal, A_{vg}, and (2) the voltage gain of the device, A_{vt}. The first term is always less than unity and represents the voltage division between the generator impedance and the input impedance of the amplifier. Referring to Fig. 3–18(a), the voltage division term can be written as

$$A_{vg} = R_i/(R_i + R_g) \tag{3–26}$$

where, for the CS amplifier,

$$R_i \simeq R_G \tag{3–27}$$

The second term represents the transistor terminal-to-terminal voltage gain, A_{vt}, and, for the CS amplifier, is

$$A_{vt} = V_o/V_i = -g_m R_{L\text{tot}} \tag{3–28}$$

Consequently, Eq. (3–20) can be written as

$$A_{vo} = (A_{vg})(A_{vt}) \tag{3–29}$$

The important relationships for the single-stage CS amplifier are summarized below:

$$\left.\begin{array}{llll} A_{vo} = (A_{vg})(A_{vt}) & & R_i = R_G \,||\, R_{ig} \simeq R_G \\ A_{vg} = R_i/(R_i + R_g) & & R_o = R_D \,||\, R_{od} \simeq R_D \\ A_{vt} = -g_m R_{L\text{tot}} \simeq -g_m R_{L\text{eq}} & & \text{signal inverted} \end{array}\right\} \tag{3–30}$$

The following example demonstrates the use of these equations.

EXAMPLE 3–7 For the CS amplifier shown in Fig. 3–20(a), determine the values of A_{vt}, A_{vg}, R_i, and R_o. The JFET is identical to the transistor used in Ex. 3–5 for large signal analysis and is biased for $V_{GS} = -0.8$ V. The values of V_P and I_{DSS} are -3.2 V and 8.3 mA as found in Ex. 3–6.

Solution: From Eq. (3–14),

$$I_D = I_{DSS}\left(1 - \frac{V_{GS}}{V_P}\right)^2 \simeq 4.7 \text{ mA} \qquad g_m = \frac{2}{V_P}\sqrt{I_D I_{DSS}} \simeq 3.9 \text{ millimhos}$$

From the circuit in Fig. 3–20,

$$R_i = R_{ig} \,||\, R_G \simeq 4.7 \text{ M} \qquad R_{L\text{eq}} = R_D \,||\, R_L \simeq 1.6 \text{ K}$$

From Eq. (3–30),

$$A_{vg} = \frac{R_i}{R_i + R_g} \simeq 1 \qquad A_{vt} \simeq -g_m R_{L\text{eq}} \simeq -6.2$$

$$A_{vo} = (A_{vg})(A_{vt}) \simeq -6.2 \qquad R_o = R_D \,||\, r_{ds} \simeq R_D = 2\text{ K}$$

The measured value of gain with 0.3 V in was −6.7 which is within 10 percent of the calculated value.

The difference between the calculated and measured values of A_{vo}, while small, is due to tolerances in resistor values and g_m. The resistor values are accurate to only ±10 percent. The value of g_m depends on the values of V_P and I_{DSS}, both of which are difficult to obtain with great accuracy. For example, in Ex. 3–5, the measured value of drain current was approximately 5.2 mA. For this value of current, g_m has a value of 4.1 millimhos. Using this value of g_m, A_{vo} is −6.6.

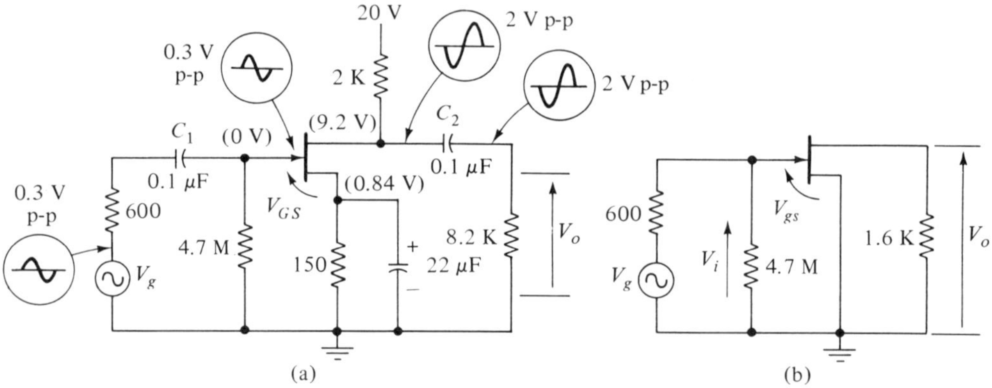

FIGURE 3–20 *CS amplifier from Ex. 3–7: (a) the complete circuit, and (b) the mid-band equivalent circuit. Measured dc voltage levels with respect to ground are given in parentheses, and the important ac waveforms are shown.*

3.7 The Common-Gate (CG) Amplifier

A second amplifier configuration using an FET is the common-gate (CG) type as shown in Fig. 3–21(a). The input signal is applied between the source and gate terminals, and the output signal is obtained between the drain and gate terminals. The objective of this section is to develop expressions for the voltage gain, input impedance, and output impedance of the CG amplifier in the mid-band frequency range. The approach taken will be similar to that used in the last section for the CS amplifier where the performance characteristics were obtained from an analysis of the mid-band equivalent circuit. The results of this analysis will show that the CG amplifier is a non-inverting, low input impedance, high output impedance amplifier.

A P-channel JFET CG amplifier is shown in Fig. 3–21(a). All circuit components perform the same bias function as they did for the CS amplifier. A gate leakage resistor, R_G, is not necessary since the gate is grounded in the CG amplifier. The mid-band equivalent circuit of this amplifier is obtained by shorting the coupling capacitors and dc supply as shown in Fig. 3–21(b). The complete mid-band equivalent circuit is obtained by replacing the JFET with its mid-band equivalent circuit as shown in Fig. 3–21(c). The analysis of the CG amplifier is simplified by reversing the polarity of the ac voltage V_{gs}. Note that the direction of the current generator $g_m V_{gs}$ must also be changed since both its

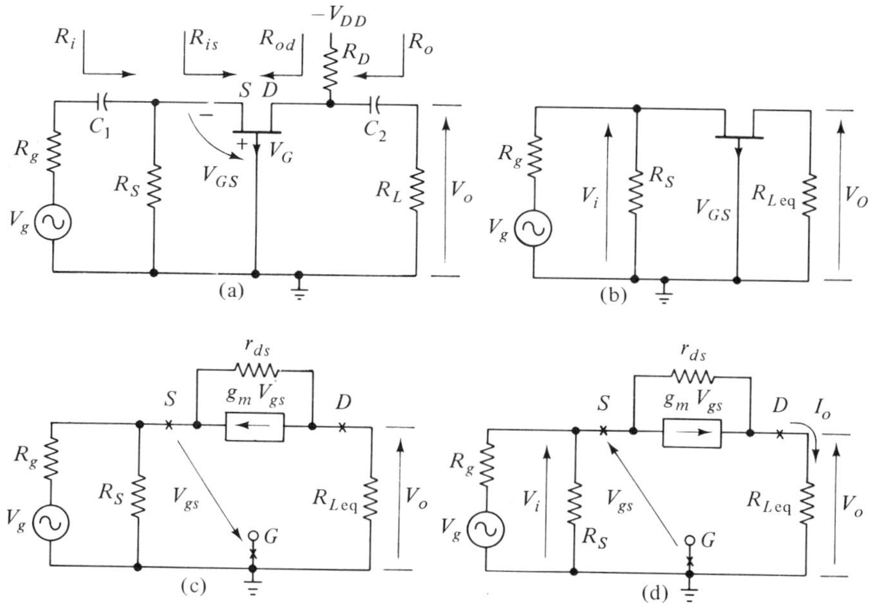

FIGURE 3–21 *A common-gate FET amplifier: (a) a complete CG amplifier, (b) a mid-band circuit for the amplifier, (c) a complete mid-band equivalent circuit, and (d) a modified equivalent circuit.*

polarity and magnitude depend on V_{gs}. These modifications produce the equivalent circuit shown in Fig. 3–21(d).

An important observation can be made from Fig. 3–21(d). A positive input signal, V_g, causes V_{gs} to be positive, and an ac current, I_o, to flow in R_{Leq}. The resulting output voltage across R_{Leq} is positive. Unlike the CS amplifier, the CG amplifier does not invert the signal.

The CG input impedance, R_{is}, can be calculated using the circuit in Fig. 3–22(a). In this figure, the circuit to the left of the source terminal in Fig. 3–21(d), which includes R_S, R_g, and V_g, has been removed and replaced by a test voltage generator, V_{ti}. The current flowing as a result of V_{ti} is

$$I_{ti} = g_m V_{gs} + \frac{V_{ti} - V_o}{r_{ds}} \tag{3–31}$$

From Fig. 3–22(a), we observe that V_{gs} equals V_{ti}, and V_o equals $I_{ti}R_{Leq}$. Consequently, Eq. (3–31) can be written as

$$I_{ti} = g_m V_{ti} + \frac{V_{ti}}{r_{ds}} - I_{ti}\frac{R_{Leq}}{r_{ds}} \tag{3–32}$$

Combining coefficients of I_{ti}, Eq. (3–32) becomes

$$I_{ti}\left[1 + \frac{R_{Leq}}{r_{ds}}\right] = V_{ti}\left[g_m + \frac{1}{r_{ds}}\right] \tag{3–33}$$

Solving for R_{is},

$$R_{is} = \frac{V_{ti}}{I_{ti}} = \left[1 + \frac{R_{Leq}}{r_{ds}}\right] \bigg/ \left[g_m + \frac{1}{r_{ds}}\right] \tag{3–34}$$

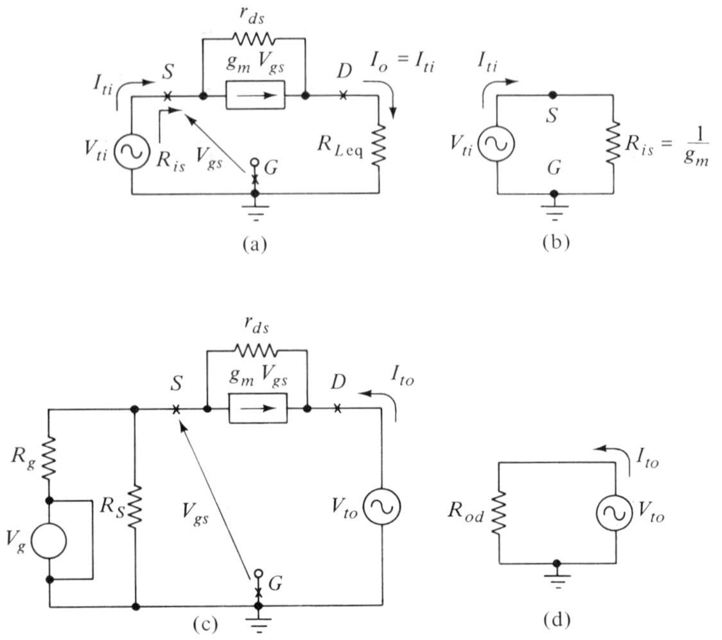

FIGURE 3–22 *Circuits used to determine the input and output impedance of a CG stage: (a) the input circuit for finding* R_{is}, *(b) the equivalent input circuit, (c) the output circuit for finding* R_{od}, *and (d) the equivalent output circuit.*

Since r_{ds} is typically much greater than either unity or R_{Leq}, R_{is} can be approximated as

$$R_{is} \simeq 1/g_m \tag{3–35}$$

This equation indicates that the FET and its load may be replaced by a resistor having a value of $1/g_m$ as shown in Fig. 3–22(b) when looking into the input terminals of a CG amplifier. With R_{is} known, the input impedance to the amplifier, R_i, can be determined from Fig. 3–21 as

$$R_i = R_S \,||\, R_{is} \simeq R_S \,||\, (1/g_m) \tag{3–36}$$

The expression for the output impedance of the CG amplifier can be derived using Fig. 3–22(c). The result is

$$R_{od} = r_{ds}\left[1 + g_m(R_S \,||\, R_g) + (R_S \,||\, R_g)/r_{ds}\right] \tag{3–37}$$

The development of this equation is left as an exercise for the reader. A comparison of the output impedance of the CS stage [Eq. (3–25)], to that of the CG stage [Eq. (3–37)] indicates that the latter has a larger value, although both can be approximated as r_{ds} for large values of r_{ds}. We would expect R_{od} to be larger for the CG stage since the source terminal is not ac grounded. With R_{od} known, the output impedance of the amplifier, R_o, can be determined from Fig. 3–21 as

$$R_o = R_D \,||\, R_{od} \simeq R_D \tag{3–38}$$

where the approximation is valid for large values of r_{ds}.

Referring to Fig. 3–21(d), the output voltage for a CG amplifier is approximately

$$V_o = I_o R_{Leq} \simeq g_m V_{gs} R_{Leq} \tag{3–39}$$

where the approximation requires r_{ds} to have a large value. The voltage V_{gs} is that portion of the generator voltage, V_g, developed across the source-gate terminals, or

$$V_{gs} = V_g R_i/(R_i + R_g) \qquad \textbf{(3–40)}$$

where R_i is given by Eq. (3–36). From Eqs. (3–39) and (3–40), the generator-to-output voltage gain is

$$A_{vo} = \frac{V_o}{V_g} = \left[\frac{R_i}{R_i + R_g}\right](g_m R_{Leq}) = (A_{vg})(A_{vt}) \qquad \textbf{(3–41)}$$

where A_{vg} is the voltage gain from the generator to the input terminals, and A_{vt} is the voltage gain from the input terminals to the output terminals.

The important relationships for the single-stage CG amplifier are summarized below.

$$\left.\begin{array}{ll} A_{vo} = (A_{vg})(A_{vt}) & R_i = R_S \,||\, R_{is} \simeq R_S \,||\, (1/g_m) \\ A_{vg} = R_i/(R_i + R_g) & R_o = R_D \,||\, R_{od} \simeq R_D \\ A_{vt} \simeq g_m R_{Leq} & \text{signal not inverted} \end{array}\right\} \qquad \textbf{(3–42)}$$

The following example demonstrates the use of the CG equations.

EXAMPLE 3–8 The CG amplifier shown in Fig. 3–23 has the following circuit values: $R_g = 1$ K; $R_S = 560$; $R_D = 4.7$ K; $R_L = 20$ M (oscilloscope input impedance); $V_{DD} = -15$ V; and $C_1 = C_2 = 0.5\ \mu$F. The P-channel JFET is characterized by $V_P = 3.5$ V, and $I_{DSS} = -3.2$ mA. The dc bias value for V_{GS} is 1 V. Find the other bias values and also the values of A_{vo}, A_{vt}, R_i, and R_o.

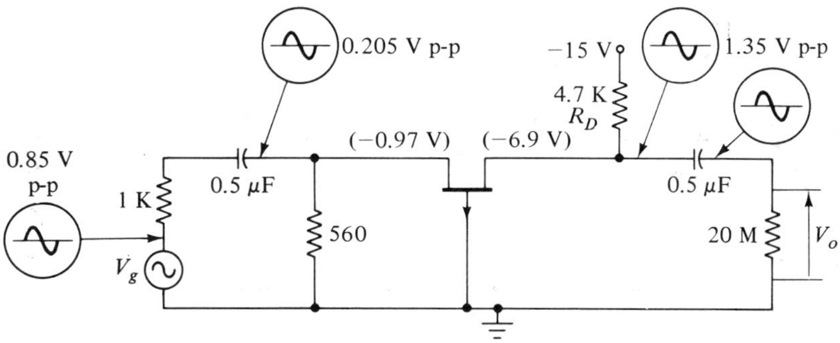

FIGURE 3–23 *The CG amplifier from Ex. 3–8. Measured dc voltage levels with respect to ground are shown in parentheses. Important ac waveforms are also shown.*

Solution: (a) Bias conditions: The dc current is

$$I_D = I_{DSS}\left(1 - \frac{V_{GS}}{V_P}\right)^2 = -3.2\left(1 - \frac{1}{3.5}\right)^2 \simeq -1.7 \text{ mA}$$

The drain-to-ground voltage is found by subtracting the dc voltage drop across R_D from the dc source voltage, or

$$V_D = -15 \text{ V} - (-1.7 \text{ mA})(4.7 \text{ K}) \simeq -7 \text{ V}$$

The drain-to-ground voltage, V_D, is the sum of voltages across R_S and V_{DS}. Consequently,

$$V_{DS} = V_D - V_S = -7 + 1 = -6 \text{ V}$$

(b) Gain and impedance calculations: From Eq. (3–12), g_m in terms of the given values is

$$g_m = \frac{2}{V_P}\sqrt{I_D I_{DSS}} = \frac{2}{3.5}\sqrt{(1.7)(3.2)10^{-6}} \simeq 1.32 \text{ millimhos}$$

The total equivalent load is R_D in parallel with R_L or

$$R_{Leq} = 4.7\text{ K} \,||\, 20\text{ M} \simeq 4.7\text{ K}$$

From the source to the equivalent load, the voltage gain is

$$A_{vt} \simeq g_m R_{Leq} = (1.32 \times 10^{-3})(4.7\text{ K}) \simeq 6.2$$

The equivalent load on the generator is

$$R_i \simeq R_S \,||\, (1/g_m) = 560 \,||\, 760 \simeq 320$$

Since the generator voltage divides across R_g and R_i, the voltage gain from the generator to the input terminals is

$$A_{vg} = R_i/(R_i + R_g) = 320/(320 + 1\text{ K}) \simeq 0.242$$

The voltage gain from the generator to the output (on V_o/V_g) is

$$A_{vo} = (A_{vg})(A_{vt}) = (0.242)(6.2) \simeq 1.5$$

If the generator resistance is decreased to zero, then A_{vo} would be equal to A_{vt} or 6.2. The resistance looking back into the output terminals is

$$R_o = R_{od} \,||\, R_D \simeq R_D = 4.7\text{ K}$$

(c) Experimental results:

$$V_S = -0.97\text{ V} \qquad V_D = -6.9\text{ V}$$

With a generator voltage (open circuit voltage) of 0.85 V, the output voltage was 1.35 V. The measured gain was

$$A_{vo} = \frac{V_o}{V_g} = \frac{1.35\text{ V}}{0.85\text{ V}} \simeq 1.6$$

Note that the gain is low because most of the generator voltage, V_g, was developed across R_g and also that the output impedance is high.

3.8 The Common-Drain (CD) Amplifier

A third useful amplifier configuration using an FET is the common-drain (CD) amplifier shown in Fig. 3–24(a). This configuration is also referred to as a "source follower" (SF). The drain terminal is usually connected directly to a power supply terminal (an ac ground). The input signal is applied between the gate and ground, and the output signal is obtained between the source and ground. The analysis of this circuit will be similar to that used for the CS and the CG amplifiers where the performance characteristics were obtained from the mid-band equivalent circuit. The results of this analysis will show that the CD amplifier is a non-inverting, high input impedance, low output impedance amplifier with a voltage gain less than unity. The CD amplifier is used primarily to match a low-impedance load to a high-impedance source.

A P-channel JFET common-drain (CD) amplifier is shown in Fig. 3–24(a). The mid-band equivalent circuit of this amplifier is obtained by shorting the coupling capacitors and ac grounding the dc power supply as shown in Fig.

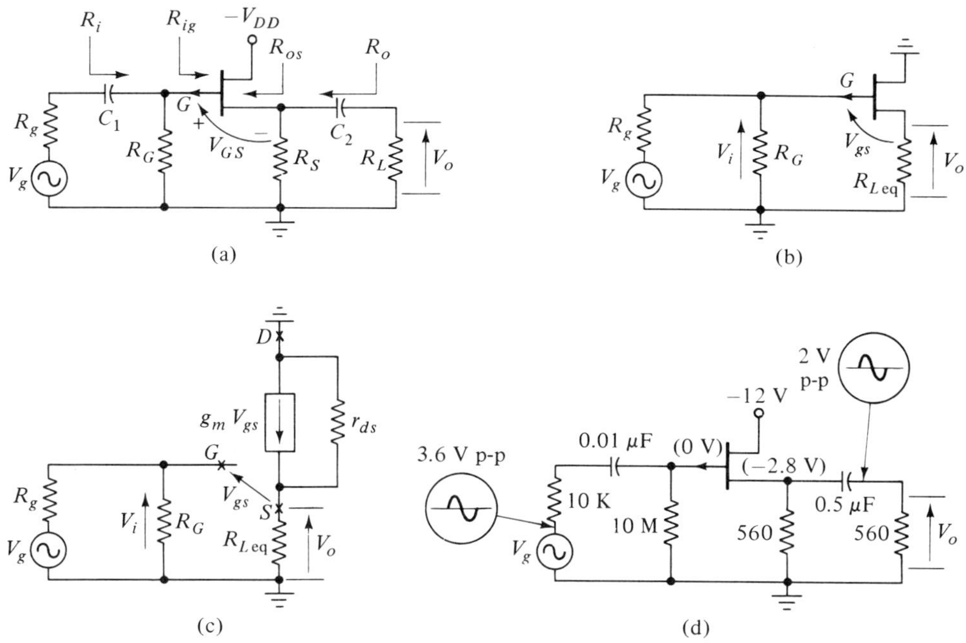

FIGURE 3–24 *A common-drain FET amplifier: (a) a complete CD amplifier, (b) a mid-band equivalent circuit for the amplifier, (c) a complete mid-band equivalent circuit, and (d) the CD amplifier from Ex. 3–9.*

3–24(b). The complete mid-band equivalent circuit is obtained by replacing the JFET with its mid-band equivalent circuit as shown in Fig. 3–24(c).

An important observation can be made from Fig. 3–24(c). A positive input signal, V_g, causes V_{gs} to be positive and causes a portion of the current $g_m V_{gs}$ to flow into R_{Leq}. The resulting output voltage is positive. Consequently, the CD amplifier does not invert the signal.

The voltage gain of the CD amplifier can be calculated using Fig. 3–24(c). Applying the current splitting theorem to r_{ds} and R_{Leq}, the output voltage is

$$V_o = (g_m V_{gs})\left(\frac{r_{ds}}{r_{ds} + R_{Leq}}\right)(R_{Leq}) \simeq g_m V_{gs} R_{Leq} \qquad \textbf{(3–43)}$$

where $g_m V_{gs}$ is the ac channel current and the approximation is valid for large values of r_{ds}. As shown in Fig. 3–24(b), the value of V_{gs} is different from V_i and is given by

$$V_{gs} = V_i - V_o \qquad \textbf{(3–44)}$$

Substituting Eq. (3–44) into (3–43) gives

$$V_o = g_m(V_i - V_o)R_{Leq} \qquad \textbf{(3–45)}$$

Combining coefficients of V_o, Eq. (3–45) becomes

$$V_o(1 + g_m R_{Leq}) = V_i g_m R_{Leq} \qquad \textbf{(3–46)}$$

Solving Eq. (3–46) for the terminal-to-terminal voltage gain of the CD stage,

$$A_{vt} = \frac{V_o}{V_i} = \frac{g_m R_{Leq}}{1 + g_m R_{Leq}} \qquad \textbf{(3–47)}$$

The voltage gain from the generator to the input terminals, A_{vg}, is also available from Fig. 3–24 if R_{ig} is assumed to be much larger than R_G. Applying voltage division to R_G and R_g,

$$A_{vg} = \frac{V_i}{V_g} = \frac{R_i}{R_i + R_g} = \frac{R_G}{R_G + R_g} \tag{3–48}$$

Combining Eqs. (3–47) and (3–48) gives the generator-to-output voltage gain

$$A_{vo} = A_{vg}A_{vt} = \left(\frac{R_G}{R_G + R_g}\right)\left(\frac{g_m R_{Leq}}{1 + g_m R_{Leq}}\right) \tag{3–49}$$

Since each term in Eq. (3–49) is less than unity, the product will be less than unity.

The input impedance of the CD amplifier can be calculated from the circuit in Fig. 3–25(a). Using the assumption that r_{gs} is much greater than $R_g \parallel R_G$, the

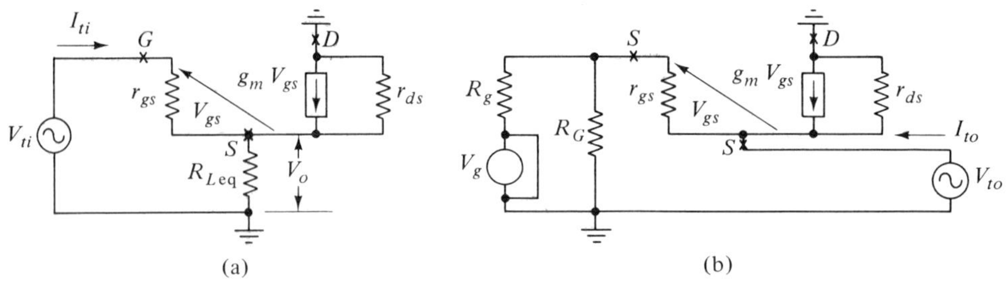

FIGURE 3–25 *Equivalent circuits used to determine the input and output impedances of a common-drain amplifier: (a) circuit for the input impedance,* R_{ig}*, and (b) circuit for the output impedance,* R_{os}*.*

result is

$$R_{ig} \simeq r_{gs}(1 + g_m R_{Leq}) \tag{3–50}$$

The development of this equation is left as an exercise for the reader. Note that R_{ig} is larger than r_{gs} because the ac load, R_{Leq}, is reflected back to the input in series with r_{gs} where the reflection factor is $g_m V_{gs}$. The CD amplifier has a higher input impedance than the CS amplifier because of this reflected impedance. With R_{ig} known, the input impedance of the amplifier, R_i, can be determined from Fig. 3–24 as

$$R_i = R_G \parallel R_{ig} \simeq R_G \tag{3–51}$$

where the approximation is valid for large values of r_{gs}.

The output impedance of the CD amplifier can be calculated using Fig. 3–25(b). The result is

$$R_{os} = \frac{1}{g_m} \parallel r_{ds} \parallel (r_{gs} + R_g \parallel R_G) \simeq \frac{1}{g_m} \tag{3–52}$$

The approximate result of R_{os} equal to $1/g_m$ should be expected since the impedance looking into the source terminal should be relatively independent of the transistor configuration. Consequently, the input impedance, (R_{is}), of the CG configuration is approximately equal to the output impedance (R_{od}) of the CD configuration since we are looking into the source terminal in each case. With

R_{os} known, the output impedance of the amplifier, R_o, can be determined from Fig. 3–24(a) as

$$R_o = R_S \,||\, R_{os} \simeq R_S \,||\, \frac{1}{g_m} \qquad \textbf{(3–53)}$$

The important relationships for the single stage CD amplifier are summarized below:

$$\left.\begin{aligned} A_{vo} &= (A_{vg})(A_{vg}) & R_i &= R_G \,||\, R_{ig} \simeq R_G \\ A_{vg} &= \frac{R_i}{R_i + R_g} & R_o &= R_S \,||\, R_{os} \simeq R_S \,||\, \frac{1}{g_m} \\ A_{vt} &\simeq \frac{g_m R_{Leq}}{1 + g_m R_{Leq}} & &\text{signal not inverted} \end{aligned}\right\} \qquad \textbf{(3–54)}$$

The following example illustrates the use of the CD equations.

EXAMPLE 3–9 The CD amplifier shown in Fig. 3–24(d) has the following circuit values: $R_g = 10$ K, $R_G = 10$ M, $R_S = 560$ ohms, $R_L = 560$ ohms, $C_1 = 0.01\ \mu$F, $C_2 = 0.5\ \mu$F, and $V_{DD} = -12$ V. The P-channel JFET is characterized by $V_P = 6$ V, and $I_{DSS} = -18$ mA. The circuit is biased for $V_{GS} = 2.8$ V. Find the other bias values, A_{vo}, A_{vt}, R_i, and R_o.

Solution: (a) Bias conditions: From Eq. (3–4), the drain current is

$$I_D = I_{DSS}\left[1 - \frac{V_{GS}}{V_P}\right]^2 = -18\text{ mA}\left[1 - \frac{2.8}{6}\right]^2 \simeq -5.1\text{ mA}$$

Since the drain voltage is V_{DD}, then

$$V_{DS} = V_{DD} - V_S = -12\text{ V} - (-2.8\text{ V}) = -9.2\text{ V}$$

(b) Gain and impedance calculations: From Eq. (3–12), the transconductance is

$$g_m = \frac{2}{V_P}\sqrt{I_D I_{DSS}} = \frac{2}{6}\sqrt{(5.1)(18)(10^{-6})} \simeq 3.2\text{ millimhos}$$

The equivalent ac load in the mid-band range is

$$R_{Leq} = R_S \,||\, R_L = 560 \,||\, 560 = 280\ \Omega$$

The gain from the gate to the equivalent ac load is

$$A_{vt} = \frac{g_m R_{Leq}}{1 + g_m R_{Leq}} = \frac{(3.2 \times 10^{-3})(280)}{1 + (3.2 \times 10^{-3})(280)} \simeq 0.48$$

The input impedance for the circuit is

$$R_i \simeq R_G \,||\, r_{gs} \simeq R_G = 10\text{ M}$$

The gain from the generator to the gate terminal is

$$A_{vg} = \frac{R_i}{R_i + R_g} \simeq \frac{10\text{ M}}{10\text{ M} + 10\text{ K}} \simeq 1$$

The gain from the generator to the load is

$$A_{vo} = A_{vg}\,A_{vt} = (1)(0.48) = 0.48$$

The output impedance is

$$R_o \simeq R_S \,||\, \frac{1}{g_m} \simeq 560 \,||\, 312 \simeq 200\ \Omega$$

Note that in this circuit, approximately half of the generator voltage is developed across the load and the other half across the gate-source terminals. Note also that the output impedance is rather low.

(c) Experimental results:

$$V_{GS} = -V_S = 2.8\text{ V} \qquad A_{vo} = 2\text{ V}/3.6\text{ V} \simeq 0.56$$

3.9 Multistage FET Amplifier Circuits

The three preceding sections were devoted to the development of the performance characteristics of three useful amplifier configurations using the FET. Many amplifiers require more than a single stage to achieve the desired characteristics. Multistage amplifiers are formed by interconnecting several single-stage amplifiers. Although a large number of multistage circuits exist, some simple techniques are available that can be used to analyze a multistage amplifier regardless of the configuration. The objective of this section is to develop these techniques. To achieve this objective, we will first summarize and compare the performance characteristics of the three single-stage amplifier circuits already considered. Next, we will study the effects of interconnecting several single-stage networks. Finally, we will study two cascaded* CS-CS amplifiers: one using ac coupling, and one using dc coupling between stages. The differential amplifier configurations will be treated in the next section.

The approximate performance characteristics of the CS, CG, and CD amplifiers are shown in Table 3–1. In each case, the FET is shown as a two-port network** in which the input and output impedances of the transistor are identified. Note that the load and bias resistors are assumed to be located external to the two-port. The general expressions for A_{vg}, A_{vt}, and A_{vo} are also given in Table 3–1.

The data in Table 3–1 can be used to compare the performance characteristics of the three configurations. Only the CS stage inverts the signal, and only the CD stage is limited to a voltage gain of less than unity. The CD stage has the highest input impedance, and the CG stage has the lowest input impedance. Comparing the output impedances, the CG configuration has the highest value, and the CD has the lowest value. Note also that the approximate impedance value is relatively independent of the configuration; i.e., the impedance looking into the gate terminal is approximately infinite (several megohms); the impedance looking into the source terminal is approximately $1/g_m$, and the impedance looking into the drain terminal is approximately r_{ds}. These approximate impedance levels will be used to simplify multistage circuit analysis.

Of the three configurations, the CS stage is probably the most widely used for two reasons; (1) the voltage gain is greater than unity, and (2) the high values of input and output impedance, R_{ig} and R_{od}, neither excessively load the generator nor limit the effective ac load impedance. By comparison, the CG stage has a low value of R_{is}, a low value of A_{vg} and, consequently, a low value of A_{vo}.

* The term "cascade" applies to multistage amplifiers in which each stage has the same configuration.

** A two-port network is any circuit having a pair of input terminals for an input port, and a pair of output terminals for an output port. A single-stage amplifier is a typical two-port network.

TABLE 3–1 *Summary of approximate performance characteristics of single-stage FET amplifiers. The quantities in parentheses indicate the typical range of parameter values.*

Configuration	A_{vt}	Stage input impedance	Stage output impedance
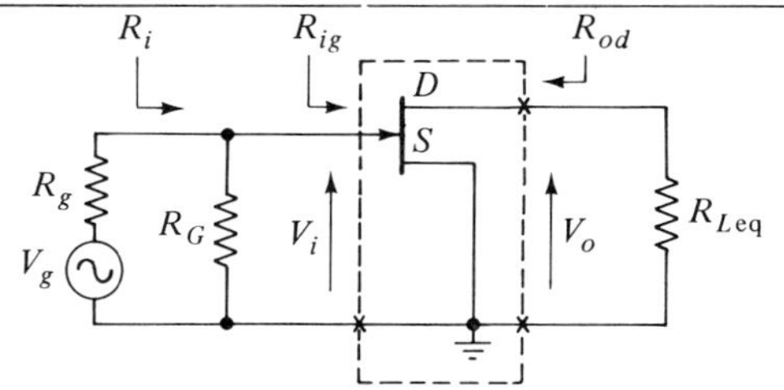 Common source (CS)	$-g_m R_{Leq}$ (signal inverted)	$R_{ig} \simeq r_{gs}$ $(\simeq \infty)$	$R_{od} \simeq r_{ds}$ (50 K to 200 K)
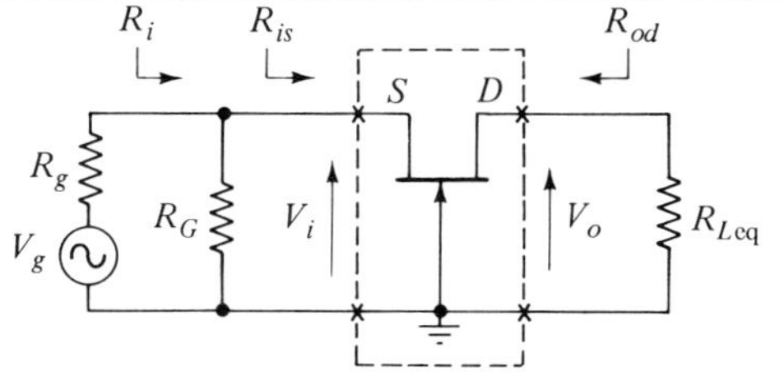 Common gate (CG)	$g_m R_{Leq}$ (signal not inverted)	$R_{is} \simeq \dfrac{1}{g_m}$ (50 to 2 K)	$R_{od} > r_{ds}$ (100 K to 10 M)
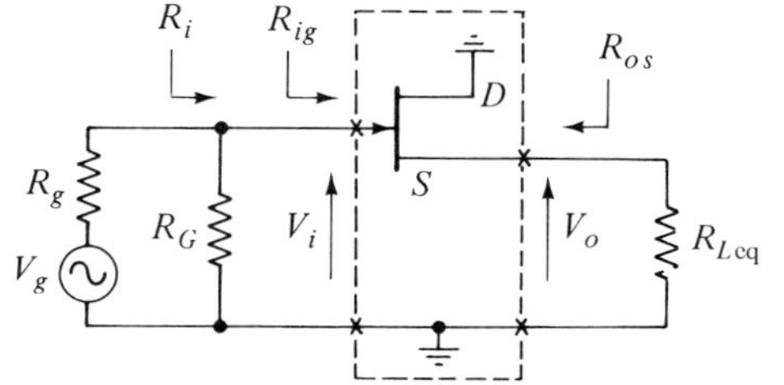 Common drain (CD)	$\dfrac{g_m R_{Leq}}{1 + g_m R_{Leq}}$ (signal not inverted)	$R_{ig} > r_{gs}$ $(\simeq \infty)$	$R_{os} \simeq \dfrac{1}{g_m}$ (50 Ω to 2 K)

$A_{vo} = V_o/V_g = A_{vg}A_{vt};\ A_{vg} = V_i/V_g = R_i/(R_i + R_g);\ A_{vt} = V_o/V_i$

Therefore, the CG stage is limited mainly to applications where a low input impedance is desirable. For example, the CG is used as the second stage of both a differential amplifier (see Sec. 3.10) and in a cascode* amplifier. The third type, the CD stage, is useful in applications requiring a very high value of input impedance, a low value of output impedance, or both. A low output impedance is often desirable when the load resistor, R_L, is small. The CD can also be used as an amplifier to isolate a small value of R_L from the input portion of a multi-stage amplifier.

An ac equivalent circuit of a multi-stage amplifier is shown in Fig. 3–26 in which each transistor is represented by a two-port network or a block with two

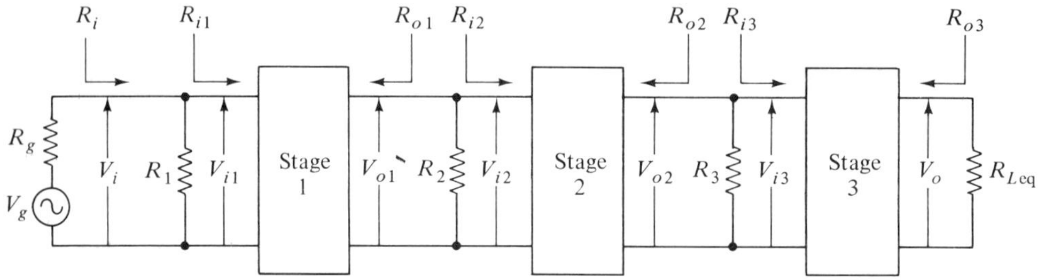

FIGURE 3–26 *A two-port representation of a three-stage amplifier.*

input and two output terminals. Resistors R_1, R_2, and R_3 correspond to the necessary bias resistors such as R_G, R_S, and R_D. The input and output impedances of each stage depend on the configuration. For example, if stage 1 is a CS stage, R_{i1} corresponds to R_{ig}, and R_{o1} corresponds to R_{od}. From this figure, the performance characteristics of the multistage amplifier in Fig. 3–26 can be expressed in terms of the performance characteristics of the individual stages. The generator-to-output voltage gain is

$$A_{vo} = \frac{V_o}{V_g} = \left(\frac{V_{i1}}{V_g}\right)\left(\frac{V_{o1}}{V_{i1}}\right)\left(\frac{V_{o2}}{V_{i2}}\right)\left(\frac{V_o}{V_{i3}}\right) = A_{vg}A_{vt1}A_{vt2}A_{vt3} \tag{3–55}$$

where each term in Eq. (3–55) is calculated in the same manner as for the single-stage amplifier. The voltage division between R_i and R_g is still given by

$$A_{vg} = V_{i1}/V_g = R_i/(R_i + R_g) \tag{3–56}$$

where R_i is the parallel combination of R_1 and R_{i1}, and R_{i1} depends on the input stage configuration. The expression for each A_{vt} also depends on the configuration. In the calculation of A_{vt}, R_{Leq} must include the input impedance of the next stage. The application of the two-port approach to multistage analysis is demonstrated in the following two examples for two-stage CS amplifiers.

The circuit in the first example contains two P-channel MOSFET units, each of which is connected as a CS linear amplifier. This circuit is shown in Fig. 3–27(a). A coupling capacitor,** C_2, is used to provide ac coupling between the drain of Q_1 and the gate of Q_2 while isolating the dc levels between these two

---

* The cascode amplifier is characterized by a high output impedance stage feeding a low input impedance stage. A possible configuration is the CS-CG amplifier. Cascode amplifiers find wide application in high-frequency circuits.

** Coupling capacitors are often called blocking capacitors since they perform dual functions.

points. Capacitors C_1 and C_3 are also coupling capacitors that allow different dc levels to exist on each side of each capacitor. The primary function of coupling capacitors is to make the ac and dc circuits compatible. This arrangement enables us to bias each stage without considering the dc levels on the input and output connections to the stage.

In the mid-frequency range, the impedance of coupling capacitors is relatively small and can be neglected or considered as an ac short. The mid-band ac circuit shown in Fig. 3–27(b) can be used with the two-port method of analysis. This circuit is obtained from Fig. 3–27(a) by replacing C_1, C_2, and C_3 with zero impedance connections (short circuits). Since the dc power sources have low ac impedances to ground, one end of R_{G1}, R_{G2}, R_{D1}, and R_{D2} is connected to ground. This simplified ac circuit has the same form as the first two stages in Fig. 3–26.

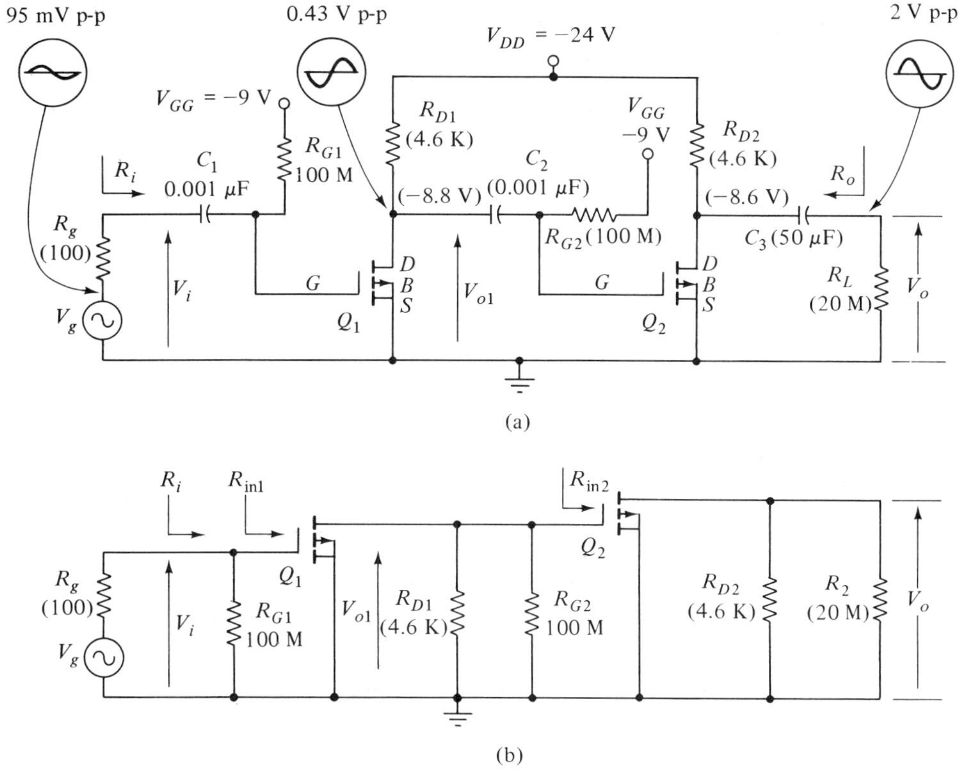

FIGURE 3–27 *A cascaded two-stage CS-CS amplifier using enhancement-mode MOSFET units: (a) a complete circuit showing coupling and bias components and dc power sources and (b) a mid-band ac circuit suitable for use with the two-port method of analysis, component and voltage values to Ex. 3–10.*

EXAMPLE 3–10 Calculate the values of A_{vo}, R_i, and R_o for the CS-CS amplifier shown in Fig. 3–27(a). Both P-channel enchancement-mode MOSFETs are biased so that both V_{GS} and V_{DS} are −9 V. For these bias values, I_D is 3.4 mA. Under these bias conditions, g_m is approximately 1 millimho.

Solution: (a) Mid-band equivalent circuit: Since the bias conditions are given, the next step is to structure the mid-band equivalent circuit of the amplifier as shown in Fig. 3–27(b). (Since the dc gate current is negligible, R_G could be connected between the gate and drain to bias the gate while providing effective ac isolation between these two points. The value of R_G is not critical but should be large. For this arrangement, the ac analysis would be more complicated, but the results would be the same.)

(b) Mid-band voltage gain: Considering Fig. 3–27(b) and neglecting r_{ds1}, the equivalent load resistance for the first stage is

$$R_{Leq1} = R_{D1} \,||\, R_{G2} \,||\, R_{in2} \simeq R_{D1} \simeq 4.6 \text{ K}$$

The terminal-to-terminal voltage gain of the first stage is

$$A_{vt1} = -g_{m1} R_{Leq1} \simeq -(1 \times 10^{-3})(4.6 \text{ K}) \simeq -4.6$$

The equivalent load resistance for the second stage is

$$R_{Leq2} = R_{D2} \,||\, R_L \simeq R_{D2} = 4.6 \text{ K}$$

The terminal-to-terminal voltage gain of the second stage is

$$A_{vt2} = -g_{m2} R_{Leq2} \simeq -(1 \times 10^{-3})(4.6 \text{ K}) \simeq -4.6$$

The input resistance of the amplifier is

$$R_i = R_{G1} \,||\, R_{in1} \simeq R_{G1} \simeq 100 \text{ M}$$

The output resistance of the amplifier is

$$R_o = R_{D2} \,||\, r_{ds2} \simeq R_{D2} \simeq 4.6 \text{ K}$$

The generator-to-input terminal voltage gain is

$$A_{vg} = R_i/(R_i + R_g) = 100 \text{ M}/(100 \text{ M} + 100) \simeq 1$$

The overall voltage gain of the amplifier is

$$A_{vo} = A_{vg}\, A_{vt1}\, A_{vt2} = (1)(-4.64)(-4.64) \simeq 21.5$$

(c) Experimental results: The measured bias levels and the observed voltage waveforms are shown in Fig. 3–27(a). The measured voltage gain with a V_g of 95 mV was 21 which is in close agreement with the calculated value.

A second example that uses a different type of coupling between stages is the two-stage dc coupled CS amplifier using JFETs shown in Fig. 3–28(a). The first transistor, Q_1, is an N-channel JFET and uses the dc voltage drop across R_{S1} to obtain the gate-to-source bias and consequently the desired drain current. The second transistor could also be an N-channel unit if V_{DD} could be made large enough to obtain the desired output voltage swing.

A better design approach is to use a P-channel JFET in the second stage as shown in Fig. 3–28(a). Note that Q_2 is "upside-down" in that the drain is on the bottom, and the source is on the top. However, a review of our previous examples using P-channel units indicates that the drain of a P-channel unit should be negative relative to the source, a condition that is provided by this "upside-down" arrangement. Since the dc gate current is very small, the voltage across R_{D1} is $I_{D1}R_{D1}$. This voltage is also the sum of V_{GS2} and the drop across R_{S2}, or

$$I_{D1}R_{D1} = V_{GS2} + I_{D2}R_{S2} \tag{3–57}$$

After biasing the first stage, V_{RD1} (or $I_{D1}R_{D1}$) is known. To bias the second stage first choose I_{D2}. Then find the value of V_{GS2} that is required for the desired I_{D2}. Using these values find R_{S2} from Eq. (3–57). The source-to-ground voltage, V_{S2}.

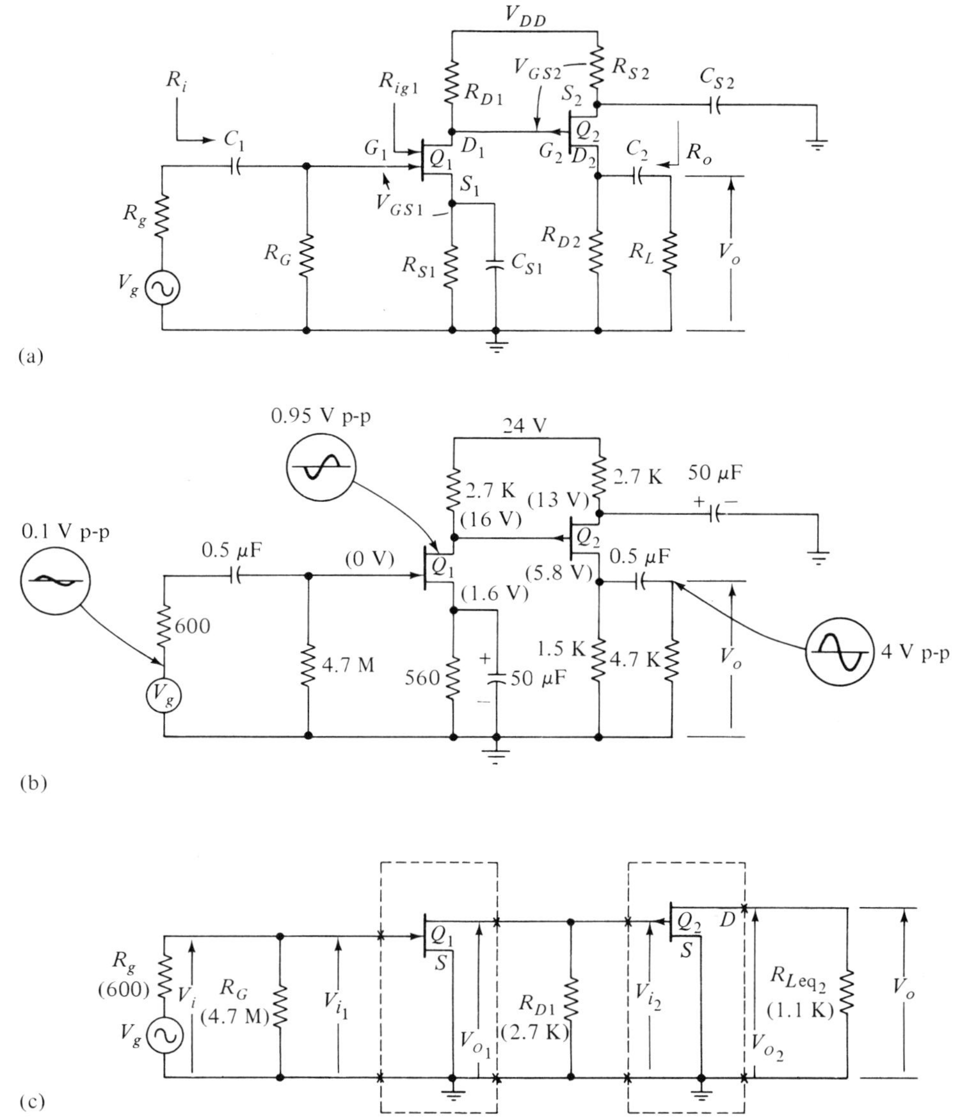

FIGURE 3–28 *A two-stage CS JFET amplifier: (a) general circuit, (b) circuit from Ex. 3–10 showing component values, measured dc voltages and ac waveforms, and (c) an ac equivalent circuit for the amplifier.*

and the drain-to-ground voltage, V_{D2}, for Q_2 are

$$V_{S2} = V_{DD} - I_{D2}R_{S2}$$

and **(3–58)**

$$V_{D2} = I_{D2}R_{D2}$$

The gate-to-source voltage for Q_2 is the difference between the gate voltage and the source voltage, or

$$V_{GS2} = V_{S2} - V_{D1} \qquad \textbf{(3–59)}$$

An equivalent circuit for the amplifier is shown in Fig. 3–28(c). This circuit can be used to calculate the characteristics of the CS cascaded amplifier. The input and output impedances of the amplifier are

$$R_i = R_{G1} \parallel R_{ig1} \simeq R_{G1} \tag{3–60}$$

and

$$R_o = R_{D2} \parallel R_{od2} \simeq R_{D2} \tag{3–61}$$

The generator-to-output voltage gain is

$$A_{vo} = A_{vg}A_{vt1}A_{vt2} \tag{3–62}$$

where

$$\left.\begin{aligned} A_{vg} &= R_i/(R_i + R_g) \qquad A_{vt1} \simeq -g_{m1}R_{Leq1} \\ A_{vt2} &\simeq -g_{m2}R_{Leq2} \qquad R_{Leq1} = R_{D1} \parallel R_{ig2} \simeq R_{D1} \\ & \qquad\qquad\qquad R_{Leq2} = R_{D2} \parallel R_L \end{aligned}\right\} \tag{3–63}$$

The complete analysis of a two-stage CS amplifier is demonstrated in the following example.

EXAMPLE 3–11 A dc coupled two-stage CS amplifier is shown in Fig. 3–28(b). The circuit component values are given in the figure, and the available data on the transistors are as follows: $V_{P1} = -3.3$ V, $I_{DSS1} = 11$ mA, $V_{P2} = 5.2$ V, and $I_{DSS2} = -22$ mA. The transistors are biased at $V_{GS1} = -1.6$ V and $V_{GS2} = 3$ V. Calculate the remaining bias conditions and the values of A_{vo}, R_i, and R_o.

Solution: (a) Bias conditions: The drain current for Q_1 is

$$I_{D1} = I_{DSS1}\left(1 - \frac{V_{GS1}}{V_{P1}}\right)^2 = 11\left(1 - \frac{1.6}{3.3}\right)^2 \simeq 2.9 \text{ mA}$$

The bias voltages are

$$V_{S1} = V_{GS1} = 1.6 \text{ V}$$
$$V_{D1} = V_{G2} = V_{DD} - I_{D1}R_{D1} \simeq 16.2 \text{ V}$$
$$V_{DS1} = V_{D1} - V_{S1} = 16.2 - 1.6 = 14.6 \text{ V}$$

The drain current of Q_2 is

$$I_{D2} = I_{DSS2}\left(1 - \frac{V_{GS2}}{V_{P2}}\right)^2 = -22\left(1 - \frac{3}{5.2}\right)^2 \simeq -3.9 \text{ mA}$$

and the bias voltages are

$$V_{S2} = V_{G2} - V_{GS2} = 13.2 \text{ V} \qquad V_{D2} = I_{D2}R_{D2} \simeq 5.9 \text{ V}$$
$$V_{DS2} = V_{D2} - V_{S2} = -7.3 \text{ V} \qquad V_{GS2} = V_{G2} - V_{S2} = 3 \text{ V}$$

(b) Gain and impedance calculations:

$$g_{m1} = \frac{2}{V_{P1}}\sqrt{I_{D1}I_{DSS1}} \simeq 3.4 \text{ millimhos} \qquad R_{Leq1} = R_{D1} \parallel R_{ig2} \simeq 2.7 \text{ K}$$

$$A_{vt1} = -g_{m1}R_{Leq1} \simeq -9.2 \qquad g_{m2} = \frac{2}{V_{P2}}\sqrt{I_{D2}I_{DSS2}} \simeq 3.6 \text{ millimhos}$$

$$R_{Leq2} = R_{D2} \parallel R_L \simeq 1.1 \text{ K} \qquad A_{vt2} = -g_{m2}R_{Leq} \simeq -4$$
$$R_i = R_{G1} \parallel R_{ig1} \simeq 4.7 \text{ M} \qquad A_{vg} = R_i/(R_i + R_g) \simeq 1$$
$$A_{vg} = R_i/(R_i + R_g) = 4.7 \text{ M}/(4.7 \text{ M} + 600) \simeq 1$$
$$A_{vo} = A_{vg}A_{vt1}A_{vt2} = (1)(-9.2)(-4) \simeq 37$$

These gain calculations apply only to small-signal conditions. If the signal levels are sufficiently large to cause unsymmetrical output signals (amplitude distortion), a graphical (or large-signal) analysis is required.

(c) Experimental results: The measured dc voltage levels are shown in Fig. 3–28(b), and they agree favorably with the calculated voltage levels. The experimental voltage gains, using peak-to-peak voltage measurements, are

$$A_{vt1} \simeq -\frac{0.95\text{ V}}{0.1\text{ V}} \simeq -9.5 \qquad A_{vt2} \simeq \frac{4.0\text{ V}}{0.95\text{ V}} \simeq -4.2 \qquad A_{vo} \simeq \frac{4.0\text{ V}}{0.1\text{ V}} \simeq 40$$

The ac waveforms are also shown in Fig. 3–28(b). Note that each CS stage inverts the signal. After two inverting stages, the output signal is in phase with the input signal.

As indicated in these examples, the analysis of multistage circuits is a simple extension of a single-stage amplifier analysis. First, the bias values are determined. Second, the value of the A_{vt} (terminal-to-terminal voltage gain) for each stage is found. The equivalent load resistance on any stage is the parallel combination of the input impedance of the next stage and any bias resistors connected to the output terminal of that stage. The value of A_{vo} for the entire amplifier is found by obtaining the product of the gains for each section. Values of A_{vg}, R_i, and R_o are calculated in the same manner as for the single-stage amplifier.

3.10 The FET Differential Amplifier

A particularly useful multistage arrangement is the differential amplifier, one form of which is shown in Fig. 3–29(a). This circuit is more complex than many multistage amplifiers since the differential amplifier can have two input signals, V_{g1} and V_{g2}, and two output signals, V_{o1} and V_{o2}. The differential circuit derives its name from the fact that both output voltages, V_{o1} and V_{o2}, depend on the differences between the two input signals, V_{i1} and V_{i2}. If only one input is present, the two output voltages are approximately equal in amplitude and are 180° out of phase with respect to each other. This condition is shown in Fig. 3–29 where the waveforms and dc voltage levels are obtained from Ex. 3–12. The differential amplifier can be analyzed using the techniques developed for both single-stage and multistage circuits.

A mid-band equivalent circuit for the differential amplifier, assuming a single input, is shown in Fig. 3–29(b). The calculation of each output voltage is simplified by dividing this equivalent circuit into two parts, one to be used in the calculation of V_{o1} and the other to be used in the calculation of V_{o2}. These two equivalent circuits are shown in Figs. 3–29(c) and (d).

The voltage gain from the gate to the drain terminal of Q_1 can be calculated using Fig. 3–29(c). This amplifier is a CS stage with an unbypassed source resistor, R_{Seq1}, where

$$R_{Seq1} = R_S \,||\, R_{is2} \simeq R_S \,||\, \frac{1}{g_{m2}} \tag{3–64}$$

The presence of R_{Seq1} reduces the CS voltage gain of Q_1 since part of the input voltage is dropped across R_{Seq1}. The calculation of the modified CS voltage gain,

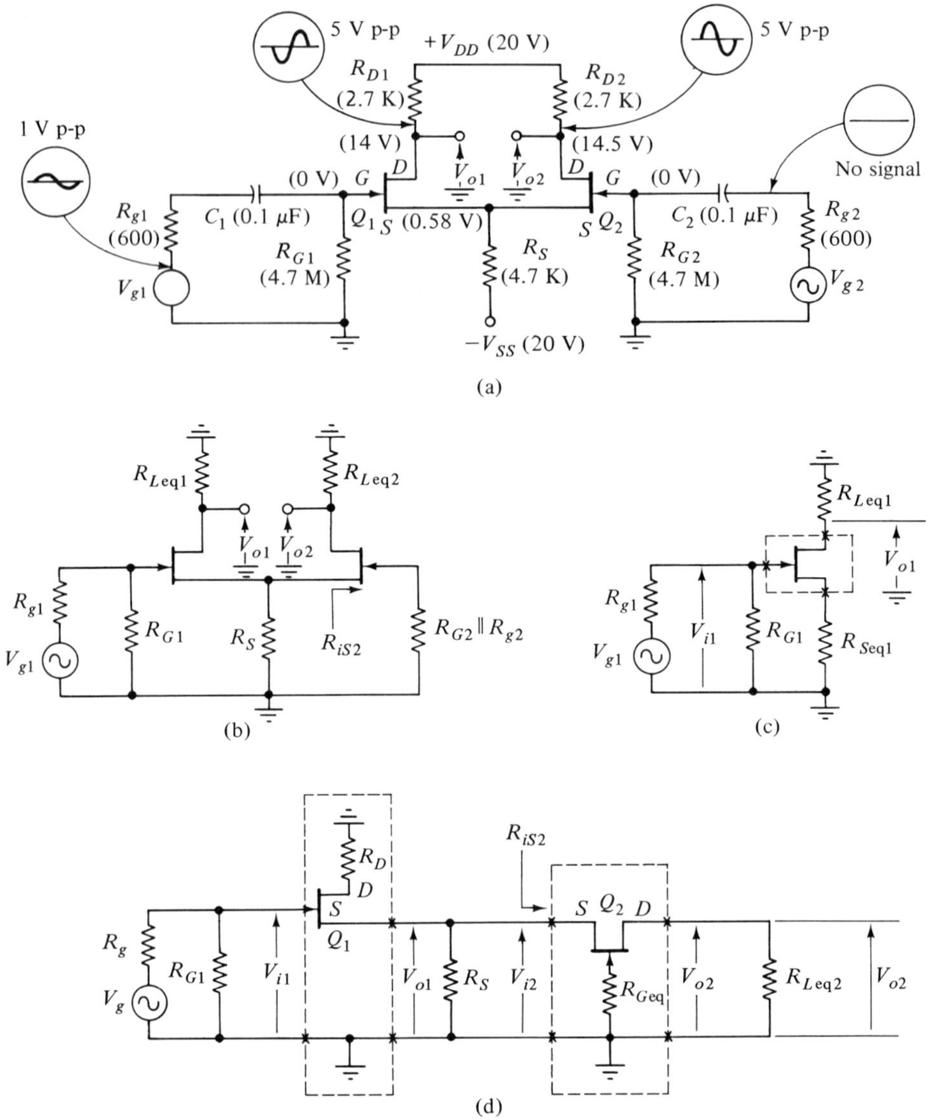

FIGURE 3–29 *An FET differential amplifier: (a) the complete circuit, (b) the amplifier ac equivalent circuit with only one input present* ($V_{g2} = 0$), *(c) the ac equivalent circuit used to calculate* V_{o1}, *and (d) the ac equivalent circuit used to calculate* V_{o2}. *The component values, dc voltages, and ac waveforms in (a) are from Ex. 3–12.*

A_{vt1}, is similar to that used in developing the voltage gain for the CD stage except that the output voltage is obtained at the drain terminal rather than the source terminal. The result of this analysis is

$$A_{vt1} = g_m R_{Leq}/(1 + g_m R_{Seq}) \tag{3–65}$$

where R_{Leq} is the ac resistance from the drain terminal to ground, and R_{Seq} is the ac resistance from the source terminal to ground. The generator-to-output voltage gain is

$$A_{vo1} = A_{vg1}A_{vt1} \tag{3–66}$$

where

$$A_{vg1} = \frac{R_{i1}}{R_{i1} + R_{g1}} \quad \text{and} \quad A_{vt1} = \frac{g_m R_{Leq1}}{1 + g_m R_{Seq}} \tag{3–67}$$

The voltage gain from the gate of Q_1 to the drain of Q_2 can be calculated using Fig. 3–29(d). In this circuit, Q_1 functions as a CD stage and Q_2 as a CG stage. The dual role of Q_1 (a CS stage when calculating V_{o1}, and a CD stage when calculating V_{o2}) should not be surprising since the type of configuration depends essentially on where the input signal is applied and where the output signal is obtained. Nothing prohibits two outputs from being obtained from a single transistor. Having recognized the configuration of the amplifier in Fig. 3–29(d), V_{o2} can be calculated as

$$V_{o2} = V_{g1}A_{vg1}A_{vt1}A_{vt2} \tag{3–68}$$

where

$$A_{vg1} = \frac{R_{i1}}{R_{i1} + R_{g1}} \qquad A_{vt1} = \frac{g_{m1} R_{Leq1}}{1 + g_m R_{Leq1}} \qquad A_{vt2} = g_{m2}R_{Leq2} \tag{3–69}$$

In Eq. (3–69),

$$R_{Leq1} = R_{Seq1} = R_S \,||\, (1/g_{m2}) \tag{3–70}$$

and

$$R_{Leq2} = R_{D2} \tag{3–71}$$

The voltage gain from the generator to each output is obtained by combining Eqs. (3–64) through (3–71). These gain expressions are

$$A_{vo1} = \frac{V_{o1}}{V_{g1}} = -\left(\frac{R_{i1}}{R_{i1} + R_{g1}}\right)\left(\frac{g_{m1}R_{Leq1}}{1 + g_{m1}(R_S \,||\, 1/g_{m2})}\right) \tag{3–72}$$

and

$$A_{vo2} = \frac{V_{o2}}{V_{g1}} = \left(\frac{R_{i1}}{R_{i1} + R_{g1}}\right)\left(\frac{g_{m1}(R_S \,||\, 1/g_{m2})}{1 + g_{m1}(R_S \,||\, 1/g_{m2})}\right)(g_{m2}R_{Leq2}) \tag{3–73}$$

Differential amplifiers are often symmetrical. This symmetry means that R_{Leq1} equals R_{Leq2}, R_{G1} equals R_{G2}, and R_{g1} equals R_{g2}. For this condition, the subscript denoting the stage can be dropped. In addition, the source resistor is usually chosen as large as possible, thereby making the parallel combination of R_S and $1/g_{m2}$ approximately equal to $1/g_{m2}$. For these conditions,

$$\left.\begin{aligned} A_{vo1} &\simeq -\left(\frac{R_i}{R_i + R_g}\right)\left(\frac{g_{m1}R_{Leq}}{1 + g_{m1}/g_{m2}}\right) \\ A_{vo2} &\simeq \left(\frac{R_i}{R_i + R_g}\right)\left(\frac{g_{m1}R_{Leq}}{1 + g_{m1}/g_{m2}}\right) \end{aligned}\right\} \tag{3–74}$$

Two important observations can be made from Eq. (3–74). First, the voltage gain from one input to each output has the same magnitude, and the two output signals are 180° out of phase as indicated by the minus sign for A_{vo1}. Second, if both input signals are equal and in phase (V_{g1} equal to V_{g2}), the ac voltage at each drain is zero. This condition occurs because the gains are approximately equal in magnitude and are 180° out of phase. At either drain, the output signal due to one input exactly cancels the output signal due to the other input. When a differential amplifier has two input signals, the output is proportional to the difference between these two signals. Note, however, that these results depend on the following two assumptions: (1) the differential amplifier is symmetrical, and (2) that R_S is very large compared to $1/g_m$. The following example demonstrates the analysis of a differential amplifier for the following two conditions: a single input signal and two input signals. The effect of a finite R_S is also included.

EXAMPLE 3–12 A JFET differential amplifier is shown in Fig. 3–29(a). The circuit component values are given in the figure, and the transistors are characterized by $V_{P1} = -1.8$ V, $I_{DSS1} = 4.5$ mA, $V_{P2} = -1.6$ V, and $I_{DSS2} = 4.7$ mA. Both transistors are biased at $V_{GS} = -0.6$ V. (a) Calculate the bias condition. (b) With $V_{g2} = 0$, calculate the voltage gain from V_{g1} to V_{o1} and V_{o2}. (c) With $V_{g1} = V_{g2} \neq 0$, calculate the voltage gain to each output.

Solution: (a) Bias conditions:

$$I_{D1} = I_{DSS1}\left(1 - \frac{V_{GS}}{V_{P1}}\right)^2 = 4.5\left(1 - \frac{0.6}{1.8}\right)^2 \simeq 2 \text{ mA}$$

$$V_{D1} = V_{DD} - I_{D1}R_{D1} = 20 - (2 \times 10^{-3})(2.7 \text{ K}) \simeq 14.6 \text{ V}$$

Since the gates are at dc ground potential ($I_G \simeq 0$) and are also 0.6 V negative relative to the sources, V_{S1} and V_{S2} are both at 0.6 V with respect to ground. Then

$$V_{DS1} = V_{D1} - V_{S1} = 14.6 - 0.6 \simeq 14 \text{ V}$$

$$I_{D2} = I_{DSS2}\left(1 - \frac{V_{GS2}}{V_{P2}}\right) = 4.7\left(1 - \frac{0.6}{1.6}\right)^2 \simeq 1.8 \text{ mA}$$

$$V_{D2} = V_{DD} - I_{D2}R_{D2} \simeq 15.1 \text{ V} \qquad V_{DS2} = V_{D2} - V_{S2} \simeq 14.5 \text{ V}$$

(b) Gain and impedance calculations with $V_{g2} = 0$:

$$g_{m1} = \frac{2}{V_{P1}}\sqrt{I_{D1}I_{DSS1}} \simeq 3.5 \text{ millimhos} \qquad g_{m2} = \frac{2}{V_{P2}}\sqrt{I_{D2}I_{DSS2}} \simeq 3.7 \text{ millimhos}$$

$$R_{Leq1} = R_{Leq2} = R_D = 2.7 \text{ K} \qquad R_{i1} \simeq 4.7 \text{ M}$$

$$R_{Seq1} \simeq R_S \,||\, (1/g_{m2}) = 4.7 \text{ K} \,||\, 270 \simeq 255\ \Omega$$

The voltage gain from V_{g1} to V_{o1} is calculated using Eq. (3–72):

$$A_{vo1} = -\left(\frac{4.7 \text{ M}}{4.7 \text{ M} + 600}\right)\left(\frac{(3.5 \times 10^{-3})(2.7 \text{ K})}{1 + (3.5 \times 10^{-3})(255)}\right) \simeq -5$$

The voltage gain from V_{g1} to V_{o2} is calculated using Eq. (3–73):

$$A_{vo2} = \left(\frac{4.7 \text{ M}}{4.7 \text{ M} + 600}\right)\left(\frac{(3.5 \times 10^{-3})(255)}{1 + (3.5 \times 10^{-3})(255)}\right)(3.7 \times 10^{-3} \times 2.7 \text{ K}) \simeq 4.7$$

Note that A_{vo1} and A_{vo2} are approximately equal. The small difference occurs because R_S is 4.7 K rather than being infinitely large.

(c) Gain and impedance calculations with $V_{g1} = V_{g2} \neq 0$: Denoting the voltage gain from V_{g1} to V_{o1} as A_{v1}, and the voltage gain from V_{g2} to V_{o1} as A_{v2},

$$V_{o1} = -V_{g1}A_{v1} + V_{g2}A_{v2} = V_g(A_{v2} - A_{v1})$$

Hence the output voltage depends on the difference between the two voltage gains. The voltage gain from V_{g2} to each output is slightly different from the corresponding values for V_{g1} since g_{m1} is not exactly equal to g_{m2}. Since g_{m1} and g_{m2} are within 6 percent of the same value, the gain differences are neglected, and

$$\frac{V_{o1}}{V_g} = \frac{V_{o2}}{V_g} \simeq (A_{v2} - A_{v1}) = 5 - 4.7 = 0.3$$

(d) Experimental results: The measured dc voltage levels are shown in Fig. 3–29(a) and agree favorably with the calculated voltage levels. The ac waveforms for a single input voltage ($V_{g2} = 0$) are also shown in Fig. 3–29(a). The experimental voltage gains were

$$A_{vo1} \simeq -5\text{ V}/1\text{ V} = -5 \qquad A_{vo2} = 5\text{ V}/1\text{ V} \simeq 5$$

With both inputs equally excited (connected to the same generator), the voltage at each output was 0.25 V with 1 V in at each input terminal.

The performance of the differential amplifier can be improved by increasing the value of R_S. The ideal component to substitute for R_S is a dc current generator with an infinite output impedance. One approach is to increase V_{SS} to a large value which allows R_S to be increased to a large value. Another approach which does not require a large value of V_{SS} is to replace R_S by a constant-current source. Such a circuit is shown in Fig. 3–30. The drain current is determined by the self-bias resistor, R_S. Since the output impedance of this stage corresponds to the output impedance of a CG stage, its value can range from 100 K to 10 M. The application of a transistor constant-current source is illustrated in the next chapter.

Several semiconductor manufacturers market dual FET transistors which consist of two FETs with closely matched values of V_P, I_{DSS}, and temperature characteristics. These dual transistors are formed either by testing and selecting individual devices and then mounting them in the same package, or by using two transistors formed in very close proximity on the same silicon chip. The ideal gain of a differential amplifier with both inputs equally excited is zero. Differential amplifiers which use matched transistors, symmetrical circuits with precision resistors, and constant-current sources in place of R_S, can have gains as low as 10^{-3} to 10^{-5} from input-to-output when both inputs are equally excited.

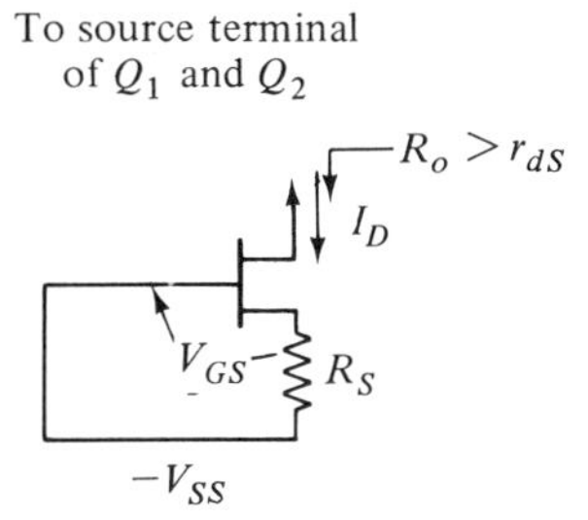

FIGURE 3–30 *An FET used as a constant-current source for a differential amplifier.*

3.11 Summary

The properties of a P-N junction and a metal-semiconductor junction were developed in Chapter 2. These properties were used to explain the physical operation of the JFET and the MOSFET devices. One physical characteristic of FET devices is that the principal current flow (source-to-drain) does not cross a junction. In general, the drain current depends on the value of V_{GS} and V_{DS}. However, for values of V_{DS} greater than $|V_P|$, the drain current becomes relatively independent of V_{DS} and is related to V_{GS} by the square-law relationship given by Eq. (3–14).

A particular FET can be characterized either by a set of static characteristics or by a set of parameters. The parameters V_P, I_{DSS}, and BV_{DGS} apply to JFET and depletion-mode MOSFET devices while the parameters V_T, I_{DO}, and BV_{DGS} apply to enhancement-mode MOSFET devices. Accurate values of these parameters can be found experimentally by calculating g_{mo} using two methods as discussed in Sec. 3.5.

An FET must be properly biased before it can be used as an active device in a linear amplifier. Proper bias means that the gate-source junction of a JFET should be reverse biased while the value of V_{GS} for a depletion-mode MOSFET can be either positive or negative. In an enhancement-mode MOSFET, V_{GS} must exceed V_T. The value of V_{DS} is chosen to ensure operation of the transistor in the saturated region.

The ac performance of an FET amplifier can be determined using either graphical techniques or small-signal techniques. Graphical analyses require an accurate set of static characteristics for the device and should be used when the input signal produces relatively large excursions from the operating or Q point. Small-signal analyses depend upon an equivalent circuit for the FET and are strictly valid only when the parameters of importance, such as g_m and r_{ds}, change very little as the input signal changes. However, small-signal analyses are often used under moderately large-signal conditions to obtain approximate solutions.

When an FET is used as a linear amplifier, it is normally used in one of three configurations: the CS, CG, or CD. For any of these configurations, the mid-band performance of an FET stage can be characterized with three terms: the terminal-to-terminal voltage gain (A_{vt}), the input impedance (Z_{in}), and the output impedance (Z_{out}). The performance characteristics of each configuration are summarized in Table 3–1. The overall mid-band response of an amplifier can also be characterized by three terms: A_{vo}, R_i, and R_o, where these terms include the effects of both the transistor and the bias network. Multistage amplifier analyses are simple extensions of single-stage analyses with the principal difference being an A_{vt} term for each stage. The equivalent load resistance on any stage is the parallel combination of the input impedance to the next stage with any bias resistors connected to the output terminal of the driving stage.

The FET is an important active device which finds wide application in electronic circuits. The detailed discussion of the physical operation and performance characteristics of the FET discussed in this chapter should aid the reader in the proper utilization of this device. The FET will be encountered often in subsequent chapters.

Problems

Section 3.1

3–1 Estimate the values of V_P, I_{DSS}, and BV_{DGS} for the P-channel JFET whose curves are given in Fig. 3–4(a).

3–2 The P-channel JFET whose curves are given in Fig. 3–4(a) was constructed using the diffusion process. What effect would interchanging the drain and source terminals have on the device curves?

3–3 Using the results of Prob. 3–1, calculate the value of V_{GS} required for a drain current of -8 mA using Eq. (3–4). Compare this value to the value of V_{GS} obtained from the curves in Fig. 3–4(a).

3–4 Using the values of V_P and I_{DSS} found in Ex. 3–1, calculate the drain current for $V_{GS} = -0.8$ V using Eq. (3–4). Compare this value with the value of I_D obtained from the curves in Fig. 3–4(b).

Section 3.2

3–5 Estimate the values of V_P and I_{DSS} for the MOSFET whose curves are given in Fig. 3–7(a).

3–6 A MOSFET unit has the following characteristics: $V_T = 3$ V and $I_{DO} = 10$ mA. What drain current will flow if V_{GS} is (a) 5 V, and (b) -5 V?

3–7 Sketch a set of characteristic curves that could apply to (a) an N-channel enhancement-mode MOSFET, and (b) a P-channel depletion-mode MOSFET. (c) How are these curves different from those given in Fig. 3–7?

3–8 An N-channel depletion mode MOSFET has the following characteristics: $V_P = -4$ V and $I_{DSS} = 6$ mA. What value of V_{GS} is required for a drain current of (a) 4 mA, and (b) 8 mA. (c) Could an N-channel JFET with the same values of V_P and I_{DSS} be substituted for the MOSFET in (a) or (b)?

Section 3.3

3–9 Repeat Ex. 3–3 for the case in which $V_{DD} = -20$ V, $R_D = 1.6$ K, and $R_S = 50$ ohms.

3–10 Design a self-bias circuit for a P-channel JFET with $V_P = 2$ V, $I_{DSS} = -8$ mA, and $BV_{DGS} = -35$ V. The desired Q point values are $I_D = -6$ mA, and $V_{DS} = -12$ V. Repeat the design using the bias circuit in Fig. 3–11(a) and also for the circuit in Fig. 3–11(b).

3–11 An N-channel JFET with $V_P = -3.5$ V and $I_{DSS} = 7$ mA is biased using the circuit in Fig. 3–11(a). The circuit values are: $-V_{GG} = -10$ V, $V_{DD} = 20$ V, $R_1 = 82$ K, $R_2 = 20$ K, and $R_D = 1.2$ K. What is the dc voltage at the drain terminal?

3–12 The MOSFET unit whose curves are given in Fig. 3–7(b) is used in the bias circuit in Fig. 3–11(e). The circuit values are: $V_{DD} = -20$ V, $R_D = 5.1$ K, and $R_G = 100$ M. Find the Q point values.

3–13 The MOSFET unit whose curves are given in Fig. 3–7(a) is used in the bias circuit in Fig. 3–11(b). The circuit values are: $R_S = 560$, $R_D = 3.9$ K, $R_1 = 91$ K, $R_2 = 10$ K, $R_G = 10$ M, and $V_{DD} = 18$ V. Find the Q point values.

Section 3.4

3–14 A circuit similar to the one shown in Fig. 3–13(a) has the following circuit values: $R_D = 1.5$ K, $V_{DD} = 18$ V, and $V_{GS} = -1$ V. The N-channel JFET curves are given in Fig. 3–13(b). (a) What value of R_S is required? (b) Sketch the output waveform if V_g is a 2-V peak-to-peak sine wave.

3–15 The CS amplifier shown in Fig. 3–10(a) is driven from a voltage generator whose internal impedance is 1 K and whose ac open circuit voltage is 2 V peak-to-peak. The FET characteristics are given in Fig. 3–10(b). The generator is connected to the gate terminal through a 0.5-μF capacitor. A 1.2-K load resistor is connected to the drain terminal through a 1-μF capacitor. (a) Sketch the complete circuit diagram. (b) Sketch the waveform of the ac voltage developed across the load resistor.

Section 3.5

3–16 Repeat Prob. 3–1, and verify the values of V_P and I_{DSS} by calculating g_{mo} using two methods.

3–17 Calculate the values of V_P and I_{DSS} for the P-channel JFET whose curves are given in Fig. 3–10(b). Verify the results by calculating g_{mo} using two methods.

3–18 Derive an expression for g_m in terms of g_{mo}, I_D, and I_{DSS}.

3–19 Verify that for an echancement-mode MOSFET,

$$g_m = \frac{2}{V_T}\sqrt{I_D I_{DO}}$$

Section 3.6

3–20 A CS amplifier of the form shown in Fig. 3–20(a) has the following circuit values: $R_g = 1$ K, $R_G = 4.7$ M, $R_D = R_L = 2$ K, $C_1 = C_2 = 1\ \mu$F, $C_S = 82\ \mu$F, and $V_{DD} = -18$ V. The JFET parameters have the following magnitudes: $|V_P| = 2$ V and $|I_{DSS}| = 6$ mA. The circuit is biased at $|I_D| = 3$ mA. (a) Is the JFET an N- or a P-channel unit? (b) Calculate the required value of R_S. (c) Sketch the complete mid-band equivalent circuit including component values. (d) Calculate the values of R_i, R_o, A_{vt}, and A_{vo}.

3–21 Using an N-channel depletion-mode MOSFET with $V_P = -4$ V and $I_{DSS} = 5$ mA, design a CS amplifier with a generator-to-output voltage gain of 5 ± 10 percent. The amplifier is to be used in a situation where $R_g = 300$ ohms and $R_L = 4.7$ K. Supply voltages of either +24 V or −24 V are available. Use only standard components.

3–22 Repeat Prob. 3–21 using an N-channel enhancement-mode MOSFET with $V_T = 2$ V and $I_{DO} = 9$ mA.

Section 3.7

3–23 A CG amplifier of the form shown in Fig. 3–21(a) has the following circuit values: $R_g = 100$, $R_D = 3.3$ K, $R_L = 5.1$ K, $C_1 = C_2 = 2.2\ \mu$F, and $V_{DD} = -20$ V. The P-channel JFET is characterized by $V_P = 4$ V and $I_{DSS} = -7$ mA. (a) Bias the circuit so $I_D = 3$ mA. (b) Sketch the complete mid-band equivalent circuit including component values. (c) Calculate the values of R_i, R_o, A_{vt}, and A_{vo}.

3–24 Design a CG amplifier with the following performance characteristics: R_i less than 500 ohms, R_o greater than 1 K, and $A_{vo} = 4 \pm 10$ percent. The circuit is to be used in a situation where $R_g = 600$ ohms and $R_L = 8.2$ K. The JFET is characterized by $V_P = -3$ V, $I_{DSS} = 7$ mA, and $BV_{DGS} = 30$ V. Use only standard components.

3–25 Verify Eq. (3–37) for R_{od} for the CG amplifier.

Section 3.8

3–26 A CD amplifier of the form shown in Fig. 3–23(a) has the following circuit values: $R_g = 5$ K, $R_G = 5.1$ M, $R_L = 470$ ohms, $C_1 = C_2 = 3.3\ \mu$F, and $V_{DD} = -15$ V. The P-channel JFET is characterized by $V_P = 3.5$ V and $I_{DSS} = -8.3$ mA. (a) Bias the circuit so that $g_m = 3$ millimhos, (b) sketch the complete mid-band equivalent circuit including component values, and (c) calculate the values of R_i, R_o, A_{vt}, and A_{vo}.

3–27 Verify Eq. (3–50) for R_{ig} for the CD amplifier.

3–28 Verify Eq. (3–52) for R_{os} for the CD amplifier.

3–29 Design a CD amplifier with A_{vo} greater than 0.5 using an enhancement-mode MOSFET with parameters $V_T = -2.5$ V and $I_{DO} = -8.2$ mA. The circuit is to be used in a situation where $R_g = 8.2$ K and $R_L = 1$ K. Supply voltages of either +15 V or −15 V are available.

Section 3.9

3–30 Calculate the bias values, A_{vo}, R_i, and R_o for the CS-CG amplifier shown in Fig. 3–31. For each JFET, $|V_P| = 2$ V, $|I_{DSS}| = 8$ mA, and $|V_{GS}| = 1$ V.

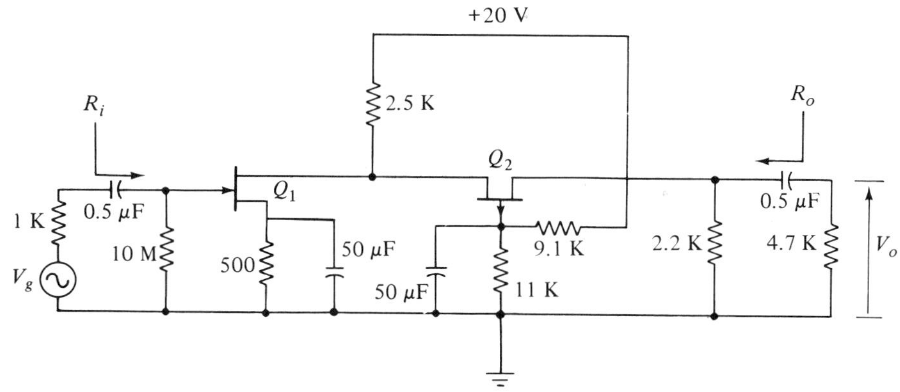

FIGURE 3–31 *Circuit for Prob. 3–30.*

3–31 Calculate the bias values, A_{vo}, R_i, and R_o for the CS-CD amplifier shown in Fig. 3–32. For each JFET, $V_P = 4$ V, $I_{DSS} = -12$ mA, and $V_{GS} = 2$ V.

3–32 Design an amplifier with the following characteristics: $A_{vo} \geq 100$, $R_i \geq 5$ megohms, and $R_o \leq 500$ ohms. The value of R_g is 600 ohms, and R_L is 500 ohms. Assume that all transistors, both N-channel and P-channel, have values of $|V_P|$ equal to 4 V and $|I_{DSS}|$ equal to 10 mA. Use only standard components.

3–33 Design an inverting amplifier with the following characteristics: $|A_{vo}| \geq 100$, $R_i \geq 5$ megohms, and $R_o \geq 1$ kilohm. The value of R_g is 600 ohms, and R_L is 5 K. Assume the same transistor characteristics as given in Prob. 3–32. Use only standard components.

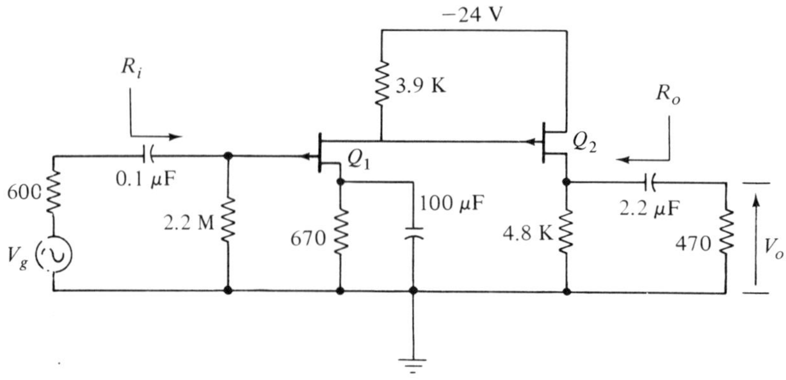

FIGURE 3–32 *Circuit for Prob. 3–31.*

Section 3.10

3–34 Calculate the value of R_S required for proper bias of the circuit shown in Fig. 3–33. For each JFET, $V_P = 2$ V, $I_{DSS} = -8$ mA, and $V_{GS} = 1$ V.

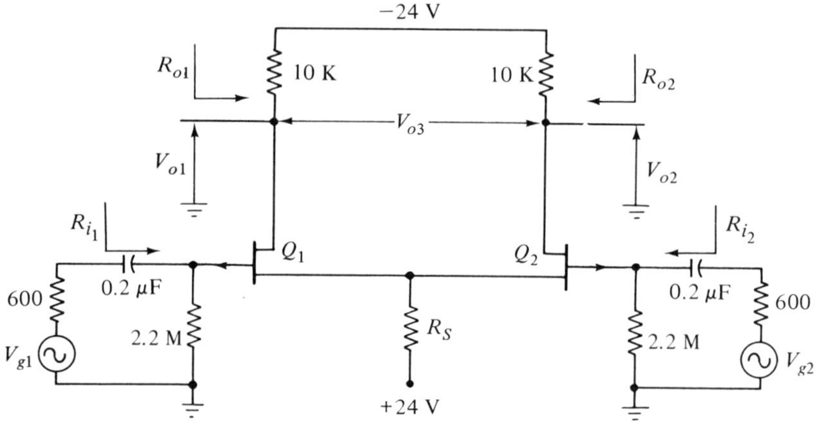

FIGURE 3–33 *Circuit for Probs. 3–34 through 3–37.*

3–35 Using the transistors parameters and the value of R_S from Prob. 3–34, calculate V_{o1}/V_{g1}, V_{o2}/V_{g1}, V_{o3}/V_{g1}, R_i, and R_o for the circuit shown in Fig. 3–33. Assume that $V_{g2} = 0$ V.

3–36 Using the transistor parameters and value of R_S from Prob. 3–34, calculate V_{o1}, V_{o2}, and V_{o3} for the circuit shown in Fig. 3–33 when $V_{g1} = 10$ mV, and $V_{g2} = 20$ mV. Assume that the two input signals are in phase.

3–37 Re-design the circuit shown in Fig. 3–33 by replacing the resistor R_S with a JFET constant-current source of the form shown in Fig. 3–30. The redesigned circuit should have the same bias conditions as those of Prob. 3–34. Specify values of V_P and I_{DSS}.

3–38 Design a JFET differential amplifier with the following characteristics: $R_i \geq 1$ M, $R_o \geq 1$ K, and the magnitude of the gain from either input to either output should be greater than 50. Assume that all transistors, either N-channel or P-channel, have values of $|V_P|$ equal to 3 V and $|I_{DSS}|$ equal to 6 mA. Use standard components.

References

Blalock, T. V., "Optimization of Semiconductor Preamplifiers for Use with Radiation Detectors." Ph.D. Thesis, The University of Tennessee, 1965.

Cobbold, Richard S. C., *Theory and Applications of Field-Effect Transistors*. New York: John Wiley & Sons, Inc., 1970.

Crawford, R. H., *MOSFET in Circuit Design*. New York: McGraw-Hill Book Company, 1968.

Dacey, G. C. and I. M. Ross, "The Field-Effect Transistor," *The Bell System Technical Journal*, 34 (November, 1955), 1149.

Early, J. M., "The Field-Effect Transistor," *Proceedings of the IRE*, 46 (1958), 1924.

Kennedy, E. J., "A Study of the Theoretical and Practical Limitations of Low-Current Amplification by Transistorized Current-Feedback DC Electrometers." Ph.D. Thesis, The University of Tennessee, 1967.

Richer, I., "The Equivalent Circuit of an Arbitrarily Doped Field-Effect Transistor," *Solid-State Electronics*, 8 (August 17, 1964), 381–93.

Sah, C. T., "Characteristics of the Metal-Oxide-Semiconductor Transistors," *IEEE Transactions of Electron Devices* (July, 1964), pp. 324–45.

Sevin, L. J., Jr., *Field-Effect Transistors*. New York: McGraw-Hill Book Company, 1965.

Shockley, W., "A Unipolar Field-Effect Transistor," *Proceedings of the IRE*, 40 (November, 1952), 1365–76.

Shockley, W. and G. L. Pearson, "Modulation of Conductance of Thin Films by Surface Charge," *Physical Review*, 74 (1948), pp. 232–33.

Wallmark, J. T. and H. Johnson, eds., *Field-Effect Transistors: Physics, Technology and Applications*. Englewood Cliffs, N. J.: Prentice-Hall, Inc., 1966.

4

Bipolar Transistors and Amplifiers

A bipolar transistor consists of two P-N junctions separated by a thin layer of lightly doped semiconductor material as shown in Fig. 4–1(b). The basic principles of operation for the bipolar transistor are radically different from those of the FET. In an FET, the main current does not cross a junction in its path through the device; however, in a bipolar transistor the main current crosses two junctions. In an FET, the control element (the gate) is used to vary the resistance of the main channel, while in a bipolar transistor the control element (the base) is used to control the current flowing across a forward-biased diode. Many other differences will become obvious in this chapter.

The objectives of this chapter are to develop the theory of operation of bipolar transistors and to use these units in practical amplifiers. Starting with the physical operation in Sec. 4.1, we will develop the static characteristics in Sec. 4.2. Any active (amplifying) device must be properly biased before it can be used as a linear amplifier. Therefore, biasing problems and bias stability are considered in Sec. 4.3. Small-signal models suitable for linear amplifier analysis are discussed in Sec. 4.4. The remaining four sections are devoted to the analysis and design of single-stage and multistage amplifiers that use bipolar transistors.

4.1 Physical Operation

The operation of a bipolar transistor depends on the interaction between two P-N junctions formed in close proximity in the same semiconductor. As shown in Chap. 2, the effects of a junction extend only a few diffusion lengths into a crystal on either side of the depletion region. If two P-N junctions are formed as shown in Fig. 4–1(a), the diodes are independent of each other and function like two diodes connected in series. However, if the depletion regions of the two junctions are separated by much less than one diffusion length, the two diodes interact and must be treated as a single unit. Bipolar transistors are fabricated by forming two P-N junctions in which both the spacing between the junctions and the doping levels on either side of the junctions are closely controlled. The principal methods of fabricating a bipolar transistor are discussed in Sec. 4.4.

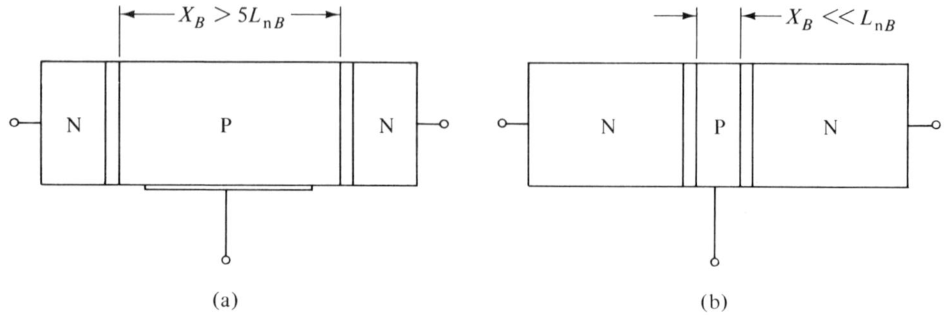

FIGURE 4–1 *Two P-N junctions formed in the same crystal bar. (a) When the distance* X_B *between depletion regions is greater than several minority carrier diffusion lengths, the diodes are independent. (b) A bipolar transistor is constructed by making* X_B *small compared with a minority carrier diffusion length so that the diodes are not independent.*

Since a bipolar transistor consists of three regions of alternating N- and P-type material, it can be fabricated as one of two types, an *NPN* or a *PNP*. The three regions are called the *emitter*, the *base*, and the *collector*. An ohmic contact is made to each region to provide for external connections. The emitter region is normally heavily doped, and the base region is lightly doped. Depending upon the type of construction, the collector may be either lightly or heavily doped. The equilibrium carrier concentrations are calculated in the same manner as for any doped semiconductor material. The important relationships from Chap. 1 are

$$\left.\begin{aligned} \sigma = \frac{1}{\rho} = q(n\mu_n + p\mu_p) \qquad np = n_i^2 \\ n + N_A = p + N_D \end{aligned}\right\} \qquad \textbf{(4–1)}$$

One set of relative carrier densities is shown in Fig. 4–2. The values shown are obtained from the following example.

EXAMPLE 4–1 Find approximate equilibrium values for the electron and hole densities in the emitter, base, and collector of a silicon PNP transistor in which the emitter resistivity is 0.05 ohm-cm, the base resistivity is 7.5 ohm-cm, and the collector resistivity is 0.5 ohm-cm.

Solution: The value of n_i^2 in silicon is approximately 2.3×10^{20} at room temperature. With μ_n and μ_p equal to 1380 and 460 cm²/volt-sec, respectively, the following densities may be calculated.

(a) Emitter hole and electron densities: Neglecting the effect of electrons on the resistivity, from Eq. (4–1),

$$p_{EO} \simeq \frac{1}{\rho q \mu_p} \simeq \frac{1}{(0.05)(460)(1.6 \times 10^{-19})} \simeq 2.7 \times 10^{17} \text{ holes/cm}^3$$

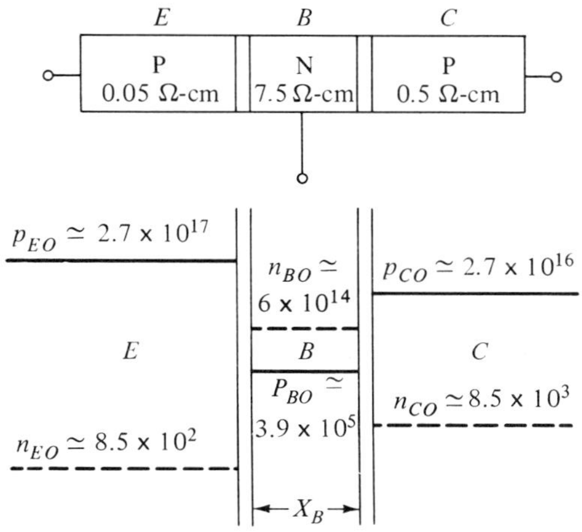

FIGURE 4–2 *Carrier densities in an unbiased PNP transistor at thermal equilibrium. The values shown were obtained from Ex. 4–1. (Not to scale)*

$$n_{EO} \simeq \frac{n_i^2}{p_{EO}} \simeq \frac{2.3 \times 10^{20}}{2.7 \times 10^{17}} \simeq 8.5 \times 10^{2} \text{ electrons/cm}^3$$

(b) Base electron and hole densities:

$$n_{BO} \simeq \frac{1}{(7.5 \times 1380 \times 1.6 \times 10^{-19})} \simeq 6 \times 10^{14} \text{ electrons/cm}^3$$

$$p_{BO} \simeq \frac{2.3 \times 10^{20}}{6 \times 10^{14}} \simeq 3.9 \times 10^{5} \text{ holes/cm}^3$$

(c) Collector hole and electron densities:

$$p_{CO} \simeq \frac{1}{(0.5)(46)(1.6 \times 10^{-19})} \simeq 2.7 \times 10^{16} \text{ holes/cm}^3$$

$$n_{CO} \simeq \frac{2.3 \times 10^{20}}{2.7 \times 10^{16}} \simeq 8.5 \times 10^{3} \text{ electrons/cm}^3$$

These carrier density values are approximate because mobility values, which are assumed to be constant here, vary with the degree of doping and compensation. As in the case of the unbiased diode, these values are approximately constant up to the edge of the depletion or space-charge region at each junction [see Fig. 2–4(a) or Fig. 4–2].

In linear amplifier circuits the emitter-base junction is biased in the forward direction, and the collector-base junction is biased in the reverse direction. Bias arrangements are shown in Fig. 4–3 for both NPN and PNP transistors. The presence of these two external bias voltages disturbs the thermal carrier equilibrium densities.* To determine the nature of the new distributions and the resulting terminal currents for normal bias voltages, apply the bias voltages to the PNP unit in Fig. 4–3(b) in two steps. First, apply the collector bias, V_{CC}, and assume that the emitter lead is open. The effect of this reverse bias is to increase the width of the base-collector depletion region and to reduce the minority carrier concentration at each edge of the depletion region to approximately zero.

---

* The reader may wish to refer to Sec. 2.2 to review the effects of forward and reverse bias on the carrier concentrations. Figure 2–4 provides a quick summary of these effects.

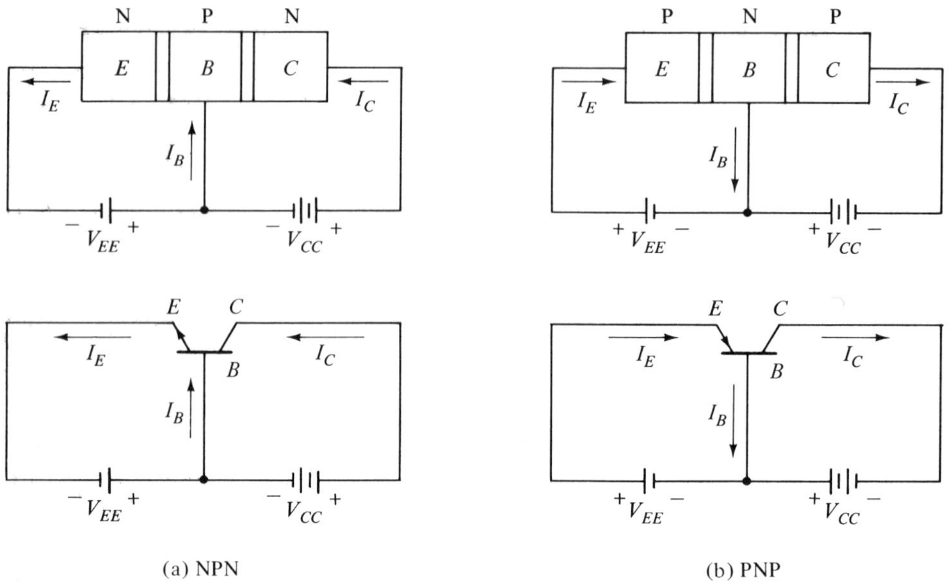

FIGURE 4–3 *Basic bias-arrangements for: (a) an NPN transistor, and (b) a PNP transistor. The arrow on the emitter indicates the direction of conventional current flow in each case.*

Also, a reverse saturation current, I_{CO}, flows. This current is similar to I_S for an isolated diode case. Minority carriers* that move into the base-collector depletion region are quickly swept to the other side of this region by the high field gradient. For V_{CC} values above a few tenths of a volt, majority carriers do not have sufficient energy to cross the barrier. As V_{CC} increases, I_{CO} increases slightly until breakdown occurs. Considering carrier densities, the important point is that the reverse bias voltage reduces the minority carrier density to approximately zero at each edge of the base-collector depletion region. This reduction is achieved by sweeping all minority carriers that arrive at the edge of this region to the other side of the junction where they become majority carriers.

As a second step, apply a small forward bias (V_{EE}) to the emitter-base junction while leaving the reverse bias on the base-collector junction. As indicated in the diode study, a forward bias reduces the height of the potential barrier and allows more majority carriers on each side of the base-emitter junction to cross to the other side of the depletion region where they become minority carriers. These added minority carriers produce an excess charge distribution on each side of the depletion region which is the same as that for the diode case. In addition, the carrier distribution on the emitter side is similar to that for a corresponding diode. However, the distribution on the base side is different. In a bipolar transistor, the minority carrier density approaches zero in one base width because of the boundary condition imposed by the collector junction, whereas in a diode, the minority carrier density approaches equilibrium in several diffusion lengths.

---

* In a PNP transistor, holes are the minority carriers in the base, and electrons are the minority carriers in the collector and emitter.

For this set of bias conditions, the minority carrier density curves in a transistor are shown in Fig. 4–4.*

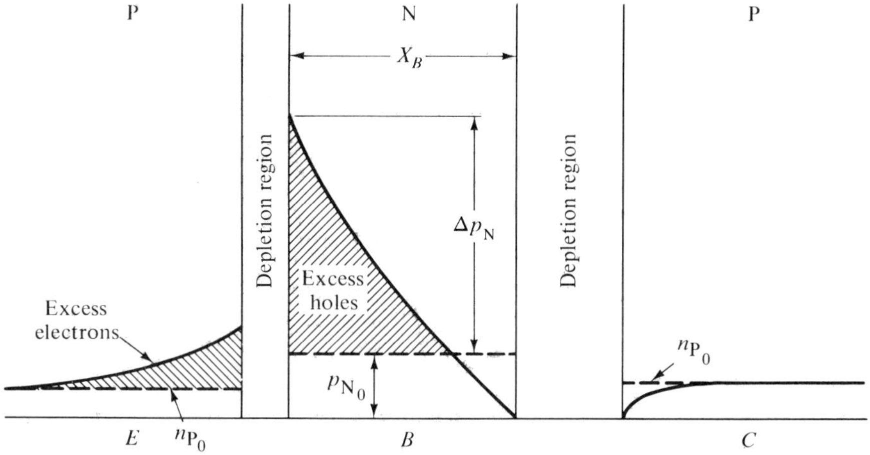

FIGURE 4–4 *Minority carrier density curves in a PNP transistor with a forward bias on the emitter-base junction and a reverse bias on the collector-base junction. The shaded areas represent excess stored minority carriers.*

Let us now consider the currents that flow inside a normally biased PNP transistor and the resulting terminal currents. Minority carriers are removed from the base in two ways. Some of the minority carriers recombine with majority carriers in the base region. Since the base width, X_B, is small compared with a diffusion length, base recombinations account for only one or two percent of the carriers that are injected into the base region from the emitter. The remainder diffuse to the edge of the base-collector depletion region and are swept into the collector by the field in this region. If recombinations in the base region are neglected, then the hole density curve in the base, as shown in Fig. 4–4, becomes a straight line with a slope of

$$\left|\frac{dp}{dx}\right| = \frac{\Delta p_n + p_{no}}{X_B} \tag{4–2}$$

when X_B is the width of the base between the two depletion regions for fixed values of V_{CC} and V_{EE}. From Eq. (1–26), the diffusion current resulting from this charge gradient is

$$|I_p| = qD_pA(dp/dx) \tag{4–3}$$

* The hole density curve in the base, as shown in Fig. 4–4, has a slight curvature due to the recombination of minority carriers (holes) in the N-type base. This curve is based on the assumption of homogeneous doping in the base. As was the case with the FET, modern bipolar transistors are often fabricated using a diffusion process (see Sec. 4.4), and the base region is not homogeneously doped. A nonhomogeneous base region radically changes the base minority carrier density curve. However, these effects will be neglected in our discussion of the physical operation of the bipolar transistor in the interest of simplicity. The interested and advanced student should refer to one of several excellent texts on transistor physics, such as the book by Lindmayer and Wrigley, listed in the bibliography of this chapter.

As the minority carriers forming this diffusion current move to the base-collector junction, they are swept into the collector region due to the reverse bias of the base-collector junction. Consequently, the collector current may be estimated for a given set of transistor parameters as indicated in the following example.

EXAMPLE 4–2 The PNP transistor in Ex. 4–1 is a uniform bar having a square cross section 0.025 cm on each side. If the net base width, X_B, between depletion regions is 10 microns (or 10×10^{-4} cm), and $L_p = 10^{-2}$ cm, estimate the dc collector current when V_{EE} is 0.6 V, and V_{CC} is 5 V. Assume that D_p is 12 and that (q/kT) is 38.8 at room temperature.

Solution: The hole density in Ex. 4–1 for the base was 3.9×10^5 holes/cm³. From Eq. (2–9), the excess hole density at the base edge of the base-emitter depletion region is

$$\Delta p_n = p_{BO}[\exp(qV/kT) - 1] = 3.9 \times 10^5[\exp(38.8 \times 0.6) - 1]$$
$$= 4.6 \times 10^{15} \text{ holes/cm}^3$$

Since this number is much larger than p_{BO}, the total density on the side of the base next to the emitter is approximately 4.6×10^{15} holes/cm³ and is zero on the side of the base next to the collector. Neglecting recombinations in the base, the slope of the hole density curve in the base is

$$\left|\frac{dp}{dx}\right| = \frac{4.66 \times 10^{15}}{10^{-3} \text{ cm}} = 4.66 \times 10^{18} \text{ holes/cm}^4$$

From Eq. (4–3), the collector current is

$$I_C = qD_p A(dp/dx) \simeq (1.6 \times 10^{-19})(12)(2.5 \times 10^{-2})^2(4.66 \times 10^{18}) \simeq 5.6 \text{ mA}$$

To obtain a better estimate for I_C, the reverse saturation current I_{CO} should be added to, and the base recombination current should be subtracted from, the above value. Since both of these quantities are small and are compensating, the above value is a reasonable estimate.

The collector current component calculated in Ex. 4–2 resulted from holes being injected* or diffused into the base from the emitter due to the forward bias of the base-emitter junction. As in the case of a diode, carriers diffuse both ways across a forward-biased junction. Consequently, electrons are injected into the emitter from the base simultaneously with the injection of holes from the emitter into the base. The electron current component can be calculated in a manner similar to that used for a diode since the minority carrier distribution on the emitter side is similar to that for a corresponding diode.** From Eq. (2–15), the electron component of the emitter current is given by

$$I_{En} = \frac{qAD_n n_{po}}{L_n}[\exp(qV/kT) - 1] \tag{4–4}$$

The total emitter current is the sum of the hole component [Eq. (4–3)] and the electron component [Eq. (4–4)] or

$$I_E = I_{Ep} + I_{En} \tag{4–5}$$

This relationship is demonstrated in the following example.

---

* The term "injection" is used to describe the diffusion of majority carriers across a forward-biased junction.

** This can be shown by comparing the excess electron distributions in Fig. 4–4 and Fig. 4–2(b).

EXAMPLE 4–3 Find the total emitter current, I_E, for the transistor in Ex. 4–2.

Solution: From Ex. 4–2, the hole component of I_E is 5.6 mA. From Ex. 4–1, n_{po} is 8.5×10^2. Assume that L_n is 10^{-2} cm and D_n is 35 cm²/sec. From Eq. (4–4),

$$I_n = \frac{AqD_n n_{po}}{L_n}[\exp(qV/kT) - 1]$$

$$= \frac{(2.5 \times 10^{-2})^2(1.6 \times 10^{-19})(35)(8.5 \times 10^2)}{10^{-2}}[\exp(39)(0.6) - 1] \simeq 3.54\ \mu\text{A}$$

Then, from Eq. (4–5),

$$I_E = 5.6\text{ mA} + 3.54\ \mu\text{A} \simeq 5.6\text{ mA}$$

The preceding example demonstrates why the emitter should be much more heavily doped than the base. For low values of base doping, n_{po}, and hence I_{En}, is small. Consequently, virtually all of the emitter current is injected into the base and will reach the collector, except for those carriers lost due to recombination in the base region. Emitter doping levels corresponding to emitter resistivity values of less than 0.001 ohm-cm are not uncommon in practice. Electron-hole recombinations in the base produce a base current component as indicated below.

With no bias on the transistor, the rate of formation of electron-hole pairs in the base is equal to the recombination rate. However, with normal bias voltages, the recombination* rate in the base increases because the minority carrier density in the base is much larger than the equilibrium value. To maintain charge neutrality in the base region, majority carriers (electrons for a PNP unit) must be supplied to the base region as fast as they are lost by recombination. The resulting component of base current, I_{BR}, may be determined from the recombination rate as

$$I_{BR} = \frac{dq}{dt} = \frac{\text{Excess stored charge in base}}{\text{Minority carrier lifetime in base}} = \frac{Q_B}{\tau_B} \tag{4–6}$$

where Q_B is the excess stored minority charge in the base, and τ_B is the average lifetime of minority carriers in the base region. The lifetime decreases slowly as the excess stored charge density increases. Therefore, this component of base current increases more rapidly than the stored charge.

Consider now the case of a thermally generated electron-hole pair that is produced near the collector junction. If produced in the base, the hole (in a PNP unit) may diffuse to the collector junction and be swept into the collector. If produced in the collector, the electron may be swept into the base. Either event provides an excess electron in the base that would flow out of the base if I_E were zero. For this condition, the base current would be I_{CO} with electrons flowing out the base lead for a PNP unit. When I_E is not zero, these electrons are used to keep the base neutral as electrons are captured by holes or as electrons move from the base into the emitter, thereby reducing the rate at which electrons must be supplied by the base lead. Therefore, the total dc base current is

$$I_B = I_{En}\text{ (electron component)} + I_{BR}\text{ (base recombination)} - I_{CO} \tag{4–7}$$

* The recombination rate includes the effects of both bulk and surface recombinations. The lifetime, τ_B, is a composite value that includes both surface and volume effects.

EXAMPLE 4–4 For the transistor parameters and bias conditions in Ex. 4–2, find the dc base current. Assume that I_{CO} is 1 μA and τ_B is 5 μsec.

Solution: The base current component due to recombination may be found from Eq. (4–6). Neglecting the equilibrium hole density, the excess charge in the base due to excess holes is

$$Q_B \simeq \left(\frac{\Delta p_n - 0}{2}\right)(\text{volume of base})(q)$$

$$= \frac{4.6 \times 10^{15}}{2}(10^{-3} \times 0.025^2)(1.6 \times 10^{-19}) = 2.33 \times 10^{-10} \text{ coulombs}$$

Then

$$I_{BR} \simeq (2.33 \times 10^{-10})/(5 \times 10^{-6}) = 46.6\ \mu\text{A}$$

The electron component of I_E was found in Ex. 4–3 as 3.54 μA. The total base current is, from Eq. (4–7),

$$I_B \simeq (46.6 + 3.54 - 1)\ \mu\text{A} = 49.1\ \mu\text{A}$$

The transistor current components calculated in Exs. 4–2, 4–3, and 4–4 are shown in Fig. 4–5. A study of the magnitudes of the various current components indicates that, in this case, approximately 99 percent of the emitter terminal current reaches the collector terminal. For a PNP transistor,* the total emitter

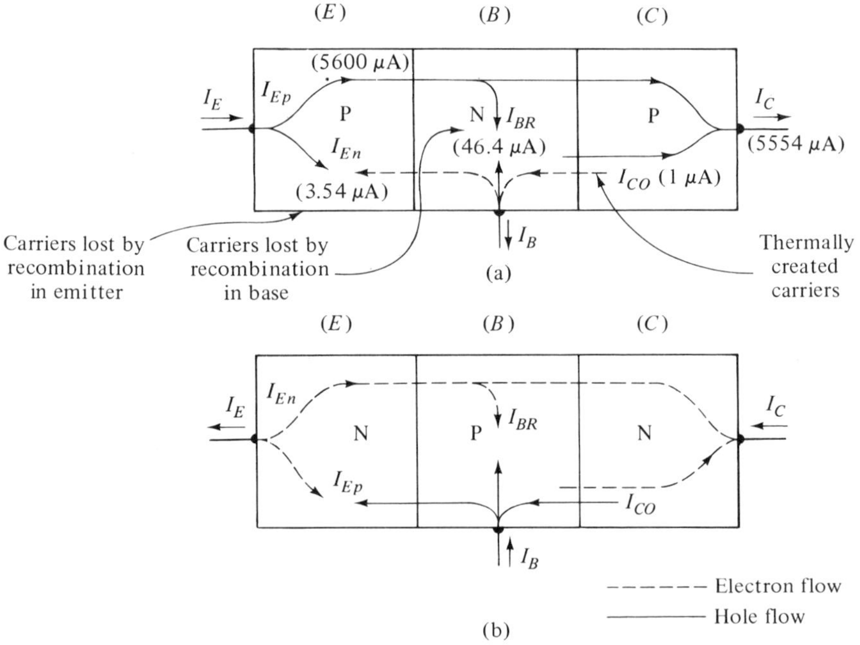

FIGURE 4–5 *(a) Current components in a PNP transistor. Values shown are from Exs. 4–2, 4–3, and 4–4. (b) Current components in an NPN transistor.*

* For an NPN transistor, the current component I_{En} is substituted for I_{Ep} in the various equations, and all current components change direction.

current is given by Eq. (4–5), the total base current by Eq. (4–7), and the total collector current is

$$I_C = I_{Ep} - I_{BR} + I_{CO} \tag{4–8}$$

One important feature of a bipolar transistor is that the emitter current, and therefore the collector current, can be changed or controlled simply by changing the emitter-base bias voltage. The function of the reverse biased collector is to collect the minority carriers that are injected into the base from the emitter. However, the base-collector bias voltage has a second-order effect on I_E. As this voltage increases, the width of the base-collector depletion region increases and causes the base width to decrease. Holding V_{BE} constant, a decrease in the effective base width causes I_E to increase because the slope of the minority carrier density curve increases [Eq. (4–3)]. This effect, known as the *Early effect*, is shown in Fig. 4–6.

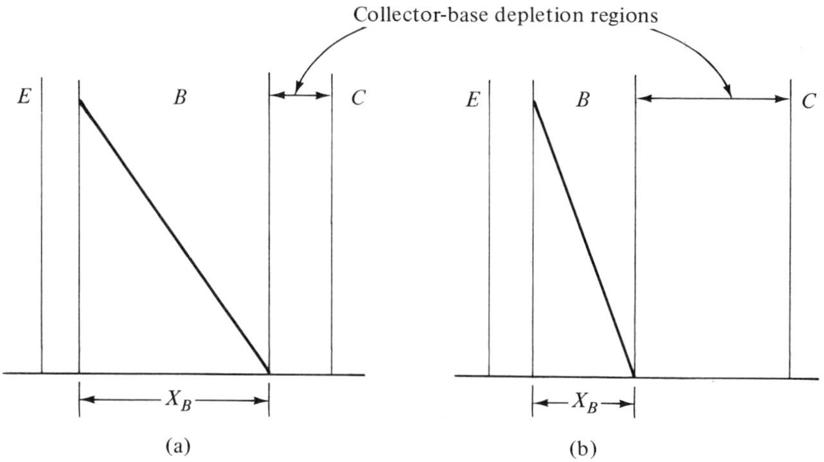

FIGURE 4–6 *The effect of an increase in collector voltage on the base width: (a) the base width with a small collector voltage, and (b) the base width with a much larger collector voltage.*

A relationship between the terminal currents in a bipolar transistor is

$$I_C = \alpha I_E + I_{CO} \tag{4–9}$$

In this equation, α is the fractional part of the emitter current that reaches the collector terminal. In small-signal work*

$$\alpha \equiv \left.\frac{\partial I_C}{\partial I_E}\right|_{V_{CB}\ \text{constant}} \simeq \left.\frac{\Delta I_C}{\Delta I_E}\right|_{V_{CB}\ \text{constant}} \tag{4–10}$$

Because α is a very important parameter and is used frequently, it should be well understood. The ideal value for α is unity, but practical values range from 0.95 to 0.995. This parameter is a product of three factors: the *emitter efficiency*, the *transport factor*, and the *collector-multiplication ratio*.

---

* The value of α is a function of frequency. Often the mid-band value of α is denoted as α_o. Since only mid-band effects will be considered in this chapter, the subscript is omitted.

The emitter efficiency, γ, is the ratio of the current injected into the base to the total emitter current. In Ex. 4–3 the emitter efficiency is

$$\gamma = \frac{I_{Ep}}{I_{Ep} + I_{En}} = \frac{5600\ \mu\text{A}}{5600\ \mu\text{A} + 3.54\ \mu\text{A}} \simeq 1 \tag{4–11}$$

This factor is usually close to unity because the emitter is doped much more heavily than the base.

The transport factor, b, is the ratio of the number of minority base carriers that arrive at the collector junction to the number that are injected into the base from the emitter. From Ex. 4–4 and Fig. 4–5, the transport factor for this particular transistor is

$$b = \frac{I_{Ep} - I_{BR}}{I_{Ep}} \simeq \frac{5600\ \mu\text{A} - 46\ \mu\text{A}}{5600\ \mu\text{A}} \simeq 0.995 \tag{4–12}$$

The transport factor approaches unity as the recombination current approaches zero. The recombination current is reduced by making the base width very small.

The collector multiplication ratio δ, is the ratio of the total current crossing the collector-base junction to the base minority carrier current that crosses this junction into the collector. The reverse saturation current, I_{CO}, is excluded from this ratio. For low collector voltages, carrier multiplication in the base-collector region is not pronounced, and δ approaches unity. For high collector-base voltages, δ may exceed unity, especially in transistors having high collector resistivity values.

In summary, the physical operation of a bipolar transistor has been discussed for normal biasing (forward bias on the *EB* junction and reverse bias on the *BC* junction) and low-level injection.* The background developed is sufficient for applications involving linear amplifiers but not for those involving switching circuits. An important concept discussed is that the base region has no net charge. If the forward bias of the base-emitter junction is changed, the emitter current changes. Also, the base current must change to maintain neutrality in the base region. If the base current is fixed at some value (by an external current source), the emitter current is fixed at the particular value required to maintain base neutrality through recombinations in the base region. Consequently, the emitter current, and hence the collector current, can be controlled either by controlling the base-emitter junction bias voltage or the base current. The important relationships of this section are summarized in Eq. (4–13) for a PNP unit. Figure 4–5 shows the direction of each current component for both the PNP and the NPN transistor.

$$\left.\begin{aligned} I_E &= I_{Ep} + I_{En} & I_B &= I_E - I_C \\ I_B &= I_{En} + I_{BR} - I_{CO} & \alpha &= \frac{\Delta I_C}{\Delta I_E} = \gamma b \delta \\ I_C &= I_{Ep} - I_{BR} + I_{CO} & I_C &= \alpha I_E + I_{CO} \\ & & I_C &\rightarrow I_E \text{ as } \alpha \rightarrow 1 \end{aligned}\right\} \tag{4–13}$$

---

* The term "low-level injection" means that the base minority carrier concentration, including the carriers injected from the emitter, is much less than the base majority carrier concentration. Several effects occur under conditions of high-level injection including a reduction in emitter efficiency and, hence, a reduction in α.

4.2 Static Characteristics

Like the FET, the bipolar transistor is a three-terminal device which can be connected in a circuit in six configurations, three of which are useful.* While the FET and the bipolar transistor have many similarities,** some important differences exist. In Chap. 3, the static characteristics of an FET were plotted as I_D versus V_{DS} for various values of V_{GS}. The value of drain current depended primarily on V_{GS} once the FET was biased in the saturation region. Hence, the FET is a voltage-controlled device. The bipolar transistor, on the other hand, is more properly described as a current-controlled device with the charge neutrality of the base region relating the base and emitter currents (and hence the base and collector currents). The most common and useful set of characteristic curves for a bipolar transistor are plotted as I_C versus V_{CE} for various values of I_B, as shown in Fig. 4–7. These static characteristics are shown for both NPN and PNP bipolar transistors connected in the common-emitter (CE) configuration.

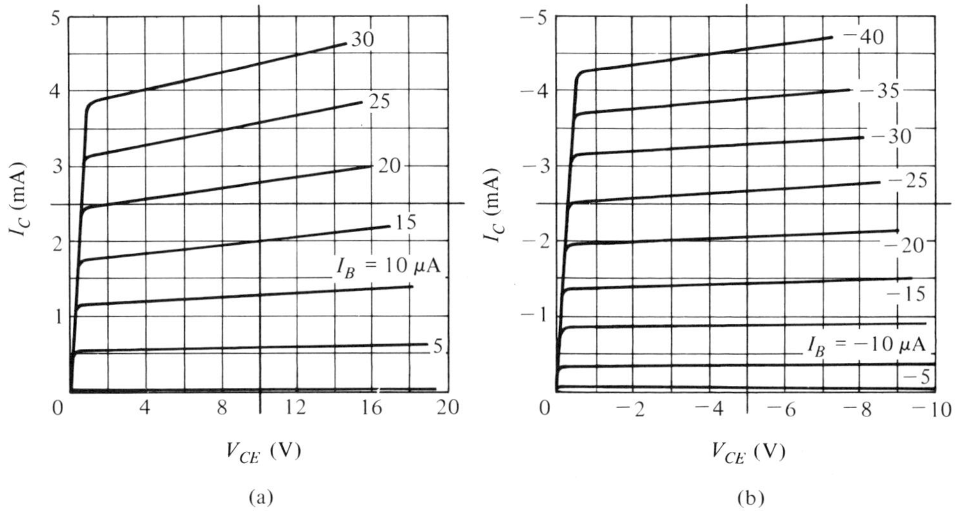

FIGURE 4–7 *Bipolar transistor common-emitter static characteristic curves: (a) curves for an NPN transistor and (b) curves for a PNP transistor.*

Although the static curves are seldom used in practice, they are discussed here to develop an understanding of the terminal characteristics of a bipolar transistor. The static curves can be generated for a PNP unit using the circuit in Fig. 4–8(a). The base voltage source, V_{BB}, and the base resistor, R_B, form an

* Three useful bipolar transistor configurations are the common-emitter (CE), common-base (CB), and the common-collector (CC) connections. These configurations correspond to the FET common-source (CS), common-gate (CG), and common-drain (CD) configurations, respectively. Like the FET, the designation is determined by the terminal that is common to both the input and output signals.

** A useful correspondence can be made between the FET and the bipolar transistors by relating the source and emitter terminals, the gate and base terminals, and the drain and collector terminals. The extent and limitations of this analogy will become apparent in the course of the chapter.

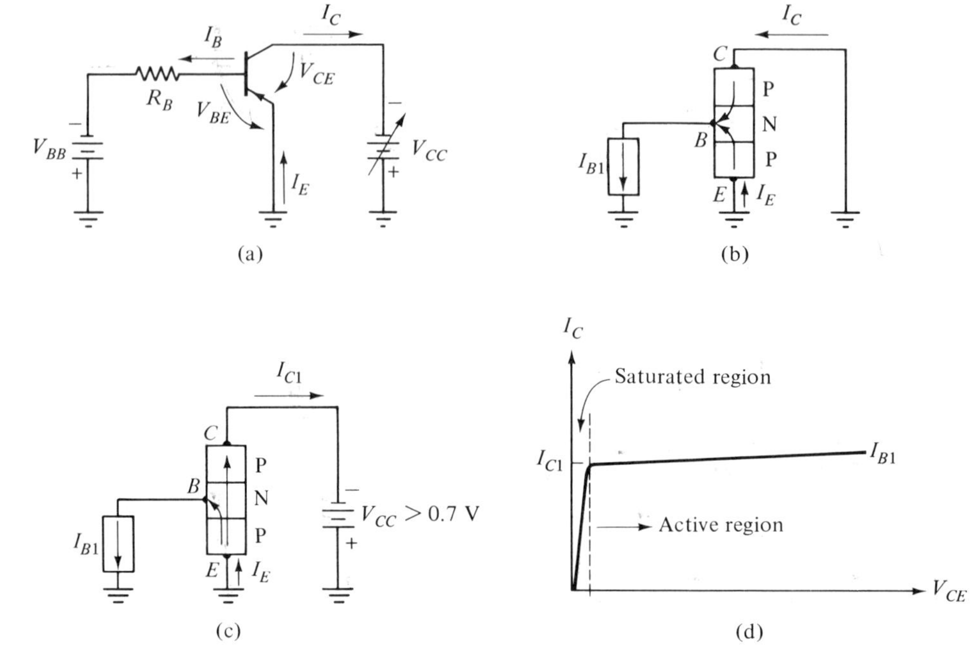

FIGURE 4–8 *(a) A common-emitter configuration with bias sources, (b) a physical equivalent for* $V_{CC} = 0$, *(c) a physical equivalent for* $V_{CC} > 0.7$ *V, and (d) one curve of the CE characteristics.*

approximate current source* if R_B has a large value. Consequently, for given values of V_{BB} and R_B, a particular value of base current, I_{B1}, will flow. With V_{CC} equal to zero, the transistor acts like two diodes with a common narrow N-region. The base current comes from both the emitter and the collector since both the *EB* and *BC* junctions are slightly forward biased. As the magnitude of V_{CC} is increased from zero volts [Fig. 4–8(c)], the *BC* junction is reverse biased. This condition produces two principal effects: (1) all of the base current comes from the emitter (neglecting I_{CO}) since only the *EB* junction is forward biased, and (2) the collector current changes direction and increases in magnitude. The second effect results from the physical operation discussed in the last section. Since a majority of the emitter current carriers reach the collector, a small base current produces a much larger emitter and collector current. For a silicon transistor, values of V_{CC} greater than about 0.7 V are sufficient to reverse bias the *BC* junction and bias the transistor in the active region, as shown in Fig. 4–8(a). Increasing V_{CC} beyond this value produces only a slight increase in the collector current with this increase being due to the Early effect. A complete family of curves can be generated by adjusting the circuit parameters for various values of I_B and plotting I_C as V_{CE} is varied. For the particular value of I_B equal to zero, the collector current is approximately $[\alpha/(1 - \alpha)]\ I_{CO}$. Since I_{CO} is similar to I_S for a diode, the value of I_{CO} is in the order of nanoamperes for a silicon transistor, and microamperes for a germanium transistor.

---

* As will be shown later, the dc input resistance to the base terminal is only a few thousand ohms, a range that is negligibly small if R_B is several hundred thousand ohms.

Two mechanisms limit the upper value of V_{CE}. The first is an avalanche breakdown between the base and collector resulting from the reverse bias of this junction as discussed in Sec. 2.7. The second mechanism is called *punch-through* and is often encountered in transistors with a very narrow base region. Punch-through results when the depletion region of the *BC* junction extends completely across the base region to the *EB* depletion region. Hence, carriers are injected directly from the emitter into the high field of the *BC* depletion region. Because of the large currents that can flow under either avalanche breakdown or punch-through conditions, a transistor can be destroyed by the resulting excessive heat. The design of a transistor determines which of these two effects occurs first.

The base and collector currents can be related by a parameter just as the emitter and collector currents are related by α. This parameter, β, is defined* as

$$\beta \equiv \left.\frac{\partial I_C}{\partial I_B}\right|_{V_{CE}=\text{constant}} \simeq \left.\frac{\Delta I_C}{\Delta I_B}\right|_{V_{CE}=\text{constant}} \tag{4–14}$$

This equation defines an ac value of β since changes in current are involved. A dc β can also be defined as

$$\beta_{DC} \equiv I_C/I_B \tag{4–15}$$

In practice, the two values of β are different but can be assumed approximately equal when making initial calculations. The same situation exists for α. Since α is approximately equal to unity in each case, the distinction between ac and dc values is usually not important.

EXAMPLE 4–5 Find the value of β and β_{DC} from the NPN transistor curves in Fig. 4–7(a) if $V_{CE} = 10$ V, and $I_C = 2$ mA.

Solution: The base current for the given bias conditions is approximately 15 μA. From Eq. (4–15),

$$\beta_{DC} = \frac{I_C}{I_B} \simeq \frac{2 \text{ mA}}{0.015 \text{ mA}} \simeq 133$$

The ac value of β is found by assuming a change in I_B and determining the corresponding change in I_C. As I_B changes from 10 μA to 20 μA on the constant V_{CE} line of 10 V,

$$\Delta I_C \simeq 2.8 \text{ mA} - 1.3 \text{ mA} \simeq 1.5 \text{ mA}$$

$$\beta = \Delta I_C/\Delta I_B \simeq 1.5 \text{ mA}/10\ \mu\text{A} \simeq 150$$

The value of β at $I_C = 1$ mA is

$$\beta = \Delta I_C/\Delta I_B \simeq (1.3 - 0.6) \text{ mA}/(10 - 5)\ \mu\text{A} \simeq 140$$

These two values of β demonstrate the dependency of β on I_C. Several manufacturers market instruments that measure β directly, usually at a frequency of 1 kHz. Using such an instrument, and with $V_{CE} = 7.5$ V and $I_E = 1$ mA, the measured value of β was approximately 130.

---

* The value of β is frequency dependent. The mid-band value of β is denoted by β_o. Since only mid-band effects will be considered in this chapter, the subscript is omitted.

The terminal currents of a bipolar transistor in terms of α and β are:

$$I_E = I_B + I_C = (\beta + 1)I_B \qquad I_C = \beta I_B = \alpha I_E + I_{CO}$$

$$I_B = I_E - I_C = (1 - \alpha)I_E - I_{CO} = \frac{I_E}{\beta + 1} - I_{CO} \tag{4–16}$$

From Eq. (4–16), and neglecting I_{CO}, the relationships between α and β are*

$$\alpha = \frac{I_C}{I_E} = \frac{\beta I_B}{(\beta + 1)I_B} = \frac{\beta}{\beta + 1} \tag{4–17}$$

$$\beta = \frac{I_C}{I_B} = \frac{\alpha I_E}{(1 - \alpha)I_E} = \frac{\alpha}{1 - \alpha} \tag{4–18}$$

As indicated by Eq. (4–18), β is strongly dependent on the value of α because α is close to unity; that is, a small change in α causes a large change in β. Also, β is a function of the emitter current. As shown in Fig. 4–9, β has a broad peak as

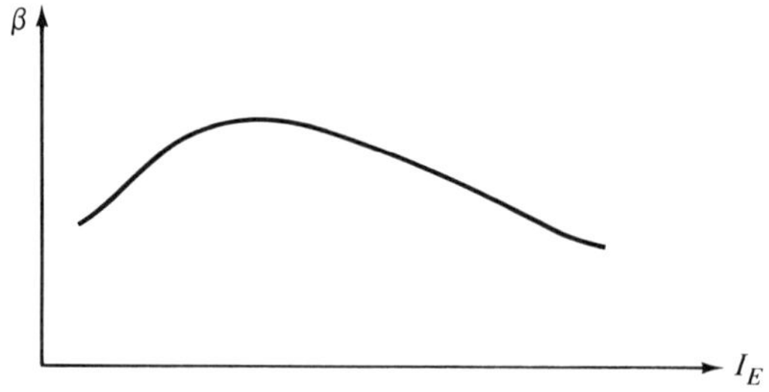

FIGURE 4–9 *Variations in β as a function of* I_E.

I_E increases. The decrease in β at low currents is due to recombinations in the *BE* junction depletion region, which increases in width as the forward bias is reduced. The decrease in β at high currents is due to a decrease in emitter efficiency (γ) since the injection of carriers from the base into the emitter increases as the current is increased. Base recombination effects are generally of secondary importance.

EXAMPLE 4–6 A PNP transistor with $\alpha = 0.98$ is biased so that I_B is 20 μA. Estimate the values of I_C and I_E.

Solution: The given value of α will be assumed to be the dc value of α. Then from Eqs. (4–16) and (4–18),

$$\beta_{DC} = \frac{\alpha}{1 - \alpha} = \frac{0.98}{1 - 0.98} \simeq 49 \qquad I_E = (\beta + 1)I_B \simeq 1 \text{ mA}$$

$$I_C = \beta I_B = (49)(20\ \mu\text{A}) \simeq 0.98 \text{ mA}$$

Two other curves are used in the dc characterization of bipolar transistors: (1) the input curve, and (2) the CB static characteristic curves. Both curves are shown in Fig. 4–10. The input curve is a plot of I_E versus V_{BE} for a constant value of V_{CB} and is similar to a diode curve. The input curve will be used in the chapter on power amplifiers. The CB curves are a plot of I_C versus V_{CB} for var-

---

* Eqs. (4–17) and (4–18) apply to both dc and ac values of α and β, but are not valid in the high-frequency range. These equations are generally written in terms of α_o and β_o.

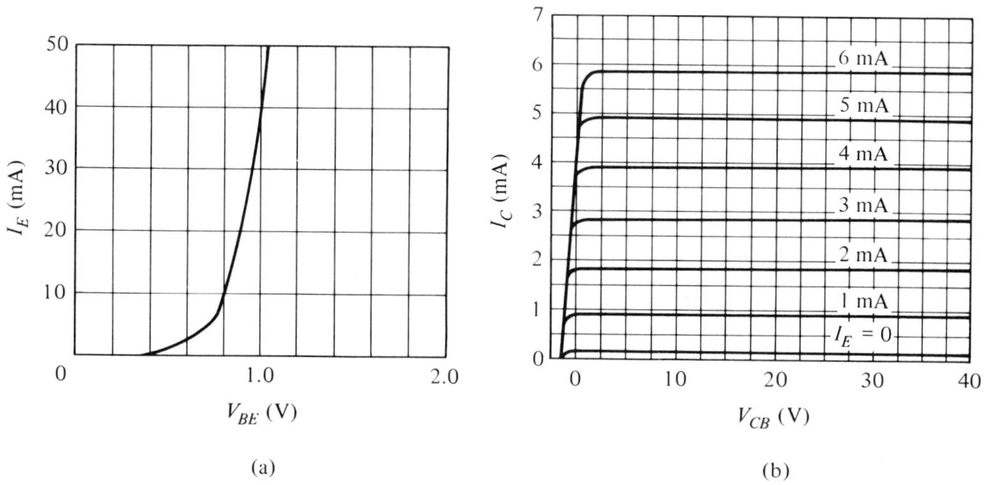

FIGURE 4–10 *Static characteristic curves: (a) the input curve for a constant value of* V_{CB}, *and (b) the common-base static characteristic curves.*

ious values of I_E. An important difference between the CE and CB curves is that the latter are more nearly horizontal indicating that the CB collector current is less dependent on the collector voltage than for the CE case.

Bipolar transistors have three important parameters that are functions of temperature: I_{CO}, V_{BE}, and β. The temperature variations of the first two parameters were discussed in Sec. 2.1 where it was noted that I_S (which corresponds to I_{CO}) will usually double for each 10°C rise in temperature, and V_{BE} must be decreased approximately 2 mV per centigrade degree rise in order to maintain a constant current. The parameter β increases with temperature due to the increase in excess charge (Δn_p or Δp_n) on the emitter side of the *BE* junction. Typically, β will double for each 50°C rise in temperature for germanium and each 100°C rise for silicon. Consequently, as the temperature increases, each of these three parameters change in a way that all three cause the collector current to increase; i.e., it is not possible for these effects to cancel each other in a single transistor. Stabilizing I_C for variations in these parameters is an important problem in circuit design. This problem will be treated later.

In this section, we studied the two principal sets of output characteristic curves of the bipolar transistor: the CE curves, the CB curves, and also the input curves. The parameter β was defined, and the difference between its dc and ac values was discussed. Of the various bipolar transistor parameters, β is probably the most important. The voltage limitations imposed by avalanche and punch-through, the relationships between the terminal currents (I_E, I_B, and I_C), and the principal temperature effects were also discussed.

4.3 DC Bias Circuits and Bias Stability

When using bipolar transistors as active elements in linear-amplifier circuits, the first problem is to bias the *EB* junction in the forward direction, and the *BC* junction in the reverse direction. These conditions must be satisfied regardless of the configuration used. In addition, these conditions must be achieved without seriously degrading the ac gain characteristics of the amplifier.

In selecting the bias point for a linear amplifier, I_C is usually the first parameter chosen. Many factors influence the choice of the value for I_C, one of the most important being the ac input impedance of the device. Since the EB junction is simply a diode modified by the presence of another junction, we would expect the ac resistance looking into the emitter terminal with the base grounded to be approximately the same as for a diode $[r_f = (26 \text{ mV})/I_E]$. Assuming that I_E is 1 mA, typical values of input impedance for the CB connection are 25 to 40 ohms. For the CE connection, the ac junction voltage is the same, but the input current, I_b, is $I_e/(\beta + 1)$. Therefore, input impedance values range from 1 to 5 kilohms. In addition, the input impedance is inversely proportional to I_E; i.e., the input impedance will approximately double if I_E is reduced from 1 mA to 0.5 mA.

Expressions for the input impedance of each configuration will be developed in this chapter. The important point here is that the value of I_E (which is approximately equal to I_C) can substantially influence the amplifier performance. In this section, we will discuss methods for obtaining the desired value of I_C, and also for evaluating the changes in I_C as transistor parameters, temperature, and circuit components change.

The next bias value chosen is usually the dc bias voltage, V_{CB}. The major limitations on this choice are breakdown (avalanche or punch-through) on the high end, and low values of β on the low end. Typical values of V_{CB} for small-signal amplifiers range from 2 to 30 V. The maximum amplitude of the output ac signal also influences the choice of V_{CB}.

When I_C and V_{CB} have been selected, the other transistor dc current and voltages are automatically determined. The circuit designer may then select the external components to obtain the desired bias conditions.

Except for the capacitors, the bias circuits in Fig. 4–11 are obtained by rearranging the same circuit as a CE, CB, or CC amplifier. The distinguishing feature of this bias circuit is a voltage dividing network in the base circuit to establish the approximate dc voltage at the base terminal. This type of bias is referred to as *base voltage bias*. A second type of bias network will be discussed later in this section.

The CE amplifier will now be analyzed in detail. First, expressions for I_E and I_C will be derived in terms of the circuit parameters and β. Then useful approximations that can often simplify the analysis will be discussed. The first step in the analysis is to replace the voltage dividing network in the base circuit (R_1 and R_2) with a Thévenin-equivalent circuit. This circuit is shown in Fig. 4–12(a) where

$$R_B = R_1 \,||\, R_2 \tag{4–19}$$

and

$$V_B = V_{CC}R_1/(R_1 + R_2) \tag{4–20}$$

Summing voltages around the base loop in Fig. 4–12(a) gives

$$V_B = I_B R_B + V_{BE} + I_E R_E \tag{4–21}$$

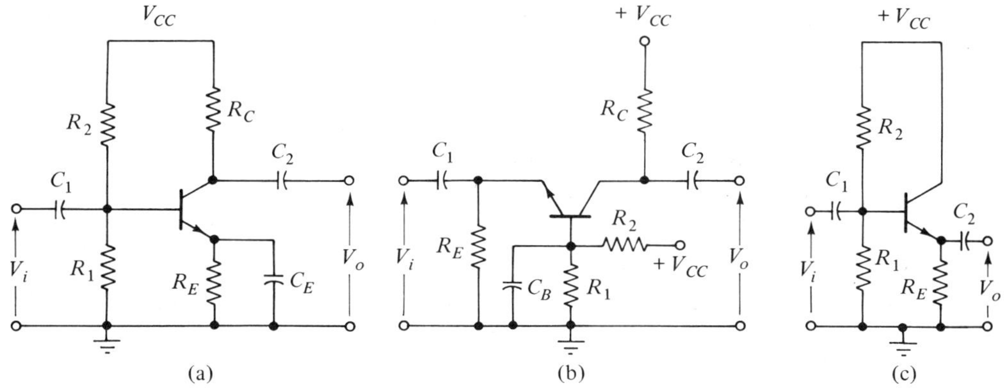

FIGURE 4–11 *Bipolar transistor base voltage bias networks: (a) a CE amplifier, (b) a CB amplifier, and (c) a CC amplifier.*

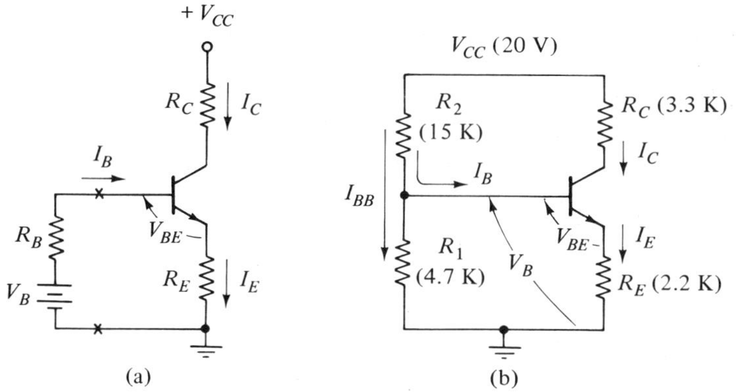

FIGURE 4–12 *The dc circuit of the CE amplifier shown in Fig. 4–11(a): (a) circuit redrawn with a Thévenin-equivalent circuit representing the base circuit, and (b) circuit showing the important bias values used in the approximate analysis. (Component values in (b) apply to Ex. 4–7.)*

Substituting the value for V_B from Eq. (4–20), and the value of I_B from Eq. (4–16) into Eq. (4–21) and solving for I_E,

$$I_E = \frac{V_{CC}(\beta + 1)R_B/R_2 - (\beta + 1)V_{BE} + (\beta + 1)I_{CO}R_B}{R_B + (\beta + 1)R_E} \tag{4–22}$$

A general equation for I_C can be obtained by substituting this equation into the equation for I_C in Eq. (4–16). After some simplification,

$$I_C = I_{CO} + \frac{V_{CC}\,\beta(R_B/R_2) - \beta V_{BE} + \beta I_{CO}R_B}{R_B + (\beta + 1)R_E} \tag{4–23}$$

Equation (4–23) is important for two reasons: (1) it may be used to find I_C for any circuit in Fig. 4–11 when the parameters are given, and (2) the new value of I_C can also be found when the parameters change. The second point will be treated in detail in the latter portion of this section.

Equation (4–21) can also be solved for I_B. Neglecting I_{CO}, I_E is equal to $I_B(\beta + 1)$. Substituting this value for I_E into Eq. (4–21), and solving for I_B yields

$$I_B = \frac{V_B - V_{BE}}{R_B + (\beta + 1)R_E} \tag{4–24}$$

This equation indicates that the emitter resistor, R_E, is effectively much larger when viewed from the base. The multiplication of R_E by $(\beta + 1)$ is due to the relationship between I_B and I_E in a normally biased transistor; i.e., a base current will cause $(\beta + 1)$ times the current's value to flow in the emitter current. This transformation of impedance from the emitter circuit to the base circuit, and vice versa, greatly simplifies circuit analysis. The results can be summarized as follows:

(1) Any impedance in the emitter circuit is multiplied by $(\beta + 1)$ when viewed from the base terminal.
(2) Any impedance in the base circuit is divided by $(\beta + 1)$ when viewed from the emitter terminal.

The above results will also apply to ac circuits since β is not restricted to dc values.

The circuit in Fig. 4–12(b) will be used to demonstrate an approximate analysis of a base voltage bias network. The current flowing in R_2 may be divided into two components: (1) I_{BB}, which flows in both R_1 and R_2, and (2) the base current, I_B, which flows in R_2. The base current usually has a low value. Hence, many bias circuits satisfy the condition that I_{BB} is much larger than I_B so that I_B can be neglected. For this condition,

$$I_{BB} \simeq V_{CC}/(R_1 + R_2) \tag{4–25}$$

Then the base-to-ground voltage, V_B, is

$$V_B \simeq V_{CC}\, R_1/(R_1 + R_2) \tag{4–26}$$

The emitter voltage (V_E) is

$$V_E = V_B - V_{BE} \tag{4–27}$$

Typical values for V_{BE} are 0.6 V for silicon and 0.15 V for germanium. The collector current, I_C, is approximately equal to I_E where

$$I_E = V_E/R_E \tag{4–28}$$

This approximate approach is generally useful if I_{BB} is greater than 10 I_B since neither the resistor values nor β are usually known to a high degree of accuracy. The approximate approach is demonstrated in the following example.

EXAMPLE 4–7 An NPN silicon transistor with $\beta = 128$ is connected in the base voltage bias network shown in Fig. 4–12(b). Calculate all bias voltages and currents.

Solution: Assume $I_{BB} \gg I_B$. Then,

$$V_B \simeq V_{CC}\, R_1/(R_1 + R_2) \simeq 4.7 \text{ V} \qquad V_E = V_B - V_{BE} = 4.1 \text{ V}$$

$$I_C \simeq I_E = V_E/R_E \simeq 1.9 \text{ mA} \qquad V_C = V_{CC} - I_C R_C \simeq 13.7 \text{ V}$$

The measured bias voltages were

$$V_B = 4.8 \text{ V} \qquad V_E = 4.2 \text{ V} \qquad V_C = 14 \text{ V}$$

which are within the 10 percent tolerance of the resistors.

A second type of bias circuit is the *base current bias* network shown in Fig. 4–13. The three networks in Fig. 4–13 are somewhat similar in that only the base current flows in the base-bias resistor.

From Fig. 4–13(b), voltages can be summed to include the voltage drops across R_B and R_E as

$$V_{CC} = I_B R_B + V_{BE} + I_E R_E = [I_E/(\beta + 1)]R_B + V_{BE} + I_E R_E \quad \text{(4–29)}$$

Solving this equation for I_E gives

$$I_E = \frac{V_{CC} - V_{BE}}{[R_B/(\beta + 1)] + R_E} \quad \text{(4–30)}$$

In this equation the division of R_B by $(\beta + 1)$ is due to the impedance transformation properties of the *EB* junction. If I_E is known, the other circuit voltages and currents can be determined as demonstrated in the following example.

EXAMPLE 4–8 A PNP silicon transistor with $\beta = 110$ is connected in the circuit shown in Fig. 4–13(b). Solve for all bias voltages and currents.

Solution: From Eq. (4–30),

$$I_C \simeq I_E = \frac{V_{CC} - V_{BE}}{[R_B/(\beta + 1)] + R_E} = \frac{-12 + 0.6}{(1\text{ M}/111) + 2.2\text{ K}} \simeq -1\text{ mA}$$

$$V_E = I_E R_E \simeq -2.2\text{ V} \qquad V_B = V_E + V_{BE} \simeq -2.8\text{ V}$$

$$V_C = V_{CC} - I_C R_C \simeq -12\text{ V} - (-1\text{ mA})(2.2\text{ K}) \simeq -9.8\text{ V}$$

The measured bias voltages were

$$V_E = -2.05\text{ V} \qquad V_B = -2.62\text{ V} \qquad V_C = -10\text{ V}$$

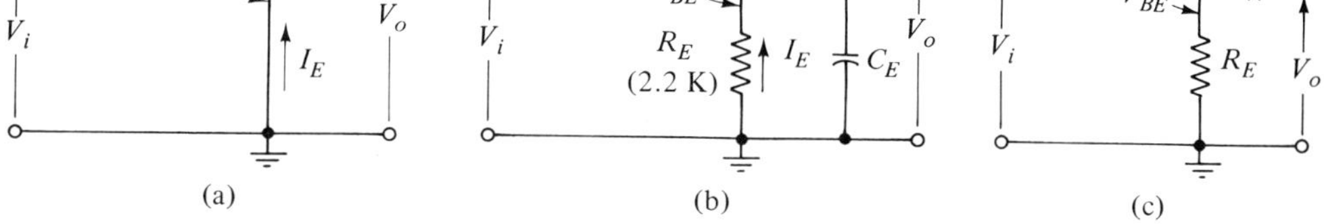

FIGURE 4–13 *Base current bias networks: (a) a CE amplifier, (b) a CE amplifier with an emitter resistor, and (c) a CC amplifier. Component values apply to Ex. 4–8.*

In design work, I_C should be held reasonably constant because a change in I_C affects the performance characteristics of an amplifier. If I_C is allowed to drift very much from the design value, reasonable ac excursions could easily drive the transistor into either a saturated or a cut-off condition. In either case, the output signal would be distorted.

If the value of I_C is not very critical, variations in the resistor values are not important. However, I_C may change drastically with changes in I_{CO}, V_{BE}, and β where variations in these parameters may be due to either unit-to-unit variations or changes in temperature. In addition, changes in the dc supply voltage (V_{CC}) may cause I_C to vary.

Given a range for each of these parameters, the change in I_C produced by a change in the parameters, either separately or jointly, may be found by multiple solutions of Eq. (4–23). However, a better understanding of the effectiveness of specific circuit elements in stabilizing I_C can be obtained from a stability factor analysis. In this approach, the effect of each of the above parameters on I_C can be isolated and evaluated. This approach is based on the method for finding the total change by using the partial derivatives.

If only small changes in the parameters are assumed, then the total change* in I_C is

$$\Delta I_C = S_I\,\Delta I_{CO} + S_V\,\Delta V_{BE} + S_\beta\,\Delta\beta + S_E\,\Delta V_{CC} \qquad \textbf{(4–31)}$$

where ΔI_{CO}, ΔV_{BE}, $\Delta\beta$, and ΔV_{CC} are the parameter changes, and S_I, S_V, S_β, and S_E are the stability factors for each of the parameters. The stability factors for the base voltage bias networks in Fig. 4–11 are

$$S_I = 1 + \frac{\beta(R_B/R_E)}{1+\beta+(R_B/R_E)} \qquad \textbf{(4–32)}$$

$$S_V = (-1/R_E)\left[\frac{\beta}{1+\beta+(R_B/R_E)}\right] \qquad \textbf{(4–33)}$$

$$S_\beta = \left[\frac{I_C - I_{CO}}{\beta}\right]\left[\frac{1+(R_B/R_E)}{1+\beta+(R_B/R_E)}\right] \qquad \textbf{(4–34)}$$

$$S_E = \left[\frac{1}{R_2}\right]\left[\frac{\beta(R_B/R_E)}{1+\beta+(R_B/R_E)}\right] \qquad \textbf{(4–35)}$$

where R_B is the parallel combination of R_1 and R_2.

The following important relationships may be obtained from Eqs. (4–32) through (4–35):

1. The lower limit for S_I is unity; i.e., ΔI_C *cannot* be less than ΔI_{CO}.
2. All the stability factors depended primarily upon the ratio of R_B/R_E. In addition, β affects the stability. However, for R_B/R_E less than 10 this dependence becomes small for the S_I, S_V, and S_E stability factors.
3. Making R_E large reduces all four stability factors, and therefore improves stability. The price for making R_E large is an increase in the dc power consumption by the circuit and low dc gain (also, low ac gain if R_E is not bypassed).
4. For a reduction in the effects of power supply variations, the value of R_2 should be as large as possible for a given value of R_B. This condition minimizes the change in V_B due to a change in V_{CC} through the voltage-divider action of R_1 and R_2.

---

* Since $I_C = f(I_{CO}, V_{BE}, \beta, V_{CC})$, then by the chain rule of differentiation the total derivative of I_C is

$$dI_C = \left(\frac{\partial I_C}{\partial I_{CO}}\right)\bigg|_{V_{BE},\beta,V_{CC}} dI_{CO} + \left(\frac{\partial I_C}{\partial V_{BE}}\right)\bigg|_{I_{CO},\beta,V_{CC}} dV_{BE} + \left(\frac{\partial I_C}{\partial \beta}\right)\bigg|_{I_{CO},V_{BE},V_{CC}} d\beta + \left(\frac{\partial I_C}{\partial V_{CC}}\right)\bigg|_{I_{CO},V_{BE},\beta} dV_{CC}$$

Equation (4–31) is obtained by replacing derivatives with delta values and designating the partial of I_C with respect to one of the variables as an S; $\partial I_C/\partial I_{CO}$ is S_I. The stability factors are evaluated in terms of the circuit parameters by substituting Eq. (4–23) for I_C into each partial derivative. See Chap. 6 of Pierce in the reference list for a detailed treatment of the stability factor analysis.

5. The I_{CO}, β, and V_{CC} effects may be reduced by using small values of R_B (low base impedance to ground). However, for good V_{BE} stability it is desirable to have a large value of R_B. Small values of S_I, S_V, and S_E require small values of β, whereas a large value of β is desired to make S_β small. Obviously, a compromise must be made for a given design. This compromise will depend upon whether silicon or germanium transistors are used and the percentage each effect contributes to the total operating-point change, or change in I_C.

Since the stability factors are based on small changes, only approximate results can be expected for the relative large changes that occur in practice. An application of stability factors is demonstrated in the following example. Note that the effect of each parameter change with an increase in temperature is translated into a change in I_C.

EXAMPLE 4–9 A germanium PNP transistor is used in a base voltage bias circuit with the following circuit values: $R_1 = 51$ K, $R_2 = 91$ K, $R_E = 5.1$ K, $R_C = 5.1$ K, and $V_{CC} = -30$ V. The following experimental data are available for the circuit:

Temperature	I_{CO}	β	I_C
30°C	2.5 μA	120	1.93 mA
60°C	37.5 μA	146	2.24 mA

Using stability factors, calculate the value of ΔI_C and compare this with the measured value.

Solution: The stability factor values are

$$S_I = 7.05 \qquad S_V = -0.185 \times 10^{-3} \qquad S_\beta = 1.01 \times 10^{-6}$$

The values of ΔI_{CO} and $\Delta\beta$ are available from the data, and ΔV_{BE} can be estimated since V_{BE} changes 2 mV for each centigrade degree rise in temperature. Thus, $\Delta I_{CO} = 35\ \mu$A, $\Delta\beta = 26$, and $\Delta V_{BE} \simeq -60$ mV. From Eq. (4–31), the total change in I_C is

$$\Delta I_C = \Delta I_{CO}\, S_I + \Delta V_{BE}\, S_V + \Delta\beta S_\beta \simeq 0.247 \text{ mA} + 0.011 \text{ mA} + 0.026 \text{ mA} \simeq 0.29 \text{ mA}$$

The measured value of ΔI_C was 0.31 mA which is within 7 percent of the calculated value. The major component of ΔI_C was due to I_{CO}. If a silicon transistor had been used, ΔI_{CO}, and hence ΔI_C, would have been much smaller over this practical temperature range.

In this section, we have treated the problem of bipolar transistor bias and bias stability. The two principal types of bias networks, base voltage bias and base current bias, were discussed. An approximate solution and also an exact solution of the bias conditions were developed for the base voltage bias case. Both types of bias networks will be encountered in the remainder of the text. However, the base-voltage bias circuit is normally used because I_C can be stabilized better against parameter changes with this circuit than with the base-current bias circuit.

4.4 Bipolar Transistor Equivalent Circuits

The purpose of a transistor equivalent circuit is to simplify the calculations required to determine the ac response of the device. When an equivalent circuit is used, we assume that the device is already biased in the active region; i.e., the *EB* junction is forward biased, and the *BC* junction is reverse biased. An equivalent circuit should be a reasonably accurate representation of the device, easy to use, and consist of components that have physical meaning. Of the many equivalent circuits available, two are particularly important and useful: the hybrid-π equivalent and the hybrid equivalent circuits. The hybrid-π circuit meets all of the above criteria. In the hybrid equivalent circuit, the parameters are not directly related to the physical processes of the transistor, but this equivalent circuit is much simpler to use in mid-band analyses. In this section, we will develop these two equivalent circuits and also study various methods used in the fabrication of bipolar transistors.

The first step in developing the hybrid-π equivalent circuit is to relate the components in this circuit with the physical operation of a transistor. To accomplish this, we will start with the cross-sectional view of a transistor shown in Fig. 4–14(a). An intermediate circuit obtained from this unit is shown in Fig. 4–14(b). The final hybrid-π circuit is developed from the intermediate circuit.

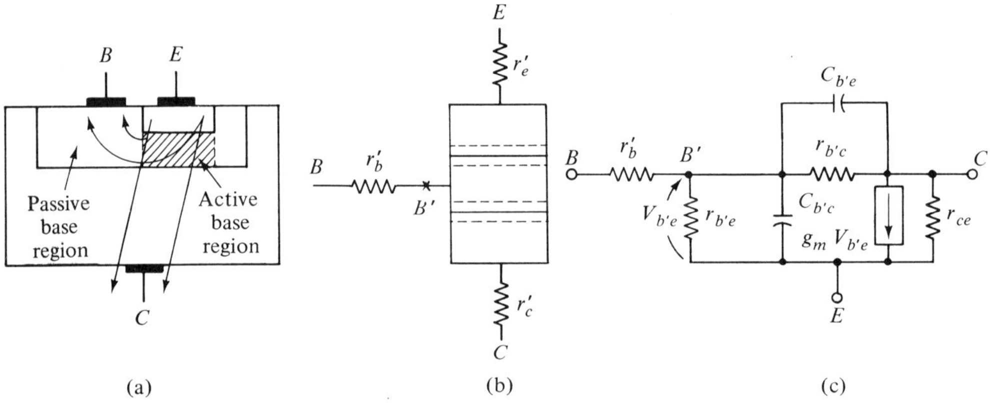

FIGURE 4–14 *Various models of a bipolar transistor: (a) cross-section of a planar constructed transistor, (b) simple model of the transistor showing* $r_{b'}$, *and (c) CE hybrid-π equivalent circuit.*

The currents flowing in each of the three external leads also flows through the resistance of the bulk material. These resistors are shown in Fig. 4–14(b) as $r_{b'}$, $r_{e'}$, and $r_{c'}$ and have values that are functions of the resistivity of the individual parts of the transistor. Since both the emitter and the collector are usually doped heavily, $r_{e'}$ and $r_{c'}$ have low values and are neglected in the equivalent circuit. However, the base is lightly doped (high resistivity material) so that $r_{b'}$ is not negligible.* Typical values of $r_{b'}$ range from 10 ohms to 1000 ohms depending on the base resistivity and path length.

After identifying the bulk resistors and discarding all but $r_{b'}$, the ideal transistor is available for analysis. The complete CE hybrid-π equivalent circuit is shown in Fig. 4–14(c). The physical significance of the remainder of the components will now be explained using the physical model in Fig. 4–14(b), and the diode equivalent circuits in Fig. 2–23.

---

* The base resistance, $r_{b'}$, does not have a constant value for at least two reasons. First the width of the base is a function of both the *EB* and *CB* voltages. Second, the resistivity of the base is a function of the charge density in the base, and hence a function of I_E. The part of $r_{b'}$ that lies in the active region is a distributed resistance called the base spreading resistance. In terms of circuit performance, the ideal value of $r_{b'}$ would be zero.

In this circuit model, the *EB* junction can be approximated as the dynamic resistance ($r_{b'e}$) of a forward-biased diode in parallel with the capacitance $C_{b'e}$. On a base-current level, the dynamic resistance at room temperature is approximately

$$r_{b'e} \simeq 26 \text{ mV}/I_B \tag{4-36}$$

Rewriting this equation in terms of the emitter current,

$$r_{b'e} \simeq \frac{26 \text{ mV}}{I_E/(\beta + 1)} = (\beta + 1)r_e \tag{4-37}$$

where r_e is the dynamic impedance of the *EB* junction when viewed from the emitter terminal and is given by

$$r_e \simeq 26 \text{ mV}/I_E \tag{4-38}$$

The significance of $r_{b'e}$ is readily appreciated by noting that the ac component of the input signal appearing across $r_{b'e}$ is actually the ac voltage across the *EB* junction. This voltage causes the emitter current I_E, and consequently I_C, to vary. The capacitance $C_{b'e}$ is the sum of the diffusion and junction capacitance of the *EB* junction. The diffusion component accounts for the stored charge in the base region and is a function of I_E. The junction component is small compared to the diffusion component except for very low values of I_E.

The *BC* junction is a reverse-biased P-N junction with a dynamic resistance $r_{b'c}$. Typical values of $r_{b'c}$ are similar to those of r_r for a reverse-biased diode, and range from 500 K to 100 M. The capacitance $C_{b'c}$ is the junction capacitance of the *BC* junction. The resistor r_{ce} represents the equivalent resistance between the emitter and collector terminals. An approximate value of r_{ce} can be determined from the inverse slope ($\Delta V/\Delta I$) of the constant I_B curve of the CE characteristics that passes through the Q point, or

$$r_{ce} = \left.\frac{\Delta V_{CE}}{\Delta I_C}\right|_{I_B=\text{constant}} \tag{4-39}$$

Typical values range from 20 to 500 K.

The remaining component in the hybrid-π circuit is the current source $g_m V_{b'e}$. The ac component of the collector current is proportional to the ac *EB* junction voltage, $V_{b'e}$, with the transconductance, g_m, being the constant of proportionality. Hence,

$$g_m = \left.\frac{\Delta I_C}{\Delta V_{BE}}\right|_{V_{CE}=\text{constant}} \tag{4-40}$$

The value of g_m cannot be obtained from the output static characteristics since V_{BE} is not a parameter in these curves (see Fig. 4–7). However, the value of g_m can be estimated* as follows. Note that in Fig. 4–14(c), I_c is $g_m V_{b'e}$ if the current in $r_{b'c}$ is neglected. The ac base current can be written in terms of $V_{b'e}$ as

$$I_b = V_{b'e}/r_{b'e} = V_{b'e}/[(\beta + 1)(r_e)] \tag{4-41}$$

The collector current resulting from I_b is

$$I_c = \beta I_b = \frac{\beta V_{b'e}}{(\beta + 1)r_e} = \left(\frac{\alpha}{r_e}\right) V_{b'e} = g_m V_{b'e} \tag{4-42}$$

---

* A more intuitive approach is to relate $V_{b'e}$ and I_c through I_e. Since $V_{b'e}$ is the ac voltage across the *EB* junction, the ac emitter current, I_e, is $V_{b'e}/r_e$ where r_e is the dynamic or ac resistance of the *EB* diode curve. If $I_e \simeq I_c$, then

$$I_e \simeq I_c \simeq V_{b'e}/r_e \simeq g_m V_{b'e}$$

Thus the small-signal value of g_m is

$$g_m = \frac{I_c}{V_{b'e}} = \frac{\alpha}{r_e} \simeq \frac{1}{r_e} \tag{4–43}$$

EXAMPLE 4–10 A PNP silicon transistor is biased at $I_C = 2$ mA and $V_{CE} = -10$ V. The parameter values are: $\beta = 85$, $r_{b'} = 100$, and $r_{ce} = 100$ K. Calculate the remaining components of the hybrid-π equivalent circuit.

Solution: Referring to Fig. 4–14(c), the remaining mid-band components are $r_{b'e}$, g_m, and $r_{b'c}$. The value of $r_{b'c}$ is typically in the high megohm range and can usually be ignored.

$$\alpha = \beta/(\beta + 1) \simeq 0.99 \qquad I_E = I_C/\alpha \simeq 2\text{ mA}$$

$$r_e = \frac{26\text{ mV}}{I_E} = \frac{26\text{ mV}}{2\text{ mA}} = 13\ \Omega \qquad g_m = \frac{\alpha}{r_e} \simeq \frac{1}{r_e} \simeq 77\text{ millimhos}$$

$$r_{b'e} = (\beta + 1)r_e = (85 + 1)(13) \simeq 1120\ \Omega$$

The remaining components of the hybrid-π equivalent circuit, $C_{b'e}$ and $C_{b'c}$, are discussed in Chap. 7.

The second equivalent circuit for a bipolar transistor, the hybrid equivalent circuit, is shown in Fig. 4–15(a). This approach uses a two-port network representation of a device.* A bipolar transistor in the CE configuration as a two-port is shown in Fig. 4–15(b) and relates the terminal voltages and currents of the general hybrid equivalent circuit (V_1, I_1, V_2, and I_2) to those of the CE stage (V_{be}, I_b, V_{ce}, and I_c). Only ac voltage and currents are considered. The interpretations of the hybrid parameters are as follows:

$$\left.\begin{aligned} h_i &\equiv V_1/I_1|_{V_2=0} = \text{input impedance in ohms} \\ h_r &\equiv V_1/V_2|_{I_1=0} = \text{reverse voltage gain} \\ h_f &\equiv I_2/I_1|_{V_2=0} = \text{forward current gain} \\ h_o &\equiv I_2/V_2|_{I_1=0} = \text{output admittance in mhos} \end{aligned}\right\} \tag{4–44}$$

where all voltages and currents are assumed to have small amplitudes. Since the units for the parameters are not all the same, the name "hybrid" is used.

Generally, the hybrid parameters have a second subscript to denote the configuration. For instance, h_{ie} is the input impedance of a CE stage, and h_{ib} is the input impedance of a CB stage. Also, h_{fe} denotes the CE forward current gain, and h_{fc} denotes the CC forward current gain. The expressions for the CE hybrid parameters will be developed in this section while those for the CB and CC configurations will be developed in later sections.

By comparing Figs. 4–15(a) and 4–14(c), the expressions for h_{ie}, h_{oe}, and h_{re} can be obtained. The value of h_{ie} is

$$h_{ie} = r_{b'} + r_{b'e} = r_{b'} + (\beta + 1)r_e \tag{4–45}$$

Hence the input impedance of a CE stage, which is the impedance between the base and emitter terminals when viewed from the base terminal, is the sum of the base spreading resistance, $r_{b'}$, and the dynamic impedance of the *EB* junction, $r_{b'e}$. If the emitter circuit contains an unbypassed resistor [shown as R_E in Fig.

---

* See Appendix A for the development of the hybrid equivalent circuit.

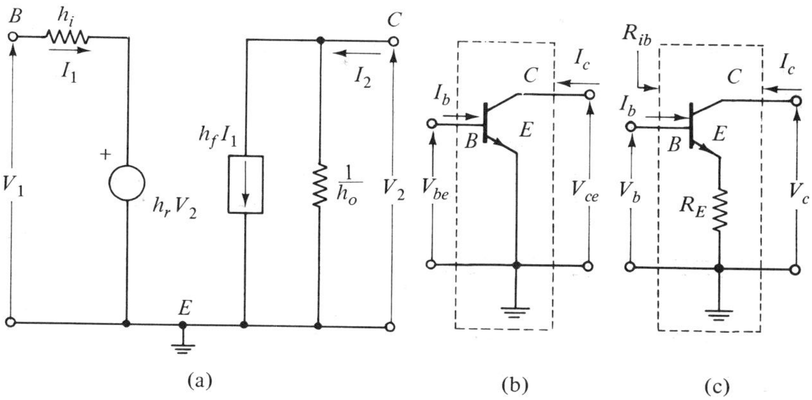

FIGURE 4–15 *A hybrid equivalent circuit for the CE case: (a) a general two-port hybrid equivalent circuit, (b) a CE stage shown as a two-port network, and (c) a CE stage with an unbypassed emitter resistor.*

4–15(c)] the input impedance, R_{ib}, is calculated by adding R_E to r_e in Eq. (4–45) giving

$$R_{ib} = r_{b'} + (\beta + 1)(r_e + R_E) = h_{ie} + (\beta + 1)R_E \qquad \textbf{(4–46)}$$

The $(\beta + 1)$ impedance transformation between the base and emitter circuits discussed in the last section is apparent in Eqs. (4–45) and (4–46).

The value of h_{oe} represents the CE output admittance and is given by

$$h_{oe} = 1 \Big/ \left(r_{ce} \Big\| \frac{r_{b'c}}{\beta + 1} \right) \simeq 1/r_{ce} \qquad \textbf{(4–47)}$$

The parameter h_{re} represents the Early effect (see Fig. 4–6) and accounts for the change in collector current due to the output ac voltage. From Fig. 4–14(c), h_{re} can be approximated by finding the division of the output voltage between $r_{b'e}$ and $r_{b'c}$ as

$$h_{re} \simeq r_{b'e}/(r_{b'e} + r_{b'e}) \qquad \textbf{(4–48)}$$

Since the Early effect is of secondary importance, particularly if the ratio V_{ce}/V_{be}* is less than 1000, the effect of h_{re} is often neglected. The fourth CE hybrid parameter (h_{fe}) represents the current gain from the base to the collector. This corresponds to our previous definition of β in Eq. (4–14), hence

$$h_{fe} = \beta \qquad \textbf{(4–49)}$$

EXAMPLE 4–11 A silicon NPN transistor connected in the CE configuration is biased at $I_E = 1$ mA and has the following hybrid parameters: $h_{ie} = 2.7$ K, $h_{re} = 10^{-5}$, $h_{fe} = 99$, and $h_{oe} = 10^{-5}$ mhos. Calculate the mid-band hybrid-π components.

Solution:

$$r_e = \frac{26 \text{ mV}}{I_E} = \frac{26 \text{ mV}}{1 \text{ mA}} = 26\ \Omega \qquad \beta = h_{fe} = 99$$

$$r_{b'e} = (\beta + 1)r_e \simeq 2.6 \text{ K} \qquad r_{b'} = h_{ie} - r_{b'e} \simeq 100\ \Omega$$

---

* The ratio V_{ce}/V_{be} corresponds approximately to A_{vt} for the CE amplifier.

$$r_{ce} \simeq 1/h_{oe} = 10^5 = 100\text{ K} \qquad g_m = \alpha/r_e \simeq 1/26 \simeq 38\text{ millimhos}$$

$$h_{re} \simeq r_{b'e}/(r_{b'e} + r_{b'c}) \simeq r_{b'e}/r_{b'c}$$

Therefore,

$$r_{b'c} = r_{b'e}/h_{re} \simeq (2.6\text{ K})/10^{-5} \simeq 260\text{ M}$$

One important difference between the hybrid-π and the hybrid equivalent circuits is the manner in which the frequency dependency of the transistor is treated. In the hybrid-π circuit, all components are treated as constants, and the frequency dependency of the transistor is accounted for by the two capacitors, $C_{b'e}$ and $C_{b'c}$. In the hybrid circuit, all four parameters are frequency dependent. Consequently, the analysis of transistor circuits as a function of frequency is simpler using the hybrid-π equivalent circuit. Techniques for finding the frequency response are discussed in Chap. 7.

The approximate CE mid-band hybrid-π equivalent circuit is shown in Fig. 4–16(a) and is obtained from Fig. 4–14(c) by neglecting both capacitors, $C_{b'e}$ and $C_{b'c}$, and also the large reverse dynamic resistance of the *BC* junction, $r_{b'c}$. The approximate CE mid-band hybrid equivalent circuit is shown in Fig. 4–16(b) and is obtained from Fig. 4–15(a) by neglecting the effect of h_{re} (which corresponds to neglecting $r_{b'c}$) and treating the other parameters as constants. The principal difference between the two equivalent circuits is the dependent current generator in the collector circuit. This current generator is voltage dependent ($g_m V_{b'e}$) in the hybrid-π circuit, and current dependent ($h_{fe} I_b$) in the hybrid circuit. Since both dependent generators account for I_c,

$$g_m V_{b'e} = I_c = \beta I_b = h_{fe} I_b \tag{4–50}$$

Although the hybrid-π equivalent circuit is useful in high-frequency studies, the hybrid circuit finds wide application in the mid- and low-frequency range. Both will be used in this chapter, with the hybrid circuit being emphasized.

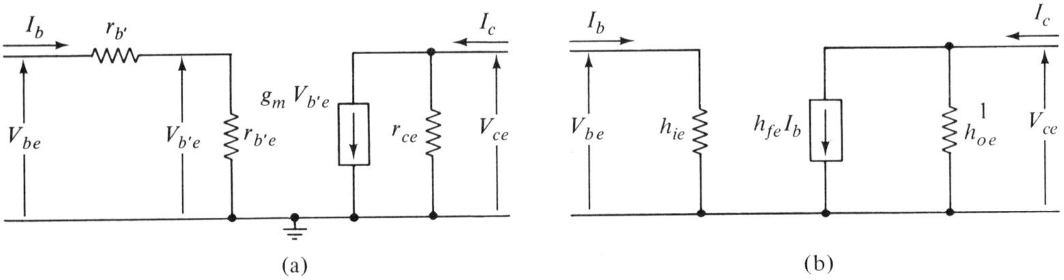

FIGURE 4–16 *The approximate CE mid-band equivalent circuits: (a) hybrid-π circuit and (b) hybrid circuit.*

At this point some of the various techniques used in the fabrication of transistors will be discussed since many of the device parameters depend directly on the manufacturing methods that are used.

Historically, the first transistor was a point-contact transistor made by Bell Laboratory in 1948. The point-contact transistor was constructed by connecting

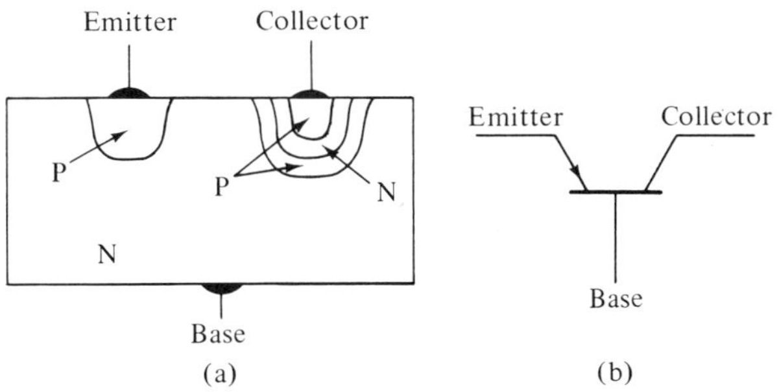

FIGURE 4–17 *A point contact transistor: (a) construction and (b) symbol.*

two metal wires to a semiconductor crystal, as shown in Fig. 4–17. The wires were attached to the semiconductor by heating the contact with a short current pulse. The exact theory of the point contact transistor has not been established primarily due to the complex collector contact. Actually, the device is a PNPN unit, and α can be greater than unity due to carrier multiplication in the region near the collector terminal. The symbol for the bipolar transistor originated from the geometry of the point-contact transistor. Since the junction bipolar transistor is superior to the point contact in many ways, very few point-contact transistors are in use today.

As indicated in Sec. 4.1, a junction bipolar transistor consists of two P-N junctions formed in close proximity. One method of achieving this arrangement is by modifying the basic crystal growing process as shown in Fig. 4–18. A monocrystalline material can be obtained from a purified, melted semiconductor by slowly withdrawing a seed crystal from the material as shown in Fig. 4–18(a). If the molten material is N-type, the resulting crystal will be N-type. During the growing process, the crystal can be changed from N- to P-type by the addition of trivalent impurities to the molten semiconductor. Then a section of P-type material is incorporated into the crystal as shown in Fig. 4–18(b). The crystal can again be made N-type by the addition of a sufficient quantity of pentavalent material as shown in Fig. 4–18(c). The result is an NPN cylindrical rod which can be sawed into many bars. A typical bar will measure 3 millimeters (mm) in length and 0.4 mm in both width and height. Ohmic contacts are made to each region, and the entire assembly is enclosed in a standard case. The grown-junction transistor is characterized by gradual junctions and poor high-frequency performance. Although this was one of the first methods of fabricating junction transistors, few modern transistors are manufactured using the grown-junction process.

The alloy-junction process is a second method for fabricating bipolar transistors. As shown in Fig. 4–19, this process consists of alloying two pellets to a semiconductor bar. The pellets should be the proper type of impurity (N or P) and should also be good conductors. Indium, which is a P-type impurity, has found wide application. To form a PNP transistor, the N-type bar, along with the two indium pellets, is heated to a temperature sufficient to cause the pellets to melt. When cooled, the semiconductor recrystallizes but now has sufficient impurities to form P-type regions for use as the emitter and collector. Ohmic

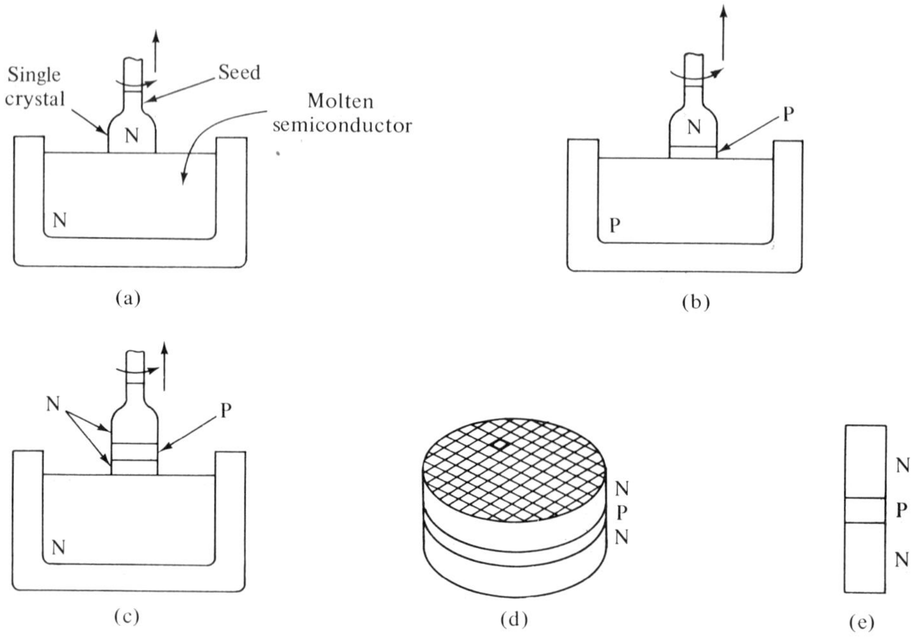

FIGURE 4–18 *A method of fabricating a grown junction transistor using the basic crystal growing process: (a) growing an N region, (b) growing a P region, (c) growing the second N region, (d) cross-section of a crystal rod before cutting, and (e) the final NPN bar used for the transistor.*

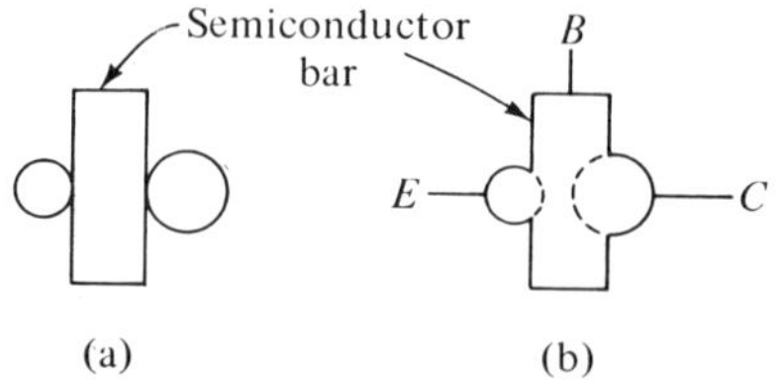

FIGURE 4–19 *A method of fabricating alloy-junction transistors: (a) prior to bonding, and (b) after bonding.*

contacts are formed, and the assembly is enclosed in a standard case. Alloy-junction transistors are characterized by abrupt junctions, poor high-frequency performance, and low values of breakdown voltage between the collector and the base terminals. The value of $r_{b'}$ can be as high as 1 K. However, alloy-junction transistors are reliable and have been used widely.

The newest technique of transistor fabrication is based on a diffusion process. Most modern transistors are manufactured using this process, one form of which is the planar, epitaxial process* shown in Fig. 4–20. This process is similar to that shown in Fig. 3-5 for the JFET. The process of forming an NPN transistor (see Fig. 4–20) begins with a slice of heavily doped N-type material. A lightly doped N-type layer is then grown on this substrate by passing an N-type gas over the slice. This layer is called an epitaxial layer [see Fig. 4–20(b)]. Using a photomasking, etching, and diffusion process similar to that used for the JFET

* This process is discussed in detail in Chap. 5.

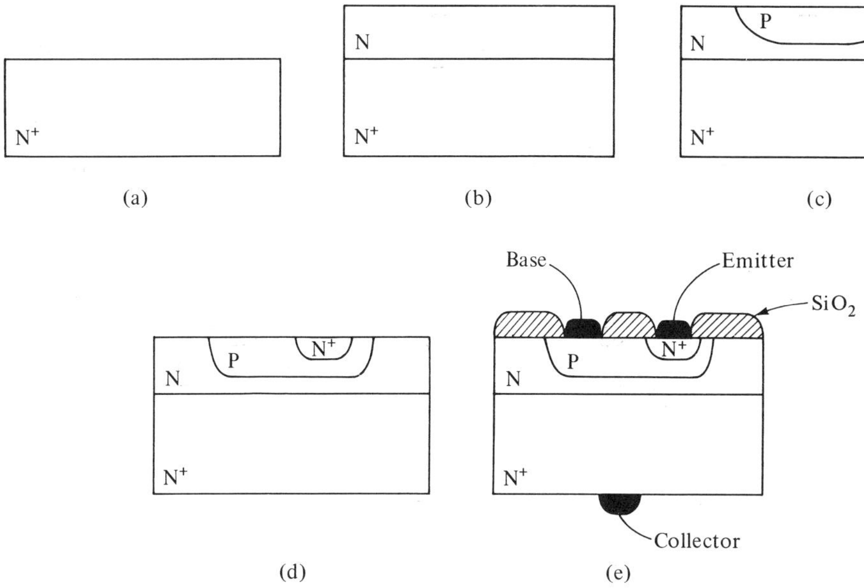

FIGURE 4–20 *Various stages in the fabrication of diffused planar, epitaxial transistor: (a) heavily doped substrate, (b) addition of lightly doped epitaxial layer, (c) addition of base region, (d) addition of emitter region, and (e) complete transistor with ohmic contacts and oxide layer.*

(see Fig. 3–5), the base region of P-type material can be formed in the epitaxial layer as shown in Fig. 4–20(c). Repeating the process, the second N-type emitter region can be formed [Fig. 4–20(d)]. The surface is then passified by growing an oxide layer over the crystal and ohmic contacts are added [Fig. 4–20(e)]. The assembly is then enclosed in a standard transistor case. Diffused transistors are characterized by nearly abrupt junctions, excellent high-frequency performance, low leakage current, and a nonhomogeneous base region. The epitaxial layer provides a high breakdown voltage between the collector and base while the bulk of the collector region is a low resistance path for the collector current. The value of $r_{b'}$ is low, typically ranging from 10 to 50 ohms, and often can be neglected.

In this section, we have discussed the two principal equivalent circuits of the bipolar transistor, the hybrid-π and the hybrid, and various methods of fabricating bipolar transistors. Our next task is to analyze the three important bipolar transistor amplifier configurations: the CE, CB, and CC.

4.5 The Common-Emitter (CE) Amplifier

Two small-signal equivalent circuits for a bipolar transistor were developed in the last section. In this section, we will use these equivalent circuits to determine the small-signal behavior of a CE amplifier. The procedure is as follows. First the bias conditions are determined. Next the mid-band equivalent circuit is structured by replacing all coupling and bypass capacitors with short circuits and replacing the transistor with its mid-band equivalent circuit. The mid-band performance of the amplifier can then be determined by standard circuit analysis techniques. The objective of this section is to develop the expressions for the voltage gain, input impedance, and output impedance of the CE amplifier in the mid-band frequency range. This section closely parallels the treatment of the CS amplifier in Sec. 3.6.

A complete CE amplifier is shown in Fig. 4–21(a), and the mid-band equivalent circuit of the amplifier is shown in Fig. 4–21(b), where

$$R_B = R_1 \,||\, R_2 \quad \text{and} \quad R_{L\text{eq}} = R_C \,||\, R_L \tag{4–51}$$

A complete CE mid-band equivalent circuit is obtained by substituting the hybrid equivalent circuit from Fig. 4–16(b) for the transistor as shown in Fig. 4–21(c).

The mid-band response of an amplifier may be characterized with three terms: (1) the generator-to-output voltage gain, $A_{vo} = V_o/V_g$, (2) the input impedance of the amplifier, R_i, and (3) the output impedance of the amplifier, R_o. In addition to these general terms, three other performance terms are usually desirable: (1) the CE terminal-to-terminal voltage gain, $A_{vt} = V_o/V_i$, (2) the CE input impedance, R_{ib}, and (3) the CE output impedance, R_{oc}.

Referring to Fig. 4–21(c), the output voltage is

$$V_o = -I_o R_{L\text{eq}} \tag{4–52}$$

where I_o is the current flowing in $R_{L\text{eq}}$ due to the current generator $h_{fe}I_b$. The current I_o may be expressed in terms of $h_{fe}I_b$ by applying the current splitting theorem to r_{ce} and $R_{L\text{eq}}$, or

$$I_o = (h_{fe}I_b)[r_{ce}/(r_{ce} + R_{L\text{eq}})] \tag{4–53}$$

The ac base current, I_b, is related to the ac base voltage, V_i, by

$$I_b = V_i/h_{ie} \tag{4–54}$$

The input impedance to the amplifier, R_i, is the parallel combination of R_B and h_{ie},

$$R_i = R_B \,||\, h_{ie} \tag{4–55}$$

The ac base voltage V_i is the ac base-to-emitter voltage and is the component of the generator voltage, V_g, that is dropped across R_i, or

$$V_i = V_g\, R_i/(R_i + R_g) \tag{4–56}$$

Combining Eqs. (4–52) through (4–56), the generator to output voltage gain, A_{vo}, is

$$A_{vo} = \frac{V_o}{V_g} = -\left(\frac{R_i}{R_i + R_g}\right)\left(\frac{h_{fe}}{h_{ie}}\right)\left(\frac{r_{ce}}{r_{ce} + R_{L\text{eq}}}\right)(R_{L\text{eq}}) \tag{4–57}$$

The negative sign in Eq. (4–57) indicates that the output signal in a CE amplifier is 180° out of phase (or inverted) relative to V_g.

The general expression for the CE input impedance, R_{ib}, was developed in the last section. From Eq. (4–46),

$$R_{ib} = r_{b'} + (\beta + 1)(r_e + R_E) = h_{ie} + (\beta + 1)R_E \tag{4–58}$$

The emitter resistor, R_E, is ac bypassed in the CE amplifier shown in Fig. 4–21. Hence the last term in Eq. (4–58) is zero. When calculating the impedance look-

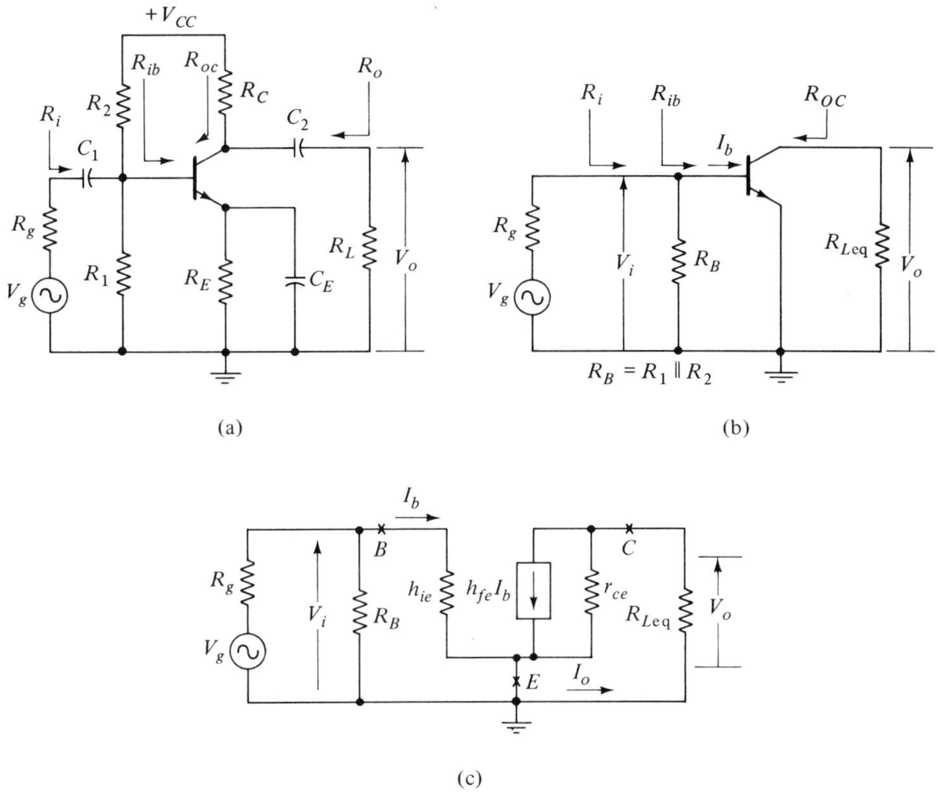

FIGURE 4–21 *A common-emitter amplifier: (a) complete CE amplifier, (b) a mid-band equivalent circuit for the amplifier, and (c) a complete mid-band equivalent circuit using the hybrid equivalent circuit to represent the transistor.*

ing into either the base or the emitter terminal, it is simpler to use the $(\beta + 1)$ transformation discussed in Sec. 4.3 rather than the test voltage generator approach used for FET circuits. As shown in Fig. 4–22(a), the results are the same. In this figure,

$$R_{ib} = V_{ti}/I_{ti} = h_{ie} \tag{4–59}$$

Figure 4–22(a) is obtained from Fig. 4–21(c) by removing all components to the left of the base terminal, including R_B, R_g, and V_g, and by placing a test voltage generator, V_{ti}, from the base terminal to ground.

The CE output impedance, R_{oc}, is obtained from the circuit shown in Fig. 4–22(b). In this figure, the circuit to the right of the collector terminal in Fig. 4–21(c), including R_{Leq}, has been removed, and a test voltage generator, V_{to}, connected across the output terminals. Also, the input generator, V_g, is replaced by a short circuit. Solving for R_{oc},

$$R_{oc} = V_{to}/I_{to} = r_{ce} \tag{4–60}$$

With R_{oc} known, the output impedance of the amplifier can be determined from Fig. 4–21(a) as

$$R_o = R_C \,||\, R_{oc} \simeq R_C \,||\, r_{ce} \simeq R_C \tag{4–61}$$

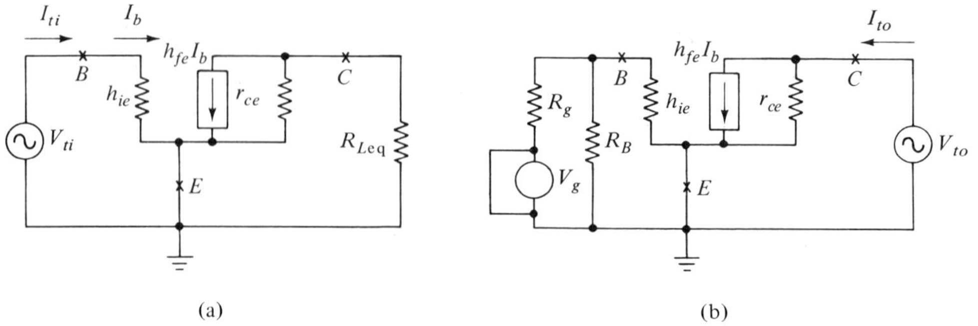

FIGURE 4–22 *Equivalent circuits used to determine the input and output impedances for the CE stage: (a) the input impedance,* R_{ib}, *and (b) the output impedance,* R_{oc}.

Neglecting r_{ce} in the calculation of R_o is generally valid only when R_C is less than about 5 K, since values for r_{ce} range from 20 K to 500 K when the transistor is biased in the active region.

The expression for A_{vo} can be rewritten in terms of the generator-to-input voltage gain, A_{vg}, and the transistor terminal-to-terminal voltage gain, A_{vt},* as

$$A_{vo} = A_{vg}A_{vt} \tag{4–62}$$

where

$$A_{vg} = V_i/V_g = R_i/(R_i + R_g) \tag{4–63}$$

$$A_{vt} = \frac{V_o}{V_i} = \frac{h_{fe}(R_{Leq} \,||\, R_{oc})}{h_{ie}} \tag{4–64}$$

The expression for A_{vt} can be generalized to include any transistor configuration. A general two-port network with a finite input impedance, R_{in}, and a total load impedance,** R_{Ltot}, is shown in Fig. 4–23. The current gain A_i of the amplifier is represented in the two-port by I_o/I_i. The terminal-to-terminal voltage gain of the two-port is

$$A_{vt} = \frac{V_o}{V_i} = \frac{-I_o\, R_{Ltot}}{I_i\, R_{in}} = -\frac{A_i\, R_{Ltot}}{R_{in}} \tag{4–65}$$

For the CE amplifier, which has a current gain (A_i) of h_{fe} (or β) and an input impedance of h_{ie}, Eq. (4–65) reduces to Eq. (4–64).

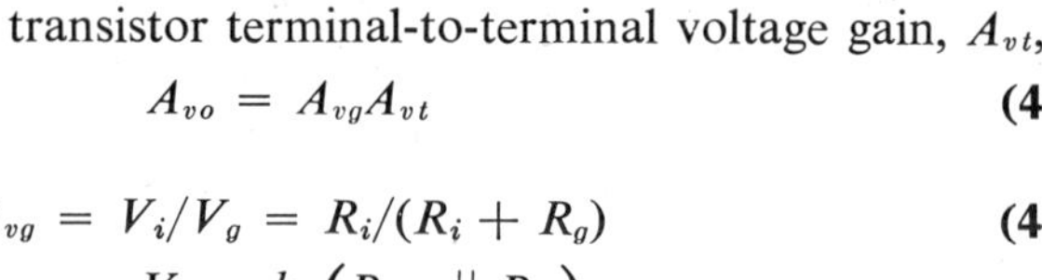

FIGURE 4–23 *A general two-port network used in the calculation of the general expression for* A_{vt}.

* Equation (4–64) for A_{vt} is often approximated as follows:

$$h_{ie} = r_{b'} + (\beta + 1)r_e \simeq (\beta + 1)r_e$$

$$A_{vt} \simeq \frac{-\beta R_{Leq}}{(\beta + 1)r_e} \simeq -\frac{R_{Leq}}{r_e}$$

This approximation for A_{vt} for the CE amplifier requires a large value of β and a small value of $r_{b'}$. These conditions are often satisfied in practice.

** The total load impedance for the CE stage is $R_{Leq}||R_{oc}$.

The important relationships for the single stage CE amplifier are summarized below:

$$\left.\begin{array}{ll} A_{vo} = (A_{vg})(A_{vt}) & R_{in} = r_{b'} + (\beta + 1)(r_e + R_E) \\ A_{vg} = R_i/(R_i + R_g) & R_i = R_B \,||\, R_{in} \\ A_{vt} = -A_i R_{L\text{tot}}/R_{in} \simeq -R_{L\text{eq}}/r_e & R_o = R_C \,||\, r_{ce} \simeq R_C \\ A_i = h_{fe} = \beta & \text{signal inverted} \end{array}\right\} \quad \textbf{(4–66)}$$

Example 4–12 illustrates the use of these equations.

EXAMPLE 4–12 For the CE amplifier shown in Fig. 4–24(a), determine the values of A_{vt}, A_{vo}, R_i, and R_o. The NPN transistor is a silicon unit with $\beta \simeq 120$ and $r_{b'} \simeq 30$ ohms.

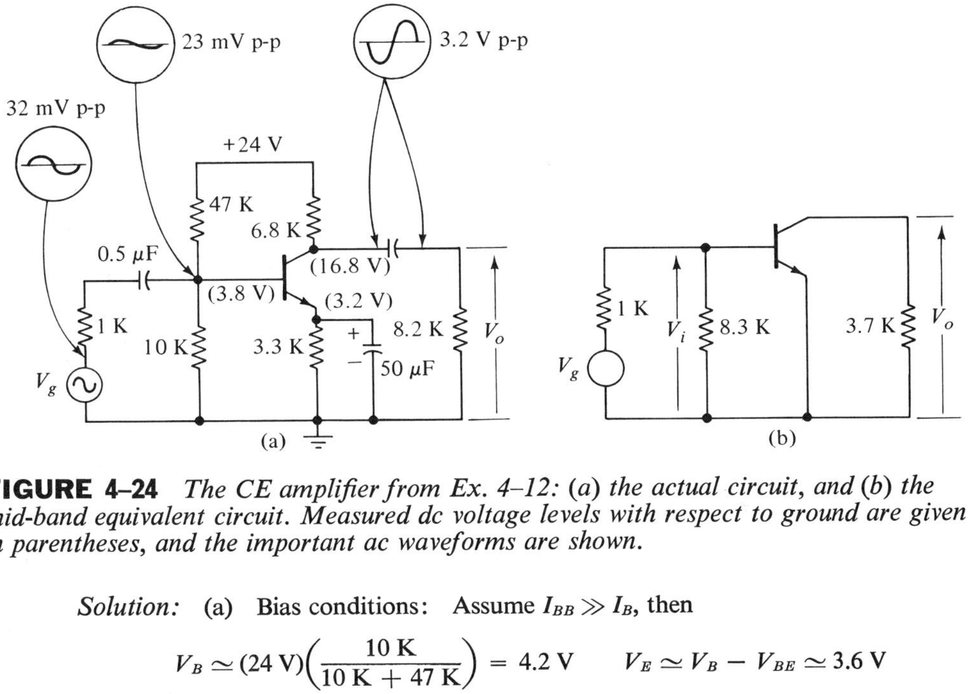

FIGURE 4–24 *The CE amplifier from Ex. 4–12: (a) the actual circuit, and (b) the mid-band equivalent circuit. Measured dc voltage levels with respect to ground are given in parentheses, and the important ac waveforms are shown.*

Solution: (a) Bias conditions: Assume $I_{BB} \gg I_B$, then

$$V_B \simeq (24\text{ V})\left(\frac{10\text{ K}}{10\text{ K} + 47\text{ K}}\right) = 4.2\text{ V} \qquad V_E \simeq V_B - V_{BE} \simeq 3.6\text{ V}$$

$$I_C \simeq I_E = V_E/R_E \simeq 1.1\text{ mA} \qquad V_C = V_{CC} - I_C R_C \simeq 16.6\text{ V}$$

$$I_{BB} = \frac{V_{CC}}{R_1 + R_2} \simeq 420\ \mu\text{A} \qquad I_B = I_C/\beta \simeq 9.2\ \mu\text{A}$$

Since I_{BB} is much greater than I_B, the approximate bias solution is valid.

(b) Gain and impedance calculations:

$$r_e = \frac{26\text{ mV}}{I_E} = \frac{26\text{ mV}}{1.1\text{ mA}} \simeq 24\ \Omega \qquad A_i = \beta = -120$$

$$R_{in} = h_{ie} = r_{b'} + (\beta + 1)(r_e) \simeq 2.9\text{ K} \qquad R_B = 10\text{ K} \,||\, 47\text{ K} \simeq 8.3\text{ K}$$

$$R_{L\text{tot}} \simeq R_{L\text{eq}} \simeq 3.7\text{ K} \qquad R_o = R_C \,||\, R_{oc} \simeq R_C = 6.8\text{ K}$$

$$R_i = R_B \,||\, R_{in} \simeq 2.15\text{ K} \qquad A_{vg} = R_i/(R_i + R_g) \simeq 0.68$$

$$A_{vt} = A_i R_{L\text{eq}}/R_{in} \simeq -154 \qquad A_{vo} = A_{vg} A_{vt} \simeq -104$$

(c) Experimental results: The measured dc voltage levels are shown in Fig. 4–24(a). The measured voltage gains were

$$A_{vt} = \frac{V_o}{V_i} \simeq -\frac{3.2\text{ V}}{0.023\text{ V}} \simeq -140 \qquad A_{vo} = \frac{V_o}{V_g} \simeq -\frac{3.2\text{ V}}{0.032\text{ V}} \simeq -100$$

These values are reasonably close to the calculated values. Note that the signal is inverted, as shown by the waveform in Fig. 4–24(a).

Since the hybrid-π equivalent circuit will be used extensively in later chapters, the equivalence between it and the hybrid circuit will be discussed briefly here. Starting with Fig. 4–21(b), and replacing the transistor with the hybrid-π equivalent circuit gives the equivalent circuit shown in Fig. 4–25. The expression for A_{vo} can be obtained from Fig. 4–25 as

$$A_{vo} = \frac{V_o}{V_g} = -\left(\frac{V_i}{V_g}\right)\left(\frac{V_{b'e}}{V_i}\right)\left(\frac{V_o}{V_{b'e}}\right) = -\left(\frac{R_i}{R_i + R_g}\right)\left(\frac{r_{b'e}}{r_{b'} + r_{b'e}}\right)(g_m R_{Ltot}) \tag{4–67}$$

Equation (4–67) can be converted into the expression for A_{vo} given in Eq. (4–57) by using the following parameter relationships:

$$h_{ie} = r_{b'} + r_{b'e} \qquad g_m r_{b'e} = \beta = h_{fe} \tag{4–68}$$

The expression for A_{vo} given in Eq. (4–67) is very similar to the expression for A_{vo} for the CS amplifier (see Table 3–1). The only difference between the two is the second term in Eq. (4–67), which accounts for the voltage division between $r_{b'}$ and $r_{b'e}$.

In summary, the hybrid equivalent circuit has been used to analyze the performance of a bipolar CE amplifier. The input impedance of this circuit is rather low (1000 to 5000 ohms). The output impedance is rather high (10 K to 500 K). The gain is reasonably high (50 to 150), and the signal is inverted. Since both the two-port approach and the hybrid-π will be used later, both were discussed briefly.

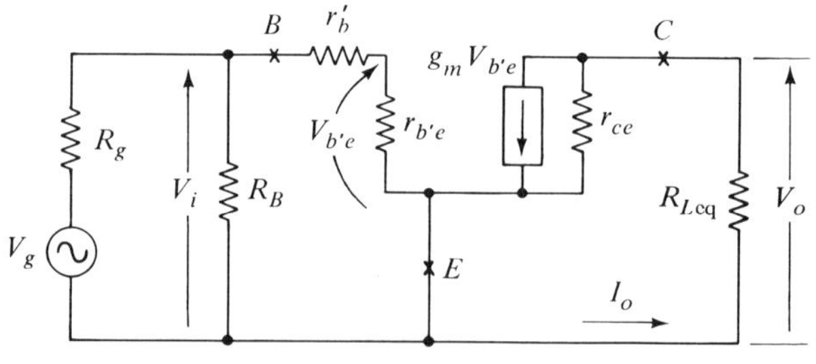

FIGURE 4–25 *The complete equivalent circuit of the CE amplifier shown in Fig. 4–21(b) using the hybrid-π equivalent circuit to represent the transistor.*

4.6 The Common-Base (CB) Amplifier

A second important amplifier configuration using a bipolar transistor is the common-base (CB) type shown in Fig. 4–26(a). The input signal is applied between the emitter and base terminals, and the output signal is obtained between the collector and base terminals. The objective of this section is to develop ex-

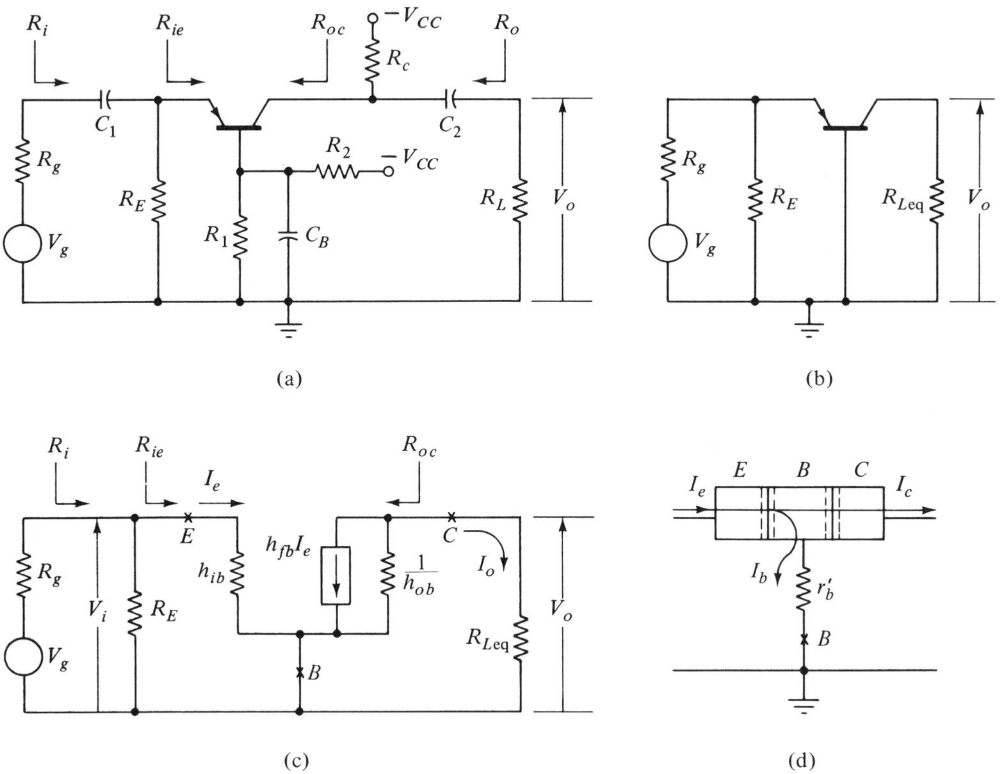

FIGURE 4–26 *A common-base amplifier: (a) a complete CB amplifier, (b) a mid-band circuit for the amplifier, (c) a complete mid-band equivalent circuit using the hybrid equivalent circuit for the transistor, and (d) a physical model of the transistor used to evaluate the CB hybrid parameters.*

pressions for the voltage gain, input impedance, and output impedance of the CB amplifier in the mid-band frequency range. The hybrid equivalent circuit will be used to represent the transistor, and the CB hybrid parameters will be developed. The results of the analysis will show that the CB amplifier is a non-inverting, low-input impedance, high-output impedance amplifier.

A PNP amplifier is shown in Fig. 4–26(a). The mid-band equivalent circuit of this amplifier is obtained by short circuiting the capacitors and dc supply as shown in Fig. 4–26(b). A complete mid-band equivalent circuit is obtained by replacing the transistor with its hybrid equivalent circuit as shown in Fig. 4–26(c).

The CB hybrid parameters can be obtained as follows. The CB current gain, h_{fb}, is the ratio I_c/I_e which corresponds to the definition of α. The sign of α, however, should receive special attention. From our previous discussion on the physical operation of a bipolar transistor operation, we know that a current entering at the emitter terminal is associated with a current leaving the collector terminal. Hence, the assumed direction* of the current generator, $h_{fb}I_e$, in Fig.

* In two-port network theory, both the input current and the output current are assumed to flow into the top terminals. For the CE case, this assumption conforms to the physical directions. However, for the CB case, the assumption is wrong and must be corrected with a sign change.

4–26(c) is opposite that indicated by the physical action. By letting h_{fb} be equal to $-\alpha$, the assumed and actual current directions are coordinated. The CB stage does not invert the signal since a positive going value of V_g produces positive going values of I_e, I_o, and V_o as shown in Fig. 4–26(c).

The CB input impedance, h_{ib}, can be obtained from Fig. 4–26(d). This figure shows the physical model of a bipolar transistor with the base lead common to both input and output circuits. The impedance obtained by looking into the emitter terminal of Fig. 4–26(d) is the *EB* junction dynamic impedance, r_e, plus a contribuiton due to $r_{b'}$. Using the $(\beta + 1)$ impedance transformation of the *EB* junction (see Sec. 4.3), the contribution of $r_{b'}$ to h_{ib} is $r_{b'}/(\beta + 1)$. The CB output impedance, $1/h_{ob}$, can be approximated by the inverse slope of the constant I_E curve of the CB characteristics which passes through the bias point or the Q point. Hence,

$$r_{cb} \simeq \frac{1}{h_{ob}} \simeq \left.\frac{\Delta V_{CB}}{\Delta I_C}\right|_{I_E=\text{constant}} \qquad \textbf{(4–69)}$$

Typical CB static curves are shown in Fig. 4–10(b). As noted in Sec. 4.2, the CB curves are more nearly horizontal than the CE curves. Consequently, r_{cb} is very large with typical values ranging from 200 K to 100 M. To summarize the CB hybrid parameters:

$$h_{fb} = -\alpha \qquad h_{ob} = 1/r_{cb} \simeq 0 \qquad h_{ib} = r_e + r_{b'}/(\beta + 1) \qquad \textbf{(4–70)}$$

The expressions for the input and output impedances of the CB amplifier can be developed using a test voltage generator as was done for the CE amplifier in the previous section. The results are already available, however, from Fig. 4–26(c) where

$$R_{ie} = h_{ib} \quad \text{and} \quad R_{oc} = 1/h_{ob} = r_{cb} \qquad \textbf{(4–71)}$$

With R_{ie} known, the input impedance of the amplifier can be calculated from Fig. 4–26(a) as

$$R_i = R_E \,||\, R_{ie} = R_E \,||\, h_{ib} \qquad \textbf{(4–72)}$$

With R_{oc} known, the output impedance of the amplifier can be calculated from Fig. 4–26(a) as

$$R_o = R_C \,||\, R_{oc} = R_C \,||\, r_{cb} \simeq R_C \qquad \textbf{(4–73)}$$

Referring to Fig. 4–26(c), the output voltage is given by

$$V_o = I_o\, R_{\text{Leq}} \simeq -h_{fb} R_{\text{Leq}}\, I_e \qquad \textbf{(4–74)}$$

where the approximation is valid for large values of r_{cb}. Rewriting Eq. (4–74) in terms of α,

$$V_o = \alpha I_e\, R_{\text{Leq}} \qquad \textbf{(4–75)}$$

The current I_e is V_i divided by h_{ib} where V_i is that portion of the generator voltage, V_g, developed across the base-emitter terminals, or

$$V_i = V_g R_i/(R_i + R_g) \qquad \textbf{(4–76)}$$

From the preceding equations, the generator-to-output voltage gain is

$$A_{vo} = \frac{V_o}{V_g} \simeq \left(\frac{R_i}{R_i + R_g}\right)\left(\frac{\alpha R_{\text{Leq}}}{h_{ib}}\right) = (A_{vg})(A_{vt}) \qquad \textbf{(4–77)}$$

where A_{vg} is the voltage gain from the generator to the input terminals, and A_{vt} is the voltage gain from the input terminals to the output terminals.

The general expression for A_{vt} can also be applied to the CB amplifier where

$$A_{vt} = -\frac{A_i R_{Leq}}{R_{in}} = \frac{\alpha R_{Leq}}{h_{ib}} \simeq \frac{R_{Leq}}{r_e} \qquad \textbf{(4–78)}$$

The important relationships* for the single-stage CB amplifier are summarized below:

$$\left.\begin{array}{ll} A_{vo} = (A_{vg})(A_{vt}) & R_{in} = r_e + \dfrac{r_{b'} + R_B}{\beta + 1} \\ A_{vg} = R_i/(R_i + R_g) & R_i = R_E \,||\, R_{in} \\ A_{vt} = -A_i R_{Leq}/R_{in} \simeq R_{Leq}/r_e & R_o = R_C \,||\, r_{cb} \simeq R_C \\ A_i = h_{fb} = -\alpha & \text{signal not inverted} \end{array}\right\} \qquad \textbf{(4–79)}$$

The following example demonstrates the use of the CB equations.

EXAMPLE 4–13 For the CB amplifier shown in Fig. 4–27, determine the values of A_{vt}, A_{vo}, R_i, and R_o. The PNP transistor is a silicon unit with $\beta = 150$ and $r_{b'} = 50$ ohms. The 20-M load resistor represents the input impedance of an oscilloscope.

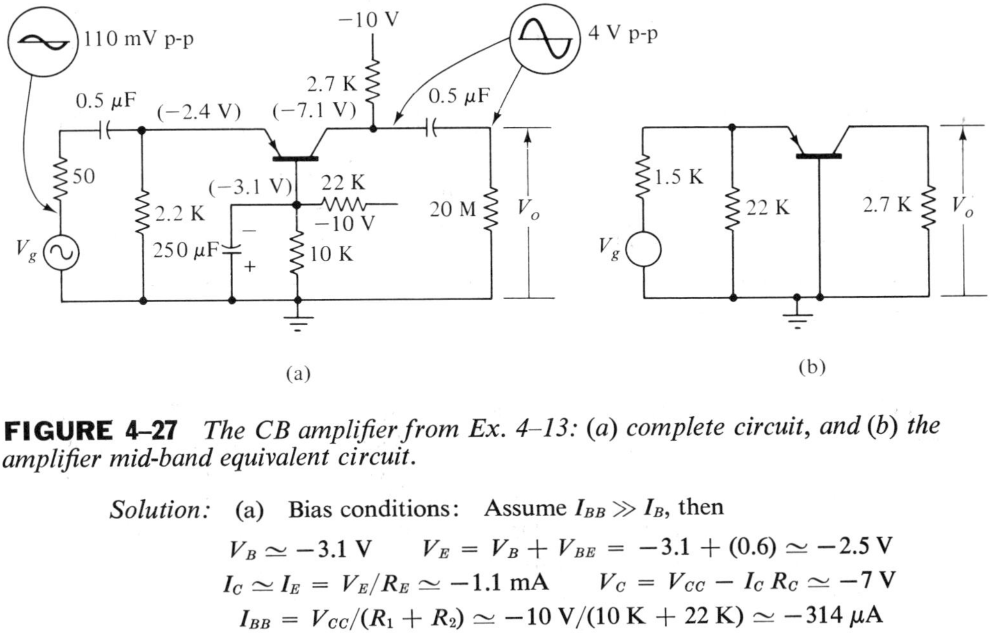

FIGURE 4–27 *The CB amplifier from Ex. 4–13: (a) complete circuit, and (b) the amplifier mid-band equivalent circuit.*

Solution: (a) Bias conditions: Assume $I_{BB} \gg I_B$, then

$$V_B \simeq -3.1\text{ V} \qquad V_E = V_B + V_{BE} = -3.1 + (0.6) \simeq -2.5\text{ V}$$

$$I_C \simeq I_E = V_E/R_E \simeq -1.1\text{ mA} \qquad V_C = V_{CC} - I_C R_C \simeq -7\text{ V}$$

$$I_{BB} = V_{CC}/(R_1 + R_2) \simeq -10\text{ V}/(10\text{ K} + 22\text{ K}) \simeq -314\ \mu\text{A}$$

$$I_B = I_C/\beta \simeq -1.1\text{ mA}/150 \simeq -7\ \mu\text{A}$$

Since I_{BB} is much greater than I_B, the approximate bias solution is valid.

(b) Gain and impedance calculations:

$$r_e = \frac{26\text{ mV}}{I_E} = \frac{26}{1.1} \simeq 24\ \Omega \qquad R_{Leq} = 2.7\text{ K} \,||\, 20\text{ M} \simeq 2.7\text{ K}$$

---

* Note that in the expression for R_{in}, R_B is the unbypassed resistor in the base leg. Usually R_B is zero.

$$R_{in} = R_{ie} = h_{ib} = r_e + r_{b'}/(\beta + 1) \simeq 24 + 50/150 \simeq 24\,\Omega$$

$$A_i = -\alpha = -\frac{\beta}{\beta+1} = -\frac{150}{150+1} \simeq -1$$

$$R_i = R_E \,||\, R_{in} = 2.2\text{ K} \,||\, 24 \simeq 24 \qquad R_o = R_C \,||\, r_{cb} \simeq 2.7\text{ K}$$

$$A_{vt} = -A_i\, R_{Leq}/R_{in} \simeq 112 \qquad A_{vg} = R_i/(R_i + R_g) \simeq 0.325$$

$$A_{vo} = A_{vg}\, A_{vt} = (0.325)(112) \simeq 36.5$$

(c) Experimental results: The measured dc voltage levels are shown in Fig. 4–27(a). With an input voltage of 0.11 V, the measured voltage gain was 36. The value of A_{vo} is within 10 percent of the calculated value. As shown in Fig. 4–27(a), the signal was not inverted.

The complete CB hybrid-π equivalent circuit is shown in Fig. 4–28. While the CE form of the hybrid-π [as shown in Fig. 4–14(c)] can be used for the CB stage, the physical interpretation is complicated. The CB hybrid-π will be used in Chap. 7 to study the frequency response of the CB amplifier.

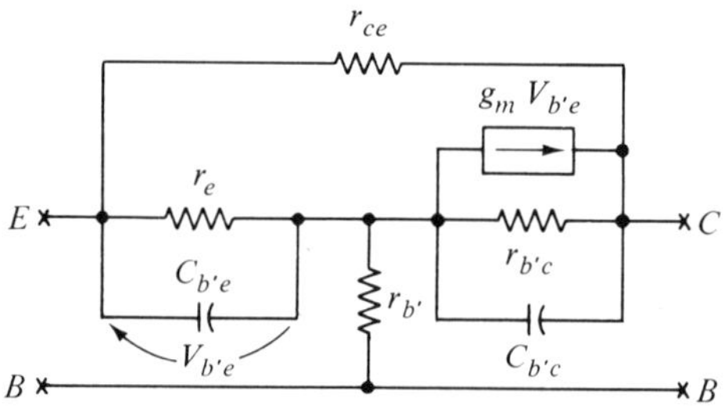

FIGURE 4–28 *The CB hybrid-π equivalent circuit.*

The CB form of the hybrid-π can be generated from the CE form as follows. First, redraw Fig. 4–14(c) with the base lead common to the input and the output terminals. Next, replace the current generator connected between the C and E terminals, $g_m V_{b'e}$, by two current generators in series having the same value, $g_m V_{b'e}$. The common point between the series current generators can be connected to node B' with no change in the circuit. The polarity of each current generator must be changed if the polarity of $V_{b'e}$ is changed. Since the $g_m V_{b'e}$ generator between E and B' has a voltage $V_{b'e}$ across it, this generator can be replaced by a resistor of value α/r_e. The development of the circuit in Fig. 4–28 is completed by combining (α/r_e) and $r_{b'e}$.

In summary, the hybrid equivalent circuit has been used to determine the small-signal performance characteristics of a bipolar common-base amplifier. This analysis indicates that (1) the input impedance is very low (10 to 100 ohms), (2) the output impedance is very high (100 K to 10 M), and (3) A_{vt} is large. The circuit gain, A_{vg}, is usually small because most of V_g is lost across R_g. The CB hybrid-π equivalent was discussed briefly only to note its mid-frequency form. The details of this circuit and its applications will be discussed in Chap. 7.

4.7 The Common-Collector (CC) Amplifier

A third useful amplifier configuration using a bipolar transistor is the common-collector (CC) amplifier shown in Fig. 4–29(a). This configuration is also referred to as an "emitter follower" (EF). The collector terminal is usually connected to a power supply (an ac ground). The input ac signal is applied between the base and ground (collector) terminals, and the output signal is obtained between the emitter and ground (collector) terminals. The performance characteristics of the CC amplifier will be developed by using the general expression for A_{vt} [Eq. (4–65)] and the $(\beta + 1)$ impedance transformation between the base and emitter terminals. The results of this analysis will show that the CC amplifier is a non-inverting, high-input impedance, low-output impedance amplifier having a voltage gain less than unity.

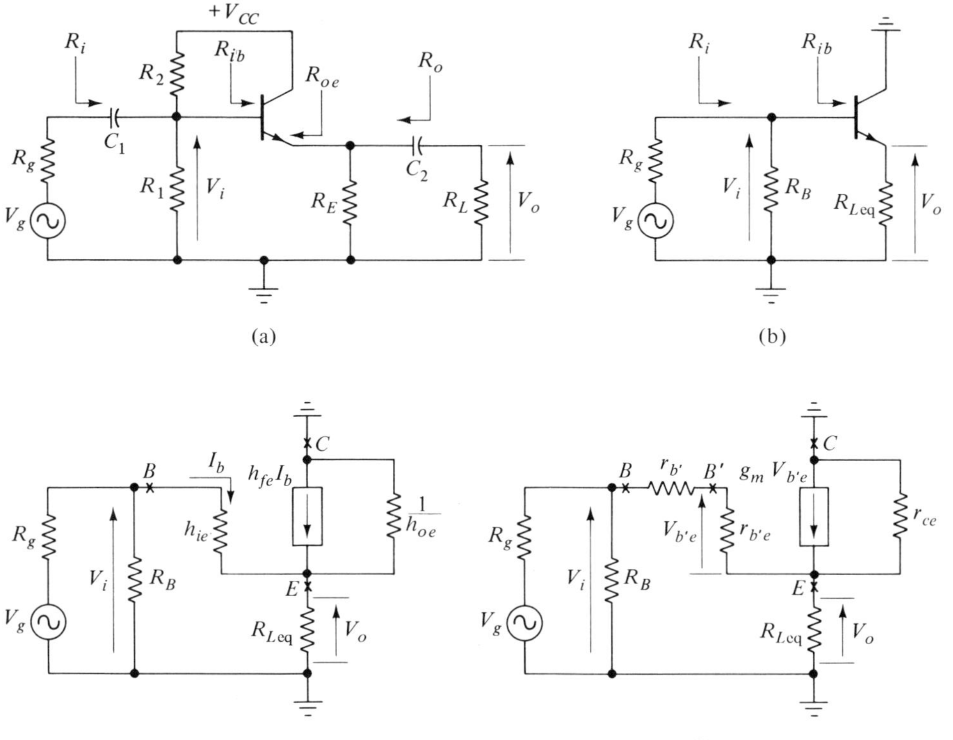

FIGURE 4–29 *A common-collector amplifier: (a) a complete CC amplifier, (b) a mid-band equivalent circuit for the amplifier, (c) a complete mid-band equivalent circuit using the hybrid circuit for the transistor, and (d) a complete mid-band equivalent circuit using the hybrid-π circuit for the transistor.*

An NPN CC amplifier is shown in Fig. 4–29(a). The mid-band equivalent circuit of this amplifier is obtained by short circuiting the coupling capacitors and replacing the dc power supply with an ac ground as shown in Fig. 4–29(b).

A complete mid-band equivalent circuit is obtained by replacing the transistor with its equivalent circuit as shown in Fig. 4–29(c) and (d). Normally the CE equivalent circuit is used in analyzing the CC amplifier. The general expression for A_{vt}, as given by Eq. (4–65), is

$$A_{vt} = -A_i R_{L\text{tot}}/R_{in} \tag{4–80}$$

From Sec. 4.1, the current gain of the CC stage is I_e/I_b or $-(\beta + 1)$ where the negative sign is necessary for the CC amplifier to correspond to the general two-port notation.* Using the *EB* junction impedance transformation, the input impedance of the CC stage is

$$R_{in} = R_{ib} = r_{b'} + (\beta + 1)(r_e + R_{L\text{eq}}) = h_{ie} + (\beta + 1)(R_{L\text{eq}}) \tag{4–81}$$

where $R_{L\text{eq}}$ is the value of the ac impedance in the emitter circuit. The total load impedance is

$$R_{L\text{tot}} = R_{L\text{eq}} \,||\, r_{ce} \simeq R_{L\text{eq}} \tag{4–82}$$

Substituting these equations into Eq. (4–80), the mid-band expression for A_{vt} is

$$A_{vt} = \frac{(\beta + 1)(R_{L\text{eq}})}{r_{b'} + (\beta + 1)(r_e + R_{L\text{eq}})} \simeq \frac{R_{L\text{eq}}}{r_e + R_{L\text{eq}}} \tag{4–83}$$

where the approximation requires that $(\beta + 1)(r_e + R_{L\text{eq}})$ be much larger than $r_{b'}$. This equation indicates that the CC voltage gain is positive (noninverting) and less than unity.

The generator-to-input voltage gain, A_{vg}, is available from Fig. 4–29 as

$$A_{vg} = V_i/V_g = R_i/(R_i + R_g) \tag{4–84}$$

where R_i is R_B in parallel with R_{ib}, and R_{ib} is given by Eq. (4–81). The generator-to-output gain, A_{vo}, is

$$A_{vo} = (V_i/V_g)(V_o/V_i) = A_{vg}A_{vt} \tag{4–85}$$

where A_{vg} is given by Eq. (4–84), and A_{vt} by Eq. (4–83). The expressions for A_{vo} and A_{vt} can also be found using the equivalent circuit for the amplifier [Fig. 4–29(c) or (d)]. The analysis is very similar to that for the CE amplifier and is left as an exercise for the reader.

The input impedance of the CC stage can be calculated using the circuit shown in Fig. 4–30(a). The impedance looking into the base terminal, R_{ib}, was found earlier and is given by Eq. (4–81). With R_{ib} known, the input impedance of the amplifier, R_i, is R_B in parallel with R_{ib}.

The output impedance of the CC stage, R_{oe}, can be calculated using Fig. 4–30(b). The impedance looking into the emitter terminal is

$$R_{oe} = r_e + \frac{r_{b'} + R_B \,||\, R_g}{\beta + 1} \tag{4–86}$$

With R_{oe} known, the output impedance of the amplifier, R_o, is R_E in parallel with R_{oe}.

* In a bipolar transistor connected as a CC amplifier, when I_b flows into the base or input terminal, I_e flows out of the emitter or output terminal, which is opposite the general two-port sign convention. Hence, a negative sign is required for the CC current gain.

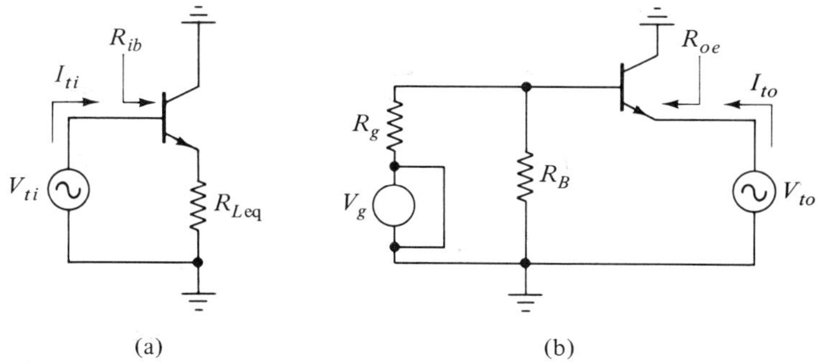

FIGURE 4–30 *Circuits used to determine (a) the input impedance, and (b) the output impedance of a CC stage.*

The important relationships for the single-stage CC amplifier are summarized below:

$$\left.\begin{aligned}
&A_{vo} = (A_{vg})(A_{vt}) \qquad R_{in} = r_{b'} + (\beta + 1)(r_e + R_{Leq}) \\
&A_{vg} = R_i/(R_i + R_g) \qquad R_i = R_B \,||\, R_{in} \\
&A_{vt} = -A_i R_{Leq}/R_{in} \qquad R_o = R_E \,||\, \left(r_e + \frac{r_{b'} + R_B \,||\, R_g}{\beta + 1}\right) \\
&A_i = h_{fc} = -(\beta + 1) \qquad \text{signal not inverted}
\end{aligned}\right\} \quad \textbf{(4–87)}$$

The following example demonstrates the use of the CC equations.

EXAMPLE 4–14 For the CC amplifier shown in Fig. 4–31, determine the values of A_{vt}, A_{vo}, R_i, and R_o. The NPN transistor is a silicon unit with $\beta = 150$ and $r_{b'} = 200$ ohms. The 100-ohm load resistor represents a terminated coaxial cable.

Solution: (a) Bias conditions: Assume $I_{BB} \gg I_B$, then

$$V_B \simeq (15\text{ V})\left(\frac{22\text{ K}}{22\text{ K} + 22\text{ K}}\right) \simeq 7.5\text{ V} \qquad V_E = V_B - V_{BE} = 6.9\text{ V}$$

$$I_C \simeq I_E = V_E/R_E = 6.9\text{ mA} \qquad I_{BB} = V_{CC}/(R_1 + R_2) \simeq 340\ \mu\text{A}$$

$$I_B \simeq I_C/\beta \simeq 6.9\text{ mA}/150 \simeq 46\ \mu\text{A}$$

Since I_{BB} is only 7.4 times greater than I_B, the more general bias analysis will be used. The general method, as outlined in Sec. 4.3 and shown in Fig. 4–12(a), is to replace the base bias network with its Thévenin equivalent circuit and then calculate I_B and I_E. The equivalent dc network of the amplifier is shown in Fig. 4–31(c) where

$$V_{B(\text{Th})} = (15\text{ V})\left(\frac{22\text{ K}}{22\text{ K} + 22\text{ K}}\right) = 7.5\text{ V} \qquad R_{B(\text{Th})} = 22\text{ K} \,||\, 22\text{ K} = 11\text{ K}$$

Summing voltages in terms of I_E,

$$7.5\text{ V} = 11\text{ K}\left(\frac{I_E}{\beta + 1}\right) + 0.6 + (1\text{ K})(I_E)$$

Solving for I_E, V_E, and V_B:

$$I_E = \frac{7.5\text{ V} = 0.6\text{ V}}{1\text{ K} + 11\text{ K}/(151)} \simeq 6.4\text{ mA} \qquad V_E = I_E R_E \simeq 6.4\text{ V}$$

$$V_B = V_E + V_{BE} = 6.4\text{ V} + 0.6\text{ V} = 7\text{ V}$$

The measured bias voltages agree very closely with the second set of calculated values.

(b) Gain and impedance calculations:

$$r_e = \frac{26\text{ mV}}{I_E} = \frac{26}{6.4} \simeq 4\,\Omega \qquad R_{L\text{eq}} = 1\text{ K} \,||\, 100 \simeq 91\,\Omega$$

$$R_{in} = r_{b'} + (\beta + 1)(r_e + R_{L\text{eq}}) = 200 + (150 + 1)(4 + 91) \simeq 14.3\text{ K}$$

$$R_i = R_B \,||\, R_{in} = 11\text{ K} \,||\, 14.3\text{ K} \simeq 6.2\text{ K}$$

$$R_o = R_E \,||\, \left(r_e + \frac{r_{b'} + R_g \,||\, R_B}{\beta + 1}\right) = 1\text{ K} \,||\, \left(4 + \frac{200 + 100 \,||\, 11\text{ K}}{151}\right) \simeq 6\,\Omega$$

$$A_i = -(\beta + 1) = -151 \qquad A_{vt} = -\frac{A_i R_{L\text{eq}}}{R_{in}} = \frac{(151)(91)}{14.3\text{ K}} \simeq 0.96$$

$$A_{vg} = R_i/(R_i + R_g) \simeq 0.99 \qquad A_{vo} = A_{vg} A_{vt} = (0.99)(0.96) \simeq 0.95$$

(c) Experimental results: The measured dc voltage levels are shown in Fig. 4–31(a). The measured voltage gain was 0.97 with $V_g = 200$ mV. The output signal was not inverted.

In summary, a bipolar CC amplifier has been analyzed by treating it as a special case of a CE amplifier. The analysis indicates that the CC amplifier has a high-input impedance, a low-output impedance, and a voltage gain of less than unity. However, the current gain is $(\beta + 1)$. This configuration is often used as an impedance matching stage to match a low-impedance load to a high-impedance generator. The CC stage could also be analyzed with the hybrid, hybrid-π, or two-port approach.

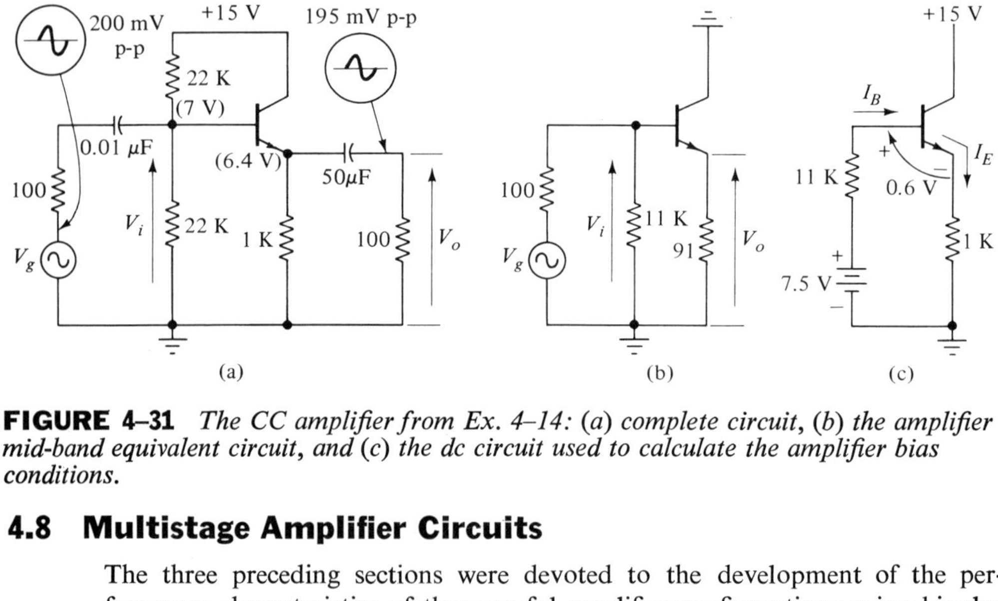

FIGURE 4–31 *The CC amplifier from Ex. 4–14: (a) complete circuit, (b) the amplifier mid-band equivalent circuit, and (c) the dc circuit used to calculate the amplifier bias conditions.*

4.8 Multistage Amplifier Circuits

The three preceding sections were devoted to the development of the performance characteristics of three useful amplifier configurations using bipolar transistors. A similar development was made for FET amplifiers in the last chapter, and the results were summarized in Sec. 3.9. Many amplifiers require more than a single stage to achieve the desired characteristics. Multistage amplifiers are formed by interconnecting several single-stage amplifiers and may consist of all bipolar transistors, all FETs, or a combination of the two types of devices.

Although a large number of multistage circuit types exist, some simple techniques are available that can be used to analyze a multistage amplifier regardless of the configuration. The objective of this section is to develop these techniques. To achieve this objective, we will first summarize the performance characteristics of the three single-stage bipolar amplifier circuits already considered in this chapter. Next, we will review the effects of interconnecting several two-port networks and apply the techniques of multistage analysis to three particular multistage circuits.

The approximate performance characteristics of the CE, CB, and CC amplifiers are shown in Table 4–1. In each case, the transistor is shown as a two-port network with the input and output impedances of the transistor identified. Note that the load and bias resistors are assumed to be located external to the two-port with two exceptions: (1) the CE configuration is shown with an unbypassed emitter resistor, R_E, and (2) the CB configuration is shown with an unbypassed base resistor, R_B. These two resistors are included to keep the expressions for both the input impedance and the approximate terminal-to-terminal voltage gain in a general form. If either R_E or R_B are ac bypassed, that parameter is zero in the corresponding expression. The general expressions for A_{vo}, A_{vg}, and A_{vt} are also given in Table 4–1.

Both the hybrid-π and the hybrid parameters are important in transistor circuit analysis. Often it is necessary to convert from one set of parameters to the other. Table 4–2 lists the important mid-band interrelationships between the two sets of parameters.

We are now in a favorable position to compare FET and bipolar amplifier stages. From Table 3–1 for the FET, and Table 4–1 for the bipolar transistor, the two stages which invert the signal are the CS stage and the CE stage. Of the three FET stages, the CS has the middle value of both input and output impedance. Of the three bipolar stages, the CE has the middle value of both input and output impedance. Consequently, the CS and the CE have similar characteristics and can be used in similar applications.

However, differences also exist between the CS and CE stages, one important difference being between their values of input impedance. The CS has a nearly infinite input impedance while the input impedance of the CE stage is only a few hundred to a few thousand ohms.

A second difference between the CS and CE stages is their available voltage gains. The expression for A_{vt} for both stages can be written in terms of g_m as

$$A_{vt} = -A_i R_{Leq}/R_{in} \simeq -\left(\frac{\beta}{(\beta+1)r_e}\right) R_{Leq} = -g_m R_{Leq} \qquad \textbf{(4–88)}$$

Most FET devices have values of g_m ranging from 1 to 25 millimhos with a few units ranging over 100 millimhos. Bipolar devices can have much higher values of g_m. Assuming the same value of load resistance, the CE amplifier is generally capable of a larger voltage gain than the CS amplifier.

TABLE 4–1 *Summary of the approximate performance characteristics of single-stage bipolar amplifiers. The quantities in parentheses indicate the typical range of parameter values.*

Configuration	Current gain (A_i)	Stage input impedance (R_{in})	Stage output impedance	Voltage gain (A_{Vt})
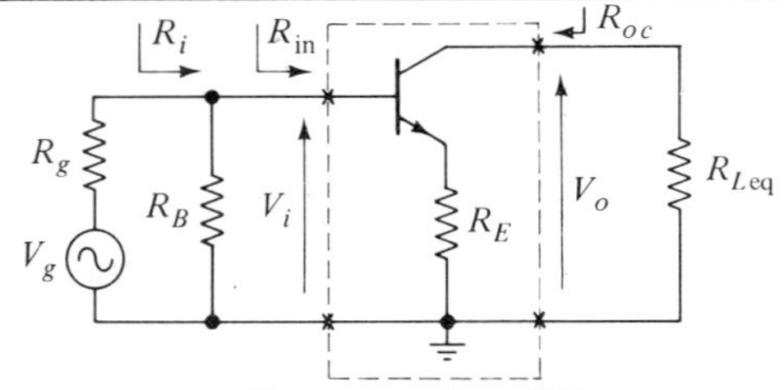 Common emitter (CE)	β (10 to 1000) [voltage signal inverted]	$r_{b'} + (\beta + 1)(r_e + R_E)$ (R_E may be ac bypassed and hence equal to zero.)	r_{ce} (20 K to 500 K)	$\dfrac{R_{Leq}}{r_e + R_E}$
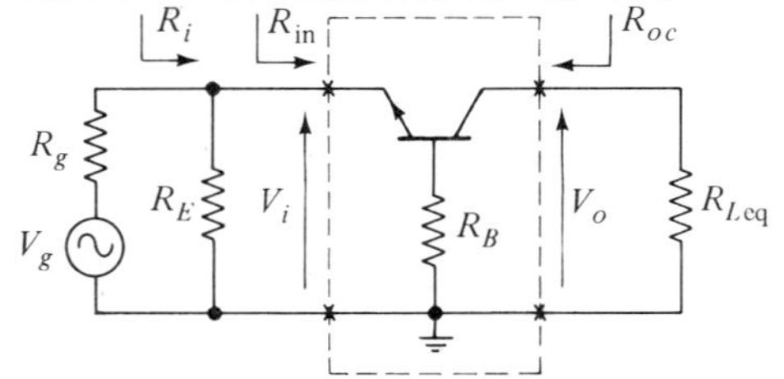 Common base (CB)	$-\alpha$ (0.9 to 0.99) [voltage signal not inverted]	$r_e + \dfrac{r_{b'} + R_B}{\beta + 1}$ (R_B may be ac bypassed and hence equal to zero.)	r_{cb} (100 K to 100 M)	$\dfrac{R_{Leq}}{r_e + \dfrac{R_B}{\beta + 1}}$
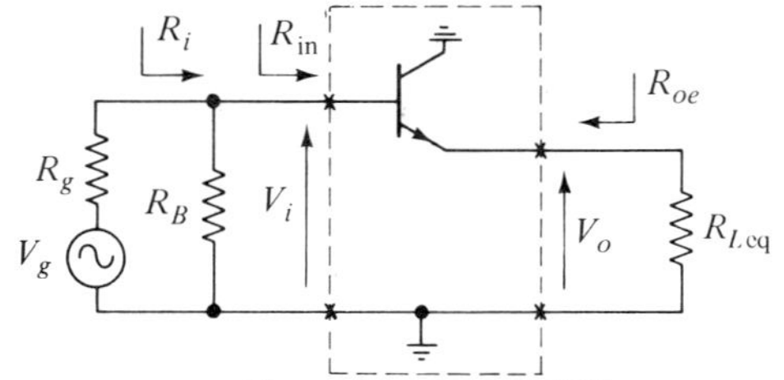 Common collector (CC)	$-(\beta + 1)$ (10 to 1000) [voltage signal not inverted]	$r_{b'} + (\beta + 1)(r_e + R_{Leq})$	$r_e + \dfrac{r_{b'} + R_g \parallel R_B}{\beta + 1}$	$\dfrac{R_{Leq}}{R_{Leq} + r_e}$

$A_{vo} = V_o / V_g = A_{vg} A_{vt}$; $A_{vg} = V_i / V_g = R_i / (R_i + R_g)$; $A_{vt} = -A_i R_{Leq} / R_{in}$

TABLE 4–2 *Approximate interrelationships for the mid-band hybrid and hybrid-π parameters*

$\beta = h_{fe} = \alpha_o/(1 - \alpha_o) \simeq g_m r_{b'e}$	$h_{ie} \simeq r_{b'} + (\beta + 1)r_e$
$g_m \simeq \alpha_o/r_e = \dfrac{\alpha_o I_E}{26\text{ mV}}$	$h_{ib} \simeq r_e + r_{b'}/(\beta + 1)$
$r_{b'e} = (\beta + 1)(r_e)$	$h_{ic} \simeq r_{b'} + (\beta + 1)r_e$
$r_e \simeq 26\text{ mV}/I_E$	$h_{oe} \simeq 1/r_{ce}$
$h_{fe} = \beta$	$h_{ob} \simeq 1/r_{cb}$
$h_{fb} = -\alpha$	$h_{oc} \simeq 1/r_{ce}$
$h_{fc} = -(\beta + 1)$	

A comparison can also be made between other configurations. For example, the CG and CB are similar since each has a low input impedance and a high output impedance. In addition, neither type inverts the signal. The CD and CC stages are similar since each has a high input impedance, low output impedance, and a voltage gain less than unity.

The bipolar transistor generally has a better high-frequency response than the FET. However, the FET can be used at very low temperatures (such as liquid nitrogen temperature, 77°K). Because of its better high-frequency response, higher gain, and greater variety of commercially available units, the bipolar transistor has found wider application than the FET. However, the FET is often used as the input stage of a multistage amplifier, in differential amplifiers, in low-temperature application, and in applications requiring a very high value of input impedance. Also, some integrated circuit fabrication techniques using MOSFET units as the active element promise economic advantages over the bipolar transistor.

In general, the value of A_{vo} for a multistage bipolar amplifier is obtained by finding the product of A_{vg} and A_{vt} for each stage, or

$$A_{vo} = A_{vg} A_{vt1} A_{vt2} A_{vt3} \ldots \tag{4–89}$$

Values of A_{vg}, R_i, and R_o are calculated in the same manner as for the single-stage amplifier. The equivalent load resistance for any stage is the parallel combination of the input impedance to the next stage and any bias resistors connected to the output terminal. The following five examples will demonstrate this method of analysis for multistage circuits.

A CE-CE NPN-PNP amplifier is shown in Fig. 4–32. This configuration corresponds to the CS-CS FET amplifier shown in Fig. 3–27 and discussed in Ex. 3–10. Base voltage bias is used for each stage. Since each stage inverts the signal, the output voltage is in phase with the input signal. The overall voltage gain can be quite high since each CE stage can have a high value of A_{vt}. An analysis of a two-stage CE amplifier is illustrated in the following example.

EXAMPLE 4–15 The CE-CE amplifier shown in Fig. 4–32(b) has silicon transistors characterized by $\beta_1 \simeq 100$, $\beta_2 \simeq 35$, and negligible values of $r_{b'}$. Calculate the values of A_{vo}, R_i, and R_o.

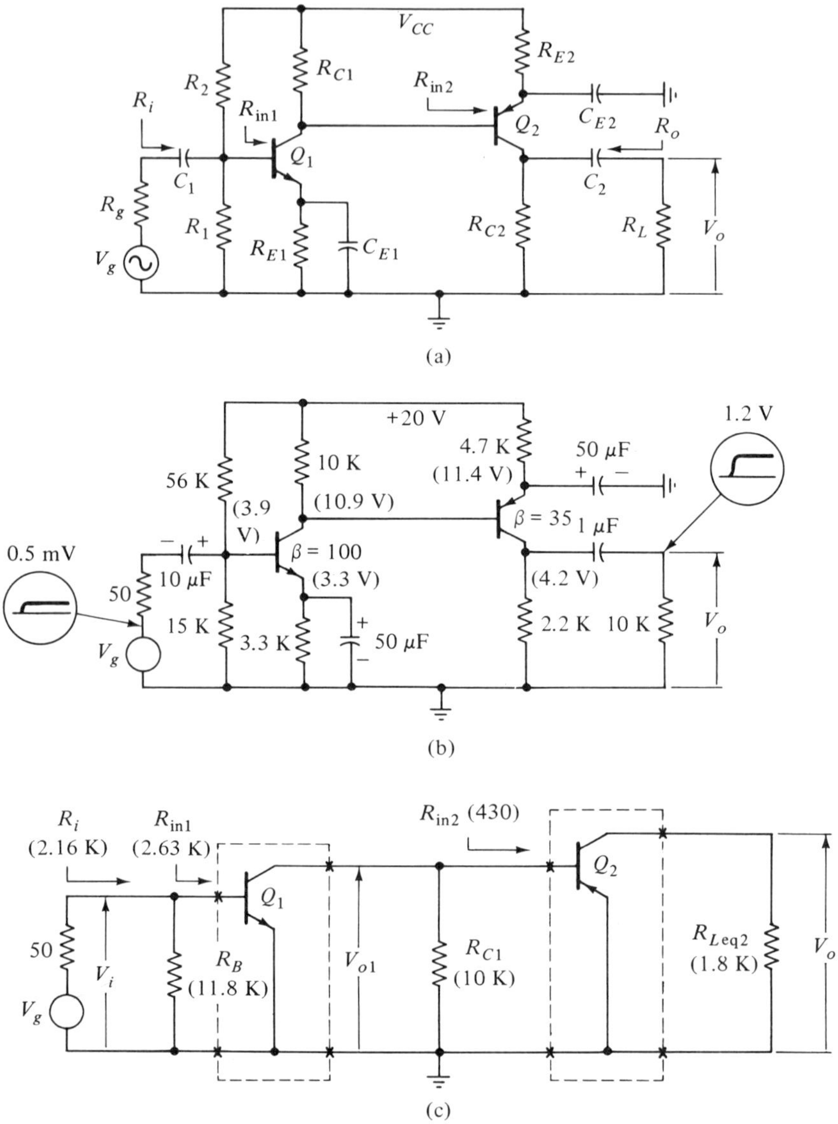

FIGURE 4–32 *A CE-CE amplifier: (a) a general circuit, (b) the circuit from Ex. 4–15 showing component values, measured dc voltage, and ac waveforms, and (c) an ac equivalent circuit for the amplifier. Measured dc voltages are given in parentheses.*

Solution: (a) Bias conditions: Using the approximate bias analysis,

$$V_{B1} = V_{CC}\left(\frac{R_1}{R_1 + R_2}\right) \simeq 4.2 \text{ V} \qquad V_{E1} = V_{B1} - V_{BE} \simeq 3.6 \text{ V}$$

$$I_{E1} = V_{E1}/R_{E1} \simeq 1.1 \text{ mA} \qquad V_{C1} \simeq V_{CC} - I_{E1} R_{C1} \simeq 9 \text{ V}$$

$$V_{E2} = V_{B2} + V_{BE2} \simeq 9.6 \text{ V} \qquad I_{E2} = (V_{CC} - V_{E2})/R_{E2} \simeq 2.2 \text{ mA}$$

$$V_{C2} = I_{C2} R_{C2} = (2.2 \text{ mA})(2.2 \text{ K}) \simeq 4.8 \text{ V}$$

(b) Gain and impedance calculations:

$$r_{e1} = 26\text{ mV}/I_{E1} \simeq 24\,\Omega \qquad R_{in1} = r_{b'1} + (\beta_1 + 1)r_{e1} \simeq 2.4\text{ K}$$

$$r_{e2} = 26\text{ mV}/I_{E2} \simeq 12\,\Omega \qquad R_{in2} = r_{b'2} + (\beta_2 + 1)r_{e2} \simeq 430\,\Omega$$

$$R_{L\text{eq}1} = R_{C1} \,||\, R_{in2} \simeq 415\,\Omega \qquad R_{L\text{eq}2} = R_{C2} \,||\, R_L \simeq 1.8\text{ K}$$

$$A_{i1} = \beta_1 \simeq 100 \qquad A_{i2} = \beta_2 \simeq 35$$

$$A_{vt1} = -\frac{A_{i1}\,R_{L\text{eq}1}}{R_{in1}} \simeq -17.3 \qquad A_{vt2} = -\frac{A_{i2}\,R_{L\text{eq}2}}{R_{in2}} \simeq -146$$

$$R_i = R_1 \,||\, R_2 \,||\, R_{in1} \simeq 2\text{ K} \qquad R_o = R_{C2} \,||\, r_{ce2} \simeq R_{C2} = 2.2\text{ K}$$

$$A_{vg} = R_i/(R_i + R_g) = 2\text{ K}/(2\text{ K} + 50) \simeq 0.98$$

$$A_{vo} = A_{vg}\,A_{vt1}\,A_{vt2} \simeq (0.98)(-17.3)(-146) \simeq 2460$$

(c) Experimental results: The measured dc voltages are shown in parentheses in Fig. 4–32(b) and compare favorably with the calculated bias values. The experimental voltage gain, using pulse amplitudes, was approximately 2400 with the amplitude of V_g equal to 0.5 mV. This gain is in excellent agreement with the calculated value. The measurement of voltage gains in excess of several hundred is difficult to accomplish without overdriving the amplifier and causing distortion. The gain of an overdriven amplifier will be less than the value calculated using small-signal analysis due to the distortion in the output signal. The generator used to test this amplifier was a precision, mercury-wetted relay pulse generator whose output can be accurately set to values less than 0.1 mV.

A CS-CB FET-bipolar amplifier is shown in Fig. 4–33. This configuration is called a *cascode* amplifier and is characterized by a stage having a high output impedance (the CS stage) driving a stage having a low input impedance (the CB stage). The first stage could also be a bipolar CE stage.

An interesting feature of the cascode amplifier is that the overall mid-band voltage gain is approximately equal to the voltage gain of the first stage assuming that it has the same equivalent load as the second stage. This feature is demonstrated as follows. From Table 3–1, the value of A_{vt} for the CS is $(-g_m R_{L\text{eq}})$. Since the second stage of a cascode amplifier is characterized by a low input impedance, the value of $R_{L\text{eq}}$ for the CS is approximately R_{in2}. Using a CB for the second stage, with α approximately unity, the overall voltage gain of the cascode amplifier is approximately

$$A_{vo} = A_{vg}A_{vt1}A_{vt2} \simeq A_{vg}\,(-g_m R_{in2})\left(\frac{\alpha R_{L\text{eq}2}}{R_{in2}}\right) \simeq A_{vg}(-g_{m1}R_{L\text{eq}2}) \quad \textbf{(4–90)}$$

Consequently, a cascode amplifier is not used solely to increase the overall voltage gain. This configuration provides excellent isolation between the input and output terminals of an amplifier in the high-frequency range and is used in many high-frequency amplifiers. The high-frequency characteristics of a cascode amplifier are treated in Chap. 8. The following example illustrates the calculation of the mid-band response of a cascode amplifier.

EXAMPLE 4–16 For the CS-CB amplifier shown in Fig. 4–33(b), calculate the values of A_{vo}, R_i, and R_o. The N-channel JFET is characterized by $I_{DSS} = 3.4$ mA, $V_P = -2.2$ V, and is biased at $V_{GS} = 0.5$ V. For the PNP silicon transistor, $\beta \simeq 130$, and $r_{b'} \simeq 50$ ohms.

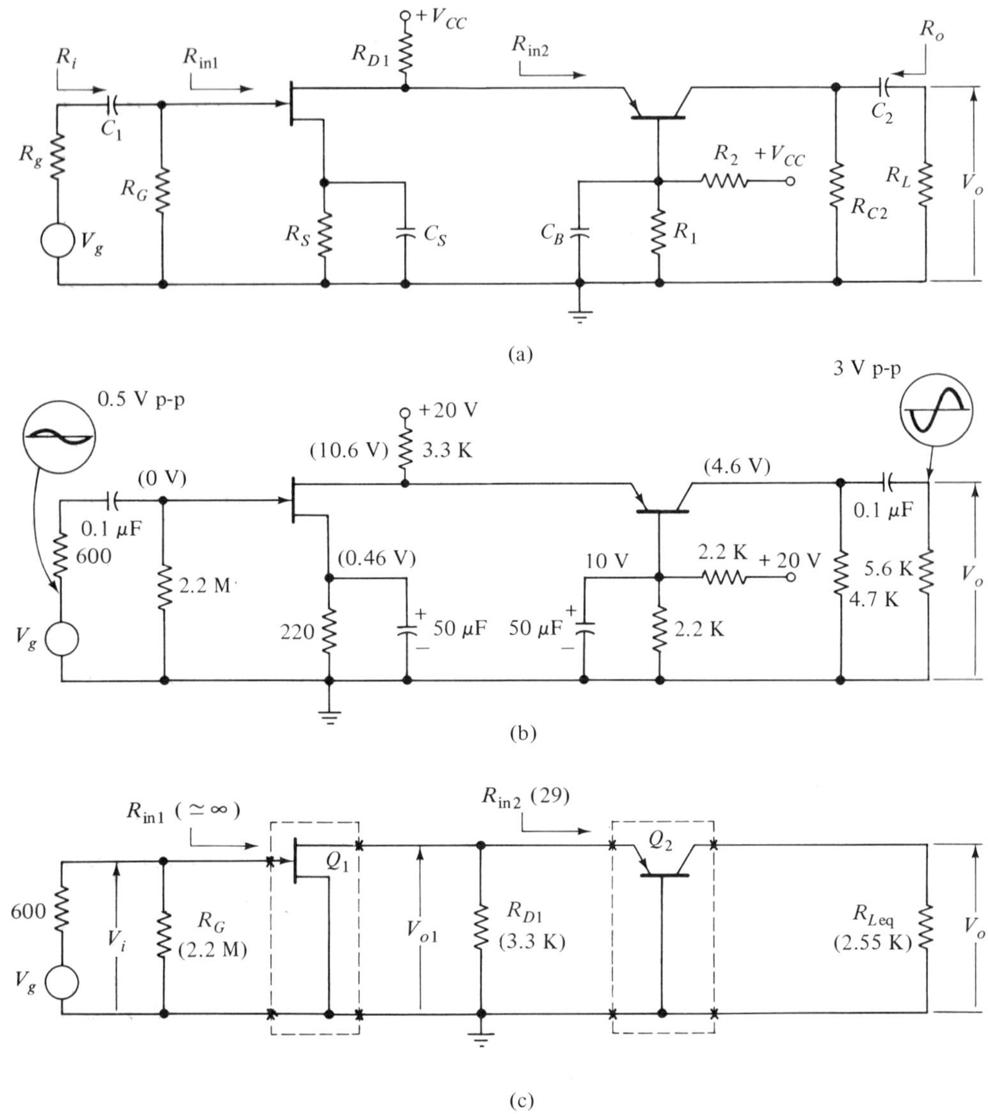

FIGURE 4–33 *A CS-CB FET-bipolar amplifier: (a) general circuit, (b) circuit from Ex. 4–16 showing component values, measured dc voltages, and ac waveforms, and (c) ac equivalent circuit for the amplifier.*

Solution: (a) Bias conditions:

$$I_{D1} = I_{DSS}\left[1 - \frac{V_{GS}}{V_P}\right]^2 \simeq 2\text{ mA} \qquad V_{S1} = I_{D1}\,R_{S1} \simeq 0.44\text{ V} \simeq -V_{GS}$$

$$V_{B2} = V_{CC}\left(\frac{R_1}{R_1 + R_2}\right) = 10\text{ V} \qquad V_{D1} = V_{E2} = V_{B2} + V_{BE2} \simeq 10.6\text{ V}$$

$$I_{E2} = (V_{CC} - V_{E2})/R_{D1} - I_{D1} \simeq (20\text{ V} - 10.6\text{ V})/3.3\text{ K} - 2\text{ mA} \simeq 0.9\text{ mA}$$

$$V_{C2} = I_{C2}\,R_{C2} \simeq I_{E2}\,R_{C2} = (0.9\text{ mA})(4.7\text{ K}) \simeq 4.2\text{ V}$$

(b) Gain and impedance calculations:

$$g_{m1} = \frac{2}{V_P}\sqrt{I_D\, I_{DSS}} \simeq 2.4 \text{ millimhos} \qquad r_{e2} = 26 \text{ mV}/I_{E2} \simeq 29\ \Omega$$

$$R_{in2} = r_{e2} + r_{b'2}/(\beta + 1) = 29 + 50/131 \simeq 29\ \Omega$$

$$R_{Leq1} = R_{D1} \,||\, R_{in2} \simeq 29\ \Omega \qquad R_{Leq2} = R_{C2} \,||\, R_L \simeq 2.55\text{ K}$$

$$A_{i2} = -\alpha = -\beta/(\beta + 1) \simeq -1 \qquad R_i = R_G \,||\, R_{in1} \simeq R_G = 2.2\text{ M}$$

$$R_o = R_{C2} \,||\, R_{o2} \simeq R_{C2} = 4.7\text{ K} \qquad A_{vt1} = -g_{m1}\, R_{Leq1} \simeq -0.07$$

$$A_{vt2} = -\frac{A_{i2}\, R_{Leq2}}{R_{in2}} \simeq 88 \qquad A_{vg} = R_i/(R_i + R_g) \simeq 1$$

$$A_{vo} = A_{vg}\, A_{vt1}\, A_{vt2} \simeq (1)(-0.07)(88) \simeq -6.1$$

Using Eq. (4–90),

$$A_{vo} \simeq A_{vg}(-g_{m1}\, R_{Leq2}) \simeq -(1)(2.4 \times 10^{-3})(2.55\text{ K}) \simeq -6.1$$

(c) Experimental results: The measured dc voltages are shown in Fig. 4–33(b) and agree closely with the calculated values. The experimental voltage gain, using peak-to-peak values, was -6 using an input voltage (V_g) of 0.5 V. This gain agrees closely with the calculated value.

A CE-CC amplifier is shown in Fig. 4–34(a). An interesting feature of this amplifier is the effect of the CC stage on the overall voltage gain of the amplifier when the load, R_L, is small. Although the CC is limited to a voltage gain of less than unity, the voltage gain of the CE stage is increased by isolating the CE from the small load. This effect is illustrated in the following example.

EXAMPLE 4–17 The CE-CC amplifier shown in Fig. 4–34(b) and (c) uses silicon transistors with $\beta_1 \simeq 35$ and $\beta_2 \simeq 22$. (a) Calculate the values of A_{vo}, R_i, and R_o. (b) Calculate the value of A_{vo} when Q_2 is removed and R_L is connected to Q_1 as shown in Fig. 4–34(d).

Solution: (a) Bias conditions: Using the approximate bias analysis, $V_{B1} \simeq 5$ V, $I_{E1} \simeq 1$ mA, $V_{B2} \simeq 14$ V, and $I_{E2} \simeq 1.3$ mA.

(b) Gain and impedance calculations for CE-CC amplifier:

$$r_{e1} = 26\text{ mV}/I_{E1} \simeq 26\ \Omega \qquad r_{e2} \simeq 20\ \Omega$$

$$R_{in1} = r_{b'1} + (\beta + 1)r_{e1} \simeq 935\ \Omega \qquad R_{Leq2} = 10\text{ K} \,||\, 100 \simeq 100\ \Omega$$

$$R_{in2} = r_{b'2} + (\beta_2 + 1)(r_e + R_{Leq2}) \simeq (22 + 1)(20 + 100) \simeq 2.75\text{ K}$$

$$R_{Leq1} = R_{C1} \,||\, R_{in2} \simeq 2.15\text{ K} \qquad R_i = R_1 \,||\, R_2 \,||\, R_{in1} \simeq 760\ \Omega$$

$$R_o \simeq R_{E2} \,||\left(r_{e2} + \frac{r_{b'2} + R_{C1} \,||\, r_{ce1}}{\beta_2 + 1}\right) \simeq 10\text{ K} \,||\left(20 + \frac{10\text{ K}}{23}\right) \simeq 455\ \Omega$$

$$A_{vt1} = -\left(\frac{A_{i1}\, R_{Leq1}}{R_{in1}}\right) \simeq -81 \qquad A_{vt2} = -\left(\frac{A_{i2}\, R_{Leq2}}{R_{in2}}\right) \simeq 0.84$$

$$A_{vg} = R_i/(R_i + R_g) \simeq 0.89 \qquad A_{vo} = A_{vg}\, A_{vt1}\, A_{vt2} \simeq -61$$

(c) Gain and impedance calculations for CE shown in Fig. 4–34(d):

$$R_{Leq1} = R_{C1} \,||\, R_L \simeq 100 \qquad A_{vt1} = -\frac{A_i\, R_{Leq}}{R_{in}} \simeq -3.8$$

$$A_{vo} = A_{vg}\, A_{vt} = (0.89)(-3.8) \simeq -3.4$$

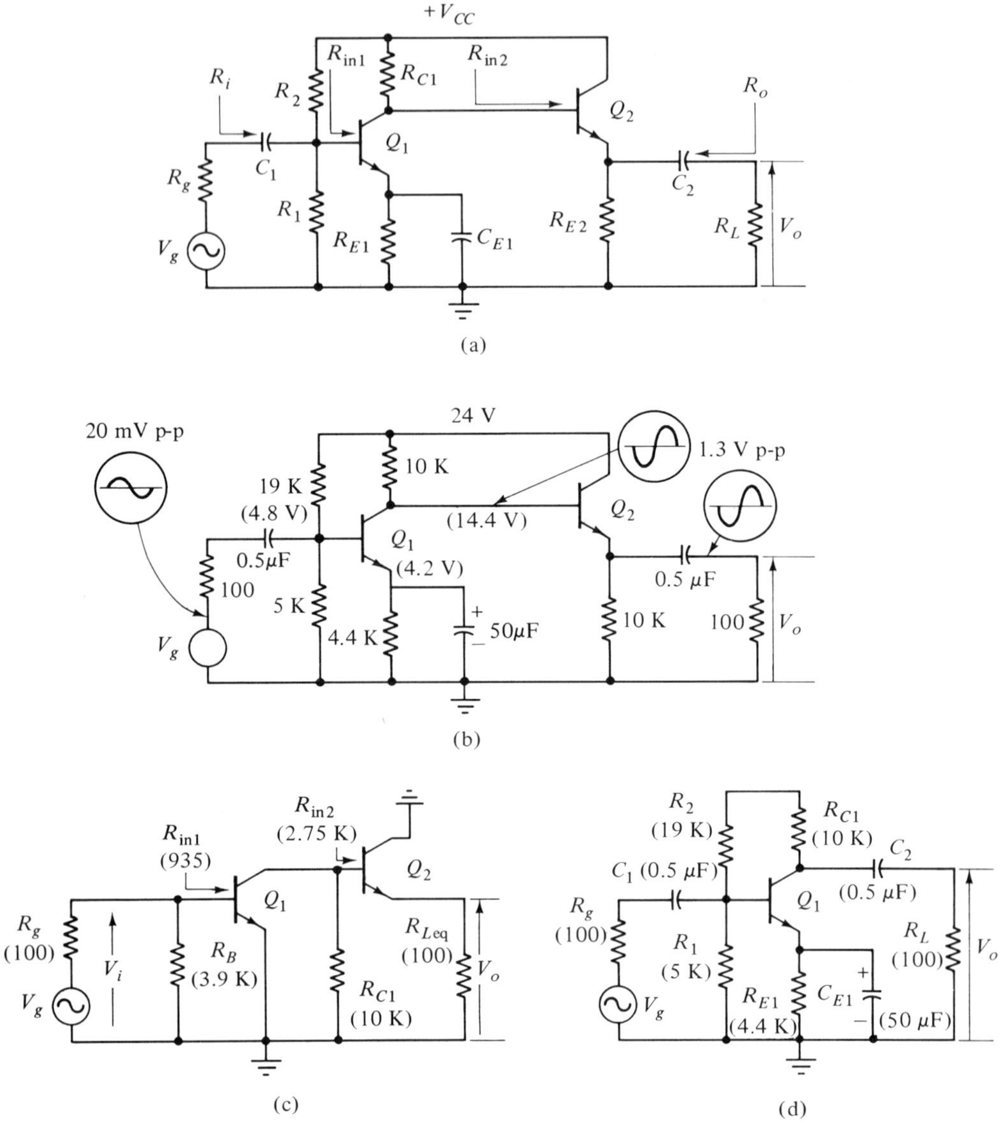

FIGURE 4–34 *A CE-CC amplifier: (a) a general circuit, (b) the circuit from Ex. 4–17 showing component values, measured dc voltages, and ac waveforms, (c) an ac equivalent circuit for the amplifier, and (d) the circuit with* R_L *connected directly to the CE stage.*

(d) Experimental results: The measured dc voltages are shown in Fig. 4–35(b). The experimental voltage gain for the CE-CC amplifier was -65 with an input voltage of 20 mV.

In this example, the CC stage increased the value of the load impedance for the CE stage (R_{Leq1}) from 100 ohms to 2150 ohms, and increased the overall voltage gain from -3.4 to -61, a factor of 18.

4.9 Differential Amplifiers

A differential amplifier using FETs was treated in Sec. 3.10. The differential amplifier is used widely in instrumentation circuits, feedback amplifiers, linear integrated circuits, and logic circuits. Because of its wide use, the differential amplifier deserves careful study.

The objective of this section is to present a detailed development of the differential amplifier. To achieve this objective, we will first analyze a bipolar transistor differential amplifier using the multistage techniques developed in the last section. Second, we will examine the basic properties of all differential amplifiers. Third, the particular characteristics of a general transistor differential amplifier will be developed and applied to two circuits. Fourth, the particular characteristics of FET differential amplifiers will be developed and applied to two circuits. Finally, the material of this section is summarized in Table 4–4.

A differential amplifier using bipolar transistors is shown in Fig. 4–35(a) and is similar to the FET differential amplifier discussed in Sec. 3.10. A constant-current source, Q_3, is used instead of an emitter resistor where the collector current of Q_3 is the sum of the emitter currents of Q_1 and Q_2.

A differential amplifier is usually symmetrical; transistors Q_1 and Q_2 are matched units, and the corresponding components associated with Q_1 and Q_2 have the same values. The differential amplifier shown in Fig. 4–35 is an example of a symmetrical circuit. Transistors Q_1 and Q_2 are usually constructed on the same substrate in very close proximity to form a dual unit. The specification sheet on a dual transistor will usually include information on the matching characteristics of the two units. For the dual transistor used in Ex. 4–18, β_1/β_2 ranges from 0.9 to 1 and $|V_{BE1} - V_{BE2}|$ is less than 3 mV. In the differential amplifier shown in Fig. 4–35, R_{12} is a variable resistor to equalize the dc currents in Q_1 and Q_2. Other methods for balancing the bias of a differential amplifier will be discussed later in this section.

The general characteristics discussed in Sec. 3–10 also apply to the bipolar form of the differential amplifier. The following example illustrates these characteristics. The approximate expressions for A_{vt}, given in Table 4–1, will be used in the calculations.

EXAMPLE 4–18 The silicon transistors used in the differential amplifier shown in Fig. 4–35 are characterized by $\beta_1 \simeq \beta_2 \simeq 450$, and $\beta_3 \simeq 100$. Calculate the bias conditions, the value of V_{o1}/V_{g1} and also V_{o2}/V_{g1} with $V_{g2} = 0$, the value of V_{o1}/V_g with $V_g = V_{g1} = V_{g2}$, and the value of R_i and R_o.

Solution: (a) Bias conditions: Assume that R_{12} is 68 K,

$$V_{B1} \simeq V_{B2} \simeq 0 \text{ V} \qquad V_{B3} \simeq -10 \text{ V}$$

$$V_{E3} \simeq V_{B3} - V_{BE3} \simeq -10.6 \text{ V} \qquad I_{C3} \simeq I_{E3} \simeq (-20 \text{ V} + 10 \text{ V})/2.2 \text{ K} \simeq 4.3 \text{ mA}$$

$$I_{E1} \simeq I_{E2} \simeq (I_{C3})/2 \simeq 2.15 \text{ mA} \qquad V_{C1} = V_{C2} \simeq 20 \text{ V} - (2.15 \text{ mA})(4.7 \text{ K}) \simeq 10 \text{ V}$$

(b) Gain V_{o1}/V_{g1} with $V_{g2} = 0$: Under these conditions, Q_1 acts as a CE stage with an unbypassed emitter resistor shown in Fig. 4–35(c).

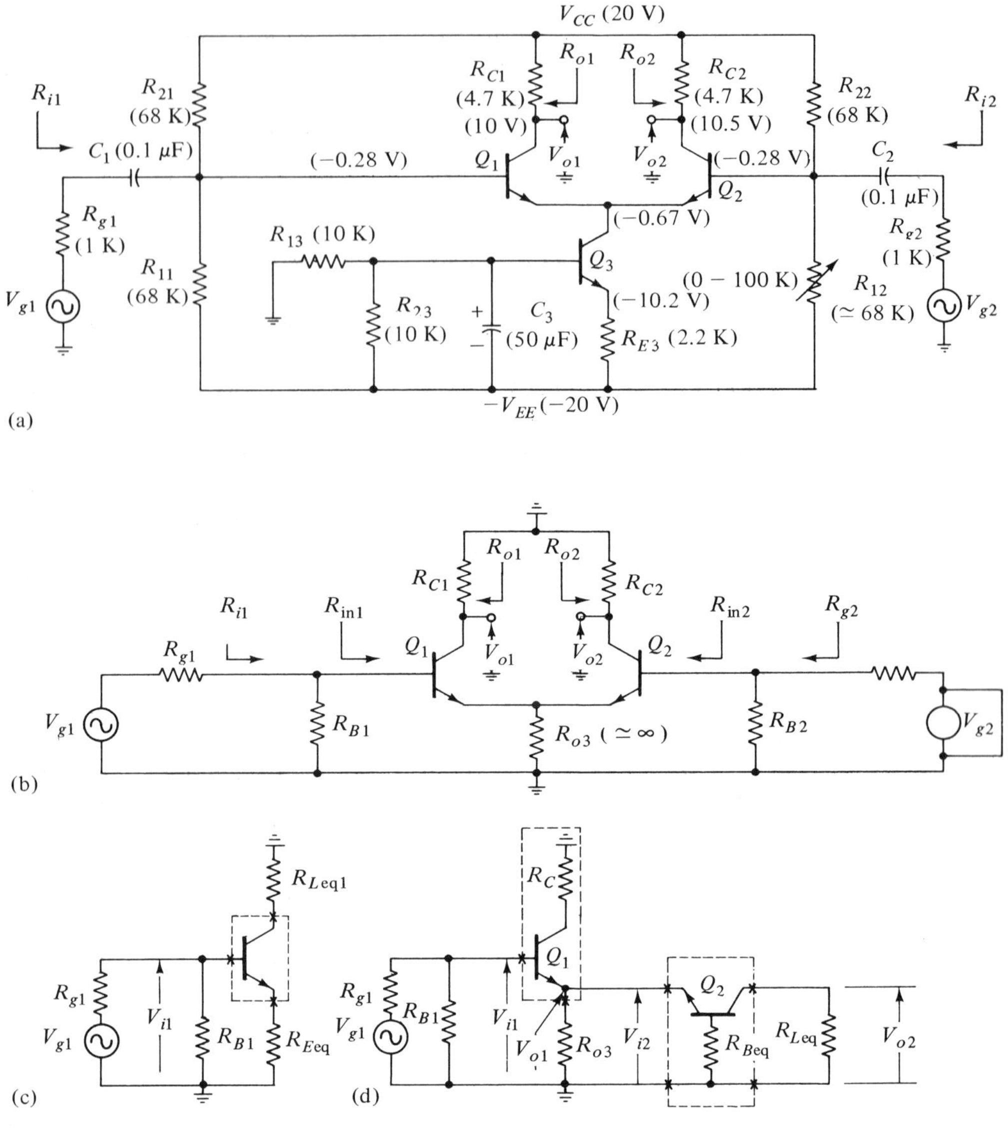

FIGURE 4–35 *A differential amplifier using bipolar transistors (a) complete circuit, (b) an ac equivalent circuit with only one input present* ($V_{g2} = 0$)*, (c) an ac equivalent circuit used to calculate* V_{o1}*, and (d) an ac equivalent circuit used to calculate* V_{o2}*. Values in parentheses were obtained from Ex. 4–18.*

$$r_{e1} \simeq r_{e2} = \frac{26\ \text{mV}}{I_{E1}} \simeq 12\ \Omega \qquad R_{ie2} = r_{e2} + \frac{R_{12} \,||\, R_{22} \,||\, R_{g2}}{\beta_2 + 1} \simeq 14\ \Omega$$

$$R_{E1} = R_{o3} \,||\, R_{ie2} \simeq 14\ \Omega \qquad R_{in1} = r_{b'1} + (\beta + 1)(r_{e1} + R_{E1}) \simeq 11.7\ \text{K}$$

$$R_{i1} = R_{11} \,||\, R_{21} \,||\, R_{in1} \simeq 8.7\ \text{K} \qquad A_{vt1(\text{CE})} \simeq -\frac{R_{Leq1}}{r_{e1} + R_{E1}} \simeq -180$$

$$A_{vg1} = R_{i1}/(R_{i1} + R_{g1}) \simeq 0.9 \qquad V_{o1}/V_{g1} = A_{vg1}\, A_{vt1(\text{CE})} \simeq -160$$

(c) Gain V_{o2}/V_{g1} with $V_{g2} = 0$: Under these conditions, Q_1 acts as a CC stage, and Q_2 acts as a CB stage.

$$A_{i1} = -(\beta + 1) = -451 \qquad A_{i2} = -\alpha = -1$$
$$R_{Leq1} = R_{ie2} \simeq 14\,\Omega \qquad R_{Leq2} = 4.7\text{ K}$$
$$R_{in1} = 11.7\text{ K} \qquad R_{ie2} = 14$$
$$A_{vt1(CC)} = \frac{R_{Leq1}}{R_{Leq1} + r_{e1}} \simeq 0.54 \qquad A_{vt2(CB)} = \frac{R_{Leq2}}{r_{e2}} \simeq \frac{4.7\text{ K}}{14} \simeq 336$$
$$A_{vg} = A_{vg1} = 0.9 \qquad V_{o2}/V_{vg1} = A_{vg1}\,A_{vt1(CC)}\,A_{vt2(CB)} \simeq 160$$

(d) Gain V_{o1}/V_g with $V_g = V_{g1} = V_{g2}$: Due to the symmetry of the circuit,

$$\frac{V_{o1}}{V_{g1}} = \frac{V_{o2}}{V_{g2}} \simeq -160 \qquad \frac{V_{o2}}{V_{g1}} = \frac{V_{o1}}{V_{g2}} \simeq 160$$

Thus, with $V_{g1} = V_{g2} = V_g$,

$$V_{o1} = V_{g1}(-160) + V_{g2}(160) = V_g(-160 + 160) = 0$$
$$V_{o2} = V_{g1}(160) + V_{g2}(-160) = V_g(160 - 160) = 0$$

and both outputs are zero when both inputs are equally excited. Also, due to symmetry,

$$R_{i1} = R_{i2} = 8.7\text{ K} \qquad R_{o1} = R_{o2} \simeq R_C = 4.7\text{ K}$$

(e) Experimental results: The measured bias voltages are shown in Fig. 4–35(a). The measured voltage gains were as follows. For $V_{g2} = 0$,

$$\frac{V_{o1}}{V_{g1}} = \frac{-3\text{ V}}{0.02\text{ V}} \simeq -150 \qquad \frac{V_{o2}}{V_{g1}} = \frac{3\text{ V}}{0.02\text{ V}} \simeq 150$$

For $V_{g1} = V_{g2} = V_g$,

$$\frac{V_{o1}}{V_g} = \frac{V_{o2}}{V_g} = \frac{0\text{ V}}{0.02\text{ V}} \simeq 0$$

The general characteristics of a balanced bipolar differential amplifier can be observed from the preceding example. First, the voltage gain from one input to either output has approximately the same magnitude, but the two output signals are 180° out of phase. Second, if the two input signals are equal and in phase, both output signals will be approximately zero. If the two input signals are unequal, the differential amplifier will amplify the differences between the two input signals (or $V_{g1} - V_{g2}$).

In general, the term "differential amplifier" does not imply a particular circuit but rather refers to a class of amplifiers which has two sets of input terminals and whose output signals depend on the difference between the two input signals. Our next goal is to derive the general equations for differential amplifiers. The block diagram of a general differential amplifier is shown in Fig. 4–36. Comparing this block diagram to the bipolar differential amplifier shown in Fig. 4–35, the input signal V_1 corresponds to the signal at the base of Q_1, the input signal V_2 corresponds to the signal at the base of Q_2, and the output signal V_o corresponds to the signal at the collector of Q_2 or V_{o2}. From Ex. 4–18 and Fig. 4–35, we know that the output signal due to V_1 is not inverted while the output signal due to V_2 is inverted. The signal inversion characteristics are indicated on the block diagram by the positive or negative sign associated with each input.

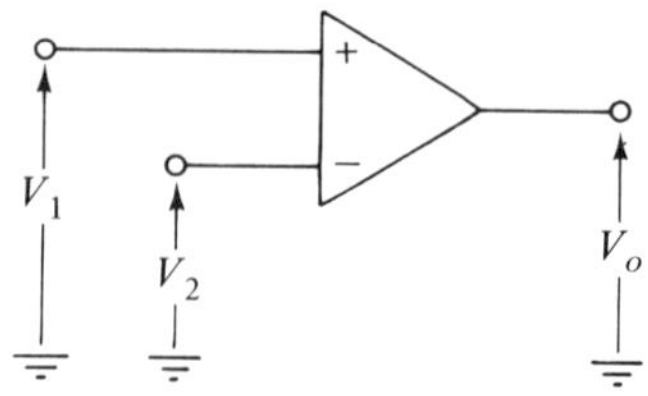

FIGURE 4–36 *A block diagram of a general differential amplifier. All signals are referenced to ground.*

The output signal of the block diagram shown in Fig. 4–36 can be written as

$$V_o = V_1A_1 + V_2A_2 \qquad \textbf{(4–91)}$$

where, for $V_2 = 0$,

$$A_1 = V_o/V_1 \qquad \textbf{(4–92)}$$

and for $V_1 = 0$,

$$A_2 = V_o/V_2 \qquad \textbf{(4–93)}$$

Note that the gain A_1 is positive since it is associated with the noninverting input, while the gain A_2 is negative since it is associated with the inverting input.

The input signal to a differential amplifier can be divided into two parts: (1) a difference-mode signal, V_d, and (2) a common-mode signal, V_c. As the name implies, the value of the difference-mode signal is equal to the difference between the two input signals or

$$V_d \equiv V_1 - V_2 \qquad \textbf{(4–94)}$$

The common-mode signal is equal to the average value of the two input signals or

$$V_c \equiv \tfrac{1}{2}(V_1 + V_2) \qquad \textbf{(4–95)}$$

By first adding and then subtracting Eqs. (4–94) and (4–95), the two input signals can be related to V_d and V_c by

$$V_1 = V_c + V_d/2 \qquad V_2 = V_c - V_d/2 \qquad \textbf{(4–96)}$$

In order to understand better the concept of difference-mode and common-mode signals, three input signal conditions are given in Table 4–3 along with the corresponding values for V_d and V_c calculated using Eqs. (4–94) and (4–95). Condition I occurs when the two input signals are equal in magnitude but are 180° out of phase. Condition II occurs when the input signal of interest (the difference signal), is part of a much larger input signal. Condition III occurs

TABLE 4–3 *Three input signal conditions for a differential amplifier resulting in the same value for the difference-mode signal but different values of the common-mode signal.*

Input Signal Condition	V_1	V_2	V_d	V_c
I	0.5 mV	−0.5 mV	1 mV	0
II	50.5 mV	49.5 mV	1 mV	50 mV
III	1 mV	0 mV	1 mV	0.5 mV

when only one input signal is present, and the other input is grounded. The value of V_d is the same for all three conditions, while the value of V_c is different for each input signal condition.

For a practical differential amplifier, the output signal will depend on the values of both the difference-mode and the common-mode input signals. Our next task is to develop the expressions for the difference-mode gain and the common-mode gain. Substituting Eq. (4–96) into (4–91), the output voltage can be written as

$$\begin{aligned} V_o &= (V_c + V_d/2)A_1 + (V_c - V_d/2)A_2 \\ &= V_c(A_1 + A_2) + (V_d/2)(A_1 - A_2) = V_cA_c + V_dA_d \end{aligned} \qquad \textbf{(4–97)}$$

where the difference-mode gain A_d is evaluated at $V_c = 0$ and is equal to

$$A_d \equiv \frac{V_o}{V_d} = \frac{1}{2}(A_1 - A_2) \qquad \textbf{(4–98)}$$

The common-mode gain A_c is evaluated at $V_d = 0$ and is equal to

$$A_c \equiv \frac{V_o}{V_c} = A_1 + A_2 \qquad \textbf{(4–99)}$$

A measure of the ability of a differential amplifier to distinguish between a difference-mode signal and a common-mode signal is the *common-mode rejection ratio* (CMRR) which is defined as

$$\text{CMRR} \equiv |A_d|/|A_c| \qquad \textbf{(4–100)}$$

If Eqs. (4–98) and (4–99) are substituted into Eq. (4–100), the CMRR can be interpreted as the ratio of the common-mode input signal to the difference-mode input signal that produces the same output signal.

The output signal can be expressed in terms of the difference-mode and common-mode parameters as

$$V_o = V_dA_d + V_cA_c = V_dA_d\left(1 + \frac{V_cA_c}{V_dA_d}\right) = V_dA_d\left(1 + \frac{V_c/V_d}{\text{CMRR}}\right) \qquad \textbf{(4–101)}$$

Ideally, the value of the CMRR should be infinite which requires that the common-mode gain, A_c, be zero. The value of A_c can be zero when the differential amplifier is perfectly symmetrical, the two active devices are perfectly matched, and the impedance of the current source (the impedance looking into the collector of Q_3 in Fig. 4–35) is infinite. Except for simple circuits, accurate calculation of the CMRR is very difficult, and its value should usually be determined experimentally.

Measurement of the CMRR requires a measurement of A_d and A_c. An experimental method for measuring A_d and A_c is shown in Fig. 4–37. If equal but opposite signals having a magnitude $V_d/2$ are applied to the two inputs of the differential amplifiers, the output signal is equal to A_dV_d as shown in Fig. 4–37(a). An approximate method for determining A_d, which requires only one input signal, is shown in Fig. 4–37(b). For small input signals and a large value of CMRR, the effect of the common-mode signal (equal to $V_d/2$) can be neglected when the input signal is applied to only one input with the other input

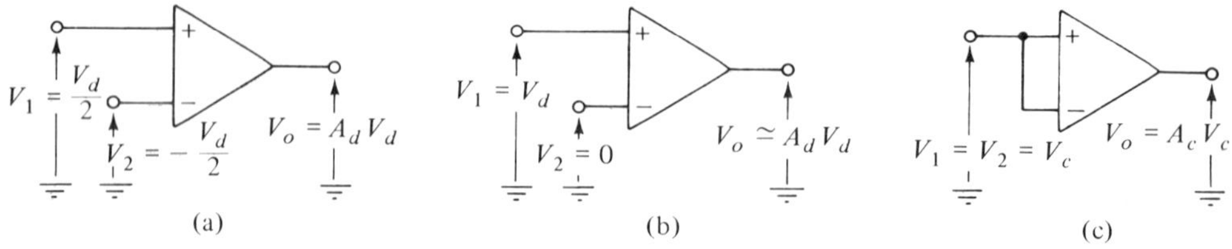

FIGURE 4–37 *An experimental method for measuring A_d and A_c to determine the value of CMRR: (a) an accurate method of measuring A_d, (b) an approximate method of measuring A_d valid for small-input signal levels, and (c) a method of measuring A_c. The value of the CMRR is given by the ratio $|A_d/A_c|$. The value of V_o should be the same for all measurements.*

grounded. The output signal is approximately A_dV_d for this condition. The value of A_c can be determined by connecting the two inputs together and applying an input signal of sufficient magnitude to produce a measurable output signal. The value of the CMRR is the ratio $|A_d/A_c|$.

Our next task is to develop the general gain expressions for the bipolar differential amplifier shown in Fig. 4–38. In this development, assume that the circuit is symmetrical and that the transistors are matched. Then,

$$\left.\begin{aligned} R_{C1} &= R_{C2} = R_C \qquad & R_{i1} &= R_{i2} = R_i \\ R_{g1} &= R_{g2} = R_g \qquad & R_{e1} &= R_{e2} = R_e \end{aligned}\right\} \tag{4–102}$$

where, for $r_{b'} = 0$,

$$R_{i1} = (\beta_1 + 1)(r_{e1} + R_E \,||\, R_{e2}) = (\beta + 1)(r_e + R_E \,||\, R_e) = R_i \tag{4–103}$$

$$R_{e1} = r_{e1} + [R_{g1}/(\beta_1 + 1)] = r_e + R_g/(\beta + 1) = R_e \tag{4–104}$$

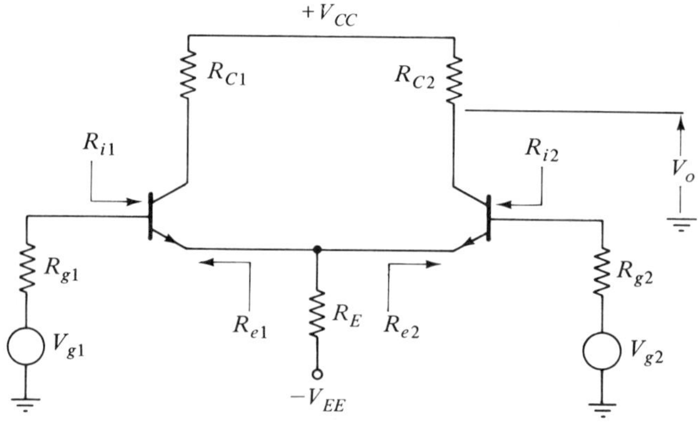

FIGURE 4–38 *A bipolar differential amplifier.*

A determination of the expressions for A_d, A_c, and CMRR requires a knowledge of the gain from each input to the output, A_1 and A_2. From Eq. (4–92) and Fig. 4–38, A_1 with $V_{g2} = 0$ is

$$A_1 \equiv \frac{V_{o2}}{V_{g1}} = A_{vg1}A_{vt1(CC)}A_{vt2(CB)} \simeq \left(\frac{R_{i1}}{R_{i1} + R_{g1}}\right)\left(\frac{R_E \,||\, R_{e2}}{R_E \,||\, R_{e2} + r_{e1}}\right)\left(\frac{R_C}{R_e}\right) \tag{4–105}$$

Using Eqs. (4–102) through (4–104), Eq. (4–105) can be simplified to

$$A_1 \simeq \frac{(R_E \parallel R_e)(R_C)}{(R_e + R_E \parallel R_e)(R_e)} \tag{4–106}$$

From Eq. (4–93) and Fig. 4–38, A_2 with $V_{g1} = 0$ is

$$A_2 \equiv \frac{V_{o2}}{V_{g2}} = A_{vg2}A_{vt2(\text{CE})} \simeq \left(\frac{R_{i2}}{R_{i2} + R_{g2}}\right)\left(\frac{-R_{C2}}{r_{e2} + R_E \parallel R_{e1}}\right) \tag{4–107}$$

Using Eqs. (4–102), (4–103), and (4–104), Eq. (4–107) can be simplified to

$$A_2 \simeq -\frac{R_C}{R_e + R_E \parallel R_e} \tag{4–108}$$

With the expressions for A_1 and A_2 available, equations for the difference-mode gain, A_d, the common-mode gain, A_c, and the CMRR can be developed. Substituting Eqs. (4–106) and (4–108) into (4–98) and simplifying,

$$A_d = \frac{1}{2}(A_1 - A_2) \simeq \frac{1}{2}\left(\frac{R_C}{R_e + R_E \parallel R_e}\right)\left(\frac{2R_E + R_e}{R_E + R_e}\right) \tag{4–109}$$

Generally, R_E is much greater than R_e, and Eq. (4–109) can be simplified to*

$$A_d \simeq \frac{R_C}{2R_e} = \frac{R_{L\text{eq}}}{2[r_e + R_g/(\beta + 1)]} \tag{4–110}$$

Substituting Eqs. (4–106) and (4–108) into (4–99) and simplifying,

$$A_c = A_1 + A_2 \simeq -\left(\frac{R_C}{R_e + R_E \parallel R_e}\right)\left(\frac{R_e}{R_E + R_e}\right) \tag{4–111}$$

If we again assume R_E is much greater than R_e, Eq. (4–111) can be simplified to

$$A_c \simeq -R_C/2R_E = -R_{L\text{eq}}/2R_E \tag{4–112}$$

Substituting Eqs. (4–111) and (4–112) into (4–100),

$$\text{CMRR} = \left|\frac{A_d}{A_c}\right| \simeq \frac{R_E}{r_e + [R_g/(\beta + 1)]} \tag{4–113}$$

Equations (4–111) and (4–113) indicate that A_c becomes very small, and the CMRR becomes very large as the value of R_E becomes very large. Consequently, a constant-current source (like Q_3 in Fig. 4–35) is normally used in place of an emitter resistor (R_E in Fig. 4–38) in practical differential amplifiers. The following example illustrates the analysis of a bipolar differential amplifier using the general equations developed above.

EXAMPLE 4–19 Calculate V_o for the circuit shown in Fig. 4–39 when $V_{g1} = 20$ mV and $V_{g2} = 0$ V. The transistors are characterized by $\beta_1 \simeq \beta_2 \simeq 450$. This circuit is identical to the differential amplifier shown in Fig. 4–35 except that the constant current source has been replaced by a resistor, R_E. The value of R_E is adjusted so that the sum of I_{E1} and I_{E2} is the same as in Ex. 4–18 or 4.3 mA.

---

* Note that the expression for A_d includes A_{vg} as indicated in Eqs. (4–105) and (4–107). Neglecting the A_{vg} effect, the magnitude of the gain from either base terminal to either collector terminal is

$$A_d \simeq \frac{R_{L\text{eq}}}{2r_e + R_g/(\beta + 1)}$$

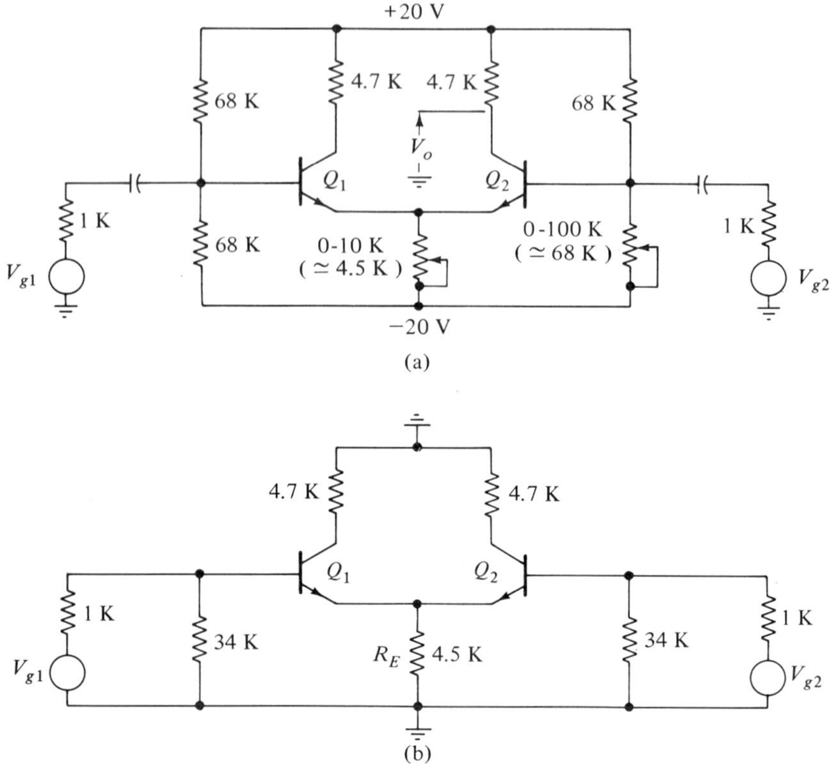

FIGURE 4–39 *The differential amplifier from Ex. 4–19: (a) the complete circuit, and (b) an ac equivalent circuit.*

Solution: (a) Bias values:

$$V_{B1} \simeq V_{B2} \simeq 0 \text{ V} \qquad I_{E1} \simeq I_{E2} \simeq \left(\frac{1}{2}\right)\left(\frac{20 \text{ V} - 0.6 \text{ V}}{4.5 \text{ K}}\right) \simeq 2.15 \text{ mA}$$

$$V_{C1} \simeq V_{C2} = 20 \text{ V} - (2.15 \text{ mA})(4.7 \text{ K}) \simeq 10 \text{ V}$$

(b) Gain values: The ac equivalent circuit shown in Fig. 4–39(b) is identical to the general differential amplifier shown in Fig. 4–30 except for the 34-K equivalent base bias resistors. However, very little error is introduced by neglecting the 34-K bias resistors, since their values are large compared to the 1-K generator resistors.* The gains are calculated as follows:

$$r_{e1} \simeq r_{e2} \simeq 26 \text{ mV}/2.15 \text{ mA} \simeq 12\ \Omega$$

$$A_d \simeq \frac{R_{Leq}}{2[r_e + R_g/(\beta + 1)]} \simeq \frac{4.7 \text{ K}}{2[12 + 1 \text{ K}/(451)]} \simeq 166$$

$$A_c \simeq -\frac{R_{Leq}}{2R_E} \simeq \frac{4.7 \text{ K}}{2(4.5 \text{ K})} \simeq -0.52$$

---

* A more accurate solution can be obtained by forming a Thévenin equivalent for the base circuit where

$$R_{Th} = 34 \text{ K} \,||\, 1 \text{ K}$$

and

$$V_{Th} = V_g\left(\frac{34 \text{ K}}{1 \text{ K} + 34 \text{ K}}\right)$$

$$\text{CMRR} \simeq \frac{R_E}{r_e + r_g/(\beta + 1)} \simeq \frac{4.5\text{ K}}{12 + 1\text{ K}/(451)} \simeq 316$$

(c) Output voltage: To calculate V_o, first find the values of the difference-mode signal, V_d, and the common-mode signal, V_c.

$$V_d = V_{g1} - V_{g2} = 20\text{ mV} - 0\text{ V} = 20\text{ mV}$$

$$V_c = \tfrac{1}{2}(V_{g1} + V_{g2}) = \tfrac{1}{2}(20\text{ mV} + 0\text{ V}) = 10\text{ mV}$$

$$V_o = V_d A_d \left(1 + \frac{V_c/V_d}{\text{CMRR}}\right) \simeq 3.3\text{ V}$$

(d) Experimental results: The measured bias levels are the same as shown in Fig. 4–35 since the value of R_E was adjusted to give these bias conditions. The measured difference-mode gain was 150 using the approximate method shown in Fig. 4–37(b) (with V_{g2} equal to zero) and an input voltage of 20 mV.

The common-mode gain was measured by connecting the two inputs together as shown in Fig. 4–37(c) resulting in $V_{g1} = V_{g2} = V_g$. With 5.5 V in, the output was -3.0 V, or a value of A_c of -0.55. From the measured values, the CMRR is $|150/(-0.55)|$ or 273. The difference between the calculated and measured values for A_c and CMRR is due primarily to asymmetry in the circuit, such as resistor tolerances and mismatch between the two transistors.

In the preceding example, the CMRR was limited primarily by the finite value of R_E. By replacing R_E with a constant-current source, as in Ex. 4–18 and Fig. 4–35, the equivalent value for R_E corresponds to the resistance looking into the collector of a CB stage, typically 1 to 10 megohms. When a constant-current source is used, the CMRR is generally limited by circuit asymmetry and transistor mismatch. For this case, a value for the CMRR should be determined experimentally. The value of the CMRR for the differential amplifier discussed in Ex. 4–18 was measured yielding the following results. For $V_{g2} = 0$,

$$A_d \simeq \frac{V_o}{V_{g1}} \simeq \frac{800\text{ mV}}{5\text{ mV}} \simeq 160$$

For $V_{g1} = V_{g2} = V_g$,

$$A_c = \frac{V_o}{V_g} \simeq \frac{800\text{ mV}}{4.1\text{ V}} \simeq 0.195 \qquad \text{CMRR} = \left|\frac{A_d}{A_c}\right| \simeq \frac{160}{0.195} \simeq 820$$

Values of CMRR ranging above 10^3 are possible in well matched differential amplifiers. When measuring A_c the large input signal should not forward bias the base-collector junction of any transistors in the circuit.

An FET differential amplifier is shown in Fig. 4–40. Equations for this circuit can be developed by following an analysis similar to that for the bipolar differential amplifier and by assuming a symmetrical circuit with matched transistors. These equations are:

$$\left.\begin{aligned} A_d = \frac{g_m R_D}{2} \qquad A_c &= -\frac{R_D}{2R_s + 1/g_m} \simeq -\frac{R_D}{2R_S} \\ \text{CMRR} &= \frac{2g_m R_s + 1}{2} \simeq g_m R_S \end{aligned}\right\} \qquad \textbf{(4–114)}$$

where, in the approximate expressions for A_c and CMRR, R_S is assumed to be much greater than $1/g_m$. As in the case of the bipolar differential amplifier, replacing R_S with a constant-current source reduces the value of A_c and in-

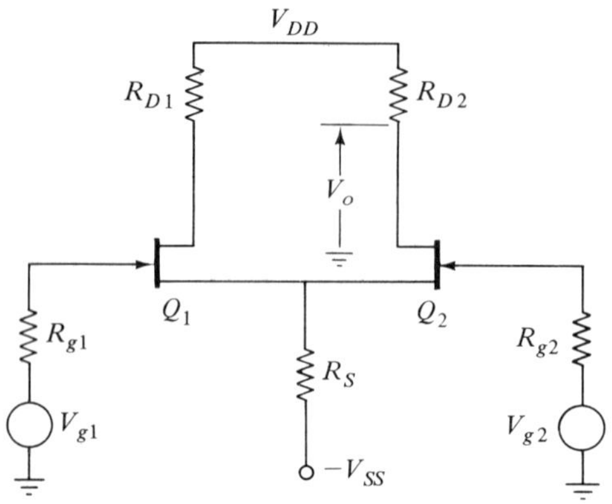

FIGURE 4–40 *An FET differential amplifier.*

creases the value of the CMRR. The following example illustrates the analysis of an FET differential amplifier.

EXAMPLE 4–20 Repeat Ex. 3–12 using Eq. (4–114).

Solution: (a) Gain values: From Ex. 3–12,

$$g_{m1} = 3.5 \text{ millimhos} \qquad R_D = 2.7 \text{ K}$$
$$g_{m2} = 3.7 \text{ millimhos} \qquad R_S = 4.7 \text{ K}$$

Using the average value of g_m,

$$A_d \simeq \tfrac{1}{2}(g_m R_D) = \tfrac{1}{2}(3.6 \times 10^{-3})(2.7 \text{ K}) \simeq 4.86$$

$$A_c \simeq -\frac{R_D}{2R_s} = -\frac{2.7 \text{ K}}{2(4.7 \text{ K})} \simeq -0.29$$

$$\text{CMRR} \simeq g_m R_s \simeq (3.6 \times 10^{-3})(4.7 \text{ K}) \simeq 16.9$$

(b) Experiment values: For $V_{g2} = 0$,

$$A_d \simeq V_{o2}/V_{g1} = 1 \text{ V}/0.2 \text{ V} \simeq 5$$

For $V_{g1} = V_{g2} = V_g$,

$$A_c = V_{o2}/V_g \simeq 1 \text{ V}/4 \text{ V} \simeq 0.25 \qquad \text{CMRR} = |A_d|/|A_c| = 5/0.25 \simeq 20$$

An FET differential amplifier with a bipolar transistor constant-current source is shown in Fig. 4–41. Except for the current source, Q_3, this amplifier is identical to the one in Exs. 3–12 and 4–20. The detailed analysis of this amplifier is left as an exercise for the reader. Using the same JFETs that were used in Exs. 3–12 and 4–20, the measured values of gain were as follows. For $V_{g2} = 0$,

$$V_{o1}/V_{g1} \simeq -5 \text{ V}/1 \text{ V} \simeq -5 \qquad V_{o2}/V_{g1} \simeq 5 \text{ V}/1 \text{ V} = 5 \qquad \textbf{(4–115)}$$

For $V_{g1} = V_{g2} = V_g$,

$$V_{o1}/V_g \simeq V_{o2}/V_g \simeq 0 \text{ V}/1 \text{ V} \simeq 0 \qquad \textbf{(4–116)}$$

The general characteristics of the bipolar and FET differential amplifiers studied in this section are summarized in Table 4–4. The effects of a potentiometer, R_{EE}, connected between the two emitter (or source) terminals has also been included. This potentiometer usually has a low value of resistance (10 ohms

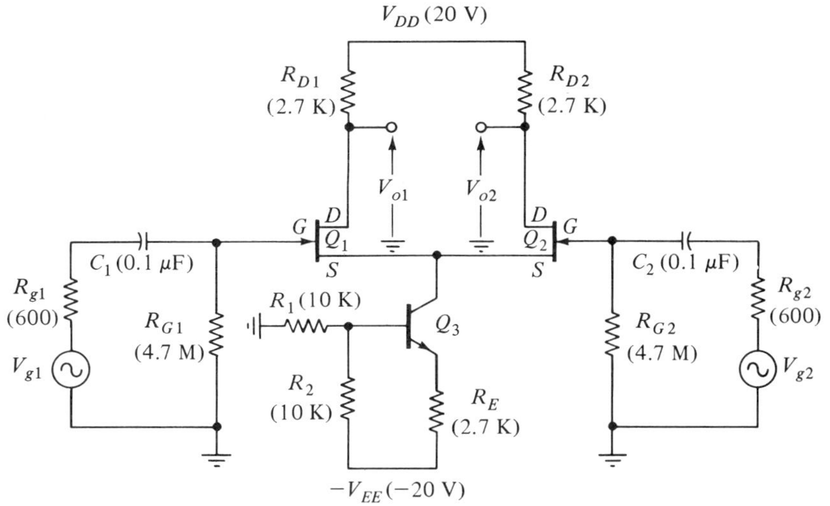

FIGURE 4–41 *A differential amplifier using FETs and a bipolar transistor constant-current source.*

to 1 K) and is used to balance the two emitter currents or the values of V_{BE}. The effects of R_{EE} (or R_{SS}) on the gain expressions can be easily determined by an analysis similar to the one developed in this section.

Table 4–4 also summarizes the signal inversions that occur in a differential amplifier. For example, if only V_{g1} is present and positive, and V_{g2} is zero, V_d is equal to V_{g1}. The output signal V_{o1} is negative or inverted relative to V_{g1}, and the output signal V_{o2} is positive (or not inverted) relative to V_{g1}. Also, if only V_{g2} is present and positive, and V_{g1} is zero, V_d is equal to $-V_{g2}$. Then the output signal V_{o1} is positive (or not inverted) relative to V_{g2}, and the output signal V_{o2} is negative (or inverted) relative to V_{g2}. When both input signals are present, the output signal magnitude and polarity can be determined using the equations given in Table 4–4.

In addition to using a constant-current bias, a well designed differential amplifier requires a symmetrical circuit and matched transistors. For bipolar transistors, the parameters that are generally matched are β and V_{BE}, while for the FET, the matched parameters are generally g_m and V_{GS}.

4.10 Summary

In this chapter, the physical operation, bias requirements, characteristics, and some applications of the bipolar transistor were discussed. Both the bipolar transistor and the FET, which was treated in Chap. 3, are important active devices that are widely used in electronic circuits. Comparing corresponding configurations of the bipolar and the FET, the bipolar transistor has a lower input resistance and a higher voltage gain. However, each circuit and active device has both advantages and disadvantages. The knowledgeable designer will choose the device and circuit which best meets the requirements of a particular application.

The physical operation of a bipolar transistor depends on the interaction between two P-N junctions. Also, the bipolar transistor is a current-controlled

TABLE 4–4 *A summary of the approximate characteristics of bipolar and FET differential amplifiers.*

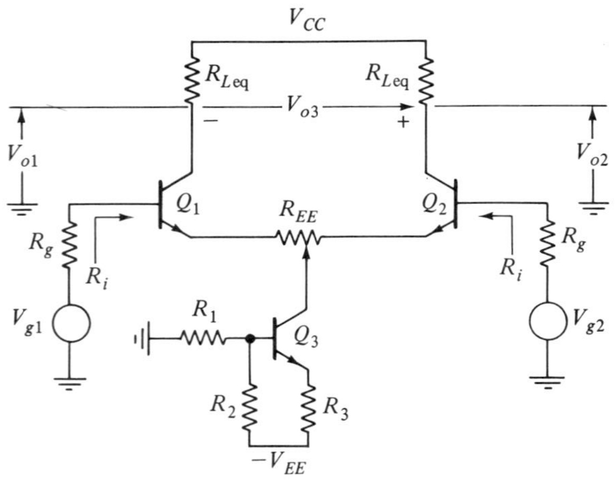

$$R_i = (\beta + 1)(2r_e + R_{EE}) + R_g$$

$$A_d \simeq \frac{R_{Leq}}{2[r_e + R_g/(\beta + 1)] + R_{EE}}$$

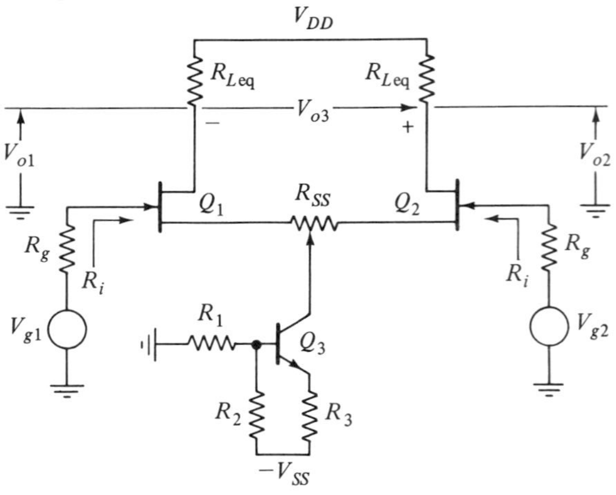

$$R_i \simeq \infty$$

$$A_d \simeq \frac{g_m R_{Leq}}{2 + g_m R_{SS}}$$

Common characteristics

$$A_c \simeq 0$$

$$\text{CMRR} \simeq \infty$$

$$V_d = V_{g1} - V_{g2}$$

$$V_c = \tfrac{1}{2}(V_{g1} + V_{g2})$$

$$\frac{V_{o1}}{V_d} = -A_d$$

$$\frac{V_{o2}}{V_d} = A_d$$

$$\frac{V_{o3}}{V_d} = 2A_d$$

device. Two important parameters of the bipolar transistor are the current gains, α and β. Both the dc and ac values of β are important.

Since a bipolar transistor consists of two P-N junctions, the bias of these two junctions is important to the operation of the device. The bias conditions are established by resistor networks external to the transistor. For linear applications, the *EB* junction is forward biased, and the *BC* junction is reverse biased.

When biased in the linear region for small-signal operation, the bipolar transistor can be modeled using either the hybrid or the hybrid-π circuit. The hybrid equivalent circuit is useful in mid-band and low-frequency analyses, and the hybrid-π equivalent circuit is useful in high-frequency and transient response analyses. Both equivalent circuits will be used in later chapters.

The bipolar transistor can be connected in six different configurations, three of which are useful as linear amplifiers: the CE, CB, and CC. Each of those configurations was analyzed in this chapter, and the important characteristics were summarized in Table 4–1. Most practical amplifiers use more than a single stage to achieve the required characteristics. Several multistage amplifiers were analyzed in Sec. 4.8, and the differential amplifier was analyzed in Sec. 4.9. The characteristics of both bipolar and FET differential amplifiers are summarized in Table 4–4. The analysis of multistage amplifiers is a simple extension of the analysis of single-stage amplifiers. The overall voltage gain of a multistage amplifier (A_{vo}) is calculated from the product of the terminal-to-terminal voltage gains (A_{vt}) of each stage and the generator-to-input voltage gain (A_{vg}). The input resistance of a multistage amplifier (R_i) depends on the first or input stage, and the output resistance of a multistage amplifier (R_o) depends on the last or output stage.

In later chapters, the bipolar transistor will be used in integrated circuits, feedback amplifiers and power amplifiers. Also, the frequency response, transient response, and thermal characteristics of bipolar transistors will be treated.

Problems

Section 4.1

4–1 A silicon NPN transistor has an emitter resistivity of 0.02 ohm-cm, a base resistivity of 2 ohm-cm, and a collector resistivity of 0.1 ohm-cm. Find the equilibrium values for the electron and hole densities in the emitter, base, and collector regions.

4–2 The transistor in Prob. 4–1 is a uniform bar having a circular cross section that is 0.1 cm in diameter. If the net base width, X_B, between depletion regions is 5 microns, and $L_n = 10^{-2}$ cm, estimate the dc collector current when V_{EE} is 0.5 V, and $V_{CC} = 10$ V. Assume $D_n = 35$ cm²/sec, and $D_p = 12$ cm²/sec.

4–3 Find the total emitter current for the transistor in Prob. 4–2. Assume that $L_p = 10^{-3}$ cm.

4–4 Find the total base current for the transistor in Prob. 4–2 if I_{CO} is 10 nA, and τ_B is 2 μsec.

Section 4.2

4–5 Find the values of β, β_{DC}, α, and α_{DC} for the PNP transistor whose curves are given in Fig. 4–7(b) if $V_{CE} = -5$ V, and $I_C = -2$ mA.

4–6 An NPN transistor with $\beta = 50$ is biased so that I_E equals 2 mA. Estimate the values of I_C and I_B.

4–7 A PNP silicon transistor has values of $I_{CO} = 5$ nA, $\beta = 100$, and is biased at $V_{BE} = 0.6$ V at 25°C. Estimate the values of I_{CO}, β, and V_{BE} at: (a) 50°C, (b) 100°C, and (c) 150°C.

Section 4.3

4–8 The CB amplifier shown in Fig. 4–11(b) uses a silicon transistor with $\beta = 150$. The component values are: $R_1 = 4.7$ K, $R_2 = 15$ K, $R_E = 2.2$ K, $R_C = 3.3$ K, and $V_{CC} = 20$ V. Calculate all bias voltages and currents.

4–9 A CC amplifier similar to the one shown in Fig. 4–11(c) uses a silicon PNP transistor with $\beta = 30$. The component values are: $R_1 = 100$ K, $R_2 = 100$ K, $R_E = 1$ K, and $V_{CC} = -20$ V. Calculate all bias voltages and currents.

4–10 An NPN silicon transistor with $\beta = 165$ is connected in a circuit of the type shown in Fig. 4–13(c) where $R_B = 2.2$ M, $R_E = 1.5$ K, and $V_{CC} = -24$ V. Calculate all bias voltages and currents.

4–11 Repeat Ex. 4–9 using a silicon transistor. Assume that all values are the same as those for the germanium transistor except that I_{CO} increases from 5 nA at 30°C to 40 nA at 60°C. Compare the results with those obtained in Ex. 4–9.

Section 4.4

4–12 A silicon NPN transistor is biased at $I_C = 4$ mA and $V_{CE} = 6$ V, and has the following component values: $\beta = 150$, $r_{b'} = 10$, and $r_{ce} = 50$ K. Calculate the remaining components of the mid-band hybrid-π equivalent circuit.

4–13 A silicon PNP transistor connected in the CE configuration is biased at $I_E = -0.5$ mA and $V_{CE} = -10$ V and has the following hybrid parameters: $h_{ie} = 4$ K (at $I_E = 1$ mA), $h_{re} = 5 \times 10^{-5}$, $h_{fe} = 150$, and $h_{oe} = 5 \times 10^{-5}$ mhos. (a) Sketch the hybrid equivalent circuit, and show the value of each component. (b) Sketch the mid-band hybrid-π equivalent circuit, and show the value of each component.

4–14 With the emitter terminal grounded, the resistance looking into the base terminal of a transistor is $h_{ie} = r_{b'} + r_{b'e}$. Write an expression for the resistance looking into the emitter terminal when the base terminal is grounded. (This expression corresponds to h_{ib}.)

Section 4.5

4–15 A silicon PNP transistor is used in a CE amplifier similar to the one shown in Fig. 4–21(a). The circuit values are: $R_g = 1$ K, $R_1 = 4.7$ K, $R_2 = 10$ K, $R_E = 2.2$ K, $R_C = R_L = 3.3$ K, $C_1 = C_2 = 0.1\ \mu$F, $C_E = 30\ \mu$F, and $V_{CC} = -15$ V. The transistor has a $\beta = 100$. (a) Sketch the circuit, (b) calculate the bias values, and (c) calculate the values of A_{vo}, R_i, and R_o.

4–16 Design a single-stage CE amplifier with the following characteristics: $|A_{vo}| \geq 150$, $R_i \geq 2$ K, and $R_o \geq 3$ K. The amplifier will be used in an application where $R_L = 10$ K and $R_g = 750$. Assume that the transistor has a β of 100 and that the available power supply voltage is $+24$ V. Use only standard components.

4–17 Repeat Ex. 4–12 with the emitter resistor bypass capacitor, $C_E = 50\ \mu F$, disconnected from the circuit. What effect does an unbypassed emitter resistor have on the values of A_{vt}, A_{vg}, R_i, and R_o?

Section 4.6

4–18 A silicon NPN transistor is used in a CB amplifier similar to the one shown in Fig. 4–26(a). The circuit values are: $R_g = 100$, $R_1 = 2.2$ K, $R_2 = 10$ K, $R_E = 1$ K, $R_C = 3.3$ K, $R_L = 5.6$ K, $C_1 = 10\ \mu F$, $C_2 = C_B = 0.1\ \mu F$, and $V_{CC} = 12$ V. The transistor has a $\beta = 175$. (a) Sketch the circuit, (b) calculate the bias values, and (c) calculate the values of A_{vo}, R_i, and R_o.

4–19 Design a single-stage CB amplifier with the following characteristics: $|A_{vo}| \geq 50$, $R_i \leq 25$ ohms, and $R_o \geq 3$ K. The amplifier will be used in an application where $R_g = 50$ ohms and $R_L = 10$ K. Assume that the transistor has a β of 200 and that the available power supply voltage is +15 V. Use only standard components.

4–20 Repeat Ex. 4–13 with the base bypass capacitor, $C_B = 250\ \mu F$, disconnected from the circuit. What effect does an unbypassed base resistor have on the values of A_{vt}, A_{vg}, R_i, and R_o?

Section 4.7

4–21 A silicon PNP transistor is used in a CC amplifier similar to the one shown in Fig. 4–29(a). The circuit values are: $R_g = 10$ K, $R_1 = 15$ K, $R_2 = 47$ K, $R_E = 2.2$ K, $R_L = 220$, $C_1 = 0.1\ \mu F$, $C_2 = 2.2\ \mu F$, and $V_{CC} = -18$ V. The transistor has a $\beta = 350$. (a) Sketch the circuit, (b) calculate the bias values, and (c) calculate the values of A_{vo}, R_i, and R_o.

4–22 Using the amplifier given in Ex. 4–14, calculate the value of A_{vo} when R_L is equal to (a) 50 ohms, (b) 500 ohms, and (c) 5 K. What effect does a CC amplifier have in insulating the previous stage against changes in R_L?

4–23 Using the amplifier given in Ex. 4–14, calculate the value of A_{vo} when R_g is equal to (a) 50 ohms, (b) 500 ohms, and (c) 1 K. What conclusions can be reached with regard to the use of the CC amplifier as an insulation or buffer stage against changes in R_g?

4–24 Design a single-stage CC amplifier with the following characteristics: $|A_{vo}| \geq 0.9$ and $R_i \geq 20$ K. The amplifier will be used in an application where $R_g = 500$ ohms and $R_L = 1$ K. Assume that the transistor has a β of 250 and that the available power supply voltage is −30 V. Use standard components.

Section 4.8

4–25 Calculate the values of R_i, R_o, and A_{vo} for the CE-CB amplifier shown in Fig. 4–42. Assume that $\beta = 100$ and $r_{b'}$ is negligible for both stages.

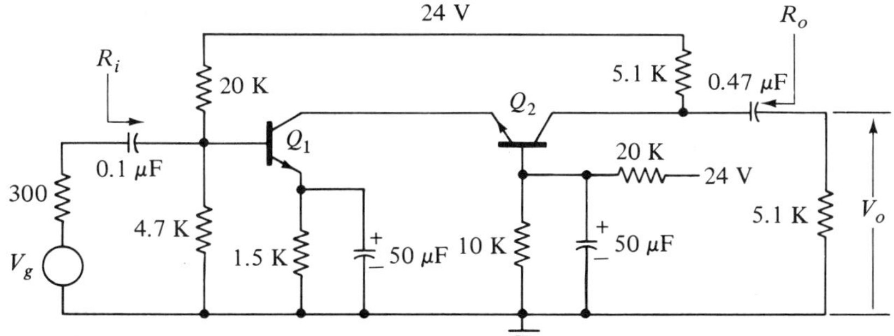

FIGURE 4–42 *Circuit for Prob. 4–25.*

4–26 The amplifier shown in Fig. 4–43 is called a "phase splitter." Calculate the voltage gains $A_{vo1} = V_{o1}/V_g$ and $A_{vo2} = V_{o2}/V_g$. The transistor has a β equal to 150.

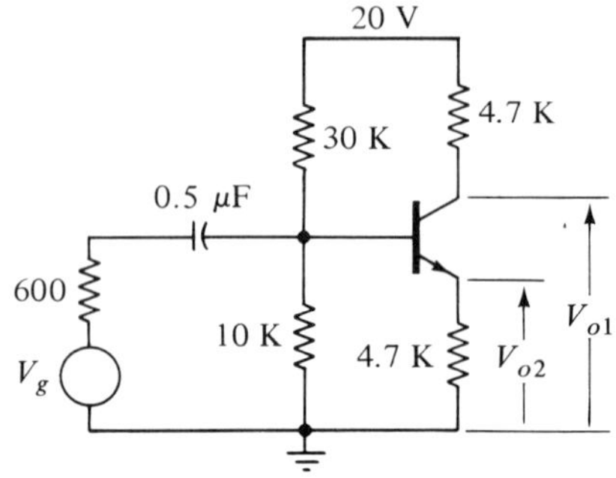

FIGURE 4–43 *Circuit for Prob. 4–26.*

4–27 Calculate the values of R_i, R_o, and A_{vo} for the CC-CE amplifier shown in Fig. 4-44. Assume that $\beta = 130$ for each transistor.

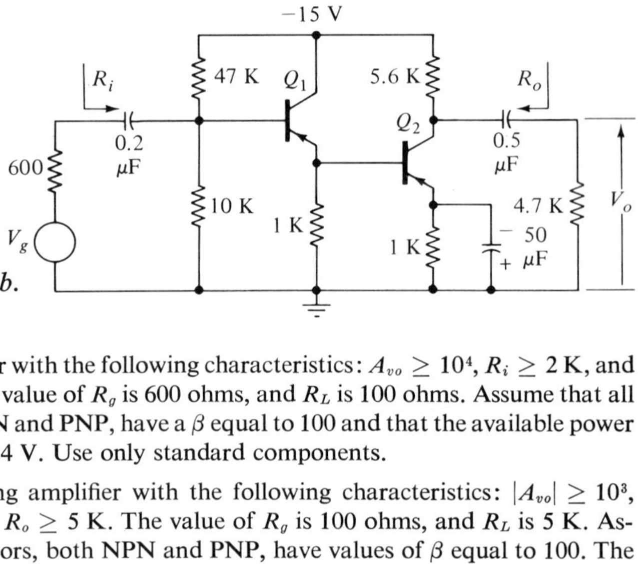

FIGURE 4–44 *Circuit for Prob. 4–27.*

4–28 Design an amplifier with the following characteristics: $A_{vo} \geq 10^4$, $R_i \geq 2$ K, and $R_o \leq 100$ ohms. The value of R_g is 600 ohms, and R_L is 100 ohms. Assume that all transistors, both NPN and PNP, have a β equal to 100 and that the available power supply voltage is +24 V. Use only standard components.

4–29 Design an inverting amplifier with the following characteristics: $|A_{vo}| \geq 10^3$, $R_i \leq 100$ ohms, and $R_o \geq 5$ K. The value of R_g is 100 ohms, and R_L is 5 K. Assume that all transistors, both NPN and PNP, have values of β equal to 100. The available power supply voltages are +15 V and −15 V. Use only standard components.

Section 4.9

4–30 Calculate V_{o1}, V_{o2}, and V_{o3} for the differential amplifier shown in Fig. 4–45. Assume that $\beta = 130$ for each transistor and that $V_{g1} = 10$ mV and $V_{g2} = 0$ mV. Repeat for the case where $V_{g1} = 0$ and $V_{g2} = 10$ mV.

4–31 Repeat Prob. 4–30 for the case where $V_{g1} = 10$ mV and $V_{g2} = 20$ mV. Assume that the two input signals are sine waves that are in phase.

4–32 Repeat Prob. 4–31 for the case where the two input signals are 180° out of phase.

4–33 Verify Eq. (4–114) for the FET differential amplifier shown in Fig. 4–40.

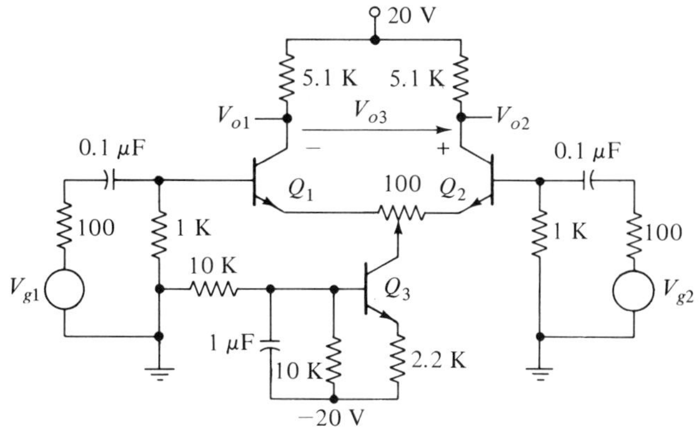

FIGURE 4–45 *Circuit for Probs. 4–30, 4–31, and 4–32.*

4–34 Calculate the values for V_{o1} and V_{o2} for the FET differential amplifier shown in Fig. 4–41. The experimental results for this amplifier are given in Eqs. (4–115) and (4–116).

4–35 Design an amplifier with a differential input having the following characteristics: $|V_o/V_g| \geq 500$, CMRR ≥ 500, $R_i \geq 10$ K, and $R_o \leq 100$ ohms. The value of R_g is 100 ohms, and R_L is 100 ohms. Assume that all bipolar transistors, both NPN and PNP, are matched and have values of β equal to 200. Assume that all JFETs, both N-channel and P-channel, are matched and have values of $|V_P|$ equal to 4 V and $|I_{DSS}|$ equal to 10 mA. The available power supply voltages are +24 V and −24 V. Use only standard components.

References

Early, J. M., "Design Theory of Junction Transistors," *The Bell System Technical Journal*, 32 (1953), 1271–1312.

Gray, P. E., D. DeWitt, A. R. Boothroyd, and J. F. Gibbons, *Physical Electronics and Circuit Models of Transistors*. New York: John Wiley & Sons, Inc., 1964.

Hyde, F. J., "The Physical Interpretation of Measurements on Transistors," *Journal of Electronics and Control*, 6, No. 4 (April, 1959), 362–64.

Lindmayer, J. and C. Y. Wrigley, *Fundamentals of Semiconductor Devices*. New York: D. Van Nostrand Company, Inc., 1965.

Millman, Jacob and C. C. Halkias, *Electronic Devices and Circuits*. New York: McGraw-Hill Book Company, 1967.

Pierce, J. F., *Semiconductor Junction Devices*. Columbus, Ohio: Charles E. Merrill Publishing Company, 1967.

Pierce, J. F., *Transistor Circuit Theory and Design*. Columbus, Ohio: Charles E. Merrill Publishing Company, 1963.

Pritchard, R. L., "Advances in the Understanding of the P-N Junction Triode," *Proc. IRE* (1958), 1130–41.

Shockley, W., *Electrons and Holes in Semiconductors*. New York: D. Van Nostrand Company, Inc., 1950.

Valdes, Leopoldo B., *The Physical Theory of Transistors*. New York: McGraw-Hill Book Company, 1961.

5

Integrated Circuits

In recent years, the technology associated with the fabrication of transistors, both FET and bipolar, has been extended to the point where entire circuits containing as many as one thousand elements can be fabricated on a chip of silicon with an area as small as 0.1 by 0.1 inch. These circuits, called *integrated circuits* or simply ICs, offer the following advantages over circuits composed of discrete elements: (1) a drastic reduction in size, volume, and weight; (2) a substantial increase in reliability; (3) a reduction in cost per individual circuit or function; and (4) an improved circuit performance in some instances. The first section will trace the development of ICs and briefly discuss their characteristics, limitations, and applications. In the remaining sections characteristics of IC components will be discussed in some detail with the eventual goal of designing simple ICs.

5.1 The Technology of Integrated Circuits

Since the announcement of the first transistor by Bell Laboratories in 1948, the size of electronic components has been reduced by a factor of approximately ten about every five years. If this trend continues, the size will be reduced by about 5 orders of magnitude, or only one ten-thousandth of the size of a 1948 electronic circuit in the near future. This miniaturization has been possible because transistors depend on crystal technology and the relative amount and type of impurities in the crystal. Consequently, the physical size of a transistor can be very small and can be measured in terms of several microns (10^{-6} m). The missile and space programs of the 1950s and 1960s provided much of the stimulus for this miniaturization. However, the effect of miniaturization has spread into many areas including computers, radio, television, and instrumentation.

In addition to the advantage of size reduction, solid state devices also require low power and low voltages. Thus a high density of devices can be achieved without exceeding rated operating temperatures. With a high density of active devices, rather complex functions can be performed in a small space. An outstanding example is the modern digital computer which was made possible by the miniaturization, lower power requirements, and reliability offered by solid state devices.

The transistor technology developed in the 1960s, particularly the silicon diffusion technology, made possible the fabrication of resistors and capacitors in a manner similar to the fabrication of transistors. An appropriately doped semiconductor of a given length and cross-section constitutes a resistor. A capacitor can be formed by placing an insulating layer of oxide between a thin metal film and a heavily doped semiconductor. The metal film can be formed at the same time that the ohmic contacts are formed at the active device terminals. Alternately, the junction capacitance of a reverse biased P-N junction can be used as a capacitor in a circuit. The only passive component that is difficult to fabricate from semiconductor material is an inductor. However, the effect of inductance can be simulated by using active devices if the requirement for an inductor cannot be circumvented.

The development of passive components using the same fabrication technology and having similar space requirements as a transistor meant that all of the circuit components could be made simultaneously and on the same silicon chip. The components can be interconnected on a chip by thin films of aluminum deposited on the surface. An example of a simple IC consisting of a transistor, resistor, and capacitor is shown in Fig. 5–1. All three components are formed in the body of the crystal, and the interconnections are made on the surface of the crystal.

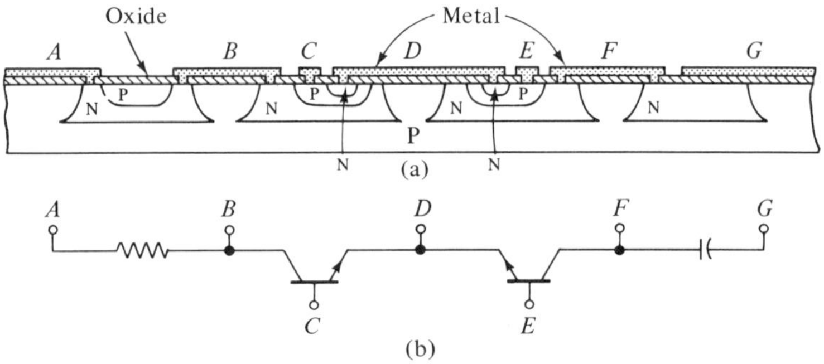

FIGURE 5–1 *A simple integrated circuit (IC): (a) a cross-sectional view of the IC, and (b) the circuit diagram of an IC.*

Very complex circuits can be fabricated by interconnecting the many individual circuits formed at the same time on the same chip. While ICs are almost impossible to repair, they can be replaced readily at low cost. An example of the impact of ICs is shown in Fig. 5–2 where one IC performs the same function as fourteen transistors, ten resistors, two capacitors, and many connections.

Along with the advantages, ICs have several inherent disadvantages. (1) The tolerance on the value of passive components may be poor ($\pm 25\%$), although the tolerance on the ratio of two resistor values can be very close. (2) The value

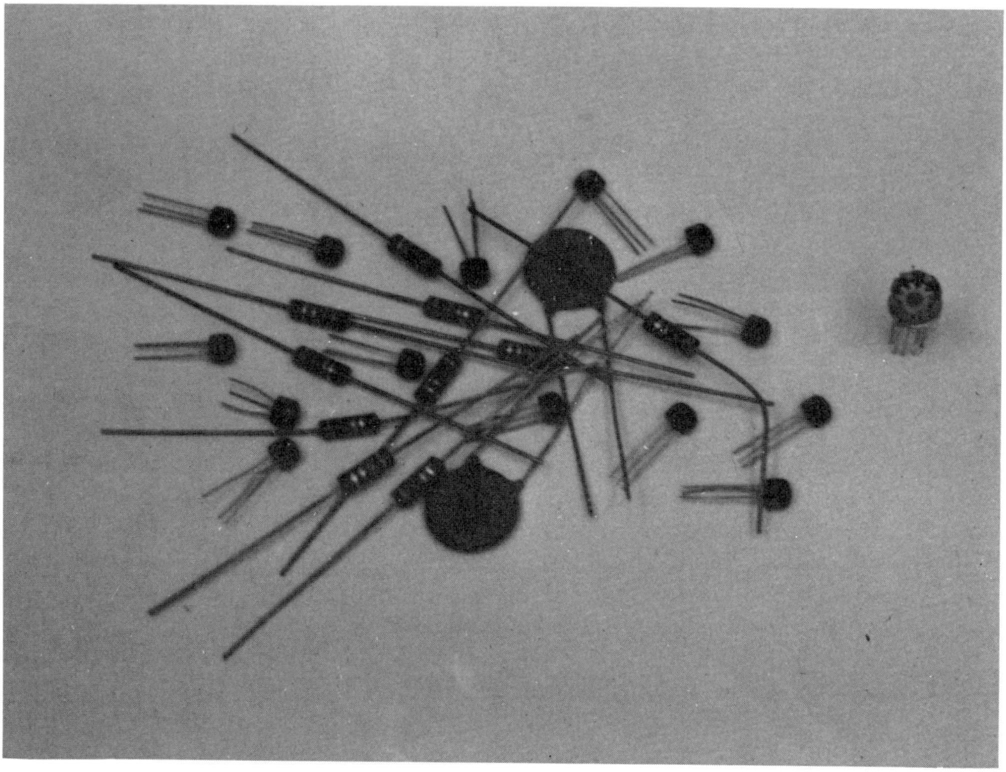

FIGURE 5–2 *A typical IC surrounded by the discrete components required to construct a similar circuit with conventional parts.*

of passive components may be voltage dependent. (3) The flexibility available in discrete component design is not always available in IC circuit design. For example, complementary active devices (NPN and PNP bipolar transistors or N-channel and P-channel FETs) of equal quality are difficult to fabricate in an IC. (4) The performance of the active devices in an IC is often poorer than the performance of the corresponding active device in discrete form. However, many of the disadvantages of ICs can be overcome by following the circuit design guides peculiar to ICs. These design guides, as well as further treatment of the characteristics of IC components, will be discussed in later sections. Before beginning this discussion we will briefly outline the basic classifications of ICs.

The basic semiconductor IC is a silicon monolithic structure in which all of the circuit components and connections are fabricated within or on the crystal. Monolithic means that the structure is a single crystal. A second type of IC is the hybrid integrated circuit. In contrast with the monolithic IC, the passive components of a hybrid IC are formed by a film process on a suitable substrate. Two film technologies are used: (1) thin-films which use vacuum deposition of various materials to form passive components, and (2) thick-films which use specially prepared inks to form passive components. The required active devices (transistors or monolithic ICs) are added separately to the film pattern.

Hybrid ICs find application mainly as a supplement to monolithic ICs since special components can be added to overcome some disadvantages of a monolithic circuit. Both types of ICs will be discussed in this chapter with the emphasis on the monolithic form.

ICs can also be grouped according to their application as: (1) *logic* or *digital* (off-on states), and (2) *linear* or *analog* (amplifiers). The initial spectacular growth of ICs was in the logic area. Logic systems utilize large numbers of similar circuits, most of which operate as switches. An IC could almost directly replace the discrete form of the logic circuit and provide substantial miniaturization. With the growing demand for computers, mass production of ICs became economically possible. Also, relatively simple demands are placed on the active devices of a digital IC. Generally, they are required to either conduct or not conduct and hence represent either the "0" or the "1" logic state. However, the speed with which a device can go from one state to the other state is an important consideration.

Linear circuits, such as amplifiers, vary widely in their configuration and application. A set of universal "standard linear circuits" has not been defined although some circuits are so widely used that they are considered basic building blocks. One basic circuit is the differential amplifier. Also, many linear circuits are high performance units, and each component is specifically chosen to meet the circuit requirements. For these and other reasons, the development of linear ICs has lagged behind the development of digital ICs. However, the decade of the 1970s promises to be a growth period for both digital and linear ICs.

The remainder of this chapter is devoted to a discussion of the basic process used in fabricating ICs and the guidelines for designing IC circuits. Only simple circuits will be treated; however, more complex ICs will be used in various applications in the remainder of the text.

5.2 Fabrication of Monolithic Circuits

While many methods have been used to fabricate monolithic circuits, most of the circuits of this type now start with a planar epitaxial process. In this process, a thin wafer 5 to 10 mils thick is sliced from a P-type silicon rod. The wafer may have an area ranging from 2 to more than 6 square inches. Since a typical circuit occupies an area of approximately 25 $\times$ 25 mils, several thousand circuits can be fabricated on one wafer. The wafer is used mainly as a substrate with the 10-mil thickness being necessary only for mechanical strength. The doping level is low, approximately 10^{14} impurity atoms per cm^3, so that later successive diffusion processes are not extremely critical.

After smoothing the surface, the next step is to grow a 5 μ to 1 mil thick epitaxial layer of N-type silicon on the surface of the wafer. This process may be carried out with the wafer in an environment of $SiCl_4$ and H_2O at 1200°C. Minute amounts of phosphorus are added to make the grown layer N-type with a doping density of approximately 1.5×10^{16} phosphorus atoms per cm^3. In this environment, the silicon crystal grows slowly on the surface of the wafer in such a way that the crystal lattice is preserved. The resulting P-N junction is rather abrupt because the phosphorus atoms do not diffuse readily at this temperature. A cross-sectional view of the wafer is shown in Fig. 5–3(a), and the P-N junction in Fig. 5–3(b).

The next step is the formation of an oxide layer approximately 0.5 μ thick on top of the epitaxial layer as shown in Fig. 5–3(b). The oxide layer, SiO_2, is grown by heating the wafer to about 1000°C in an oxygen atmosphere. The oxide layer is then covered by a film of photoresist as shown in Fig. 5–3(d). The photoresist is soluble in certain solvents unless it is polymerized by being exposed to ultraviolet light. In this way, certain portions of the resist can be removed to allow selective diffusion of impurities into the epitaxial layer. Application of the photoresist layer completes the epitaxial step.

The circuit being fabricated is composed of alternate regions of N- and P-type material. Designing the geometry of the various components is part of the layout problem and will be discussed in Sec. 5.6. Once the geometry of an IC is determined, a series of master drawings is made on a large scale. These drawings are photographically reduced in size and are used to make transparencies called *masks*. These masks contain the exact sizes of all the components being fabricated.

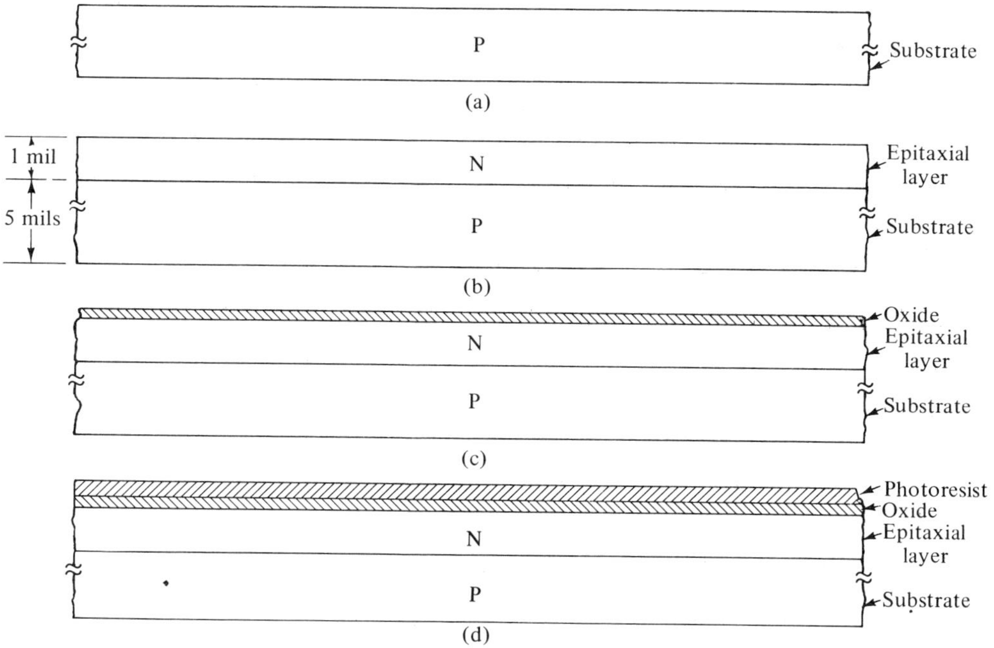

FIGURE 5–3 *First monolithic process step: (a) the P-type slice of silicon which forms the substrate, (b) the epitaxial layer, (c) the oxide layer, and (d) the photoresist layer.*

The second step is shown in Fig. 5–4. The first photographic mask contains the isolation pattern of the circuit. Although the various parts of the IC are physically separated, they would be electrically connected by the epitaxial layer if isolation were not used. The isolation mask is placed over the photoresist as shown in Fig. 5–4(a), and the entire assembly is exposed to ultraviolet light. The exposed photoresist is polymerized. The unexposed photoresist and underlying

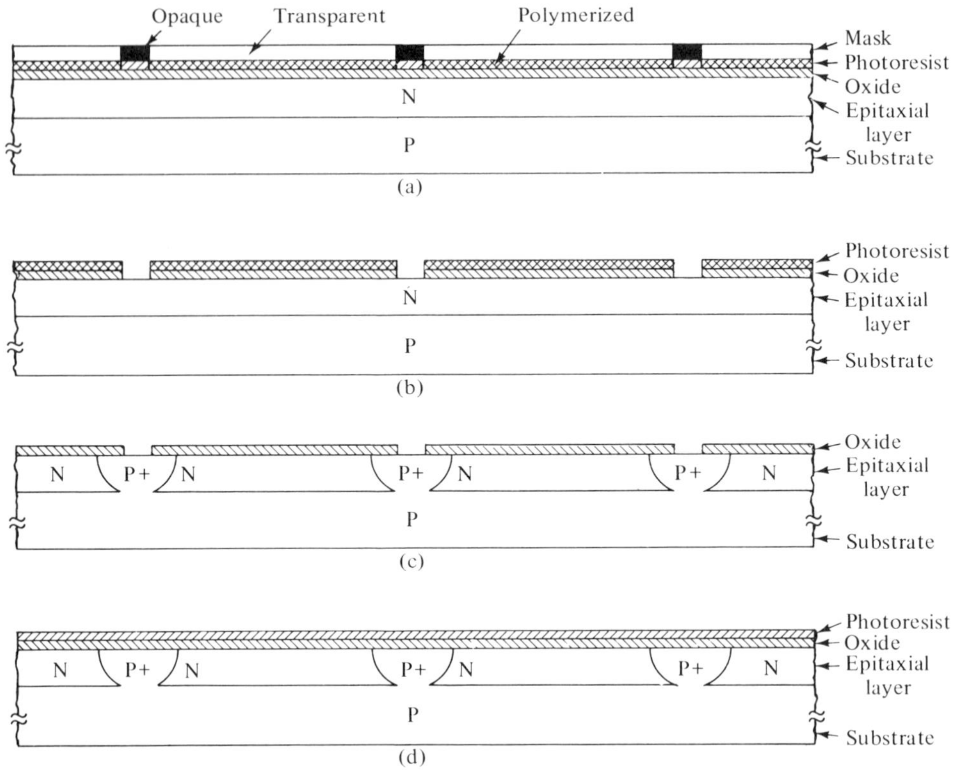

FIGURE 5–4 *The second step in making a monolithic circuit:* (*a*) *first or isolation mask and effect of being exposed to ultraviolet light,* (*b*) *results of first etching,* (*c*) *results of first or isolation diffusion, and* (*d*) *second oxide and photoresist layer.*

oxide layer are then removed chemically by an etching solution, leaving the pattern shown in Fig. 5–4(b).

Once the openings in the oxide have been made, impurities can be diffused into the epitaxial layer to form N- or P-type regions. Diffusion is a complex process and will not be treated in detail here. Basically the chip is heated in an atmosphere of gas containing the desired impurities.* The results of the isolation diffusion are shown in Fig. 5–4(c). Since diffusion is a three-dimensional process, the impurity atoms move under the oxide approximately as far as they move into the chip. The resulting isolation junction is formed under the oxide layer which protects it against surface contamination. The effect of the isolation diffusion is to form islands of N-type material completely surrounded by P-type material. By reverse biasing the resulting P-N junction, the N-type regions are electrically isolated from each other by the reverse resistance and capacitance of a diode. After the isolation diffusion, the oxide layer is formed again and covered with photoresist as shown in Fig. 5–4(d).

---

* The impurity gas may be of unlimited supply or may be restricted in supply. Each condition results in a different impurity profile or impurity concentration level as a function of distance from the surface into the chip. Generally, the impurity profile is described by an error function (*erf*) or a complimentary error function (*erfc*).

The third step in making a monolithic circuit is shown in Fig. 5–5. The second or base region mask, which outlines all the regions to receive P-type impurities, is placed over the wafer shown in Fig. 5–4(d). The entire assembly is exposed to ultraviolet light which results in a pattern of polymerized material as shown in Fig. 5–5(a). After exposure, the surface is again etched as shown in 5–5(b). P-type impurities are diffused through the windows that are etched in the photoresist to form the base region of transistors, the P-type side of diodes, and the resistors shown in Fig. 5–5(c). Then the surface is covered with an oxide layer and photoresist as shown in Fig. 5–5(d).

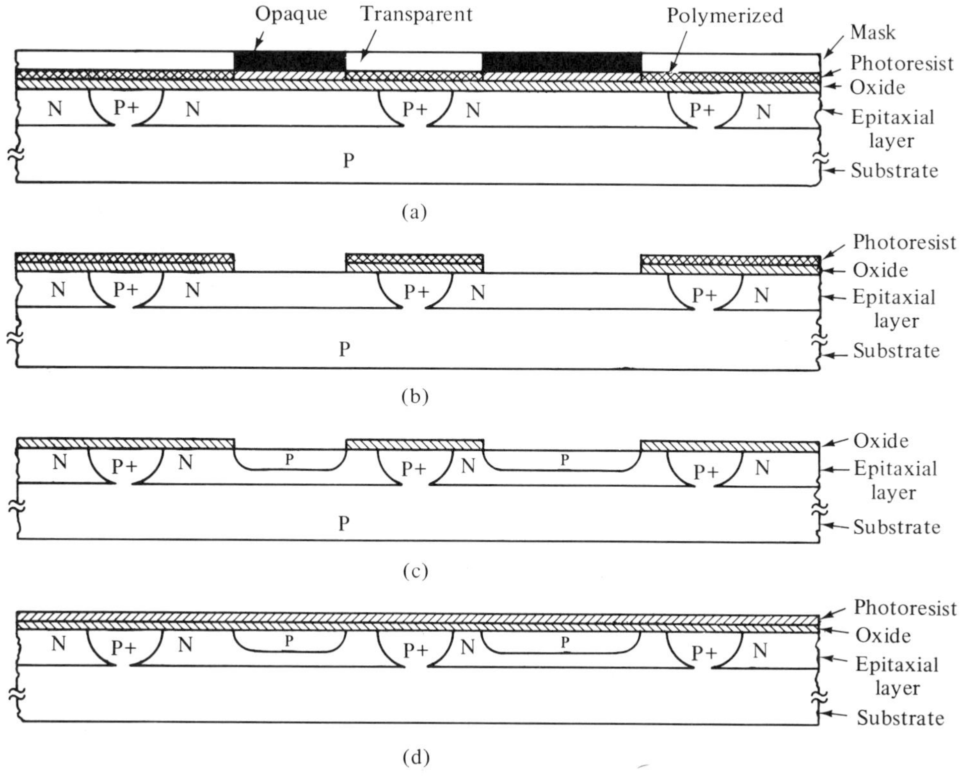

FIGURE 5–5 *Third monolithic process step: (a) second or base region mask and effect of being exposed to ultraviolet light, (b) results of second etching, (c) results of second or base diffusion, and (d) third oxide and photoresist layer.*

The fourth monolithic process step is shown in Fig. 5–6. The third or emitter region mask, which outlines the areas to receive N^+ impurities, is placed over the wafer shown in Fig. 5–5(d). The assembly is then exposed to ultraviolet light yielding the polymerized pattern shown in Fig. 5–6(a). After the third etch, the slice appears as shown in Fig. 5–6(b). The third diffusion forms the emitter regions of the transistors, N-sides of the diodes, and the collector contact regions as shown in Fig. 5–6(c). Then the slice is again covered with oxide and photoresist as shown in Fig. 5–6(d).

The fifth monolithic process step is shown in Fig. 5–7. The fourth or ohmic contact mask, which outlines the areas where contacts are to be made to each

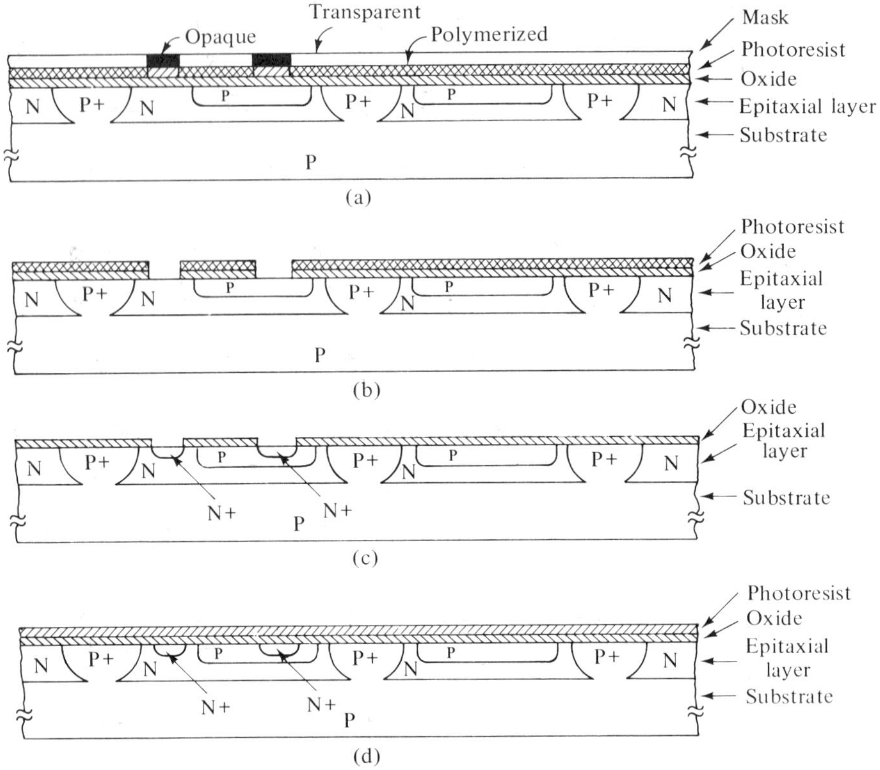

FIGURE 5–6 *Fourth monolithic process step: (a) third or emitter region mask and effect of being exposed to ultraviolet light, (b) results of third etching, (c) results of third or emitter diffusion, and (d) fourth oxide and photoresist layer.*

device, is placed over the wafer as shown in Fig. 5–6(d). The result of exposing this assembly to ultraviolet light is shown in Fig. 5–7(a). After exposure and etching, the slice appears as shown in Fig. 5–7(b). Then a metal, such as aluminum, is deposited or evaporated in a thin film over the entire slice as shown in Fig. 5–7(c). Photoresist is again spread over the surface [see Fig. 5–7(d)].

The sixth step is shown in Fig. 5–8. The fifth or interconnection mask, which outlines the unwanted regions of the metallic layer, is placed on the wafer shown in Fig. 5–7(d). The results of exposing this assembly to ultraviolet light are shown in Fig. 5–8(a). After exposure, the unwanted regions of the metallic layer are etched off leaving the final circuit [see Fig. 5–8(b)]. The circuit diagram of the IC is shown in Fig. 5–8(c).

With the fabrication of the IC completed, each circuit on the chip is tested. Then the individual circuits are broken apart along lines scribed with a diamond tool. Finally, each chip is encapsuled and fitted with leads.

Two important features result from the monolithic process. First, the dimensions of the various devices can be very small since photographic techniques were used to determine the precise geometry. Simultaneous fabrication of a large number of devices can be made on a small chip area. Second, the various P-N junctions are formed under an oxide layer which protects them from surface contamination. Such P-N junctions have low leakage currents and high reliability.

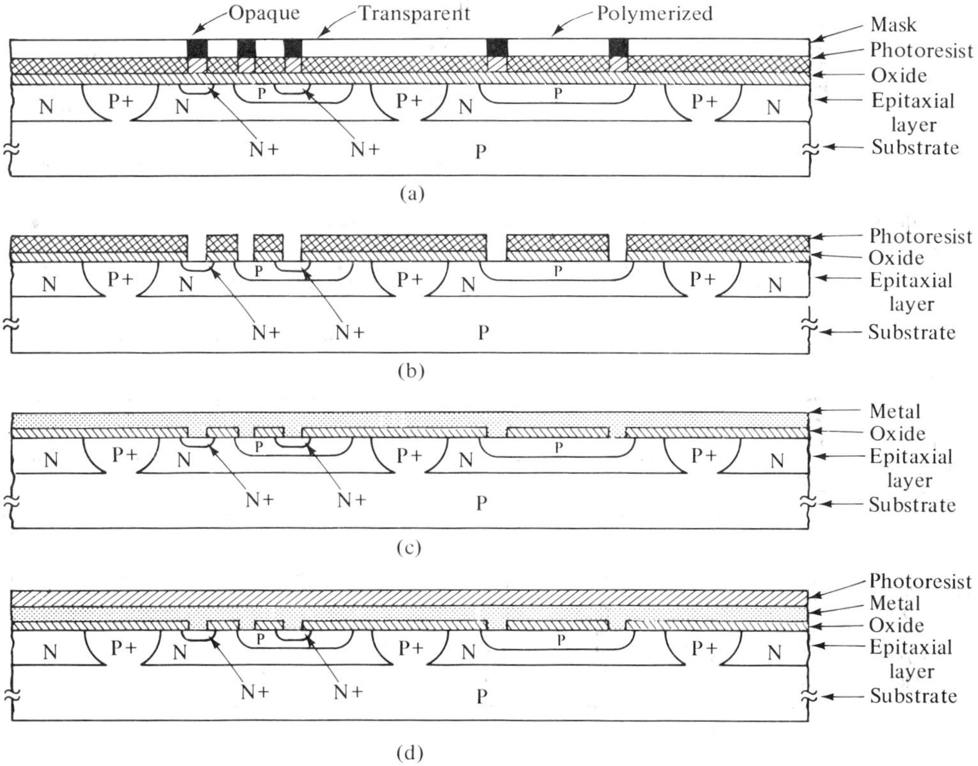

FIGURE 5–7 *Fifth monolithic process step: (a) fourth or ohmic contact mask and effect of being exposed to ultraviolet light, (b) results of fourth etching, (c) metallic film, and (d) photoresist layer.*

In the following two sections, we will examine the characteristics of integrated active and passive devices when bipolar transistors are used for the active device. MOSFET devices, both active and passive, are discussed in Sec. 5.7.

Now that a typical fabrication process has been described, let us study some of the problems of multiple diffusion processes. As indicated in the preceding discussion, the epitaxial layer is grown first and forms the collector regions of bipolar transistors. A typical impurity concentration is 1.2 to 1.8×10^{16} impurity atoms per cm^3 and is uniform throughout the 1 mil thick epitaxial layer as indicated by the straight dashed line in Fig. 5–9. After the isolation diffusion, the P-type base region was diffused into the N-type islands in the epitaxial layer. This diffusion is obtained by holding the surface concentration of the P-type dopant, usually boron, at 0.4 to 0.8×10^{18} atoms/cm^3 while maintaining the entire assembly at a high temperature. As shown by the dashed curve in Fig. 5–9, the resulting P-type impurity distribution is nonlinear. The base-collector junction occurs where the P-type and N-type impurity concentrations are equal; i.e., where the two dashed lines intersect at a depth of approximately 2 to 2.5 μ.

The next diffusion forms the emitter and must start with a higher surface impurity concentration than the base diffusion. Phosphorus is usually used in this case since it has a solubility in silicon of 1.2×10^{21} atoms/cm^3. The emitter

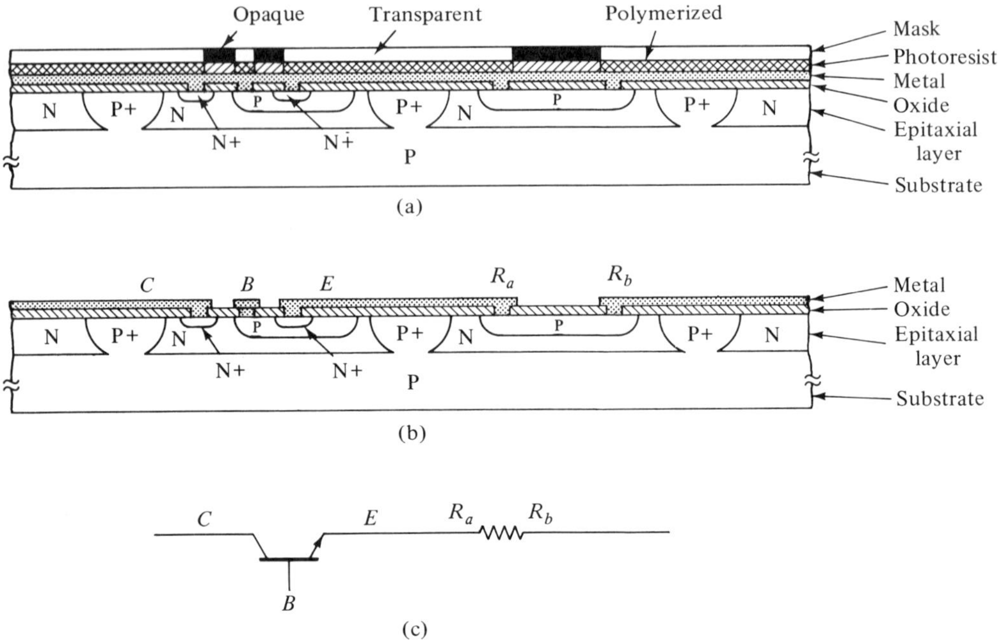

FIGURE 5–8 *Sixth monolithic process step: (a) Fifth or interconnection mask and effect of being exposed to ultraviolet light, (b) results of fifth etching and completed circuit, and (c) circuit diagram.*

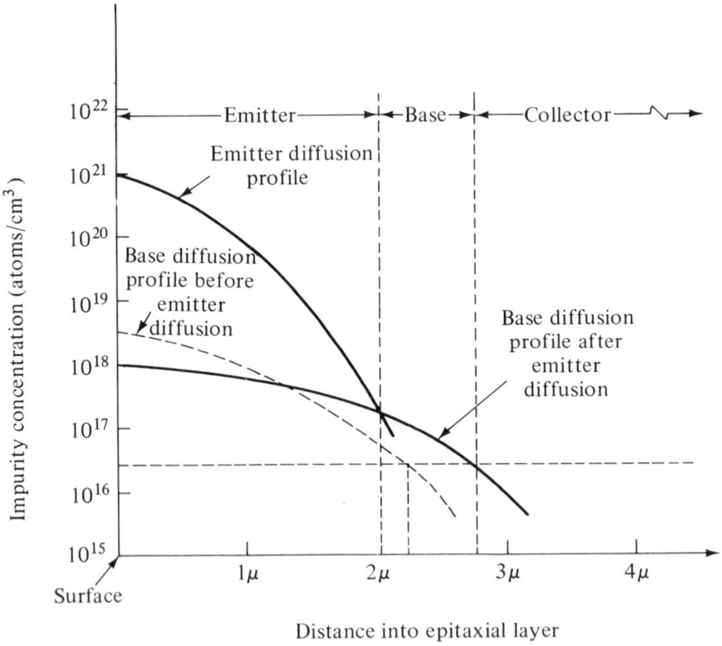

FIGURE 5–9 *The emitter, base, and collector impurity concentrations as a function of distance into the epitaxial layer. Note how the base diffusion profile was modified by the emitter diffusion process.*

diffusion profile is the top curve in Fig. 5–9. Note that while the emitter diffusion was occuring, the base diffusion profile changed because the boron atoms diffused further into the material under the temperature conditions that allowed phosphorus to diffuse into the silicon. In Fig. 5–9, the base collector junction occurred at 2.3 μ before the emitter diffusion and moved to 2.7 μ during the emitter diffusion. Junction drift is one of the main reasons for keeping the number of diffusion steps to a minimum. Of course the same effect occurs when a transistor or diode is drastically overheated.

The emitter-base junction occurs where the N-type emitter doping is equal to the P-type base doping, or at the intersection of the two solid curves in Fig. 5–9. Note that the resulting base width is only 0.6 to 0.8 μ wide. A narrow base width is associated with a high value of β.

In this section, some of the techniques used in fabricating a monolithic transistor have been briefly discussed. As indicated in later sections, other common integrated circuit components (diodes, resistors, and capacitors) involve some of the same steps used to fabricate a transistor.

5.3 Monolithic Active Devices

Having outlined the processing steps required to fabricate a typical monolithic transistor, we will continue the discussion of device formation to include a typical layout for a transistor. Some device properties will also be discussed, including the junction breakdown voltages and the common-emitter current gain, β. Other device properties, including the capacitance of various junctions, will be treated in the next section on monolithic passive devices.

The active device most commonly used in a monolithic IC is the NPN bipolar transistor. The top and side views of a typical low-current NPN transistor are shown in Fig. 5–10. This geometry is referred to as a "single-base-strip" geometry. The IC designer has considerable freedom in the layout since the diffusion patterns are established by a photographic process. However, the width of the emitter strip is a limiting dimension. Emitter widths as narrow as 0.25 mil have been achieved, but a 1-mil width represents a typical value. The length of the emitter can be as long as necessary to handle the required emitter current. The area of the emitter in Fig. 5–10(a) is 1 mil $\times$ 1.5 mils. The metallic film forming the emitter ohmic contact is 0.5 mil wide, leaving a 0.25-mil clearance on either side. The clearance for the junctions and the contacts is 0.5 mil.

Once the emitter dimensions and the device geometry are established, the resulting dimensions can be calculated using 0.5-mil clearance for each junction contact. The width of the base region is obtained by adding 0.5 mil to each side of the emitter length of 1.5 mils, giving a base width of 2.5 mils. The length of the base is obtained in a similar manner. For a metallic base strip of 0.5 mil, three 0.5-mil clearances, and a 1-mil emitter width, the total base width is 3 mils. The base area is then 2.5 $\times$ 3 mils as shown in Fig. 5–10.

The width of the collector contact area is 1.5 mils and is obtained by using a 0.5-mil metallic strip with a 0.5-mil clearance on either side. The length of the collector contact is the same as the width of the base region, or 2.5 mils.

In order to calculate the area of either the isolation island or the isolation mask, we must first examine the side view given in Fig. 5–10(b). The isolation

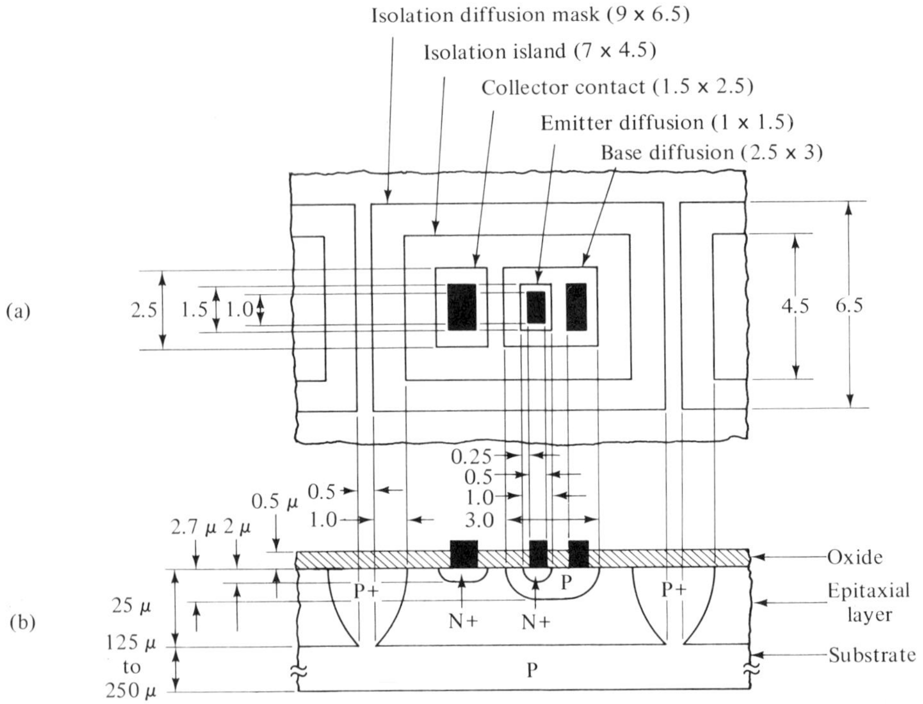

FIGURE 5–10 *A layout, with typical dimensions, of a low-current NPN monolithic transistor: (a) top view, and (b) side view. All dimensions are mils unless otherwise noted. (1 mil = 10^{-3} inches and 25.4 μ = 1 mil.)*

island was formed by diffusing P+ impurities from the surface to a depth approximately equal to the depth of the epitaxial layer on 1 mil. Diffusion, however, is a three-dimensional process. Consequently, the P+ impurities will also diffuse under the oxide layer a distance of approximately 1 mil as shown in Fig. 5–10(b). Therefore, the size of the isolation island will be two mils smaller in each dimension than the size of the isolation mask. By allowing a 1-mil clearance between the edge of the isolated island and either the base area or the collector contact area, the area of the island is 7 mils × 4.5 mils. The inside dimensions of the required isolation mask are 9 mils by 6.5 mils.

A typical set of depth dimensions for an NPN monolithic transistor is given in Fig. 5–10(b). For a base diffusion of 2.7 μ and an emitter diffusion of 2 μ, the distance across the base region is approximately 0.7 μ.

The single-base-strip geometry discussed above is only one of many possible geometries for a bipolar transistor. Several other possible geometries are shown in Fig. 5–11. The double-base-strip layout shown in Fig. 5–11(b) is one form often used in low-power ICs. As shown in Figs. 5–11(c) and (d), the double-base- and double-collector-strip geometry and the multiple emitter geometry have an elongated emitter strip to provide for higher current-carrying capability. The multiple emitter form is used in some logic circuits.

One important limitation of a P-N junction is its reverse breakdown. Voltage breakdown can result from either avalanche effects or Zener effects, both of which are discussed in Sec. 2.7. Many P-N junctions are characterized by one

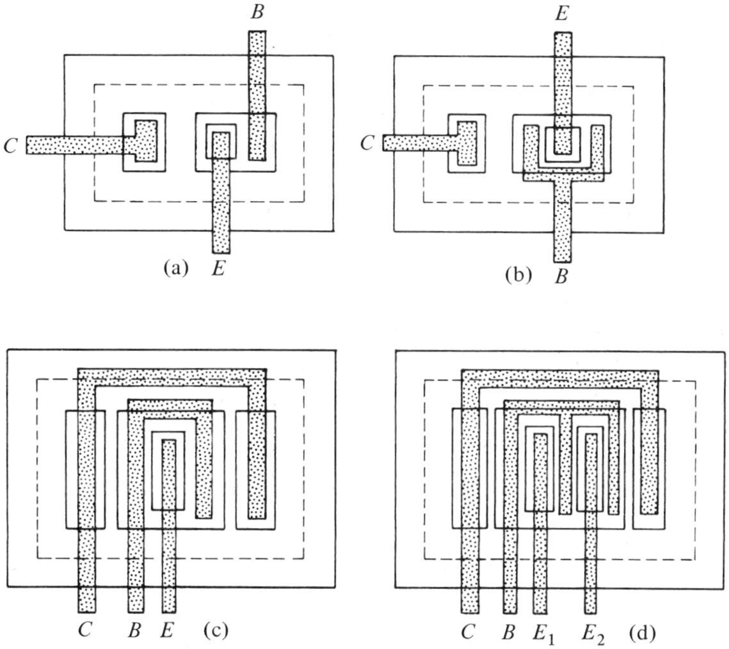

FIGURE 5–11 *Typical NPN monolithic transistor geometries: (a) single-base-strip, (b) double-base-strip, (c) double-base- and double-collector-strip, and (d) multiple emitters. The current-carrying capability of all geometries can be increased by lengthening the emitter strip as shown in (c) and (d).*

side being more heavily doped than the other side. For this type of junction, the breakdown voltage is controlled by the impurity level of the lighter doped side. A plot of the breakdown voltage as a function of the impurity concentration of the lighter doped material is given in Fig. 5–12. This curve can be used to estimate the breakdown voltages of the various junctions present in the monolithic transistor as shown by the following example.

EXAMPLE 5–1 An NPN transistor has the following approximate doping levels:

emitter $= 10^{20}$ atoms/cm^3 collector $= 10^{16}$ atoms/cm^3

base $= 10^{18}$ atoms/cm^3 substrate $= 10^{15}$ atoms/cm^3

Estimate the value of the following breakdown voltages: collector-base, emitter-base, collector-substrate, and emitter-collector.

Solution: The collector-base breakdown voltage is controlled by the lighter or collector impurity level, which is 10^{16} atoms/cc. From Fig. 5–12, BV_{CBO} is approximately 60 V. The emitter-base breakdown voltage is controlled by the base impurity level, which is 10^{18} atoms/cc. From Fig. 5–12, BV_{EBO} is approximately 5 V.

The collector-substrate breakdown voltage is not controlled by the substrate because the P-type impurities used in the isolation diffusion contain a high impurity concentration. Consequently, BV_{CSO} is controlled by the collector impurity level and is approximately the same as BV_{CBO} or 60 V.

The collector-emitter breakdown voltage cannot be found from Fig. 5–12 since the collector and emitter do not form a junction. However, the value of BV_{CEO} can be estimated as

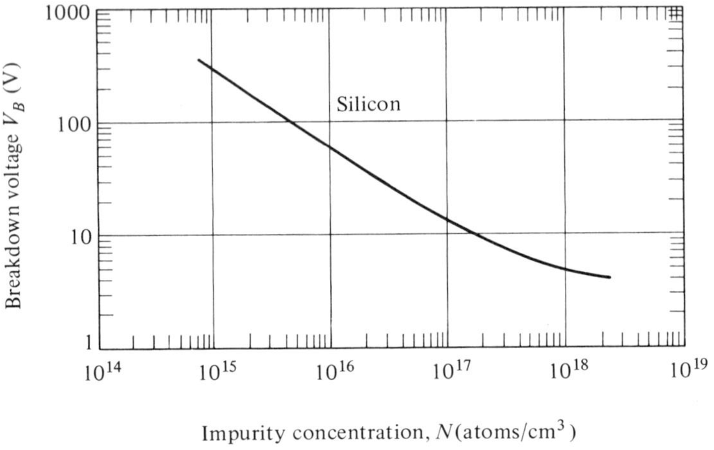

FIGURE 5–12 *Breakdown voltage of a silicon P-N junction, with one side lightly doped, as a function of the impurity level of the lightly doped side. (Adapted from L. Stern,* Fundamentals of Integrated Circuits, *Hayden Book Co., Inc., 1968.)*

$$BV_{CEO} \simeq BV_{CBO}/\sqrt[4]{\beta}$$

Using this equation and a β of 50, BV_{CEO} is approximately 22.5 V.

A second parameter of importance is the common emitter current gain, β. From Eq. (4–18), the value of β in terms of the common base current gain, α, is

$$\beta = \alpha/(1 - \alpha) \qquad \textbf{(5–1)}$$

Since α has a value approaching unity, any slight change in the value of α will produce a large change in β. From Eq. (4–13), α is

$$\alpha = \gamma b \delta \qquad \textbf{(5–2)}$$

where: (1) γ is the emitter efficiency and represents the ratio of the current injected into the base to the total emitter current; (2) b is the base transport factor and represents the ratio of the number of minority carriers that reach the collector to the number that are injected into the base from the emitter; and (3) δ is the collector multiplication ratio and is generally equal to unity.

The value of γ depends on the impurity concentration of the emitter region and the base region at the edge of the emitter-base junction. From Fig. 5–9, it is apparent that the emitter and base regions are not homogeneously doped. The impurity concentration changes with the distance from the surface. The exact value of impurity concentrations at the edges of the emitter-base junction is difficult to predict and, consequently, the exact value of γ is difficult to predict. Any uncertainty in the value of γ is associated with the same degree of uncertainty in the value of α and a much higher degree of uncertainty in the value of β.

The value of b is a measure of the percentage of carriers lost in the base region due to electron-hole recombinations. The number of carriers that recombine depends on the time required by the minority carriers to move across the base region. In Chap. 4, we assumed a homogeneously doped region, and the carriers moved across the base region only by diffusion. However, when the base is nonhomogeneously doped, the electric field developed in the base region aids the movement of carriers across the base region.* Thus the movement of base minority carriers has both a drift component and a diffusion component. Because of the nonlinear impurity concentration in the base region, it is difficult to predict the exact value of b. As was the case with γ, any uncertainty in the value of b produces the same uncertainty in the value of α and a much higher uncertainty in the value of β.

The preceding discussion indicates some of the difficulties encountered when attempting to calculate a value of β for a nonhomogeneously doped transistor of the type found in a monolithic IC. Development of an accurate expression for β is also difficult because of the many nonlinearities present. Consequently, we will only examine values of β that are typical for monolithic ICs.

Values of β for an NPN monolithic transistor range from approximately 15 to more than 1000 and in some cases as high as 10,000. The higher values of β are obtained with a very narrow base width and high impurity concentrations. These conditions also produce low values of breakdown voltage, in the order of 3 V. Typical values of β range around 100.

As in discrete transistors, β is a function of collector current. By adjusting the length of the emitter strip (see Fig. 5–11), the current at which β reaches a peak value can be changed. An estimate of the value of collector current (in mA) at which β will peak is given by

$$I_C \text{ for } \beta_{\max} \simeq (4)(\text{effective emitter perimeter in mils}) \qquad \textbf{(5–3)}$$

The effective emitter perimeter depends on the geometry of the transistor. Only those edges of the emitter close to a base strip are considered in determining the effective emitter perimeter. For example, a single base-strip geometry with an emitter area 1 mil × 1.5 mils has an effective emitter perimeter of 1.5 mils, since only one side of the emitter is adjacent to the base. For this geometry, β has a maximum value at a collector current of approximately 6 mA. The peak in the curve of β versus I_C is fairly broad as shown in Fig. 5–13.

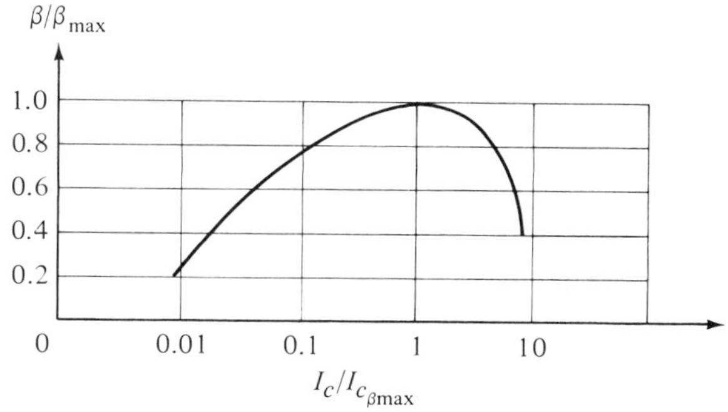

FIGURE 5–13 *Beta variations as a function of collector current.*

A diode is fabricated in a monolithic IC by connecting an NPN transistor in one of the configurations shown in Fig. 5–14. The reverse breakdown characteristics of a diode are determined by the junction used to form the diode. For example, the diodes shown in Figs. 5–14(a) and (b) have a breakdown voltage

* A detailed treatment of transistors with a nonhomogeneous impurity distribution, called "drift" or "graded-base" transistors, can be found in Chapter 4 of Lindmayer, J. and C. Y. Wrigley, *Fundamentals of Semiconductor Devices* (see references at the end of this chapter).

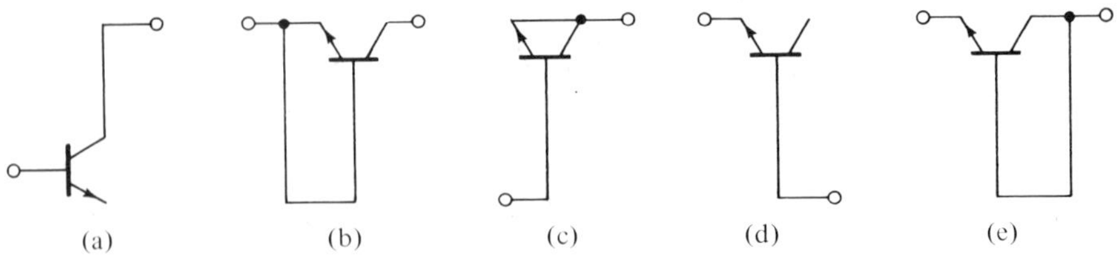

FIGURE 5–14 *Various diode connections.*

similar to BV_{CBO} or approximately 60 V. The remaining three diodes have a breakdown voltage similar to BV_{EBO} or approximately 7 V. These latter configurations are also useful as Zener or reference diodes.

A monolithic PNP transistor can be formed in several ways, all of which either involve additional diffusion steps or an inferior geometry. For example, the emitter, base, and collector of a PNP can be diffused by separate diffusion steps independent of the formation of an NPN transistor. A second and more common method is to form a "lateral" PNP transistor as shown in Fig. 5–15. In this geometry, the N-type epitaxial layer is used for the base region, and the emitter and collector regions are formed during the base diffusion of an NPN transistor or during the isolation diffusion.

In addition to the basic NPN and lateral PNP geometries discussed above, numerous other active device geometries are possible in monolithic ICs. Two geometries that show promise are the "buried layer" and the "TRIM" structures discussed below.

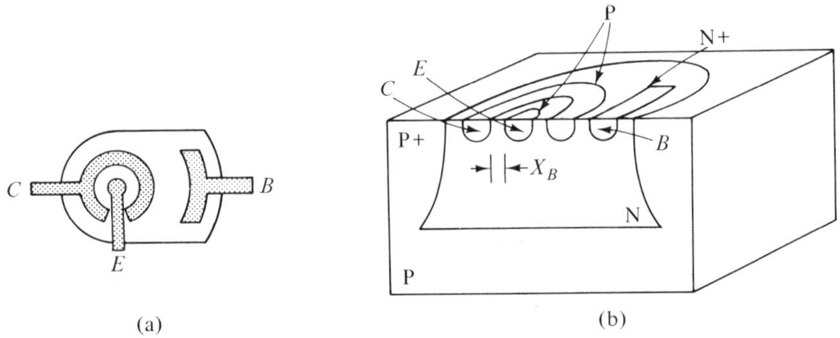

FIGURE 5–15 *A lateral PNP transistor: (a) top view, and (b) side view.*

Consider the monolithic NPN transistor whose side view is shown in Fig. 5–16. This geometry is identical to that shown in Figs. 5–8(b) and 5–10 except for the N^+ layer located between the epitaxial layer and the substrate. Since the epitaxial layer is lightly doped (10^{16} to 10^{17} atoms/cc), the series resistance of the collector can be relatively large. A large series collector resistance is generally undesirable, particularly in transistors used in switching or logic circuits. The heavily doped N^+ buried layer effectively short circuits this series collector resistance. The buried layer is diffused into the substrate prior to the growth of the epitaxial layer.

The top view of a TRIM structure transistor is shown in Fig. 5–17(a) and the side view in Fig. 5–17(b). Fabrication of a TRIM transistor starts with a lightly

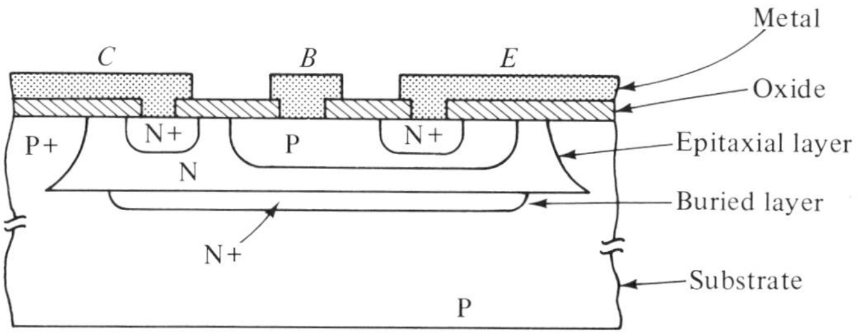

FIGURE 5–16 *Side view of a monolithic NPN transistor using a buried layer to reduce the collector series resistance.*

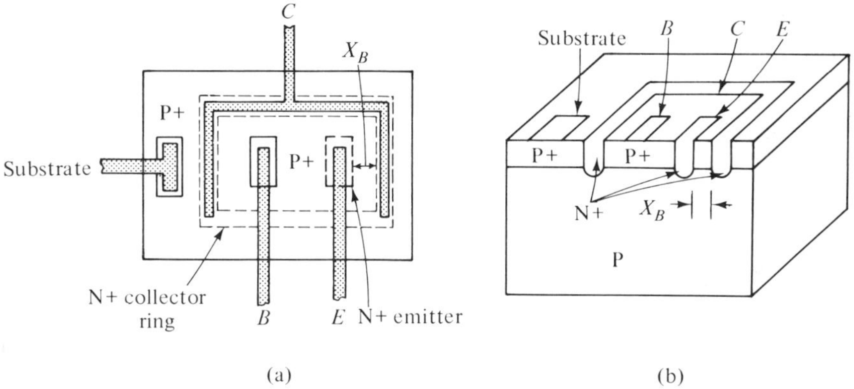

FIGURE 5–17 *A TRIM structure transistor: (a) top view, and (b) side view.*

doped P-type substrate into which a P+ layer is diffused. Following the same process of oxide growth, photoresist application, mask placement, and N+ diffusion discussed in Sec. 5.2, the emitter and collector regions are formed simultaneously in the P+ surface layer. These N+ regions diffuse completely through the P+ regions. Consequently, the P+ base region is isolated from the surrounding P+ substrate. The ohmic contacts and interconnection patterns are formed in the same manner discussed in Sec. 5.2. The TRIM structure requires only three masks: (1) the emitter and collector region mask, (2) the ohmic contact mask, and (3) the interconnection pattern mask. Since only three masks are used, the name "Tri-Mask" or "TRIM" has been applied to transistors constructed in this manner. In contrast, the conventional NPN fabrication technique discussed in Sec. 5.2 requires five masks.

A comparison between Figs. 5–15 and 5–17 shows that the TRIM structure is very similar to the lateral PNP structure. In each case, the principal current flow is parallel to the surface. TRIM structure transistors have found their widest application in digital ICs.

The basic types and characteristics of bipolar transistors used in monolithic integrated circuits have been reviewed in this section. Because of its superior performance, the NPN transistor is normally used. However, other forms of the bipolar transistor, such as the TRIM structure, may be developed in the future. In the next section, passive devices used in monolithic ICs will be discussed. Later sections treat the problems of monolithic circuit layout.

5.4 Monolithic Passive Components

In addition to transistors and diodes, most electronic circuits require passive components such as resistors, capacitors, and inductors. The resistor is the most common monolithic passive component and will be discussed first.

Resistors used in monolithic ICs are fabricated using thin layers of semiconductor material which are formed simultaneously with the formation of the base or emitter region. A term used in the calculation of the resistance of thin layers is "sheet resistance" which can be defined with the aid of the model shown in Fig. 5–18. The equivalent resistance between opposite sides of a square as shown in Fig. 5–18 is given by

$$R = \frac{\rho l}{A} = \frac{\rho l}{ld} = \frac{\rho}{d} \tag{5–4}$$

where ρ is the average resistivity of the sample, l is the length and width of the sample, and d is the depth of the sample. From Eq. (5–4) it is apparent that the equivalent resistance between opposite sides of a square is independent of the surface area of the square. This relationship gives rise to the term "sheet resistance" which has units of ohms per square. A plot of resistivity as a function of impurity concentration is shown in Fig. 5–19. Since the resistivity and depth of a particular semiconductor layer are generally known, the equivalent resistance of a thin layer of semiconductor material can be calculated by the simple process of counting squares. This procedure is illustrated in the following example.

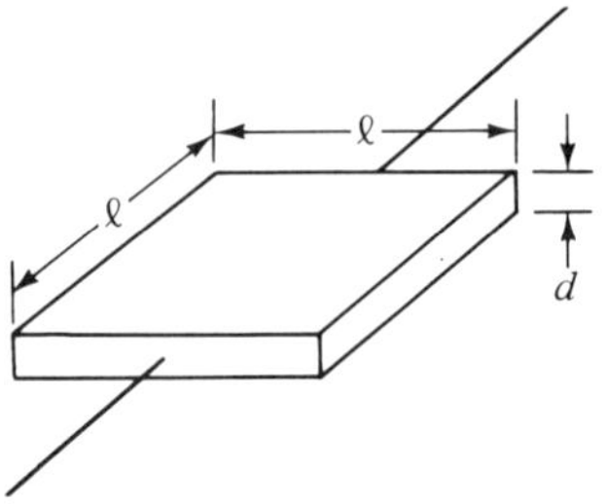

FIGURE 5–18 *Model showing dimensions for the calculation of sheet resistance.*

EXAMPLE 5–2 Calculate the resistance of a film of semiconductor material measuring 1 mil × 5 mils formed during the base diffusion of the transistor whose characteristics are given in Ex. 5–1.

Solution: From Ex. 5–1, the base impurity concentration is 10^{18} atoms/cc. From Fig. 5–19, $\rho_B = 7 \times 10^{-2}$ ohm-cm. The depth of the base diffusion was 3 μ. From Eq. (5–4), the sheet resistance of the base region is

$$R_{SB} = \rho_B / X_B = (7 \times 10^{-2} \text{ ohm-cm})/3\ \mu \simeq 235 \text{ ohm/sq}$$

A sample 1 mil × 5 mils consists of 5 equivalent squares with dimensions of 1 mil × 1 mil. Consequently, the total resistance of the 1 mil × 5 mils strip is

$$R = (5)(R_{SB}) = (5)(235) = 1175\ \Omega$$

Monolithic resistors can be formed in any geometry which will best utilize the available surface area of the chip. Regardless of the configuration, however, regions or pads are often formed at each end of the resistor to allow metallization and contact to the other components of the IC. An example of a contact pad is

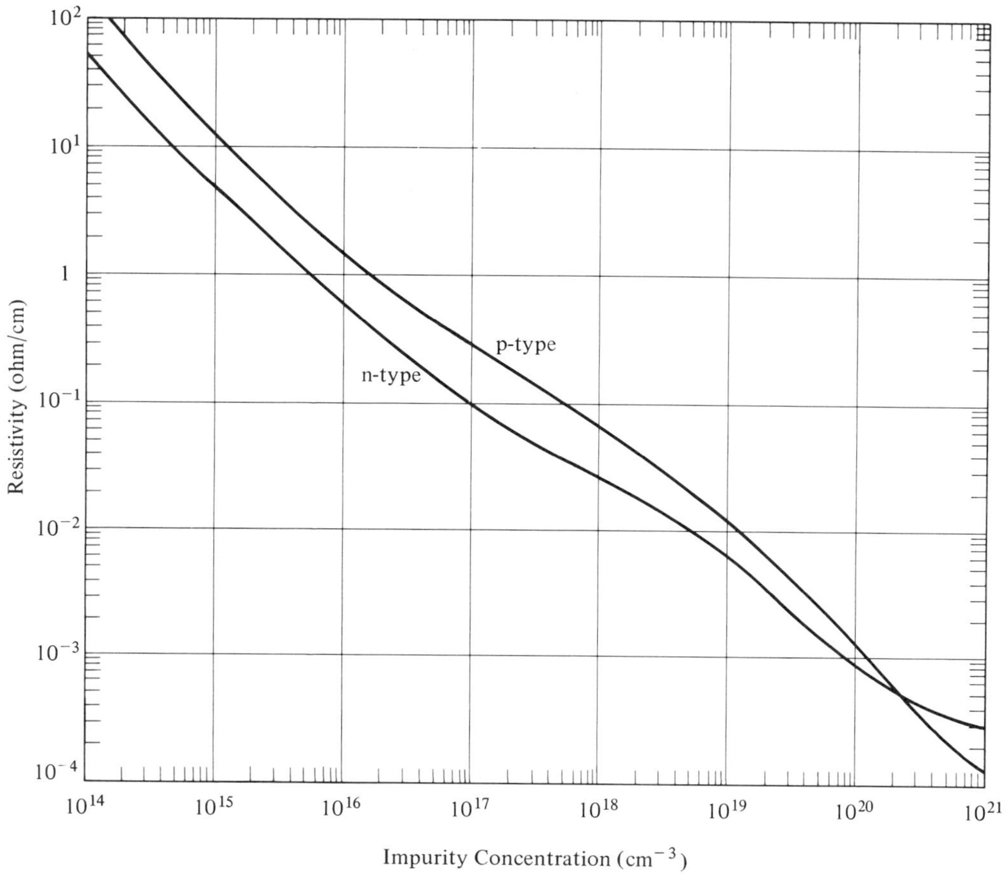

FIGURE 5–19 *Resistivity as a function of impurity concentration. (J. C. Irwin, "Resistivity of Bulk Silicon and of Diffused Layers in Silicon," BSTJ, 41 (March, 1962), 388. Copyright, 1962, The American Telephone and Telegraph Co., reprinted by permission.)*

shown in Fig. 5–20(a). The pad width is larger than the metallization to insure that the metallization does not short circuit the resistor to the substrate. The equivalent resistance between the end of the resistor and the metal contact is approximately equal to that of 0.65 square.*

Resistor geometries other than the straight-line resistors are often required in monolithic ICs to best use the available chip area. The two most common geometries are the quarter-bend shown in Fig. 5–20(b) and the half-bend shown in Fig. 5–20(c). Normally, square corners are avoided because of the high current densities that they produce. The equivalent resistance between points X and *Y* for the quarter-bend is approximately 2.45 squares, and that for a half-bend is approximately 3.88 squares.* The following example illustrates the calculation of the equivalent resistance of a resistor formed using quarter- and half-bend geometries.

EXAMPLE 5–3 Calculate the resistance between contact *X* and contact *Y* of the monolithic resistor shown in Fig. 5–21. The sheet resistance is 200 ohms/sq.

---

* See Camenzind in the references at the end of the chapter.

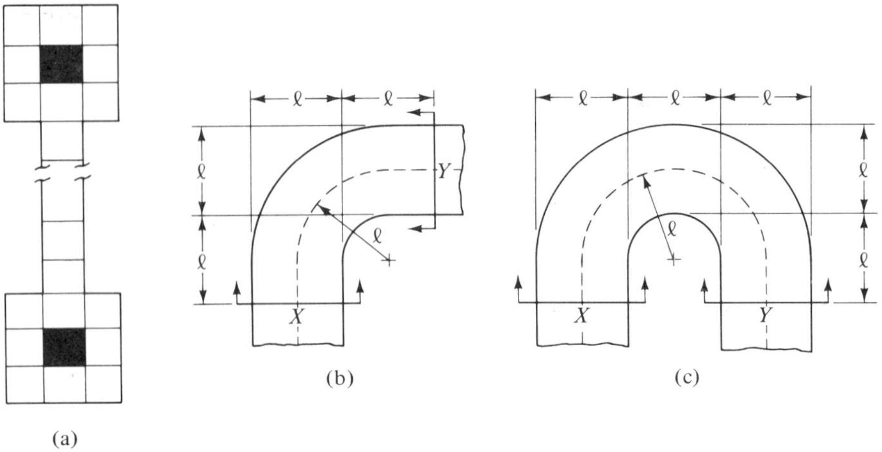

FIGURE 5–20 *Various resistor geometries encountered in monolithic ICs: (a) contact pad, (b) quarter-bend, and (c) half-bend. (Adapted from H. R. Camenzind,* Circuit Design for Integrated Electronics, *1968, Addison-Wesley, Reading, Mass.*

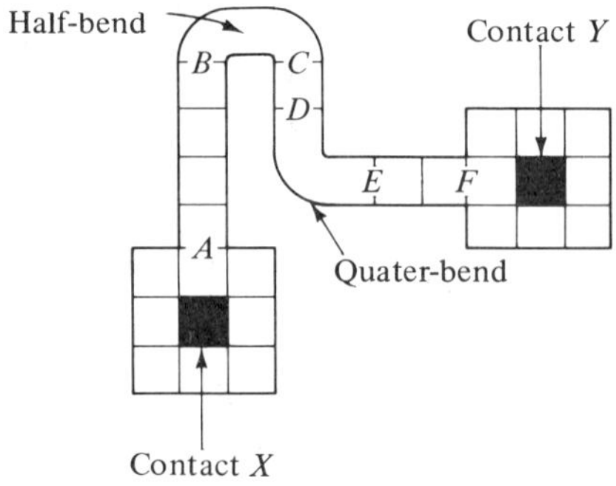

FIGURE 5–21 *The monolithic resistor used in Example 5–3.*

Solution: The equivalent resistance is calculated by dividing the resistor into various parts which correspond to the diagrams shown in Fig. 5–20. Then the total resistance is

$$R = R_{XA} + R_{AB} + R_{BC} + R_{CD} + R_{DE} + R_{EF} + R_{FY}$$

Using the factors given above, the resistance can be calculated as

$$R = (0.65)(200) + (4)(200) + (3.88)(200) + (1)(200) + (2.45)(200) + (2)(200) + (0.65)(200) \simeq 2930\ \Omega$$

The second most common passive component used in monolithic ICs is the capacitor. Monolithic capacitors can be considered as parallel plate capacitors whose capacitance is given by

$$C = K\varepsilon_o A/d \text{ Farads} \quad \textbf{(5–5)}$$

where K is the relative dielectric constant of the insulating material; ε_o is the permittivity of free space (8.85×10^{-14} F/cm); A is the area in cm^2; and d is the distance between the plates in cm. For the metal-insulator-semiconductor (MOS) capacitor shown in Fig. 5–22, the dielectric is silicon dioxide which has a relative dielectric constant ranging from 2.7 to 4.2. Assuming the highest value, the capacitance is

$$C = 9.4 \times 10^{-3}\ A/d \text{ pF} \quad \textbf{(5–6)}$$

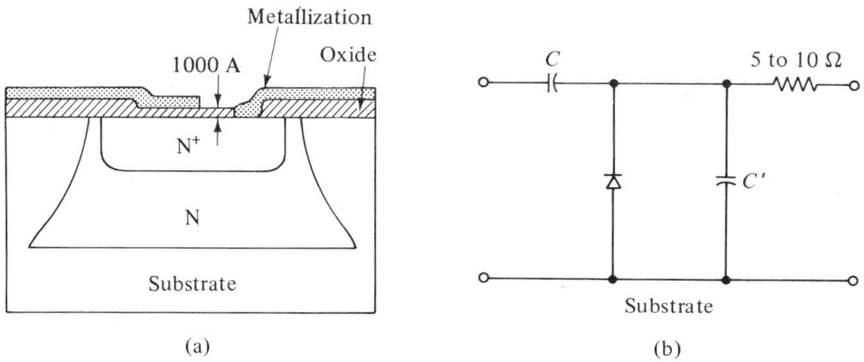

FIGURE 5–22 *A metal-insulator-semiconductor capacitor: (a) a cross-sectional view, and (b) an approximate equivalent circuit.*

where A is given in square mils and d in mils. For a typical oxide thickness of 1000 A, the capacitance per unit area is approximately 0.25 pF/mil^2. One advantage of this type of capacitor is that it is nonpolar; i.e., voltages of either polarity can be applied. However, a capacitor having a reasonable value requires a large surface area. For example, using the value found above, a 100 pF capacitor requires 400 mil^2. Note that the thickness of the SiO_2 dielectric is reduced in the capacitor area. A thickness of 1000 A has a breakdown voltage of approximately 50 V.

Unfortunately, several parasitic components are associated with a MOS capacitor as shown in Fig. 5–22(b). A junction diode and C' represent the collector-substrate junction. If this junction is reverse biased with 8 to 10 V, then C is 6 to 8 times larger than C'. The resistor R represents the ac contact resistance plus the resistance of the N^+ layer.

A second type of capacitor used in monolithic ICs is the junction capacitance, C_T, associated with a reverse-biased P-N junction. A cross-sectional view of a capacitor of this type is shown in Fig. 5–23(a). The P region can be formed either during the isolation diffusion or during the base diffusion. From Eq. (2–28),

$$C_T = K(V_T)^{-1/2} = A(q\varepsilon KN/2V_T)^{1/2} \tag{5–7}$$

where N is the doping level of the lighter doped material, and V_T is the total reverse bias of the junction. As indicated in Chap. 2, the total reverse bias is equal to the applied voltage, V, plus the barrier voltage, V_B (approximately 0.6 V for silicon at room temperature). The relationship between C_T and V_T/N is given in Fig. 5–23(b).

When calculating the capacitance associated with the reverse-biased junction, the area of the side walls of the junction must also be considered. The junction formed by the side walls is characterized by nonhomogeneous doping which complicates the calculation. However, an estimate of the capacitance can be made by using the approximation of an abrupt junction with homogeneous doping. The following example illustrates the calculation of the collector-to-substrate capacitance of the single-base-strip geometry given in Fig. 5–9.

EXAMPLE 5–4 Calculate the collector-to-substrate capacitance of the single-base-strip geometry shown in Fig. 5–10 when the substrate is reverse biased with 3.8 V. The doping levels are given in Ex. 5–1.

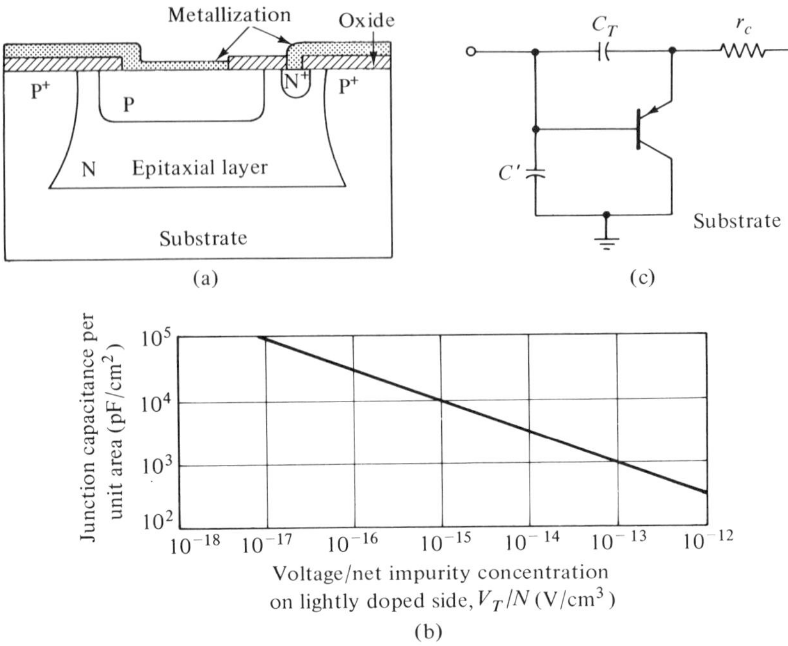

FIGURE 5–23 *A monolithic junction capacitor: (a) cross section, (b) a plot of* C_T *as a function of* V_T/N *where N is the impurity concentration of the lighter doped material, and* V_T *is the total junction voltage, and (c) an approximate equivalent circuit.*

Solution: From Fig. 5–10, the area of the bottom of the isolation island is 6.5 mils × 9 mils. This area also corresponds to the collector area. The substrate (except for the P^+ isolation diffusion) is more lightly doped than the collector and has an impurity concentration of 10^{15} atoms/cm³. The ratio V_T/N is

$$V_T/N = (3.8 + 0.6)\ \text{V}/10^{15} \simeq 4.4 \times 10^{-15}$$

From Fig. 5–23(b),

$$C_T/A \simeq 5 \times 10^3\ \text{pF/cm}^2$$

The bottom area of the collector is

$$A_C = 6.5\ \text{mils} \times 9\ \text{mils} \simeq 58.5\ \text{mils}^2$$

The component of capacitance due to the collector bottom is

$$C_{CS(B)} \simeq (5 \times 10^3\ \text{pF/cm}^2)(6.45 \times 10^{-6}\ \text{cm}^2/\text{mils}^2)(58.5\ \text{mils}^2) \simeq 1.9\ \text{pF}$$

The side wall component of the collector substrate capacitance is calculated in a similar manner except that in this case the collector region is the lighter doped region. The side wall area is

$$A_S = (1)(6.5 + 9 + 6.5 + 9)\ \text{mils}^2 \simeq 31\ \text{mils}^2$$

The ratio V_T/N is

$$V_T/N = 4.4\ \text{V}/10^{16} \simeq 4.4 \times 10^{-16}$$

From Fig. 5–23(b),

$$C_T/A \simeq 1.5 \times 10^4\ \text{pF/cm}^2$$

The side wall component of the capacitance is

$$C_{CS(S)} \simeq (1.5 \times 10^4 \text{ pF/cm}^2)(6.45 \times 10^{-6} \text{ cm}^2/\text{mils}^2)(31 \text{ mils})^2 \simeq 3.0 \text{ pF}$$

The total capacitance between the collector and the substrate is the sum of the bottom component and the side wall component, or

$$C_{CS} \simeq C_{CS(B)} + C_{CS(S)} \simeq (1.9 + 3.0) \text{ pF} \simeq 4.9 \text{ pF}$$

The parasitic components associated with a junction type capacitor are shown in Fig. 5–23(c). These components include the capacitance across the reverse biased collector-substrate junction (C'), the contact resistance (r_c), and a low β PNP transistor. Because of these parasitic components, a junction type capacitor should not be used unless absolutely necessary. In some applications, this type may be used as a bypass capacitor but almost never as a coupling capacitor.

The remaining passive component to be considered is the inductor. While a monolithic inductor can be fabricated, the Q of the inductor is very low. Consequently, inductors are usually connected to the circuit external to the monolithic chip. Alternately, an inductor can be simulated using capacitors and IC amplifiers. One example is the capacitively loaded gyrator which can simulate an inductor with a Q as high as 1000.

5.5 Monolithic Circuit Design Guides

The goal of any design is to meet a set of performance specifications with the required circuit reliability and at a minimum cost. The objectives of this section are to review the characteristics of monolithic components and to establish some general guides for the design of circuits which are to be fabricated in monolithic form.

Monolithic transistors are similar to discrete transistors since both types are fabricated by the same process. The principal difference is that PNP active devices increase the cost of an IC and usually result in low β devices. The monolithic circuit designer is generally limited to using NPN transistors.* The use of lateral PNP transistors should be minimized since their characteristics are generally inferior to those of the NPN transistor. This limitation is not serious, however, since good circuit design can circumvent the need for a large number of complimentary devices. An example is the "level shift" circuit discussed later in this section.

The principal difference between monolithic and discrete components occurs with the passive components. High quality resistors and capacitors whose tolerance and temperature coefficients are small are readily available in discrete form. Monolithic passive components, on the other hand, are characterized by poor tolerances and poor temperature coefficients. Resistance values may have tolerances of ± 30 percent in absolute value and are voltage dependent. However, the ratio of two resistance values can be less than ± 5 percent since both resistors are formed at the same time and in close proximity. Large resistance values are

---

* Complementary MOS devices are not difficult to fabricate. However, these devices are normally used in logic circuits (see Sec. 5.7).

difficult to obtain due to the chip space that they would occupy.* Similarly, large capacitance values are unattainable in monolithic ICs. Using silicon dioxide as the insulator, thin film capacitance values in the order 0.25 to 0.30 pF/mil^2 can be obtained. Typical junction capacitance values, assuming a 5-V reverse bias, are: collector-substrate approximately 0.05 pF/mil^2, collector-base approximately 0.1 pF/mil^2, and base-emitter approximately 0.4 pF/mil^2. Assuming a value of 0.2 pF/mil^2, a 10-pF capacitor would require 50 mil^2, and a 1000 pF capacitor would require 5000 mil^2 of chip area. Consequently, large capacitors must either be avoided or be provided external to the monolithic chip.

Monolithic components, however, do have certain advantages over their discrete counterparts. The foremost advantage is the availability of well matched active devices which require very little chip space. Circuits using balancing techniques that depend on matched transistors can often replace circuits requiring precision passive components. One important example is in the biasing of a CE amplifier. Bias stability is achieved in discrete component circuits by using a large emitter resistor. The ac gain of the CE is maintained by bypassing the emitter resistor with a large capacitor. The requirement for a large bypass capacitor can be avoided in monolithic circuits by using one of the bias schemes shown in Fig. 5–24. These circuits are discussed below.

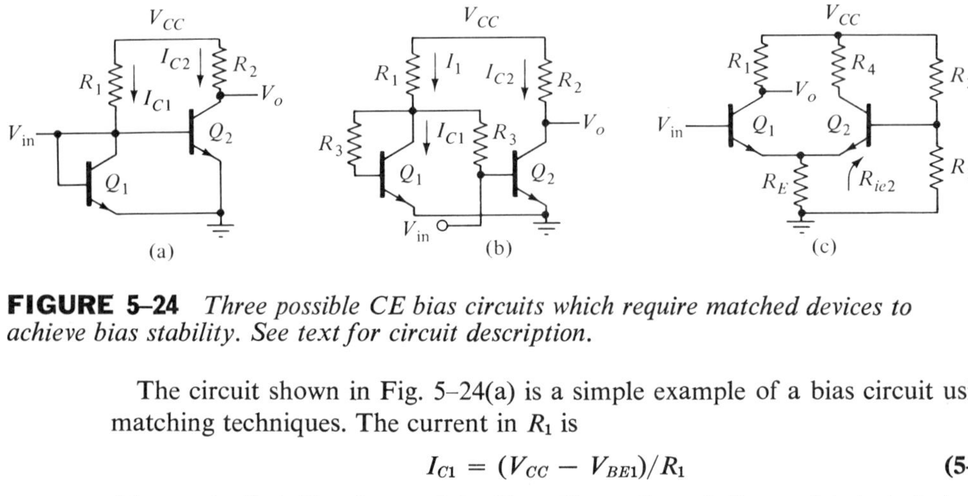

FIGURE 5–24 *Three possible CE bias circuits which require matched devices to achieve bias stability. See text for circuit description.*

The circuit shown in Fig. 5–24(a) is a simple example of a bias circuit using matching techniques. The current in R_1 is

$$I_{C1} = (V_{CC} - V_{BE1})/R_1 \tag{5–8}$$

Also, note that V_{BE1} is equal to V_{BE2}. Since Q_1 and Q_2 are fabricated simultaneously and in very close proximity, they can be treated as matched transistors. Matched transistors with the same BE bias voltage will have the same collector current. Consequently,

$$I_{C1} \simeq I_{C2} \tag{5–9}$$

* An exception to this is the "pinch" resistor which uses the emitter diffusion to reduce the thickness of a resistor formed during the base diffusion. From Eq. 5–4, it is apparent that reducing the thickness will increase the sheet resistance allowing formation of larger resistance values in a given area. Sheet resistance values in excess of 10 kilohms/square are possible. However, resistors formed by this method are nonlinear and have breakdown voltages near BV_{EBO}, or about 5 to 10 V.

The circuit shown in Fig. 5–24(b) is a modification of the circuit shown in Fig. 5–24(a). The collector currents of each transistor will be equal since their base currents are equal. For either transistor the collector current is

$$I_C = \frac{V_{CC} - V_{BE}}{R_1} - I_B\left(2 + \frac{R_3}{R_1}\right) \simeq \frac{V_{CC} - V_{BE}}{R_1} \tag{5–10}$$

where the approximation is valid for very small base currents. The dc voltage at the collector of Q_2 is

$$V_{C2} = V_{CC} - I_{C2}R_2 \simeq V_{CC}(1 - R_2/R_1) \tag{5–11}$$

where the approximation is valid if V_{CC} is much larger than V_{BE}. When R_1 is equal to $2R_2$, V_{C2} is equal to $V_{CC}/2$ regardless of the value of V_{CC}. Verification of Eqs. (5–10) and (5–11) is left as an exercise for the reader.

Possibly the most versatile single configuration is the differential amplifier, one form of which is shown in Fig. 5–24(c). In this circuit, Q_1 is connected as a CE stage with an emitter bias resistor, R_E. The emitter resistor is effectively bypassed by Q_2 since the resistance looking into the emitter of Q_2 is

$$R_{ie2} \simeq r_{e2} + [(R_2 \parallel R_3)/(\beta_2 + 1)] \tag{5–12}$$

Consequently, Q_2 performs the same function as an emitter bypass capacitor in a discrete component circuit. However, the voltage gain A_{vt1} is reduced in any differential amplifier. An approximate expression for A_{vt1} is*

$$A_{vt1} \simeq -R_1/(r_{e1} + R_{ie2}) \tag{5–13}$$

The differential amplifier can also be used to provide a positive voltage gain (no signal inversion) by taking the output from the collector of Q_2. The availability of both positive and negative voltage gains, the ability to amplify the difference between two input signals, and the inherent balance of the differential amplifier account for its wide use in monolithic ICs. The performance of a differential amplifier is usually improved by replacing R_E with a constant-current source which can be formed easily in an IC.

A bias problem occurs when all transistors are NPN types. This problem, called the "voltage buildup" problem, can be illustrated using Fig. 5–25(a). For proper circuit operation, the *BC* junction of Q_1 and Q_2 must be reversed biased by at least an amount equal to the peak voltage swing that will occur at the collector terminal. Consequently, for the circuit shown in Fig. 5–25(a), V_{C2} must be greater than V_{C1} which, in turn, must be greater than V_{IN}. The dc power supply voltage, V_{CC}, effectively limits the number of stages that can be connected in this manner.

The voltage buildup problem can be avoided in discrete component circuits by alternating NPN and PNP transistors as shown in Fig. 5–25(b). For proper bias, the collector of a PNP must be at a lower voltage than the base. For the circuit shown in Fig. 5–25(b), both V_{IN} and V_{C2} may be near the same dc voltage level. A second method of avoiding voltage buildup in discrete component circuits is to use a coupling capacitor between stages as shown in Fig. 5–25(c). In this circuit, the bias conditions of Q_1 and Q_2 are independent, and voltage

* See Sec. 4.9 and Table 4–4.

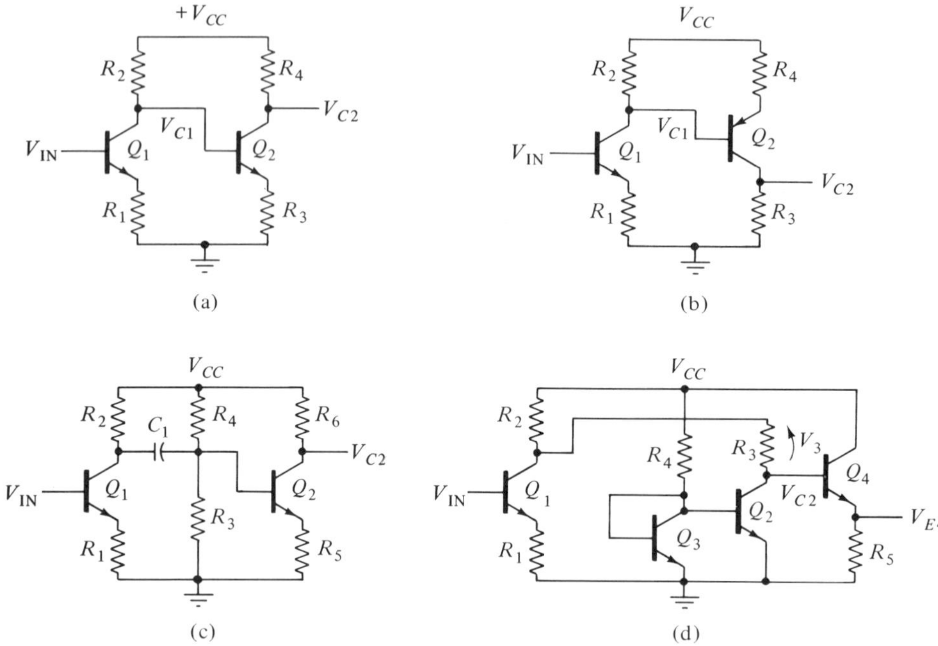

FIGURE 5–25 *Circuits used to illustrate the dc buildup problem. See text for circuit descriptions.*

buildup is not a problem. However, as indicated earlier, neither NPN-PNP combinations nor capacitive coupling are readily available in monolithic ICs.

One solution to the voltage buildup problem in a monolithic IC is the level-shift circuit shown in Fig. 5–25(d). In this circuit, Q_1 functions as a CE stage, Q_2 as constant-current source, and Q_4 as a CC stage. The bias circuit for Q_2 is the same circuit as shown in Fig. 5–24(a). The voltage across R_3 provides the desired dc voltage shift. Using this voltage drop, it is possible to design a circuit so that V_{IN} and V_{E4} are near the same dc voltage level. The ac signal gain from the collector of Q_1 to the emitter of Q_4 can be approximately unity since the resistance looking into the base of Q_4 or into the collector of Q_2 can be large compared to R_3. The level shift circuit is one example of several types of circuits that are often used to circumvent the shortcomings of monolithic components.

In this section, we have reviewed the area requirements and the characteristics of monolithic components. Area requirements are particularly important since the cost per circuit depends directly on the chip area occupied by the circuit. For this reason, large resistance values and large capacitance values should be avoided in monolithic ICs, because they require large areas. Also, complementary NPN and PNP transistors are not always available in the same monolithic circuit. On the other hand, matched active devices of high quality can be fabricated easily in small areas. These characteristics determine the general guides for the circuits which are fabricated in monolithic form. Stated simply, these design guides are:

(1) Minimize the number of resistors and capacitors used rather than number of transistors.

(2) Use circuits based on matched components.
(3) Minimize the use of PNP transistors, since the characteristics of both NPNs and PNPs cannot be optimized on the same chip.
(4) Avoid capacitors when possible since they may require an additional process and require large surface areas.
(5) Large resistors (> 20 K) should be avoided because they require large surface areas.
(6) Inductors should not be used.

5.6 Monolithic Circuit Layout Guides

Using the general guidelines that were discussed in the last section, the first step in designing an IC is to develop a design on paper. With the completion of the design on paper, the next step is to build up or "breadboard" the circuit using discrete components. This step will check the validity of the design. Once satisfactory results are obtained from the discrete model, the circuit should be breadboarded a second time using equivalent monolithic components. These components can be obtained from the manufacturer who may ultimately fabricate the IC. The second model should be extensively tested to insure that it meets the design specifications particulary with respect to temperature variations. Once satisfactory results are obtained from the second model, the layout of the IC can be started. In this section, we will examine general layout guides for monolithic circuits. The procedures will be illustrated with a simple differential amplifier. For the remainder of the section, we will assume that the initial design with two breadboard models has met successfully the desired specifications.

The geometry of the various components is the ultimate factor governing the layout of a monolithic IC. The geometry of the active device is generally fixed by constraints other than layout. Resistors and capacitors, on the other hand, are more flexible and can be laid out in a manner to minimize the overall IC area. Metallization strips for interconnections can cross any resistor since resistors are covered by an oxide layer. Bonding pads that are necessary for the connections of an external lead should be arranged around the periphery of the chip.

The first step in the layout of a circuit is to determine the approximate area required by each component. The spacing and area requirements for a typical monolithic layout are listed in Table 5–1. The values in this table are conservative and do not represent ultimate limits. Most of these dimensions are being reduced as the fabrication processes are improved.

The second step in the layout procedure is to determine the number and sizes of the required isolation regions. The function of an isolation region is to electrically isolate the various components from each other. Isolation is obtained by reverse biasing the P-N junction formed by N-type isolation islands and the P-type substrate. Connecting the substrate to the most negative voltage in the circuit would insure the proper bias for isolation.

With few exceptions, each transistor requires a separate isolation region. The only exception in linear circuits is the CC stage whose collector terminal (for an NPN) is connected to the most positive voltage in the circuit.* The transistors

* Transistors whose collectors are connected together can be formed in the same isolation region. This situation is often encountered in logic circuits.

TABLE 5–1 *Typical dimensions and parameter values used in monolithic circuits*

Components	Typical Value
Transistors:*	
single-base-strip area	9×6.5 mils²
double-base-strip area	10×6.5 mils²
Resistors:	
width	0.5 mil
length	value dependent
spacing between resistors	0.5 mil
base sheet resistance	200 ohms/square
emitter sheet resistance	2.5 ohms/square
Capacitors:	
area	value dependent
SiO_2 thin film	0.25 pF/mil²
emitter-base**	0.4 pF/mil²
collector-base**	0.1 pF/mil²
collector-substrate**	0.05 pF/mil²
Bonding pads:	
area	4×4 mils²
spacing between pads	2 mils
Isolation regions:	
spacing to components	1 mil
spacing between isolation regions	0.5 mil
Metallization:	
width	0.5 mil
spacing between metallization	0.5 mil

* 1 mil × 1.5 mils emitter area

** 5 V reverse bias

associated with these CC stages, along with all the resistors, can be placed in the same isolation region.* Electrical isolation between each resistor is obtained by connecting the N-type isolation island to the most positive voltage in the circuit. Each capacitor requires a separate isolation region.

Once the component area requirements and the required isolation regions are determined, an actual layout can be attempted. This is a trial-and-error procedure. Several atttempts are usually necessary before a layout is found which minimizes the chip area and provides the necessary interconnections without metallization crossovers. The layout generally proceeds from the circuit diagram to a geometric layout in which the components are grouped in their respective isolation regions. The required interconnections are drawn and checked for crossovers. Once an acceptable layout is found, the circuit is drawn to scale using the actual geometry of each component.

To illustrate the procedures discussed in the first part of this section, we will develop a monolithic layout for the differential amplifier shown in Fig. 5–26(a). Assume that both the discrete component breadboard and the monolithic component breadboard have satisfied the design specifications. The differential

* One exception is the "crossover" resistor formed by a short N^+ region which is used to prevent metallization crossover. Each crossover resistor requires a separate isolation region.

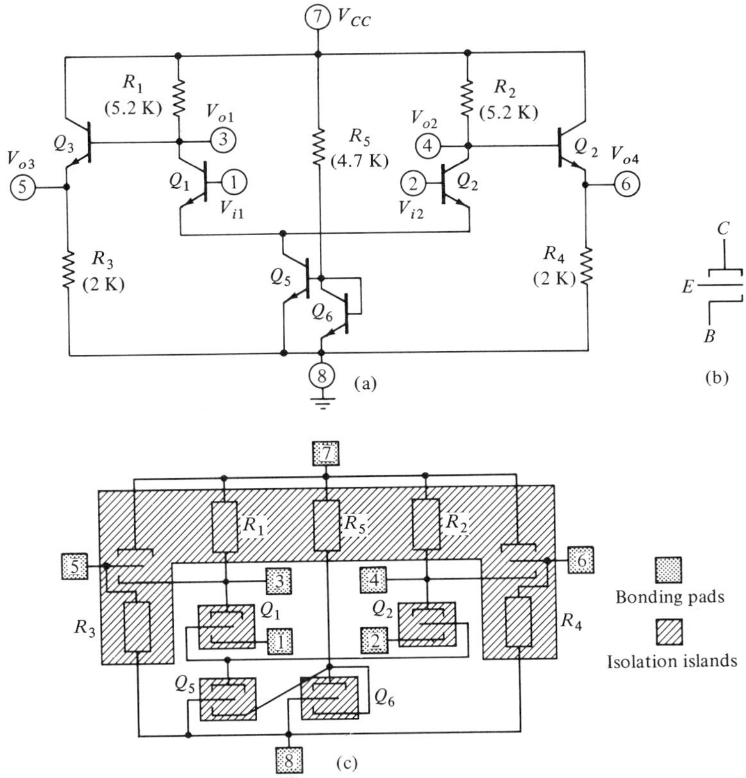

FIGURE 5–26 *A differential amplifier used to illustrate monolithic layout principles: (a) the complete circuit, (b) the geometric transistor symbol, and (c) a geometric representation of the circuit.*

amplifier has two high-resistance outputs (V_{o1} and V_{o2}) and two low-resistance outputs (V_{o3} and V_{o4}). Assuming that the transistors have a minimum β of 100, the amplifier is characterized by an input impedance greater than 5 K, an output resistance of less than 70 ohms, and a voltage gain $|A_{vt}|$ from either input to any output of approximately 100. Verification of these values is left as an exercise for the reader. Assume that V_{CC} equals 10 V.

The initial step in developing a monolithic layout is to determine the approximate area required by each component. For this calculation, we will assume that the transistors are of the single-base-strip geometry. From Table 5–1, or from Fig. 5–10, the area required by each transistor is approximately 9 $\times$ 6.5 mils^2, giving a total area for all six transistors of approximately 350 mils^2.

A calculation of the area needed by each resistor requires a knowledge of the value of sheet resistance for the material used to form resistors as well as the width of the resistors. From Table 5–1, the typical width of a resistor is 0.5 mil. Assuming that the resistors are formed during the base diffusion, which has a typical sheet resistance of 200 ohms/square, the approximate length and area required by each resistor can be calculated. For R_1, the required number of squares is

$$\frac{5.2 \text{ K}}{200 \text{ ohms/square}} = 26 \text{ squares} \tag{5–14}$$

assuming R_1 is 0.5 mil wide, the required length of R_1 is 13 mils, and the required area is 7.5 mils^2. The geometries of the various resistors are summarized in Table 5–2.

TABLE 5–2 *Summary of resistor characteristics for the differential amplifier shown in Fig. 5–26.*

Resistor	Value	Width (mil)	Length (mil)	Area (mil^2)
R_1	5.2 K	0.5	13	6.5
R_2	5.2 K	0.5	13	6.5
R_3	2 K	0.5	5	2.5
R_4	2 K	0.5	5	2.5
R_5	4.7 K	0.5	12	6.0

The total area required by all the resistors is approximately 24 mils^2. The area required by the eight bonding pads is calculated by using the typical dimensions given in Table 5–1.

$$\text{Bonding pad area} = 8(4 \text{ mils} \times 4 \text{ mils}) = 128 \text{ mils}^2 \qquad \textbf{(5–15)}$$

The total area required by all the components is

$$A_T = (350 + 24 + 128)\text{mils}^2 = 502 \text{ mils}^2 \qquad \textbf{(5–16)}$$

Note that the transistors account for approximately 70 percent of the total component area.

The next step in the layout procedure is to determine the number and size of the isolation regions. Transistors Q_1, Q_2, Q_5, and Q_6 require individual isolation regions. Since Q_3 and Q_4 are CC stages whose collector terminals connect to the most positive circuit voltage, they can be included in the resistor isolation region. Consequently, a total of five isolation regions are required.

We can now begin the actual layout of the circuit with a trial-and-error procedure. Generally, a good approach is to redraw the circuit diagram by simply replacing each component with a geometric symbol as shown in Fig. 5–26(c) where the geometric representation for the transistor is shown in Fig. 5–26(b). Also, the required isolation regions and bonding pads can be indicated. Provisions should be made for contact to the substrate and to the resistor isolation region for proper electrical isolation.

Next, attempt to convert the geometric circuit diagram into a simple layout, and check for metallization crossovers. Note that the resistors can be placed in such a manner as to ease the crossover problem. One such layout for the differential amplifier is shown in Fig. 5–27.

Once a rough layout appears satisfactory, a scale drawing of the actual layout can be attempted. In the scale drawing, the required spacing between the various components, metallization strips, and isolation regions must be observed. The final scale drawing for the differential amplifier is shown in Fig. 5–28. Note that the extent of the isolation diffusion is shown in dashed lines and the metallization strips by solid lines. The total area required for this amplifier is 32.5 mils × 27.5 mils or 894 mils^2 (excluding metallization extensions).

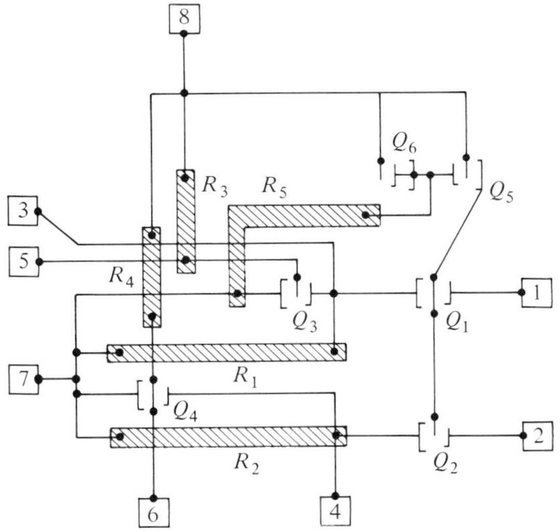

FIGURE 5–27 *Initial rough layout for the differential amplifier shown in Fig. 5–25.*

When an acceptable scale drawing of the layout has been completed, the various masks can be prepared. The dimensions for the masks can be obtained from the scale drawing. A separate mask must be prepared for: (1) the isolation diffusion, (2) the base region and resistor diffusion, (3) the emitter region and collector contact region diffusion, (4) the openings for ohmic or metallization contacts, and (5) the interconnection pattern. An additional mask may be required if capacitors are used in the circuit. The complete set of five masks required for the differential amplifier just discussed is shown in Fig. 5–29.

The layout of a monolithic IC has been discussed in this section. Some simple layout guides were developed. Typical dimensions and parameters are given in Table 5–1. However, the basic rule for monolithic layout can be stated quite simply: minimize the chip area in a manner consistent with a high manufacturing yield.

5.7 MOS Integrated Circuits

The MOSFET, first introduced in Sec. 3.2, is an active device with three terminals: gate, source, and drain. A fourth terminal, the substrate, is sometimes available but is generally connected to the source terminal to insure proper bias. Four types of MOSFET devices are available: (1) N-channel depletion-mode, (2) P-channel depletion-mode, (3) N-channel enhancement-mode, and (4) P-channel enhancement-mode.

The principal application of MOS integrated circuits is in the logic or digital area. The enhancement-mode device is particularly suited for digital circuits and dominates the MOS IC field. The P-channel enhancement-mode type is the most common active device, although both P-channel and N-channel units can be fabricated on the same chip. Because of its dominance, the P-channel enhancement-mode unit will form the basis for most of the discussion in this section.

The objective of this section is to develop an understanding of MOS IC technology. To achieve this objective, the material is presented in the following order: (1) a brief review of the fabrication and geometry of MOS devices, (2) a

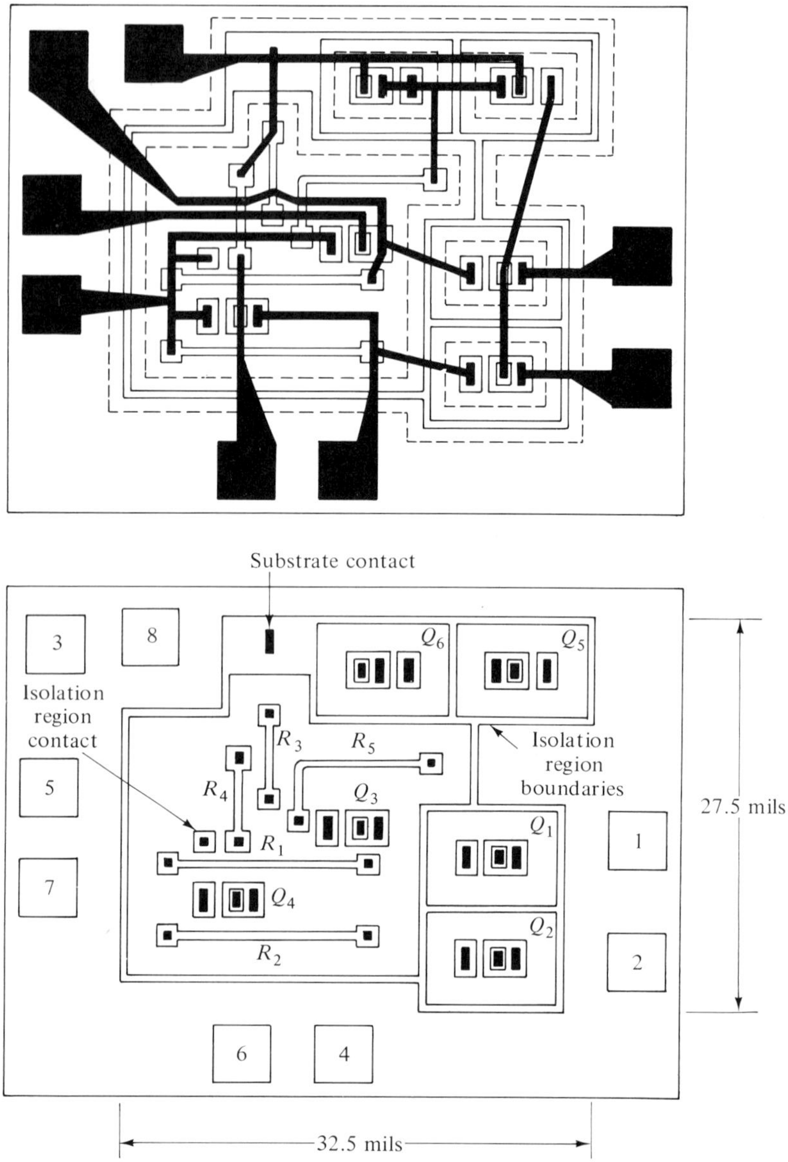

FIGURE 5–28 *Final scale drawings of the layout for the differential amplifier shown in Figs. 5–26 and 5–27. The extent of the isolation diffusion is shown by the dashed lines.*

discussion of the characteristics of MOS devices used as active devices, resistors, and constant-current sources, (3) an examination of the layout of simple MOS circuits, and (4) a comparison of MOS IC with bipolar monolithic ICs studied earlier in this chapter.

One method of fabricating a P-channel enhancement-mode MOS device begins with a substrate of lightly doped N-type silicon. After oxidation, the chip appears as shown in Fig. 5–30(a). As was the case for bipolar transistors, openings are made in the oxide layer using photoresist exposed to ultraviolet

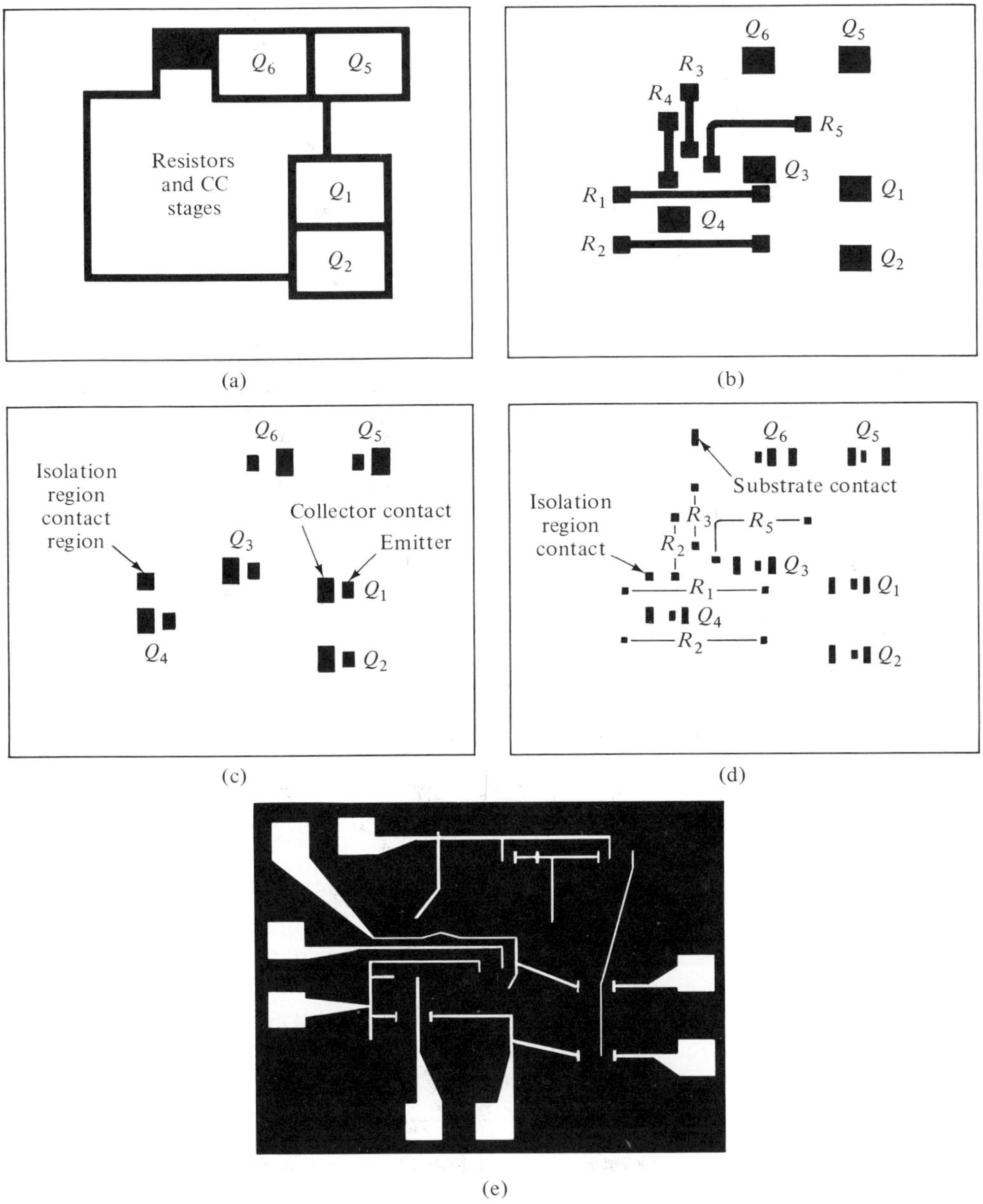

FIGURE 5–29 *Masks for the circuit shown in Fig. 5–28: (a) isolation diffusion mask, (b) base region mask, (c) emitter region mask, (d) ohmic contact mask, and (e) interconnection mask.*

light through a mask. The ultraviolet light polymerizes the exposed photoresist. The unexposed photoresist and the underlying oxide layer are removed chemically with an etching solution. The source and drain regions are formed by diffusing P^+ impurities through the oxide openings. The results of this diffusion are shown in Fig. 5–30(b). In a similar manner, openings are made in a second

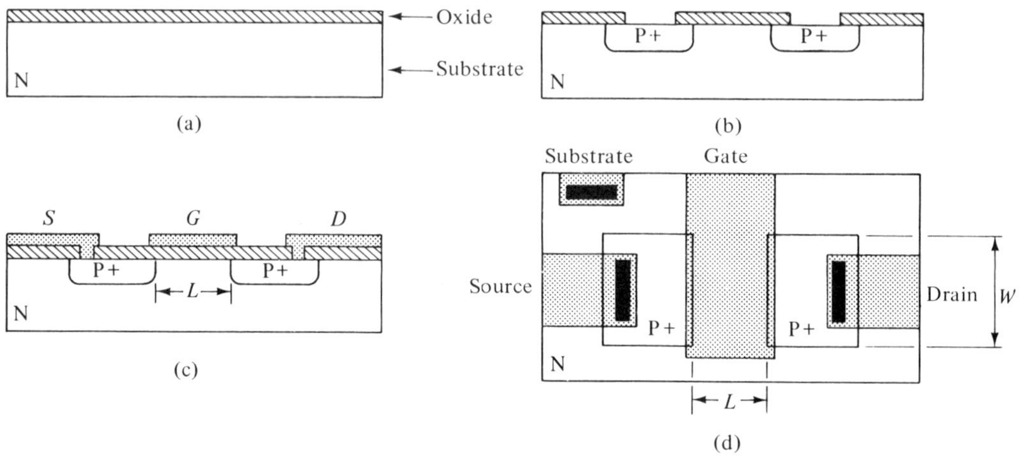

FIGURE 5–30 *Fabrication sequence for a P-channel enhancement-mode MOS device: (a) N-type substrate with oxide layer, (b) results of source and drain diffusion, (c) side view of completed device, and (d) top view of completed device. The dimension* W *indicates the width of the channel, and the dimension* L *indicates the length of the channel.*

layer to provide ohmic contacts to the source, drain, and substrate. Then the entire surface is covered with a thin film of metal. When the unwanted regions of the metallic layer are etched away, the metal forms the interconnection pattern. A side view of the completed device is shown in Fig. 5–30(c), and a top view in Fig. 5–30(d). The dimension W indicates the width of the channel, and the dimension L indicates the length of the channel.

Fabrication of the MOS device required three masks: (1) the source and drain region mask, (2) the ohmic contact mask, and (3) the interconnection pattern mask. This number compares favorably with a minimum of five masks for the bipolar device discussed in Sec. 5.2. Reducing the number of diffusion steps results in a higher manufacturing yield. Consequently, the cost per MOS device is lower than for a bipolar device.

In addition to a simple fabrication process, the MOS device is ideally suited for integrated circuits because it can be used as an amplifier, a resistor, or a constant-current source. A simple CS amplifier of the type studied in Sec. 3.6 is shown in Fig. 5–31(a). From Table 3–1, the gain of this amplifier is $-g_m R_L$. An MOS device connected as a resistor is shown in Fig. 5–31(b). In this configuration, the gate and drain are usually connected together, and the substrate is connected to ground. The resistance looking into the source terminal is given in Table 3–1 as $1/g_m$. By combining the circuits shown in Figs. 5–31(a) and (b), a CS amplifier can be formed using only MOS devices as shown in Fig. 5–31(c). In this figure, Q_1 is the CS stage, and Q_2 is the load resistor.* The gain of this amplifier is $-g_{m1}/g_{m2}$ since R_{Leq1} is equal to $1/g_{m2}$. When the CS amplifier in Fig. 5–31(c) is fabricated as an IC, the substrate of each device will be common, and both must be connected to the same potential, ground in this case.

* This configuration can also be compared to a CS-CG amplifier with the output taken from the drain of the CS stage.

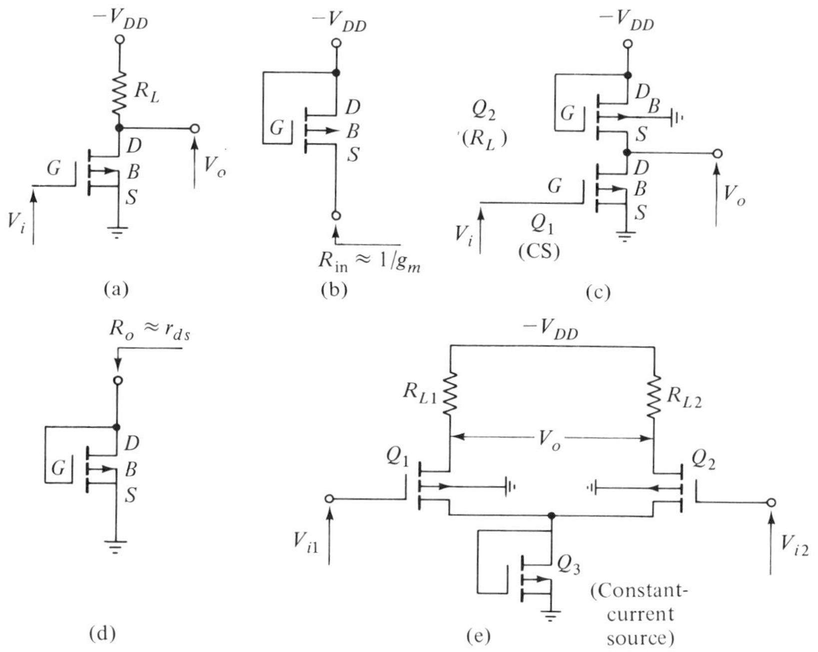

FIGURE 5–31 *Various circuit configurations using MOS devices: (a) a CS amplifier, (b) a resistor with a value of $1/g_m$, (c) a CS amplifier with an MOS device as a load resistor, (d) a constant-current source, and (e) a differential amplifier with an MOS constant-current source.*

In Fig. 5–31(d), the MOS device is connected as a constant-current source. This configuration is identical to that shown in Fig. 5–31(b) except that the resistance looking into the drain terminal is now the important quantity. From Table 3–1, the resistance looking into the drain terminal is r_{ds} (50 K to 200 K). A constant-current source is often used in a differential amplifier, one form of which is shown in Fig. 5–31(e). In this circuit, Q_1 and Q_2 form the differential pair, and Q_3 functions as a constant-current source. Note that all three substrate terminals are connected to the same potential (ground). An all-MOS differential amplifier can be made by replacing R_{L1} and R_{L2} with MOS resistors [Fig. 5–31(b)]. Both the all-MOS CS amplifier, Fig. 5–31(c), and the all-MOS differential amplifier will be designed in integrated circuit form later in this section.

Capacitors are also available using the MOS configuration shown in Fig. 5–22. However, capacitors are seldom needed in MOS ICs because the gate is electrically insulated from the channel, a condition that allows stages to be directly coupled.

To design an all-MOS IC, first determine the relationship between the device performance parameters and the physical dimensions of the device. The MOSFET was characterized in Sec. 3.2 in terms of its static characteristics. From Eq. (3–6),

$$I_D = I_{DO}\,[1 - (V_{GS}/V_T)]^2 \tag{5–17}$$

Equation (5–17) can be rewritten as

$$I_D = (I_{DO}/V_T^2)(V_{GS} - V_T)^2 \tag{5–18}$$

The constant I_{DO}/V_T^2 can be expressed in terms of the physical parameters of the device as $K(W/L)$ where K is a device constant with a typical value of 2 micromhos/volt,* W is the channel width, and L is the channel length. The ratio (W/L) is also called the *aspect ratio* and governs many of the properties of an MOS device. The drain current can now be expressed as

$$I_D \simeq (2 \times 10^{-6})(W/L)(V_{GS} - V_T)^2 \qquad \textbf{(5–19)}$$

This equation gives the drain current in the saturated region, i.e., for bias voltages that satisfy the conditions

$$|V_{DS}| > |V_{GS} - V_T| \quad \text{and} \quad |V_{GS}| > |V_T| \qquad \textbf{(5–20)}$$

Often, an MOS is biased with V_{GS} equal to V_{DS}, a condition that insures operation in the saturation region of the characteristic curves.

A second parameter of importance is the transconductance, g_m. From Eq. (3–14),

$$g_m = \frac{2}{V_T}\sqrt{I_D I_{DO}} = 2(I_{DO}/V_T^2)^{1/2}\sqrt{I_D} = 2\sqrt{2}(W/L)^{1/2}\sqrt{I_D} \qquad \textbf{(5–21)}$$

Consider the all-MOS CS amplifier shown in Fig. 5–32(a). Our immediate task is to design the integrated circuit layout for this amplifier and calculate the required power supply voltage. The gain of a CS amplifier is given in Table 3–1 as $-g_{m1}R_{Leq1}$. For the circuit shown in Fig. 5–32(a),

$$R_{Leq1} = R_{out1} \,||\, R_{in2} \simeq r_{ds1} \,||\, 1/g_{m2} \simeq 1/g_{m2} \qquad \textbf{(5–22)}$$

Then

$$A_{vt1} \simeq -g_{m1}R_{Leq1} = -g_{m1}/g_{m2} \qquad \textbf{(5–23)}$$

Combining the above expressions and noting that I_{D1} is equal to I_{D2}, the gain is

$$A_{vt1} \simeq -\left(\frac{W_1/L_1}{W_2/L_2}\right)^{1/2} \simeq -\left(\frac{\text{Aspect ratio of } Q_1}{\text{Aspect ratio of } Q_2}\right)^{1/2} \qquad \textbf{(5–24)}$$

Consequently, the gain of the CS amplifier shown in Fig. 5–32 depends primarily on the dimensions of the channels of the two MOS devices. The value of the gain can be adjusted by varying the aspect ratio of either device.

The relative values of the bias voltages for the two MOS devices in Fig. 5–32(a) are also dependent on the ratio of the aspect ratios. With equal drain currents and the bias conditions given in Eq. (5–20),

$$(W_1/L_1)(V_{GS1} - V_{T1})^2 = (W_2/L_2)(V_{GS2} - V_{T2})^2 \qquad \textbf{(5–25)}$$

Consequently,

$$\frac{V_{GS2} - V_{T2}}{V_{GS1} - V_{T1}} = \left(\frac{W_1/L_1}{W_2/L_2}\right)^{1/2} = |A_{vt}| \qquad \textbf{(5–26)}$$

The application of Eq. (5–26) to determine the bias conditions for an all-MOS CS amplifier is illustrated later in Ex. 5–5.

Before the design of an MOS IC can be completed, we need to examine the typical spacing and area requirements for the various components. As with the bipolar monolithic IC, the resolution of the photographic and diffusion processes places a lower limit on the size of an integrated component. Since the source and drain regions are simply regions into which P+ impurities have been

---

* See R. S. C. Cobbold in the references at the end of the chapter.

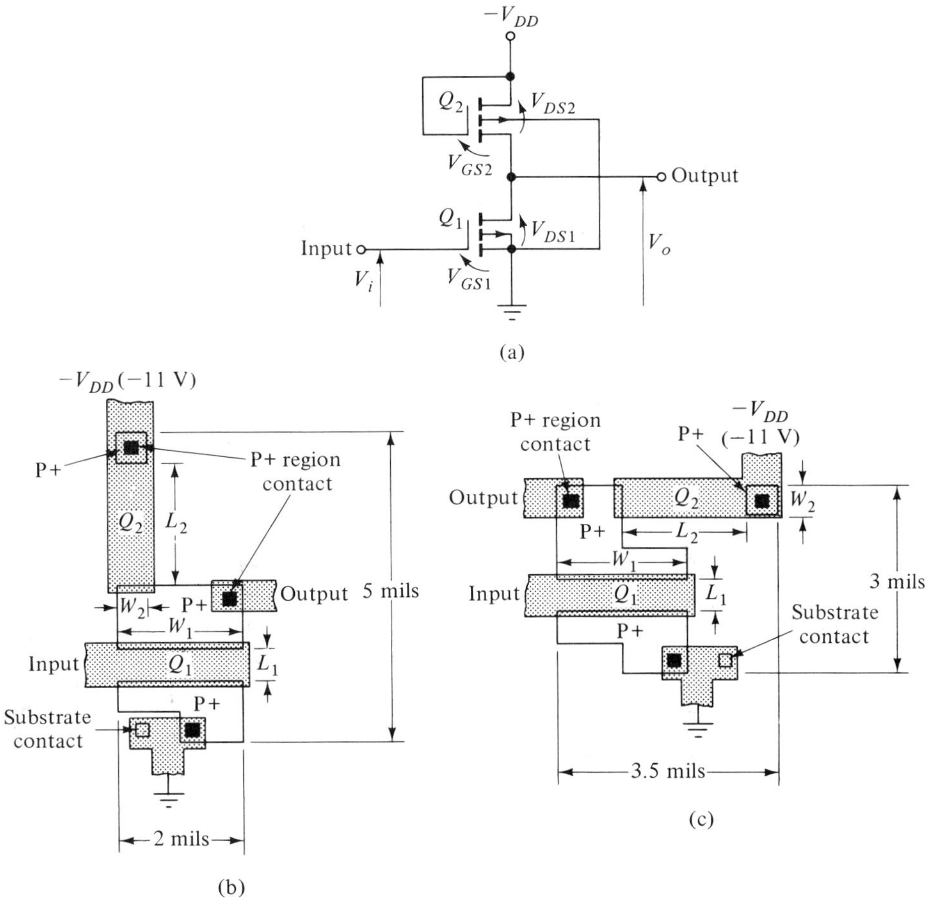

FIGURE 5–32 *An all-MOS CS amplifier: (a) the circuit diagram, (b) the top view of an IC layout, and (c) an alternate layout. Dimension values apply to Ex. 5–5.*

diffused, they resemble the diffused resistor discussed in Sec. 5.4. From Table 5–1, the typical width of a diffused resistor was 0.5 mil, and the typical spacing between resistors was 0.5 mil. Using these same criteria, the minimum dimension of the source or drain region is 0.5 mil, and the minimum spacing between P+ regions is 0.5 mil. Also, from Table 5–1, the typical width of a metallization strip and the spacing between metallization strips is 0.5 mil. Using these criteria, the minimum channel dimensions, either length (L) or width (W) is 0.5 mil. The typical dimensions for MOS devices used in integrated circuits are summarized in Table 5–3.

EXAMPLE 5–5 Design an all-MOS CS amplifier of the type shown in Fig. 5–32(a) having a gain of −4. Assume that V_T is −3 V for both devices.

Solution: (a) Circuit layout: From Eq. (5–24),

$$A_{vt1} \simeq -\left(\frac{W_1/L_1}{W_2/L_2}\right)^{1/2} = -4 \quad \text{or} \quad W_1/L_1 = 16(W_2/L_2)$$

TABLE 5–3 *Typical dimensions of MOS devices used in integrated circuits.**

Component	Typical Value
Source and drain region	
smallest dimensions	0.5 mil
spacing between regions	0.5 mil
Metallization	
width	0.5 mil
spacing between metallization	0.5 mil
Channel	
smallest dimensions (W or L)	0.5 mil
Gate overlap of source and drain regions	0.1 mil
Ohmic contact openings	0.2×0.2 mil²
Spacing between components	0.5 mil

* The values given in Table 5–3 should not be considered as state-of-the-art. They correspond to the values given in Table 5–1 for bipolar integrated devices and are used to allow easy comparison between the relative sizes of the two types of integrated components. These dimensions have already been reduced to 0.1 those given in Table 5–3 for Si-gate MOS units with further reductions highly probable.

If we choose W_1/L_1 equal to 4, then

$$W_1/L_1 = 4 \qquad W_2/L_2 = 1/4$$
$$W_1 = 4L_1 \qquad L_2 = 4W_2$$

From Table 5–3, the smallest channel dimension is typically 0.5 mil. Using this value,

$$L_1 = W_2 = 0.5 \text{ mil} \qquad W_1 = 4L_1 = 2 \text{ mils} \qquad L_2 = 4W_2 = 2 \text{ mils}$$

Once the length and width of each channel have been determined, the layout can be completed. Two possible layouts are shown in Figs. 5–32(b) and (c). The spacing requirements given in Table 5–3 have been observed in both layouts. Note in each layout that the drain of Q_1 and the source of Q_2 are formed by the same P^+ region. Also, the gate and drain of Q_2 are connected together by simply extending the gate metallization of Q_2 over the ohmic contact opening of the drain region. The substrate is connected to the source of Q_1, and both are connected to ground. This arrangement insures a reverse bias between each P^+ region and the substrate for a diode.

(b) Power supply requirement: The magnitude of V_{GS1} is chosen greater than the magnitude of V_T to insure that the device is conducting. Since V_T is equal to -3 V, V_{GS1} can be chosen as -4 V. Then,

$$V_{GS2} - V_T = |A_{vt}|(V_{GS1} - V_T) = (4)(-4 + 3) = -4 \text{ V}$$
$$V_{GS2} = -4 \text{ V} - 3 \text{ V} = -7 \text{ V}$$

Each MOS device is biased in the saturation region if $V_{GS} = V_{DS}$. Consequently, choose $V_{DS1} = V_{GS1} = -4$ V and $V_{DS2} = V_{GS2} = -7$ V. Proper bias is insured if the supply voltage is

$$V_{DD} = V_{DS1} + V_{DS2} = -4 \text{ V} - 7 \text{ V} = -11 \text{ V}$$

The gate terminal of Q_1 can be biased at -4 V either by the input generator or by connecting a large resistor (several megohms) between the gate and drain of Q_1 as shown in Fig. 3–11(e).

(c) Device parameters: From Eq. (5–19),

$$I_{D1} \simeq (2 \times 10^{-3})(W_1/L_1)(V_{GS1} - V_T)^2 \text{ mA} \simeq 8\ \mu\text{A}$$

From Eq. (5–21),

$$g_{m1} \simeq 2\sqrt{2}\,(W_1/L_1)^{1/2}\sqrt{I_D} \text{ millimhos} \simeq 16 \text{ micromhos}$$
$$g_{m2} = 4 \text{ micromhos}$$

The preceding example illustrates the design of an MOS IC. The total area of the layout shown in Fig. 5–32(b) is 2 mils × 5 mils or 10 mils², excluding the metallization extensions. The total area of the layout in Fig. 5–32(c) is 3 mils × 3.5 mils or 10.5 mils². This total area is only slightly larger than that of the area of the base region of the bipolar transistor shown in Fig. 5–4 which has an area of 2.5 mils × 3 mils or 7.5 mils². Consequently, a complete CS amplifier can be fabricated as an MOS IC using a chip area only slightly larger than the area required for the base region of a bipolar transistor. This comparison illustrates the size advantage of MOS ICs. However, note that the gain, A_{vt1}, was only -4, and that g_m for the CS stage was only 16 micromhos. Low values for g_m are very undesirable in linear circuits, but are acceptable in digital or logic circuits. Consequently, most MOS ICs are used in logic circuits.

As a second example of an MOS IC, consider the differential amplifier shown in Fig. 5–33(a). If a differential gain of -4 is desired with the output signal obtained between the two output terminals as shown in Fig. 5–33(a), the CS

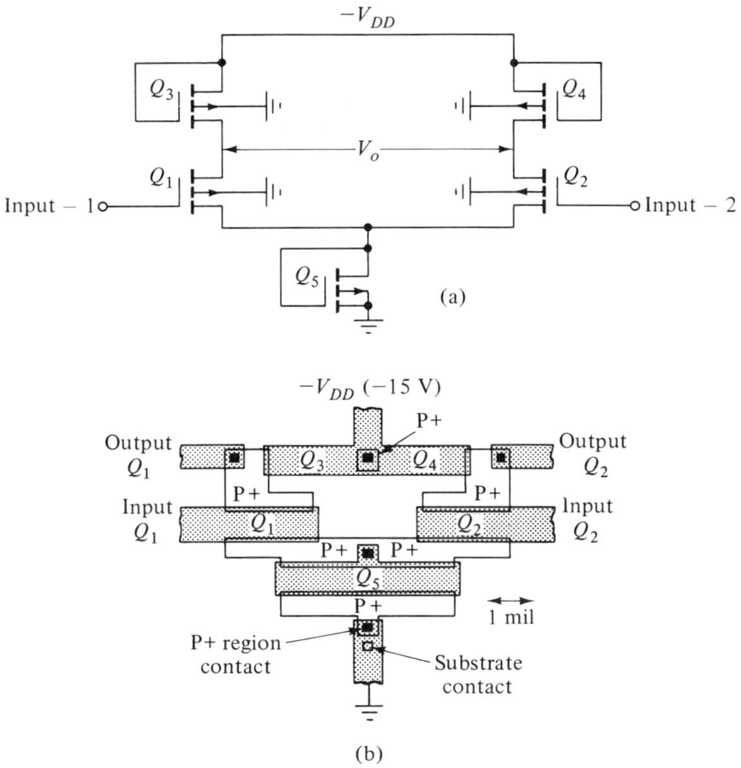

FIGURE 5–33 *An all-MOS differential amplifier: (a) the circuit diagram, and (b) the top view of an IC layout of the circuit.*

amplifier shown in Fig. 5–32 can be used for each side of the differential amplifier. The constant-current source, Q_5, is the only remaining component to be designed. Note that I_{D5} is equal to $(I_{D1} + I_{D2})$ or $2\,I_{D1}$ for a balanced differential amplifier. If V_{GS5} is -4 V and V_T is -3 V as before, the aspect ratio of Q_5 can be calculated from Eq. (5–19) as 8. Letting L_5 equal 0.5 mil, W_5 is 4 mils. The supply voltage must also be changed to -15 V to provide the additional -4 V for V_{DS5}. One possible layout for the differential amplifier is shown in Fig. 5–33(b) where the CS amplifier layout shown in Fig. 5–32(c) has been used to form each side of the differential amplifier. The total area required for this amplifier is 6.5 mils × 4.6 mils or 29.9 mils² (excluding metallization extensions) compared with 894 mils² required by the bipolar differential amplifier shown in Fig. 5–28. Verification of the design of the differential amplifier shown in Fig. 5–33 is left as an exercise for the reader.

An extension of the CS or inverter stage discussed earlier in this section is the complementary circuit* shown in Fig. 5–34(a). A complementary circuit gener-

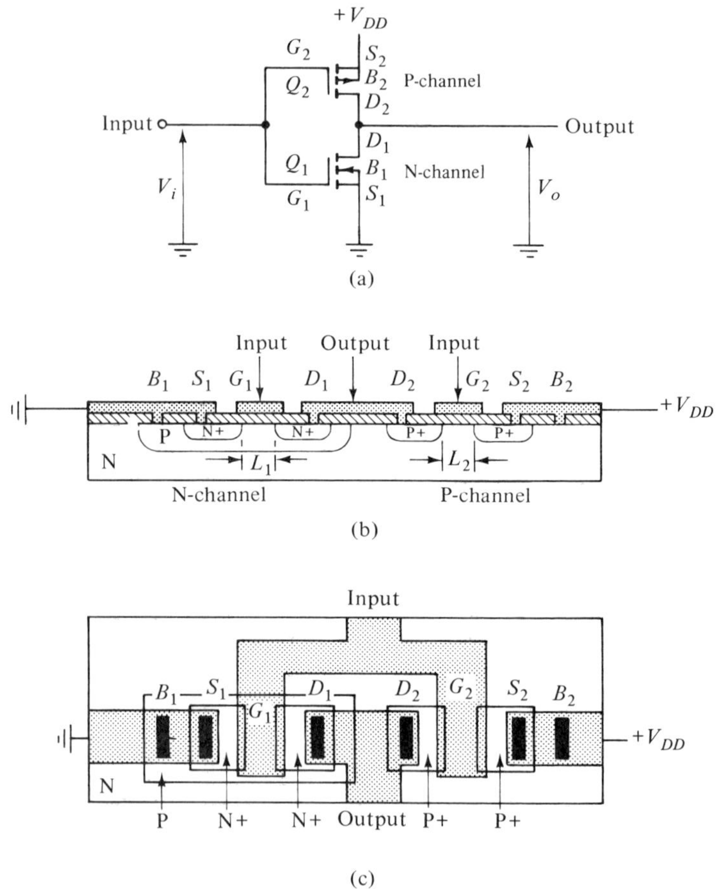

FIGURE 5–34 *A complementary MOS CS amplifier: (a) a circuit diagram, (b) a side view, and (c) a top view of a circuit layout.*

* A complementary circuit is characterized by the use of N-channel and P-channel devices (or NPN and PNP) in the same circuit.

ally has a higher speed of response than the more conventional circuit shown in Fig. 5–32. Since the input signal drives the gate of both devices, it is meaningless to term one device the CS stage and the other device the load resistor. Side and top views of a circuit layout are shown in Figs. 5–34(b) and (c). In addition to the P^+ diffusion required to form the P-channel device, two other diffusions are required to form the N-channel device: one to form the P-type substrate and the second to form the N^+ source and drain regions.

In this section, we have examined the fabrication, geometry, and characteristics of MOS devices used in integrated circuits. The basic MOS device can serve as an amplifier, a resistor, or a constant-current source. Since the gate is electrically insulated from the channel, no coupling capacitor is required between stages. Consequently, an MOS circuit can be fabricated without using conventional monolithic passive components. MOS integrated circuits find their widest application in logic or digital circuits where their advantages of small size, high input resistance, high component density, and low power requirements are particularly attractive. Since the MOS IC requires fewer diffusion steps, has a smaller size, and produces a higher yield, the cost per circuit can be less than that for the bipolar IC studied earlier. The bipolar IC, however, does have a higher speed of response than the MOS IC. Like the discrete MOSFET, MOS ICs are seldom used in linear circuits because of their low gain. However, some integrated amplifiers do use FETs in combination with bipolar devices to take advantage of the desirable characteristics of each. Additional improvements and applications of MOS devices in both linear and digital ICs will probably occur in the future.

5.8 MSI and LSI

The preceding sections of this chapter have described the characteristics, fabrication, design, and layout of circuits using either bipolar or MOS devices and their associated passive components. The next step in the integration process is to fabricate many circuits and their interconnections on the same chip to form a complete electronic subsystem. These completely integrated subsystems are referred to as "Medium Scale Integration (MSI)" and "Large Scale Integration (LSI)." The objective of this section is to discuss briefly the following points: (1) the meaning of the terms MSI and LSI, (2) the design approaches to MSI and LSI, and (3) some of the applications of MSI and LSI.

Most MSI and LSI circuits are used in logic or digital applications. To develop definitions for MSI and LSI, we need to examine briefly some characteristics of logic circuits. The design of logic circuits can be traced to George Boole, a nineteenth century mathematician, who showed that all logic operations can be performed using a binary system and the three operations, AND, OR, and NOT. The binary system and the AND and OR operations were examined in Sec. 2.6. Any number can be expressed in the binary number system using only two digits, 0 and 1. The output of an AND circuit will assume the 1 state only when all inputs are in the 1 state while the output of an OR circuit will assume the 1 state when one or more of the inputs are in the 1 state. The NOT operation is implemented using an inverting amplifier (CE or CS); i.e., when the input is high, the output is low, and when the input is low, the output is high. Although the AND and OR functions were implemented using diodes in Sec. 2.6, they can also be implemented using transistors. Consequently, the three basic logic functions can all be performed with transistors. Circuits which perform the AND, OR, or NOT operations are called *gates*.

The basic distinction between an IC, MSI, and LSI is the complexity of the circuit. Using the number of gates as a criterion, an IC generally refers to a circuit having the complexity associated with 1 to 10 gates. MSI refers to a circuit with 10 to 99 equivalent gates, and LSI refers to a circuit with over 100 equivalent logic gates. However, the above definition is arbitrary, and some writers prefer to define LSI simply as a single chip containing a large number of components connected in such a way as to perform several functions. The latter definition will be used in our discussion.

One of the major problems encountered in manufacturing LSI units is the low percentage yield of good chips. When several thousand components are fabricated on the same chip, the probability of finding one or more defective components is fairly high. In a *fixed wiring* approach, the circuit is fabricated as designed. Consequently, many (and sometimes all) of the components on the chip are used in a circuit. If a defective component is used, the circuit may not perform properly. Consequently, the entire chip would have to be discarded.

Several approaches have been and are being developed to maximize the yield and minimize the cost. One proposed method is the discretionary wiring approach. This approach begins with a silicon chip that is typically 1 to 2 inches in diameter. A large number of small-area circuits are fabricated on the chip, using the methods discussed earlier in this chapter. Since these small circuits can be of different types, all of the basic logic circuits can be fabricated on the same chip. Each individual circuit is tested to determine which circuits are good, and the results are stored in a computer. The tested LSI chip then becomes part of an inventory available for future processing.

The next step is to determine the particular logic function to be performed. Generally, this information is provided by the user in the form of a logic circuit diagram.* A computer is used to generate the required interconnection pattern, which will use only the good circuits on an LSI chip in the inventory. The metallization masks are made using a precision cathode ray tube controlled by a computer.

In the discretionary wiring pattern approach described above, an LSI circuit is fabricated using an existing chip in the inventory. The fixed wiring pattern approach results in a higher density of used components. However, additional design effort is required, and the manufacturing yield may be lower.

The fixed wiring pattern approach is a good example of computer-aided design.** The design begins with a basic logic diagram which is programmed into a computer. The computer also has a library of standard logic circuits. Additional programs are used to: (1) determine the location and orientation of each element, (2) arrange the interconnection pattern to eliminate metallization crossovers, and (3) determine the size and actual geometry of each element. Component densities approaching 200,000 devices per square inch have been achieved using this method.

---

* A logic circuit diagram is a block diagram showing the interconnections of the basic logic function blocks (AND, OR, NOT, etc.) and does not include the details of each basic circuit (transistors, resistors, etc.).

** See H. R. Camenzind and F. G. Heath in the references listed at the end of this chapter.

The two design approaches to LSI discussed in this section can be combined by using MSI circuits connected by a discretionary wiring pattern. This combined approach has achieved component densities approaching 100,000 devices per square inch and does provide some leeway for failure of an MSI unit.

5.9 Hybrid, Compatible, and Monolithic Breadboard Integrated Circuits

The preceding sections have treated several aspects of monolithic integrated circuits in which all components were fabricated in or on a single silicon chip. In this section, we will examine three types of ICs: the hybrid, the compatible, and the monolithic breadboard IC. These forms of integrated circuits are distinguished from the single chip circuits studied earlier in this chapter by one or both of the following characteristics: (1) some or all of the passive components are fabricated using thick-film or thin-film techniques, and (2) some or all of the interconnections within the circuit are formed using wire bonding techniques or film techniques. Our immediate task is to examine the technology of thick-film and thin-film components.

The material used in the fabrication of thick-film conductors, resistors, and dielectrics are special inks or paints. These inks are deposited on a ceramic substrate using a silk screen technique. The inks used for conductors and resistors contain a mixture of powdered metal (often gold or platinum) and glass suspended in an organic binder. The inks used for dielectrics are a mixture of powdered dielectrics suspended in an organic binder. Once the inks are screened and dried on the substrate, they are fired in a high-temperature furnace (500° to 1200°C) to burn off the organic binder and reflow the glass-metal mixture. The resulting material is called a *cermet*.

The characteristics of thick-film resistors can be adjusted several ways. Sheet resistance values ranging from 10 ohms/square to 1 megohm/square can be obtained by varying the composition of the ink. Resistance values can also be varied by adjusting the shape of the resistor. The power rating of a resistor can be increased by increasing the surface area.

Thick-film capacitors can be fabricated with capacitance values in excess of 0.1 pF/mil^2. Although this value is similar to the capacitance per unit area for monolithic capacitors (Table 5–1), thick-film capacitors have lower loss and better voltage breakdown characteristics.

Thin-film components are fabricated by two processes, vacuum deposition and sputtering. In the vacuum deposition or evaporation process, resistance materials, such as chromium and nichrome, are heated in a high vacuum, and the resulting metal vapor is deposited through openings in a mask onto the substrate. Sputtering, on the other hand, depends on the low pressure ionization of a gas. In this process, a potential of several kilovolts is applied between two electrodes in the presence of an inert gas. The negative electrode, called the cathode, is made of the material to be deposited on the substrate. One example is tantalum. The inert gas ionizes due to the high electric field, and the heavy positive gas ions bombard the negative electrode, releasing ions from the cathode material. These ions are then deposited on the substrate. Both of the thin-film fabrication techniques are more complicated than the thick-film techniques discussed above.

Thin-film resistors can be formed with sheet resistance values ranging from 10 to 250 K ohms/square with tolerances of ±5 percent to ±10 percent. Thin-film capacitors are formed by depositing a metal layer, a dielectric layer, and then a second metal layer. The dielectric may be formed from a variety of materials including silicon monoxide, tantalum oxide, and ceramic. Capacitance values up to several pF/mil² can be obtained.

Active devices can also be fabricated using thin-film techniques. The side view of a thin-film transistor* (TFT) is shown in Fig. 5–35. The metal, semiconductor, and oxide layers can all be formed using the thin-film techniques discussed above. Normally, the semiconductor layer is polycrystalline as opposed to the monocrystalline structure used in discrete and monolithic transistors. Since the substrate of a TFT is an insulator, electrical isolation between various components is easily obtained. However, TFTs suffer from instability in performance parameters and are not well understood. Consequently, they have not progressed beyond the laboratory stage. Laboratory models of the TFT have produced characteristic curves similar to those of the MOS transistor.

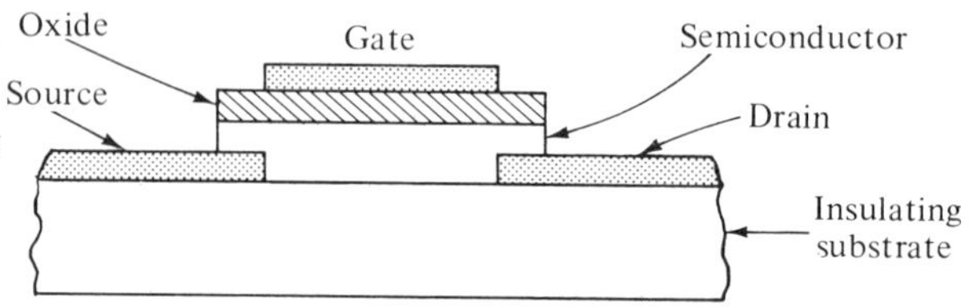

FIGURE 5–35 *Side view of a thin-film transistor (TFT). (Adapted from Paul K. Weimer, "Thin-Film Transistors" in J. Torkel Wallmark and Harwick Johnson, Eds.,* Field-Effect Transistors: Physics, Technology and Applications, *© 1966, Prentice-Hall, Inc., p. 218.)*

The principal advantages of thick-film or thin-film passive components when compared to monolithic passive components are their closer tolerance, better voltage breakdown characteristics, and better thermal characteristics (smaller variations in value with temperature changes). The principal disadvantage of the film techniques is that reliable active devices are not available. Normally, active devices must be separately attached to a substrate to form a hybrid integrated circuit.

A hybrid-IC generally consists of several separate components mounted on the same substrate and connected internally by thin-film, thick-film, or wire bonding techniques. The separate components can be individual transistors, complete monolithic ICs, and thick- or thin-film passive components. Since hybrid ICs are not restricted to one technology, they can be designed to take advantage of the desirable characteristics of each technology. Hybrid ICs are generally used in applications requiring high voltage or high current values. Also, hybrid ICs may be more economical in some low-volume, special purpose circuits. Since most active devices and monolithic ICs can be obtained as individual chips or dies, the hybrid IC offers great flexibility to the circuit designer.

The fabrication of a hybrid IC generally begins with a ceramic substrate which is carefully cleaned. The desired thick- or thin-film passive components are then formed on this substrate. Active devices or monolithic ICs are then mounted on the substrate, and the entire structure is enclosed in a suitable package.

* The thin-film transistor is discussed by the R. S. C. Cobbold and Lindmayer and Wrigley references listed at the end of the chapter.

A variation of the hybrid structure is the compatible IC. In this case, a complete monolithic IC is fabricated and then covered with an oxide layer. The close tolerance passive components are fabricated on the top of this oxide layer using thin-film techniques. In this way, the advantages of the monolithic structure and the accuracy of thin-film passive components are combined to improve circuit performance. Interconnections between the monolithic circuit and thin-film components are made through openings in the oxide layer. Compatible ICs are very useful in negative-feedback amplifiers where the gain is determined by the ratio of two resistors.*

The final type of IC to be discussed is the monolithic breadboard. In this case, a complete monolithic IC is fabricated, except for the interconnection pattern, and mounted in an open top package. More than one monolithic chip can be mounted if desired. The resulting structure then forms a breadboard on which any number of circuits can be connected by simply interconnecting the various components using wire bonding techniques. The layout of one such monolithic breadboard is shown in Fig. 5–36, and a schematic diagram of the circuit is shown in Fig. 5–37. The monolithic breadboard IC is especially useful when only a few ICs are required, or as a means of checking the characteristics of a circuit to be fabricated in monolithic form.**

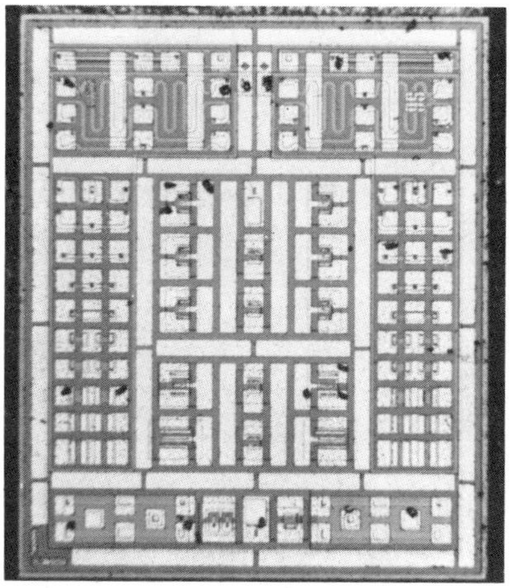

FIGURE 5–36 *A monolithic breadboard. The circuit is a complete monolithic IC except that no interconnection pattern has been fabricated. (Courtesy Silicon General, Inc.)*

The typical parameters for the transistors on the monolithic breadboard are listed in Table 5–4. The value of f_T, a transistor parameter useful in frequency-response analysis,*** is included for completeness. The values of C_{CB} and C_{CS} are also useful in frequency-response analysis. Transistors Q_1–Q_6 are small-signal NPN transistors, and Q_9–Q_{10} are medium current NPN transistors, the principal difference being the length of the emitter region. Transistors Q_7–Q_8 are lateral PNP transistors, and Q_{11} is a substrate PNP transistor which can be used only as a CC stage since its collector (the integrated circuit substrate) must be connected to the lowest potential in the circuit.

* Negative-feedback amplifiers are discussed in Chap. 10.

** The use of a monolithic breadboard for the evaluation of an IC design is discussed in Sec. 5.6.

*** Frequency-response analysis begins in Chap. 7.

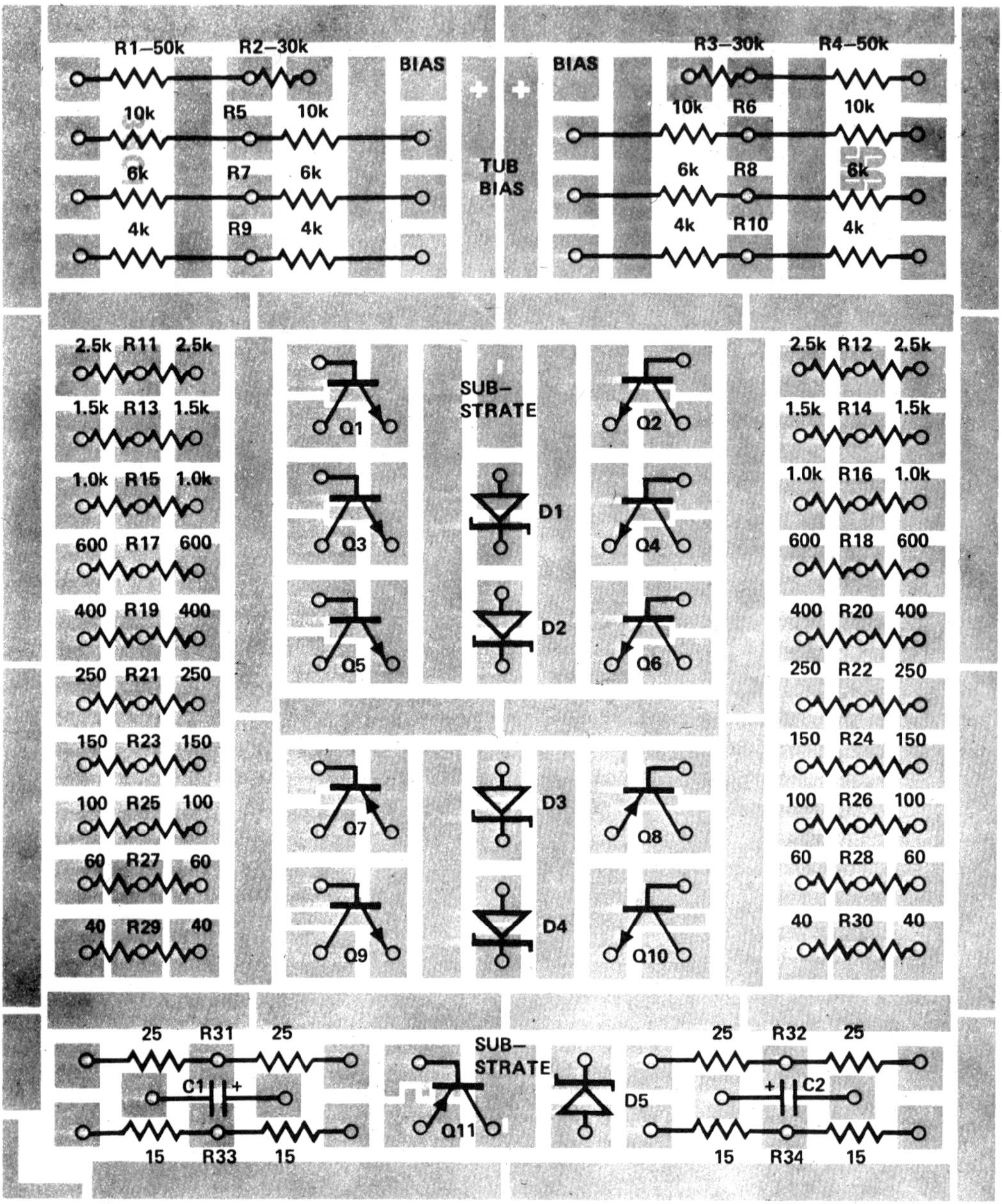

FIGURE 5–37 *Circuit diagram for the monolithic breadboard shown in Fig. 5–36. (Courtesy of Silicon General, Inc.)*

The isolation regions for $R_{31} - R_{33}$ and $R_{32} - R_{34}$ are formed by large area base diffusion and can be used as a P-N junction capacitor. Capacitors C_1 and C_2 in Fig. 5–37 are formed in this manner, and their typical values are given in Table 5–5. Note that C_1 and C_2 are polarized and that their value is voltage dependent. The breakdown voltage for these capacitors is equal to BV_{CBO} or about 50 V.

TABLE 5–4 *Typical transistor characteristics for the monolithic breadboard shown in Fig. 5–37. Values of* f_T, C_{CB}, *and* C_{CS} *are included for completeness and are used later in the text. (Courtesy Silicon General, Inc.)*

Parameter	Test I_C	Conditions V_{CE}	Q_1–Q_6 NPN	Q_7–Q_8 Lateral PNP	Q_9–Q_{10} NPN	Q_{11} Substrate PNP	Units
BV_{CSO}	10 μA		80	50	80	—	V
BV_{CBO}	10 μA		50	50	50	80	V
BV_{CEO}	100 μA		30	25	30	80	V
BV_{EBO}	10 μA		6.2	50	6.2	6.2	V
β	10 μA	5 V	70	15	50	40	
β	100 μA	5 V	80	10	70	40	
β	1 mA	5 V	80	3	80	30	
β	10 mA	5 V	65	—	80	5	
f_T	100 μA	5 V	200	4	100	5	MHz
f_T	1 mA	5 V	600	1	500	5	MHz
f_T	10 mA	5 V	800	—	1100	—	MHz
C_{CB}		5 V	0.5	1.0	1.3	5.0	pF
C_{CS}		5 V	2.0	1.0	4.0	—	pF

TABLE 5–5 *Typical capacitor characteristics for the monolithic breadboard shown in Fig. 5–37. (Courtesy Silicon General, Inc.)*

Junction Voltage	Capacity
0 V	20 pF
2 V	13 pF
5 V	10 pF
10 V	8 pF

Five diodes are included in the monolithic breadboard shown in Fig. 5–37. In addition to their usual diode functions, these diodes can be used as low-voltage reference diodes. Diode parameters are given in Table 5–6.

TABLE 5–6 *Typical diode characteristics for the monolithic breadboard shown in Fig. 5–37. (Courtesy Silicon General, Inc.)*

Parameter	Test Conditions	$D1$–$D4$	$D5$	Units
V_Z	$I_{ZT} = 1$ mA	6.2	5.4	volts
Z_Z	$I_{ZT} = 1$ mA	25	50	ohms
TC	$I_{ZT} = 1$ mA	0.04	0.02	%/°C
V_F	$I_F = 1$ mA	0.72	0.83	volts

Three different types of resistors are included in the monolithic breadboard shown in Fig. 5–37. Resistors $R_1 - R_4$ are "pinch" type resistors and have a tolerance of +100 percent to −50 percent and a breakdown voltage of 6.2 V. Resistors $R_5 - R_{30}$ are formed during the base diffusion process and have a sheet resistance of approximately 150 ohms/square with a tolerance of ±20 percent in absolute value. However, the ratio of two resistors is within ±5 percent. The breakdown voltage for $R_5 - R_{30}$ is 30 V. Resistors $R_{31} - R_{34}$ are formed during the emitter diffusion and have the same tolerance as $R_5 - R_{30}$ and a breakdown voltage of only 6.2 V. The nominal value of each resistor and its capacity to the substrate is given in Table 5–7. Resistors $R_5 - R_{34}$ are center-

TABLE 5–7 *Typical resistor characteristics for the monolithic breadboard shown in Fig. 5–37. (Courtesy Silicon General, Inc.)*

Resistor Designation	Nominal Value (ohms)	Capacitance (picofarads)
*R*1, *R*4	50 K	
*R*2, *R*3	30 K	3.8
*R*5, *R*6	20 K	5.6
*R*7, *R*8	12 K	3.4
*R*9, *R*10	8.0 K	2.4
*R*11, *R*12	5.0 K	1.8
*R*13, *R*14	3.0 K	1.2
*R*15, *R*16	2.0 K	0.8
*R*17, *R*18	1.2 K	0.6
*R*19, *R*20	800	0.8
*R*21, *R*22	500	1.0
*R*23, *R*24	300	1.0
*R*25, *R*26	200	1.3
*R*27, *R*28	120	1.5
*R*29, *R*30	80	1.8
*R*31, *R*32	50	7.5
*R*33, *R*34	30	9.0

tapped so that half of each value can be obtained; however, the capacity to the substrate will be the same.

As an example of the use of a monolithic breadboard, consider the amplifier* shown in Fig. 5–38(a). The required interconnection pattern is constructed using the breadboard circuit diagram shown in Fig. 5–37. One form of this pattern is shown in Fig. 5–38(b). Using metalized strips on the circuit, the need for wiring cross-overs is reduced and the reliability is increased. The interconnections are made using a wire bonding machine and wire that is typically 0.7 to 1.5 mils in diameter. The complete amplifier in integrated circuit form is shown in Fig. 5–38(c).

Three types of integrated circuits have been discussed in this section: the hybrid, the compatible, and the monolithic breadboard. In addition, we have examined the technology associated with the thick-film and thin-film components. The techniques discussed in this section generally augment the monolithic integrated circuits discussed earlier and can be considered as variations of the basic monolithic technology. Since the technology for producing integrated circuits is changing rapidly, the interested reader should consult the current literature for the latest developments.

* This circuit is a negative-feedback amplifier. Negative feedback is discussed in Chap. 10.

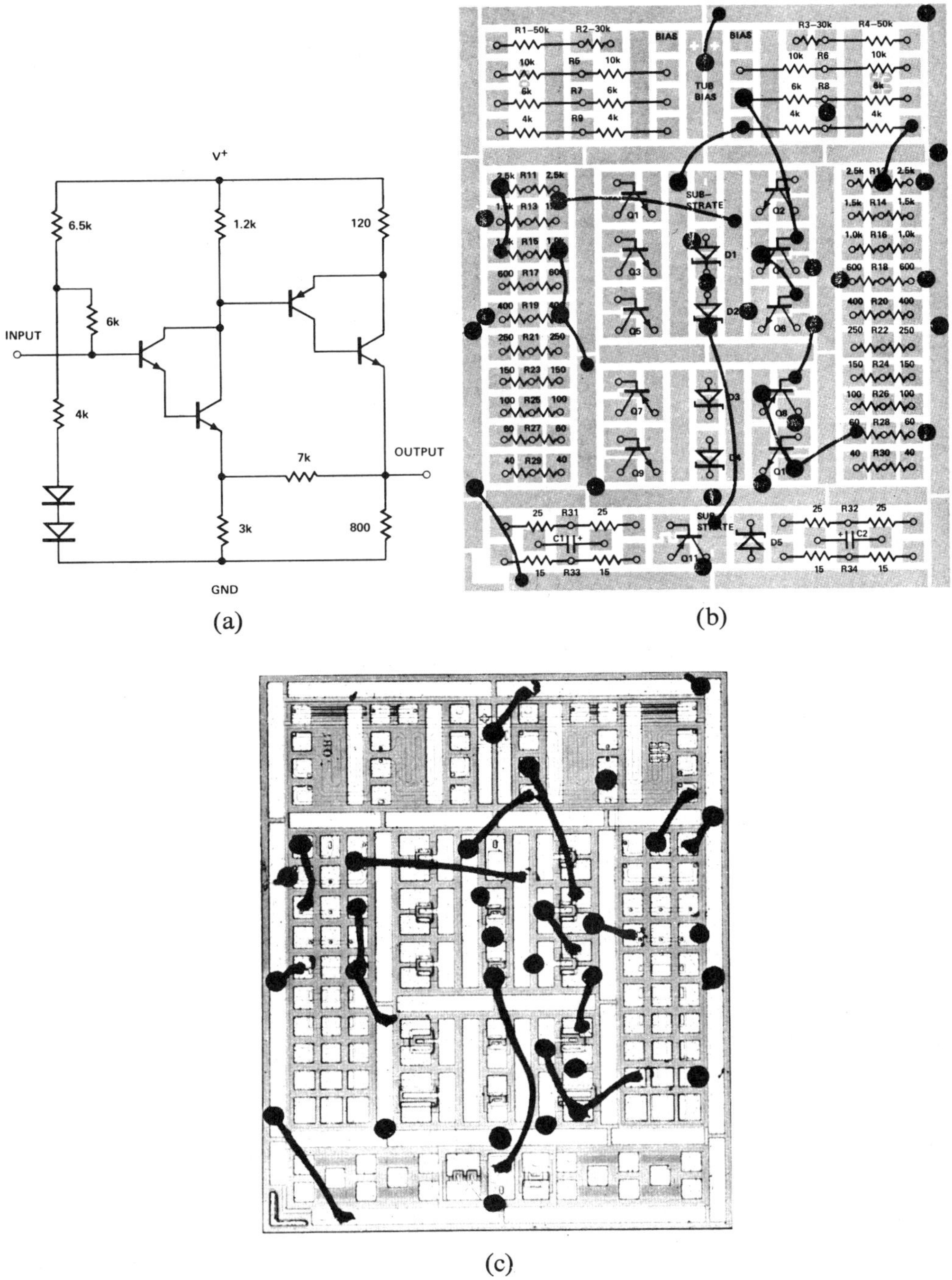

(a) (b) (c)

FIGURE 5–38 *An example of the use of a monolithic breadboard: (a) the circuit diagram of an amplifier, (b) the layout of required interconnections using the breadboard circuit diagram shown in Fig. 5–37, and (c) the completed amplifier in integrated circuit form. (Courtesy Silicon General, Inc.)*

5.10 Summary

In this chapter, we have introduced the concept of an integrated circuit, some IC fabrication techniques, and some guides for the design of integrated circuits. Integrated circuits have several advantages over comparable discrete circuits. For example, they are smaller, more reliable, less expensive, and often perform better than discrete circuits.

The most common method of IC fabrication is based on a photolithographic process whereby selective diffusions are made into a slice of silicon. Both bipolar and FET active devices can be fabricated in this manner. Integrated circuits using bipolar transistors and their associated passive components are used in both linear and digital applications while ICs using MOS devices (as both active devices and passive devices) are used primarily in digital circuits. Generally, an all-MOS circuit requires a smaller chip area resulting in a higher density of components than a bipolar transistor IC. However, bipolar transistor ICs generally have a higher gain and a better speed of response.

The principal disadvantage of monolithic ICs lies in the performance of the passive components. Both the resistors and the capacitors of a monolithic IC have poor tolerances and are generally voltage dependent. In addition, large values of either resistance or capacitance use an excessive amount of chip area and are not desirable. Thick-film and thin-film techniques, on the other hand, can be used to fabricate excellent passive components but not reliable active devices. Two types of ICs, the hybrid and the compatible IC, have been developed to combine the advantages of monolithic and film technologies. While these methods can be used to produce circuits having acceptable performance, additional processing steps are required which result in higher cost.

In addition to the basic bipolar and MOS technologies, other forms of IC fabrication were also discussed. These devices include the TRIM structure transistor, the thin-film transistor (TFT), complementary MOS circuits, and monolithic breadboard ICs. Integrated circuits using some of these techniques are already commercially available. Other techniques will certainly be developed. The interested reader must consult the current literature to keep abreast of this fast changing field. The application of ICs as well as some additional IC characteristics will be introduced in later chapters.

Problems

Section 5.1

5–1 List three advantages and three disadvantages of ICs when compared with discrete component circuits.

5–2 Discuss why hybrid ICs may find wide application in linear amplifier circuits.

Section 5.2

5–3 A slice of P-type silicon 2 inches in diameter is used to form the substrate of many monolithic ICs. Each IC occupies an area of approximately 25 mils × 25 mils. The expected yield from the manufacturing process is 30 percent. Assuming that the total manufacturing cost is one thousand dollars, estimate the cost of each acceptable circuit.

5–4 In order to insure isolation between various parts of an IC, should the substrate be connected to the most positive or the most negative voltage in the circuit? Why?

5–5 Why is the epitaxial layer in an IC doped lighter than the base region?

5–6 What processes limit the base width of a monolithic IC transistor?

5–7 What processes limit the size of monolithic IC components?

5–8 After an IC circuit was overheated, a number of transistors no longer functioned. What is the one most probable cause of this failure?

Section 5.3

5–9 Using the same dimensions as given for the emitter strip and clearances of the single-base-strip geometry, calculate the dimensions of a double-base-strip geometry NPN transistor.

5–10 An NPN transistor has the following approximate doping levels:

emitter—10^{21} atoms/cc	base—5×10^{18} atoms/cc
collector—5×10^{16} atoms/cc	substrate—5×10^{15} atoms/cc

Estimate the values of breakdown voltage between: (a) the collector and base, (b) the emitter and base, (c) the collector and substrate, and (d) emitter and collector. Assume that β is 100.

5–11 An NPN monolithic transistor has a β_{max} of 100. Estimate the value of β at a collector current of 0.4 mA assuming that the transistor is of the single-base-strip geometry with an emitter area of 2 mils $\times$ 1 mil.

5–12 Repeat Prob. 5–11 for a double-base-strip geometry.

5–13 Sketch the top and side view of a monolithic IC consisting of an NPN and a lateral PNP transistor where the collector of the NPN is connected to the base of the PNP. The remaining terminals of the two devices should be available for external connection.

Section 5.4

5–14 Repeat Ex. 5–2 for a resistor formed during the emitter diffusion. The doping level of the emitter region is approximately 10^{20} atoms/cm^3.

5–15 Assuming a sheet resistance of 200 ohms/square, estimate the total resistance between the two terminals of the monolithic resistor shown in Fig. 5–39.

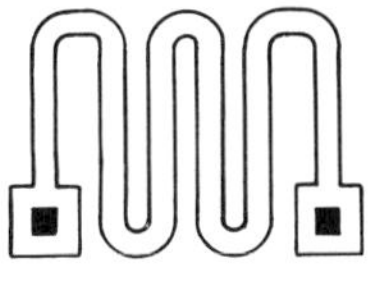

FIGURE 5–39 *The monolithic resistor from Prob. 5–15.*

5–16 Repeat Ex. 5–4 for the double-base-strip geometry calculated in Prob. 5–9 using the doping levels given in Prob. 5–10.

Section 5.5

5–17 Derive Eqs. (5–10) and (5–11).

5–18 The circuit shown in Fig. 5–25(a) has the following component values: $R_1 = 600$, $R_2 = 4.4$ K, $R_3 = 1$ K, $R_4 = 3$ K, and $V_{CC} = 6$ V. What dc value of V_{IN} will cause the base collector junction of Q_2 to become forward biased?

5–19 A level-shift circuit of the form shown in Fig. 5–25(d) has the following component values: $R_1 = 400$, $R_2 = 4$ K, $R_3 = 5.2$ K, $R_4 = 14.8$ K, $R_5 = 800$, $V_{CC} = 8$ V, and $\beta = 100$. Assuming $I_{E1} = I_{E2}$, calculate: (a) all bias voltages and currents, and (b) the ac voltage gain from the base of Q_1 to the emitter of Q_4.

Section 5.6

5–20 Draw the circuit diagram for the IC shown in Fig. 5–40.

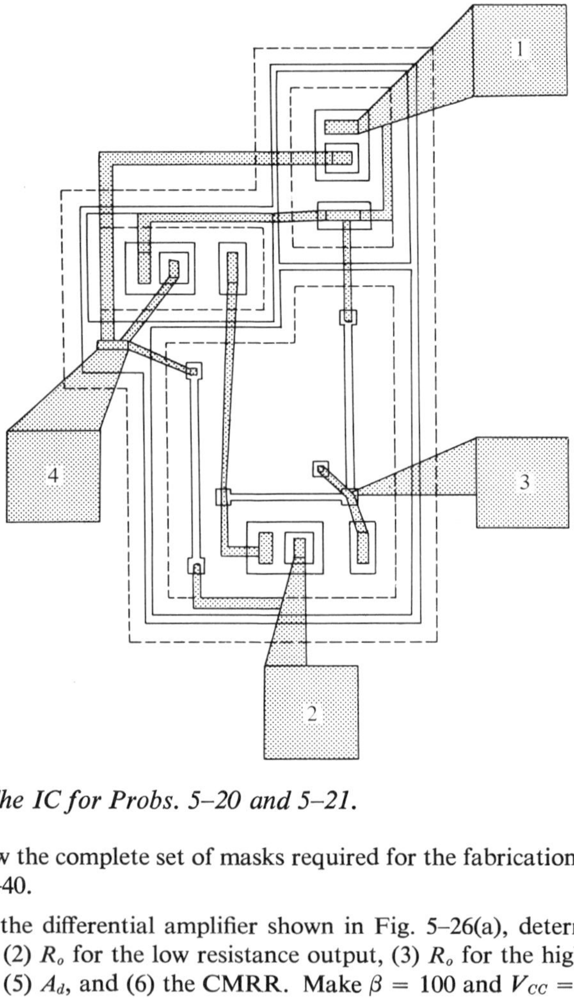

FIGURE 5–40 *The IC for Probs. 5–20 and 5–21.*

5–21 Draw the complete set of masks required for the fabrication of the IC shown in Fig. 5–40.

5–22 For the differential amplifier shown in Fig. 5–26(a), determine the values of: (1) R_i, (2) R_o for the low resistance output, (3) R_o for the high resistance output, (4) A_c, (5) A_d, and (6) the CMRR. Make $\beta = 100$ and $V_{CC} = 10$ V.

5–23 Draw the circuit diagram for the IC shown in Fig. 5–41.

5–24 Draw the complete set of masks required for the fabrication of the IC shown in Fig. 5–41.

5–25 What purpose does the separate transistor in the lower left-hand corner of the IC shown in Fig. 5–41 perform?

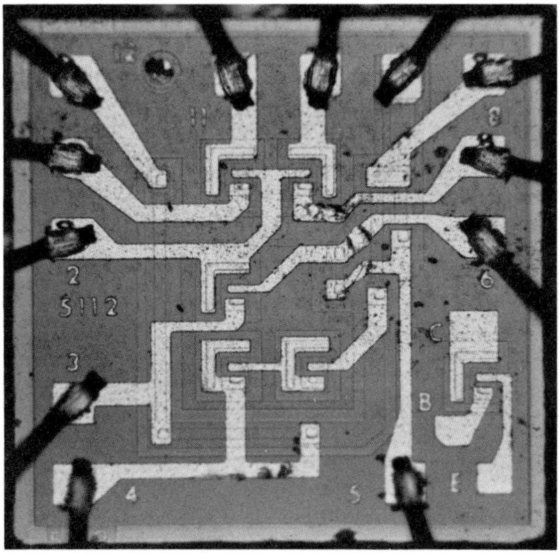

FIGURE 5–41 *The IC for Probs. 5–23, 5–24, and 5–25. (Courtesy Radio Corporation of America.)*

5–26 For the amplifier shown in Fig. 5–42, calculate all bias voltages and currents and the value of A_d.

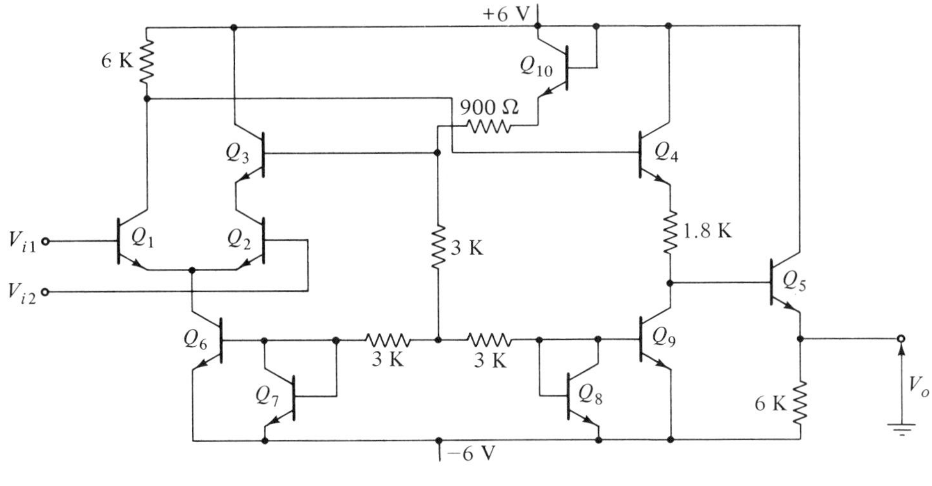

FIGURE 5–42 *Circuit for Probs. 5–26 and 5–27.*

5–27 Construct a scale drawing for a monolithic layout for the circuit shown in Fig. 5–42. The following components are not in the IC: R_g and R_L.

5–28 Construct a scale drawing for a monolithic layout for the circuit shown in Fig. 5–25(d). Use the component values given in Prob. 5–18.

5–29 Design, lay out, and construct the required masks for a monolithic IC to satisfy the following requirements: $A_{vo} = -50$, $R_i \geq 10$ K, and $R_o \leq 100$ ohms. The IC will be used in an application where R_g is 1 K and R_L is 100 ohms. Assume that each NPN transistor has a β of 200, each lateral PNP transistor has a β of 50, and that the available power supply voltages are +9 V and −9 V. R_g and R_L are not on the chip.

5–30 Design, lay out, and construct the required masks for a monolithic IC to satisfy the following requirements: $A_{vo} \geq 200$, $R_i \geq 2$ K, and $R_o > 1$ K. The IC will be used in an application where R_g is 300 ohms and R_L is 1 K. Use the same transistor and power supply specifications as given in Prob. 5–29.

Section 5.7

5–31 For the all-MOS differential amplifier shown in Fig. 5–33, verify that the required power supply is -15 V and that the value of A_d is 4.

5–32 Draw a complete set of masks for the all-MOS differential amplifier shown in Fig. 5–33.

5–33 Draw the circuit diagram for the MOS IC shown in Fig. 5–43 and estimate the values of A_{vt}, R_i, and R_o.

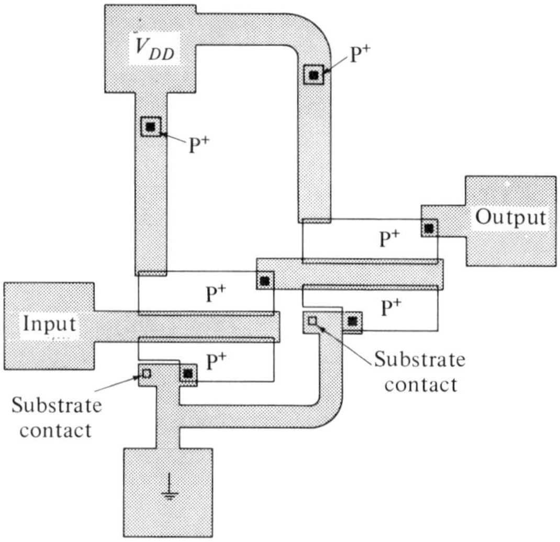

FIGURE 5–43 *The MOS IC for Probs. 5–33 and 5–34.*

5–34 Draw a complete set of masks for the MOS IC shown in Fig. 5–43.

5–35 Design, lay out, and draw the required masks for an all-MOS IC having a gain, A_{vt}, of -50. The value of R_L external to the IC can be assumed infinite (open circuit output).

5–36 Design, lay out, and draw the required masks for an all-MOS differential amplifier having the following specifications: $A_d \geq 30$, $|A_c| \leq 0.1$, CMRR ≥ 300, $R_i \geq 1$ M, and $R_o \leq 1$ K. The value of R_L external to the IC can be assumed infinite.

Section 5.8

5–37 Give one reason why MSI and LSI circuits have had little impact in the linear circuit field.

5–38 Why does the fixed wiring pattern result in a higher density of components than the discretionary wiring pattern of LSI fabrication? Which scheme is more flexible?

Section 5.9

5–39 List two advantages and two disadvantages of the hybrid IC as compared to the monolithic IC discussed earlier in this chapter.

5–40 Connect the various components of the monolithic breadboard IC shown in Figs. 5–36 and 5–37 to fabricate the differential amplifier shown in Fig. 5–26(a).

5–41 Repeat Prob. 5–40 for the level-shift circuit shown in Fig. 5–25(d).

5–42 Using the monolithic breadboard IC shown in Figs. 5–36 and 5–37, design an amplifier to satisfy the specifications given in Prob. 5–29.

5–43 Using the monolithic breadboard IC shown in Figs. 5–36 and 5–37, design an amplifier to satisfy the specifications given in Prob. 5–30.

References

Camenzind, H. R., *Circuit Design for Integrated Electronics.* Reading, Mass.: Addison-Wesley Publishing Co., Inc., 1968.

Cobbold, R. S. C., *Theory and Applications of Field-Effect Transistors.* New York: Wiley-Interscience, John Wiley & Sons, Inc., 1970.

Heath, F. G., "Large-Scale Integration in Electronics," *Scientific American*, 222, No. 2 (February, 1970), 22–31.

Irvin, J. C., "Resistivity of Bulk Silicon and of Diffused Layers in Silicon," *The Bell System Technical Journal*, 41, No. 2 (March, 1962), 387–410.

Lindmayer, J. and C. Y. Wrigley, *Fundamentals of Semiconductor Devices.* New York: D. Van Nostrand Company, Inc., 1965.

Stern, L., *Fundamentals of Integrated Circuits.* New York: Hayden Book Co., Inc., 1968.

Thun, R. E., "Thick Film or Thin," *IEEE Spectrum*, 6, No. 10 (October, 1969), 73–79.

Wallmark, J. T. and H. Johnson, eds., *Field-Effect Transistors: Physics, Technology and Applications.* Englewood Cliffs, N. J.: Prentice-Hall, Inc., 1966.

Warner, R. M., Jr., ed., *Integrated Circuits: Design Principles and Fabrication.* New York: McGraw-Hill Book Company, 1965.

6

Vacuum Tubes

Vacuum tubes were developed in the early part of this century and were used exclusively by the electronic industry until bipolar transistors became commercially available in the early 1950s. Transistors have many advantages over vacuum tubes, not the least of which is a lower cost, and have largely replaced tubes. However, tubes are still used in many special applications. For example, transmitters that require large amounts of signal power use tubes. In addition, television picture tubes and cathode ray tubes belong in the vacuum tube family. Because of several special applications and the large amount of vacuum tube equipment that is still being used, a short treatment of tubes is justified at this point.

Two problems that are basic to any active device are: (1) to make a sufficient but limited number of current carriers available, and (2) to control the flow of these carriers with an electrical signal. In solid-state devices, a reasonable number of carriers is made available at room temperature by doping a semiconductor with an impurity. In a field-effect device, control is exercised by changing the width of a channel; while in a bipolar transistor, control is achieved by changing the height of a potential barrier (*EB* junction). This chapter is devoted to the methods used in obtaining carriers in a vacuum, ways of controlling these carriers, and applications of vacuum tubes.

6.1 Electron Emission

A vacuum tube is constructed by placing two or more metal electrodes or elements in a metal or glass envelope, pumping as much air as practical from the envelope, and then sealing it. As a result, the space between the electrodes or elements contains very few atoms of gas. To simplify our study, we will usually assume that a perfect vacuum exists inside the envelope. Leads to the elements are brought out through airtight glass-to-metal seals.

The only carriers of any consequence in a vacuum tube are electrons.* The element that provides the electrons is called the cathode. A cathode will emit electrons, however, only if sufficient energy is supplied to it. The study of electron emission consists of two parts: (1) determining the amount of energy required to release an electron, and (2) evaluating methods of supplying energy to the cathode.

In Chap. 1, we noted that at a temperature of 0°K all of the electrons in a material occupied the lowest possible energy levels. However, the lowest energy level (first orbitals) accommodates only two electrons per atom while the number of electrons that can be accommodated increases as the orbital number increases. A plot of the distribution of electrons as a function of energy for a metal is shown in Fig. 6–1(a) for three different temperatures as follows: T_0 for 0°K, T_R

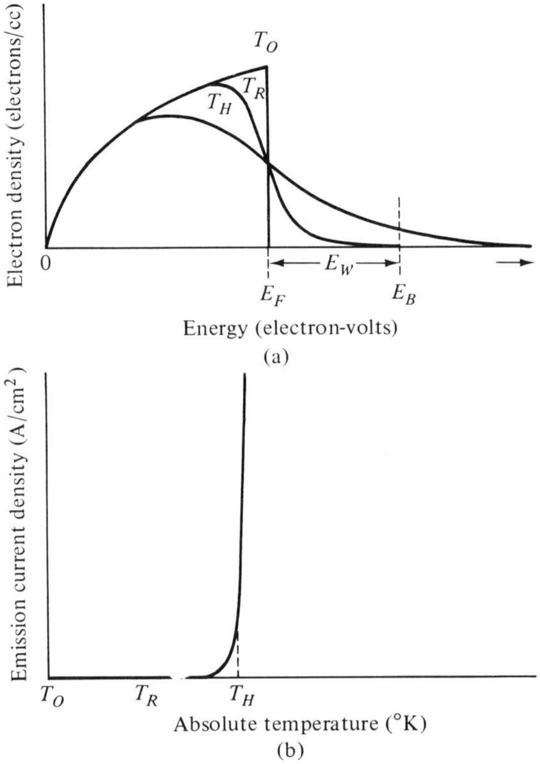

FIGURE 6–1 *Energy distribution of electrons and the emitted current for a metal: (a) electron distribution as a function of energy for three different temperatures and (b) the current density of the electrons that are emitted as a function of temperature. (Adapted from R. H. Krackhardt,* Vacuum Tube Electronics, *Charles E. Merrill Publishing Company, 1966.)*

for room temperature, and T_H for a high temperature. In a metal at 0°K (or T_0) the energy distribution breaks sharply at the Fermi level, E_F, which is the highest energy that an electron can have in a metal at 0°K.

* In gas tubes, a small amount of gas is released inside the envelope after the air has been pumped out. In these tubes, gas atoms are ionized by collisions, and the ionized gas molecules become important carriers.

As the temperature increases, some of the electrons are excited into levels above E_F as indicated in Chap. 1. However, at room temperature (or T_R) virtually none of the electrons have sufficient energy to completely escape from the surface of the metal. For example, tungsten, which is one of the metals used in cathodes, emits one electron per cm^2 each 10^{41} years (on an average) at room temperature (300°K). However, at a temperature of 3000°K, the same material can emit 13 amperes per cm^2. This functional relationship is shown in Fig. 6–1(b).

To escape from the surface of a metal, an electron must have a minimum energy of E_B associated with a velocity perpendicular to the surface of the metal. The perpendicular to the surface of the metal is called the x direction. The number of electrons per unit volume, dn_x (electrons/cc), having a velocity in the x direction, v_x (cm/sec), is*

$$\frac{dn_x}{dv_x} = \frac{4\pi\, m^2\, kT}{h^3}\, ln\left(1 + \exp\frac{E_F - (1/2)mv_x^2}{kT}\right) \tag{6–1}$$

where h is Planck's constant, k is Boltzmann's constant, and m is the mass of an electron. This equation indicates that at 0°K, the number of electrons at the Fermi level having a component of v_x is zero. As the temperature increases, the number of electrons having a velocity of v_x or an energy of $(1/2)\,(mv_x^2)$ in the x direction increases.

To escape from the surface of a metal, an electron must obtain from some source an energy equal to $(E_F - E_B)$. This difference is called the work function,** E_W, and is different for different metals. The number of electrons having an energy equal to or greater than the work function for a velocity component in the x direction can be obtained as a function of temperature from Eq. (6–1). Since all of these electrons escape from the surface of a metal, the resulting equation is a universal theoretical expression for the maximum current available from a metal as a function of temperature. This equation is

$$I = AC_e T^2 \exp(-E_W/kT) \text{ Amperes} \tag{6–2}$$

where A is the surface area (m^2), C_e is a proportionality constant having a theoretical value of 120×10^4 (amperes/m^2)/T^2, k is Boltzmann's constant (1.38×10^{-23} joules/°K), E_W is the work function in joules, and T is the temperature (°K). This equation is known as the Richardson-Dushman equation. A general plot of this function is shown in Fig. 6–1(b).

Many factors such as surface roughness and contamination are not included in Eq. (6–2). Consequently, while the general form is satisfactory, the value of C_e is not. However, this value can be determined experimentally. Values for C_e and also the work function, E_W, are given in Table 6–1 for a few common cathode materials.

Only a few materials can be used in practical cathodes because most metals melt before a reasonable rate of emission is achieved. The only pure element that has been used commercially is tungsten. Although the work function for this element is high, its melting point is 3640°K. At a temperature of 2500 to 2600°K, tungsten emits a reasonable number of electrons per second.

---

* See pages 137–39 in Millman, *Vacuum-Tube and Semiconductor Electronics* in the references at the end of the chapter.

** Work functions are designated with a W in the literature on vacuum tubes. The symbol E is used here to correspond with the symbols used in Chap. 1.

TABLE 6–1 *Values of proportionality constant and work function for several materials*

Material	C_e (Amps/m^2)/T^2	E_W (eV)
Cesium	16×10^4	1.93
Copper	65×10^4	4.45
Nickel	30×10^4	4.96
Thorium	60×10^4	3.4
Tungsten	60×10^4	4.54
Thoriated tungsten	3×10^4	2.6

Tungsten is difficult to shape because it is brittle. Usually this material is formed into a wire. The wire is then placed inside a tube to form a filament. By passing a current through this filament, the resulting I^2R loss (Watts) heats the filament to the desired temperature of 2500°K. At this temperature, the filament emits directly 1 to 10 mA for each watt of power dissipated in the wire with the actual value depending on the design of the filament. The current emitted per Watt of dissipation is called the *emission efficiency*. This quantity is used as a basis of comparison for various materials. While the emission efficiency of tungsten is comparatively low, this material is relatively immune to damage from high velocity ions* and finds wide application in high-voltage tubes such as x-ray tubes and high-voltage transmitter tubes.

One way to reduce the work function of tungsten is to add a small amount of thorium oxide to the material. When heated, the thorium diffuses to the surface and forms a monatomic layer on the surface of the tungsten. The presence of this layer of thorium increases the emission efficiency to 50 to 200 mA/W at a temperature of 1800 to 2000°K. In this temperature range, the thorium evaporates slowly whereas at 2500°K the evaporation rate would be too large to be practical. A layer of barium on tungsten also produces a much higher emission efficiency than that for pure tungsten.

For circuit voltages less than 1000 V, a more delicate type of emitter construction can be used because the damage associated with positive ion bombardment is usually negligible. A very efficient emitter can be made by coating a sleeve of base material, such as nickel, with a mixture of barium and strontium oxides. The oxide coating has a work function of only 1.0 to 1.5 eV and can be used to provide a current density as high as 100 mA/cm^2 at a cathode temperature of only 1000°K. This temperature can easily be achieved by placing a filament inside the sleeve and heating the filament. Almost all of the cathodes for the smaller tubes are made in this manner.

Thermal energy is usually supplied to emitters or cathodes made from tungsten, thoriated tungsten, or oxides by passing a current through a wire. Two widely used arrangements are shown in Fig. 6–2. A directly heated cathode is shown in (a) and an indirectly heated cathode in (b). In designing a filament

* The normal vacuum in a tube is in the neighborhood of 10^{-5} mm of Hg. At this level many gas molecules are floating around in the tube. As we will see later, when these atoms become ionized, the heavy positive ions move to the cathode and can cause considerable damage to some types of emitters because the impact energy is sufficient to dislodge pieces of the surface.

for this purpose, an appropriate voltage* is first selected. Then a filament is designed to supply the power required to obtain the desired maximum emission current; i.e., a given number of mA for a given emitter efficiency. Indirect heating has the advantage that the entire cathode is at the same potential; whereas in a directly heated filament the current produces a voltage drop along the filament. For an ac filament voltage, the emission of the filament varies as a function of the instantaneous current. While this effect is smoothed by the thermal capacity of a filament, the variation is very noticeable as noise in a high-gain amplifier.

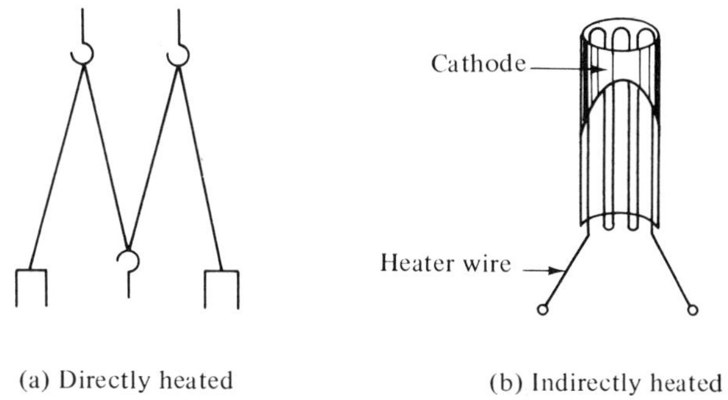

FIGURE 6–2 *Cathode construction (a) directly heated and (b) indirectly heated. (Adapted from R. H. Krackhardt,* Vacuum Tube Electronics, *Charles E. Merrill Publishing Company, 1966.)*

This chapter is concerned primarily with thermionic emission in which energy is supplied to the emitter by heating a filament. However, three other types of emission are worth noting: photoemission, secondary emission, and high-field emission. If the work function of a material is sufficiently low, light energy can be transferred to surface electrons and cause emission.** This effect is called photoemission. Quantum theory indicates that electromagnetic energy is composed of small units of energy called photons. The energy, E_p, of a photon is

$$E_p = hf \tag{6–3}$$

where h is Planck's constant, and f is the frequency of the radiation in Hertz (cycles per second). If the energy of a photon is sufficient to enable an electron to free itself from the surface (equal to the work function or greater), then emission occurs. For this case the number of free electrons is almost directly proportional to the light intensity. However, if a photon has an energy less than the work function, emission will not occur even when the light intensity is increased. Levels of current that can be obtained by photoemission are very small, usually in the microampere range. Consequently, applications of photoemissive devices are usually confined to light detecting and light measuring devices.

Another way to release electrons is by secondary emission. When an atomic particle strikes the surface of a metal, a small part of the energy of the primary particle is transferred to electrons on or near the surface, and the remainder is transferred to thermal energy. A few electrons gain sufficient impact energy in the proper direction to move away from the metal surface and become free

---

* Standard voltages are 1.5, 3.0, 5, 6.3, and 12.6 V with a few special purpose tubes requiring 30, 50, and 117 V.

** Materials used in photocells include cesium and rubidium.

electrons for a short period of time. These free electrons are called *secondary electrons.*

Almost any type of atomic particle can be used to release secondary electrons. Secondary emission can be useful or it can be a nuisance. Useful applications include the following: (1) certain gas tubes which rely on the positive ion bombardment of the cathode to release the desired electrons, (2) photomultipliers in which each primary electron releases several electrons at each of several successive cathodes (dynodes), and (3) cathode ray tubes in which the secondary electrons released from the screen are collected and thereby complete the current circuit for the primary electrons in the beam. Secondary emission is undesirable in a multigrid tube when the primary electrons strike the plate and release several secondary electrons. This problem will be discussed later in this chapter.

A fourth form of emission occurs when the electric field is so high near the surface of a material that electrons are literally pulled out of the material. This type is called high field emission. The field strength required is in the order of 10^6 V/cm and can easily be obtained at a sharp protrusion on the surface of a metal. This form of emission has several useful applications, the best known of which is in the electron microscope. In this application, high field emission occurs at the point of a needle that is located at the center of a portion of a hollow sphere. The emitted electrons are collected on the inside surface of the sphere and transfer the image of the point of the needle to a much larger surface. The area transformation produces a magnification that is much larger than can be obtained with optical microscopes. High field emission is generally a nuisance in electric circuits because the emission that may occur at sharp corners can cause corona and breakdown. Consequently, sharp corners should be avoided, even in low-voltage circuits.

In this section we have looked briefly at four ways of obtaining electron emission, the most important of which is thermionic emission. This type is found in almost all vacuum tubes. Thermionic emission is obtained in a vacuum tube by heating a cathode with an electric current. In the next section, a two-element tube called a diode will be considered.

6.2 Vacuum Diodes

A vacuum diode is constructed by placing a second metal electrode in the same evacuated glass or metal envelope with a cathode. The second electrode is called a plate or anode and usually surrounds the cathode. A typical arrangement is shown in Fig. 6–3(a). Using the cathode voltage as a reference, the volt-ampere characteristics of a diode are shown in Fig. 6–3(b). A discussion of the physical operation of a vacuum diode and a few typical diode applications are included in this section.

The diode curve in Fig. 6–3(b) has three interesting regions. First, a small amount of current flows in the plate circuit when the voltage between the plate and cathode is zero. Consequently, a small negative voltage on the plate is required to cause the plate current to approach zero. Second, as the plate voltage increases in the positive direction, the plate current also increases but in a nonlinear manner. Third, when the plate current reaches a relatively high value, a further increase in plate voltage does not cause the plate current to increase very much; i.e., current saturation occurs. Each of these regions will now be discussed.

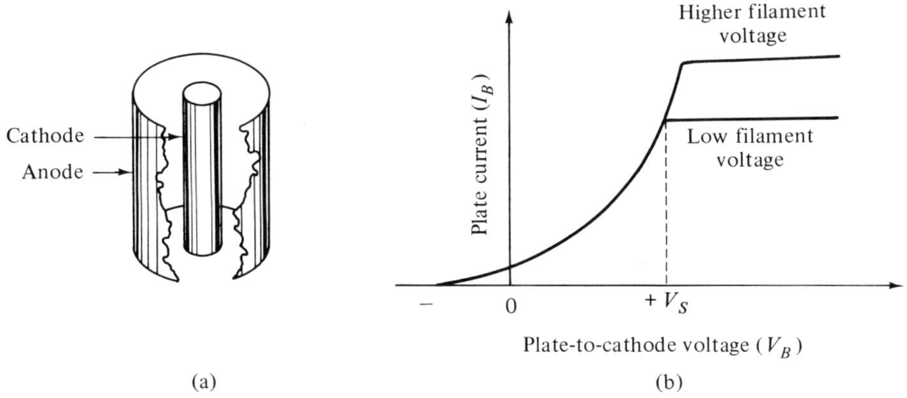

FIGURE 6–3 *(a) Construction of a diode with cylindrical electrodes and (b) a typical volt-ampere characteristic curve of a diode. (Adapted from R. H. Krackhardt,* Vacuum Tube Electronics, *Charles E. Merrill Publishing Company, 1966.)*

When a cathode is heated to a high temperature, T_H, electrons are emitted as indicated in Sec. 6.1. Most of these electrons leave the metal surface with very little excess energy and form an electron cloud around the cathode. However, a few are very energetic and travel some distance away from the cathode. Some of the most energetic electrons strike the plate and are captured. When an electron is captured, the plate becomes negative, and the cathode becomes positive. If the two elements are connected electrically, the charge equalizes by means of current flow through the external circuit. This current is shown for V_B equal to zero in Fig. 6–3(b). A negative voltage on the plate reduces the current below this value. A small negative voltage, typically -1 V, reduces the plate current to approximately zero. In this region, the characteristic curve is approximately exponential in form.

In the region from V_B slightly above zero to saturation, the plate current* is given by

$$I_P = KV_B^{3/2}/X^2 \tag{6–4}$$

where V_B is the voltage across the diode, X is the distance between the anode and cathode, and K is a constant for a given diode. This equation is known as

---

* This equation can be derived for 2 large parallel plane electrodes as follows: Let J be the magnitude of the current density. Combining the energy equation ($qV = \frac{1}{2}mv^2$), Poisson's equation ($d^2V/dx^2 = qn/\varepsilon_o$), and the current density-charge density relationship ($J = qnv$), we obtain

$$\frac{d^2V}{dx^2} = \frac{J}{\varepsilon_o}\sqrt{\frac{m}{2qV}}$$

Multiplying both sides of this equation by 2 (dV/dx) and integrating twice we obtain

$$\frac{4\ V^{3/4}}{3} = \left(\sqrt{\frac{4J}{\varepsilon_o}}\sqrt{\frac{m}{2q}}\right)x$$

Both constants of integration are assumed to be zero because the potential at the cathode is zero. Solving for J,

$$J = \left(\frac{4\varepsilon_o}{9\sqrt{\dfrac{m}{2q}}}\right)\frac{V^{3/2}}{X^2} = \frac{K\ V^{3/2}}{X^2}$$

The current associated with this current density is given by Eq. (6–4).

the three-halves power law, the Langmuir-Child law, or the space-charge equation. The three-halves law for vacuum diodes is due primarily to the current limiting features of the charge in transit between the two electrodes. For an overly simplified explanation of this effect, assume that each electric flux line starts on a positive charge on the plate and terminates on an electron as shown in Fig. 6–4. In Fig. 6–4(a), the cathode is cold, and all flux lines terminate on the cathode. The voltage gradient is V_B/X. In Fig. 6–4(b), the cathode is hot and

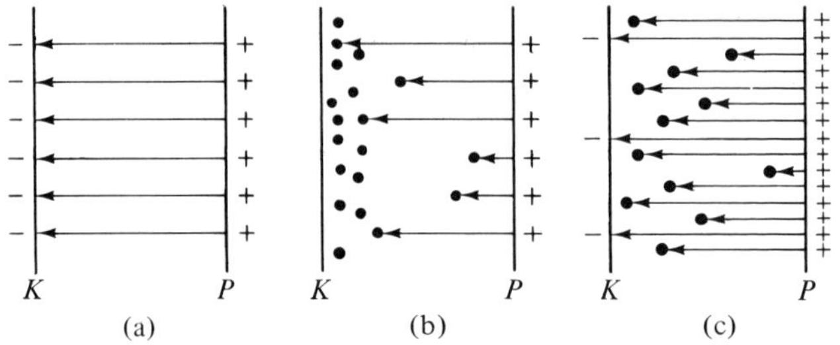

FIGURE 6–4 *Electric flux lines for three space-charge conditions: (a) zero space-charge with all flux lines terminating on the cathode, (b) all flux lines terminating on electrons in flight so that the current is limited by the charge in the interelectrode space, and (c) saturated conditions in which some flux lines terminate on cathode while waiting for more electrons to be emitted.*

is surrounded by a cloud of electrons. All of the electric flux lines terminate on electrons in the space between the electrodes. The current that flows at a positive voltage V_B is the result of these flux lines pulling electrons to the plate. However, the current is limited because the transit time for an electron is finite (a few nanoseconds). A given flux line cannot return to the cathode region for another electron until the one it terminates on reaches the plate. Thus for a given voltage and electrode separation, the number of flux lines is finite, and the transit time is finite. Consequently, if the rate of emission is sufficient, the current that flows is limited by the electrons in transit or the space charge. However, the current can be increased by increasing the number of flux lines; i.e., by increasing the voltage. As indicated by Eq. (6–4), the current increases more rapidly than the voltage.

At some value of plate voltage, V_S, the current saturates or ceases to increase with an increase in voltage because all of the electrons are pulled away from the cathode as fast as they are emitted. This condition is predicted by Eq. (6–2) and is shown in Fig. 6–4(c). Note that some of the flux lines terminate on the cathode while waiting for electrons to be emitted. Saturation can occur at a low voltage if the cathode temperature is low (low emission rate) or at a high voltage if the cathode temperature is high. Consequently, this condition is called temperature or emission saturation. Although an increase in V_B beyond the saturation voltage, V_S, produces very little increase in the plate current, a slight increase occurs because electrons that have almost enough energy to become free can be assisted by the pull of an electric flux line. This field assisted emission is called the *Schottky effect.*

EXAMPLE 6–1 A diode has a cylindrical tungsten emitter with a diameter of 1 mm and a length of 20 cm. The cathode temperature is 2500°K. If the Schottky effect is neglected, calculate the maximum current that can be supplied by the emitter.

Solution: From the Richardson-Dushman equation, Eq. (6–2),

$$I = AC_eT^2 \exp(-E_W/kT) \text{ Amperes}$$

Neglecting the end areas of the emitter, the area is

$$A = \pi dl = (3.14)(1 \text{ mm})(20 \text{ cm}) = 6.28 \times 10^{-4} m^2$$

From Table 6–1: $C_e = 60 \times 10^4 \text{ A}/(m^2T^2)$

$$E_W = 4.54 \text{ eV} = (4.54 \text{ eV})(1.6 \times 10^{-19} J/\text{eV}) = 7.27 \times 10^{-19} \text{ joules}$$

$$I = (6.28 \times 10^{-4} m^2)(60 \times 10^4 A/m^2T^2)(2500°\text{K})^2$$

$$\exp\left(\frac{-7.27 \times 10^{-19} \text{ joules}}{(1.38 \times 10^{-23} J/°\text{K})(2500°\text{K})}\right)$$

$$= (2.36 \times 10^9)e^{-21.2} = (2.36 \times 10^9)(7.5 \times 10^{10}) \text{ A} = 1.77 \text{ A}$$

Normally, the emission-limited current is very high compared with the rated operating current; i.e., several amperes for a rated current of 100 mA. Consequently, if V_B were increased to obtain the saturated current level, the plate dissipation would exceed the rated plate dissipation, and the plate would be destroyed* rapidly. When trying to determine the emission-limited current, pulsed operation is highly recommended.

In addition to the limitation on the maximum current due to the maximum emission rate, vacuum diodes have two other physical limitations, a maximum plate dissipation and a maximum signal frequency. As previously indicated, the energy that an electron has when it strikes the plate is converted to heat. Since a vacuum diode contains very few gas particles, the plate dissipates most of this heat by infrared radiation and the remainder by conduction through the plate leads and support post. An equilibrium condition must be achieved at a reasonable plate temperature which may range from 100°C for small receiving tubes to 800°C for large transmitting tubes with special anodes. Thus each tube type has a rated plate dissipation that should not be exceeded.

The power dissipated on a plate can be found by summing the energies of the electrons that strike the plate over a unit period of time. However, the power is also the product of the rms plate current** and rms plate-to-cathode voltage. For example, if the dc plate-to-cathode voltage is 10 V and the diode dc current is 100 mA, the plate dissipation is (10 V × 100 mA) or 1 W. For this steady-state condition, the temperature of the plate would rise until it could radiate 1 W.

The maximum signal frequency limitation is imposed by the time required for an electron to travel from the cathode to the plate. For good operation, the signal voltage should not change appreciably during the transit period of an

* This condition produces a monode. As yet, no one has found a useful application for a monode.

** rms means root-mean-square and is the dc equivalent on a heating basis.

electron. To find the transit time between two large parallel-plates,* let X be the distance between plates as shown in Fig. 6–5, and let the cathode be the origin. The force acting on an electron is

$$f = (-q)(-V_B/X) \tag{6–5}$$

where f is in Newtons, q is the charge on an electron (1.6×10^{-19} coulombs), and V_B/X is the electric field intensity. The minus sign on q indicates that the

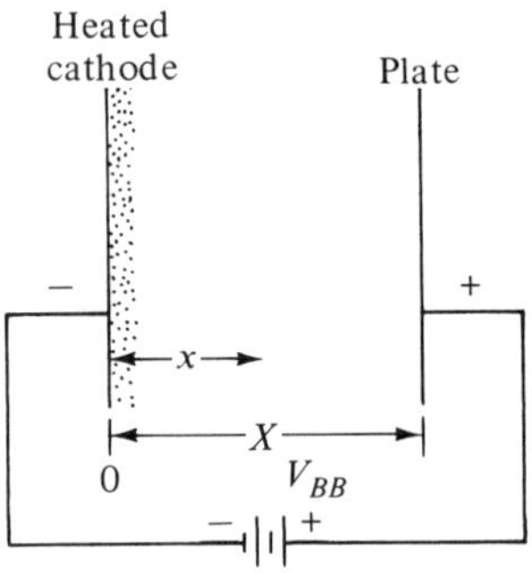

FIGURE 6–5 *A planar diode.*

charge on an electron is negative, while the minus sign on V_B/X indicates that the electric field direction is from right to left. In an electric field, an electron is accelerated in the direction of the positive electrode. From Newton's second law,

$$f = ma \tag{6–6}$$

where m is the mass of an electron, and a is the acceleration or the change in velocity, dv/dt. Combining these two equations and solving for the acceleration,

$$a = \frac{qV_B}{mX} = \frac{dv}{dt} \text{ meters/sec}^2 \tag{6–7}$$

Assuming that the electron moves only in the x direction and that the electron starts with zero velocity at the cathode,

$$v_x = \int \frac{qV_B}{mX}\,dt + C_1 = \frac{qV_Bt}{mX} \text{ meters/sec} \tag{6–8}$$

where C_1 is zero because v_x is zero at $t = 0$.

The x position at any time t can be found by integrating the expression for v_x, or

$$x = \frac{qV_Bt^2}{2mX} + x_o \text{ meters} \tag{6–9}$$

Using the cathode as the origin, x_o is zero at $t = 0$. If $x = X$, then the transit time between the electrodes, T, is

$$T = X\sqrt{\frac{2m}{qV_B}} \text{ sec} \tag{6–10}$$

where T is in seconds, m is in Kg, and X is in meters. This equation assumes that V_B is constant and that the velocity is small compared with the speed of light.

---

* This case is the easiest one to analyze because of the simple geometry and the absence of edge effects near the central area.

EXAMPLE 6–2 The cathode and plate of a diode are 0.5 cm apart. If a dc voltage of 10 V is impressed across the diode, find (a) the transit time of an electron, (b) the velocity on impact, and (c) the energy of the electron just before impact. Use the MKS system of units and assume that the electron starts at the cathode with zero velocity.

Solution: (a) $X = 0.5\text{ cm} = 5 \times 10^{-3}\text{ m}$; $m = 9.1 \times 10^{-31}\text{ Kg}$. From Eq. (6–10),

$$T = X\sqrt{\frac{2m}{qV_B}} = 5 \times 10^{-3}\sqrt{\frac{2 \times 9.1 \times 10^{-31}}{1.6 \times 10^{-19} \times 10}} \simeq 5.3\text{ nsec}$$

(b) At the instant just prior to impact, $t \simeq T$. From Eq. (6–8), the impact velocity, v_X, is

$$v_X = \frac{qV_BT}{mX} = \frac{(1.6 \times 10^{-19})(10)(5.3 \times 10^{-9})}{(9.1 \times 10^{-31})(5 \times 10^{-3})} \simeq 1.86 \times 10^6\text{ m/sec}$$

(c) Kinetic energy is $(1/2)mv^2$ where in this case $v = v_X$. Therefore, the impact energy is

$$(1/2)(mv_X^2) = (1/2)(9.1 \times 10^{-31})(1.86 \times 10^6)^2 \simeq 1.6 \times 10^{-18}\text{ joule}$$

An alternate approach is to use qV where V is the voltage difference (10 V in this case) and q is the charge on an electron. Thus, the impact energy is

$$qV = (1.6 \times 10^{-19})(10) = 1.6 \times 10^{-18}\text{ joule}$$

Vacuum diodes are rarely used in low-voltage circuits for two reasons: (1) they are more expensive than semiconductor diodes and (2) they require a filament source, usually a transformer. High-voltage power supplies sometimes exceed the peak inverse voltage (PIV) ratings of semiconductor diodes. Although semiconductor diodes can be placed in series to increase the PIV rating, high-voltage vacuum tubes may be preferable in some cases. A typical full-wave rectifier circuit with a filter is shown in Fig. 6–6. Note that the rectifier arrangement is the same as that shown in Fig. 2–12, where the arrow on the semiconductor diode corresponds to the plate or anode on the vacuum diode. The filter section is similar to that shown in Fig. 2–16(c). The following example provides some insight into the performance of a circuit of this type.

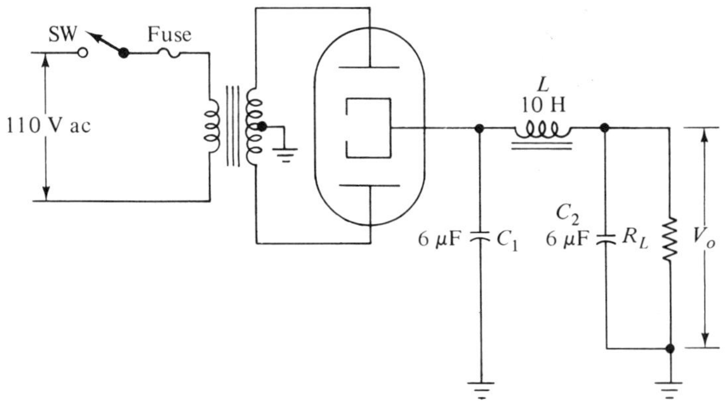

FIGURE 6–6 *A dc power supply using a full-wave vacuum diode rectifier circuit and a π-type filter.*

EXAMPLE 6–3 The circuit shown in Fig. 6–6 is a full-wave rectifier with a π-type filter. With 110 V rms 60-Hz sine wave input to the power transformer, the secondary voltage was approximately 400 V rms on each side of the secondary center tap (or ground point) or 800 V rms across the total secondary. Assuming that R_L was infinite and that the capacitors had no appreciable leakage current, the output dc voltage, V_O, was $400\sqrt{2}$ or approximately 560 V. As R_L decreased from infinity, V_O dropped rapidly, and the ac ripple on V_O increased. For a 3000-ohm load, the measured value of V_O was 400 V with a 3-V peak-to-peak ripple. The load current was 133 mA. With a 1500-ohm load, V_O decreased to 340 V with a 3.8-V peak-to-peak ripple. The load current was 226 mA. At this current level, the dc voltage drop across the coil* was 18 V, indicating that the coil had a resistance of approximately 80 ohms.

A decrease of 60 V in V_O as the current increases from 133 mA to 226 mA is typical for this circuit. Part of this drop occurs in the resistance of the coil, part in the tube, part in the transformer windings, and part due to the inability of the input capacitor to maintain a reasonably high voltage over a full half-cycle period. The regulation could be improved, and the ripple reduced, by increasing the values of C_1, C_2, and L and also by decreasing the resistance of the windings.

6.3 The Vacuum Triode

Vacuum diodes discussed in the previous section are limited to switching and rectifying applications. However, by placing a properly designed third electrode between the cathode and plate, the resulting tube can be used as an amplifier. This third electrode is called a grid, and is usually made of fine wire attached to two support posts as shown in Fig. 6–7(a).

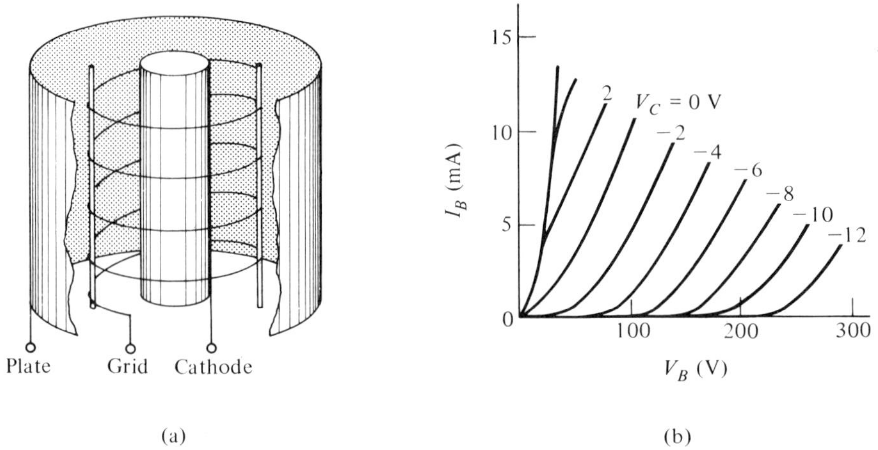

FIGURE 6–7 *A vacuum triode: (a) typical construction and (b) static characteristic curves. (Adapted from R. H. Krackhardt,* Vacuum Tube Electronics, *Charles E. Merrill Publishing Company, 1966.)*

The purpose of the grid is to place a potential between the plate and cathode. This potential is usually negative relative to the cathode. The function of the grid is to vary the plate current as the grid-to-cathode voltage is varied, i.e., to

* The term "choke" is often used to designate the large coil or inductor used in power supplies.

control the rate of flow of electrons to the plate. If the grid is made more negative than the cathode (−0.5 V or more), electrons are repelled from the grid wires so that the grid current is extremely small or essentially zero. However, electrons can move to the positive plate by going between the grid wires.

The effect of the grid-to-cathode voltage on the plate current is shown in Fig. 6–7(b). This set of curves, called the static plate characteristic curves, is the type usually found in the tube manuals and may be used to determine both the dc bias points and the ac characteristics. Note that the plate current is a function of both the negative grid voltage and the positive plate voltage. For example, in Fig. 6–7(b) a plate current of 5 mA can be obtained with an infinite number of combinations of plate voltage, V_B, and grid voltage, V_C. However, as V_C is made more negative, the value of V_B required to obtain the same current must be made more positive. In addition, since the grid structure is much closer to the cathode than the plate, a large positive change must be made in the plate voltage, V_B, to compensate for a small negative change in grid voltage, V_C.

As with any active or amplifying device, the triode must be properly biased before using it to amplify signals. While the choice of a bias point is usually not critical, a few guidelines should be observed. First, the dc supply voltage, V_{BB}, should not exceed the rated breakdown voltage of the tube unless special circuits are incorporated around the tube to protect it in the event that the plate current, I_B, is interrupted. If the supply voltage is too low, the signal will be distorted, even on small excursions around the operating point. In small- and medium-signal amplifiers, V_{BB} ranges from 100 to 250 V and may be as high as 400 V for rather large output signals.

Second, the resistor R_P in the plate circuit should be large enough to obtain a reasonable voltage gain but not so large that excessive distortion results. Typical values range from 10 K to 100 K. Third, the grid bias voltage, V_C, should place the operating point in the linear range of the characteristic curves for the expected amplitude of ac grid voltage. A typical bias circuit for a triode is shown in Fig. 6–8. Note that this circuit is basically the same as the one normally used for a JFET. Since the grid is not usually biased in the positive region, only one resistor is necessary in the grid circuit. Resistor R_G provides a path from grid to ground for the few electrons and ionized atoms that strike the grid wires. However, the resistance of R_G should be reasonably large so that the ac signal is not shorted to ground. Typical values range from 100 K to 1 M. Since the dc grid current is extremely small, the grid is virtually at ground potential if R_G is not excessively large; that is, the grid-to-ground voltage is essentially zero. Consequently, the dc plate current flowing through R_K produces a voltage drop that is equal to the grid-to-cathode bias voltage. The following example illustrates a practical approach to the bias problem.

EXAMPLE 6–4 A tube manual gives the following data for the tube shown in Fig. 6–8:

Plate voltage: 300 V max, 250 V typical

Grid bias: −6 V typical

Plate dissipation: 2.5 W

Using a supply voltage of 250 V, find the bias resistors, plate current, and dc circuit voltages.

Solution: With only these restrictions, a unique solution does not exist. The operating point, or Q point, should be on the -6-V grid line and somewhere in the linear operating region. The point shown in Fig. 6–8(b) is chosen rather arbitrarily. At this point, I_P is approximately 4 mA, and the plate to cathode voltage, V_{PK}, is approximately 165 V. The value of R_K is

$$R_K = 6\ \text{V}/4\ \text{mA} = 1500\ \Omega$$

The dc load line is drawn from the supply voltage, V_{BB}, through the Q point and on to the current axis where the intercept is 12 mA. The total external dc resistance is determined by the inverse slope of this dc load line or

$$R_K + R_P = 250\ \text{V}/12\ \text{mA} = 20.8\ \text{K}$$

Therefore, the value of R_P is

$$R_P = 20.8\ \text{K} - 1.5\ \text{K} = 19.3\ \text{K} \simeq 20\ \text{K}$$

The dc circuit voltages are shown in Fig. 6–8(a). The plate dissipation is 165 V $\times$ 4 mA or 0.66 W.

A current intercept of 10 mA or 14 mA might also have been chosen, giving different Q points on the -6-V bias line and, consequently, different values of plate current. The choices in this range are not critical.

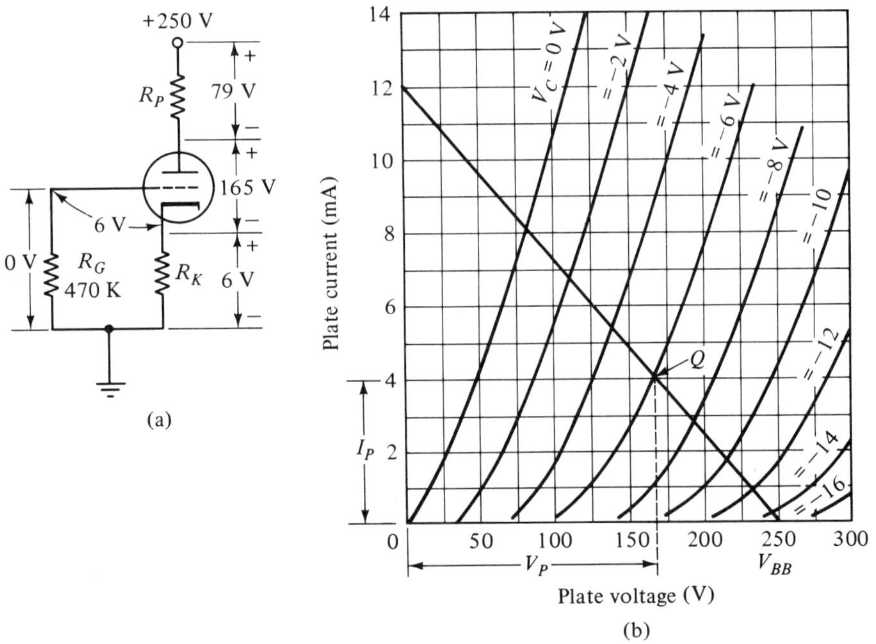

FIGURE 6–8 *Graphical analysis of a triode bias circuit: (a) bias circuit and (b) graphical solution of circuit voltages.*

6.4 Analysis of Large-Signal Triode Amplifiers

When properly biased, a triode can be used as an amplifier. Like the devices discussed in Chaps. 3 and 4, three configurations are practical: (1) the common cathode (CK), (2) the common plate (CP), and the common grid (CGr).* The

* The CK corresponds to the CS and CE, the CP corresponds to the CD and CC, and the CGr corresponds to the CG and CB configurations.

major emphasis is placed on the CK stage since it is the one most often encountered in practice.

Although the approaches used to analyze vacuum tube amplifiers are similar to those used for JFET and bipolar transistor amplifiers, a few differences exist because a tube does not operate physically like either type of transistor. One obvious difference is the insensitivity of the plate current in a tube to changes in ambient temperature compared with the corresponding currents in transistors. As a result, the ordinary triode amplifier does not require stabilization against changes in temperature.

Another difference is the magnitude of the dc supply voltages. The usual supply voltage is above 100 V. Since the output signal can be a significant fraction of the dc source voltage, the output voltage can range up to one or two orders of magnitude higher than that usually associated with transistor amplifiers.

The large-signal analysis of a CK stage is very similar to the large-signal analysis of a CS stage, since each requires an accurate set of static characteristic curves for the active device. The basic steps in large-signal analysis are: (1) construction of dc and ac load lines, and (2) construction of the output signal waveform from the input signal waveform and the ac load line. The following example illustrates this type of analysis for a CK amplifier.

EXAMPLE 6–5 A single-stage common-cathode (CK) amplifier is shown in Fig. 6–9(a). If V_g is a sine wave with a 4-V peak value, find the mid-frequency output voltage. Use the bias circuit designed in Ex. 6–4 and the characteristic curves shown in Fig. 6–8(b).

Solution: The impedance looking into the grid terminal of the CK stage is several tens of megohms in the mid-frequency range and is determined largely by the external leakage around the base of the tube and tube socket. Consequently, the ac voltage from grid to ground is

$$V_i = V_g R_G/(R_G + R_g) = (4\text{ V})(470\text{ K})/(10\text{ K} + 470\text{ K}) \simeq 4\text{ V peak}$$

To find the output voltage, we must first construct an ac load line for $R_{L\text{eq}}$ where

$$R_{L\text{eq}} = R_P \parallel R_L = 20\text{ K} \parallel 33\text{ K} \simeq 12.4\text{ K}$$

One point on the ac load line is the Q point that was determined for the bias circuit. This point is shown in Fig. 6–9(b) on the dc load line. In the mid-frequency range, the ac load line is straight because the circuit is assumed to contain only resistive elements. Consequently, only one additional point is needed. A second point may be found by assuming a change in current (or voltage) from the Q point and finding the corresponding change in voltage (or current) that is produced across $R_{L\text{eq}}$. Assume an increase of 8 mA above the 4-mA bias current. The decrease in plate voltage produced by an increase of 8 mA is

$$|\Delta V_B| = 8\text{ mA} \times 12.4\text{ K} \simeq 100\text{ V}$$

The second point* is found by starting at the Q point (165 V) and moving to the left along the bias current (4 mA) line to (165 V − 100 V), or 65 V, as shown in Fig. 6–9(b). Then move vertically to (4 mA + 8 mA), or 12 mA, to locate the second point at 65 V and 12 mA. The ac load line is constructed by drawing a straight line through this point and the Q point. The ac load line is the locus of all possible instantaneous combinations of plate current, plate voltage, and grid-voltage in the mid-frequency region.

---

* The second point can also be located using the method described in Ex. 3–5.

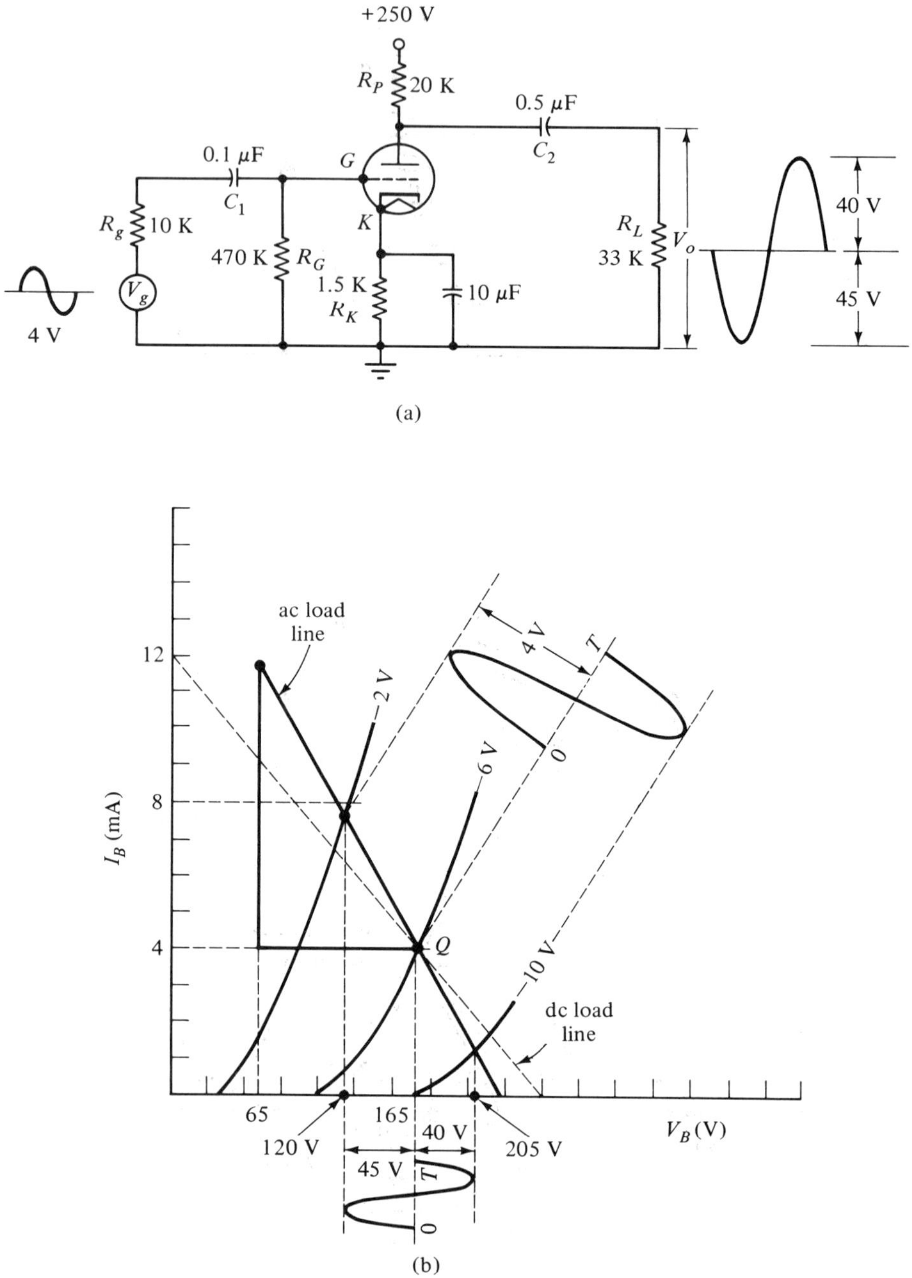

FIGURE 6–9 *A large-signal triode amplifier: (a) a circuit diagram and (b) a graphical analysis.*

With a grid voltage swing of 4 V peak, the instantaneous operating point moves along the ac load line from a grid voltage of −6 V to −2 V, back to −6 V and then to −10 V as shown in Fig. 6–9(b). One cycle of the input voltage wave is completed by returning to −6 V. The −2-, −6-, and −10-V grid voltage curves from Fig. 6–8(b) are shown in Fig. 6–9(b) for clarity. By projecting these critical points down to the plate voltage axis, the output ac voltage may be sketched as shown in this figure. Note that as the plate current increases, the plate voltage decreases; i.e., as the grid becomes less negative, the plate becomes less positive. This relationship is the familiar 180° phase shift between the input and output ac voltage of a CK (or a CE) amplifier in the mid-frequency range. The plate voltage swings from 165 V at the Q point to 120 V on the low-voltage side and to 205 V on the high-voltage side. The output ac voltage is slightly distorted as indicated by an amplitude of 45 V peak on one side and 40 V peak on the other. The average amplitude of the output signal is

$$V_o = \frac{V_{\max} - V_{\min}}{2} \simeq \frac{205 - 120}{2} = \frac{85}{2} = 42.5 \text{ V peak}$$

The gain is then

$$A_{vo} = V_o/V_g = 42.5 \text{ V}/4 \text{ V} \simeq 10.6$$

The graphical procedure is enlightening because an ac swing around the operating point can be visualized readily as rapid variations in the dc values along a straight-line locus. The ac load line is different from the dc load line because the coupling capacitor C_2 is assumed to be a short circuit to the ac signal. This condition is achieved when the voltage across the capacitor does not change appreciably during the period of one-half cycle of the voltage wave.

6.5 Small-Signal Analysis of Triode Amplifiers

When the input signal is small, a graphical analysis is not appropriate because the reading errors become too large relative to the signal amplitudes. Consequently, ac equivalent circuits will now be introduced to use with small-signal cases. As in the FET and bipolar circuits, we assume that the triode is biased properly and consider only the small ac excursions around the operating point.

Two mid-frequency ac equivalent circuits* are shown in Fig. 6–10, one with a dependent current source ($g_m V_{gk}$) and one with a dependent voltage source (μV_{gk}). Three triode parameters are included in the equivalent circuits, g_m, r_p, and μ. The parameter g_m is a transconductance and is the ratio of a small change in plate current to a corresponding change in grid voltage when the plate voltage is held constant, or

$$g_m \simeq \left.\frac{\Delta I_B}{\Delta V_C}\right|_{V_B=\text{constant}} \tag{6–11}$$

---

* These equivalent circuits may be derived as follows. Assuming that the grid current is zero, the plate current is a function of V_B and V_C, or

$$I_B = f(V_B, V_C)$$

Then

$$\Delta I_B \simeq dI_B = \frac{\partial I_B}{\partial V_B} dV_B + \frac{\partial I_B}{\partial V_C} dV_C$$

For small ac signals, let $dV_B = V_p$ and $dV_C = V_{gk}$. Then

$$I_p = g_p V_p + g_m V_g = \frac{1}{r_p} V_p + g_m V_g$$

This equivalent circuit is shown in Fig. 6–10(a). The equivalent circuit shown in Fig. 6–10(b) can be obtained by making a source transformation where

$$\mu = g_m r_p$$

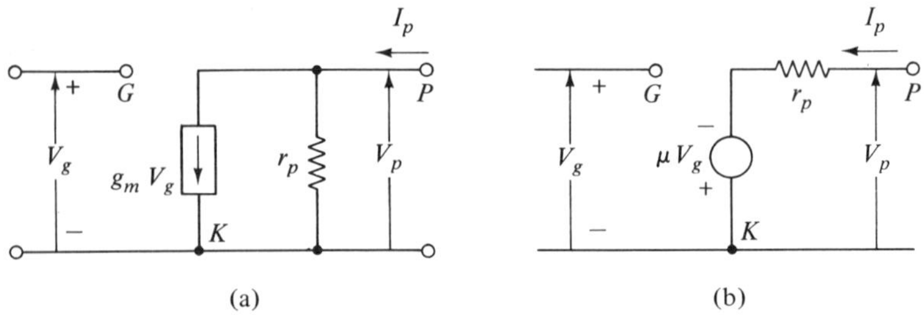

FIGURE 6–10 *Two small-signal mid-frequency equivalent circuits for triodes: (a) a circuit with a dependent current generator, and (b) a circuit with a dependent voltage generator.*

The value of g_m can be estimated from the plate characteristic curves as follows: (1) draw a vertical line through the operating point, (2) select a convenient change in grid voltage (say one or two volts) with the Q point in the center of this increment, (3) find the corresponding change in plate current, and (4) find the ratio of the change in plate current to the selected change in grid voltage.

The parameter r_p is the incremental plate resistance and is the ratio of a change in plate voltage to a change in plate current with the grid voltage held constant, or

$$r_p \simeq \left.\frac{\Delta V_B}{\Delta I_B}\right|_{V_C=\text{constant}} \tag{6–12}$$

This parameter may also be estimated from the plate characteristic curves. To simulate a constant grid voltage, draw a tangent to the constant grid voltage line at the operating point. Along this line, select an arbitrary increment of plate voltage (say 50 to 100 V) and find the corresponding change in plate current. Then r_p is the ratio of the increment of plate voltage and the corresponding change in plate current.

The parameter μ is the ratio of a change in plate voltage to a change in grid voltage with the plate current held constant, or

$$\mu \simeq \left.\frac{\Delta V_B}{\Delta V_C}\right|_{I_B=\text{constant}} \tag{6–13}$$

To obtain an estimated value of μ from the plate characteristic curves, first draw a horizontal straight line through the Q point and then select a small increment of grid voltage (say 1 to 4 V) along the line with the Q point centered in this increment. For this increment find the corresponding change in plate voltage. The value of μ is the ratio of these two corresponding increments.

From practical considerations, μ and r_p are the easiest coefficients to determine graphically from the plate characteristic curves. Then g_m may be obtained from

$$g_m = \mu/r_p \tag{6–14}$$

These parameters change if the Q point is moved. In general, μ and g_m have low values for low plate currents and both increase for increasing values of plate current. However, the relationships are nonlinear and cannot be normalized.

A typical small-signal triode amplifier is shown in Fig. 6–11(a). If the circuit is already designed, the first problem in analyzing the circuit is to locate the Q

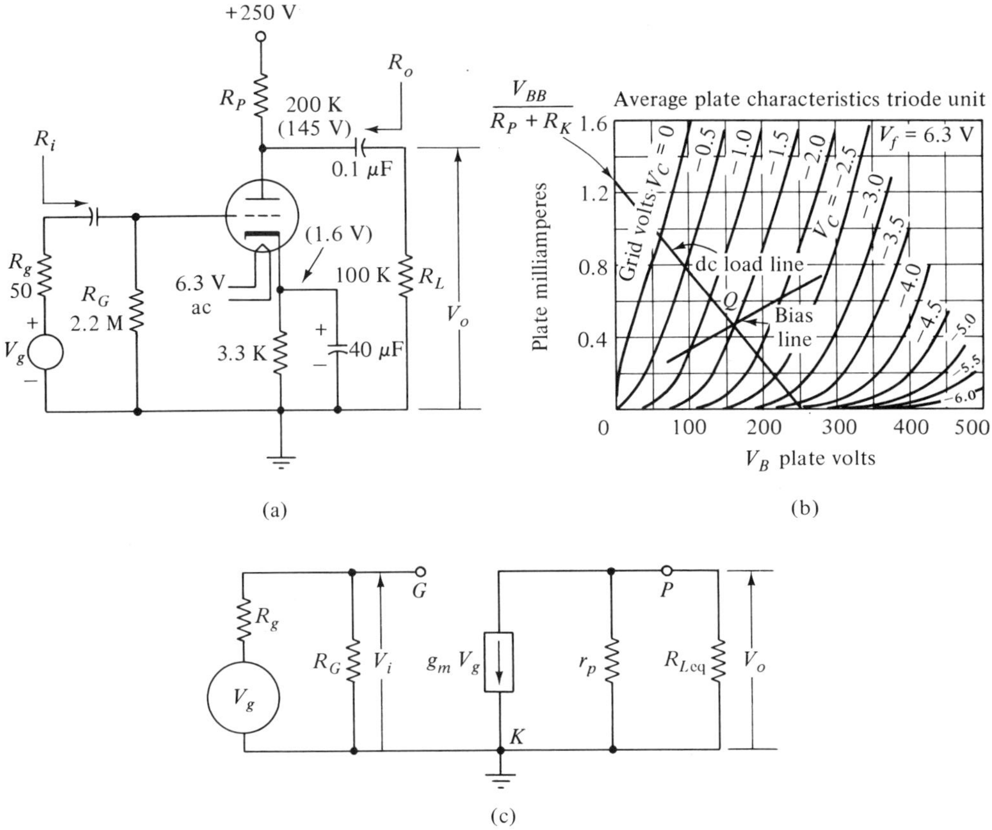

FIGURE 6–11 *A small-signal CK triode amplifier: (a) amplifier circuit, (b) graphical construction to locate the Q point, and (c) the complete equivalent circuit. Measured dc voltages with respect to ground are shown in parentheses.*

point. This point can be located by drawing in the dc load line and the bias line. The Q point lies at the intersection of these two lines. The voltage intercept of the dc load line is V_{BB}, and the current intercept is $V_{BB}/(R_P + R_K)$. The bias line in this circuit is obtained by selecting a grid voltage and dividing this voltage by R_K for one point on the line. This step is then repeated for other values of grid voltages until a line can be drawn that intersects the dc line. When the Q point is located, the triode parameters can be calculated and incorporated into an equivalent circuit as shown in Fig. 6–11(c). With a complete equivalent circuit, the voltage gains, input impedance, and output impedance are calculated in the same way that these performance parameters were calculated for bipolar and FET amplifiers.* These equations are summarized in Table 6–2. The following example illustrates the application of these equations to a CK triode amplifier.

EXAMPLE 6–6 For the triode amplifier circuit shown in Fig. 6–11(a), find (a) the operating point, (b) g_m and r_p, and (c) the performance parameters. Use the plate characteristic curves shown in Fig. 6–11(b).

* See Sec. 3.6 and especially Eq. 3–30.

TABLE 6–2 *Summary of approximate performance characteristics of single-stage vacuum-tube amplifiers. The quantities in parentheses indicate typical ranges of parameter values.*

Configuration	A_{vt}	Stage input impedance	Stage output impedance
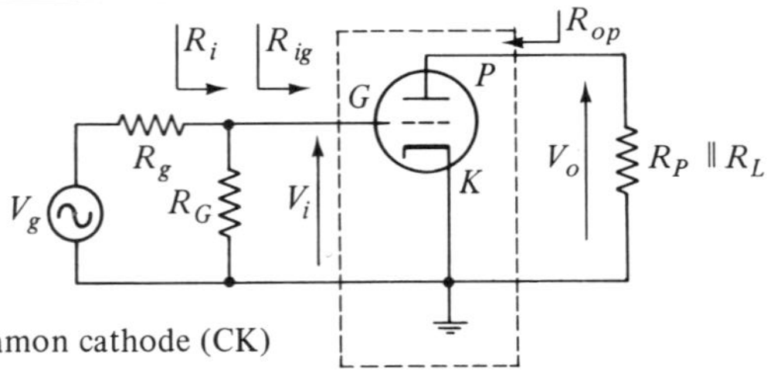 Common cathode (CK)	$-g_m(R_P \parallel R_L \parallel r_p)$ (signal inverted)	$R_{ig} \simeq r_{gk}$ $(\simeq \infty)$	$R_{op} \simeq r_p$ (triode – 10 K to 100 K) (pentode – 100 K to 1 M)
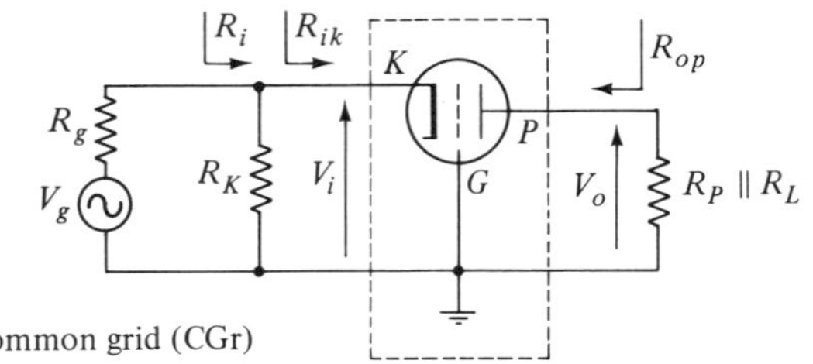 Common grid (CGr)	$g_m(R_P \parallel R_L \parallel r_p)$ (signal not inverted)	$R_{ik} \simeq \dfrac{r_p + R_L \parallel R_P}{g_m r_p + 1}$ (20 ohms to 2 K)	$R_{op} = r_p + (g_m r_p + 1)(R_K \parallel R_g)$ (triode – 15 K to 150 K) (pentode – 100 K – 1 M)
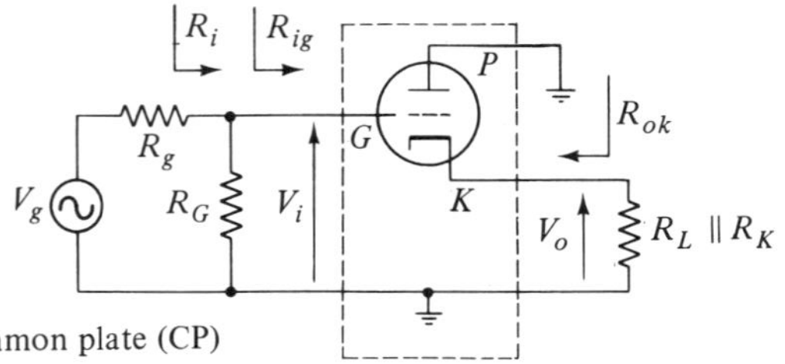 Common plate (CP)	$\dfrac{g_m(r_p \parallel R_K \parallel R_L)}{1 + g_m(r_p \parallel R_K \parallel R_L)}$ (signal not inverted)	$R_{ig} > r_{gk}$ $(\simeq \infty)$	$R_{ok} = \dfrac{1}{g_m} \parallel r_p$ (50 Ω to 2 K)

$A_{vo} = V_o/V_g = A_{v_g}A_{vt}$; $A_{v_g} = V_i/V_g = R_i/(R_i + R_g)$; $A_{vt} = V_o/V_i$

Solution: (a) Bias conditions: The dc load line and the bias line are drawn on the characteristic curves as shown in Fig. 6–11(b). The intersection of the dc load line and the bias line determines the Q point. The bias values at the Q point are

$$I_B \simeq 0.45 \text{ mA} \qquad V_{PK} \simeq 160 \text{ V} \qquad V_{GK} \simeq -1.5 \text{ V}$$

The plate voltage with respect to ground can also be found from

$$V_P = V_{BB} - I_B R_P \simeq 250 \text{ V} - (0.45 \text{ mA})(200 \text{ K}) \simeq 160 \text{ V}$$

(b) Tube parameters: The values of g_m and r_p can be approximated from the characteristic curves at the Q point as

$$g_m \simeq \frac{\Delta I_B}{\Delta V_{GK}} \simeq \frac{1 \text{ mA} - 0.1 \text{ mA}}{-1 \text{ V} - (-2 \text{ V})} \simeq 0.9 \text{ millimho}$$

$$r_p \simeq \frac{\Delta V_{PK}}{\Delta I_B} \simeq \frac{250 \text{ V} - 120 \text{ V}}{1.4 \text{ mA} - 0} \simeq 93 \text{ K}$$

(c) Gain and impedance calculations:

$$R_{L\text{eq}} \simeq R_P \,||\, R_L \,||\, r_p \simeq 200 \text{ K} \,||\, 100 \text{ K} \,||\, 93 \text{ K} \simeq 39 \text{ K}$$

$$A_{vt} \simeq -g_m R_{L\text{eq}} \simeq -(0.9 \times 10^{-3})(39 \text{ K}) \simeq -35$$

$$R_i \simeq R_G \simeq 2.2 \text{ M}$$

$$R_o \simeq R_P \,||\, r_p \simeq 200 \text{ K} \,||\, 93 \text{ K} \simeq 63 \text{ K}$$

$$A_{vg} = R_i/(R_i + R_g) \simeq 2.2 \text{ M}/(2.2 \text{ M} + 50) \simeq 1$$

$$A_{vo} = A_{vg} A_{vt} \simeq (1)(-35) \simeq -35$$

(d) Experimental results: The measured dc voltage levels are shown in Fig. 6–11(a) and are within 20 percent of the calculated values. The measured voltage gain was -36.5 with an input voltage of 0.55 V.

As indicated by a comparison between the above example and Ex. 3–7, an FET common-source amplifier and a triode common-cathode have much in common. The mid-frequency equivalent circuits are the same, and the input resistance for both devices is very high. The value of g_m is usually determined with an equation for an FET and graphically for a triode. However, the calculations for the performance parameters of an amplifier are similar.

While the common-cathode amplifier is the type most frequently used, two other types are sometimes found in practice: the common-grid amplifier which is similar to the common-gate FET amplifier, and the common-cathode amplifier (or the cathode follower) which is similar to the common-drain (or source-follower) FET amplifier. The analysis of these two types will parallel the two corresponding FET cases.

A common-grid (CGr) amplifier is shown in Fig. 6–12(a). The input resistance to the triode in this connection, R_{ik}, can be found from the partial equivalent circuit in Fig. 6–12(b). Using a voltage source, V_{ti}, to drive this partial circuit, and referring to the FET case,

$$R_{ik} = \frac{V_{ti}}{I_{ti}} = \frac{1 + [(R_P \,||\, R_L) r_p]}{g_m + (1/r_p)} = \frac{r_p + R_P \,||\, R_L}{\mu + 1} \simeq \frac{1}{g_m} \tag{6–15}$$

where $\mu = g_m r_p$.

From Fig. 6–12(a), the input resistance is

$$R_i = R_K \,||\, R_{ik} \tag{6–16}$$

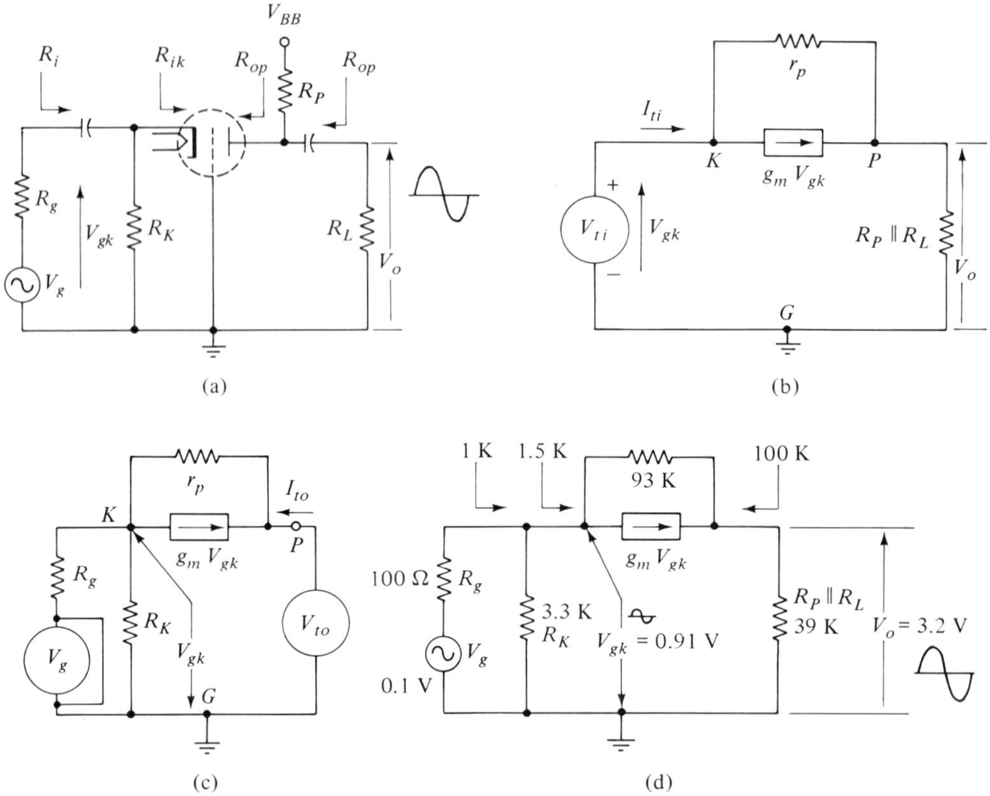

FIGURE 6–12 *A triode connected as a common-grid amplifier: (a) complete circuit, (b) circuit for determining* R_{ik}, *(c) circuit for determining* R_{op}, *and (d) complete equivalent circuit with values from Ex. 6–7.*

and the gain from the source to the cathode (input) terminal is

$$A_{vg} = R_i/(R_i + R_g) \tag{6–17}$$

The source-to-output voltage gain* is

$$A_{vo} = V_o/V_g = [A_{vg}][A_{vt}] \simeq [R_i/(R_i + R_g)]\ [(g_m)(R_P \,||\, R_L \,||\, r_p)] \tag{6–18}$$

An expression for the output resistance of this stage may be obtained from the partial circuit shown in Fig. 6–12(c). In this circuit, V_g has been shorted, and the output is driven by a voltage source, V_{to}. The output resistance is

$$R_{op} = \frac{V_{to}}{I_{to}} = r_p\left(1 + g_m\,(R_K \,||\, R_g) + \frac{R_K \,||\, R_g}{r_p}\right) = r_p + (\mu + 1)\,(R_K \,||\, R_g) \tag{6–19}$$

Since the values for g_m for different tube types range from 500 to 50,000 μmhos, approximations can be made only after a particular tube type has been selected.

---

* An accurate expression for A_{vt} is

$$A_{vt} = \frac{(\mu + 1)(R_L \,||\, R_P)}{(R_L \,||\, R_P) + r_p}$$

Since μ is not known accurately, the second factor in Eq. (6–18) is a reasonable approximation.

However, the following example will provide some perspective for this configuration.

EXAMPLE 6–7 Using the same bias conditions specified in Ex. 6–6 and the same values of R_K, R_P, and R_L, find the mid-frequency performance parameters and the output voltage when the input signal, V_g, is 0.1 V and R_g is 100 ohms.

Solution: From Eq. (6–15) the input resistance at the cathode is

$$R_{ik} = \frac{r_p + R_{Leq}}{\mu + 1} = \frac{93\text{ K} + 39\text{ K}}{(0.9 \times 10^{-3} \times 93 \times 10^3) + 1} = \frac{132\text{ K}}{85} \simeq 1.5\text{ K}$$

The load on the generator is

$$R_i = R_K \,||\, R_{ik} = 3.3\text{ K} \,||\, 1.5\text{ K} \simeq 1\text{ K}$$

The gain from the generator to the cathode is

$$A_{vg} = 1\text{ K}/(1\text{ K} + 100) \simeq 0.91$$

Consequently, the cathode-to-ground voltage, V_i, is 0.091 V. The gain from the cathode to the plate is

$$A_{vt} \simeq g_m R_{Leq} = 0.9 \times 10^{-3} \times 39 \times 10^3 \simeq 35$$

The output voltage is

$$V_o \simeq 0.1\text{ V}\,(A_{vg})(A_{vt}) \simeq 0.1(0.91)(35) \simeq 3.2\text{ V}$$

The output signal is in phase with V_g. From Eq. (6–19) the output resistance, R_o, is

$$R_{op} = r_p\left(1 + g_m(R_K \,||\, R_g) + \frac{R_K \,||\, R_g}{r_p}\right) = 93\text{ K}\left(1 + 0.9 \times 10^{-3}(100) + \frac{100}{93\text{ K}}\right)$$

$$\simeq 93\text{ K} \times 1.1 \simeq 100\text{ K}$$

These calculated values are shown in Fig. 6–12(d). Note that some of the approximations made in the CG FET amplifier are not valid here, especially if R_{Leq} is large.

As indicated by the preceding example, the common-grid amplifier has a low input impedance and a high output impedance. As with any low input impedance amplifier, the generator impedance must be low, and the equivalent load impedance must be reasonably large in order to obtain a reasonable generator-to-output voltage gain.

A third type of useful triode amplifier is the common-plate (CP) or cathode follower (CF) amplifier. As indicated in Fig. 6–13, the plate is connected directly to the dc power supply and, consequently, is ac grounded. The input signal is applied between the grid and ground while the output signal is obtained between the cathode and ground. Thus, the output signal, V_o, in the mid-frequency region is always less than the input signal, V_i, because part of the input voltage must be used to drive the triode to obtain an output voltage. From Fig. 6–13,

$$V_i = V_{gk} + V_o \tag{6–20}$$

An ac equivalent circuit for a CP amplifier is shown in Fig. 6–13(b). From this figure, the output voltage, V_o, is

$$V_o = (g_m)(V_{gk})\,(r_p \,||\, R_K \,||\, R_L) \tag{6–21}$$

Since

$$V_{gk} = V_i - V_o \tag{6–22}$$

then from Eq. (6–21),

$$V_o = g_m(V_i - V_o)(r_p \,||\, R_K \,||\, R_L) \tag{6–23}$$

Solving for A_{vt},

$$A_{vt} = \frac{V_o}{V_i} = \frac{g_m(r_p \,||\, R_K \,||\, R_L)}{1 + g_m(r_p \,||\, R_K \,||\, R_L)} \tag{6–24}$$

From this equation, we see that if $g_m(r_p \,||\, R_K \,||\, R_L) \gg 1$, then A_{vt} is approximately unity.

The grid-to-cathode impedance, r_{gk}, is always very large (several tens of megohms), and the effect of the current generator and $(R_K \,||\, R_L)$ is to multiply r_{gk} by a factor of $[1 + g_m(R_K \,||\, R_L)]$. Consequently, R_{ig} can be approximated as infinitely large compared with the grid leak resistor, R_G. Therefore,

$$R_i \simeq R_G \tag{6–25}$$

Then the gain from the generator to the grid terminal is

$$A_{vg} = R_G/(R_G + R_g) \tag{6–26}$$

From Eqs. (6–24) and (6–26), A_{vo} is

$$A_{vo} = A_{vg}A_{vt} = \left(\frac{R_G}{R_G + R_g}\right)\left(\frac{g_m(r_p \,||\, R_K \,||\, R_L)}{1 + g_m(r_p \,||\, R_K \,||\, R_L)}\right) \tag{6–27}$$

The output resistance may be found by placing a voltage generator, V_{to}, from the cathode to ground and shorting V_g as shown in Fig. 6–13(c). Note that when these changes are made, the grid is ac grounded, and the cathode swing is V_{to}. Therefore, V_{gk} is equal to $-V_{to}$ and

$$I_{to} = I_{rp} + g_m V_{to} \tag{6–28}$$

Then R_{ok} is

$$R_{ok} = \frac{V_{to}}{I_{to}} = r_p \,||\, (1/g_m) = \frac{r_p}{\mu + 1} \tag{6–29}$$

Combining R_{ok} in parallel with R_K, the impedance looking back from the load terminals is

$$R_o = \frac{r_p}{\mu + 1} \,||\, R_K \tag{6–30}$$

These relationships indicate that the cathode follower has a gain of less than unity and that the output signal is in phase with the input signal. In addition, the input impedance is large, and the output impedance is small. This type of amplifier is often used between a high-impedance source and a low-impedance load. One important consideration when designing a CP stage to drive a low-impedance load is the available current range. The bias point should be selected so that the available current swing on each side of the Q point is greater than the peak ac current required. For example, the peak current required for a 5-V peak sine wave across a 1000-ohm ac load is 5 mA. If the ac load is reduced to 100 ohms, the Q point must be selected to allow a 50-mA peak current on each side of the Q point without distortion. Another important application of a cathode follower is to form part of a differential amplifier.

One form of a triode differential amplifier is shown in Fig. 6–14(a). The similarity between the vacuum tube and the FET allows us to use the performance equations developed for the FET differential amplifier for the vacuum tube

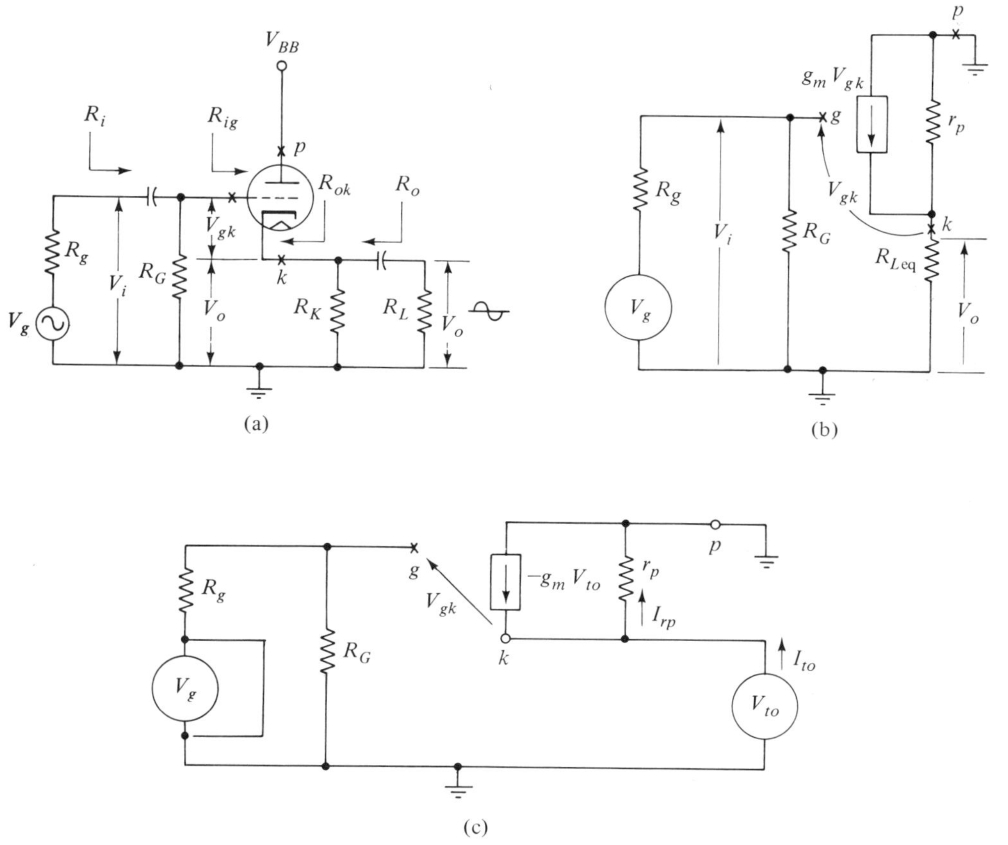

FIGURE 6–13 *A triode common-plate amplifier: (a) circuit diagram, (b) ac equivalent circuit, and (c) partial circuit for finding* R_{ok}.

differential amplifier with only minor modifications. The expressions given in Eq. (4–114) are appropriate for the differential amplifier with a fixed resistor bias [R_K in Fig. 6–14(a)]. These expressions, modified for triodes, are

$$A_d = g_m R_{Leq}/2 \qquad A_c \simeq -R_{Leq}/2R_K \qquad \text{CMRR} \simeq g_m R_K \tag{6–31}$$

The first problem is to determine expressions for R_{Leq} for each triode stage in Fig. 6–14(a). In general, R_{op} cannot be neglected for a triode when calculating the effective load. Each tube in the differential amplifier is different from a simple CK configuration because of the presence of an unbypassed cathode resistor. From Eq. (6–19) the resistance looking into the plate terminal can be calculated by analogy with the resistance looking into the drain terminal of a common gate (CG) stage, or

$$R_{op} = r_p[1 + g_m(R_K \parallel R_g) + (R_K \parallel R_g)/r_p] \tag{6–32}$$

To apply Eq. (6–32) to a triode differential amplifier, replace $R_K \parallel R_g$ with the equivalent resistance in the cathode circuit, R_{Keq}. Then for a triode,

$$R_{op} = r_p[1 + g_m R_{Keq} + R_{Keq}/r_p] \tag{6–33}$$

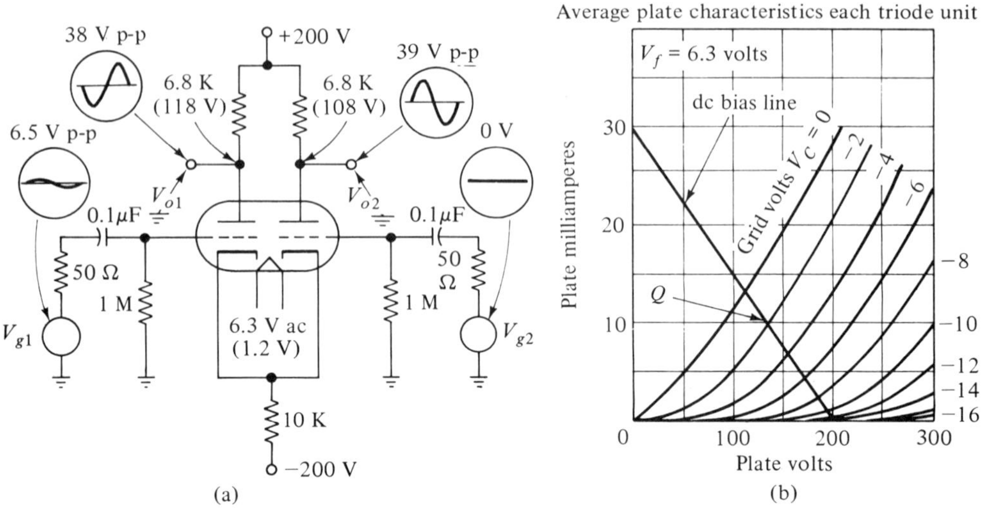

FIGURE 6–14 *A triode differential amplifier: (a) the actual circuit, and (b) the static characteristics of each tube. (Measured dc voltage levels with respect to ground from Ex. 6–9 are shown in parentheses. The important ac waveforms are also shown.)*

The value of $R_{K\text{eq}}$ for the triode differential amplifier shown in Fig. 6–14(a) is

$$R_{K\text{eq}} \simeq R_K \,||\, (1/g_m) \simeq 1/g_m \tag{6–34}$$

Substituting into Eq. (6–33),

$$R_{op} \simeq r_p[1 + g_m/g_m + g_m/r_p] \simeq 2r_p \tag{6–35}$$

The effective load impedance for each stage of the triode differential amplifier is

$$R_{L\text{tot}} = R_P \,||\, R_{op} \simeq R_P \,||\, 2r_p \tag{6–36}$$

An analysis of a triode differential amplifier is illustrated in the following example.

EXAMPLE 6–8 For the triode differential amplifier shown in Fig. 6–14(a), determine the following: (a) the bias values; (b) the difference-mode gain, A_d; (c) the common-mode gain, A_c; and (d) the common-mode rejection ratio, CMRR. The static characteristic curves for each tube are given in Fig. 6–14(b).

Solution: (a) The calculation of the bias conditions is complicated by the fact that the two triodes are directly coupled. However, the Q point for each tube can be estimated by assuming that the triodes are matched and that the grid-cathode voltage, V_{GK}, is small or that the cathodes are at ground potential. By neglecting V_{GK}, the dc load line for each tube can be constructed as shown in Fig. 6–8(b). The dc load line terminates on the voltage axis at 200 V and on the current axis at

$$V_{BB}/R_P = 200\text{ V}/6.8\text{ K} \simeq 29.5\text{ mA}$$

The current in R_K is the sum of the two plate currents or

$$I_{B1} + I_{B2} = I_{RK} = \frac{V_{GK} - V_{KK}}{R_K} \simeq \frac{200\text{ V}}{10\text{ K}} \simeq 20\text{ mA}$$

For matched triodes,

$$I_{B1} \simeq I_{B2} = I_{RK}/2 \simeq 10\text{ mA}$$

The intersection of the dc load line with the 10-mA plate current determines the Q point as shown in Fig. 6–14(b). The Q point values are

$$I_B \simeq 10 \text{ mA} \qquad V_{GK} \simeq -2 \text{ V} \qquad V_{PK} \simeq 130 \text{ V}$$

(b) Tube parameters: From Fig. 6–14(b), the values of g_m and r_p at the Q point are

$$g_m = \frac{\Delta I_P}{\Delta V_{GK}} \simeq \frac{17 \text{ mA} - 5 \text{ mA}}{0 - (-4 \text{ V})} \simeq \frac{12 \text{ mA}}{4 \text{ V}} \simeq 3 \text{ millimhos}$$

$$r_p = \frac{\Delta V_{PK}}{\Delta I_P} \simeq \frac{200 \text{ V} - 100 \text{ V}}{25 \text{ mA} - 5 \text{ mA}} \simeq \frac{150 \text{ V}}{20 \text{ mA}} \simeq 7.5 \text{ K}$$

(c) Gain and impedance calculations: From Eq. (6–36),

$$R_{Leq} \simeq R_P \,||\, 2r_p = 6.8 \text{ K} \,||\, 2(7.5 \text{ K}) \simeq 4.7 \text{ K}$$

From Eq. (6–31),

$$A_d = g_m R_{Leq}/2 = (3 \times 10^{-3})(4.7 \text{ K})/2 \simeq 7$$

$$A_c \simeq -R_{Leq}/2R_K = -(4.7 \text{ K})/2(10 \text{ K}) \simeq -0.24$$

$$\text{CMRR} \simeq g_m R_K = (3 \times 10^{-3})(10 \text{ K}) \simeq 30$$

For each stage,

$$R_i \simeq R_G = 1 \text{ M}$$

$$R_o = R_P \,||\, R_{op} \simeq R_P \,||\, 2r_p = 6.8 \text{ K} \,||\, 2(7.5 \text{ K}) \simeq 4.7 \text{ K}$$

(d) Experimental results: The measured dc voltage levels are shown in Fig. 6–14(a). Note that the two dc plate voltages are slightly different. This difference indicates that the differential amplifier is not perfectly symmetrical. The values of A_d and A_c were measured using the arrangement shown in Fig. 4–37 with the following results:

$$A_d \simeq 39 \text{ V}/6.5 \text{ V} = 6 \quad A_c = -10 \text{ V}/25 \text{ V} \simeq -0.4 \quad \text{CMRR} = \left|\frac{A_d}{A_c}\right| = \frac{6}{0.4} = 15$$

The measured CMRR was only half the calculated value which further indicates that the differential amplifier is not perfectly symmetrical. The tubes could have been balanced by placing a 100-ohm potentiometer at the junction of R_K and the two cathodes to make the two bias voltages slightly different.

The triode has two significant handicaps. First, the plate current is a function of both the grid voltage and the plate voltage. When a change in grid voltage causes the plate current to increase in a CK amplifier, this increase in plate current causes the plate voltage to decrease. A decrease in plate voltage partially offsets the increase in current. Consequently, the gain is less than that which would be possible if the plate current were independent of the plate voltage. A second and a far more serious handicap is the capacitance between the grid and plate, C_{gp}. This capacitance is due to the finite areas of the grid and plate and the small distance between them (parallel plate capacitance). The range of C_{gp} is 10 to 100 pF. However, the capacitance is multiplied by a factor of A_{vt} when looking into the grid terminal of a CK amplifier.* Consequently, in the high-frequency region, the signal source is loaded by a low value of reactance. The capacitor C_{gp} also couples the input and output circuits and, under proper condi-

---

* This multiplication of C_{gp} (or C_{gd} for an FET and $C_{b'c}$ for a bipolar unit) is called the Miller effect and is discussed in Sec. 7.3.

tions, can allow the output signal to drive the amplifier. Oscillation results and makes the amplifier useless as a linear amplifier. To reduce the effects of these handicaps, additional grids are placed in a tube. Tubes with more than one grid are considered in the next section.

6.6 Multigrid Tubes

The capacitance between the grid and plate, C_{gp}, in a vacuum tube can be reduced by placing an electrostatic shield between the grid and plate. This second grid is sometimes made from a wire mesh and is called a screen grid or G_2. The tube now has four electrodes and is called a *tetrode*. By placing a positive voltage on the screen grid with respect to the cathode, an electric field is established in the tube similar to that in a triode. In fact, the screen grid may be considered as a plate containing many holes. Most of the electrons reaching the screen grid go through the holes and strike the plate. The screen grid wires intercept approximately 20 to 25 percent of the electrons. These electrons cause a screen grid current to flow. If the space current is varied by placing an ac signal on the control grid, both the screen and the plate currents vary. To hold the screen voltage constant, a rather large bypass capacitor is connected from the screen grid to either the cathode or ground. The ac current flows through this bypass capacitor to ground and produces very little voltage drop in the mid- and high-frequency ranges. Sometimes the screen grid is connected directly to the power supply so that the capacitors in the power supply serve as a bypass capacitor.

Holding the screen voltage constant makes the electric field in the control grid-cathode region almost independent of the plate voltage; i.e., the plate voltage can swing over a wide range without changing the space current appreciably. This effect is shown in Fig. 6–15 for three different plate voltages and for constant bias voltages on both grids. Note that the potential in the control grid-cathode region does not change. Consequently, the space current remains constant. Since plate voltage variations do not disturb the area around the control grid, the screen acts as a good electrostatic shield between the plate and control grid; i.e., the capacitance between these two electrodes is reduced to a very low value. A typical value for C_{gp} in a tetrode is 0.001 pF.

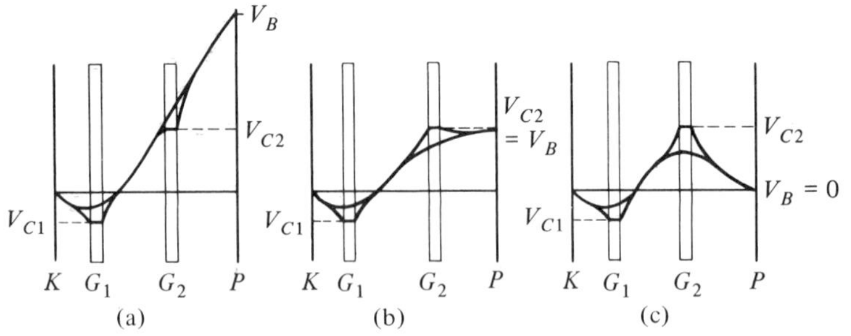

FIGURE 6–15 *Potential distribution in a tetrode for constant bias values on* G_1 *and* G_2 *and three different plate voltages: (a) plate voltage,* V_B*, larger than the screen voltage,* V_{C2}*, (b)* $V_B \simeq V_{C2}$*; and (c)* $V_B < V_{C2}$*. Note that the slope of the potential curve does not change in the vicinity of* G_1*. (Adapted from R. H. Krackhardt,* Vacuum Tube Electronics*; Charles E. Merrill Publishing Company, 1966.)*

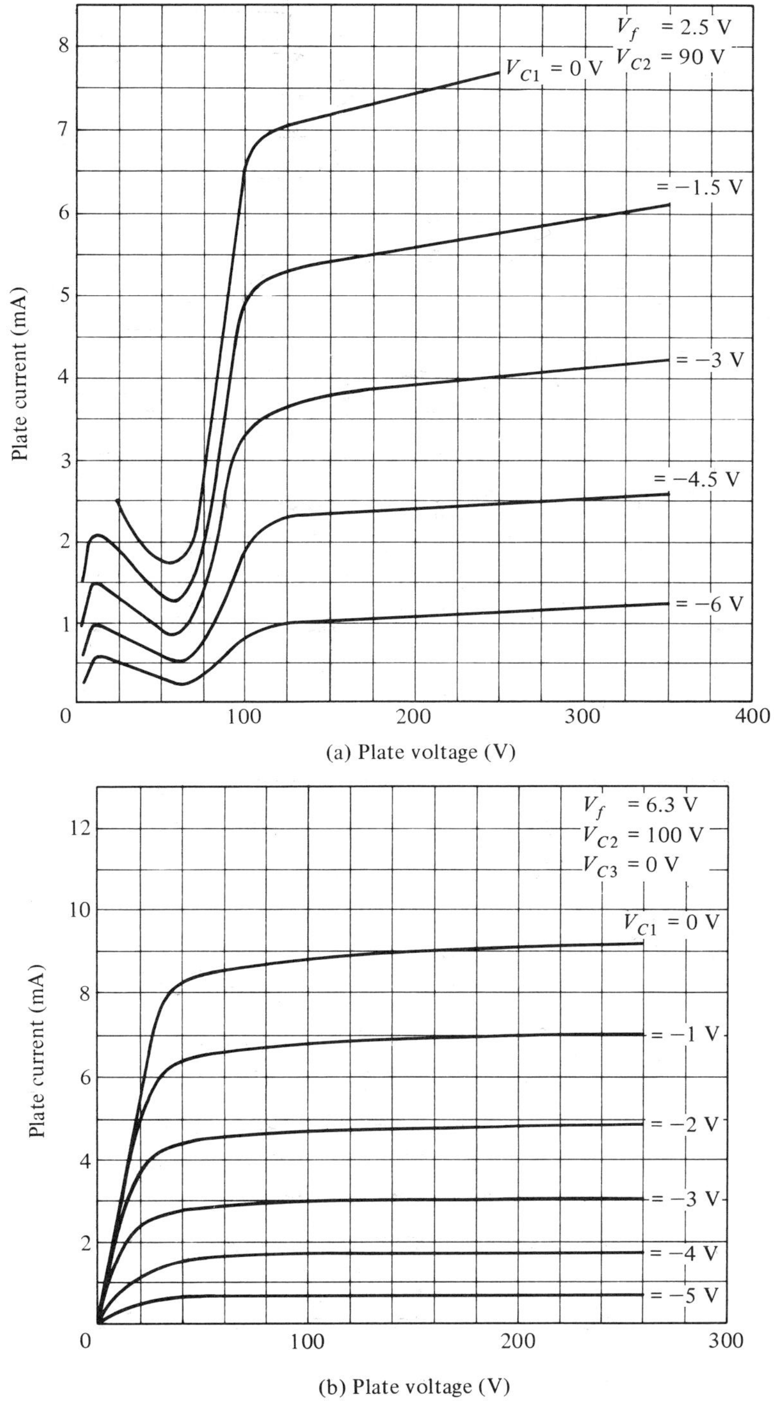

FIGURE 6–16 *Plate characteristics for two multigrid tubes: (a) curves for a tetrode showing the effects of secondary emission in the low plate voltage region and (b) curves for a pentode.*

While the screen grid removes the problem areas for a triode, it introduces a new problem. As indicated earlier, secondary emission occurs when high velocity electrons strike the plate. In a triode, these electrons return to the plate because the plate is biased at the highest positive voltage in the tube. However, in a tetrode the screen voltage can be, and often is, higher than the plate voltage as indicated in Fig. 6–15(c). For this condition, many of the low-energy secondary electrons go to the screen, thereby increasing the screen current and decreasing the plate current. This effect is so pronounced in some tetrodes that the plate current decreases for an increase in plate voltage over a small range of plate voltage. The net result of secondary emission in a tetrode is to produce undesirable deviations from a smooth curve in the V_B-I_B characteristic curves as shown in Fig. 6–16(a). A tetrode should not be operated in the nonlinear region for linear amplifier applications. This condition limits the lowest minimum instantaneous plate voltage to values above the screen bias, V_{C2}.

The problem created by secondary electrons can be eliminated by placing a grid between the screen and plate. This third grid, called a suppressor grid or G_3, is biased with a voltage that is always negative relative to the plate. The tube now has five electrodes and is called a *pentode*. Since the plate voltage is usually positive relative to the cathode, the suppressor grid is usually connected to the cathode. High velocity electrons that pass through the screen experience little difficulty in going through the suppressor grid to reach the plate. The secondary electrons that are emitted by the plate have relatively low velocities. Consequently, these low-energy electrons quickly lose their kinetic energy as they move from the plate in the direction of the suppressor. As soon as the kinetic energy of a secondary electron reaches zero, it stops and is then accelerated in the direction of the plate by the field between the plate and suppressor. A secondary electron returns to the plate with a velocity which is approximately equal to its initial velocity. The important point is that the suppressor grid prevents the secondary electrons from reaching the screen while allowing the primary electrons to reach the plate. The effect on the plate characteristic curves is shown in Fig. 6–16(b). Note that the suppressor grid has removed the undesirable deviations in the low-voltage part of the plate curves.

Special purpose tubes may have one or more grids in addition to the three found in a pentode. One example is a pentagrid converter which has five grids: two control grids, two screen grids, and a suppressor grid. A converter is often used in the front end of a receiver to mix the incoming signal with a signal from a local oscillator to produce a difference signal.* However, the pentode is the type most often used in amplifier work. Pentode amplifiers will be discussed in the next section.

6.7 Pentode Amplifiers

Except for the bias circuit associated with the screen, pentode amplifiers are analyzed in the same way that triode amplifiers were analyzed. The small-signal equivalent circuits and performance equations are the same for both. Consequently, the equations in Table 6–2 can be used to analyze and design pentode amplifiers.

---

* See F. E. Terman in the references listed at the end of this chapter.

The screen grid is biased by connecting the screen to the dc source, V_{BB}, through a voltage dropping resistor, R_{C2}, as shown in Fig. 6–17(a). The value of R_{C2} is

$$R_{C2} = (V_{BB} - V_{C2})/I_{C2} \tag{6–37}$$

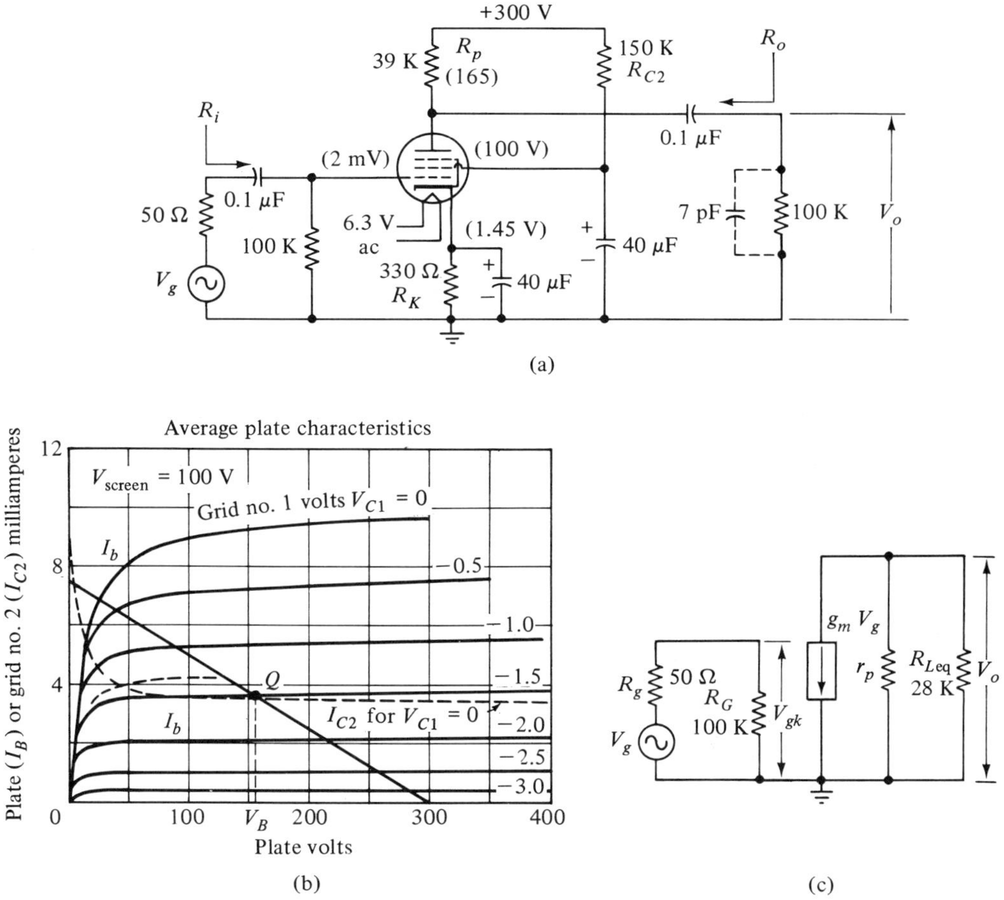

FIGURE 6–17 *The CK amplifier from Ex. 6–3: (a) the complete circuit, (b) the static characteristic curves of the pentode showing the dc load line, and (c) the mid-band equivalent circuit.*

where I_{C2} is the screen current for a given screen voltage, V_{C2}, and control grid voltage, V_{C1}. For example, the tube in Fig. 6–16(b) is normally operated with 100 V on the screen and −3 V on the control grid. For this condition, the plate current is approximately 3 mA, and the screen current is approximately 0.9 mA, or a total space current of 3.9 mA. Thus, slightly less than 25 percent of the space current is collected by the screen. Since the plate current can be found easily from the plate characteristic curves, the screen current can be estimated for other bias conditions using this same relationship for this tube type. However, as indicated by the dashed curve in Fig. 6–17(b), the current division does not remain constant for low plate voltages. The screen bias circuit also contains a bypass capacitor connected between the screen and ground. The reactance of this capacitor should be small at the lowest frequency that is to be amplified so that the ac voltage developed on the screen is very small over the entire operating frequency range. The requirements for effective bypassing are discussed in Chap. 7.

In the remainder of this section, two examples will be used to illustrate the analysis of typical pentode amplifiers. The first one is a CK single-stage amplifier.

EXAMPLE 6–9 For the CK pentode amplifier shown in Fig. 6–17(a), determine the values of A_{vt}, A_{vo}, R_i, and R_o. The static characteristic curves of the pentode are shown in Fig. 6–17(b).

Solution: (a) Bias conditions: The dc load line is constructed on the plate characteristic curves as shown in Fig. 6–17(b). The load line intersects the voltage axis at V_{BB} or 300 V and the current axis at $V_{BB}/(R_P + R_K)$ or 7.6 mA. A bias line cannot be constructed on the characteristics of a pentode as easily as for a triode because the total current through R_K is the sum of the plate and screen currents while the current coordinate is only the plate current. Since the screen current, I_{C2}, is 20 to 30 percent of the total current, a significant error would be introduced by neglecting this component of the total current and assuming that I_K is equal to I_B. The bias circuit may be designed in the following manner. Select a grid bias voltage, V_{GK}, in the mid-region of the dc load line, say -1.5 V. The Q point is located at the intersection of the -1.5 V grid line and the dc load line. The coordinates of this point are $I_B \simeq 3.8$ mA and $V_B \simeq 155$ V. To estimate the percentage of I_K that reaches the plate, note that on the plate characteristic curves, $I_{C2} \simeq 3.8$ mA and $I_B \simeq 9.4$ mA for $V_{C1} = 0$, $V_{C2} = 100$ V, and $V_B \simeq 150$ V. For these voltages,

$$I_K \simeq (9.4 + 3.8) \text{ mA} \simeq 13.2 \text{ mA}$$

Then

$$I_B \simeq (9.4 \text{ mA}/13.2 \text{ mA})I_K \simeq 0.72 I_K$$

For $V_{C1} = -1.5$ V and $V_{C2} = 100$ V,

$$I_K \simeq I_B/0.72 \simeq (3.8/0.72) \text{ mA} \simeq 5.3 \text{ mA}$$

$$I_{C2} \simeq I_K - I_B \simeq (5.3 - 3.8) \text{ mA} \simeq 1.5 \text{ mA}$$

The values of R_K and R_{C2} are

$$R_K = V_K/I_K \simeq 1.5 \text{ V}/5.3 \text{ mA} \simeq 290\ \Omega$$

$$R_{C2} \simeq (V_{BB} - V_{C2})/I_{C2} \simeq (300 - 100)/1.5 \text{ mA} \simeq 133 \text{ K}$$

Using standard resistor values of 330 ohms and 150 K, the measured values were as follows: V_K was 1.45 V, V_B was 165 V, and V_{C2} was 100 V.

(b) Tube parameters: The values of g_m and r_p can be estimated from the plate characteristic curves at the Q point:

$$g_m \simeq \frac{\Delta I_B}{\Delta V_{GK}} \simeq \frac{1.8 \text{ mA}}{0.5 \text{ V}} \simeq 3.6 \text{ millimhos}$$

$$r_p = \frac{\Delta V_{PK}}{\Delta I_B} \simeq \frac{400 \text{ V}}{0.4 \text{ mA}} \simeq 1000 \text{ K}$$

(c) Gain and impedance calculations: Using Table 6–2,

$$R_{Leq} \simeq R_P \,||\, R_L \,||\, r_p \simeq 39 \text{ K} \,||\, 100 \text{ K} \,||\, 1000 \text{ K} \simeq 28 \text{ K}$$

$$A_{vt} \simeq -g_m R_{Leq} \simeq -(3.6 \times 10^{-3})(28 \text{ K}) \simeq -100$$

$$R_i \simeq R_G \simeq 100 \text{ K} \qquad R_o \simeq R_P \,||\, r_p \simeq 39 \text{ K}$$

$$A_{vg} = R_i/(R_i + R_g) \simeq 1 \qquad A_{vo} = (A_{vg})(A_{vt}) \simeq -100$$

(d) Experimental results: The measured dc voltage levels are shown in Fig. 6–3(b) and are within 20 percent of the calculated values. The measured voltage gain was −95 with $V_g = 0.25$ V.

The preceding example indicated that the gain in a pentode CK amplifier is reasonably large if R_{Leq} is large. In most cases, a 1-megohm value for r_p is too large to determine graphically and can usually be neglected in the performance equations.

If the load on a pentode CK amplifier is low, the gain is also low. Consequently, a common plate (CP) amplifier is usually placed between a CK stage and a low impedance load to maintain the level of gain available from a pentode. The next circuit consists of a CK stage driving a CP stage which in turn drives a low impedance load.

EXAMPLE 6–10 For the CK-CP amplifier shown in Fig. 6–18(a), determine the values of A_{vt}, A_{vo}, R_i, and R_o. The static characteristics for T_1 are shown in Fig. 6–17(b).

Solution: (a) Bias conditions: The Q points for T_1 and T_2 were determined experimentally and are as follows:

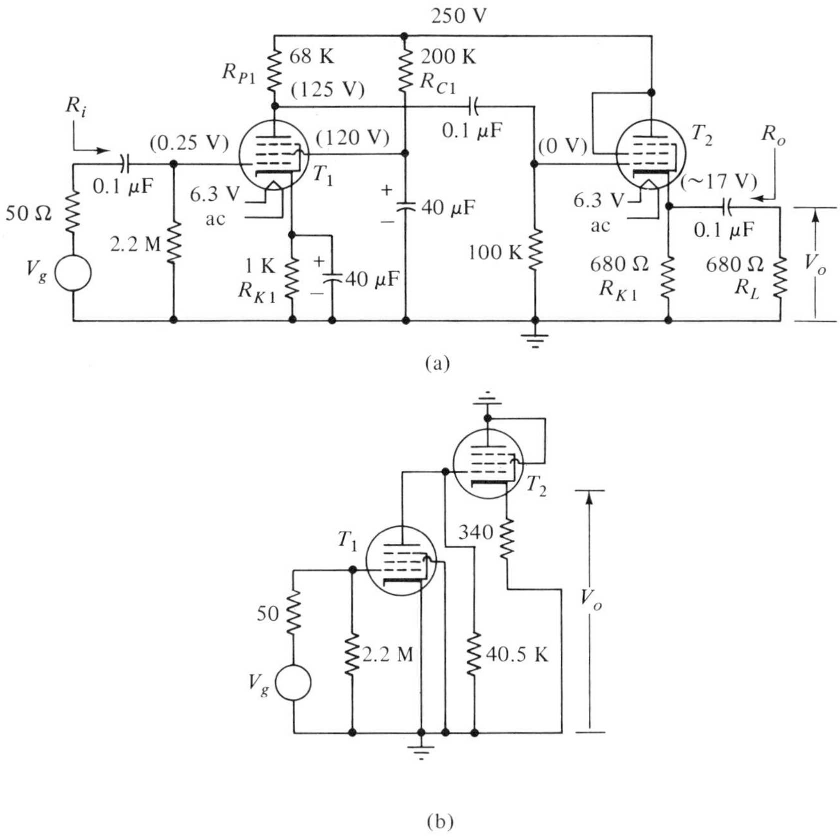

FIGURE 6–18 *The CK-CP amplifier from Ex. 6–10: (a) the complete circuit, (b) the mid-band equivalent circuit. The static characteristics for* T_1 *are shown in Fig. 6–17(b).*

$$I_{B1} \simeq 2 \text{ mA} \qquad I_{B2} \simeq 28 \text{ mA}$$
$$V_{GK1} \simeq -2.1 \text{ V} \qquad V_{GK2} \simeq -20 \text{ V}$$

(b) Gain and impedance calculations: From the curves, $g_{m1} \simeq 2.6$ millimhos; $g_{m2} \simeq 2$ millimhos, and $r_{p2} \simeq 110$ K. Using Table 6–2,

$$R_{Leq1} \simeq R_{P1} \,||\, R_{G2} \,||\, r_{p1} \simeq 40.5 \text{ K} \qquad R_i \simeq R_{G1} = 2.2 \text{ M}$$
$$A_{vt1} = -g_{m1}R_{Leq1} \simeq -106 \qquad R_{Leq2} \simeq R_{K2} \,||\, R_L = 340\ \Omega$$
$$R_o \simeq R_{K2} \,||\, (1/g_{m2}) \simeq 290\ \Omega \qquad A_{vt2} = g_{m2}R_{Leq2}/(1 + g_{m2}R_{Leq2}) \simeq 0.4$$
$$A_{vg} = R_1/(R_i + R_g) \simeq 1 \qquad A_{vo} = (A_{vg})(A_{vt1})(A_{vt2}) \simeq -43$$

(c) Experimental results: The measured dc voltage levels are shown in Fig. 6–4(b). The measured voltage gains were

$$A_{vt1} = -11 \text{ V}/0.1 \text{ V} \simeq -110 \qquad A_{vt2} \simeq 0.45$$
$$A_{vo} = V_{o2}/V_g \simeq -5 \text{ V}/0.1 \text{ V} \simeq -50$$

6.8 Cathode-Ray Tubes

Perhaps the most important application of vacuum tube principles is found in the cathode-ray tube (CRT). As shown in Fig. 6–19, a CRT has three major sections: an electron gun, a deflection system, and a screen. The purpose of an electron gun is to provide a highly columnated beam of electrons to the deflection system.

The deflection system consists of two sets of two parallel plates each. Electrical signals applied to the deflection plates cause the beam to move away from its central position. The vertical plates are used to produce a vertical movement of the beam while the horizontal plates are used to deflect the beam in the horizontal direction.

The interior surface of the screen is coated with a translucent fluorescent material. When high-speed electrons strike this material, light is produced at the point of impact. The persistence of the fluorescent material makes the path of the beam easy to follow, especially when the same path is traversed repeatedly.

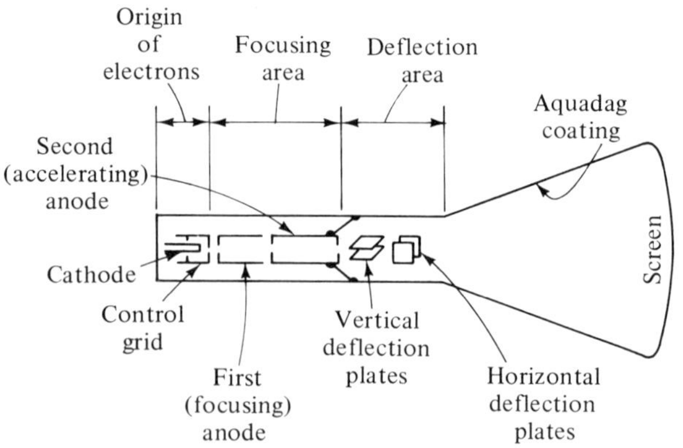

FIGURE 6–19 *A cathode-ray tube showing the various functional parts.*

As shown in Fig. 6–19, the electron gun has four electrodes: a cathode, a grid, a focusing anode, and an accelerating anode. The first three electrodes perform the same function as the electrodes in a triode; i.e., the cathode supplies the electrons, the grid controls the number of electrons that reach the anode, and the anode voltage accelerates the electrons. However, the anode has a hole in its center. Some of the electrons pass through this hole without striking the anode material and are then further accelerated by the high positive voltage on the accelerating electrode. In addition, both the anode and accelerating electrode are cylindrical sleeves that are close together. This physical arrangement, together with the voltages that are applied to these electrodes, form an electrostatic lens that focuses the electrons like a glass lens focuses light. The beam of electrons leaves the second electrode through a small aperture so that electrons which are not properly focused are intercepted by the metal.

The focused beam of high-velocity electrons then passes successively between the vertical and horizontal deflection plates. The point at which the beam strikes the screen is determined by the voltages on the deflection plates and the period of time that the electrons are in the electric field between the two plates. Our immediate objective is to determine the amount of deflection that is produced by a given voltage on the deflection plates.

As indicated above, the deflection is determined by the electrostatic field strength and the period of time that an electron is in the field. Considering only the vertical deflection plates in Fig. 6–20, the electric field strength between the plates, $\mathcal{E}_y$, is

$$\mathcal{E}_y = V_D/d \quad \text{V/m} \tag{6–38}$$

where V_D is the voltage between the vertical deflection plates, and d is the distance between the plates. The time, t, that an electron remains between the plates is

$$t = l/v \text{ sec} \tag{6–39}$$

where l is the length of the plates, and v is the horizontal velocity of an electron when it leaves the accelerating electrode. From Ex. 6–2 and Eq. (6–8)

$$v_x = \left(\frac{qV_A}{mX}\right)(T) = \left(\frac{qV_A}{mX}\right)\left(\frac{X\sqrt{2m}}{\sqrt{qV_A}}\right) = \left(\frac{2qV_A}{m}\right)^{1/2} \tag{6–40}$$

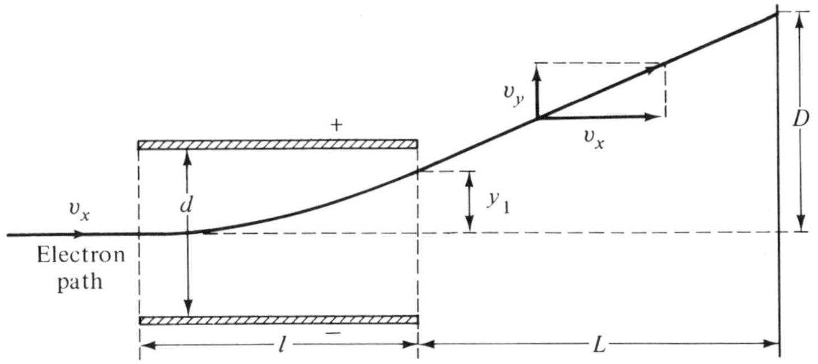

FIGURE 6–20 *The path of an electron that has been subjected to an electrostatic field while passing between the vertical deflection plates. The top plate is positive relative to the bottom plate. (Adapted from R. H. Krackhardt,* Vacuum Tube Electronics, *Charles E. Merrill Publishing Company, 1966.)*

Since q/m is the charge-to-mass ratio for an electron, this quantity is a constant having a value of 1.76×10^{11} coulombs/kg. Therefore, v_x is a function of the accelerating voltage only and is

$$v_x = 5.9 \times 10^5 \sqrt{V_A} \quad \text{m/sec} \tag{6–41}$$

Assuming that the vertical velocity is zero as the electrons enter the space between the vertical deflection plates, then the vertical velocity, v_y, as the electrons leave the plates is

$$v_y = \frac{q}{m} \mathcal{E}_y t = 1.76 \times 10^{11} \mathcal{E}_y \left(\frac{l}{v_x}\right) \quad \text{m/sec} \tag{6–42}$$

In Fig. 6–20, the value of v_x is unchanged, and v_y is also constant after an electron leaves the space between the plates. The amount of vertical deflection, D, at the screen is

$$D = L(v_y/v_x) + y_1 \quad \text{m} \tag{6–43}$$

where y_1 is the vertical deviation that an electron has already achieved while between the plates. The vertical deflection sensitivity is the deflection produced on the screen per volt on the deflection plates. The following example illustrates some of the major points discussed above.

EXAMPLE 6–11 A cathode-ray tube has an accelerating potential of 1600 V. The vertical deflection plates are 2 cm long and 0.5 cm apart. If the distance L is 20 cms, find the vertical deflection produced on the screen by a deflection voltage of 200 V.

Solution: From Eq. (6–38), the electric field intensity is

$$\mathcal{E}_y = V_D/d = 200 \text{ V}/(0.5 \times 10^{-2}) = 4 \times 10^4 \text{ V/m}$$

From Eq. (6–41), the horizontal velocity is

$$v_x = 5.9 \times 10^5 \sqrt{V_A} = 5.9 \times 10^5 \times 40 = 23.9 \times 10^6 \text{ m/sec}$$

At the right-hand edge of the deflection plates, the vertical velocity, from Eq. (6–42), is

$$v_y = \frac{1.76 \times 10^{11} \times \mathcal{E}_y l}{v_x} = \frac{1.76 \times 10^{11} \times 4 \times 10^4 \times 2 \times 10^{-2}}{2.39 \times 10^7} = 5.88 \times 10^6 \text{ m/sec}$$

From Eq. (6–9), the vertical displacement at the right-hand edge of the plates is

$$y_1 = \frac{qV_D}{2md} t^2 = \frac{1.76 \times 10^{11}}{2} \times 4 \times 10^4 \times \left(\frac{2 \times 10^{-2}}{23.9 \times 10^6}\right)^2 \simeq 0.25 \text{ cm}$$

From Eq. 6–43, the vertical displacement at the screen is

$$D = \frac{Lv_y}{v_x} + y_1 = \frac{0.2 \times 5.88 \times 10^6}{2.37 \times 10^7} + 0.25 \times 10^{-2} \simeq 5.15 \text{ cm}$$

The vertical deflection sensitivity is 5.15 cms/200 V or 2.57×10^{-2} cm/V.

Since the deflection voltage is usually obtained from linear amplifiers, the deflection sensitivity should be as high as practical so that the maximum required deflection voltage can be as low as possible. The deflection sensitivity can be increased by decreasing the acceleration voltage, lengthening the deflection plates, decreasing the distance between the deflection plates, and increasing L.

However, each of these changes produces an undesirable result if carried to an extreme. In general, several compromises must be made in the overall design of a CRT.

High velocity electrons are necessary to produce sharp clear traces. One way to achieve high values of deflection sensitivity and also high velocity electrons is to use a low value of V_A (say 300 V) and then to use an accelerating anode, called an intensifying electrode, close to the screen. This accelerating electrode is a conducting ring around the inside of the tube. The ring is connected to an accelerating voltage of several thousand volts.

One simple application of a CRT is shown in Fig. 6–21. A sine wave is applied to the vertical deflection plates, and a sawtooth sweep voltage having the same period is applied to the horizontal deflection plates. Starting at the left side of

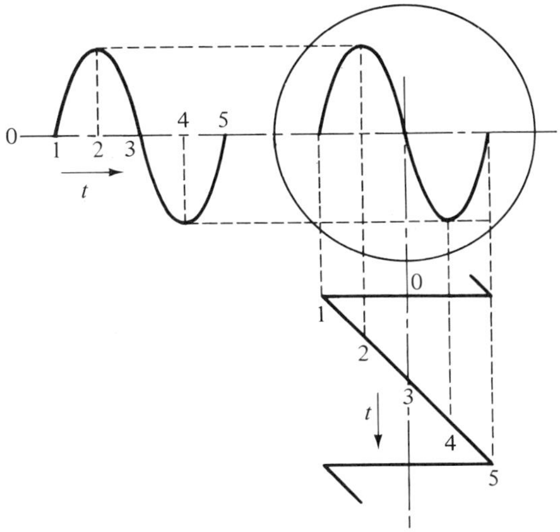

FIGURE 6–21 *A trace on the screen of a CRT obtained by applying a linear sweep voltage on the horizontal deflection plates and a sine wave on the vertical deflection plates. (Adapted from John D. Ryder,* Electronic Fundamentals and Applications, *3rd ed., © 1964. Reprinted by permission of Prentice-Hall, Inc., Englewood Cliffs, N. J.)*

the tube face, the sawtooth wave applied to the horizontal deflection plates moves the beam across the tube as a linear function of time. As the beam moves from left to right, the sine wave applied to the vertical deflection plates moves the beam up to a peak value, down to a negative peak value, and then back to the center position during the time that the sawtooth sweep voltage goes from a negative to a positive peak value. Assuming that the sawtooth wave can return to its initial value instantaneously, the entire process is repeated. In this way, the phosphor is excited along the same path a number of times each second equal to the repetition rate of the sweep voltage. Thus, even for phosphors having short persistence periods, the trace is clearly visible as a solid line plot to the human eye because of the persistence of the eye.

In most CRTs used in television sets, magnetic deflection is used instead of electric deflection because the tube can be made shorter. In a magnetic deflection system a coil is placed on each side of the neck of the tube. A magnetic field is produced by passing a current through the coils. The magnetic flux is usually perpendicular to the path of the electron beam as shown in Fig. 6–22. As indicated in this figure, an electron entering this field with an initial velocity v_x

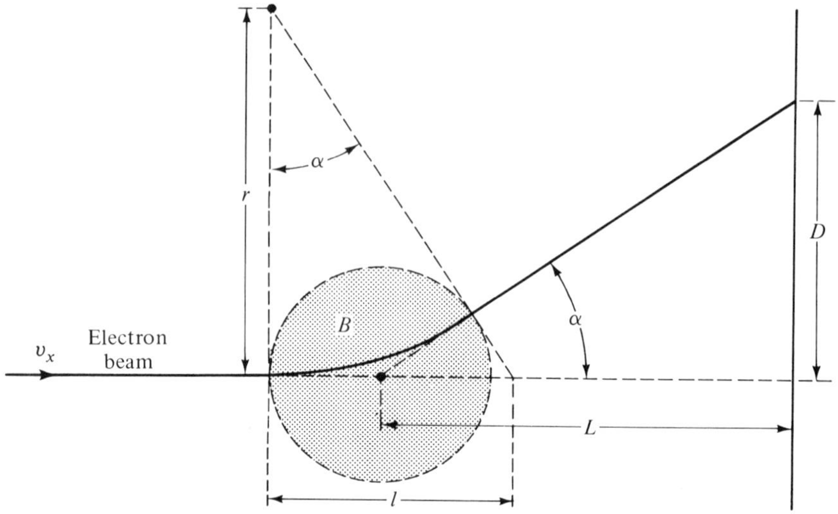

FIGURE 6–22 *The path of an electron that has a velocity,* v_x, *as it moves through a constant magnetic field,* B, *that is directed into the page. (Adapted from R. H. Krackhardt,* Vacuum Tube Electronics, *Charles E. Merrill Publishing Company, 1966.)*

is deflected along the arc of a circle. If the magnetic field is constant, the exit velocity is also $|v_x|$ because the kinetic energy of a particle can not be altered by a steady* magnetic field. Our objective is to find an expression for the deflection distance, *D*.

Assuming that the magnetic flux density, *B*, is uniform and directed into the page in Fig. 6–22, the force on an electron that enters the field with a velocity v_x is

$$f = Bqv_x \sin\theta \tag{6–44}$$

where θ is the angle between the flux lines and the electron path. In our case, θ is 90°. Consequently, the force is constant, and the electron moves upward along an arc of a circle while in the magnetic field. The radius of the circle may be found by equating the centripetal and centrifugal forces, or

$$Bvq = mv_x^2/r \tag{6–45}$$

Solving for the radius,

$$r = mv_x/(Bq) \tag{6–46}$$

Constructing perpendiculars to the electron path at the entrance and exit points, the angle α is the angle subtended by the arc path of an electron. By similar triangles,

$$\tan\alpha = l/r = D/L \tag{6–47}$$

Then the deflection, *D*, is

$$D = \frac{Ll}{r} = \frac{LlBq}{mv_x} = LlB\left(\frac{q}{2mV_A}\right)^{1/2} \tag{6–48}$$

The following example illustrates the application of these equations.

---

* The assumption is made in practice that the flux density of the magnetic field does not change appreciably while an electron moves through it.

EXAMPLE 6–12 In a CRT using magnetic deflection, B is 1×10^{-3} Weber per square meter over a region whose diameter, l, is 3 cm. If L is 16 cm, and V_A is 1600 V, find the deflection of the beam at the screen.

Solution: Using Eq. (6–48) and substituting the numerical value for (q/m)

$$D = LlB\left(\frac{1.76 \times 10^{11}}{2V_A}\right)^{1/2} \text{ m}$$

$$= (16 \times 10^{-2})(3 \times 10^{-2})(10^{-3})\frac{(8.8 \times 10^{10})^{1/2}}{(1600)^{1/2}} \simeq 3.5 \text{ cm}$$

The fluorescent coating on the inside face of a CRT is rather thin so that the light produced by the impact of high-velocity electrons can be observed on the outside of the tube. Several different types of phosphors may be used in the coating material to provide a certain persistence characteristic for a given application. These phosphors are designated by number, $P1$, $P2$, . . . $P25$. However, a higher number does not denote a shorter or longer persistence. Type $P1$ phosphor has a medium short persistence and is used for general purpose oscillography. Type $P4$, which has a short persistence and a white trace, is used in television picture tubes. The persistence ranges from less than 30 μsec for type $P5$ to several seconds for a type $P7$ phosphor.

If the electrons were not removed from the screen, negative charge would accumulate on the screen and repel incoming electrons. However, secondary electrons are emitted by the screen. These electrons are then attracted to a conducting graphite coating on the interior of the walls of the tube because this coating is connected to a positive voltage.

An alternate approach in removing the electrons is to deposit an extremely thin layer of aluminum on the inside surface of the fluorescent material. High velocity electrons can easily go through a thin aluminum coating. However, the excess electrons are readily conducted away from the screen when the aluminum coating is connected to a positive voltage. This coating also has several other advantages. Heavy negative ions that are normally accelerated along with the electrons will usually produce a dark spot in the center of the tube when magnetic focusing is used. These ions (charged atoms) do not penetrate the aluminum screen easily so that the damage to the fluorescent screen is greatly reduced. In addition, the aluminum layer increases the light output by reflecting the backward radiation to the front of the tube.

In this section, we have looked briefly at the operation of cathode ray tubes. Many variations of the electron gun arrangement and also of the deflection system may be found in practice. For example, an accelerating grid, similar in function to a screen grid, may be added to the electron gun. Many configurations of the coils may be found in magnetic deflection systems. The many methods used for focusing and deflecting electron beams are highly specialized, and a complete treatment is beyond the scope of this chapter.

6.9 Summary

While the primary objective of this chapter was to describe the construction, operation, and applications of various types of vacuum tubes, a significant amount of the material was devoted to electron ballistics in electrostatic and also in magnetic fields. Starting with the energy level concepts developed in Chap. 1, four types of electron emission were discussed with emphasis placed on thermal emission. After determining the characteristics of an emitter (or a cathode) in a vacuum, a second electrode called a plate or an anode was introduced. The composite structure, called a diode, had gross characteristics similar to the solid-state diode. Although the fundamentals of operation were different for the two devices, both conducted easily in only one direction. Applications of the two devices were essentially the same, as rectifiers and logic circuit elements.

By placing a grid between the cathode and anode, the vacuum diode was converted into an active device called a triode. The analysis of the triode closely paralleled the analysis of the FET. First, bias circuits were discussed. Then the large-signal case was treated. This case clearly demonstrated the effect of the grid voltage variations on the plate current. For the small-signal case, equivalent circuits were developed and incorporated into amplifier circuits.

Three amplifier configurations were considered: the common cathode, the common grid, and the common plate. The performance characteristics of amplifiers using a triode in these three configurations are summarized in Table 6–2.

Triodes have several disadvantages, two of which are the high input capacitance and the dependence of the plate current on plate voltage. Adding a screen grid between the control grid and the plate corrected these disadvantages but added another problem. Since the screen grid was always positive and physically near the plate, secondary electrons from the plate were attracted to the screen when the plate voltage was less than the screen voltage. This problem was corrected by placing a suppressor between the screen and plate and connecting the suppressor to the cathode or to ground. With five electrodes (three grids, a plate and a cathode), the resulting tube structure was called a pentode. Pentode amplifiers were discussed and illustrated.

Since the cathode-ray tube is one of the most important special applications of vacuum tube principles, the construction and operation of typical CRTs were discussed. Both electrostatic and magnetic deflection principles were considered in some detail.

While the material in this chapter is not an extensive treatment of vacuum tubes, basic principles have been emphasized. This introduction should provide a firm basis for further study.

Problems

Section 6.1

6–1 Find the longest wave length of light that can be used to release electrons from a cesium emitter.

6–2 Find the work function at the surface of a metal which emits 5000 times more electrons at 2300°K than it does at 2000°K.

6–3 Find the percentage decrease in the maximum rate of electron emission from tungsten when the temperature is decreased from 2500°K to 2450°K.

6–4 Find the approximate temperature at which each of the following materials will emit a current of one ampere per square centimeter: (a) tungsten, (b) thoriated tungsten, and (c) cesium.

Section 6.2

6–5 The current flowing in a diode may be limited by the space charge or by the rate at which electrons are emitted by the cathode. For a cylindrical plate diode, the space charge limited current is

$$I = \frac{(14.7 \times 10^{-6})(V_B^{3/2})(\text{length of anode in cm})}{K_1\ (\text{radius of anode in cm})}$$

where K_1 is a function of the ratio of the radius of the anode to that of the cathode. If the tungsten filament of a diode has a diameter of 0.03 cm, a length of 6 cm, and an anode diameter of 2 cm, will the current be temperature or space-charge limited when V_B is 100 V and the filament temperature is 2500°K? Assume that $K_1 \simeq 1$ for ratios greater than 50.

6–6 A parallel-plate diode has a spacing of 0.2 cm. Find the transit time of an electron that starts from the cathode with zero initial velocity if the plate-to-cathode voltage is (a) 10 V and (b) 100 V.

6–7 The current in a particular diode is 20 mA when the anode-to-cathode voltage, V_B, is 30 V. Find the current when V_B is 100 V.

6–8 Design a dc power supply that uses a half-wave rectifier and an RC filter circuit. When connected directly to the 110 V, 60 Hz ac lines (no transformer), this power supply must deliver 40 mA of dc current at 90 V dc to operate a small table model radio. The ripple voltage must not exceed 0.2 V at the load. Specify the tube type, the values and wattage ratings of all resistors, and the value(s) and breakdown voltage rating(s) of all capacitors. Use Figs. 2–10 and 2–11.

6–9 Design a 350 V, 200 mA dc power supply using a full-wave, vacuum diode rectifier and an RC filter section. The output ripple should be less than 0.5 V peak-to-peak. Specify the transformer turns ratio, the required peak inverse voltage rating for the tube, and the values and ratings of the filter components.

Section 6.3

6–10 Repeat Ex. 6–4 for a grid bias of–4V.

6–11 Find the bias point, the.dc plate current, and plate-to-cathode voltage for the circuit in Fig. 6–23(b). Use the plate characteristics in Fig. 6–23(a) for the tube.

Section 6.4

6–12 Find the output ac voltage for the triode amplifier circuit in Fig. 6–23(b) by using the large signal (graphical) approach.

6–13 Repeat Ex. 6–5 for the case where V_{BB} is 300 V, R_P is 22 K, and R_K is 2 K.

6–14 Using a triode whose plate characteristics are shown in Fig. 6–23(a), design a single-stage CK amplifier with the following characteristics: $|A_{vo}| \geq 15$, $R_i \geq 500$ K, and $R_o \geq 10$ K. The amplifier output voltage should be capable of a 100-V peak-to-peak swing without driving the grid positive. The amplifier will be used in an application where R_L is 51 K and R_g is 1 K. Use only standard components.

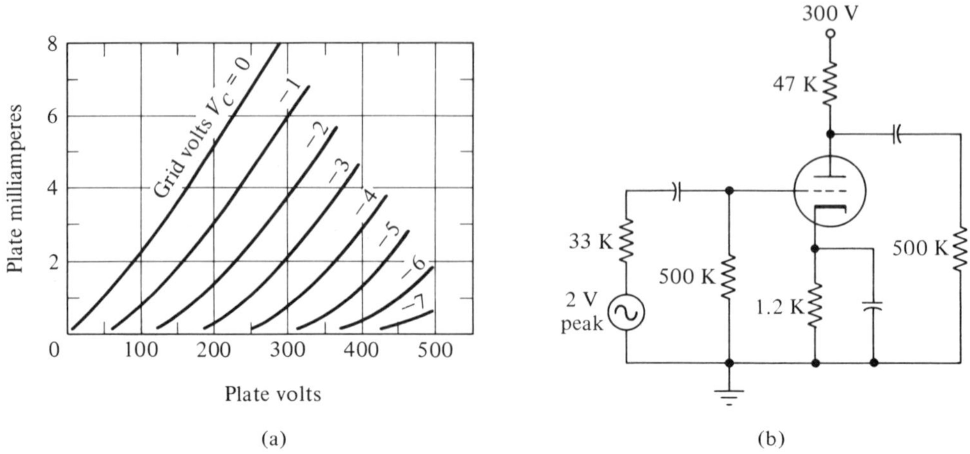

FIGURE 6–23 *Triode characteristics and circuit diagram for Prob. 6–11.*

Section 6.5

6–15 Estimate the values of g_m, r_p, and μ for the triode whose characteristic curves are shown in Fig. 6–23(a) when the operating point is located at 200 V and 3 mA.

6–16 Repeat Prob. 6–12 using small signal analysis. Explain the difference in the amplitude of the output signal obtained with the two methods.

6–17 A triode whose plate characteristics are shown in Fig. 6–23(a) is used in a CGr amplifier similar to the one shown in Fig. 6–12(a). The circuit values are: $V_{BB} = 300$ V, $R_P = 47$ K, $R_K = 680$ ohms, $R_L = 500$ K, and $R_g = 1$ K. Calculate the bias values and values for A_{vo}, R_i, and R_o.

6–18 A triode whose plate characteristic curves are shown in Fig. 6–23(a) is to be used in a CP amplifier similar to the one shown in Fig. 6–13(a). The circuit values are: $V_{BB} = 350$ V, $R_K = 56$ K, $R_G = 220$ K, $R_g = 1$ K, and $R_L = 1$ K. Calculate the bias conditions and values of A_{vo}, R_i, and R_o.

6–19 For the triode differential amplifier shown in Fig. 6–14, the left-hand input signal is 5 V p-p and the right-hand input signal is 4.9 V p-p. Calculate the output voltages, V_{o1} and V_{o2}.

6–20 The triode differential amplifier shown in Fig. 6–14(a) is to be modified by replacing the 10-K cathode resistor with a triode constant-current source. Design the constant-current source so that the current in each side of the differential amplifier remains the same as in Ex. 6–8. Use a triode whose static characteristics are shown in Fig. 6–14(b). Also calculate the CMRR of the modified amplifier.

Section 6.6

6–21 Estimate the value of the dynamic plate resistance (r_p) of the tetrode whose plate characteristics are shown in Fig. 6–16(a) for values of plate voltage between 25 and 50 V.

6–22 Estimate the values of r_p, g_m, and μ for the tetrode whose plate characteristics are shown in Fig. 6–16(a) with bias conditions of 250 V and 4 mA.

6–23 Estimate the values of g_m, r_p, and μ for the pentode whose plate characteristics are shown in Fig. 6–16(b) with bias conditions of 200 V and 7 mA.

Section 6.7

6–24 Repeat Ex. 6–9 for the case where V_{BB} equals 400 V, R_P equals 27 K, and R_K equals 470 ohms.

6–25 The CK-CP amplifier shown in Fig. 6–18(a) is changed to a single-stage CK amplifier by capacitively coupling the 680-ohm load directly to the plate of T_1 and by removing T_2. Calculate the value of A_{vo} for this CK amplifier and compare it with the value of A_{vo} obtained in Ex. 6–10. Briefly discuss the effect of the CP stage on the overall performance of the CK-CP amplifier shown in Fig. 6–18(a).

6–26 Using the pentode characteristic curves shown in Fig. 6–24, design a small-signal amplifier with an overall gain, A_{vo}, of at least 100 when R_g is 50 K and R_L is 100 K. Use a bias point located at a grid voltage of -2 V and a plate-to-cathode voltage of 250 V. Bias the screen at 150 V. Use cathode bias. Specify R_K, R_P, V_{BB}, and R_G. Draw the complete circuit and show all dc voltages and currents. If V_g is 0.1 V, show the magnitude and phase of the ac signal at the control grid and also across R_L.

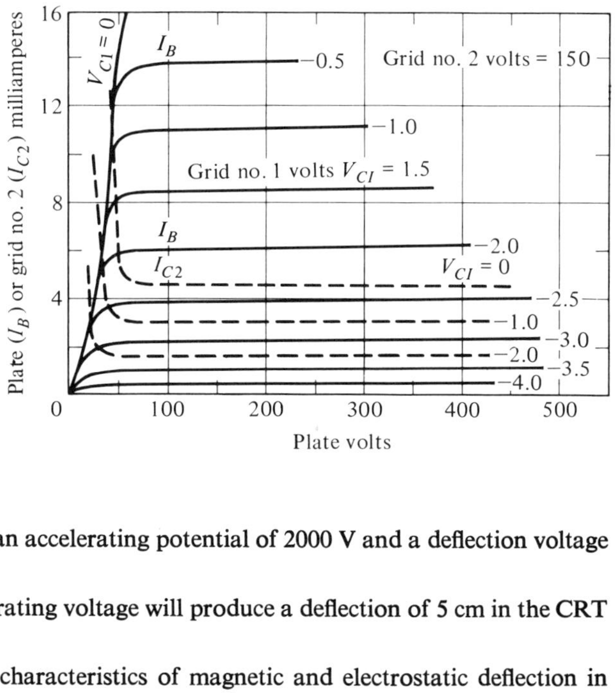

FIGURE 6–24 *Plate characteristic curves for the tube in Prob. 6–26.*

Section 6.8

6–27 Repeat Ex. 6–11 for an accelerating potential of 2000 V and a deflection voltage of 100 V.

6–28 What value of accelerating voltage will produce a deflection of 5 cm in the CRT described in Ex. 6–12.

6–29 Briefly compare the characteristics of magnetic and electrostatic deflection in a CRT.

References

Krackhardt, R. H., *Vacuum Tube Electronics*. Columbus, Ohio: Charles E. Merrill Publishing Company, 1966.

Millman, J., *Vacuum-Tube and Semiconductor Electronics*. New York: McGraw-Hill Book Company, 1958.

Millman, J. and S. Seely, *Electronics*, 2d ed. New York: McGraw-Hill Book Company, 1951.

Ryder, J. D., *Electronic Fundamentals and Applications*, 3rd ed. Englewood Cliffs, N. J.: Prentice-Hall, Inc., 1964.

Terman, F. E., *Electronic and Radio Engineering*, 4th ed. New York: McGraw-Hill Book Company, 1955.

7

Frequency and Transient Response of Single-Stage Amplifiers

In the previous four chapters, the characteristics of several types of amplifiers were developed for the mid-frequency range. In this approach, we assumed that the voltage and current gains, the input impedance, and the output impedance of an amplifier remained constant as the frequency varied over the mid-frequency range. If an amplifier contained no reactive circuit elements (capacitors and inductors) and if the parameters of the active device used were not functions of frequency, the mid-frequency range would extend from zero frequency (dc) to an infinite frequency.

However, the presence of blocking and bypass capacitors causes the parameters of the amplifier to vary in the low-frequency range. This effect should be expected since the reactance of a capacitor approaches infinity (an open circuit) as the signal frequency approaches zero (dc). In addition, the parameters of an amplifier are affected in the high-frequency range by the stray capacitance to ground and by the physical limitations of the device. These effects can also be appreciated by noting that the reactance of a capacitor approaches zero as the signal frequency approaches infinity.

The objective of this chapter is to determine the effects of capacitive type circuit elements on the frequency and transient response of single-stage amplifiers. The problems encountered in finding the frequency response of an amplifier are discussed in Sec. 7.1, and, the material in Sec. 7.2 describes the high-frequency limitations inherent in active devices. The analysis of inverting type amplifiers presents special problems because of the effect of the gain on the input and output capacitance. These problems are treated in Sec. 7.3. Using the material in the first three sections, the frequency response of an input section of an amplifier is discussed in Sec. 7.4, and the corresponding transient or time response is examined in Sec. 7.5. Building on this foundation, the response of complete single-stage amplifiers is treated in the remaining three sections. Three functional types of amplifiers are considered: (1) inverting amplifiers (Sec. 7.6), (2) low-input impedance amplifiers (Sec. 7.7), and (3) low-output impedance amplifiers (Sec. 7.8).

7.1 Frequency-Response Problems

The mid-frequency equivalent circuits discussed in the preceding four chapters consisted only of resistive elements and current and voltage sources. These circuits were relatively easy to analyze because they usually had only two nodes. When the blocking, bypass, and stray capacitive elements are included in an equivalent circuit, the number of nodes increases. Consequently, the difficulty of analyzing a circuit increases. Some appreciation of this problem may be obtained by studying the equivalent circuit of the common-source amplifier shown in Fig. 7–1.

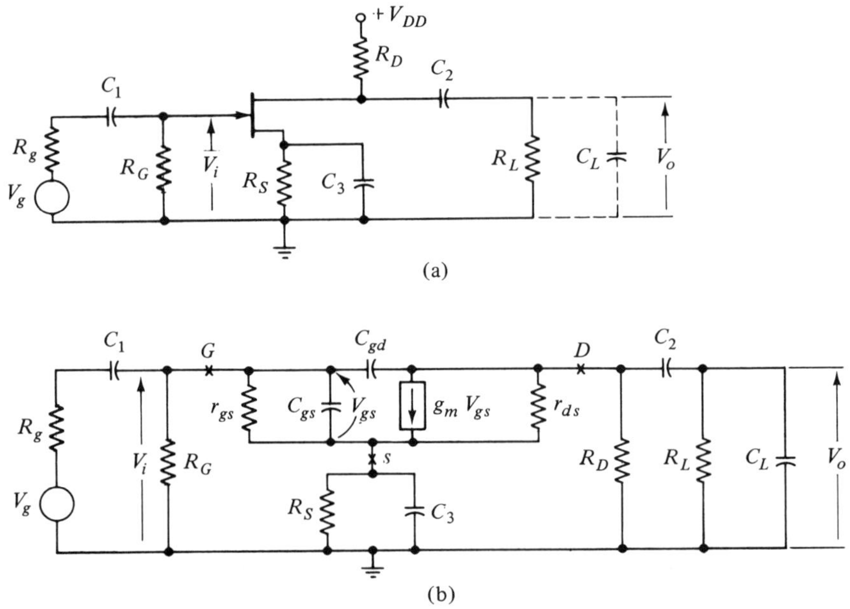

FIGURE 7–1 *A CS amplifier: (a) complete circuit, and (b) complete equivalent circuit.*

Since this circuit was discussed in Chap. 3, the mid-frequency components may be easily recognized. Several capacitors have been added to complete the circuit. The stray capacitance from the output terminal to ground is represented by C_L and usually has a value in the picofarad range. The circuit contains two blocking (or coupling) capacitors, C_1 and C_2, and a bypass capacitor, C_3. These three capacitors have values in the 0.1 μF to 100 μF range.

A relatively complete equivalent circuit is shown in Fig. 7–1(b) and is obtained by replacing the FET with its equivalent circuit (see Fig. 3–15). The parasitic capacitors C_{gs} and C_{gd} are in the device and have values in the picofarad range.

The frequency-response problem is one of finding a low frequency and also a high frequency at which the gain is 0.707 of that in the mid-frequency range. These specific frequencies are designated as f_1 and f_2, or f_{lo} and f_{hi} respectively, and correspond to the -3-db frequencies. These are reference frequencies and, as shown later, are very useful in design work. Since the circuit in Fig. 7–1(b) has six nodes, the problem of finding f_{lo} and f_{hi} is cumbersome unless a digital computer is used. Even when a computer is used, a circuit designer should be able to check the results by making approximate calculations.

The objective at this point is to simplify the complete circuit of Fig. 7–1(b) so that f_1 and f_2 can be easily calculated. As a first step in this direction, allow the frequency of the input signal, V_g, to start at zero (dc) and increase to a very high frequency while observing the effects of the decreasing reactances* on the equivalent circuit.

At a frequency of zero, all reactances in the circuit are infinite, and all capacitors can be replaced by open circuits. As shown in Fig. 7–2(a), none of the input signal is available to drive the gate of the FET, and the output voltage, V_o, is zero. Even with dc coupling on the input, V_o would still be zero because the output signal path is open.

The reactance values of C_1, C_2, and C_3 decrease as the frequency increases above zero frequency. For audio amplifiers, these three capacitors are selected so that the reactance values decrease to the level of the circuit resistance values in the frequency range of 10 to 40 Hz. The reactance values of C_{gd}, C_{gs}, and C_L in this frequency range are still in the hundreds or thousands of megohms. Consequently, these last three capacitors can be omitted from the low-frequency equivalent circuit as shown in Fig. 7–2(b). This simplified circuit is used to determine f_{lo} or f_1.

A further increase in the signal frequency causes the reactances of C_1, C_2, and C_3 to become negligibly small compared with circuit resistance values. In this frequency range, the circuit can be simplified as shown in Fig. 7–2(c). This range is the mid-frequency case discussed in Chap. 3.

A further increase in frequency causes the reactances of C_{gd}, C_{gs}, and C_L to decrease to the magnitude of the circuit resistances. Consequently, these capacitors must be included in the equivalent circuit as shown in Fig. 7–2(d). This is the high-frequency circuit that determines f_{hi} (or f_2).

As the frequency approaches infinity, the reactances of C_{gd}, C_{gs}, and C_L decrease to zero and short the associated circuit resistances. For this case, the output voltage approaches zero, and the equivalent circuit reduces to that shown in Fig. 7–2(e).

Since the mid-frequency case in Fig. 7–2(c) was considered in Chap. 3, this chapter will cover the low- and high-frequency circuits in Figs. 7–2(b) and (d). The approximate approach for finding f_1 and f_2 is to isolate each dominant capacitor with its associated resistive network and find the response for that part of

---

* The magnitude of the reactance of a capacitor is $|X| = \left|\frac{1}{2\pi fC}\right|$ and consequently decreases as frequency increases.

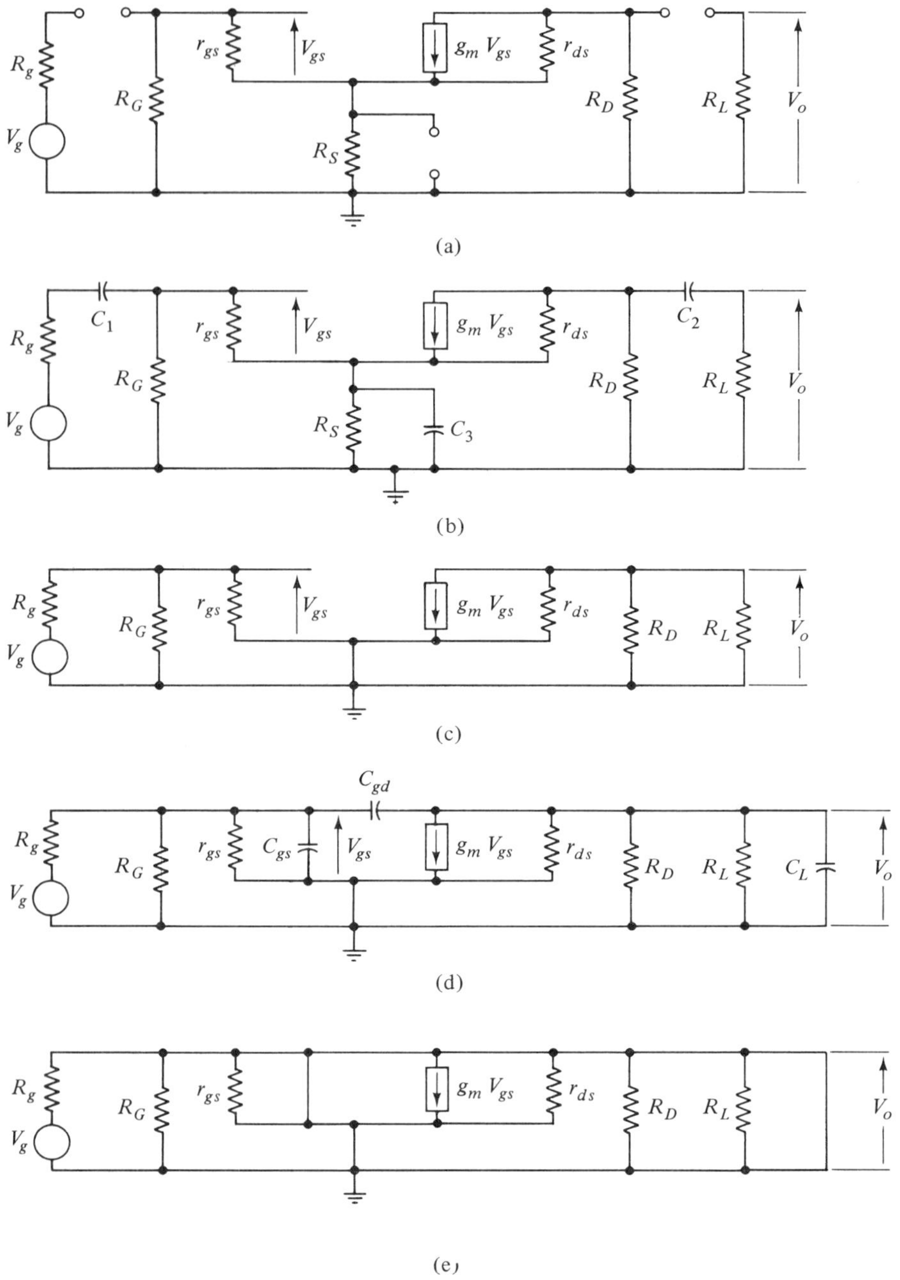

FIGURE 7–2 *The effect of varying the frequency of the input signal on the equivalent circuit shown in Fig. 7–1(b): (a) circuit for a dc input signal, (b) circuit for a low-frequency input signal, (c) circuit for a mid-band frequency input signal, (d) circuit for a high-frequency input signal, and (e) circuit for a very high-frequency input signal.*

the circuit. Then the responses of the individual *RC* circuits are combined into an overall low- or high-frequency response. This approach will be treated in detail later in this chapter. Before proceeding with the frequency-response analysis, the frequency limitations imposed by the active devices must be well understood. These limitations will be considered in the next section.

7.2 High-Frequency Limitations of Active Devices

The physical operation and equivalent circuits of FETs, bipolar transistors, and vacuum tubes were discussed in previous chapters. These discussions indicated that active devices can amplify signals in the mid-frequency range. Active devices can also amplify signals in the low-frequency range down to and including dc. A dc input signal simply corresponds to a new bias condition. However, a different condition exists in the high-frequency range. The objective of this section is to gain an intuitive understanding of the performance of transistors and vacuum tubes* at high frequencies. This objective will be accomplished by examining: (1) the simplified high-frequency equivalent circuits of each active device, and (2) the physical operation of each active device.

Consider the common-emitter hybrid-π equivalent circuit shown in Fig. 7-3(a). As the frequency of the input signal increases, the magnitudes of the capacitive reactances decrease. For the CE circuit, this effect reduces the input impedance $(r_{b'e} \,||\, 1/j\omega C_{b'e})$ and tends to short the base and emitter terminals. The voltage drop across the *EB* junction is reduced which in turn reduces $g_m V_{b'e}$ and therefore V_o.

While the reduction in gain is apparent from the approximate high-frequency equivalent circuits, the actual physical mechanism by which this effect occurs can be quite complicated. Electron transit times, junction capacitances, interelectrode capacitances, stored charge effects, lead inductances, and bulk resistances all influence the frequency response of active devices. These effects on the high-frequency performance are usually characterized by cutoff frequencies, time constants, and equivalent elements (*R*, *L*, and *C*) derived from the fundamental physical operation of the device. We will briefly review these aspects of device physics for the bipolar transistor, the FET, and the vacuum tube in that order.

The physical operation of a bipolar transistor depends on the injection of carriers from the emitter into the base and the movement** of these carriers across the base region to the collector. One limit on the high-frequency performance is the *base transit time*, τ_B, which is the time required for a carrier to

---

* This treatment is limited to vacuum triodes, tetrodes and pentodes. Microwave tubes such as the magnetron operate on different principles and are not discussed in this chapter.

** The physical operation of a bipolar transistor, as presented in Sec. 4.1, assumed a homogeneously doped base region. In such a transistor, carriers move across the base region only by diffusion. However, as pointed out in Sec. 4.4, most modern transistors are fabricated by a diffusion process which results in an inhomogeneously doped base region. The base is more heavily doped at the emitter side than at the collector side. One result is a built-in electric field in the base region which causes the injected carriers to move across the base by drift as well as by diffusion. This condition significantly reduces the base transit time and improves the high-frequency performance.

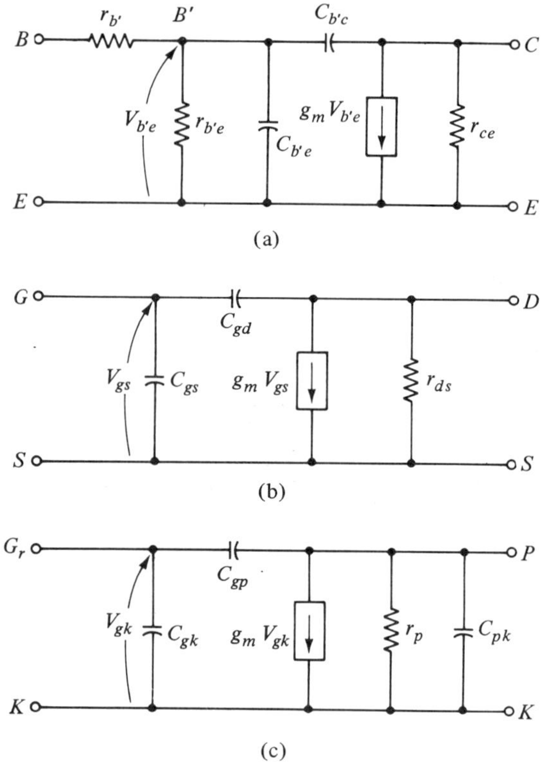

FIGURE 7–3 *Simplified high-frequency equivalent circuits for active devices connected as inverting amplifiers: (a) the common-emitter (CE) hybrid-π (see Fig. 4–14), (b) the common-source (CS) FET (see Fig. 3–15), and (c) the common-cathode (CK) vacuum tube (see Fig. 6–10).*

move across the base region. The base transit time can be written in terms of the stored charge in the base region (see Fig. 4–4) and the dc emitter current as *

$$\tau_B = Q_s/I_E \tag{7–1}$$

When the period of an input voltage and τ_B are of the same order of magnitude, recombinations in the base and changing boundary conditions decrease the amplitude of the ac variations on I_E; i.e., the ac current that flows at high frequencies is not as large as that in the mid-frequency range for the same input voltage magnitude.

In addition to the base transit time, the junction capacitance of the *EB* and *BC* junctions also limits the high-frequency performance of a bipolar transistor. An ac base voltage varies the width of the depletion region and, consequently, the number of uncovered impurity atoms or charges within the depletion region.

---

* Equation (7–1) is derived for an NPN transistor as follows. The dc emitter current is approximately equal to the electron current, I_n, injected into the base. This current can be written, from Eq. (1–19), as

$$I_e \simeq I_n = nq\,A\,v_n$$

where v_n is the velocity of an electron in the base region. The incremental transit time across a distance dx is

$$d\tau_B = dx/v_n$$

Integrating from the emitter to the collector, the time is

$$\tau_B = \int_E^C dx/v_n = \frac{1}{I_n}\int_E^C nq\,A\,dx = \frac{Q_s}{I_n} \simeq \frac{Q_s}{I_E}$$

A finite time is required to effect this change in charge. This period is characterized by a time constant which depends on the junction capacitance.

The physical limitations listed above are incorporated into the hybrid-π equivalent circuit in the following manner. The capacitor $C_{b'e}$ is the sum of the junction and diffusion capacitances of the *EB* junction. Hence, the effect of changing the *EB* junction bias is included in $C_{b'e}$. Since τ_B depends on the stored charge in the base region, which in turn determines the diffusion capacitance, base transit time effects are also included in $C_{b'e}$. The resistive components in the hybrid-π circuit describe the paths internal to the transistor through which the charging currents for the capacitors must flow.

Most device manufacturers do not specify values for $C_{b'e}$. Rather, they specify a value for either the β cutoff frequency, f_β, or for the CE *current gain-bandwidth product*, f_T, from which $C_{b'e}$ may be computed. Expressions for these two characteristic frequencies can be developed from the hybrid-circuit. The CE gain, β, is defined as*

$$\beta = \left.\frac{I_c}{I_b}\right|_{V_{ce}=0} \tag{7-2}$$

Hence, the frequency dependence of β can be calculated by placing an ac short circuit across the output terminals of the hybrid-π and driving the base terminal with a current generator as shown in Fig. 7–4(a). With the output shorted, the

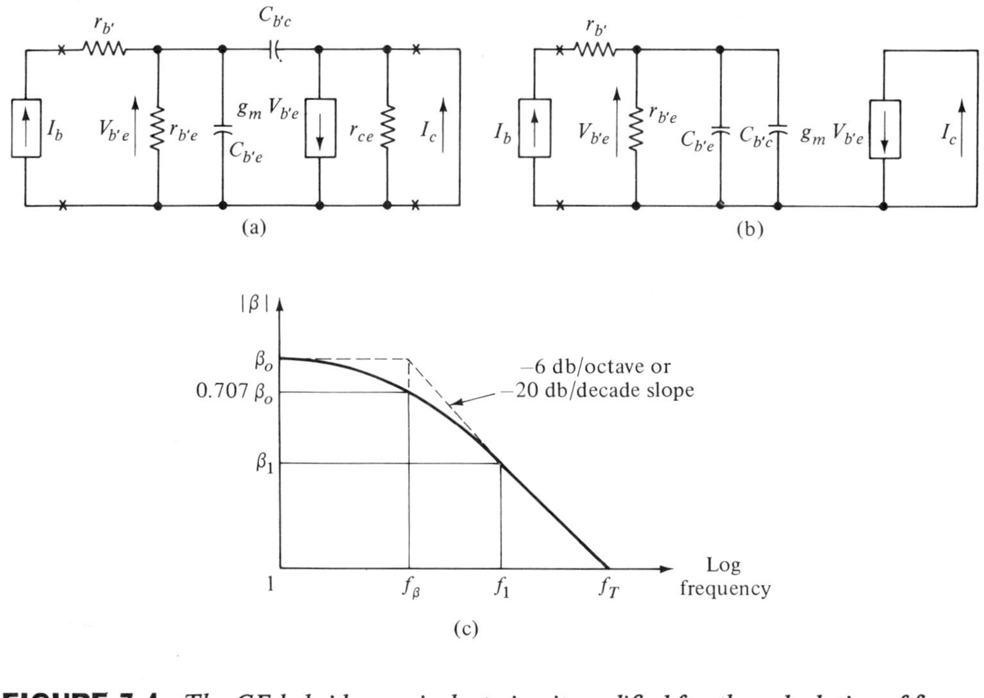

FIGURE 7–4 *The CE hybrid-π equivalent circuit modified for the calculation of* f_β *(a) the complete circuit, (b) a simplified form of (a), and (c) the dependency of β on frequency.*

* See Sec. 4.2.

circuit can be simplified as shown in Fig. 7–4(b). From this figure, $V_{b'e}$ is

$$V_{b'e} = I_b\left(r_{b'e} \,||\, \frac{1}{j\omega(C_{b'e} + C_{b'c})}\right) = I_b\left(\frac{r_{b'e}}{1 + j\omega r_{b'e}(C_{b'e} + C_{b'c})}\right) \quad \textbf{(7–3)}$$

Then I_c is

$$I_c = g_m V_{b'e} = I_b\left(\frac{g_m r_{b'e}}{1 + j\omega r_{b'e}(C_{b'e} + C_{b'c})}\right) \quad \textbf{(7–4)}$$

Solving this equation for I_c/I_b,

$$\beta = \frac{I_c}{I_b} = \frac{g_m r_{b'e}}{1 + j\omega r_{b'e}(C_{b'e} + C_{b'c})} \quad \textbf{(7–5)}$$

Equation (7–5) can be rewritten as

$$\beta = \frac{\beta_o}{1 + j\dfrac{\omega}{\omega_\beta}} = \frac{\beta_o}{1 + j\dfrac{f}{f_\beta}} \quad \textbf{(7–6)}$$

where $\beta_o = g_m r_{b'e}$ and is the mid-band value of β, and where

$$f_\beta \equiv \frac{1}{2\pi[r_{b'e}(C_{b'e} + C_{b'c})]} \quad \textbf{(7–7)}$$

and

$$\omega_\beta \equiv 2\pi f_\beta \quad \textbf{(7–8)}$$

Equation (7–6) is plotted in Fig. 7–4(c) with β either given in db or plotted on a log scale, and frequency plotted on a log scale.* The value of $|\beta|$ at the frequency f_β is

$$|\beta|_{f_\beta} = \left|\frac{\beta_o}{1 + j1}\right| = \frac{\beta_o}{\sqrt{2}} = 0.707\,\beta_o \quad \textbf{(7–9)}$$

Hence f_β corresponds to the 0.707 or -3-db frequency of $|\beta|$ and is called the *β cutoff frequency*. A general plot of Eq. (7–6) has a slope of -6 db/octave (or -20 db/decade) for frequencies much greater than f_β. This slope means that $|\beta|$ decreases by an order of magnitude for each order of magnitude increase in frequency. Along the -6 db/octave slope, the product of any particular value of $|\beta|$ and its corresponding frequency is a constant, and is equal to the so-called *current gain-bandwidth product*, f_T, for the CE. Hence, from Fig. 7–4(c),

$$\beta_o f_\beta = |\beta_1| f_1 = f_T \quad \textbf{(7–10)}$$

where f_T is the frequency at which the straight line approximation for $|\beta|$ is unity.

Semiconductor device manufacturers usually specify a value for $C_{b'c}$. This value is often listed on a specification sheet as C_{ob}, less often as C_{cb}, and seldom as $C_{b'c}$. Also given is a value for f_T, either as f_T directly, or as a value for $|h_{fe}|$ at some test frequency. The latter case corresponds to $|\beta_1| f_1$ as shown in Fig. 7–4(c). Some specification sheets include plots of C_{ob} and f_T as functions of I_C and V_{CE}. A value of $C_{b'e}$ can be calculated from f_T using Eqs. (7–7) and (7–10). Since the capacitance of a forward-biased junction is generally greater than that of a corresponding reverse-biased junction, f_β can be approximated as

$$f_\beta \simeq 1/(2\pi r_{b'e} C_{b'e}) \quad \textbf{(7–11)}$$

---

* This curve is commonly called a "Bode Plot" and is discussed in Appendix B.

Then,

$$f_T = \beta_o f_\beta \simeq \frac{\beta_o}{2\pi r_{b'e} C_{b'e}} \simeq \frac{\beta_o}{2\pi(\beta_o + 1) r_e C_{b'e}} \simeq \frac{1}{2\pi r_e C_{b'e}} \tag{7–12}$$

Solving Eq. (7–12) for $C_{b'e}$ gives

$$C_{b'e} \simeq \frac{1}{2\pi r_e f_T} \tag{7–13}$$

The values for $C_{b'e}$ and $C_{b'c}$, along with the mid-band components of the hybrid-π, complete the high-frequency equivalent circuit of a bipolar transistor. If all stray capacitance associated with the transistor case, socket, and circuit are included, the hybrid-π can accurately represent the transistor for frequencies approaching f_T. Second-order effects, such as the emitter and collector bulk resistance and lead inductance, can generally be neglected for frequencies less than f_T. Typical values of f_T range from as low as 100 kHz for high-power units to over 5000 MHz for some high-frequency units. Values of $C_{b'e}$ range from less than 1 pF to over 1000 pF. The values of $C_{b'c}$ and f_T are dependent on bias conditions. For a particular device, the corresponding specification sheet should be consulted to determine the optimum bias conditions for the best high-frequency performance.

Let us now consider the parameters that limit the frequency response of an FET. As discussed in Chap. 3, the basic physical operation of a JFET depends on the input voltage changing the width of the depletion region of a reverse-biased P-N junction which, in turn, changes the conductance of the channel. A change in the width of the depletion region means that the number of uncovered impurity ions (charge) in the depletion region is changing (see Sec. 2.4). Since a finite time is required to charge and discharge the junction capacitance, a limitation is imposed on the frequency that can be amplified by a JFET. Although the junction capacitance is distributed along the channel, it can be approximated by two discrete capacitors, C_{gs} and C_{gd}, as shown in Fig. 7–3(b). Values for C_{gs} range from 60 pF for a low-frequency unit to less than 4 pF for a high-frequency unit while C_{gd} can be less than 1 pF for a high-frequency unit. Specification sheets may give a value for C_{gd} as C_{rss}, and a value for C_{gs} as C_{iss}. In general, cutoff frequencies and gain-bandwidth products are not defined for a JFET. Occasionally, specification sheets include a value for g_m at some high frequency, or give plots of the input impedance, output impedance, and g_m as functions of frequency. Such plots are useful when the amplifier is operated at a very high frequency, and the device is characterized as a two-port network whose parameters are frequency dependent. Second-order effects, such as carrier transit time through the channel and also the bulk resistance of the channel outside the depletion region, are usually negligible. For this study the equivalent circuit shown in Fig. 7–3(b) can be used to determine the approximate high-frequency performance of a JFET amplifier.

Although the construction of a MOSFET is different from that of a JFET, the two devices are similar in electrical characteristics. Both have a significant gate-to-channel capacitance. However, in the MOSFET, this capacitance is associated with a conductor on each side of the thin oxide insulating layer rather than with a P-N junction. A finite time is required to charge and discharge the capacitance as the input ac voltage changes the voltage across the oxide, and this time imposes a limitation on the high-frequency range over which a signal can be amplified by a MOSFET. Specifications for the MOSFET are generally similar to those for the JFET. The approximate high-frequency equivalent circuit shown in Fig. 7–3(b) will be used for the MOSFET amplifier analysis. Emphasis will be placed on the JFET in discrete component circuits since they are more readily available commercially in a variety of N- and P-channel units. The MOSFET has its greatest potential in the field of digital integrated circuits, particularly in medium- and large-scale integrated circuits (MSI and LSI).

Since a vacuum tube is a mechanical structure (as opposed to the crystal structure of a transistor) the interelectrode capacitance arises from the spacing and area of the elements that make up the tube. Typical values for these capacitors are in the low picofarad range for a high-frequency vacuum tube. As discussed in Chap. 6, the pentode has a lower plate-to-control grid capacitance than the triode due to the electrostatic shield provided by the screen grid. For this reason, pentodes are generally used in high-frequency applications. Other high-frequency effects are also present. Since vacuum tubes are physically larger than transistors, electron transit times are larger. In addition, the inductance of the cathode reduces the effective g_m in the high-frequency range. However, these two effects are usually of secondary importance and will be neglected in our study.

The purpose of this section was to develop an understanding of the high-frequency limitations of transistors and vacuum tubes. Both the approximate high-frequency equivalent circuits and the underlying physical processes were reviewed briefly. For the remainder of this chapter, these limitations will be incorporated into equivalent circuits. These equivalent circuits will then be used to determine the frequency-response of electronic amplifiers.

7.3 Input and Output Impedance of Inverting Amplifiers

In the high-frequency range, the general expression for the input and output impedances, Z_i and Z_o, includes a reactive component due to the capacitive components in the active devices. The objective of this section is to calculate the input and output capacitances of inverting amplifiers. These capacitances, along with the resistive components, determine the input and output impedances and also the voltage gain of an amplifier as a function of frequency.

A CS amplifier is shown in Fig. 7–5(a), and its ac equivalent circuit is shown in Fig. 7–5(b). The high-frequency equivalent circuit is obtained as follows: (1) replace the circuit coupling and bypass capacitors with short circuits, (2) ac ground the dc power supply terminal,* and (3) replace the JFET with its equivalent circuit [Fig. 7–3(b)]. The input capacitance can be calculated using Fig. 7–5(c) where circuit components to the left of the gate terminal, including R_G, R_g, and V_g, have been replaced by a test voltage generator, V_{ti}, and where r_{ds} and R_D have been combined as R_{Leq}. The current flowing in V_{ti} is given by

$$I_{ti} = \frac{V_{ti}}{1/j\omega C_{gs}} + \frac{V_{ti} - V_o}{1/j\omega C_{gd}} = V_{ti}(j\omega C_{gs}) + V_{ti}(1 - V_o/V_{ti})(j\omega C_{gd}) \quad \textbf{(7–14)}$$

The ratio (V_o/V_{ti}) is equal to the terminal-to-terminal voltage gain of the amplifier, A_{vt}, which is equal to $-g_m R_{Leq}$. Consequently, Eq. (7–14) can be

* A dc power supply usually has a low output impedance to an ac wave. Typical values of ac impedance for good power supplies are less than one ohm.

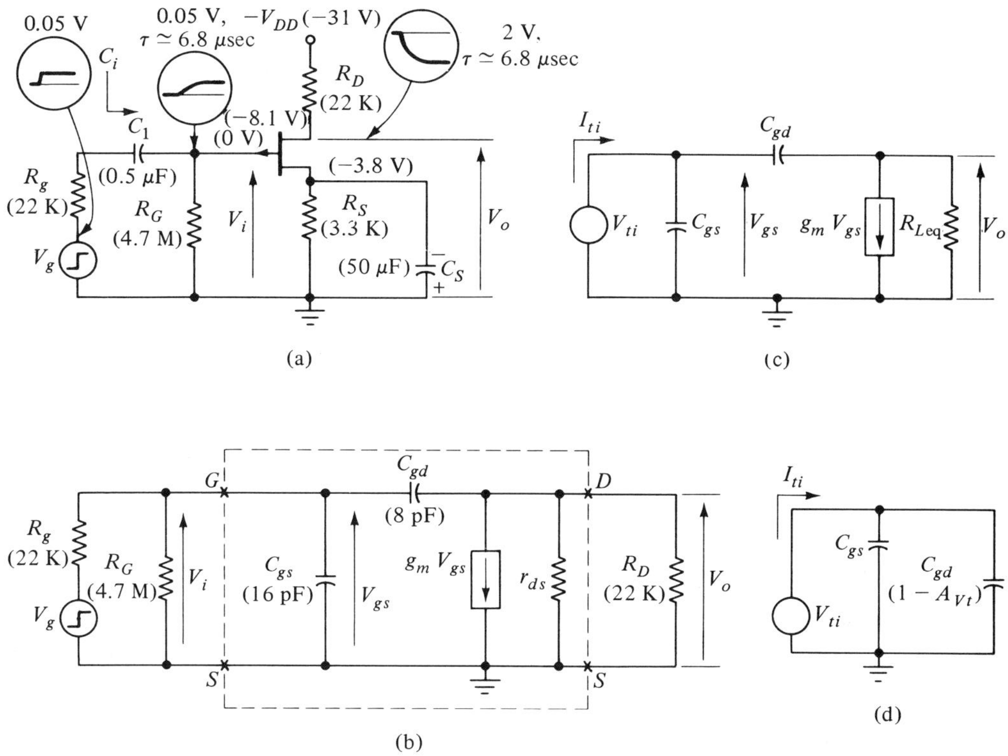

FIGURE 7–5 *A CS amplifier circuit: (a) the complete circuit, (b) an ac high-frequency equivalent circuit, (c) a circuit used to calculate the input capacitance, and (d) a simplified high-frequency input circuit. (Component values, voltages, and waveforms were obtained from Ex. 7–1.)*

written as

$$I_{ti} = V_{ti}\, j\omega[C_{gs} + C_{gd}(1 - A_{vt})] \tag{7–15}$$

Solving for Z_{in},

$$Z_{\text{in}} = \frac{V_{ti}}{I_{ti}} = \frac{1}{j\omega[C_{gs} + C_{gd}(1 - A_{vt})]} \tag{7–16}$$

Hence, the input capacitance of the CS amplifier is

$$C_{\text{in(CS)}} = C_{gs} + C_{gd}(1 - A_{vt}) = C_{gs} + C_{gd}(1 + g_m R_{\text{Leq}}) \tag{7–17}$$

Equation (7–17) shows that the input capacitance of the CS amplifier can be large due to the multiplication of C_{gd} by $(1 - A_{vt})$ where A_{vt} is the gate-to-drain voltage gain of the amplifier. This multiplication effect occurs because the total ac voltage developed across C_{gd} is the sum of the input and output voltages.*

Since the high-frequency equivalent circuits of the three inverting amplifiers are similar (see Fig. 7–3), the results for the CS amplifier can be extended to the

* Because of the amplification and phase inversion, the total ac voltage across C_{gd} is $V_{ti} - (-V_o) = V_{ti} + V_o = V_{ti}(1 - A_{vt})$. Since C_{gd} is to be replaced by another capacitor which is across the input as shown in Fig. 7–5(d), the new capacitor must be larger than C_{gd} if the input current remains constant. The voltage across the new capacitor is V_{ti}, and the voltage across C_{gd} is $V_{ti}(1 - A_{vt})$. Therefore, the new capacitor is $C_{gd}(1 - A_{vt})$.

CK and CE amplifiers. By direct analogy, the input capacitance of the CK amplifier is

$$C_{\text{in(CK)}} = C_{gk} + C_{gp}(1 - A_{vt}) \tag{7–18}$$

Since $r_{b'}$ is small and can be neglected for most modern bipolar transistors, the input capacitance of the CE amplifier is approximately

$$C_{\text{in(CE)}} \simeq C_{b'e} + C_{b'c}(1 - A_{vt}) \tag{7–19}$$

The following example demonstrates the calculation of input capacitance.

EXAMPLE 7–1 Calculate the value of the input capacitance of the CS amplifier shown in Fig. 7–5(a). The P-channel JFET is characterized by $V_P = 4.8$ V, $I_{DSS} = -18.5$ mA, $C_{gs} = 16$ pF, and $C_{gd} = 8$ pF. The JFET is biased such that $g_m = 2$ millimhos.

Solution: (a) Since the value of g_m is known, the bias conditions need not be calculated. The mid-band value of A_{vt} is (see Table 3–1)

$$A_{vt} = -g_m R_{Leq} = -(2 \times 10^{-3})(22\text{ K}) \simeq -44$$

From Eq. (7–17),

$$C_{\text{in(CS)}} \simeq C_{gs} + C_{gd}(1 - A_{vt}) \simeq 16 + 8(1 + 44) \simeq 376\text{ pF}$$

(b) Experimental results: The measured bias voltages are shown in Fig. 7–5(a). The input capacitance can be estimated using the following method. The voltage generator, V_g, supplies a very fast step in voltage. The voltage at the gate terminal will rise on an exponential curve whose time constant is determined by the product of the resistance and capacitance between the gate terminal and ground. From Fig. 7–5(b), this time constant is

$$\tau = (R_g \,||\, R_G)C_{\text{in}} = (22\text{ K} \,||\, 4.7\text{ M})(376\text{ pF}) \simeq 8.3\ \mu\text{sec}$$

The observed waveform at the gate (and drain) terminal is shown in Fig. 7–5(a) and has a time constant of 6.8 μsec, which agrees closely with the calculated value. Since the resistive component of τ is known, the measured value of τ verifies the calculation of C_{in}. The measured mid-band voltage gain was $A_{vo} = V_o/V_i = -2\text{ V}/0.05\text{ V} \simeq -40$.

As demonstrated by the analysis and example above, the input impedance of the CS amplifier depends on the voltage gain of the amplifier. This effect can be generalized to include any inverting type amplifier whose input and output terminals are connected by an impedance. Such an amplifier is shown in Fig. 7–6(a) where the basic amplifier can be any configuration and is characterized by values for A_{vt}, Z_{in}, and Z_{out}. The input and output terminals of the amplifier are connected together through an impedance, Z. Our objective is to calculate the network input and output impedances, Z_i and Z_o.

The network in Fig. 7–6(a) can be transformed to the one shown in Fig. 7–6(b) provided that the terminal conditions (V_i, I_i, V_o, and I_o) are unchanged by the transformation. From Fig. 7–6(a),

$$\left.\begin{aligned} I_1 &= \frac{V_i - V_o}{Z} = \frac{V_i(1 - V_o/V_i)}{Z} = \frac{V_i(1 - A_{vt})}{Z} \\ I_2 &= \frac{V_o - V_i}{Z} = \frac{V_o(1 - V_i/V_o)}{Z} = \frac{V_o(1 - 1/A_{vt})}{Z} \end{aligned}\right\} \tag{7–20}$$

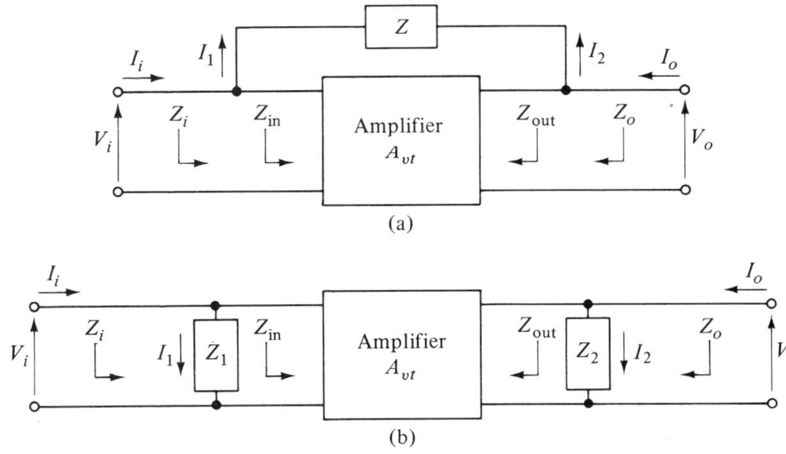

FIGURE 7–6 *(a) A network containing an amplifier whose input and output terminals are connected by an impedance, Z. (b) An equivalent network resulting from the application of Miller's Theorem.*

From Fig. 7–6(b),

$$I_1 = V_i/Z_1 \qquad I_2 = V_o/Z_2 \tag{7–21}$$

Consequently,

$$Z_1 = \frac{Z}{(1 - A_{vt})} \qquad Z_2 = \frac{Z}{(1 - 1/A_{vt})} \tag{7–22}$$

The network input and output impedances can be calculated from Fig. 7–6(b) and Eq. (7–22) as

$$Z_i = Z_{\text{in}} \,||\, \frac{Z}{(1 - A_{vt})} \qquad Z_o = Z_{\text{out}} \,||\, \frac{Z}{(1 - 1/A_{vt})} \tag{7–23}$$

The results expressed in Eq. (7–22) are often called *Miller's theorem* or *Miller's effect* and will find extensive application in this and later chapters. For the special case of the CS input capacitance,

$$\left.\begin{aligned} Z &= 1/(j\omega C_{gd}) \qquad & Z_i &= \frac{1}{j\omega C_{gs}} \,||\, \frac{1}{(j\omega C_{gd})(1 - A_{vt})} \\ Z_{\text{in}} &= 1/(j\omega C_{gs}) \qquad & C_i &= C_{gs} + C_{gd}(1 - A_{vt}) \end{aligned}\right\} \tag{7–24}$$

where the expression for C_i agrees with Eq. (7–17).

The output capacitance of the inverting amplifiers can be calculated using Eq. (7–22) for Z_2, yielding

$$\left.\begin{aligned} C_{o(\text{CS})} &= C_{gd}(1 - 1/A_{vt}) \simeq C_{gd} \\ C_{o(\text{CK})} &= C_{pk} + C_{gp}(1 - 1/A_{vt}) \simeq C_{pk} + C_{gp} \\ C_{o(\text{CE})} &= C_{b'c}(1 - 1/A_{vt}) \simeq C_{b'c} \end{aligned}\right\} \tag{7–25}$$

where the approximations are valid for large values of A_{vt}.

The results of this section show that the input and output impedances of an inverting type amplifier depend on the voltage gain of the amplifier. We will use this effect in the calculation of the frequency and time response of amplifiers.

7.4 Frequency Response of an Amplifier Input Section

In Sec. 7.1, the frequency response of a CS amplifier was discussed qualitatively. The objective of this section is to develop the techniques and approximations useful in determining both the low- and high-frequency responses of the input section of a CS amplifier. The techniques and approximations developed will then be generalized and used to determine the complete response of an amplifier.

The CS amplifier from Ex. 7–1 is shown in Fig. 7–7(a). The input section of the amplifier can be represented by the *RC* bandpass* network shown in Fig. 7–7(b) where the amplifier circuit to the right of the coupling capacitor, C_1, has been replaced by the amplifier input resistance, R_i, and input capacitance, C_i. The output voltage of the amplifier, V_o, depends directly on the input voltage, V_i. Consequently, the frequency response of V_i determines to a large degree the frequency response of the amplifier. As indicated in Sec. 7.1, we would expect V_i to be reduced at low frequencies due to the blocking effect of C_1, and to be reduced at high frequencies due to the shorting effect of C_i. Our immediate goal is to determine the voltage gain of the *RC* network in Fig. 7–7(b) as a function of frequency.

The voltage gain of the *RC* network shown in Fig. 7–7(b) could be found using standard circuit analysis techniques. However, the analysis can be simplified by dividing the network into three equivalent circuits: one for the mid-band range, one for the low-frequency range, and one for the high-frequency range.** In previous chapters, the mid-band equivalent circuit of an amplifier was formed by replacing the coupling and bypass capacitors with short circuits and replacing the active device with its mid-band equivalent circuit.*** For the *RC* network,

FIGURE 7–7 *A common source amplifier: (a) the complete circuit and (b) the ac equivalent input circuit for the circuit in (a).*

* A bandpass network is one that will pass signals only in a certain range or band of frequencies, f_{lo} to f_{hi}. Signals whose frequencies are lower than approximately f_{lo} or higher than approximately f_{hi} are attenuated significantly by the network.

** Dividing the *RC* network into three equivalent circuits corresponds to assuming that C_1 is much greater than C_i, an approximation which is generally valid. This division also assumes that there is no intersection between the two capacitors. The problem of interaction between various capacitors will be treated in the next chapter.

*** In mid-band equivalent circuits of the FET, the active device capacitances, C_{gs} and C_{gd}, are open circuits. This simplification results in a purely resistive input impedance. Consequently, the input capacitance of the amplifier is treated as an open circuit in the mid-band range.

this corresponds to replacing C_1 with a short circuit and replacing C_i with an open circuit as shown in Fig. 7–8(a). For frequencies below the mid-band range, the effect of C_1 must be included, but C_i can still be neglected. The low-frequency equivalent circuit is shown in Fig. 7–8(b). For frequencies above mid-band, C_1 can be neglected, but the effect of C_i must be included as shown in Fig. 7–8(c).

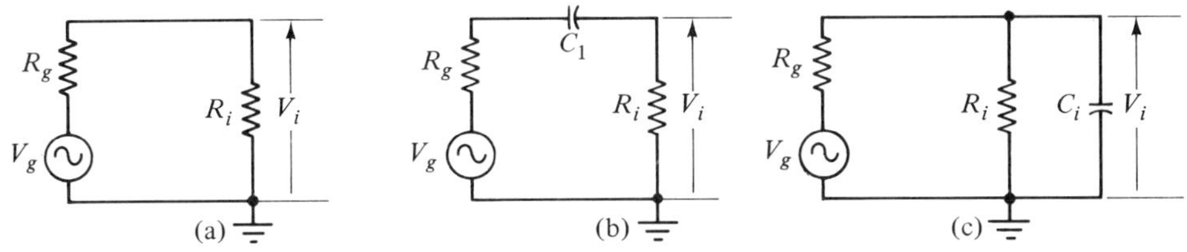

FIGURE 7–8 *Simplified versions of the ac equivalent circuit in Fig. 7–7(b): (a) mid-frequency equivalent, (b) low-frequency equivalent, and (c) high-frequency equivalent.*

With the input network divided into three equivalent circuits, we can calculate its overall frequency response by analyzing each equivalent circuit separately. The mid-band voltage gain is available from Fig. 7–8(a). Using voltage division between R_i and R_g,

$$\left.\frac{V_i}{V_g}\right|_{\text{mid}} = \frac{R_i}{R_i + R_g} \tag{7–26}$$

Equation (7–26) can be recognized as the generator-to-input voltage gain of the amplifier, A_{vg}, obtained in Chap. 3.

The low-frequency gain is available from Fig. 7–8(b) as

$$\left.\frac{V_i}{V_g}\right|_{\text{lo}} = \frac{R_i}{R_g + 1/j\omega C_1 + R_i} \tag{7–27}$$

Equation (7–27) can be written in terms of the complex variable,* s, where, for real frequencies, s is equal to $j\omega$. Substituting s for $j\omega$ in Eq. (7–27) gives

$$\left.\frac{V_i}{V_g}\right|_{\text{lo}} = \frac{R_i}{R_g + 1/sC_1 + R_i} = \frac{R_i(sC_1)}{1 + s(R_i + R_g)C_1} \tag{7–28}$$

Equation (7–28) actually includes the mid-band as well as the low-frequency response of the *RC* network. Multiplying the numerator and denominator of Eq. (7–28) by $(R_i + R_g)$ gives

$$\left.\frac{V_i}{V_g}\right|_{\text{lo}} = \left(\frac{R_i}{R_i + R_g}\right)\left(\frac{s(R_i + R_g)C_1}{1 + s(R_i + R_g)C_1}\right) \tag{7–29}$$

This equation shows the voltage gain divided into a mid-band term and a low-frequency term. Before discussing the physical significance of the low-frequency term, we will complete the analysis by calculating the high-frequency response of the *RC* network.

---

* The variable s is a complex variable equal to $\sigma + j\omega$ which is used in Laplace transform analysis. For real frequencies (frequencies that can be measured) s is equal to $j\omega$. The use of s reduces an equation from one involving complex quantities to a regular algebraic equation which greatly simplifies the analysis. Reactive terms are expressed in terms of s as

$$X_L = j\omega L = sL \qquad X_C = 1/j\omega C = 1/sC$$

The high-frequency voltage gain is available from Fig. 7–8(c) as

$$\left.\frac{V_i}{V_g}\right|_{\text{hi}} = \frac{(R_i) \,||\, (1/sC_i)}{R_g + (R_i) \,||\, (1/sC_i)} = \frac{R_i}{R_i + R_g + sC_i(R_i R_g)} \qquad \textbf{(7–30)}$$

Equation (7–30) can be written in terms of the mid-band gain by factoring out $(R_i + R_g)$ from the denominator giving

$$\left.\frac{V_i}{V_g}\right|_{\text{hi}} = \left(\frac{R_i}{R_i + R_g}\right)\left(\frac{1}{1 + s(R_i \,||\, R_g)C_i}\right) \qquad \textbf{(7–31)}$$

This equation shows the voltage gain divided into a mid-band term and a high-frequency term.

A complete expression for the voltage gain for the input *RC* network as a function of frequency can be written as*

$$\frac{V_i}{V_g} = \left(\frac{R_i}{R_i + R_g}\right)\left(\frac{s(R_i + R_g)C_1}{1 + s(R_i + R_g)C_1}\right)\left(\frac{1}{1 + s(R_i \,||\, R_g)C_i}\right) \qquad \textbf{(7–32)}$$

Equation (7–32) can be simplified by using the following relationships:

$$A_{\text{mid}} \equiv R_i/(R_i + R_g) \qquad \textbf{(7–33)}$$

$$f_{\text{lo}} \equiv 1/[2\pi(R_i + R_g)C_1] \qquad f_{\text{hi}} \equiv 1/[2\pi(R_i \,||\, R_g)C_i] \qquad \textbf{(7–34)}$$

Using these definitions, and replacing s by $j\omega$, or $j2\pi f$, Eq. (7–32) can be written as

$$\frac{V_i}{V_g} = (A_{\text{mid}})\left(\frac{jf/f_{\text{lo}}}{1 + jf/f_{\text{lo}}}\right)\left(\frac{1}{1 + jf/f_{\text{hi}}}\right) \qquad \textbf{(7–35)}$$

Equation (7–35) is the transfer function**of the input *RC* bandpass network written in its most convenient form. The transfer function consists of four terms: (1) a constant, A_{mid}, (2) a zero at zero frequency, jf/f_{lo}, representing the blocking action of C_1 to dc, (3) a pole, $1/(1 + jf/f_{\text{lo}})$, at the frequency f_{lo}, which cancels the zero for frequencies above f_{lo}, thereby indicating that C_1 passes signals above zero frequency, and (4) a pole, $1/(1 + jf/f_{\text{hi}})$, at the frequency f_{hi}, indicating that C_i short circuits V_i at high frequencies. An asymptotic Bode plot*** of the magnitude of each term in Eq. (7–35) is plotted in Fig. 7–9(a), and the approximate phase plot of each term is shown in Fig. 7–9(b). The composite Bode plot of the magnitude of the transfer function is shown in Fig. 7–9(c), and the phase plot is shown in Fig. 7–9(d). These composite plots show that the analysis can be conveniently divided into three regions: (1) the low-frequency region below f_{lo} where the effect of C_1 is important, (2) the mid-band region between f_{lo} and f_{hi} where the effects of C_1 and C_i can be neglected, and (3) the high-frequency region above f_{hi} where the effect of C_i is important. The extent of the three regions is found by calculating the corner frequencies, f_{lo} and f_{hi}. At these two

---

* Equation (7–32) assumes that the time constant $(R_i + R_g)C_1$ is much larger than the time constant $(R_i \,||\, R_g)C_i$.

** The transfer function of a network relates the output signal to the input signal for an arbitrary input signal. The mid-band voltage gains developed in previous chapters are also examples of transfer functions. A complete transfer function is frequency dependent and can always be expressed as the ratio of two polynomials. The roots of the denominator are called "poles," and the roots of the numerator are called "zeros." The transfer function in Eq. (7–35) has already been factored. Transfer functions are discussed further in Appendix B.

*** Refer to Appendix B for a discussion of the method for constructing Bode plots.

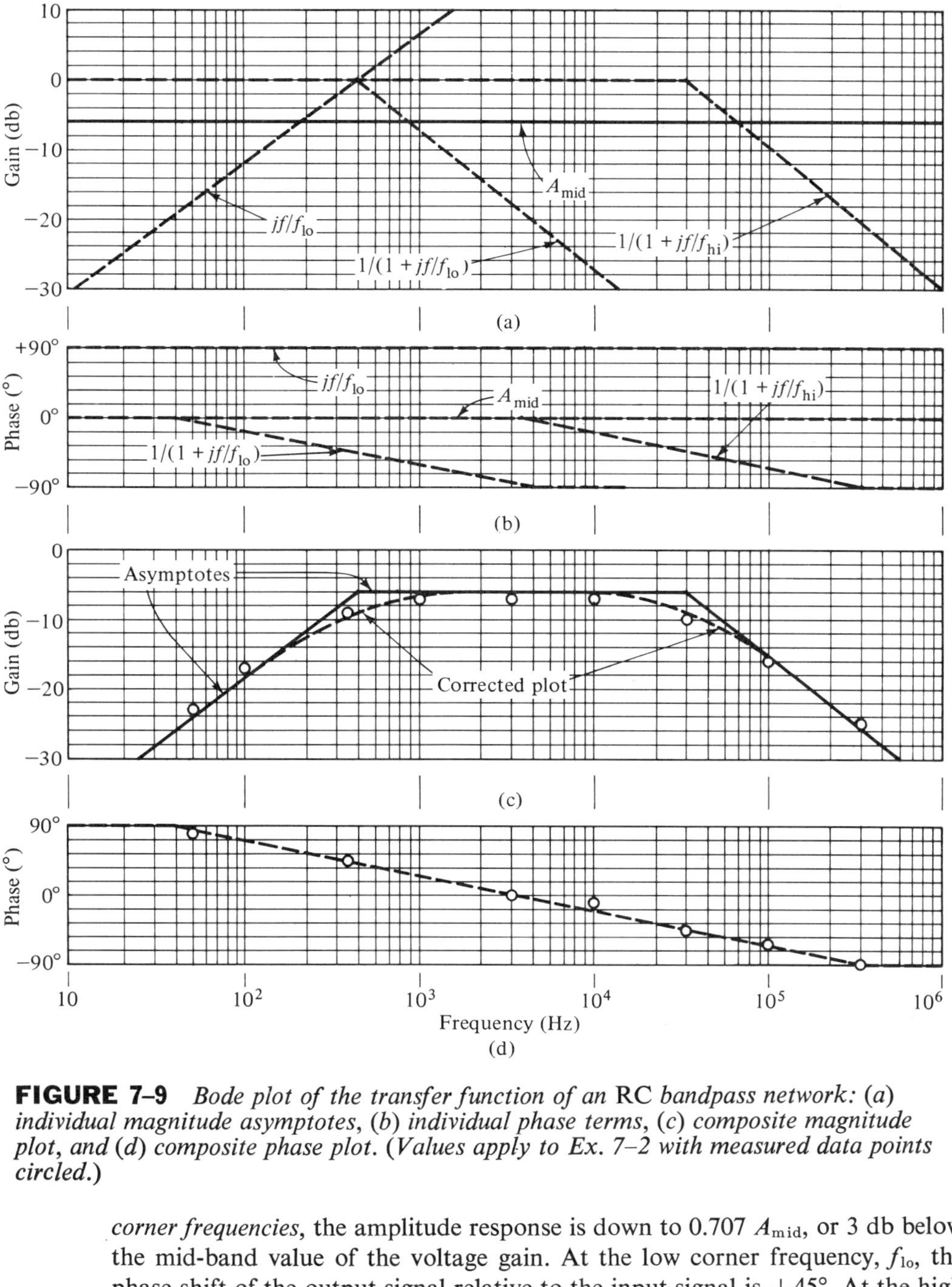

FIGURE 7–9 *Bode plot of the transfer function of an* RC *bandpass network: (a) individual magnitude asymptotes, (b) individual phase terms, (c) composite magnitude plot, and (d) composite phase plot. (Values apply to Ex. 7–2 with measured data points circled.)*

corner frequencies, the amplitude response is down to 0.707 A_{mid}, or 3 db below the mid-band value of the voltage gain. At the low corner frequency, f_{lo}, the phase shift of the output signal relative to the input signal is $+45°$. At the high corner frequency, f_{hi}, the phase shift is $-45°$.

Now that the general form of the response of the *RC* network is known, the actual calculations for a particular network can be simplified. The network voltage gain or transfer function, as given in Eq. (7–35), is characterized by three constants: (1) the mid-band voltage gain, A_{mid}, (2) the low corner frequency,

f_{lo}, and (3) the high corner frequency, f_{hi}.* The value of A_{mid} can be calculated from the mid-band equivalent circuit. The low corner frequency can be found from the low-frequency equivalent circuit by calculating the *RC* time constant associated with the coupling capacitor, C_1. The value of this time constant, τ_{lo}, is equal to the product of C_1 and the total resistance shunting C_1. Referring to Fig. 7–8(b),

$$\tau_{\text{lo}} = (R_i + R_g)C_1 \qquad \textbf{(7–36)}$$

The time constant is related to the corner frequency by

$$f = 1/(2\pi\tau) \qquad \textbf{(7–37)}$$

Hence,

$$f_{\text{lo}} = 1/(2\pi\tau_{\text{lo}}) = 1/[2\pi(R_i + R_g)C_1] \qquad \textbf{(7–38)}$$

which is in agreement with Eq. (7–34). A high corner frequency, f_{hi}, can be obtained from the high-frequency equivalent circuit in Fig. 7–8(c). An expression for τ_{hi} is

$$\tau_{\text{hi}} = (R_i \,||\, R_g)C_i \qquad \textbf{(7–39)}$$

and

$$f_{\text{hi}} = 1/(2\pi\tau_{\text{hi}}) = 1/[2(R_i \,||\, R_g)C_i] \qquad \textbf{(7–40)}$$

This equation is in agreement with Eq. (7–34). The following example demonstrates this method of frequency response analysis.

EXAMPLE 7–2 Calculate the complete frequency response of the *RC* bandpass network shown in Fig. 7–7(b) if $R_i = R_g = 2$ K, $C_1 = 0.1\ \mu$F, and $C_i = 0.005\ \mu$F.

Solution: The analysis can be divided into three parts, one for each frequency region.

(a) Mid-band response: In the mid-band region, the impedance of C_1 is negligibly small and can be approximated as a short circuit. The impedance of C_i is very large and can be approximated as an open circuit. From Eq. (7–26) or (7–33), the mid-band voltage gain is

$$A_{\text{mid}} = \left.\frac{V_i}{V_g}\right|_{\text{mid}} = \frac{R_i}{R_i + R_g} = \frac{2\text{ K}}{2\text{ K} + 2\text{ K}} = 0.5$$

Expressing A_{mid} in db,

$$A_{\text{mid}} = 20 \log |0.5| = -20 \log |2| = -6 \text{ db}$$

(b) Low-frequency response: From Eq. (7–36), the low-frequency time constant is

$$\tau_{\text{lo}} = (R_i + R_g)C_1 = (2\text{ K} + 2\text{ K})(0.1\ \mu\text{F}) = 0.4 \text{ msec}$$

The low corner frequency is

$$f_{\text{lo}} = 1/(2\pi\tau_{\text{lo}}) = 0.159/0.4 \text{ msec} \simeq 400 \text{ Hz}$$

The asymptotic low-frequency response drops off at a slope of 6 db/octave (20 db/decade) for the frequencies below f_{lo} as shown in Fig. 7–9(a). The approximate phase response is shown in Fig. 7–9(b).

(c) High-frequency response: From Eq. (7–39), the high-frequency time constant is

$$\tau_{\text{hi}} = (R_i \,||\, R_g)C_i = (2\text{ K} \,||\, 2\text{ K})(0.005\ \mu\text{F}) \simeq 5\ \mu\text{sec}$$

---

* Since the *RC* bandpass network represents the input section of an amplifier, we can expect a similar situation in amplifier analysis.

The high corner frequency is

$$f_{\text{hi}} = 1/(2\pi\tau_{\text{hi}}) = 0.159/5\ \mu\text{sec} \simeq 32\ \text{kHz}$$

The asymptotic high-frequency response drops off at a slope of −6 db/octave (−20 db/decade) for frequencies above f_{hi} as shown in Fig. 7–9(a). The approximate phase response is shown in Fig. 7–9(b).

(d) Composite magnitude response: The asymptotic Bode plot of the complete magnitude response is shown in Fig. 7–9(c). The actual response deviates from the asymptotes as follows: −3 db at the corner frequencies, −1 db one octave on either side of the corner frequencies, and approaches the asymptotes a decade on either side of the corner frequencies. The complete magnitude response is plotted in Fig. 7–9(c).

(e) Composite phase response: The low-frequency phase response can be approximated as a straight line passing through 45° at f_{lo}, 90° at $0.1f_{\text{lo}}$, and 0° at $10f_{\text{lo}}$.* This corresponds to a slope of −45°/decade. In a similar manner, the high-frequency phase response can be approximated as a straight line passing through −45° at f_{hi} with a slope of −45°/decade. The composite phase response is obtained by combining the phase response of each term as shown in Fig. 7–9(d).

(f) Experimental results: The complete frequency response was plotted by taking readings from 50 Hz to 400 kHz. Data points are circled in Fig. 7–9. The mid-band voltage gain was measured at 5 kHz as

$$A_{\text{mid}} = V_i/V_g = 2.0\ \text{V}/4.4\ \text{V} \simeq 0.45$$

Converting to db,

$$A_{\text{mid}} = 20 \log |0.45| \simeq -7\ \text{db}$$

Since the shape of the frequency response curve is known, it is usually sufficient to measure only the corner frequencies and the mid-band gain.

The results of this section indicate that the voltage gain of an amplifier as a function of frequency can be divided into three parts: (1) the mid-band gain, (2) the low-frequency response, and (3) the high-frequency response. The mid-band gain can be calculated using the techniques discussed in previous chapters. The low-frequency response is calculated by finding the time constant associated with each circuit coupling and bypass capacitor. The high-frequency response is calculated by finding the time constant associated with the input and output capacitances of each stage of the amplifier. Each time constant is associated with a pole or zero of the transfer function. The transfer function can be plotted as a Bode plot using the corner frequency associated with each pole and zero. When more than one low or high corner frequency is present, the overall corner frequencies (−3 db frequencies) can be found from the composite Bode plot. The approximate analysis discussed in this section will be used extensively in this and later chapters.

7.5 Time Response of an Amplifier Input Section

The fundamentals of frequency response analysis were developed in the last section using the input circuit of a CS amplifier as the basis for the discussion. The time response of an amplifier is also important. In this section, we will

---

* A more accurate phase plot is given in Appendix B.

examine the time response of the input section of a CS amplifier.* This analysis will provide some insight into the time response of any amplifier. The objectives of this section are: (1) to develop the techniques and approximations useful in determining the time response of amplifiers, and (2) to develop the relationships between the time response and the frequency response of amplifiers. In the last section, an *RC* bandpass network was used to represent the input section of a CS amplifier. The same *RC* network shown in Fig. 7–7(b) will be used in this analysis and is repeated in Fig. 7–10(b). Again we will assume that values of C_1 are much greater than C_i.

Often the time response of an amplifier is given in terms of how well the amplifier can follow a perfectly square-wave input signal. The square wave is chosen because it is a combination of the most abrupt voltage change available and the slowest possible voltage change. The input waveform, V_g, is shown in Fig. 7–10(a). The output waveform, V_i, is shown in Fig. 7–10(c). As indicated in Fig. 7–10(c), the *RC* network distorts the square wave. This distortion is due to C_1 and C_i, and is discussed below.

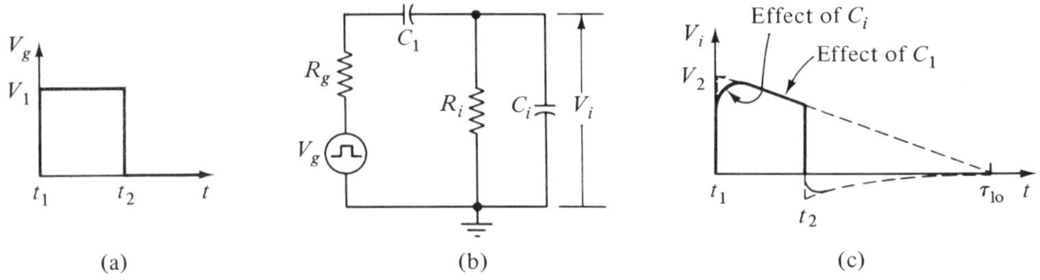

FIGURE 7–10 *An* RC *bandpass network representing the input section of a CS amplifier [see Fig. 7–7(b)]: (a) a square wave input signal, (b) the network, and (c) the output waveform of the circuit,* V_i.

A study of the time response of an *RC* network to sudden changes in voltage is based on two principles of circuit theory: (1) the voltage across a capacitor cannot change instantaneously,** and (2) the voltage across a capacitor will change according to an exponential curve with a time constant that depends on the circuit components. Referring to Fig. 7–10(a), V_g changes suddenly from 0 to $+V_1$ at time t_1. Since the voltages across C_1 and C_i cannot change instantaneously, the output voltage, V_i, cannot follow this abrupt change in input voltage. Also, since the voltage across C_i is the output voltage, the output voltage can change only as fast as the voltage on C_i changes. During the time interval that the voltage on C_i is changing, the voltage on C_1 remains relatively constant*** because C_1 is much greater than C_i. Consequently, C_1 can be replaced by an ac short circuit when calculating the effect of C_i on the output waveform. This simplification reduces the *RC* network to the high-frequency equivalent circuit shown

* An example of time response is contained in Ex. 7–1 where the voltage waveforms at the gate and drain terminals of the FET were exponential functions of time when the input signal, V_g, was a step in voltage.

** An instantaneous change in voltage across a capacitor would require an infinite current.

*** For two capacitors in series, the voltage division is inversely proportional to the capacitance since the change in charge is the same. Usually C_1 is 10^3 to 10^5 times larger than C_i. Therefore, the initial drop across C_1 is negligible.

in Fig. 7–8(c). The effect of C_i on V_i is to "round off" the sudden changes in the input voltage as shown in Fig. 7–10(c).

After the sudden change in V_g at time t_1, the output voltage builds up to a limiting or final value, V_2, as shown in Fig. 7–10(c). The value of V_2 is determined by R_i, R_g, and V_1 as

$$V_2 = V_1 R_i/(R_i + R_g) \qquad \textbf{(7–41)}$$

The ratio V_2/V_1 corresponds to the mid-band voltage gain of the RC network since

$$A_{\text{mid}} = \left.\frac{V_i}{V_g}\right|_{\text{mid}} = \frac{R_i}{R_i + R_g} = \frac{V_2}{V_1} \qquad \textbf{(7–42)}$$

However, the output voltage does not remain at a value of V_2 with increasing time. If the input voltage were to remain at $+V_1$, the voltage across C_1 would eventually build up to a value equal to V_1, thereby causing the output voltage to decay to zero volts. Hence, the effect of C_1 is to cause the output voltage to decay appreciably during the time interval from t_1 to t_2 as shown in Fig. 7–10(c). Since C_1 is much greater than C_i, C_i can be neglected when calculating the effect of C_1 on the output waveform. Neglecting C_i corresponds to replacing it with an open circuit. This simplification reduces the RC network to the low-frequency equivalent circuit shown in Fig. 7–8(b).

The preceding discussion indicates that the analysis of the time response of the RC network representing the input section of a CS amplifier can be divided into three parts. First, the mid-band gain is equal to the ratio of the maximum possible value of the output voltage to the actual maximum value of the input voltage. Second, the transient response occurring when the input suddenly changes for 0 to $+V_1$ is controlled by the high-frequency equivalent of the RC network. Third, the long-term response is controlled by the low-frequency equivalent circuit of the RC network. Expressions for the transient and long-term response are developed below.

The high-frequency equivalent circuit of the RC network was given in Fig. 7–8(c) and is repeated in Fig. 7–11(b). The leading edge of the input square wave signal is shown in Fig. 7–11(a), and the output waveform in Fig. 7–11(c). From circuit theory, the expression for the output voltage is*

$$V_i = 1 - \exp(-t/\tau_{\text{hi}}) \qquad \textbf{(7–43)}$$

where the network time constant, τ_{hi}, is

$$\tau_{\text{hi}} = (R_i \,||\, R_g)C_i \qquad \textbf{(7–44)}$$

---

* Equation (7–43) can be derived from the transfer function of the network in Fig. 7–11(a) which is given in Eq. (7–31) as

$$\left.\frac{V_i(s)}{V_g(s)}\right|_{\text{hi}} = \left(\frac{R_i}{R_i + R_g}\right)\left(\frac{1}{1 + s\,\tau_{\text{hi}}}\right)$$

where τ_{hi} is given by Eq. (7–39). For a unit step voltage input,

$$V_g(s) = 1/s$$

and the normalized output is

$$V_i(s) = \left(\frac{1}{s}\right)\left(\frac{1}{s + \tau_{\text{hi}}}\right)$$

Equation (7–43) is the inverse Laplace transform of $V_i(s)$.

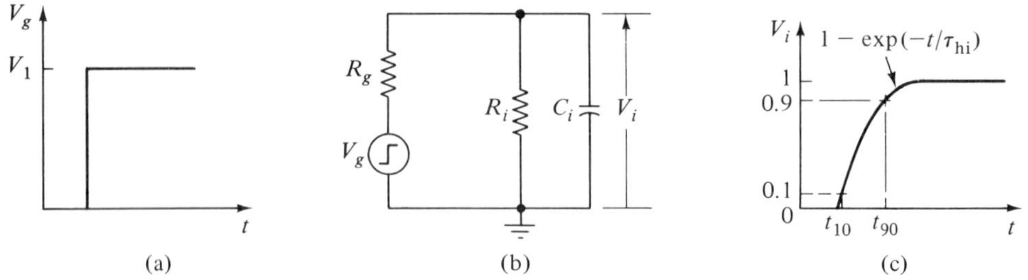

FIGURE 7–11 *A high-frequency equivalent circuit for the* RC *bandpass: (a) the leading edge of the square wave input, (b) the network, and (c) the output transient response.*

Often the transient response is described in terms of the *rise time*, τ_r, of the output. The rise time is the time required for the signal to change from 10 percent to 90 percent of its final value. From Fig. 7–11(c), the rise time is the difference between t_{90} and t_{10}. The value of t_{10} can be found by substituting V_i equal to 0.1 into Eq. (7–43) yielding

$$t_{10} = 0.1\ \tau_{\rm hi} \qquad \textbf{(7–45)}$$

Substituting V_i equal to 0.9 into Eq. (7–43) yields

$$t_{90} = 2.3\ \tau_{\rm hi} \qquad \textbf{(7–46)}$$

The 10 to 90 percent rise time can now be expressed as

$$\tau_r = t_{90} - t_{10} = 2.2\ \tau_{\rm hi} \qquad \textbf{(7–47)}$$

From Eq. (7–40), $\tau_{\rm hi}$ can be expressed in terms of the high corner frequency, $f_{\rm hi}$, as

$$\tau_{\rm hi} = 1/(2\pi f_{\rm hi}) \qquad \textbf{(7–48)}$$

Consequently, the rise time and high corner frequency are related by

$$\tau_r = 2.2/2\pi f_{\rm hi} = 0.35/f_{\rm hi} \qquad \textbf{(7–49)}$$

This equation indicates that the product of the rise time, τ_r, and the high corner frequency, $f_{\rm hi}$, has a constant value* of 0.35. This important relationship is used frequently to convert from one parameter to the other.

The low-frequency equivalent circuit of the *RC* network was given in Fig. 7–8(b) and is repeated in Fig. 7–12(b). The input square wave is shown in Fig. 7–12(a), and the output waveform in Fig. 7–12(c). With the input held at a value of $+V_1$, the output voltage is** approximately

$$V_i = V_2 \exp(-t/\tau_{\rm lo}) \qquad \textbf{(7–50)}$$

---

* Equation (7–49) is exact only for amplifiers with a single or dominant high corner frequency. However, it is often used as an approximation for more complicated amplifiers.

** Equation (7–50) can be derived from the transfer function of the network in Fig. 7–12(b). From Eq. (7–29), this transfer function is

$$\left.\frac{V_i(s)}{V_g(s)}\right|_{\rm lo} = \left(\frac{R_i}{R_i + R_g}\right)\left(\frac{s\,\tau_{\rm lo}}{1 + s\,\tau_{\rm lo}}\right)$$

where $\tau_{\rm lo}$ is given by Eq. (7–36). For a unit step voltage input, $V_g(s)$ is $1/s$, and the output is

$$V_i(s) = V_2\left(\frac{1}{s}\right)\left(\frac{s\,\tau_{\rm lo}}{1 + s\,\tau_{\rm lo}}\right) = \frac{V_2\,\tau_{\rm lo}}{1 + s\,\tau_{\rm lo}}$$

Equation (7–50) is the inverse Laplace transform of $V_i(s)$.

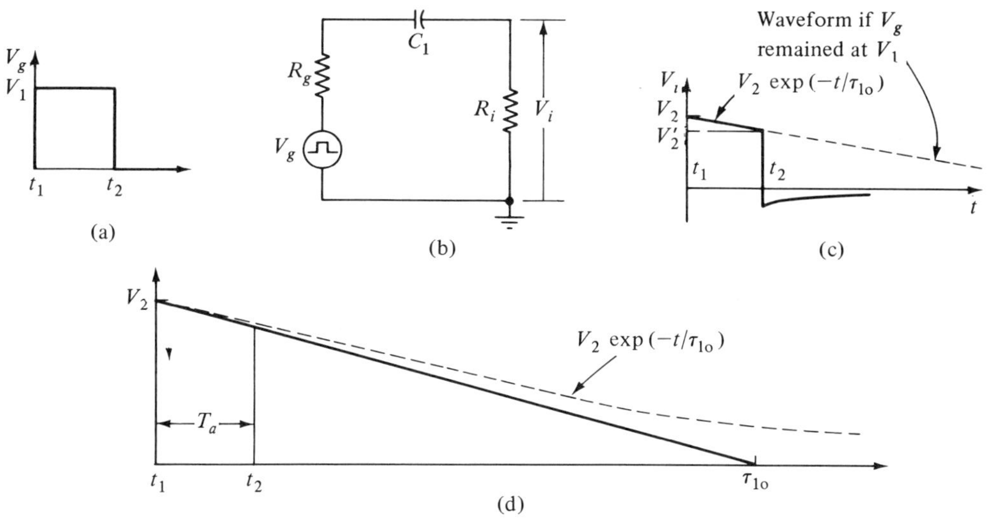

FIGURE 7–12 *A low-frequency equivalent circuit for the* RC *bandpass network: (a) the input signal, (b) the network, (c) the output waveform, and (d) a simplified high-frequency input circuit.*

where the network time constant, τ_{lo}, is

$$\tau_{lo} = (R_i + R_g)C_1 \tag{7-51}$$

During the time interval from t_1 to t_2, the output voltage will decay from V_2 to V_2'. Our goal is to describe the percentage change in V_i during this time interval. This change is usually called the tilt.

The *percentage tilt*, *P*, of the output waveform is defined as

$$P \equiv [(V_2 - V_2')/V_2] \times 100\% \tag{7-52}$$

Assuming that the voltage across C_i reaches a value of V_2 at t_1, then V_2', the voltage at t_2, can be found from a decaying exponential curve having a time constant of τ_{lo}, or, from Eq. (7–50),

$$V_2' = V_2[\exp(-T_a/\tau_{lo})] \tag{7-53}$$

where T_a is the duration of the input rectangular wave. This equation gives V_2' for any ratio of $(t_2 - t_1)/\tau_{lo}$. However, if the period of the pulse is small compared to the time constant τ_{lo}, the solution is much easier. As shown in Fig. 7–12(c), if the initial slope of the decaying exponential is extended to the x-axis, the point of intersection is one time constant, τ_{lo}, away from the initial time; i.e., if the initial rate of decay continued until the voltage across R_i reached zero, the time required would be one time constant. If T_a is small compared to τ_{lo}, then only a small error is introduced by assuming that the exponential curve follows the initial slope as shown in Fig. 7–12(d). For this case, V_2' can be obtained from similar triangles as

$$V_2' = V_2(1 - T_a/\tau_{lo}) \tag{7-54}$$

Then the percentage tilt is

$$P = \frac{V_2 - V_2(1 - T_a/\tau_{lo})}{V_2} \times 100\% = \frac{T_a}{\tau_{lo}} \times 100\% \tag{7-55}$$

During the period that the input voltage is present, an output voltage is produced by a charging current flowing through R_i. This charging current also develops a voltage, v_c, across C_1 that increases as a function of time and cancels part of the input voltage. Therefore, the current at any instant is

$$i = (V_g - v_c)/(R_g + R_i) \tag{7–56}$$

From this equation, we see that as v_c becomes larger, the current becomes smaller. Consequently, the output voltage decays since it is simply iR_i. When the input returns to zero, the voltage v_c causes the current to reverse and produces an undershoot in the output voltage as shown in Fig. 7–12(c). In this figure, the amount of decay, $V_2 - V_2'$, is equal to the undershoot. This negative output voltage then approaches zero on an exponential function with a time constant of τ_{lo} as C_1 discharges.

Only a single rectangular pulse was considered in the preceding discussion for simplicity. However, this case can easily be extended to the more useful case of a repetitive square wave input. In this case, the period of the wave (positive and negative) is T or $2T_a$ for a symmetrical square wave. Substituting this relationship into Eq. (7–55), the percent tilt in terms of T is approximately*

$$P = (T/2\tau_{lo}) \times 100\% \tag{7–57}$$

From Eq. (7–38), the value of τ_{lo} in terms of the low corner frequency is

$$\tau_{lo} = 1/(2\pi f_{lo}) \tag{7–58}$$

Consequently, the tilt and the low corner frequency are related by

$$P = (T/2)(2\pi f_{lo}) = T\pi f_{lo} \tag{7–59}$$

This equation indicates that the amplifier's response to the flat portion of the square wave and its low-frequency response are closely related. Equation (7–59) is accurate only for values of tilt less than approximately 15 percent because the linear approximation is only good over this range. An approach to use in experimental work is to adjust the period of the square wave until the tilt is approximately 10 percent. The experimental values obtained for P and T can then be used in Eq. (7–59) to find f_{lo}.

The time-response techniques developed in this section find wide application in the experimental evaluation of amplifier circuits. With the proper choice of the square wave period, the complete response of an amplifier is easily observed on an oscilloscope by simply changing the time scale to show first the rise time and then the tilt. The mid-band gain is available from the ratio of the peak output signal to the peak input signal. Values for the high and low corner frequencies are available from the values of τ_r and P using Eqs. (7–49) and (7–59). The following example illustrates the use of square-wave testing to determine the response of a network.

EXAMPLE 7–3 Calculate the rise time, percent tilt, and mid-band gain for the RC bandpass network shown in Fig. 7–10(b) if $R_i = R_g = 2$ K, $C_1 = 0.1\ \mu$F, and $C_i = 0.005\ \mu$F. The period of the input square wave is 0.1 msec. Using the values of τ_r and P, calculate the low and high corner frequencies of the network, and compare the results with those obtained in Ex. 7–2.

---

* The approximation arises because the voltage that accumulates on C_1 during the negative part of the wave produces some overshoot on the positive swing.

Solution: (a) Rise time: From Eqs. (7–44) and (7–47), the high-frequency time constant and the 10-to-90 percent rise time are

$$\tau_{\mathrm{hi}} = (R_i \,||\, R_g)C_i = (2\text{ K} \,||\, 2\text{ K})(0.005\ \mu\text{F}) \simeq 5\ \mu\text{sec}$$
$$\tau_r = 2.2\tau_{\mathrm{hi}} = (2.2)(5\text{ msec}) \simeq 11\ \mu\text{sec}$$

From Eq. (7–49), the corresponding high corner frequency is

$$f_{\mathrm{hi}} = 0.35/\tau_r = 0.35/11\ \mu\text{sec} = 32\text{ kHz}$$

(b) Tilt: From Eqs. (7–51) and (7–57), the low-frequency time constant and the tilt are

$$\tau_{\mathrm{lo}} = (R_i + R_g)C_1 = (2\text{ K} + 2\text{ K})(0.1\ \mu\text{F}) = 0.4\text{ msec}$$
$$P = (T/2\tau_{\mathrm{lo}}) \times 100\% = 12.5\%$$

From Eq. (7–59), the corresponding low corner frequency is

$$f_{\mathrm{lo}} = P/(\pi T) = 0.125/[\pi(0.1\text{ msec})] = 400\text{ Hz}$$

(c) Mid-band gain: From Eq. (7–42), the mid-band gain is

$$A_{\mathrm{mid}} = R_i/(R_i + R_g) = 2\text{ K}/(2\text{ K} + 2\text{ K}) = 0.5$$

(d) Experimental results: The measured results were

$$A_{\mathrm{mid}} = V_i/V_g = 2.0/4.4 \simeq 0.45 \qquad \tau_r = 12\ \mu\text{sec} \qquad P = 10\%$$

These values correspond closely with the calculated values.

7.6 Response of Inverting Amplifiers (CE, CS, CK)

The preceding five sections treated the fundamentals of frequency and time response analysis, and indicated that the gain, input impedance, and output impedance of an amplifier were functions of frequency. The gain is reduced in the low-frequency range by the effects of the coupling and bypass capacitors, and in the high-frequency range by the limitations of the active device and also by the parasitic capacitance. Both frequency and time response analyses are based on the calculation of the various *RC* time constants in the circuit. These time constants can be calculated using the ac equivalent circuit of an amplifier.

The frequency response of an amplifier is obtained by calculating the following quantities: (1) the bias conditions, (2) the mid-band gain, (3) the high-frequency response from the *RC* time constants associated with the input and output capacitances of the active device, and (4) the low-frequency response from the *RC* time constants associated with the coupling and bypass capacitors. Each low- and high-frequency time constant is associated with a corner frequency. The overall frequency response of an amplifier can be found by combining the corner frequencies on a Bode plot. This method of analysis can be applied to both single and multi-stage amplifiers and will be used extensively in the remainder of this chapter and also in the next chapter.

This procedure will give only approximate results due to the interaction between the various circuit time constants. However, the overall high and low corner frequencies of the amplifier will be reasonably accurate. The deviation from the approximate response due to the interactions of the various time constants is generally confined to frequencies above the overall high corner frequency and frequencies below the overall low corner frequency. The effects of interactions are treated in Chap. 8.

The objective of this section is to determine the response of single-stage inverting amplifiers: the CE, the CS, and the CK. Both the CS and the CE stages will be treated in detail. The CK case will be discussed by analogy with the other two types.

A CS amplifier is shown in Fig. 7–13(a) with the input impedance of an oscilloscope acting as a load. The input capacitance of the oscilloscope is included as part of the load impedance* because this stray capacitance will influence the high-frequency response. When analyzing an amplifier, all important stray capacitors should be included. In addition, the impedance of the generator, R_g. should be considered. The frequency response of an amplifier can be quite different under different test conditions.

The bias and mid-band analysis of the CS is discussed in Sec. 3.6 and will not be repeated here. The FET amplifier summary found in Table 3–1 can be used as a reference for the mid-band gain and impedance expressions. The complete high-frequency equivalent circuit of the amplifier is shown in Fig. 7–13(b). This equivalent circuit is obtained by considering the coupling and bypass capacitors and also the dc power supply as short circuits. In addition, the FET is replaced with its equivalent circuit shown in Fig. 7–3(b). Miller's theorem is used to divide the equivalent circuit into an input section as shown in Fig. 7–13(c). The effect of Miller's theorem on the input capacitance is given in Eq. (7–17), and the effect on the output capacitance is given in Eq. (7–25).

The high-frequency response of the CS amplifier is calculated using the time constants associated with the input capacitance and with the output capacitance. Referring to Fig. 7–13(c), the time constant associated with the input capacitance, C_i, is

$$\tau_{Ci} = [R_g \parallel R_G][C_{gs} + C_{gd}(1 - A_{vt})] = (R_g \parallel R_i)(C_i) \tag{7–60}$$

The corresponding high corner frequency of the input section is

$$f_{Ci} = 1/(2\pi\tau_{Ci}) \tag{7–61}$$

The time constant associated with the output capacitance, C_o, is

$$\tau_{Co} = [R_{Leq}][C_{gd}(1 - 1/A_{vt}) + C_L] \simeq (R_{Leq})(C_o + C_L) \tag{7–62}$$

The corresponding high corner frequency of the output section is

$$f_{Co} = 1/(2\pi\tau_{Co}) \tag{7–63}$$

If the two high corner frequencies are separated by as much as a factor of 10 (a decade in frequency), the lower of the two will determine the overall high-frequency response, and a Bode plot is not necessary. When this condition exists, the amplifier has a dominant high-frequency pole, or a dominant high corner frequency.

To determine the low-frequency response of the CS amplifier, we will calculate the time constant associated with each coupling and bypass capacitor. Since the circuit contains three capacitors, three low-frequency equivalent circuits can be drawn as shown in Figs. 7–13(d), (e), and (f). In each case, the equivalent resistance shunting the capacitor is found from the amplifier circuit in Fig. 7–13(a). For example, the resistance shunting the input coupling capacitor, C_1, is (R_g + R_i), where R_i is the resistive component of the input impedance, Z_i. For the CS

---

* The value of 70 pF is typical for a ×1 oscilloscope probe, and a value of 7 pF is typical of a ×10 probe.

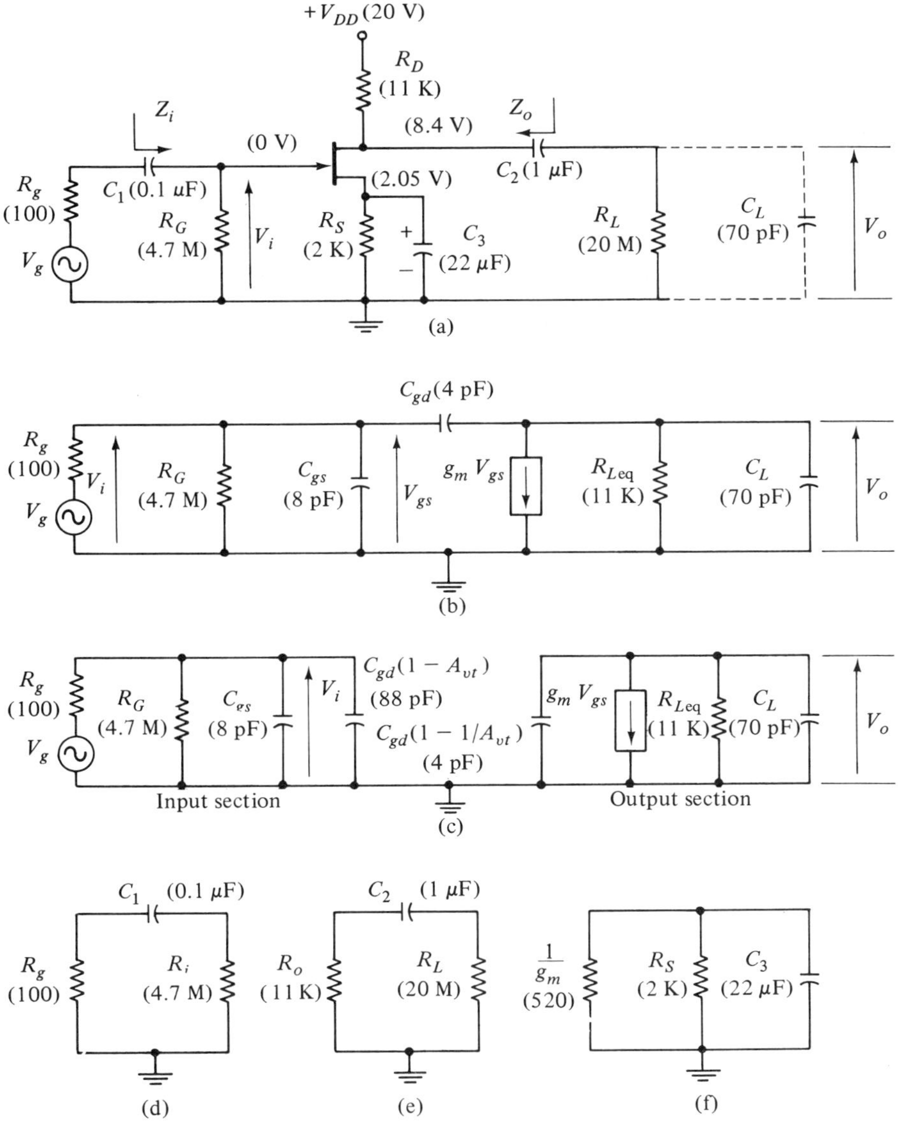

FIGURE 7–13 *A CS amplifier: (a) the complete circuit, (b) a complete high-frequency equivalent circuit, (c) the separation of the high-frequency equivalent circuit into an input section and an output section using Miller's theorem, (d) the low-frequency equivalent circuit for* C_1, *(e) the low-frequency equivalent circuit for* C_2, *and (f) the low-frequency equivalent circuit for* C_3. *(Values in parentheses apply to Ex. 7–4.)*

amplifier, R_i is approximately equal to R_G. From Figs. 7–13(d) and (e), the time constant and low corner frequencies associated with C_1 and C_2 are

$$\begin{aligned} \tau_{C1} &= (R_i + R_g)C_1 & f_{C1} &= 1/(2\pi\tau_{C1}) \\ \tau_{C2} &= (R_o + R_L)C_2 = (R_D + R_L)C_2 & f_{C2} &= 1/(2\pi\tau_{C2}) \end{aligned} \quad \textbf{(7–64)}$$

The low-frequency equivalent circuit for the source bypass capacitor, C_3, is shown in Fig. 7–13(f). When calculating the resistance shunting C_3, the resistance looking into the source terminal, $(1/g_m)$, must be included. The time constant and low corner frequency associated with C_3 are

$$\tau_{C3} = [(1/g_m) \parallel R_S]C_3 \qquad f_{C3} = 1/(2\pi\tau_{C3}) \tag{7–65}$$

If the highest low corner frequency is a factor of 10 greater than the next corner frequency, the highest one will determine the overall low corner frequency, making a Bode plot unnecessary. When this condition exists, the amplifier has a dominant low-frequency pole, or a dominant low corner frequency.

The analysis of the frequency response of the CS amplifier was divided into four parts: (1) bias conditions, (2) mid-band gain, (3) high-frequency response, and (4) low-frequency response.* A fifth step can be added if the time response is required. The following example illustrates the analysis of a CS amplifier.

EXAMPLE 7–4 For the CS amplifier shown in Fig. 7–13, calculate the bias conditions, mid-band gain, input and output impedance, and overall high and low corner frequencies. The N-channel JFET has values** of $V_P = -3$ V, $I_{DSS} = 8.4$ mA, $C_{gs} = 8$ pF (max), and $C_{gd} = 4$ pF (max), and is biased at $V_{GS} = -2$ V.

Solution: (a) Bias conditions:

$$I_D = I_{DSS}\left(1 - \frac{V_{GS}}{V_P}\right)^2 \simeq 1\text{ mA} \qquad V_D \simeq 9\text{ V}$$

(b) Mid-band gain and impedances:

$$g_m = \frac{2}{V_P}\sqrt{I_D I_{DSS}} \simeq 1.93\text{ millimhos} \qquad R_i \simeq R_G = 4.7\text{ M}$$

$$R_{Leq} \simeq R_D = 11\text{ K} \qquad A_{vt} = -g_m R_{Leq} \simeq -21$$

$$A_{vg} = R_i/(R_i + R_g) \simeq 1 \qquad A_{vo} = A_{vg}A_{vt} = (1)(-21) \simeq -21$$

(c) Input and output impedances: The resistive components are given in (b). The capacitive components are

$$C_i = C_{gs} + C_{gd}(1 - A_{vt}) \simeq [8 + 4(1 + 21)]\text{ pF} \simeq 96\text{ pF}$$

$$C_o = C_{gd}(1 - 1/A_{vt}) \simeq 4\text{ pF}$$

(d) High-frequency response: From Fig. 7–13(c) and Eqs. (7–60) through (7–63),

$$\tau_{Ci} = (R_g \parallel R_i)C_i \simeq 9.6\text{ nsec} \qquad f_{Ci} = 1/(2\pi\tau_{Ci}) \simeq 16.6\text{ MHz}$$

$$\tau_{Co} = R_{Leq}(C_o + C_L) \simeq 810\text{ nsec} \qquad f_{Co} = 1/(2\pi\tau_{Co}) \simeq 200\text{ kHz}$$

Since f_{Ci} is greater than $10f_{Co}$, the overall high corner frequency is approximately f_{Co} or 200 kHz.

(e) Low-frequency response: The effects of C_1 and C_2 are calculated using Figs. 7–13(d) and (e):

$$\tau_{C1} = (R_g + R_i)C_1 \simeq 0.47\text{ sec} \qquad f_{C1} = 1/(2\pi\tau_{C1}) \simeq 0.34\text{ Hz}$$

$$\tau_{C2} = (R_o + R_L)C_2 \simeq 20\text{ sec} \qquad f_{C2} = 1/(2\pi\tau_{C2}) \simeq 0.008\text{ Hz}$$

The effects of C_3 are calculated using Fig. 7–13(f):

$$\tau_{C3} = [(1/g_m) \parallel R_S](C_3) = (520 \parallel 2\text{ K})(22\ \mu\text{F}) \simeq 9.1\text{ msec}$$

$$f_{C3} = 1/(2\pi\tau_{C3}) = 0.159/9.1 \times 10^{-3} \simeq 17.5\text{ Hz}$$

---

* The overall transfer function for the CS amplifier can be formed by analogy to Eq. (7–35), Sec. 7.4, using the mid-band gain (A_{mid}), the three low corner frequencies (f_{C1}, f_{C2}, and f_{C3}), and the two high corner frequencies (f_{Ci} and f_{Co}).

** The values for V_P and I_{DSS} are measured values, and the values of C_{gs} and C_{gd} were obtained from the device specification sheet.

The amplifier has a dominant low corner frequency at f_{C3}. Hence, the overall low corner frequency, f_{lo}, is f_{C3} or 17.5 Hz.

(f) Experimental results: The measured bias voltages are shown in Fig. 7–13(a). The measured mid-band gain, A_{vg}, with $V_g = 0.15$ V was -20. Using a sine wave generator, the measured corner frequencies were

$$f_{\text{hi}} \simeq 240 \text{ kHz} \qquad f_{\text{lo}} \simeq 15 \text{ Hz}$$

Several observations can be made concerning the CS amplifier in Ex. 7–4. The amplifier is characterized by a dominant high corner frequency, f_{Co}, and a dominant low corner frequency, f_{C3}.* The high corner frequency of the input section occurred at a very high frequency (16.6 MHz) primarily due to the low value of generator resistance ($R_g = 100$ ohms). The capacitive component of the output section of the high-frequency equivalent circuit was dominated by the stray capacitance. If an oscilloscope probe with 7 pF stray capacitance (×10 probe) had been used, the output high corner frequency, f_{Co}, would have been approximately 1.3 MHz. Consequently, the high-frequency response of the CS amplifier was very dependent on the test conditions. The dominant low corner frequency was associated with C_3 even though C_3 was much larger than either C_1 or C_2. This condition occurred primarily due to the low resistance looking into the source terminal.

The response of a CE amplifier, as shown in Fig. 7–14(a), will now be considered. The bias and mid-band analysis of the CE stage is treated in Sec. 4.5 and will not be repeated here. The bipolar amplifier summary given in Table 4–1 can be used as a reference for the mid-band gain and impedance expressions. The complete high-frequency equivalent circuit is shown in Fig. 7–14(b) where the hybrid-π equivalent circuit is used to represent the transistor [see Fig. 7-4(a)]. Miller's theorem is used to divide the equivalent circuit into an input section and an output section. The low-frequency equivalent circuits associated with each coupling and bypass capacitor are shown in Figs. 7–14(d), (e), and (f).

A comparison between Fig. 7–14 for the CE and Fig. 7–13 for the CS shows that the low-frequency equivalent circuits have the same form. No basic differences exist between the CE and the CS with respect to the low-frequency analysis. The output sections of the high-frequency equivalent circuits of the CE and the CS also have the same form. The form of the input sections of the high-frequency equivalent circuits is different only by the inclusion of $r_{b'}$ in the CE hybrid-π equivalent circuit. Most modern bipolar transistors are fabricated by a diffusion process. Values of $r_{b'}$ typically range from 10 to 50 ohms for this type of transistor with the exact value being difficult to determine. Neglecting $r_{b'}$ introduces only a small error into the high-frequency analysis of the input section of the CE amplifier if R_g is much larger than $r_{b'}$. When $r_{b'}$ is neglected, the form of the input section of the CE has the same as that for the CS case.

This discussion indicates that the equivalent circuits used in the calculation of the frequency response of the CE and the CS amplifiers are of the same form.

---

* The transfer function of the CS amplifier in Ex. 7–4 can be written in terms of the dominant corner frequencies as

$$A_{vo} \simeq -21\left(\frac{jf/17.5\text{ Hz}}{1 + jf/17.5\text{ Hz}}\right)\left(\frac{1}{1 + jf/200\text{ kHz}}\right)$$

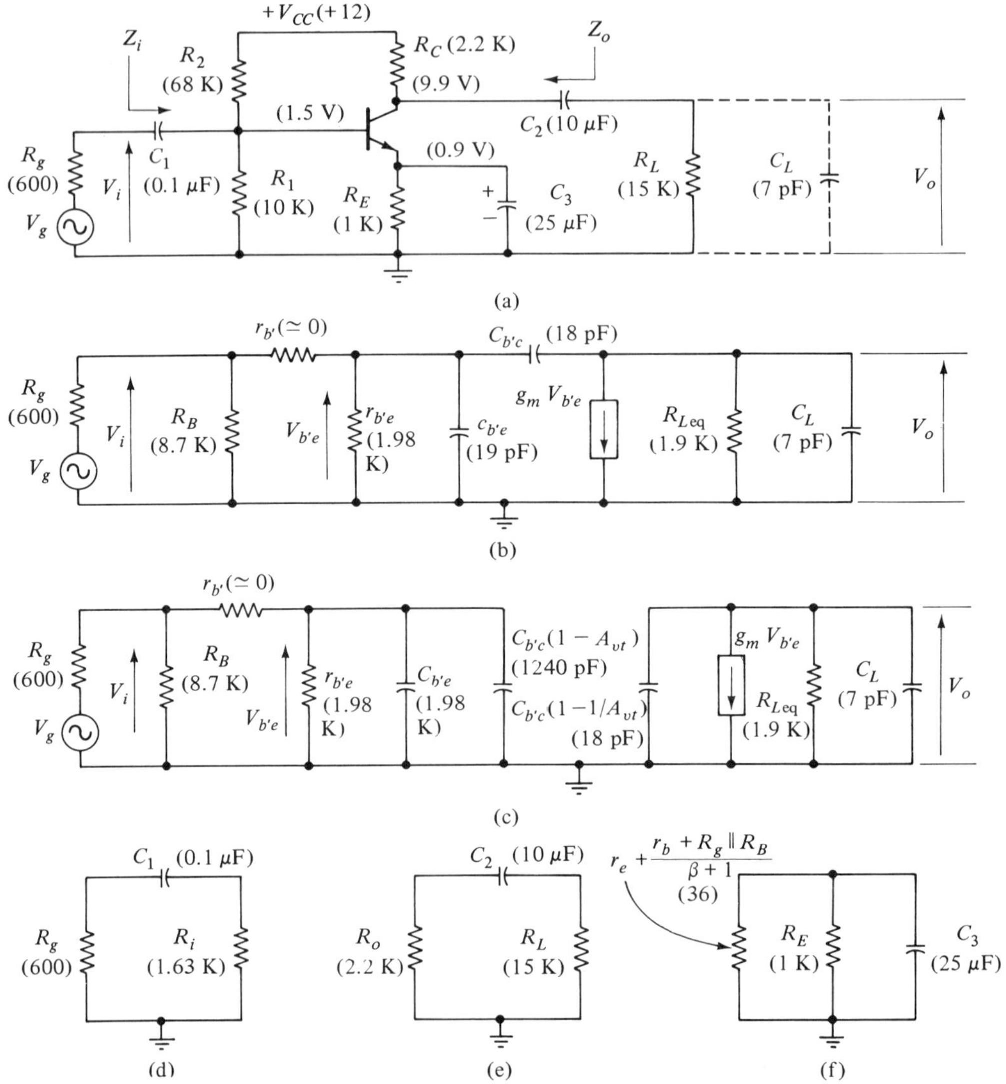

FIGURE 7–14 *A CE amplifier: (a) the complete circuit, (b) a complete high-frequency equivalent circuit, (c) the separation of the high-frequency equivalent circuit into an input section and an output section using Miller's theorem, (d) the low-frequency equivalent circuit for* C_1, *(e) the low-frequency equivalent circuit for* C_2, *and (f) the low-frequency equivalent circuit for* C_3. *(Values in parentheses apply to Ex. 7–5.)*

Consequently, the techniques used in the calculation of the frequency response of the two amplifiers are the same. Only the designation and the magnitude of the corresponding components are different. The following example illustrates an analysis of a CE amplifier as a function of frequency.

EXAMPLE 7–5 For the CE amplifier shown in Fig. 7–14, calculate the bias conditions, mid-band gain, input and output impedances, and overall high and low corner frequencies. The NPN silicon transistor has values* of $\beta = 70$, $f_T = 300$ MHz (typical), and $C_{ob} = 18$ pF.

Solution: (a) Bias conditions:

$$V_B = V_{CC}\left(\frac{R_1}{R_1 + R_2}\right) \simeq 1.5 \text{ V} \qquad I_E \simeq 0.9 \text{ mA} \qquad V_C \simeq 10 \text{ V}$$

(b) Mid-band gain and impedances:

$$r_e = 26 \text{ mV}/I_E \simeq 28\ \Omega \qquad R_o \simeq R_C = 2.2 \text{ K}$$

$$R_i = R_1 \,||\, R_2 \,||\, (r_{b'} + r_{b'e}) \simeq 68 \text{ K} \,||\, 10 \text{ K} \,||\, 1.96 \text{ K} \simeq 1.63 \text{ K}$$

$$A_{vt} = -\frac{A_i R_{\text{Leq}}}{R_{\text{in}}} \simeq -68 \qquad A_{vg} = R_i/(R_i + R_g) \simeq 0.73$$

$$A_{vo} = A_{vg}A_{vt} = (0.73)(-68) \simeq -49.5$$

(c) Input and output impedances: The resistive components are given in (b). The capacitive components are

$$C_{b'e} \simeq 1/(2\pi r_e f_T) \simeq 57 \text{ pF} \qquad C_o = C_{b'c}(1 - 1/A_{vt}) \simeq 18 \text{ pF}$$

$$C_i = C_{b'e} + C_{b'c}(1 - A_{vt}) = 57 + 18(1 + 68) \simeq 1299 \text{ pF}$$

(d) High-frequency response:

$$\tau_{Ci} = (R_g \,||\, R_i)C_i = 0.57\ \mu\text{sec} \qquad f_{Ci} = 1/(2\pi\tau_{Ci}) \simeq 280 \text{ kHz}$$

$$\tau_{Co} = R_{\text{Leq}}(C_o + C_s) = 47.5 \text{ nsec} \qquad f_{Co} = 1/(2\pi\tau_{Co}) \simeq 3.3 \text{ MHz}$$

Since f_{Co} is greater than $10f_{Ci}$, the overall high corner frequency is 280 kHz.

(e) Low-frequency response:

$$\tau_{C1} = (R_g + R_i)C_1 \simeq 0.22 \text{ msec} \qquad f_{C1} = 1/(2\pi\tau_{C1}) \simeq 720 \text{ Hz}$$

$$\tau_{C2} = (R_o + R_L)C_2 = 172 \text{ msec} \qquad f_{C2} = 1/(2\pi\tau_{C2}) \simeq 0.92 \text{ Hz}$$

$$\tau_{C3} = \left[R_E \,||\, \left(r_e + \frac{r_{b'} + R_1 \,||\, R_2 \,||\, R_g}{\beta + 1}\right)\right](C_3) \simeq 0.9 \text{ msec}$$

$$f_{C3} = 1/(2\pi\tau_{C3}) = 0.159/0.9 \times 10^{-3} \simeq 176 \text{ Hz}$$

The overall low-frequency response is due to both f_{C1} and f_{C3} since these two low corner frequencies are not separated by a factor of 10. A Bode plot can be used to determine the overall low corner frequency. However, since the circuit capacitors have a tolerance of ± 20 percent, the low corner frequency can be estimated as being greater than f_{C1} and less than $10f_{C3}$, or approximately 900 Hz.

(f) Experimental results: The measured bias voltages are shown in Fig. 7–14(a). The measured mid-band gain, A_{vo}, was -48 with V_g equal to 10 mV. The measured corner frequencies were

$$f_{\text{hi}} \simeq 300 \text{ kHz} \qquad f_{\text{lo}} \simeq 1 \text{ kHz}$$

A CK amplifier is shown in Fig. 7–15(a). The bias and mid-band analyses of the CK are treated in Chap. 6 and are not repeated here. The vacuum tube amplifier summary given in Table 6–2 may be used as a reference for the mid-band gain and impedance expressions. The complete high-frequency equivalent circuit for the amplifier is shown in Fig. 7–15(b) where the equivalent circuit shown in Fig. 7–4(c) is used to represent the pentode. Miller's theorem is used

---

* The values of β and C_{ob} were measured. The value of f_T was obtained from the device specification sheet. The value of C_{ob} can be estimated with reasonably accurate results for $C_{ob} > 10$ pF by using a Tektronix Type 575 curve tracer.

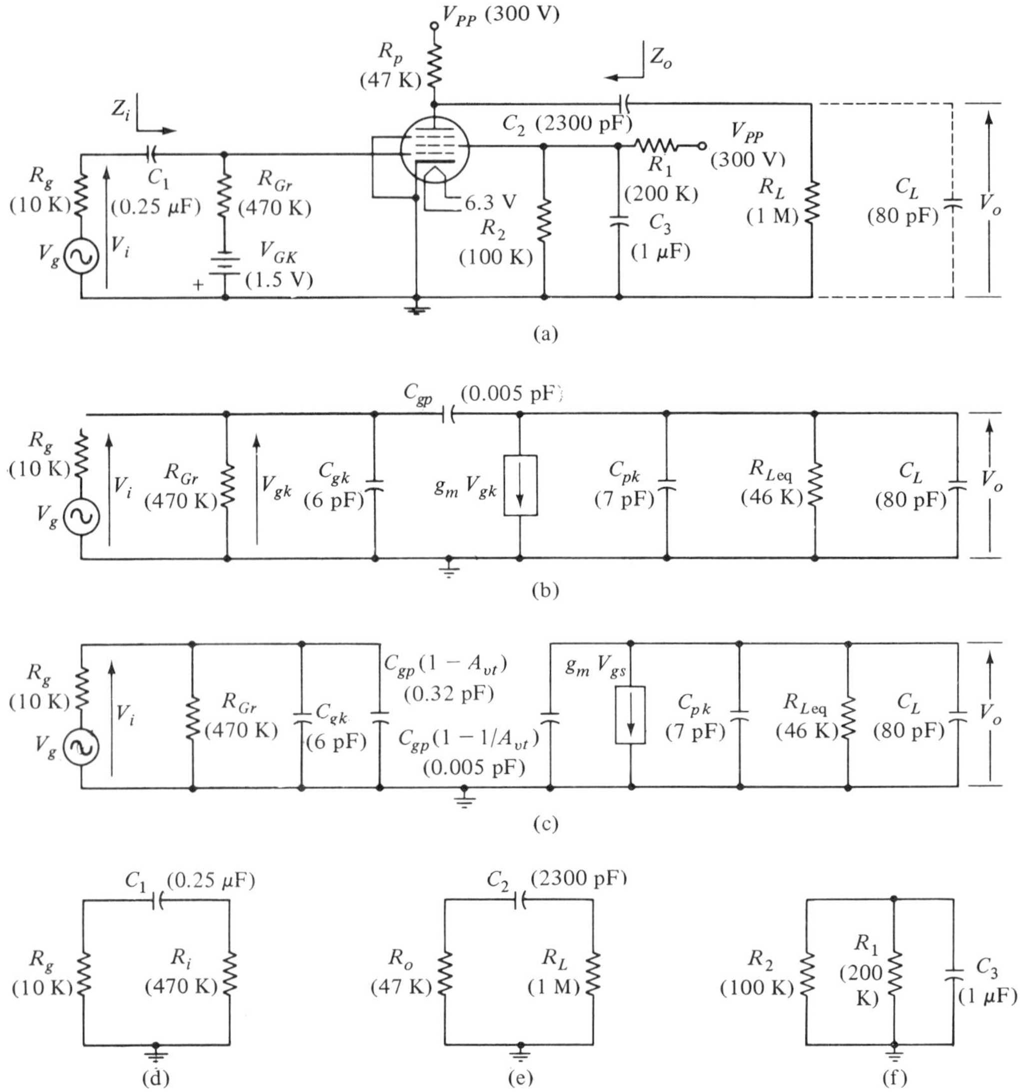

FIGURE 7–15 *A CK amplifier: (a) the complete circuit, (b) a complete high-frequency equivalent circuit, (c) the separation of the high-frequency equivalent circuit into an input section and an output section using Miller's theorem, (d) the low-frequency equivalent circuit for* C_1*, (e) the low-frequency equivalent circuit for* C_2*, and (f) the low-frequency equivalent circuit for* C_3*. (Values in parentheses apply to chapter problem.)*

to divide the equivalent circuit into an input section and an output section. The low-frequency equivalent circuits associated with each coupling and bypass capacitor are shown in Figs. 7–15(d), (e), and (f).

A comparison of the various equivalent circuits for the CK shown in Fig. 7–15 with the corresponding equivalent circuits for the CS shown in Fig. 7–13 indicates that there is no basic difference between the two equivalent circuits. Consequently, the frequency response analysis of the CK will be analogous to the analysis for the CS. The detailed analysis of the CK amplifier shown in

Fig. 7–15 is left as an exercise for the reader. Using a pentode characterized by $g_m = 1.3$ millimhos, $C_{gk} = 6$ pF, $r_p = 10$ M, $C_{gp} = 0.005$ pF, and $C_{pk} = 7$ pF, the measured gain at a frequency of 1 kHz was -62 with a grid voltage of 1 V. The high corner frequency was 36 kHz, and the low corner frequency was 80 Hz.

In this section we have determined the frequency response of inverting amplifiers. Comparing the results of Exs. 7–4 and 7–5, we find that the dominant high corner frequency occurred at the input for the CE and at the output for the CS. The dominant low corner frequency was associated with the input coupling capacitor, C_1, for the CE and with the bypass capacitor, C_3, for the CS. Hence, no general rule can be used for determining which corner frequency will dominate the response of an inverting amplifier. However, the method of analysis was similar for each type of active device.

7.7 Response of Low Input Impedance Amplifiers (CB, CG, CGr)

A second class of useful amplifiers are those with relatively low values of input impedance: the CB, CG, and CGr. The objective of this section is to determine the frequency and transient response of this class of amplifiers. The same techniques that were used for inverting amplifiers are also used in this study. However, the form of the high-frequency equivalent circuits is different.

The first task is to develop the CB hybrid-π equivalent circuit from the CE form of the hybrid-π. The resulting equivalent circuit can then be used to determine the frequency response of a single-stage CB amplifier. The CG and CGr high-frequency equivalent circuits will be developed by analogy with the CB equivalent circuit.

The CB hybrid-π equivalent circuit can be developed from the CE form as shown in Fig. 7–16. The first step is to redraw the CE hybrid-π, Fig. 7–16(a), with the base lead common, as shown in Fig. 7–16(b). Next, the current generator $g_m V_{b'e}$ is replaced by two current generators in series, each generator having the same value. The common point between the two generators is connected to the B' node. This reconnection does not change the equivalent circuit since any current flowing into node B' due to the right-hand current generator is removed from node B' by the left-hand generator. Consequently, the equivalent circuits shown in Figs. 7–16(a), (b), and (c) are exactly equivalent. The current generator connected between B' and E can be replaced by a resistor having a value of $1/g_m$ as shown in Fig. 7–16(d) since the current flowing in this generator depends directly on the voltage across the generator ($V_{b'e}$). Replacing the parallel combination of $1/g_m$ and $r_{b'e}$ by its equivalent resistance, r_e, completes the development of the CB hybrid-π equivalent circuit as shown in Fig. 7–16(e). For most modern bipolar transistors, $r_{b'}$ is so small that it can be neglected. The resulting simplified high-frequency equivalent circuit is shown in Fig. 7–16(f).

The expressions for the input and output impedance of the CB stage can be found from the CB equivalent circuit shown in Fig. 7–16(e) by assuming that $r_{b'}$ is negligible.* The resistive and capacitive components of the CB input and

---

* Neglecting $r_{b'}$ amounts to neglecting a pole and zero in the expression for the input impedance of the CB stage. The exact form of the input impedance expression is derived in Sec. 8.4.

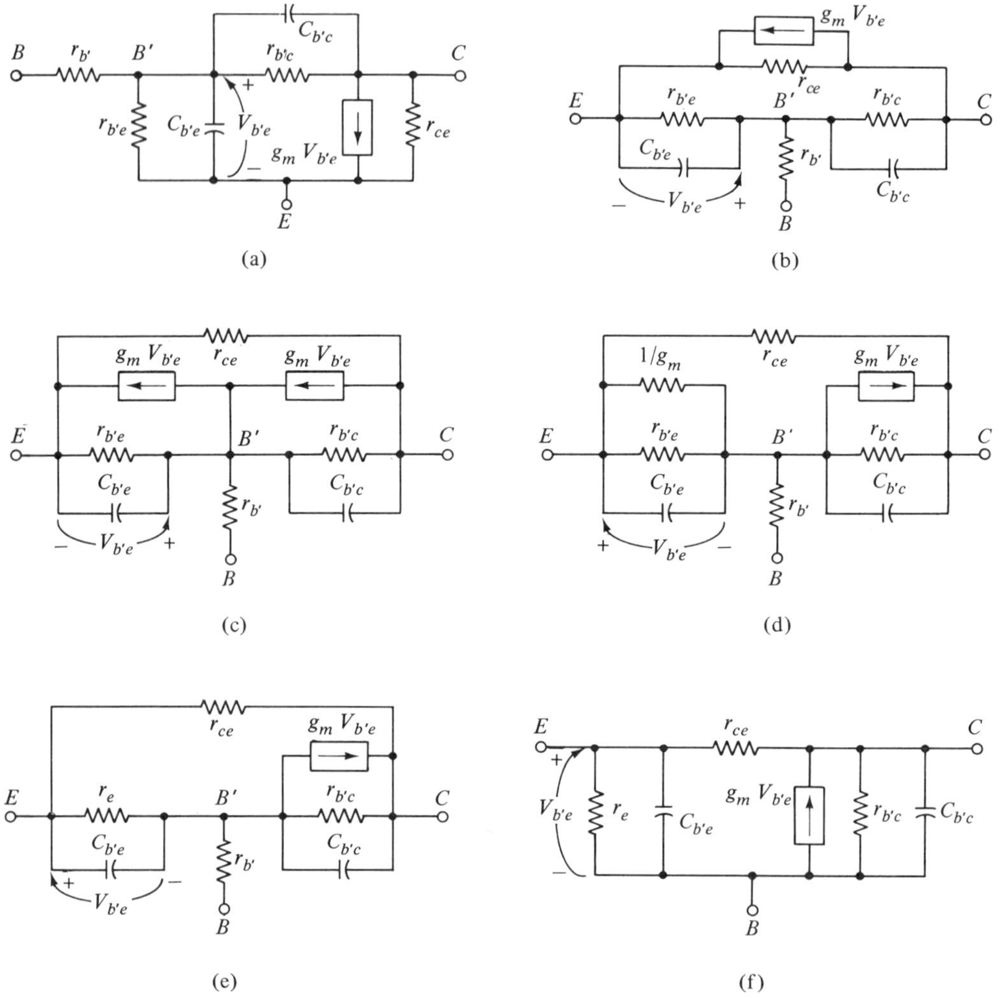

FIGURE 7–16 *Development of the CB hybrid-π equivalent circuit: (a) the CE hybrid-π, (b) the CE hybrid-π redrawn with the base lead common, (c) reconnection of the current generator,* $g_m V_{b'e}$, *(d) replacement of the input current generator with a resistor, (e) the complete CB hybrid-π equivalent circuit, and (f) a simplifier CB high-frequency equivalent circuit neglecting* $r_{b'}$.

output impedance are:

$$R_{\text{in(CB)}} \simeq r_e \qquad C_{\text{in(CB)}} \simeq C_{b'e}$$

$$R_{\text{out(CB)}} \simeq r_{cb} \simeq r_{b'c} \qquad C_{\text{out(CB)}} \simeq C_{b'c} \simeq C_{ob} \simeq C_{bc} \tag{7–66}$$

The expression for $R_{\text{out(CB)}}$ given in Eq. (7–66) may not be apparent from Fig. 7–16(f), since r_{ce} is also connected to the output terminal. However, a close examination of Fig. 7–16(f) indicates that any current flowing from the collector to the emitter terminal through r_{ce} also flows in r_e. The current in r_e develops a voltage $V_{b'e}$ which, in turn, causes a current $g_m V_{b'e}$ to flow into the collector terminal. This current cancels most of the original current flowing in r_{ce}. Consequently, very little net current flows from the collector back to the emitter so that r_{ce} has little effect on the CB output resistance.*

A CB amplifier is shown in Fig. 7–17(a). The bias and mid-band analysis of the CB stage was treated in Sec. 4.6 and is not repeated here. The complete high-frequency equivalent circuit for the amplifier is shown in Fig. 7–17(b). An approximate equivalent circuit of the amplifier which may be used in calculating the high-frequency response is shown in Fig. 7–17(c). The low-frequency response of the amplifier is calculated using the equivalent circuits associated with each coupling and bypass capacitor shown in Figs. 7–17(d), (e), and (f). The following example illustrates the procedure for determining the response of a CB amplifier as a function of frequency.

EXAMPLE 7–6 For the CB amplifier shown in Fig. 7–17, calculate the bias conditions, mid-band gain, input and output impedances, and overall high and low corner frequencies. The PNP silicon transistor has a β of 40, an f_T of 100 MHz, and a C_{ob} of 12 pF.

Solution: (a) Bias conditions:

$$V_B = V_{CC}\left(\frac{R_1}{R_1 + R_2}\right) \simeq -4.3\text{ V} \qquad I_E \simeq -1.7\text{ mA} \qquad V_C \simeq -10\text{ V}$$

(b) Mid-band gain and impedances:

$$r_e = \frac{26\text{ mV}}{I_E} \simeq 15\ \Omega \qquad R_i = R_E \,||\left(r_e + \frac{r_{b'}}{\beta + 1}\right) \simeq 15\ \Omega$$

$$R_o \simeq R_C = 8.25\text{ K} \qquad R_{\text{in}b} = r_{b'} + (\beta + 1)(r_e + R_E \,||\, R_g) \simeq 4.5\text{ K}$$

$$A_{vt} = -\frac{A_i R_{\text{Leq}}}{R_{\text{in}}} \simeq -\frac{(-1)(6.5\text{ K})}{15} \simeq 430 \qquad A_{vg} = R_i/(R_i + R_g) \simeq 0.13$$

$$A_{vo} = A_{vg}A_{vt} \simeq (0.13)(430) \simeq 56$$

(c) Input and output impedances: The resistive components are given in (b). The capacitive components are

$$C_i = C_{b'e} \simeq 1/(2\pi r_e f_T) \simeq 100\text{ pF} \qquad C_o \simeq C_{b'c} = 12\text{ pF}$$

(d) High-frequency response:

$$\tau_{Ci} = (R_g \,||\, R_i)C_i \simeq 1.3\text{ nsec} \qquad f_{Ci} = 1/(2\pi\tau_{Ci}) \simeq 122\text{ MHz} > f_T$$

$$\tau_{Co} = (R_{\text{Leq}})(C_o + C_L) \simeq 530\text{ nsec} \qquad f_{Co} = 1/(2\pi\tau_{Co}) \simeq 300\text{ kHz}$$

Since f_{Ci} is greater than $10f_{Co}$, the overall high corner frequency* and rise time values are

$$f_{\text{hi}} \simeq f_{Co} \simeq 300\text{ kHz} \qquad \tau_r = 0.35/f_{\text{hi}} \simeq 1.2\ \mu\text{sec}$$

(e) Low-frequency response:

$$\tau_{C1} = (R_g + R_i)C_1 \simeq 0.38\text{ msec} \qquad f_{C1} = 1/(2\pi\tau_{C1}) \simeq 420\text{ Hz}$$

$$\tau_{C2} = (R_o + R_L)(C_2) \simeq 18\text{ msec} \qquad f_{C2} = 1/(2\pi\tau_{C2}) \simeq 8.8\text{ Hz}$$

$$\tau_{C3} = (R_{\text{in}b} \,||\, R_1 \,||\, R_2)(C_3) \simeq 137\text{ msec} \qquad f_{C3} = 1/(2\pi\tau_{C3}) \simeq 1.2\text{ Hz}$$

---

* This condition is one in which the generator impedance, R_g, is high compared to r_e so that essentially all of the current flowing through r_{ce} also flows through r_e to produce $V_{b'e}$. If R_g is very low, then $V_{b'e}$ is low, and the cancellation effect is reduced. If $(R_g + r_{b'})$ is zero, $V_{b'e}$ is zero, and R_o is the parallel combination of r_{ce} and $r_{b'c}$.

* No significant interaction occurs between the high-frequency time constants of the CB stage.

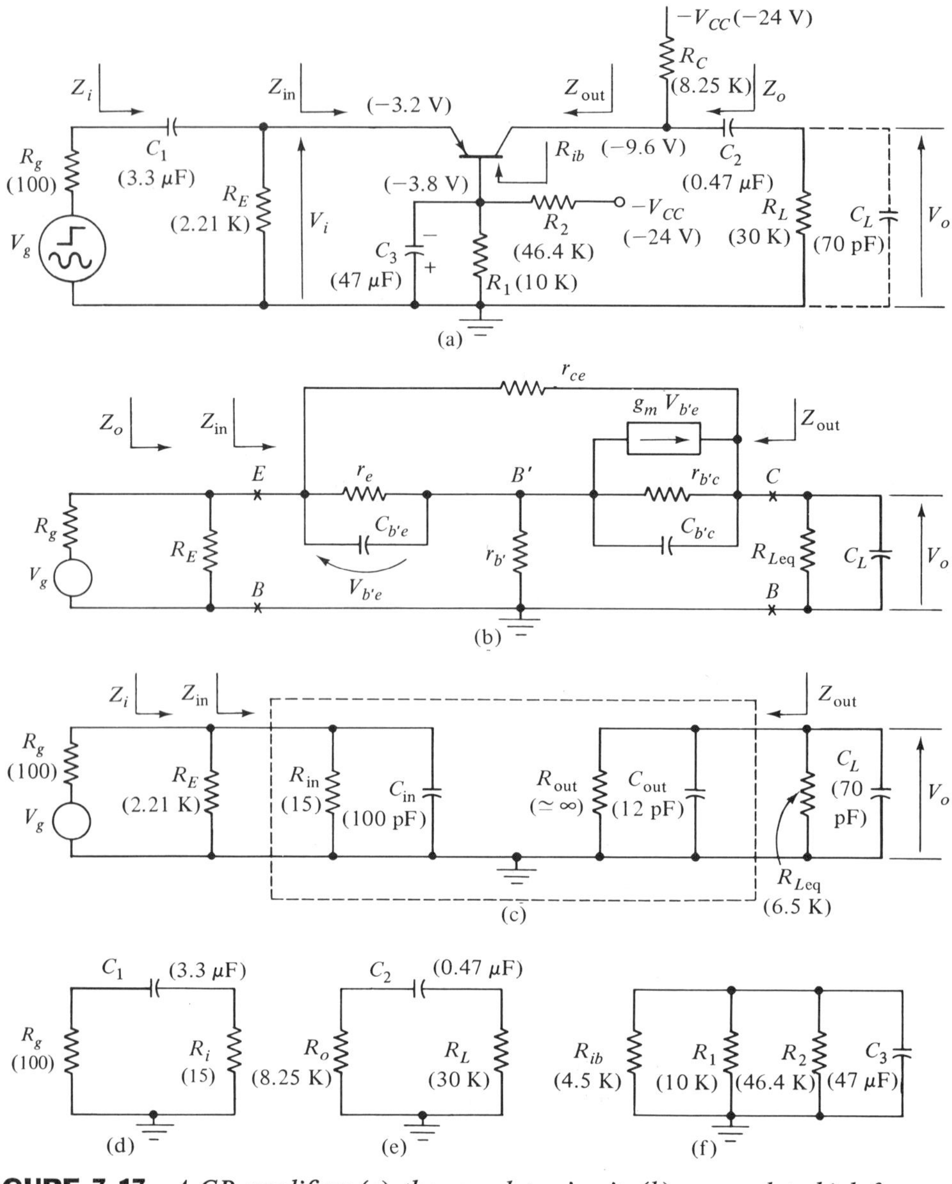

FIGURE 7–17 *A CB amplifier: (a) the complete circuit, (b) a complete high-frequency equivalent circuit, (c) an approximate high-frequency equivalent circuit showing the input and output impedances of the amplifier, (d) the low-frequency equivalent circuit for* C_1, *(e) the low-frequency equivalent circuit for* C_2, *and (f) the low-frequency equivalent circuit for* C_3. *(Values in parentheses apply to Ex. 7–6.)*

Since f_{C1} is the dominant low corner frequency, f_{lo} is equal to f_{C1} or 420 Hz.

(f) Experimental results: The measured bias voltages are shown in Fig. 7–17(a). The measured mid-band gain was

$$A_{vo} = V_o/V_g = 1.2\text{ V}/20\text{ mV} \simeq 60$$

Using a sine wave generator, the measured corner frequencies were 500 Hz and 320 kHz. The measured rise time was 1.1 μsec with a ×1 (70 pF) probe and 0.26 μsec with a ×10 (7 pF) probe.

Three important observations can be made concerning the frequency response of the CB amplifier. First, the high corner frequency of the input section of a CB stage is in the neighborhood of f_T, or $1/(2\pi r_e C_{b'e})$, since the time constant, τ_{Ci}, is $r_e C_{b'e}$ [from Fig. 7–17(c)]. The high corner frequency occurring at the emitter terminal of the CB stage can generally be nelgected in frequency response analyses.

Second, the high corner frequency occurring at the collector terminal of the common base stage is very dependent on the stray capacitance from the collector terminal-to-ground. This dependency is demonstrated in Ex. 7–6 by the different measured values for rise time for different values of stray capacitance. If the stray capacitance had been neglected, the calculated rise time would have been

$$\tau_r = 2.2\,\tau_{\text{hi}} = 2.2\,R_{L\text{eq}}C_{b'c} = 2.2\,(6.5\text{ K})(12\text{ pF}) \simeq 170\text{ nsec} \qquad \textbf{(7–67)}$$

which deviates significantly from the measured rise time.

Third, the bandwidth of a CB amplifier can extend into the very high frequency range if the transistor used has a high value for f_T and a low value for C_{ob} because the capacitive multiplication is not significant. Ex. 7–6 was repeated using a transistor with $f_T = 700$ MHz, $C_{ob} = 1.7$ pF, and $C_L = 7$ pF. Using these values, the calculated rise time was

$$\tau_r = 2.2\,R_{L\text{eq}}(C_{ob} + C_L) = 2.2\,(6.5\text{ K})[(1.7 + 7)\text{ pF}] = 124\text{ nsec} \qquad \textbf{(7–68)}$$

When a $\times 10$ (7 pF) probe was used, the measured rise time was 130 nsec.

The CB stage is the bipolar form of a low input impedance amplifier. The corresponding FET stage is the CG, and the corresponding vacuum tube stage is the CGr. A simplified high-frequency equivalent circuit for each of the three active devices connected as low input impedance amplifiers is shown in Fig. 7–18. The CB form, Fig. 7–18(a), is identical to Fig. 7–16(f) and is repeated for comparison. The CG high-frequency equivalent circuit can be developed from

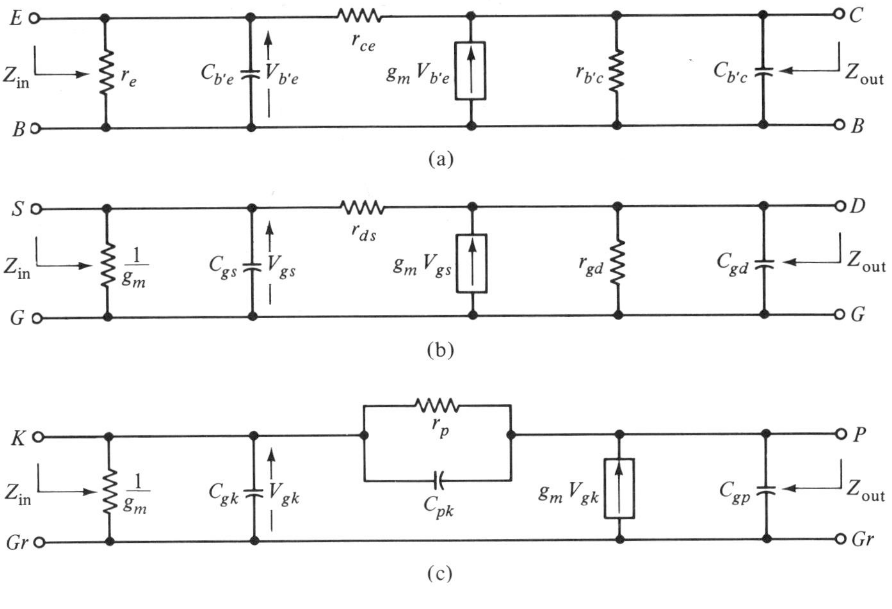

FIGURE 7–18 *Simplified high-frequency equivalent circuits for active devices connected as low input impedance amplifiers: (a) the CB, (b) the CG, and (c) the CGr.*

the CS high-frequency equivalent circuit in a manner analogous to the development of the CB hybrid-π from the CE hybrid-π. Also the CGr high-frequency equivalent circuit can be developed from the CK high-frequency equivalent circuit.

From Fig. 7–18(b), the resistive and capacitive components of the CG input and output impedance are:

$$R_{\text{in(CG)}} \simeq 1/g_m \qquad C_{\text{in(CG)}} \simeq C_{gs}$$

$$R_{\text{out(CG)}} \simeq r_{gd} \qquad C_{\text{out(CG)}} \simeq C_{gd} \qquad \textbf{(7–69)}$$

A CG amplifier is shown in Fig. 7–19(a). The bias and mid-band analysis of the CG stage is treated in Sec. 3.7 and is not repeated here. The high-frequency equivalent circuit for the amplifier can be drawn using the CG equivalent circuit shown in Fig. 7–18(b). However, a simplified circuit suitable for an approximate high-frequency analysis is shown in Fig. 7–19(b). In this circuit, the transistor has been replaced by a two-port network with input and output impedances equal to those of the transistor. Expressions for the resistive and capacitive components of both the input and output impedances are given in Eq. (7–69).

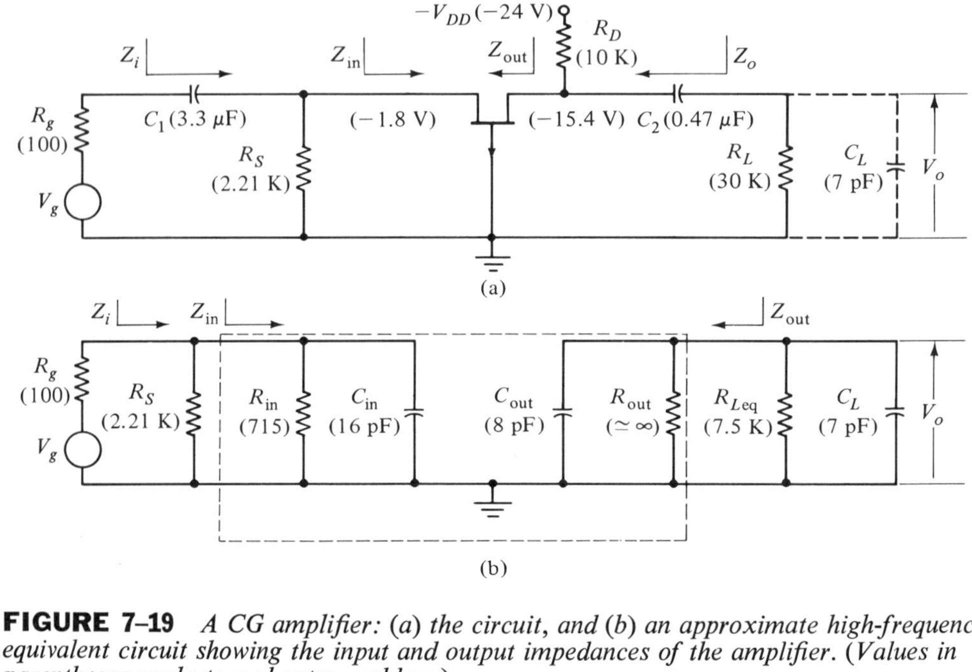

FIGURE 7–19 *A CG amplifier: (a) the circuit, and (b) an approximate high-frequency equivalent circuit showing the input and output impedances of the amplifier. (Values in parentheses apply to a chapter problem.)*

An analysis of the CG amplifier shown in Fig. 7–19 follows the same form used in the analysis of the CB stage. Details of such an analysis are left as an exercise for the reader. Using a JFET having a g_m of 1.4 millimhos, C_{gs} of 16 pF, and C_{gd} of 8 pF, the measured characteristics of the amplifier were as follows:

$$A_{vo} = 9.5 \qquad f_{\text{lo}} = 75\text{ Hz} \qquad \tau_r = 248\text{ nsec} \qquad \textbf{(7–70)}$$

The CGr high-frequency equivalent circuit is shown in Fig. 7–18(c). The similarity between the CGr and the other two high-frequency equivalent circuits should be noted. The principal difference between the CGr and either the CB or the CG is the presence of an input-to-output capacitor, C_{pk}, in the equivalent circuit. This capacitance occurs because the grid structure is not solid and also because the connecting wires to the tube elements are parallel. Miller's theorem can be applied to give an expression for the input and output capacitance. From Fig. 7–18(b), the resistive and capacitive components of the CGr input and output impedance are:*

$$R_{\text{in(CGr)}} \simeq 1/g_m \qquad C_{\text{in(CGr)}} \simeq C_{gk} + C_{pk}(1 - A_{vt})$$
$$R_{\text{out(CGr)}} \gg r_p \qquad C_{\text{out(CGr)}} \simeq C_{gp} + C_{pk}(1 - 1/A_{vt}) \tag{7–71}$$

The CGr amplifier is similar to both the CB and CG amplifiers. An approximate representation of the amplifier suitable for a high-frequency analysis is a two-port network whose input and output impedances are equal to those of the CGr stage as given in Eq. (7–71). The resulting equivalent circuit is similar to that shown for the CG in Fig. 7–19(b). An example of the CGr stage is included in Chap. 8 as part of a two-stage amplifier.

The results of this section indicate that the methods used in analyzing a low input impedance amplifier are very similar to those used for analyzing an inverting amplifier. The high-frequency response is governed by the *RC* time constants associated with the input section and output section of the amplifier. In general, the high corner frequency at the input terminal (emitter, source, or cathode) occurs at a very high frequency and can be neglected in the overall response. Stated differently, a dominant high corner frequency generally occurs at the output terminal (collector, drain, or plate) of a low input impedance amplifier. The low-frequency response was calculated using the time constant associated with each coupling and bypass capacitor in the circuit. Often the dominant low corner frequency is associated with the input coupling capacitor due to the low input resistance of this type of amplifier. The general characteristics of low input impedance amplifiers outlined above will be useful in the analysis of multistage amplifiers.

7.8 Response of Low Output Impedance Amplifiers (CC, CD, CP)

A third class of useful amplifiers are those with relatively low values of output impedance: the CC, CD, and CP. The objective of this section is to determine the frequency and transient response of these low output impedance amplifiers. The analysis of this class of amplifiers is identical to the analysis of the inverting and low input impedance amplifiers considered in the two preceding sections. However, the forms of the high-frequency equivalent circuits are different.

---

* Equation (7–71) indicates that a problem can arise due to the CGr input capacitance. Since the CGr amplifier is a non-inverting amplifier, A_{vt} is a positive number, and the value of $C_{\text{in(CGr)}}$ can be negative. A similar situation exists for the CB and CG stages where the input-to-output capacitance is due to the stray capacitance in the circuit. A negative input capacitance for any configuration can give rise to a right-half plane, high-frequency pole which can cause oscillation. While this situation can occur, it is complicated by the interaction between the various high-frequency time constants. A complete analysis, similar to the one used in Sec. 8.4, is often required to completely characterize the response of a low input impedance amplifier. The circuits of low input impedance amplifiers should be planned so that the input-to-output capacitance is at a minimum.

Consider the CC amplifier shown in Fig. 7–20(a). A standard form of the hybrid-π equivalent circuit is not necessary for the CC case because the CE hybrid-π can be adapted easily to this case. A complete high-frequency equivalent circuit for the CC stage is shown in Fig. 7–20(b) where the CE hybrid-π has been used to represent the transistor. Our immediate goal is to simplify the equivalent circuit into the form shown in Fig. 7–20(c). In this figure, the transistor is represented by a two-port network whose input and output impedance are equal to those of the transistor.*

The resistive component of the input impedance is given by Eq. (4–81) as

$$R_{\text{in(CC)}} = r_{b'} + (\beta + 1)(r_e + R_{L\text{eq}}) \tag{7–72}$$

To simplify the development of the expression for the input capacitance, neglect $r_{b'}$ and any stray output capacitance, C_L. The input capacitance consists of two components. First, from Fig. 7–20(b), $C_{b'c}$ appears in parallel with the input terminals since the collector terminal is grounded. Second, the effect of $C_{b'e}$ on the input capacitance can be obtained using Miller's theorem. From this theorem, the complete expression for the input capacitance becomes

$$C_{\text{in(CC)}} = C_{b'c} + C_{b'e}[1 - A_{vt\text{(CC)}}] \tag{7–73}$$

The component of the input capacitance due to $C_{b'e}$ is often designated as $C_{b'g}$, the equivalent capacitance from the terminal B' to ground. A simplified expression for $C_{b'g}$ can be obtained as follows. From Eq. (7–13), $C_{b'e}$ is $1/(2\pi r_e f_T)$, and from Eq. (4–83) $A_{vt\text{(CC)}}$ is $R_{L\text{eq}}/(r_e + R_{L\text{eq}})$. $C_{b'g}$ can now be expressed as

$$C_{b'g} = C_{b'e}[1 - A_{vt\text{(CC)}}] \simeq \frac{1}{2\pi(r_e + R_{L\text{eq}})f_T} \tag{7–74}$$

Equation (7–74) could have been anticipated since $R_{L\text{eq}}$ can be considered in series with r_e. The resistive component of the output impedance is given by Eq. (4–86) as

$$R_{\text{out(CC)}} \simeq r_e + \frac{r_{b'} + R_B \,||\, R_g}{\beta + 1} \tag{7–75}$$

An expression for the output capacitance of a CC stage is not readily apparent from the equivalent circuit shown in Fig. 7–20(b). However, an approximate expression for the output capacitance can be obtained by noting the analogy between the CC output high corner frequency and the CB input high corner frequency. Since the CC output and the CB input both occur at the emitter terminal, the two corner frequencies will be nearly the same; i.e., both occur in the neighborhood of f_T. Thus the RC time constant of the CC output impedance is controlled by f_T or

$$R_{\text{out(CC)}}\, C_{\text{out(CC)}} \simeq 1/(2\pi f_T) \tag{7–76}$$

Substituting Eq. (7–75) into Eq. (7–76) and solving for $C_{\text{out(CC)}}$ yields

$$C_{\text{out(CC)}} \simeq \frac{1}{2\pi\left(r_e + \dfrac{R_b}{\beta + 1}\right)f_T} \tag{7–77}$$

---

* The two-port representation for the CC stage that is developed in this section must be treated as very approximate since the interaction between the input and output time constants is neglected. The complete expressions for the input and output impedances of the CC stage are developed in Sec. 8.4.

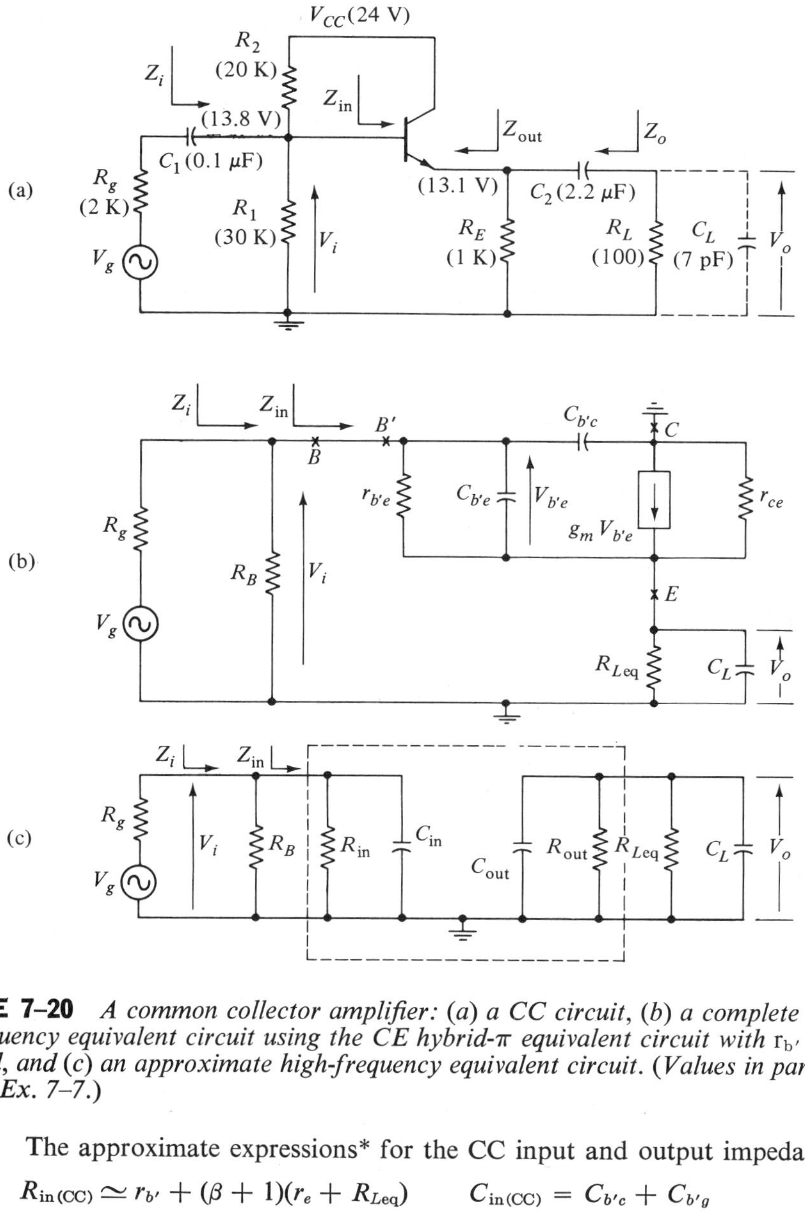

FIGURE 7–20 *A common collector amplifier: (a) a CC circuit, (b) a complete high-frequency equivalent circuit using the CE hybrid-π equivalent circuit with* $r_{b'}$ *neglected, and (c) an approximate high-frequency equivalent circuit. (Values in parentheses apply to Ex. 7–7.)*

The approximate expressions* for the CC input and output impedances are

$$R_{\text{in(CC)}} \simeq r_{b'} + (\beta + 1)(r_e + R_{L\text{eq}}) \qquad C_{\text{in(CC)}} = C_{b'c} + C_{b'g}$$

$$R_{\text{out(CC)}} \simeq r_e + \frac{r_{b'} + R_b}{\beta + 1} \qquad C_{\text{out(CC)}} \simeq \frac{1}{2\pi\left(r_e + \dfrac{R_b}{\beta + 1}\right) f_T} \tag{7–78}$$

The following example illustrates the use of the impedance expressions in the analysis of a CC amplifier.

EXAMPLE 7–7 For the CC amplifier shown in Fig. 7–20, calculate the bias conditions, mid-band gain, input and output impedances, and overall high and low corner

---

* The complete expressions for the input and output impedances of a CC stage are developed in Sec. 8.4.

frequencies. The NPN silicon transistor has a β of 150, an f_T of 50 MHz minimum, and a C_{ob} of 20 pF maximum.

Solution: (a) Bias conditions:

$$V_B \simeq 14.4\ \text{V} \qquad I_E \simeq 13.8\ \text{mA}$$

(b) Mid-band gain and impedances:

$$r_e = 26\ \text{mV}/I_E \simeq 2\ \Omega \qquad R_{Leq} = R_L \,||\, R_E \simeq 91\ \Omega$$

$$R_{in} = r_{b'} + (\beta + 1)(r_e + R_{Leq}) \simeq 14\ \text{K} \qquad R_i = R_1 \,||\, R_2 \,||\, R_{in} \simeq 6.5\ \text{K}$$

$$R_o = R_E \,||\left(r_e + \frac{r_{b'} + R_1 \,||\, R_2 \,||\, R_g}{\beta + 1}\right) \simeq 1\ \text{K} \,||\, (2 + 12) \simeq 14\ \Omega$$

$$A_{vt} = -\frac{A_i R_{Leq}}{R_{in}} \simeq \frac{R_{Leq}}{r_e + R_{Leq}} \simeq 0.99 \qquad A_{vg} = \frac{R_i}{R_i + R_g} \simeq 0.77$$

$$A_{vo} = A_{vg}A_{vt} \simeq 0.76$$

(c) Input and output impedances: The resistive components are given in (b). The capacitive components are

$$C_{b'g} = 1/[2\pi(r_e + R_{Leq})f_T] \simeq 35\ \text{pF} \qquad C_{in} \simeq C_{b'c} + C_{b'g} = 55\ \text{pF}$$

$$C_{out} \simeq 1 \bigg/ \left[2\pi\left(r_e + \frac{R_b}{\beta + 1}\right)f_T\right] \simeq 0.159/[(14)(50 \times 10^6)] \simeq 228\ \text{pF}$$

(d) High-frequency response:

$$\tau_{Ci} = (R_g \,||\, R_i)C_i \simeq 84\ \text{nsec} \qquad f_{Ci} = 1/(2\pi\tau_{Ci}) \simeq 1.9\ \text{MHz}$$

$$\tau_{Co} = (R_L \,||\, R_o)(C_o + C_L) \simeq 2.9\ \text{nsec} \qquad f_{Co} = 1/(2\pi\tau_{Co}) = 55\ \text{MHz}$$

Since $f_{Co} \gg f_{Ci}$, the overall high corner frequency, f_{hi}, is f_{Ci} or 1.9 MHz. The rise time, τ_r, is $0.35/f_{hi}$ or 185 nsec.

(e) Low-frequency response:

$$\tau_{C1} = (R_g + R_i)C_1 \simeq 0.85\ \text{msec} \qquad f_{C1} = 1/(2\pi\tau_{C1}) \simeq 187\ \text{Hz}$$

$$\tau_{C2} = (R_o + R_L)C_2 \simeq 0.25\ \text{msec} \qquad f_{C2} = 1/(2\pi\tau_{C2}) = 640\ \text{Hz}$$

Since neither of the low corner frequencies dominates the response, the overall low corner frequency for the amplifier may be obtained from a Bode plot. Alternately, the overall low corner frequency can be roughly approximated as the sum of the two low corner frequencies or

$$f_{lo} \simeq f_{C1} + f_{C2} = 187 + 640 \simeq 827\ \text{Hz}$$

(f) Experimental results: The measured bias voltages are shown in Fig. 7–20(a). The measured mid-band gain, A_{vo}, was 0.8 with an input of 1 V. The measured rise time, τ_r, was 170 nsec. A square wave with a period, T, of 30 μsec had an 8 percent tilt. The corresponding low frequency was

$$f_{lo} = P/\pi T \simeq 0.08/(\pi)(30 \times 10^{-6}) \simeq 850\ \text{Hz}$$

Two important observations can be made concerning the frequency response of the CC amplifier. First, the high corner frequency at the output or emitter terminal occurs at a very high frequency and can be neglected in frequency response analysis unless the load capacitor, C_L, is very large. Second, the bandwidth of a CC amplifier can extend into the high frequency range if the transistor has a high value for f_T and a low value for C_{ob}. The Miller effect reduces the value of $C_{b'e}$ to $C_{b'g}$. The reduced value of input capacitance increases the value of the input high corner frequency. A CC stage can often be used in a multistage amplifier to improve the overall high-frequency response with little loss in overall gain.

Consider the CD amplifier shown in Fig. 7–21(a). A complete high-frequency equivalent circuit for the CD stage is shown in Fig. 7–21(b) where the CS high-frequency equivalent circuit has been used to represent the FET. The similarity between the CD and the CC high-frequency equivalent circuits allows us to write the expression for the CD input capacitance by inspection as [see Eq. (7–73)]

$$C_{\text{in(CD)}} = C_{gd} + C_{gs}\,[1 - A_{vt(\text{CD})}] \tag{7–79}$$

From Table 3–1,

$$A_{vt(\text{CD})} \simeq \frac{g_m R_{L\text{eq}}}{1 + g_m R_{L\text{eq}}} \tag{7–80}$$

Substituting this expression for $A_{vt(\text{CD})}$ into Eq. (7–79),

$$C_{\text{in(CD)}} = C_{gd} + \frac{C_{gs}}{1 + g_m R_{L\text{eq}}} \tag{7–81}$$

The resistance looking into the gate terminal of an FET is very large and can be approximated as infinitely large.

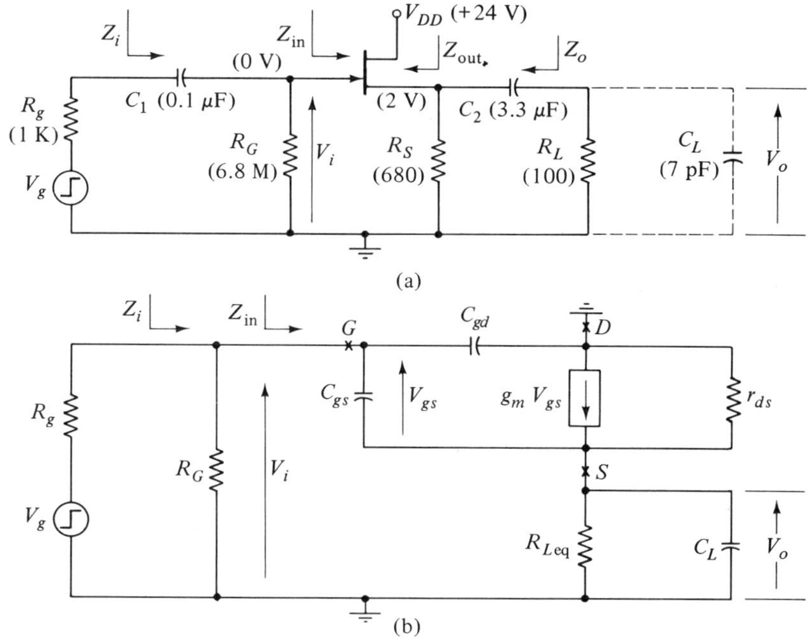

FIGURE 7–21 *A CD amplifier: (a) the complete circuit, and (b) a complete high-frequency equivalent circuit using the CS high-frequency equivalent circuit to represent the transistor. (Values in parentheses refer to a chapter problem.)*

Since the output of the CD and the input of the CG both occur at the source terminal, the input impedance of the CG stage can be used to approximate the output impedance of the CD stage.* Consequently, from Eq. (7–69),

* The exact expression for the output impedance of a CD stage is given in Sec. 8.4.

$$R_{\text{out(CD)}} \simeq R_{\text{in(CG)}} \simeq 1/g_m$$
$$C_{\text{out(CD)}} \simeq C_{\text{in(CG)}} \simeq C_{gs} \qquad \textbf{(7–82)}$$

The approximate expression for the resistive and capacitive components of the CD input and output impedance can be summarized as follows:

$$\left.\begin{array}{ll} R_{\text{in(CD)}} \simeq \infty & R_{\text{out(CD)}} \simeq 1/g_m \\ C_{\text{in(CD)}} \simeq C_{gd} + C_{gs}/(1 + g_m R_{L\text{eq}}) & C_{\text{out(CD)}} \simeq C_{gs} \end{array}\right\} \qquad \textbf{(7–83)}$$

The complete analysis of the CD amplifier shown in Fig. 7–21 is left as an exercise for the reader. Using a JFET characterized by a V_p of -4 V, an I_{DSS} of 12.3 mA, a C_{gs} of 6 pF maximum, a C_{gd} of 2 pF maximum, and biased at V_{GS} equal -2 V, the measured characteristics of the amplifier were as follows:

$$A_{vo} = 0.22 \qquad f_{\text{lo}} = 237 \text{ Hz} \qquad \tau_r = 19 \text{ nsec} \qquad \textbf{(7–84)}$$

A third low output impedance amplifier is the CP amplifier. Since the approximate high-frequency equivalent circuits of the FET and the vacuum tube are very similar (see Fig. 7–3), the approximate expression for the input and output impedances of the CP stage will be similar to those for the CD stage. The only difference is that the capacitor C_{pk} also appears across the output terminal (cathode-to-ground) of the CP stage. By analogy with Eq. (7–83), the resistive and capacitive components of the input and output impedances of the CP stage can be written as

$$\begin{array}{ll} R_{\text{in(CP)}} \simeq \infty & R_{\text{out(CP)}} \simeq 1/g_m \\ C_{\text{in(CP)}} \simeq C_{gp} + C_{gk}/(1 + g_m R_{L\text{eq}}) & C_{\text{out(CP)}} \simeq C_{pk} + C_{gk} \end{array} \qquad \textbf{(7–85)}$$

The results of this section indicate that, in general, the high corner frequency at the output terminal (emitter, source, or cathode) occurs at a very high frequency and can often be neglected in the overall response. Stated differently, a dominant high corner frequency generally occurs at the input terminal (base, gate, or grid) of a low output impedance amplifier. Often the dominant low corner frequency is associated with the output coupling capacitor due to the low output resistance of this type of amplifier. The general characteristics of the low output impedance amplifiers outlined above will be useful in the analysis and design of multistage amplifiers.

7.9 Summary

In this chapter, the techniques for making an approximate frequency response analysis of single-stage amplifiers have been developed. Electronic amplifiers are frequency dependent in the low-frequency region because of the presence of circuit coupling and bypass capacitors. In the high-frequency region, the active devices usually limit the response of an amplifier.

The approximate analysis of an amplifier is accomplished by calculating the following: (1) the bias conditions, (2) the mid-band gain, (3) the high-frequency response from the input and output time constants, and (4) the low-frequency response from the time constants associated with each coupling and bypass capacitor. A fifth step can be added if the time response is required.

Usually one corner frequency dominates the response at each end of the band. However, if one frequency is not dominant, the overall response of the amplifier can be determined from a Bode plot.

TABLE 7–1 *Summary of the approximate expressions for the resistive and capacitive components of the input and output impedances of single-stage amplifiers*

Configuration		Input Resistance (R_{in})	Input Capacitance (C_{in})	Output Resistance (R_{out})	Output Capacitance (C_{out})
Bipolar	CE	$r_{b'} + (\beta + 1)r_e$	$C_{b'e} + C_{b'c}(1 - A_{vt})$	r_{ce}	$C_{b'c}$
	CB	$r_e + r_{b'}/(\beta + 1)$	$C_{b'e}$	$r_{b'c}$	$C_{b'c}$
	CC	$r_{b'} + (\beta + 1)(r_e + R_{Leq})$	$C_{b'c} + C_{b'g}$	$r_e + \frac{R_B}{\beta + 1}$	$\frac{1}{2\pi[r_e + R_b/(\beta + 1)]f_T}$
FET	CS	∞	$C_{gs} + C_{gd}(1 - A_{vt})$	r_{ds}	C_{gd}
	CG	$1/g_m$	C_{gs}	r_{ds}	C_{gd}
	CD	∞	$C_{gd} + C_{gs}/(1 + g_m R_{Leq})$	$1/g_m$	C_{gs}
Vacuum Tube	CK	∞	$C_{gk} + C_{gp}(1 - A_{vt})$	r_p	$C_{pk} + C_{gp}$
	CGr	$1/g_m$	$C_{gk} + C_{pk}(1 - A_{vt})$	r_p	$C_{gp} + C_{pk}$
	CP	∞	$C_{gp} + C_{gk}/(1 + g_m R_{Leq})$	$1/g_m$	$C_{pk} + C_{gk}$

The frequency and time response of an amplifier are both controlled by the same time constants. Consequently, these two responses are closely related. The product of f_{hi} and τ_r is 0.35. The low-frequency response, f_{lo}, is equal to the tilt, P, divided by (πT).

The analyses presented in this chapter do not include the effects of time constant interaction. Consequently, the solutions must be treated as approximate. However, these approximate solutions are generally adequate for determining the overall high and low corner frequency of an amplifier. Deviations between the approximate solutions and a more accurate solution generally occur at frequencies above the overall high corner frequency and below the overall low corner frequency. The effects of time constant interaction are discussed in the next chapter.

The high-frequency response of an amplifier depends to a large degree on the input and output impedances of the particular configuration. Table 7–1 provides a convenient summary of these impedances as well as a means of comparing the various configurations. Inverting amplifiers (CE, CS, and CK) have a high gain, but also have a high value of input capacitance due to the Miller effect. The high-frequency response of the low input impedance amplifiers (CB, CG, and CGr) is generally governed by the output high corner frequency, and the high-frequency response of low output impedance amplifiers (CC, CD, and CP) is generally limited by the input high corner frequency. Stated differently, the high corner frequency at a low resistance terminal (emitter, source, or grid) will generally occur at a very high frequency and can be neglected in the overall response of an amplifier.

The techniques discussed in this chapter, although approximate, are very useful in the design and analysis of practical amplifiers and are extended to multistage amplifiers in the next chapter.

Problems

Section 7.1

7–1 A cutoff frequency can be defined as the frequency where $|R|$ is equal to $|X|$. At the cutoff frequency, what are the values of (a) $|Z|$ and (b) $\underline{/Z}$ (phase of Z) if $Z = R \,||\, X$ and $X = 1/(j\omega C)$?

7–2 In Fig. 7–2(b), what is the expression for the equivalent resistance in parallel with C_1 if C_3 is assumed to be a short circuit?

7–3 In Fig. 7–2(d), what are the expressions for (a) the equivalent resistance, and (b) the equivalent capacitance from the output node to ground if C_{gs} is assumed to be a short circuit?

Section 7.2

7–4 An NPN transistor is biased at $I_C = 2$ mA and is characterized by $\beta_o = 200$, $f_T = 200$ MHz, and $C_{ob} = 10$ pF. Calculate the frequency at which the magnitude of $r_{b'e}$ is equal to the magnitude of the reactance of $C_{b'e}$.

7–5 A PNP transistor is biased at $I_C = 1.3$ mA and is characterized by $\beta_o = 100$, $f_T = 400$ MHz, and $C_{cb} = 2$ pF. Calculate the $|\beta|$ in db at (a) 0.4 MHz, (b) 2 MHz, (c) 4 MHz, (d) 8 MHz, and (e) 40 MHz.

7–6 The results of testing a certain bipolar transistor indicate that $|h_{fe}| = 6$ at a frequency of 100 MHz when biased at $I_C = 1$ mA. Estimate the values of (a) f_T and (b) $C_{b'e}$.

7–7 A silicon NPN transistor has the following characteristics when biased at $I_C = 1$ mA and $V_{CE} = 10$ V: $h_{ie} = 2.7$ K, $h_{re} = 10^{-4}$, $h_{fe} = 99$, $h_{oe} = 10^{-5}$ mhos, $f_T = 300$ MHz, and $C_{ob} = 10$ pF. Sketch the CE hybrid-π equivalent circuit, and calculate the value of every component.

7–8 A P-channel JFET is characterized by $V_P = 5$ V, $I_{DSS} = -10$ mA, $C_{iss} = 10$ pF, and $C_{rss} = 4$ pF, and is biased at $V_{GS} = 1$ V. Sketch the high-frequency equivalent circuit, and calculate or estimate the value of each component.

7–9 A P-channel MOSFET (enhancement mode) is characterized by $V_T = -6$ V, $I_{DO} = -3.2$ mA, $C_{iss} = 16$ pF, and $C_{rss} = 6$ pF, and is biased at $V_{GS} = -10$ V. Sketch the high-frequency equivalent circuit, and calculate or estimate the value of each component.

7–10 What effect will the lead inductance of the emitter circuit have on the value of $V_{b'e}$ of a bipolar transistor at high frequency?

Section 7.3

7–11 Verify Eq. (7–25) for $C_{o(\text{CE})}$.

7–12 Calculate the resistive and capacitive components of Z_i for the CE amplifier shown in Fig. 7–22. The NPN silicon transistor has values of $\beta = 100$, $f_T = 100$ MHz, and $C_{ob} = 10$ pF.

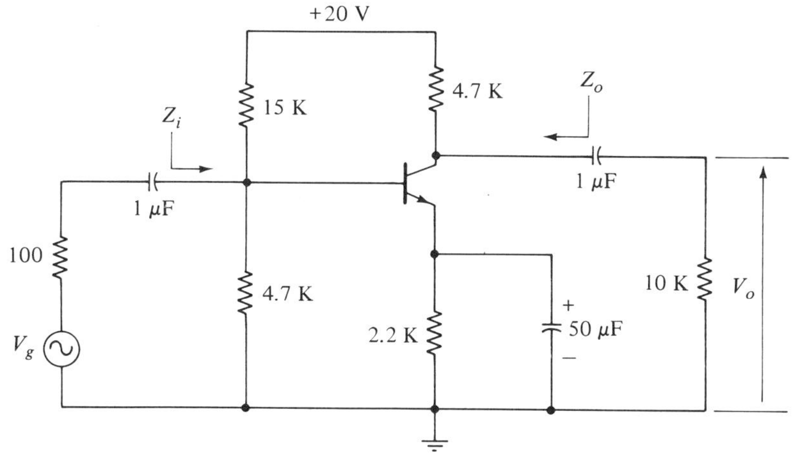

FIGURE 7–22 *Circuit for Probs. 7–12 and 7–16.*

7–13 Using Miller's theorem, calculate the approximate input resistance and capacitance for the network shown in Fig. 7–23. Assume that the amplifier has an infinite input resistance and negligible input capacitance.

7–14 (a) Sketch the output waveform of the network shown in Fig. 7–24 for the time interval 0 to 1 sec. (b) What property of the input signal does the output signal

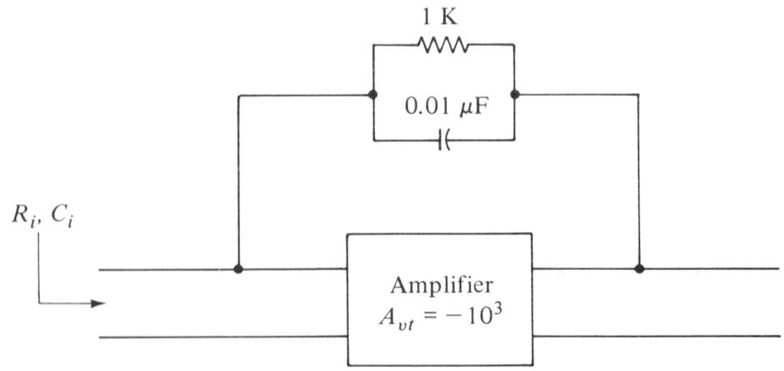

FIGURE 7–23 *Network for Prob. 7–13.*

represent? Assume that the amplifier has an infinite input resistance and negligible input capacitance.

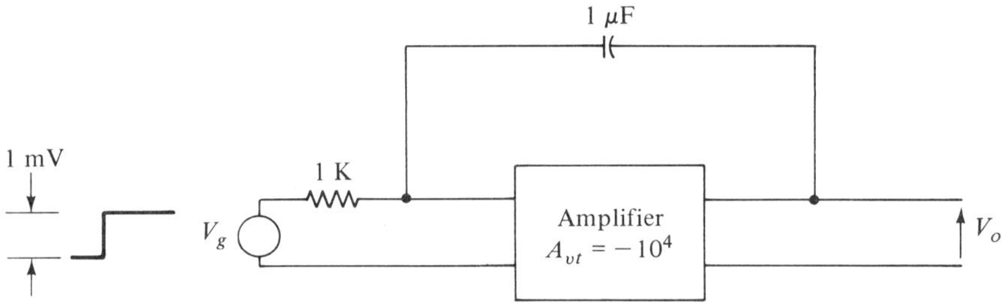

FIGURE 7–24 *Network for Prob. 7–14.*

Section 7.4

7–15 An *RC* network similar to the one shown in Fig. 7–7(b) has values of $R_i = 470$, $R_g = 4.7$ K, $C_1 = 1\ \mu$F, and $C_i = 100$ pF. Calculate the mid-band voltage gain, f_{hi}, and f_{lo}.

7–16 (a) Sketch the *RC* bandpass network which represents the input section of the CE amplifier shown in Fig. 7–22. (b) Calculate the values for $A_{mid} = A_{vg}$, f_{lo}, and f_{hi} for the input section obtained in (a).

7–17 Express the transfer function of the *RC* network in Ex. 7–2 in the form of Eq. (7–35).

Section 7.5

7–18 A 1-V, 200-Hz square wave is applied to the *RC* network described in Prob. 7–15. Sketch the output voltage waveform.

7–19 Calculate the rise time of the amplifier input section obtained in Prob. 7–16.

7–20 A particular circuit has a measured tilt of 10 percent for an input square wave frequency of 100 Hz. Estimate the low corner frequency of this circuit.

Section 7.6

7–21 Calculate the frequency response of the CS amplifier shown in Fig. 7–5. The transistor data are given in Ex. 7–1, and $C_L = 7$ pF.

7–22 A silicon PNP transistor is used in a CE amplifier similar to the one shown in Fig. 7–14(a). The circuit values are: $R_g = 1$ K, $R_1 = 4.7$ K, $R_2 = 10$ K, $R_E = 2.2$

K, $R_C = R_L = 3.3$ K, $C_1 = C_2 = 0.1\ \mu$F, $C_E = 30\ \mu$F, and $V_{CC} = -15$ V. The transistor has values of $\beta = 100$, $C_{ob} = 10$ pF, and $f_T = 300$ MHz. Calculate the mid-band gain, rise time, and low corner frequency of the amplifier. (*Note:* the mid-band analysis of this amplifier was determined in Prob. 4–22.)

7–23 Calculate the frequency response of the CK amplifier shown in Fig. 7–15. The pentode is characterized by $g_m = 1.3$ millimhos, $C_{gk} = 6$ pF, $r_p = 10$ M, $C_{pk} = 7$ pF, and $C_{gp} = 0.005$ pF. The experimental results are given in the text.

7–24 Design an inverting amplifier with the following characteristics: $|A_v| > 10$, $R_i > 5$ M, $R_o > 2$ K, $f_{\text{hi}} > 200$ kHz, and $f_{\text{lo}} < 200$ Hz. The amplifier is to be used in an application where $R_g = 500$ ohms, and $R_L = 10$ K. Use standard component values.

7–25 Design an inverting amplifier with the following characteristics: $|A_v| > 100$, $R_i > 2$ K, $R_o > 5$ K, $f_{\text{hi}} > 300$ kHz, and $f_{\text{lo}} < 100$ Hz. The amplifier is to be used in an application where $R_g = 600$ ohms, and $R_L = 10$ K. Use standard component values.

Section 7.7

7–26 Show that the parallel combination of $r_{b'e}$ and $(1/g_m)$ is equal to r_e.

7–27 Starting with the CS high-frequency equivalent circuit shown in Fig. 7–3(b), generate the CG high-frequency equivalent circuit shown in Fig. 7–18(b).

7–28 A silicon NPN transistor is used in a CB amplifier similar to the one shown in Fig. 7–17(a). The circuit values are: $R_g = 600$ ohms, $R_E = R_1 = R_C = R_L = 4.7$ K, $R_2 = 8.2$ K, $C_1 = C_2 = 1\ \mu$F, $C_3 = 10\ \mu$F, and $V_{CC} = 15$ V. The transistor has values of $\beta = 90$, $C_{ob} = 12$ pF, and $f_T = 200$ MHz. Calculate the mid-band gain, rise time, and low corner frequency of the amplifier. Assume $C_L = 7$ pF.

7–29 Calculate the gain and frequency responses of the CG amplifier shown in Fig. 7–19.

7–30 Design a low input impedance amplifier with the following characteristics: $|A_v| > 50$, $R_i \simeq 25$ ohms, $R_o > 5$ K, $\tau_r < 100$ nsec, and $f_{\text{lo}} < 500$ Hz. The amplifier is to be used in an application where $R_g = 25$ ohms, $R_L = 10$ K, and $C_L = 7$ pF. Use standard component values.

Section 7.8

7–31 A silicon PNP transistor is used in a CC amplifier similar to the one shown in Fig. 7–20(a). The circuit values are: $R_g = 1$ K, $R_1 = 47$ K, $R_2 = 82$ K, $R_E = 3.3$ K, $R_L = 470$, $C_1 = C_2 = 1\ \mu$F, $C_L = 7$ pF, and $V_{CC} = -18$ V. The transistor has values of $\beta = 300$, $C_{ob} = 8$ pF, and $f_T = 290$ MHz. Calculate the mid-band gain, rise time, and low corner frequency.

7–32 The CC amplifier in Fig. 7–20(a) is modified so that $R_E = 4.7$ K, and a resistor, $R_C = 1$ K, is connected between the collector terminal and V_{CC}. For this modified circuit, calculate the mid-band gain, input capacitance, and high corner frequency.

7–33 Calculate the gain and frequency responses of the CD amplifier shown in Fig. 7–21.

7–34 Design a low output impedance amplifier with the following characteristics: $|A_v| > 0.8$, $R_i > 50$ K, $R_o < 50$ ohms, $\tau_r < 100$ nsec, and $f_{\text{lo}} < 500$ Hz. The amplifier is to be used in an application where $R_g = 5$ K, $R_L = 100$ ohms, and $C_L = 10$ pF. Use standard component values.

References

Abraham, R. R., "Transistor Behavior at High Frequencies," *Trans. IRE*, ED-7, No. 1 (1960), 59–70.

Cobbold, Richard S. C., *Theory and Applications of Field-Effect Transistors*. New York: John Wiley & Sons, Inc., 1970.

Hoffman, T. R., "Effect of Emitter Bypass Capacitance on the Frequency Response of a Grounded-Emitter Transistor Amplifier Stage," *Semiconductor Prods.* (July, 1970), 27.

Lindmayer, J. and C. Y. Wrigley, *Fundamentals of Semiconductor Devices*. New York: D. Van Nostrand Company, Inc., 1965.

Pierce, J. F., *Transistor Circuit Theory and Design*. Columbus, Ohio: Charles E. Merrill Publishing Company, 1963.

Salaman, R. G., "Designing Transistorized Video Amplifiers, I.," *Electronic Inds.*, 18, No. 5 (May, 1959), 79–83.

Searle, C. L., A. R. Boothroyd, E. J. Angelo, Jr., P. E. Gray, and D. O. Pederson, *Elementary Circuit Properties of Transistors*. New York: John Wiley & Sons, Inc., 1964.

Terman, Frederick E., *Electronic and Radio Engineering*. New York: McGraw-Hill Book Company, 1955.

Thomas, E. E., "Some Design Considerations for High-Frequency Transistor Amplifiers," *The Bell System Technical Journal*, 38, No. 6 (November, 1959), 1551–80.

Thornton, R. D., C. L. Searle, D. O. Pederson, R. B. Adler, and E. J. Angelo, Jr., *Multistage Transistor Circuits*. New York: John Wiley & Sons, Inc., 1965.

8

Frequency and Transient Response of Multistage Amplifiers

The characteristics of three types of single-stage amplifiers were discussed in the previous chapter. In practice, more than one stage is generally required to produce the necessary gain at the desired impedance levels. The resulting coupling circuits between stages often present special problems. The objective of this chapter is to extend the techniques developed in the previous chapter to the analysis and design of multistage amplifiers.

In the first section, the techniques developed in Chap. 7 are extended to multistage amplifiers. Seven useful multistage circuits are analyzed to illustrate typical problems in discrete circuits. Special problems created by integrated circuits are treated in Sec. 8.2. With this background, the design approach to multistage amplifiers is discussed in Sec. 8.3. Interactions between incompletely isolated *RC* circuits are considered in Sec. 8.4.

8.1 Some Useful Multistage Amplifier Circuits

As indicated in the preceding chapter, the low-frequency response of a single-stage amplifier can be determined from the *RC* time constants associated with the coupling and bypass capacitors in the circuit. Each low-frequency time constant is associated with a low corner frequency. The overall low corner frequency of an amplifier is determined from the dominant low corner frequency or from a Bode plot. The low-frequency analysis of a multistage amplifier is similar to that for a single-stage amplifier and is complicated only by the possible presence of additional coupling and bypass capacitors.

The high-frequency analysis of multistage amplifiers is more complicated than that for single-stage circuits because of the presence of additional circuit nodes. For example, a two-stage amplifier has three principal nodes:* one at the input of the first stage, one between the two stages, and one at the output of the second stage. These nodes are designated as the input node, the interstage node, and the output node. The interstage node was not present in a single-stage circuit. As indicated in Chap. 7, the value of the high-frequency time constant associated with a particular node is the product of the equivalent resistance and capacitance from that node to ground. The overall high corner frequency of a multistage amplifier can be found from the dominant high corner frequency or from a Bode plot.

The remainder of this section consists of seven examples of multistage amplifier frequency response analysis. These seven amplifiers are: CE-CE, CS-CS, CE-CB, CS-CB, CK-CGr, CC-CE, and the bipolar differential amplifier. Four of the examples include a detailed analysis while in the other three, only the circuits that determine the dominant corner frequencies are considered. A discussion accompanies each example.

EXAMPLE 8–1 Calculate the frequency response of the CE-CE amplifier shown in Fig. 8–1. Both transistors are silicon and are characterized by

Transistor	β	$f_{T\,(\text{min})}$	$C_{ob\,(\text{max})}$
Q_1	100	350 MHz	5 pF
Q_2	35	100 MHz	20 pF

Solution: (a) The bias and mid-band characteristics were evaluated in Ex. 4–15. The important results required for the frequency-response analysis are:

$$r_{e1} = 24\,\Omega \qquad R_o = 2.2\text{ K} \qquad A_{vt1} = -17.3$$
$$r_{e2} = 12\,\Omega \qquad R_{Leq1} = 415\,\Omega \qquad A_{vt2} = -146$$
$$R_i = 2\text{ K} \qquad R_{Leq2} = 1.8\text{ K} \qquad A_{vo} = 2460$$

(b) High-frequency analysis: The high-frequency response depends on three RC time constants, one associated with each principal node in the amplifier as shown in Fig. 8–1(b). The values of these time constants depend on the values of the input and output capacitances of each stage. Using Table 7–1,

$$C_{\text{in1}} = C_{b'e1} + C_{b'c1}(1 - A_{vt1}) = [18 + 5(1 + 17.3)]\text{ pF} \simeq 110\text{ pF}$$
$$C_{\text{out1}} \simeq C_{b'c1} = 5\text{ pF} \qquad C_{\text{out2}} \simeq C_{b'c2} = 20\text{ pF}$$
$$C_{\text{in2}} = C_{b'e2} + C_{b'c2}(1 - A_{vt2}) = [132 + 20(1 + 146)]\text{ pF} \simeq 3000\text{ pF}$$

The three high-frequency time constants are

$$\tau_{\text{in}} = (R_i \,||\, R_g)(C_{\text{in1}}) = (2\text{ K} \,||\, 50)(110\text{ pF}) \simeq 5.5\text{ nsec}$$
$$\tau_{\text{int}} = (R_{\text{out1}} \,||\, R_{C1} \,||\, R_{\text{in2}})(C_{\text{out1}} + C_{\text{in2}}) = (R_{Leq1})(C_{\text{out1}} + C_{\text{in2}}) \simeq 1.25\,\mu\text{sec}$$
$$\tau_{\text{out}} = (R_{\text{out2}} \,||\, R_L)(C_{\text{out2}} + C_L) = (R_{Leq2})(C_{\text{out2}} + C_L) \simeq 49\text{ nsec}$$

Since τ_{int} is the dominant time constant, the overall rise time and high corner frequency values are:

$$\tau_r \simeq 2.2\tau_{\text{int}} \simeq 2.75\,\mu\text{sec} \qquad f_{\text{hi}} = 0.35/\tau_r \simeq 128\text{ kHz}$$

---

* The node internal to the hybrid-π equivalent circuit, B', is not included in the principal nodes of the amplifier.

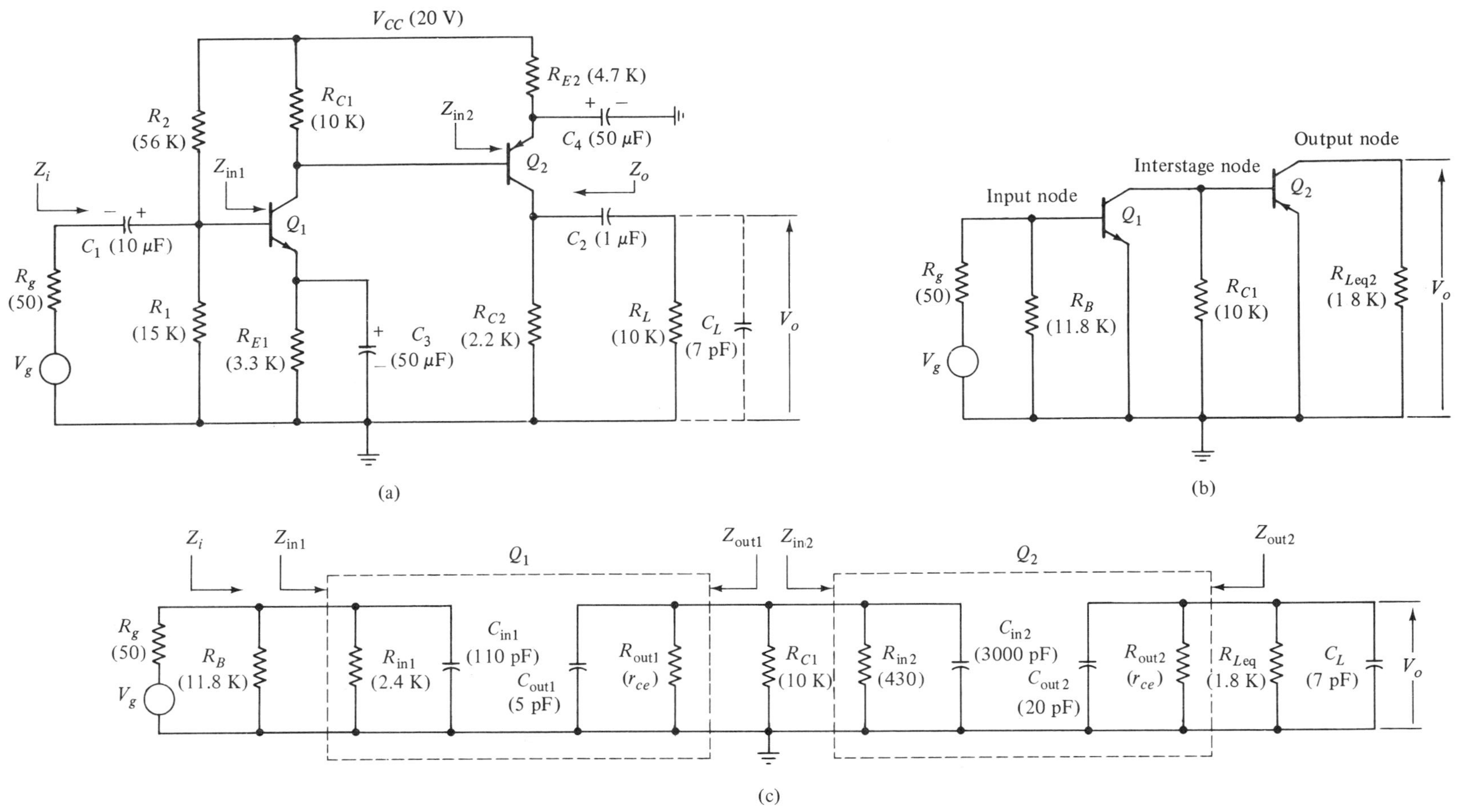

FIGURE 8–1 *The CE-CE from Ex. 8–1: (a) the circuit diagram, (b) a mid-band equivalent circuit, and (c) a high-frequency equivalent circuit showing the input and output impedance of each stage.*

(c) Low-frequency response: The low-frequency response depends on four *RC* time constants, one associated with each circuit coupling and bypass capacitor. From Fig. 8–1(a),

$$\tau_{C1} = (R_g + R_i)C_1 \simeq 20 \text{ msec} \qquad \tau_{C2} = (R_o + R_L)C_2 \simeq 12.2 \text{ msec}$$

$$\tau_{C3} = \left[R_{E1} \,||\, \left(r_{e1} + \frac{R_B \,||\, R_g}{\beta_1 + 1}\right)\right] C_3 \simeq 1.25 \text{ msec}$$

$$\tau_{C4} = \left[R_{E2} \,||\, \left(r_{e2} + \frac{R_{C1}}{\beta_2 + 1}\right)\right] C_4 = [4.7\text{ K} \,||\, (12 + 10\text{ K}/36)](50\ \mu\text{F}) \simeq 14.5 \text{ msec}$$

Since τ_{C3} is the dominant low-frequency time constant, the overall low corner frequency of the amplifier is

$$f_{lo} \simeq 1/(2\pi\tau_{C3}) = 0.159/(1.25 \times 10^{-3}) \simeq 128 \text{ Hz}$$

(d) Experimental results: The measured bias voltages were given in Ex. 4–15. The measured gain and frequency responses were

$$A_{vo} \simeq \frac{1.2 \text{ V}}{0.5 \text{ mV}} \simeq 2400 \qquad \tau_r \simeq 1.5\ \mu\text{sec} \qquad f_{lo} \simeq 105 \text{ Hz}$$

The measured rise time is less than the calculated value of 2.75 μsec. This type of discrepancy often results when the high-frequency analysis is based on the minimum value of f_T and the maximum value of C_{ob} that are given on a specification sheet. If the value of C_{ob2} is assumed as 10 pF rather than 20 pF, the calculated rise time becomes 1.37 μsec.

The results of Ex. 8–1 are typical of the response of an amplifier consisting of two inverting stages. The interstage node often dominates the high-frequency response of such an amplifier because both the resistance and the capacitance to ground are large. Compared to the interstage node, the input node usually has a lower resistance to ground (due to R_g), and the output node has a lower capacitance to ground. Either condition results in a smaller *RC* product and, consequently, a higher corner frequency.* The following example illustrates the same type of amplifier using JFETs as the active devices.

EXAMPLE 8–2 Calculate the mid-band gain and the frequency response of the CS-CS amplifier shown in Fig. 8–2. The transistor characteristics are:

Transistor	V_P	I_{DSS}	V_{GS}	$C_{gs(\max)}$	$C_{gd(\max)}$
Q_1	−3 V	6 mA	−0.8 V	20 pF	6 pF
Q_2	−4.2 V	14 mA	−1 V	8 pF	4 pF

Solution: (a) Bias conditions: Within the accuracy of the circuit parameters, the drain current can be calculated from the ratio V_{GS}/R_S yielding

$$I_{D1} \simeq 3.8 \text{ mA} \qquad I_{D2} \simeq 10.4 \text{ mA}$$

(b) Mid-band gain and impedances:

$$g_{m1} \simeq 3.2 \text{ millimhos} \qquad g_{m2} \simeq 5.75 \text{ millimhos}$$

$$A_{vt1} = -g_{m1}R_{Leq1} = -7.7 \qquad A_{vt2} = -g_{m2}R_{Leq2} \simeq -4.7$$

$$A_{vg} \simeq 1 \qquad A_{vo} = A_{vg}A_{vt1}A_{vt2} \simeq 36$$

---

* The interaction between the three high-frequency time constants increases the dominance of the interstage pole. The effects of this interaction are discussed in Sec. 8.4.

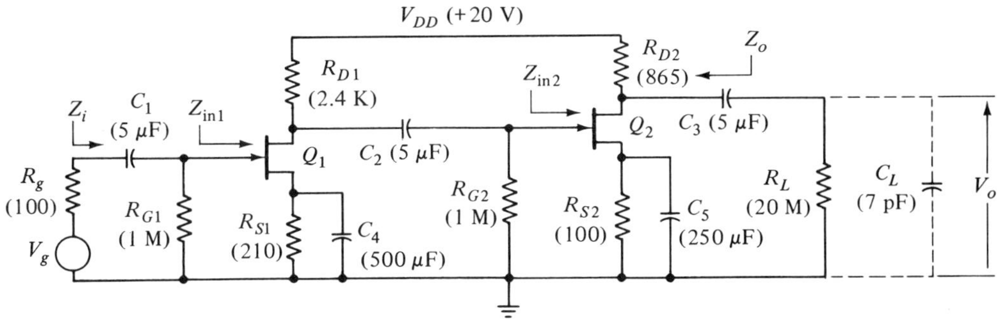

FIGURE 8–2 *The CS-CS amplifier from Ex. 8–2.*

(c) High-frequency response: A detailed analysis shows that the interstage node dominates the response. For the interstage high corner frequency,

$$C_{\text{out1}} \simeq C_{gd1} = 6\text{ pF}$$

$$C_{\text{in2}} = C_{gs2} + C_{gd2}(1 - A_{vt2}) = [8 + 4(1 + 4.7)]\text{ pF} \simeq 31\text{ pF}$$

$$\tau_{\text{int}} = (R_{L\text{eq1}})(C_{\text{out1}} + C_{\text{in2}}) \simeq 89\text{ nsec} \qquad f_{\text{hi}} = 1/(2\pi\tau_{\text{int}}) \simeq 1.8\text{ MHz}$$

(d) Low-frequency response: A detailed analysis shows that the dominant low-frequency time constant is associated with C_5.

$$\tau_{C5} = \left(R_{S2} \,\|\, \frac{1}{g_{m2}}\right)(C_5) \simeq 15.8\text{ msec} \qquad f_{\text{lo}} \simeq 1/(2\pi\tau_{C5}) \simeq 10\text{ Hz}$$

(e) Experimental results:

$$A_{vo} = V_o/V_g \simeq 6.4\text{ V}/0.2\text{ V} \simeq 32 \qquad f_{\text{hi}} \simeq 1.75\text{ MHz} \qquad f_{\text{lo}} \simeq 8.4\text{ Hz}$$

The high-frequency response of an inverting amplifier is usually limited by its large input capacitance. For example, a CE stage with a large voltage gain, A_{vt}, has a large input capacitance because $C_{b'c}$ is multiplied by $(1 - A_{vt})$. An amplifier with an improved high-frequency response and approximately the same gain as a CE stage can be obtained with a two-stage amplifier consisting of an inverting stage followed by a low input impedance stage (such as a CE-CB amplifier). This circuit is called a *cascode* amplifier and is illustrated in the next three examples. An alternate approach is to drive an inverting stage with a low output impedance stage. The CC-CE amplifier is an example of this configuration and is discussed in Ex. 8–6.

EXAMPLE 8–3 Calculate the frequency response of the CE-CB cascode amplifier shown in Fig. 8–3. Both transistors are silicon and are characterized by

Transistor	I_E	β	$f_{T(\text{min})}$	$C_{ob(\text{max})}$
Q_1	1 mA	40	100 MHz	12 pF
Q_2	1 mA	38	100 MHz	12 pF

Solution: (a) Mid-band gain:

$$r_{e1} \simeq r_{e2} \simeq 26\text{ mV}/1\text{ mA} \simeq 26\ \Omega \qquad R_{\text{in1}} = r_{b'} + (\beta_1 + 1)(r_{e1}) \simeq 1\text{ K}$$

$$R_i = R_1 \,\|\, R_2 \,\|\, R_{\text{in1}} \simeq 360\ \Omega \qquad R_o \simeq R_{C2} = 4.7\text{ K}$$

$$A_{vt1} = -\frac{A_{vi}\,R_{L\text{eq1}}}{R_{\text{in1}}} \simeq -\frac{(\beta_1)(r_{e2})}{(\beta_1 + 1)(r_{e1})} \simeq -1 \qquad A_{vt2} = -\frac{A_{i2}R_{L\text{eq2}}}{R_{\text{in2}}} \simeq 90$$

$$A_{vg} = R_i/(R_i + R_g) \simeq 0.78 \qquad A_{vo} = A_{vg}A_{vt1}A_{vt2} \simeq -70.4$$

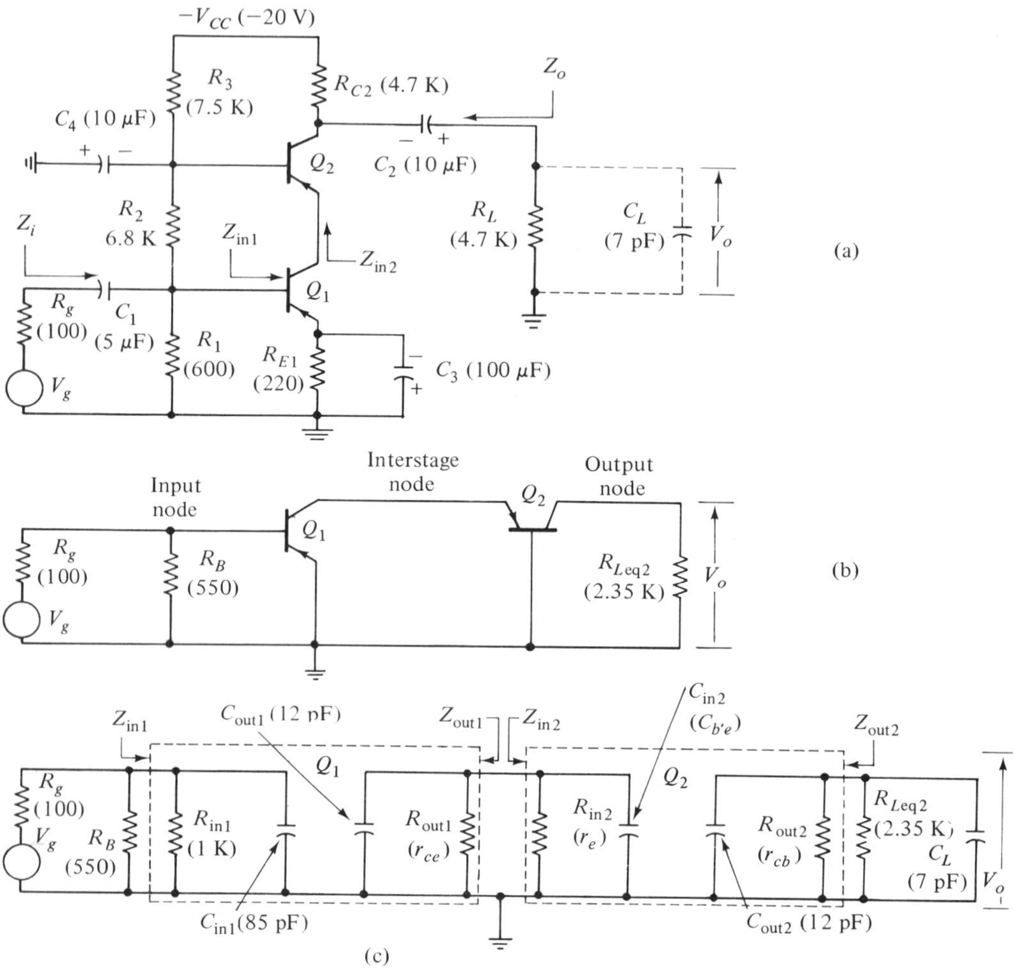

FIGURE 8–3 *The CE-CB cascode amplifier from Ex. 8–3: (a) the complete circuit, (b) a mid-band equivalent circuit, and (c) a high-frequency equivalent circuit showing the input and output impedances of each stage.*

(b) High-frequency response: The high corner frequency of the interstage can be neglected since it occurs at the input of a CB stage and will be in the neighborhood of f_T. The remaining high corner frequencies depend on the value of either C_{in1} or C_{out2}. From Table 7–1,

$$C_{in1} = C_{b'e1} + C_{b'c1}(1 - A_{vt1}) \simeq [61 + 12(1 + 1)] \text{ pF} \simeq 85 \text{ pF}$$
$$C_{out2} \simeq C_{b'c2} \simeq 12 \text{ pF}$$

The corresponding time constants are

$$\tau_{in} = (R_i \,||\, R_g)(C_{in1}) = (360 \,||\, 100)(85 \text{ pF}) \simeq 6.8 \text{ nsec}$$
$$\tau_{out} = (R_{Leq2})(C_{out2} + C_L) \simeq (2.35 \text{ K})[(12 + 7) \text{ pF}] \simeq 45 \text{ nsec}$$

The overall high corner frequency of the amplifier is slightly less than the high corner frequency associated with τ_{out}. Using only τ_{out},

$$f_{hi} < 1/(2\pi\tau_{out}) = 0.159/45 \times 10^{-9} \simeq 3.5 \text{ MHz}$$

Since the effect of the other two time constants is to reduce f_{hi}, an estimated value for f_{hi} is 3 MHz.

(c) Low-frequency response: A detailed analysis indicates that the dominant low corner frequency is associated with C_1. Therefore, τ_{C1} and f_{lo} are

$$\tau_{C1} = (R_i + R_g)(C_1) \simeq 2.3 \text{ msec} \qquad f_{lo} \simeq f_{C1} = 1/(2\pi\tau_{C1}) \simeq 70 \text{ Hz}$$

(d) Experimental results: The measured characteristics of the amplifier were

$$A_{vo} = V_o/V_g \simeq 7 \text{ V}/0.1 \text{ V} = 70 \qquad f_{hi} \simeq 2.2 \text{ MHz} \qquad f_{lo} \simeq 100 \text{ Hz}$$

The results of Ex. 8–3 are typical of a cascode amplifier. The overall voltage gain of the amplifier, A_{vo}, is approximately equal to that of a single CE stage if the equivalent load resistance of the CE is equal to that of the CB stage. The distinguishing characteristic of a cascode amplifier is that the gain of the first stage is approximately unity. Consequently, the voltage gain is obtained from the low input impedance stage and not from the inverting stage. The input capacitance is reduced since, for a CE first stage, $C_{b'c}$ is multiplied by a factor of two and not by the voltage gain of the entire amplifier. The dominant high corner frequency generally occurs at the output node. The high corner frequency of the interstage node can generally be neglected since it occurs at the input of a low input impedance amplifier. The high corner frequency of the input node usually occurs at a high frequency due to the low values of R_g and C_{in1}.

The cascode amplifier in Ex. 8–3 used two PNP transistors. An alternate form of the cascode amplifier, sometimes called the *folded cascode*,* is shown in Fig. 8–4 where an N-channel JFET is followed by a PNP transistor. In this circuit, the PNP transistor provides a drop in dc voltage which can be used to reduce the dc voltage buildup problem. Either form of the cascode amplifier has an improved frequency response over a single-stage inverting amplifier. An analysis of a folded cascode amplifier is illustrated in the following example.

EXAMPLE 8–4 Calculate the frequency response of the CS-CB folded cascode amplifier shown in Fig. 8–4. The transistor characteristics are:

Transistor	g_m	$C_{gs(max)}$	$C_{gd(max)}$	β	$f_{T(min)}$	$C_{ob(max)}$
Q_1	2.4 millimhos	4 pF	2 pF	—	—	—
Q_2	—	—	—	130	350 MHz	10 pF

Solution: (a) The bias and mid-band characteristics were evaluated in Ex. 4–16.

(b) High-frequency analysis: The voltage gain of the first stage is low, thereby causing the input capacitance to be small (approximately equal to $C_{gs1} + C_{gd1}$). Since R_g is also small, the time constant of the input node is small and can be neglected. The time constant at the interstage node can also be neglected since it occurs at the input of a CB stage which has a low input impedance. The time constant of the output node dominates the response and is

$$\tau_{out} = (R_{Leq2})(C_{out2} + C_L) \simeq (R_{Leq2})(C_{ob2} + C_L) = (2.55 \text{ K})(17 \text{ pF}) \simeq 43.5 \text{ nsec}$$

$$\tau_r = 2.2\tau_{out} = (2.2)(43.5 \text{ nsec}) \simeq 96 \text{ nsec}$$

---

* A folded cascode is different from a regular cascode in that two types of devices are used: an NPN-PNP arrangement, some comparable circuit using FETs, or an FET and bipolar combination.

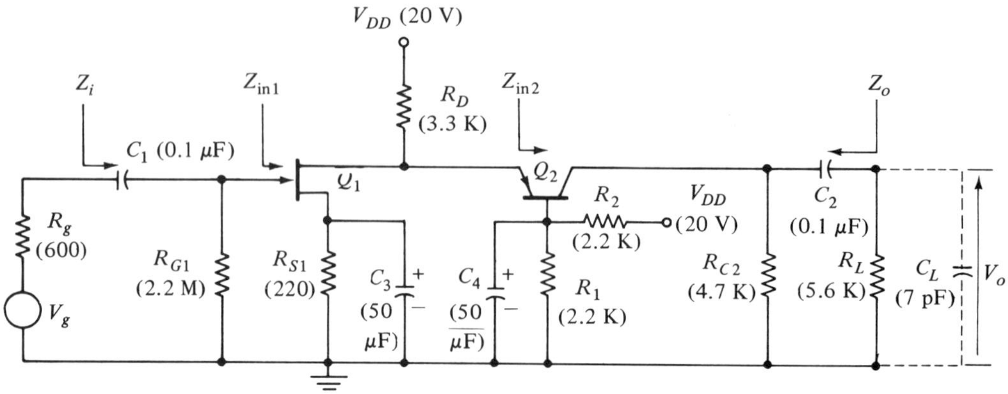

FIGURE 8–4 *The CS-CB amplifier from Ex. 8–4.*

(c) Low-frequency response: A detailed analysis indicates that the dominant low-frequency time constant is associated with C_2. This time constant and the associated value for f_{lo} are

$$\tau_{C2} = (R_o + R_L)C_2 = (4.7\text{ K} + 5.6\text{ K})(0.1\ \mu\text{F}) \simeq 1\text{ msec}$$
$$f_{lo} = 1/(2\pi\tau_{C2}) = 0.159/10^{-3} = 159\text{ Hz}$$

(d) Experimental results: The measured bias voltages were given in Ex. 4–16. The measured gain and frequency response values were

$$A_{vo} = -3\text{ V}/0.5\text{ V} \simeq -6 \qquad \tau_r = 100\text{ nsec} \qquad f_{lo} \simeq 150\text{ Hz}$$

The techniques of multistage analysis discussed in this section also apply to vacuum tube amplifiers. The following example illustrates the analysis of a vacuum tube cascode amplifier.

EXAMPLE 8–5 Calculate the frequency response of the CK-CGr cascode amplifier shown in Fig. 8–5. Both triodes are in the same glass envelope and form a dual unit. The specification sheet values for each triode are:

Triode	g_m	r_p	C_{gp}	C_{gk}	C_{pk}
T_1, T_2	2 millimhos	6.7 K	4 pF	2.6 pF	1 pF

Solution: (a) Mid-band gain: Since T_2 is a common-grid stage, the impedance looking into its plate terminal, R_{op2}, is much greater than r_{p2}. The value of R_{op2} can be determined from Eq. (6-19) as

$$R_{op2} = r_{p2}[1 + g_{m2}R_{Keq2} + R_{Keq2}/r_{p2}] \simeq r_{p2}[1 + g_{m2}r_{p1} + r_{p1}/r_{p2}] \simeq 103\text{ K}$$

The equivalent load resistance for T_2 is

$$R_{Leq2} = R_{op2} \| R_{P2} \| R_L \simeq 103\text{ K} \| 20\text{ K} \| 1\text{ M} \simeq 16.8\text{ K}$$
$$A_{vt2} = g_{m2}R_{Leq2} \simeq (2 \times 10^{-3})(16.8\text{ K}) \simeq 33.6$$
$$A_{vt1} = -g_{m1}R_{Leq1} \simeq -g_{m1}/g_{m2} \simeq -1$$
$$A_{vg} = R_i/(R_i + R_g) \simeq 20\text{ K}/(20\text{ K} + 100) \simeq 1$$
$$A_{vo} = A_{vg}A_{vt1}A_{vt2} \simeq (1)(-1)(33.6) \simeq -33.6$$

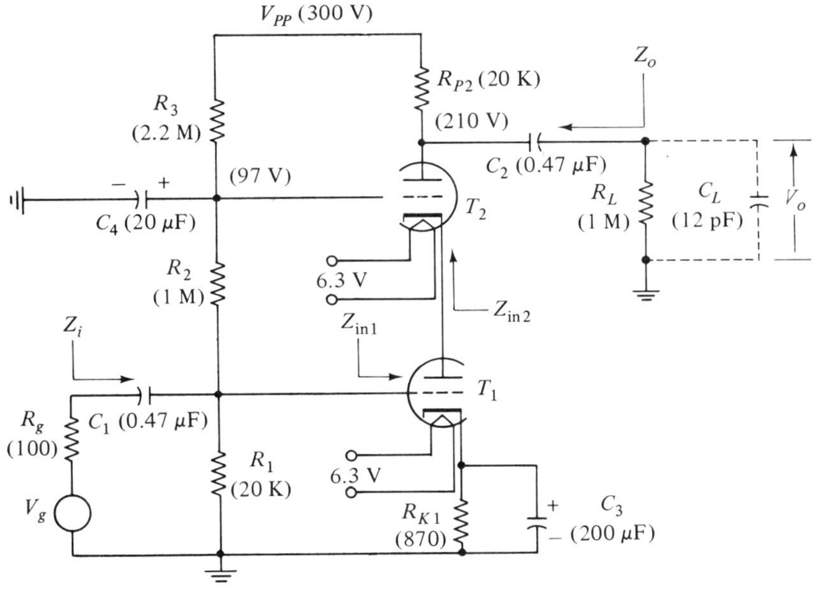

FIGURE 8–5 *The CK-CGr amplifier from Ex. 8–5.*

(b) High-frequency analysis: The time constant of the output node dominates the response. This time constant is

$$\tau_{\text{out}} = (R_{L\text{eq}2})(C_{\text{out}2} + C_L) = (R_{L\text{eq}2})(C_{gp} + C_{pk} + C_L) \simeq 285 \text{ nsec}$$

$$f_{\text{hi}} \simeq f_{\text{out}} = 1/(2\pi\tau_{\text{out}}) = 0.159/(285 \times 10^{-9}) \simeq 560 \text{ kHz}$$

(c) Low-frequency response: A detailed analysis shows that the dominant low-frequency time constant is associated with C_1 and has a value of

$$\tau_{C1} = (R_g + R_i)(C_1) \simeq (100 + 20\text{ K})(0.47\ \mu\text{F}) \simeq 9.4 \text{ msec}$$

The corresponding frequency is

$$f_{\text{lo}} \simeq f_{C1} = 1/(2\pi\tau_{C1}) = 0.159/(9.4 \times 10^{-3}) \simeq 16.9 \text{ Hz}$$

(d) Experimental results: The measured bias voltages are given in parentheses in Fig. 8–5. The measured gain and frequency response values were:

$$A_{vo} = -70\text{ V}/2\text{ V} = -35 \qquad f_{\text{hi}} \simeq 500 \text{ kHz} \qquad f_{\text{lo}} \simeq 16 \text{ Hz}$$

The cascode amplifier is one method of improving the frequency response of an inverting amplifier. Alternately, the frequency response of an inverting amplifier can be improved by using a circuit in which the inverting amplifier is preceded by a low output impedance amplifier such as the CC stage. In this arrangement, the high input capacitance of an inverting amplifier appears in parallel with the low output impedance of the first stage. Consequently, the value of the interstage *RC* time constant is reduced.* A CC-CE amplifier, which is an example of this configuration, is analyzed in the following example.

---

* A differential amplifier using both a CC input stage and a CB cascode arrangement to improve circuit performance is shown in Fig. 8–26.

EXAMPLE 8–6 Calculate the gain and frequency responses of the CC-CE amplifier shown in Fig. 8–6. Both transistors are silicon and are characterized by

Transistor	β	$f_{T(\text{min})}$	$C_{ob(\text{max})}$
Q_1	180	390 MHz	5 pF
Q_2	70	390 MHz	5 pF

Solution: (a) Bias conditions:

$$V_{B1} \simeq 3.3\text{ V} \qquad I_{E1} \simeq 2.7\text{ mA} \qquad I_{E2} \simeq 0.95\text{ mA}$$

(b) Mid-band gain and impedance:

$$r_{e1} \simeq 26\text{ mV}/2.7\text{ mA} \simeq 10\ \Omega \qquad r_{e2} \simeq 26\text{ mV}/0.95\text{ mA} \simeq 27\ \Omega$$

$$R_{\text{in}2} \simeq (\beta_2 + 1)(r_{e2}) \simeq 1.9\text{ K} \qquad R_{L\text{eq}2} = R_{C2} = 10\text{ K}$$

$$R_{L\text{eq}1} = R_{E1} \,||\, R_{\text{in}2} \simeq 0.66\text{ K} \qquad R_{\text{in}1} \simeq (\beta_1 + 1)(r_{e1} + R_{L\text{eq}1}) \simeq 12.1\text{ K}$$

$$R_i \simeq R_1 \,||\, R_2 \,||\, R_{\text{in}1} = R_B \,||\, R_{\text{in}1} \simeq 8.5\text{ K} \,||\, 12.1\text{ K} \simeq 5\text{ K}$$

$$R_{\text{out}1} \simeq r_{e1} + (R_g \,||\, R_B)/(\beta_1 + 1) \simeq 10 + (2\text{ K} \,||\, 8.5\text{ K})/181 \simeq 19\,\Omega$$

$$R_o \simeq R_{C2} = 10\text{ K} \qquad A_{vt1} \simeq R_{L\text{eq}1}/(R_{L\text{eq}1} + r_{e1}) \simeq 0.99$$

$$A_{vt2} \simeq -R_{L\text{eq}2}/r_{e2} \simeq -370 \qquad A_{vg} = R_i/(R_i + R_g) \simeq 0.71$$

$$A_{vo} = A_{vg}A_{vt1}A_{vt2} \simeq (0.71)(0.99)(-370) \simeq -260$$

(c) High-frequency response:

$$C_{b'g1} \simeq 1/[2\pi(r_{e1} + R_{L\text{eq}1})f_{T1}] \simeq 0.159/[(10 + 660)(390 \times 10^6)] \simeq 0.6\text{ pF}$$

$$\tau_{\text{in}} = (R_i \,||\, R_g)(C_{\text{in}1}) = (R_i \,||\, R_g)(C_{b'c1} + C_{b'g1}) = (2\text{ K} \,||\, 5\text{ K})(5.6\text{ pF}) \simeq 8\text{ nsec}$$

$$C_{b'e2} \simeq 1/(2\pi r_{e2} f_{T2}) \simeq 15\text{ pF} \qquad C_{\text{in}2} \simeq C_{b'e2} + C_{b'c2}(1 - A_{vt2}) \simeq 1870\text{ pF}$$

$$C_{\text{out}1} \simeq 1/2\pi[r_e + (R_B \,||\, R_g)/(\beta_1 + 1)]f_T \simeq 0.159/(19)(390 \times 10^6) \simeq 21\text{ pF}$$

$$\tau_{\text{int}} \simeq (R_{\text{out}1} \,||\, R_{L\text{eq}1})(C_{\text{out}1} + C_{\text{in}2}) \simeq (19 \,||\, 660)[(21 + 1870)\text{ pF}] \simeq 35.5\text{ nsec}$$

$$\tau_{\text{out}} \simeq (R_{L\text{eq}2})(C_{\text{out}2} + C_L) \simeq (10\text{ K})(5\text{ pF} + 7\text{ pF}) \simeq 120\text{ nsec}$$

The largest high-frequency time constant is τ_{out}. The interaction between τ_{int} and τ_{out} will cause τ_{out} to dominate the high-frequency response, and a Bode plot is not necessary.* Consequently, the rise time of the amplifier is

$$\tau_r \simeq 2.2\tau_{\text{out}} = (2.2)(120\text{ nsec}) \simeq 264\text{ nsec}$$

(d) Low-frequency response:

$$\tau_{C1} = (R_i + R_g)C_1 \simeq 3.5\text{ msec} \qquad f_{C1} = 1/(2\pi\tau_{C1}) \simeq 45\text{ Hz}$$

$$\tau_{C2} = \left[R_{E2} \,||\, \left(r_{e2} + \frac{R_{E1} \,||\, R_{\text{out}1}}{(\beta_2 + 1)}\right)\right] C_2$$

$$= \left[2.2\text{ K} \,||\, \left(27 + \frac{1\text{ K} \,||\, 19}{71}\right)\right](50\ \mu\text{F}) \simeq 1.3\text{ msec}$$

$$f_{C2} = 1/(2\pi\tau_{C2}) = 0.159/(1.3 \times 10^{-3}) \simeq 122\text{ Hz}$$

The overall low corner frequency, f_{lo}, obtained from a Bode plot was 135 Hz.

(e) Experimental results:** The measured bias voltages are shown in Fig. 8–6(a).

* Time constant interactions are discussed in Sec. 8.4.

** A potential problem with the CC-CE amplifier is a negative input resistance resulting from the capacitive loading of the CC stage by $C_{\text{in}2}$. This problem is discussed in Sec. 8.4.

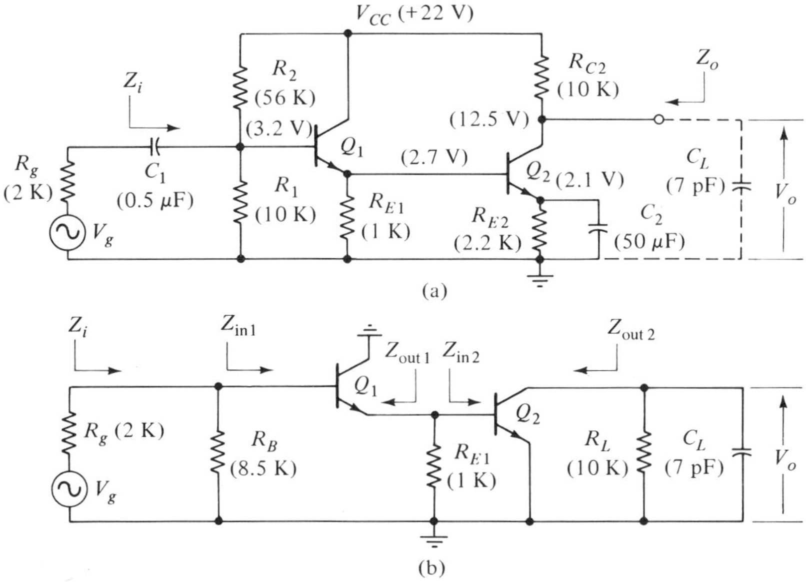

FIGURE 8–6 *The CC-CE amplifier from Ex. 8–6: (a) the circuit diagram and (b) a mid-band equivalent circuit.*

The measured characteristics of the amplifier were:

$$A_{vo} = 0.5\text{ V}/1.8\text{ mV} \simeq -280 \qquad \tau_r \simeq 270\text{ nsec} \qquad f_{lo} \simeq 110\text{ Hz}$$

One of the most versatile multistage configurations is the differential amplifier which finds wide application both in discrete component circuits and in integrated circuits. The following example illustrates the frequency response analysis of a bipolar differential amplifier.

EXAMPLE 8–7 Calculate the frequency response of the bipolar differential amplifier shown in Fig. 8–7. The transistors are silicon and are characterized by

Transistor	β	$f_{T(\min)}$	$C_{ob(\max)}$
Q_1, Q_2 (dual)	450	60 MHz	6 pF
Q_3	100	40 MHz	4 pF

Solution: (a) The bias and mid-band characteristics were evaluated in Ex. 4–18. The important values that are required for a frequency response analysis are:

$$r_{e1} = r_{e2} \simeq 12\,\Omega \qquad R_{Leq1} = R_{Leq2} = 4.7\text{ K}$$
$$R_{i1} = R_{i2} = 8.7\text{ K} \qquad A_{vt(CE)} = -180$$
$$R_{ie2} \simeq 14\,\Omega$$

(b) High-frequency response: Because of the symmetry of the circuit, only the high-frequency poles associated with one input node and one output node need be calculated. The pole occurring at the junction of the emitter terminals of Q_1 and Q_2

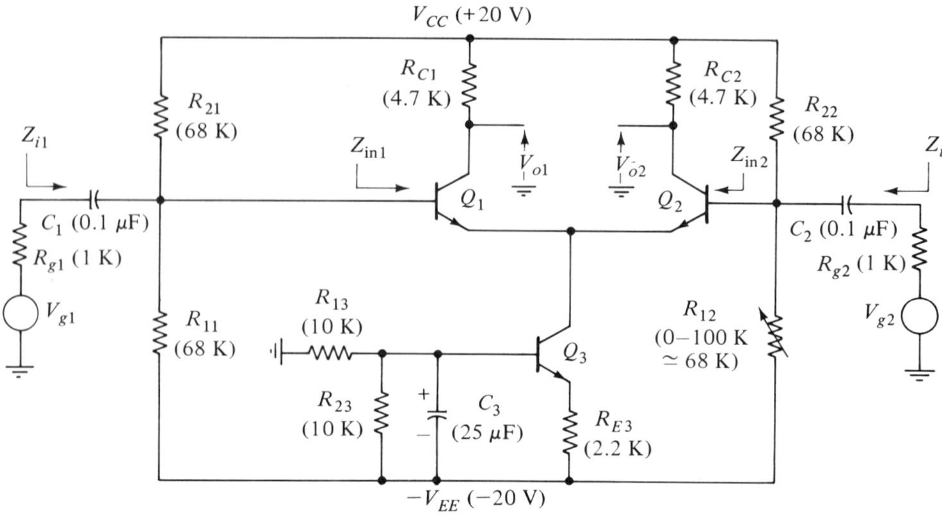

FIGURE 8–7 *The bipolar transistor differential amplifier from Ex. 8–7.*

will occur at a very high frequency and can be neglected. For the input and output poles,

$$C_{b'g1} \simeq 1/[2\pi(r_{e1} + R_{ie2})f_T] = 0.159/(12 + 14)(60 \times 10^6) \simeq 102 \text{ pF}$$
$$C_{\text{in1}} = C_{b'g1} + C_{b'c1}(1 - A_{vt1}) = [102 + 6(1 + 180)] \text{ pF} \simeq 1200 \text{ pF}$$
$$C_{\text{out1}} \simeq C_{b'c1} = 6 \text{ pF} \qquad \tau_{\text{in1}} = (R_g \,\|\, R_i)(C_{\text{in1}}) \simeq 1.1 \,\mu\text{sec}$$
$$\tau_{\text{out1}} = (R_{Leq1})(C_{\text{out1}} + C_{L1}) = (4.7 \text{ K})(6 \text{ pF} + 7 \text{ pF}) \simeq 61 \text{ nsec}$$

Since τ_{in1} is the dominant time constant, the overall high corner frequency of the amplifier is

$$f_{\text{hi}} = 1/(2\pi\tau_{\text{in1}}) = 0.159/(1.1 \times 10^{-6}) \simeq 145 \text{ kHz}$$

(c) Low-frequency response: The low corner frequency associated with C_3 can be neglected since Q_3 is not in the signal path. Also, since the circuit is symmetrical, the time constant associated with only one input coupling capacitor need be calculated. For this pole,

$$\tau_{C1} = (R_i + R_g)C_1 \simeq 0.97 \text{ msec} \qquad f_{\text{lo}} = 1/(2\pi\tau_{C1}) \simeq 154 \text{ Hz}$$

(d) Experimental results: The measured bias voltages were given in Ex. 4–18. The measured characteristics of the amplifier were:

$$|A_{vo}| = \frac{3 \text{ V}}{20 \text{ mV}} = 150 \qquad f_{\text{hi}} \simeq 325 \text{ kHz} \qquad f_{\text{lo}} \simeq 210 \text{ Hz}$$

The frequency response characteristics of multistage amplifiers have been examined in this section. First, the response of an amplifier consisting of two inverting stages was examined. For this type of amplifier, the interstage node often governed the high-frequency response. The overall high-frequency response of such an amplifier was limited due to the high input capacitance of an inverting amplifier. Next, two methods of improving the frequency response of inverting amplifiers were examined. Finally the frequency response of the differential amplifier was examined. Using the characteristics of the circuits discussed in this section as a base, the design approach to multistage amplifiers will be developed in Sec. 8.3.

8.2 Frequency Response of Integrated Circuits

In the last section we examined the frequency response of multistage amplifiers. Each transistor circuit discussed was constructed using individual transistors, resistors, and capacitors. Many of these circuits, or their equivalent, can also be fabricated as integrated circuits. In general, the frequency response analysis of an integrated circuit multistage amplifier follows the same form as that of a discrete component circuit. However, a few differences exist between integrated and discrete component circuits which affect the frequency response. The objective of this section is to examine these differences. We will also examine the typical frequency response data available for an integrated circuit.

The principal difference between discrete components and monolithic components is that the latter are distributed in nature. For example, consider the monolithic 4-K resistor shown in Fig. 8–8.* The resistor is fabricated from

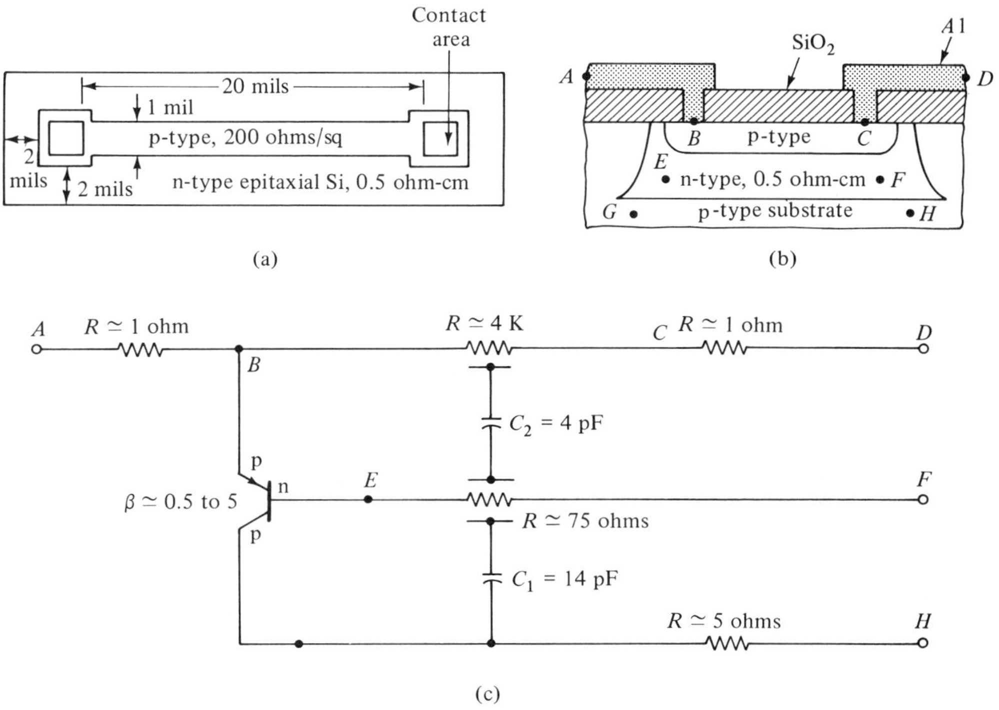

FIGURE 8–8 *A monolithic 4-K resistor: (a) top view, (b) side view, and (c) approximate equivalent circuit. (Courtesy of Motorola.)*

P-type material that is diffused into an N-type epitaxial layer. Consequently, a P-N junction is formed between the body of the resistor and the epitaxial layer. The transition capacitance of this P-N junction is a distributed capacitance along the length of the resistor. A parasitic PNP transistor is formed by the body of the resistor, the epitaxial layer, and the substrate. The epitaxial layer and also the substrate have a distributed resistance. The P-N junction formed by the epitaxial layer and the substrate has a distributed capacitance. Consequently, a

---

* Monolithic resistors are discussed in Sec. 5.4.

monolithic resistor has a complicated equivalent circuit, an approximate form of which is shown in Fig. 8–8(c).

The principal difference between monolithic and discrete transistors is that the collector contact is formed on the same side as the emitter contact for monolithic transistors, and on the opposite side for discrete transistors (shown in Fig. 8–9). In the monolithic transistor, the P-N junction between the epitaxial layer and the substrate adds additional capacitance at the collector terminal. This added capacitance degrades the high-frequency response relative to that of a similar discrete unit. However, the small size of an IC reduces the values of stray capacitance, stray inductance, and signal propagation delay, so that the response of an IC is not necessarily inferior to that of a discrete unit.

The IC manufacturers seldom give specifications for the individual components in an IC. Rather, the specifications indicate the performance of the entire unit. Data may be given on the response of an IC in typical applications or as a set of curves relating the circuit performance to power supply variations, temperature variations, input signal levels, and other parameters.

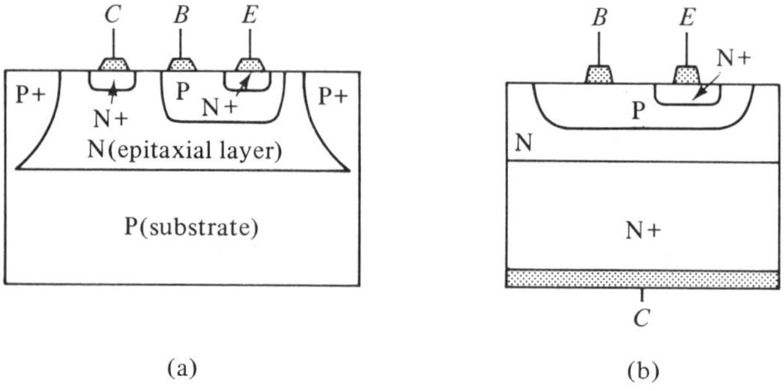

FIGURE 8–9 *Comparison of the side views of (a) a monolithic, and (b) a discrete NPN transistor.*

A general purpose monolithic differential amplifier* is shown in Fig. 8–10(a), and the actual circuit layout in Fig. 8–10(b). Typical of the many performance curves provided by the manufacturer are those shown in Figs. 8–10(c), (d), (e), and (f). The voltage transfer characteristic [Fig. 8–10(c)] and gain characteristics [Figs. 8–10(d) and (e)] are given for the differential output case in which the output voltage is obtained between the two output terminals (either between terminals 2 and 4 or between terminals 5 and 7). The overall frequency response of the circuit is shown in Fig. 8–10(f) and indicates a high corner frequency of 1.5 MHz. The characteristics of the amplifier can also be estimated using the techniques developed earlier in this chapter. The following example illustrates such a calculation.

EXAMPLE 8–8 Estimate the differential gain and high corner frequency of the differential amplifier shown in Fig. 8–10(a). Assume a supply voltage of +12 V, a 50-ohm source impedance, and an infinite load impedance.

---

* The general characteristics of differential amplifiers are discussed in Sec. 4.9 and are summarized in Table 4–4.

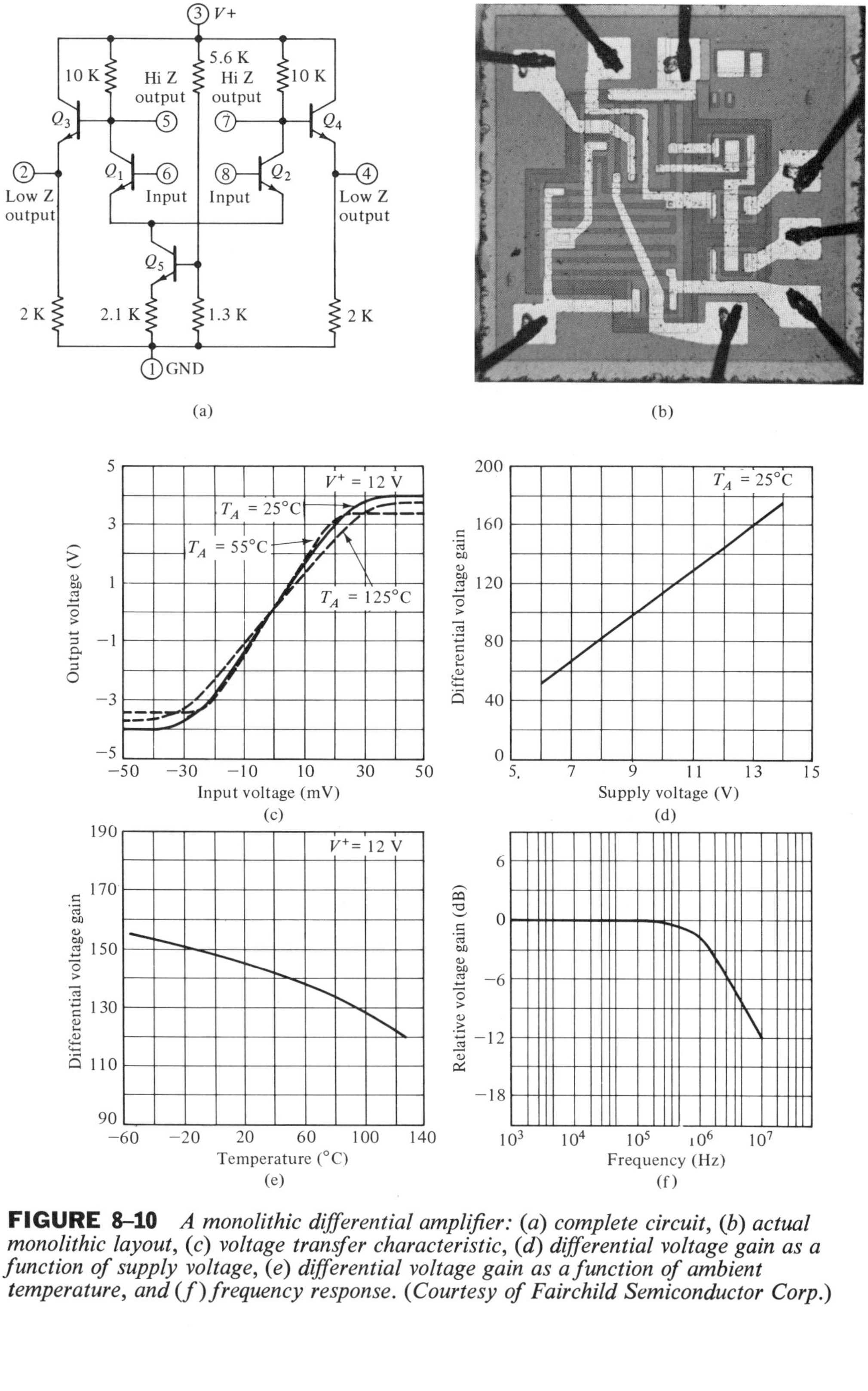

FIGURE 8–10 *A monolithic differential amplifier: (a) complete circuit, (b) actual monolithic layout, (c) voltage transfer characteristic, (d) differential voltage gain as a function of supply voltage, (e) differential voltage gain as a function of ambient temperature, and (f) frequency response. (Courtesy of Fairchild Semiconductor Corp.)*

Solution: (a) Bias conditions:

$$V_B \simeq (12\text{ V})\left(\frac{1.3\text{ K}}{1.3\text{ K} + 5.6\text{ K}}\right) \simeq 2.25\text{ V}$$

$V_{E5} \simeq 2.25\text{ V} - 0.6\text{ V} \simeq 1.65\text{ V}$ $\quad$ $I_{E5} \simeq 1.65\text{ V}/2.1\text{ K} \simeq 0.79\text{ mA}$

$I_{E1} \simeq I_{E2} \simeq I_{E5}/2 \simeq 0.4\text{ mA}$ $\quad$ $V_{C1} \simeq V_{C2} \simeq 12\text{ V} - (0.4\text{ mA})(10\text{ K}) \simeq 8\text{ V}$

$V_{E3} \simeq V_{E4} \simeq 8\text{ V} - 0.6\text{ V} \simeq 7.4\text{ V}$ $\quad$ $I_{E3} \simeq I_{E4} \simeq 7.4\text{ V}/2\text{ K} \simeq 3.7\text{ mA}$

(b) Mid-band gain and impedances:

$$r_{e1} \simeq r_{e2} \simeq 26\text{ mV}/0.4\text{ mA} \simeq 65\Omega \qquad r_{e3} \simeq r_{e4} \simeq 26\text{ mV}/3.7\text{ mA} \simeq 7\Omega$$

Assuming unity gain for each CC, the voltage gain from either input to any output is

$$|A_{vt}| \simeq R_{Leq1}/(r_{e1} + r_{e2}) \simeq 10\text{ K}/(65 + 65) \simeq 77$$

The differential gain is twice the gain for a single output or 154.

(c) High-frequency response: The dominant high corner frequency is determined by the high impedance output, or at the collector of Q_1 or Q_2. Since the CC stages have an equivalent emitter resistance of 2 K, the value of $C_{b'g}$ is very small.

$$C_{b'g} = 1/[2\pi(r_e + R_E)f_T] \simeq 0.159/[(2\text{ K})f_T] \simeq 0\text{ pF}$$

Consequently, the capacitance-to-ground at the dominant node is approximately

$$C \simeq C_{ob1} + C_{ob3} \simeq 2\,C_{ob}$$

Values for C_{ob} for a small-signal monolithic transistor range from 1 to 10 pF with 5 pF being a reasonable first estimate. Using this assumption, the values of the dominant high-frequency time constant and corner frequency are

$$\tau_{hi} \simeq (10\text{ K})(5\text{ pF} + 5\text{ pF}) \simeq 100\text{ nsec} \qquad f_{hi} \simeq 1/(2\pi\tau_{hi}) \simeq 1.59\text{ MHz}$$

(d) Comparison with manufacturer's data: From Fig. 8–10(d), the differential voltage gain is approximately 145 which agrees closely with the calculated value of 154. From Fig. 8–10(f), the high corner frequency is approximately 1.5 MHz indicating that the assumed value of C_{ob} equal to 5 pF was reasonable.

As a second example of the frequency response of an integrated circuit, consider the amplifier shown in Fig. 8–11(a). Assume that this was constructed using the monolithic breadboard shown in Figs. 5–36 and 5–37. The characteristics of the various components are given in Tables 5–4 through 5–7. Since the circuit is fabricated using monolithic techniques, additional parasitic components are present. Two such components that influence the high-frequency response of the circuit are the capacitance between the collector and the substrate (C_{CS}), and the distributed capacitance between the resistor body and the epitaxial layer (C_R). Since the substrate is generally connected to the lowest potential in the circuit, C_{CS} generally appears as an additional capacitance to ground at the collector terminal. The resistor isolation region is generally connected to the most positive potential in the circuit which corresponds to an ac ground. As a first order approximation, the distributed capacitance associated with each resistor can be represented by a capacitor to ground connected at each end of the resistor. The value of each of these equivalent capacitors is approximately equal to one half of the total distributed capacitance associated with the resistor. The effects of C_{CS} and C_R are illustrated in the partial circuit diagram shown in Fig. 8–11(b). The total capacitance from the collector of Q_7 to ground is approximately equal to the sum of C_{CB7}, C_{CS7}, $C_{R7}/2$, C_{CB9}, and

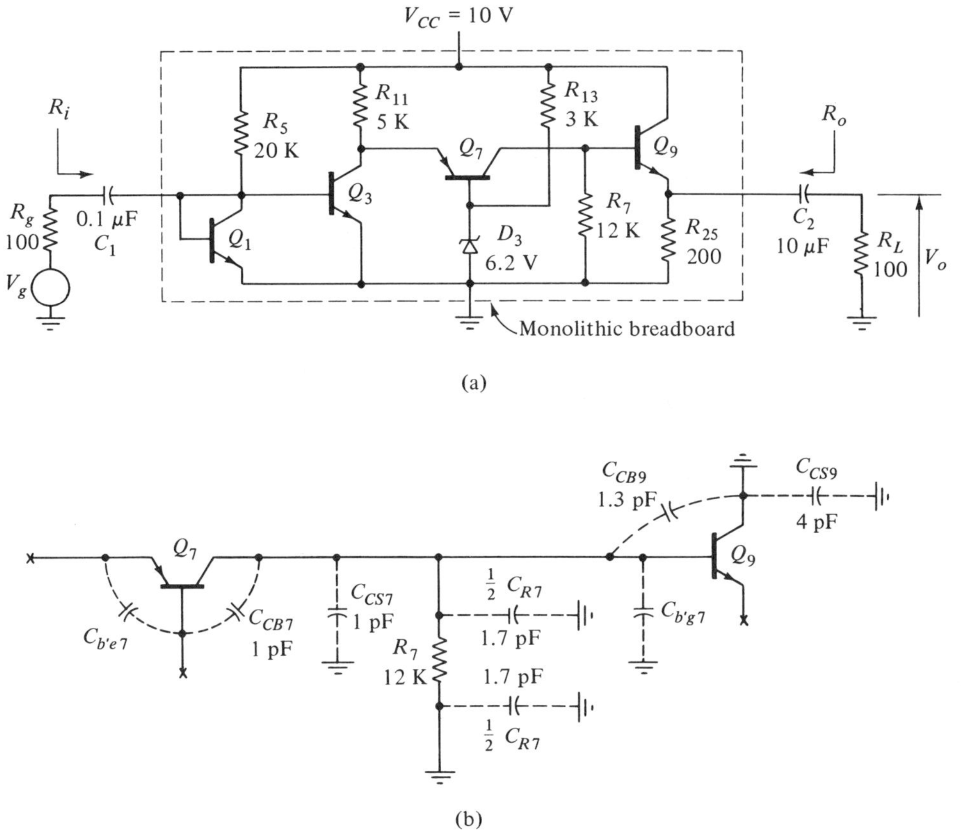

FIGURE 8–11 *An amplifier constructed using the monolithic breadboard shown in Figs. 5–36 and 5–37: (a) the complete circuit, and (b) a partial equivalent circuit showing the stray components present. The component numbers* (R5, Q3, *etc.*) *identify the devices used in Fig. 5–37.*

$C_{b'g9}$.* Calculation of the performance characteristics of the amplifier shown in Fig. 8–11 is left as an exercise for the reader.

In this section, we have compared circuits composed of discrete components with those fabricated as monolithic ICs. The principal difference is the presence of additional parasitic components in the monolithic circuit. IC manufacturers usually provide performance specifications for the entire circuit rather than data on the individual components. Methods of improving frequency response, such as the cascode circuit and the use of CC stages, find wide application in ICs as well as in discrete component circuits. Consequently, much of the discussion on discrete component circuits can be directly applied to integrated circuits.

8.3 Design of Multistage Amplifiers

The material on single-stage and multistage amplifiers, coupled with the discussion of the mid-band analysis given in previous chapters, forms the necessary background for amplifier design studies. To some degree, the design of a multi-

* C_{CB} is another symbol used for the device capacitor $C_{b'c}$ or C_{ob}.

stage amplifier is a trial-and-error procedure whereby some basic configuration is modified until the desired specifications are achieved. However, a judicious choice of a configuration, based on the characteristics of each stage, greatly simplifies the design procedure. The objective of this section is to develop methods for designing multistage amplifiers. To achieve this objective, we will study two design problems. First, the specifications for an amplifier will be developed. Then an amplifier will be designed to meet these specifications. The solutions are generally not unique. However, the approaches used in these designs can be applied to other situations.

As a first example of multistage amplifier design, consider the following situation. For safety reasons, an experiment is being conducted in an open field some distance from the data processing equipment. The monitoring equipment for the experiment is connected to the data processing unit with a long coaxial cable. The cable has a characteristic impedance, Z_o, of 100 ohms. Since the cable attenuates the signal, a "booster amplifier" is to be placed midway in the cable. The signals of interest are sine waves ranging in frequency from 1 kHz to 100 kHz and have amplitudes ranging from 10 mV to 30 mV. The gain of the booster amplifier should lie between 10 and 40. The exact value of the gain is not critical since the entire system (monitor, coaxial cable, booster amplifier, and data processing equipment) will be calibrated prior to use. A sketch of the experimental setup and its equivalent circuit is shown in Fig. 8–12.

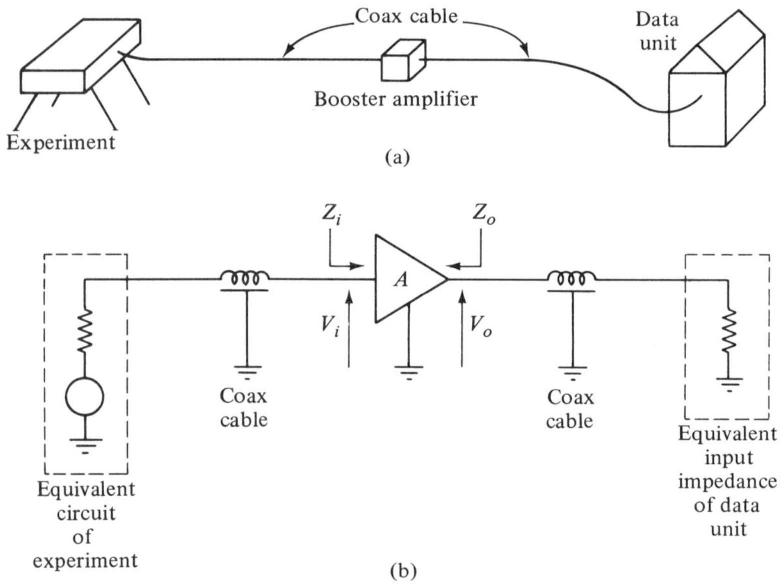

FIGURE 8–12 *An outdoor experimental setup: (a) a sketch of the actual experimental layout, and (b) the electrical equivalent circuit for the experimental layout.*

The first task is to develop the specifications for the amplifier including the overall gain, frequency response, input impedance, and output impedance. Since the signals will be sine waves and no restrictions were placed on signal inversion, either an inverting or a noninverting amplifier is acceptable. A gain of -25 will be used as the initial design goal since the actual value is not critical.

One important parameter of a coaxial cable is its characteristic impedance, Z_o. When a lossless cable* is terminated or loaded by a resistor whose value is equal to Z_o, the impedance looking into the cable from the other end is purely resistive and equal to Z_o. Also, a proper termination reduces signal reflections in the cable and provides better signal transmission. Consequently, it is desirable for the input and output impedances of the amplifier to be resistive and equal to Z_o or 100 ohms.

The high-frequency response of the amplifier must be sufficient to amplify the 100-kHz signal. If the value of f_{hi} is chosen a factor of 10 greater than the highest signal frequency, the amplifier will operate well within its mid-band range for the frequencies of interest. Consequently, the design value for the overall high corner frequency of the amplifier is chosen as 1 MHz or greater.

The low-frequency response of the amplifier must be sufficient to amplify the 1-kHz signal. Consequently, the low corner frequency design value is chosen as 100 Hz or lower.

One final consideration is necessary before the specifications are complete. Since the amplifier will be operated in an open field, an ac line may not be available to supply power to a dc power supply. In this case, batteries may be used so that the amplifier is completely self-contained and portable. Assume that the dc power will be provided by 6-V batteries.

The amplifier specifications discussed above are summarized as follows.

$$\begin{array}{lll} A_{vo} \simeq -25 & f_{lo} \simeq 100\ \text{Hz} & R_i = 100\ \Omega \\ f_{\text{hi}} \simeq 1\ \text{MHz} & V_{\text{dc}} = \pm 6\ \text{V} & R_o = 100\ \Omega \end{array} \quad \textbf{(8–1)}$$

The specifications for A_{vo} and f_{lo} are not particularly difficult to satisfy. The value for f_{hi} can be achieved by careful attention to the various high-frequency time constants and by using high-frequency circuits such as the cascode configuration. Designing an amplifier with 100-ohm input and output impedances, however, requires additional consideration. This restriction involves the choice of the input stage and the output stage of the amplifier which, in general, are the first decisions that must be made in designing a multistage amplifier. Four possible combinations will now be discussed.

A block diagram of one method of terminating a coaxial cable, called "series termination," is shown in Fig. 8–13(a). If the basic amplifier has a very low value of input resistance, the desired value of R_i (100 ohms) can be obtained by placing a resistor in series with the amplifier input. In a similar manner, the desired value of R_o (100 ohms) can be obtained by placing a resistor in series with a low output impedance amplifier. A CB stage has a low value of input impedance and would be a suitable choice for an input stage. The CC stage has a low value of output impedance and would be a suitable output stage as indicated in Fig. 8–13(b).** The values of R_1 and R_2 are chosen to give the required values for R_i and R_o.

A block diagram of an alternate method of terminating a coaxial cable, called "shunt termination," is shown in Fig. 8–14(a). In this method, the terminating resistor is placed in parallel with a high input impedance amplifier. Since the FET has the highest input impedance of all semiconductor active devices, this

---

* Most signal-handling coaxial cables have characteristic impedances in the range of 50 to 200 ohms and can generally be treated as lossless.

** Negative feedback can be used advantageously to determine the input impedance and output impedance of an amplifier. Negative feedback is discussed in Chap. 10.

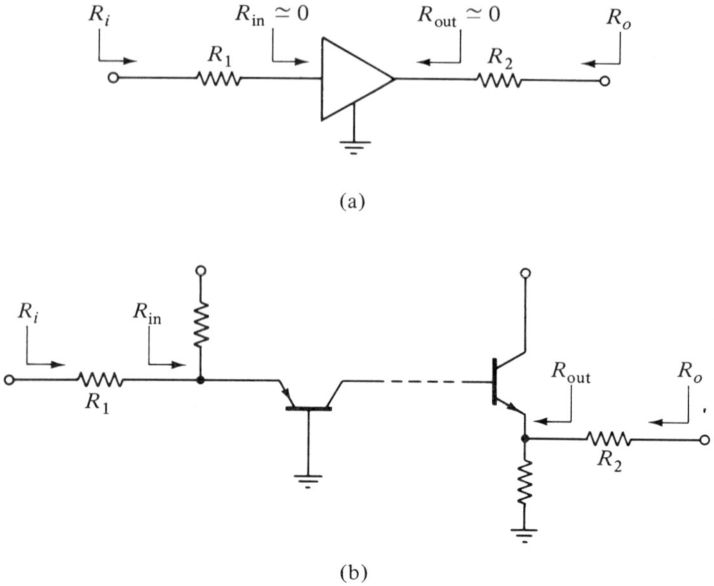

FIGURE 8–13 *Series method of terminating a coaxial cable: (a) a block diagram, and (b) a partial circuit diagram.*

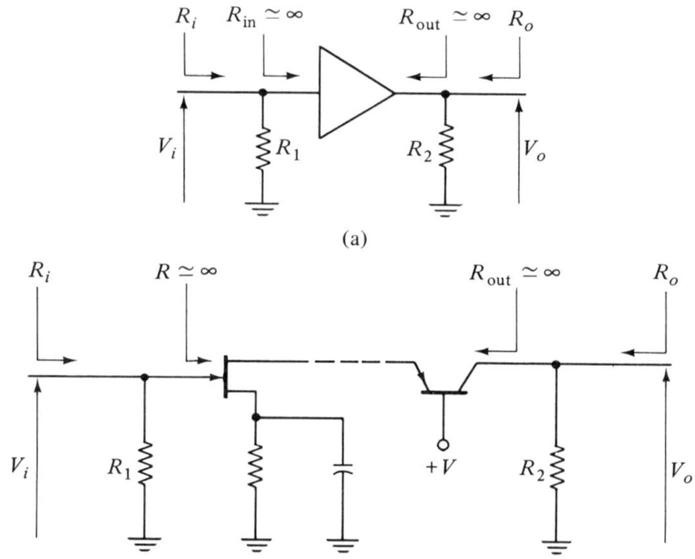

FIGURE 8–14 *Shunt method of terminating a coaxial cable: (a) a block diagram, and (b) a partial circuit diagram.*

device is an ideal choice for the input stage. A low input impedance amplifier has the highest value of output impedance and would be a suitable choice for the output stage. A partial circuit diagram using a CS as the input stage and a CB as the output stage is shown in Fig. 8–14(b).

Two other circuit configurations can also be used for the termination of coaxial cables. One is obtained by combining a series input termination with a shunt output termination and the other by combining a shunt input termination with a series output termination. While any of the four possible input-output arrangements discussed above can be used, we will use the shunt-input series-output scheme for this design example. A partial circuit diagram of this configuration is shown in Fig. 8–15(a).

The advantage of an FET input stage shunted by a resistor having a value equal to Z_o is that the resistive portion of the input impedance can be precisely controlled. The input capacitance can be minimized by using a cascode configuration for the first two stages. Minimizing the input capacitance is also

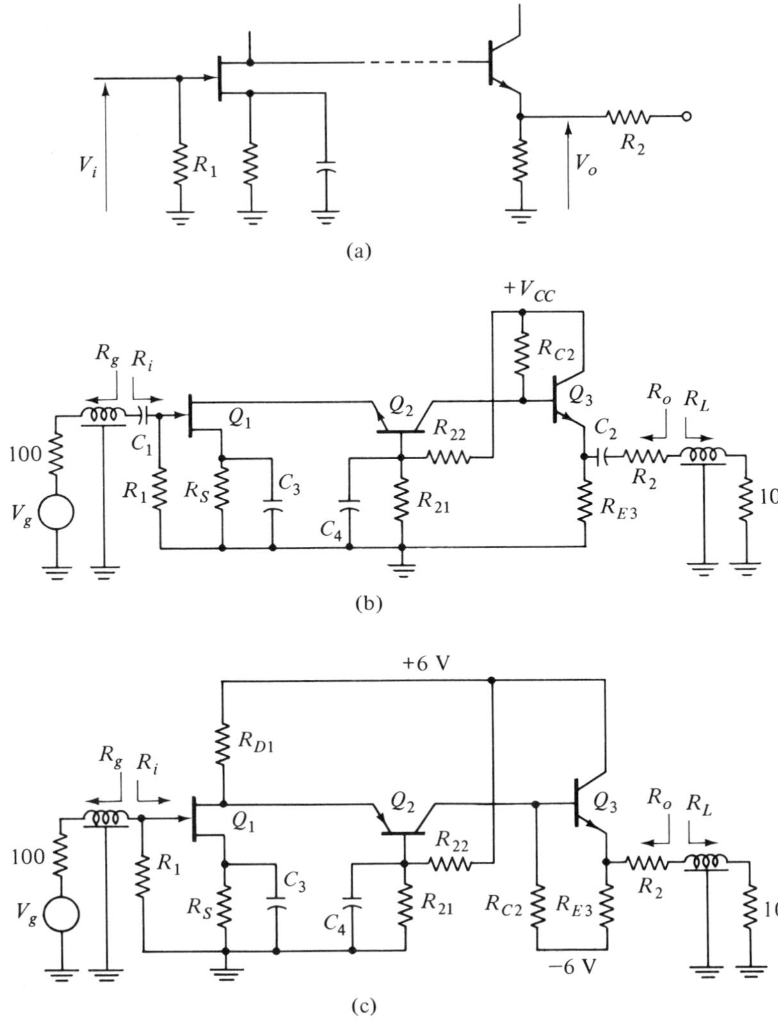

FIGURE 8–15 *Various stages in the design of an amplifier to satisfy the specifications given in Eq. (8–1).*

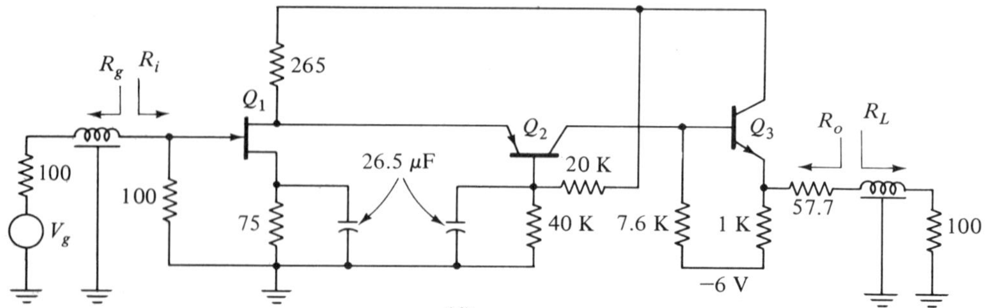

(d)

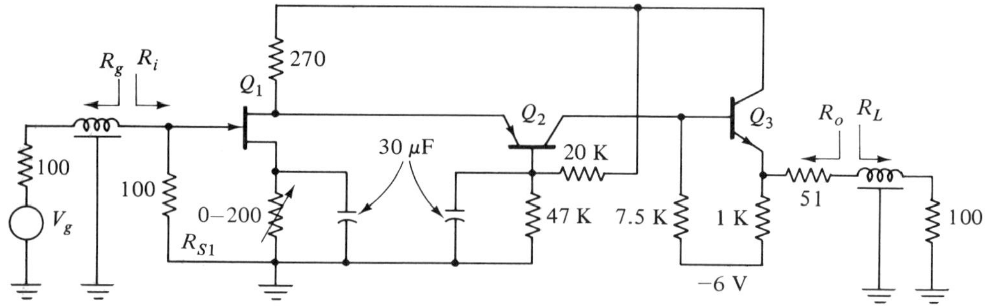

(e)

FIGURE 8–15 (cont.)

important for proper termination of the coaxial cable. The selection of a series output termination is based mainly on the advantages of a CC output stage. Most useful amplifiers, when loaded with a low impedance load (such as a coaxial cable), use a CC as the output stage.* Among other advantages, the CC stage transforms the low impedance load into a much higher value, thus preserving the gain of the preceding stage.

Having chosen a CS as the input stage and a CC as the output stage, we can proceed with the remainder of the design. A CB configuration has several advantages when used as a second stage. First, the cascode circuit has an excellent high-frequency characteristic and will help satisfy the 1-MHz requirement. Second, the cascode circuit reduces the value of the input capacitance.

At this point in the design, we have a CS-CB input configuration and a CC output stage. By combining these three stages as shown in Fig. 8–15(b), we have a configuration that might satisfy the specifications given in Eq. (8–1). Also note in Fig. 8–15(b) that both R_g and R_L are assumed to be 100 ohms. This assumption requires that the coaxial cable be terminated by its characteristic impedance at the experimental location and also at the data processing unit. As discussed earlier in this section, a coaxial cable terminated by its characteristic impedance appears as a resistor with a value of Z_o.

---

* When large-signal levels of both plus and minus polarity are encountered, the output circuit will usually be some variation of the single CC stage. Useful output stages for large signals are discussed in Chap. 12.

Our next task is to evaluate the CS-CB-CC configuration. Assuming that the impedance looking into the base of the CC stage (Q_3) is large, the equivalent load impedance for Q_2 is approximately R_{C2}. From the previous discussions of the cascode circuit, the combined voltage gain of Q_1 and Q_2 is approximately

$$A_{vt1}A_{vt2} \simeq g_{m1}R_{Leq2} \simeq g_{m1}R_{C2} \qquad \textbf{(8–2)}$$

Since the CC stage usually has a voltage gain of slightly less than unity, the overall voltage gain of the CS-CB-CC amplifier shown in Fig. 8–15(b) is approximately

$$A_{vt1}A_{vt2}A_{vt3} \simeq (g_{m1}R_{C2})(1) \simeq g_{m1}R_{C2} \qquad \textbf{(8–3)}$$

The dominant high corner frequency for the CS-CB-CC amplifier is located at the collector of Q_2 and is governed by the time constant

$$\tau_{\text{hi}} \simeq R_{C2}(C_{ob2} + C_{ob3} + C_{b'g3}) \qquad \textbf{(8–4)}$$

Before discussing the actual parameter values of the CS-CB-CC amplifier, we will make two changes which will simplify the design problem. First, replace the regular cascode circuit shown in Fig. 8–15(b) with a folded cascode* circuit shown in Fig. 8–15(c). While this change does not materially affect the gain, it allows the currents of Q_1 and Q_2 to be adjusted independently. Also, the value of R_{C2} is not restricted by the value of I_{D1}. Second, use both a +6-V and a −6-V dc power supply so that the dc voltage at the output terminal (emitter of Q_3) can be adjusted to zero volts. This condition will eliminate the need for an output coupling capacitor. No input coupling capacitor is required since the dc voltage at the gate terminal of Q_1 is already very close to zero. Both changes are shown in Fig. 8–15(c).

The next task is the choice of active devices. The actual choice is usually restricted by availability, cost, etc. For this design, we will assume typical values without specifying the actual devices by number. However, the reader should realize that no circuit can be built and tested until the choice of active devices by number has been made.

The assumed parameters for the JFET are: $V_P = -3$ V, $I_{DSS} = 5.6$ mA, $C_{gs} = 8$ pF, and $C_{gd} = 4$ pF. These values are reasonable for a small-signal, plastic encapsulated N-channel unit. For both bipolar transistors, β is 200, f_T is 300 MHz, and C_{ob} is 5 pF. No specifications are given for the breakdown voltages, since the total supply voltage is only 12 V.

Having made the initial choice of active devices, we can complete the design of the bias circuits.** If we choose to operate the JFET at $V_{GS} = -0.33$ V, the circuit values would be:

$$I_D = I_{DSS}\left(1 - \frac{V_{GS}}{V_P}\right)^2 = 5.6 \text{ mA}\left(1 - \frac{0.33 \text{ V}}{3 \text{ V}}\right)^2 \simeq 4.4 \text{ mA}$$

$$g_m = \frac{2}{V_P}\sqrt{I_D I_{DSS}} \simeq \frac{2}{3}\sqrt{(4.4)(5.6)(10^{-6})} \simeq 3.3 \text{ millimhos}$$

---

* The distinction between the regular and the folded cascode circuits is discussed in the paragraph preceding Ex. 8–4.

** If only bipolar units are used, the bias conditions can be and often are determined prior to the choice of active devices. Since an FET is used in the design, the values of V_P and I_{DSS} are required before the bias conditions can be determined. One exception is the case where negative feedback is used to establish and stabilize the bias conditions regardless of the FET parameter values. A second exception occurs when a bias adjustment is incorporated into the design.

$$R_{S1} = V_{GS}/I_D = (0.33 \text{ V})/(4.4 \text{ mA}) \simeq 75\ \Omega$$
$$R_{C2} = A_{vt}/g_{m1} \simeq 25/(3.3 \times 10^{-3}) \simeq 7.6 \text{ K}$$
$$I_{C2} \simeq (6.6 \text{ V})/(R_{C2}) \simeq (6.6 \text{ V})/(7.6 \text{ K}) \simeq 0.87 \text{ mA}$$

The value of V_{DS1} should be greater than V_P or 3 V to ensure that Q_1 operates in the pentode or saturated region. Adjusting the dc voltage at the base of Q_2 to 4 V will set the voltage at the drain of Q_1 to 4.6 V. Then V_{DS1} is approximately 3.6 V. Choosing R_{21} as 40 K and R_{22} as 20 K, the base voltage of Q_2 is 4 V.

The emitter current of Q_3 is selected as 6 mA.* Consequently,

$$R_{E3} = 6 \text{ V}/6 \text{ mA} = 1 \text{ K} \qquad r_{e3} = 26 \text{ mV}/I_{E3} \simeq 4.3\ \Omega$$

$$R_{D1} = \frac{V_{CC} - V_{D1}}{I_{D1} + I_{E2}} \simeq \frac{6 \text{ V} - 4.6 \text{ V}}{(4.4 + 0.87)\text{mA}} \simeq 265\ \Omega$$

The two remaining resistive components are the terminating resistors for the coaxial cable, R_1 and R_2. Since the impedance looking into the gate terminal of Q_1 is very large, R_1 will limit the value of R_i. Consequently, R_1 is chosen as 100 ohms to satisfy the specification for R_i. The value of R_2 is chosen so that R_o will equal 100 ohms. From Fig. 8–15(c),

$$R_2 = R_o - R_{E3} \,||\, [r_{e3} + R_{C2}/(\beta_3 + 1)] \simeq 100 - 42.3 \simeq 57.7\ \Omega$$

The low-frequency response is determined by the values of C_3 and C_4. Assuming equal values for C_3 and C_4, the time constant associated with C_3 will dominate the low-frequency response. The minimum value for C_3 is

$$C_3 = 1/(2\pi[R_{S1} \,||\, (1/g_{m1})]f_{\text{lo}}) \simeq 0.159/(75 \,||\, 300)(100 \text{ Hz}) \simeq 26.5\ \mu\text{F}$$

If C_3 and C_4 are each 26.5 μF, the low-frequency specification of 100 Hz is satisfied.

The dominant high-frequency time constant occurs at the collector of Q_2. From Figs. 8–15(c) and (d), the equivalent load resistor for Q_3 is

$$R_{L\text{eq}3} = R_{E3} \,||\, (R_2 + R_L) \simeq 1 \text{ K} \,||\, (57.7 + 100) \simeq 136\ \Omega$$

The value of $C_{b'g3}$ is

$$C_{b'g3} = 0.159/[(r_{e3} + R_{L\text{eq}3})f_{T3}] \simeq 3.8 \text{ pF}$$

The value of $R_{L\text{eq}2}$ is

$$R_{L\text{eq}2} = R_{C2} \,||\, R_{\text{in}3} \simeq 7.6 \text{ K} \,||\, [(201)(4.3 + 136)] \simeq 6 \text{ K}$$

and the values of the dominant time constant and high corner frequency are

$$\tau_{\text{hi}} = (R_{L\text{eq}2})(C_{ob2} + C_{ob3} + C_{b'g3}) \simeq 83 \text{ nsec}$$
$$f_{\text{hi}} = 0.159/\tau_{\text{hi}} \simeq 0.159/(83 \times 10^{-9}) \simeq 1.9 \text{ MHz}$$

This value exceeds the specified high corner frequency of 1 MHz.

The preceding analysis indicates that the CS-CB-CC configuration shown in Fig. 8–15(d) meets or exceeds the design specifications given in Eq. (8–1). The next step in the design is to convert the calculated values for the various components into practical or standard values. Figure 8–15(e) shows a circuit using 5 percent tolerance carbon composition resistors. Normally, it is not good design

* The actual choice of I_{E3} depends on the magnitude of the negative-going signal at the output. The maximum value of the negative-going signal is limited to $I_{E3}R_{L\text{eq}3}$ since the emitter current of Q_3 can be reduced by an amount no greater than the emitter bias current. The design and analysis of large-signal amplifiers is discussed in Chap. 12.

practice to combine resistors in various series and parallel combinations to obtain a value equal to the calculated value. However, a potentiometer can be included in the circuit to provide the necessary adjustment. For this circuit, the potentiometer will be adjusted to provide 0 V at the output or emitter of Q_3.

The most important step in the design is to ensure that all specifications are satisfied by the final design configuration. Consequently, the final circuit must be analyzed in a manner similar to the multistage analysis discussed in Sec. 8.1. Only when this step has been completed is the design finished (except for experimental verification). The analysis of the circuit shown in Fig. 8–15(e) is left as an exercise for the reader.

As a second design example, consider the following situation. A manufacturing process is equipped with a quality control monitor which measures some characteristic of the article being manufactured. The monitor has an output voltage consisting of pulses which have peak amplitudes of 1 mV and pulse durations ranging from 5 μsec to 5 msec. The minimum time between pulses is 5 msec. The output resistance of the monitor is approximately 100 ohms. The pulses are to be recorded on a wideband instrument having an input impedance of 100 ohms. The peak signal amplitude into the recorder should be 10 V. The recorded pulses provide a record of the manufacturing process. The problem is to design an amplifier to place between the system monitor and the recording instrument. The amplifier should preserve the pulse shapes of the signals from the monitor and cause a peak indication on the recording instrument corresponding to the peak signal from the monitoring unit.

The first problem is to develop the specifications for the amplifier including the overall gain, frequency response, input impedance, and output impedance. The voltage gain is available from the ratio of the peak recorder input to the peak monitor output signal:

$$A_{vo} \simeq 10 \text{ V}/1 \text{ mV} = 10^4$$

The required rise time for the amplifier is determined by the smallest pulse width or 5 μsec. To ensure that the output of the amplifier reaches the peak value during the 5 μsec pulse, the dominant high-frequency time constant of the amplifier must be less than one-fourth of the pulse width since the output pulse reaches 98 percent of its final value in four time constants. Thus,

$$\tau_{\text{hi}} < 5 \ \mu\text{sec}/4 = 1.25 \ \mu\text{sec}$$

The rise time of the amplifier must be less than

$$\tau_r = 2.2\tau_{\text{hi}} = 2.2(1.25 \ \mu\text{sec}) = 2.75 \ \mu\text{sec}$$

The high corner frequency must be greater than

$$f_{\text{hi}} = 1/(2\pi\tau_{\text{hi}}) = 0.159/1.25 \ \mu\text{sec} \simeq 127 \text{ kHz}$$

The high corner frequency is chosen as 200 kHz to preserve the important features of the pulse shape from the monitor

The required low-frequency response of the amplifier is determined by the longest pulse width or 5 msec. To have less than 10 percent tilt, the amplifier low corner frequency should be less than

$$f_{\text{lo}} = P/\pi T \simeq 0.1/(\pi 5 \times 10^{-3}) \simeq 6.4 \text{ Hz}$$

and is chosen as 3 Hz.

The input impedance of the amplifier is not critical since the output impedance of the monitor is only 100 ohms. However, the value of R_i should be greater than 1 K so that at least 90 percent of the output signal from the monitor is available to drive the input of the amplifier, or $A_{vg} \geq 0.9$.

The output impedance of the amplifier should be low since the load on the amplifier is only 100 ohms (due to the input impedance of the recorder). Consequently, the value of R_o should be in the range of 100 ohms or less.

The amplifier specifications are summarized as follows:

$A_{vo} = 10^4$	$R_i \geq 1$ K
$f_{\mathrm{hi}} \geq 200$ kHz	$R_o \leq 100\ \Omega$
$f_{\mathrm{lo}} \leq 3$ Hz	Signal not inverted

The amplifier will be designed to meet these specifications. Only bipolar transistors will be used in the design because of their high gain. An FET is not required for the input stage since R_i can be as low as 1 K. For the purpose of this design, the following characteristics are assumed for the bipolar transistors:

$$\beta = 200 \qquad C_{ob} = 5 \text{ pF} \qquad f_T = 300 \text{ MHz}$$

Supply voltages of both plus and minus 24 V dc are also assumed to be available.

In the first design attempt, we will try to meet the mid-band specifications. Since a CE stage can easily have a voltage gain greater than 10^2, two CE stages, or their equivalent, should satisfy the gain requirement. By using both plus and minus power supply voltages, the collector resistors can be made reasonably large and, consequently, the voltage gain can be increased. The low output impedance suggests that a CC stage should be used for the output stage. Hence, a CE-CE-CC amplifier of the form shown in Fig. 8–16(a) should satisfy the mid-band specifications.

Although the CE-CE-CC circuit will satisfy the mid-band specifications, this arrangement probably will not satisfy the frequency response specifications. The low corner frequency specification of 3 Hz requires very large values for the emitter bypass capacitors and output coupling capacitor. The Miller effect, particularly for Q_2, produces a large capacitance from the base of Q_2 to ground which limits the high-frequency response. Consequently, the circuit in Fig. 8–16(a) must be modified to meet the frequency response specifications.

The low-frequency response can be improved by eliminating all the coupling and bypass capacitors from the circuit as shown in Fig. 8–16(b). The coupling capacitors are eliminated by adjusting the dc voltages at the base of Q_1 and the emitter of Q_3 to zero volts. This condition can be achieved with the two potentiometers. The bypass capacitors are eliminated by using a differential amplifier instead of a CE stage. For example, Q_1 in Fig. 8–16(a) is replaced by a differential pair, Q_1 and Q_4, in Fig. 8–16(b). While the gain is reduced slightly by this change, a gain of greater than 10^4 can be obtained from the amplifier shown in Fig. 8–16(b).

The high-frequency response of the amplifier in Fig. 8–16(b) will be similar to the high-frequency response of the original circuit shown in Fig. 8–16(a) since the capacitive multiplication due to the CE stages, Q_1 and Q_2, is still large. The high-frequency response can be improved by using a cascode circuit for Q_2, as shown in the partial circuit diagram of Fig. 8–16(c). An alternate approach would be to use a CC stage between Q_1 and Q_2 as shown in the partial circuit diagram in Fig. 8–16(d). Both approaches reduce the value of the high-frequency time constant at the base of Q_2.

Another method of improving the high-frequency response is to use the non-inverting output of each differential amplifier as shown in Fig. 8–16(e). Since the signal path does not contain inverting type amplifiers, the capacitive multiplication due to the Miller effect is negligible. The two important high-frequency time constants occur at the base Q_2 and at the base Q_3. The mid-frequency response extends to 0 Hz since the circuit has no coupling or bypass capacitors. The complete analysis of the circuit shown in Fig. 8–16(e) is left as an exercise for the reader. Using transistors with the characteristics given above, the amplifier has a response of

$$A_{vo} \simeq 2.6 \times 10^4 \qquad f_{hi} \simeq 900 \text{ kHz} \qquad f_{lo} = 0 \text{ Hz}$$

These values exceed the design specifications.

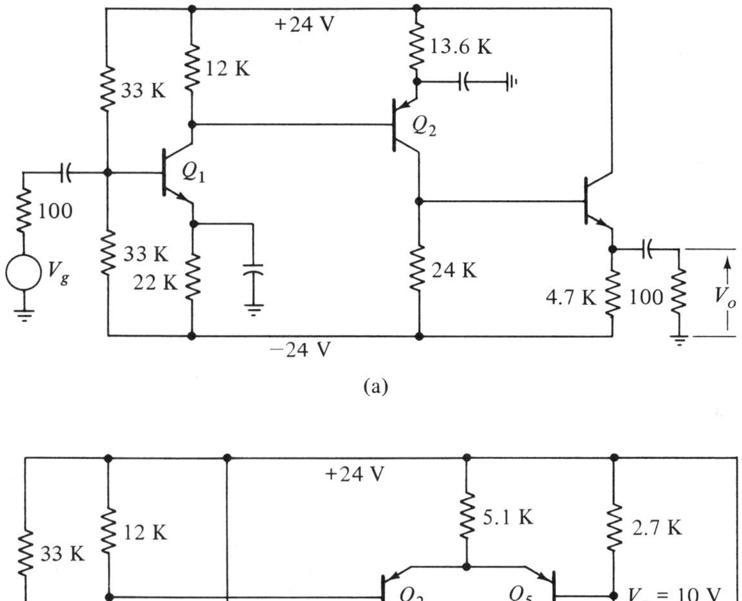

(a)

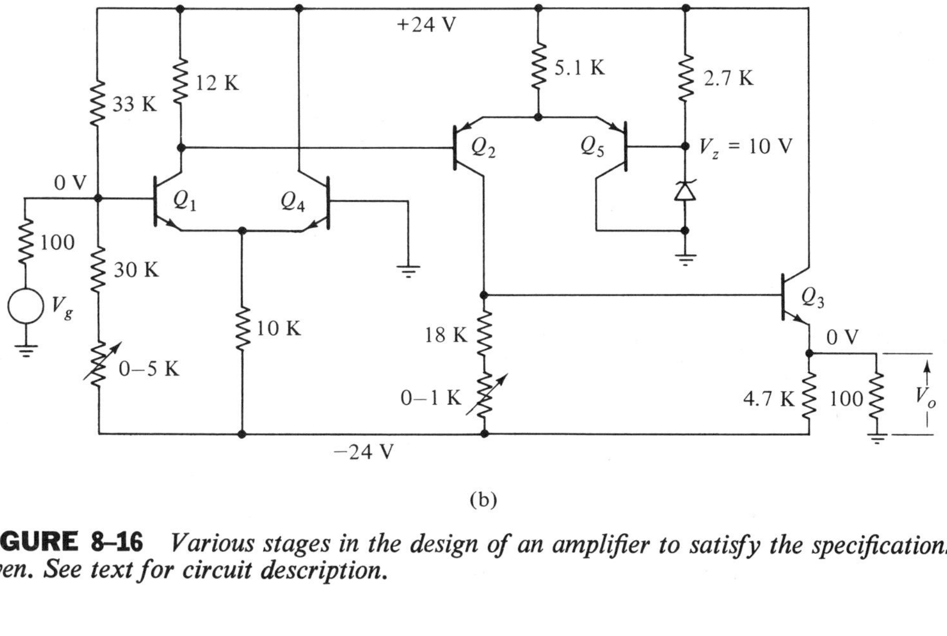

(b)

FIGURE 8–16 *Various stages in the design of an amplifier to satisfy the specifications given. See text for circuit description.*

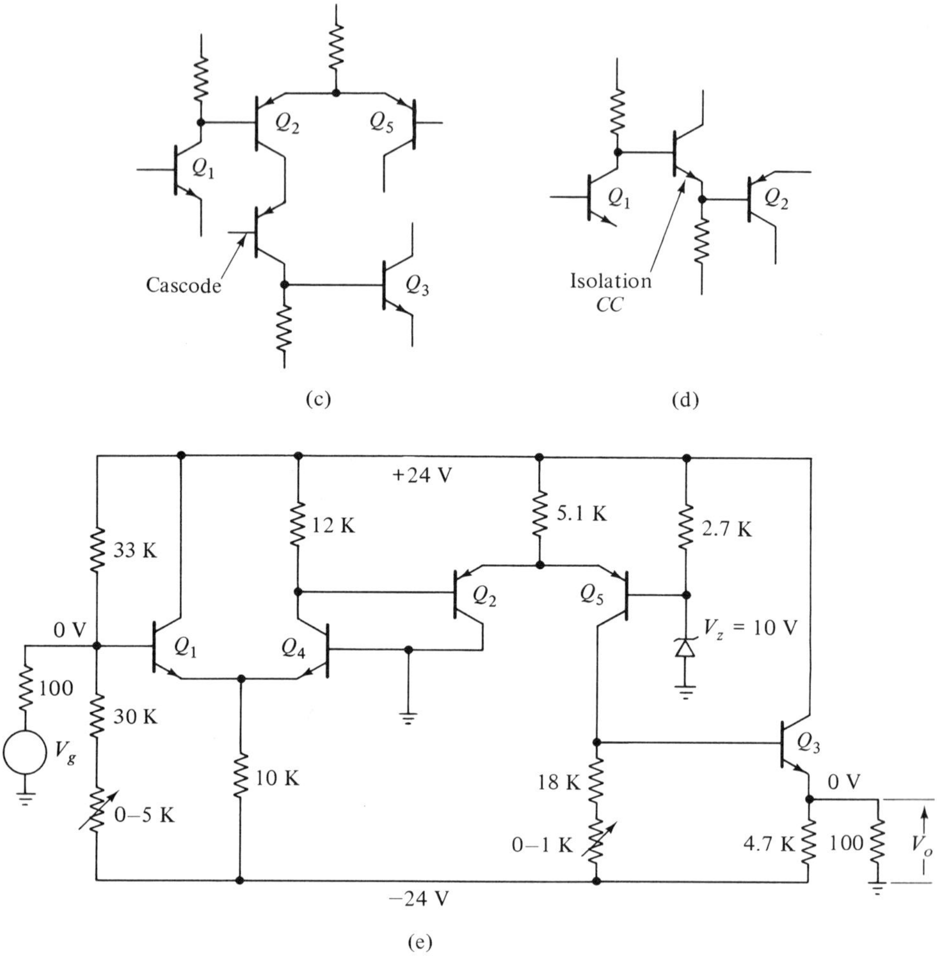

FIGURE 8–16(cont.)

In this section, we have developed two amplifiers to satisfy particular requirements. As is often the case in practice, the first task was to develop the specifications from the general requirements. Once the specifications are known, a design can be started. Usually, a design approach consists of a trial-and-error procedure in which a basic circuit is modified until the desired specifications are achieved.

8.4 Some Important Second Order Effects in Amplifier Frequency Response

The high-frequency response analysis discussed in this and the previous chapter is based on the assumption that the various high-frequency time constants do not interact. Stated differently, we have assumed that the value of the time constant associated with each node in the circuit is independent of the values of the other time constants in the circuit. While an approximate analysis gives accurate results for the overall high corner frequency of an amplifier, the high-frequency response for those frequencies above the overall −3-db frequency is not always accurately described by this approach. An accurate evaluation of the high-frequency response is necessary for the calculation of the frequency response of feedback amplifiers and oscillators. The objective of this section is to review briefly some of the results that can be obtained from a more accurate analysis of the various amplifier configurations. The details of this analysis can be found in Appendix C. The topics to be discussed are: (1) high-frequency time constant interaction in CE and CC stages, (2) negative input resistance effects in CC stages, (3) expressions for the input impedance for the CB stage, and (4) expressions for the output impedance of CB and the CC stages. First, however, we will examine a method of simplifying the mathematical expressions that are often obtained in the detailed analysis of a circuit.

Nodal analysis is the primary tool used in the complete analysis of a circuit. The expressions that result from nodal analysis, such as transfer functions, input and output impedance, etc., often involve polynomials of second or higher order. The important characteristics of these polynomials are most apparent when the polynomials are written in factored form. Factoring can be accomplished easily if the roots of the polynomial are widely separated. For example, a second order polynomial can be written as

$$c_1s^2 + c_2s + c_3 = (s\tau_1 + 1)(s\tau_2 + 1) = s^2(\tau_1\tau_2) + s(\tau_1 + \tau_2) + 1 \quad \textbf{(8–5)}$$

where c_1, c_2, and c_3 are coefficients, and τ_1 and τ_2 are time constants. Equating coefficients of like powers of s gives

$$c_1 = \tau_1\tau_2 \qquad c_2 = \tau_1 + \tau_2 \qquad c_3 = 1 \quad \textbf{(8–6)}$$

If the roots are widely separated, τ_1 is much greater than τ_2, and the values of τ_1 and τ_2 can be determined from the coefficients of the polynomials as*

$$\begin{aligned} c_1/c_2 &= \tau_1\tau_2/(\tau_1 + \tau_2) \simeq \tau_1\tau_2/\tau_1 \simeq \tau_2 \\ c_2/c_3 &= (\tau_1 + \tau_2)/1 \simeq \tau_1 \end{aligned} \quad \textbf{(8–7)}$$

This approximate but simple method of polynomial factoring finds wide application in later chapters as well as in the development of the material summarized in this section.

The effect of time constant interaction is particularly important in an inverting amplifier. For the CE stage, the approximate analysis gave the high-frequency response as

$$A_{vo} \simeq (A_{\text{mid}})\left(\frac{1}{1 + s\tau_{\text{in}}}\right)\left(\frac{1}{1 + s\tau_{\text{out}}}\right) \quad \textbf{(8–8)}$$

where

$$\tau_{\text{in}} \simeq (R_g \parallel R_i)[C_{b'e} + C_{b'c}(1 - A_{vt})] \quad \textbf{(8–9)}$$

$$\tau_{\text{out}} \simeq R_{L\text{eq}}(C_{b'c} + C_L) \quad \textbf{(8–10)}$$

and A_{mid} is the mid-band value of A_{vo}. A complete nodal analysis of the single-stage CE circuit (see Appendix C) gives the high-frequency response as**

---

* These results can be extended to higher order polynomials where the roots are still approximately equal to the ratio of adjacent coefficients provided that the roots are widely separated.

** The approximate signs in Eqs. (8–11), (8–12), and (8–13) are due to the use of the approximate method of factoring a polynomial outlined in Eqs. (8–5) through (8–7).

$$A_{vo} \simeq \frac{(A_{\text{mid}})(1 - sC_{b'c}/g_m)}{(1 + s\tau_1)(1 + s\tau_2)} \tag{8–11}$$

where

$$\tau_1 \simeq (R_g \parallel R_i)[C_{b'e} + C_{b'c}(1 - A_{vt})] + R_{L\text{eq}}(C_{b'c} + C_L) \tag{8–12}$$

$$\tau_2 \simeq \frac{(R_g \parallel R_i)(C_{b'e}C_L + C_{b'c}C_L + C_{b'e}C_{b'c})}{(R_g \parallel R_i)[C_{b'e} + C_{b'c}(1 - A_{vt})] + R_{L\text{eq}}(C_{b'c} + C_L)} \tag{8–13}$$

and A_{mid} is again the mid-band value of A_{vo}.

Before examining the poles of the transfer function in Eq. (8–11), the location of the zero will be discussed briefly. Since $1/g_m$ is approximately equal to r_e, the zero has a time constant equal to $r_eC_{b'c}$. Since $C_{b'c}$ is generally less than $C_{b'e}$, and the time constant associated with f_T is $r_eC_{b'c}$, the zero will generally occur at some frequency greater than f_T. Consequently, we can neglect the effect of the zero in subsequent analyses.

The two time constants, τ_1 and τ_2, are the actual time constants of the amplifier. Note that the amplifier has two high corner frequencies, one associated with τ_1 and the other associated with τ_2 where τ_1 is associated with the lower high corner frequency, and τ_2 is associated with the upper high corner frequency. The values of τ_1 and τ_2 depend on the values of τ_{in} and τ_{out}. Comparing Eq. (8–12) with Eq. (8–9), we see that τ_1 is equal to the sum of the two approximate time constants or

$$\tau_1 = \tau_{\text{in}} + \tau_{\text{out}} \tag{8–14}$$

Since adding τ_{in} and τ_{out} is similar to combining their associated high corner frequencies on a Bode plot, the results of the approximate analysis are identical with those of the nodal analysis for the lowest high corner frequency or overall −3-db frequency.

The value of τ_2 can be calculated directly in Eq. (8–13). However, consider the case in which one time constant (τ_{in} or τ_{out}) is much larger than the other time constant. When τ_{in} is much larger than τ_{out}, the implication in Eqs. (8–9) and (8–10) is that A_{vt} is large, and C_L is small. Under these conditions, Eq. (8–13) reduces to

$$\tau_2 \simeq r_eC_{b'e} \tag{8–15}$$

and the high corner frequency associated with τ_2 becomes

$$f_{\text{hi2}} \simeq 1/(2\pi\tau_2) \simeq 1/(2\pi r_eC_{b'e}) \simeq f_T \tag{8–16}$$

Thus when τ_{in} dominates the response of a CE stage, the second high corner frequency is in the neighborhood of f_T. This result will be used often in later chapters.

When τ_{out} is much larger than τ_{in}, the implication from Eqs. (8–9) and (8–10) is that A_{vt} is small, and C_L is large. Under these conditions, Eq. (8–13) reduces to

$$\tau_2 \simeq (R_g \parallel R_i)(C_{b'e} + C_{b'c}) \tag{8–17}$$

When τ_{out} dominates the response of a CE stage, the second high corner frequency is similar to τ_{in} except that $C_{b'c}$ is no longer multiplied by $(1 - A_{vt})$.

We are now in a favorable position to summarize the effects of high-frequency time constant interaction in inverting amplifiers. The time constants τ_{in} and τ_{out} result from the approximate analysis and can be identified with the input and output nodes of the single-stage inverting amplifier. The time constants, τ_1 and

τ_2, on the other hand, result from a complete nodal analysis and may not be associated with a particular node in the circuit. In a two-pole system, τ_1 is always associated with the lower high corner frequency regardless of the relative values of τ_{in} and τ_{out}. The mathematical analysis can be extended to a three-pole system* with mathematical time constants τ_1, τ_2, and, τ_3 and circuit time constants τ_{in}, τ_{int}, and τ_{out}. In a three-pole system, τ_1 is associated with the lowest high corner frequency; τ_2 is associated with the second lowest high corner frequency, and τ_3 is associated with the highest high corner frequency. The results of high-frequency time constant interactions for the single-stage CE circuit are summarized in Table 8–1, and the use of this table is illustrated in Ex. 8–9 for a CE-CE amplifier.

TABLE 8–1 *Summary of the effects of interaction between high-frequency time constants for the single-stage CE amplifier.*

	Dominant Time Constant τ_1	Second Time Constant τ_2
$\tau_{\text{in}} \gg \tau_{\text{out}}$	$\tau_{\text{in}} + \tau_{\text{out}}$	$1/2\pi f_T$
$\tau_{\text{out}} \gg \tau_{\text{in}}$	$\tau_{\text{in}} + \tau_{\text{out}}$	$(R_g \parallel R_i)(C_{b'e} + C_{b'c})$
$\tau_{\text{in}} \simeq \tau_{\text{out}}$	$\tau_{\text{in}} + \tau_{\text{out}}$	Use Eq. (8–13)

$$\tau_{\text{in}} = [R_g \parallel R_i][C_{b'e} + C_{b'c}(1 - A_{vt})]$$
$$\tau_{\text{out}} = R_{L\text{eq}}(C_{b'c} + C_L)$$

EXAMPLE 8–9 For the CE-CE amplifier discussed in Ex. 8–1 and shown in Fig. 8–1, calculate the complete high-frequency response including the effects of interaction between the high-frequency time constants.

Solution: From Ex. 8–1, the three high-frequency time constants are

$$\tau_{\text{in}} = 5.5 \text{ nsec} \qquad \tau_{\text{int}} = 1.25\ \mu\text{sec} \qquad \tau_{\text{out}} = 49 \text{ nsec}$$

The interstage time constant corresponds to the output time constant for the first stage, τ_{out1}, and to the input time constant for the second stage, τ_{in2}. Since, τ_{int} has the largest value, it will dominate the response. From Table 8–1, the dominant time constant for the first stage, τ_{11},** is

$$\tau_{11} = \tau_{\text{out1}} = \tau_{\text{int}} = 1.25\ \mu\text{sec}$$

and the dominant time constant for the second stage, τ_{12}, is

$$\tau_{12} = \tau_{\text{in2}} = \tau_{\text{int}} = 1.25\ \mu\text{sec}$$

Since τ_1 is the dominant time constant for the entire amplifier,

$$\tau_1 = \tau_{11} = \tau_{12} = \tau_{\text{int}} = 1.25\ \mu\text{sec}$$

* A CE-CE amplifier is an example of a three-pole system with time constants at the input, interstage, and output nodes.

** The time constants τ_1, τ_2, and τ_3 refer to the overall mathematical time constants of the amplifier. The time constants τ_{11} and τ_{21} refer to the first stage mathematical time constants while τ_{12} and τ_{22} refer to the second stage mathematical time constants where the second subscript is used to designate the stage.

The first stage is characterized by

$$\tau_{\text{out1}} \gg \tau_{\text{in1}}$$

From Table 8–1, the value of the second time constant for the first stage, τ_{21}, is given by

$$\tau_{21} = (R_g \,||\, R_i)(C_{b'e} + C_{b'c}) \simeq (50 \,||\, 2\text{ K})[(18 + 5)\text{ pF}] \simeq 0.115\text{ nsec}$$

which is much smaller than the value of 5.5 nsec that was obtained by the approximate solution.

The second stage is characterized by

$$\tau_{\text{in2}} \gg \tau_{\text{out2}}$$

From Table 8–1, the value of the second time constant for the second stage, τ_{22}, is

$$\tau_{22} = 1/(2\pi f_T) = 0.159/(100 \times 10^6) \simeq 1.59\text{ nsec}$$

This value is much smaller than that of 49 nsec obtained by the approximate solution.

The three mathematical time constants for the amplifier can be written as

$$\tau_1 = \tau_{11} = \tau_{12} = 1.25\ \mu\text{sec} \qquad \tau_2 = \tau_{22} = 1.59\text{ nsec}$$
$$\tau_3 = \tau_{21} = 0.115\text{ nsec}$$

The high-corner frequencies are related to the high-frequency time constant by $f_{\text{hi}} = 1/(2\pi\tau)$. A comparison between the results of the approximate solution and the solution including the interaction effects is listed below:

High corner frequencies	f_{hi1}	f_{hi2}	f_{hi3}
Approximate solution	128 kHz	3.2 MHz	28 MHz
Table 8–1 solution	128 kHz	100 MHz	138 MHz

The complete high-frequency response is plotted in Fig. 8–17 for both the approximate solution and the solution including the effects of time constant interaction. Although the two curves deviate considerably above the -3 db frequency, they have the same value for the overall high corner frequency.

Time constant interaction also occurs in CC stages. As shown in Appendix C, the transfer function of a CC stage resulting from a nodal analysis is

$$A_{vo} \simeq \frac{(A_{\text{mid}})(1 + sr_eC_{b'e})}{(1 + s\tau_1)(1 + s\tau_2)} \tag{8–18}$$

where A_{mid} is the mid-band value of A_{vo}. Note that the zero in Eq. (8–18) occurs at f_T. As was the case with the CE stage, the dominant time constant, τ_1, is

$$\tau_1 \simeq \tau_{\text{in}} + \tau_{\text{out}} \tag{8–19}$$

where

$$\tau_{\text{in}} \simeq (R_i \,||\, R_g)(C_{b'c} + C_{b'g}) \tag{8–20}$$

$$\tau_{\text{out}} \simeq (R_{\text{out(CC)}} \,||\, R_{L\text{eq}})(C_{\text{out(CC)}} + C_L) \tag{8–21}$$

As shown in Appendix C, the second time constant, τ_2, is associated with a high corner frequency that occurs in the neighborhood of f_T regardless of the relative values of τ_{in} and τ_{out}. Since the zero of the transfer function also occurs at f_T, the CC amplifier can be characterized as a single time constant amplifier. Thus Eq. (8–18) can be rewritten as

$$A_{vo} \simeq \frac{A_{\text{mid}}}{[1 + s(\tau_{\text{in}} + \tau_{\text{out}})]} \tag{8–22}$$

A potential problem can arise with a CC stage if the emitter circuit consists of a parallel RC network. For this case the input impedance of the CC amplifier

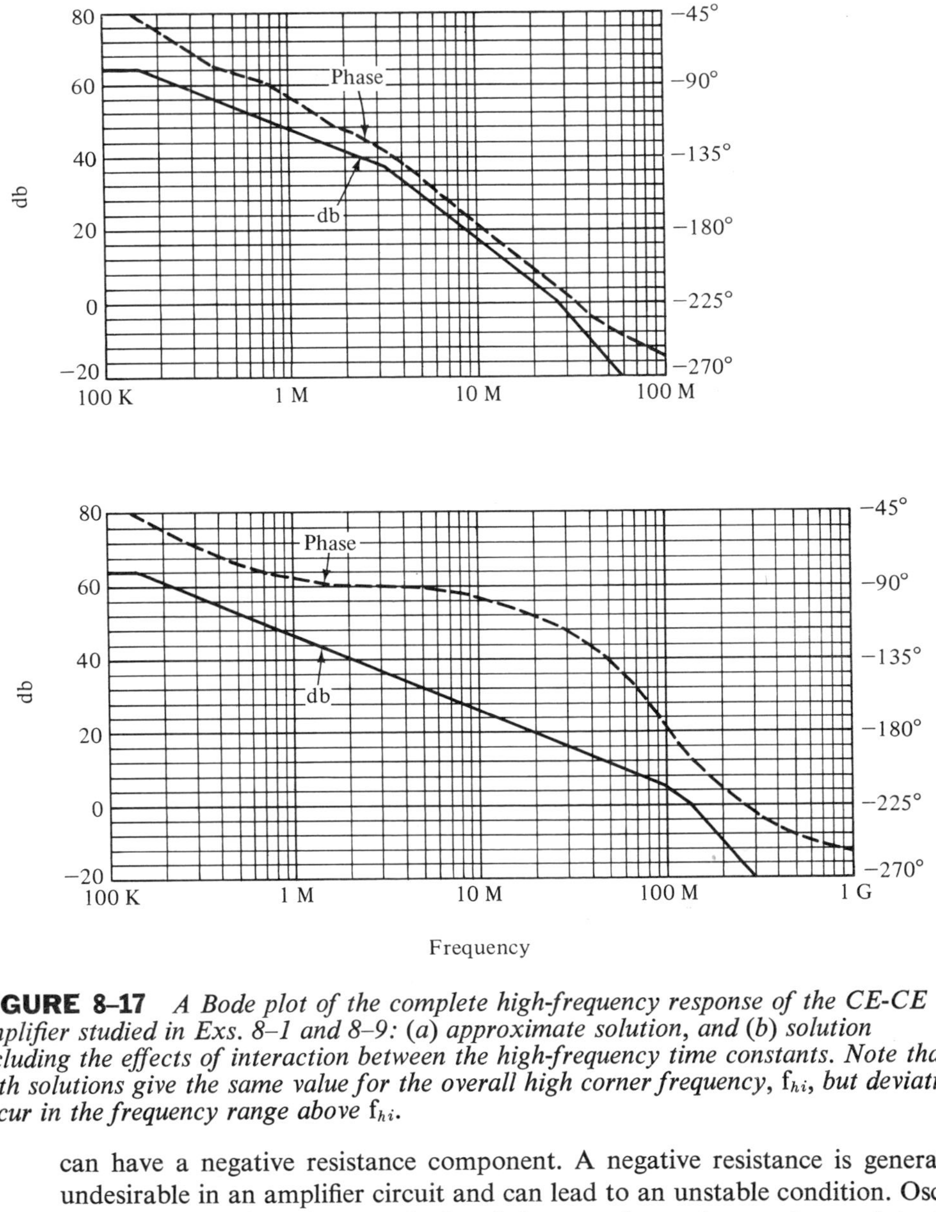

FIGURE 8–17 *A Bode plot of the complete high-frequency response of the CE-CE amplifier studied in Exs. 8–1 and 8–9: (a) approximate solution, and (b) solution including the effects of interaction between the high-frequency time constants. Note that both solutions give the same value for the overall high corner frequency,* f_{hi}*, but deviations occur in the frequency range above* f_{hi}.

can have a negative resistance component. A negative resistance is generally undesirable in an amplifier circuit and can lead to an unstable condition. Oscillation results when the magnitude of the negative resistance is equal to the magnitude of the equivalent generator resistance, causing the circuit to become lossless. Then the input capacitance of the CC stage and any stray inductance in the base circuit form an *LC* resonant circuit. The frequency of the oscillations cannot be predicted with any degree of accuracy due to the problem of evaluating the stray inductance and also the nonlinear nature of a negative resistance oscillator.*

* Many simple CC output stage circuits have been observed that were stable (no oscillations) when the output signal was monitored using a ×10 probe (7 pF), and were unstable when the output signal was monitored using a ×1 probe (70 pF).

Expressions for the maximum value of the negative resistance, NR_{max}, and the frequency at which the maximum negative resistance occurs, $f_{m(NR)}$, are developed in Appendix C and summarized below:

$$\left.\begin{aligned} NR_{max} &= r_{b'} + \frac{a_1 - \omega_m^2(a_1 b_1 - a_2 b_2)}{(1 - \omega_m^2 b_1)^2 + \omega_m^2 b_2^2} \\ \omega_m &= 2\pi f_{m(NR)} \\ f_{m(NR)} &\simeq f_\beta[M + (M+1)\sqrt{M}]^{1/2} \\ f_\beta &= f_T/\beta \\ M &= (r_{b'e}C_{b'e})/(R_E C_E) \\ a_1 &= (\beta+1)(r_e + R_E) \\ a_2 &= r_{b'e}R_E(C_{b'e} + C_E) \\ b_1 &= r_{b'e}R_E(C_{b'c}C_{b'e} + C_{b'c}C_E + C_{b'e}C_E) \\ b_2 &= (\beta+1)(r_e + R_E)C_{b'c} + R_E C_E + r_{b'e}C_{b'e} \end{aligned}\right\} \qquad (8\text{–}23)$$

The following example illustrates the calculation of the negative input resistance of a CC amplifier.

EXAMPLE 8–10 Estimate the maximum value of negative resistance and the frequency at which this maximum value occurs for a CC amplifier with the following parameters: $I_E = 1$ mA, $V_{CE} = 7.6$ V, $r_{b'} = 75$ ohms, $\beta = 101$, $f_T = 30$ MHz, $C_{b'c} = 18.5$ pF, $R_E = 1$ K, and $C_E = 680$ pF.

Solution: (a) Calculations:

$$r_e \simeq 26\text{ mV}/1\text{ mA} \simeq 26\,\Omega \qquad r_{b'e} = (\beta+1)r_e \simeq 2.65\text{ K}$$

$$f_\beta = f_T/\beta \simeq 300\text{ kHz} \qquad C_{b'e} \simeq 1/2\pi r_e f_T \simeq 203\text{ pF}$$

$$M = r_{b'e}C_{b'e}/R_E C_E = (2.65\text{ K})(203\text{ pF})/(1\text{ K})(680\text{ pF}) \simeq 0.79$$

$$f_{m(NR)} \simeq f_\beta[M + (M+1)\sqrt{M}]^{1/2} = (300\text{ kHz})[0.79 + (1.79)\sqrt{0.79}]^{1/2} \simeq 470\text{ kHz}$$

$$\omega_m = 2\pi f_{m(NR)} = 2.95 \times 10^6$$

$$a_1 = (\beta+1)(r_e + R_E) \simeq 1.05 \times 10^5 \qquad a_2 = r_{b'e}R_E(C_{b'e} + C_E) \simeq 2.35 \times 10^{-3}$$

$$b_1 = r_{b'e}R_E(C_{b'c}C_{b'e} + C_{b'c}C_E + C_{b'e}C_E) \simeq 0.41 \times 10^{-12}$$

$$b_2 = (\beta+1)(r_e + R_E)C_{b'c} + R_E C_E + r_{b'e}C_{b'e} \simeq 3.16 \times 10^{-6}$$

$$NR_{max} = r_{b'} + \frac{a_1 - \omega_m^2(a_1 b_1 - a_2 b_2)}{(1 - \omega_m^2 b_1) + \omega_m^2 b_2^2} \simeq -2.2\text{ K}$$

(b) Experimental results: The measured value of the maximum negative resistance was approximately -2.7 K occurring at approximately 700 kHz.*

Several steps can be taken to reduce the possibility of oscillations in a CC amplifier. Two obvious precautions are: (1) minimize the capacitance from the emitter terminal to ground, and (2) minimize the stray inductance in the base circuit. The positive part of the input resistance can be increased by adding a resistor either in series with the base lead or in series with the emitter lead (between the emitter terminal and the $R_E C_E$ load). The latter method is often preferable since a small emitter resistor appears multiplied by $(\beta + 1)$ when viewed from the base terminal.

---

* The measurement techniques are described in DeWitt and Pierce in the references at the end of this chapter. The method consisted essentially of measuring the magnitude and phase of the voltages on either side of a series base resistor.

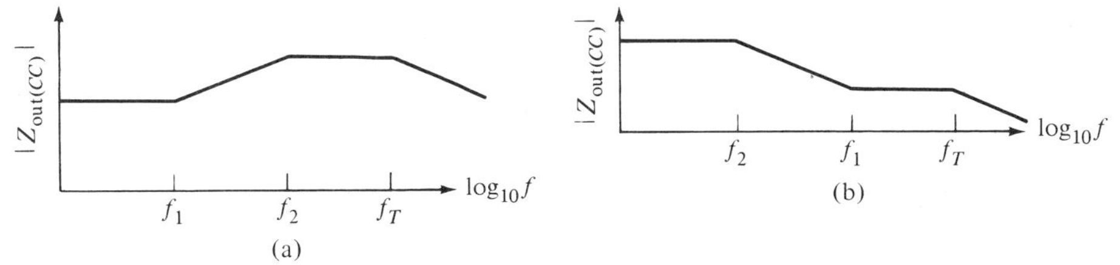

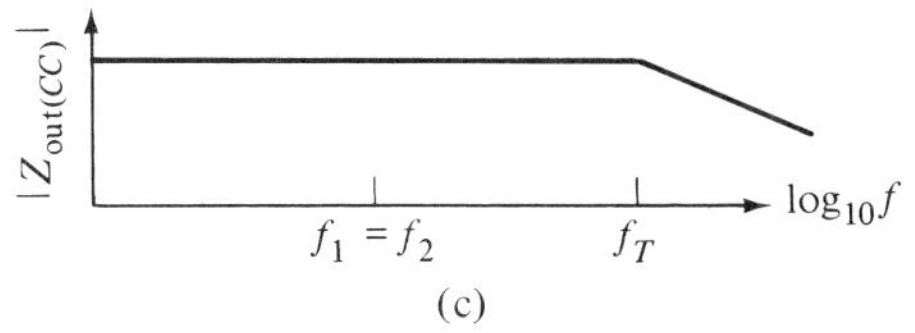

FIGURE 8–18 *Plot of* Z_{out} *as a function of frequency for: (a)* $f_1 < f_2$, *(b)* $f_1 > f_2$, *and (c)* $f_1 = f_2$.

An exact expression for the output impedance of a CC stage (see Appendix C) is

$$Z_{out(CC)} = \left[r_e + \frac{R_b}{\beta + 1}\right]\left[\frac{1 + jf/f_1}{(1 + jf/f_2)(1 + jf/f_T)}\right] \tag{8–24}$$

where

$$\left.\begin{array}{l} f_1 \equiv 1/[2\pi(R_b \,||\, r_{b'e})(C_b + C_{b'e})] \\ f_2 \equiv 1/(2\pi R_b C_b) \qquad f_T \simeq 1/(2\pi r_e C_{b'e}) \end{array}\right\} \tag{8–25}$$

and R_b and C_b represent the equivalent resistance and capacitance (including C_{ob}) from the base terminal to ground. The expression for $Z_{out(CC)}$ given by Eq. (8–24) shows that the output impedance of a CC stage can not be represented by a simple *RC* network. The mid-band value of $|Z_{out(CC)}|$, however, has the same value that was given in Table 4–1: r_e plus any resistance in the base circuit divided by $(\beta + 1)$. In the high-frequency region, the output impedance can appear as a complex impedance with a resistive component and either an inductive or a capacitive component depending on the relative values of f_1 and f_2. For example, if f_1 is less than f_2, then the zero in Eq. (8–24) occurs first, and $Z_{out(CC)}$ has an inductive component over the frequency range between f_1 and f_2 as shown in Fig. 8–18(a). When f_1 is greater than f_2, then $Z_{out(CC)}$ has a capacitive component over the frequency range between f_1 and f_2 as shown in Fig. 8–18(b). When f_1 and f_2 have approximately the same value, their effects tend to cancel, and $Z_{out(CC)}$ remains resistive for frequencies less than f_T. This last condition is shown in Fig. 8–18(c) and corresponds to the case treated in Sec. 7.8. Assuming that f_1 and f_2 are approximately equal, the resistive and capacitive components of $Z_{out(CC)}$ are the same as given in Table 7–1, or

$$\left.\begin{array}{l} R_{out(CC)} \simeq r_e + R_b/(\beta + 1) \\ C_{out(CC)} \simeq 1/2[r_e + R_b/(\beta + 1)]f_T \end{array}\right\} \tag{8–26}$$

Following a procedure similar to that used in the CC analysis, the following expression for the output impedance of a CD stage can be derived:

$$Z_{out(CD)} = \left[\frac{1}{g_m}\right]\left[\frac{1 + sR_g(C_g + C_{gs})}{(1 + sR_gC_g)(1 + sC_{gs}/g_m)}\right] \tag{8–27}$$

where R_g and C_g represent the equivalent resistance and capacitance (including C_{gd}) from the gate terminal to ground.

The input impedance of the CB stage is directly related to the output impedance of the CC stage since, in each case, we are looking into the emitter terminal. Consequently, Eq. (8–24) represents $Z_{\text{in(CB)}}$ as well as $Z_{\text{out(CC)}}$. As in the case of $Z_{\text{out(CC)}}$, $Z_{\text{in(CB)}}$ can appear as a complex impedance with a resistive component and either an inductive or a capacitive component depending on the circuit values. The inductance of a high-frequency low-power silicon transistor when $Z_{\text{in(CB)}}$ appears inductive is approximately equal to $r_{b'}/2\pi f_T$.* Some integrated circuits utilize this effect when small values of inductance are required.

The output impedance of a CB is of interest for both the CB amplifier and the bipolar constant-current source. From the derivation in Appendix C,

$$R_{\text{out(CB)}} = r_{b'c} \parallel \left[\frac{r_{ce} + R_e \parallel R_{\text{ine}}}{1 - [\beta/(\beta+1)][R_e/(R_e + R_{\text{ine}})]}\right] \tag{8–28}$$

where

$$R_{\text{ine}} = r_e + [R_b/(\beta + 1)] \tag{8–29}$$

R_e is the total ac resistance in the emitter circuit, and R_b is the total ac resistance in the base circuit. This expression for $R_{\text{out(CB)}}$ can often be simplified. For example, when R_e is completely bypassed, the value of $R_{\text{out(CC)}}$ reduces to $R_{\text{out(CE)}}$ or simply $r_{b'c} \parallel r_{ce}$. When R_b is completely bypassed, and the value of r_{ce} is much greater than r_e, the value of $R_{\text{out(CC)}}$ reaches its largest value or $r_{b'c} \parallel (\beta + 1)r_{ce}$. When neither R_b nor R_e is bypassed, an intermediate result is obtained, and the value of $R_{\text{out(CB)}}$ depends on the values of R_b and R_e. Consequently, the output resistance of a CB stage (or a bipolar transistor constant-current source) increases when the base is ac grounded, and the equivalent emitter resistance is large. Table 8–2 summarizes the important results.

TABLE 8–2 *Summary of output resistance expressions for CB stage.*

Circuit Condition	$R_{\text{out(CB)}}$
General	Eq. (8–28)
$R_e = 0$	$r_{b'c} \parallel r_{ce}$
$R_b = 0,\ R_e \neq 0$	$r_{b'c} \parallel (\beta + 1)(r_{ce})$
$R_b \neq 0,\ R_e \neq 0$	$r_{b'c} \parallel \dfrac{(\beta + 1)(r_{ce})}{1 + R_b/R_e}$

Developing an expression for the output capacitance of a CB stage requires a solution of a three-node circuit plus the reference node. However, the effect of an unbypassed base resistor on the value of $C_{\text{out(CB)}}$ can be approximated as**

$$C_{\text{out(CB)}} \simeq C_{ob}(1 + R_b/R_e) \tag{8–30}$$

Equation (8–30) shows that the output capacitance of a CB stage (or a bipolar transistor constant-current source) has a minimum value when the base is ac grounded.

In this section, some of the important second order effects encountered in transistor circuits have been introduced. The details of the various derivations

---

* See Kennedy listed in the references at the end of this chapter.

** Equation (8–30) has been indirectly verified experimentally in Hill and Albritton in the references listed at the end of this chapter.

are contained in Appendix C. A simplified method of factoring a polynomial when the roots are widely separated was presented in Eq. (8–5) through Eq. (8–7). The effects of time constant interactions were examined for the CE and CC stages.* A negative resistance component of the input impedance of a CC stage can occur when the emitter circuit contains a parallel RC network. In both the output impedance in a CC stage and the input impedance of a CB stage we found that either an inductive component or a capacitive component may be present, depending on the values of the various circuit components. Finally, the output impedance of a CB stage or a bipolar constant-current source was examined. The various effects summarized in this section will be used in later chapters.

8.5 Summary

In this chapter, we have discussed the analysis and design of multistage amplifier circuits. The seven multistage amplifier examples discussed in Sec. 8.1 provided some insight into the performance of various configurations. For example, an amplifier consisting of two inverting stages can have a high value of gain but often has a relatively low overall high corner frequency. The overall high corner frequency of such an amplifier is limited by the high input capacitance of an inverting amplifier.

Because of its versatility, the differential amplifier configuration is often used even when only one input signal is present. This circuit can have an excellent low-frequency response since a bypass capacitor is not usually required. When operated as a non-inverting amplifier, the differential configuration corresponds to the CC-CB configuration and has a relatively high overall high corner frequency if the collector resistor of the CC stage is omitted. When operated as an inverting amplifier, the differential configuration corresponds to the CE stage with an unbypassed emitter resistor approximately equal to r_{e2}. An unbypassed emitter resistor reduces the voltage gain from the base-to-collector and, consequently, reduces the value of the input capacitance since C_{ob} is multiplied by a smaller number. Also, the effect of $C_{b'e}$ is reduced due to the presence of r_{e2}.

Often the high-frequency response of a differential amplifier is improved by either using a CB stage at each output to form cascode circuits, or by using a CC stage at each input to reduce the input high-frequency time constant. The monolithic amplifier shown in Fig. 8–26 used both techniques to improve the high-frequency response of a basic differential amplifier.

The frequency response of an intergrated circuit amplifier is generally given in terms of the overall circuit performance rather than in terms of the individual transistors. An analysis is complicated by the distributed parasitic capacitances associated with both the active devices and the passive components used in the integrated circuit. However, the response of many integrated circuits can be estimated using the techniques of multistage analysis and reasonable estimates for the various capacitors in the circuit.

A natural extension of the analysis of multistage circuits is the design of multistage circuits. Although a design procedure is often a trial-and-error process, a judicious choice of the configuration, based on the characteristics of

* Since the input and output nodes of the CB stage are well isolated, time constant interaction is not as important.

each stage, greatly simplifies the procedure. The first step in the design of an amplifier is to determine the specifications. Once the specifications are known, the next step is to select the input and output stages based on the required input and output impedances. Then the remainder of the amplifier is designed with both the gain and frequency response in mind. The final circuit configuration must be analyzed to ensure that all specifications are satisfied. If possible, the final design should use only standard components that are readily available at low cost. The final step in the design sequence is to build and test the circuit to check its performance experimentally.

Problems

Section 8.1

8–1 Calculate the mid-band gain and frequency response of the bipolar transistor folded cascode circuit shown in Fig. 8–19. Both transistors are silicon and have values of $\beta \simeq 100$, $f_T \simeq 300$ MHz, and $C_{ob} \simeq 5$ pF.

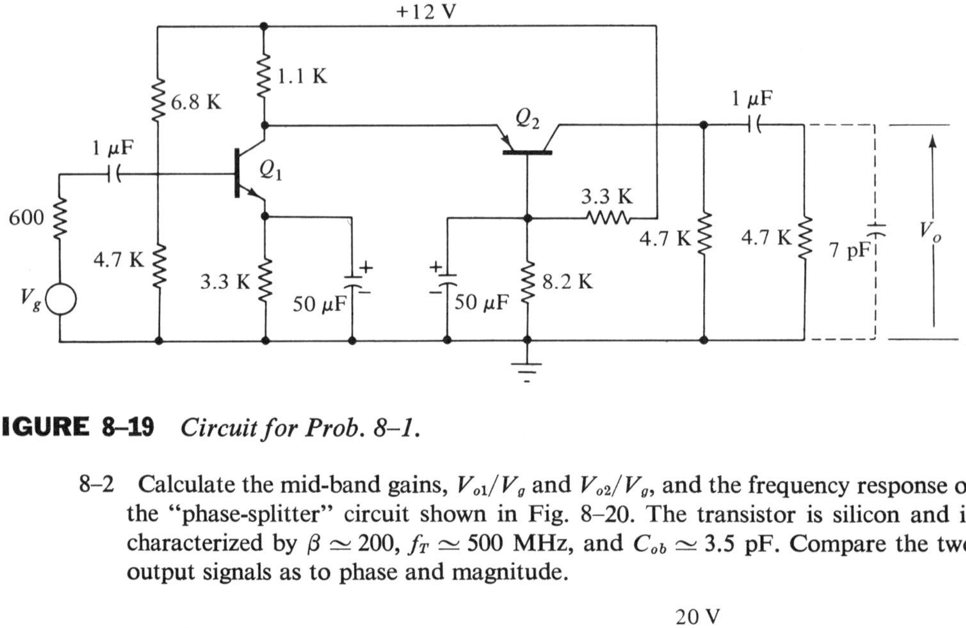

FIGURE 8–19 *Circuit for Prob. 8–1.*

8–2 Calculate the mid-band gains, V_{o1}/V_g and V_{o2}/V_g, and the frequency response of the "phase-splitter" circuit shown in Fig. 8–20. The transistor is silicon and is characterized by $\beta \simeq 200$, $f_T \simeq 500$ MHz, and $C_{ob} \simeq 3.5$ pF. Compare the two output signals as to phase and magnitude.

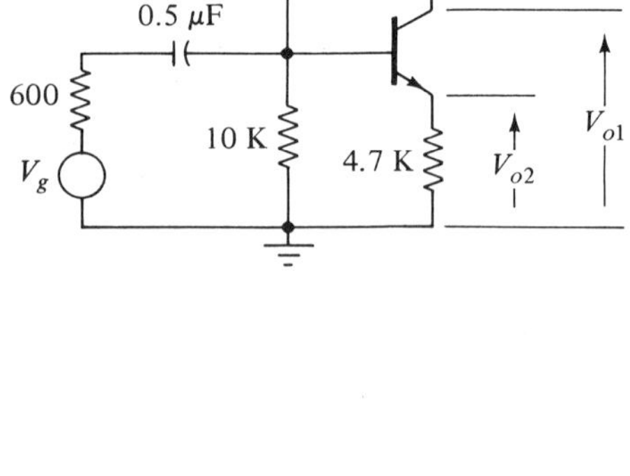

FIGURE 8–20 *Circuit for Prob. 8–2.*

8–3 Calculate the mid-band gain and frequency response of the CB-CC amplifier shown in Fig. 8–21. Both transistors are silicon and have values of $\beta \simeq 75$, $f_T \simeq 400$ MHz, and $C_{ob} \simeq 8$ pF.

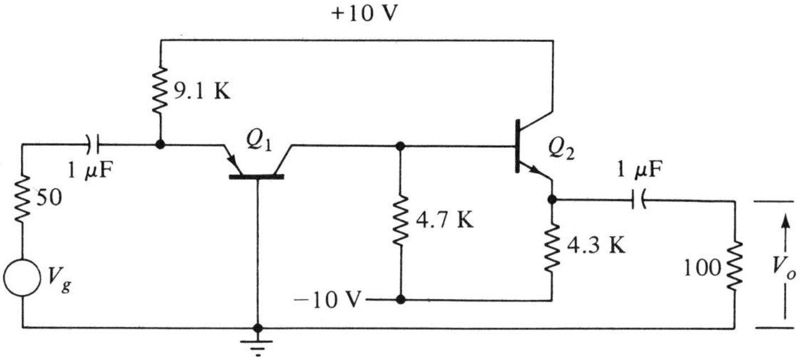

FIGURE 8–21 *Circuit for Prob. 8–3.*

8–4 Calculate the mid-band gain and frequency response of the FET differential amplifier shown in Fig. 8–22. The N-channel JFETs are characterized by

Transistor	V_P	I_{DSS}	V_{GS}	C_{gs}	C_{gd}
Q_1	−1.8 V	4.5 mA	−0.6 V	8 pF	4 pF
Q_2	−1.6 V	4.7 mA	−0.6 V	8 pF	4 pF

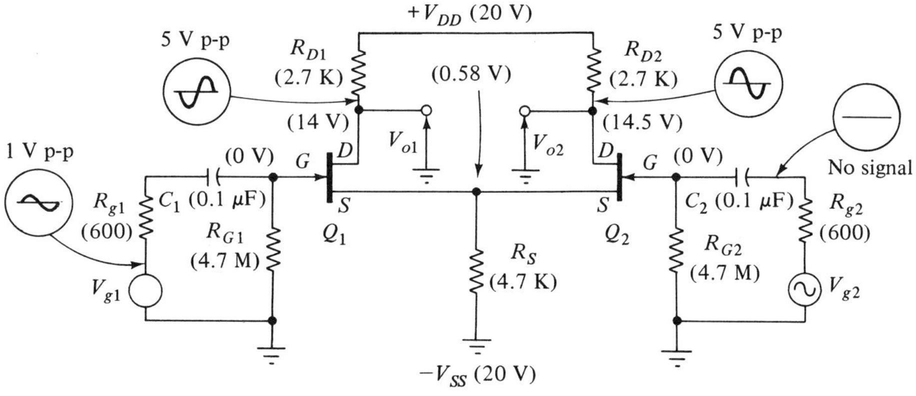

FIGURE 8–22 *Circuit for Prob. 8–4.*

8–5 Calculate the mid-band gain and rise time of the amplifier shown in Fig. 8–23. All transistors are silicon and are characterized by

Transistor	β	f_T	C_{ob}
Q_1	110	400 MHz	6 pF
Q_2	58	600 MHz	1.7 pF
Q_3	76	400 MHz	8 pF

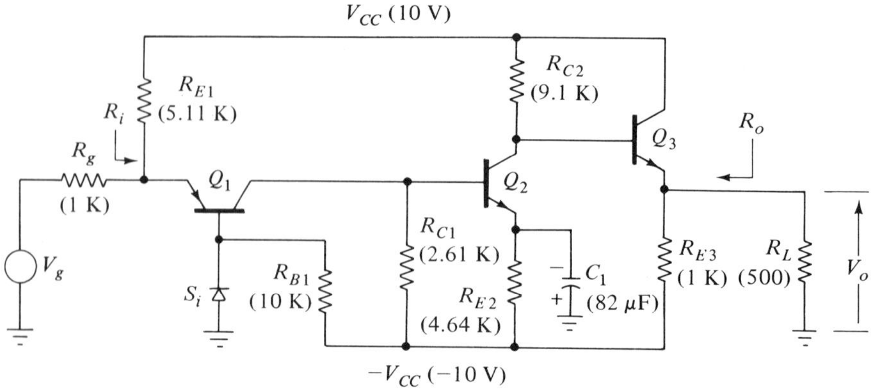

FIGURE 8–23 *Circuit for Prob. 8–5.*

8–6 Calculate the mid-band gain and rise time of the amplifier shown in Fig. 8–24. All transistors are silicon and are characterized by

Transistor	β	f_T	C_{ob}
Q_1, Q_2 (dual)	800	60 MHz	3 pF
Q_3	130	200 MHz	6 pF
Q_4	200	200 MHz	8 pF
Q_5	100	500 MHz	3 pF

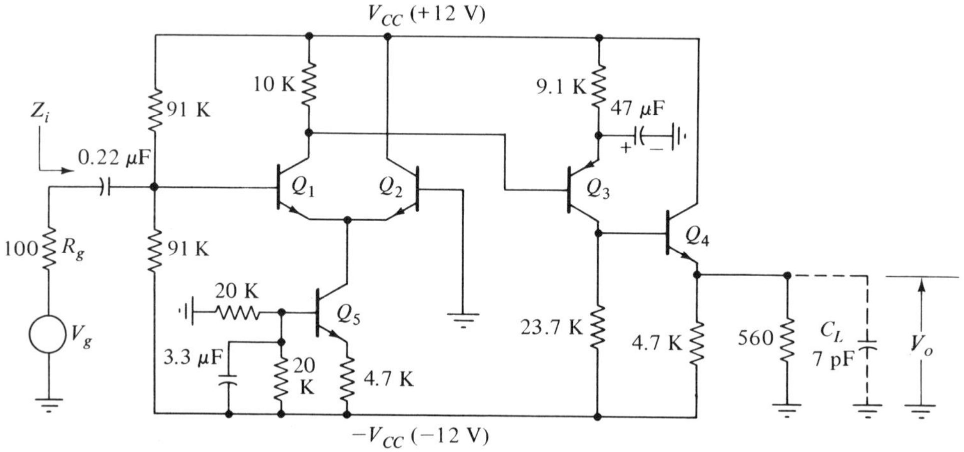

FIGURE 8–24 *Circuit for Prob. 8–6.*

8–7 Calculate the voltage gain $V_o/(V_{g1} - V_{g2})$, and estimate the rise time of the triode differential amplifier shown in Fig. 8–25. The two triodes are matched and are characterized by values of $g_m \simeq 3000$ μmhos, $r_p \simeq 10$ K, $C_{gk} \simeq 2$ pF, $C_{gp} \simeq 5$ pF, and $C_{pk} \simeq 3$ pF when biased at a plate current of 3 mA and $V_{GK} = -2$ V.

Section 8.2

8–8 The monolithic differential amplifier shown in Fig. 8–10 is to be operated at 100°C using a supply voltage of 9 V. Estimate the values of: (a) the single-ended voltage gain, (b) the high-frequency response, and (c) the total power dissipation of the IC.

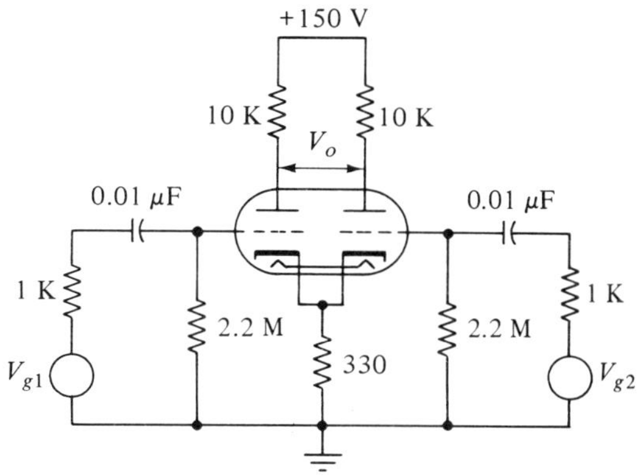

FIGURE 8–25 *Circuit for Prob. 8–7.*

8–9 For the monolithic amplifier shown in Fig. 5–42, calculate (a) all bias voltages and currents, (b) the difference-mode gain, (c) the common-mode gain, and (d) estimate the high-frequency response.

8–10 Repeat Prob. 8–9 using supply voltages of +9 V and −9 V.

8–11 Repeat Prob. 8–9 for the circuit shown in Fig. 8–26.

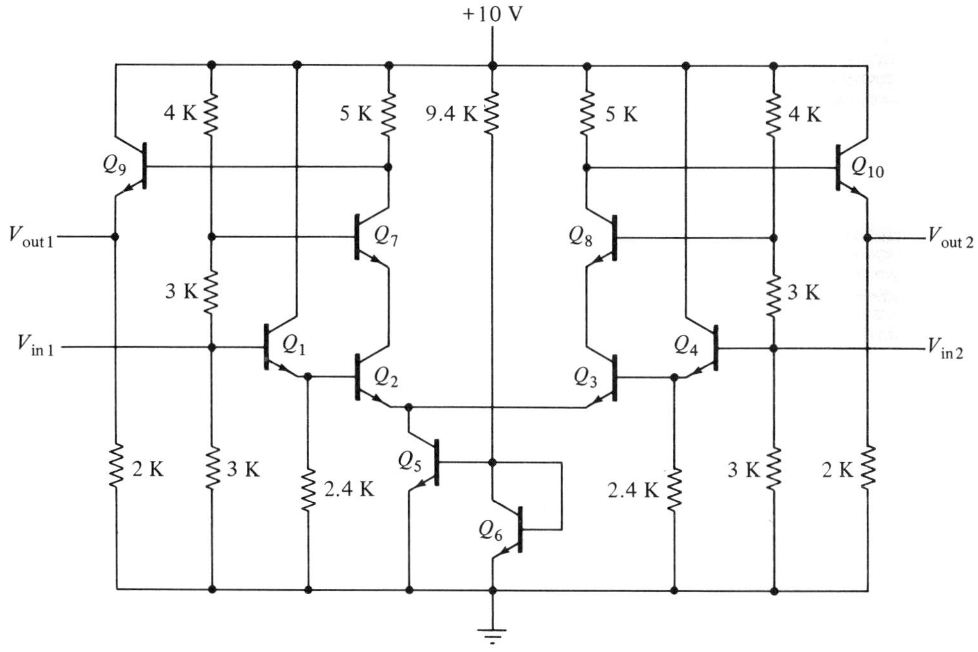

FIGURE 8–26 *Circuit for Prob. 8–11.*

8–12 Draw an ac equivalent circuit for the monolithic breadboard amplifier shown in Fig. 8–11(a). In the equivalent circuit, represent each monolithic resistor with a resistor shunted at each end by a capacitor having a value $\frac{1}{2}C_R$. The diode $D3$ can be represented by its equivalent resistance. Each transistor should be represented by its circuit symbol with the various device and substrate capacitors connected external to the transistor symbol. The values of the various components depend on the bias conditions and are given in Tables 5–4 through 5–7. [A partial equivalent circuit of this amplifier is shown in Fig. 8–11(b).]

8–13 Using the results of Prob. 8–12, determine the values of A_{vo}, f_{hi}, and f_{lo} for the amplifier shown in Fig. 8–11(a).

Section 8.3

8–14 For the amplifier shown in Fig. 8–15(e), determine the value of R_{S1} required to set the dc voltage at the emitter of Q_3 equal to zero. Using this value of R_{S1}, verify that this amplifier satisfies the design specifications given in Eq. (8–1). The JFET parameters are $V_P = -3$ V, $I_{DSS} = 5.6$ mA, $C_{gs} = 8$ pF, and $C_{gd} = 4$ pF. The bipolar transistor parameters are: $\beta = 200$, $f_T = 300$ MHz, and $C_{ob} = 5$ pF.

8–15 Design an amplifier to satisfy the specifications given in Eq. (8–1) using a nominal voltage gain of $+25$. For this design, use the JFET and bipolar parameters from Prob. 8–14.

8–16 Repeat Prob. 8–15 using a nominal voltage gain of -200.

8–17 Repeat Prob. 8–15 using a nominal voltage gain of $+200$.

8–18 Verify that the circuit shown in Fig. 8–16(e) satisfies the following design specifications: $A_{vo} \geq 10^4$, $R_i \geq 1$ K, $R_o \leq 100$ ohms, $f_{hi} \geq 200$ kHz, and $f_{lo} \leq 3$ Hz. The transistor parameters are given in Prob. 8–14.

8–19 Design an amplifier to satisfy the specifications given in Prob. 8–18 except for a gain of -10^4. For this design, use the bipolar transistor parameters given in Prob. 8–14.

8–20 Repeat Prob. 8–19 using an input impedance of greater than 10 M and a voltage gain of $+10^3$.

8–21* Design an amplifier to meet the following specifications:

$$A_{vo} = -10^6, \pm 10\% \qquad R_i > 1\text{ K}$$
$$f_{hi} > 1\text{ kHz} \qquad R_o < 1\text{ K}$$
$$f_{lo} < 10\text{ Hz}$$

The amplifier is to be used in an application where R_L is 10 K and R_g is 100 ohms. Supply voltages of both $+24$ V and -24 V are available. Use only standard components, and specify the active devices by number.

8–22 Design an amplifier to meet the following specifications:

$$A_{vo} = +500, \pm 10\% \qquad R_i = 50\ \Omega$$
$$f_{hi} \geq 2\text{ MHz} \qquad R_o = 50\ \Omega$$
$$f_{lo} \leq 10\text{ Hz}$$

The amplifier is to be used in an application where R_g is 50 ohms and R_L is 50 ohms. Supply voltages of both $+12$ V and -12 V are available. Use only standard components, and specify the active devices by number.

8–23 Repeat Prob. 8–21 using the monolithic breadboard shown in Figs. 5–36 and 5–37. Component characteristics are given in Tables 5–4 through 5–7.

8–24 Repeat Prob. 8–22 using the monolithic breadboard shown in Figs. 5–36 and 5–37. Component characteristics are given in Tables 5–4 through 5–7.

---

* For Probs. 8–21, 8–22, and 8–25, the specification of the actual devices by number is required. For analysis and design, use typical or average values of β, V_P, and I_{DSS}; the maximum values of C_{ob}, C_{gs}, and C_{gd}, and the minimum value of f_T. The use of these values will generally yield a conservative design.

8–25 A certain transducer is characterized by an output resistance of 1 M and an output capacitance of 10 pF. The signal appearing at the terminals of the transducer when the terminals are short circuited consists of pulses which have peak amplitudes ranging from 1 to 20 μA, durations ranging from 200 to 10 μsec, and times between pulses ranging from 1 to 10 msec. The pulses will be monitored on an instrument which has an input resistance of 200 ohms and an input capacitance of 50 pF. Design an amplifier to place between the transducer and the monitoring instrument which will essentially preserve the shape of the pulses and produce a peak output signal ranging up to 1 V.

Section 8.4

8–26 Repeat Ex. 7–5 to include the effects of high-frequency time constant interaction.

8–27 For the two-stage CE amplifier shown in Fig. 8–27, calculate the mid- and high-frequency response including the effects of high-frequency time constant interactions. Both transistors are characterized by a β of 100, a C_{ob} of 2 pF, and an f_T of 500 MHz.

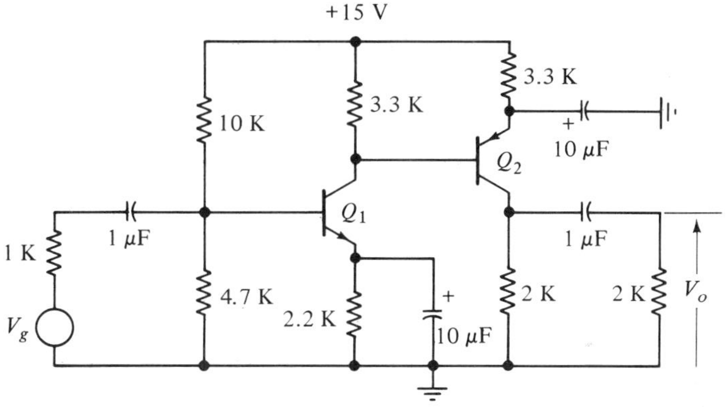

FIGURE 8–27 *Circuit for Prob. 8–27.*

8–28 Repeat Ex. 8–6 to include the effects of high-frequency time constant interaction.

8–29 The CC stage of the CC-CE amplifier discussed in Ex. 8–6 is loaded with a parallel *RC* circuit where the load capacitance is equal to the input capacitance of the CE stage. If this CC stage has a negative input resistance, determine the frequency at which the maximum value of negative resistance occurs.

8–30 Assuming that r_{ce} is 50 K, calculate the resistance and capacitance looking into the collector terminal of the transistor described in Prob. 8–2 and shown in Fig. 8–20.

8–31 Calculate the high-frequency time constant at node "A" indicated on the partial circuit diagram shown in Fig. 8–28. All transistors are characterized by a β of 100, a C_{ob} of 5 pF, an r_{ce} of 20 K, and an f_T of 300 MHz. Note that the base terminals are not ac grounded.

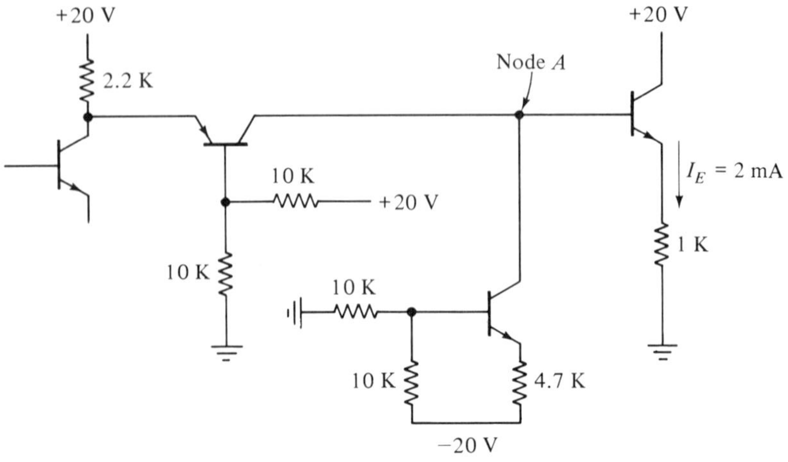

FIGURE 8–28 *The partial circuit diagram for Prob. 8–31.*

References

DeWitt, B. F. and J. F. Pierce, "Input Impedance of Emitter Followers," *Scientific Report*, No. 15, Contract No. W7405, Eng. —26(2160), Oak Ridge National Laboratory, Oak Ridge, Tennessee, 1964.

Goshell, R. E., *Engineering Mathematics*. New York: The Dryden Press, Inc., 1958.

Hill, N. W. and W. P. Albritton, "An Input-Capacity-Insensitive, Charge-Sensitive Preamplifier for Simultaneous Use in Vacuum with Fast Amplifiers for Charged-Particle Studies with Semiconductor Detectors," *Nuclear Instruments and Methods*, 75 (1969), 18–28.

Kennedy, E. J., Unpublished laboratory notes. Oak Ridge National Laboratory, Oak Ridge, Tennessee.

Pierce, J. F., *Transistor Circuit Theory and Design*. Columbus, Ohio: Charles E. Merrill Publishing Company, 1963.

Searle, C. L., A. R. Boothroyd, E. J. Angelo, Jr., P. E. Gray, and D. O. Pederson, *Elementary Circuit Properties of Transistors*. New York: McGraw-Hill Book Company, 1955.

9

Tuned Amplifiers

The amplifiers discussed in the preceding chapters are often called wide-band amplifiers because the ratio of the high -3-db frequency to the low -3-db frequency is several thousand to one or larger. These amplifiers are used in audio, video, and instrumentation systems.

Amplifiers used in radio frequency applications have low values for the ratio of the high to the low -3-db frequencies, usually less than two to one. Because of this low ratio, they are called narrow-band amplifiers. While narrow-band amplifiers can be made with resistors and capacitors,* radio frequency (RF) amplifiers generally use inductors and capacitors in tuned circuit arrangements. The objectives of this chapter are to review briefly the fundamentals of tuned or resonant circuits and then to study narrow-band tuned amplifiers by combining tuned circuits and active devices.

The characteristics of tuned circuits are summarized in Sec. 9.1. These tuned circuits are used as loads on single-stage amplifiers in Sec. 9.2. Multistage amplifiers are developed in Sec. 9.3. Since tuned amplifiers of the inverting type tend to be unstable, Sec. 9.4 includes a short discussion on the cause and cure of instability. A short discussion of IC tuned amplifiers is included in Sec. 9.5.

9.1 Tuned Circuits

Tuned circuits can be subdivided into two broad catagories: (1) series resonant circuits, and (2) parallel resonant circuits. The objective of this section is to review briefly the performance characteristics of tuned circuits.**

A simple series resonant circuit is shown in Fig. 9–1(a). The resistor, R_c, represents the effective resistance of the coil. At very low frequencies, the equivalent impedance, Z, has a capacitive component and a leading phase angle. At

---

* Narrow-band amplifiers using *RC* components include bridge circuits and twin-T filters.

** For a more detailed treatment of resonant circuits, see C. M. Close and K. Y. Tang in the references listed at the end of the chapter or any other good texts on ac circuit theory.

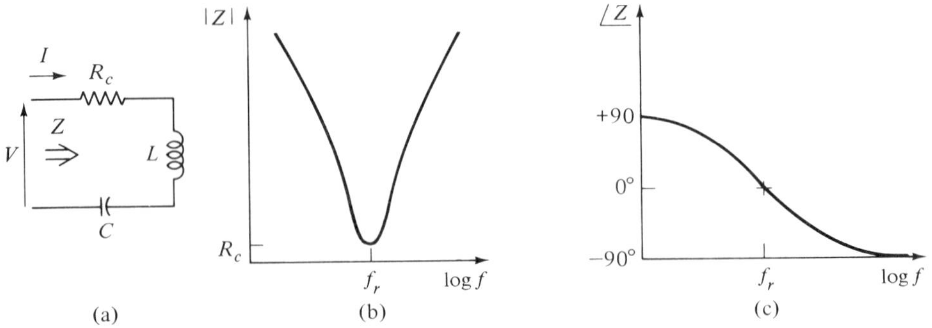

FIGURE 9–1 *Series resonant circuits: (a) circuit diagram, (b) magnitude response of the equivalent impedance, and (c) phase response of the equivalent impedance.*

high frequencies, Z has an inductive component and a lagging phase angle. The frequency at which the magnitudes of the capacitive reactance and inductive reactance are equal is called the resonant frequency, f_r, of the circuit. This frequency is given by

$$f_r = 1/(2\pi\sqrt{LC}) \tag{9–1}$$

At the resonant frequency the following approximate conditions exist: (1) Z is purely resistive and is equal to R_c, (2) the current has a maximum value and is limited by R_c, and (3) the phase angle is zero.

Generally, the performance of a resonant circuit is expressed in terms of the "multiplication factor," "quality factor," or "Q" of the circuit. Physically, the circuit Q is

$$Q = \frac{(2\pi)(\text{Peak energy stored})}{\text{Energy lost per cycle}} \tag{9–2}$$

In terms of the circuit components,

$$Q = \frac{2\pi f_r L}{R_c} = \frac{1}{R_c}\sqrt{\frac{L}{C}} = \frac{1}{2\pi f_r C R_c} \tag{9–3}$$

Values of Q less than approximately 5 are usually termed "low" while values greater than approximately 10 are termed "high." Practical values of Q often exceed 100 in electronic circuits.

The normalized current response of a series resonant circuit as a function of frequency is shown in Fig. 9–2. The selectivity of a tuned circuit is often given in terms of the −3-db bandwidth, BW, where

$$BW_{3\text{ db}} \simeq f_r/Q \tag{9–4}$$

A parallel resonant circuit is shown in Fig. 9–3(a). This circuit consists of a coil in parallel with an ideal capacitor.* The coil has a resistance of R_c. At very low frequencies, the equivalent impedance, Z, has an inductive component and a lagging phase angle. At very high frequencies, Z has a capacitive component and a leading phase angle. Like the series tuned circuit, the frequency at which the input impedance appears purely resistive is called the resonant frequency,

* An ideal capacitor is one with no leakage. A non-ideal capacitor has leakage that is usually represented by a parallel resistor. Most capacitors used in electronic tuned circuits have negligible leakage.

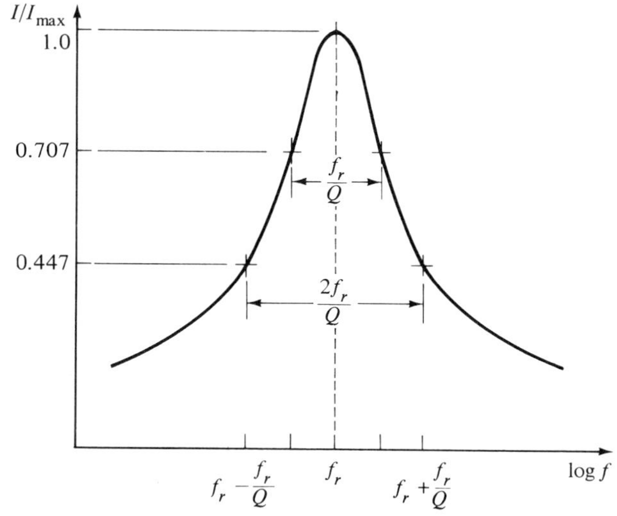

FIGURE 9–2 *Normalized current response in a series resonant circuit as a function of frequency.*

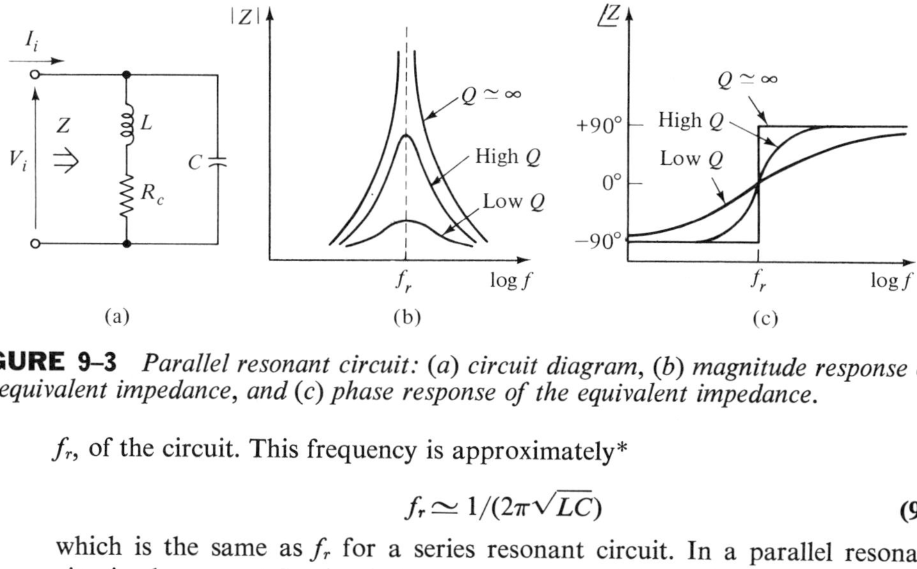

FIGURE 9–3 *Parallel resonant circuit: (a) circuit diagram, (b) magnitude response of the equivalent impedance, and (c) phase response of the equivalent impedance.*

f_r, of the circuit. This frequency is approximately*

$$f_r \simeq 1/(2\pi\sqrt{LC}) \tag{9–5}$$

which is the same as f_r for a series resonant circuit. In a parallel resonance circuit, the current flowing in the branches is called the circulating current. At the resonant frequency, the circulating current is Q times larger than the line or input current. In addition, the input impedance at f_r is resistive and is approximately equal to Q times the reactance of one of the branches. Defining the equivalent input resistance at f_r as R_{op}, then

$$R_{op} \simeq V_i/I_i \simeq Q(2\pi f_r L) \simeq Q/(2\pi f_r C) \tag{9–6}$$

Solving for QR_c in terms of f_r and L in Eq. (9–3) and substituting into Eq. (9–6) gives

$$R_{op} \simeq Q^2 R_c \tag{9–7}$$

---

* This approximation is valid only when $Q \geq 10$, a criterion that is usually met in practice.

When tuned circuits are used with other circuits, the choice of a series or parallel type depends upon the nature of the other circuits. In the series case, the value of Q depends on the total equivalent series resistance or

$$Q_{\text{series}} = 2\pi f_r L / R_{\text{series}} \tag{9–8}$$

To obtain a reasonable value of Q, R_{series} must be small. In the parallel case, however, the value of Q depends on the total resistance shunting or loading the tuned circuit.

$$Q_{\text{parallel}} = R_{\text{parallel}}/(2\pi f_r L) \tag{9–9}$$

To obtain a reasonable value of Q, R_{parallel} must be large. When the shunting resistance is infinite, then the value of R_{parallel} is simply equal to the equivalent resistance of the tuned circuit, R_{op}. Both expressions for Q are based on the more fundamental definition given by Eq. (9–2). Therefore, in parallel tuned circuits, the principal loading effects are due to the equivalent resistance and capacitance placed across the terminals of the tuned circuit. A resistive load lowers the effective Q of the circuit. A capacitive load lowers the resonant frequency, f_r.

The normalized response of a parallel tuned circuit in terms of its equivalent impedance is shown in Fig. 9–4. This curve has the same shape as the curve shown in Fig. 9–2. Figure 9–4 is often referred to as the *universal resonant curve* since it can be used to estimate the response of any tuned circuit having a Q of 10 or larger. The following example illustrates both the selectivity of a tuned circuit and the effects of loading.

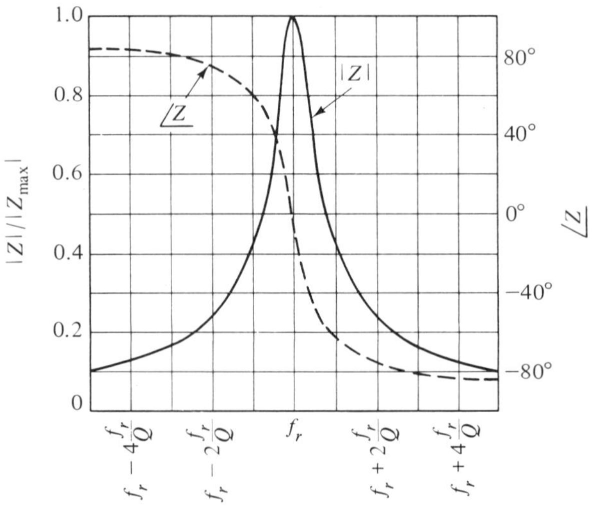

FIGURE 9–4 *A normalized impedance curve for a parallel tuned circuit.*

EXAMPLE 9–1 A parallel tuned circuit with a 455-kHz resonant frequency is constructed using a 100-μH coil having an effective resistance of approximately 3.2 ohms at 455 kHz.* (a) Find the value of the tuning capacitance. (b) Calculate the values

---

* The measured dc resistance of the coil was 1.6 ohms. The difference in resistance at 0 Hz and 455 kHz is due to the "skin effect." The skin effect results from a high-frequency field which causes the current to be concentrated in the outer part of the conductor. This effect reduces the effective cross-sectional area of a conductor in which current flows and increases the resistance.

of Q_L, Q_U, R_{op}, and the -3-db bandwidth. (c) What effect does a 10-K resistor connected in parallel with the tuned circuit have on these values?

Solution: (a) Value of tuning capacitance: From Eq. (9–1),

$$C = (0.159)^2/(455^2 \times 10^6 \times 100 \times 10^{-3}) \simeq 1225 \text{ pF}$$

(b) Unloaded characteristics of the tuned circuit: From Eqs. (9–3), (9–4), and (9–6),

$$Q_U = 2\pi f_r L/R_c \simeq (2\pi)(455 \times 10^3)(10^{-4})/3.2 \simeq 89$$

$$BW_{3\text{ db}} \simeq f_r/Q_U \simeq (455 \text{ kHz})/89 \simeq 5.1 \text{ kHz}$$

$$R_{op} \simeq Q_U^2 R_c \simeq (89)^2(3.2) \simeq 25.4 \text{ K}$$

(c) Loaded characteristics of the tuned circuit: The effect of the 10-K resistor connected in parallel with the tuned circuit is to reduce the value of the circuit Q from Q_U to Q_L. The equivalent resistance for the circuit is

$$R_{\text{eq}} = 10 \text{ K} \,||\, R_{op} \simeq 10 \text{ K} \,||\, 25.4 \text{ K} \simeq 7.2 \text{ K}$$

From Eqs. (9–9) and (9–4),

$$Q_L \simeq \frac{R_{\text{eq}}}{2\pi f_r L} \simeq \frac{7.2\ K}{(2\pi)(455 \times 10^3)(10^{-4})} \simeq 25$$

$$BW_{3\text{ db}} \simeq f_r/Q_L \simeq (455 \text{ kHz})/25 \simeq 18.2 \text{ kHz}$$

(d) Experimental results: The resonant frequency of the unloaded tuned circuit was found by connecting the tuned circuit to a sinusoidal signal generator through a 1-M resistor. Since R_{op} is much less than 1 M, this arrangement did not significantly load the tuned circuit. The signal across the tuned circuit was monitored using an $\times 10$ oscilloscope probe and a sensitive oscilloscope. Using a digital frequency meter, the measured resonant frequency was 447.8 kHz for the unloaded case. The *BW* was approximately 5 kHz. The resistor connecting the generator to the tuned circuit was changed to 10 K to measure the loaded characteristics of the tuned circuit. With this 10-K load, the measured value of f_r was 441.5 kHz, and the *BW* was approximately 21 kHz.

In this section, we have examined some of the characteristics of tuned circuits. Both series and the parallel tuned circuits are used in the narrow-band amplifiers discussed in the remainder of this chapter.

9.2 Single-Tuned Amplifiers

Tuned amplifiers are different from the wide-band amplifiers discussed in Chaps. 7 and 8 because of the frequency selective characteristics of the tuned circuits. The voltage gain of a tuned amplifier generally has a maximum value at f_r and decreases rapidly as the frequency deviates from f_r. The objective of this section is to study the characteristics of a simple single-stage tuned amplifier.

Some typical tuned circuits used in narrow-band amplifier applications are shown in Fig. 9–5. Transformer coupling is used in each case.* The circuits shown in Fig. 9–5(a) and (b) are called *single-tuned* circuits since only the primary circuit is tuned to f_r. Tuning is often accomplished by adjusting a moveable powdered iron slug. This arrangement is indicated by double arrows between the transformer windings. The circuits in Figs. 9–5(a) and (b) are different in that

---

* Capacitive coupling and other forms of inductive coupling are also used in some narrow band amplifiers operating at very high frequencies. One type is discussed later in this section.

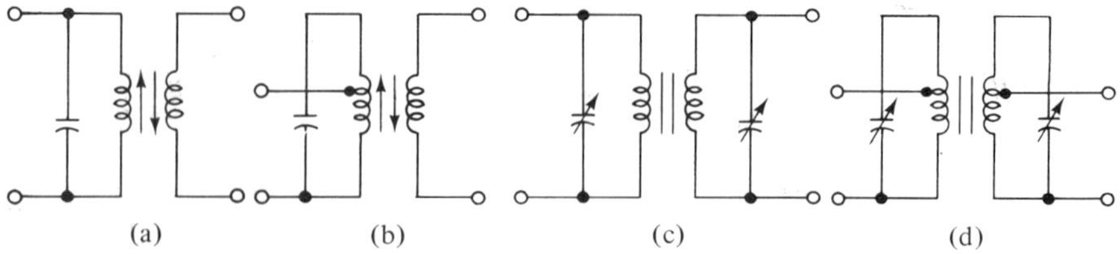

FIGURE 9–5 *Some typical tuned circuits: (a) single-tuned, (b) single-tuned with tapped primary, (c) double-tuned, and (d) double-tuned with tapped primary and tapped secondary.*

the latter has a tapped primary. The tapped primary arrangement reduces the loading effects of the equivalent primary resistance and preserves the high Q of the tuned circuit. The tuned circuit shown in Fig. 9–5(b) is considered later in this section.

The tuned circuits shown in Figs. 9–5(c) and (d) are called *double-tuned* circuits since both the primary and the secondary circuits form resonant circuits. The response of a double-tuned circuit depends on the mutual coupling between the two coils as well as the properties of each tuned circuit. The amount of coupling between the two circuits is given in terms of the *coefficient of coupling* or k, where k can take on values between 0 and 1.* In terms of the circuit parameters,

$$k = M\sqrt{L_p L_s} \qquad \textbf{(9–10)}$$

where M is the mutual inductance between the primary and secondary, L_p is the self-inductance of the primary circuit, and L_s is the self-inductance of the secondary circuit. The response of a double-tuned circuit for various values of k is shown in Fig. 9–6(a). The maximally flat response occurs when k equals k_c, the *critical coefficient of coupling*, where

$$k_c = 1/\sqrt{Q_p Q_s} \qquad \textbf{(9–11)}$$

With k equal to k_c, the double-tuned circuit has a more desirable bandpass characteristic than the single-tuned circuit as shown in Fig. 9–6(b).

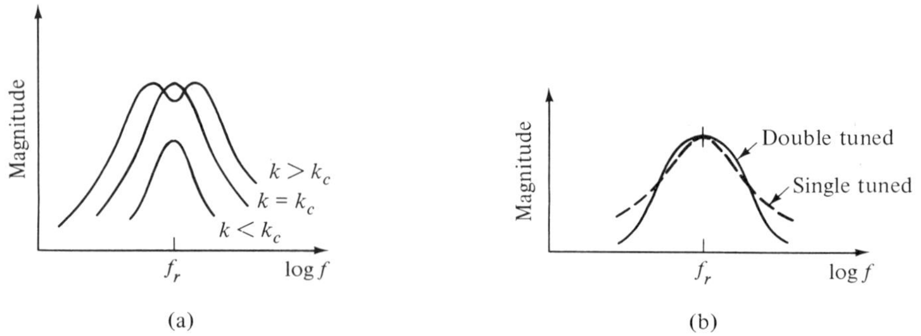

FIGURE 9–6 *Comparison of the bandpass characteristics of tuned circuits: (a) double-tuned circuits with various values of* k, *and (b) double-tuned critically coupled tuned circuit and single-tuned circuit.*

---

* When $k = 0$, the two coils are completely isolated from each other. When $k = 1$, all the flux lines link all the turns of both the primary and the secondary.

Although a double-tuned circuit with critical coupling has a very desirable bandpass characteristic, it is not used in all applications. For example, a two-stage bipolar transistor amplifier using single-tuned circuits can be adjusted to have approximately the same bandpass characteristic as the double-tuned circuit. This arrangement, called *stagger tuning*, is discussed in the next section.

The type of tuned circuit used in a narrow-band amplifier depends on many constraints, one of the more important being the type of active device used in the amplifier. Generally, parallel-tuned circuits are placed at a high impedance point in the circuit to maintain a high value for Q_L. Since a bipolar transistor in the CE configuration has a relatively high value of output resistance (r_{ce}) and a relatively low value of input resistance ($r_{b'e}$), tuned circuits are often placed in the collector circuit. The single-tuned circuit is used widely in narrow-band amplifiers using bipolar transistors. On the other hand, FETs and vacuum tubes have very high values of input resistance and moderate to high values of output resistance. Double-tuned circuits are often used with FETs and vacuum tubes.

Consider the single-stage, single-tuned CE amplifier shown in Figs. 9–7(a) and (b). The circuit is biased like any other linear amplifier studied in previous chapters. The filter network $R_D C_D$ is used to decouple or isolate the tuned circuit from the power supply. Also, the tuned circuit is generally housed in a metal shield that is connected to the circuit ground. One of the main objectives of an analysis of the circuit shown in Fig. 9–7 is to determine the value of the loaded Q of the circuit, Q_L. Normally the first step is to reflect both the load impedance (R_L and C_L) and the transistor output impedance (R_o and C_o) across the primary by using the transformer turns ratio squared. The tuned circuit itself can be represented by the parallel components R_{op}, L_p, and C_t, as shown in Fig. 9–7(c). The components $R_{o'}$ and $C_{o'}$ represent the loading effects of the transistor output impedance on the tuned circuit and are given by

$$R_{o'} = R_o\left(\frac{N_1 + N_2}{N_1}\right)^2 \qquad C_{o'} = C_o\left(\frac{N_1}{N_1 + N_2}\right)^2 \tag{9–12}$$

The components $R_{L'}$ and $C_{L'}$ represent the effects of the load impedance on the tuned circuit and are given by

$$R_{L'} = R_L\left(\frac{N_1 + N_2}{N_3}\right)^2 \qquad C_{L'} = C_L\left(\frac{N_3}{N_1 + N_2}\right)^2 \tag{9–13}$$

Once the value of each component of the equivalent circuit shown in Fig. 9–7(c) is known, values for Q_L, BW, and f_r can be calculated.

With the loaded characteristics of the tuned circuit known, the overall voltage gain of the amplifier can be calculated. From Fig. 9–7(b),

$$A_{vo} = \frac{V_o}{V_g} = \left(\frac{V_i}{V_g}\right)\left(\frac{V_{oQ}}{V_i}\right)\left(\frac{V_o}{V_{oQ}}\right) = A_{vg}A_{vt}A_{vX} \tag{9–14}$$

where A_{vg} is the familiar generator-to-input voltage division term, A_{vt} is the familiar terminal-to-terminal voltage gain of the transistor, and A_{vX} is the voltage gain of the output transformer. For the amplifier shown in Fig. 9–7,

$$A_{vg} = \frac{R_i}{R_i + R_g} \qquad A_{vt} \simeq -\frac{R_{L\text{eq}}}{r_e} \qquad A_{vX} = \frac{N_3}{N_1} \tag{9–15}$$

where, at the resonant frequency,

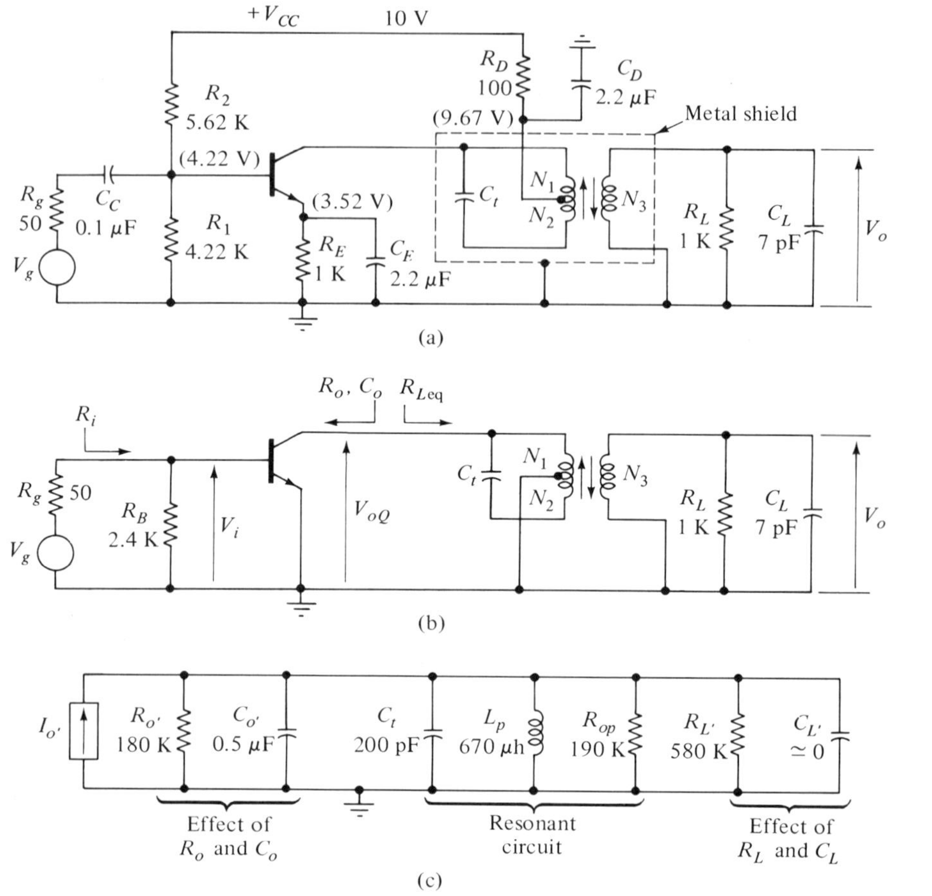

FIGURE 9–7 *A single-stage, single-tuned CE amplifier: (a) a circuit diagram, (b) an ac equivalent circuit, and (c) an equivalent circuit for the tuned circuit showing the effects of loading. Component values apply to Exs. 9–2 and 9–3.*

$$R_{Leq} = (R_{o'} \parallel R_{op} \parallel R_{L'})\left(\frac{N_1}{N_1 + N_2}\right)^2 \tag{9–16}$$

The following example illustrates the analysis of the amplifier shown in Fig. 9–7.

EXAMPLE 9–2 Calculate the voltage gain V_o/V_g and the 3-db bandwidth of the amplifier shown in Fig. 9–7(a). The transistor characteristics are: $\beta \simeq 85$, $r_{ce} \simeq 20$ K, $f_T \simeq 500$ MHz, and $C_{ob} \simeq 5$ pF. The tuned circuit parameters are:

$$N_1/N_3 = 8 \qquad C_t = 200 \text{ pF} \qquad f_r \simeq 455 \text{ kHz}$$
$$(N_1 + N_2)/N_1 = 3 \qquad R_c \simeq 19\ \Omega \qquad L_p \simeq 670\ \mu\text{H (tunable)}$$

Solution: (a) Bias conditions:

$$V_B \simeq 4.3 \text{ V} \qquad V_E \simeq 3.7 \text{ V} \qquad I_E \simeq 3.7 \text{ mA} \qquad r_e \simeq 7.1\ \Omega$$

(b) Unloaded characteristics of the tuned circuit: Values of C_t and L_p are given. From Eqs. (9–3) and (9–7),

$$Q_U = 2\pi f_r L/R_c \simeq (2\pi)(455 \times 10^3)(670 \times 10^{-6})/19 \simeq 100$$
$$R_{op} \simeq Q_U^2 R_c \simeq (100)^2(19) \simeq 190 \text{ K}$$

(c) Loaded characteristics of the tuned circuit: From Eq. (9–12) the effect of the output impedance of the transistor is

$$R_{o'} = R_o\left(\frac{N_1 + N_2}{N_1}\right)^2 \simeq (r_{ce})(3)^2 \simeq (20\text{ K})(9) \simeq 180\text{ K}$$

$$C_{o'} = C_o\left(\frac{N_1}{N_1 + N_2}\right)^2 \simeq (C_{o'})(1/3)^2 \simeq 5\text{ pF}/9 \simeq 0.5\text{ pF}$$

The effect of the load impedance can be found using Eq. (9–13) as follows:

$$R_{L'} = R_L\left(\frac{N_1 + N_2}{N_3}\right)^2 = (R_L)\left(\frac{N_1 + N_2}{N_1}\right)^2\left(\frac{N_1}{N_3}\right)^2 = (1\text{ K})(3)^2(8)^2 \simeq 580\text{ K}$$

$$C_{L'} = C_L\left(\frac{N_3}{N_1 + N_2}\right)^2 = (C_L)\left(\frac{N_1}{N_1 + N_2}\right)^2\left(\frac{N_3}{N_1}\right)^2 = (7\text{ pF})(1/3)^2(1/8)^2 \simeq 0$$

The component values are shown in Fig. 9–7(c). The equivalent resistance shunting the tuned circuit is

$$R_{Leq'} = R_{op} \,||\, R_{L'} \,||\, R_{o'} \simeq 190\text{ K} \,||\, 580\text{ K} \,||\, 180\text{ K} \simeq 80\text{ K}$$

From Eq. (9–9), the loaded Q is

$$Q_L \simeq \frac{R_{Leq'}}{2\pi f_r L} \simeq \frac{80\text{ K}}{(2\pi)(455 \times 10^3)(670 \times 10^{-6})} \simeq 42$$

From Eq. (9–4), the bandwidth of the amplifier is

$$BW_{3\text{ db}} \simeq f_r/Q_L \simeq (455\text{ kHz})/42 \simeq 10.8\text{ kHz}$$

(d) Voltage gain at resonant frequency:

$$R_{\text{in}} \simeq (\beta + 1)(r_e) \simeq (86)(7.1) \simeq 610\ \Omega$$

$$R_i = R_1 \,||\, R_2 \,||\, R_{\text{in}} \simeq 4.22\text{ K} \,||\, 5.62\text{ K} \,||\, 610 \simeq 500\ \Omega$$

$$A_{vg} = R_i/(R_i + R_g) \simeq 500/(500 + 50) \simeq 0.91$$

From Eq. (9–16), the load on the collector of the transistor is

$$R_{Leq} = (R_{Leq'})\left(\frac{N_1}{N_1 + N_2}\right)^2 \simeq (80\text{ K})(1/3)^2 \simeq 8.9\text{ K}$$

$$A_{vt} \simeq -R_{Leq}/r_e \simeq -8.9\text{ K}/7.1 \simeq -1250$$

Since the generator is not coupled to the amplifier by a tuned circuit, the effect of the input high corner frequency must be considered. At the input,

$$\tau_{\text{hi}} = (R_i \,||\, R_g)[C_{b'e} + C_{ob}(1 - A_{vt})] \simeq 286\text{ nsec}$$

$$f_{\text{hi}} \simeq 0.159/\tau_{\text{hi}} \simeq 550\text{ kHz}$$

At 455 kHz, the effect of the pole at 550 kHz is to reduce the value of A_{vg} from 0.91 to 0.70. From Eq. (9–14),

$$A_{vo} = A_{vg}A_{vt}A_{vX} \simeq (0.70)(-1250)(1/8) \simeq -110$$

(e) Experimental results: The measured bias values are shown in Fig. 9–7(a.) The measured values of gain and bandwidth were:

$$V_o/V_g \simeq 1\text{ V}/10\text{ mV} \simeq -100 \qquad BW \simeq 11.3\text{ kHz}$$

The *BW* value is in close agreement with the calculated value of 10.8 kHz. Note that a 1-V peak-to-peak output signal corresponds to approximately $(N_1/N_3)V_o$ or 8 V peak-to-peak at the collector terminal.

In addition to restricting the bandwidth, a tuned circuit can also be used to provide conditions for maximum power transfer. This condition occurs when the effective generator resistance is equal to the effective load resistance at f_r. Referring to Fig. 9–7(c), maximum power will be transferred to the load when

$R_{o'} \parallel R_{op}$ is equal to $R_{L'}$.* Assuming that the values of $R_{o'}$ and R_{op} are fixed, the value of $R_{L'}$ can be changed to provide for maximum power transfer either by changing the value of R_L or the value of the transformer turns ratio N_1/N_3. The following example illustrates these calculations.

EXAMPLE 9–3 For the amplifier shown in Fig. 9–7 and studied in Ex. 9–2, calculate the values of (a) R_L and (b) N_1/N_3 which will provide for maximum power transfer to the load.

Solution: (a) Maximum power transfer occurs when $R_{o'} \parallel R_{op}$ is equal to $R_{L'}$. From Ex. 9–2,

$$R_{o'} \parallel R_{op} \simeq 180 \text{ K} \parallel 190 \text{ K} \simeq 93.5 \text{ K}$$

The value of R_L that will cause $R_{L'}$ to equal 93.5 K is

$$R_L = R_{L'}\left(\frac{N_3}{N_1}\right)^2\left(\frac{N_1}{N_1 + N_2}\right)^2 = (93.5 \text{ K})(1/8)^2(1/3)^2 \simeq 160\ \Omega$$

(b) With R_L equal to 1 K, maximum power transfer can be achieved if the turns ratio N_3/N_1 is adjusted such that $R_{L'}$ equals 93.5 K. From Eq. (9–13),

$$\frac{R_{L'}}{R_L} = \left(\frac{N_1 + N_2}{N_3}\right)^2 = \left(\frac{N_1 + N_2}{N_1}\right)^2\left(\frac{N_1}{N_3}\right)^2$$

Solving for (N_1/N_3),

$$\frac{N_1}{N_3} = \left[\left(\frac{R_{L'}}{R_L}\right)\left(\frac{N_1}{N_1 + N_2}\right)^2\right]^{1/2} \simeq \left[\left(\frac{93.5 \text{ K}}{1 \text{ K}}\right)\left(\frac{1}{9}\right)\right]^{1/2} \simeq 3.22$$

In the high-frequency ranges, the impedance levels become so low that capacitive coupling can be used advantageously. This approach is especially attractive for the input circuit in a radio receiver because the resonant frequency of this circuit must be varied over a wide range of frequencies. A capacitively coupled circuit is shown in Fig. 9–8(a). In Fig. 9–8(b), the transistor input impedance is represented by a parallel RC network. The values of R_i and C_i are generally specified for the particular frequency of interest and are obtained from specification sheets or by direct measurement using an admittance bridge. Capacitor C_2 is a coupling capacitor and isolates the tuned circuit from the input capacitance C_i since C_2 and C_i are effectively in series. Capacitor C_1 is used to tune the resonant circuit over the desired frequency range. The coil, L, is tapped so that the generated resistance, R_g, does not load the tuned circuit excessively.

An approximate equivalent circuit for the loaded resonant circuit is shown in Fig. 9–8(c). Tapping the coil or placing R_g across part of the coil has the effect of transforming R_g to $R_{g'}$ across the entire coil where

$$R_{g'} = R_g\left(\frac{N_1 + N_2}{N_2}\right)^2 \qquad \textbf{(9–17)}$$

The effects of C_2, R_i, and C_i can be determined by deriving an expression for Z as indicated in Figs. 9–8(b) and (c). This analysis** results in

---

* An alternate condition that results in maximum power transfer from the transistor occurs when $R_{o'}$ is equal to $R_{op} \parallel R_{L'}$. If R_{op} is very large and assumed infinite, then the conditions for maximum power transfer from the transistor and maximum power transfer to the load coincide; i.e., $R_{o'}$ equals $R_{L'}$.

** In Eq. (9–19), the expressions for $R_{i'}$ and $C_{i'}$ are based on the following assumptions:

$$\left.\begin{aligned}[\omega R_i(C_i + C_2)]^2 &\gg 1\\ \omega^2 R_i^2 C_i(C_i + C_2) &\gg 1\end{aligned}\right\} \qquad \textbf{(9–18)}$$

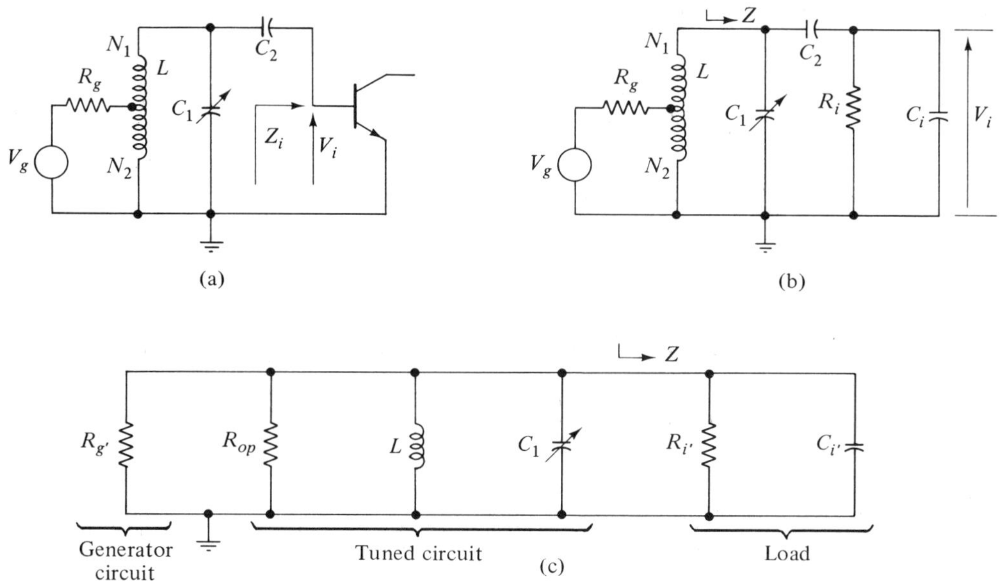

FIGURE 9–8 *An RF amplifier using capacitive coupling at the input: (a) a partial circuit diagram, (b) an equivalent circuit representation, and (c) an approximate equivalent circuit for the tuned circuit.*

$$R_{i'} \simeq R_i \left(\frac{C_i + C_2}{C_2}\right)^2 \qquad C_{i'} \simeq \frac{C_i C_2}{C_i + C_2} \tag{9–19}$$

In Eq. (9–19) the ratio $[(C_i + C_2)/C_2]$ effectively performs the same resistance transformation as the turns ratio of a transformer. Often an equivalent "turns ratio" is defined for the capacitively coupled circuit as

$$n \equiv (C_i + C_2)/C_2 \tag{9–20}$$

The following example will illustrate the response of a capacitively coupled circuit.

EXAMPLE 9–4 The amplifier shown in Fig. 9–8(a) has the following component values: $L = 0.22\ \mu\text{H}$, $R_i = 200$ ohms, $C_i = 50$ pF, and $R_g = 300$ ohms. The coil is assumed to be lossless. Calculate the values of C_1 and C_2 for the circuit to operate with maximum power transfer at 100 MHz with a bandwidth of 1 MHz.

Solution:

$$Q_L = f_r/BW = (100\text{ MHz})/1\text{ MHz} = 100$$

The equivalent parallel resistance of the tuned circuit is

$$R_{Leq'} = Q(2\pi f_r L) \simeq (100)(2\pi)(10^8)(0.22 \times 10^{-6}) \simeq 13.8\text{ K}$$

From Fig. 9–8(c) maximum power transfer occurs when $R_{g'}$ equals $R_{i'}$. Since $R_{Leq'}$ equals $R_g \,||\, R_{i'}$,

$$R_{g'} = R_{i'} = 2R_{Leq'} \simeq (2)(13.8\text{ K}) \simeq 27.6\text{ K}$$

From Eq. (9–17),

$$(N_1 + N_2)/N_1 = (R_{g'}/R_i)^{1/2} = (27.6\text{ K}/300)^{1/2} \simeq 9.6$$

Thus N_2 equals 8.6 N_1. From Eq. (9–19),

$$(C_1 + C_2)/C_2 = (R_{i'}/R_i)^{1/2} \simeq (27.6\text{ K}/200)^{1/2} \simeq 11.75$$

$$C_2 \simeq C_i/10.75 \simeq 50\text{ pF}/10.75 \simeq 4.65\text{ pF}$$

A commercially available value for C_2 is 4.7 pF. From Eq. (9–18),

$$C_{i'} = \frac{C_i C_2}{C_i + C_2} \simeq \frac{(50\text{ pF})(4.7\text{ pF})}{54.7\text{ pF}} \simeq 4.3\text{ pF}$$

The total capacitance required for resonance at 100 MHz with L equal to 0.22 μH is approximately 11.6 pF. Thus the value of C_1 is

$$C_1 \simeq 11.6\text{ pF} - C_{i'} \simeq 11.6\text{ pF} - 4.3\text{ pF} \simeq 7.3\text{ pF}$$

A commercially available variable capacitor that can be used for C_1 has the nominal range of 5 to 25 pF.

In this section we have looked briefly at some of the characteristics of tuned amplifiers. The design of the input and output coupling networks forms an important part of the overall design. Maximum power transfer results when the transistor input and output impedances are conjugate-matched to the source and load impedance, respectively. Both capacitive coupling and inductive coupling are used in these matching networks. Experimental evaluation of the final design of a tuned amplifier is particularly important since stray inductance and capacitance can drastically affect the circuit operation.

9.3 Multistage Tuned Amplifiers

Many tuned amplifiers require more gain or different bandwidth characteristics than can be provided by a single-stage tuned amplifier. In this section we will study the frequency characteristics of multistage amplifiers together with some multistage amplifier configurations.

One of the most important properties of a tuned amplifier is its ability to reject signals in adjacent channels; i.e., to provide selectivity. The figure of merit normally used to compare tuned amplifiers is the shape factor or the ratio of the bandwidth at 60 db down on the response curve to the bandwidth at 6 db down. The ideal shape factor is unity. With this value for the shape factor, signals in adjacent channels would not be amplified. As shown in Fig. 9–9, the bandwidth of a single-tuned circuit increases rapidly as the frequency deviates from the center frequency. The relative response, m, for any frequency, f, reasonably close to f_r is approximately

$$m \simeq \frac{1}{\sqrt{1 + \left(\dfrac{f_r - f}{f_r - f_{-3\text{ db}}}\right)^2}} \tag{9–21}$$

where $f_{-3\text{ db}}$ is the half-power frequency or the frequency for which the relative response is 0.707 of that at mid-frequency. The value of m is unity when f and f_r are equal. The fraction under the radical is the ratio of two bandwidths. Solving Eq. (9–21) for this ratio, we obtain

$$\frac{\text{bandwidth for power loss of } 1/m^2}{\text{bandwidth for power loss of } 1/(0.707)^2} = \sqrt{(1/m^2) - 1} \tag{9–22}$$

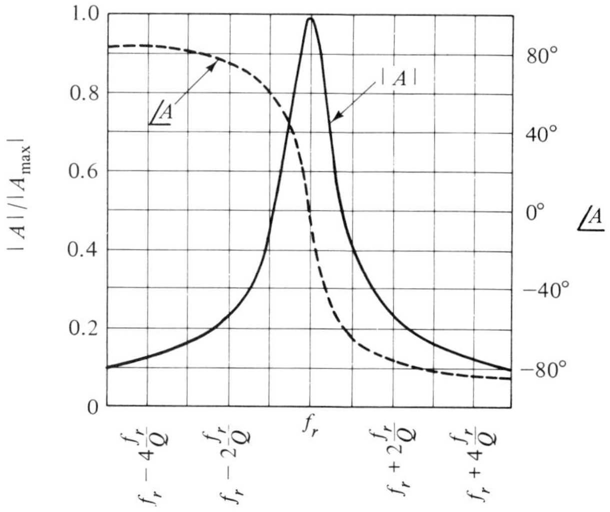

FIGURE 9–9 *A normalized response curve for an amplifier loaded with a single-tuned circuit.*

Knowing that the half-power bandwidth is f_r/Q_L, this equation can be used to find the approximate bandwidth for other values of relative response.

EXAMPLE 9–5 The single-tuned coupling circuit of a tuned amplifier has a Q_L of 30. If the center frequency is 450 kHz, find the ratio of the bandwidth at −60 db and at −6 db.

Solution: The half-power bandwidth is 450 kHz/30 or 15 kHz. The relative response at −6 db is 0.5 and that for −60 db is 0.001 or 10^{-3}. Therefore, from Eq. (9–22),

$$\text{Bandwidth at } -6 \text{ db} = 15 \text{ kHz } [(1/0.5)^2 - 1]^{1/2} \simeq 26 \text{ kHz}$$

$$\text{Bandwidth at } -60 \text{ db} = 15 \text{ kHz } (10^6 - 1)^{1/2} \simeq 15 \text{ MHz}$$

$$\text{Shape factor} = \frac{BW_{60}}{BW_6} = \frac{15 \text{ MHz}}{26 \text{ kHz}} \simeq 577$$

The tuned circuit in Ex. 9–5 does not offer the circuit designer much selectivity. When selectivity is a dominant factor in the design problem, crystal filters may be used at certain frequencies. Ceramic filters are available with center frequencies of 455 kHz and bandwidths (at −6 db) ranging from 4 kHz to 20 kHz. The shape factors for these crystals are especially good. A typical value for the 60/6-db shape factor is 1.6/1 with a 16-kHz bandwidth at −6 db.* A typical circuit using a ceramic transformer is shown in Fig. 9–10.

Frequently one tuned stage does not have sufficient gain for the required bandwidth or sufficient bandwidth for the required gain. In either case single-tuned stages can be cascaded to produce the desired results. One important problem encountered when stages are cascaded is that the bandwidth shrinks. Let us consider first the case of cascaded identical stages.

---

* Impedance matching with these ceramic transformers is obtained by placing the electrodes properly with respect to the voltage waveform on a piece of polarized ceramic. If the ceramic is in the form of a circular disk that vibrates at the first overtone of its circular mode, the input electrode may be placed in the center of the disk to provide a relatively high impedance of 5 to 20 kilohms. If the output electrode is placed close to the edge it will provide a smaller impedance of 1 to 5 kilohms.

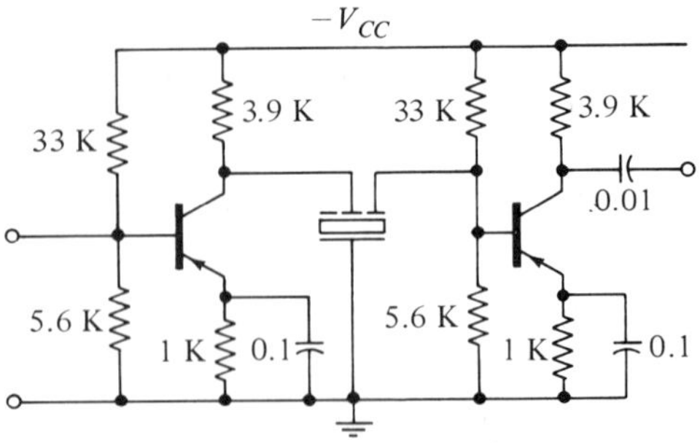

FIGURE 9–10 *A tuned amplifier using a ceramic transformer.*

Since the gains of cascaded stages are multiplied to obtain the overall gain, a relative gain of 0.707 (or −3 db) for one stage becomes $(0.707)^2$ or 0.5 gain for two stages. Thus for two stages, the overall half-power bandwidth is less than that for one stage. The overall half-power bandwidth for n identical stages in terms of the single-stage bandwidth* is

$$BW_n = BW_1\sqrt{2^{1/n} - 1} \simeq BW_1/(1.2n) \qquad \textbf{(9–23)}$$

The off-channel selectivity for n identical stages is improved relative to the single-stage response, because each stage adds 6 db to the slope of the overall response curve. The overall bandwidth of n identical single stages for a relative response (m) is obtained by modifying Eq. (9–22) for n stages to obtain

$$\frac{BW_n \text{ for power loss of } 1/m^2}{BW_1 \text{ for power loss of } 1/(0.707)^2} = \sqrt{\left(\frac{1}{m^2}\right)^{1/n} - 1} \qquad \textbf{(9–24)}$$

Consider the two-stage, single-tuned CE amplifier shown in Fig. 9–11. The analysis of this circuit is similar to the single-stage case illustrated in the last section. First the bias of each stage is determined. Next the loaded characteristics of each tuned circuit are determined. Note that the input impedance of Q_2 serves as the load impedance of tuned circuit B. A similar condition exists for tuned circuit A. Once the loaded characteristics of each tuned circuit are known, the gain and bandwidth of each stage can be found. For identical stages, the effect of bandwidth shrinkage can be calculated using Eq. (9–23). The following example illustrates the calculation of the response of a multistage tuned amplifier.

EXAMPLE 9–6 Calculate the overall voltage gain at resonance and the bandwidth of the amplifier shown in Fig. 9–11. The resonant frequency of each tuned circuit is adjusted to 455 kHz. The transistor parameters are

Transistor	β	r_{ce}	f_T	C_{ob}
Q_1	85	20 K	500 MHz	5 pF
Q_2	80	20 K	500 MHz	5 pF

The tuned circuit parameters are

Tuned Circuit	$\frac{N_1}{N_3}$	$\frac{N_1 + N_2}{N_1}$	L_p (μh)	R_c (ohms)
A	8.8	2.1	610	17.8
B	8.0	3.0	670	19.0
C	3.7	1.5	475	13.6

* Equation (9–23) is valid only for single-tuned circuits all tuned to the same frequency or synchronously tuned. Double-tuned circuits and ceramic transformers have less bandwidth shrinkage.

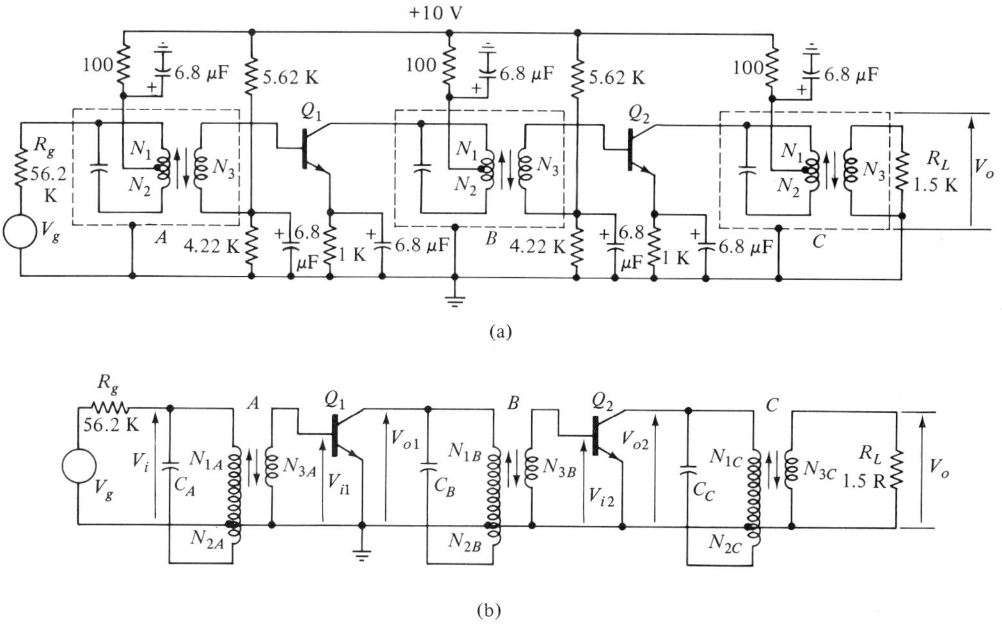

FIGURE 9–11 *A two-stage, single-tuned CE amplifier: (a) the circuit diagram, and (b) an ac equivalent circuit. (Component values apply to Ex. 9–6.)*

Solution: (a) Bias and impedance values:

$$V_{E1} \simeq V_{E2} \simeq 3.7\text{ V} \qquad r_{e1} \simeq r_{e2} \simeq 7.1\ \Omega$$

$$R_{\text{in1}} \simeq (\beta_1 + 1)r_{e1} \simeq 610\ \Omega \qquad R_{\text{in2}} \simeq (\beta_2 + 1)r_{e2} \simeq 575\ \Omega$$

(b) Loaded characteristics of tuned circuit C:

$$Q_{UC} = \frac{2\pi f_r L}{R_c} \simeq 100 \qquad R_{opC} \simeq (Q_{UC})^2 R_C \simeq 136\text{ K}$$

$$R_{L'C} = R_L\left(\frac{N_{1C} + N_{2C}}{N_{3C}}\right)^2 \simeq (1.5\text{ K})(1.5)^2(3.7)^2 \simeq 46\text{ K}$$

$$R_{o'C} = R_{o2}\left(\frac{N_{1C} + N_{2C}}{N_{1C}}\right)^2 \simeq r_{ce2}(1.5)^2 \simeq (20\text{ K})(1.5)^2 \simeq 45\text{ K}$$

$$R_{\text{Leq}'C} = R_{L'C} \parallel R_{opC} \parallel R_{o'C} \simeq 46\text{ K} \parallel 136\text{ K} \parallel 45\text{ K} \simeq 13.6\text{ K}$$

$$Q_{LC} = \frac{R_{\text{Leq}'C}}{2\pi f_r L_C} \simeq \frac{13.6\text{ K}}{(2\pi)(455\text{ kHz})(475\ \mu\text{h})} \simeq 10$$

$$BW_C \simeq f_r/Q_{LC} \simeq (455\text{ kHz})/10 \simeq 45.5\text{ kHz}$$

$$R_{\text{Leq2}} \simeq (R_{\text{Leq}'C})\left(\frac{N_{1C}}{N_{1C} + N_{2C}}\right)^2 \simeq (13.6\text{ K})/(1.5)^2 \simeq 6\text{ K}$$

(c) Loaded characteristics of tuned circuit B: Using the same procedure as in (b),

$$Q_{UB} \simeq 100 \qquad R_{opB} \simeq 190\text{ K} \qquad R_{L'B} \simeq 330\text{ K} \qquad R_{o'B} \simeq 180\text{ K}$$

$$R_{\text{Leq}'B} \simeq 72.5\text{ K} \qquad Q_{LB} \simeq 38 \qquad BW_B \simeq 12\text{ kHz} \qquad R_{\text{Leq1}} \simeq 8\text{ K}$$

(d) Loaded characteristics of tuned circuit A: Using the same procedure as in (b),

$$Q_{UA} \simeq 100 \qquad R_{opA} \simeq 136\text{ K} \qquad R_{L'A} \simeq 212\text{ K} \qquad R_{g'A} \simeq 250\text{ K}$$

$$R_{\text{Leq}'A} \simeq 62\text{ K} \qquad Q_{LA} \simeq 35.5 \qquad BW_A \simeq 12.8\text{ kHz}$$

The equivalent resistance across the terminals of coil N_{1A} is $R_g \parallel R_i$ or

$$R_g \parallel R_i = R_{\text{eq}'A}\left(\frac{N_{1A}}{N_{1A} + N_{2A}}\right)^2 \simeq (62\text{ K})\left(\frac{1}{2.1}\right)^2 \simeq 14.1\text{ K}$$

Solving for R_i with R_g equal to 56.2 K,

$$R_i = \frac{R_g(R_g \parallel R_i)}{R_g - (R_g \parallel R_i)} \simeq \frac{(56.2\text{ K})(14.1\text{ K})}{56.2\text{ K} - 14.1\text{ K}} \simeq 18.8\text{ K}$$

(e) Voltage gain at 455 kHz:

$$A_{vg} = R_i/(R_i + R_g) \simeq (18.8\text{ K})/(18.8\text{ K} + 56.2\text{ K}) \simeq 0.25$$

$$A_{vt1} \simeq -R_{Leq1}/r_{e1} \simeq -8\text{ K}/7.1 \simeq -1120$$

$$A_{vt2} \simeq R_{Leq2}/r_{e2} \simeq -6\text{ K}/7.1 \simeq -850$$

From Fig. 9–11(b),

$$A_{vo} = \left(\frac{V_i}{V_g}\right)\left(\frac{V_{i1}}{V_i}\right)\left(\frac{V_{o1}}{V_{i1}}\right)\left(\frac{V_{i2}}{V_{o1}}\right)\left(\frac{V_{o2}}{V_{i2}}\right)\left(\frac{V_o}{V_{o2}}\right)$$

$$= (A_{vg})\left(\frac{N_{3A}}{N_{1A}}\right)(A_{vt1})\left(\frac{N_{3B}}{N_{1B}}\right)(A_{vt2})\left(\frac{N_{3C}}{N_{1C}}\right)$$

$$\simeq (0.25)\left(\frac{1}{8.8}\right)(-1120)\left(\frac{1}{8}\right)(-850)\left(\frac{1}{3.7}\right) \simeq 910$$

(f) Overall bandwidth of the amplifier:

$$BW_A \simeq 12.8\text{ kHz} \qquad BW_B \simeq 12\text{ kHz} \qquad BW_C \simeq 45.5\text{ kHz}$$

The responses of tuned circuits *A* and *B* will control the overall responses of the amplifier. Assuming that the two stages are identical, a very approximate estimate of the overall *BW* can be found from Eq. (9–23).

$$BW_t \simeq BW_1/(1.2 \times 2) \simeq 12.4\text{ kHz}/2.4 \simeq 5.2\text{ kHz}$$

The other tuned circuit will increase this value slightly.

(g) Experimental results: The measured response was

$$A_{vo} = V_o/V_g \simeq 5.1\text{ V}/5\text{ mV} \simeq 850 \qquad BW \simeq 7.2\text{ kHz}$$

Cascading identical or nearly identical stages produces an increase in gain and a decrease in bandwidth relative to a single stage. Both gain and bandwidth can be increased relative to that of a single stage by adjusting the center frequency of the various tuned circuits to slightly different values. The swap of some overall gain for some overall bandwidth is usually determined by the required bandwidth and the desired shape of the overall response curve. The general case is rather involved and will not be treated here. One interesting special case is that in which the overall response curve has a maximally flat top. For maximal flatness in a two-stage amplifier the stages should be tuned to frequencies f_{r1} and f_{r2} as follows:

$$f_{r1} = f_r - 0.35\ BW_t \qquad f_{r2} = f_r + 0.35\ BW_t \tag{9–25}$$

where f_r is the overall center frequency, and BW_t is the half-power bandwidth for the overall amplifier. In addition, the half-power bandwidth for each stage should be

$$BW_1 = 0.71\ BW_t \tag{9–26}$$

This arrangement allows the individual curves to intersect at their half-power frequencies, which occur at the center frequency f_r, as shown in Fig. 9–12. This combination of two tuned stages is called a *staggered double* and may be treated as a single stage for further cascading.

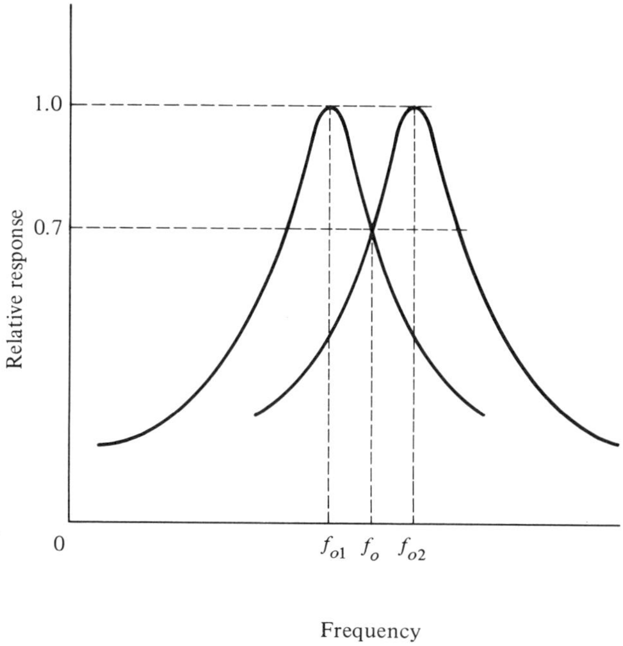

FIGURE 9–12 *Two single-tuned circuit responses arranged to produce a maximally flat overall response.*

EXAMPLE 9–7 For the amplifier in Ex. 9–6, calculate f_r, BW, and Q_L for tuned circuits A and B if the amplifier is stagger tuned with an overall BW of 20 kHz and an overall resonant frequency of 455 kHz.

Solution: From Eq. (9–25),

$$f_{rA} \simeq 455\text{ kHz} - (0.35)(20\text{ kHz}) \simeq 448\text{ kHz}$$
$$f_{rB} \simeq 455\text{ kHz} + (0.35)(20\text{ kHz}) \simeq 462\text{ kHz}$$

From Eq. (9–26),

$$BW_A = BW_B \simeq (0.71)(20\text{ K}) \simeq 14.2\text{ kHz}$$

From Eq. (9–4),

$$Q_{LA} = f_{rA}/BW_A \simeq 448\text{ kHz}/14.2\text{ kHz} \simeq 31.5$$
$$Q_{LB} = f_{rB}/BW_B \simeq 462\text{ kHz}/14.2\text{ kHz} \simeq 33.5$$

In this section, the responses of some simple multistage tuned amplifiers have been considered. When each tuned circuit in a cascaded amplifier has the same resonant frequency, the multistage amplifier has a larger gain and a smaller bandwidth relative to the single stage. A more desirable response is often provided by stagger tuning the stages. The response of the staggered double is somewhat similar to the response of the double-tuned amplifier studied in the last section.

9.4 Tuned Amplifier Stability

Much of the analysis studied in this chapter deals with the response of tuned circuits at resonance. At the resonant frequency, tuned circuits appear resistive, and little difficulty is encountered in the analysis. At frequencies other than the resonant frequency, however, the tuned circuit appears either inductive or capacitive. The input impedance of an amplifier can have a negative resistance component for certain reactive type loads. The CC stage, when loaded with a parallel *RC* load in the emitter circuit, is an example that was considered in Chap. 8. In an inverting type amplifier, $C_{b'c}$ (or C_{ob}) provides a signal path between the output and input circuits of the amplifier. If the load on the output contains an inductive component, the impedance reflected across the input circuit because of $C_{b'c}$ may contain a negative-resistance component. The objective of this section is to study this negative-resistance component and methods of reducing its effect on circuit performance.

Consider the single-stage, tuned amplifier shown in Fig. 9–13(a). In previous sections we studied methods of reflecting the various impedances in the circuit to the primary of the tuned circuit. In this study, however, we will assume that the various impedances have been reflected across the N_1 terminals and form the load impedance of the transistor. Since a resonant condition is not assumed, the equivalent load impedance is represented by a parallel *RLC* circuit as shown in Fig. 9–13(b). The expression for Z_L in terms of the complex variable s is

$$Z_L = R_L \,||\, \frac{1}{sC} \,||\, sL = \frac{sL}{s^2LC + s(L/R_L) + 1} \tag{9–27}$$

The equivalent circuit of the input section can be found using the hybrid-π equivalent circuit and Miller's theorem. The resulting circuit is shown in Fig. 9–13(c). The components $r_{b'}$, $r_{b'e}$, and $C_{b'e}$ are identical to the CE hybrid-π circuit while the components C_{ob} and $g_mZ_LC_{ob}$ result from the application of Miller's theorem. With a purely resistive load, Z_L reduces to R_L, and the effect of the load is simply to add capacitance in parallel with $C_{b'e}$, or $C_{ob}(1 + g_mR_L)$. When the load impedance is resistive and inductive, the added phase shift may produce a negative-resistance component across $C_{b'e}$.

Our next task is to find an equivalent circuit to represent the component $g_mZ_LC_{ob}$ shown in Fig. 9–13(c). The equivalent impedance Z' indicated in Fig. 9–13(c) is

$$Z' = 1/(sg_mZ_LC_{ob}) \tag{9–28}$$

Substituting Eq. (9–27) for Z_L into Eq. (9–28) and replacing s by $j\omega$ yields

$$Z' = \frac{\omega^2LC - 1}{\omega^2g_mLC_{ob}} + \frac{1}{j\omega g_mRC_{ob}} = R' - jX' \tag{9–29}$$

The real term in Eq. (9–29) can also be written in terms of the resonant frequency so that

$$Z' = \frac{(f/f_r)^2 - 1}{(2\pi f)^2g_mLC_{ob}} + \frac{1}{j(2\pi f)g_mR_LC_{ob}} \tag{9–30}$$

From Eq. (9–30) we can see that Z' consists of an *RC* series combination. The value of the capacitor is the same as that for a resistive load or $g_mR_LC_{ob}$. The resistance component is of particular interest since it is negative for all values of frequency below the resonant frequency, f_r. Just as in the CC case in Sec. 8.4, a negative resistance component can cause self-sustained oscillations. An alternate equivalent circuit for the transistor impedance is shown in Fig. 9–13(d) to illustrate the various components present.

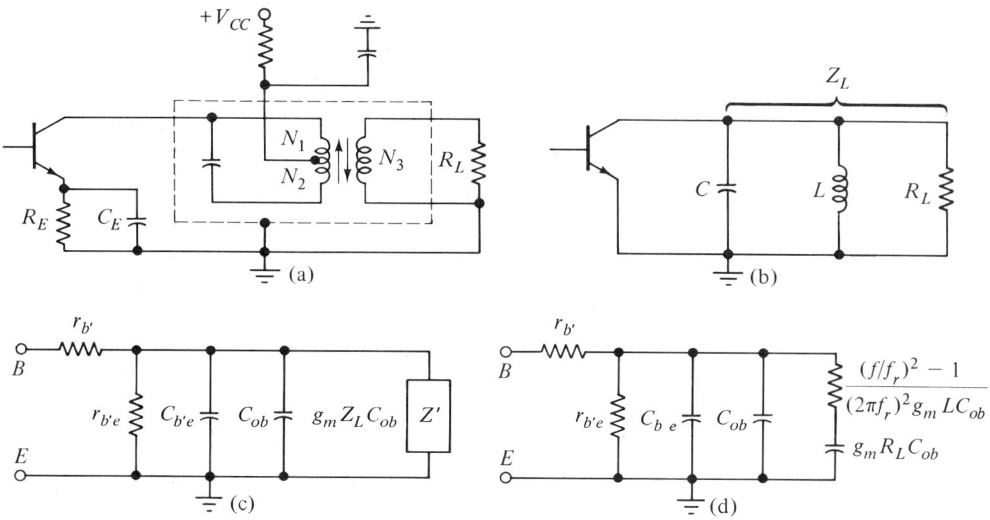

FIGURE 9–13 *Circuit diagrams used in the development of instability effects in tuned amplifiers: (a) a partial circuit diagram, (b) an ac equivalent circuit, (c) an equivalent circuit for the input impedance, and (d) an alternate form of the equivalent circuit for the input impedance.*

As indicated previously, the negative resistance component in the input impedance of a tuned amplifier occurs because of a signal path from the output of the amplifier back to the input of the amplifier through C_{ob} and because the load is inductive. We will briefly examine three methods of reducing the effect of C_{ob}: (1) neutralization, (2) the cascode amplifier, and (3) the non-inverting differential amplifier.

One approach to neutralization consists of adding a signal path from some point in the amplifier back to the input of the amplifier to reduce or cancel the signal fed back through C_{ob}. If the network exactly cancels the signal fed back through the transistor, the amplifier is said to be *unilateralized.* When the signal is only partially cancelled (the more general case), the term *neutralization* is used. One arrangement is shown in Fig. 9–14 in which a single-stage, single-tuned amplifier of the type studied in Sec. 9.2 is shown. The unneutralized circuit, Fig. 9–14(a), is modified by adding C_N in Fig. 9–14(b). Since the center top of the

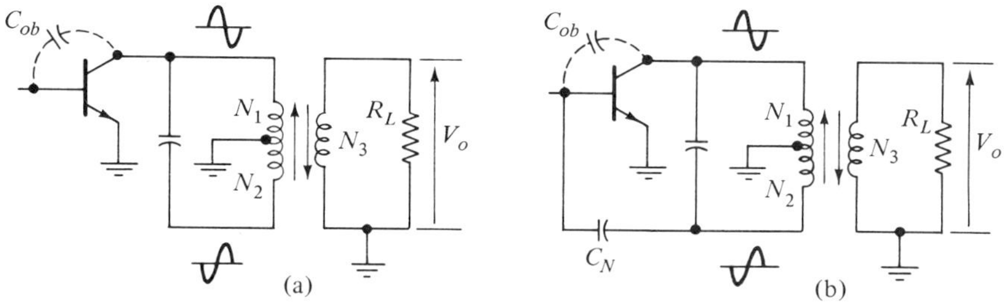

FIGURE 9–14 *The effect of neutralization: (a) circuit diagram without neutralization, and (b) circuit diagram with neutralization provided by* C_N.

transformer primary is grounded, the signals at each end of the transformer are 180° out of phase. If N_1 and N_2 are equal, C_N would be chosen approximately equal to C_{ob}. Other forms of neutralization used include a series RC network or a parallel RC network in place of C_N as shown in Fig. 9–14(b).

The principal cause of instability results from the signal fed back to the input through C_{ob}. Rather than simply cancel this signal, as is done in neutralization, we can reduce its amplitude and thus reduce the instability problem. Two familiar circuits that can be used for this purpose are the cascode circuit and one form of the differential amplifier. A simple form of each circuit is shown in Fig. 9–15. Basically, the voltage gain is obtained from a CB stage rather than a CE stage. Stated differently, the isolation between the input and the output is greatly improved due to the inherent input-output isolation of the CB stage.

In this section we have examined both the origin and some possible cures for the problem of instability in tuned amplifiers. The instability is due to the phase characteristic of the tuned circuit and the output-to-input signal path through C_{ob}. This effect can be reduced by neutralization. Alternately, the isolation between the input and the output can be improved using circuits other than the single-stage CE.

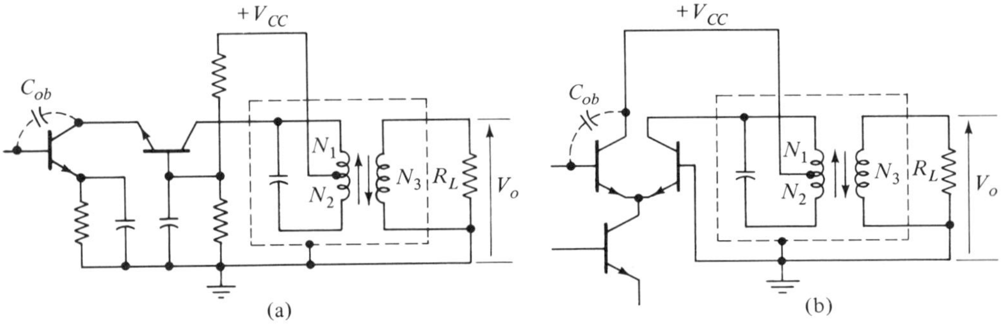

FIGURE 9–15 *Tuned circuit arrangements used to reduce the instability in tuned amplifiers: (a) the cascode circuit, and (b) one form of the differential amplifier with the tuned circuit at the non-inverting output terminal.*

9.5 IC Tuned Amplifiers

Most tuned amplifiers use few components other than tuned circuits and gain stages. As noted in Sec. 5.4, fabrication of a high-Q coil in integrated form is not practical. Hence, ICs find their main application in tuned amplifiers as gain stages. The objective of this section is to examine briefly the advantages of ICs in tuned amplifiers and some typical circuits encountered in practice.

One advantage of using an IC in a tuned amplifier is the elimination of many biasing resistors and coupling and bypassed capacitors. Also, improved stability can be achieved by using a multistage circuit such as the cascode circuit or the differential amplifier shown in Fig. 9–15. Thus, the advantages of small size and simplicity of circuit construction common in IC circuits are partially available in tuned amplifiers. The multistage tuned amplifier shown in Fig. 9–11 is shown in Fig. 9–16 using ICs for the gain stages.

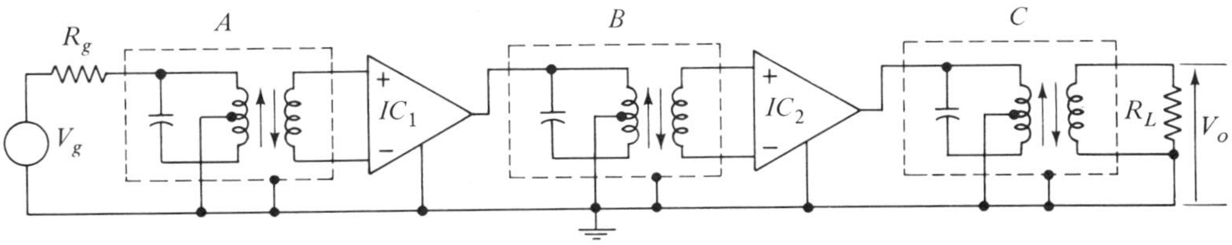

FIGURE 9–16 *A two-stage tuned amplifier using ICs for the gain stages. The bias circuit is not included.*

A typical IC used in tuned amplifiers and RF applications is shown in Fig. 9–17. Basically, the circuit consists of a differential pair, Q_1 and Q_2, biased with a constant-current source, Q_3. A tuned circuit load can be connected from the collector terminal of either Q_1 or Q_2 to ac ground, or connected between the two collector terminals. Also, Q_3 and either Q_1 or Q_2 can be used to form a cascode circuit. Simple ICs of the type shown in Fig. 9–17 are often very high frequency units capable of operation at frequencies greater than 100 MHz.

In this section we have briefly examined the use of ICs in tuned amplifiers. With the exception of the tuned circuit, especially the inductor, most tuned amplifiers can be fabricated using ICs. The principal advantages of the use of ICs in this application are small size, simplicity, and improved stability.

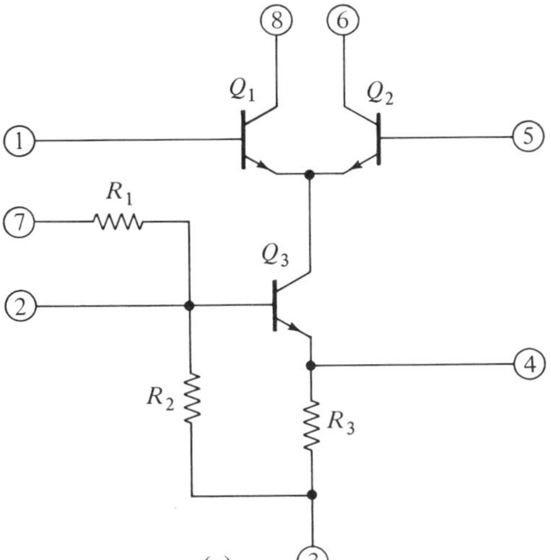

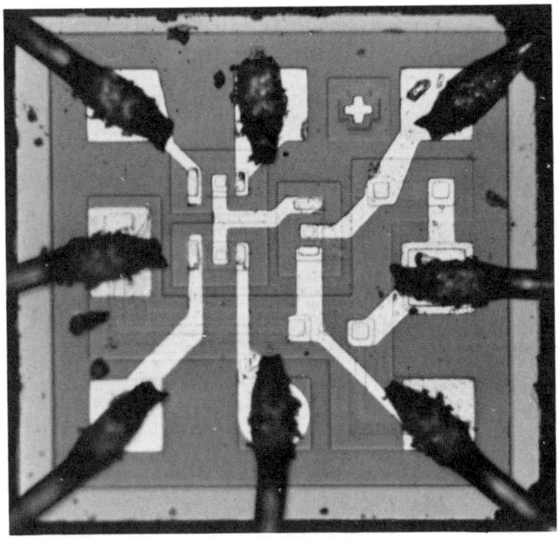
(b)

FIGURE 9–17 *A typical IC used in tuned amplifier and RF application: (a) circuit diagram and (b) photomicrograph of the actual layout. (Courtesy of Radio Corporation of America.)*

9.6 Summary

Tuned circuits are used as frequency sensitive loads and driving circuits in narrow-band amplifiers. These amplifiers are used extensively in communications equipment to obtain gain over a narrow band of frequencies while rejecting frequencies on either side of the desired band. For both series and parallel tuned circuits, the resonant frequency, f_r, is approximately $1/[2\pi\sqrt{(LC)}]$. The half-power (−3 db) bandwidth of a tuned circuit is f_r/Q where Q is the quality factor of the circuit. This quality factor is determined by the loss associated with the circuit. Consequently, two values of Q are usually specified: Q_U which considers only the loss in the components of the tuned circuit, and Q_L which considers the loss associated with the entire circuit. Along with the bandwidth, the shape factor is used to compare the selectivity of tuned amplifiers.

The analysis of both single-stage and multistage amplifiers has been considered. Depending upon the impedance levels in the circuit, a stage may use a single-tuned circuit or a double-tuned circuit. Single-tuned circuits are normally used with bipolar transistors while double-tuned circuits are used with the higher impedance devices. Cascading identical stages increases the gain relative to a single stage but decreases the bandwidth. Stagger tuning is sometimes used to obtain a desired bandwidth at the expense of some overall gain.

Inverting type amplifiers are normally used in tuned stages. Inductive loads on inverting type amplifiers can reflect sufficient negative resistance to the input to make the amplifier unstable. Since a parallel tuned circuit is inductive for frequencies below f_r, high-gain tuned amplifiers often need some form of neutralization to maintain stability.

Integrated amplifiers are often used in narrow-band amplifiers. However, coils are not easily fabricated in integrated circuit form. Consequently, narrow-band amplifiers are usually hybrid circuits with the tuned circuit being connected external to an IC amplifier.

Problems

Section 9.1

9–1 For the circuit shown in Fig. 9–1(a), $R_c = 2$ ohms, $L = 100\ \mu$h, and $C = 100$ pF. Calculate f_r, Q, and BW.

9–2 Repeat Ex. 9–1 for a 10.7-MHz resonant frequency.

9–3 What value of resistance must be connected in parallel with the tuned circuit studied in Ex. 9–1 to give a 10-kHz bandwidth?

9–4 Estimate the frequencies at which the magnitude of the equivalent impedance of the unloaded tuned circuit studied in Ex. 9–1 is equal to 5 kilohms.

Section 9.2

9–5 The value of R_g in Fig. 9–7(a) is changed from 50 ohms to 1 K. What two effects will this have on the circuit gain? (*Hint:* One effect is due to the circuit frequency response.)

9–6 In the amplifier studied in Ex. 9–2, the tap on the primary of the tuned circuit is to be moved so that the overall bandwidth is 8 kHz. Calculate the new values of $(N_1 + N_2)/N_2$ and A_{vo}.

9–7 For the amplifier shown in Fig. 9–7 and discussed in Ex. 9–2, calculate the values of (a) R_L and (b) N_1/N_3 which will provide for maximum power transfer from the transistor. Compare the results with those obtained in Ex. 9–3.

9–8 Derive Eq. (9–19).

9–9 Repeat Ex. 9–4 for the case where the circuit is to operate with maximum power transfer at 50 MHz.

Section 9.3

9–10 Repeat Ex. 9–5 for $Q_L = 100$.

9–11 Calculate the shape factor of a single-tuned amplifier consisting of two cascaded stages with each stage having a Q_L of 50 and an f_r of 455 kHz.

9–12 The amplifier studied in Ex. 9–6 is to be modified to provide for maximum power transfer from the generator and from each transistor by modifying the turns ratio of each tuned circuit. (a) Calculate the required values of N_1/N_3 and $(N_1 + N_2)/N_1$ for tuned circuits *A*, *B*, and *C*. (b) Calculate the overall voltage gain of the modified amplifier.

9–13 Modify the amplifier shown in Fig. 9–11 to provide for the stagger tuned characteristics calculated in Ex. 9–7.

Section 9.4

9–14 Derive Eq. (9–30) using Fig. 9–13.

9–15 The amplifier shown in Fig. 9–11 is to be unilateralized using the method shown in Fig. 9–14(b). Estimate the value of C_N required for each transistor. Component data are given in Ex. 9–6.

Section 9.5

9–16 For the IC shown in Fig. 9–17, identify the terminals of the actual layout in terms of the circuit diagram.

9–17 The amplifier shown in Fig. 9–16 is made up of the tuned circuits described in Ex. 9–6 and the IC shown in Fig. 9–17. Other component values are: V_{CC} is ± 10 V, R_g is 10 K, and R_L is 1 K. (a) Sketch a circuit diagram complete with bias components. (b) With β equal to 100, calculate V_o/V_g at f_r equal to 455 kHz.

References

Close, C. M., *The Analysis of Linear Circuits*. New York: Harcourt, Brace & World, Inc., 1966.

Giles, J. N., *Fairchild Semiconductor Linear Integrated Circuits Applications Handbook*. New York: Fairchild Publications, Inc., 1967.

Pierce, J. F., *Transistor Circuit Theory and Design*. Columbus, Ohio: Charles E. Merrill Publishing Company, 1963.

Ryder, J. D., *Electronic Fundamentals and Applications*, Fourth Ed. New York: Prentice-Hall, Inc., 1970.

Tang, K. Y., *Alternating-Current Circuits*, Third Ed. Scranton, Pa.: International Textbook Co., 1962.

10

Characteristics of Negative-Feedback Amplifiers in the Mid-Band Frequency Range

As indicated in the preceding seven chapters, the gain, input impedance, output impedance, and response characteristics are all functions of the generator impedance, load impedance, power supply voltage, and parameter tolerances. Differences between calculated and experimental results as large as 20 percent are not unusual even when most of the parameters are known accurately. In many applications, a potentiometer is used to adjust the gain to a desired value, and the frequency or time response characteristics are conservatively designed. In these cases, the performance characteristics are not critical.

Other applications require that the performance characteristics of an amplifier must remain relatively constant over a wide range of generator and load impedance values, circuit parameter values, and dc voltages. This condition can be achieved by incorporating negative feedback into the amplifier designs to make the gain rather insensitive to many parameter variations. However, the price paid for this desirable feature is always a reduction in gain and sometimes a tendency to oscillate. The purposes of this chapter are to study negative-feedback amplifiers in the mid-frequency region and to compare the sensitivity of the performance characteristics of feedback and non-feedback amplifiers to changes in circuit parameters. In the first section, the effects of parameter variations on the performance characteristics will be illustrated for an amplifier that has no ac feedback. The dramatic changes in gain and other performance characteristics produced by changes in the circuit parameters should provide insight into the problem of maintaining a constant gain. The second and third sections contain two specific examples of feedback circuits and are intended to develop both a perspective and an intuitive approach to the analysis of feedback circuits.

A rather general approach to the analysis of feedback amplifiers is introduced in Secs. 10.4 and 10.5. Approximate methods of analysis are treated in Sec. 10.6 and further simplifications are introduced in Sec. 10.7. The detailed analysis of negative-feedback amplifiers is discussed in Secs. 10.8 through 10.13.

One difficulty with negative-feedback amplifiers is a tendency for the circuit to be unstable or to oscillate. This problem can be overcome by properly designed compensation networks. The types of compensation and values of the components used depend on the characteristics of the active devices and other circuit parameters. Most of the amplifiers considered in this chapter require some type of high-frequency compensation to ensure stable operation. The problems of oscillation and compensation are treated in the next chapter.

10.1 Effect of Circuit Parameter Variations on Amplifier Performance

The objective of this section is to illustrate how various performance parameters of a specific amplifier change for changes in the values of various circuit parameters. The specific amplifier chosen for this study is the CE-CC circuit that was analyzed in Ex. 4–17 and shown in Fig. 4–34. This circuit is also shown in Fig. 10–1(a). The approach used in this study was to change one circuit parameter at a time and then to compare the new performance values with those found in

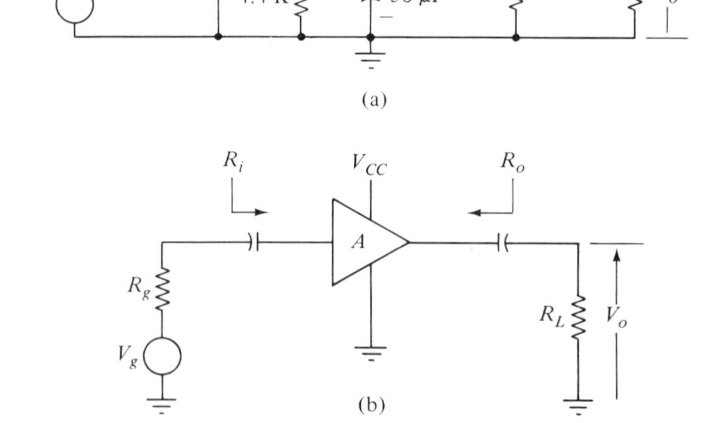

FIGURE 10–1 *A CE-CC amplifier: (a) the actual circuit, and (b) a general amplifier circuit diagram used to represent the CE-CC amplifier.*

Ex. 4–17. The circuit parameters to be changed are the first five parameters listed in the first column of Table 10–1. The performance values from Ex. 4–17 are shown in the second column and are used as reference values.

Consider first a change in the β values of the transistors. This parameter increases by a factor of two for each 50°C rise in the temperature of a germanium transistor and for each 100°C rise for a silicon transistor. In addition, the ratio of the highest value to the lowest value within the same type may range from 4/1 to 10/1, depending upon the selection process. Consequently, increasing both βs by a factor of two is not at all unreasonable. The results of this increase are summarized in the third column of Table 10–1. With this change, A_{vo} increased approximately 70 percent, R_i increased approximately 65 percent, and R_o decreased to 53 percent of its reference value.

TABLE 10–1 *Summary of the amplifier performance changes for variations in circuit parameters for the CE-CC amplifier studied in Ex. 4–17 and shown in Fig. 10–1. (The circuit parameter which was changed is indicated in bold type.)*

		Ex. 4–17 Results	β Doubled	R_L Increased	R_g Increased	V_{CC} Decreased
Circuit Parameters	β_1	35	**70**	35	35	35
	β_2	22	**44**	22	22	22
	R_L	100	100	**1 K**	100	100
	R_g	100	100	100	**1 K**	100
	V_{CC}	+24 V	+24 V	+24 V	+24 V	**+12 V**
Performance Characteristics	R_i	760	1.26 K	760	760	1.4 K
	R_o	455	242	455	455	472
	A_{vt1}	−81	−134	−261	−81	−39.5
	A_{vt2}	0.84	0.93	0.89	0.43	0.93
	A_{vg}	0.89	0.93	0.89	0.43	0.93
	A_{vo}	−61	−104	−228	−29.2	−27
	% change in $\|A_{vo}\|$	—	+70%	+275%	−52%	−56%

The value of R_L can change widely from one application to another. Changing the βs back to their original values and increasing R_L from 100 ohms to 1000 ohms produced the performance values listed in the fourth column of Table 10–1. This change resulted in an increase in $|A_{vo}|$ by a factor of 2.75. On the other end of the circuit, if R_g is increased from 100 ohms to 1000 ohms, the value of $|A_{vo}|$ is reduced to approximately 0.5 of its reference value. The results of this change are listed in column five.

Supply voltage variations can result from power line voltage changes, poor dc supply design,* etc. While these variations are usually small, if the value of

* Generally, most quality dc power supplies use negative feedback to stabilize the value of the dc voltage.

V_{CC} for the CE-CC amplifier shown in Fig. 10–1(a) is reduced from 24 V to 12 V, the value of $|A_{vo}|$ is reduced to approximately 56 percent of its reference value as shown in Table 10–1. Also, the values of R_i and R_o change. Consequently, variations in supply voltage can cause significant variations in the performance characteristics of an amplifier.

In this section, some of the effects of circuit parameter variations on the values of amplifier performance characteristics have been illustrated. The use of negative-feedback circuits can reduce these changes considerably. The remainder of this chapter is devoted to a study of feedback circuits.

10.2 Introduction to the Effects of Negative Feedback in Amplifier Circuits

The objectives of this section are to show how negative feedback can be used to reduce the magnitude of the changes in the performance characteristics discussed in the previous section, and to introduce some of the basic terminology associated with negative-feedback amplifiers. To achieve these objectives, we will analyze and experimentally characterize a negative-feedback amplifier. The results will show that the mid-band gain of a negative-feedback amplifier can be made to depend primarily on the ratio of resistance values and remain relatively constant with changes in circuit components. Also, we will see that the input and output impedances of a negative-feedback amplifier are changed by the use of negative feedback.

Consider the feedback* amplifier shown in Fig. 10–2(a). The ac equivalent circuit of this amplifier is shown in Fig. 10–2(b). This amplifier is different from the amplifiers studied previously because R_{f1} and R_{f2} provide a path between the output and the input of the amplifier. To understand how the amplifier operates, first examine the signal path from the input to the output and then from the output to the input.

Referring to the ac equivalent circuit shown in Fig. 10–2(b), the generator signal V_g produces an input signal V_i at the base terminal of Q_1. This voltage drives Q_1 and causes a signal V_{o1} at the collector terminal of Q_1, which in turn causes a signal V_o at the collector of Q_3. The signal path from the input to the output is referred to as the *forward signal path* of the amplifier.

Negative feedback is obtained by using a fractional part of the output signal to drive the base of Q_2. This signal, called V_f, is the second input signal to the Q_1-Q_2 differential pair.** Neglecting the resistance looking into the base of Q_2 as large ($R_{\text{in}2} \gg R_{f1}$), the value of V_f is

$$V_f = V_o[R_{f1}/(R_{f1} + R_{f2})] \quad \textbf{(10–1)}$$

The signal path from the output terminal back to the input circuit is referred to as the *feedback signal path*, and the signal V_f is called the *feedback signal*. Note that the signals V_i and V_f are in phase, a relationship that causes the output voltage of Q_1, V_{o1}, to be less than that for the case of no feedback. Stated differently, the differential pair Q_1-Q_2 effectively subtracts the feedback signal, V_f, from the input signal, V_i.

---

* For simplicity, we will refer to negative-feedback amplifiers simply as feedback amplifiers in this and later sections.

** Differential amplifiers are discussed in Sec. 4.9.

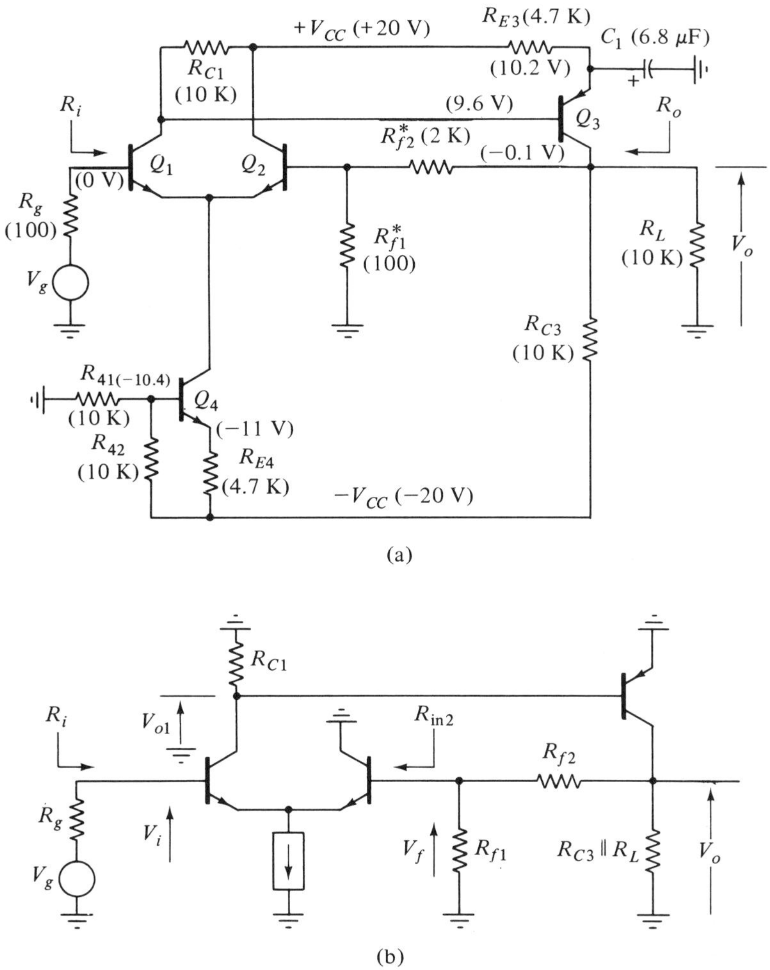

FIGURE 10–2 *A non-inverting negative-feedback amplifier: (a) the complete circuit, and (b) an ac equivalent circuit. Resistors marked with (*) indicate 1 percent values.*

Before discussing the response of the circuit shown in Fig. 10–2, we will briefly summarize the feedback characteristics discussed above. The amplifier shown in Fig. 10–2 consists of a "forward path" (Q_1, Q_2, and Q_3) and a "feedback path" (R_{f1} and R_{f2}). The feedback network effectively *samples* the output voltage, and the differential pair (Q_1-Q_2) combines or *sums* the *input signal* with the *feedback signal.* Since the summing operation is negative,* the amplifier is said to use *negative feedback.*

The performance characteristics of the amplifier shown in Fig. 10–2 will be obtained by using the experimental results listed in Table 10–2. One important

---

* If V_i and V_f were 180° out of phase, the differential pair Q_1-Q_2 would effectively add V_i and V_f since, from Table 4–4,

$$V_{o1} = A_{d1-2}(V_i - V_f)$$

When the input signal and the feedback signal are added, the result is *positive feedback* which can, as indicated in Chap. 11, lead to self-sustained oscillations.

TABLE 10–2 *Experimental values of the voltage gain of the amplifier shown in Fig. 10–2 for various values of* R_g, R_L, V_{CC}, *and* β. *(The circuit parameter which was changed is indicated in bold type.)*

	Initial Results	Effect of Changing β	Effect of Changing V_{CC}	Effect of Changing R_g			Effect of Changing R_L		
β_1	170	**450**	170	170	170	170	170	170	170
β_2	170	**450**	170	170	170	170	170	170	170
β_3	90	**250**	90	90	90	90	90	90	90
V_{CC}	±20 V	±20 V	**±10 V**	±20 V	±20 V	±20 V	±20 V	±20 V	±20 V
R_g	100	100	100	**1 K**	**10 K**	**100 K**	100	100	100
R_L	10 K	10 K	10 K	10 K	10 K	10 K	∞	**1 K**	**100**
V_o/V_i	$\frac{1\text{ V}}{48\text{ mV}}$	$\frac{1\text{ V}}{48\text{ mV}}$	$\frac{1\text{ V}}{48\text{ mV}}$	$\frac{1\text{ V}}{48\text{ mV}}$	$\frac{1\text{ V}}{48\text{ mV}}$	$\frac{0.88\text{ V}}{48\text{ mV}}$	$\frac{1\text{ V}}{48\text{ mV}}$	$\frac{1\text{ V}}{48\text{ mV}}$	$\frac{0.1\text{ V}}{5\text{ mV}}$
$A_v = V_o/V_i$	21	21	21	21	21	18.3	21	21	20*

* With R_L equal to 100 Ω, the maximum negative output voltage is limited to the value $I_{E3}R_{Leq3}$ or approximately 0.2 V. Consequently, the gain was measured at a lower output signal level to eliminate distortion.

performance characteristic of this amplifier is that the voltage gain changes very little for large changes in R_g, R_L, V_{CC}, and β. A constant voltage gain is in sharp contrast with the gain variations observed in the non-feedback amplifier in Sec. 10.1.

The value of the gain with feedback depends primarily on the value of the feedback network components, R_{f1} and R_{f2}. For the amplifier shown in Fig. 10–2, the gain of the feedback path is

$$\frac{V_f}{V_o} \simeq \frac{R_{f1}}{R_{f1} + R_{f2}} \simeq \frac{100}{100 + 2\text{ K}} \simeq \frac{1}{21} \qquad \textbf{(10–2)}$$

From the experimental results given in Table 10–2, the voltage gain of the amplifier, V_o/V_i, is the reciprocal of the gain of the feedback network or

$$V_o/V_i \simeq (R_{f1} + R_{f2})/R_{f1} \simeq 21 \qquad \textbf{(10–3)}$$

The derivation and the limitations of Eq. (10–3) are discussed in later sections.

A third effect of feedback on the performance characteristics of the amplifier shown in Fig. 10–2 was to reduce the value of the output impedance, R_o. This effect can be observed from the experimental data in Table 10–2 and with the aid of the Thévenin equivalent circuit* of the output of the amplifier shown in Fig. 10–3. From Fig. 10–3, we see that the output voltage, V_o, is a fractional part of the Thévenin voltage, V_{Th}, and is determined by the voltage division between R_{Th} and R_L. Neglecting the effects of feedback,

$$\begin{aligned} R_o &\simeq r_{ce3} \,||\, R_{C3} \,||\, (R_{f2} + R_{f1} \,||\, R_{\text{in}2}) \\ &\simeq 10\text{ K} \,||\, (2\text{ K} + 100) \simeq 1.84\text{ K} \end{aligned} \qquad \textbf{(10–4)}$$

Although the value of R_L was varied from an open circuit condition ($R_L = \infty$) to as low as 100 ohms, the gain remained nearly constant. Consequently, the value of R_{Th}, and hence the value of R_o, is less than 100 ohms, a much smaller value than 1.84 K as calculated in Eq. (10–4) by neglecting the effects of feedback. In a later section, we will show that in feedback amplifiers where the output voltage is sampled, the value of R_o is generally reduced to a small value.

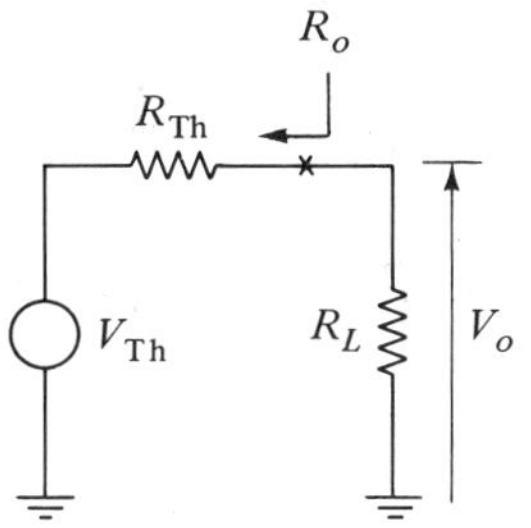

FIGURE 10–3 *A Thévenin equivalent circuit for the output of the amplifier shown in Fig. 10–2.*

A fourth effect of the feedback on the performance characteristics of the amplifier in Ex. 10–1 was to increase the value of the input impedance, R_i. This effect can be observed from the experimental data given in Table 10–2 and with the aid of the equivalent circuit of the input section of the amplifier shown in

---

* A Thévenin equivalent circuit of the output is found by disconnecting R_L and placing the looking back impedance of the amplifier, R_{Th}, in series with the open circuit output voltage, V_{Th}.

Fig. 10–4. From Fig. 10–4 we see that the input voltage to the amplifier is the generator voltage, V_g, reduced by the voltage division between R_g and R_i.

Neglecting the effects of feedback,

$$\begin{aligned} R_i &\simeq R_{\text{in}1} \simeq (\beta_1 + 1)(r_{e1} + r_{e2}) + R_{f1} \,||\, (R_{f2} + R_{C3} \,||\, R_L) \\ &\simeq (171)(52) + 100 \,||\, (2\text{ K} + 10\text{ K} \,||\, 10\text{ K}) \simeq 8.8\text{ K} \end{aligned} \tag{10–5}$$

If feedback did not change the value of R_i, the gain would be reduced to 0.5 when R_g and R_i were equal. From Table 10–2, however, the gain was not changed from the reference value with R_g equal to 10 K and was reduced only 13 percent (to 18.3) with R_g equal to 100 K. Consequently, the value of R_i was increased significantly with feedback. In a later section, we will show that in feedback amplifiers where voltages are summed or combined at the input, the value of R_i generally increases.

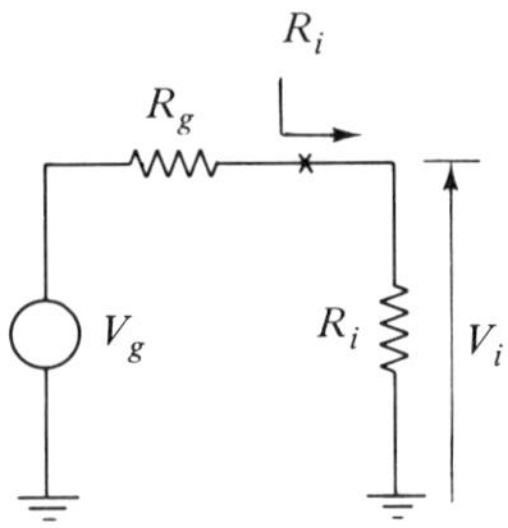

FIGURE 10–4 *An equivalent circuit of the input section of the amplifier shown in Fig. 10–2.*

The only difference between this amplifier and the amplifiers studied in previous chapters was the presence of the feedback network, R_{f1} and R_{f2}. The input signal to the feedback network was the output voltage of the amplifier V_o, and the output signal of the feedback network was the fedback signal V_f. The input signal, V_i, and the fedback signal, V_f, were summed or combined through the action of the differential pair, Q_1-Q_2. This type of feedback amplifier is called a *voltage sample-voltage sum* feedback amplifier. Assuming proper design, the performance characteristics of a voltage sample-voltage sum feedback amplifier can be summarized as follows: (1) the voltage gain is determined by the feedback network; (2) the value of R_o is reduced, and the value of R_i is increased; and (3) the voltage gain remains relatively constant for changes in circuit parameters.

10.3 A Second Example of the Effects of Negative Feedback in Amplifier Circuits

For the specific circuit studied in the preceding section, we found that negative feedback stabilized the voltage gain against changes in generator impedance and load impedance. Also, the input resistance was increased, and the output resistance was decreased. However, the feedback circuit used is only one of several types available. The objective of this section is to show that the effect of feedback on the performance characteristics of an amplifier depends on the type of feedback used. To achieve this objective, we will examine the performance of a second type of feedback amplifier.

Consider the feedback amplifier shown in Fig. 10–5(a). An ac equivalent circuit for the amplifier is shown in Fig. 10–5(b), and a simplified equivalent circuit is shown in Fig. 10–5(c). This amplifier is different from the one con-

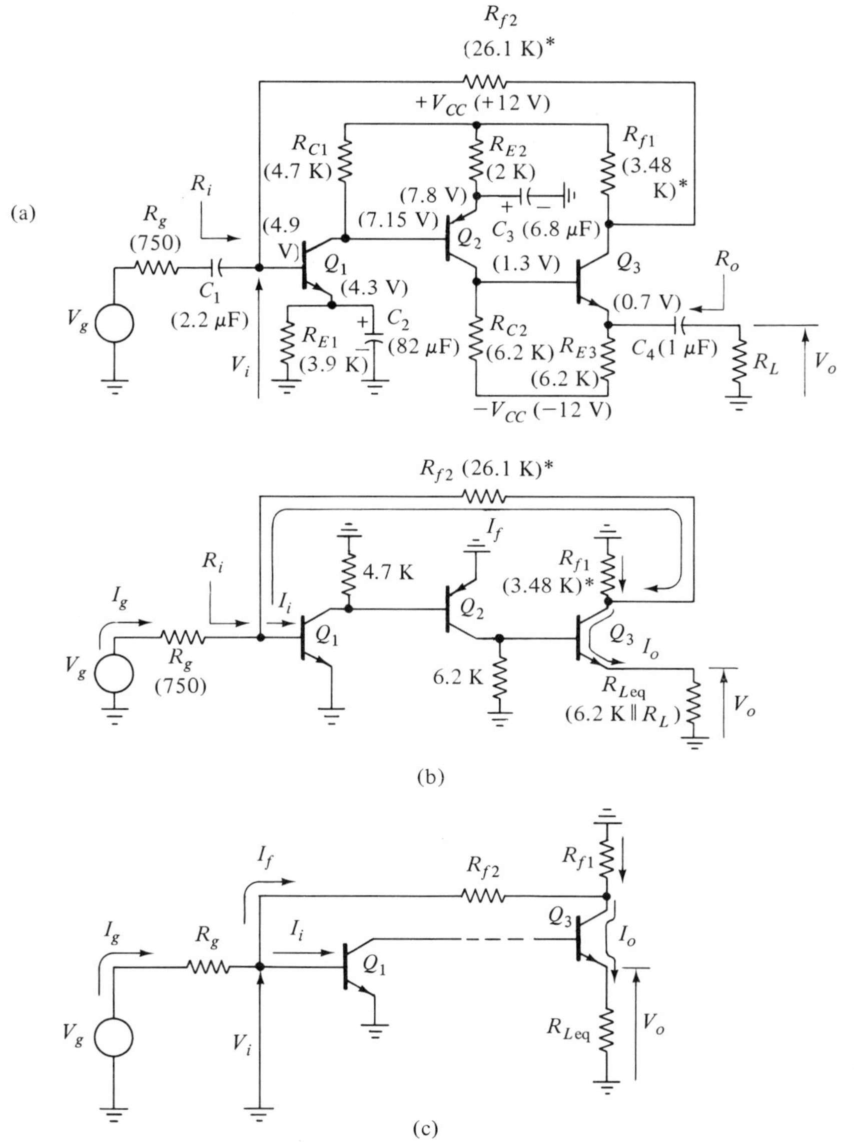

FIGURE 10–5 *A second type of feedback amplifier: (a) the actual circuit, (b) an ac equivalent circuit, and (c) a simplified ac equivalent circuit. (Resistors marked with (*) indicate 1 percent values.)*

sidered in the previous section in two important respects. First, the feedback network (R_{f1} and R_{f2}) is not connected at the same point where the output voltage is obtained as in the previous case. The output voltage appears at the emitter terminal of Q_3, and the feedback network is *sampling* the output current, not the output voltage. The output current, I_o, which flows in R_{Leq}, also flows in the feedback network (R_{f1} and R_{f2}). Second, the feedback resistor, R_{f2}, and the generator impedance, R_g, are connected at the same node, the base terminal of Q_1. This point serves as the *summing* node for the generator current, I_g, and the feedback current, I_f. The difference between these two components is the actual input current to the amplifier, I_i. These two points suggest that the feedback amplifier shown in Fig. 10–5 should be called a *current sample-current sum* feedback amplifier.

The performance characteristics of the amplifier shown in Fig. 10–5 will be studied using the experimental results listed in Table 10–3. One important experimental result apparent from Table 10–3 is that the voltage gain, A_v, varies over a wide range for changes in R_g and R_L. However, the current gain, $A_i = I_o/I_g$, remained relatively constant for changes in R_g and R_L. The output current is related to the output voltage by*

$$I_o = V_o/R_{Leq} \tag{10–6}$$

and the generator current is related to the generator voltage by

$$I_g = V_g/(R_g + R_i) \tag{10–7}$$

As will be shown later in this section, the effect of summing currents at the input is to reduce the value of R_i, usually to a very small value. Thus, I_g is approximately related to V_g by

$$I_g \simeq V_g/R_g \tag{10–8}$$

Combining Eqs. (10–6) and (10–8), the current gain, A_i, is

$$\frac{I_o}{I_g} \simeq \frac{V_o/R_{Leq}}{V_g/R_g} \simeq \frac{V_o}{V_g}\frac{R_g}{R_{Leq}} \tag{10–9}$$

The experimental data given in Table 10–3 indicates that the current gain of the amplifier shown in Fig. 10–5 remains relatively constant for changes in R_g and R_L. Consequently, we should expect that this type of feedback network produces a stable current gain rather than a stable voltage gain as was the case in the last section. The value of the stable gain, however, is still determined by the feedback network. From Fig. 10–6, the fedback signal, I_f, is related to the output signal, I_o, by the current division between R_{f1} and R_{f2} as

$$I_f = I_o[R_{f1}/(R_{f1} + R_{f2})] \tag{10–10}$$

Solving Eq. (10–10) for I_o/I_f,

$$\frac{I_o}{I_f} = \frac{R_{f1} + R_{f2}}{R_{f1}} = \frac{3.48\text{ K} + 26.1\text{ K}}{3.48\text{ K}} \simeq 8.5 \tag{10–11}$$

Note that the current gain calculated in Eq. (10–11) is in agreement with the experimental values of current gain given in Table 10–3. Just as the voltage gain of the voltage sample-voltage sum feedback amplifier in the last section was determined by the values of the feedback resistors, the current gain of the current sample-current sum feedback amplifier in this section is also determined by the values of the feedback resistors.

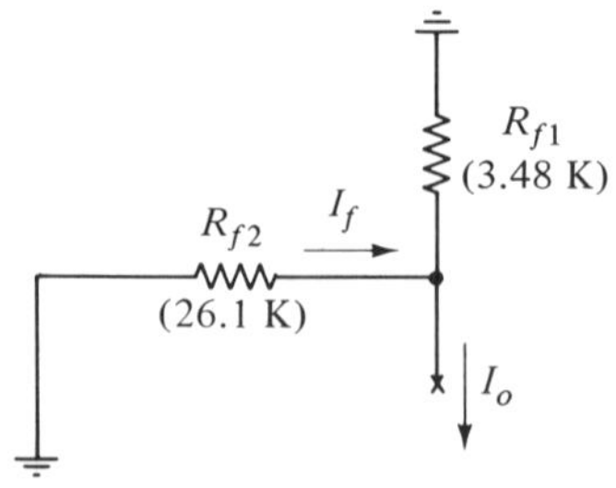

FIGURE 10–6 *The feedback network for the amplifier shown in Fig. 10–5 assuming that the equivalent resistance from the base terminal of* Q_1 *to ground is approximately zero ohms.*

* From Fig. 10–5, the value of R_{Leq} is $R_{E3} \parallel R_L$.

TABLE 10–3 *Experimental values of the voltage gain, A_v, and the current gain, A_i, for the amplifier shown in Fig. 10–5 for various values of R_g and R_L. (The circuit parameter which was changed is indicated by bold type.)*

	Initial Values	Effect of Changing R_g			Effect of Changing R_L		
R_g	750	**1 K**	**2 K**	**10 K**	750	750	750
R_L	1 K	1 K	1 K	1 K	**100**	**6.2 K**	∞
R_{Leq}	860	860	860	860	**97**	**3.1 K**	**6.2 K**
V_o/V_g	$\frac{0.98\text{ V}}{0.1\text{ V}}$	$\frac{0.74\text{ V}}{0.1\text{ V}}$	$\frac{0.38\text{ V}}{0.1\text{ V}}$	$\frac{0.074\text{ V}}{0.1\text{ V}}$	$\frac{0.11\text{ V}}{0.1\text{ V}}$	$\frac{1.75\text{ V}}{0.05\text{ V}}$	$\frac{3.5\text{ V}}{0.05\text{ V}}$
$A_v = V_o/V_g$	9.8	7.4	3.8	0.74	1.1	35	70
$A_i = I_o/I_g$	8.6	8.6	8.8	8.6	8.5	8.5	8.5

Another effect of the feedback on the performance characteristics of the amplifier shown in Fig. 10–5 was to reduce the value of the input impedance, R_i. This effect can be observed by connecting a resistor, R_x, between the base terminal of Q_1 and ground as shown in Fig. 10–7. Neglecting the effects of feedback, the value of R_i is*

$$R_i \simeq (R_{f2} + R_{f1}) \,||\, R_{\text{in}1} \simeq (R_{f2} + R_{f1}) \,||\, (\beta_1 + 1)(r_{e1}) \simeq (26.1\text{ K} + 3.48\text{ K}) \,||\, (100 + 1)(23.6) \simeq 2.2\text{ K} \qquad \textbf{(10–12)}$$

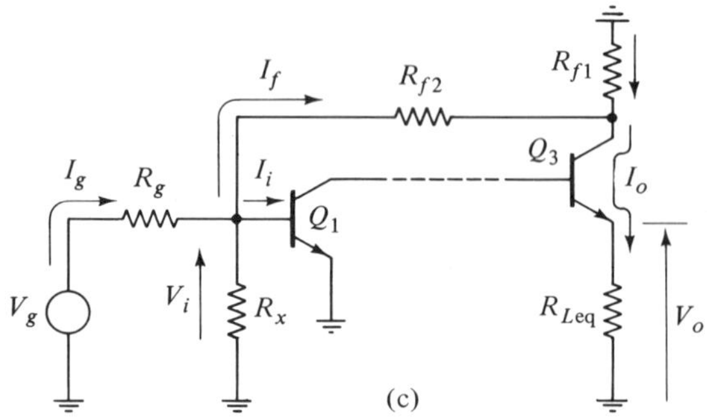

FIGURE 10–7 *Modification of the simplified ac equivalent circuit in Fig. 10–5(c) showing the presence of a resistor from the base terminal of* Q_1 *and ground.* (R_x *was coupled to the base of* Q_1 *through a 6.8-μF capacitor.*)

If the feedback circuit did not reduce the value of R_i, placing a small resistor R_x in parallel with R_i should reduce the gain by diverting a portion of I_g away from the amplifier's input. The experimental results for various values of R_x are given in Table 10–4. Note that the value of the current gain, and consequently the voltage gain, remained constant for values of R_x as low as 10 ohms. Therefore, the value of R_i is much less than 10 ohms and certainly less than the value calculated in Eq. (10–12), by neglecting the effects of feedback. In a later section, we will show that in negative-feedback amplifiers where currents are summed at the input, the value of R_i is generally reduced.

TABLE 10–4 *Experimental values of the voltage gain,* A_v, *and the current gain,* A_i, *of the amplifier shown in Fig. 10–7 for various values of* R_x.

R_x	R_g	R_L	R_{Leq}	$A_v = V_o/V_g$	$A_i = I_o/I_g$
∞	750	1 K	860	0.98 V/0.1 V = 9.8	8.6
1 K	750	1 K	860	0.98 V/0.1 V = 9.8	8.6
100	750	1 K	860	0.98 V/0.1 V = 9.8	8.6
10	750	1 K	860	0.98 V/0.1 V = 9.8	8.6

The effect of feedback on the output impedance of the amplifier shown in Fig. 10–5 can be determined from the experimental data given in Table 10–3. A partial ac equivalent circuit of the amplifier output section is shown in Fig. 10–8(a). The corresponding Thévenin equivalent circuit is shown in Fig. 10–8(b). Neglecting the effects of feedback, the value of R_o is**

* The value of r_{e1} is available from the bias voltages shown in Fig. 10–5(a). The value of β_1 was measured.

** The value of r_{e3} is available from the bias voltages shown in Fig. 10–5(a). The value of β_3 was measured.

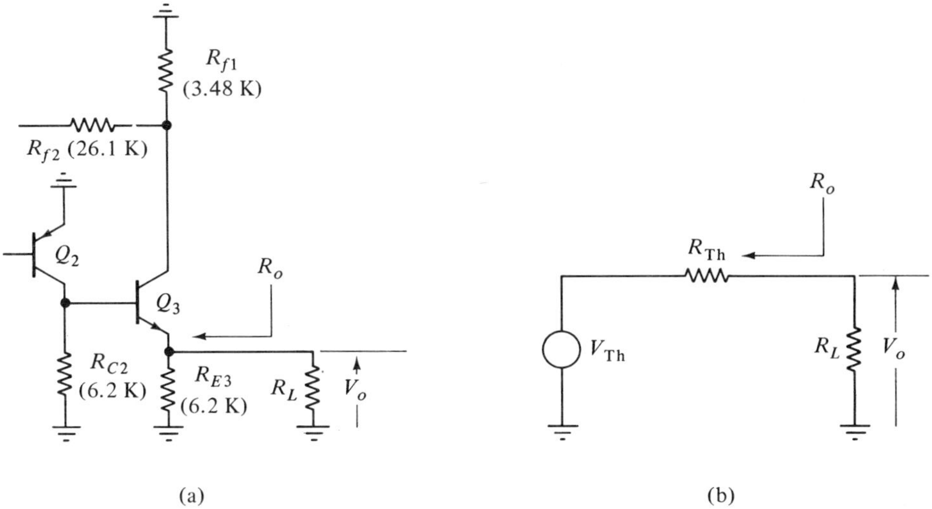

FIGURE 10–8 *Circuit used for the calculation of* R_o *of the amplifier shown in Fig. 10–5: (a) a partial ac equivalent circuit, and (b) a Thévenin equivalent circuit.*

$$R_o \simeq R_{E3} \,||\, \left(r_{e3} + \frac{R_{C2}}{\beta_3 + 1}\right) \simeq 6.2 \text{ K} \,||\, \left(12.7 + \frac{6.2 \text{ K}}{151}\right) \simeq 54\ \Omega \quad \textbf{(10–13)}$$

From Fig. 10–8(b), the output voltage is that part of the open circuit voltage, V_{Th}, that appears across R_L by simple voltage division between R_{Th} and R_L. If the value of R_o calculated in Eq. (10–13) is correct, the output voltage should be reduced to 50 percent of the open circuit value when R_L is approximately 54 ohms. From Table 10–3, however, the value of V_o was reduced 50 percent when R_L was changed from an infinite value (open circuit) to 6.2 K. Consequently, the apparent output impedance is approximately 6.2 K, a value much higher than that calculated in Eq. (10–13). As indicated in Fig. 10–8(a), R_{E3} is 6.2 K and places an upper limit on the value of R_o. The effect of this type of feedback was to greatly increase the value of the resistance looking into the emitter terminal of Q_3. In a later section, we will show that in negative-feedback amplifiers where the output current is sampled, the value of R_o is generally increased.

In this section, we have studied the characteristics of a current sample-current sum feedback amplifier. Assuming proper design, the effects of feedback on the performance characteristics of this type of feedback amplifier can be summarized as follows: (1) the current gain is determined by the feedback network, (2) the value of R_i is reduced, (3) the value of R_o is increased, and (4) the current gain remains relatively constant for changes in circuit parameters. The performance characteristics listed above are valid provided that the product of the gain of the forward signal path and the feedback signal path is much greater than unity. A general method of calculating this gain product is discussed in a later section.

10.4 Characteristics of Ideal Amplifiers

The feedback amplifiers studied in the two preceding sections had three salient characteristics: (1) a stable current or voltage gain, (2) either a very high or a very low value of input impedance, and (3) either a very high or a very low value of output impedance. Amplifiers with these characteristics are often termed *ideal amplifiers*. In this section, we will discuss the characteristics and types of ideal amplifiers because they are very useful in the analysis and design of feedback amplifiers and in the analysis and design of large systems composed of a series of amplifiers and other electronic systems.

A general block diagram of an amplifier is shown in Fig. 10–9. The amplifier is characterized by its input impedance R_i, its output impedance R_o, and its gain A, where A is the output signal divided by the input signal. While the input and output signals may be voltages, currents, or power, only voltage and current signals will be discussed in this chapter.

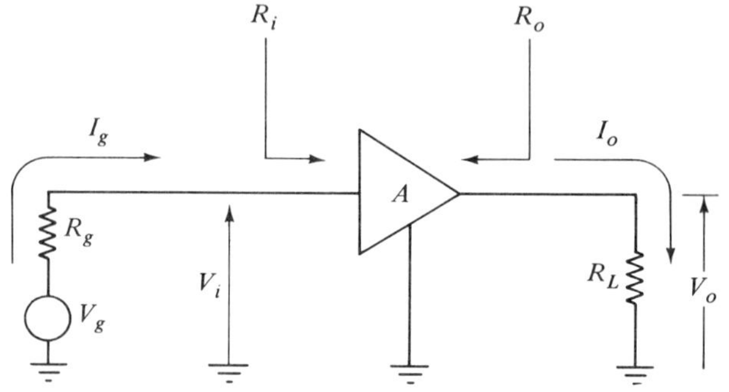

FIGURE 10–9 *A basic block diagram of an amplifier.*

Gain expressions and calculations were developed in previous chapters for voltage, A_{vt}, A_{vg}, and A_{vo}. Voltage gain was also stressed in the first two sections of this chapter. In Sec. 10.3, however, the current gain of an amplifier was stabilized by feedback. To expand our study of feedback, we must be able to obtain the following four gain relationships: (1) V_o/V_g, (2) I_o/I_g, (3) I_o/V_g, and (4) V_o/I_g. Negative feedback can be used to stabilize any one of these gains in an amplifier. In the remainder of this section, we will consider the characteristics of ideal amplifiers associated with each of these four expressions. Also, we will discuss a method for converting from one gain expression to another for a given amplifier. Feedback circuits that will stabilize these gains are discussed in later sections.

An ideal *voltage amplifier* should provide a constant-voltage gain for large changes in the values of R_g and R_L. To satisfy this condition, a voltage amplifier should have a large value for R_i (ideally infinite) and a small value of R_o (ideally zero). When the value of R_i is much greater than the value of R_g, the value of A_{vg} is essentially unity. When the value of R_o is much less than the value of R_L, the output voltage changes very little as the value of R_L changes. The characteristics of an ideal voltage amplifier are summarized in Table 10–5. One example of a nearly ideal voltage amplifier is the voltage sample-voltage sum feedback amplifier discussed in Sec. 10.2.

An ideal *current amplifier* should provide a constant-current gain for large changes in the values of R_g and R_L. A current amplifier should have a small value of R_i (ideally zero) and a large value of R_o (ideally infinite). The characteristics of an ideal current amplifier are summarized in Table 10–5. One example

TABLE 10–5 *Summary of ideal amplifier characteristics.*

Amplifier type	Stable gain	Input impedance	Output impedance	Block diagram
Voltage	$A_v = V_o/V_g$	$R_i >> R_g$ $R_i \simeq \infty$	$R_o << R_L$ $R_o \simeq 0$	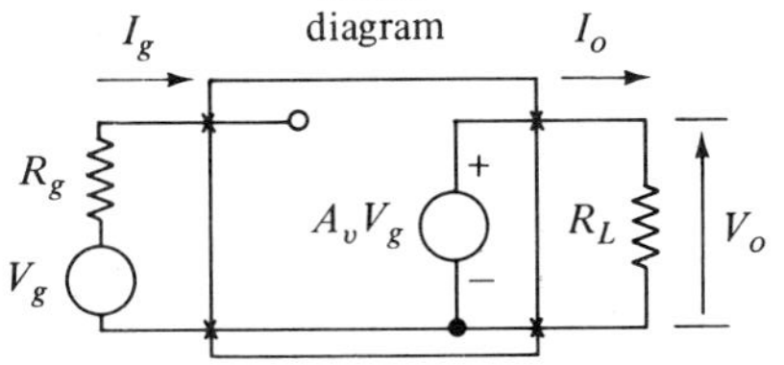
Current	$A_i = I_o/I_g$	$R_i << R_g$ $R_i \simeq 0$	$R_o >> R_L$ $R_o \simeq \infty$	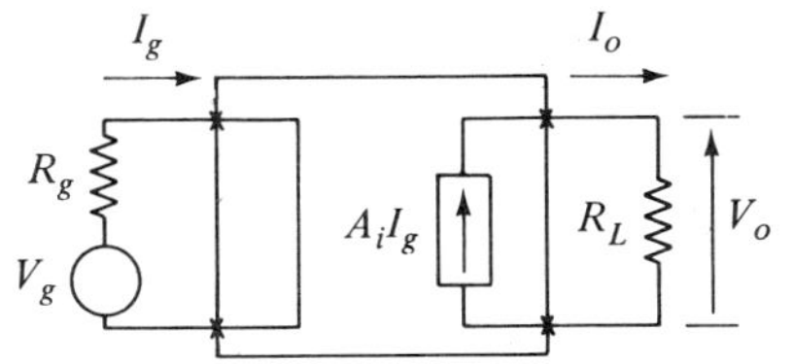
Transconductance	$A_g = I_o/V_g$	$R_i >> R_g$ $R_i \simeq \infty$	$R_o >> R_L$ $R_o \simeq \infty$	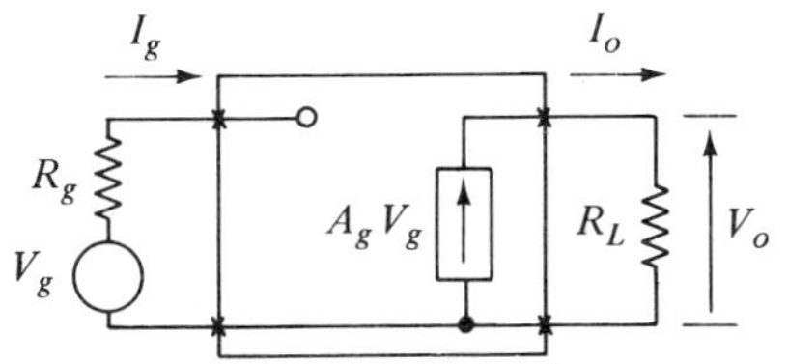
Transresistance	$A_r = V_o/I_g$	$R_i << R_g$ $R_i \simeq 0$	$R_o << R_L$ $R_o \simeq 0$	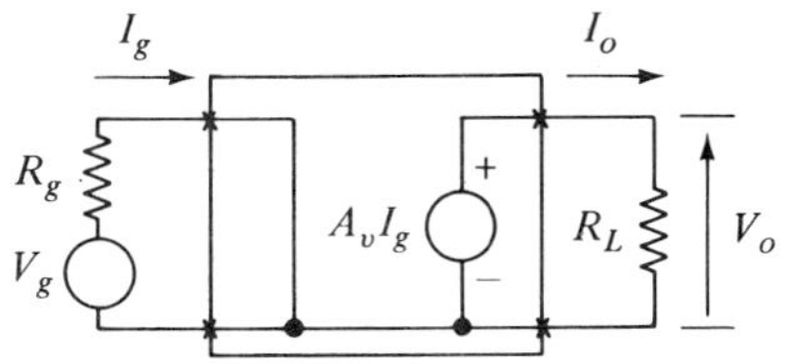

of a nearly ideal current amplifier is the current sample-current sum feedback amplifier discussed in Sec. 10.3. In addition, a CB stage with a low value of r_e ($r_e \ll R_g$) approximates an ideal current amplifier with a current gain of α, or approximately unity.

By combining the characteristics of the ideal voltage amplifier and the ideal current amplifier, we can determine the characteristics of the other two types of ideal amplifiers. For example, an ideal *transconductance amplifier* should provide an output current directly proportional to the input voltage. The name of the transconductance amplifier is derived from the units of the gain expression, amperes per volt, or conductance. A transconductance amplifier should have a large value of R_i, and ideally R_o should be infinite. The characteristics of an ideal transconductance amplifier are summarized in Table 10–5. One example of a nearly ideal transconductance amplifier is a MOSFET connected as a CS configuration with a large value of r_{ds} ($R_{Leq} \ll r_{ds}$). For this amplifier, the value of the transconductance gain is equal to the g_m of the device.

The fourth class of ideal amplifier is the *ideal transresistance amplifier* where the output voltage is directly proportional to the input current. The name of the transresistance amplifier is derived from the units of the gain expression, volts per ampere, or resistance. A transresistance amplifier should have small values of both R_i and R_o, ideally zero. The characteristics of a transresistance amplifier are summarized in Table 10–5.

An important effect of feedback is to stabilize one gain of an amplifier. Note that negative feedback stabilized the voltage gain of the amplifier in Sec. 10.2 and stabilized the current gain of the amplifier in Sec. 10.3. Depending on the type of feedback, the stable gain may be A_v, A_i, A_g, or A_r. Often, it is desirable to convert from one gain expression to another in a given amplifier. This conversion is illustrated in the following example.

EXAMPLE 10–1 The amplifier shown in Fig. 10–10 approximates an ideal current amplifier with a stable current gain of 8.5. When R_g is 750 ohms and R_{Leq} is 3.1 K, determine the values of A_v, A_g, and A_r for the amplifier.

Solution: (a) Voltage gain, A_v: The voltage gain can be written in terms of A_i, R_{Leq}, and R_g as follows:

$$A_v = \frac{V_o}{V_g} = \frac{I_o R_{Leq}}{I_g(R_g + R_i)} \simeq A_i(R_{Leq}/R_g)$$

For the amplifier shown in Fig. 10–10,

$$A_v \simeq A_i(R_{Leq}/R_g) \simeq (8.5)(3.1\text{ K}/750) \simeq 35$$

(b) Transconductance gain A_g: The transconductance gain can be written in terms of A_i, R_{Leq}, and R_g as follows:

$$A_g = \frac{I_o}{V_g} = \frac{I_o}{I_g(R_g + R_i)} \simeq \frac{A_i}{R_g}$$

For the amplifier shown in Fig. 10–10,

$$A_g \simeq A_i/R_g \simeq 8.5/750 \simeq 11.3 \text{ milliohms}$$

(c) Transresistance gain A_r: The transresistance gain can be written in terms of A_i, R_{Leq}, and R_g as follows:

$$A_r = \frac{V_o}{I_g} = \frac{I_o R_{Leq}}{I_g} = A_i R_{Leq}$$

For the amplifier shown in Fig. 10–10,

$$A_r = A_i R_{Leq} \simeq (8.5)(3.1\text{ K}) \simeq 26.4\text{ K}$$

(d) Experimental results: The current amplifier in this example is the same as the current sample-current sum feedback amplifier considered in the last section. The experimental voltage gain is given in Table 10–3 as

$$A_v = V_o/V_g = 1.75\text{ V}/0.05\text{ V} \simeq 35$$

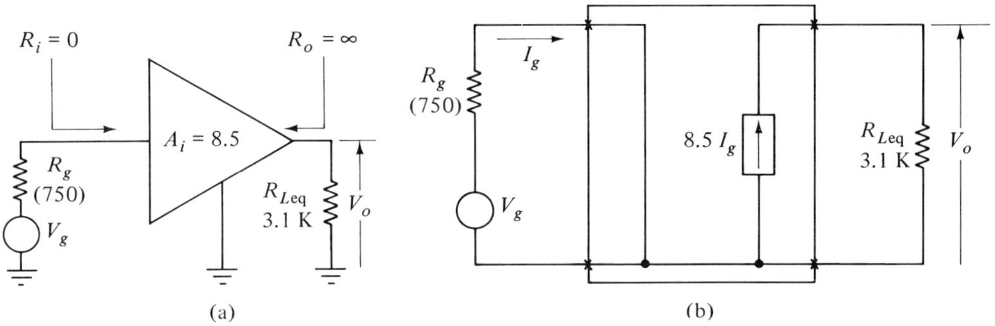

FIGURE 10–10 *The circuit diagram from Ex. 10–1: (a) a general block diagram for the amplifier, and (b) use of the ideal current amplifier block diagram to represent the amplifier.*

In this section, we have discussed the characteristics of ideal amplifiers and summarized them in Table 10–5. Also, we illustrated how to convert from one gain into another using the values of R_g and R_L. In the next two sections, we will discuss methods of determining the stable gain of a feedback amplifier. Once the stable gain is known, any other gain can be found using the material in this section.

10.5 Basic Concepts in Negative-Feedback Amplifiers

Much of the treatment of feedback amplifiers presented in the previous sections was based on experimental data. This approach avoided unnecessary analytical complications while the concept of feedback was still new. We have now reached a point, however, where a more basic treatment of feedback is appropriate. The objectives of this section are: (1) to study the arrangement and characteristics of the basic parts of a feedback amplifier, and (2) to derive the gain expression for an amplifier with feedback. Additional feedback characteristics are discussed later in this chapter.

A block diagram showing the basic parts of a feedback amplifier is shown in Fig. 10–11. In this figure, the feedback amplifier consists of four basic parts: (1) a high gain amplifier, A; (2) a feedback network, B; (3) a sampling network; and (4) a summing network. As indicated in the last three sections, the important input or output signal in an amplifier can be either a voltage or a current. Consequently, to initiate a general treatment, we will use the symbol X to represent the signals in a feedback amplifier.

The generator circuit in Fig. 10–11 produces a signal, X_i, at the input terminals of the summing network. The second input signal to the summing network is

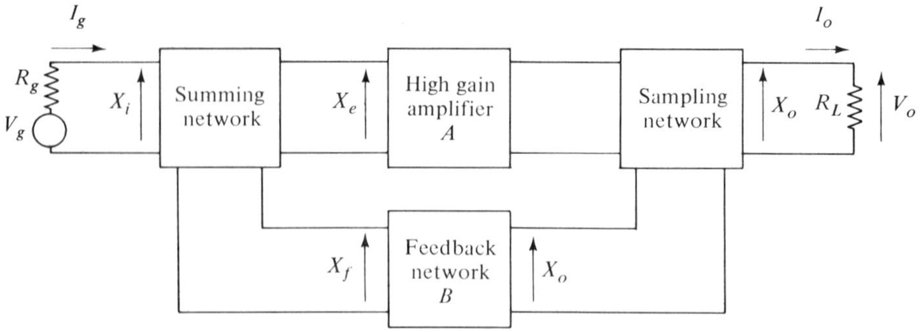

FIGURE 10–11 *A block diagram showing the basic parts of a feedback amplifier.*

the feedback signal, X_f. The output of the summing network is the *error signal,* X_e, and is equal to the difference between X_i and X_f in a negative-feedback amplifier. The error signal is the actual input signal to the high gain amplifier, A. The output signal of the amplifier, X_o, is also the input signal to the feedback network, B. The feedback network is connected between the sampling network and the summing network and is the feedback signal path. The output signal of the feedback network is X_f. The gain of the feedback network is X_f/X_o; that is, B is X_f/X_o.

Our next task is to discuss briefly the characteristics of each part of the feedback amplifier shown in Fig. 10–11. The high-gain amplifier, A, is generally a multistage amplifier which can have a variety of configurations. Many of the multistage amplifiers discussed in previous chapters are high-gain amplifiers. One important requirement is that the high-gain amplifier be nearly *unilateral;* i.e., the signal gain from the input to the output should be much greater than the signal gain from the output back to the input. This requirement is generally satisfied in a multistage amplifier. The ideal gain of the amplifier, A, is infinity.

The function of the sampling network is to sample the output signal, X_o. Two possible methods of sampling the output signal are shown in Fig. 10–12. In Fig. 10–12(a), the signal being sampled is the output voltage V_o, and the sampling network is simply the common connection between the output of the amplifier and the input to the feedback network, B. In Fig. 10–12(b), the signal being sampled is the output current I_o, and the sampling network is actually part of the amplifier, A. The signal being sampled at the output is important, since this determines the type of gain stabilized in a feedback amplifier. Generally, the problem of determining which output signal is being sampled (V_o or I_o) is relatively simple. If the feedback network and the output terminal have a common connection, V_o is the signal being sampled. If the feedback network and the output terminal do not have a common connection, I_o is generally the signal being sampled.

Negative feedback tends to hold the sampled signal constant for changes in circuit components. If V_o is sampled, the value of V_o remains relatively constant against component changes, and the output of the amplifier appears as a voltage source. The resistance of an ideal voltage source is zero. Thus, voltage sampling at the output reduces the output impedance of an amplifier, an effect observed from experimental data in Sec. 10.2.

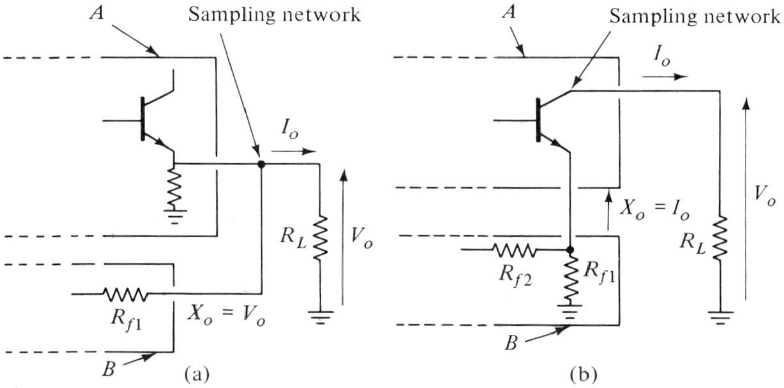

FIGURE 10–12 *Two possible methods of sampling the output signal: (a) voltage sampling, and (b) current sampling.*

If I_o is sampled, the value of I_o remains relatively constant while component values change, and the amplifier appears as a current source. The resistance of an ideal current source is infinite. Thus, current sampling at the output increases the output impedance of an amplifier, an effect also observed in Sec. 10.3.

The function of the summing network is to sum the signal fed back, X_f, and the input signal, X_i. Two possible methods of summing signals at the input of an amplifier are shown in Fig. 10–13. In this figure, the summing network consists of a simple connection common to the generator circuit, the feedback network, and the amplifier input. Since only currents can be summed at a node, the important signals at the input of the amplifier shown in Fig. 10–13(a) must be currents. The input signal X_i corresponds to the generator current I_g; the feedback signal X_f corresponds to the current in the B network, I_f; and the error signal X_e is the difference between I_g and I_f.

A second method of summing signals, shown in Fig. 10–13(b), is different from that shown in Fig. 10–13(a) because the feedback network, the amplifier

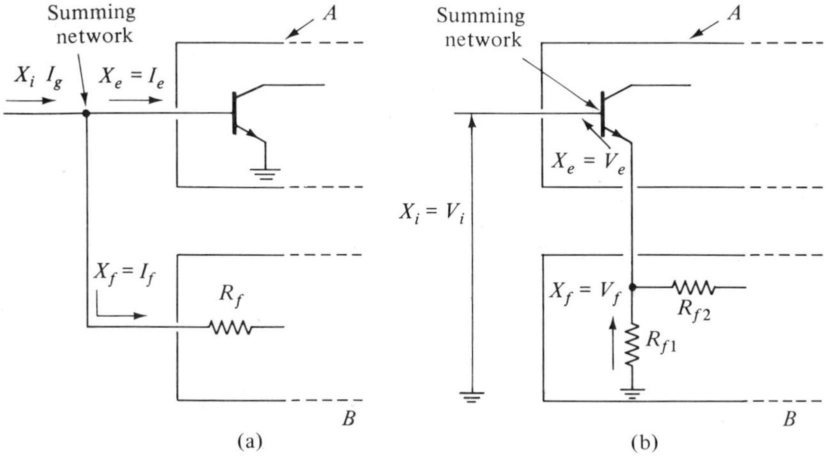

FIGURE 10–13 *Two possible methods of summing signals at the input of a feedback amplifier: (a) current summing, and (b) voltage summing.*

input, and the generator circuit do not have a common connection. This circuit illustrates one method of summing voltages at the input of a feedback amplifier. The input signal X_i corresponds to the input voltage V_i; the feedback signal X_f corresponds to V_f, the voltage across R_{f1}. The error signal X_e is the difference between V_i and V_f and corresponds to the actual signal voltage that appears across the base-emitter junction of the input transistor. A second method of summing V_i and V_f at the input of a feedback amplifier is to use a differential pair. This method was illustrated in Fig. 10–2. The method of signal summing at the input is important since it is one of the ways used to determine the stable gain of a feedback amplifier, an effect considered later in this section.

Negative feedback tends to idealize the load impedance presented to the generator circuit. When the input current is the important signal, as is the case with current summing, the ideal input impedance is zero. Thus, current summing at the input of a feedback amplifier should reduce the value of R_i, an effect observed in Sec. 10.3. In voltage summing, the ideal input impedance is infinite. Thus, voltage summing at the input of a feedback amplifier should increase the value of R_i. This effect was observed in Sec. 10.2.

The function of the feedback network is to provide a signal path from the output back to the input. The input signal to the B network is the output signal, X_o. The output signal of the B network is the feedback signal, X_f, which must be either a voltage, V_f, or a current, I_f, depending on the type of summing used. Consequently, one function of the B network is to convert X_o into a suitable form to allow summing of X_i and X_f.

From the preceding discussion, the following expression for the gain of the feedback or B network is appropriate:

$$B \equiv \frac{X_f}{X_o} = \frac{\text{Signal fedback for summing at the input}}{\text{Signal sampled at the output}} \tag{10–14}$$

Since X_o can correspond to either V_o or I_o, and X_f can correspond to either V_f or I_f, there are four possible expressions for B:

$$B = V_f/V_o;\ \text{voltage sample-voltage sum} \tag{10–15}$$

$$B = I_f/I_o;\ \text{current sample-current sum} \tag{10–16}$$

$$B = V_f/I_o;\ \text{current sample-voltage sum} \tag{10–17}$$

$$B = I_f/V_o;\ \text{voltage sample-current sum} \tag{10–18}$$

Note that the four possible gain expressions for B correspond to both of the following: (1) the four possible configurations of the feedback network, and (2) the four types of ideal amplifiers discussed in the last section. The gain of the B network is of particular importance since the overall gain of a feedback amplifier depends on the B network, an effect observed experimentally in Secs. 10.2 and 10.3.

One final point concerning the B network is that it is assumed to be unilateral; i.e., the only signal transmission is from the output back to the input of the amplifier. This assumption is generally satisfied in a well designed and constructed feedback amplifier.

Our next task is to derive the general expression for the gain of an amplifier with feedback. A simplified block diagram of a feedback amplifier is shown in Fig. 10–14. The block marked A_{ol} represents a high gain amplifier, and A_{ol} is the

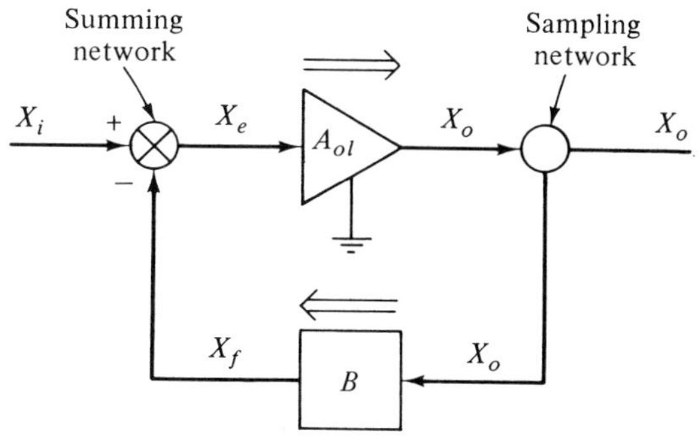

FIGURE 10–14 *A simplified block diagram of a negative-feedback amplifier. The double lined arrows indicate the direction of unilateral signal flow.*

value of the *open-loop gain* or the gain without feedback. The double lined arrows indicate the direction of the unilateral signal flow. Three gain expressions are important in a feedback amplifier and can be defined as follows:

$$\text{Gain with feedback:} \quad A_f \equiv X_o/X_i \tag{10–19}$$

$$\text{Open loop gain:} \quad A_{ol} \equiv X_o/X_e \tag{10–20}$$

$$\text{Gain of } B \text{ network:} \quad B \equiv X_f/X_o \tag{10–21}$$

The gain with feedback, A_f, is also termed the *closed loop gain.*

The summing network in Fig. 10–14 has a "+" and "−" sign to indicate the type of feedback and polarity of the amplifier gain. From this figure,

$$X_e = X_i - X_f \tag{10–22}$$

Substituting Eq. (10–22) into (10–20) gives

$$X_o = A_{ol}X_i - A_{ol}X_f \tag{10–23}$$

Substituting Eq. (10–21) into (10–23) yields

$$X_o = A_{ol}X_i - A_{ol}BX_o \tag{10–24}$$

Solving Eq. (10–24) for A_f,

$$A_f = \frac{X_o}{X_i} = \frac{A_{ol}}{1 + A_{ol}B} \tag{10–25}$$

Equation (10–25) is the classic expression for the gain of a negative-feedback system.* An equivalent expression more useful for our purposes is obtained by multiplying the numerator and the denominator of Eq. (10–25) by B yielding

$$A_f = \left(\frac{A_{ol}}{1 + A_{ol}B}\right)\left(\frac{B}{B}\right) = \left(\frac{A_{ol}B}{1 + A_{ol}B}\right)\left(\frac{1}{B}\right) \tag{10–26}$$

From Eq. (10–26), we observe that when the value of $A_{ol}B$ is much greater than unity, then

$$A_f \simeq 1/B \tag{10–27}$$

The product $A_{ol}B$ appears so often in the analysis and design of feedback amplifiers that a separate symbol will be used to represent this quantity. From

* If the indicated summation of the summing network had been chosen as positive, then X_e would be equal to $(X_i + X_f)$, and Eq. (10–25) would be changed to

$$X_o/X_i = A_{ol}/(1 - A_{ol}B) \tag{10–28}$$

For our purpose, we will use Eq. (10–25) while keeping in mind that the assumed summation was negative.

the block diagram shown in Fig. 10–14, the physical significance of the product $A_{ol}B$ is that it represents the signal gain completely around the feedback loop. This gain is called the *loop transmission* or simply the *loop gain* and is represented by the symbol T, or

$$T \equiv A_{ol}B \tag{10–29}$$

Using this definition for T, the expression for A_f becomes

$$A_f = \left(\frac{T}{1+T}\right)\left(\frac{1}{B}\right) \tag{10–30}$$

When $T \gg 1$, A_f reduces to $1/B$ as indicated in Eq. (10–27).

To gain additional insight into the operation and characteristics of a negative-feedback amplifier, the significance of the various results derived above will be discussed briefly. First, the sign of T is always negative in a negative-feedback amplifier. Otherwise, any slight variation or change in the amplifier could propagate around the loop and add to the original disturbance in such a way as to increase its value and cause the amplifier to become unstable. Actually, a negative sign for T was assumed in the derivation of the gain with feedback, A_f, when the indicated summation at the input was assumed negative as shown in Eq. (10–22). In order to avoid confusion, the expressions for A_f [Eqs. (10–30) and (10–26)], will be written using the absolute value of T, or

$$A_f = \left(\frac{|T|}{1+|T|}\right)\left(\frac{1}{B}\right) \simeq \frac{1}{B} \tag{10–31}$$

The circuit designer must still determine, however, that the loop gain, and hence the feedback, is negative in a given amplifier.

A second point that should be noted is that the loop gain has no units; T is a pure number. This effect is observed from the denominator of Eq. (10–30) where T is added to the number 1. If T had units, then the indicated addition would be meaningless. As a direct consequence of the unitless nature of T, the units of A_{ol} and B must be reciprocal because the product of A_{ol} and B is equal to T [Eq. (10–29)]. Since the units of B are determined by the method of signal sampling at the output and the method of signal summing at the inputs, the units of both B and A_{ol} are determined by the configuration of the feedback network. Also, the gain with feedback, A_f, has the same units as A_{ol}, an effect apparent from Eq. (10–25) where the numerator is A_{ol}, and the denominator, $(1 + A_{ol}B)$, has no units.

To summarize, the loop gain T is unitless, and A_f, A_{ol}, and $1/B$ all have the same units. Once the gain and units of the feedback network are determined, the gain and units of the entire feedback amplifier are known, assuming that $|T|$ is much greater than unity. Stated differently, if $|T|$ can be assumed to be much greater than unity, the performance of a feedback amplifier can be determined by the method of signal sampling, the method of signal summing, and the ratio X_f/X_o or B. Consequently, the gain of a feedback amplifier and also the effects of feedback on the input and output impedances can be determined from the feedback network configuration.

The effects of the feedback network configuration on the performance characteristics of a feedback amplifier are summarized in Table 10–6. By combining the results summarized in Table 10–6 with the analysis of ideal amplifiers in the last section, the gain of a feedback amplifier can be determined easily. This approximate method of feedback analysis is illustrated by the series of examples in the next section.

TABLE 10–6 *Summary of the effects of the feedback configuration on the performance characteristics of a feedback amplifier. The second subscript of* A_f *indicates the gain stabilized by feedback. The values of* R_i *and* R_o *for* $|T|$ *equal to infinity are given in parentheses.*

Signal sampled Signal summed	Voltage Voltage	Current Current	Voltage Current	Current Voltage
Stable gain	$A_{fv} = V_o/V_i$	$A_{fi} = I_o/I_i$	$A_{fr} = V_o/I_i$	$A_{fg} = I_o/V_g$
B network gain	$B = V_f/V_o$	$B = I_f/I_o$	$B = I_f/V_o$	$B = V_f/I_o$
R_i	Increased ($\simeq \infty$)	Decreased ($\simeq 0$)	Decreased ($\simeq 0$)	Increased ($\simeq \infty$)
R_o	Decreased ($\simeq 0$)	Increased ($\simeq \infty$)	Decreased ($\simeq 0$)	Increased ($\simeq \infty$)

In this section, we have studied the arrangement and characteristics of the basic parts of a feedback amplifier. Also, we derived the gain of an amplifier with feedback and discussed the significance of the gain expressions. The most significant results of this section are summarized in Table 10–6. These results, coupled with the concept of ideal amplifiers, provide a simple method of determining the approximate performance characteristics of a feedback amplifier. In the following section, we will illustrate this method of analysis with a series of examples.

10.6 An Approximate Analysis of Negative-Feedback Amplifiers

The objectives of this section are (1) to illustrate the approximate analysis of feedback amplifiers, and (2) to develop block diagrams to show the functional parts of feedback amplifiers.

The approximate analysis is based on the assumption that $|T|$ is much greater than unity. For this case, the following steps can be used to evaluate the approximate response of a feedback amplifier:

1. Determine the signal sampled, X_o
2. Determine the signal fed back for summing, X_f
3. Determine the gain of the feedback network, $B = X_f/X_o$
4. Determine the value of the gain stabilized by feedback using

$$A_f = X_o/X_1 \simeq 1/B \qquad \textbf{(10–32)}$$

5. Estimate the values of R_i and R_o
6. Determine the signal gain for the amplifier using ideal amplifier analysis

The five examples in this section will illustrate this type of analysis.

EXAMPLE 10–2 For the feedback amplifier shown in Fig. 10–15, determine the feedback network configuration and the type of gain stabilized by feedback. Also, estimate the values of R_i, R_o, and A_{vo} assuming that $|T| \gg 1$.

Solution: (a) Feedback network configuration: From Fig. 10–15(b), the output signal being sampled is V_o, since the feedback network and the output voltage terminal have a common connection. Voltages are summed at the input, since the feedback network and the generator circuit do not have a common connection. Hence the amplifier is a voltage sample-voltage sum type.

(b) Gain of the feedback network: The feedback network is shown separately in Fig. 10–15(c). The resistance looking into the source terminal of Q_1 is increased by feedback and can be neglected in comparison with R_{f1}. The value of B, from voltage division, is

$$B = \frac{X_f}{X_o} = \frac{V_f}{V_o} = \frac{R_{f1}}{R_{f1} + R_{f2}} = \frac{50}{50 + 460.5} = \frac{1}{10.2}$$

(c) Gain stabilized by feedback: From Table 10–6, the stable gain of a voltage sample-voltage sum feedback amplifier is A_{fv} where

$$A_{fv} = V_o/V_i \simeq 1/B = (R_{f1} + R_{f2})/R_{f1} = 10.2$$

(d) Input impedance: Voltage summing at the input increases the value of the impedance looking into the input terminal of the first stage. Ideally, R_{in1} is infinite, and

$$R_i = 10^9 \,||\, R_{\text{in1}} \simeq 10^9\ \Omega$$

(e) Output impedance: Voltage sampling at the output reduces the value of R_o. Ideally, R_o is zero.

(f) Voltage gain, A_{vo}: Since the stable gain of this amplifier is A_{fv}, the voltage gain A_{vo} is different from A_{fv} only by the voltage division between R_i and R_g at the input.

$$A_{vg} = \frac{V_i}{V_g} = \frac{R_i}{R_i + R_g} \simeq \frac{10^9}{10^9 + 50} \simeq 1$$

$$A_{vo} = \frac{V_o}{V_g} = \left(\frac{V_i}{V_g}\right)\left(\frac{V_o}{V_i}\right) = A_{vg}A_{fv} \simeq (1)(10.2) \simeq 10.2$$

(g) Experimental results: The measured dc bias voltages are shown in Fig. 10–15(a). With an input voltage, V_g, of 30 mV, the measured voltage gain was 10.3.

A block diagram that can be used to represent the voltage sample-voltage sum feedback amplifier in Ex. 10–2 is shown in Fig. 10–15(d). This block diagram is a modified form of the differential amplifier block diagram shown in Fig. 4–36. The input signal is applied to the non-inverting or "+" input, since V_o is in phase with V_g, or the gain V_o/V_g is positive. The feedback signal, V_f, is applied to the inverting or "−" input to obtain negative feedback. The open loop gain, A_v, is also positive. Since the feedback network and the generator circuit do not have a common connection, voltages are summed at the input. Since the feedback network and the output voltage terminal have a common connection, the output voltage is the sampled signal. Thus the block diagram contains all the essential information concerning the feedback characteristics of the amplifier shown in Fig. 10–15(a). The block diagram can be used for the approximate analysis provided that $|T|$ is much greater than unity.

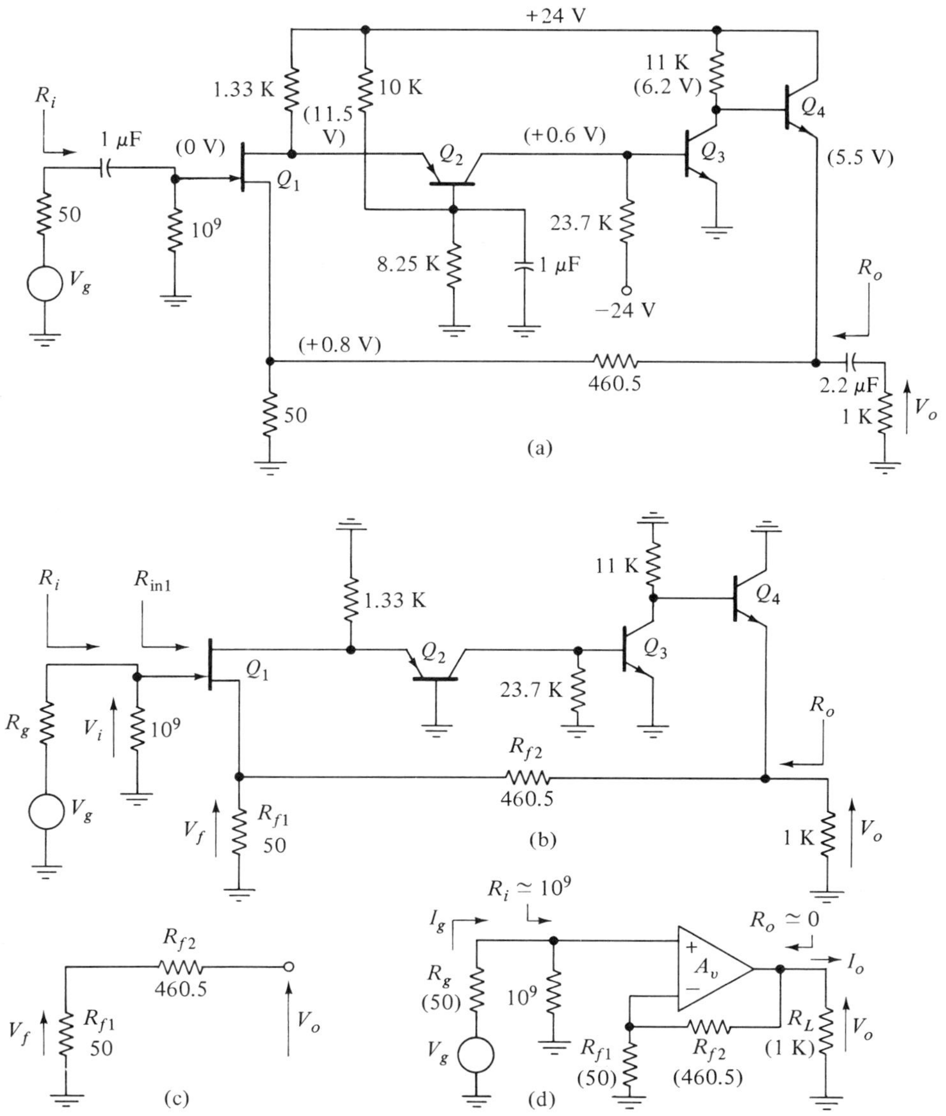

FIGURE 10–15 *The feedback amplifier from Ex. 10–2: (a) a complete circuit, (b) an ac equivalent circuit, (c) the feedback network shown separately, and (d) an equivalent block diagram representation.*

EXAMPLE 10–3 For the feedback amplifier shown in Fig. 10–16, determine the feedback network configuration and the gain stabilized by feedback. Estimate the values of R_i, R_o, and A_{vo} assuming that $|T| \gg 1$.

Solution: (a) Feedback network configuration: From Fig. 10–16(a), the output signal being sampled is I_o, since the feedback network and the output voltage terminal do not have a common connection. Currents are summed at the input, since the feedback network and the generator circuit have a common connection. Hence the amplifier is a current sample-current sum type.

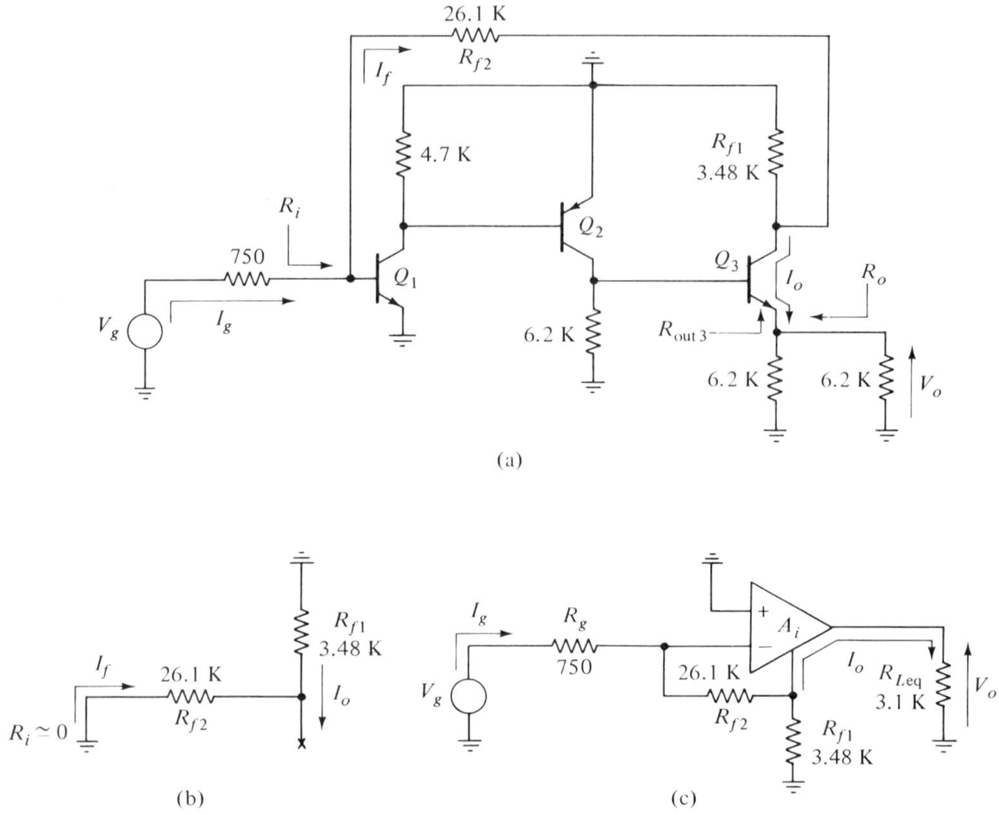

FIGURE 10–16 *The feedback amplifier from Ex. 10–3: (a) an ac equivalent circuit, (b) the feedback network shown separately, and (c) an equivalent block diagram. The complete circuit diagram is shown in Fig. 10–5(a).*

(b) Gain of the feedback network: The feedback network is shown separately in Fig. 10–16(b). The effect of current summing is to cause the resistance between the summing node and ground to be very small. Hence the terminal of R_{f2} connected to the summing node can be connected to an ac ground when calculating the value of B. From Fig. 10–16(b),

$$B = \frac{X_f}{X_o} = \frac{I_f}{I_o} = \frac{R_{f1}}{R_{f1} + R_{f2}} = \frac{3.48\text{ K}}{3.48\text{ K} + 26.1\text{ K}} = \frac{1}{8.5}$$

(c) Gain stabilized by feedback: From Table 10–6, the stable gain of a current sample-current sum feedback amplifier is A_{fi} where

$$A_{fi} = I_o/I_g \simeq 1/B = (R_{f1} + R_{f2})/R_{f1} = 8.5$$

(d) Input impedance: Current summing at the input reduces the value of the impedance looking into the summing node. Ideally, R_i should be zero. However, the actual value may be a few ohms.

(e) Output impedance: Current sampling at the output increases the value of the impedance looking into the output terminals of the last stage. Ideally, R_{out3} is infinite, and

$$R_o = 6.2\text{ K} \parallel R_{out3} \simeq 6.2\text{ K}$$

(f) Voltage gain A_{vo}: Since the stable gain of this amplifier is a current gain, we can convert to a voltage gain using the method of ideal amplifier analysis.

$$A_{vo} = \frac{V_o}{V_g} = \frac{I_o R_{Leq}}{I_g(R_g + R_i)} \simeq A_{fi}\left(\frac{R_{Leq}}{R_g}\right) = 8.5\left(\frac{3.1\text{ K}}{750}\right) \simeq 35$$

(g) Experimental results: The experimental performance of this amplifier is given in Sec. 10.2. The measured voltage gain was 35 with an input voltage, V_g, of 50 mV.

A block diagram that can be used to represent the current sample-current sum feedback amplifier in Ex. 10–3 is shown in Fig. 10–16(c). The feedback network and the output voltage terminal do not have a common connection indicating current sampling at the output. The feedback network and the generator circuit have a common connection indicating current summing. The summing node occurs at the "−" input indicating negative feedback. Also, the open-loop current gain, A_i, is negative. However, the voltage gain is positive, which is consistent with the general notation used with two-port networks. Thus the block diagram contains all the essential information concerning the feedback characteristics of the amplifier whose ac equivalent circuit is shown in Fig. 10–16(a). The actual circuit diagram for this amplifier is shown in Fig. 10–5(a).

In Ex. 10–3, we calculated the voltage gain, A_{vo}, and the gain stabilized by feedback. Generally, the most easily measured gain of an amplifier is its voltage gain since the instruments usually used are oscilloscopes or voltmeters. Also, it is general practice to show voltage waveforms in a circuit rather than current waveforms. The current gain was calculated in Ex. 10–3 because it is the gain stabilized by feedback. Once the current gain is known, the voltage gain is easily calculated for any specific value of R_g and R_L.

EXAMPLE 10–4 For the feedback amplifier shown in Fig. 10–17, determine the feedback network configuration and the type of gain stabilized by feedback. Assuming that $|T| \gg 1$, estimate the values of R_i, R_o, and A_{vo}.

Solution: (a) Feedback network configuration: From Fig. 10–17(b), the output signal being sampled is V_o, since the feedback network is connected to the output voltage terminal. Currents are summed at the input since the feedback network and the generator circuit have a common connection. Therefore, the amplifier is of the voltage sample-current sum type.

(b) Gain of the feedback network: The effect of current summing is to cause the resistance between the summing node and ground to be very small. Hence the terminal of R_f that is connected to the summing node can be connected to an ac ground when calculating the value of B. From Fig. 10–17(c),

$$B = X_f/X_o = I_f/V_o = -1/R_f = -1/100\text{ K}$$

(c) Gain stabilized by feedback: From Table 10–6, the stable gain of a voltage sample-current sum feedback amplifier is A_{fr} where

$$A_{fr} = V_o/I_g \simeq 1/B = R_f = -100\text{ K}$$

(d) Input impedance: Current summing at the input reduces the value of the impedance looking into the summing node. Ideally, R_i is zero.

(e) Output impedance: Voltage sampling at the output reduces the value of R_o. Ideally, R_o is zero.

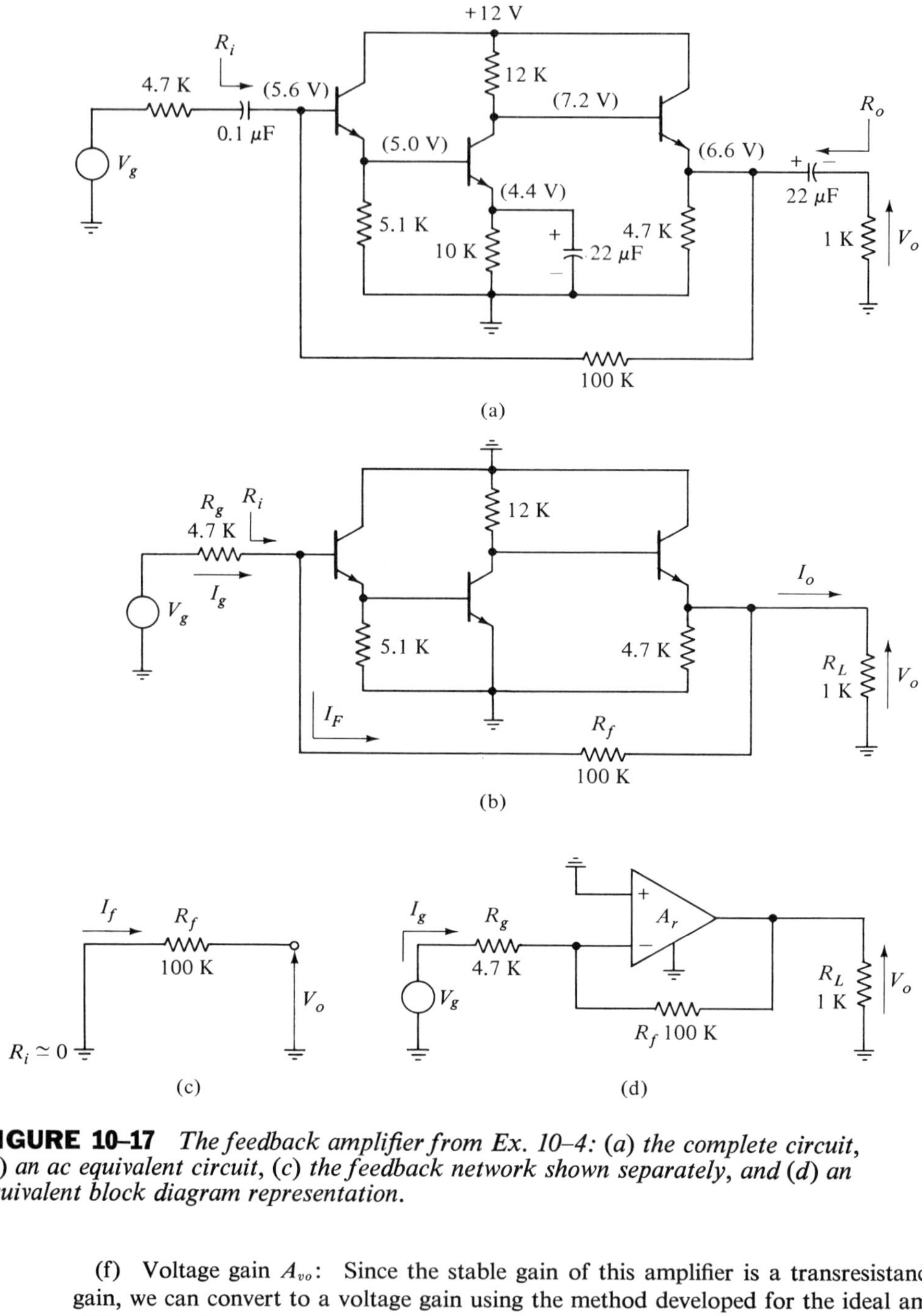

FIGURE 10–17 *The feedback amplifier from Ex. 10–4: (a) the complete circuit, (b) an ac equivalent circuit, (c) the feedback network shown separately, and (d) an equivalent block diagram representation.*

(f) Voltage gain A_{vo}: Since the stable gain of this amplifier is a transresistance gain, we can convert to a voltage gain using the method developed for the ideal amplifier analysis, or

$$A_{vo} = \frac{V_o}{V_g} = \frac{V_o}{I_g(R_g + R_i)} \simeq A_{fr}\left(\frac{1}{R_g}\right) \simeq -\frac{R_f}{R_g} = -\frac{100\ \text{K}}{4.7\ \text{K}} \simeq -21.3$$

(g) Experimental results: The measured dc bias voltages are shown in Fig. 10–17(a). With an input signal, V_g, of 48 mV, the measured voltage gain was -21.3.

A block diagram that can be used to represent the voltage sample-current sum feedback amplifier in Ex. 10–4 is shown in Fig. 10–17(d). The feedback network and the output voltage terminal have a common connection, indicating voltage sampling. The feedback network and the generator circuit also have a common connection, indicating current summing. The summing node occurs at the "–" input to indicate the negative feedback. Also, the open loop gain, A_r, is negative because a positive current entering the summing node (produced by a positive generator voltage) causes a negative voltage at the output. The voltage gain is also negative. Thus the block diagram contains all the essential information concerning the feedback characteristics of the amplifier shown in Fig. 10–17(a). The block diagram can be used for the approximate analysis provided that $|T|$ is much greater than unity.

EXAMPLE 10–5 For the feedback amplifier shown in Fig. 10–18, determine the feedback network configuration and the gain stabilized by feedback. Assuming that $|T| \gg 1$, estimate the values of R_i, R_o, and A_{vo}.

Solution: (a) Feedback network configuration: From Fig. 10–18(b), the output signal being sampled is I_o since the feedback network and the output voltage terminal do not have a common connection. Voltages are summed at the input since the feedback network and the generator circuit do not have a common connection. Hence the amplifier is the current sample-voltage sum type.

(b) Gain of the feedback network: The feedback network is shown separately in Fig. 10–18(c). The resistance looking into the emitter terminal of Q_1 is increased by the feedback and can be neglected in comparison to R_{f3}. The value of B can be calculated from Fig. 10–18(c) as

$$B = \frac{X_f}{X_o} = \frac{V_f}{I_o} = -\left(\frac{R_{f1}}{R_{f1} + R_{f2} + R_{f3}}\right)(R_{f3}) = \frac{(10\text{ K})(220)}{(10\text{ K} + 2.2\text{ K} + 220)} \simeq -\frac{10^3}{5.65}$$

(c) Gain stabilized by feedback: From Table 10–6, the stable gain for a current sample-voltage sum feedback amplifier is A_{fg} where

$$A_{fg} = \frac{I_o}{V_i} \simeq \frac{1}{B} = -\frac{R_{f1} + R_{f2} + R_{f3}}{R_{f1}R_{f3}} \simeq -5.65 \times 10^{-3}$$

(d) Input impedance: Voltage summing at the input increases the value of the impedance looking into the input terminal of the first stage. Ideally, $R_{\text{in}1}$ is infinite, and

$$R_i = 33\text{ K} \,||\, 47\text{ K} \,||\, R_{\text{in}1} \simeq 19.5\text{ K}$$

(e) Output impedance: Current sampling at the output increases the value of the impedance looking into the output terminal of the last stage. Ideally, $R_{\text{out}3}$ is infinite, and since $R_o = R_{\text{out}3}$, R_o is also infinite.

(f) Voltage gain, A_{vo}: Since the stable gain of this amplifier is a transconductance gain, we can convert to a voltage gain using the method of ideal amplifier analysis.

$$A_{vo} = \frac{V_o}{V_g} = \left(\frac{V_i}{V_g}\right)\left(\frac{V_o}{V_i}\right) = \left(\frac{V_i}{V_g}\right)\left(\frac{I_oR_{L\text{eq}}}{V_i}\right) = A_{vg}A_{fg}R_{L\text{eq}}$$

$$A_{vg} = R_i/(R_i + R_g) = 19.5\text{ K}/(19.5\text{ K} + 1\text{ K}) \simeq 0.95$$

$$A_{vo} = A_{vg}A_{fg}R_{L\text{eq}} \simeq (0.95)(-5.65 \times 10^{-3})(5.6\text{ K}) \simeq -30$$

(g) Experimental results: The measured dc bias voltages are shown in Fig. 10–18(a). The measured voltage gain, A_{vo}, was -11.5 with an input voltage, V_g, of 100 mV. Note that the measured voltage gain is much less than the calculated value. This discrepency is discussed below.

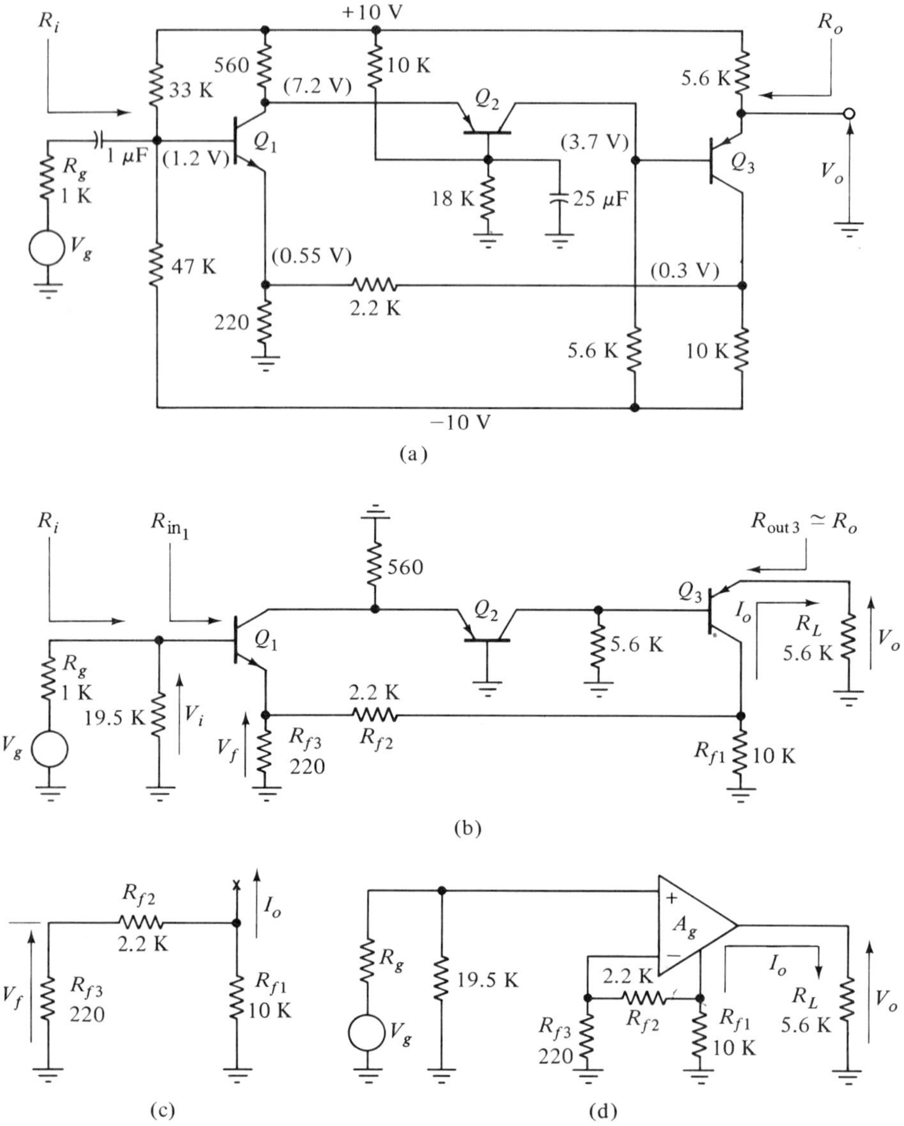

FIGURE 10–18 *The feedback amplifier from Ex. 10–5: (a) the complete circuit, (b) an ac equivalent circuit, (c) the feedback network shown separately, and (d) an equivalent block diagram representation.*

The basic assumption made when using the approximate analysis is that $|T|$ is much greater than unity. This is not the case for the circuit shown in Fig. 10–18(a) since the value of $|T|$ is approximately 0.7. Example 10–5 is repeated below using the actual value of $|T|$. Methods of calculating $|T|$ are discussed in Sec. 10.8.

EXAMPLE 10–6 Estimate the value of A_{vo} for the amplifier from Ex. 10–5, assuming β_1 is equal to 100, and $|T|$ is equal to 0.7.

Solution: From Ex. 10–5, the expression for A_{vo} is

$$A_{vo} = \frac{V_o}{V_g} = \left(\frac{V_i}{V_g}\right)\left(\frac{V_o}{V_i}\right) = A_{vg}A_{fg}R_{Leq}$$

Since $|T|$ is less than unity, the feedback has very little effect on the impedance looking into the base terminal of Q_1. Neglecting the effects of feedback,

$$\begin{aligned} R_{in1} &\simeq [\beta_1 + 1][r_{e1} + R_{f3} \,||\, (R_{f2} + R_{f1})] \\ &\simeq [100 + 1][8 + 220 \,||\, (2.2\text{ K} + 10\text{ K})] \simeq 23\text{ K} \\ R_i &= 33\text{ K} \,||\, 47\text{ K} \,||\, R_{in1} \simeq 33\text{ K} \,||\, 47\text{ K} \,||\, 23\text{ K} \simeq 10.5\text{ K} \\ A_{vg} &= R_i/(R_i + R_g) \simeq 10.5\text{ K}/(10.5\text{ K} + 1\text{ K}) \simeq 0.92 \end{aligned}$$

Using the general expression for A_f from Eq. (10–31),

$$A_{fg} = \left(\frac{0.7}{1 + 0.7}\right)(-5.65 \times 10^{-3}) \simeq -2.32 \times 10^{-3}$$

The voltage gain is

$$A_{vo} = A_{vg}A_{fg}R_{Leq} \simeq (0.92)(-2.32 \times 10^{-3})(5.6\text{ K}) \simeq -11.8$$

Note that the calculated value of -11.8 for the condition of $|T|$ approximately equal to 0.7 is in close agreement with the measured value of -11.5 given in Ex. 10–5.

The results obtained in Exs. 10–5 and 10–6 indicate the importance of $|T|$, particularly whether or not $|T|$ is much greater than unity. When $|T|$ is much greater than unity, the approximate analysis of the performance of a feedback amplifier yields accurate results as indicated by Exs. 10–2 through 10–4.

A block diagram that can be used to represent the current sample-voltage sum feedback amplifier studied in Exs. 10–6 and 10–7 is shown in Fig. 10–18(d). Although $|T|$ was actually less than unity, we will assume $|T|$ is much greater than unity in the following discussion. The feedback network and the output voltage terminal do not have a common connection indicating current sampling. Also, the feedback network and the generator circuits do not have a common connection indicating voltage summing. The feedback signal, V_f, is applied to the inverting or "$-$" input to indicate negative feedback. The open loop gain A_g is positive because a positive voltage at the input produces a current entering the output terminal of the amplifier and produces a negative output voltage. The positive value of A_g is indicated by connecting the generator to the "+" input. The voltage gain is negative. Thus the block diagram contains all the essential information concerning the feedback characteristics of the amplifier shown in Fig. 10–18(a). The block diagram can be used for the approximate analysis, provided that $|T|$ is much greater than unity.

In this section, we have illustrated the approximate analysis of all four types of feedback amplifiers. Also, we have discussed a block diagram representation for each type of feedback amplifier which contains all of the essential information on a feedback amplifier. Both the type of gain stabilized by feedback and the voltage gain of each amplifier were calculated and experimentally verified. A summary of the block diagram representation, gain stabilized by feedback, and voltage gain expression for the four types of feedback amplifiers, assuming $|T|$ is much greater than unity, is given in Table 10–7. The effect of a small value of $|T|$ was illustrated in Exs. 10–5 and 10–6. In this case, the approximate anal-

TABLE 10–7 *Summary of the block diagram representations, gain stabilized by feedback, and voltage gain of the four types of feedback amplifiers assuming |T| is much greater than unity.*

Configuration	Block diagram	Gain stabilized by feedback	Voltage gain
Voltage sample Voltage sum	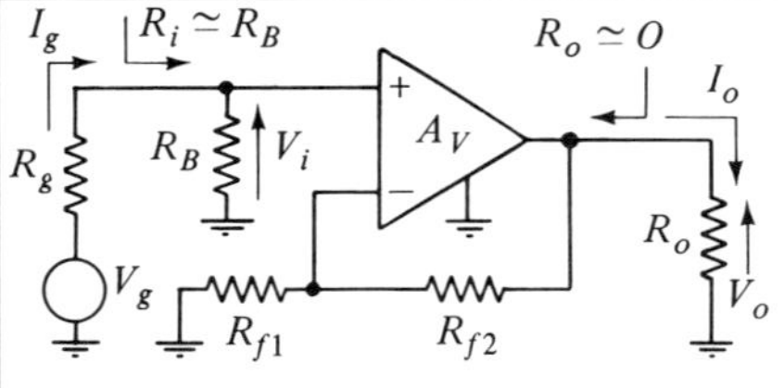	$A_{fv} = \frac{V_o}{V_i} = \frac{R_{f1} + R_{f2}}{R_{f1}}$	$A_{vo} = \frac{V_o}{V_g} = A_{vg} A_{fv}$
Current sample Current sum	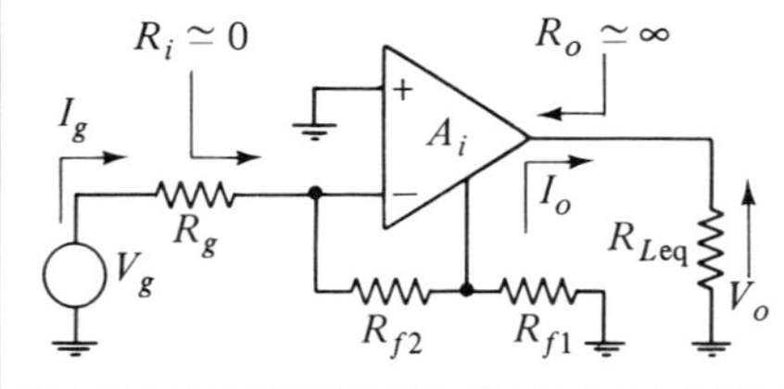	$A_{fi} = \frac{I_o}{I_g} = \frac{R_{f1} + R_{f2}}{R_{f1}}$	$A_{vo} = \frac{V_o}{V_g} = A_{fi} \frac{R_{Leq}}{R_g}$
Voltage sample Current sum	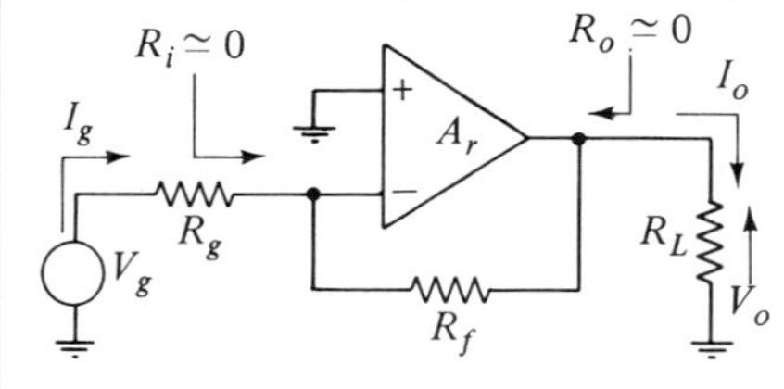	$A_{fr} = \frac{V_o}{I_g} = -R_f$	$A_{vo} = \frac{V_o}{V_g} = -\frac{R_f}{R_g}$
Current sample Voltage sum	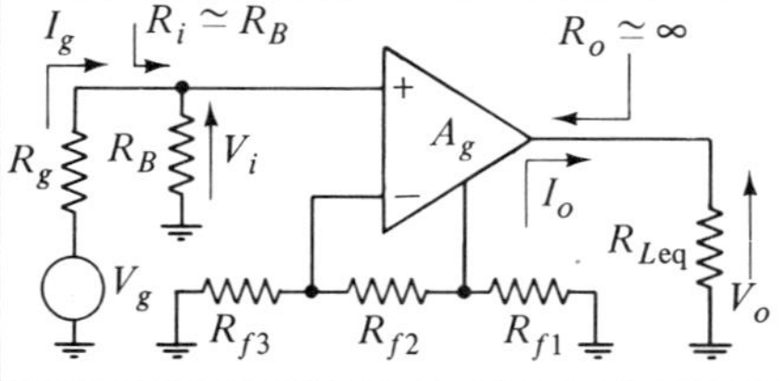	$A_{fg} = \frac{I_o}{V_i} = \frac{R_{f1} + R_{f2} + R_{f3}}{R_{f1} R_{f3}}$	$A_{vo} = \frac{V_o}{V_g} = A_{vg} A_{fg} R_{Leq}$

ysis was not valid. Consequently, the value of $|T|$ must be known before the amplifier can be successfully characterized. Methods of calculating the value of $|T|$ are discussed in Sec. 10.8. Of the four types of feedback amplifiers treated in this section, two of the more useful types will be discussed further in the next section.

10.7 Additional Examples of the Approximate Analysis of Negative-Feedback Amplifiers Using Voltage Sampling

In terms of application, the most important types of feedback amplifiers are those using voltage sampling at the output. Voltage sampling generally reduces the value of R_o and makes the gain of the amplifier relatively independent of the load impedance. Using voltage sampling at the output, an amplifier can be designed with either a positive voltage gain or a negative voltage gain depending on the manner in which signals are summed at the input. From Table 10–7, the voltage sample-voltage sum configuration has a positive value of A_{vo} and can have a large value of R_i. This configuration is also called a *voltage amplifier** since the important input signal is a voltage. The voltage sample-current sum configuration has a negative value of A_{vo} and can have a low value of R_i. This configuration is also called a *current amplifier* since the important input signal is a current.** Note that in either case, the important output signal is V_o, and the value of R_o is reduced by the voltage sampling.

The objective of this section is to further illustrate the approximate analysis of feedback amplifiers. We will restict this study to the "voltage amplifiers" and "current amplifiers" defined above.

EXAMPLE 10–7 Assuming that $|T| \gg 1$ for the amplifier shown in Fig. 10–19(a), estimate the values of R_i, R_o, and A_{vo}.

Solution: (a) Analysis: The amplifier shown in Fig. 10–19(a) is a voltage amplifier since the feedback configuration is of the voltage sample-voltage sum type. Assuming that $|T| \gg 1$, then from Table 10–7,

$$R_i = 1.96\text{ K} \,||\, R_{\text{in}1} \simeq 1.96\text{ K} \qquad R_o \simeq 0$$

$$A_{vg} = \frac{V_i}{V_g} = \frac{R_i}{R_i + R_g} \simeq 0.975 \qquad A_{fv} = \frac{V_o}{V_i} \simeq \frac{R_{f1} + R_{f2}}{R_{f1}} \simeq 14.1$$

$$A_{vo} = \frac{V_o}{V_g} = A_{vg}A_{fv} \simeq (.975)(14.1) \simeq 13.7$$

(b) Experimental results: The measured bias values are shown in Fig. 10–19(a). The measured voltage gain was 13.4 when V_g was 50 mV.

Although the voltage amplifier shown in Fig. 10–19(a) is fairly complex, the analysis is simple using the methods developed in the last section for the approximate analysis of feedback amplifiers. The output portion of the amplifier, Q_6 and Q_7, is called a *complementary emitter follower* circuit since both the NPN

---

* This type is also called a *voltage feedback amplifier* or a *series feedback amplifier*.

** This type is also called a *current feedback amplifier* or a *shunt feedback amplifier*.

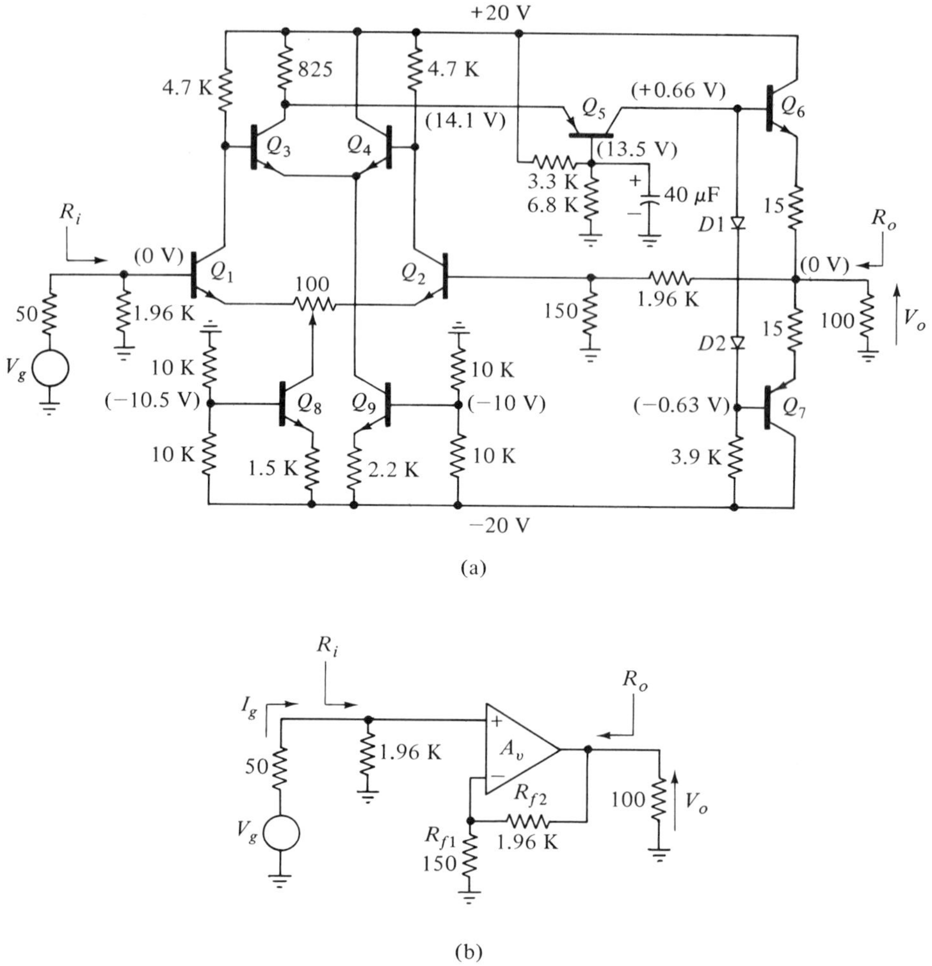

FIGURE 10–19 *The voltage amplifier from Ex. 10–7: (a) the complete circuit, and (b) an equivalent block diagram representation.*

and PNP transistors are connected as CC stages. The diodes, D_1 and D_2, provide the necessary bias voltages between the bases of Q_6 and Q_7 and are essentially short circuits to the ac signal. The complementary emitter follower circuit is useful when both polarities of output signal are required, and the load impedance is relatively small. This circuit is considered in Chaps. 11 and 12.

EXAMPLE 10–8 Assuming that $|T| \gg 1$ for the amplifier shown in Fig. 10–20(a), estimate the values of R_i, R_o, and A_{vo}.

Solution: (a) Analysis: The amplifier shown in Fig. 10–20(a) is a current amplifier since the feedback configuration is of the voltage sample-current sum type. From Table 10–7 and assuming $|T| \gg 1$,

$$R_i \simeq 0 \qquad R_o \simeq 0$$

$$A_{vo} = \frac{V_o}{V_g} \simeq -\frac{R_f}{R_g} = -\frac{30\text{ K}}{1\text{ K}} = -30$$

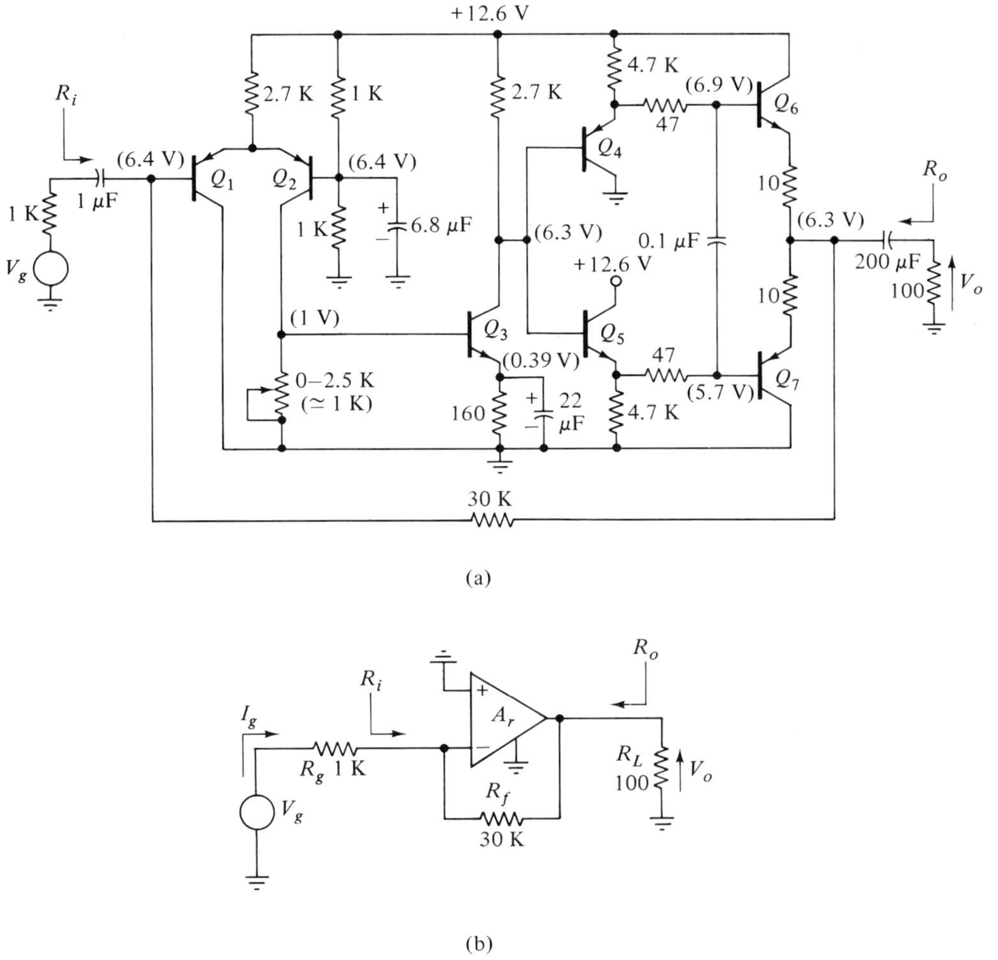

FIGURE 10–20 *The current amplifier from Ex. 10–8: (a) the complete circuit, and (b) an equivalent block diagram representation.*

(b) Experimental results: The measured bias values are shown in Fig. 10–20(a). The measured voltage gain was −29.7 when V_g was 370 mV. Since R_f and R_g were 5 percent resistors, exact agreement was not expected.

Two comments should be made concerning the configuration of the current amplifier shown in Fig. 10–20(a). First, the variable resistor in the collector circuit of Q_2 was used to adjust the output dc voltage at half the supply voltage or 6.3 V. This arrangement allows the maximum peak-to-peak value of the output voltage to be obtained with a minimum value of distortion. Second, the four output transistors, Q_4 through Q_7, are a modification of the complementary emitter follower output circuit shown in Fig. 10–19(a). This configuration, often called the *diamond driver*, is useful when large output voltage and current swings are encountered with low values of load impedance.

EXAMPLE 10–9 Assuming that $|T| \gg 1$ for the amplifier shown in Fig. 10–21(a), estimate the values of R_i, R_o, and A_{vo}.

Solution: (a) Analysis: The amplifier shown in Fig. 10–21(a) is a voltage amplifier. From Table 10–7,

$$R_i \simeq 1\ \text{M} \qquad R_o \simeq 0$$

$$A_{vg} = \frac{V_o}{V_g} \simeq \frac{R_i}{R_i + R_g} \simeq 1 \qquad A_{fv} = \frac{V_o}{V_i} \simeq \frac{R_{f1} + R_{f2}}{R_{f1}} \simeq 7.7$$

$$A_{vo} = \frac{V_o}{V_g} = A_{vg}A_{fv} = (1)(7.7) = 7.7$$

(b) Experimental results: The measured bias values are shown in Fig. 10–21(a). The measured voltage gain was 7.7 with $V_g = 1.3$ V.

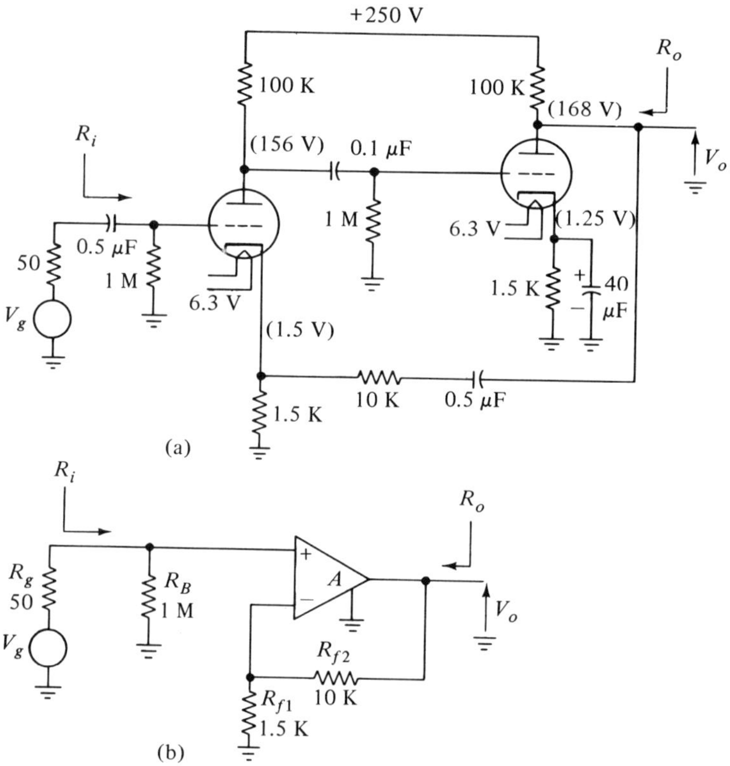

FIGURE 10–21 *The voltage amplifier from Ex. 10–9: (a) the complete circuit, and (b) an equivalent block diagram representation.*

The vacuum tube voltage amplifier studied in Ex. 10–9 further illustrates that the gain with feedback can be relatively independent of the active devices. A second example of a vacuum tube feedback amplifier is the current amplifier shown in Fig. 10–22. The measured voltage gain for this amplifier was

$$A_{vo} = V_o/V_g \simeq -25\ \text{V}/10\ \text{V} = -2.5$$

The analysis of this vacuum tube current amplifier is left as an exercise for the reader.

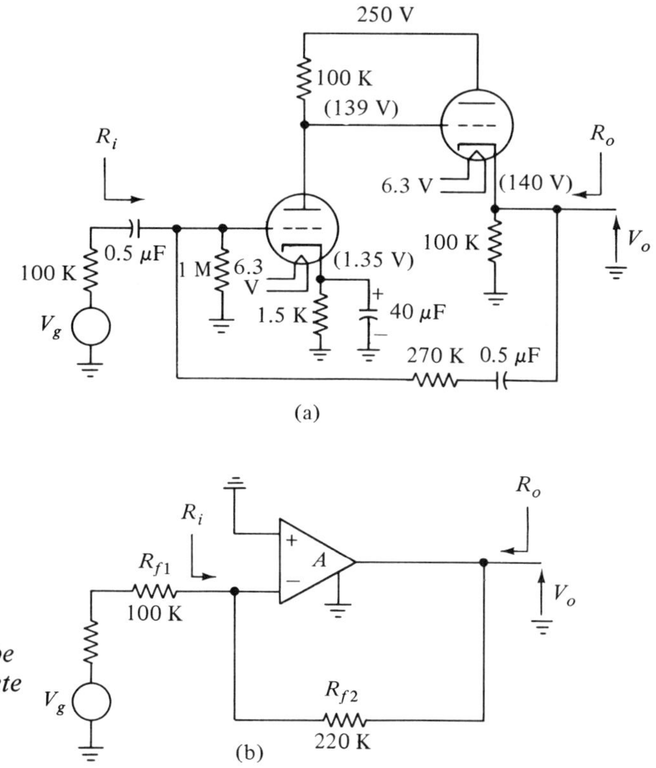

FIGURE 10–22 *A vacuum tube current amplifier: (a) the complete circuit, and (b) an equivalent block diagram representation.*

The advent of high-performance linear integrated circuits has made the block diagram approach to feedback amplifiers particularly useful. In many linear ICs, the actual circuit configuration is not particularly important. The manufacturer's data sheet usually gives sufficient information on the required dc voltages and frequency compensation networks.* The equivalent block diagram can be used to design the feedback network to achieve desired gain.

The circuit diagram of a high performance linear IC is shown in Fig. 10–23(a), and a top view of the physical layout of the chip is shown in Fig. 10–23(b). Neglecting the power supply and frequency compensation terminals, the multistage IC is characterized by an inverting input terminal, a noninverting input terminal, and an output terminal. Hence, the block diagram shown in Fig. 10–23(c) can be used to represent the IC. Note that this block diagram is identical to the block diagram used to represent a feedback amplifier except that the feedback network has not been connected.

The linear IC shown in Fig. 10–23 can be connected as a current amplifier by arranging the feedback network and the generator circuit as shown in Fig. 10–24(a).** For the resistor values shown, and assuming $|T| \gg 1$,

---

* Most high-performance feedback amplifiers require frequency compensation to avoid high-frequency oscillations. The general topic of instability as well as methods of frequency response compensation are discussed in the next chapter. For the amplifier shown in Fig. 10–23(a), compensation networks would be connected between nodes *A* and *B* and as part of the feedback network.

** Note the slight change in symbolism for R_g to R_{f1} in Fig. 10–24(a). While both symbols are used in practice, the use of R_{f1} is more appropriate in the discussion of the circuits shown in Fig. 10–24.

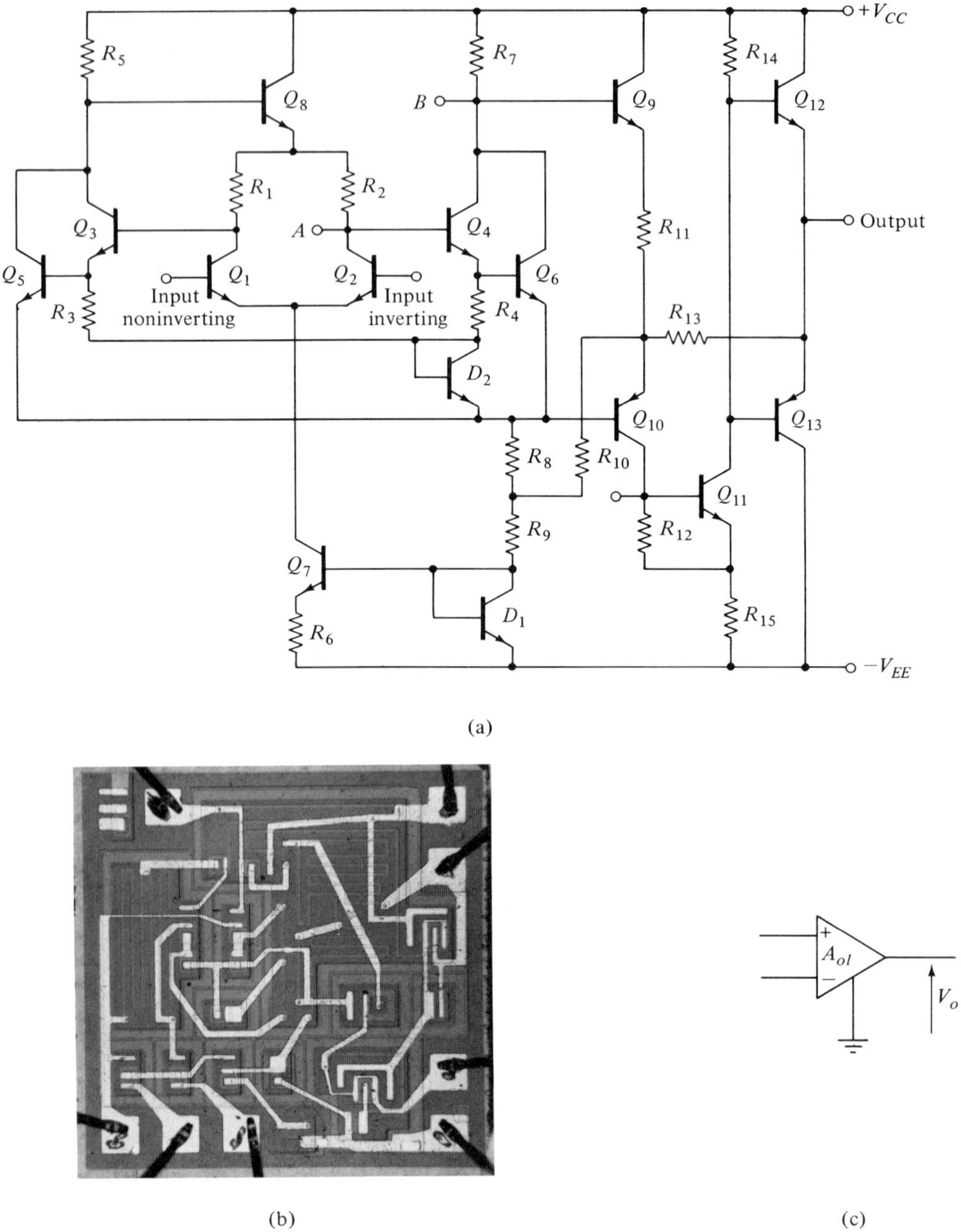

FIGURE 10–23 *A high performance linear integrated circuit: (a) a circuit diagram, (b) a photomicrograph of the circuit, and (c) an equivalent block diagram representation. (Courtesy of Fairchild Semiconductor Corp.)*

$$A_{vo} = V_o/V_g \simeq -R_{f2}/R_{f1} = -10\text{ K}/1\text{ K} = -10 \tag{10–33}$$

The measured voltage gain for this circuit was

$$A_{vo} = V_o/V_g \simeq -300\text{ mV}/30\text{ mV} = -10 \tag{10–34}$$

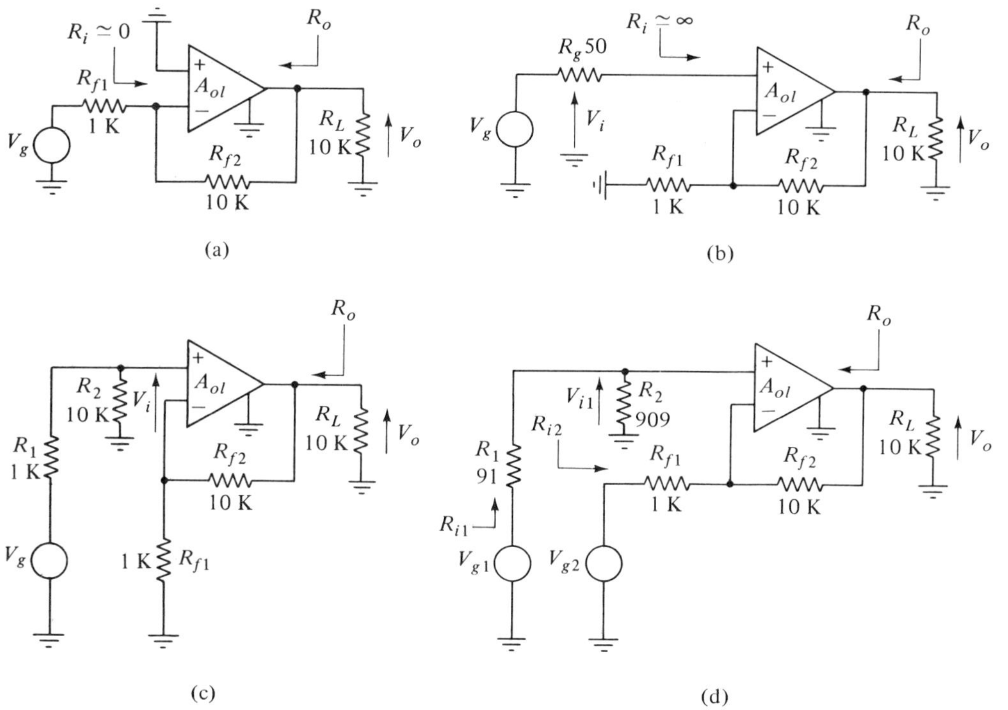

FIGURE 10–24 *The linear IC shown in Fig. 10–23 connected in four configurations: (a) a current amplifier, (b) a voltage amplifier, (c) a modified voltage amplifier, and (d) a differential amplifier.*

The linear IC can also be connected as a voltage amplifier by arranging the feedback network and the generator circuit as shown in Fig. 10–24(b). For the resistor values shown,

$$A_{vo} = \frac{V_o}{V_g} = A_{vg}A_{fv} \simeq (1)\left(\frac{R_{f1} + R_{f2}}{R_{f1}}\right) = \frac{11\text{ K}}{1\text{ K}} = 11 \qquad \textbf{(10–35)}$$

The measured voltage gain for this circuit with $V_g = 30$ mV was 11.

A modified form of the voltage amplifier using the same linear IC is shown in Fig. 10–24(c). A voltage divider network has been added at the input to slightly reduce the value of A_{vo}. For the resistor value shown, and assuming $|T| \gg 1$,

$$A_{vo} \simeq \left(\frac{R_2}{R_1 + R_2}\right)\left(\frac{R_{f1} + R_{f2}}{R_{f1}}\right) = \left(\frac{10\text{ K}}{1\text{ K} + 10\text{ K}}\right)\left(\frac{1\text{ K} + 10\text{ K}}{1\text{ K}}\right) = 10 \qquad \textbf{(10–36)}$$

The measured gain for this circuit with $V_g = 30$ mV was 10.

By comparing Figs. 10–24(a) and (c) and Eqs. (10–33) and (10–36), it is apparent that the negative gain and the positive gain have equal magnitudes for the same feedback network resistor values. Equal gain values are achieved when the ratio of R_2/R_1 is equal to the ratio R_{f2}/R_{f1}. A circuit having two inputs and one output with the gain from either input to the output equal except for the sign, suggests that the circuit contains a differential amplifier.* The linear IC shown

* Differential amplifiers are discussed in Sec. 4.9.

in Fig. 10–23 is connected as a differential amplifier by arranging the feedback network and the two generator circuits as shown in Fig. 10–24(d). The values of R_1 and R_2 have been adjusted so that R_{i1} and R_{i2}, as defined in Fig. 10–24(d), are equal. Thus,

$$R_{i2} \simeq R_{f1} = 1\text{ K} \tag{10–37}$$

$$R_{i1} \simeq R_1 + R_2 = 91 + 909 = 1\text{ K} \tag{10–38}$$

The two voltage gain expressions are

$$A_{vo1} = \frac{V_o}{V_g} \simeq \left(\frac{R_2}{R_1 + R_2}\right)\left(\frac{R_{f1} + R_{f2}}{R_{f1}}\right)$$

$$= \left(\frac{909}{91 + 909}\right)\left(\frac{1\text{ K} + 10\text{ K}}{1\text{ K}}\right) = +10 \tag{10–39}$$

$$A_{vo2} \simeq V_o/V_g \simeq -R_{f2}/R_{f1} = -10\text{ K}/1\text{ K} = -10 \tag{10–40}$$

where the gain from one input to the output is calculated with the other input short circuited. Using the methods shown in Fig. 4–37 for measuring the gains in a differential amplifier, the measured values for the amplifier shown in Fig. 10–24(d) were

$$A_d = \frac{300\text{ mV}}{30\text{ mV}} = 10 \qquad A_c = \frac{10\text{ mV}}{2.5\text{ V}} = 4 \times 10^{-3}$$

$$\text{CMRR} = \left|\frac{A_d}{A_c}\right| \simeq \frac{10}{4 \times 10^{-3}} \simeq 2.5 \times 10^3 \tag{10–41}$$

In this section, we have considered additional examples of the approximate analysis of feedback amplifiers. The circuits studied are often classified as voltage amplifiers and current amplifiers to indicate the important input signal. In both cases the important output signal was V_o, and the value of R_o was reduced. Also, a differential amplifier whose gain was stabilized by feedback was discussed. Although the amplifiers were generally more complex than those shown in earlier sections, the feedback network configuration could be used to determine the principal mid-band characteristics of the amplifiers.

10.8 Methods of Calculating the Loop Gain

The approximate analysis of feedback amplifiers discussed in the last two sections was based on the assumption that $|T|$ was much greater than unity. Examples 10–5 and 10–6 illustrated that the approximate analysis is not valid for small values of $|T|$. Consequently, the condition that $|T| \gg 1$ must be verified before approximate methods can be used.

The actual value of T is important for at least three reasons. First, the value of T is a general indication of how insensitive the performance parameters of an amplifier are to changes in the values of circuit components. Second, T can be used in the calculations of the input and output impedances of a feedback amplifier. These calculations are discussed in the next section. Third, the closed loop frequency response depends on the magnitude and frequency responses of T. Frequency response calculations are discussed in Chap. 11.

The objective of this section is to develop methods for finding the value of T. The techniques developed for the analysis of nonfeedback amplifiers will be used in the methods for finding T.

The concept of a loop transmission gain (T), was introduced in Sec. 10.5. The definition of T was given by Eq. (10–29) as

$$T \equiv A_{ol}B \tag{10–42}$$

where A_{ol} is the open loop gain, and B is the gain of the feedback network. Also, we found that T is dimensionless, and that in a negative-feedback amplifier the sign of T is always negative. The most obvious method for finding T is to use Eq. (10–42). This approach requires that A_{ol} be found first. In our previous discussion, B was defined in such a way that $1/B$ gave the approximate value of the gain that was stabilized by feedback. Consequently, B was different for the various types of feedback. This choice of the definition of B was an essential part of the approximate analysis. Due to the dimensionless nature of T, A_{ol} must have the same units as $1/B$. Also, the value of A_{ol} must reflect the loading effects of the sampling network and the summing network. Consequently, a calculation of the proper form and value of A_{ol} for use in Eq. (10–42) depends upon the type of feedback, generator impedance, and load impedance. This approach complicates the calculation of A_{ol} because B was defined for each type of feedback. Consequently, this method for finding T is not recommended.*

Rather than using Eq. (10–42) to calculate T, a direct calculation will be made. This approach is much simpler and allows a direct application of the analytical methods discussed in earlier chapters. With the values of T and B known, the value of A_{ol} can be found using Eq. (10–42). This direct method for finding T will be illustrated using the voltage amplifier from Sec. 10.2. The complete circuit of this amplifier is shown in Fig. 10–2(a), and an ac equivalent circuit is shown in Fig. 10–2(b). The ac equivalent circuit is repeated in Fig. 10–25(a) for convenience. The first step in calculating the loop gain is to break the loop at some point without altering the gain of the stages. One convenient point in this circuit is at the collector of Q_3 as shown below.

The ac equivalent circuit shown in Fig. 10–25(a) can be modified by replacing any of the transistors with its equivalent circuit. In Fig. 10–25(b) Q_3 has been replaced by its hybrid equivalent circuit.** From circuit theory we know that the current from an ideal current generator is not changed when it is short circuited. Also, the voltage across an ideal voltage generator does not change when it is open circuited. In Fig. 10–25(c), the dependent current generator, $\beta_3 I_{b3}$, in the hybrid equivalent circuit of Q_3 has been replaced by a test current generator, I_t, and the dependent current generator, $\beta_3 I_{b3}$ has been short circuited. An equivalent version of Fig. 10–25(c) is shown in Fig. 10–25(d) to illustrate that the loop has actually been opened. Note that shorting the current generator $\beta_3 I_{b3}$ and replacing it by a test current generator, I_t, has not changed any impedance values in the circuit; i.e., the values of R_{in} and $R_{L\text{eq}}$ for each transistor have not been changed.

The loop gain can be considered as the gain *once around the loop*. Using this concept, the loop gain can be calculated from Fig. 10–25(c) or (d) as the gain from the test current generator, I_t, to the shorted current generator, $\beta_3 I_{b3}$. From Fig. 10–25(d),

---

* One important case where $A_{ol}B$ can be used to evaluate T occurs in some operational amplifier circuits. Operational amplifiers are discussed in Secs. 11.9 and 11.10.

** See Chap. 4 for hybrid equivalent circuits.

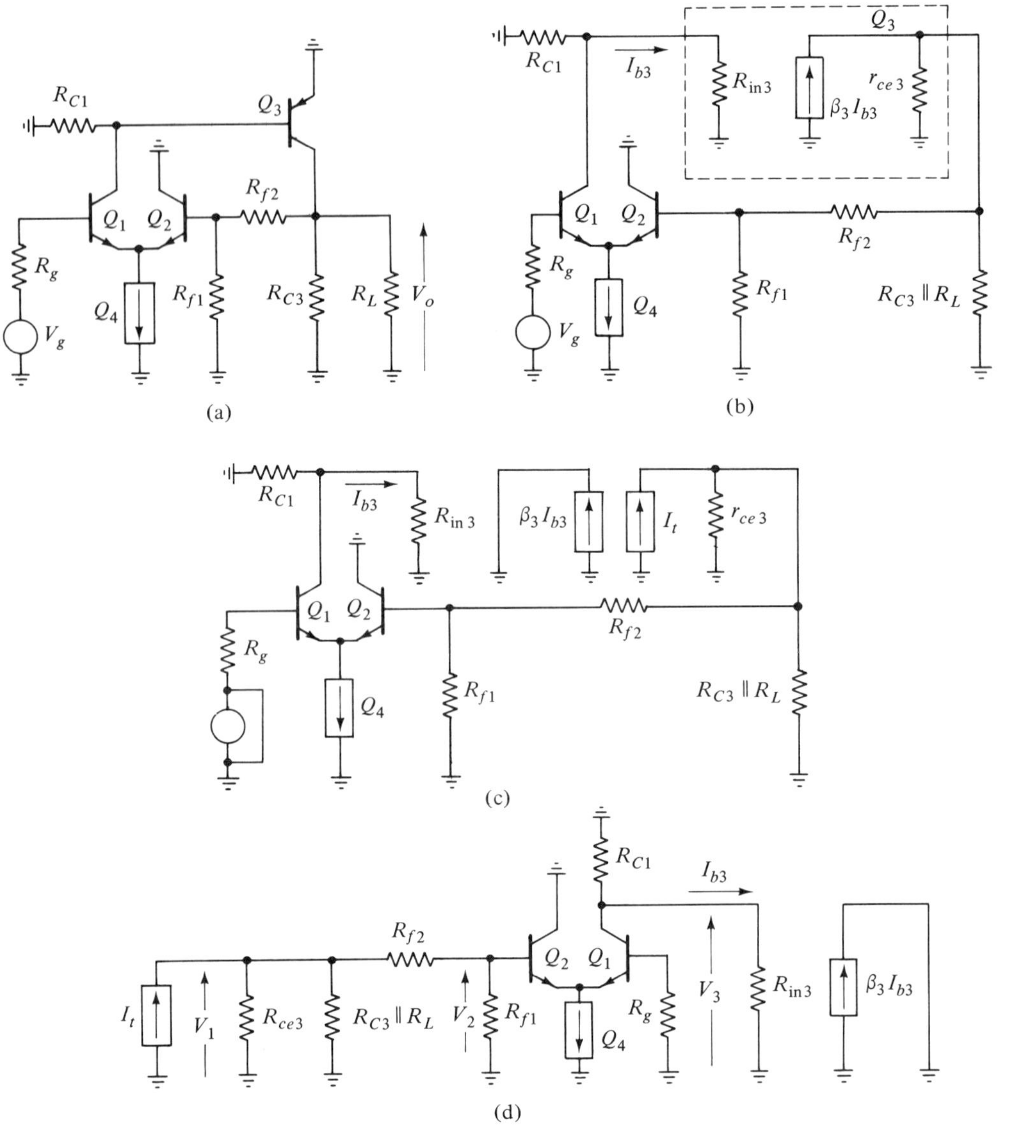

FIGURE 10–25 *The voltage amplifier used to illustrate the calculation of* T*: (a) an ac equivalent circuit, (b) an ac equivalent circuit in which* Q_3 *is replaced by its hybrid equivalent circuit, (c) a modified ac equivalent circuit showing the test generator* I_t*, and (d) an equivalent representation of (c). The complete circuit diagram is shown in Figs. 10–2(a) and 10–28(a).*

$$\frac{V_1}{I_t} = r_{ce3} \parallel R_{C3} \parallel R_L \parallel (R_{f2} + R_{f1} \parallel R_{in2}) = R_{Leq3} \tag{10–43}$$

$$\frac{V_2}{V_1} = \frac{R_{f1} \parallel R_{in2}}{R_{f1} \parallel R_{in2} + R_{f2}} \tag{10–44}$$

$$\frac{V_3}{V_2} = A_{d1-2} \simeq \frac{R_{C1} \parallel R_{in3}}{r_{e1} + r_{e2}} = \frac{R_{Leq1}}{r_{e1} + r_{e2}} \tag{10–45}$$

$$\frac{I_{b3}}{V_3} = \frac{1}{R_{\text{in3}}} \tag{10–46}$$

$$\frac{\beta_3 I_{b3}}{I_{b3}} = \beta_3 \tag{10–47}$$

The loop gain is equal to the ratio $\beta I_{b3}/I_t$ or

$$T = \frac{\beta_3 I_{b3}}{I_t} = \left(\frac{\beta_3 I_{b3}}{I_{b3}}\right)\left(\frac{I_{b3}}{V_3}\right)\left(\frac{V_3}{V_2}\right)\left(\frac{V_2}{V_1}\right)\left(\frac{V_1}{I_t}\right) \tag{10–48}$$

Substituting Eqs. (10–43) through (10–47) into (10–48) and simplifying gives

$$T \simeq \left(\frac{\beta_3 R_{Leq3}}{R_{\text{in3}}}\right)\left(\frac{R_{Leq1}}{r_{e1} + r_{e2}}\right)\left(\frac{R_{f1} \,||\, R_{\text{in2}}}{R_{f1} \,||\, R_{\text{in2}} + R_{f2}}\right) \tag{10–49}$$

The first two terms of Eq. (10–49) correspond to the voltage gains A_{vt3} and A_{d1-2}. Therefore, Eq. (10–49) can be rewritten as

$$T = A_{vt3} A_{d1-2} \left(\frac{R_{f1} \,||\, R_{\text{in2}}}{R_{f1} \,||\, R_{\text{in2}} + R_{f2}}\right) \tag{10–50}$$

From Eq. (10–50) and Fig. 10–25, the loop gain is equal to the signal gain from the base of Q_3, around the loop and back to the base of Q_3. A second approach in calculating T is to redraw the ac equivalent circuit in Fig. 10–25(a) as a continuous repetitive circuit without feedback as shown in Fig. 10–26. In this circuit the loop gain is the ratio of voltages at similar points in the circuit. Thus,

$$T = \frac{V_{b3'}}{V_{b3}} = A_{vt3} \left(\frac{R_{f1} \,||\, R_{\text{in2}}}{R_{f1} \,||\, R_{\text{in2}} + R_{f2}}\right) A_{d1-2} \tag{10–51}$$

and also

$$T = \frac{V_{b2'}}{V_{b2}} = A_{d1-2} A_{vt3} \left(\frac{R_{f1} \,||\, R_{\text{in2}}}{R_{f1} \,||\, R_{\text{in2}} + R_{f2}}\right) \tag{10–52}$$

Note that Eqs. (10–49), (10–50), (10–51), and (10–52) are all equivalent. Consequently, if we assume a signal is present at some point in the circuit inside the loop, then the loop gain is the signal gain from that point around the loop and back to that same point or the signal gain *once around the loop.*

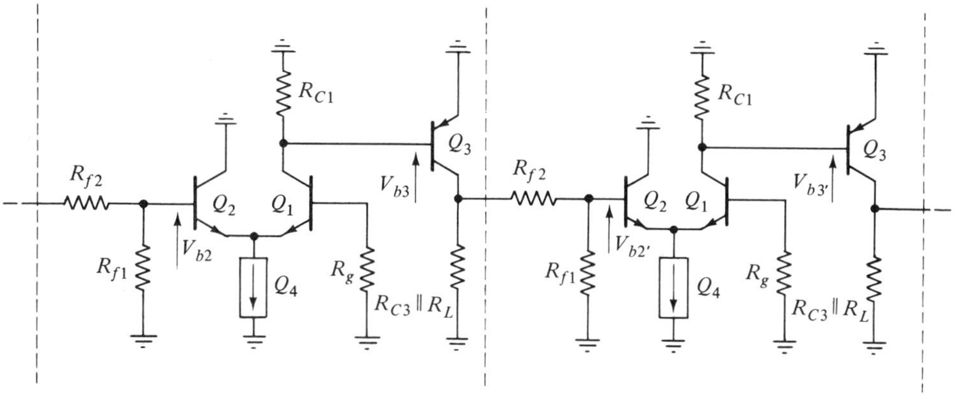

FIGURE 10–26 *The ac equivalent circuit shown in Fig. 10–25(a) redrawn as a continuous repetitive circuit with no feedback.*

Several points should be mentioned concerning the calculation of T. First, T is dimensionless in Eqs. (10–49) through (10–52), since each term is the ratio of similar quantities. Second, for the calculation of T, Q_3 appears as a CE stage, Q_2 appears as a CC stage, and Q_1 appears as a CB stage.* Thus a single inversion occurs in the loop gain, due to Q_3 (the CE stage), and the sign of T is negative. Third, the effects of feedback on impedance levels were not considered in the calculation of T. For example, feedback causes the value of R_o to be small, but when calculating T, a short circuit was not placed at the collector of Q_3. Also, the value of $R_{\text{in}2}$ was not neglected as being very large as in the calculation of B. To summarize, the effects of feedback are not considered when calculating the value of T since the loop has been opened. The calculation of T is based on the concept of opening the loop in such a way as to not change the bias values, load impedance, input impedance, or gain of any stage. Analytically, one way to achieve this is by constructing the circuit diagram as a continuous repetitive (or partially repetitive) amplifier, as shown in Fig. 10–26.

Now that the concept of T has been discussed, the procedure for finding T can be simplified. The ac equivalent circuit of the voltage amplifier studied in this section is shown in Fig. 10–27. Assume that a *disturbance signal*, V_δ, is applied at the base of Q_3. The origin of this disturbance signal is not important since we are using it simply as an analytical tool. However, such disturbance signals are important in amplifiers and may be present due to incident radiation, minute changes in power supply voltages, noise generators, and a wide variety of other sources. We will briefly outline the effects of the feedback in reducing the disturbance signal.

The disturbance signal will cause a voltage at the collector of Q_3. This voltage then produces a voltage at the base of Q_2, which then causes a *correction signal*, V_ψ, at the base of Q_3. Since the polarity of V_ψ is opposite that of V_δ, the effect of the feedback is to reduce V_δ to a low value at which point V_ψ is also small. The correcting signal, V_ψ, is equal to the disturbance signal, V_δ, multiplied by the loop gain. Starting at the base of Q_3,

$$V_\psi = V_\delta(A_{vt3})\left(\frac{R_{\text{in}2} \,||\, R_{f1}}{R_{\text{in}2} \,||\, R_{f1} + R_{f2}}\right)(A_{d1-2}) \tag{10–53}$$

Solving for the loop gain, or V_ψ/V_δ

$$T = \frac{V_\psi}{V_\delta} = (A_{vt3})\left(\frac{R_{\text{in}2} \,||\, R_{f1}}{R_{\text{in}2} \,||\, R_{f1} + R_{f2}}\right)(A_{d1-2}) \tag{10–54}$$

Comparing Eqs. (10–54) and (10–51), we see that the ratio V_ψ/V_δ is equal to the loop gain. Also note that we effectively assumed that the loop was open when we calculated V_ψ/V_δ.

To summarize, there are several methods that can be used to calculate the value of T. First, we can determine the value of A_{ol} and use the product $A_{ol}B$ to determine T. This method can be confusing since both A_{ol} and B cannot be defined arbitrarily. Second, we can use a current generator (or a voltage generator), at some point inside the loop, usually in the equivalent circuit of an active device, by short circuiting the original current generator and replacing it with a test current generator. An accurate expression for T can always be obtained by this second method. Third, we can construct the circuit diagram of an amplifier

* In the forward signal path, Q_1 and Q_2 both appear as CE stages.

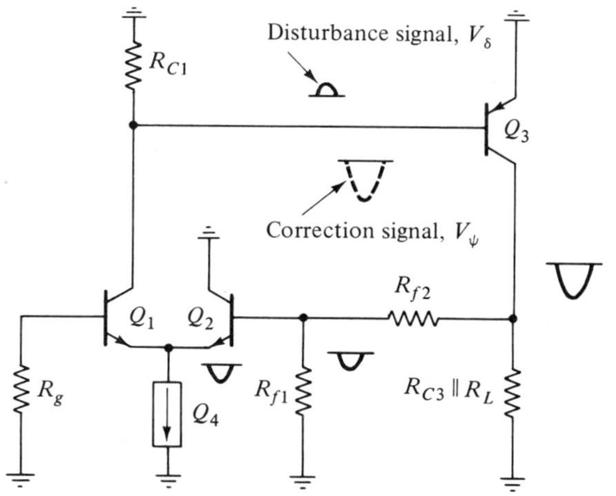

FIGURE 10–27 *The ac equivalent circuit of the voltage amplifier studied in this section showing the disturbance signal, V_δ, and the correction signal, V_ψ.*

as a continuous, repetitive circuit and calculate the signal gain from one point to that same point later in the circuit as shown in Fig. 10–26. Fourth, we can assume a disturbance signal at some point inside the loop and calculate the correcting signal at that same point as illustrated in Fig. 10–27. In this last method, we neglect the effects of feedback on the various impedance values since we have *mentally* opened the loop. Generally, the disturbance signal method is the easiest one to apply since no additional construction is required and since the analytical methods discussed in previous chapters can be used. The following example illustrates the disturbance signal method of calculating T.

EXAMPLE 10–10 Calculate the value of T for the voltage amplifier shown in Fig. 10–28(a). The transistors are characterized by the following values of β: β_1 and $\beta_2 = 170$, $\beta_3 = 90$, and $\beta_4 = 100$.

Solution: (a) Bias values:

$$I_{E4} \simeq (-V_{CC} - V_{E4})/(R_{E4}) \simeq [-20\text{ V} - (-10.6)\text{ V}]/4.7\text{ K} \simeq 2\text{ mA}$$
$$I_{E1} \simeq I_{E2} \simeq I_{E4}/2 \simeq 1\text{ mA} \qquad I_{E3} = (V_{CC} - V_{E3})/R_{E3} \simeq 2\text{ mA}$$

The bases of Q_1 and Q_2 are at approximately 0 V due to the 100 ohms connected between the base terminals and ground. Assuming that little or no current flows through R_{f2},

$$V_{C3} \simeq -V_{CC} + I_{C3}R_{C3} \simeq -20\text{ V} + (2\text{ mA})(10\text{ K}) \simeq 0\text{ V}$$

Since V_{B2} and V_{C3} are both approximately 0 V, the assumption of negligible current in R_{f2} is verified.

(b) Gain and impedance values:

$$r_{e1} \simeq r_{e2} \simeq 26\ \Omega \qquad r_{e3} \simeq 13\ \Omega$$
$$R_{\text{in}2} \simeq (\beta_2 + 1)(r_{e1} + r_{e2}) + R_g \simeq (171)(52) + 100 \simeq 9\text{ K}$$
$$R_{\text{in}3} \simeq (\beta_3 + 1)(r_{e3}) \simeq 1.18\text{ K} \qquad R_{L\text{eq}1} = R_{C1} \| R_{\text{in}3} \simeq 1.06\text{ K}$$
$$R_{L\text{eq}3} = R_L \| R_{C3} \| (R_{f2} + R_{f1} \| R_{\text{in}2}) \simeq 10\text{ K} \| 10\text{ K} \| (2\text{ K} + 100 \| 9\text{ K}) \simeq 1.48\text{ K}$$

From Table 4–4,

$$A_{d1-2} \simeq \frac{R_{L\text{eq}1}}{2[r_{e1} + R_g/(\beta_1 + 1)]} \simeq \frac{1.06\text{ K}}{2[26 + 100/171]} \simeq 20.2$$

From Table 4–1,

$$A_{vt3} \simeq -R_{L\text{eq}3}/r_{e3} \simeq -1.48\text{ K}/13 \simeq -114$$

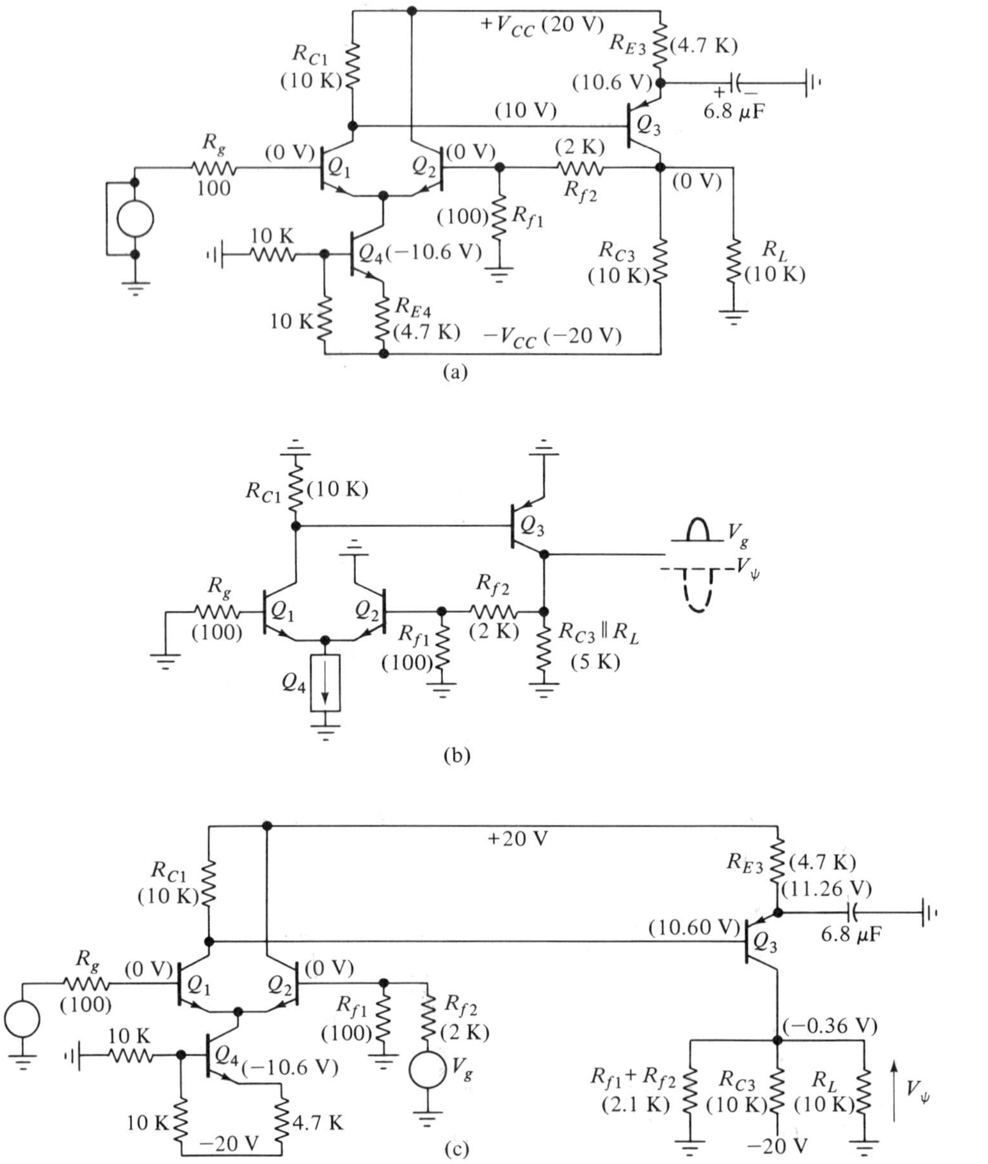

FIGURE 10–28 *The voltage amplifier from Ex. 10–10: (a) the complete circuit, (b) an ac equivalent circuit showing the disturbance signal and the correction signal, and (c) an experimental setup used to measure the loop gain.*

(c) Loop gain: The loop gain can be calculated from the ratio V_ψ/V_δ as

$$T = \frac{V_\psi}{V_\delta} = \left(\frac{R_{\text{in}2} \,\|\, R_{f1}}{R_{\text{in}2} \,\|\, R_{f1} + R_{f2}}\right)(A_{d1\text{–}2})(A_{vt3})$$

$$\simeq \left(\frac{9\text{ K} \,\|\, 100}{9\text{ K} \,\|\, 100 + 2\text{ K}}\right)(20.2)(-114) \simeq -110$$

(d) Experimental results: The loop gain can be measured because of the inherent dc balance of the voltage amplifier shown in Fig. 10–28(a). The experimental setup for measuring T is shown in Fig. 10–28(c) where V_δ is applied to one terminal of R_{f2}, and V_ψ is measured at the collector of Q_3. The loading effects of the feedback network have been included by placing a resistor equal to $(R_{f1} + R_{f2})$ or 2.1 K from the collector of Q_3 to ground. The effects of opening the loop on the bias values are indicated by comparing the dc voltages given in Figs. 10–28(a) and (c). The measured loop gain was

$$T = V_\psi / V_\delta \simeq 2.6 \text{ V}/25 \text{ mV} \simeq -104$$

In this section we have studied various methods of calculating the loop gain, T. Generally, the simplest method is the disturbance signal method in which a disturbance signal is assumed at some point in the loop, and the resulting correction signal is calculated by neglecting the effects of feedback on impedance values. The direct experimental determination of the loop gain illustrated in Ex. 10–10 was possible only because of the inherent dc balance in the circuit. In many circuits T cannot be measured because the loop cannot be opened. However, the value of T can be measured indirectly either by measuring the input or output impedance, or by measuring the closed-loop frequency response. Additional examples illustrating the steps used in calculating T are presented in later sections.

10.9 Calculation of the Input and Output Impedance of a Negative-Feedback Amplifier

Experimental data, introduced in Secs. 10.2 and 10.3, indicated that the input impedance of a negative-feedback amplifier was increased by summing voltages at the input and reduced by summing currents at the input. These data also indicated that the output impedance was reduced by sampling the output voltage and increased by sampling the output current. In this section, methods for calculating the values of the input and output impedances of a feedback amplifier will be introduced. While the emphasis will be placed on the calculation of the input and output resistance values, the method is general and can be applied to reactive components as well.

Numerous techniques are available from circuit theory that can be used to determine the value of the impedance looking into a pair of terminals. One useful method is based on Blackman's theorem and is discussed below.

Consider the block diagram shown in Fig. 10–29(a). The input impedance of an amplifier with feedback is denoted by Z_{if}. The value of the impedance Z_{T1} looking into the terminals A_1-B_1 is the sum of R_g and Z_{if} or

$$Z_{T1} = R_g + Z_{if} \tag{10–55}$$

Blackman's theorem states that Z_{T1} is also equal to

$$Z_{T1} = Z_{ol}\left(\frac{1 - T_{sc}}{1 - T_{oc}}\right) = Z_{ol}\left(\frac{1 + |T_{sc}|}{1 + |T_{oc}|}\right) \tag{10–56}$$

where Z_{ol} is $(Z_i + R_g)$ or the impedance looking into terminals A_1-B_1 neglecting the effects of feedback; Z_i is the input impedance of the amplifier neglecting the effects of feedback; T_{sc} is the value of the loop gain with terminals A_1-B_1 short circuited; and T_{oc} is the value of the loop gain with terminals A_1-B_1 open circuited. The values of R_g and Z_i can generally be found from the circuit diagram of an amplifier and the characteristics of the active devices. The values

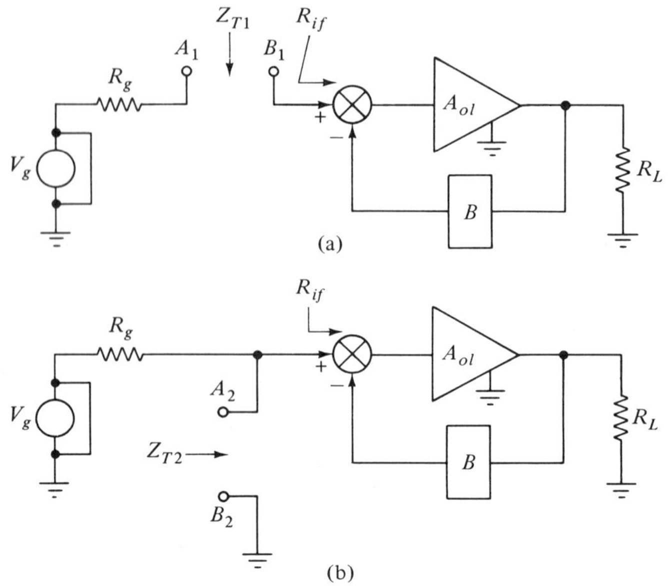

FIGURE 10–29 *Block diagram of a feedback amplifier used in the study of Blackman's theorem: (a) finding the equivalent series impedance, and (b) finding the equivalent parallel impedance.*

of T_{sc} and T_{oc} are calculated in the manner described in the last section. Normally, either T_{sc} or T_{oc} is equal to zero, and the other is generally equal to the normal loop gain of the amplifier. Combining Eqs. (10–55) and (10–56),*

$$Z_{if} + R_g = Z_{ol}\left(\frac{1 + |T_{sc}|}{1 + |T_{oc}|}\right) \qquad \textbf{(10–57)}$$

Since R_g, Z_{ol}, T_{sc}, and T_{oc} can be calculated, the value of Z_{if} can be determined from Eq. (10–57). This form of Blackman's theorem, called the series form, is useful when the feedback tends to increase the value of the input impedance, or when voltages are summed at the input. For this case, Z_{if} is generally much greater than R_g, and Z_{if} is approximately equal to the right hand side of Eq. (10–57). Also, the series form of Blackman's theorem is useful in calculating the output impedance of a current sampled amplifier since current sampling tends to increase the value of R_o. Before illustrating the use of Eq. (10–57), we will consider an alternate form of Blackman's theorem called the parallel form.

Consider the block diagram shown in Fig. 10–29(b). The value of Z_{T2}, the impedance looking into the terminals A_2-B_2, is the parallel combination of R_g and Z_{if}.

$$Z_{T2} = R_g \,||\, Z_{if} \qquad \textbf{(10–58)}$$

Blackman's theorem states that Z_{T2} is also

$$Z_{T2} = Z_{ol}\left(\frac{1 - T_{sc}}{1 - T_{oc}}\right) = Z_{ol}\left(\frac{1 + |T_{sc}|}{1 + |T_{oc}|}\right) \qquad \textbf{(10–59)}$$

* From Eq. (10–57), it might appear that Z_{if} is a function of R_g. Actually, this is not the case although the value of the loop gain (either T_{oc} or T_{sc}) is a function of R_g. Blackman's theorem is structured in such a manner that Z_{if} is independent of R_g. This effect is illustrated in Ex. 10–12.

where Z_{ol} is $Z_i \,||\, R_g$ or the impedance looking into the terminals A_2-B_2 neglecting the effects of feedback; Z_i is the input impedance of the amplifier neglecting the effects of feedback; T_{sc} is the value of the loop gain with terminals A_2-B_2 shorted; and T_{oc} is the value of the loop gain with terminals A_2-B_2 open. As in the series form of Blackman's theorem, the value of either T_{sc} or T_{oc} is usually zero, and the other is generally equal to the normal value of the loop gain of the amplifier. Combining Eqs. (10–58) and (10–59) gives

$$R_g \,||\, Z_{if} = Z_{ol}\left(\frac{1 + |T_{sc}|}{1 + |T_{oc}|}\right) \tag{10–60}$$

Since R_g, Z_{ol}, T_{sc}, and T_{oc} can be calculated, the value of Z_{if} can be calculated from Eq. (10–60). The parallel form of Blackman's theorem is useful when the feedback tends to reduce the input impedance or when currents are summed at the input. Then Z_{if} is generally much less than R_g and dominates the parallel combination of R_g and Z_{if}. Under these conditions, Z_{if} is approximately equal to the right-hand side of Eq. (10–60). Also, the parallel form of Blackman's theorem is useful in calculating the output impedance of a voltage sampled amplifier since voltage sampling tends to reduce the value of R_o.

EXAMPLE 10–11 Calculate the values of T, R_{if}, and R_{of} for the amplifier shown in Fig. 10–30(a). The transistor parameters are given in Ex. 10–10.

Solution: (a) Bias values: The only difference between the amplifier in Fig. 10–30(a) and the one in Ex. 10–10 is the value of R_g. Neglecting the $I_{B1}R_g$ voltage drop, the bias values are the same as calculated in Ex. 10–10.

(b) Impedance values: From Ex. 10–10,

$$r_{e1} \simeq r_{e2} \simeq 26\ \Omega \qquad r_{e3} \simeq 13\ \Omega$$

$$R_{Leq1} = R_{C1} \,||\, R_{in3} = 10\text{ K} \,||\, 1.18\text{ K} \simeq 1.06\text{ K}$$

$$R_{in2} \simeq (\beta_2 + 1)(r_{e1} + r_{e2}) + R_g \simeq (171)(52) + 10\text{ K} \simeq 18.9\text{ K}$$

$$R_{Leq2} \simeq r_{e1} + R_g/(\beta_1 + 1) \simeq 26 + 10\text{ K}/171 \simeq 84.5\ \Omega$$

$$R_{Leq3} = R_L \,||\, R_{C3} \,||\, (R_{f2} + R_{f1} \,||\, R_{in2}) \simeq 10\text{ K} \,||\, 10\text{ K} \,||\, (2\text{ K} + 100 \,||\, 18.9\text{ K}) \simeq 1.48\text{ K}$$

(c) Loop gain: The loop gain can be calculated from Fig. 10–30(b) using the disturbance signal method. For this calculation, Q_1 and Q_2 will not be considered as a differential pair because the equivalent base resistances are unequal, causing the circuit to be unsymmetrical. For the calculation of T, Q_1 acts as a CB stage, and Q_2 acts as a CC stage. From Table 4–1,

$$A_{vt3(\text{CE})} \simeq -R_{Leq3}/r_{e3} \simeq -1.48\text{ K}/13 \simeq -114$$

$$A_{vt2(\text{CC})} \simeq \frac{R_{Leq2}}{R_{Leq2} + r_{e2}} \simeq \frac{84.5}{84.5 + 26} \simeq 0.76$$

$$A_{vt1(\text{CB})} \simeq \frac{R_{Leq1}}{r_{e1} + R_g/(\beta_1 + 1)} \simeq \frac{1.06\text{ K}}{84.5} \simeq 12.6$$

$$T = \frac{V_\psi}{V_\delta} = A_{vt3}\left(\frac{R_{f1} \,||\, R_{in2}}{R_{f1} \,||\, R_{in2} + R_{f2}}\right) A_{vt2}A_{vt1} \simeq -52$$

Note that the product $A_{vt2}A_{vt1}$ is 9.6 as compared to the value of 20.2 calculated in Ex. 10–10 when R_g was 100 ohms.

(d) Input impedance: Since voltage summing tends to increase the input impedance, the series form of Blackman's theorem will be used to calculate R_{if}. From Eq. (10–57),

$$R_{if} + R_g = R_{ol}\left(\frac{1 + |T_{sc}|}{1 + |T_{oc}|}\right)$$

From Fig. 10–30(c), the resistance looking into the terminals A_1-B_1, neglecting the effects of feedback, is

$$R_{ol} = R_g + R_{\text{in}1} \simeq R_g + (\beta_1 + 1)(r_{e1} + r_{e2}) + R_{f1} \,||\, (R_{f2} + R_{C3} \,||\, R_L)$$
$$\simeq 10\text{ K} + (171)(52) + 100 \,||\, (2\text{ K} + 10\text{ K} \,||\, 10\text{ K}) \simeq 19\text{ K}$$

With terminals A_1-B_1 short circuited, the equivalent circuits shown in Figs. 10–30(b) and (c) are identical and have the same value of T. Thus,

$$T_{sc} = -52$$

With the terminals A_1-B_1 open circuited, the effective resistance in the base circuit of Q_1 is infinite. The gain of Q_1 becomes

$$A_{vt1} = \frac{R_{L\text{eq}1}}{r_{e1} + R_g/(\beta_1 + 1)} = \frac{R_{L\text{eq}1}}{\infty} = 0$$

With A_{vt1} equal to zero, the value of T_{oc} is also equal to zero. Substituting these values into Eq. (10–57),

$$R_{if} + 10\text{ K} = (19\text{ K})\left(\frac{1 + 52}{1 + 0}\right) \simeq 1\text{ M}$$

$$R_{if} \simeq 1\text{ M} - 10\text{ K} \simeq 990\text{ K}$$

(e) Output impedance: Since voltage sampling tends to reduce the output impedance, the parallel form of Blackman's theorem will be used to calculate R_{of}. From Eq. (10–60),

$$R_{of} \,||\, R_L = R_{ol}\left(\frac{1 + |T_{sc}|}{1 + |T_{oc}|}\right)$$

From Fig. 10–30(d), the resistance looking into the terminals A_2-B_2, neglecting the effects of feedback, is

$$R_{ol} = R_L \,||\, R_{C3} \,||\, r_{ce3} \,||\, (R_{f2} + R_{f1} \,||\, R_{\text{in}2}) \simeq 1.48\text{ K}$$

With terminals A_2-B_2 short circuited, $R_{L\text{eq}3}$ is zero, causing A_{vt3} to be zero. Consequently, $T_{sc} = 0$. With terminals A_2-B_2 open circuited, the equivalent circuits shown in Figs. 10–30(b) and (d) are identical and have the same value of T. Thus, $T_{oc} = -52$. Substituting these values into Eq. (10–60),

$$R_{of} \,||\, 10\text{ K} = (1.48\text{ K})\left(\frac{1 + 0}{1 + 52}\right) \simeq 28\ \Omega$$

Since the parallel combination of R_{of} and R_L is much less than R_L, $R_{of} \simeq 28$ ohms.

(f) Experimental value of R_{if}: The value of R_{if} can be measured by measuring the voltage drops across R_g and across R_{if} since

$$(V_g - V_i)/V_i = R_g/R_{if}$$

Measuring the voltage difference between V_i and V_g is difficult because this difference is small and also because R_{if} is large. Using an oscilloscope with a differential plug-in unit having an input impedance of 10 M, the measured values were

$$V_g - V_i \simeq 1\text{ mV} \qquad V_i \simeq 98\text{ mV}$$

Thus

$$R_{if} \simeq R_g V_i/(V_g - V_i) \simeq (10\text{ K})(98\text{ mV})/(1\text{ mV}) \simeq 980\text{ K}$$

(g) Experimental value of R_{of}: The value of R_{of} can be determined from the open circuit voltage and the short circuit current at terminals A_2-B_2 by using Thévenin's theorem, or

$$R_{\text{Th}} = \frac{V_{\text{open circuit}}}{I_{\text{short circuit}}}$$

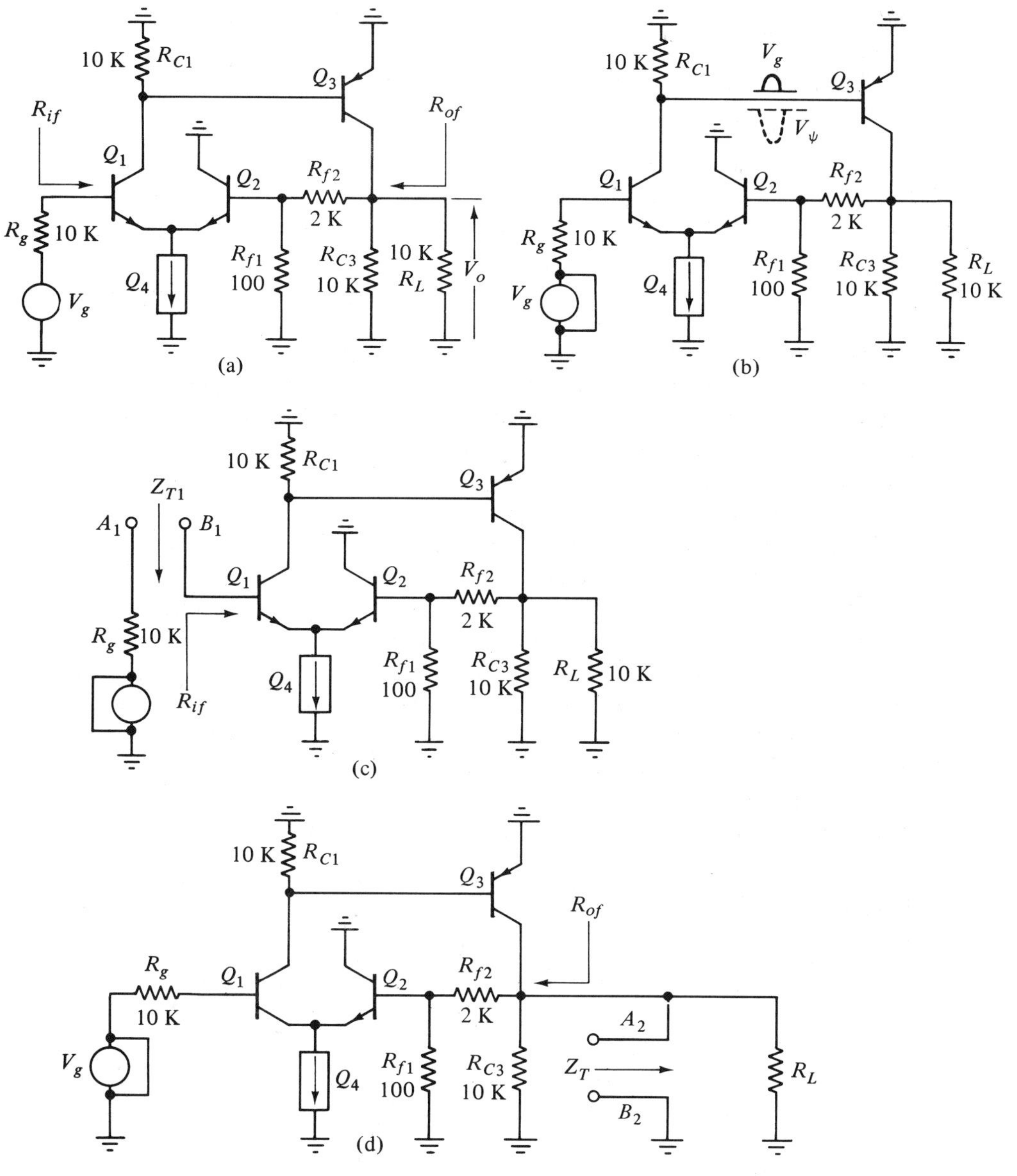

FIGURE 10–30 *The voltage sample-voltage sum amplifier from Ex. 10–11: (a) an ac equivalent circuit, (b) the equivalent circuit used in the calculation of the loop gain, (c) the equivalent circuit used in the calculation of* Z_{if}, *and (d) the equivalent circuit used in the calculation of* Z_{of}. *(The complete circuit diagram of this amplifier is shown in Fig. 10–28(a), except that* R_g *has been changed from 100 Ω to 10 K.)*

The short-circuit current was measured by measuring the voltage across a 1-ohm resistor connected in series with a 1000-μF capacitor between the terminals A_2-B_2. In making these measurements, precautions were taken to ensure that the capacitor was an effective short circuit at the signal frequency, and that the 1-ohm resistor did not distort the output. Using a 26-kHz sine wave input signal, the peak-to-peak signals were

$$I_{sc} \simeq 2\text{ mV}/1\,\Omega \simeq 2\text{ mA peak-to-peak}$$

$$V_{oc} \simeq 67\text{ mV peak-to-peak}$$

$$R_{of} \parallel R_L \simeq R_{of} \simeq V_{oc}/I_{sc} \simeq 67\text{ mV}/2\text{ mA} \simeq 34\,\Omega$$

Note that the dc collector current of Q_3 was 2 mA which is greater than the peak value of the short circuit current (1 mA) indicating that the output was not distorted during the measurement of I_{sc}.

The preceding example verifies the experimental results discussed in Sec. 10.2 in that voltage summing increases the input resistance, and voltage sampling decreases the output resistance. The factor by which the input impedance or the output impedance changes is approximately equal to $(1 + |T|)$. An experimental evaluation of R_{if} and R_{of} can be a formidable task since the amplifier must remain linear during the measurements. In addition, the impedance of the instruments used must not interact significantly with the circuit.

From Ex. 10–11, it would appear that the value of R_{if} increases without limit as the value of T increases. However, the resistance $r_{b'c1}$ of the reverse-biased base-collector junction of Q_1 is effectively in parallel with the input and places an upper limit on R_{if}. By a proper choice of the input transistor (assuming a bipolar transistor is used and not an FET), values of R_{if} in excess of 10 M can be achieved. The value of R_{of}, however, can approach zero as T increases. Values of R_{of} in the milliohm range can be achieved without great difficulty by using a low output impedance configuration, such as the CC amplifier, as the output stage of the feedback amplifier. Also, the value of R_{if} is not a function of R_g. This independence is illustrated in the following example.

EXAMPLE 10–12 Rework Ex. 10–11 with R_g equal to 100 ohms. The loop gain value from Ex. 10–10 corresponds to this condition.

Solution: (a) Input impedance: From Exs. 10–10 and 10–11,

$$R_g = 100\,\Omega \qquad R_{ol} = R_g + R_{\text{in}1} \simeq 100 + 9\text{ K} \simeq 0.1\text{ K}$$

$$T_{sc} \simeq -110 \qquad T_{oc} = 0$$

$$R_{if} = (R_{ol})\frac{1 + |T_{sc}|}{1 + |T_{oc}|} - R_g \simeq (9.1\text{ K})\frac{1 + 110}{1 + 0} - 100 \simeq 1\text{ M}$$

(b) Output impedance: From Exs. 10–10 and 10–11,

$$R_{ol} = R_L \parallel R_{C3} \parallel r_{ce3} \parallel (R_{f1} + R_{f2} \parallel R_{\text{in}2}) \simeq 1.48\text{ K}$$

$$T_{sc} = 0 \qquad T_{oc} = -110$$

$$R_{of} \parallel R_L = R_{ol}\left(\frac{1 + |T_{sc}|}{1 + |T_{oc}|}\right) \simeq 13.4\,\Omega$$

$$R_{of} \simeq 13.4\,\Omega$$

(c) Experimental results: Using the measurement method described in Ex. 10–11, the value of R_{of} was

$$R_{of} \simeq V_{oc}/I_{sc} \simeq 30\text{ mV}/2\text{ mV} \simeq 15\,\Omega$$

In Exs. 10–11 and 10–12, we found that the value of R_{if} was approximately 1 M when R_g was either 10 K or 100 ohms. Thus the value of R_{if} does not depend on the value of R_g, a result consistent with circuit theory. However, the value of R_{if} is a function of R_L. In addition, the value of R_{of} is independent of R_L but is a function of R_g. Comparing the results of Exs. 10–11 and 10–12, R_{of} was approximately 29 ohms when R_g was 10 K, and approximately 13 ohms when R_g was 100 ohms.

In this section, we have studied methods for calculating the input and output impedances of a feedback amplifier. The analysis was based on Blackman's theorem which includes the effects of R_g and R_L in the calculation of R_{if} and R_{of}. The calculations required a knowledge of the loop gain and the input (or output) impedance without feedback. Both quantities can be calculated in a straightforward manner. Later sections include additional examples of the calculations of input and output impedances.

10.10 Forward Gain of the Feedback Network

A basic requirement for the application of the general closed-loop gain expression [Eq. (10–25)] is that both the amplifier and the feedback network are unilateral. Generally, this assumption is valid for the forward gain of a multistage amplifier. The feedback network, however, may or may not be unilateral. The worst case occurs with the voltage sample-current sum configuration, since the feedback network connects the output of an amplifier directly to the input. The effect of a passive signal transfer from the input directly to the output through the feedback network will now be discussed in some detail.

A voltage sample-current sum feedback amplifier can be represented by the block diagram given in Table 10–7, which is repeated in Fig. 10–31(a). To analyze this circuit, we will replace the basic amplifier with a hybrid equivalent circuit and represent the generator circuit with a Norton equivalent. These changes are reflected in Fig. 10–31(b). Also, for simplicity we will assume that R_g is very large compared with R_{in}, and the generator will be simply represented by I_g. From Fig. 10–31(b),

$$B = -1/R_f \tag{10–61}$$

and

$$T = -\frac{(R_o \parallel R_L)A_i}{R_f + R_{\text{in}} + R_o \parallel R_L} \tag{10–62}$$

Using these equations, the closed loop or *active* gain can be expressed as

$$A_{fr} = \left.\frac{V_o}{I_g}\right|_{\text{active}} = \frac{T}{B}\frac{1}{1+T} = -\frac{(R_o \parallel R_L)A_iR_f}{R_f + R_{\text{in}} + (R_o \parallel R_L)(1 + A_i)} \tag{10–63}$$

The term $1/(1 + T)$ will be useful later and can be found from Eq. (10–62) as

$$\frac{1}{1+T} = \frac{R_f + R_{\text{in}} + (R_o \parallel R_L)}{R_f + R_{\text{in}} + (R_o \parallel R_L)(1 + A_i)} \tag{10–64}$$

The second path that the signal can take from the input to the output is through the feedback path, R_f. This path, called the *passive* signal path, is obtained from Fig. 10–31(b) by neglecting the current generator A_iI_{in}, which repre-

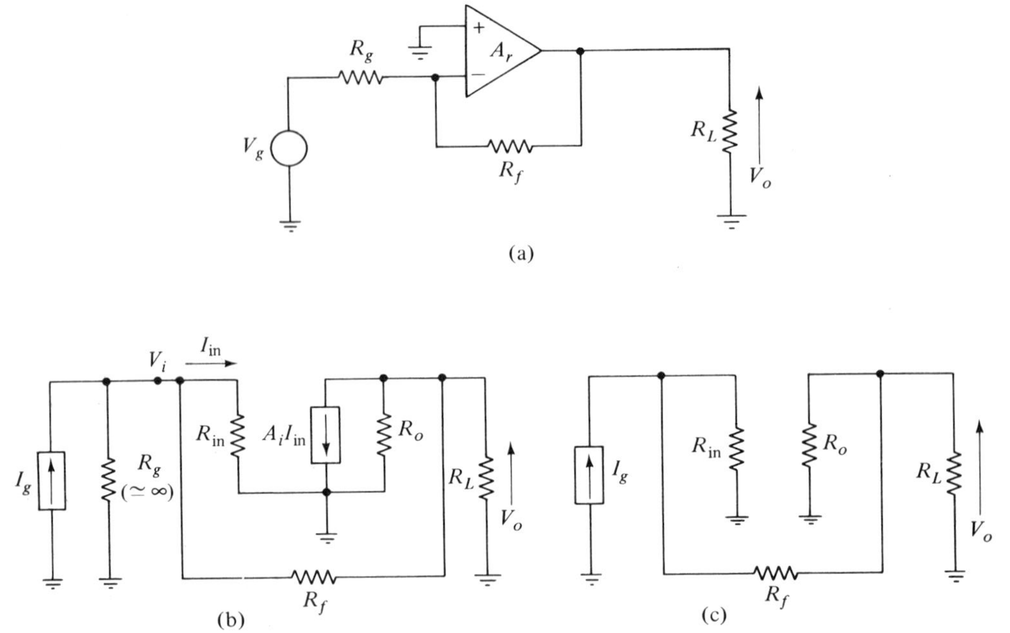

FIGURE 10–31 *Circuits used in the study of passive signal transfer through the feedback network: (a) block diagram of a voltage sample-current sum feedback amplifier, (b) circuit diagram with the basic amplifier represented by a hybrid equivalent circuit, and (c) circuit for calculating the passive signal gain of the amplifier.*

sents the active gain of the amplifier. This path is shown in Fig. 10–31(c). From this figure, the passive signal gain is

$$\left.\frac{V_o}{I_g}\right|_{\text{passive}} = \frac{(R_o \parallel R_L)R_{\text{in}}}{R_{\text{in}} + R_f + (R_o \parallel R_L)} \tag{10–65}$$

The actual gain of the amplifier, including the effects of the forward signal path and the feedback signal path, can be found from Fig. 10–31(b) using nodal analysis. The node equations for this circuit are

$$\left.\begin{aligned} I_g &= V_i\left(\frac{1}{R_{\text{in}}} + \frac{1}{R_f}\right) - V_o\left(\frac{1}{R_f}\right) \\ -\frac{A_iV_i}{R_{\text{in}}} &= -V_i\left(\frac{1}{R_f}\right) + V_o\left(\frac{1}{R_o} + \frac{1}{R_L} + \frac{1}{R_f}\right) \end{aligned}\right\} \tag{10–66}$$

Using any standard technique, such as Cramer's rule, Eq. (10–66) can be solved for the gain V_o/I_g. One form of the solution is

$$\frac{V_o}{I_g} = -\frac{A_iR_f(R_o \parallel R_L)}{R_{\text{in}} + R_f + (R_o \parallel R_L)(1 + A_i)} + \frac{R_{\text{in}}(R_o \parallel R_L)}{R_{\text{in}} + R_f + (R_o \parallel R_L)(1 + A_i)} \tag{10–67}$$

Comparing Eqs. (10–67) and (10–63), we see that the first term in Eq. (10–67) is the active signal gain of the amplifier. The second term in Eq. (10–67) corresponds to the passive signal gain.

Multiplying the numerator and the denominator of the second term in Eq. (10–67) by $(R_f + R_{\text{in}} + R_o \parallel R_L)$, we obtain

$$\frac{R_{\text{in}}(R_o \,\|\, R_L)}{R_{\text{in}} + R_f + (R_o \,\|\, R_L)(1 + A_i)}$$

$$= \left[\frac{R_{\text{in}}(R_o \,\|\, R_L)}{R_{\text{in}} + R_f + (R_o \,\|\, R_L)}\right]\left[\frac{R_{\text{in}} + R_f + (R_o \,\|\, R_L)}{R_{\text{in}} + R_f + (R_o \,\|\, R_L)(1 + A_i)}\right] \tag{10–68}$$

Note that the first term in Eq. (10–68) is the passive gain of the amplifier as given in Eq. (10–65). The second term in Eq. (10–68) is equal to $1/(1 + T)$ as given in Eq. (10–64).

The results of the nodal analysis can be written as

$$\frac{V_o}{I_g} = \left[\frac{T}{1 + T}\right]\frac{1}{B} + \frac{\text{Passive Gain}}{1 + T} \tag{10–69}$$

Two observations can be made from this equation. First, the assumption of a unilateral signal flow in the feedback network is generally valid for large values of loop gain since any passive signal gain is reduced by $(1 + T)$. Second, expressions for T, A_{ol}, or B may be difficult to identify from expressions obtained using nodal analysis since the results of a nodal analysis include the effects of passive signal transfer.

The effect of passive gain in a voltage sample-voltage sum feedback amplifier can be analyzed in a parallel manner to that for the voltage sample-current sum feedback amplifier. The basic circuit diagrams needed for the derivation are shown in Fig. 10–32. Development of the expressions which are similar to Eqs. (10–67) and (10–69) for the voltage amplifier are left as an exercise for the reader.

In this section we briefly examined the effect of the forward gain of the feedback network. As long as the value of $|T|$ is large, this effect can be neglected.

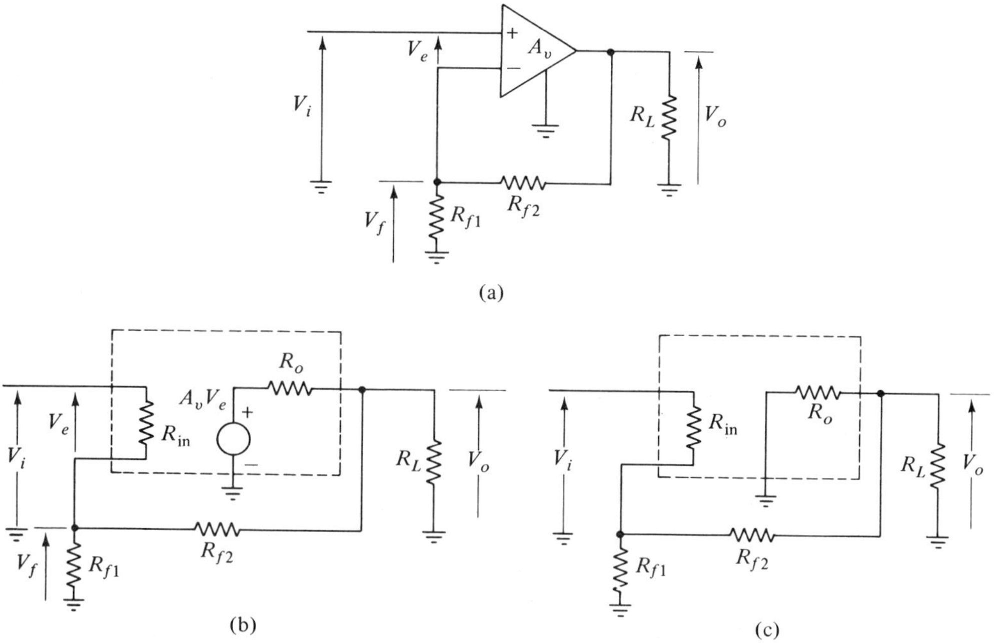

FIGURE 10–32 *Circuits used in the study of passive signal transfer through the feedback network of a voltage amplifier: (a) block diagram of a voltage amplifier, (b) circuit diagram with the basic amplifier represented by an equivalent circuit, and (c) circuit for calculating the passive signal gain of the amplifier.*

10.11 Analysis of the Mid-Band Response of a Voltage Sample-Voltage Sum Feedback Amplifier

The material presented in the preceding ten sections covered various aspects of the performance characteristics of feedback amplifiers. The objective of this section is to illustrate the complete analysis of the mid-band response of a voltage sample-voltage sum feedback amplifier. In this analysis, the following values will be determined: (1) bias conditions; (2) approximate values of A_f, A_{vo}, R_i, and R_o; (3) the value of the loop gain; and (4) the actual values of A_f, A_{vo}, R_i, and R_o. We will also study the problem of input and output dc *offset* voltages and several methods of reducing their effects.

EXAMPLE 10–13 Calculate the complete mid-band response of the amplifier shown in Fig. 10–33(a). The transistors have the following values of β: β_1 and β_2 are 220, β_3 is 170, β_4 is 58, and β_5 is 90. The 90-ohm potentiometer between the emitter terminals of Q_1 and Q_2 is adjusted so that the dc voltage at the output terminal is zero.*

Solution: (a) Approximate analysis: From the ac equivalent circuit shown in Fig. 10–33(b), the amplifier is a voltage sample-voltage sum type. Assuming that $|T| \gg 1$, then from Tables 10–6 and 10–7,

$$B = \frac{X_f}{X_o} = \frac{V_f}{V_o} = \frac{R_{f1}}{R_{f1} + R_{f2}} = \frac{1\text{ K}}{1\text{ K} + 9.09\text{ K}} = \frac{1}{10.09}$$

$$A_{fv} = \frac{V_o}{V_i} \simeq \frac{1}{B} \simeq \frac{R_{f1} + R_{f2}}{R_{f1}} \simeq 10 \qquad R_{if} \simeq \infty$$

$$A_{vg} = \frac{V_i}{V_g} = \frac{R_{if}}{R_{if} + R_g} \simeq 1 \qquad R_{of} \simeq 0$$

$$A_{vo} = \frac{V_o}{V_g} = A_{vg}A_{fv} \simeq (1)(10) \simeq 10$$

(b) Bias values: Since the output dc voltage has been adjusted to 0 V, we will begin the bias calculations at this terminal. From Fig. 10–33(a),

$$V_{E4} \simeq 0\text{ V (adjusted)} \qquad I_{E4} \simeq 24\text{ V}/4.64\text{ K} \simeq 5.2\text{ mA}$$

$$V_{C3} \simeq 0.6\text{ V} \qquad I_{E3} \simeq I_{C3} \simeq 24.6\text{ V}/12.1\text{ K} \simeq 2\text{ mA}$$

$$V_{E3} \simeq 24\text{ V} - I_{E3}R_{C3} \simeq 14.7\text{ V} \qquad V_{B3} \simeq V_{E3} - V_{BE3} \simeq 14.1\text{ V}$$

$$I_{E1} \simeq I_{C1} \simeq (24\text{ V} - 14.1\text{ V})/10\text{ K} \simeq 1\text{ mA} \qquad V_{B5} \simeq -12\text{ V}$$

$$V_{E5} \simeq -12.6\text{ V} \qquad I_{C5} \simeq I_{E5} \simeq (-24\text{ V} + 12.6\text{ V})/5.62\text{ K} \simeq 2\text{ mA}$$

$$I_{E2} \simeq I_{C5} - I_{E1} \simeq 2\text{ mA} - 1\text{ mA} \simeq 1\text{ mA}$$

Since I_{C1} and I_{C2} are approximately equal, the potentiometer is adjusted to its mid-point value where V_{E4} is equal to 0 V.

(c) Impedance values:

$$r_{e1} \simeq r_{e2} \simeq 26\ \Omega \qquad r_{e3} \simeq 13\ \Omega \qquad r_{e4} \simeq 5\ \Omega$$

$$R_{\text{in2}} \simeq (\beta_2 + 1)(r_{e1} + r_{e2} + R_{EE}) + R_g \simeq (221)(142) + 100 \simeq 31.5\text{ K}$$

$$R_{L\text{eq}4} \simeq R_L \,||\, R_{E4} \,||\, (R_{f2} + R_{f1} \,||\, R_{\text{in2}})$$
$$\simeq 1\text{ K} \,||\, 4.64\text{ K} \,||\, (9.09\text{ K} + 1\text{ K} \,||\, 31.5\text{ K}) \simeq 760\ \Omega$$

$$R_{\text{in4}} \simeq (\beta_4 + 1)(r_{e4} + R_{L\text{eq}4}) \simeq (59)(765) \simeq 45\text{ K}$$

---

* The potentiometer was rated at 100 ohms, but measured approximately 90 ohms. The value of r_{ce3} is 35 K.

$$R_{Leq3} \simeq R_{C3} \,||\, R_{in4} \,||\, r_{ce3} \simeq 7.5\text{ K} \qquad R_{in3} \simeq (\beta_3 + 1)r_{e3} \simeq 2.2\text{ K}$$

$$R_{Leq1} \simeq R_{C1} \,||\, R_{in3} \simeq 10\text{ K} \,||\, 2.22\text{ K} \simeq 1.81\text{ K}$$

$$R_{in1} \simeq (\beta_1 + 1)(r_{e1} + r_{e2} + R_{EE}) + R_{f1} \,||\, \left[R_{f2} + \left(r_{e4} + \frac{R_{C3}}{\beta_4 + 1}\right) \,||\, R_L \,||\, R_{E4}\right] \simeq 31.5\text{ K}$$

(d) Loop gain: Using the disturbance signal method, from Fig. 10–33(b),

$$T = \frac{V_\psi}{V_\delta} = A_{vt3}A_{vt4}\left(\frac{R_{f1} \,||\, R_{in2}}{R_{f1} \,||\, R_{in2} + R_{f2}}\right)A_{d1-2}$$

$$A_{vt3} \simeq -\frac{R_{Leq3}}{r_{e3}} \simeq -580 \qquad A_{vt4} \simeq \frac{R_{Leq4}}{R_{Leq4} + r_{e4}} \simeq 0.99$$

$$\frac{R_{f1} \,||\, R_{in2}}{R_{f1} \,||\, R_{in2} + R_{f2}} \simeq \frac{1\text{ K} \,||\, 31.5\text{ K}}{1\text{ K} \,||\, 31.5\text{ K} + 9.09\text{ K}} \simeq \frac{1}{10}$$

$$A_{d1-2} \simeq \frac{R_{Leq1}}{r_{e1} + r_{e2} + R_g/(\beta_1 + 1) + R_{EE}} \simeq \frac{1.81\text{ K}}{142} \simeq 12.8$$

$$T \simeq (-580)(0.99)(1/10)(12.8) \simeq -735$$

(e) Input impedance: Using the series form of Blackman's theorem, Eq. (10–57), and Fig. 10–33(d),

$$R_{ol} = R_g + R_{in1} \simeq 100 + 31.5\text{ K} \simeq 31.6\text{ K}$$

$$|T_{sc}| \simeq 735 \qquad |T_{oc}| = 0$$

$$100 + R_{if} \simeq (31.6\text{ K})\left(\frac{1 + 735}{1 + 0}\right) \simeq 23.3\text{ M}$$

$$R_{if} \simeq 23.3\text{ M} - R_g \simeq 23.3\text{ M}$$

(f) Output impedance: Using the parallel form of Blackman's theorem, Eq. (10–60), and Fig. 10–33(e),

$$R_{ol} = R_L \,||\, R_{E4} \,||\, \left(r_{e4} + \frac{R_{C3}}{\beta_4 + 1}\right) \,||\, (R_{f2} + R_{f1} \,||\, R_{in2}) \simeq 190\ \Omega$$

$$|T_{sc}| = 0 \qquad |T_{oc}| \simeq 735$$

$$1\text{ K} \,||\, R_{of} \simeq (190)\left(\frac{1 + 0}{1 + 735}\right) \simeq 0.26\ \Omega$$

$$R_{of} \simeq 0.26\ \Omega$$

(g) Actual gain values:

$$A_{fv} = \frac{V_o}{V_i} = \left(\frac{|T|}{1 + |T|}\right)\left(\frac{1}{B}\right) = \left(\frac{735}{736}\right)(10.09) \simeq 10$$

$$A_{vg} = V_i/V_g = R_{if}/(R_{if} + R_g) = 23.3\text{ M}/(23.3\text{ M} + 100) \simeq 1$$

$$A_{vo} = V_o/V_g = A_{vg}A_{fv} \simeq (1)(10) \simeq 10$$

(h) Experimental results: The measured dc bias values are shown in parentheses in Fig. 10–33(a). Varying the potentiometer R_{EE} from one extreme to the other caused the dc voltage at the output terminal to vary from +0.6 V to −0.98 V. The value of T was measured using the method shown in Fig. 10–28(c). The measured value of T was

$$T = V_\psi/V_\delta \simeq -7.5\text{ V}/10\text{ mV} \simeq -750$$

The measured voltage gain, A_{vo}, was 10 with an input voltage, V_g, of 0.1 V.

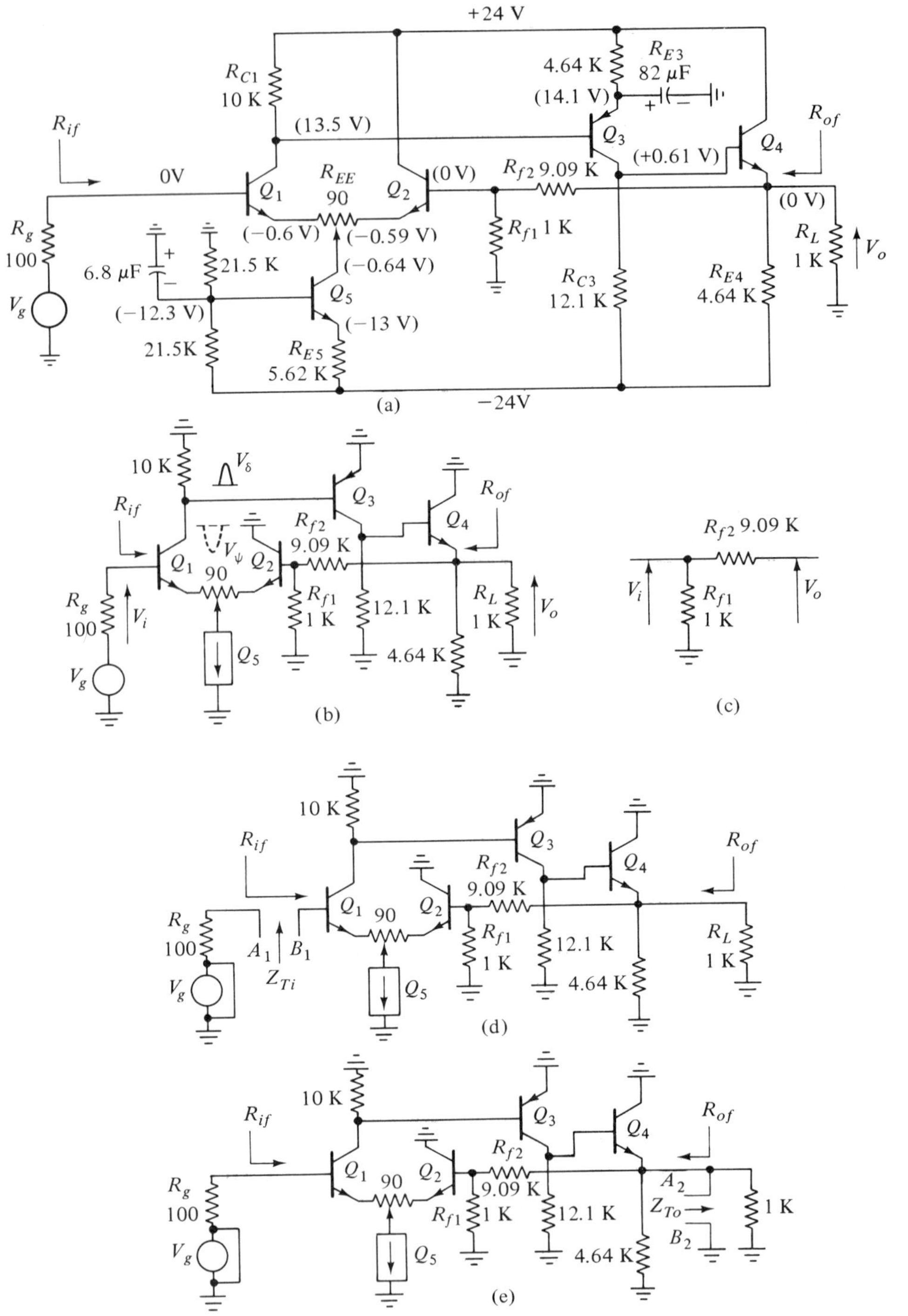

FIGURE 10–33 *The amplifier from Ex. 10–13: (a) the complete circuit, (b) an ac equivalent circuit, (c) the feedback network, (d) the equivalent circuit used for the calculation of* R_{if}, *and (e) the equivalent circuit used for the calculation of* R_{of}. *All resistor values have a 1 percent tolerance.*

A basic condition that must be satisfied in any amplifier is proper bias of all active components. One bias condition that has largely been ignored in our discussion is the requirement for a dc path for the base current of a bipolar transistor, the gate leakage current for an FET, and the grid leakage current for a vacuum tube. In the amplifier shown in Fig. 10–33(a), the base current of Q_1 flows in the generator circuit. If the generator is disconnected, this path is broken, and the amplifier will not operate in a linear manner.* In the remainder of this section we will briefly examine a bias problem called *offset voltage*.

Consider the circuit shown in Fig. 10–34(a). The Thévenin equivalent of the base bias circuit of Q_1 is shown in the partial circuit diagram in Fig. 10–34(b). Using the values from Ex. 10–13,

$$I_{B1} = I_{E1}/(\beta_1 + 1) \simeq 1\text{ mA}/221 \simeq 4.5\ \mu\text{A} \tag{10–70}$$

and

$$V_{B1} \simeq -I_{B1}R_B \simeq -(4.5\ \mu\text{A})(45.5\text{ K}) \simeq -0.205\text{ V} \tag{10–71}$$

Since the amplifier is dc coupled, the dc voltage V_{B1} will appear at the output terminal multiplied by the dc voltage gain of the amplifier.** The measured values of dc voltages are given in Fig. 10–34(a). Note that the dc values of both the input voltage and the output voltage are different from zero. These voltages are referred to as *offset voltages* and are generally undesirable in amplifiers which are to be directly coupled to the generator or load circuit.

Several methods can be used to reduce the value of an offset voltage. First, a potentiometer can be connected between the emitter terminals of Q_1 and Q_2 as shown in Fig. 10–33(a). This method is advisable only when the output offset voltage is small. Using a 100-ohm potentiometer, the dc voltage at the output terminal could be varied from +0.6 V to −1.9 V. Alternately, either R_{11} or R_{12} can be made variable.

Probably the best method of reducing input offset voltage is to use FETs (or vacuum tubes) as input devices. This method is shown in Fig. 10–34(c). With a proper choice of the FETs, gate leakage currents as low as 10^{-12} A (for JFETs) to 10^{-18} A (for MOSFETs) can be obtained. These low values of leakage current allow large values of R_g, often ranging as high as 10^9 and 10^{14} ohms, thus preserving the high input impedance of both the FET and the voltage summing amplifier. The disadvantage of using FETs is their low gain. However, the required magnitude of loop gain can be obtained with a good amplifier design.

In this section, an analysis of the mid-band response of a voltage sample-voltage sum feedback amplifier has been made. An analysis of the mid-band response of any feedback amplifier can be made in a similar manner. Also, we have introduced the problem of offset voltages and the various methods of reducing their effects.

---

* When the generator circuit was disconnected from the base terminal of Q_1, the output dc voltage changed to −4.6 V. This value is determined by the breakdown voltage characteristics of the transistors in the circuit.

** The dc voltage gain of the amplifier may be less than the mid-band voltage gain of the amplifier since the value of T may be less in the low-frequency region.

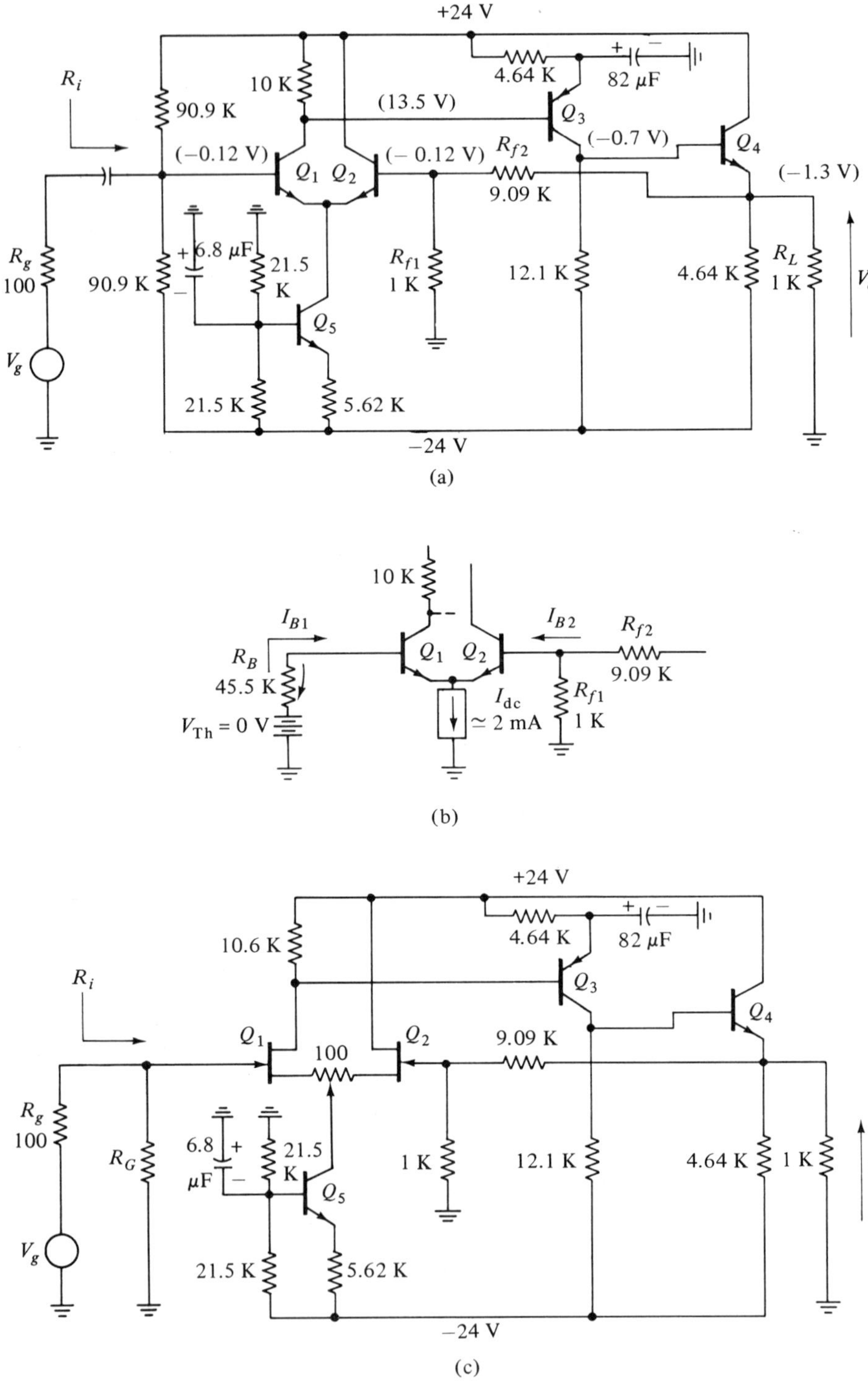

FIGURE 10–34 *Variations of the amplifier shown in Fig. 10–33(a) used in the study of offset voltages: (a) the use of base voltage bias for* Q_1*, (b) the Thévenin equivalent circuit for the bias network of* Q_1*, and (c) the use of JFETs for the input differential pair.*

10.12 Analysis of the Mid-Band Response of a Voltage Sample-Current Sum Feedback Amplifier

Feedback amplifiers using voltage sampling are widely used because the voltage gain is relatively independent of the load impedance. The voltage sample-voltage sum amplifier studied in the last section is one example of this type of amplifier. Another type is the voltage sample-current sum amplifier. In this section, we will illustrate the complete analysis of this second type of feedback amplifier. In addition, we will discuss a simple way of increasing the value of the loop gain in a direct-coupled amplifier using a constant-current source as the load resistor for a CE or CB stage.

EXAMPLE 10–14 Calculate the mid-band response of the amplifier shown in Fig. 10–35(a) if β_1 is 110, β_2 is 58, and β_3 is 76.

Solution: (a) Approximate analysis: From the ac equivalent circuit shown in Fig. 10–35(b), the amplifier is a voltage sample-current sum type. Assuming that $|T| \gg 1$, then from Tables 10–6 and 10–7,

$$B = \frac{X_f}{X_o} = \frac{I_f}{V_o} = -\frac{1}{R_f} = -\frac{1}{10\text{ K}} \qquad R_{if} \simeq 0$$

$$A_{fr} = \frac{V_o}{I_g} \simeq \frac{1}{B} \simeq -10\text{ K} \qquad R_{of} \simeq 0$$

$$A_{vo} = \frac{V_o}{V_g} = \frac{V_o}{I_g(R_g + R_{if})} \simeq -\frac{R_f}{R_g} \simeq -\frac{10\text{ K}}{1\text{ K}} \simeq -10$$

(b) Bias values: The bias values can be determined either from a set of dc equations, or by a trial-and-error process. The latter method is often simpler and will be used in this example. The silicon diode in the base circuit of Q_1 holds the base at approximately 0.6 V below ground. Thus, $V_{B1} \simeq -0.6$ V, and the emitter voltage of Q_1 is

$$V_{E1} \simeq V_{B1} + V_{BE1} \simeq -0.6\text{ V} + 0.6\text{ V} \simeq 0\text{ V}$$

As a first trial, assume that little or no current flows in R_f.* Then,

$$V_{E1} \simeq V_{E3} \simeq 0\text{ V}$$

$$I_{E1} \simeq (V_{CC} - V_{E1})/R_{E1} \simeq (10\text{ V} - 0\text{ V})/5.11\text{ K} \simeq 1.96\text{ mA}$$

$$V_{C1} = V_{B2} = -V_{CC} + I_{C1}R_{C1} \simeq -10\text{ V} + (1.96\text{ mA})(2.61\text{ K}) \simeq -4.9\text{ V}$$

$$V_{E2} \simeq V_{B2} - V_{BE2} \simeq -4.9\text{ V} - 0.6\text{ V} \simeq -5.5\text{ V}$$

$$I_{E2} \simeq (-V_{CC} - V_{E2})/R_{E2} \simeq (-10\text{ V} + 5.5\text{ V})/4.6\text{ K} \simeq 0.97\text{ mA}$$

$$V_{C2} \simeq V_{CC} - I_{C2}R_{C2} \simeq 10\text{ V} - (0.97\text{ mA})(9.09\text{ K}) \simeq 1.2\text{ V}$$

$$V_{E3} \simeq V_{C2} - V_{BE3} \simeq 1.2\text{ V} - 0.6\text{ V} \simeq 0.6\text{ V}$$

$$I_{E3} \simeq (-V_{CC} + V_{E3})/R_{E3} + V_{E3}/R_L \simeq 10.6\text{ V}/1\text{ K} + 0.6\text{ V}/500 \simeq 11\text{ mA}$$

Since the calculated value of V_{E3} is only 0.6 V different from the assumed value of 0 V, and R_f is 10 K, the calculated bias values are reasonably accurate.

(c) Impedance values:

$$r_{e1} \simeq 13\ \Omega \qquad r_{e2} \simeq 26\ \Omega \qquad r_{e3} \simeq 2.4\ \Omega$$

$$R_{Leq1} \simeq R_{C1} \,||\, R_{in2} \simeq 970\ \Omega \qquad R_{Leq2} \simeq R_{C2} \,||\, R_{in3} \simeq 6.7\text{ K}$$

$$R_{Leq3} \simeq R_L \,||\, R_{E3} \,||\, (R_f + r_{e1} \,||\, R_g \,||\, R_{E1}) \simeq 500 \,||\, 1\text{ K} \,||\, 10\text{ K} \simeq 325\ \Omega$$

---

* This assumption is also reasonable since both the generator circuit and the load resistor are directly coupled to the amplifier.

(d) Loop gain: Using the disturbance signal method, then from Fig. 10–35(b),

$$A_{vt2} \simeq -\frac{R_{Leq2}}{r_{e2}} \simeq -257 \qquad A_{vt3} \simeq \frac{R_{Leq}}{R_{Leq3} + r_{e3}} \simeq 0.99$$

$$\frac{r_{e1} \,||\, R_g \,||\, R_{E1}}{r_{e1} \,||\, R_g \,||\, R_{E1} + R_f} \simeq \frac{1}{780} \qquad A_{vt1} \simeq \frac{R_{Leq1}}{r_{e1}} \simeq 75$$

$$T = \frac{V_\psi}{V_\delta} = A_{vt2}A_{vt3}\left(\frac{r_{e1} \,||\, R_g \,||\, R_{E1}}{r_{e1} \,||\, R_g \,||\, R_{E1} + R_f}\right) A_{vt1} \simeq -25$$

(e) Input impedance: Using the parallel form of Blackman's theorem, Eq. (10–60), and Fig. 10–35(d),

$$R_g = 1\text{ K} \qquad R_{ol} \simeq R_g \,||\, R_{E1} \,||\, R_f \,||\, r_{e1} \simeq 13\ \Omega$$

$$|T_{sc}| = 0 \qquad |T_{oc}| \simeq 25$$

$$1\text{ K} \,||\, R_{if} \simeq (13)\left(\frac{1+0}{1+25}\right) \simeq 0.5\ \Omega$$

Since $(R_g \,||\, R_{if})$ is much less than R_g, then R_{if} is approximately 0.5 ohms.

(f) Output impedance: Using the parallel form of Blackman's theorem, Eq. (10–60), and Fig. 10–35(e),

$$R_L = 500 \qquad R_{ol} \simeq R_L \,||\, R_f \,||\, R_{E3} \,||\left(r_{e3} + \frac{R_{C2}}{\beta_3 + 1}\right) \simeq 87\ \Omega$$

$$|T_{sc}| = 0 \qquad |T_{oc}| \simeq 25$$

$$R_{of} \,||\, 500 \simeq (87)\left(\frac{1+0}{1+25}\right) \simeq 3.3\ \Omega$$

Since $R_{of} \,||\, R_L$ is much less than R_L, then $R_{of} \simeq 3.3$ ohms.

(g) Actual gain values:

$$A_{fr} = \frac{V_o}{I_g} = \left(\frac{|T|}{1+|T|}\right)\left(\frac{1}{B}\right) = \left(\frac{25}{26}\right)(-10\text{ K}) \simeq -9.6\text{ K}$$

$$A_{vo} = \frac{V_o}{V_g} = \frac{V_o}{I_g(R_g + R_{if})} \simeq -\frac{9.6\text{ K}}{1\text{ K}} \simeq -9.6$$

(h) Experimental results: The measured dc bias values are shown in Fig. 10–35(a) in parentheses. The value of T was determined from the frequency response of the amplifier as approximately -28.* The measured voltage gain was -9.8 with an input signal, V_g, of 100 mV.

The magnitude of the loop gain in the preceding example was only 25. Our next task is to develop methods for increasing the value of $|T|$. Since Q_3 is a CC stage which has a gain of less than unity, increasing $|T|$ requires that the value of A_{vt1}, or A_{vt2}, or both, be increased. For both the CB and the CE stages, the value of A_{vt} depends directly on the value of R_{Leq}. Therefore, increasing the value of either R_{Leq1} or R_{Leq2} would increase the value of $|T|$.

The value of R_{Leq1} is limited by R_{in2} to 1.53 K. Consequently, increasing R_{C1} will have little effect on the value of either A_{vt1} or $|T|$. The value of R_{Leq2} is

$$R_{Leq2} = r_{ce2} \,||\, R_{C2} \,||\, R_{in3} \tag{10–72}$$

---

* The frequency response of a feedback amplifier and its dependence on the magnitude of T is discussed in the next chapter.

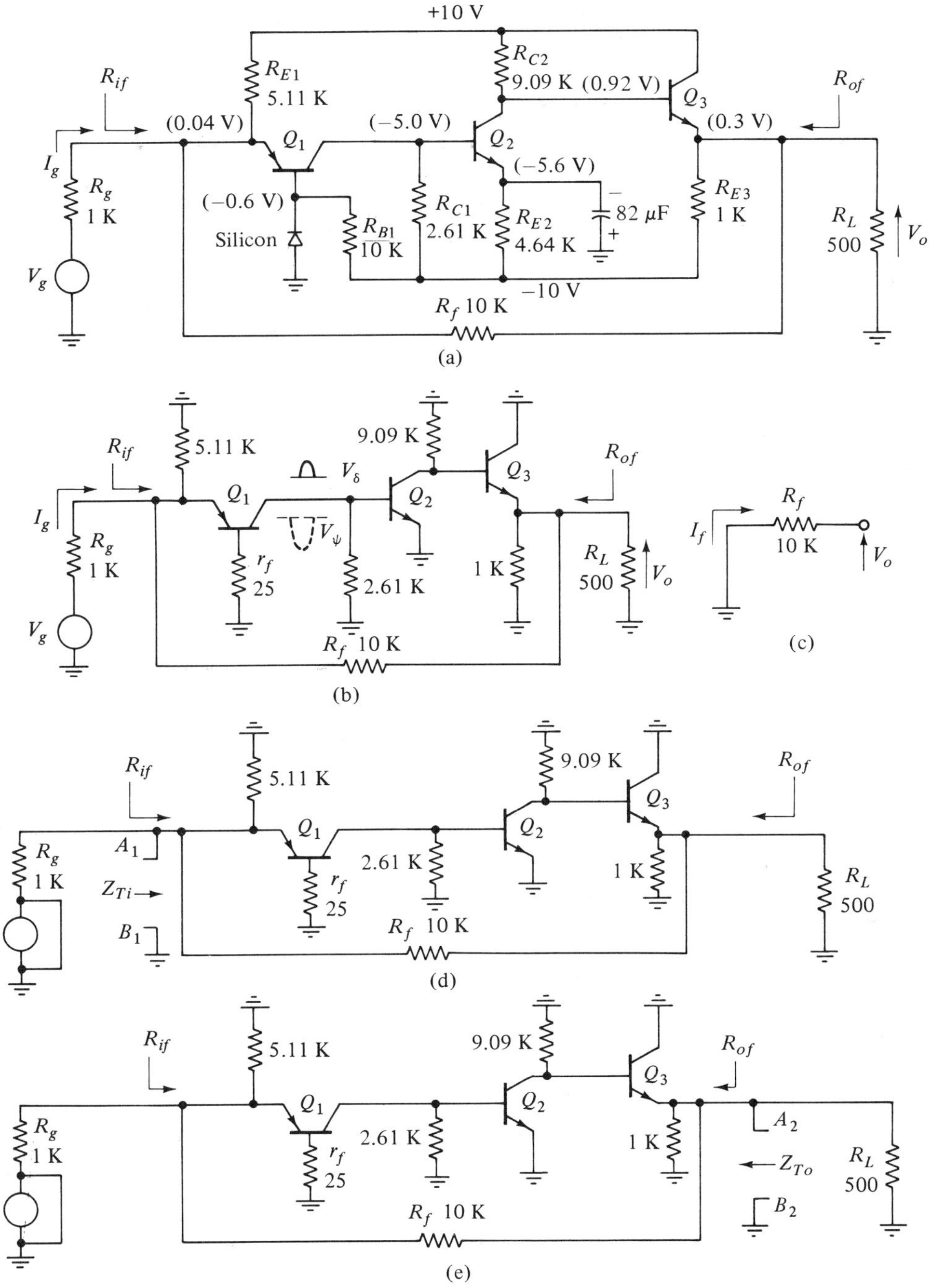

FIGURE 10–35 *The amplifier from Ex. 10–14: (a) the complete circuit, (b) an ac equivalent circuit, (c) the feedback network, (d) the equivalent circuit used for the calculation of* R_{if}*, and (e) the equivalent circuit used for the calculation of* R_{of}*. The value of all resistors have a 1 percent tolerance.*

For this discussion, assume that r_{ce2} is 50 kilohms. Using the values from Ex. 10–14,*

$$R_{Leq2} = 50\text{ K} \parallel 9.09\text{ K} \parallel 25\text{ K} \simeq 5.9\text{ K} \quad \textbf{(10–73)}$$

If the value of R_{C2} could be made very large, then R_{Leq2} would become

$$R_{Leq2} \simeq 50\text{ K} \parallel \infty \parallel 25\text{ K} \simeq 16.7\text{ K} \quad \textbf{(10–74)}$$

Using this value of R_{Leq2} in Ex. 10–14, A_{vt2} increases from −257 to −640, and T increases from −25 to −62.

One method of increasing R_{C2} to a very high value is to replace it with a constant-current source as shown in Fig. 10–36(a). The impedance looking into the

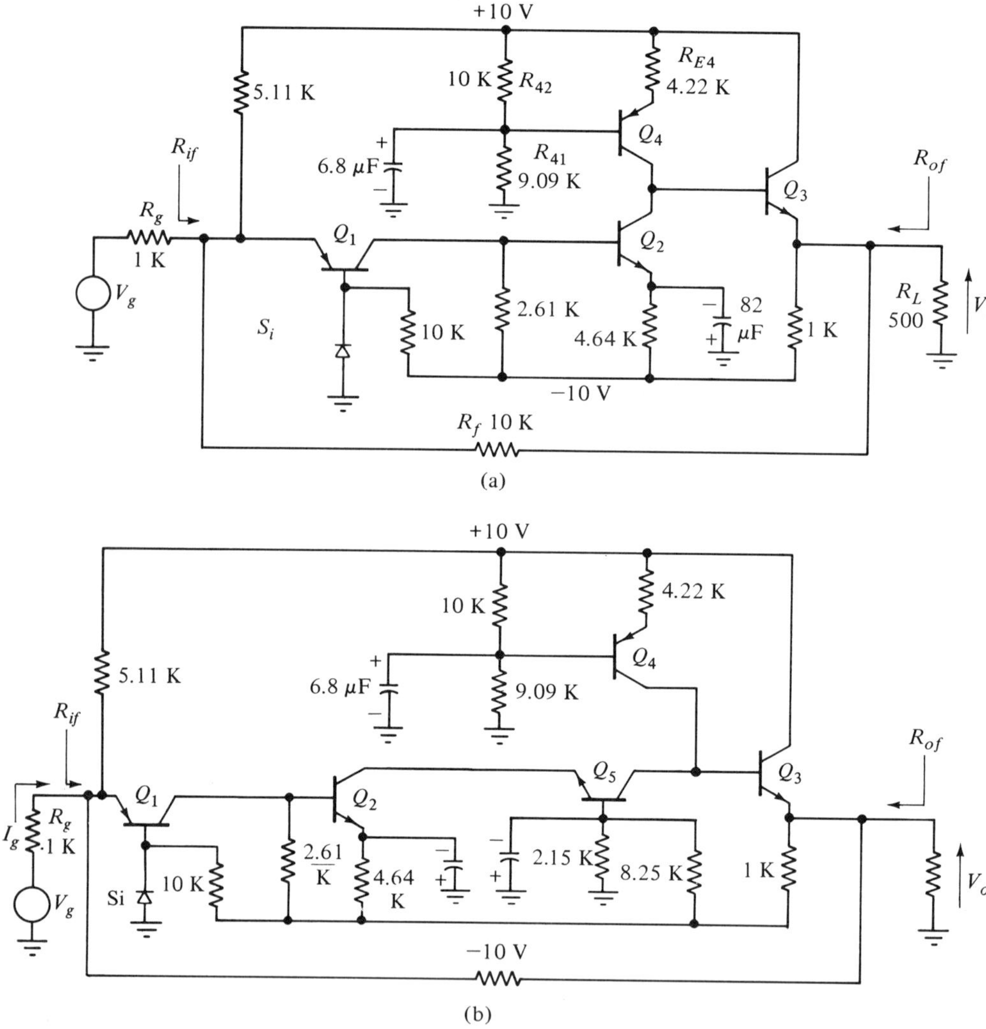

FIGURE 10–36 *Two methods of increasing the value of the loop gain of the amplifier in Ex. 10–14: (a) a constant-current source used as the load resistor for Q_2, and (b) a CB stage loaded with a constant-current source.*

* This value of R_{Leq2} is slightly less than the value of 6.7 K calculated in Ex. 10–14 where we neglected the effect of r_{ce2}.

collector of Q_4 in Fig. 10–36(a) corresponds to the output impedance of a CB stage and can be in the megohm range. However, the currents I_{C4} and I_{C2} must be equal. The current I_{C4} is determined by the base voltage bias network and also R_{E4}. The dc voltage at the collector terminals of Q_2 and Q_4 is established and stabilized by the overall dc feedback. Generally a constant-current source of the type shown in Fig. 10–36(a) should be used as a load only in a dc coupled feedback amplifier.

To increase the value of R_{Leq2} even more, the values of R_{in3} and/or r_{ce2} must be increased. Rather than trying to find a transistor with a higher value of r_{ce}, we can simply add a CB stage to the circuit as shown in Fig. 10–36(b). The CE-CB combination is referred to as a *cascode circuit.** One advantage of this arrangement is that the output impedance of Q_5 is much greater than r_{ce}. The load impedance on Q_5 is

$$R_{Leq5} \simeq r_{b'c5} \,||\, r_{b'c4} \,||\, R_{in3} \simeq R_{in3} \simeq 25 \text{ K} \tag{10–75}$$

Using the transistor parameters given in Ex. 10–14, the value of T for the amplifier shown in Fig. 10–36(b) is approximately -93. If β_3 is increased from 76 to 200, T increases to approximately -245. Verification of these values of T is left as an exercise for the reader.

In this section we have illustrated the complete mid-band analysis of a voltage sample-current sum amplifier. Also, we have studied the use of constant-current sources as load resistors. The use of a constant-current source in this manner is one way of increasing the value of the loop gain. Generally, constant-current sources can only be used as loads in dc coupled feedback amplifiers where the collector voltages can be stabilized. A particularly useful circuit is the CB stage with a constant-current source used as a load and followed by a CC stage. Transistors Q_5, Q_4, and Q_3 in Fig. 10–36(b) illustrate this configuration. The equivalent load impedance for the CB stage can exceed 100 kilohms if the circuit values are properly chosen. With loads in this range, the voltage gain of a CB stage can approach 10^4. Many high performance feedback amplifiers use this basic circuit, with variations, to achieve a very high value of loop gain.

10.13 Additional Examples of Negative-Feedback Amplifiers Using Current Sampling

The emphasis in this chapter has been placed on feedback amplifiers which use voltage sampling of the output signal. This approach is appropriate since the majority of the feedback amplifiers found in practice utilize voltage sampling. However, the current sampling type is also found in important applications. The objectives of this section are to make a complete mid-band analysis of a feedback amplifier which uses current sampling and to discuss several applications in which current sampling is useful.

---

* Cascode circuits were discussed in Exs. 4–16, 8–3, 8–4, and 8–5.

EXAMPLE 10–15 Calculate the complete mid-band response of the amplifier shown in Fig. 10–37(a) assuming that β_1 is 85, β_2 is 200, and β_3 is 95.

Solution: (a) Approximate analysis: Using an analysis similar to that in Ex. 10–3,

$$A_{fi} = \frac{I_o}{I_g} \simeq \frac{1}{B} \simeq \frac{10\text{ K} + 5.1\text{ K}}{5.1\text{ K}} \simeq 2.96 \qquad R_{if} \simeq 0$$

$$R_{of} \simeq \infty \,||\, 2.7\text{ K} \simeq 2.7\text{ K} \qquad A_{vo} = \frac{V_o}{V_g} \simeq \frac{I_o R_{Leq3}}{I_g(R_g + R_{if})} \simeq 2.16$$

(b) Bias values: The bias values can be estimated by assuming that very little current flows in R_{f2}. Then,

$$V_{C3} \simeq V_{B1} \simeq V_{BE1} + V_{BE2} \simeq 1.2\text{ V}$$

$$I_{C3} \simeq (V_{CC} - V_{C3})/R_{f1} \simeq 2.1\text{ mA} \qquad V_{E3} = -V_{CC} + I_{C3}R_{E3} \simeq -6.3\text{ V}$$

$$I_{E1} \simeq I_{E2} \simeq (-V_{CC} + V_{B3})/R_{C2} \simeq (-12\text{ V} + 5.7\text{ V})/6.2\text{ K} \simeq 1\text{ mA}$$

(c) Impedance values:

$$r_{e1} \simeq r_{e2} \simeq 26\,\Omega \qquad r_{e3} \simeq 12.4\,\Omega$$

$$R_{\text{in}1} \simeq (\beta_1 + 1)(r_{e1} + r_{e2}) \simeq (86)(52) \simeq 4.5\text{ K}$$

$$R_{\text{in}3} \simeq (\beta_3 + 1)(r_{e3} + R_{E3} \,||\, R_L) \simeq (96)(12.4 + 730) \simeq 71\text{ K}$$

(d) Loop gain: From Fig. 10–37(b),

$$A_{vt1} \simeq \frac{r_{e2}}{r_{e2} + r_{e1}} \simeq 0.5 \qquad A_{vt2} \simeq \frac{R_{C2} \,||\, R_{\text{in}3}}{r_{e2}} \simeq \frac{5.7\text{ K}}{26} \simeq 220$$

$$A_{vt3(\text{CE})} \simeq -\frac{R_{f1} \,||\, (R_{f2} + R_g \,||\, R_{\text{in}1})}{r_{e3} + R_{E3} \,||\, R_L} \simeq -\frac{3.48\text{ K}}{742} \simeq -4.7$$

$$T = \frac{V_\psi}{V_\delta} = A_{vt3}\left(\frac{R_{\text{in}1} \,||\, R_g}{R_{\text{in}1} \,||\, R_g + R_{f2}}\right) A_{vt1}A_{vt2} \simeq -39$$

(e) Input impedance: Applying the parallel form of Blackman's theorem and Eq. (10–60) to the circuit in Fig. 10–37(d) gives

$$R_g \,||\, R_{if} = R_{ol}\left(\frac{1 + |T_{sc}|}{1 + |T_{oc}|}\right) \simeq (R_g \,||\, R_{\text{in}1}) \,||\, (R_{f2} + R_{f1})\left(\frac{1 + |T_{sc}|}{1 + |T_{oc}|}\right) \simeq 19.4\,\Omega$$

$$R_{if} \simeq 19.4\,\Omega$$

(f) Output impedance: The actual value of the output impedance is $R_{of'} \,||\, R_{E3}$. Blackman's theorem can be used to find $R_{of'}$. Using the series form, Eq. (10–57), and the circuit in Fig. 10–37(e) gives

$$R_{of'} + R_{E3} \,||\, R_L = R_{ol}\left(\frac{1 + |T_{sc}|}{1 + |T_{oc}|}\right) \simeq \left(R_{E3} \,||\, R_L + r_{e3} + \frac{R_{C2}}{\beta_3 + 1}\right)\left(\frac{1 + |T_{sc}|}{1 + |T_{oc}|}\right)$$

$$R_{of'} \simeq (730 + 12.4 + 64.6)\left(\frac{1 + 39}{1 + 0}\right) - 730 \simeq 31.5\text{ K}$$

From Fig. 10–37(b),

$$R_{of} = R_{of'} \,||\, R_{E3} \simeq 31.5\text{ K} \,||\, 2.7\text{ K} \simeq 2.5\text{ K}$$

(g) Actual gain values:

$$A_{fi} = \frac{I_o}{I_g} = \left(\frac{|T|}{1 + |T|}\right)\left(\frac{1}{B}\right) \simeq \left(\frac{39}{40}\right)(2.96) \simeq 2.9$$

$$A_{vo} = \frac{V_o}{V_g} = \frac{I_o R_{Leq3}}{I_g(R_g + R_i)} \simeq (2.9)\left(\frac{730}{1\text{ K} + 19.4}\right) \simeq 2.1$$

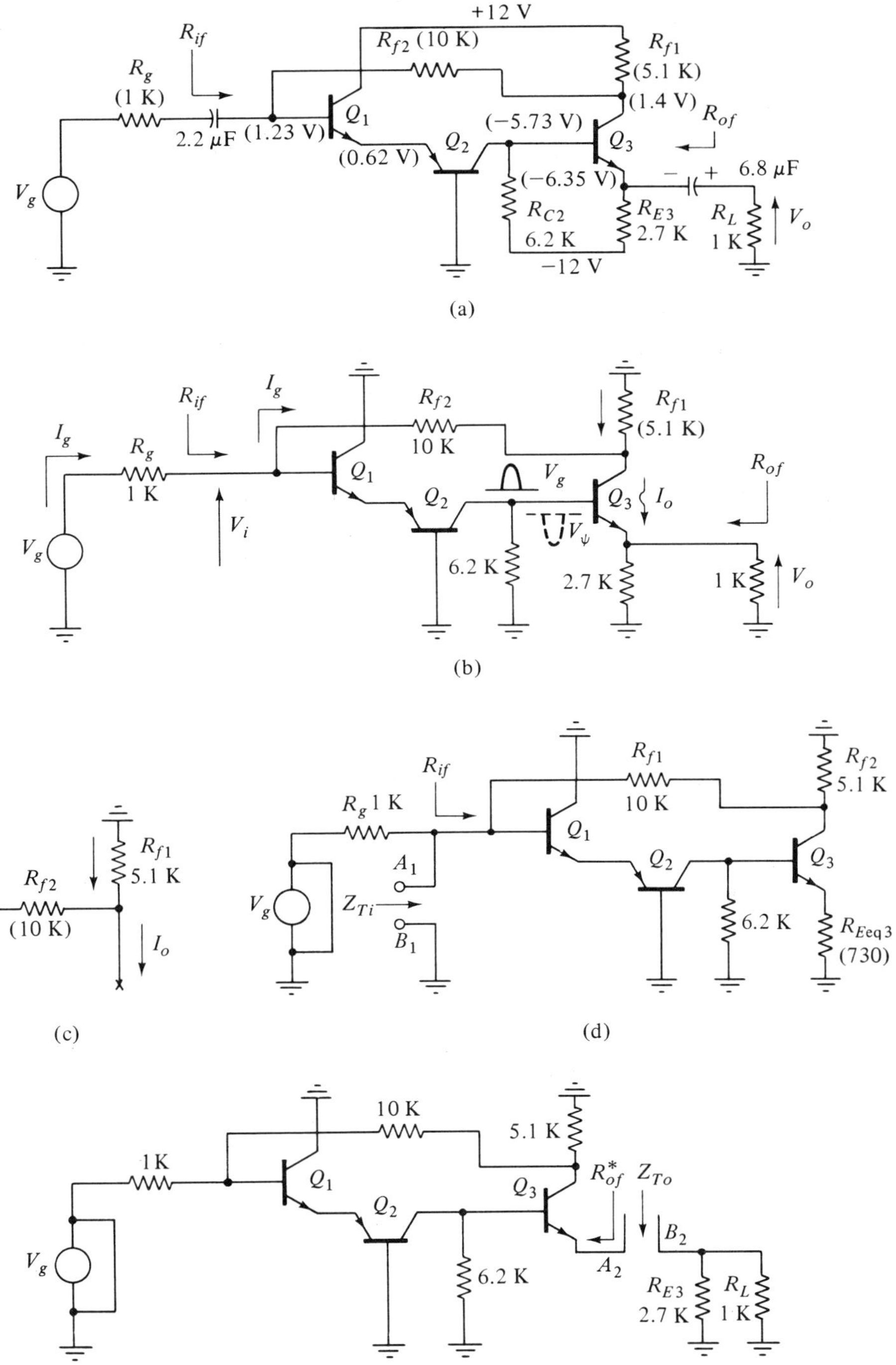

FIGURE 10–37 *The current sample-current sum feedback amplifier from Ex. 10–15: (a) the complete circuit, (b) an ac equivalent circuit, (c) the feedback network, (d) the equivalent circuit used for the calculation of* R_{if}*, and (e) the equivalent circuit used for the calculation of* R_{of}*. (All resistors have 5 percent tolerances.)*

(h) Experimental results: The measured dc bias values are shown in Fig. 10–37(a). The measured value of input impedance was

$$R_{if} = \frac{V_i}{I_g} = \frac{V_i}{(V_g - V_i)/R_g} \simeq \frac{5\text{ mV}}{(240\text{ mV} - 5\text{ mV})/1\text{ K}} \simeq 21.2\ \Omega$$

The value of R_{of} can be determined from the output open circuit voltage and short circuit current as discussed in Ex. 10–11. With R_L (equal to 1 K) disconnected,

$$R_{of} \simeq R_{\text{Th}} \simeq \frac{V_{oc}}{I_{sc}} \simeq \frac{602\text{ mV}}{0.225\text{ mA}} \simeq 2.7\text{ K}$$

The measured voltage gain, A_{vo}, was 2.09 with an input signal, V_g, of 0.24 V.

The preceding example illustrates the complete analysis of the mid-band response of a current sample-current sum feedback amplifier. In the remainder of this section we will briefly examine three applications in which current sampling is useful.

One common form of a current sample feedback amplifier is the single-stage CE circuit with an unbypassed emitter resistor shown in Fig. 10–38(a). An analysis of this circuit can be made easily by using the techniques discussed in Chap.

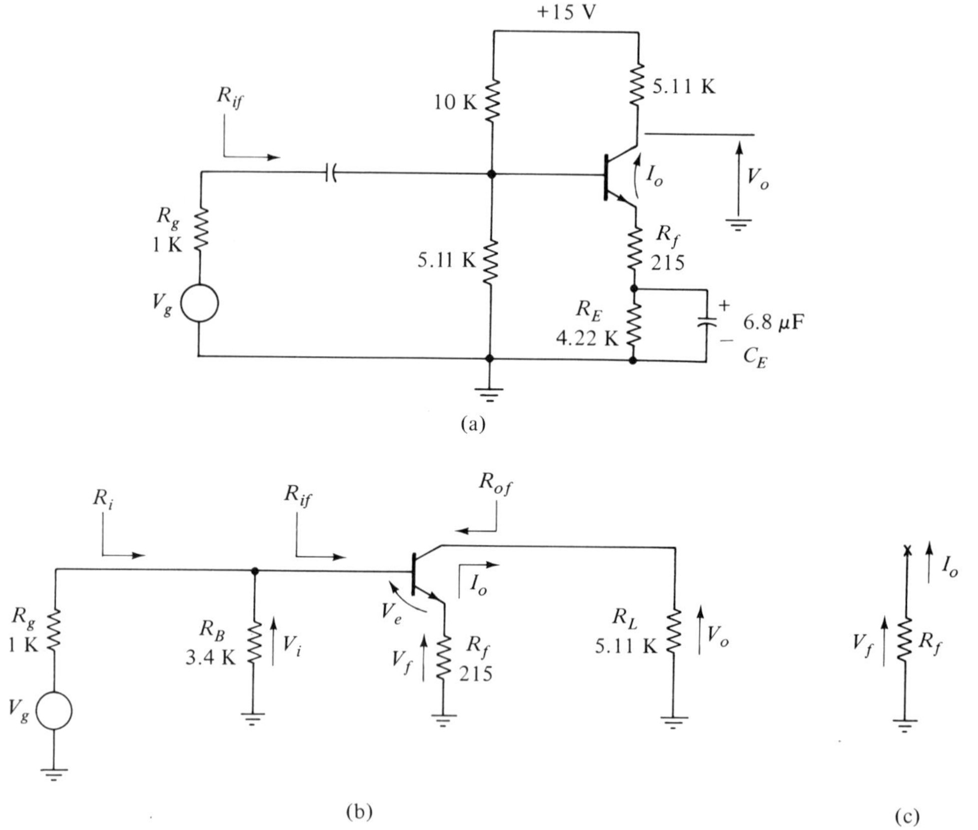

FIGURE 10–38 *A CE amplifier with an unbypassed emitter resistor: (a) the complete circuit, (b) an ac equivalent circuit, and (c) the feedback network.*

4. However, the operation of this circuit can be appreciated by using a feedback approach. This circuit is interesting because it contains both ac and dc feedback with different loop gains.

In this circuit, the resistor R_f forms the ac feedback network. From Fig. 10–38(c),

$$B = X_f/X_o = V_f/I_o = -R_f \quad \textbf{(10–76)}$$

The feedback loop is obscured because the sampling network and the basic amplifier are formed by the same active device. However, the ac open loop gain can be calculated as

$$A_{ol} = X_o/X_e = I_o/V_e = g_m \quad \textbf{(10–77)}$$

The ac loop gain is

$$T = A_{ol}B = -g_m R_f \simeq -R_f/r_e \quad \textbf{(10–78)}$$

From Eq. (10–78), the amount of negative feedback present for ac signals depends upon the ratio of the unbypassed emitter resistor, R_f, and r_e. For the circuit shown in Fig. 10–38 in which I_E is approximately 1 mA,

$$T \simeq -R_f/r_e \simeq -215/26 \simeq -8.3 \quad \textbf{(10–79)}$$

At zero frequency or dc, the bypass capacitor, C_E, is an open circuit. Consequently, the total emitter resistance is the sum of R_f and R_E or approximately 4.4 K. The loop gain for dc is

$$T_{\text{dc}} \simeq -\frac{R_E + R_f}{r_e} \simeq -\frac{4.4\text{ K}}{26} \simeq -160 \quad \textbf{(10–80)}$$

This large value of loop gain indicates that the amplifier has good bias stability. This result is consistent with the result obtained from the bias stability analysis in Sec. 4.3 which indicated that all four stability factors were reduced (improved) by making R_E large.

A basic building block used in many multistage amplifier designs is the cascode circuit. One form of this circuit is shown in Fig. 10–39(a). A major advantage of this circuit is that it has a low value of input capacitance resulting from the low value of voltage gain of Q_1 which is due to the low input impedance of Q_2. The approximate voltage gain of this circuit is

$$A_{vt1}A_{vt2} \simeq g_{m1}R_{C2} \quad \textbf{(10–81)}$$

The magnitude of the voltage gain given in Eq. (10–81) can be small because the value of g_m for an FET can be small.

A modified form of the CS-CB cascode circuit is shown in Fig. 10–39(b) in which a single-stage current sample-current sum feedback amplifier has been added between the CS and the CB stages.* Since this type of feedback amplifier has a low value of input resistance, A_{vt1} is small. Consequently, C_{in1} is as small as that of a regular cascode circuit. However, the current gain associated with the feedback amplifier stage will effectively increase the value of g_{m1} and the voltage gain, V_o/V_i.

The single-stage feedback amplifier is shown separately in Fig. 10–39(c). Assuming that β_3 is 200 and that I_{E2} and I_{E3} are each 2 mA, the feedback charac-

* This circuit is part of a charge-sensitive preamplifier discussed in T. V. Blalock in the references listed at the end of the chapter. The reported value of A_{fi}, calculated in Eq. (10–84), was approximately 16.

teristics are as follows:

$$B = \frac{R_{C3}}{R_f + R_{C3}} = -\frac{1\text{ K}}{20\text{ K} + 1\text{ K}} = -\frac{1}{21} \qquad \textbf{(10–82)}$$

$$T = \frac{V_\psi}{V_\delta} \simeq \left(-\frac{R_{Leq}}{r_{e2} + r_{e3}}\right)\left(\frac{R_d \,||\, R_{in3}}{R_d \,||\, R_{in3} + R_f}\right)$$

$$\simeq \left(-\frac{960}{26}\right)\left(\frac{3.48\text{ K} \,||\, 5.2\text{ K}}{3.48\text{ K} \,||\, 5.2\text{ K} + 20\text{ K}}\right) \simeq -3.4 \qquad \textbf{(10–83)}$$

and

$$A_{fi} = \left(\frac{|T|}{1 + |T|}\right)\left(\frac{1}{B}\right) \simeq \left(\frac{3.4}{4.4}\right)(-21) \simeq -16.2 \qquad \textbf{(10–84)}$$

Assuming that g_{m1} is 4.3 millimhos, the overall voltage gain of the amplifier shown in Fig. 10–39(b) is

$$V_o/V_i \simeq g_{m1}A_{fi}R_{Leq2} \simeq (4.3 \times 10^{-3})(-16.2)(12\text{ K}) \simeq -840 \qquad \textbf{(10–85)}$$

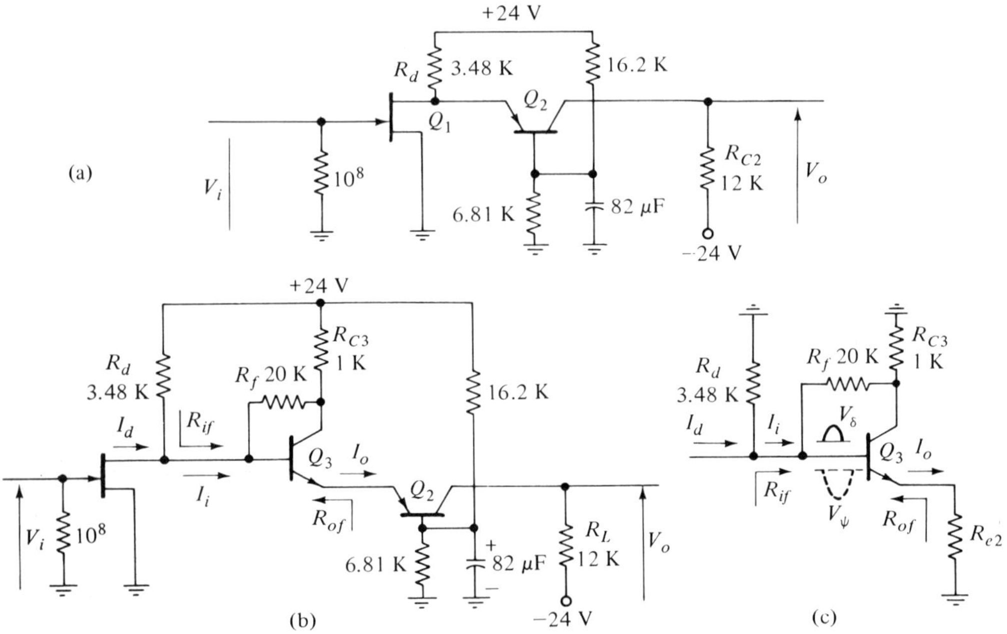

FIGURE 10–39 *A cascode amplifier that uses current sample-current sum feedback to increase the gain: (a) the original cascode circuit, (b) the modified cascode circuit, and (c) the added amplifier shown separately. (Adapted from T. V. Blalock, "Wide-Band Low-Noise Charge Sensitive Preamplifier,"* IEEE Transactions on Nuclear Science, *June, 1966.)*

A third application of current sampling is illustrated by the constant-current source shown in Fig. 10–40.* The output current, I_o, flows through a temperature sensitive resistor called a *sensistor*. If I_o is independent of temperature, then V_o will vary directly with the value of the sensistor. The result is a voltage whose

* This circuit is part of an electronic digital temperature converter discussed in Duncan and Blalock in the references listed at the end of the chapter.

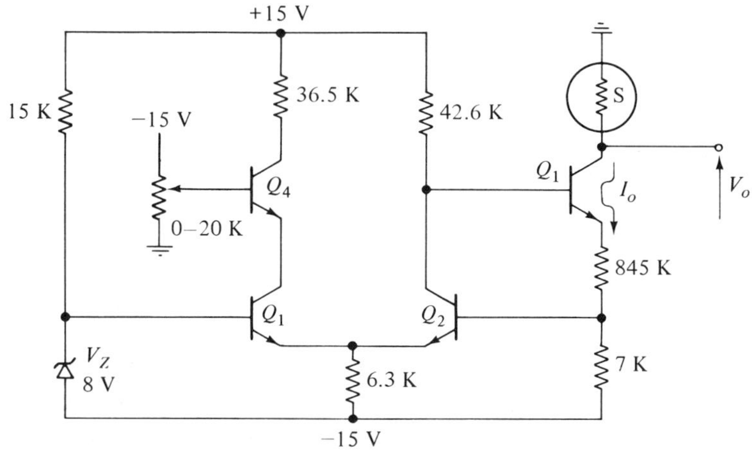

FIGURE 10–40 *A constant-current source that is stabilized by negative feedback.*

value is directly related to temperature and, consequently, is an electronic thermometer. The amplifier shown in Fig. 10–40 is a current sample-voltage sum type with the input voltage being the voltage drop across the zener diode.** By using a zener diode with temperature coefficient of approximately zero, I_o can be made nearly independent of temperature changes. Also, any changes in I_o due to changes in A_{ol} resulting from temperature changes are reduced by the factor $(1 + |T|)$. The reported value of T for the circuit shown in Fig. 10–40 was approximately −400, assuming that each transistor had a β of approximately 150.

In this section, we have illustrated the mid-band analysis of a feedback amplifier using current sampling. In addition three applications of this type of feedback amplifier were discussed. Although not as widely used as voltage sampling, current sampling is found in applications where high output impedances and stable output currents are required.

10.14 Summary

In this chapter we have examined the operation, characteristics, and analysis of negative-feedback amplifiers in the mid-band frequency range. The first three sections served to introduce the topics of negative feedback. The experimental results studied in Secs. 10.2 and 10.3 indicate that negative feedback can stabilize the gain of the amplifier and drastically alter its input and output impedances.

The second major topic studied in this chapter was an approximate analysis of feedback amplifiers. The type of sampling (voltage or current) coupled with the type of summing (voltage or current) established the approximate characteristics of four types of feedback amplifiers. Their characteristics are summarized in Table 10–6. The block diagram and gain expressions for the four types

** The 8-V zener diode is actually a series combination of a 6.8-V zener with a +3 mV/°C temperature coefficient and two silicon diodes with temperature coefficients of −1.7 mV/°C. This combination produces an 8-V zener diode with a temperature coefficient of −0.4 mV/°C. The zener was chosen to have a slightly negative temperature coefficient to correct for thermal drift in later portions of the system.

of feedback network configurations are summarized in Table 10–7. Once the type of gain that is stabilized by feedback is known, any other gain can be found using the ideal amplifier analysis from Sec. 10.4.

The third major topic studied in this chapter was the detailed analysis of feedback amplifiers. This analysis was primarily concerned with the calculation of the loop gain and its effect on circuit operation. The *disturbance signal* method is one simple way to calculate T. Blackman's theorem is one simple way of calculating the input and output impedances of a feedback amplifier. In general, the effect of feedback was to increase or decrease the original impedance by a factor of $(1 + |T|)$. Also, signal transfer from the input through the feedback network to the output is generally reduced by the factor $(1 + |T|)$. A nodal analysis includes the passive signal path as part of the circuit.

The fourth major topic included in this chapter was the complete analysis of the mid-band response of feedback amplifiers. The examples in Secs. 10.11 through 10.13 illustrated the complete analysis and served as a review of the first part of the chapter. These three sections also introduced the topics of offset voltage, active or constant-current source loads, and some applications using current sampling.

The frequency response, transient response, and stability analysis of a feedback amplifier requires some knowledge of the mid-band response of that amplifier. Consequently, the material in this chapter will find extensive application in the next chapter. Also, most high quality audio power amplifiers use negative feedback to improve their performance. Thus the material in this chapter will also be useful in Chap. 12.

Problems

Section 10.1

10–1 Verify the values given in Table 10–1 for the change in R_L.

10–2 In Fig. 10–1(a), the 4.4-K emitter resistor of Q_1 is replaced by a 3.9-K and 500-ohm series combination. The 50-μF capacitor is used to bypass only the 3.9-K resistor. For this modified circuit, determine the percentage change in $|A_{vo}|$ when β_1 and β_2 are doubled from their values in Ex. 4–17. (The unbypassed 500-ohm resistor represents one type of negative feedback.)

Section 10.2

10–3 Move resistor R_{C1} in Fig. 10–2 from the collector circuit of Q_1 to the collector circuit of Q_2. Then connect the collector terminal of Q_1 to $+V_{CC}$ and the base terminal of Q_3 to the collector terminal of Q_2. Will the modified amplifier work as a negative-feedback amplifier? What is the value of V_o/V_g?

10–4 Calculate the bias values for the circuit shown in Fig. 10–2 and compare them with the measured values.

10–5 The value of R_{f2} in Fig. 10–2 is reduced from 2 K to 1 K. Estimate the value of V_o/V_g.

10–6 The value of R_{f1} in Fig. 10–2 is increased from 100 ohms to 1 K. Estimate the value of V_o/V_g.

Section 10.3

10–7 In the ac equivalent circuit shown in Fig. 10–5(b), the CE stage Q_2 is replaced by a CC stage. Assuming that the circuit can be properly biased, will the modified amplifier work as a negative-feedback amplifier? What is the value of I_o/I_g?

10–8 The value of R_{f2} in Fig. 10–5 is reduced from 26.1 K to 15 K. Estimate the values of I_o/I_g and V_o/V_g.

10–9 The value of R_{f1} in Fig. 10–5 is reduced from 3.48 K to 2.7 K. Estimate the values of I_o/I_g and V_o/V_g.

Section 10.4

10–10 Calculate the following gains for the ideal amplifier system shown in Fig. 10–41: (a) V_o/V_g, (b) I_o/V_g, (c) V_o/I_g, and (d) I_o/I_g.

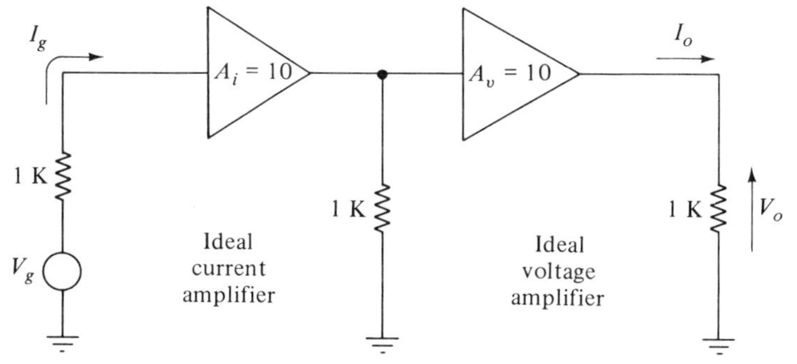

FIGURE 10–41 *Circuit diagram for Prob. 10–10.*

10–11 The current sample-current sum amplifier shown in Fig. 10–5 can be approximated as an ideal current amplifier. However, the experimental values given in Table 10–3 indicate that R_o is not infinite but approximately 6.2 K. Modify the ideal current amplifier block diagram shown in Table 10–5 to include the effect of this finite output resistance.

10–12 Often ideal amplifiers are used to obtain fixed but non-zero (and non-infinite) values of R_i and R_o. (a) Show how an ideal voltage amplifier can be used to obtain R_i equal to 100 ohms and R_o equal to 100 ohms. (b) Repeat (a) using an ideal current amplifier. (c) Give one application requiring fixed, non-zero values of R_i and R_o.

Section 10.5

10–13 In the derivation of Eq. (10–25) what two basic assumptions are made about the signal flow in the feedback amplifier?

10–14 List the type of ideal amplifier that corresponds to each type of feedback configuration.

10–15 Give one reason why active devices are seldom used in the feedback network.

10–16 A certain feedback amplifier has a value of B equal to 0.1. What value of loop gain would be required if the closed loop gain is to be within 1 percent of the value set by the feedback network? What is the required value of open loop gain?

10–17 Derive Eq. (10–26).

Section 10.6

10–18 For the feedback amplifier shown in Fig. 10–42, determine the feedback network configuration and the type of gain stabilized by feedback. Also, estimate the values of R_i, R_o, and A_{vo} assuming that $|T| \gg 1$.

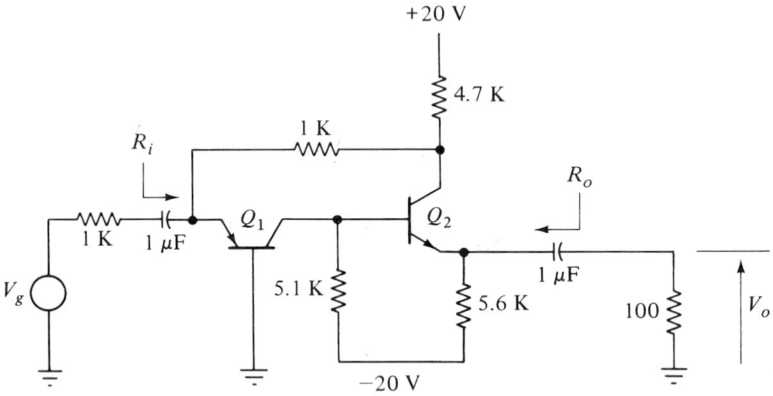

FIGURE 10–42 *Circuit for Prob. 10–18.*

10–19 Repeat Prob. 10–18 for the amplifier shown in Fig. 10–43.

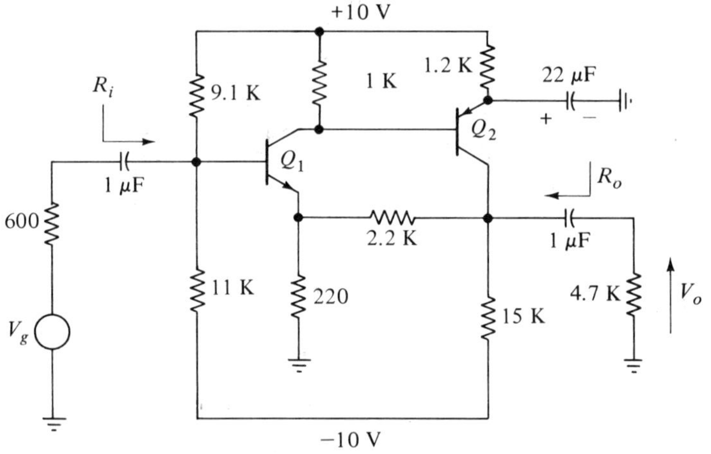

FIGURE 10–43 *Circuit for Prob. 10–19.*

10–20 Repeat Prob. 10–18 for the amplifier shown in Fig. 10–44.

Section 10.7

10–21 The feedback amplifier shown in Fig. 10–18(a) is modified by taking the output voltage from the collector terminal of Q_3 and ac bypassing the 5.6-K emitter resistor of Q_3. For the modified amplifier, determine the feedback configuration and the gain stabilized by feedback. Assuming that $|T| \gg 1$, estimate the new values of R_i, R_o, and A_{vo}.

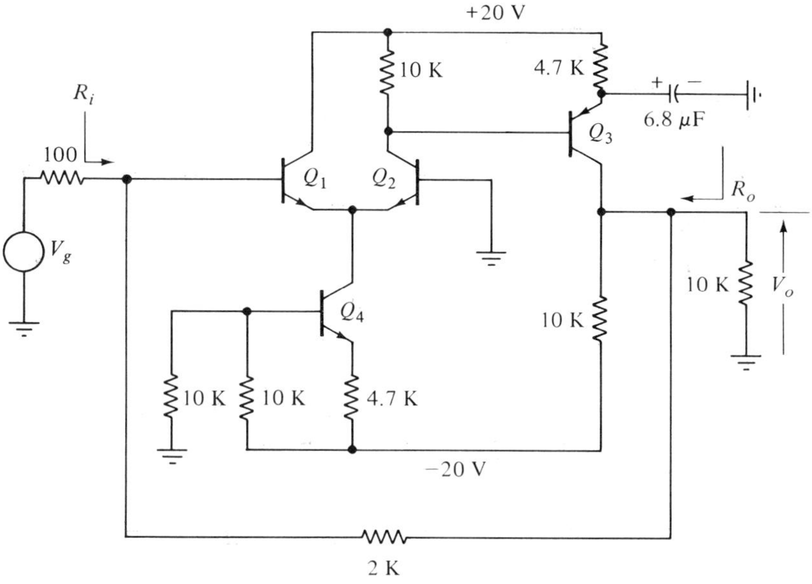

FIGURE 10–44 *Circuit for Prob. 10–20.*

10–22 The feedback amplifier shown in Fig. 10–5(a) is modified by connecting C_4 and R_L to the collector terminal of Q_3 and ac bypassing R_{E3}. For the modified amplifier, determine the feedback configuration and the gain stabilized by feedback. Assuming that $|T| \gg 1$, estimate the new values of R_i, R_o, and A_{vo}.

10–23 Convert the feedback amplifier shown in Fig. 10–19(a) into a current amplifier.

10–24 Convert the feedback amplifier shown in Fig. 10–20(a) into a voltage amplifier.

10–25 Numerous examples of voltage amplifiers are illustrated in Chap. 10 including those shown in Figs. 10–2(a), 10–15(a), 10–19(a), and 10–21(a). From a study of these and other voltage amplifier circuits what is the polarity of the voltage gain V_o/V_g of this type of negative-feedback amplifier?

10–26 Numerous examples of current amplifiers are illustrated in Chap. 10 including those shown in Figs. 10–17(a), 10–20(a), and 10–22(a). From a study of these and other current amplifier circuits, what is the polarity of the voltage gain V_o/V_g of this type of negative-feedback amplifier?

10–27 Repeat Ex. 10–9 for the amplifier shown in Fig. 10–22(a).

10–28 The IC shown in Fig. 10–23(a) has an internal feedback network provided by R_{13}. Identify this internal feedback amplifier, and estimate the voltage gain from the emitter terminal of Q_9 to the output.

10–29 Make the necessary changes in the amplifier shown in Fig. 10–19(a) so that it corresponds to the block diagram shown in Fig. 10–24(d).

Section 10.8

10–30 Calculate the value of T for the amplifier shown in Fig. 10–44. Compare this value with the results of Ex. 10–10. The transistors have the following values of β: β_1 and β_2 are 170, β_3 is 90, and β_4 is 100.

10–31 Calculate the value of T for the amplifier shown in Fig. 10–43. Each transistor has a β of 100.

10–32 Calculate the value of T for the vacuum tube amplifier shown in Fig. 10–21(a). Each triode has an r_p of 5400 ohms and a μ of 65.

10–33 Calculate the value of T for the vacuum tube amplifier shown in Fig. 10–22(a). Use the values for μ and r_p given in Prob. 10–32.

10–34 Calculate the value of T for the amplifier shown in Fig. 10–18(a). Each transistor has a β of 150.

10–35 Calculate the value of T for the amplifier shown in Fig. 10–17. Each transistor has a β of 200.

Section 10.9

10–36 Calculate the values of R_{if} and R_{of} for the amplifier shown in Fig. 10–44. The transistors have the following values of β: β_1 and β_2 are 170, β_3 is 90, and β_4 is 100.

10–37 Calculate the values of R_{if} and R_{of} for the amplifier shown in Fig. 10–43. Each transistor has a β of 100.

10–38 Using Fig. 10–30, discuss why R_{if} depends on the value of R_L.

Section 10.10

10–39 Calculate the component of the output voltage of the amplifier shown in Fig. 10–44 that is due to passive signal transfer. What percentage of the total output voltage is due to passive signal transfer? Use β values from Prob. 10–30.

10–40 Using Fig. 10–32, derive the expression for the voltage gain of a voltage sample-voltage sum feedback amplifier including the effect of passive signal transfer through the feedback network.

10–41 Repeat Prob. 10–39 for the amplifier shown in Fig. 10–17. β values are 200.

Section 10.11

10–42 Repeat Ex. 10–13 for the amplifier shown in Fig. 10–45. Assume that each transistor has a β of 100 and an r_{ce} of 50 K. The dc voltage at the output terminal is adjusted to 0 V.

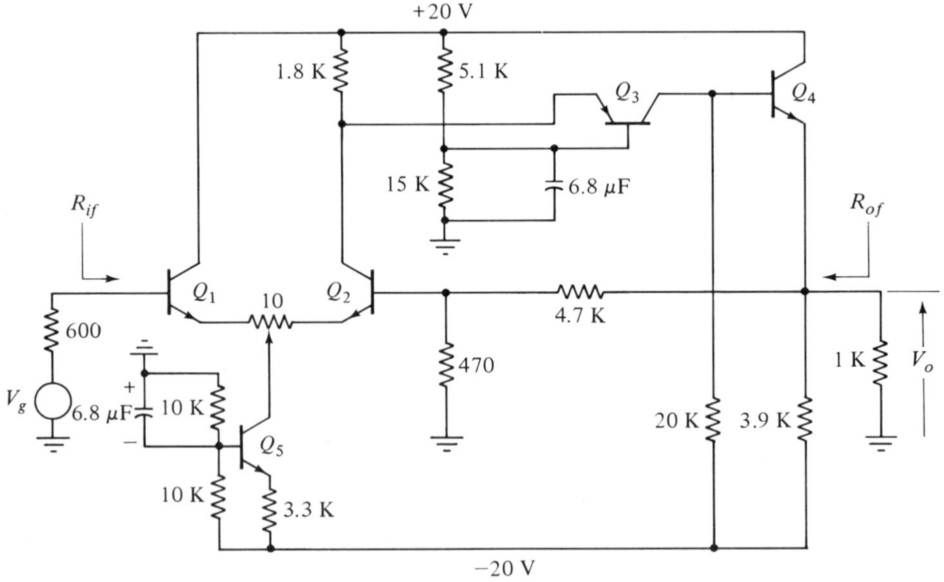

FIGURE 10–45 *Circuit for Prob. 10–42.*

10–43 Modify the circuit shown in Fig. 10–45 to allow the generator circuit to be ac coupled to the input. For the modified amplifier, the value of R_{if} should exceed 20 K, the input offset voltage should be less than 10 mV, and the output offset voltage should be adjustable to 0 V.

10–44 Repeat Ex. 10–13 for the vacuum tube amplifier shown in Fig. 10–21(a). Each triode has an r_p of 5400 ohms and a μ of 65.

10–45 Convert the non-feedback amplifier shown in Fig. 8–16(e) into a voltage sample-voltage sum feedback amplifier with a gain, V_o/V_g, of 20.

10–46 Design a voltage sample-voltage sum feedback amplifier with the following characteristics: V_o/V_g equal to 15, $|T|$ greater than 500, R_{if} greater than 100 K, and R_{of} less than 1 ohm. The generator circuit will be directly connected to the amplifier, and R_g will be in the range of 10 ohms to 1 K. Also, R_L will be directly connected to the amplifier and will be in the range of 100 ohms to 10 K. Use only standard components and a ±24-V power supply.

Section 10.12

10–47 Repeat Ex. 10–14 for the amplifier shown in Fig. 10–46. Assume that each transistor has a β of 150 and an r_{ce} of 30 K. The dc voltage at the output terminal is adjusted to 0 V.

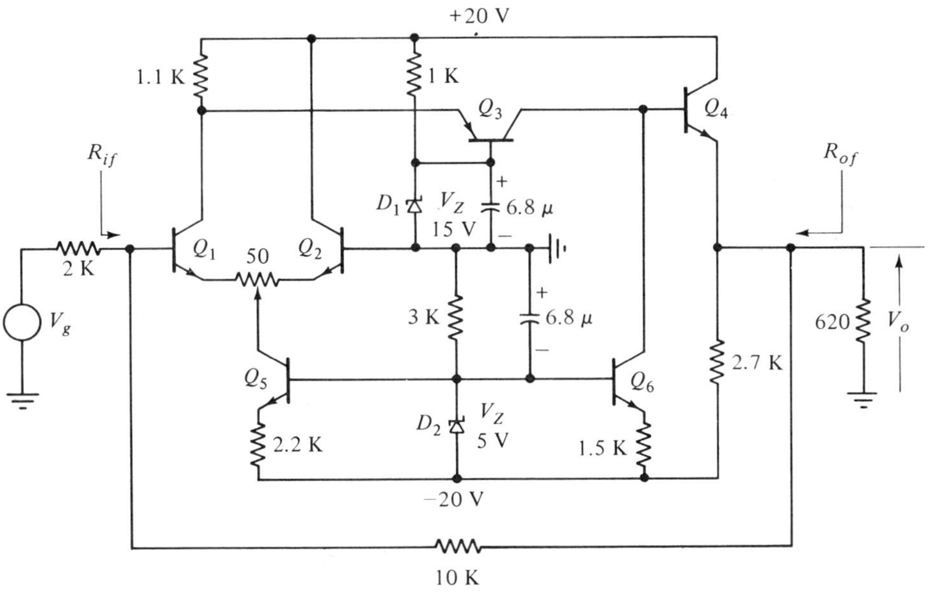

FIGURE 10–46 *Circuit for Prob. 10–47.*

10–48 Repeat Ex. 10–14 for the amplifier shown in Fig. 10–36(b) with β_3 equal to 200.

10–49 Convert the non-feedback amplifier shown in Fig. 8–15(e) into a voltage sample-current sum feedback amplifier with a gain V_o/V_g equal to −2. Terminate each coaxial cable at each end in its characteristic impedance of 100 ohms.

10–50 Repeat Ex. 10–14 for the amplifier shown in Fig. 10–17(a). Assume that each transistor has a β of 100.

10–51 Repeat Ex. 10–14 for the vacuum tube amplifier shown in Fig. 10–22(a). Each triode has an r_p of 5400 ohms and a μ of 65.

10–52 Design a voltage sample-current sum feedback amplifier with the following characteristics: V_o/V_g equal to -10, $|T|$ greater than 10^3, R_{if} less than 1 ohm, and R_{of} less than 1 ohm. The generator circuit will be directly coupled to the amplifier, and R_g is 600 ohms. Also, R_L will be directly connected to the amplifier and will be in the range of 1 K to 10 K. Use only standard components and a $\pm$15-V power supply.

Section 10.13

10–53 Repeat Ex. 10–15 for the circuit shown in Fig. 10–42. Assume that each transistor has a β of 100.

10–54 Repeat Ex. 10–15 for the circuit shown in Fig. 10–5(a). Assume that each transistor has a β of 100.

10–55 In the circuit shown in Fig. 10–38(a), the value of R_f is increased from 215 ohms to 510 ohms, and R_E is decreased from 4.22 K to 3.3 K. Estimate the new values of the ac loop gain, dc loop gain, and the voltage gain, V_o/V_g.

10–56 Using a single stage current sample-current sum feedback amplifier of the type shown in Fig. 10–39(c), modify the cascade circuit shown in Fig. 8–4 so that the voltage gain is -15. Assume that the parameters of the added transistor are equal in magnitude to those of the bipolar transistor in the original circuit. Component data are given in Ex. 8–4.

References

Black, H. S., "Wave Translation System," U. S. Patent No. 2,102,671, Dec. 21, 1937.

Blackman, R. B., "Effect of Feedback on Impedance," *The Bell System Technical Journal,* 22, No. 3 (October, 1943), 269–77.

Blalock, T. V., "Wide-Band Low-Noise Charge Sensitive Preamplifier," *IEEE Transactions on Nuclear Science* (June, 1966), 457–67.

Bode, H. W., *Network Analysis and Feedback Amplifier Design.* New York: D. Van Nostrand Company, Inc., 1945.

Cherry, E. M. and D. E. Hooper, *Amplifying Devices and Low-Pass Amplifier Design.* New York: John Wiley & Sons, Inc., 1968.

Duncan, W. R. and T. V. Blalock, "The Development of an Electronic Digital Temperature Converter for the Measurement of Body Temperature," Thesis, University of Tennessee, June, 1970.

Kuo, B. C., *Automatic Control Systems,* 2nd Edition. Englewood Cliffs, N. J.: Prentice-Hall, Inc., 1967.

Savant, C. J., Jr., *Basic Feedback Control System Design.* New York: McGraw-Hill Book Company, 1958.

11

Response, Stability, and Compensation of Negative-Feedback Amplifiers

The performance of an amplifier is not completely defined until its frequency response or its time response is determined. A determination of the frequency or time response generally starts with the mid-band response. Thus the material in Chap. 10 serves as a base for the frequency and time response analysis of feedback amplifiers. The material in Chaps. 7 and 8 will also be used extensively in this study. Starting with this foundation, the objectives of this chapter are to develop methods for finding the frequency and transient responses of feedback amplifiers and methods for improving these responses.

The sections in this chapter can be divided into five general groups. Secs. 11.1 through 11.3 introduce methods for finding the response and stability of closed-loop feedback amplifiers. An approximate but very useful method of analysis is introduced in Secs. 11.4 and 11.5 for finding both the frequency response and the transient response. Techniques for improving the response of feedback amplifiers by the addition of compensation networks are developed in Secs. 11.6 through 11.8. The performance characteristics of an important class of amplifiers called *operational amplifiers* are included in Secs. 11.9 and 11.10. The design of feedback amplifiers is discussed in Sec. 11.11.

The general emphasis in this chapter is on small-signal, very high-frequency or wide-band amplifiers. While an analysis of the complete response of audio power, or large-signal, amplifiers is considered in the next chapter, most of the methods developed in this chapter can also be applied to high-quality audio power amplifiers.

11.1 Frequency Response of Feedback Amplifiers with Dominant High and Low Corner Frequencies

In practice, many amplifiers have essentially one dominant high and one dominant low corner frequency. The objective of this section is to determine the frequency response of this type of amplifier when the feedback loop is closed around the circuit. The closed-loop gain of a feedback amplifier is

$$A_f = \frac{1}{B}\left(\frac{T}{1+T}\right) \tag{11-1}$$

where A_f is the gain with feedback, B is the gain of the feedback path, and T is the loop gain. Both B and T can be functions of frequency. However, in this section, we will examine only circuits in which the B network is purely resistive.* Consequently, the frequency response of A_f is due to the frequency dependence of T. Our immediate objective is to calculate the frequency response characteristics of T.

The value of A_{vo} calculated in Chaps. 3, 4, and 6 corresponded to the mid-band value of gain, A_{mid}, in the frequency response analysis. The frequency characteristics of the gain were given as a product of the mid-band gain and the poles and zeros of the gain. For example, from Chap. 7 the transfer function of an amplifier with a dominant high corner frequency, f_{hi}, and a dominant low corner frequency, f_{lo}, is

$$A = A_{\text{mid}}\left(\frac{jf/f_{\text{lo}}}{1 + jf/f_{\text{lo}}}\right)\left(\frac{1}{1 + jf/f_{\text{hi}}}\right) \tag{11-2}$$

A similar situation exists for the loop gain. The frequency response of T can be expressed as the product of T_{mid} and the poles and zeros of T.

A calculation of the frequency dependence of T is analogous to the calculation of the frequency dependence of the gain of the non-feedback amplifier. In Chap. 10 the value of T_{mid} was calculated by assuming that the feedback loop was open or that no feedback effects were present. The same approach is used for the calculation of the frequency dependence of T as illustrated in the following example.

EXAMPLE 11–1 Find the frequency response of the loop gain of the feedback amplifier shown in Fig. 11–1(a). The transistor parameters are:

Transistor	β	$f_{T(\text{min})}$	$C_{ob(\text{max})}$
Q_1	110	400 MHz	6 pF
Q_2	58	600 MHz	1.7 pF
Q_3	76	400 MHz	8 pF

Solution: (a) Mid-band value of T: The mid-band analysis of the amplifier was given in Ex. 10–14. The following results are useful in the frequency response analysis:

$$r_{e2} \simeq 26\ \Omega \qquad R_{Leq1} \simeq 970\ \Omega$$

$$A_{vt2} \simeq -257 \qquad T_{\text{mid}} \simeq -25$$

(b) High-frequency response: The amplifier has four principal nodes with each node being associated with a high corner frequency. However, the high corner frequencies associated with the input circuit of Q_1 and the output circuit of Q_3 are very

---

* The use of reactive components in the B network is an important type of compensation. The discussion of this type of feedback starts with Sec. 11.7.

high since they occur at emitter terminals and can be neglected. Also, the high corner frequency at the output or collector terminal of Q_2 can be neglected due to the interaction of the time constants for the CE stage. The remaining high corner frequency is associated with the input of Q_2 and dominates the high-frequency response of T. This frequency is calculated as follows:

$$C_{b'e2} = 1/(2\pi r_{e2} f_{T2}) \simeq 10 \text{ pF} \qquad C_{\text{out}1} \simeq C_{ob1} \simeq 6 \text{ pF}$$
$$C_{\text{in}2} = C_{b'e2} + C_{b'c2}(1 + A_{vt2}) \simeq [10 + 1.7(1 + 257)] \text{ pF} \simeq 450 \text{ pF}$$
$$\tau_{\text{in}2} = R_{Leq1}(C_{\text{out}1} + C_{\text{in}2}) \simeq (970)(6 \text{ pF} + 450 \text{ pF}) \simeq 440 \text{ nsec}$$
$$f_{\text{hi}} = 1/(2\pi\tau_{\text{in}2}) = 0.159/(440 \times 10^{-9}) \simeq 360 \text{ kHz}$$

(c) Low-frequency response: The dominant low-frequency response is associated with C_1 and is

$$\tau_1 = \left[R_{E2} \,||\, \left(r_{e2} + \frac{R_{C1}}{\beta_2 + 1}\right)\right] C_1 \simeq 5.7 \text{ msec}$$
$$f_{\text{lo}} = 1/(2\pi\tau_1) = 0.159/(5.7 \times 10^{-3}) \simeq 28 \text{ Hz}$$

(d) Loop gain expression: Combining the mid-band gain, high-frequency response, and low-frequency response, the loop gain can be expressed as

$$T \simeq (-25)\left(\frac{jf/28 \text{ Hz}}{1 + jf/28 \text{ Hz}}\right)\left(\frac{1}{1 + jf/360 \text{ kHz}}\right)$$

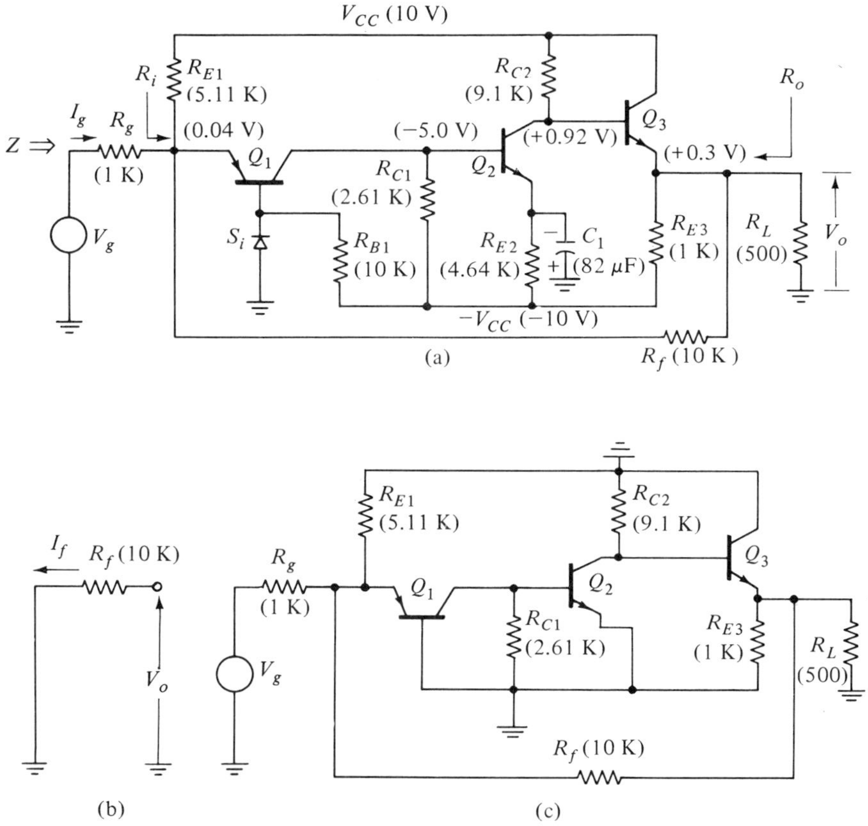

FIGURE 11–1 *The amplifier from Ex. 11–1: (a) complete circuit, (b) feedback or* B *network, and (c) mid-band equivalent circuit.*

Our next task is to develop an expression for the frequency response of the closed-loop gain, A_f. The loop gain calculated in Ex. 11–1 had a dominant high and a dominant low corner frequency. For this condition, the loop gain can be expressed in terms of s, τ_{lo} and τ_{hi} as*

$$T = T_{mid}\left(\frac{s\tau_{lo}}{1 + s\tau_{lo}}\right)\left(\frac{1}{1 + s\tau_{hi}}\right) \tag{11–3}$$

The frequency response of A_f can be found by substituting the expression for T [Eq. (11–3)] into the expression for A_f [Eq. (11–1)].

$$A_f = \frac{1}{B} \times \frac{\dfrac{|T_{mid}|s\tau_{lo}}{(1 + s\tau_{lo})(1 + s\tau_{hi})}}{1 + \dfrac{|T_{mid}|s\tau_{lo}}{(1 + s\tau_{lo})(1 + s\tau_{hi})}} \tag{11–4}$$

After some algebraic manipulation, Eq. (11–4) can be written as

$$A_f = (A_{f(mid)}) \frac{s\tau_{lo}}{\dfrac{s^2\tau_{lo}\tau_{hi}}{1 + |T_{mid}|} + s\left(\tau_{lo} + \dfrac{\tau_{hi}}{1 + |T_{mid}|}\right) + \dfrac{1}{1 + |T_{mid}|}} \tag{11–5}$$

where

$$A_{f(mid)} = \frac{|T_{mid}|}{1 + |T_{mid}|}\frac{1}{B} \tag{11–6}$$

Equation (11–5) contains the frequency response information for the closed-loop gain. However, the results are more apparent when the denominator is factored. This polynomial can be factored using the approximate method discussed in Sec. 8.4. The larger time constant, τ_1, is associated with the low-frequency response and can be approximated as the ratio of the coefficients of the s^1 term and the s^0 term as

$$\tau_1 \simeq \frac{\tau_{lo} + \tau_{hi}/(1 + |T_{mid}|)}{1/(1 + |T_{mid}|)} \simeq \tau_{lo}(1 + |T_{mid}|) \tag{11–7}$$

where the second approximation is made because τ_{lo} is much larger than τ_{hi}.

The smaller time constant, τ_2, is associated with the high-frequency response and can be approximated as the ratio of the coefficients of the s^2 terms and the s^1 term as

$$\tau_2 \simeq \frac{\tau_{lo}\tau_{hi}/(1 + |T_{mid}|)}{\tau_{lo} + \tau_{hi}/(1 + |T_{mid}|)} \simeq \frac{\tau_{hi}}{1 + |T_{mid}|} \tag{11–8}$$

Rewriting Eq. (11–5) in terms of the approximate roots, τ_1 and τ_2, gives

$$A_f \simeq A_{f(mid)} \frac{s\tau_{lo}}{[1 + s\tau_{lo}(1 + |T_{mid}|)][1 + s\tau_{hi}/(1 + |T_{mid}|)]} \tag{11–9}$$

Replacing s with $j\omega$ and τ with $1/(2\pi f)$ in Eq. (11–9), the expression A_f becomes

$$A_f \simeq A_{f(mid)}\left(\frac{jf/f_{lo(f)}}{1 + jf/f_{lo(f)}}\right)\left(\frac{1}{1 + jf/f_{hi(f)}}\right) \tag{11–10}$$

where

$$f_{lo(f)} = f_{lo}/(1 + |T_{mid}|) \tag{11–11}$$

$$f_{hi(f)} = f_{hi}\,(1 + |T_{mid}|) \tag{11–12}$$

* In this study, $s = j\omega$ and $\tau = 1/(2\pi f)$.

From Eq. (11–11), the low corner frequency with feedback is a factor of $(1 + |T_{\text{mid}}|)$ lower than the low corner frequency of T. Also, the high corner frequency with feedback is a factor of $(1 + |T_{\text{mid}}|)$ higher than the high corner frequency of T. The price paid for this improvement in frequency response is a reduction in the mid-frequency gain by the factor $(1 + |T_{\text{mid}}|)$.

The closed-loop gain, A_f, open-loop gain, A_{ol}, and loop gain, T, are plotted in Fig. 11–2 for a particular set of values and illustrate the effect of feedback on the frequency response of an amplifier with dominant time constants. In Fig. 11–2, note that the closed-loop corner frequencies are approximately equal to the frequencies where the value of $|T|$ is 0 db or unity. This result occurs because the slope of $|T|$ for frequencies less than f_{lo} and also for frequencies greater than f_{hi} has a magnitude of 6 db/octave. For this slope, the product $(|T_{\text{mid}}|)f_{\text{hi}}$ is approximately equal to $f_{\text{hi}(f)}$, and the product $(|T_{\text{mid}}|)f_{\text{lo}(f)}$ is approximately equal to f_{lo}.* The following example illustrates the calculation of the frequency response of a feedback amplifier which has dominant time constants.

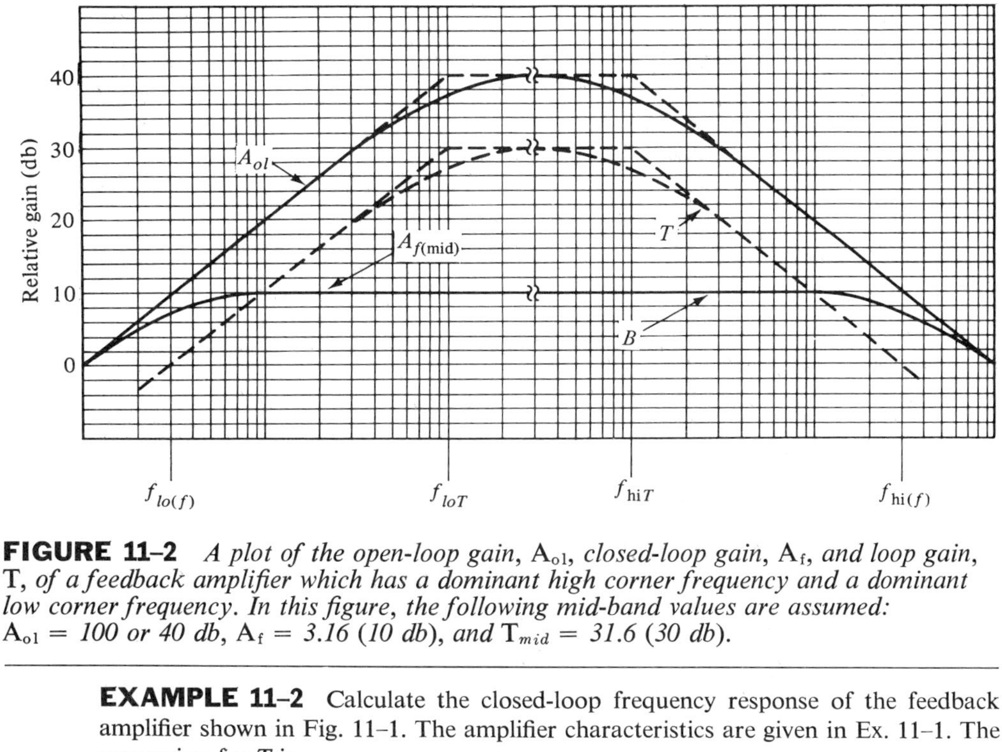

FIGURE 11–2 *A plot of the open-loop gain,* A_{ol}*, closed-loop gain,* A_f*, and loop gain,* T*, of a feedback amplifier which has a dominant high corner frequency and a dominant low corner frequency. In this figure, the following mid-band values are assumed:* A_{ol} = *100 or 40 db,* A_f = *3.16 (10 db), and* T_{mid} = *31.6 (30 db).*

EXAMPLE 11–2 Calculate the closed-loop frequency response of the feedback amplifier shown in Fig. 11–1. The amplifier characteristics are given in Ex. 11–1. The expression for T is

$$T \simeq -25\left(\frac{jf/28\text{ Hz}}{1 + jf/28\text{ Hz}}\right)\left(\frac{1}{1 + jf/360\text{ kHz}}\right)$$

Solution: (a) Closed-loop frequency response with feedback: From Eqs. (11–11) and (11–12),

$$f_{\text{lo}(f)} = f_{\text{lo}}/(1 + |T_{\text{mid}}|) \simeq 28/(1 + 25) \simeq 1.1\text{ Hz}$$
$$f_{\text{hi}(f)} = f_{\text{hi}}(1 + |T_{\text{mid}}|) = (360\text{ kHz})(1 + 25) \simeq 9.4\text{ MHz}$$

* This situation is analogous to the f_β, β_o, and f_T relationships in Chap. 7.

The corresponding rise time with feedback is

$$\tau_{r(f)} = 0.35/f_{\text{hi}(f)} = 0.35/9.4 \times 10^6 \simeq 37 \text{ nsec}$$

(b) Experimental results: The measured low corner frequency was approximately 1.5 Hz. The measured rise time was approximately 10 nsec for an output signal of 0.1 V, and approximately 30 nsec for an output signal of 1.0 V. Although undesirable, the rise time of a fast amplifier is often a function of the output signal level. The lower value for rise time, 10 nsec, is a more accurate value for the small-signal case and indicates that the value of C_{ob} for the CE stage is closer to 0.5 pF than 1.7 pF. Calculations based on the minimum value of f_T and the maximum value of C_{ob} often give conservative results.

In this section, we have studied the frequency response of the loop gain and the closed-loop gain of a feedback amplifier having a dominant high corner frequency and a dominant low corner frequency. In each case, the closed-loop frequency response was extended by the factor $(1 + |T_{\text{mid}}|)$. For a high corner frequency to be dominant, the next high corner frequency of T should occur at least a decade above the closed-loop high corner frequency. Stated differently, if T has two high corner frequencies, f_{hi1} and f_{hi2}, then f_{hi2} should be greater than $10f_{\text{hi}(f)}$ or, from Eq. (11–12), f_{hi2} should be greater than $10(1 + |T_{\text{mid}}|)f_{\text{hi1}}$. Low corner frequencies should also be separated by the factor $10(1 + |T_{\text{mid}}|)$ before the feedback amplifier can be classified as having a dominant low corner frequency. The meaning of dominant corner frequencies is discussed further in the next section. Although Eqs. (11–11) and (11–12) are strictly valid only for cases of single dominant corner frequencies, they are often used as first order approximations for more complex amplifiers.

11.2 Closed-Loop Frequency Response of Feedback Amplifiers with Multiple High Corner Frequencies

The analysis discussed in the last section was based on the assumption that the feedback amplifier has a dominant high and a dominant low corner frequency. Many feedback amplifiers do not meet these specifications, especially at the high-frequency end, because the high corner frequencies of T are generally determined by the parameters of the active devices. On the other hand, the low-frequency response can be extended to dc by eliminating all circuit coupling and bypass capacitors. The objective of this section is to determine the closed-loop, high-frequency response of a feedback amplifier having multiple high corner frequencies. Two methods of analysis are studied: (1) an analytical method valid for a second-order system, and (2) a graphical method valid for any order system.

The loop gain of an amplifier having two important high corner frequencies can be written as

$$T = T_{\text{mid}}\left(\frac{1}{1 + jf/f_{\text{hi1}}}\right)\left(\frac{1}{1 + jf/f_{\text{hi2}}}\right) \tag{11–13}$$

The high-frequency response of the closed-loop gain, A_f, can be found by substituting Eq. (11–13) into Eq. (11–1) yielding

$$A_f = \left(\frac{1}{B}\right)\left(\frac{T_{\text{mid}}/[(1 + jf/f_{\text{hi1}})(1 + jf/f_{\text{hi2}})]}{1 + T_{\text{mid}}/[1 + jf/f_{\text{hi1}})(1 + jf/f_{\text{hi2}})]}\right) \tag{11–14}$$

After some algebraic manipulation, Eq. (11–14) can be rewritten as

$$A_f = \frac{A_{f(\text{mid})}}{(jf)^2/[(f_{\text{hi1}}f_{\text{hi2}})(1 + T_{\text{mid}})] + (jf)(1/f_{\text{hi1}} + 1/f_{\text{hi2}})/(1 + T_{\text{mid}}) + 1} \tag{11–15}$$

where $A_{f(\text{mid})}$ is the mid-frequency gain with feedback as given by Eq. (11–6).

The denominator of this expression for A_f is a quadratic whose roots can be either real or complex. Consequently, the approximate method of polynomial factoring cannot be applied. In terms of the general quadratic form, Eq. (11–15) becomes

$$A_f = \frac{A_{f(\text{mid})}}{[(jf)^2/f_n^2] + [(jf)2\xi/f_n] + 1} \tag{11–16}$$

where ξ is the damping ratio, and f_n is the undamped natural frequency. By comparing the coefficients of like powers of (jf) in Eqs. (11–15) and (11–16),

$$f_n = \sqrt{f_{\text{hi1}}f_{\text{hi2}}(1 + |T_{\text{mid}}|)} \tag{11–17}$$

and

$$\xi = (f_{\text{hi1}} + f_{\text{hi2}})/(2f_n) \tag{11–18}$$

To determine the high-frequency response of a feedback amplifier with two important corner frequencies, the following values must be calculated: T_{mid}, f_{hi1}, and f_{hi2}. Using these values, f_n and ξ can be determined from Eqs. (11–17) and (11–18). The magnitude and phase of the quadratic term are plotted in Appendix B as a function of ξ and f/f_n and can be used to plot the closed-loop high-frequency response. The following example illustrates this type of analysis.

EXAMPLE 11–3 Calculate the closed-loop, high-frequency response of the feedback amplifier shown in Fig. 11–3. The transistor parameters are:

Transistor	β	$f_{T(\text{min})}$	$C_{ob(\text{max})}$
Q_1, Q_2 (dual)	220	100 MHz	6 pF
Q_3	170	200 MHz	7 pF
Q_4	58	250 MHz	8 pF
Q_5	90	300 MHz	5 pF

Solution: (a) Mid-band value of T: The mid-band response of the amplifier was found in Ex. 10–13. The following results are useful in the high-frequency response analysis:

$$r_{e1} \simeq r_{e2} \simeq 26\ \Omega \qquad r_{e3} \simeq 13\ \Omega \qquad r_{e4} \simeq 5\ \Omega$$

$$R_{Leq1} = R_{C1} \,||\, R_{\text{in}3} \simeq 10\text{ K} \,||\, 2.22\text{ K} \simeq 1.81\text{ K} \qquad R_{\text{in}2} \simeq 31.5\text{ K}$$

$$A_{vt3} \simeq -580 \qquad T_{\text{mid}} \simeq -735 \qquad A_{fv} \simeq 10$$

(b) High-frequency response of T: The circuit has six principal nodes and six corresponding high corner frequencies that must be considered when finding the frequency response of T.* Of the six corner frequencies, the pole at the base of Q_3 should occur at the lowest frequency because of the Miller effect in Q_3. The second lowest corner frequency should occur at the base of Q_2. The corner frequencies at the emitters of Q_1, Q_2, and Q_4 should occur at very high frequencies and will be neglected. The corner frequency at the collector of Q_3 will also be neglected because of the time constant interaction. The high corner frequency at the base of Q_3 is calculated as follows:

---

* The high corner frequency at the base of Q_1 is essentially outside the feedback loop.

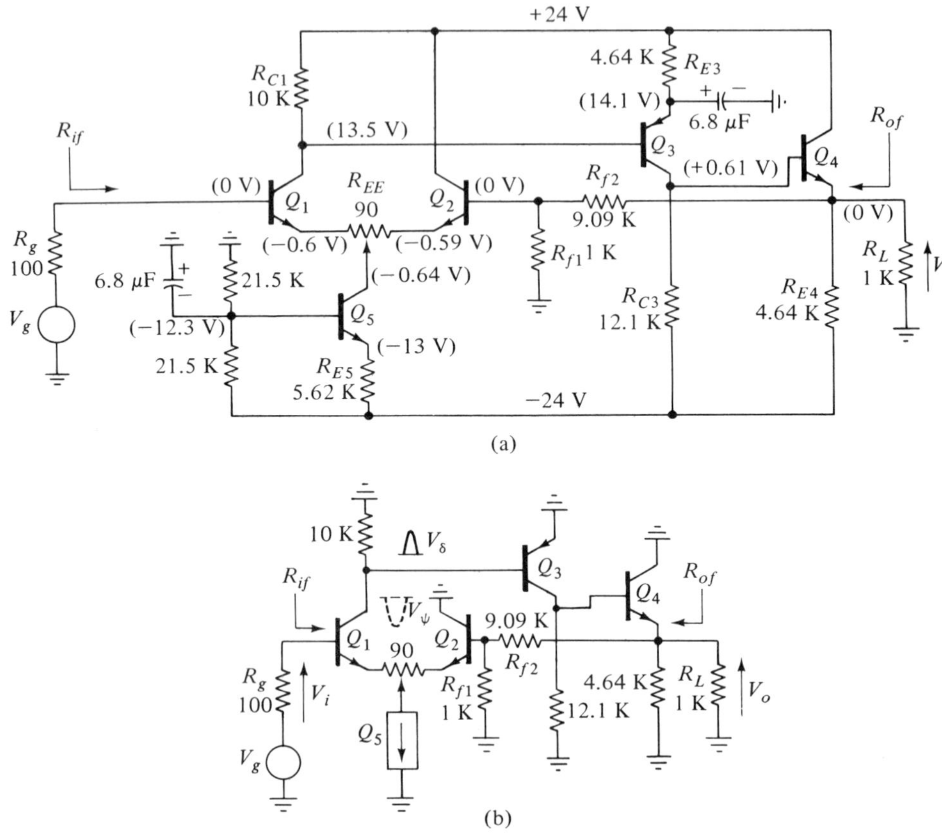

FIGURE 11–3 *The amplifier from Ex. 11–3: (a) the circuit diagram and (b) an ac equivalent circuit. [This figure is the same as Figs. 10–33(a) and (b).]*

$$R_{L\text{eq}1} \simeq 1.81\text{ K} \qquad C_{\text{out}1} \simeq C_{ob1} \simeq 6\text{ pF}$$
$$C_{b'e3} = 1/(2\pi r_{e3} f_{T3}) \simeq 0.159/[(13)(200 \times 10^6)] \simeq 61\text{ pF}$$
$$C_{\text{in}3} = C_{b'e3} + C_{b'c3}(1 - A_{vt3}) \simeq [61 + 7(1 + 580)]\text{ pF} \simeq 4130\text{ pF}$$
$$\tau_{\text{hi}1} = R_{L\text{eq}1}(C_{\text{out}1} + C_{\text{in}3}) \simeq (1.81\text{ K})(4130\text{ pF}) \simeq 7.5\ \mu\text{sec}$$
$$f_{\text{hi}1} = 1/(\pi \tau 2_{\text{hi}1}) \simeq 0.159/(7.5 \times 10^{-6}) \simeq 21\text{ kHz}$$

The high corner frequency at the base of Q_2 is calculated as follows:

$$R_2 = R_{\text{in}2} \,||\, R_{f1} \,||\, \left\{ R_{f2} + R_{E4} \,||\, R_L \,||\, \left(r_{e4} + \frac{R_{C3}}{\beta_4 + 1} \right) \right\} \simeq 900\ \Omega$$
$$C_{b'g2} \simeq 1/[2\pi(r_{e1} + r_{e2} + R_{EE}) f_{T2}] \simeq 0.159/[(142)(100 \times 10^6)] \simeq 11.2\text{ pF}$$
$$C_{\text{in}2} = C_{b'g2} + C_{b'c2} = 11.2\text{ pF} + 6\text{ pF} \simeq 17.2\text{ pF}$$
$$\tau_{\text{hi}2} = R_2 C_{\text{in}2} \simeq (900)(17.2\text{ pF}) \simeq 15.5\text{ nsec}$$
$$f_{\text{hi}2} = 1/(2\pi\tau_{\text{hi}2}) = 0.159/(15.5 \times 10^{-9}) \simeq 10\text{ MHz}$$

(c) Closed-loop high-frequency response: From Eqs. (11–17) and (11–18),

$$f_n = \sqrt{f_{\text{hi}1} f_{\text{hi}2}(1 + |T_{\text{mid}}|)} = \sqrt{(21 \times 10^3)(10 \times 10^6)(736)} \simeq 12.5\text{ MHz}$$
$$\xi = (f_{\text{hi}1} + f_{\text{hi}2})/(2 f_n) \simeq (21 \times 10^3 + 10 \times 10^6)/(2 \times 12.5 \times 10^6) \simeq 0.4$$

The particular curve for ξ equal to 0.4 is repeated in Fig. 11–4(a). The closed-loop, mid-band gain in db is 20 log (10) or 20 db. The closed-loop, high-frequency plot will have approximately the same shape as the curve for which ξ is equal to 0.4. The normalized point $f/f_n = 1$ occurs at

$$f = f_n \simeq 12.5 \text{ MHz}$$

The closed-loop, high-frequency response is plotted in Fig. 11–4(b). From Fig. 11–4(b), the closed-loop, high corner frequency, which occurs when the gain is 3 db below its mid-band value of 20 db, is approximately

$$f_{\text{hi}(f)} \simeq 18 \text{ MHz}$$

(d) Experimental results: The measured high-frequency data points are circled in Fig. 11–4(b). Note that the experimental curve has a slightly higher peak than the calculated curve. The difference occurs because the effects of the high-frequency poles were neglected. The measured value of the closed-loop, high corner frequency was approximately 18.5 MHz. This value compares favorably with the calculated value of 18 MHz. The peak value of gain was 17.3 (24.6 db) and occurred at 10 MHz. The 4.6-db peak in gain above the mid-band value of 20 db is associated with a ξ of approximately 0.35. The measured value of T_{mid} obtained in Ex. 10–13 was -750. The measured value of the lowest high corner frequency of T was approximately 20 kHz. The close agreement between the calculated and experimental results in this example is due in part to the use of measured transistor parameters including all values of β, C_{ob1}, C_{ob2}, C_{ob3}, and r_{ce3}. Other parameter values were obtained from data sheets.

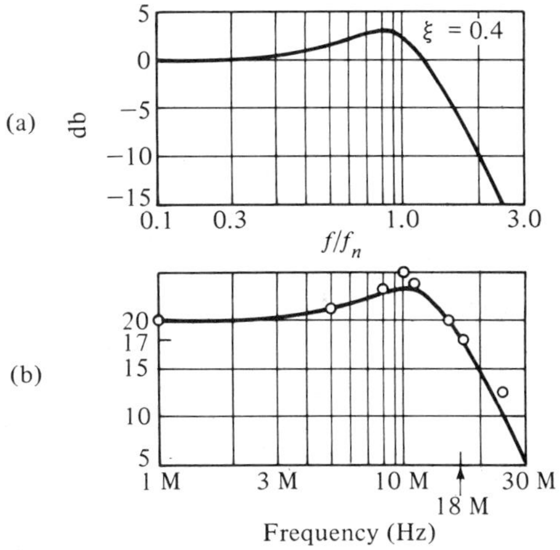

FIGURE 11–4 *The high-frequency response of the amplifier from Ex. 11–3: (a) the normalized $\xi = 0.4$ curve, and (b) the high-frequency Bode plot. Measured data points are circled in (b).*

The basic difficulty encountered in determining the closed-loop response of a feedback amplifier is in converting the expression for T into $T/(1 + T)$. For systems higher than second order, the resulting numerator and denominator polynomials of $T/(1 + T)$ are often of third or higher order and cannot be factored easily. Then the method illustrated in Ex. 11–3 cannot be applied. However, the Nichols chart, shown in Fig. 11–5, can be used to sketch a Bode plot of $T/(1 + T)$ using data from the Bode plot of T. Before discussing this procedure, we will briefly examine the structure of the Nichols chart.

A Nichols chart* is shown in Fig. 11–5 and has two sets of coordinates that are superimposed: (1) a rectangular or linear coordinate system, and (2) a

* The Nichols chart is discussed and derived in many texts on control theory as in the text by D'Azzo and Houpis listed at the end of the chapter.

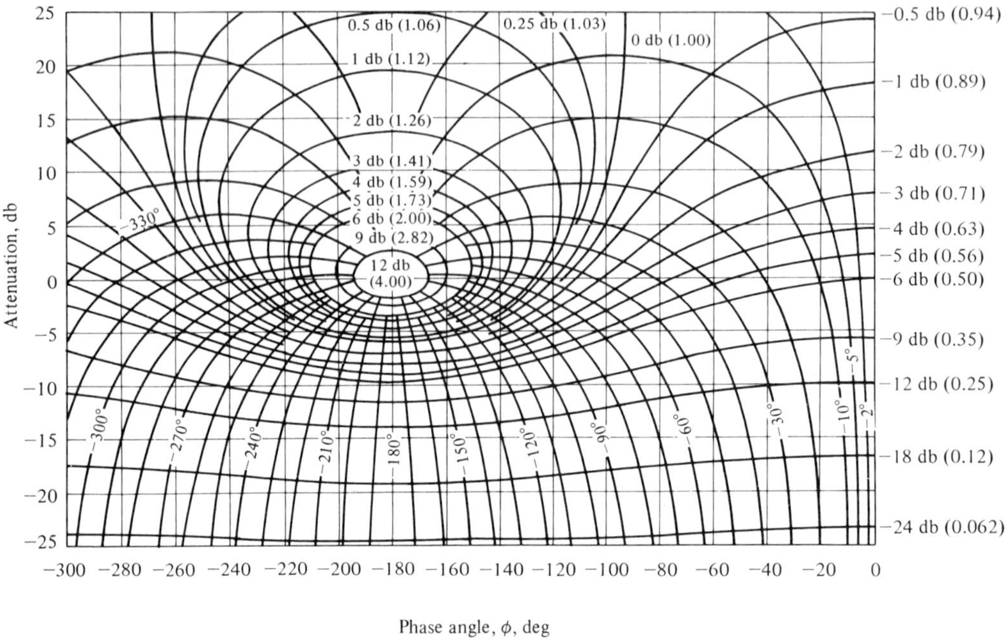

FIGURE 11–5 *A Nichols chart. (From* Feedback Control System Analysis and Synthesis, *2nd ed., by John J. D'Azzo and C. H. Houpis. Copyright 1966 by McGraw-Hill Book Company. Used with permission of McGraw-Hill Book Company.)*

curved coordinate system. The linear coordinates have db values on the vertical axis and degree values on the horizontal axis. The curved coordinates also have values in db and degrees. We will use the Nichols chart only as a conversion chart whereby the values of $|T|$ and $\underline{/T}$ are converted into the corresponding values of $|T/(1 + T)|$ and $\underline{/T/(1 + T)}$.

The graphical solution of the closed-loop frequency response of a feedback amplifier using a Nichols chart consists of the following steps:

(1) Construct the Bode plot of $|T|$ and $\underline{/T}$.

(2) Using the data from (1), construct the plot of $|T|$ versus $\underline{/T}$ on the linear coordinates of the Nichols chart with frequency as a parameter.

(3) Read the values of $|T/(1 + T)|$ and $\underline{/T/(1 + T)}$ from the curved coordinates of the Nichols chart.

(4) Using the data from (3), construct the Bode plot of $|T/(1 + T)|$ and $\underline{/T/(1 + T)}$.

(5) Combine the Bode plots of $T/(1 + T)$ and $(1/B)$ to form the Bode plot of A_f.

This graphical method is illustrated in the following example.

EXAMPLE 11–4 Repeat Ex. 11–3 using the Nichols chart. The expression for T in Ex. 11–3 was

$$T \simeq -735\left(\frac{1}{1 + jf/21 \text{ kHz}}\right)\left(\frac{1}{1 + jf/10 \text{ MHz}}\right)$$

and the value of $(1/B)$ was 10.

Solution: (a) Bode plot of T: The straight-line approximation* of this Bode plot is shown in Fig. 11–6(a) where $|T_{\text{mid}}|$ is approximately 58 db.

(b) Nichols plot: Construct a plot of $|T|$ versus $\underline{/T}$ using the linear coordinates on the Nichols chart as shown in Fig. 11–6(b). Data for this plot are obtained from the Bode plot of T. For example, at f_4 or 8.1 MHz, $|T| \simeq 5$ db and $\underline{/T} \simeq -131°$.

(c) Values of $T/(1+T)$: Values for $|T/(1+T)|$ and $\underline{/T/(1+T)}$ can be obtained from the curved coordinates of the Nichols chart. For example, at f_4 or 8.1 MHz, $|T/(1+T)| \simeq 2.2$ db and $\underline{/T/(1+T)} \simeq -33°$. The data obtained from the Nichols chart are tabulated in Table 11–1.

TABLE 11–1 *Summary of the data from the Bode plot and the Nichols chart from Fig. 11–6 and Ex. 11–4.*

	Frequency	$\|T\|$	$\underline{/T}$	$\|T/(1+T)\|$	$\underline{/T/(1+T)}$
f_1	0.9 MHz	25 db	−90°	0 db	−3°
f_2	2.5 MHz	15 db	−109°	0.3 db	−10°
f_3	4.7 MHz	10 db	−121°	1 db	−18°
f_4	8.1 MHz	5 db	−131°	2.2 db	−33°
f_5	12 MHz	0 db	−139°	3 db	−68°
f_6	16 MHz	−5 db	−144°	−1 db	−110°
f_7	21 MHz	−10 db	−149°	−7.5 db	−134°
f_8	37 MHz	−20 db	−161°	−20 db	−158°

(d) Bode plot of $T/(1 + \text{T})$: Construct the Bode plot of $T/(1+T)$. Data for this plot are obtained from the Nichols chart and are tabulated in Table 11–1. The Bode plot of $|T/(1+T)|$ is shown in Fig. 11–6(c).

(e) Bode plot of A_f: Construct the Bode plot of A_f where

$$A_f = \left(\frac{T}{1+T}\right)\left(\frac{1}{B}\right)$$

The Bode plot of $T/(1+T)$ is combined with the Bode plot of $\left(\frac{1}{B}\right)$ where

$$\left(\frac{1}{B}\right) = 20 \log |10| = 20 \text{ db}$$

The complete Bode plot of $|A_f|$ is shown in Fig. 11–6(d).

(e) Experimental results: The measured high-frequency response for this amplifier was given in Fig. 11–4(b) and is repeated in Fig. 11–6(d). The measured data points are circled. The calculated results of this example agree favorably with both the calculated results of Ex. 11–3 and the experimental results.

In this section we have studied two methods for determining the closed-loop frequency response of a feedback amplifier with multiple corner frequencies. The first or analytical method is valid only for second-order systems. The second or graphical method is based on the Nichols chart which can be used to convert the Bode plot of T into the Bode plot of $T/(1+T)$. Although an approximate method for determining the closed-loop gain characteristics directly from the

* The straight-line approximation of a Bode plot is described in Appendix B.

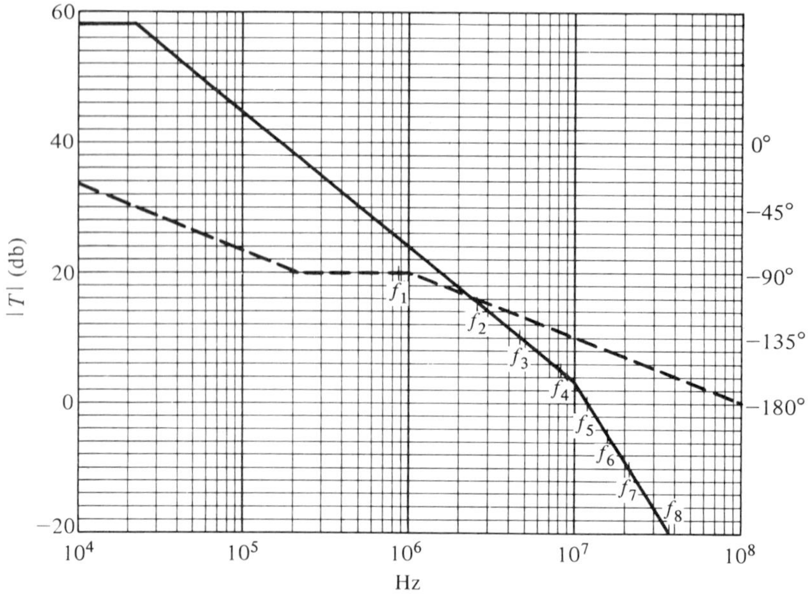

(a)

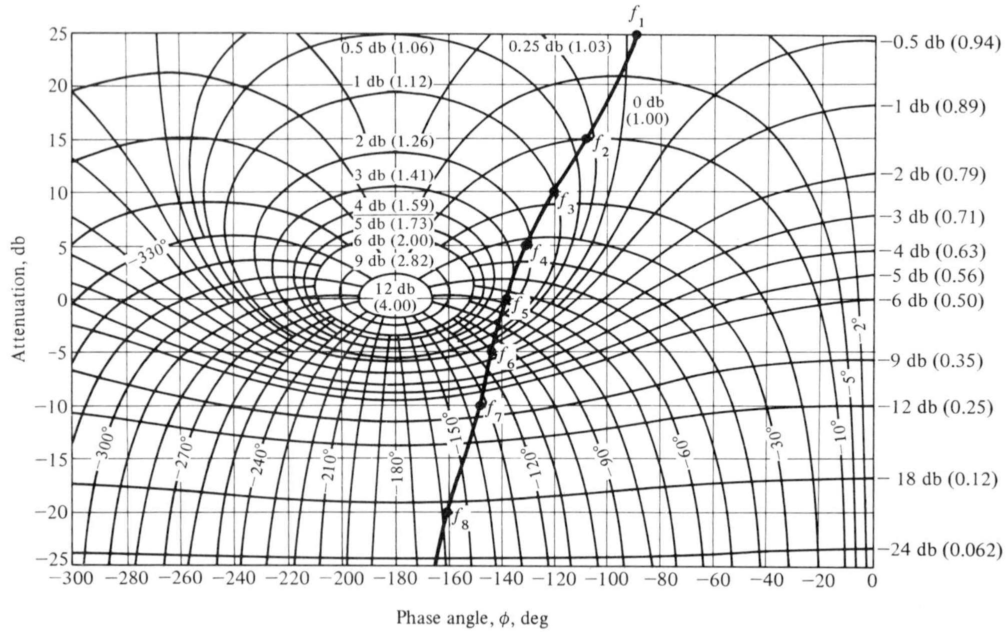

(b)

FIGURE 11–6 *The graphical construction from Ex. 11–4:* (*a*) *Bode plot of* T, (*b*) *the Nichols chart construction,* (*c*) *the Bode plot of* $|T/(1 + T)|$, *and* (*d*) *the Bode plot of* $|A_f|$. *Experimental data points are circled in* (*d*).

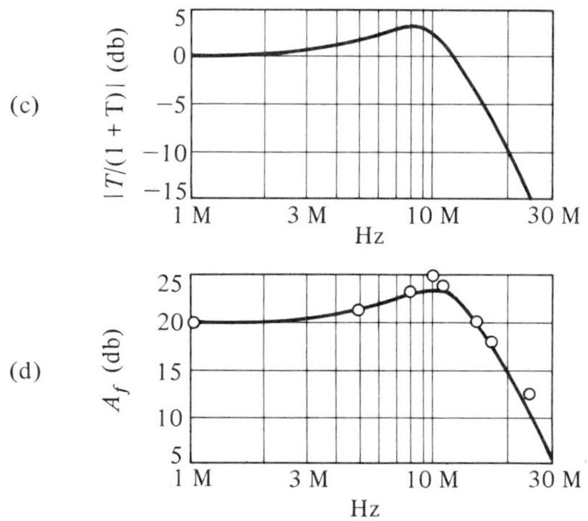

FIGURE 11–6 (cont.)

loop-gain characteristics is developed in Sec. 11.4, the graphical method studied in this section will be the basic technique for finding the closed-loop response of a complex feedback amplifier.

11.3 Stability of Feedback Amplifiers

Negative feedback exists when the signal fed back from the output combines or sums with the input signal in such a way as to reduce the output signal. This condition is satisfied at mid-band if the loop gain is negative. From the previous section, we know that T changes in both magnitude and phase as a function of frequency. If the phase shift of T exceeds 180° relative to the mid-band phase shift, the feedback becomes positive. Then the signal fed back from the output has an in-phase component that adds to the input signal, resulting in an increase in the output signal. An increase in the output signal increases the input signal, and the process may be self-sustaining. The result is an oscillating output signal that is limited in amplitude only by the non-linearities of the active devices. The objective of this section is to study the conditions which can cause instability.

The concept of system stability is treated extensively in texts on feedback systems and control theory. Since an electronic negative-feedback amplifier is a type of feedback system, many of the results of control theory can be applied to the analysis of a feedback amplifier. System stability is often examined using the Nyquist stability criteria. In simple terms, the Nyquist stability criterion states that a negative-feedback system is stable if the magnitude of the loop gain is less than unity when the phase shift of the loop gain reaches 180°.* This condition can be portrayed graphically in a number of ways, some of which are shown in Fig. 11–7. The Nyquist or polar plot is shown in Fig. 11–7(a). An unstable condition exists when the polar plot encloses the $-1 + j0$ point. Similar information is available from the Bode plot of T as shown in Fig. 11–7(b). Normally, the plots of $|T|$ and $\underline{/T}$ are drawn so that the 0 db line of $|T|$ and the $-180°$ line of $\underline{/T}$ correspond. Then it is a simple matter to determine if $|T|$

* We exclude the case of a marginally stable system.

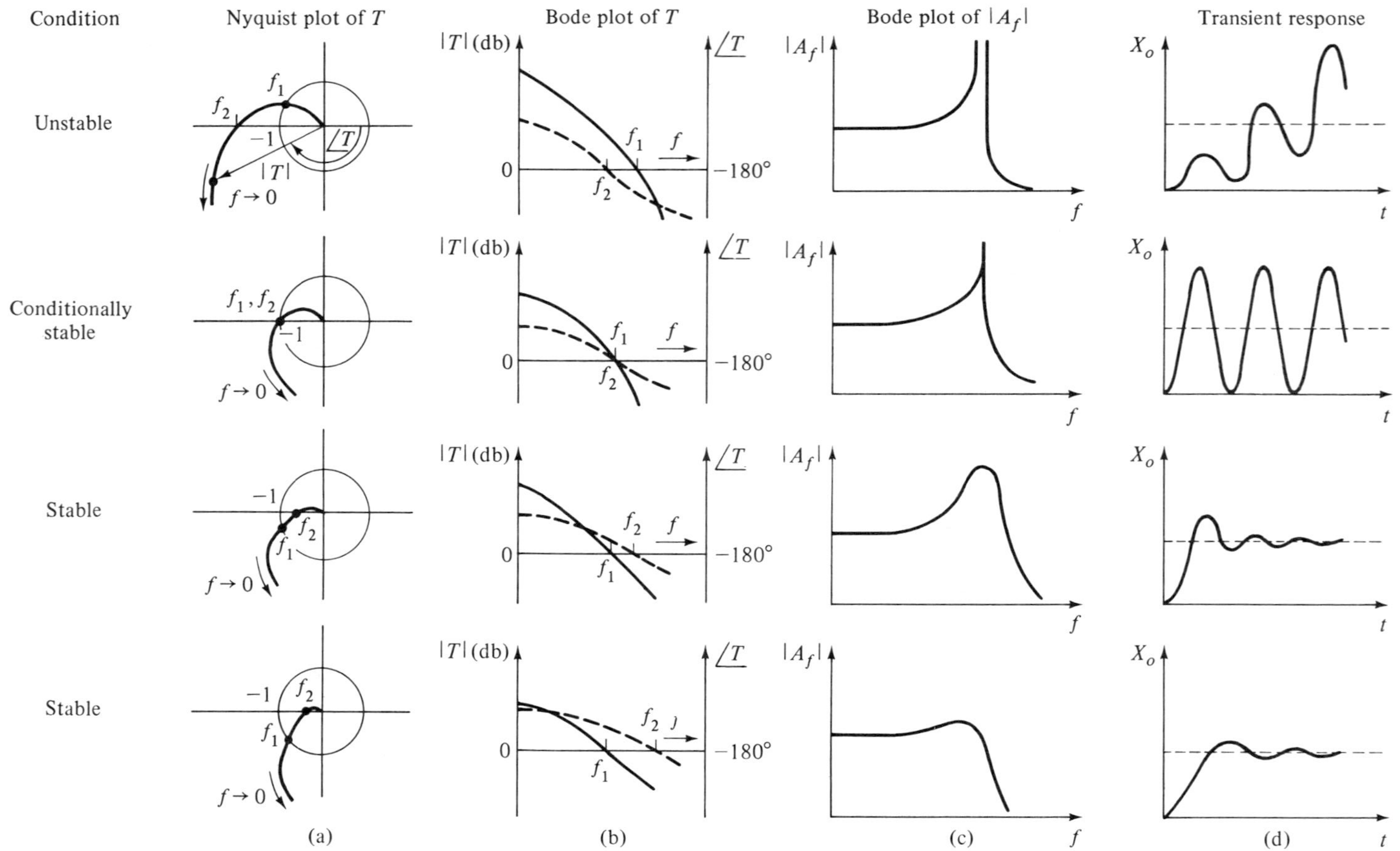

FIGURE 11–7 *A comparison between various plots used in the study of system stability: (a) Nyquist or polar plot of* T, *(b) Bode plot of* T, *(c) closed-loop Bode plot of* $|A_f|$, *and (d) the transient or time response of the output signal.*

reaches 0 db (or unity) before or after $\underline{/T}$ reaches 180° as frequency increases (or decreases) from the mid-band range.

The closed-loop, high-frequency response of $|A_f|$ is shown in Fig. 11–7(c) for various conditions of stability. The peak in the plot of $|A_f|$ is indicative of the degree of stability of the system. The transient or time response is shown in Fig. 11–7(d) for comparison and will be discussed in a later section. The following example illustrates the analysis of an amplifier to determine its stability against self oscillation.

EXAMPLE 11–5 Determine whether or not the amplifier shown in Fig. 11–8(a) is stable. The transistor parameters are:

Transistor	β	f_T	C_{ob}
Q_1	40	350 MHz	5 pF
Q_2	160	200 MHz	14 pF
Q_3	90	350 MHz	6 pF

Solution: (a) Bias values: The bias values can be determined either from the solution of simultaneous bias equations or by trial and error. Using the latter method, the bias currents are approximately

$$I_{E1} \simeq 1\text{ mA} \qquad I_{E2} \simeq 2.1\text{ mA} \qquad I_{E3} \simeq 2\text{ mA}$$

(b) Mid-band value of T:

$$A_{vt1} \simeq -\frac{R_{Leq1}}{r_{e1}} \simeq -54 \qquad A_{vt2} \simeq -\frac{R_{Leq2}}{r_{e2}} \simeq -460$$

$$A_{vt3(\text{CE})} \simeq -\frac{R_{Leq3}}{r_{e3} + R_{Eeq3}} \simeq -3.5 \qquad \frac{R_g \,||\, R_{\text{in}1}}{R_g \,||\, R_{\text{in}1} + R_{f2}} \simeq \frac{1}{62}$$

$$T_{\text{mid}} = \frac{V_\psi}{V_\delta} \simeq A_{vt2}A_{vt3(\text{CE})}\left(\frac{R_g \,||\, R_{\text{in}1}}{R_g \,||\, R_{\text{in}1} + R_{f2}}\right)A_{vt1} \simeq -1400$$

(c) High-frequency response of T: The amplifier shown in Fig. 11–8 consists of three CE stages when calculating the loop gain. Hence, the problem of interactions between high-frequency time constants must be considered.* First calculate the approximate time constant associated with each node in the circuit. These time constants are designated as τ_A through τ_D in Fig. 11–8(b). The actual value of the amplifier time constants can then be determined with the aid of Table 8–1.

$$\begin{aligned}
\tau_A &\simeq [R_g \,||\, R_{\text{in}1} \,||\, (R_{f1} + R_{f2})][C_{b'e1} + C_{b'c1}(1 - A_{vt1})] \\
&\simeq [750 \,||\, 1.06\text{ K} \,||\, (26.1\text{ K} + 3.48\text{ K})][17\text{ pF} + (5\text{ pF})(54)] \simeq 120\text{ nsec} \\
\tau_B &\simeq [R_{Leq1}][C_{b'c1} + C_{b'e2} + C_{b'c2}(1 - A_{vt2})] \simeq 9.1\ \mu\text{sec} \\
\tau_C &\simeq [R_{Leq2}][C_{b'c2} + C_{b'g3} + C_{b'c3}(1 - A_{vt3})] \simeq 233\text{ nsec} \\
\tau_D &\simeq [R_{Leq3}][C_{b'c3}] \simeq [3.1\text{ K}][6\text{ pF}] \simeq 18.6\text{ nsec}
\end{aligned}$$

The actual high-frequency time constants of the amplifier shown in Fig. 11–8(b) will be designated τ_1, τ_2, τ_3, and τ_4. The values of τ_1 through τ_4 can be determined from the relative values of τ_A through τ_D using Table 8–1. The largest time constant, τ_B, corresponds to $\tau_{\text{out}1}$ or $\tau_{\text{in}2}$. This time constant is not affected by interaction. Thus,

$$\tau_1 \simeq \tau_B \simeq 9.1\ \mu\text{sec}$$

Since $\tau_{\text{out}1}$ is much greater than $\tau_{\text{in}1}$ ($\tau_B \gg \tau_A$), the second time constant associated with Q_1 becomes

$$\tau_2 \simeq [R_g \,||\, R_{\text{in}1} \,||\, (R_{f2} + R_{f1})][C_{b'e1} + C_{b'c1}] \simeq 9.5\text{ nsec}$$

---

* Time constant interaction is discussed in Sec. 8.4.

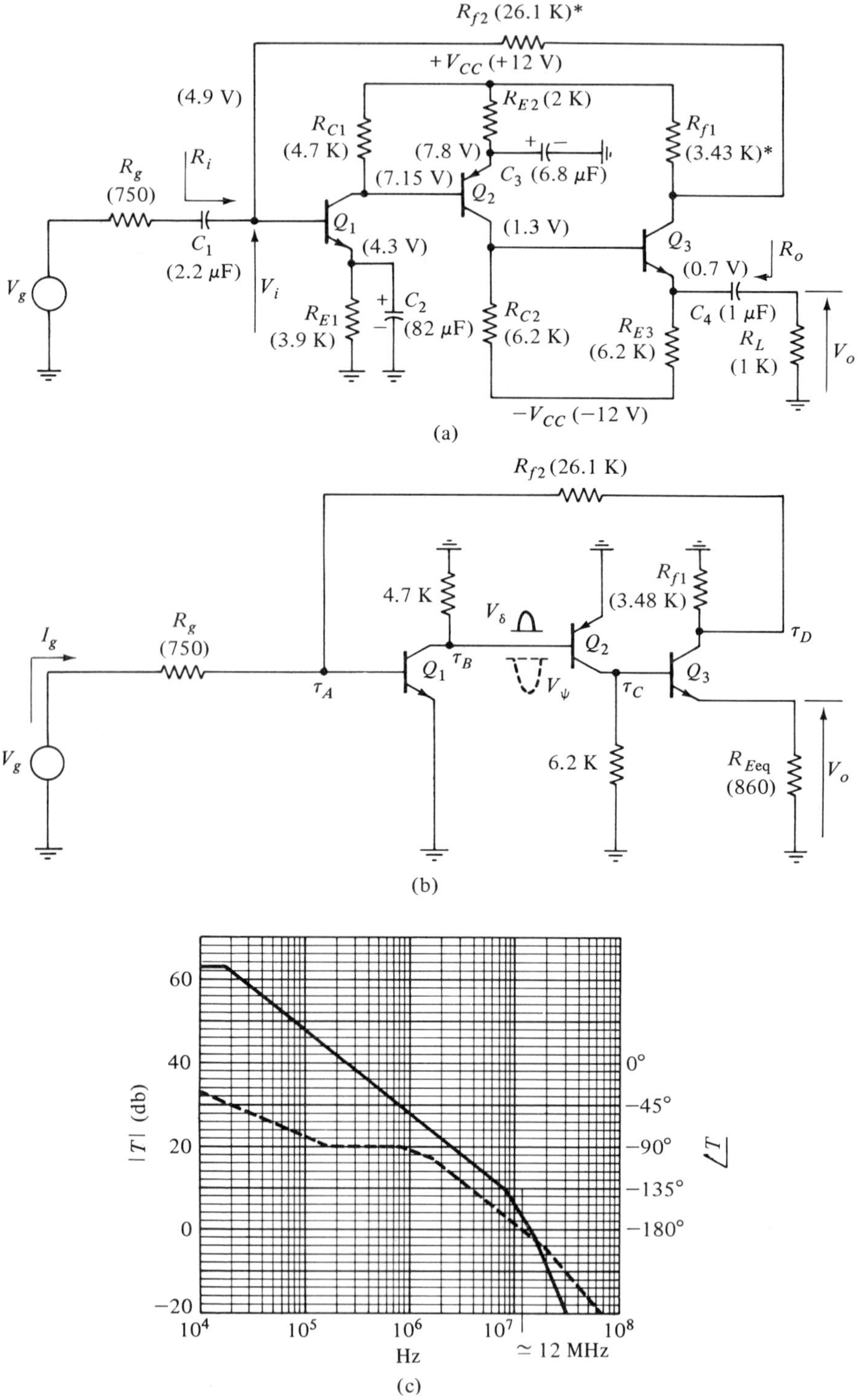

FIGURE 11–8 *The amplifier from Ex. 11–5: (a) the complete circuit, (b) an ac equivalent circuit, and (c) the Bode plot of T.*

Since τ_{in2} is much greater than τ_{out2} ($\tau_B \gg \tau_C$), the second time constant associated with Q_2 becomes

$$\tau_3 \simeq 1/(2\pi f_{T2}) \simeq 0.8 \text{ nsec}$$

The fourth time constant in the circuit is associated with Q_3. The value of τ_{in3} may appear to be equal to τ_C but is actually equal to τ_3 due to the interaction effects in Q_2. Consequently, τ_{out3} is much greater than τ_{in3} ($\tau_D \gg \tau_3$), and

$$\tau_4 \simeq \tau_D \simeq 18.6 \text{ nsec}$$

The four high corner frequencies of the amplifier are

$$f_{\text{hi1}} \simeq 1/(2\pi\tau_1) \simeq 17.5 \text{ kHz} \qquad f_{\text{hi2}} = 1/(2\pi\tau_2) \simeq 16.8 \text{ MHz}$$
$$f_{\text{hi3}} \simeq 1/(2\pi\tau_3) \simeq 200 \text{ MHz} \qquad f_{\text{hi4}} \simeq 1/(2\pi\tau_4) \simeq 8.6 \text{ MHz}$$

(d) Stability: Using the results from parts (b) and (c),

$$T \simeq -1400\left(\frac{1}{1+jf/17.5 \text{ kHz}}\right)\left(\frac{1}{1+jf/8.6 \text{ MHz}}\right)\left(\frac{1}{1+jf/16.8 \text{ MHz}}\right)\left(\frac{1}{1+jf/200 \text{ MHz}}\right)$$

The value of $|T_{\text{mid}}|$ in db is

$$|T_{\text{mid}}| = 20 \log |-1400| \simeq 63 \text{ db}$$

The Bode plot of T is shown in Fig. 11–8(c). From this plot, we see that $\underline{/T}$ reaches 180° before $|T|$ reaches 0 db. The phase crossover frequency (the frequency where $\underline{/T} = -180°$) is approximately 12 MHz. Consequently, the amplifier is unstable and will tend to oscillate at a frequency in the neighborhood of 12 MHz.

(e) Experimental results: The amplifier oscillated at approximately 14.3 MHz.

In addition to the problem of closed-loop oscillation studied in the example above, other and more subtle forms of instability can occur in an amplifier. Low-frequency oscillations, often called *motor-boating*, can result from inadequate filtering in the dc circuits. High-frequency oscillation, often called *parasitic oscillation*, can also disrupt circuit operation and even change the bias conditions.

Parasitic oscillations often occur because of coupling from one part of the circuit to another part through a common power supply lead in such a way as to provide a positive feedback path. This cause can be removed by inserting filters between the circuit and the power supply leads. A simple filter for this purpose is shown in Fig. 11–9. This circuit consists of a small series resistor, usually 10 to 50 Ω, and a bypass capacitor which can range typically from 0.01 μF to 50 μF. Since electrolytic capacitors often tend to become inductive in the

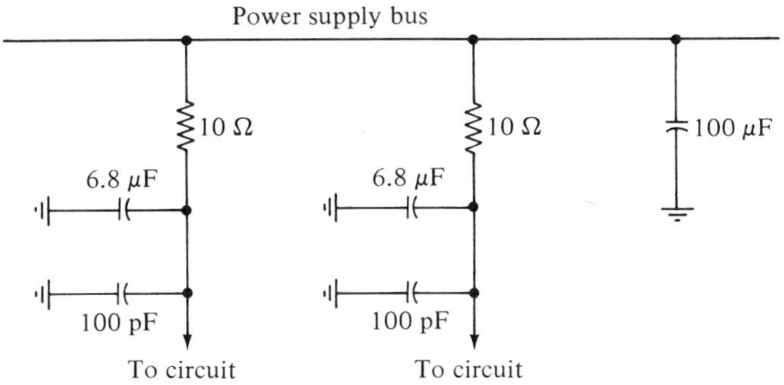

FIGURE 11–9 *A simple* RC *decoupling filter.*

frequency range above 1 MHz, a small high-frequency capacitor (typically 50 to 1000 pF) is often connected in parallel with the larger bypass capacitor. Also, complete feedthrough filters are commercially available which usually consist of a ferrite core inductor and bypass capacitors.

Another cause of parasitic oscillation is the interaction between the lead inductance in the circuit and any negative resistance present. Such oscillation often occurs at frequencies approaching f_T and is difficult to detect. The addition of a resistor having a low value (10 to 50 ohms) in series with the base lead of the associated transistor often stops the oscillation.

Amplifier stability against oscillation has been discussed in this section. Oscillation results if the $|T|$ is greater than unity (or 0 db) when the $\underline{/T}$ is $-180°$. More subtle types of oscillation can also result from inadequate power supply filtering, inadequate circuit decoupling, and stray circuit parameters. These latter types of oscillation are difficult to analyze and are usually corrected experimentally.

11.4 An Approximate Analysis of the Closed-Loop Frequency Response

In the preceding section, feedback amplifiers were divided into two classes, stable and unstable. To be useful as a linear amplifier, the first requirement is that a feedback amplifier be stable. In a linear amplifier, peaks in the frequency response curve are undesirable because signals in the frequency range where the peaks occur are amplified more than those in the mid-band range. This condition produces a form of signal distortion. Thus, a second requirement for a useful amplifier is that the frequency response curve be relatively free of peaks.

As indicated in Sec. 11.3, the criterion for stability was that the magnitude of the loop gain, $|T|$, fall to 0 db before the phase of the loop gain, $\underline{/T}$, shifts 180° relative to the mid-band phase. Peaks in the frequency response curves of feedback amplifiers occur when two or more important poles are present because the additional pole(s) produce relatively large amounts of phase shift for very little attenuation.* A peak occurs** in the frequency response curve of a feedback amplifier when the phase of T has shifted close to but less than 180° for a frequency at which $|T|$ is 0 db. If the phase shift is 170°, the peak is very high, and a phase shift of 90° produces almost no peaking. In the preceding section, two of the four conditions shown in Fig. 11–7 were stable.*** However, the amount of peaking in the frequency response curve of $|A_f|$ (and also the amount of overshoot in the transient response) was different. The amount of peaking is directly related to the $\underline{/T}$ when $|T|$ is 0 db.

In this section, we will relate the closed-loop high-frequency response of an amplifier to the frequency response of the loop gain. The objectives of developing this relationship are: (1) to be able to approximate the closed-loop response

---

* For example, a system with two poles at the same frequency has 90° of phase shift for an attenuation of only 6 db while a system with only one pole has the same phase shift with an infinite amount of attenuation.

** Peaking can occur at either end of the band. However, peaking at the high-frequency end is more prevalent and also more difficult to control. Consequently, only the high-frequency case is considered in this section.

*** Also compare the frequency response curves in Exs. 11–2 and 11–4.

without using a Nichols chart, and (2) to be able to translate the design specifications for a feedback amplifier into specifications for the loop gain. In this section only second-order systems will be considered; i.e., T will have two important high corner frequencies, f_{hi1} and f_{hi2}. Many stable systems having more than two important high corner frequencies can be approximated as a second-order system since $\underline{/T}$ is less than 180° when $|T|$ reaches 0 db for all stable systems including second-order systems.* The first step in this study is to develop several special symbols and terms for a second-order system.

Let M be the ratio of A_f at any frequency to the mid-frequency value of A_f. Assuming a second-order system, the normalized closed-loop gain, M, can be expressed using Eq. (11–16) as

$$M \equiv \frac{A_f}{A_{f(\text{mid})}} = \frac{1}{(jf)^2/f_n^2 + (jf)2\xi/f_n + 1} \qquad \textbf{(11–19)}$$

where the expressions for f_n and ξ are given by Eqs. (11–17) and (11–18). A curve of M is plotted in Fig. 11–10(a) showing the definitions of the various terms used in the following discussion. The peak value of M, M_{pf}, is related to ξ by

$$M_{\text{pf}} = 1/(2\xi\sqrt{1 - \xi^2}) \qquad \textbf{(11–20)}$$

and the frequency at which M reaches its peak, f_p, is related to f_n and ξ by**

$$f_p = f_n\sqrt{1 - 2\xi^2} \qquad \textbf{(11–21)}$$

Values for ξ and f_n can be calculated using Eqs. (11–17) and (11–18). However, their values can also be estimated from a Bode plot of T. A generalized Bode plot of T is shown in Fig. 11–10(b). An important characteristic of the Bode plot of T is the phase margin, P.M., which is defined as the difference between $\underline{/T}$ and $-180°$ when $|T|$ is 0 db, or

$$\text{P.M.} \equiv 180° - \underline{/T} \text{ when } |T| = 0 \text{ db} \qquad \textbf{(11–22)}$$

The phase margin of a second order system is related to ξ by

$$\text{P.M.} \simeq \tan^{-1}\left(\frac{4\xi^2}{(4\xi^4 + 1)^{1/2} - 2\xi^2}\right)^{1/2} \qquad \textbf{(11–23)}$$

and the frequency (f_c) where $T = 0$ db is related to f_n and ξ by***

---

* Conditionally stable systems are excluded.

** For the derivation of Eqs. (11–20) and (11–21), refer to a text on control theory such as the text by Kuo, pp. 392–393, noted in the list of references at the end of this chapter.

*** For the derivation of Eqs. (11–23) and (11–25), refer to the text by Savant, Appendix VIII, noted in the list of references at the end of this chapter. The derivation consists of: (1) approximating the plot for T as shown in Fig. 11–10(c) or

$$T \simeq \frac{f_n^2}{jf(jf + 2\xi f_n)} \qquad \textbf{(11–24)}$$

(2) setting the magnitude of Eq. (11–24) equal to 1 to determine the expression for f_c, (3) substituting the expression for f_c into Eq. (11–24) to determine the phase of T at $f = f_c$, and (4) using Eq. (11–22) to determine the expression for P.M. The approximation for T given in Eq. (11–24) is based on the assumption that f_{hi1} occurs at a low frequency; i.e., the value of T is inversely proportional to frequency for frequencies less than f_{hi2}. This assumption essentially eliminates the terms f_{hi1} and T_{mid} from the derivation. However, note that in the high-frequency region near the crossover frequency, f_c, the Bode plots of T shown in Figs. 11–10(b) and (c) are identical.

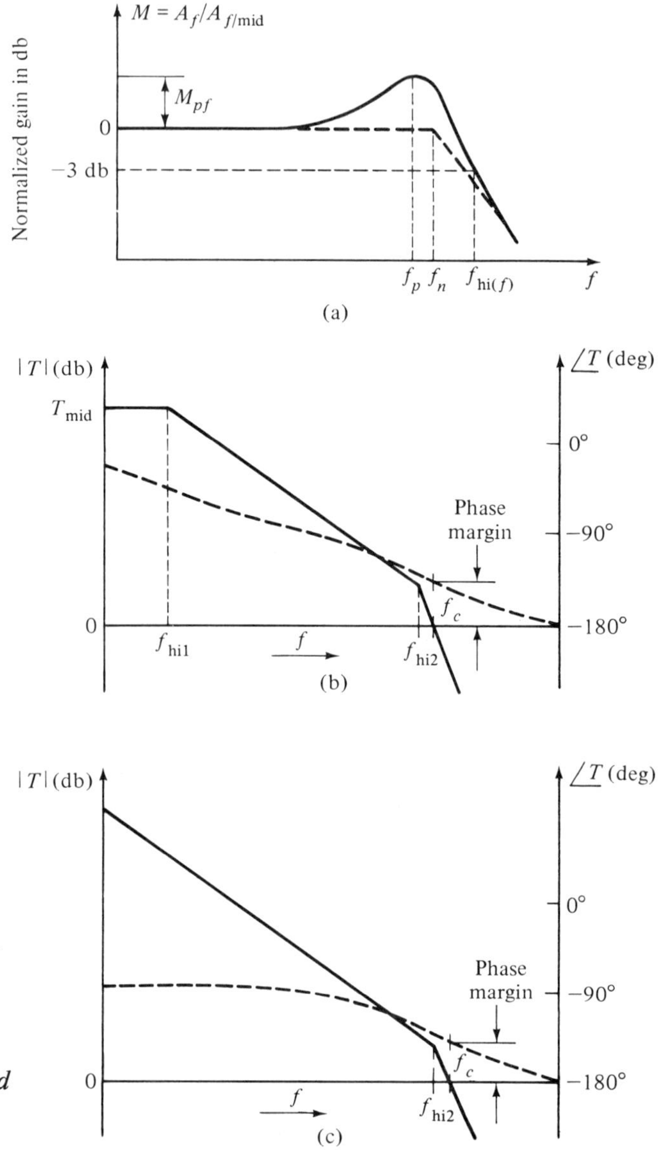

FIGURE 11–10 *Curves illustrating the various terms used in the analysis of a second-order feedback amplifier: (a) a Bode plot of the normalized closed-loop gain, (b) a Bode plot of* T*, and (c) the approximation in the plot of* T *required for the development of Eq. (11–24).*

$$f_c \simeq f_n\sqrt{(4\xi^4 + 1) - 2\xi^2} \qquad \textbf{(11–25)}$$

For the values of P.M. ranging from 0° to 60°, and values of ξ ranging from 0 to 0.6, Eq. (11–23) can be approximated as

$$\text{P.M.} \simeq 100\xi \qquad \textbf{(11–26)}$$

The closed-loop high corner frequency, $f_{\text{hi}(f)}$, is related to f_n and ξ by*

$$f_{\text{hi}(f)} \simeq f_n[1 - 2\xi^2 + \sqrt{2(2\xi^4 - 2\xi^2 + 1)}]^{1/2} \qquad \textbf{(11–27)}$$

---

* The derivation of Eq. (11–27) is similar to the derivation of Eqs. (11–21) and (11–25). The ratio $f_{\text{hi}(f)}/f_n$ is found by setting the magnitude of Eq. (11–19) equal to 0.707, which corresponds to the value of $|M|$ at $f_{\text{hi}(f)}$.

The closed-loop high-frequency response of a second-order negative-feedback amplifier is described by Eqs. (11–20), (11–21), (11–23), (11–25), and (11–27). Using these equations, the closed-loop characteristics, M_{pf}, f_p, ξ, f_n, and $f_{hi(f)}$, can be related to the loop gain characteristics, f_c and P.M. These relationships are tabulated in Table 11–2, and plotted in Fig. 11–11.

TABLE 11–2 *Summary of the relationships between the closed-loop high-frequency response and the loop gain high-frequency response for a second-order system.*

ξ	P.M.	f_n/f_c	f_p/f_c	$f_{hi(f)}/f_c$	M_{pf}, db
0	0°	1.00	1.00	1.55	∞
0.1	11.4°	1.01	1.00	1.56	14.02
0.2	22.6°	1.04	1.00	1.57	8.13
0.3	33.4°	1.09	0.99	1.59	4.85
0.4	43.3°	1.17	0.97	1.61	2.68
0.5	52.0°	1.27	0.90	1.62	1.25
0.6	59°	1.40	0.74	1.60	0.36
0.7	65°	1.54	0.22	1.59	0.08
0.707	66°	1.55	0	1.55	0
0.8	70°	1.70	—	1.48	—
0.9	74°	1.88	—	1.40	—
1.0	77°	2.06	—	1.32	—

Figure 11–11 and Table 11–2 provide a convenient method for approximating the closed-loop high-frequency response of a feedback amplifier. Although this method is developed for a second-order system (an amplifier completely characterized by two high corner frequencies), it can also be applied to more complex systems. The approximation for a higher-order stable system is based on the assumption that all stable systems have less than a 180° phase shift at f_c, the crossover frequency of $|T|$. The analysis of a feedback amplifier using Fig. 11–11 or Table 11–2 consists of the following steps:

(1) Construct a Bode plot for the loop gain, T.
(2) Determine the values of f_c and P.M. from this Bode plot.
(3) From Fig. 11–11(a), estimate the value of ξ corresponding to this value of P.M.
(4) Using this value of ξ along with Figs. 11–11(b) and (c), determine the values of f_n/f_c, f_p/f_c, $f_{hi(f)}/f_c$, and M_{pf}.
(5) Using the results from (4) and the value of f_c from (2), determine the values of f_n, f_p, $f_{hi(f)}$, and M_{pf}.
(6) Using the values from (5), sketch the Bode plot of A_f.

The following example illustrates the calculation of the response of a feedback amplifier using Fig. 11–11 (or Table 11–2).

EXAMPLE 11–6 Repeat Ex. 11–3 using Fig. 11–11. The amplifier in Ex. 11–3 is shown in Fig. 11–3. The Bode plot of T for this amplifier is shown in Fig. 11–6(a). The value of $A_{f(\mathrm{mid})}$ is 20 db.

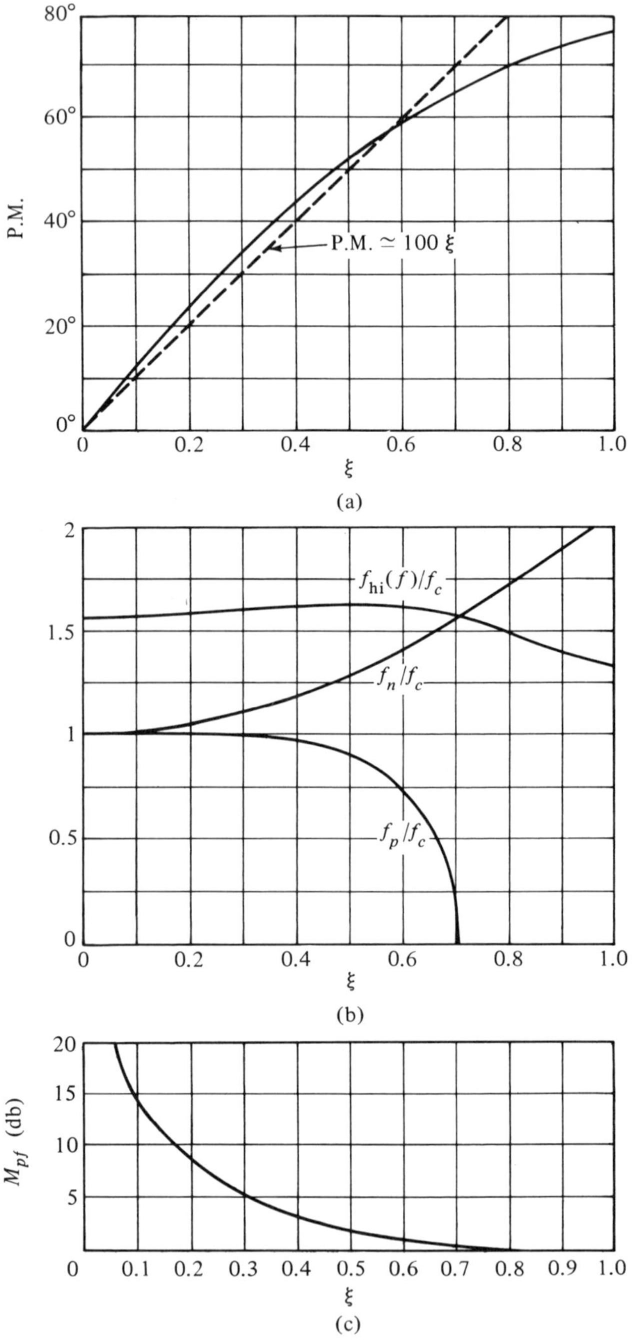

FIGURE 11–11 *Plot of the closed-loop characteristics of a second-order negative-feedback amplifier in terms of the loop gain characteristics. The interpretation of the various terms is shown in Figs. 11–10(a) and (b).*

Solution: (a) Loop gain characteristics: From Fig. 11–6(a),

$$f_c \simeq 12 \text{ MHz} \qquad \text{P.M.} \simeq 41°$$

(b) Closed-loop characteristics: From Fig. 11–11 or Table 11–2,

$$\xi \simeq 0.37 \qquad f_p/f_c \simeq 0.97$$
$$f_n/f_c \simeq 1.15 \qquad f_{\text{hi}(f)}/f_c \simeq 1.6$$
$$M_{\text{pf}} \simeq 3 \text{ db}$$

Solving for the closed-loop characteristics,

$$f_p \simeq (0.97)(12 \text{ MHz}) \simeq 11.6 \text{ MHz}$$
$$f_n \simeq (1.15)(12 \text{ MHz}) \simeq 13.8 \text{ MHz}$$
$$f_{\text{hi}(f)} \simeq (1.6)(12 \text{ MHz}) \simeq 19.2 \text{ MHz}$$

(c) Bode plot of $|A_f|$: The slope of the straight line approximation of the Bode plot of $|A_f|$ is zero for frequencies below f_n, and -40 db/decade for frequencies above f_n or 13.8 MHz. The mid-band value of $|A_f|$ is 20 db, and the peak value is 3 db greater or 23 db. The peak value occurs at the frequency f_p or 11.6 MHz. The value of $|A_f|$ at $f_{\text{hi}(f)}$ or 19.2 MHz is 3 db below the mid-band value of $|A_f|$ of 17 db. Using these values, the Bode plot of $|A_f|$ can be sketched as shown in Fig. 11–12.

(d) Experimental results: The experimental results for this amplifier were given in Ex. 11–3 as

$$M_{\text{pf}} \simeq 4.6 \text{ db} \qquad f_p \simeq 10 \text{ MHz} \qquad f_{\text{hi}(f)} \simeq 18.5 \text{ MHz}$$

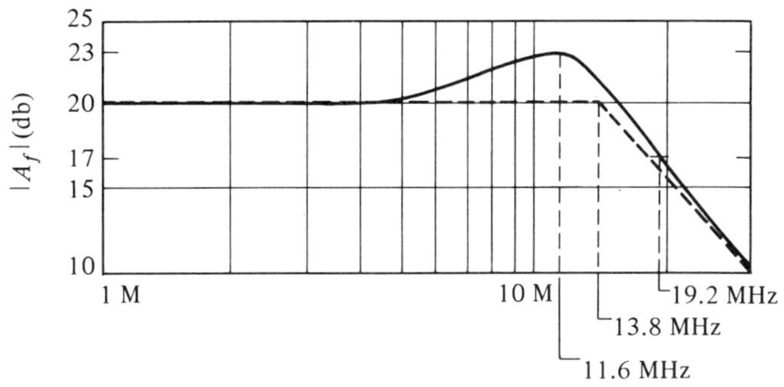

FIGURE 11–12 *An approximate Bode plot of the closed-loop high-frequency response of the amplifier from Ex. 11–6.*

In this section, the closed-loop high-frequency response, A_f, of an amplifier has been related to the high-frequency response of its loop gain, T. The important parameters of T are the crossover frequency, f_c, and the phase margin, P.M. If these values are known, the values of ξ, f_p, f_n, $f_{\text{hi}(f)}$, and M_{pf} can be estimated using either Table 11–2 or Fig. 11–11. Although the analysis is strictly valid only for a second-order system, the response of higher-order stable systems can be approximated using the same procedure. These data are also useful in the design of compensation networks and in the overall design of feedback amplifiers.

11.5 An Approximate Analysis of the Closed-Loop Transient Response

In the preceding sections, a technique was developed for finding the frequency response of a negative-feedback amplifier. The objective of this section is to establish the relationship between the transient and frequency responses of a second-order system. Also, we will relate the transient response to the frequency response of the loop gain using the magnitude crossover frequency, f_c, and the phase margin, P.M., of the loop gain.*

The closed-loop gain of a second-order feedback amplifier was given by Eq. (11–16) as

$$A_f = \frac{X_o}{X_i} = \frac{A_{f(\text{mid})}}{(jf)^2/f_n^2 + (jf)2\xi/f_n + 1} \quad \textbf{(11–28)}$$

Rearranging this equation, the output signal can be expressed as

$$X_o = \frac{X_i A_{f(\text{mid})} f_n^2}{(jf)^2 + 2\xi f_n (jf) + f_n^2} \quad \textbf{(11–29)}$$

The normalized time response of the output signal is available from this equation by using Laplace Transform techniques. Assuming a step input signal, the time response of the normalized output signal becomes**

$$X_o(t) = 1 - \frac{e^{-2\pi\xi f_n t}}{\sqrt{1-\xi^2}} \sin\left[(2\pi f_n\sqrt{1-\xi^2})t + \cos^{-1}\xi\right] \quad \textbf{(11–30)}$$

The general form of $X_o(t)$ is shown in Fig. 11–13. In addition, the interpretation of the terms used in the following discussion is also included in this figure.

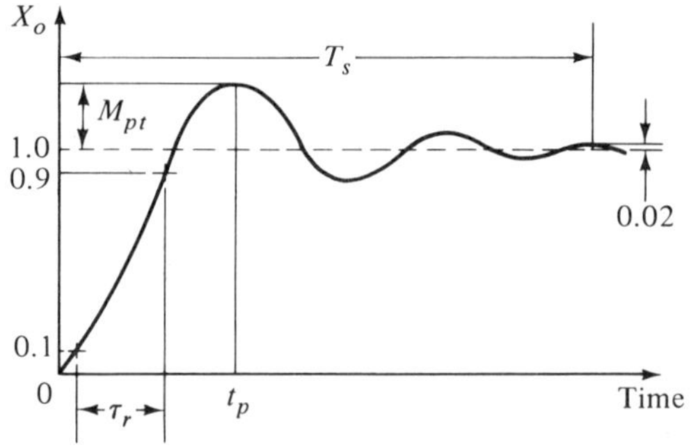

FIGURE 11–13 *General form of the transient response of a second-order system for a step input signal showing the interpretation of the rise time, τ_r, percentage overshoot, M_{pt}, time to first peak, t_p, and settling time, T_s.*

In the last section, we found that the closed-loop frequency response can have a peak in the high-frequency region. This type of peak corresponds to an overshoot in the time response of an amplifier assuming a step input signal. The percentage overshoot, M_{pt}, is related to the damping ratio, ξ, by

$$M_{pt} = 100e^{-\pi\xi/\sqrt{1-\xi^2}} \quad \textbf{(11–31)}$$

The peak value of the first overshoot occurs at time t_p given by***

$$t_p = 1/(2f_n\sqrt{1-\xi^2}) \quad \textbf{(11–32)}$$

* For additional treatment of the time response of a second-order system, refer to a text on control theory such as the text by Kuo, pp. 231–234, noted in the list of references at the end of this chapter.

** Eq. (11–30) is the inverse Laplace Transform of Eq. (11–29) assuming a unit step input.

*** Equations (11–31) and (11–32) are obtained by setting the derivative of Eq. (11–30) equal to zero.

The exponential term in Eq. (11–30) governs the decay of the oscillating portion of the output signal. This term is characterized by the time constant $1/(2\pi\xi f_n)$. The settling time T_s is defined as the time required for the signal to reach 2 percent of its final value and is

$$T_s \equiv 4/(2\pi\xi f_n) = 2/(\pi\xi f_n) \tag{11–33}$$

The rise time of the output signal as a function of f_n and ξ is also available from Eq. (11–30). The simplest way to determine the relationship between τ_r and f_n and ξ is to plot Eq. (11–30) using a computer. For a given value of ξ,

$$f_n\tau_r = f_nt_2 - f_nt_1 = f_n(t_2 - t_1) \tag{11–34}$$

where f_nt_2 causes X_o to equal 0.9, and f_nt_1 causes X_o to equal 0.1.

The closed-loop transient response may be determined when the following four values are known: the rise time, τ_r; percentage overshoot, M_{pt}; the time to the first peak, t_p; and the settling time, T_s. Equations (11–31) through (11–34) relate these characteristics to ξ and f_n. The frequency f_c where $|T|$ is unity or zero db is related to f_n by Eq. (11–24), and the phase margin of the loop gain is related to ξ by Eq. (11–23). Consequently, the transient response characteristics of a feedback amplifier can also be related to its loop-gain characteristics. These relationships are tabulated in Table 11–3 and are plotted* in Fig. 11–14.

TABLE 11–3 *Summary of the relationships between the closed-loop transient response and the loop gain high-frequency response for a second-order system.*

ξ	P.M.	f_ct_p	$f_{\text{hi}(f)}\tau_r$	$f_c\tau_r$	M_{pt}	T_sf_c
0	0°	0.50	—	—	—	∞
0.1	11.4°	0.49	0.28	0.18	73%	6.30
0.2	22.6°	0.47	0.29	0.18	53%	3.06
0.3	33.4°	0.44	0.29	0.19	37%	1.94
0.4	43.3°	0.40	0.32	0.20	25%	1.36
0.5	52°	0.36	0.34	0.21	16%	1.00
0.6	59°	0.32	0.35	0.22	9%	0.76
0.7	65°	0.29	0.36	0.23	5%	0.59
0.8	70°	0.29	0.34	0.23	2%	0.47
0.9	74°	0.33	0.34	0.24	0.2%	0.38
1.0	77°	—	—	—	0	0.31

A convenient method of approximating the transient response of a feedback amplifier can be developed using Fig. 11–14 and Table 11–3. Although this method is developed for a second-order system (an amplifier completely characterized by two high corner frequencies), it can also be applied to more complex stable systems. The approximation of a higher-order stable system as a second-order one is based on the assumption that all stable systems have less than a 180° phase shift at f_c, the crossover frequency of $|T|$. This approach to the transient analysis of a feedback amplifier using Fig. 11–14 or Table 11–3 consists of the following steps:

(1) Construct a Bode plot of T.
(2) Determine the values of f_c and P.M. from this Bode plot.

* The relationship between $f_{\text{hi}(f)}\tau_r$ and ξ was obtained by combining the curves $f_c\tau_r$ versus ξ and $f_{\text{hi}(f)}/f_c$ versus ξ. The latter relationship is plotted in Fig. 11–11(b).

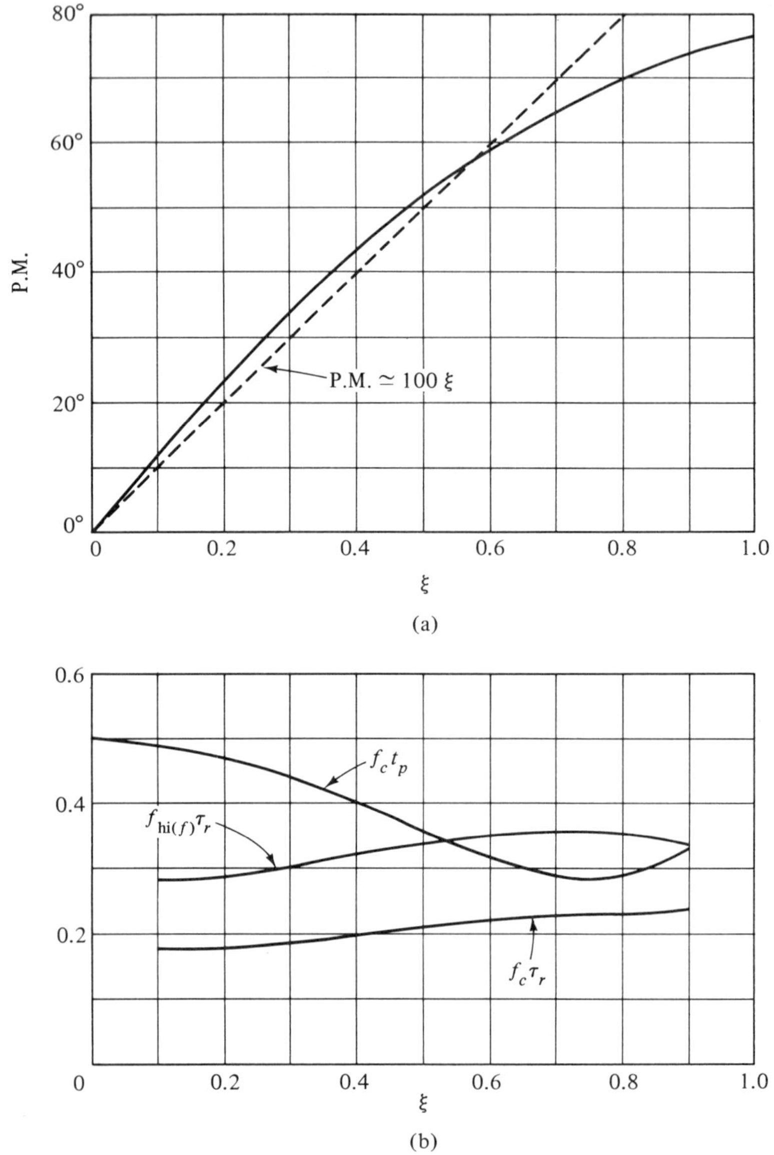

FIGURE 11–14 *Plot of the closed-loop transient response characteristics of a second-order negative-feedback amplifier in terms of the loop gain characteristics.*

(3) From Fig. 11–14(a), estimate the value of ξ corresponding to the value of P.M. determined in (2).

(4) Using this value of ξ along with Figs. 11–14(b), (c), and (d), determine the values of $f_c t_p$, $f_c \tau_r$, M_{pt}, and $T_s f_c$.

(5) Using the results of (4) and the value of f_c from (2), determine the values of t_p, τ_r, and T_s.

The following example illustrates the use of Fig. 11–14 (or Table 11–3) in the transient analysis of a feedback amplifier.

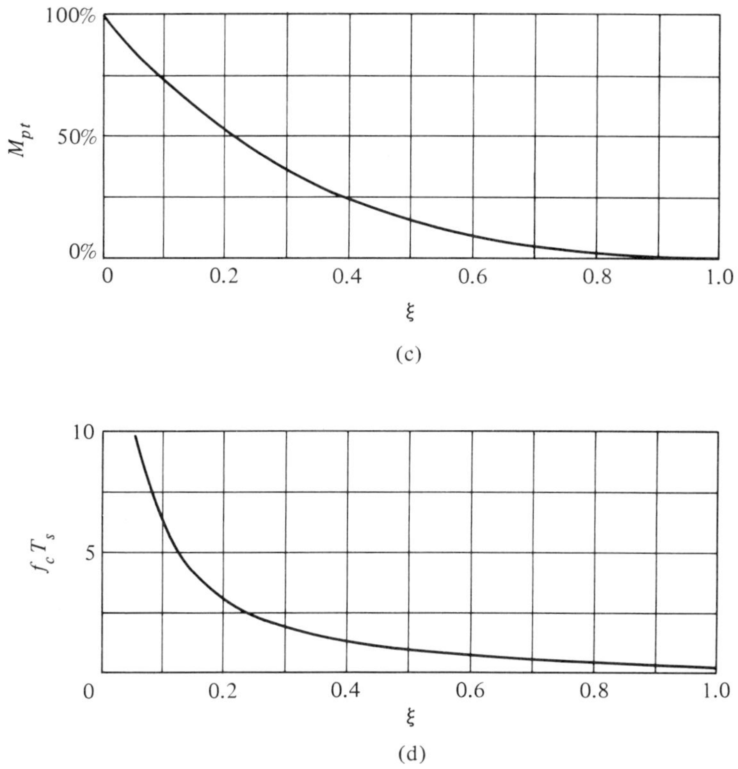

FIGURE 11–14 (cont.)

EXAMPLE 11–7 Calculate the transient response of the voltage sample-voltage sum amplifier shown in Fig. 11–3. The Bode plot of the loop gain for this amplifier is given in Fig. 11–6(a).

Solution: (a) Loop gain characteristics: From the Bode plot of T in Fig. 11–6(a),

$$f_c \simeq 12 \text{ MHz} \qquad \text{P.M.} \simeq 41°$$

(b) Closed-loop characteristics: From Fig. 11–14 or Table 11–3,

$$\xi \simeq 0.37 \qquad f_c t_p \simeq 0.41 \qquad f_{\text{hi}(f)} \tau_r \simeq 0.31$$
$$f_c \tau_r \simeq 0.19 \qquad M_{pt} \simeq 28\% \qquad f_c T_s \simeq 1.5$$

Solving for the closed-loop transient characteristics,

$$t_p \simeq 0.41/(12 \times 10^6) \simeq 34 \text{ nsec} \qquad \tau_r \simeq 0.19/(12 \times 10^6) \simeq 15.8 \text{ nsec}$$
$$T_s \simeq 1.5/(12 \times 10^6) \simeq 125 \text{ nsec}$$

(c) Experimental results: With an output signal level of 1 V, the measured response was

$$M_{pt} \simeq 28\% \qquad t_p \simeq 42 \text{ nsec}$$
$$\tau_{r(\text{input})} \simeq 10 \text{ nsec} \qquad \tau_{r(\text{output})} \simeq 20 \text{ nsec}$$

Since the input signal had a rise time of 10 nsec, the rise time for the amplifier is less than the 20 nsec rise time value measured for the output signal. The actual value of the rise time for the amplifier is approximately 15.2 nsec. The effect of an input signal with a finite rise time is discussed below.

Often in testing high speed amplifiers having rise times of approximately 100 nsec or less, the effects of the rise times of the generator and measuring circuit must be considered. Such a test setup is shown in Fig. 11–15 where each unit has a finite rise time. The combined effect of the generator and oscilloscope can be determined experimentally by measuring the rise time of the generator with the oscilloscope. For the test setup shown in Fig. 11–15, the observed rise time of the output signal is related to the rise times of the various units (within 10 percent) by*

$$\tau_{r(\text{output})} \simeq 1.1\sqrt{\tau_{r(\text{gen})}^2 + \tau_{r(\text{amplifier})}^2 + \tau_{r(\text{scope})}^2} \tag{11–35}$$

For the amplifier considered in Ex. 11–7, the combined rise time of the generator and the oscilloscope was 10 nsec. The observed rise time of the output signal was 20 nsec. Substituting these values into Eq. (11–35), we find that the rise time of the amplifier alone is 15.2 nsec.

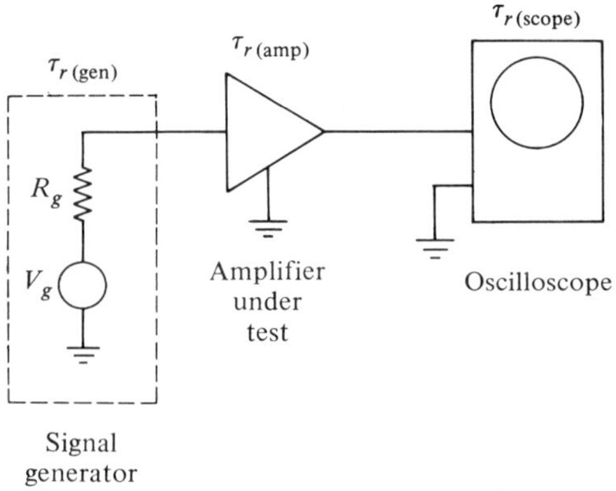

FIGURE 11–15 *An experimental setup used to measure the rise time of an amplifier where the signal generator, amplifier under test, and oscilloscope have finite rise times.*

In this section we have discussed the transient response of feedback amplifiers that can be described as second-order systems. If a feedback amplifier has only one dominant high corner frequency, then the amplifier is a first-order system, and Fig. 11–14 and Table 11–3 do not apply.** A first-order system has no overshoot, and the rise time is related to the closed-loop high corner frequency by

$$\tau_r \simeq 0.35/f_{\text{hi}(f)} \tag{11–36}$$

For a second-order system, or a system that can be approximated as second order, Fig. 11–14 and Table 11–3 can be used to estimate the transient response of a feedback amplifier. Also, we considered the effect of a signal generator and an oscilloscope having finite rise times on the rise time of the output signal. The material in this section is useful in the analysis, design, and testing of feedback amplifiers.

11.6 High-Frequency Cascade Compensation

The two preceding sections were concerned with simple methods for determining the closed-loop response of feedback amplifiers having multiple high corner

---

* Derivation of Eq. (11–35) is discussed in Millman and Taub in the references listed at the end of the chapter.

** Equation (11–30) which relates the output, X_o, to ξ and f_n, is valid only for values of ξ less than unity.

frequencies. The closed-loop, high-frequency response had a peak in the high-frequency region. The closed-loop transient response was characterized by an overshoot and damped oscillation or ringing. Most high performance feedback amplifiers have various passive networks added to the basic amplifier to reduce the undesirable peaking in the closed-loop frequency response and overshoot in the transient response. These passive networks, called compensation networks, can be grouped by location into two types: (1) cascade compensation, and (2) feedback compensation. Cascade compensation networks are added to the forward signal path of an amplifier and change the response characteristics of A_{ol} and T. Feedback compensation networks are added to the feedback path and change the response characteristics of B and T. Many amplifiers use both types of compensation. The objective of this section is to study three types of cascade compensation: (1) gain compensation, (2) dominant pole compensation, and (3) RC lag compensation. Feedback compensation is studied in the next section.

Before beginning the study of cascade compensation, we will review briefly some of the desirable characteristics of feedback amplifiers and the relationship between the closed-loop response and the loop-gain response. Three desirable characteristics of the closed-loop high-frequency response of feedback amplifiers are: (1) a large value of $|T_{\text{mid}}|$, (2) a high value of $f_{\text{hi}(f)}$, and (3) a small or zero value of M_{pf}. Three desirable characteristics of the closed-loop transient response are: (1) a large value of $|T_{\text{mid}}|$, (2) a small value of τ_r, and (3) a zero or small value of M_{pt}. Our study in this chapter shows that these closed-loop characteristics can be related to the loop gain characteristics, f_c and P.M. A high value of $f_{\text{hi}(f)}$, or a low value of τ_r, requires a high value of f_c. A low value of M_{pf}, or a low value of M_{pt}, requires a large value of P.M. As we will see in this section, increasing the value of P.M. generally reduces the value of f_c. Consequently, the design of a network for compensation generally requires judicious compromises.

Reducing the value of M_{pf} or M_{pt} generally requires that the value of the loop gain P.M. be increased. One way to increase the value of the P.M. is to reduce the value of $|T_{\text{mid}}|$ without changing any of the poles of T. This type of cascade compensation is called *gain compensation*, and its effect is illustrated in Fig. 11–16. The uncompensated amplifier is assumed to have a loop gain characterized by f_c and P.M. as shown in Fig. 11–16(a). The closed-loop transient response is characterized by τ_r and M_{pt} as shown in Fig. 11–16(b). By shifting the $|T|$ curve downward, the value of P.M. can be increased. However, the value of τ_r is increased since the value of f_c is reduced as shown in Figs. 11–16(c) and (d). The compensated values are denoted by the superscript c. The effects of gain compensation on the closed-loop response of the amplifier can be summarized as follows: (1) the value of P.M. is increased, and the values of M_{pt} and M_{pf} are reduced, (2) the values of f_c and $f_{\text{hi}(f)}$ are reduced, and the value of τ_r is increased, and (3) the value of $|T_{\text{mid}}|$ is reduced.

In some feedback amplifiers the values of $|T_{\text{mid}}|$ and f_c are sufficiently high so that their values can be decreased without compromising the design goal of the amplifier. The feedback amplifier shown in Fig. 11–3 and studied in several previous examples is a case in point. The circuit of this amplifier is shown in Fig. 11–17. The potentiometer between the emitter terminals of Q_1 and Q_2 effectively reduces the value of $|T_{\text{mid}}|$. Increasing the value of R_{EE} would further reduce the value of $|T_{\text{mid}}|$ and would also increase the value of the high corner

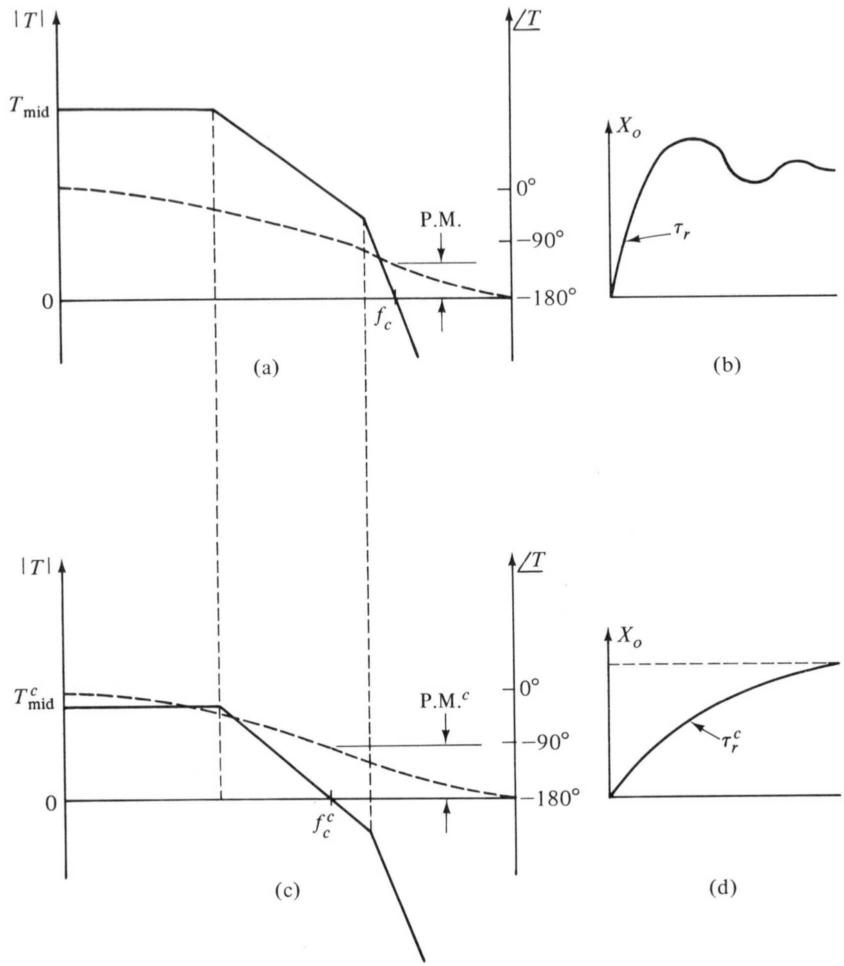

FIGURE 11–16 *The effect of gain compensation: (a) the Bode plot of* T *for the uncompensated amplifier, (b) the transient response of the uncompensated amplifier, (c) the Bode plot of* T *for the compensated amplifier, and (d) the transient response of the compensated amplifier.*

frequency at the base of Q_2 by reducing the value of $C_{b'g2}$. Verification of this effect is left as an exercise for the reader. When R_{EE} is increased from 90 ohms to 500 ohms, the measured value of τ_r increased from 15.2 nsec to 50 nsec, and the measured value of M_{pt} decreased from 28 percent to approximately 0.1 percent.

True gain compensation does not always result when the value of $|T_{\text{mid}}|$ is reduced. For example, reducing the value of $|T_{\text{mid}}|$ by lowering the value of A_{vt3} will have little effect on the closed-loop transient response.* This result occurs because the value of $C_{\text{in}3}$ depends on the value of C_{ob3} $(1 - A_{vt3})$. As the value of A_{vt3} is reduced, the pole at the base of Q_3 moves to a higher frequency. The

* The value of A_{vt3} can be easily lowered by ac bypassing part of R_{C3} or by not bypassing part of R_{E3}.

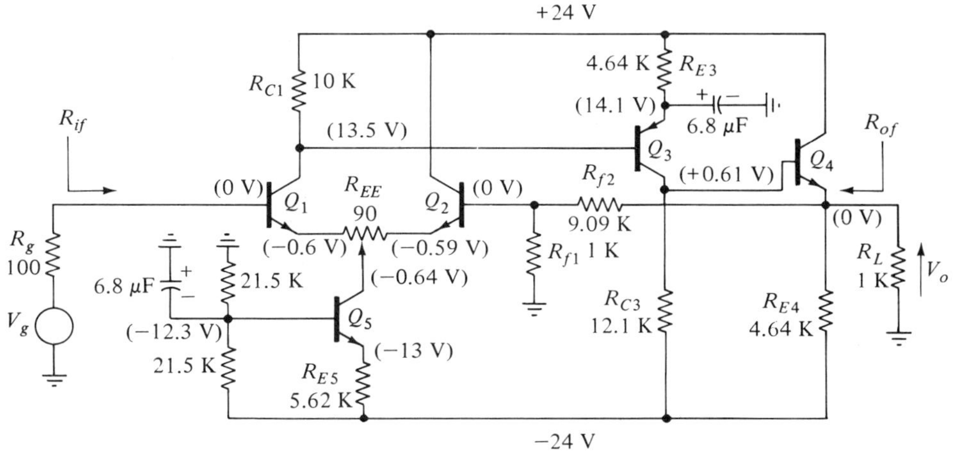

FIGURE 11–17 *An example of gain compensation using the amplifier shown in Fig. 11–3.*

result is that the loop-gain characteristics, f_c and P.M., are nearly the same as the original amplifier. Since the changes in f_c and P.M. are small, the change in the closed-loop response is also small.

As noted earlier, a reduction in the value of M_{pf} or M_{pt} generally requires an increase in the value of the phase margin of the loop gain. The essential effect of dominant pole compensation is to reduce the value of the lowest high corner frequency of T (f_{hi1}) to a new value (f_{hi1}^c) to produce the desired value of P.M. Comparing the uncompensated Bode plots of T [Fig. 11–18(a)] and the compensated Bode plot of T [Fig. 11–18(c)], we can see that the value of f_c was reduced from f_c to f_c^c, but the value of the phase margin was increased from P.M. to P.M.^{c}. The effect of compensation on the transient response is shown in Figs. 11–18(b) and (d). The value of τ_r increased while the value of M_{pt} decreased.

Generally, the lowest high corner frequency of T is reduced from f_{hi1} to f_{hi1}^c by simply increasing the capacitance to ground at the node which contributes the dominant pole. As shown in Fig. 11–19, additional capacitance can be added directly to ground, or the Miller effect of an inverting stage can be used. The latter method [shown in Fig. 11–19(b)], is generally preferable since smaller values of capacitance are required.

The following procedure can be used to calculate the value of capacitance required for dominant pole compensation:

(1) Construct the Bode plot of $|T|$ and $\underline{/T}$.
(2) Determine the required value of ξ from the closed-loop specifications for M_{pf} or M_{pt} using Fig. 11–11(c) or Fig. 11–14(c).
(3) Determine the required value of P.M. from the value of ξ found in (2) using Fig. 11–11(a) or Fig. 11–14(a).
(4) Use the Bode plot constructed in (1) to determine the frequency at which $\underline{/T}$ is equal to 180° minus the required value of P.M. determined in (3). This frequency corresponds to f_c^c.
(5) Determine the required value of f_{hi1}^c from the value of f_c^c.
(6) Determine the required value of C_{comp} from the value of f_{hi1}^c.

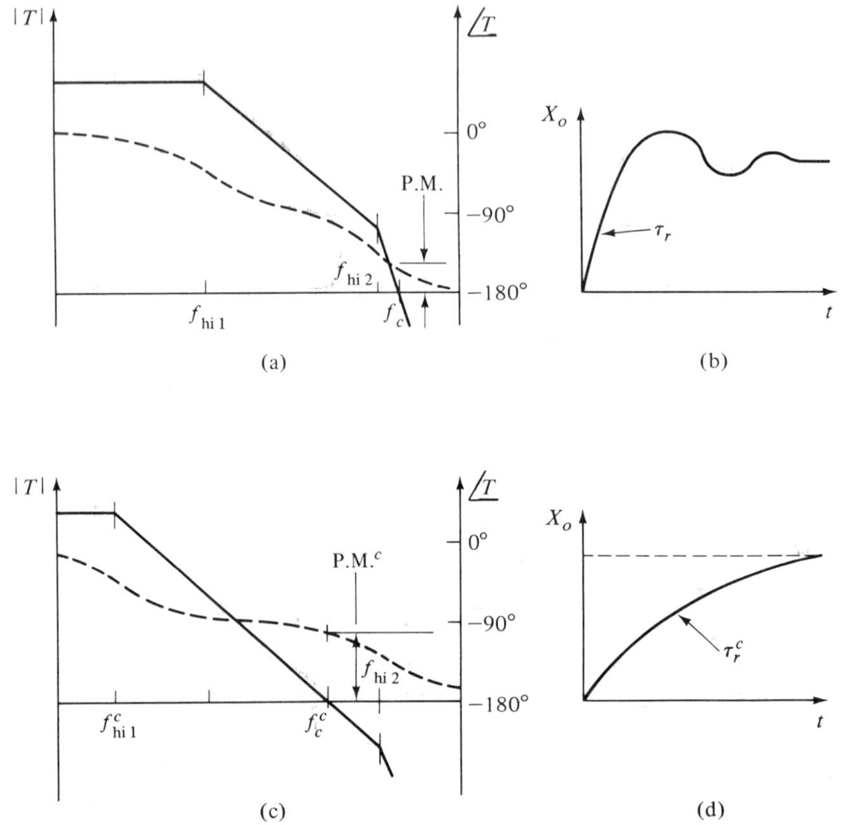

FIGURE 11–18 *The effect of dominant pole compensation: (a) the Bode plot of* T *and (b) the transient response of the uncompensated amplifier, (c) the Bode plot of* T *and (d) the transient response of the compensated amplifier.*

The following example illustrates the use of dominant pole compensation.

EXAMPLE 11–8 The amplifier shown in Fig. 11–17 has a loop gain that is characterized by the Bode plot shown in Fig. 11–20(a). Determine the value of capacitance that must be connected between the base and collector terminals of Q_3 to reduce the value of M_{pt} to 5 percent. This amplifier was analyzed in Ex. 11–3 with the following results:

$$T_{\text{mid}} \simeq -735 \qquad A_{vt3} \simeq -580$$

Base of Q_3:

$$\tau_{\text{hi1}} = R_1C_1 \simeq (1.81\text{ K})(4130\text{ pF}) \simeq 7.5\ \mu\text{sec}$$
$$f_{\text{hi1}} = 1/(2\pi\tau_{\text{hi1}}) \simeq 21\text{ kHz}$$

Base of Q_2:

$$\tau_{\text{hi2}} = R_2C_2 \simeq (900)(17.2\text{ pF}) \simeq 15.5\text{ nsec}$$
$$f_{\text{hi2}} = 1/(2\pi\tau_{\text{hi2}}) \simeq 10\text{ MHz}$$

Solution: (a) Characteristics of the uncompensated amplifier: The Bode plot of T is shown in Fig. 11–20(a). From this curve

$$f_c \simeq 12\text{ MHz} \qquad \text{P.M.} \simeq 41°$$

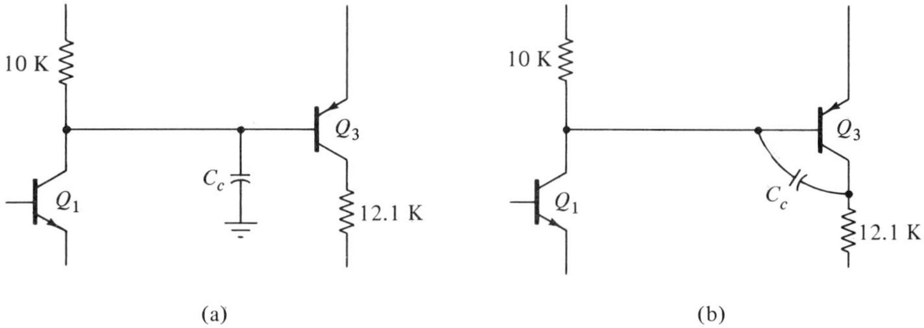

FIGURE 11–19 *Two methods of achieving dominant pole compensation: (a) addition of a capacitor to ground, and (b) the use of the Miller effect in an inverting stage.*

From the closed-loop transient response in Ex. 11–7,

$$\tau_r \simeq 15.8 \text{ nsec} \qquad M_{pt} \simeq 28\%$$

(b) Loop gain characteristics of the compensated amplifier: The desired value of M_{pt} is 5 percent or less. From Fig. 11–14(c), this value of M_{pt} corresponds to ξ equal to 0.7. From Fig. 11–14(a), the corresponding value of P.M. is approximately 65°. From the Bode plot of T, a P.M. of 65° occurs at approximately 3.5 MHz. Thus, the desired loop gain characteristics are

$$f_c^c \simeq 3.5 \text{ MHz} \qquad \text{P.M.} \simeq 65°$$

(c) Required value of $f_{\text{h1}1}^c$: The value of $f_{\text{h1}1}^c$ must be chosen so that the desired values determined in (b) are obtained. The value of $f_{\text{h1}1}^c$ can be found by constructing a new Bode plot of T in the high-frequency region through the point 0 db and 3.5 MHz. The intersection of this new plot with the horizontal line, $|T_{\text{mid}}| = 58$ db, gives the value of $f_{\text{h1}1}^c$. From Fig. 11–20(b), $f_{\text{h1}1}^c$ is 5 kHz. However, since f_c^c occurs at a lower frequency than $f_{\text{hi}2}$ (3.5 MHz < 10 MHz), the Bode plot of $|T|$ in the frequency range from $f_{\text{h1}1}^c$ to f_c^c has a constant slope of -6 db/octave (or -20 db/decade). Thus,

$$f_{\text{h1}1}^c \simeq f_c^c/|T_{\text{mid}}| \simeq 3.5 \text{ MHz}/735 \simeq 4.76 \text{ kHz}$$

(d) Value of C_{comp}: The time constant associated with $f_{\text{h1}1}^c$ is

$$\tau_{\text{h1}1}^c = 1/(2\pi f_{\text{h1}1}^c) \simeq 0.159/4.7 \text{ kHz} \simeq 33.4 \ \mu\text{sec}$$

This time constant is the product of the equivalent resistance and capacitance from this node to ground or

$$\tau_{\text{h1}1}^c = R_1 C_1^c \simeq (1.8 \text{ K})(C_1^c)$$
$$C_1^c = \tau_{\text{h1}1}^c/R_1 \simeq 33.4 \ \mu\text{sec}/1.81 \text{ K} \simeq 18{,}400 \text{ pF}$$

The original capacitance to ground at this node was approximately 4130 pF. Consequently, the additional capacitance needed is 14,270 pF. The capacitance that must be added between the base and the collector terminals of Q_3 is

$$C_{ob3}^c = C_{\text{comp}}/(1 - A_{vt3}) \simeq 14{,}270 \text{ pF}/(1 + 580) \simeq 25 \text{ pF}$$

(e) Experimental results: With C_{ob3}^c equal to 25 pF, the measured response was

$$M_{pt} \simeq 1\% \qquad \tau_r \simeq 76 \text{ nsec}$$

Using Fig. 11–14, the corresponding values of ξ, P.M., and f_c are

$$\xi \simeq 0.85 \qquad \text{P.M.} \simeq 72° \qquad f_c\tau_r \simeq 0.24$$
$$f_c \simeq 0.24/\tau_r \simeq 0.24/76 \text{ nsec} \simeq 3.15 \text{ MHz}$$

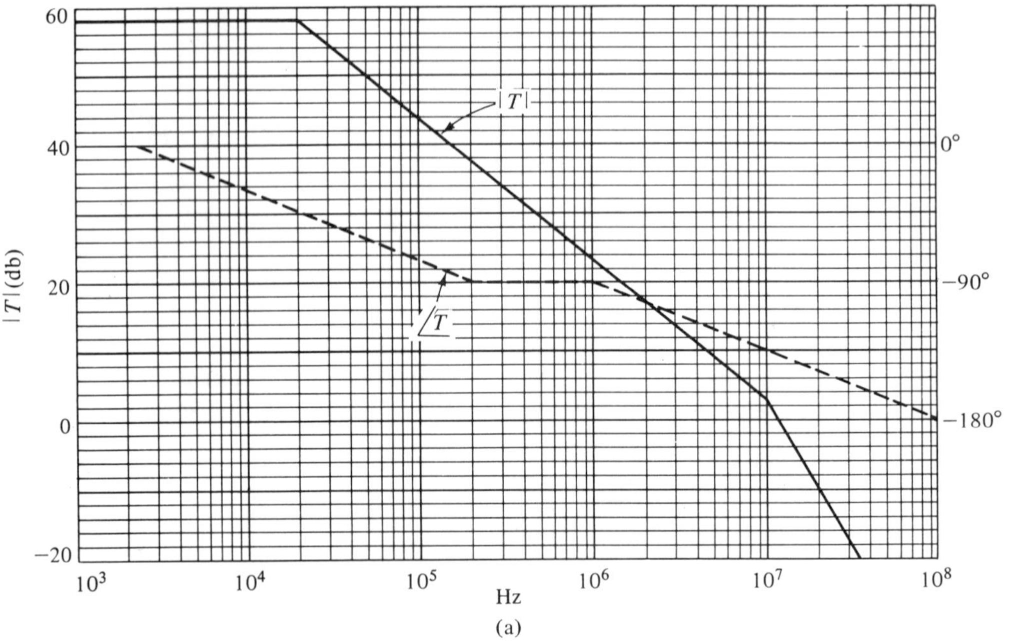

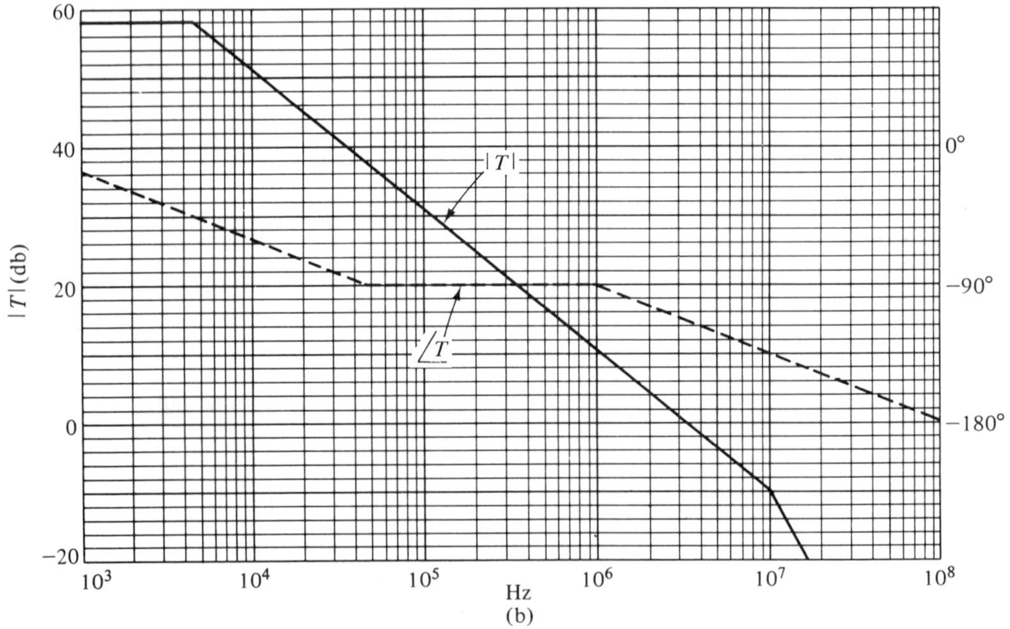

FIGURE 11–20 *A Bode plot of* T *for the amplifier studied in Ex. 11–8: (a) uncompensated Bode plot and (b) compensated Bode plot.*

(f) An alternate solution: The error between the desired value of 5 percent and the measured value of 1 percent for M_{pt} is largely due to the use of the straight-line approximation for the phase plot. When the more accurate phase response curve given in Appendix B is used, the following results are obtained:

$$f_c^c \simeq 5 \text{ MHz for P.M.} \simeq 65°$$
$$f_{\text{hl1}}^c \simeq f_c^c/|T_{\text{mid}}| \simeq 5 \text{ MHz}/735 \simeq 6.8 \text{ kHz}$$
$$\tau_{\text{hl1}}^c \simeq 1/(2\pi f_{\text{hl1}}^c) \simeq 0.159/6.8 \text{ kHz} \simeq 23.4\ \mu\text{sec}$$
$$C_1^c \simeq \tau_{\text{hl1}}^c/R_1 \simeq 23.4\ \mu\text{sec}/1.81 \text{ K} \simeq 12{,}900 \text{ pF}$$
$$C_{\text{comp}}^c \simeq C_1^c - C_1 \simeq 12{,}900 \text{ pF} - 4130 \text{ pF} \simeq 8770 \text{ pF}$$
$$C_{ob3}^c \simeq C_{\text{comp}}^c/(1 - A_{vt3}) \simeq 8770 \text{ pF}/(1 + 580) \simeq 15 \text{ pF}$$

With C_{ob3}^c equal to 15 pF, the measured response was

$$M_{pt} \simeq 5\% \qquad \tau_r \simeq 50 \text{ nsec}$$

Using Fig. 11–14, the corresponding values of ξ, P.M., and f_c are

$$\xi \simeq 0.7 \qquad \text{P.M.} \simeq 65°$$
$$f_c\tau_r \simeq 0.23 \qquad f_c \simeq 0.23/\tau_r \simeq 4.6 \text{ MHz}$$

Consequently, the value of C_{ob3}^c based on the more accurate phase plot of $\angle T$ gave more accurate results. However, rather than abandon the simpler straight line Bode plots of $|T|$ and $\angle T$, we will interpret the results as only approximate. In general, the exact value of C_{ob3}^c should be determined experimentally with the analysis indicating the approximate size of capacitor needed.

RC lag compensation differs from gain and dominant pole compensation in that a zero is added to the loop gain expression. The phase shift introduced by this zero can be used to increase the loop gain P.M. One form of an *RC* lag compensation network is shown in Fig. 11–21(a). The compensation network consists of the series combination R_1 and C_1 which is connected from one node in the amplifier to ground. An alternate form of an *RC* lag network is shown in Fig. 7–21(b) where the series network is connected between the base and collector terminals of an inverting stage. The two arrangements shown in Fig. 11–21 (a) and (b) are essentially the same since R_2 and C_2 [Fig. 11–21(b)], can be replaced by an equivalent network connected between node A and ground using Miller's theorem. The effect of applying Miller's theorem to R_2 and C_2 is shown in Fig. 11–21(c). An ac equivalent circuit which can be used to study the response of an *RC* lag network is shown in Fig. 11–21(d) where $R_{L\text{eq}1}$ and $C_{L\text{eq}1}$ represent the equivalent impedance from Node A to ground with the *RC* lag network disconnected. The current source, I_s, represents the ac signal coming from the collector terminal of Q_1. Our next task is to develop the transfer function, V_a/I_s, of the equivalent circuit shown in Fig. 11–21(d).

With the R_1C_1 network disconnected,

$$V_a/I_s = Z_L = R_{L\text{eq}1} \,||\, (1/sC_{L\text{eq}1}) = R_{L\text{eq}1}/(1 + sR_{L\text{eq}1}C_{L\text{eq}1}) \qquad \textbf{(11–37)}$$

The corner frequency associated with the pole in this equation occurs at

$$f_{p1} = 1/(2\pi R_{L\text{eq}1}C_{L\text{eq}1}) \qquad \textbf{(11–38)}$$

With the R_1C_1 lag network connected,

$$V_a/I_s = Z_L^c = Z_L \,||\, (R_1 + 1/sC_1) \qquad \textbf{(11–39)}$$

$$Z_L^c = \frac{R_{L\text{eq}1}(1 + sR_1C_1)}{s^2R_{L\text{eq}1}R_1C_{L\text{eq}1}C_1 + s[R_{L\text{eq}1}(C_{L\text{eq}1} + C_1) + R_1C_1] + 1} \qquad \textbf{(11–40)}$$

The component values of a useful *RC* lag network are generally such that

$$R_{L\text{eq}1}(C_{L\text{eq}1} + C_1) \gg R_1C_1 \qquad \textbf{(11–41)}$$

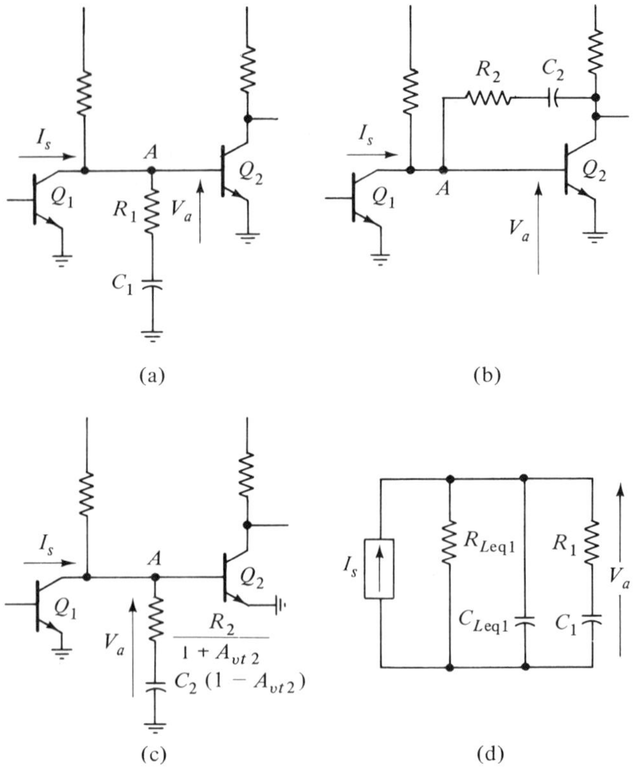

FIGURE 11–21 RC *lag compensation: (a) a partial circuit diagram showing one connection of an* RC *lag network, (b) a partial circuit diagram showing a second connection of an* RC *lag network, (c) the effect of applying Miller's theorem to the network in (b), and (d) an ac equivalent circuit used to represent (a), (b), or (c).*

Using Eq. (11–41) and the approximate method of factoring a polynomial, Eq. (11–40) can be rewritten as

$$\frac{V_a}{I_s} \simeq \frac{R_{Leq1}(1 + sR_1C_1)}{[1 + sR_{Leq1}(C_{Leq1} + C_1)]\left[1 + sR_1\left(\dfrac{C_{Leq1}C_1}{C_{Leq1} + C_1}\right)\right]} \tag{11–42}$$

By comparing Eqs. (11–37) and (11–42), we can see that the effect of *RC* lag compensation is to replace the single pole with a pole-zero-pole complex. From Eq. (11–42), the corner frequencies of the compensated system are

$$f^c_{p1} \simeq 1/[2\pi R_{Leq1}(C_{Leq1} + C_1)] \qquad f^c_z = 1/(2\pi R_1C_1)$$
$$f^c_{p2} \simeq 1 \bigg/ \left[2\pi R_1\left(\frac{C_{Leq1}C_1}{C_{Leq1} + C_1}\right)\right] \tag{11–43}$$

Note that the lowest pole of the compensated system, f^c_{p1}, occurs at a frequency lower than the pole of the uncompensated system [Eq. (11–38)] due to the addition of C_1. Also, from Eqs. (11–41) and (11–43), the zero occurs at a frequency between f^c_{p1} and f^c_{p2}.

The effect of an *RC* lag network is shown in Fig. 11–22 where the Bode plot of the magnitude and phase response of a system with and without the *RC* lag

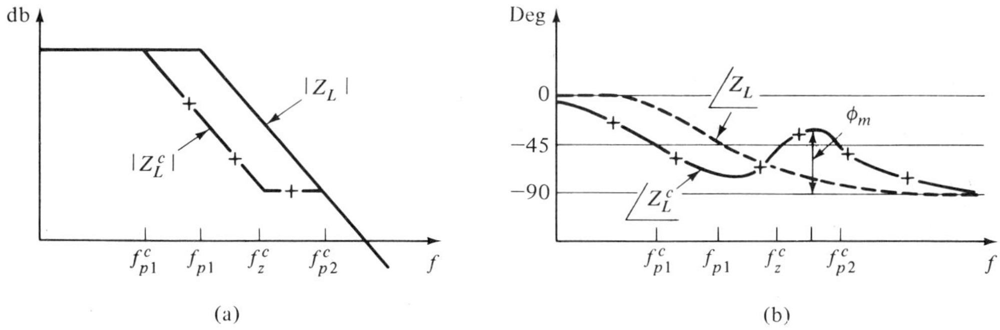

FIGURE 11–22 *Bode plot showing the effect of an* RC *lag compensation network: (a) magnitude plot, and (b) phase plot. The compensated curves are indicated by* Z_L^c.

network are compared. Note that the magnitude response of the compensated and uncompensated systems are the same for frequencies below f_{p1}^c and for frequencies above f_{p2}^c; i.e., the magnitude response of the compensated system is bounded by the magnitude response of the uncompensated system

The phase response [shown in Fig. 11–22(b)] is of particular interest. For frequencies below f_{p1}^c, the phase shift of the compensated system is greater than the phase shift of the uncompensated system.* However, the phase shift of the compensated system decreases due to the zero at f_z^c and reaches a relative minimum at some frequency between f_z^c and f_{p2}^c. The minimum phase shift is approximately equal to $(-90° + \phi_m)$ where ϕ_m is the maximum value of the positive phase shift introduced by the *RC* lag network. This relative minimum in phase shift is useful in increasing the loop gain phase margin of a feedback amplifier.

The effect of *RC* lag compensation on the loop-gain characteristics of a feedback amplifier is illustrated in Fig. 11–23. The uncompensated loop gain has two important high corner frequencies, $f_{\mathrm{hi}1}$ and $f_{\mathrm{hi}2}$, and is characterized by values of f_c and P.M. The *RC* lag network replaces the pole at $f_{\mathrm{hi}1}$ with the pole-zero-pole complex f_{p1}^c, f_z^c, and f_{p2}^c. This complex produces a new crossover frequency, f_c^c, for the loop gain and a new value for the phase margin, P.M.^{c}. Note that the value of f_c^c is less than f_c, and that the value of P.M.^{c} is greater than P.M. In terms of the closed-loop high-frequency response, the effect of compensation is to reduce the values of $f_{\mathrm{hi}(f)}$ and M_{pf}. In terms of the closed-loop transient response, the effect of compensation is to increase the value of τ_r and to reduce the value of M_{pt}.

The design goal for an *RC* lag network is to provide the necessary phase shift to obtain the desired loop gain P.M. and to have this phase shift reach its maximum value at the new loop-gain crossover frequency. From Fig. 11–23(b), this condition means that ϕ_m should occur** at f_c^c.

---

* The initial lagging phase shift of the compensated system relative to the uncompensated system is the basis for the name *lag network* or lag compensation. Strictly speaking, a lag network is one where the lowest frequency pole occurs at a lower frequency than the lowest frequency zero. A lead network has the opposite arrangement of the lowest frequency pole and zero. Lead networks are discussed in the next section.

** An excellent treatment of cascade compensation using both lead and lag networks is given in Chap. 10 of the text by Kuo.

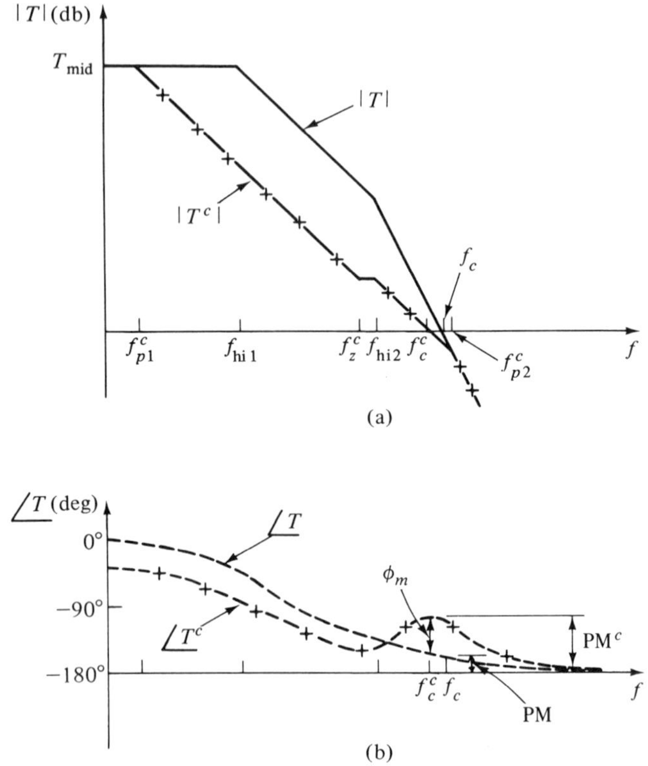

FIGURE 11–23 *The effects of* RC *lag compensation on the loop gain characteristics of a feedback amplifier:* (*a*) *magnitude response and* (*b*) *phase response.*

As noted above, the effect of the *RC* lag network was to replace the single pole, f_{p1}, with a pole-zero-pole complex, f^c_{p1}, f^c_z, and f^c_{p2}. These various corner frequencies are interdependent and can be related to each other by

$$f^c_{p1} = f_{p1}/a \qquad f^c_{p2} = af^c_z \tag{11–44}$$

where a is a parameter of the *RC* lag network. The value of a is related to the maximum phase shift that is introduced by the *RC* lag network, ϕ_m, by

$$\sin \phi_m = \frac{a-1}{a+1} \tag{11–45}$$

For optimum results, the frequency where ϕ_m occurs should coincide with f^c_c. This condition results if f^c_c is chosen as that frequency where the uncompensated value of $|T|$ is

$$|T| = 10 \log a \quad \text{db} \tag{11–46}$$

The value of ϕ_m is equal to the difference between the value of P.M. and P.M.^{c}. Actually, since f^c_c is decreased slightly due to *RC* lag compensation (see Fig. 11–23), the value of P.M. is low, resulting in a slightly high value of ϕ_m. Generally, a better result is obtained by using the value of ϕ_m given by

$$\phi_m \simeq \text{P.M.}^c - \text{P.M.} - 3° \tag{11–47}$$

where the 3° is an arbitrary value to account for the increase in the P.M. of the uncompensated system due to the reduction of f_c to f^c_c.

Once values for ϕ_m and f_c^c are determined, the value for a can be determined. The values of R_1 and C_1 can then be found from the following relationships:

$$C_1 = (a - 1)C_{Leq1} \qquad R_1 = \sqrt{a}/(2\pi f_c^c C_1) \tag{11–48}$$

The important equations for the design of RC lag compensation networks may be summarized as follows:

$$\left.\begin{aligned} \phi_m &\simeq \text{P.M.}^c - \text{P.M.} - 3^\circ & C_1 &= (a-1)C_{Leq1} \\ \sin\phi_m &= \frac{a-1}{a+1} & R_1 &= \sqrt{a}/(2\pi f_c^c C_1) \\ f_c^c &= \text{frequency where } |T| = 10\log a & & \end{aligned}\right\} \tag{11–49}$$

The expressions for the various corner frequencies of the RC lag network may be summarized as follows:

$$\begin{aligned} f_{p1} &= 1/(2\pi R_{Leq1}C_{Leq1}) & f_{p1}^c &= f_{p1}/a \\ f_z^c &= 1/(2\pi R_1 C_1) & f_{p2}^c &= a f_z^c \end{aligned} \tag{11–50}$$

The following procedure can be used to calculate the values of R_1 and C_1 required for RC lag compensation:

(1) Construct the Bode plot of T and $\underline{/T}$, and determine the values of P.M. and f_c.
(2) Determine the required value of ξ from the closed-loop specifications for M_{pf} or M_{pt} using Fig. 11–11(c) or Fig. 11–14(a).
(3) Determine the value of P.M.^{c} corresponding to the value of ξ found in (2) using either Fig. 11–11(a) or Fig. 11–14(a).
(4) Determine the values of ϕ_m, a, f_c^c, R_1, and C_1 from Eq. (11–49).
(5) Determine the values of f_{p1}^c, f_z^c, and f_{p2}^c from Eq. (11–43) or Eq. (11–50).
(6) If the alternate form of RC lag compensation shown in Fig. 11–21(b) is to be used, determine the values of R_2 and C_2 using the values of R_1 and C_1 and Miller's theorem.
(7) Construct the Bode plot of the compensated loop gain to insure that all design specifications are satisfied.

Before considering an example illustrating the use of RC lag compensation, we will briefly discuss a third connection of an RC lag network that is useful in differential amplifiers. As shown in Fig. 11–24, the compensation network is connected between the collector terminals of a differential amplifier.* The zero of this compensation network also occurs at f_z^c as given by Eq. (11–50). The poles of this network, however, are more difficult to locate due to the changing gain of the differential amplifier. For frequencies much less than f_z^c, the load resistance for each stage is approximately R_C. For frequencies much greater than f_z^c, the equivalent load resistance is approximately $R_C \,||\, R_2$. Consequently, the component of the input capacitance due to C_{ob} is changing since A_{vt} is changing. Appropriate values of R_2 and C_2 should be chosen experimentally. A good experimental procedure is to pick values of R_2 and C_2 such that f_z^c is approximately equal to $f_c/2$. Choosing C_2 as a fixed capacitor (in values ranging from 10 pF to 0.01μF) and R_2 as a good quality multiturn potentiometer, R_2 can be varied to obtain the desired response.

---

* Often RC lag compensation in a differential amplifier is used when one differential amplifier is connected to a second differential amplifier. Such a circuit is shown in Fig. 10–19(a) where the RC lag network would be connected between the collector terminals of Q_1 and Q_2.

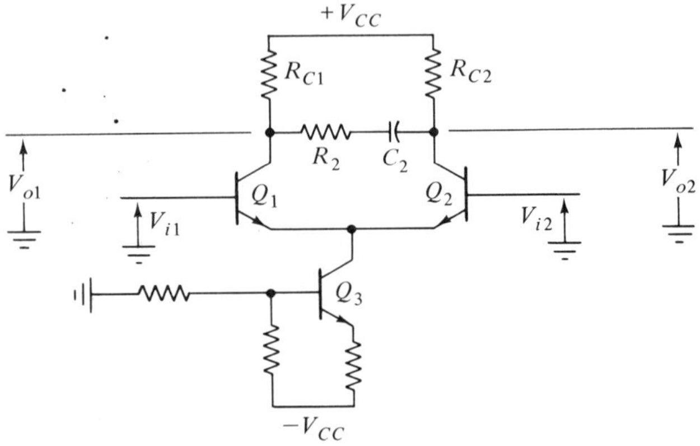

FIGURE 11–24 RC *lag compensation for a differential amplifier.*

EXAMPLE 11–9 The amplifier shown in Fig. 11–17(a) has a loop gain that is characterized by the Bode plot shown in Fig. 11–20(a). The dominant pole compensation of this amplifier was studied in Ex. 11–8. Determine the values of R_2 and C_2 of an *RC* lag network that must be connected between the base and collector terminals of Q_3 to reduce the value of M_{pt} to 5 percent. From Ex. 11–8, some of the important characteristics of this amplifier are

$$T_{\text{mid}} \simeq -735 \qquad A_{vt3} \simeq -580$$

Base of Q_3:

$$\tau_{\text{hi1}} = R_{Leq1}C_{Leq1} \simeq (1.81\text{ K})(4130\text{ pF}) \simeq 7.5\ \mu\text{sec}$$
$$f_{\text{hi1}} = 1/(2\pi\tau_{\text{hi1}}) \simeq 21\text{ kHz}$$

Base of Q_2:

$$f_{\text{hi2}} \simeq 10\text{ MHz}$$

Solution: (a) Characteristics of the uncompensated amplifier: The Bode plot of T is shown in Fig. 11–20(a). From this curve,

$$f_c \simeq 12\text{ MHz} \qquad \text{P.M.} \simeq 41°$$

(b) Loop-gain characteristics of the compensated amplifier: The desired value of M_{pt} is 5 percent or less. From Fig. 11–14, this value of M_{pt} corresponds to a ξ of 0.7 and a P.M.^{c} of approximately 65°.

(c) Calculation of R_1 and C_1: From Eq. (11–49),

$$\phi_m \simeq \text{P.M.}^c - \text{P.M.} - 3° \simeq 65° - 45° - 3° \simeq 21°$$

$$\sin\phi_m = \frac{a-1}{a+1} = \sin 21° \simeq 0.358$$

The corresponding value of a is 2.1. The new crossover frequency, f_c^c, occurs at that frequency where

$$|T| = 10\log a = 10\log 2.1 \simeq 3.2\text{ db}$$

From Fig. 11–20(a), $|T|$ is 3.2 db at approximately 10 MHz. Thus, f_c^c is 10 MHz. From Eq. (11–49),

$$C_1 = (a-1)C_{Leq1} \simeq (2.1-1)(4130\text{ pF}) \simeq 4550\text{ pF}$$
$$R_1 = \sqrt{a}/(2\pi f_c^c C_1) \simeq (\sqrt{2.1})(0.159)/(10\text{ MHz})(4550\text{ pF}) \simeq 5.1\ \Omega$$

(d) Values of the *RC* lag network components when connected between the base and collector terminals of Q_3: For this example, the *RC* lag network shown in Fig. 11–21(b) will be used. Using Miller's theorem,

$$R_2 = R_1(1 - A_{vt3}) \simeq (5.1\ \Omega)(581) \simeq 2.96\text{ K}$$
$$C_2 = C_1/(1 - A_{vt3}) \simeq (4550\text{ pF})/581 \simeq 7.8\text{ pF}$$

The series combination of R_2 and C_2 connected between the base and collector terminals of Q_3 should give the same results as the series combination of R_1 and C_1 connected from the base terminal of Q_3 to ground.

(e) Corner frequencies of the compensated amplifier: The uncompensated amplifier had two high corner frequencies,

$$f_{\text{hi}1} \simeq 21\text{ kHz} \qquad f_{\text{hi}2} \simeq 10\text{ MHz}$$

The corner frequency $f_{\text{hi}1}$ is replaced by f_{p1}^c, f_z^c, and f_{p2}^c. From Eq. (11–50),

$$f_{p1}^c = f_{p1}/a = f_{\text{hi}1}/a \simeq 21\text{ kHz}/2.1 \simeq 10\text{ kHz}$$
$$f_z^c = 1/(2\pi R_1C_1) = 1/(2\pi R_2C_2) \simeq 0.159/(2.96\text{ K})(7.8\text{ pF}) \simeq 6.9\text{ MHz}$$
$$f_{p2} = af_z^c \simeq (2.1)(6.9\text{ MHz}) \simeq 14.5\text{ MHz}$$

(f) Experimental results: The amplifier was tested using a pulse generator and oscilloscope having a combined rise time of approximately 10 nsec. With R_2 equal to 2.7 K and C_2 equal to 8 pF (values approximately equal to those based on the straight-line Bode plot of T), the measured output response was

$$M_{pt} \simeq 0 \qquad \tau_r \simeq 35\text{ nsec}$$

The measured rise time contains the rise time of the amplifier and also the rise time of the instruments. Using Eq. (11–35), $\tau_{r(\text{amp})}$ is approximately 30 nsec. Using Fig. 11–14, the corresponding loop-gain characteristics are

$$\xi \simeq 1 \qquad \text{P.M.}^c \simeq 75° \qquad f_c^c \simeq 0.24/\tau_{r(\text{amp})} \simeq 8\text{ MHz}$$

The calculated values were

$$\xi \simeq 0.7 \qquad \text{P.M.}^c \simeq 65° \qquad f_c^c \simeq 10\text{ MHz}$$

With R_2 equal to 3.83 K, and C_2 equal to 6.8 pF, the measured output response parameters were:

$$M_{pt} \simeq 5\% \qquad \tau_r \simeq 32\text{ nsec} \qquad \tau_{r(\text{amp})} \simeq 27.2\text{ nsec}$$

Using Fig. 11–14, the corresponding loop gain characteristics are

$$\xi \simeq 0.7 \qquad \text{P.M.}^c \simeq 65° \qquad f_c^c \simeq 0.23/\tau_{r(\text{amp})} \simeq 8.5\text{ MHz}$$

The experimental closed-loop transient response of the amplifier is discussed below.

The amplifier studied in Ex. 11–9 was characterized by three high-frequency poles and one high-frequency zero. While an accurate calculation of the closed-loop frequency response can be obtained by using the Nichols chart, conversion of the results into the transient response is a difficult mathematical task. A calculation of the transient response using the second-order system curves given in Fig. 11–14 will give only approximate results. However, these results are generally indicative of the performance. Three of the observed transient responses at the output of the amplifier in Ex. 11–9 are shown in Fig. 11–25. The response shown in Fig. 11–25(c) corresponds to the experimental results given in Ex. 11–9. Note that the percentage overshoot is perhaps better described as a percentage undershoot. The other transient responses shown in Fig. 11–25 were obtained experimentally.

In this section, three methods of cascade compensation were discussed. The gain compensation method reduces the value of T_{mid} while the dominant pole method reduces the value of the lowest high corner frequency of T to obtain

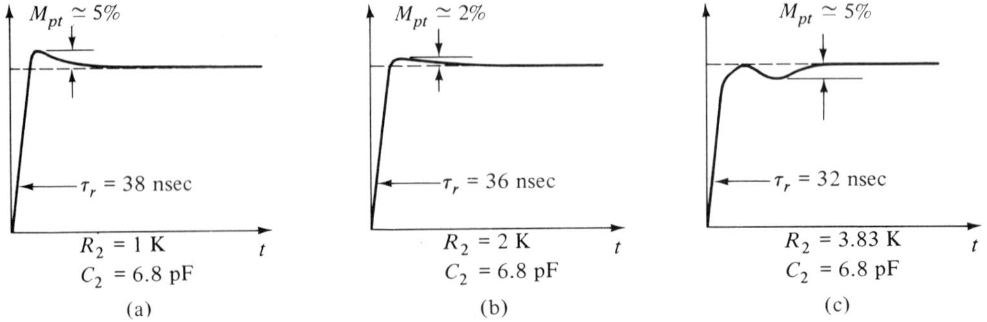

FIGURE 11–25 *Three of the observed transient response waveforms of the amplifier studied in Ex. 11–9. See text for description.*

the desired value of P.M. (also M_{pf} and M_{pt}). The RC lag method converts a pole of T into a pole-zero-pole complex where the zero is used to improve the P.M. When compared to dominant pole compensation, RC lag compensation has the advantage of a smaller rise time for the same value of M_{pt}. Since the same basic amplifier was used in both cases, a direct comparison can be made. From Ex. 11–8, the measured rise time for 5 percent overshoot using dominant pole compensation was 50 nsec. From Ex. 11–9, the measured rise time for 5 percent overshoot using RC lag compensation was 27 nsec.

11.7 The Use of Feedback Compensation in Current Amplifiers

Methods of improving the high-frequency and transient response of feedback amplifiers using cascade compensation were discussed in Sec. 11.6. This section and also the next one will be concerned with the second principal type of compensation, feedback compensation. As noted at the beginning of Sec. 11.6, feedback compensation changes the expression for both T and B. Consequently, both terms in the expression for the closed-loop gain, Eq. (11–1), are affected by feedback compensation. Depending on the circuit values, either $[T/(1+T)]$ or $(1/B)$ can determine the closed-loop high-frequency and transient responses. When $(1/B)$ controls the response, both the closed-loop mid-band gain and the closed-loop high-frequency response can be found quickly from the values of the components in the feedback network. The objective of this section is to study the characteristics of feedback compensation in a current amplifier.* A parallel study of feedback compensation in a voltage amplifier is given in the next section.

A block diagram of a current amplifier is shown in Fig. 11–26. The most simple and most often used type of feedback compensation consists of a capacitor in parallel with the feedback resistor, R_{f2}, as shown in Fig. 11–26(b). The basic amplifier has a voltage gain, A_v; a finite value of input resistance, R_{in}; and a low value of output resistance, R_{out}. Initially, we will assume that the basic amplifier has an infinite bandwidth so that the feedback network alone determines the frequency response.

* In Sec. 10.7, a voltage sample-current sum feedback amplifier was defined as a current amplifier.

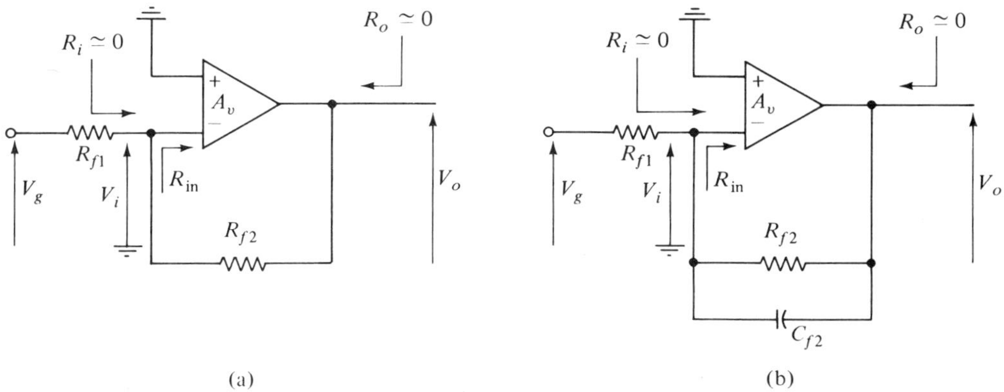

FIGURE 11–26 *A block diagram of a current amplifier: (a) without feedback compensation, and (b) with feedback compensation.*

With C_{f2} disconnected as shown in Fig. 11–26(a), the characteristics of the amplifier correspond to the mid-band values and are

$$\left.\begin{aligned} A_v &= -V_o/V_i & B_{\text{mid}} &= -1/R_{f2} \\ T_{\text{mid}} &= \left[\frac{R_{f1} \,||\, R_{\text{in}}}{R_{f1} \,||\, R_{\text{in}} + R_{f2}}\right][A_v] & A_{fv} &= \frac{V_o}{V_g} \simeq -\frac{R_{f2}}{R_{f1}} \\ A_{f(\text{mid})} &= \left[\frac{|T_{\text{mid}}|}{1 + |T_{\text{mid}}|}\right]\left[\frac{1}{B_{\text{mid}}}\right] \simeq -R_{f2} \end{aligned}\right\} \quad \textbf{(11–51)}$$

With C_{f2} connected as shown in Fig. 11–26(b), the expression for R_{f2} changes to Z_{f2} where

$$Z_{f2} = R_{f2} \,||\, (1/sC_{f2}) \quad \textbf{(11–52)}$$

Replacing R_{f2} with Z_{f2} in Eq. (11–51) gives

$$B^c = 1/Z_{f2} = B_{\text{mid}}(1 + sR_{f2}C_{f2}) \quad \textbf{(11–53)}$$

$$T^c = \left[\frac{R_{f1} \,||\, R_{\text{in}}}{R_{f1} \,||\, R_{\text{in}} + Z_{f2}}\right][A_v] = [T_{\text{mid}}]\left[\frac{1 + sR_{f2}C_{f2}}{1 + s(R_{f1} \,||\, R_{f2} \,||\, R_{\text{in}})C_{f2}}\right] \quad \textbf{(11–54)}$$

where B_{mid} and T_{mid} are given by Eq. (11–51). An expression for the compensated closed-loop gain, A_f^c, is found by substituting the expressions for the B^c and T^c into the closed-loop gain expression, Eq. (11–1). Then by assuming that T_{mid} has a large value, this equation reduces to

$$A_f^c \simeq -R_{f2}/(1 + sR_{f2}C_{f2}) \quad \textbf{(11–55)}$$

We are now in a favorable position to examine the first-order effects* of feedback compensation. As shown by Eq. (11–53), the presence of C_{f2} adds a zero to the expression for B^c. This zero occurs at the corner frequency associated with the $R_{f2}C_{f2}$ time constant. This same zero appears in the expression for the compensated loop gain, T^c, as shown by Eq. (11–54). However, the compensation also adds a pole to the expression for T^c. Since we assumed that the basic amplifier had an infinite bandwidth, the pole and zero in the expression for T^c are due only to the compensation network. Finally, the expression for the compensated loop gain, A_f^c, given by Eq. (11–55), has only a single pole. For a large

* Since the reactance of a capacitor decreases as frequency increases, the effect of C_{f2} is to increase the fed back quantity as frequency increases. For negative feedback, an increase in the feedback causes a decrease in the output. For $f = \infty$, C_{f2} is a short circuit, and $V_i = V_o$.

value of T_{mid}, this pole in A_f^c occurs at the same frequency as the zero in B^c, the corner frequency associated with the $R_{f2}C_{f2}$ time constant. This last result could have been anticipated since, for large values of T_{mid},

$$A_f^c \simeq \frac{1}{B^c} \simeq -\frac{1}{Z_{f2}} = -\frac{R_{f2}}{(1 + sR_{f2}C_{f2})} \tag{11–56}$$

The material in Chap. 10 indicated that in a properly designed feedback amplifier, the closed-loop mid-band value of gain was determined by the feedback network. Equation (11–56) shows that in a properly designed feedback amplifier, the closed-loop, high-frequency response can also be determined by the feedback network. The limitations imposed by the basic amplifier having a finite bandwidth will now be considered.

For an amplifier having a finite bandwidth and using feedback compensation, assume that the uncompensated amplifier has two important high corner frequencies, f_1 and f_2. Then the loop gain can be expressed as

$$T = T_{\text{mid}}\left(\frac{1}{1 + jf/f_1}\right)\left(\frac{1}{1 + jf/f_2}\right) \tag{11–57}$$

From Eq. (11–54) the effect of feedback compensation is to add a zero and a pole in the loop-gain expression so that

$$T^c = T_{\text{mid}}\left(\frac{1}{1 + jf/f_1}\right)\left(\frac{1}{1 + jf/f_2}\right)\left(\frac{1 + jf/f_z^c}{1 + jf/f_p^c}\right) \tag{11–58}$$

Often, however, feedback compensation modifies a pole that is already present rather than introducing a pole at f_p^c. Under these conditions,

$$T^c = T_{\text{mid}}\left(\frac{1}{1 + jf/f_1}\right)\left(\frac{1}{1 + jf/f_2^c}\right)(1 + jf/f_z^c) \tag{11–59}$$

where f_2^c is the modified pole. An example later in this section will illustrate the modified pole effect. For this development, assume that Eq. (11–58) represents the compensated loop gain.

The expression for B^c [Eq. (11–53)] has a zero at the frequency

$$f_z^c = 1/(2\pi R_{f2}C_{f2}) \tag{11–60}$$

Consequently, an expression for $(1/B^c)$ will have a pole of the same frequency, or

$$\frac{1}{B^c} = \left(\frac{1}{B_{\text{mid}}}\right)\left(\frac{1}{1 + jf/f_z^c}\right) \tag{11–61}$$

The effects of feedback compensation are shown in Fig. 11–27. The five curves used to represent the amplifier response are $|T|$, $|1/B|$, $|T/(1 + T)|$, $|A_f|$, and the transient response, X_o. The five curves shown in Fig. 11–27(a) represent the uncompensated response of the amplifier. Note that the closed-loop, high-frequency response has a pronounced peak, and the transient response has overshoot and ringing. Figures 11–27(b) and (c) represent two conditions that are discussed below.

If the value of C_{f2} is chosen such that f_z^c occurs at a frequency much less than the original closed-loop high corner frequency ($f_{\text{hi}(f)}$), the set of curves shown in Fig. 11–27(b) is obtained. The compensated loop gain, T_1^c, has one more pole and zero, f_p^c and f_z^c, than the uncompensated closed loop gain. Also, T_1^c has a higher crossover frequency and a larger P.M. However, the zero in T_1^c corre-

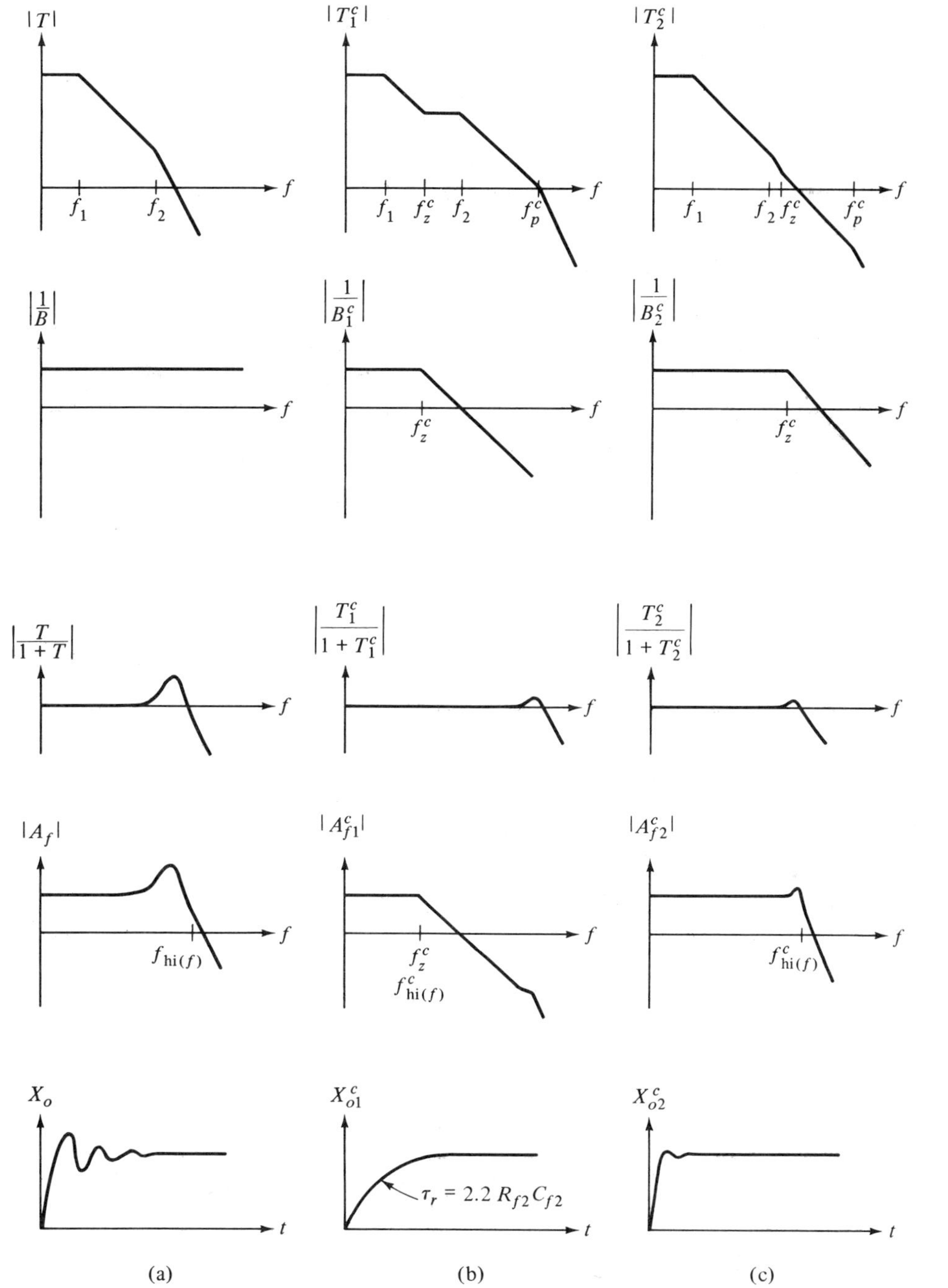

FIGURE 11–27 *A comparison of various responses of an amplifier with feedback compensation: (a) no compensation, (b) conditions when the feedback network determines the closed-loop response, and (c) conditions when the feedback network is adjusted for minimum rise time and minimum overshoot.*

sponds to a pole in $(1/B_1^c)$. When $|T_1^c/(1 + T_1^c)|$ and $|1/B_1^c|$ are combined to form A_{f1}^c, the pole in $1/B_1^c$ dominates the closed-loop high-frequency response. Under these conditions,

$$f_{\text{hi}(f)}^c \simeq f_z^c = 1/(2\pi R_{f2}C_{f2}) \qquad \tau_f^c \simeq 2.2\, R_{f2}C_{f2} \tag{11–62}$$

To summarize briefly, when C_{f2} is chosen such that f_z^c is much less than the uncompensated value of $f_{\text{hi}(f)}$, then the time constant of the feedback network, $R_{f2}C_{f2}$, controls the closed-loop high corner frequency and the closed-loop rise time. This condition is illustrated in an example later in this section.

A second choice for the value of C_{f2} is that value which will reduce the values of M_{pf} and M_{pt} while maintaining $f_{\text{hi}(f)}^c$ at a high value and τ_r^c at a low value. For this condition, f_z^c is chosen near the value of the original loop gain crossover frequency, f_c. Lowering the value of f_z^c has two effects: (1) the loop gain P.M. is increased, thus reducing the values of M_{pf} and M_{pt}, and (2) the value of $f_{\text{hi}(f)}$ is reduced due to $(1/B_2^c)$. Increasing the value of f_z^c produces the opposite effect. Consequently, the exact value of C_{f2} should be determined experimentally. This effect is illustrated by an example later in this section.

Regardless of the value of C_{f2}, the stability of the amplifier should be investigated. Certain values of C_{f2} can cause an otherwise stable amplifier to become unstable as described in Sec. 11.3.

The current amplifier shown in Fig. 11–28 will be used to illustrate the effects of feedback compensation. Before beginning this study, however, we will briefly examine the mid-band characteristics of this amplifier. The input stage, Q_1, forms one-half of a differential pair, Q_1-Q_2.* Transistors Q_1 and Q_3 form a folded cascode arrangement, and Q_4 serves as a constant-current type load for Q_3. The use of a constant-current source as a load resistor is discussed in Sec. 10.12 using Fig. 10–36. Transistors Q_5 and Q_6 form a complementary emitter follower, a circuit that was introduced in Sec. 10.7 using Fig. 10–19. This circuit is discussed in greater detail below. Both the +24-V and −24-V power supplies are connected to the amplifier through a filter circuit consisting of 10-ohm resistors and 6.8-μF capacitors. This type of filter circuit is discussed in Sec. 11.3 using Fig. 11–9. The potentiometer between the emitter terminals of Q_1 and Q_2 is used to adjust the output dc voltage to zero. This potentiometer is shunted by an 11-ohm resistor to increase the value of the loop gain. However, this emitter resistance can still be considered as a form of cascade or gain compensation.

The bias conditions of the complementary emitter-follower circuit can be examined with the aid of Fig. 11–29(a). If the output dc voltage is adjusted to 0 V, then I_{E5} is equal to I_{E6}. Assuming that the diode junctions match the *BE* junctions of Q_5 and Q_6, the emitter current in Q_5 and Q_6 will be slightly less than the diode current due to the voltage drop across R_{E5} and R_{E6}. The use of matched junctions to establish the values of I_{E5} and I_{E6} is similar to the bias arrangement used in the IC shown in Fig. 5–24(a).

One reason for using a complementary emitter-follower circuit is its ability to respond to both positive and negative signals. As shown in Fig. 11–29(b), when a positive ac voltage is applied to the bases of Q_5 and Q_6, an ac current flows into both base terminals. The resulting emitter ac currents are $(\beta + 1)$

* The base of Q_2 could also be used as an input terminal, resulting in a voltage amplifier. This arrangement is treated in the next section.

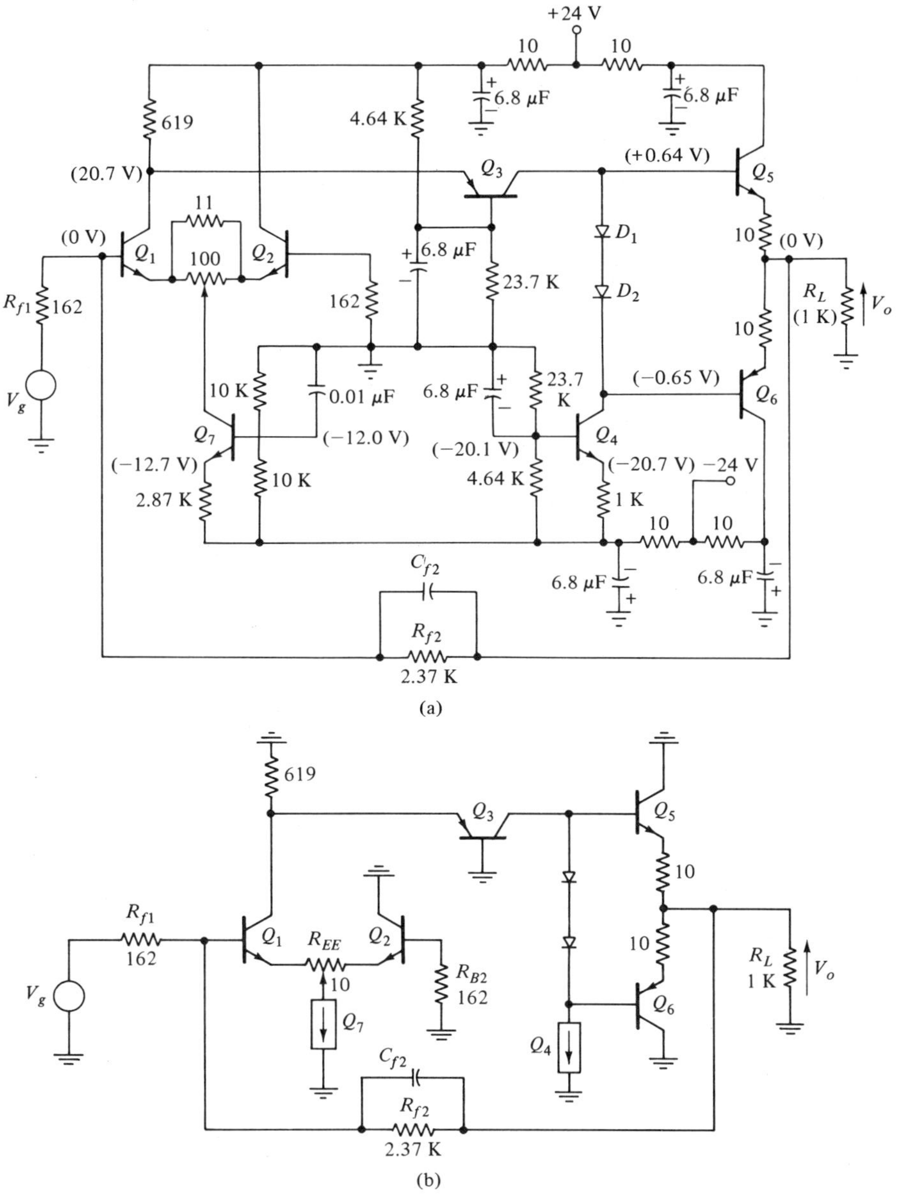

FIGURE 11–28 *The current amplifier used in Exs. 11–10, 11–11, and 11–12: (a) the complete circuit diagram, (b) an ac equivalent circuit, (c) the Bode plot of the uncompensated loop gain, and (d) the Bode plot of the loop gain with* C_{f2} *equal to 100 pF.*

times larger than the base currents. Note that I_{e5} increases the current in Q_5, and that I_{e6} reduces the current flowing in Q_6. Consequently, the output current required for a positive output voltage is furnished primarily by Q_5. Under these conditions, Q_5 is being turned on, and Q_6 is being turned off. A transistor responds much faster when it is turning on. When a negative voltage is applied to the bases of Q_5 and Q_6, the opposite conditions result as shown in Fig. 11–29(c).

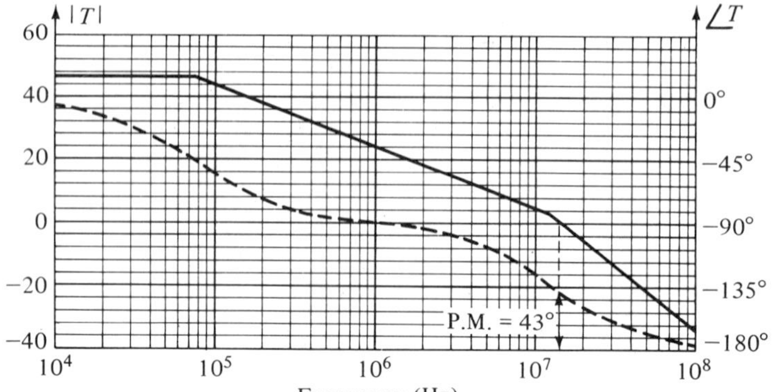

(c)

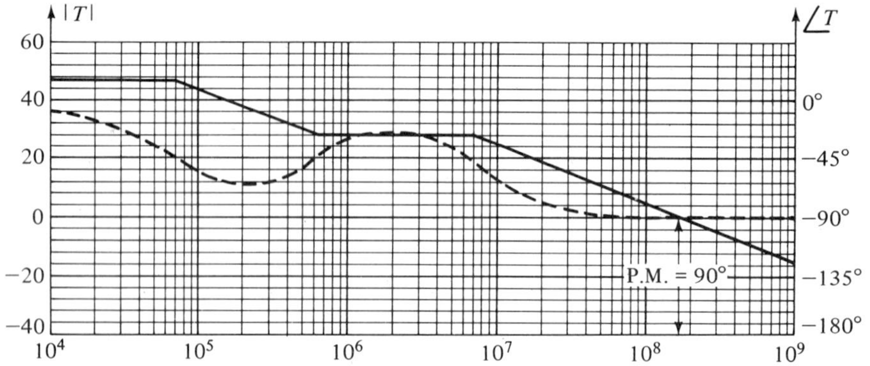

(d)

FIGURE 11–28 (cont.)

Before incorporating a complementary emitter follower into an amplifier, we need an expression for the impedance looking into the base terminals of Q_5 and Q_6. This impedance is indicated by $Z_{i(\text{CEF})}$ in Fig. 11–29(a). The expression for $Z_{i(\text{CEF})}$ is greatly simplified if we assume that either Q_5 or Q_6 is turned on and that the other transistor is turned off.* For a positive signal at the base of Q_5,

$$R_{i(\text{CEF})} \simeq (\beta_5 + 1)(r_{e5} + R_E + R_{L\text{eq}})$$
$$C_{i(\text{CEF})} \simeq C_{b'g5} + C_{ob5} + C_{ob6} \qquad \textbf{(11–63)}$$

Results similar to Eq. (11–63) are obtained when a negative voltage is applied at the base of Q_5. Since Q_5 and Q_6 are normally chosen as complementary transistors, the average values of the various transistor parameters can be used to calculate the impedance looking into the base terminals of Q_5 and Q_6. In the remainder of this section, we will examine the performance of the current amplifier shown in Fig. 11–28 with and without feedback compensation. The following example illustrates the calculation of the response of the amplifier without feedback compensation.

* The complementary emitter follower circuit is treated in detail in Barrett and Blalock in the references listed at the end of this chapter. Equation (11–63) is a reasonable approximation for $Z_{i(\text{CEF})}$ even when both Q_5 and Q_6 are conducting.

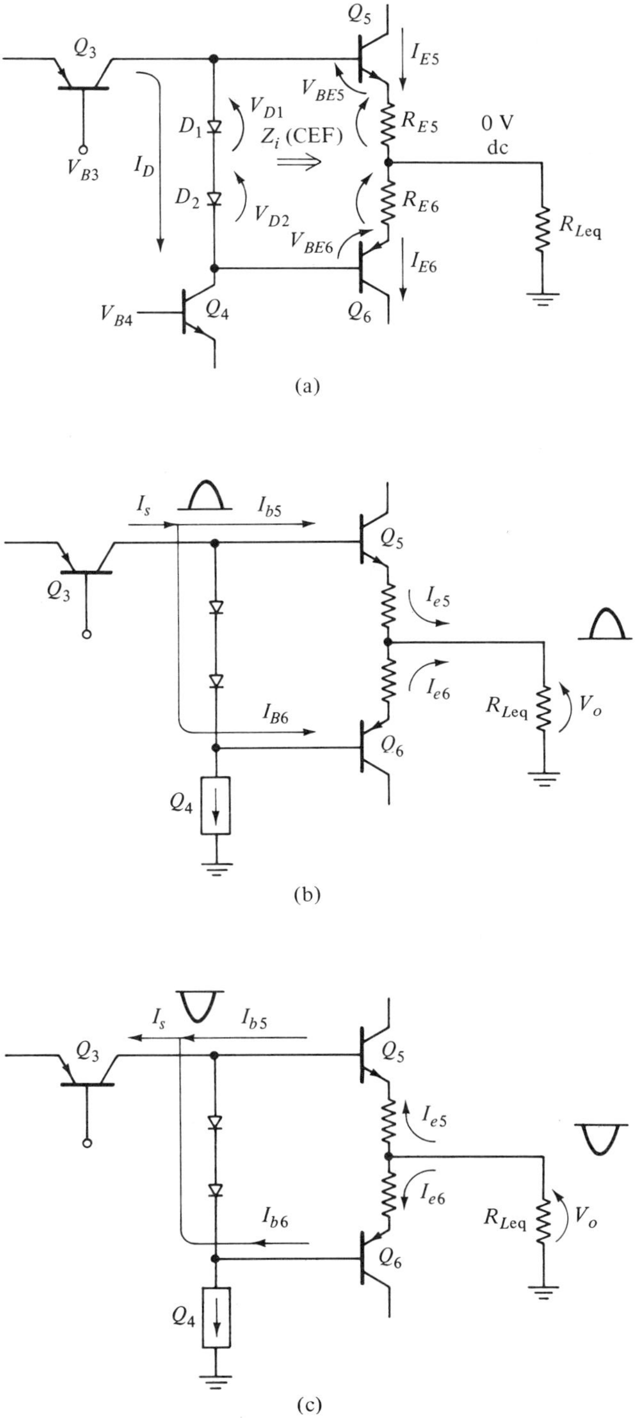

FIGURE 11–29 *A partial circuit diagram showing the complementary emitter follower in Fig. 11–28: (a) bias conditions, (b) circuit conditions for a positive output voltage, and (c) circuit conditions for a negative output voltage.*

EXAMPLE 11–10 Calculate the closed-loop transient response of the feedback amplifier shown in Fig. 11–28(a). The output dc voltage is adjusted to 0 V. The transistor parameters are

Transistor	β	f_T	C_{ob}
Q_1, Q_2	170	60 MHz	8 pF
Q_3, Q_6	225	200 MHz	5 pF
Q_4, Q_5	75	250 MHz	5 pF
Q_7	180	200 MHz	8 pF

Solution: (a) Bias conditions:

$$I_{E5} \simeq I_{E6} \simeq I_{E3} \simeq I_{E4} \simeq 3.4 \text{ mA}$$
$$I_{E1} \simeq [(24\text{ V} - V_{E3})/619] - I_{E3} \simeq 2.1 \text{ mA}$$
$$I_{E2} \simeq I_{E7} - I_{E1} \simeq 1.9 \text{ mA}$$

Since I_{E1} and I_{E2} are both approximately 2 mA, the potentiometer between the emitter terminals of Q_1 and Q_2 is adjusted to its mid-position for 0 V dc at the output.

(b) Value of T_{mid}: Using the average value of β_5 and β_6 or 150,

$$R_{L\text{eq}3} \simeq [\beta_{5,6} + 1][r_{e5,6} + R_E + R_L \,||\, (R_{f2} + R_{f1} \,||\, R_{\text{in}1})]$$
$$\simeq [151][7.6 + 10 + 720] \simeq 110 \text{ K}$$

$$A_{vt1}A_{vt3} \simeq -\frac{R_{L\text{eq}3}}{r_{e1} + r_{e2} + R_{EE} + R_{B2}/(\beta_2 + 1)} \simeq \frac{110\text{ K}}{13 + 13 + 10 + 1} \simeq -3000$$

$$T_{\text{mid}} = (A_{vt1}A_{vt3})(A_{vt5,6})\left(\frac{R_{f1} \,||\, R_{\text{in}1}}{R_{f1} \,||\, R_{\text{in}1} + R_{f2}}\right) \simeq (-3000)(1)\left(\frac{160}{160 + 2.37\text{ K}}\right) \simeq -192$$

The corresponding value of T_{mid} in db is

$$T_{\text{mid}} = 20 \log |-192| \simeq 46 \text{ db}$$

(c) High-frequency response of T: The two principal high corner frequencies of T occur at the collector terminals of Q_3 and at the base terminal of Q_1. Neglecting the effect of $C_{b'g}$ in the complementary emitter-follower circuit due to the large equivalent emitter resistance,

$$\tau_{\text{hi}1} \simeq R_{L\text{eq}3}C_{L\text{eq}3} \simeq R_{L\text{eq}3}(C_{ob3} + C_{ob4} + C_{ob5} + C_{ob6}) \simeq 2.2\ \mu\text{sec}$$
$$f_{\text{hi}1} \simeq 1/(2\pi\tau_{\text{hi}1}) \simeq 72 \text{ kHz}$$

At the base of Q_1,

$$C_{b'g1} = 0.159/[r_{e1} + r_{e2} + R_{EE} + R_{B2}/(\beta_2 + 1)]f_{T1}] \simeq 72 \text{ pF}$$
$$A_{vt1} \simeq -r_{e3}/[r_{e1} + r_{e2} + R_{EE} + R_{B2}/(\beta_2 + 1)] \simeq 7.6/37 \simeq -0.2$$
$$C_{\text{in}1} = C_{b'g1} + C_{ob1}(1 - A_{vt1}) \simeq [72 + 8(1.2)] \text{ pF} \simeq 82 \text{ pF}$$

The resistance looking into the emitter terminal of either Q_5 or Q_6 is large due to the large value of resistance looking into the collector terminals of Q_3 and Q_4. Thus,

$$\tau_{\text{hi}2} \simeq [R_{f1} \,||\, R_{\text{in}1} \,||\, (R_{f2} + R_L)](C_{\text{in}1}) \simeq (162 \,||\, 6.3\text{ K} \,||\, 2.37)(82\text{ pF}) \simeq 12.3 \text{ nsec}$$
$$f_{\text{hi}2} \simeq 1/(2\pi\tau_{\text{hi}2}) \simeq 13 \text{ MHz}$$

(d) Closed-loop transient response: Using the data calculated in (b) and (c), the Bode plot of T can be constructed as shown in Fig. 11–28(c). From this Bode plot,

$$f_c \simeq 13.5 \text{ MHz}, \qquad \text{P.M.} \simeq 43°$$

Using Fig. 11–14,

$$\xi \simeq 0.4 \qquad \tau_r \simeq 0.2/f_c \simeq 15 \text{ nsec} \qquad M_{pt} \simeq 25\%$$

(e) Experimental results: The measured bias voltages are shown in parentheses in Fig. 11–28(a). The measured output response at a signal level of 1 V was

$$\tau_r \simeq 22 \text{ nsec} \qquad M_{pt} \simeq 28\%$$

When the effects of the 10-nsec response of the generator and oscilloscope are removed, the amplifier rise time is 17 nsec. Using a 1-kHz sine wave, the maximum output signal obtained without noticeable distortion was approximately ±19.5 V or 39 V peak to peak.

The results of the preceding example indicate that the transient response of the uncompensated amplifier is characterized by a rise time of approximately 15 nsec with a 25 percent overshoot. Our next task is to apply feedback compensation to this amplifier to improve the closed-loop transient response. An improved response can imply many conditions, two of which are: (1) the closed-loop rise time is predictable from the feedback network and reproducible in many amplifiers of the same type, and (2) the closed-loop rise time has a minimum value for negligible overshoot. The first condition is important in the production of a large number of amplifiers whose transient response must meet precise specifications. This condition is illustrated in Ex. 11–11. The second condition is important when a particular amplifier is to be adjusted to obtain the optimum response. This condition is illustrated in Ex. 11–12.

EXAMPLE 11–11 For the amplifier considered in Ex. 11–10, calculate the value of C_{f2} required to produce an output rise time of 500 nsec. Also, determine the effect of the compensation on the loop-gain response.

Solution: (a) Value of C_{f2}: Since the compensated rise time, τ_r^c, is to be 500 nsec, and the uncompensated rise time is only 15 nsec, the value of C_{f2} can be determined from Eq. (11–62) as

$$C_{f2} = \tau_r^c/2.2R_{f2} = (500 \text{ nsec})/(2.2)(2.37 \text{ K}) \simeq 96 \text{ pF}$$

A practical fixed value for C_{f2} is 100 pF. Then,

$$\tau_r^c \simeq 2.2R_{f2}C_{f2} \simeq (2.2)(2.37 \text{ K})(100 \text{ pF}) \simeq 520 \text{ nsec}$$

(b) Effect of C_{f2} on the loop-gain response: The principal effect of C_{f2} on the loop gain is to add a zero at the frequency f_z^c and change the value of f_{hi2}. The zero occurs at

$$f_z^c = 1/(2R_{f2}C_{f2}) \simeq 670 \text{ kHz}$$

The new time constant associated with the pole at the base of Q_1 is

$$\tau_{\text{h12}}^c \simeq (R_{f1} \,||\, R_{\text{in1}} \,||\, R_{f2})(C_{\text{in1}} + C_{f2}) \simeq (150)(182 \text{ pF}) \simeq 27.3 \text{ nsec}$$
$$f_{\text{h12}}^c \simeq 1/(2\pi\tau_{\text{h12}}^c) \simeq 5.8 \text{ MHz}$$

The expression for the compensated loop gain becomes

$$T^c \simeq -192\left(\frac{1}{1 + jf/72 \text{ kHz}}\right)(1 + jf/670 \text{ kHz})\left(\frac{1}{1 + jf/5.8 \text{ MHz}}\right)$$

The Bode plot of the compensated loop gain is shown in Fig. 11–28(d). The new value of f_c^c is approximately 150 MHz, and the P.M.^{c} is approximately 90°. These values are unrealistic since f_c^c is of the same order of magnitude as f_T for the various transistors. The numerous higher order poles and zeros that have been neglected are now important. Because of the circuit complexity, the stability of the amplifier should be determined experimentally or using a computer aided analysis.

(b) Experimental results: With C_{f2} equal to 100 pF, the measured rise time was 540 nsec with no overshoot.

The important result of Ex. 11–11 is that the closed-loop high-frequency response or the closed-loop transient response can be predicted accurately from the time constant of the feedback network when the rise time associated with this time constant is sufficiently larger than the rise time of the uncompensated amplifier. When this condition is satisfied, along with a large value of T_{mid}, both the closed-loop gain and the closed-loop transient response are relatively constant for changes in transistors, power supply voltages, generator resistance, and load resistance. Since the transient response is controlled by Z_{f2}, the amplifier can be approximated by a single pole. Often, however, the problem is to obtain the minimum rise time possible (or maximum bandwidth) for a low value of overshoot from a particular amplifier. Then the exact value of C_{f2} should be determined experimentally. This condition is illustrated in the following example.

EXAMPLE 11–12 For the amplifier in Ex. 11–10, estimate the value of C_{f2} required to give a minimum value of rise time with no overshoot.

Solution: (a) Initial choice of C_{f2}: The initial choice of C_{f2} should be such that the zero, f_z^c, occurs near the original crossover frequency of the loop gain, f_c.

$$f_z^c = 1/(2\pi R_{f2}C_{f2}) \simeq f_c \simeq 13.5 \text{ MHz}$$

Solving for C_{f2},

$$C_{f2} \simeq 0.159/(2.37 \text{ K})(13.5 \text{ MHz}) \simeq 5 \text{ pF}$$

A trimmer capacitor with a range on each side of 5 pF should be chosen for C_{f2}. Trimmer capacitors available in this range are 2-8 pF and 5.5-18 pF. These values are only nominal and can vary $\pm 20\%$ at each extreme.

(b) Experimental results: Using a 5.5-18 pF capacitor, the minimum output rise time without noticeable overshoot was 34 nsec. Since the instrument system had a 10-nsec rise time, the amplifier rise time was approximately 28 nsec.

Adjusting the trimmer capacitor to one extreme gave an output rise time of 28 nsec with 5 percent overshoot. Adjusting the trimmer capacitor to the other extreme gave an output rise time of 66 nsec with no overshoot.

In this section the effects of feedback compensation in a current amplifier have been considered. Two important results were obtained. First, the value of the closed-loop, high corner frequency or the closed-loop transient response can be determined primarily by the feedback network under certain conditions. This effect was illustrated in Ex. 11–11. Second, feedback compensation can be used to provide an optimum response by adjusting the value of C_{f2} for minimum overshoot and minimum rise time. This condition was illustrated in Ex. 11–12. Also, some of the important characteristics of the complementary emitter-follower output or driver circuit were discussed. The results of this section will be used in the remainder of this chapter and in the next chapter.

11.8 The Use of Feedback Compensation in Voltage Amplifiers

The effects of feedback compensation in current amplifiers were considered in the last section. In this section, we will study the effects of feedback compensation in voltage amplifiers. As in the last section, our study will begin with a block diagram so that the effects of feedback compensation in voltage and current amplifiers can be compared.

A block diagram of a voltage amplifier is shown in Fig. 11–30. The simplest and most often used type of feedback compensation consists of adding a capacitor in parallel with the feedback resistor, R_{f2}, as shown in Fig. 11–30(b). The basic amplifier may be characterized by the following parameters: a voltage gain, A_v, equal to V_o/V_i; a finite value of resistance, R_{in}, looking into the input terminal; and a low value of output resistance, R_{out}. Initially, we will also assume that the basic amplifier has an infinite bandwidth so that the feedback network alone determines the frequency response.

With C_{f2} disconnected, as shown in Fig. 11–30(a), the amplifier characteristics correspond to their mid-band values and:

$$A_v = \frac{V_o}{V_i} \qquad T_{\text{mid}} = \frac{R_{f1} \,||\, R_{\text{in}}}{R_{f1} \,||\, R_{\text{in}} + R_{f2}}(A_v)$$

$$B_{\text{mid}} = \frac{R_{f1}}{R_{f1} + R_{f2}} \qquad A_{f(\text{mid})} = \left(\frac{T_{\text{mid}}}{1 + T_{\text{mid}}}\right)\left(\frac{1}{B_{\text{mid}}}\right) \simeq \frac{R_{f1} + R_{f2}}{R_{f1}} \tag{11–64}$$

With C_{f2} connected, as shown in Fig. 11–30(b), R_{f2} becomes Z_{f2} where

$$Z_{f2} = R_{f2} \,||\, (1/sC_{f2}) = R_{f2}/(1 + sR_{f2}C_{f2}) \tag{11–65}$$

Replacing R_{f2} with Z_{f2} in Eq. (11–64),

$$B^c = [B_{\text{mid}}]\left[\frac{1 + sR_{f2}C_{f2}}{1 + s(R_{f1} \,||\, R_{f2})C_{f2}}\right] \tag{11–66}$$

$$T^c = T_{\text{mid}}\left[\frac{1 + sR_{f2}C_{f2}}{1 + s(R_{f1} \,||\, R_{f2} \,||\, R_{\text{in}})C_{f2}}\right] \tag{11–67}$$

where B_{mid} and T_{mid} are given by Eq. (11–64).

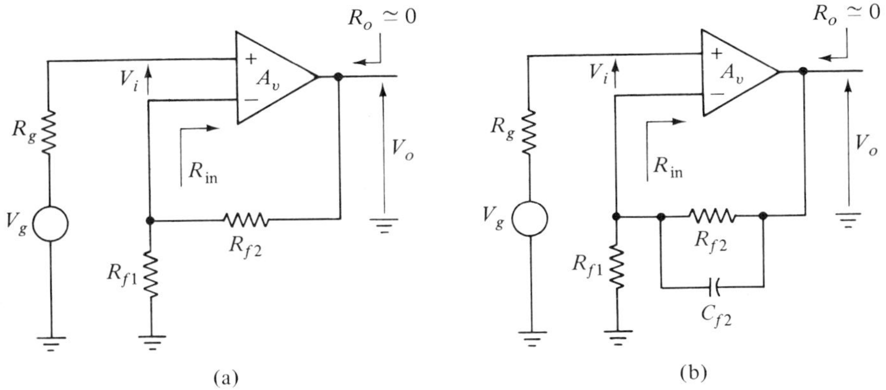

FIGURE 11–30 *A block diagram representation of a voltage amplifier (a) without feedback compensation, and (b) with feedback compensation.*

An expression for the compensated closed-loop gain is found by substituting the expression for B^c and T^c into the closed-loop gain expression, Eq. (11–1). Then by assuming a large value of T_{mid}, the compensated gain with feedback is

$$A_f^c \simeq A_{f(\text{mid})}\left[\frac{1 + s(R_{f1} \parallel R_{f2})C_{f2}}{1 + sR_{f2}C_{f2}}\right] \tag{11–68}$$

We are now in a favorable position to compare the results of feedback compensation in a voltage amplifier with the effects of feedback compensation in a current amplifier. By comparing Eqs. (11–54) and (11–67), we can see that the expression for T^c for both a current amplifier and a voltage amplifier have an additional pole and zero due to the presence of C_{f2}.* Since the zero occurs at a lower frequency than the pole, the zero can introduce a leading or positive phase shift to improve the loop-gain phase margin. Consequently, the general results of the last section for a current amplifier can also be applied to a voltage amplifier.

Although the expression for T^c has the same form for both the current and the voltage amplifier, the expressions for B^c and A_f^c are different. In a current amplifier, the expression for B^c [Eq. (11–54)] had a single zero, and the expression for A_f^c [Eq. (11–56)] had a single pole. The corner frequency associated with both the zero in B^c and the pole in A_f^c was f_z^c where,

$$f_z^c = 1/(2\pi R_{f2}C_{f2}) \tag{11–69}$$

In a voltage amplifier, however, the expression for B^c [Eq. (11–66)] has both a zero and a pole where the corner frequency of the zero is given by Eq. (11–69), and the corner frequency of the pole is

$$f_p^c = 1/[2\pi(R_{f1} \parallel R_{f2})C_{f2}] \tag{11–70}$$

Combining Eqs. (11–66), (11–69), and (11–70), an expression for B^c is

$$B^c = \left[\frac{R_{f1}}{R_{f1} + R_{f2}}\right]\left[\frac{1 + jf/f_z^c}{1 + jf/f_p^c}\right] \tag{11–71}$$

In a similar manner, we can combine Eqs. (11–68) through (11–70) and express A_f^c as

$$A_f^c \simeq \left[\frac{R_{f1} + R_{f2}}{R_{f1}}\right]\left[\frac{1 + jf/f_p^c}{1 + jf/f_z^c}\right] \tag{11–72}$$

Equations (11–71) and (11–72) illustrate one important effect of feedback compensation. The zero of B^c becomes the pole of A_f^c, and the pole of B^c becomes the zero of A_f^c. This effect could have been anticipated since A_f^c is approximately equal to $1/B^c$. Since we assumed in the block diagram analysis that the basic amplifier had an infinite bandwidth, the pole and zero in the expression for A_f^c [Eq. (11–72)] are due only to the compensation network.

Using Eqs. (11–64), (11–69), and (11–70), the expressions for f_p^c and A_f^c can be written as

$$f_p^c = \frac{1}{2\pi\left[\dfrac{R_{f1}R_{f2}}{R_{f1} + R_{f2}}\right]C_{f2}} = \left[\frac{R_{f1} + R_{f2}}{R_{f1}}\right]\left[\frac{1}{2\pi R_{f2}C_{f2}}\right] = A_{f(\text{mid})}f_z^c$$

$$A_f^c \simeq A_{f(\text{mid})}\left[\frac{1 + jf/A_{f(\text{mid})}f_z^c}{1 + jf/f_z^c}\right] \tag{11–73}$$

* As noted in the last section, the feedback compensation does not always introduce a new pole. Sometimes the only effect is to modify a pole that is already there.

From this last equation the Bode plot of the closed-loop gain can be constructed as shown in Fig. 11–31(a). Note that the gain changes from its mid-band value of f_z^c to zero db (unity) at f_p^c. The value of A_f^c will not go below zero db until the response of the basic amplifier causes a reduction in T^c. If the time constant of the feedback network, $R_{f2}C_{f2}$, controls the closed-loop, high-frequency response, then the pole and zero of A_f^c should be widely separated. Since f_p^c and f_z^c are separated in frequency an amount equal to $A_{f(\text{mid})}$, feedback compensation in voltage amplifiers is most effective when the value of $A_{f(\text{mid})}$ is greater than approximately 10. Also, the high corner frequency associated with $R_{f2}C_{f2}$ should be much less than the value of $f_{\text{hi}(f)}$ without feedback compensation. This effect is shown in Fig. 11–31(b).

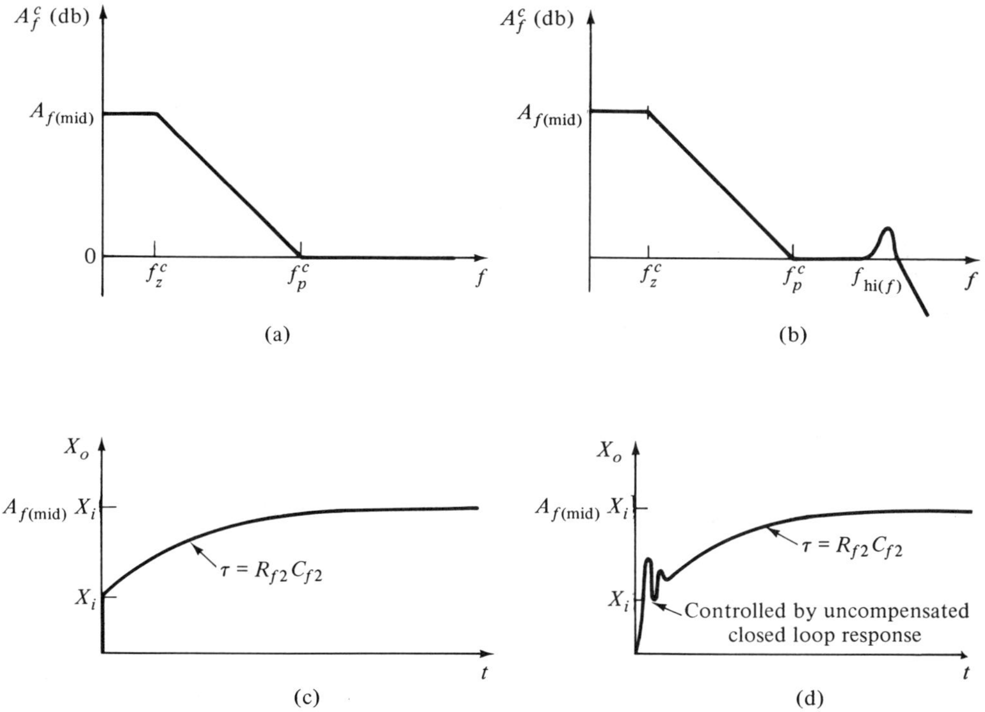

FIGURE 11–31 *Closed-loop characteristics of a voltage amplifier using feedback compensation: (a) Bode plot of the high-frequency response when the basic amplifier has an infinite bandwidth, (b) Bode plot of the high-frequency response when the basic amplifier has a finite bandwidth, (c) transient response when the basic amplifier has a zero rise time, and (d) transient response when the basic amplifier has a finite rise time with overshoot.*

The pole and zero in A_f^c also will affect the closed-loop transient response. Assuming a step input, the expression for X_o corresponding to the expression for A_f^c in Eq. (11–73) is*

$$X_o = X_i\{A_{f(\text{mid})} - [A_{f(\text{mid})} - 1][\exp(-t/R_{f2}C_{f2})]\} \qquad \textbf{(11–74)}$$

At t equal to zero, X_o is equal to X_i. The general form of the transient response for a step input is shown in Fig. 11–31(c). Note that the output signal consists of a sudden jump equal to X_i, followed by an exponential rise to a final value

* Equation (11–74) is obtained by forming the inverse Laplace transform of Eq. (11–73) assuming a step function input.

which is equal to $A_{f(\text{mid})}X_i$. Actually, the initial jump in X_o is not a perfect step. The characteristics of the initial jump (rise time and overshoot) correspond to those of the amplifier with R_{f2} shorted since, at very high frequencies, C_{f2} is essentially a short circuit. Figure 11–31(d) shows the transient response when the basic amplifier has a finite rise time with overshoot. The exponential portion of the curve is controlled by the time constant of the feedback network. If the value of $A_{f(\text{mid})}$ is greater than 10, then the initial jump in X_o is less than 10 percent of the final value of the output signal, and its effect on the transient response can be neglected. Under these conditions, the closed-loop rise time for the compensated system is 2.2 $R_{f2}C_{f2}$ provided that the value of τ_r^c is much greater than the value of τ_r without feedback compensation.

EXAMPLE 11–13 Calculate the value of C_{f2} that is required to produce a 500-nsec rise time for the amplifier shown in Fig. 11–32. The transistor parameters are given in Ex. 11–10.

Solution: (a) Characteristics of the uncompensated loop gain: When calculating the loop gain, the ac equivalent circuit shown in Fig. 11–32(b) is identical to that shown in Fig. 11–28(b). Consequently, the results of Ex. 11–10 can be used to characterize the uncompensated loop gain. From Ex. 11–10,

$$T_{\text{mid}} \simeq -192 \qquad f_c \simeq 13.5\text{ MHz} \qquad \text{P.M.} \simeq 43°$$

(b) Characteristics of the uncompensated closed-loop response: The mid-band value of A_f is

$$A_f \simeq 1/B = (R_{f1} + R_{f2})/R_{f1} = (162 + 2.37\text{ K})/162 \simeq 15.6$$

Using Fig. 11–14 and the loop-gain characteristics from (a), the closed-loop transient response can be estimated as

$$\xi \simeq 0.4 \qquad \tau_r \simeq 0.2/f_c \simeq 15\text{ nsec} \qquad M_{pt} \simeq 25\%$$

(c) Value of C_{f2}: Since the value of τ_r^c is much greater than τ_r (500 nsec ≫ 15 nsec), and the value of $A_{f(\text{mid})}$ is greater than 10, the value of C_{f2} is

$$C_{f2} \simeq \tau_r^c/2.2R_{f2} \simeq (500\text{ nsec})/(2.2)(2.37\text{ K}) \simeq 96\text{ pF}$$

A practical fixed value of C_{f2} is 100 pF. Then,

$$\tau_r^c \simeq 2.2R_{f2}C_{f2} = (2.2)(2.37\text{ K})(96\text{ pF}) \simeq 520\text{ nsec}$$

(d) Effect of initial jump in V_o: The amplitude of the initial jump in V_o is equal to the amplitude of V_i. Since the final value of V_o is $15.6V_i$, the initial jump in V_o is only 6.5 percent of the final value of V_o. Consequently, the initial jump will have little effect on the closed-loop transient response.

(e) Experimental results: The measured bias voltages are shown in Fig. 11–32(a). The mid-band gain was measured at 10 kHz. With V_g equal to 0.64 V, V_o was 10 V, giving a gain, $A_{f(\text{mid})}$, of 15.6. At an output signal level of 1 V, the measured transient response of the uncompensated amplifier was

$$\tau_r \simeq 24\text{ nsec} \qquad M_{pt} \simeq 28\%$$

When the effects of the 10-nsec rise time of the generator and oscilloscope are removed, the measured response of the amplifier becomes

$$\tau_{r(\text{A})} \simeq 19.4\text{ nsec} \qquad M_{pt} \simeq 28\%$$

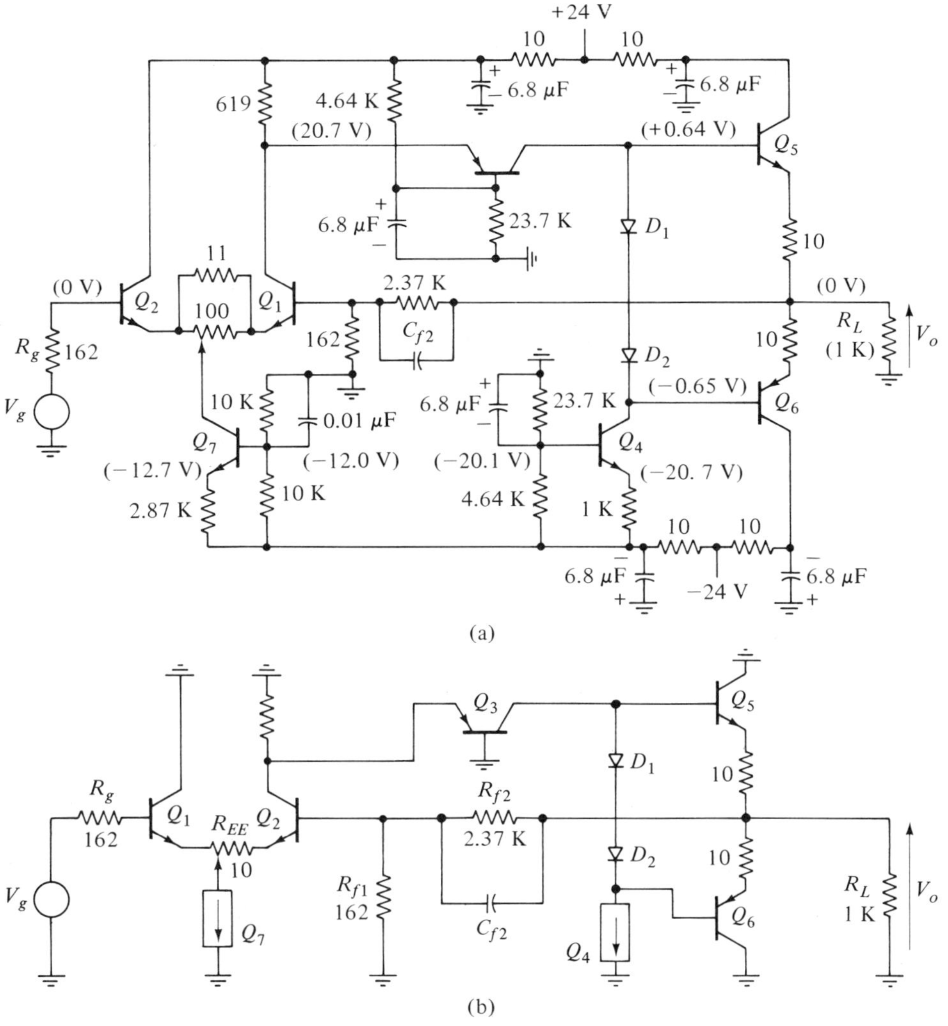

FIGURE 11–32 *The feedback amplifier from Exs. 11–13 and 11–14: (a) the complete circuit, and (b) an ac equivalent circuit. This circuit is identical to the one shown in Fig. 11–28 except that the input signal is applied to the base of* Q_2. *The functions of the various active devices in the circuit are discussed in Sec. 11.7.*

The measured output response of the compensated amplifier with C_{f2} equal to 100 pF was

$$\tau_r^c \simeq 550 \text{ nsec}$$

There was no overshoot in the step response of the compensated amplifier. The initial jump in V_o was obscured by the finite rise time of the amplifier without compensation.

The important result of Ex. 11–13 is that the closed-loop, high-frequency response or the closed-loop, transient response can be predicted from the time constant of the feedback network when the rise time associated with this time constant is much larger than the rise time of the uncompensated amplifier. When this condition is satisfied along with a large value of T_{mid} and an $A_{f(mid)}$ greater than 10, both the closed-loop gain and the closed-loop transient response are relatively constant for changes in transistors, power supply, generator and load resistors. By comparing the results of Exs. 11–11 and 11–13, we can see that the transient response of a current amplifier and a voltage amplifier are very similar when using the same feedback network components. The polarity and value of the mid-band gains, however, are different.

Often it is desirable to obtain the minimum rise time possible (or maximum bandwidth) for a low value of overshoot from a particular amplifier. For this condition, f_z^c is chosen near the value of the uncompensated loop gain crossover frequency, f_c. If the value of $A_{f(mid)}$ is greater than 10, the approach used with current amplifiers and illustrated in Ex. 11–12 can also be used with voltage amplifiers. This point is illustrated in the following example.

EXAMPLE 11–14 For the amplifier in Ex. 11–13, estimate the value of C_{f2} that is required to give a minimum value of rise time with no overshoot.

Solution: (a) Initial choice of C_{f2}: The initial choice of C_{f2} should be such that the zero, f_z^c, occurs near the original crossover frequency of the loop gain, f_c.

$$f_z^c = 1/(2\pi R_{f2}C_{f2}) \simeq f_c \simeq 13.5 \text{ MHz}$$

Solving for C_{f2},

$$C_{f2} \simeq 0.159/(2.37 \text{ K} \times 13.5 \text{ MHz}) \simeq 5 \text{ pF}$$

A trimmer capacitor having a range around 5 pF should be chosen for C_{f2}.

(b) Experimental results: Using a 5.5-18-pF capacitor, the minimum output rise time without noticeable overshoot was 40 nsec. Since the measuring system had a 10-nsec rise time, the amplifier rise time was approximately 35 nsec.

Adjusting the trimmer capacitor to one extreme gave an output rise time of 34 nsec with 5 percent overshoot. Adjusting the trimmer capacitor to the other extreme gave an output rise time of 69 nsec with no overshoot. A small initial jump in V_o was observed in this last setting of C_{f2} indicating that the output was not a simple exponential curve.

Our final task is to illustrate the effect of a small value of $A_{f(mid)}$ on the transient response of a voltage amplifier. A modified form of the amplifier shown in Fig. 11–32 is shown in Fig. 11–33 where R_{f2} has been changed from 2.37 K to 162 ohms. When the modified amplifier is connected as a voltage amplifier as shown in Fig. 11–33(b), V_o/V_g is approximately 2, and the initial jump in V_o will equal V_i or 50 percent of the final value of V_o. The rise time and percent overshoot of the initial jump can be calculated with C_{f2} replaced by a short circuit. The $R_{f2}C_{f2}$ time constant controls the response of V_o between the output signal levels of V_i and $A_{f(mid)}V_i$. This exponential portion of V_o has a rise time of 2.2 $R_{f2}C_{f2}$ or approximately 200 nsec. The observed transient response of the voltage amplifier is shown in Fig. 11–33(b). The two portions of the response are easily identified. The initial jump had a measured rise time of 15 nsec with 35 percent overshoot. The measured rise time of the exponential portion of V_o was 200 nsec. The rise time of the measuring circuit was approximately 10 nsec.

The modified amplifier can also be connected as a current amplifier as shown in Fig. 11–33(c). Under these conditions, V_o/V_g is approximately -1, and the

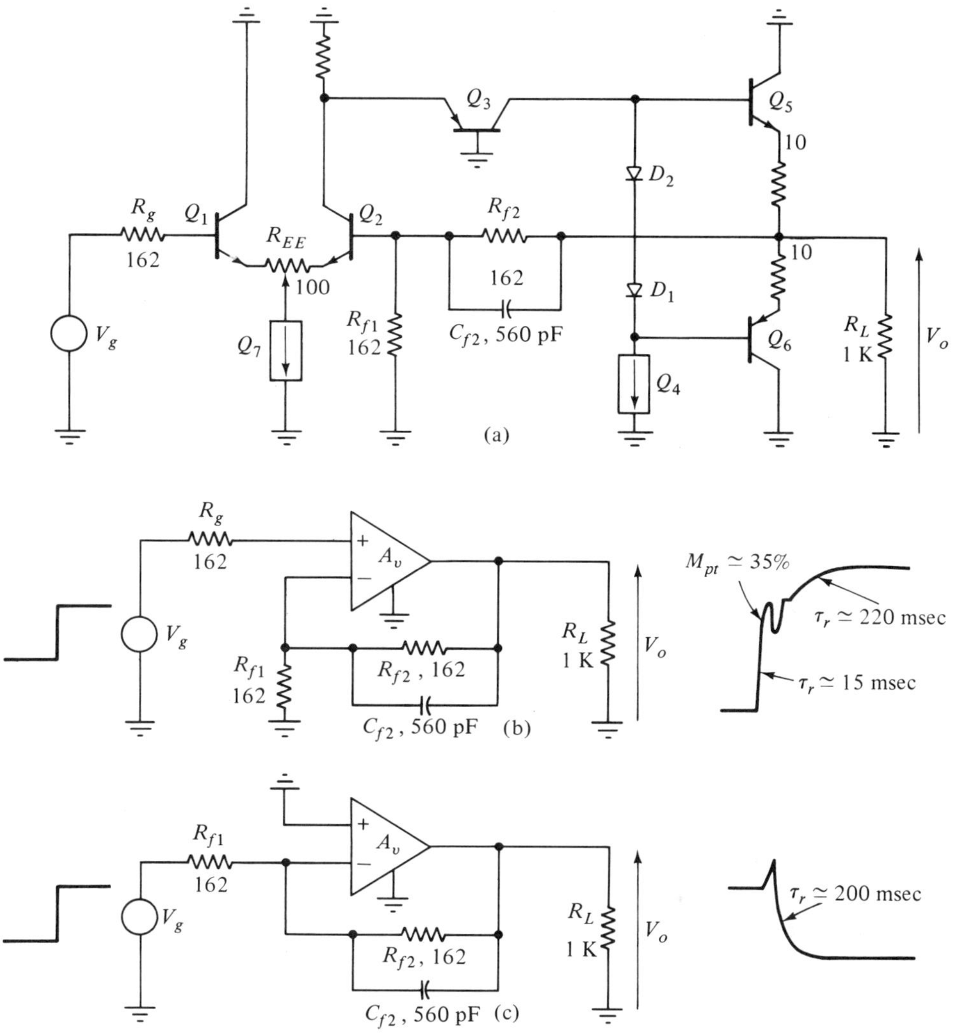

FIGURE 11–33 *A modified form of the amplifier shown in Fig. 11–32: (a) an ac equivalent circuit, (b) a block diagram of the amplifier connected as a voltage amplifier, and (c) a block diagram of the amplifier connected as a current amplifier. See text for circuit description.*

rise time is controlled by 2.2 $R_{f2}C_{f2}$ or approximately 200 nsec. The observed transient response is shown in Fig. 11–33(c). The initial positive spike in the output of the current amplifier is due to the passive feedthrough of the feedback network.* The measured rise time of the current amplifier was 220 nsec.

In this section we have examined the effects of feedback compensation in voltage amplifiers. When the value of $A_{f(\mathrm{mid})}$ is greater than approximately 10, the effects of feedback compensation in voltage amplifiers are similar to the

* Calculation of the initial spike requires a knowledge of the time delay of the forward signal path of the amplifier. While calculation of this transient effect is beyond the scope of this text, the mid-band effects of passive signal transfer in current amplifiers are included in Sec. 10.10.

effects of feedback compensation in current amplifiers. The feedback compensation can control the closed-loop transient response when τ_r^c is sufficiently larger than τ_r as illustrated in Ex. 11–13. Alternately, the feedback compensation can be used to obtain a nearly optimum step response for a given amplifier as illustrated in Ex. 11–14. When the value of $A_{f(\text{mid})}$ is much less than 10, the pole and zero of A_f tend to cancel each other, and the value of feedback compensation is reduced as illustrated in Fig. 11–33.

11.9 Characteristics of Operational Amplifiers

Negative-feedback amplifiers with large values of loop gain and certain input and output characteristics are called *operational amplifiers.* The objective of this section is to examine some of the important characteristics of operational amplifiers. Several applications of the operational amplifier are introduced in the next section.

A block diagram of an operational amplifier is shown in Fig. 11–34(a). Note that this diagram corresponds to the familiar voltage sample-current sum feedback configuration or current feedback amplifier. The gain block can be represented by the approximate equivalent circuit shown in Fig. 11–34(b) where R_{in} is the differential input resistance; R_o is the output resistance, and A_{vol} is the open-loop voltage gain of the amplifier. The ideal characteristics of an operational amplifier are: (1) infinite voltage gain, (2) infinite input resistance, (3) zero output resistance, (4) infinite bandwidth, and (5) zero input and output offset voltage and current.

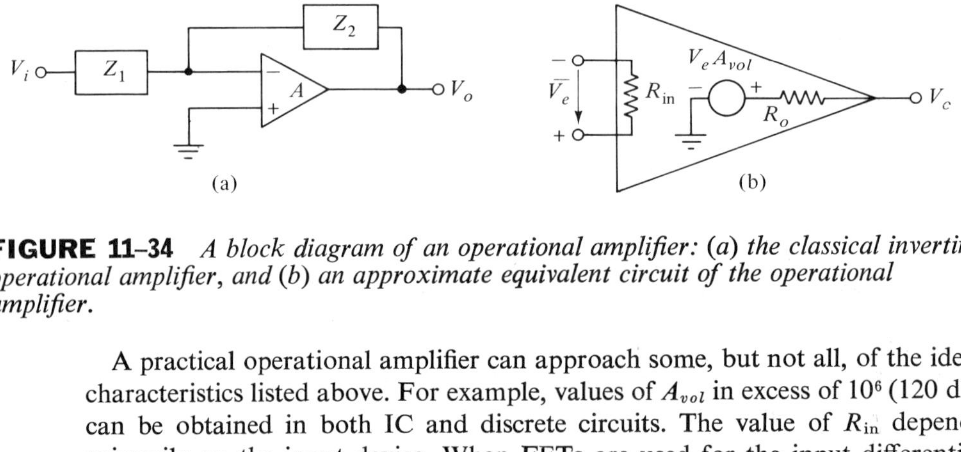

FIGURE 11–34 *A block diagram of an operational amplifier: (a) the classical inverting operational amplifier, and (b) an approximate equivalent circuit of the operational amplifier.*

A practical operational amplifier can approach some, but not all, of the ideal characteristics listed above. For example, values of A_{vol} in excess of 10^6 (120 db) can be obtained in both IC and discrete circuits. The value of R_{in} depends primarily on the input device. When FETs are used for the input differential pair, a very large value of R_{in} can be obtained. Values of R_o are often low since some type of CC output stage is generally used in an operational amplifier. Offset voltages and currents are usually small, and their values depend on many factors. Data sheets usually list values for A_{vol}, R_{in}, R_o, offset voltage, offset current, CMRR, maximum output voltage swing, and maximum power dissipation.

Calculation of the frequency response of an operational amplifier circuit is complicated by several factors. First, the frequency response of an operational amplifier corresponds to the frequency response of the open-loop gain and does not include the effects of the feedback network. Second, high-frequency compensation is generally required to obtain closed-loop stability for the overall feedback amplifier. Third, the output voltage of the amplifier is limited to some maximum rate of change, called the *slew rate*, which imposes both frequency and transient response limitations on the overall amplifier. Each of these three topics is considered below.

Data sheets for operational amplifiers use a variety of methods to describe the frequency response. Often a Bode plot of A_{vol}, both magnitude and phase, is provided. Many operational amplifiers have external leads that connect to various points internal to the amplifier to allow external cascade compensation.* The effects of some typical cascade compensation networks, usually dominant pole or *RC* lag networks, are often shown in terms of the compensated open-loop gain [see Fig. 11–35(b)].

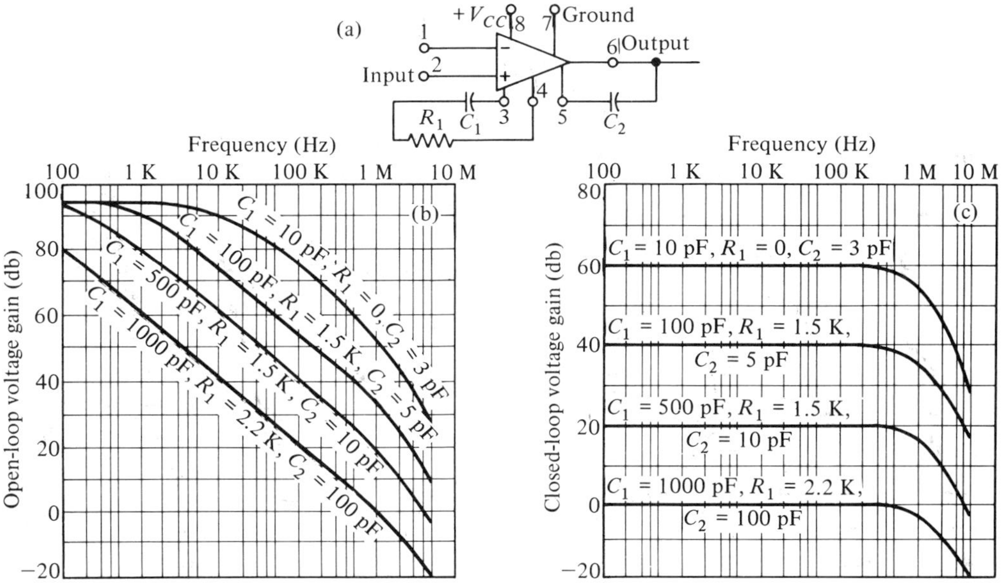

FIGURE 11–35 *Typical frequency response data provided by operational amplifier manufacturers: (a) Block diagram showing pin connection for external compensation networks, (b) effect of various compensation networks on the open-loop gain, and (c) effect of various compensation networks on the closed-loop gain. (Courtesy Fairchild Semiconductor Corp.)*

In addition to showing the effects of various compensation networks on the open-loop gain of the operational amplifier, device manufacturers often provide data on the closed-loop frequency response for various values of closed-loop gain. One set of curves on the closed-loop response is shown in Fig. 11–35(c). The compensation networks recommended by the manufacturer are generally good first choices of component values for a particular operational amplifier application. A parameter often given is the closed-loop, high corner frequency when Z_1 equals Z_2 (the unity gain condition).

---

* Terminals *A* and *B* of the IC shown in Fig. 10–23 are provided for external cascade compensation.

In the small-signal, high-frequency analysis used in this and the previous chapters we assumed that the rate of change in voltage at any node was determined by the effective RC time constant at that node. This type of circuit can be represented by the circuit shown in Fig. 11–36(a). Note that the only restraint on the maximum value of I is that imposed by V_g/R. Often, however, the value of current available to charge a capacitor is limited by the bias values in the amplifier. When the maximum charging current is limited, the RC circuit with current drive shown in Fig. 11–36(b) is a better model for the amplifier. Neglecting R as being very large in Fig. 11–36(b),

$$V_o = V_C \simeq \frac{1}{C} \int I \, dt \tag{11–75}$$

The maximum rate of change of the output voltage is defined as the slew rate, or

$$\text{Slew Rate} \equiv \text{Maximum} \frac{dV_C}{dt} \simeq \frac{I}{C} \tag{11–76}$$

The units of slew rate are typically V/μsec.

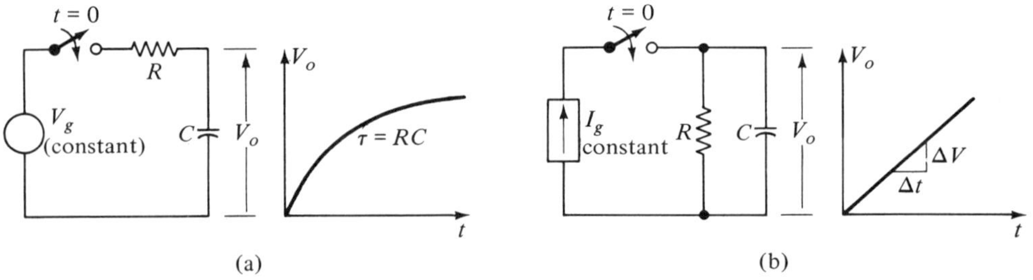

FIGURE 11–36 *Two circuits used in the discussion of slew rates: (a) an* RC *circuit with voltage drive, and (b) an* RC *circuit with current drive.*

The amplifiers discussed in the last two sections can serve to illustrate the slew rate problem. The complete circuit diagram is shown in Fig. 11–32(a). The ac equivalent circuit shown in Fig. 11–32(b) is repeated in Fig. 11–37(a) for convenience. To illustrate how slewing effects can occur, assume that a negative-going step input is applied to the input terminal. This signal causes a negative-going signal at the emitter of Q_3 which tends to cut-off Q_3. If the signal at the emitter of Q_3 is sufficiently large, then Q_3 is completely cut-off, and the feedback loop is effectively opened. Using the partial equivalent circuit shown in Fig. 11–37(b) to approximate the circuit at node A, we see that the constant-current source, Q_4, will charge the equivalent capacitance from node A to ground and produce a negative-going ramp in voltage. The output voltage also tends to be a negative-going ramp, a portion of which is fed back in the base of Q_2. As the feedback signal increases in amplitude, the net voltage at the emitter of Q_3 is reduced, and Q_3 again conducts. This condition effectively reestablishes the feedback loop, and linear analysis can again be used.

The preceding discussion leads to several important observations. First, a stage internal to the amplifier can be cut-off by a sufficiently fast input signal of the proper amplitude, effectively disconnecting the feedback loop. During the period that this condition exists, the node voltages of the amplifier may bear no

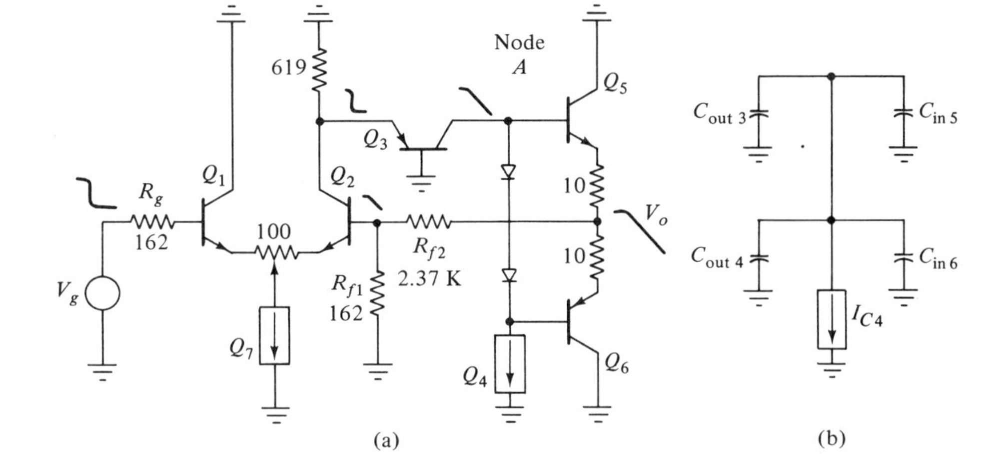

FIGURE 11–37 *The feedback amplifier from Fig. 11–32(a): (a) an ac equivalent circuit, and (b) a partial equivalent circuit under slewing conditions.*

relationship to the input voltage. Second, slewing can occur when the current available to charge a capacitor is limited to some finite value. Third, slew effects are non-linear and difficult to analyze. In general, the slewing characteristics obtained with a positive-going signal are different from those obtained with a negative-going signal.

Some typical waveforms illustrating the effects of slew rate are shown in Fig. 11–38. The ideal square-wave output is shown in Fig. 11–38(a). If the slew rate completely controls the response, the output may be a triangular waveform as shown in Fig. 11–38(b). A non-symmetrical waveform resulting from different slewing effects from positive-going signals and negative-going signals is shown in Fig. 11–38(c). In Fig. 11–38(c), a positive-going signal is linear for the first portion of the response, and slewing controls the remainder of the response as well as the entire negative-going response. The negative overshoot is also due to slewing effects.

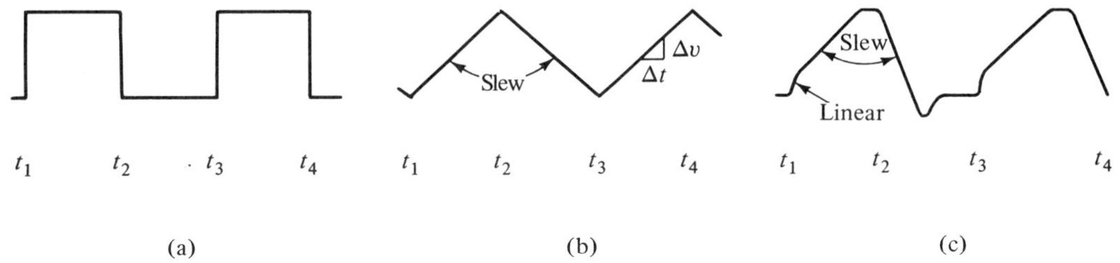

FIGURE 11–38 *Some typical waveforms resulting from slewing effects: (a) ideal square wave output, (b) symmetrical slewing effects and (c) non-symmetrical slewing effects.*

Some operational amplifier manufacturers provide slew-rate information in terms of a maximum frequency for full power operation. To determine this relationship, assume that the output waveform is a sinusoidal signal given by

$$V_o = V_M \sin (2\pi f_M)t \tag{11–77}$$

where f_M is the maximum frequency for full power operation. Differentiating Eq. (11–77) gives

$$dV_o/dt = 2\pi f_M V_M \cos(2\pi f_M)t \tag{11–78}$$

The maximum rate of change occurs as the waveform crosses zero. Thus, from Eq. (11–76),

$$\text{Slew Rate} = 2\pi f_M V_M \tag{11–79}$$

In this section we have examined some of the basic characteristics of operational amplifiers. Several examples of operational amplifier applications are given in the next section.

11.10 Operational Amplifier Applications

The objective of this section is to illustrate some of the applications of operational amplifiers. Actually, several of the amplifiers already studied in this and the last chapter can be described as operational amplifiers. Our first step will be to treat several of these circuits as operational amplifiers. Next we will study the basic computational circuits which perform the operations of addition, subtraction, multiplication by a constant, integration, and differentiation. This study leads to the analog computer, which is briefly discussed. Finally, we will examine some of the practical difficulties encountered in the use of IC operational amplifiers.

Table 11–4 shows the block diagram and transfer function of some of the more important operational amplifier circuits encountered in previous sections. The non-inverting amplifier corresponds to the voltage sample-voltage sum configuration. Two examples of practical operational amplifiers of this type are shown in Figs. 10–19 and 10–24. The inverting amplifier shown in Table 11–4 corresponds to the voltage sample-current sum configuration. Two practical examples of this type of operational amplifier are shown in Figs. 10–20 and 10–24. The differential amplifier is a logical extension of the inverting and non-inverting amplifier. From Table 11–4, the inverting gain and the non-inverting gain of a differential amplifier will be equal if $R_1 = R_3$ and $R_2 = R_4$.

A low-pass filter is one that will pass (amplify) low-frequency signals and attenuate high-frequency signals. The low-pass filter can be implemented using feedback compensation with a voltage feedback amplifier or a current feedback amplifier. The latter is preferred since the non-inverting low-pass filter has an initial output jump for a step input. For voltage gains greater than approximately 10, however, both inverting and non-inverting amplifiers give good results when used as low-pass filters.

An important operational amplifier circuit is the inverting, summing amplifier shown in Fig. 11–39(a). The block diagram indicates current summing at the input which results in an impedance from the input node (node A) to ground of approximately zero. Thus, the current flowing in R_1 depends only on V_1 and not on V_2 or V_3. The total current entering node A through resistors R_1, R_2, and R_3 will flow in the feedback resistor R_4. Since node A is an ac ground, the output voltage is

$$V_o = -(I_1 + I_2 + I_3)R_4 \tag{11–80}$$

TABLE 11–4 *Block diagram and transfer function of some operational amplifier circuits from previous sections.*

Block diagram	Transfer function ($\lvert T \rvert >> 1$)	Figure reference
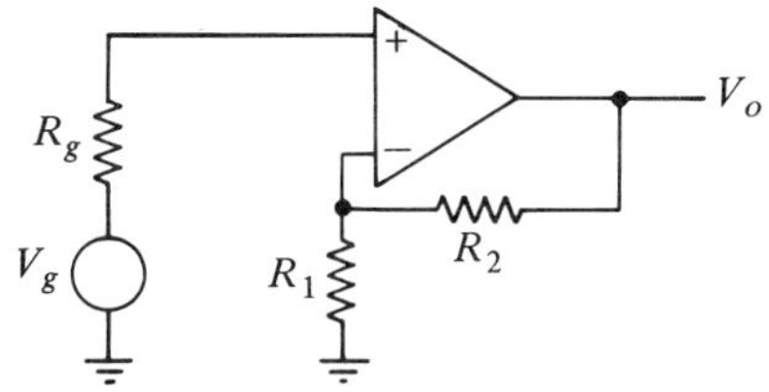	Noninverting amplifier $\frac{V_o}{V_g} \simeq \frac{R_1 + R_2}{R_1}$	Figure 10-19 Figure 10-24
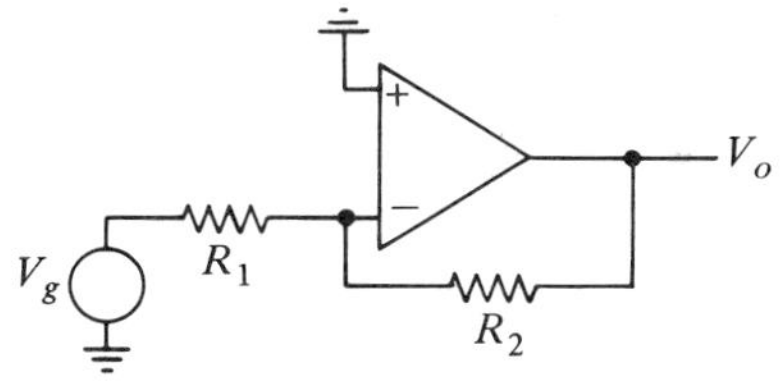	Inverting amplifier $\frac{V_o}{V_g} \simeq -\frac{R_2}{R_1}$	Figure 10-20 Figure 10-24
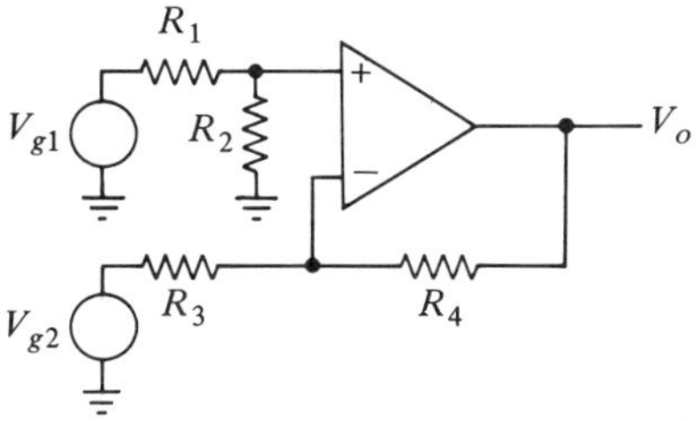	Differential amplifier $\frac{V_o}{V_{g1}} \simeq \left(\frac{R_2}{R_1 + R_2}\right)\left(\frac{R_3 + R_4}{R_3}\right)$ $\frac{V_o}{V_{g2}} \simeq \frac{R_4}{R_3}$	Figure 10-24
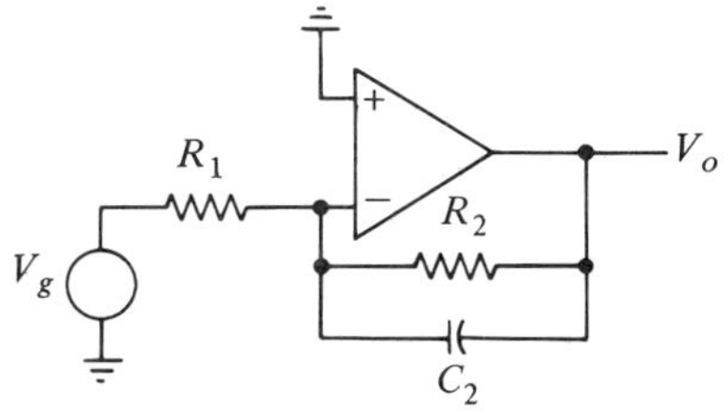	Inverting low pass filter $\frac{V_o}{V_g} \simeq \left(-\frac{R_2}{R_1}\right)\left(\frac{1}{1 + sR_2C_2}\right)$	Figure 11-28
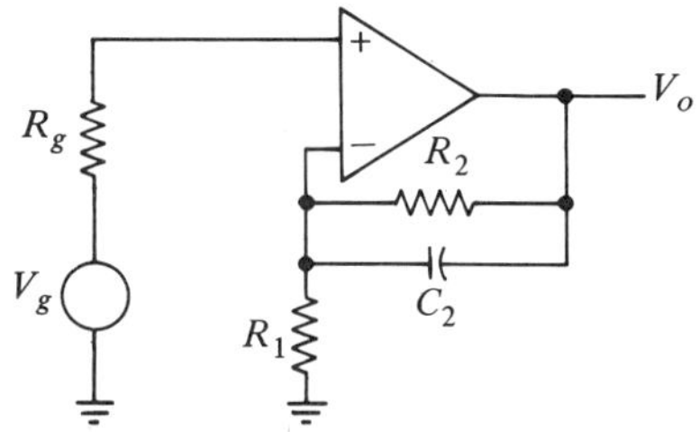	Noninverting low pass filter $\frac{V_o}{V_g} \simeq \left(\frac{R_1 + R_2}{R_1}\right)\left(\frac{1}{1 + sR_2C_2}\right)$ $R_2 >> R_1$	Figure 11-32

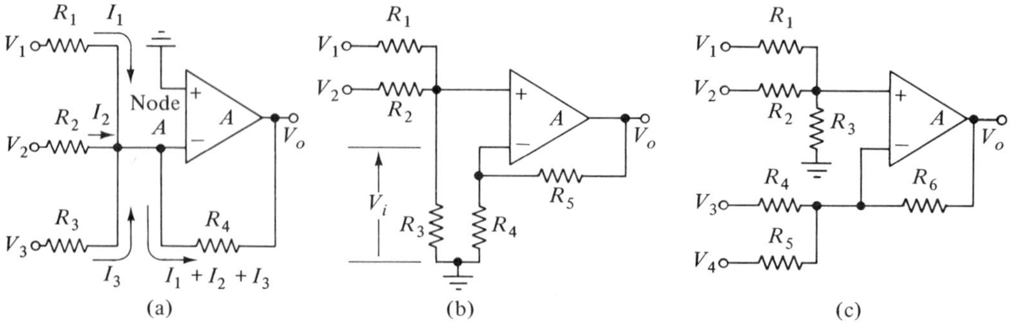

FIGURE 11–39 *Some basic computational circuits using operational amplifiers: (a) inverting summing amplifier, (b) non-inverting summing amplifier, and (c) addition-subtraction amplifier.*

In terms of the input voltage, V_o becomes*

$$V_o = -\left(V_1 \frac{R_4}{R_1} + V_2 \frac{R_4}{R_2} + V_3 \frac{R_4}{R_3}\right) \qquad \textbf{(11–81)}$$

Additional input signals can be summed by connecting additional resistors to node A.

Several special cases of the inverting, summing amplifier shown in Fig. 11–39(a) are worthy of consideration. For example, if all of the external resistors are equal ($R_1 = R_2 = R_3 = R_4$), the output is simply the algebraic sum of the inverted input voltages. When the resistors are not equal, each input can be given a different weighting factor. Finally, when R_1, R_2, and R_3 are all equal to $3R_4$, the output is the inverted average value of the input signals.

The summing function can also be performed using a non-inverting amplifier as shown in Fig. 11–39(b). The actual input voltage of the amplifier V_i, can be calculated from superposition as

$$V_i = V_1[(R_2 \,||\, R_3)/(R_1 + R_2 \,||\, R_3)] + V_2[(R_1 \,||\, R_3)/(R_2 + R_1 \,||\, R_3)] \qquad \textbf{(11–82)}$$

The output voltage is

$$V_o = V_i(R_4 + R_5)/R_4 \qquad \textbf{(11–83)}$$

When R_1, R_2, and R_3 are all equal to R, Eq. (11–82) becomes

$$V_i = (1/3)\,(V_1 + V_2) \qquad \textbf{(11–84)}$$

Adjusting the gain V_o/V_i to 3 will provide for the addition of the two input signals. Additional input signals can be summed by connecting additional resistors to the V_i node in Fig. 11–39(b). However, the input voltage division term [Eq. (11–82)] will be different, and the value of R_4 must be changed if V_o is to equal exactly the sum of the input signals.

An arrangement that combines addition and subtraction is shown in Fig. 11–39(c). Comparing this figure with the differential amplifier configuration in Table 11–4 shows that the only difference is the number of input signals. The circuit in Fig. 11–39(c) can also be viewed as a combination of the inverting and

---

* Note that Eq. (11–81) reduces to the familiar gain expression $-R_{f2}/R_{f1}$ when only one input signal is considered.

non-inverting summing amplifiers in Figs. 11–39(a) and (b). When resistors R_1 through R_6 are all equal, the output is

$$V_o = V_1 + V_2 - V_3 - V_4 \tag{11–85}$$

The general form of an important class of operational amplifier circuit is shown in Fig. 11–40(a). In this circuit, the resistive components of the feedback amplifier are replaced by general impedances, Z_i and Z_f. The voltage gain of this circuit is

$$V_o/V_i = -Z_f/Z_i \tag{11–86}$$

Numerous mathematical operations can be approximated by the proper choice of Z_f and Z_i. We will now consider two of the basic operations. Additional applications are given in the chapter problems.

In Fig. 11–40(b), Z_i is a capacitive reactance, and Z_f is a resistor. The output of this circuit is approximately equal to the derivative of the input signal. From Eq. (11–86),

$$V_o = V_i(-Z_f/Z_i) = V_i[-R/(1/sC)] = s(-RCV_i) \tag{11–87}$$

In the time domain,*

$$V_o = -RC\, dV_i/dt \tag{11–88}$$

Thus if V_i is a ramp given by kt, then V_o will be the derivative of the ramp or a step function having a value of $-kRCV_i$. When V_i is a step in voltage, V_o will be a negative-going pulse of very short duration as shown in Fig. 11–40(b).

In Fig. 11–40(c), Z_i is a resistor, and Z_f is a capacitor. The output of this circuit is approximately equal to the integral of the input signal since, from Eq. (11–86),

$$V_o = V_i(-Z_f/Z_i) = V_i[(1/sC)/R] = (1/s)(-RCV_i) \tag{11–89}$$

In the time domain,**

$$V_o = -RCV_i\, dt \tag{11–90}$$

When V_i is a step in voltage, V_o will be a negative going ramp as shown in Fig. 11–40(c). Often a resistor is connected in parallel with the integrating capacitor

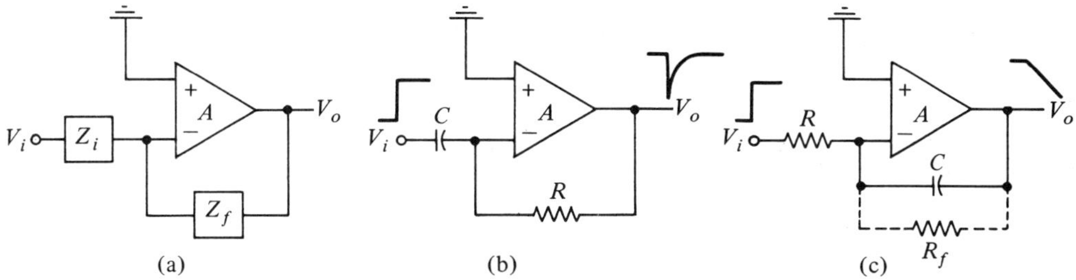

FIGURE 11–40 *Additional computational circuits using an inverting operational amplifier: (a) a general case, (b) a differentiator, and (c) an integrator.*

---

* Multiplication by s in the frequency domain corresponds to differentiation with respect to t in the time domain.

** Division by s in the frequency domain corresponds to an integration with respect to t in the time domain.

to obtain closed-loop stability by effectively limiting the voltage gain to $-(R_f/R)$. Then the output signal will be an exponential function with a time constant given by R_fC, and the output will approximate the integral for values of time less than about $0.15R_fC$.*

One important application of operational amplifiers is in analog simulation.** As an illustration of analog simulation, we will solve a second-order linear equation using the operational amplifier circuits discussed above. Specifically, assume that the differential equation describing the input/output relationship of a linear system is

$$X_o'' + 2X_o' + 3X_o = X_i \tag{11–91}$$

The first step in an analog simulation is to solve for the highest order derivative or

$$X_o'' = X_i - 2X_o' - 3X_o \tag{11–92}$$

A simulation can begin by assuming that X_i, X_o, and X_o' are available. The configuration shown in Fig. 11–41(a) uses the inverting summing amplifier of Fig. 11–39(a) to obtain X_o''. Once X_o'' is available, the integration configuration shown in Fig. 11–40(c) can be used to obtain X_o' and X_o as shown in Figs. 11–41(b) and (c). The final simulation for the solution of the differential equation given by Eq. (11–91) is shown in Fig. 11–41(d). Note that two additional inverting amplifiers are required, one to change X_o' to $-X_o'$, and the other to change $-X_o$ to X_o.***

The availability of high performance IC operational amplifiers has led to their wide application in analog simulation circuits. These ICs should be used with care, however, since they are generally capable of very high-frequency operation. Good high-frequency techniques should be used in the circuit layout, even though the signals of interest may be in the low-frequency range, because high-frequency parasitic oscillation can result from improper layout. Some construction guides are: (1) use RC decoupling filters of the type shown in Fig. 11–9 at each power supply point; (2) keep all leads between the IC and the RC filters as short as possible; (3) keep input and output leads short, and use terminated coaxial cable if necessary; (4) use a good ground connector; and (5) use sufficient cascade compensation to obtain closed-loop stability. Good results are usually obtained when the circuit is constructed on copper-clad board or used with printed circuits.

In this section we have considered some applications of operational amplifiers in computational circuits. The few circuits studied here should serve to illustrate the wide range of circuits available to the design engineer. Other circuits are given in the chapter problems.

---

* This time range corresponds to the region where the exponential function is approximately linear. The linear region of the exponential function is also discussed with percentage tilt in Sec. 7.5.

** Our objective here is to briefly illustrate analog simulation rather than to treat the topic in a comprehensive manner. A much more detailed treatment of analog simulation can be found in Korn and Korn in the references listed at the end of the chapter.

*** Although the analog simulation configuration shown in Fig. 11–41(d) can be used to solve Eq. (11–92), it may not be a practical circuit. Due to dynamic range problems, the amplitude of the various terms may need scaling. Also, the time variable is often scaled. Both magnitude and time scaling are discussed in most texts on analog simulation.

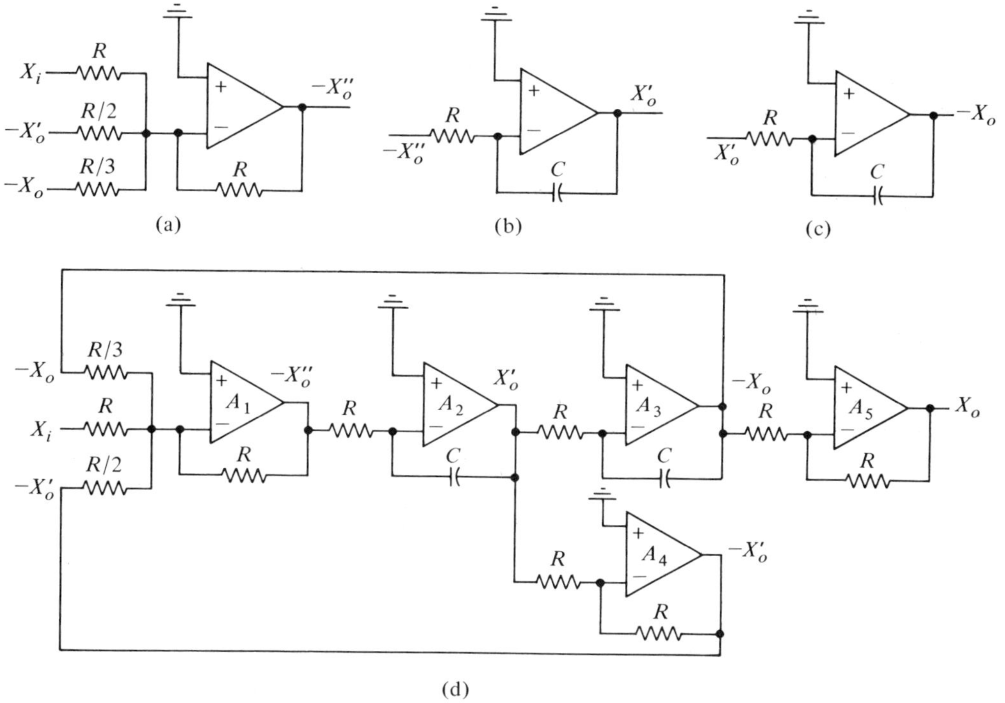

FIGURE 11–41 *Analog simulation of a second-order differential equation: (a) summation to obtain* X_o'', *(b) integration to obtain* X_o', *(c) integration to obtain* X_o, *and (d) complete analog simulation.*

11.11 Design of Negative-Feedback Amplifiers

In the preceding sections we were concerned with various aspects of the analysis of feedback amplifiers. The objective of this section is to outline briefly an approach to the design of feedback amplifiers. In this approach a given set of closed-loop specifications is converted into loop gain specifications. Also, the actual configuration of an amplifier will be studied. The set of specifications used in this design exercise is given in Table 11–5.

TABLE 11–5 *Design specifications for the amplifier in Sec. 11.11.*

1. $A_{vf} = 10$, $R_g = 100$ ohms, $f_{hi(f)} \geq 20$ MHz, $f_{lo(f)} = 0$ Hz.
2. Input impedance: greater than 1 M is parallel with less than 5-pF capacitance over the frequency range of 0 Hz to 2 MHz.
3. Drive: ±3 V into a 100-ohms load with slew rate of less than 150 V/μsec; ±10 V into 1-K load with less than 10 percent overshoot for a square wave input.
4. Power supply: ±15 V, standby current less than 20 mA, and power supply rejection ratio greater than 40 db.
5. Offset voltages: for zero input signal, both the input and the output dc voltages should be 0 V ± 20 mV.
6. Gain stability: 100 percent change in open-loop gain should produce less than a 0.5 percent change in closed-loop gain.

The first step is to select a suitable feedback network configuration. The drive requirements given in Table 11–5 suggest a low output impedance. The high input impedance requirement indicates voltage summing. Thus a voltage sample-voltage sum feedback configuration (a voltage feedback amplifier) is selected. Also, this arrangement has a non-inverting voltage gain that is consistent with the design specifications. A block diagram of the amplifier is shown in Fig. 11–42(a).

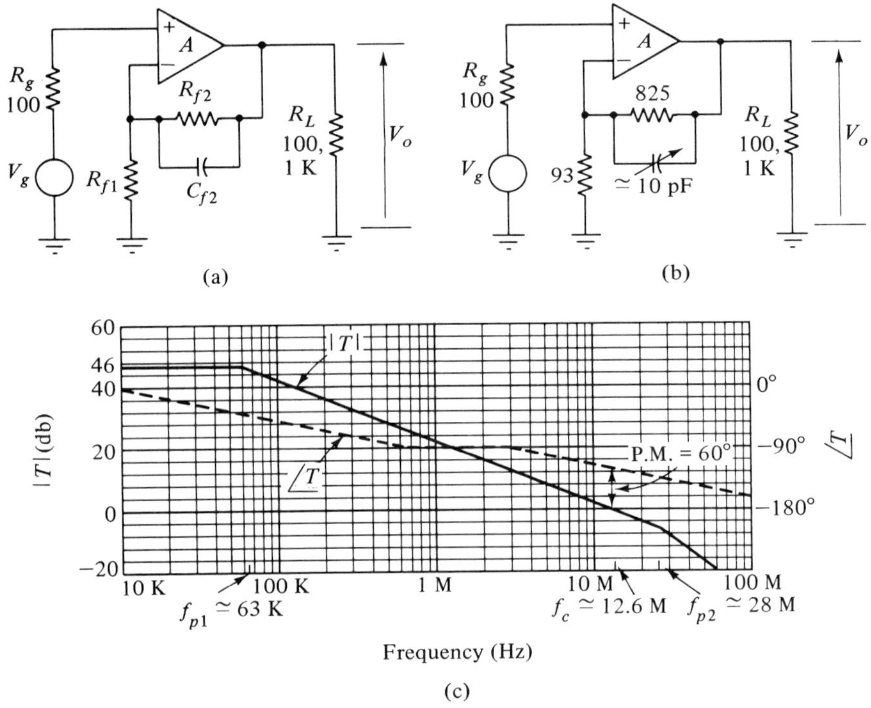

FIGURE 11–42 *Initial stages of the design of an amplifier whose specifications are given in Table 11–5: (a) basic block diagram showing component values, and (b) Bode plot of the approximate loop-gain characteristics.*

Before choosing values of R_{f1} and R_{f2}, we should examine the effects of the 20-MHz high-frequency requirement. The feedback compensation normally used in wide-band feedback amplifiers also effectively limits the high-frequency response of the amplifier to approximately $1/(2\pi R_{f2}C_{f2})$. In attempting to optimize the high-frequency response, C_{f2} is normally chosen as a variable capacitor. Typical trimmer capacitors are 2–8 pF, 5.5–18 pF, 5–25 pF, etc. For this design, assume a nominal value of 10 pF for C_{f2}. Then the value of R_{f2} is

$$R_{f2} \simeq 1/(2\pi f_{\mathrm{hi}(f)} C_f) \simeq (0.159)/(20\ \mathrm{MHz})(10\ \mathrm{pF}) \simeq 800\ \Omega \qquad \textbf{(11–93)}$$

A practical 1 percent tolerance resistor is 825 ohms. For a closed-loop gain of 10, R_{f1} is $(R_{f2}/9)$ or 92 ohms. The nearest 1 percent resistor is 93 ohms. The resulting feedback network is shown in Fig. 11–42(b).

The next step is to establish boundary conditions on the loop gain response. The minimum value of T_{mid} can be found from the sixth specification in Table 11–5 as follows.

$$A_f = A_{ol}/(1 + T) \tag{11–94}$$

A change in A_{ol}, ΔA_{ol}, causes a change in A_f, ΔA_f, given by

$$\Delta A_f = \Delta A_{ol}/(1 + T) \tag{11–95}$$

Solving for $1 + T$,

$$1 + T = \Delta A_{ol}/\Delta A_f \tag{11–96}$$

Thus, the design specifications require that

$$T_{\text{mid}} \geq 100\%/0.5\% = 200 \text{ or } 46 \text{ db} \tag{11–97}$$

The required frequency response of T can be approximated with the aid of Fig. 11–14. Assuming a second-order system, the 10 percent overshoot specification corresponds to a ξ of 0.6, a phase margin of 60°, a product of $f_{\text{hi}(f)}\tau_r$ equal to 0.35, and a product of $f_c\tau_r$ equal to 0.22. Thus, the loop-gain crossover frequency, f_c, should be

$$f_c \simeq \frac{0.22}{\tau_r} \simeq \left(\frac{0.22}{0.35}\right) f_{\text{hi}(f)} \simeq \left(\frac{0.22}{0.35}\right)(20 \text{ MHz}) \simeq 12.6 \text{ MHz} \tag{11–98}$$

Assuming that the amplifier is a second-order system, the Bode plot of the required loop gain is shown in Fig. 11–42(c) where f_{p1} is determined by f_c/T_{mid}, and f_{p2} is determined by the 60° phase margin.

A few comments should be made concerning the interpretation of the Bode plot of T shown in Fig. 11–42(c). This plot is valid only for a second-order system, and very few high performance amplifiers can be represented exactly as second-order systems. For example, the zero at 20 MHz due to the $R_{f2}C_{f2}$ network has not been included in this Bode plot. The principal advantage of the Bode plot in Fig. 11–42(c) is to show the approximate minimum loop-gain response required. The loop-gain response of a practical amplifier configuration can then be compared with this approximate response characteristic to determine if the amplifier can possibly satisfy the closed-loop specifications.

The required response of A_{ol} can also be specified. Since T equals $A_{ol}B$, and A_f is approximately equal to $1/B$, then A_{ol} is approximately equal to A_fT or 2000 for the amplifier under study. Unity gain crossover frequency can be approximated as $A_f f_{\text{hi}(f)}$ or 200 MHz.

Once the desired loop gain characteristics are known, the design of a feedback amplifier is very similar to the design of a multistage amplifier, a topic in Sec. 8.3. The initial approach generally deals with the input and output circuit configurations. For this amplifier, the input circuit must have a high value of input impedance and allow for voltage summing. The JFET differential amplifier will perform these two functions very well. Since the amplifier must provide both positive- and negative-going output signals (± 3 V into 100 ohms, ± 10 V into 1 K), the complementary emitter-follower circuit appears to be a good choice for the output circuit. A partial circuit diagram of the input and output circuits is shown in Fig. 11–43. The potentiometer R_{SS} between the two source terminals provides a means for adjusting the dc output voltage to zero.

The constant-current type of bias network for Q_1 and Q_2 has several interesting features. The base voltage of Q_5 is determined by the Zener diode D_3. Since most Zener diodes operate under avalanche conditions, their currents have many minute fluctuations called noise. Capacitor C_1 and the R_2C_2 filter are

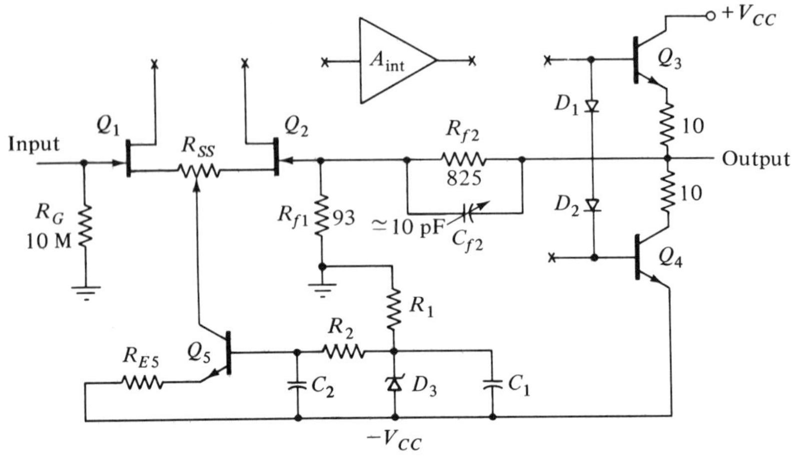

FIGURE 11–43 *A partial circuit diagram showing one possible configuration for input and output circuits.*

added to reduce the effects of the noise introduced by D_3. Diode D_3 is connected between the base of Q_5 and the negative power supply. This arrangement tends to hold the voltage across R_{E5} constant with changes in power supply voltage. The current in Q_1 and Q_2 as well as in Q_5 remains constant for variations in $-V_{CC}$, giving a high value of power supply rejection ratio. Power supply rejection ratio (PSRR) is similar to CMRR and can be defined as

$$\text{PSRR} = \left(\frac{\Delta V_{\text{supply}}}{\Delta V_{\text{out}}}\right) \tag{11–99}$$

The final part of the design is to determine the circuit configuration between the JFET input circuit and the complementary emitter-following output circuit. This portion of the amplifier is denoted as A_{int} in Fig. 11–43. Numerous configurations are possible. If A_{int} connects to the drain terminal of Q_1, one net inversion must occur in order to have negative feedback. If A_{int} is connected to the drain terminal of Q_2, no inverting stages are required for negative feedback. Also, A_{int} can have a differential input which allows connections to be made to the drain terminals of both Q_1 and Q_2. In the remainder of this section we will briefly examine one design solution.

One design solution for the amplifier whose specifications are given in Table 11–5 is shown in Fig. 11–44(a). The function of the input stages (Q_1, Q_2) has already been discussed. The 220-ohm resistor in series with the 100-ohm potentiometer serves to extend the range of the potentiometer in adjusting the output dc voltage to zero. The interstage amplifier consists of Q_6 through Q_{11} and is a cascoded CE stage (Q_6, Q_8) with a constant-current type load (Q_9). The high load impedance of Q_8 is maintained by isolating the complementary emitter-follower output stages from Q_8 by a CC stage (Q_{10}) biased with a constant-current source (Q_{11}). Note that each constant-current source (Q_5, Q_9, Q_{11}) is biased by the same Zener diode, D_3. Transistor Q_7 is used to bypass R_{E6} and maintain direct coupling throughout the amplifier. Diode D_4 determines the base voltage of Q_6 and Q_7. Connecting D_4 between the power supply and the base of Q_4 improves the PSRR. Diodes D_5 and D_6 ensure that the base-collector junction of Q_3 is reverse biased by the voltage drop across one diode.

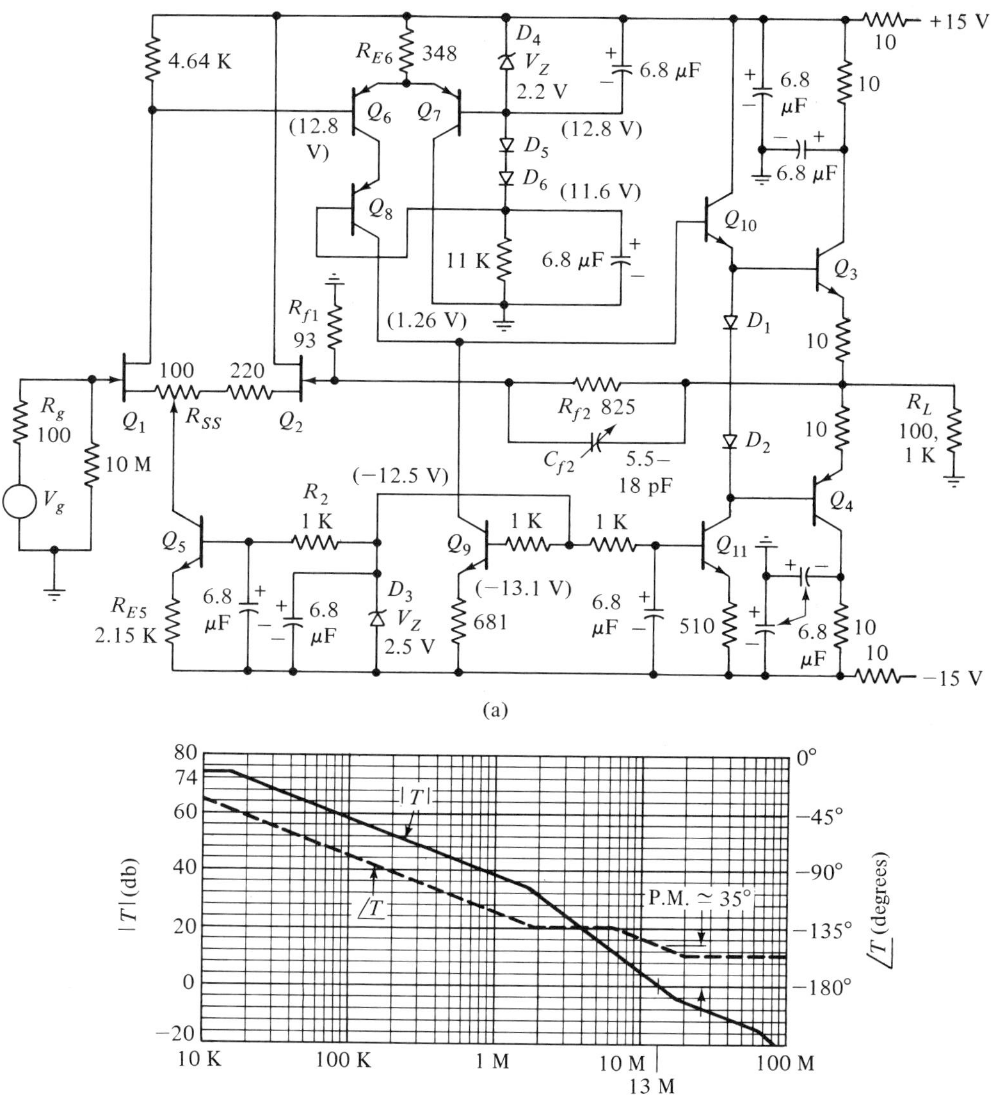

FIGURE 11–44 *One design solution to the amplifier whose specifications are given in Table 11–5: (a) circuit diagram, and (b) Bode plot of* T.

The requirement of ±10 V output voltage using a ±15 V power supply places several constraints on the design of the bias circuit. Since Q_{10}, Q_3, and Q_4, are all CC stages having a voltage gain of approximately unity, the full ±10 V output swing will also occur at the following terminals: E_{10}, C_{11}, C_8, and C_9. The bias circuit design must ensure that all base-collector junctions remain reverse-biased under peak signal swings. This requirement, coupled with the desire to connect the diodes between the power supply and the bias point, required the small values of Zener voltages that were used in the circuit.

The characteristics of the active devices used in the amplifier shown in Fig. 11–44 are listed in Table 11–6. These data can be used to calculate the response

TABLE 11–6 *Characteristics of the active devices used in the amplifier shown in Fig. 11–44.*

Transistor	V_P	I_{DSS}	C_{gd}	C_{gs}
Q_1, Q_2	2.1 V	5.5 mA	4 pF	6 pF

Transistor	β	r_{ce}	f_T	C_{ob}
Q_3	110	100 K	390 MHz	5 pF
Q_4	200	14 K	260 MHz	6 pF
Q_5	90	160 K	390 MHz	5 pF
Q_6	330	12 K	320 MHz	4 pF
Q_7	400	8 K	320 MHz	4 pF
Q_8	200	11 K	450 MHz	4.5 pF
Q_9	600	22 K	100 MHz	4 pF
Q_{10}	100	50 K	390 MHz	5 pF
Q_{11}	850	24 K	100 MHz	4 pF

of the amplifier, an exercise left for the reader. The calculated loop gain is shown in Fig. 11–44(b) and should be compared with the approximate design goal shown in Fig. 11–42(c).

The final design of the amplifier was determined experimentally. The bypass capacitor at the base of Q_9 was removed to reduce the percentage overshoot.* Also the dc current in Q_{11} was increased to improve the slew-rate response, especially for negative-going output signals. The experimental bias values are shown in Fig. 11–44(a), and other measured data are given in Table 11–7.

TABLE 11–7 *Experimental step response data for the amplifier shown in Fig. 11–44.*

V_o	R_L	M_{pt}	$T_{r(\text{output})}$	$T_{r(\text{amplifier})}$
+1.2 V	100	10%	14 nsec	8 nsec
−1.2 V	100	15%	14 nsec	8 nsec
+3 V	100	8%	17 nsec	10 nsec
−3 V	100	20%	15 nsec	9 nsec
+10 V*	1 K	0	32 nsec	28 nsec
−10 V*	1 K	0	32 nsec	28 nsec

* slew rate $\simeq$ 8 V/32 nsec $\simeq$ 250 V/μsec

In this section we have examined some aspects of the design of negative-feedback amplifiers. The first task was to determine the feedback network configuration and component values. Next, the characteristics of T and/or A_{ol} were determined from the closed-loop response specifications. Finally, the actual configuration of the amplifier was determined beginning with the input and the output sections. While a design problem has no unique solution, the procedures outlined here have achieved the design response with a minimum expenditure of time. Any complex design should be checked experimentally before the design is considered complete.

* As shown in Sec. 8.4, when the base of a constant-current source is unbypassed, the output resistance is reduced, and the output capacitance is increased.

11.12 Summary

The analysis and design of negative-feedback amplifiers are considerably more difficult than that for amplifiers without feedback, because of the possibility of oscillation in the high- or low-frequency range. The basic design problem becomes one of optimizing the desirable characteristics obtained with negative feedback while achieving acceptable degrees of stability against regeneration and oscillation.

When the frequency response of a negative-feedback amplifier is dominated by a single pole at each end of the band, regeneration is no problem. The frequency response is extended at each end of the band by a factor of $(1 + |T_{\text{mid}}|)$. In this case, negative feedback reduces the mid-frequency gain and increases the bandwidth by approximately the same factor.

For second- or higher-order systems (multiple corner frequencies), the methods for finding the response characteristics is more difficult. Section 11.2 includes two methods for finding the frequency response of second-order systems, an analytical method and a graphical method using the Nichols chart. The latter method can be used to approximate the response of higher-order systems. In these systems, self-oscillation results if $|T| > 1$ when $\underline{/T} = 180°$.

The undesirable peaks in the frequency response and the overshoot and ringing in the transient response of a second- or higher-order system can be removed to some degree by compensation. Two types of compensation are considered in detail: (1) cascade compensation, in which the compensating network is added to the forward signal path; and (2) feedback compensation, in which the compensating network is added to the feedback path. In general, the net effect of compensation is to increase the phase margin at the cross-over frequency so that the desired amount peaking or overshoot is achieved. Several methods of achieving the desired conditions are considered. Some of these methods are then applied to current and voltage feedback amplifiers.

Since operational amplifiers are so extremely useful, their general characteristics were discussed in Sec. 11.9 with special emphasis on the topic of slew rate. Several applications of operational amplifiers are reviewed in Sec. 11.10. These applications include the functions of inverting, summing, subtraction, integration, differentiation, and analog simulation.

An approach to the design of feedback amplifiers is considered in Sec. 11.11. Given a set of specifications, the step by step process of arriving at an amplifier design is discussed in detail. This procedure is not unique; however, it does allow the reader to achieve a satisfactory design in a reasonable period of time.

Problems

Section 11.1

11-1 For the amplifier shown in Fig. 11–45, calculate $f_{\text{hi}(f)}$ and $f_{\text{lo}(f)}$. The JFET is biased at $V_{GS} = -2$ V and has the following characteristics: $V_P = -4$ V, $I_{DSS} = 8$ mA, $C_{gd} = 4$ pF, and $C_{gs} = 8$ pF. The bipolar transistor has the following characteristics: $\beta = 200$, $f_T = 200$ MHz, and $C_{ob} = 10$ pF.

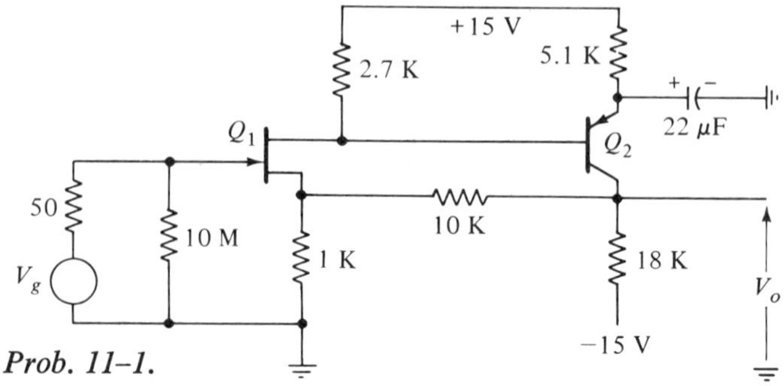

FIGURE 11–45 *Circuit for Prob. 11–1.*

11–2 The loop gain of a certain feedback amplifier has the following characteristics:

$$T_{mid} = -100 \qquad f_{hi1} = 10 \text{ kHz} \qquad f_{hi2} = 1 \text{ MHz}$$

Does this amplifier have a dominant high corner frequency? If not, what can be changed to produce dominant high corner frequency operation?

11–3 The two feedback amplifiers shown in Figs. 10–42 and 10–43 have input and output coupling capacitors. Does the negative feedback increase or decrease the low corner frequency associated with these capacitors?

Section 11.2

11–4 Calculate the closed-loop, high-frequency response for the amplifier shown in Fig. 11–46. Assume that each transistor has a β of 200, a C_{ob} of 6 pF, an f_T of 200 MHz, and an r_{ce} of 50 K.

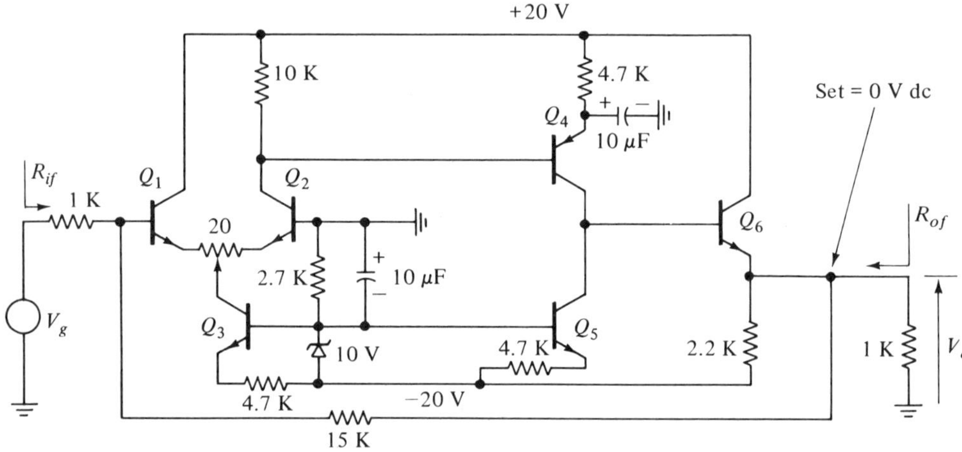

FIGURE 11–46 *Circuit for Prob. 11–4.*

11–5 Calculate the closed-loop, high-frequency response of the amplifier whose characteristics are given in Prob. 11–2.

11–6 The circuit shown in Fig. 11–3 is modified by adding a 10-K resistor between the collector terminal of Q_2 and the +24-V power supply. Repeat Ex. 11–3 for the modified amplifier.

11–7 Repeat Prob. 11–4 using the Nichols chart.

11–8 For phase shifts greater than $-90°$, what is the range of values for $|T|$ corresponding to the condition that $|T/(1 + T)|$ is -3 db? Can this information be used to estimate the value of $f_{\text{hi}(f)}$ directly from the Bode plot of $|T|$?

Section 11.3

11–9 The amplifier in Ex. 11–5 can be made stable if the value of $|T_{\text{mid}}|$ is reduced. Modify the circuit shown in Fig. 11–8(a) so that the value of $|T_{\text{mid}}|$ is reduced, and verify that the modified amplifier is stable.

11–10 The amplifier in Ex. 11–5 can be made stable if the value of f_{hi1} is reduced. Modify the amplifier shown in Fig. 11–8 so that the value of f_{hi1} is reduced, and verify that the modified amplifier is stable.

11–11 The amplifier shown in Fig. 11–3 is modified by the addition of a CB stage between Q_3 and Q_4. This forms a cascode amplifier to replace the CE stage. Will the modified circuit be more stable or less stable?

Section 11.4

11–12 Repeat Prob. 11–4 using Fig. 11–11.

11–13 Calculate the closed-loop, high-frequency response of the amplifier shown in Fig. 10–45. The mid-band response of this amplifier was found in Prob. 10–42. Assume that each transistor has a β of 100, an r_{ce} of 50 K, a C_{ob} of 5 pF, and an f_T of 200 MHz. The dc voltage at the output terminal is adjusted to 0 V.

11–14 Calculate the closed-loop, high-frequency response of the amplifier shown in Fig. 10–46. The mid-band response of this amplifier was found in Prob. 10–47. Assume that all transistors have the same characteristics as given in Prob. 11–13 and that the dc output voltage is adjusted to 0 V.

11–15 The specifications for a certain feedback amplifier include

$$T_{\text{mid}} = 50 \text{ db} \qquad f_{\text{hi}(f)} = 20 \text{ MHz} \qquad M_{pf} = 2 \text{ db}$$

Assuming that the amplifier can be represented as a second-order system, find the two high corner frequencies of the loop gain.

Section 11.5

11–16 Calculate the closed-loop transient response of the amplifier shown in Fig. 11–46. The transistor characteristics are given in Prob. 11–4.

11–17 Calculate the closed-loop transient response of the amplifier shown in Fig. 10–45. The transistor characteristics are given in Prob. 11–13.

11–18 Calculate the closed-loop transient response of the amplifier shown in Fig. 10–46. The transistor characteristics are given in Prob. 11–13.

11–19 The specifications for a certain feedback amplifier include

$$T_{\text{mid}} = 60 \text{ db} \qquad \tau_r = 15 \text{ nsec} \qquad M_{pt} = 5\%$$

Assuming that the amplifier can be represented as a second-order system, find the two high corner frequencies of the loop gain.

Section 11.6

11–20 Calculate the values of τ_r and M_{pt} for the amplifier shown in Fig. 11–17 when R_{EE} is increased from 90 ohms to 500 ohms.

11–21 The amplifier shown in Fig. 11–17 is modified by changing R_{E3} from 4.64 K to a series combination of 3.9 K and 680 ohms. Only the 3.9-K resistor is ac bypassed by the 6.8-μF capacitor. Calculate the new values of f_c and P.M., and compare the results with the values given in Ex. 11–6.

11–22 Calculate the value of R_{EE} for the amplifier shown in Fig. 11–46 that will cause M_{pt} to be 2 percent. What is the value of τ_r^c? The high-frequency response of this amplifier was calculated in Prob. 11–4.

11–23 Repeat Prob. 11–22 using dominant pole compensation. The compensating capacitor should be connected between the base and the collector terminals of Q_4.

11–24 Calculate the value of R_{EE} for the amplifier shown in Fig. 10–45 that will cause M_{pf} to be 1 db. What is the value of τ_r^c? The high-frequency response of this amplifier was calculated in Prob. 11–13.

11–25 Repeat Prob. 11–24 using dominant pole compensation. The compensating capacitor should be connected between the collector terminal of Q_3 and ground.

11–26 Calculate the value of R_{EE} for the amplifier shown in Fig. 10–46 that will cause M_{pt} to be 2 percent. What is the value of τ_r^c? The high-frequency response of this amplifier was calculated in Prob. 11–14.

11–27 Repeat Prob. 11–26 using dominant pole compensation. The compensating capacitor should be connected between the collector terminal of Q_3 and ground.

11–28 Repeat Prob. 11–22 using *RC* lag compensation. Compare the resulting value of τ_r^c with the values found in Probs. 11–22 and 11–23.

11–29 Repeat Prob. 11–24 using *RC* lag compensation. Compare the resulting value of τ_r^c with the values found in Probs. 11–24 and 11–25.

11–30 Repeat Prob. 11–26 using *RC* lag compensation. Compare the resulting value of τ_r^c with the values found in Probs. 11–26 and 11–27.

Section 11.7

11–31 Derive Eq. (11–54) directly from Fig. 11–26(b).

11–32 Modify the amplifiers shown in Figs. 11–46, 10–45, and 10–46 so that the output CC stage is replaced by a complementary emitter-follower circuit.

11–33 Repeat Exs. 11–11 and 11–12 for the amplifier shown in Fig. 11–46. The high-frequency response of this amplifier was calculated in Prob. 11–4.

11–34 Repeat Exs. 11–11 and 11–12 for the amplifier shown in Fig. 10–46. The high-frequency response of this amplifier was calculated in Prob. 11–14.

Section 11.8

11–35 Derive Eq. (11–67) directly from Fig. 11–30(b).

11–36 Repeat Exs. 11–13 and 11–14 for the amplifier shown in Fig. 10–45. The high-frequency response of this amplifier was calculated in Prob. 11–13.

Section 11.9

11–37 An operational amplifier has the following specifications: Slew rate $\simeq$ 35 V/μsec, $R_{\text{out}} \simeq 4$ K, input offset voltage $\simeq$ 2 mV, $R_{\text{in}} \simeq 300$ K, $A_{vol} \simeq 100{,}000$, and $V_o \simeq \pm 10$ V into 10 K. Can this amplifier be used as a voltage feedback amplifier with the following specifications: $A_{vf} \simeq +30$, $R_{\text{out}(f)} < 1$ ohm, $V_o \simeq \pm 10$ V into 10 K at 10 kHz, and $R_{\text{in}(f)} > 1$ M?

11–38 In Fig. 11–37(a), the dc collector current of Q_4 is approximately 3.4 mA, and the value of C_{ob} for Q_3, Q_4, Q_5, and Q_6 is approximately 5 pF. Estimate the slew rate for this circuit.

Section 11.10

11–39 Express V_o in terms of the input signals V_1 through V_5 for the circuit shown in Fig. 11–47.

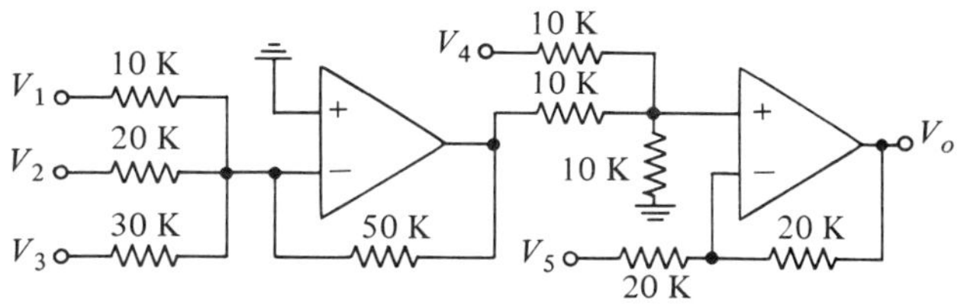

FIGURE 11–47 *Circuit for Prob. 11–39.*

11–40 Using as few operational amplifiers as possible, devise a method of combining 8 linear signals (V_1 through V_8) according to the relationship $V_o = V_1 + 2V_2 + (V_3 + V_4 + V_5)/3 - V_6 - (V_7 + 3V_8)/2$.

11–41 Design an analog simulation system to solve the following equation for X_o: $X_o'' + 5X_o' - 3X_o = X_i$.

11–42 Sketch the output waveform of the amplifier shown in Fig. 11–48(a) for a sine wave input signal. What function does this circuit perform?

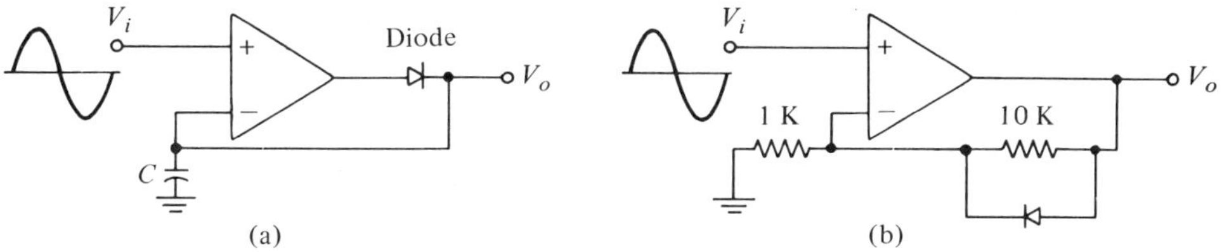

FIGURE 11–48 *Circuit for (a) Prob. 11–42 and (b) Prob. 11–43.*

11–43 Repeat Prob. 11–42 for the circuit shown in Fig. 11–48(b).

Section 11.11

11–44 Calculate the closed-loop, high-frequency response of the amplifier shown in Fig. 11–44(a). The active device characteristics are given in Table 11–6.

11–45 Estimate the value of input capacitance of the amplifier shown in Fig. 11–44(a).

11–46 Design an amplifier with the following specifications: $R_{i(f)} = 100$ ohms $\pm$ 1 percent and $R_{o(f)} = 100$ ohms $\pm$ 1 percent for $f < 100$ kHz, $V_o/V_g = -20$, $\tau_r \leq 20$ nsec, $M_{pt} \leq 5$ percent, $R_L = 100$ ohms, and $f_{lo(f)} = 0$ Hz. The closed-loop gain of the amplifier should change less than 0.5 percent for a 50 percent change in the open-loop gain. Supply voltages of $\pm$24 V are available. Use only standard components, and specify the active devices by number.

11–47 Repeat Prob. 11–46 using an input impedance greater than 10 megohms and a voltage gain of +20.

References

Barrett, J. G. and T. V. Blalock, "A Wideband, Linear, Transistorized Cable Driver," University of Tennessee, Electrical Engineering Dept., Report No. 22, ORNL Contract No. W7405 Eng. 26(2160), 1966.

Cherry, E. M. and D. E. Hooper, *Amplifying Devices and Low-Pass Amplifier Design.* New York: John Wiley & Sons, Inc., 1968.

D'Azzo, John J. and C. H. Houpis, *Feedback Control System Analysis and Design.* New York: McGraw-Hill Book Company, 1966.

"Handbook of Operational Amplifier Applications." Tucson, Arizona: Burr-Brown Research Corporation, 1963.

Korn, G. A. and T. M. Korn, *Electronic Analog and Hybrid Computers.* New York: McGraw-Hill Book Company, 1964.

Kuo, B. C., *Automatic Control Systems.* Englewood Cliffs, N. J.: Prentice-Hall, Inc., 1967.

McCollum, P. A. and B. F. Brown, *Laplace Transform Tables and Theorems.* New York: Holt, Rinehart & Winston, Inc., 1965.

Millman, J. and H. Taub, *Pulse, Digital, and Switching Waveforms.* New York: McGraw-Hill Book Company, 1965.

Savant, C. J., Jr., *Basic Feedback Control System Design.* New York: McGraw-Hill Book Company, 1958.

Thornton, R. D., C. L. Searle, D. O. Pederson, R. B. Adler, and E. J. Angelo, Jr., *Multistage Transistor Circuits.* New York: John Wiley & Sons, Inc., 1965.

12

Audio Power Amplifiers

Many applications require large amounts of audio power. One familiar example is the power required to drive the speakers in sound reproducing systems which may require from one to several hundred watts, depending upon the type of system.

Several new problems are introduced when active devices are used to convert large amounts of dc power to ac power. For example, active devices are notoriously nonlinear when operated over a large dynamic range. Consequently, the large swings in current and voltage that are required to produce several watts of ac power can also produce large amounts of distortion. Since distortion is undesirable, methods for quantifying and minimizing it will be considered. Some of the feedback circuits introduced in Chap. 11 will be used to keep the distortion low.

Another important problem is introduced by the relatively large amount of heat that must be dissipated by a power transistor. This heat is generated largely at the base-collector junction and must be radiated into the air. Methods for calculating the size of a heat sink to keep the junction temperature below a safe operating temperature will be discussed. In addition, methods for improving the dc to ac conversion efficiency will be discussed.

This study is initiated with a short discussion of sound reproduction systems in Sec. 12.1. Several of the power amplifier requirements are developed in this discussion. In the next section, the characteristics and limitations of power transistors are discussed. This discussion provides some insight into the nonlinearity problems and also the thermal problems.

After a brief discussion and evaluation of the operation and efficiency of ideal amplifiers in Sec. 12.3, the distortion problem is considered in Sec. 12.4. While three types of distortion are introduced, only amplitude distortion is analyzed.

Various types of practical power amplifiers are discussed in Secs. 12.5, 6, and 7, and examples are developed in Sec. 12.8. Integrated circuit power amplifiers are considered briefly in Sec. 12.9 and regulated dc power supplies in Sec. 12.10. Throughout this discussion, emphasis is placed on the practical aspects of the analysis and design of high quality audio power amplifiers.

12.1 Elements of a Sound Reproduction System

Prior to studying the characteristics of audio power amplifiers, we will briefly examine the various parts of a sound reproduction system. These systems, generally referred to as audio systems, are used widely in home entertainment sets (radio, TV, and stereo), public address systems, movie sound tracks, etc. Sound reproduction dates back to 1877 and Edison's talking machine. These first machines, as well as modern ones, provide a means for the direct recording, storage, and reproduction of sound. The objective of this section is to outline the principal characteristics of a sound reproduction system.

A block diagram of a sound reproduction system may be divided into five sections as shown in Fig. 12–1. The first or input section might be a microphone, a phonograph cartridge, or a magnetic tape head. The function of the input section is to convert recorded sound or the original sound into electrical signals. Generally, the signal level at the output of the first section is in the microvolt to millivolt range. The second section, or the preamplifier, has a voltage gain which can be as high as 10^3. The third section contains the control circuits which provide a means of adjusting the volume and/or the frequency response of the system. The fourth section is a power amplifier which increases the power level of the audio signal to the required level for driving the speaker system. The last section is a speaker system which converts the electrical output signal from the power amplifier into an acoustical signal, thus completing the process of sound reproduction. In the remainder of this section, we will study the desirable characteristics of the power amplifier, the preamplifier, and the control circuits.

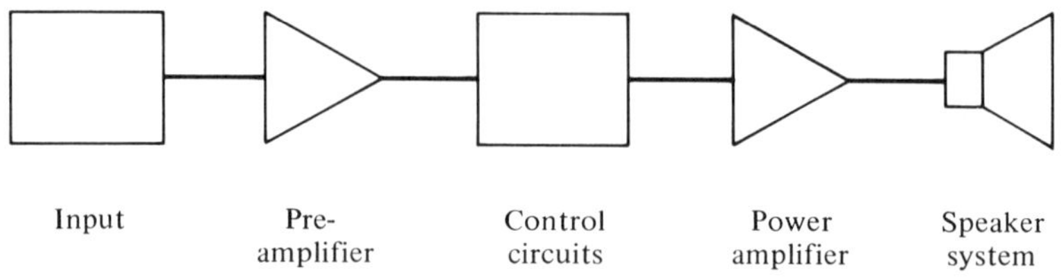

FIGURE 12–1 *A block diagram representation of a sound reproduction system.*

The characteristics of an amplifier are based on the sound intensity requirements of the system. Some typical sounds and their associated intensity levels are listed in Table 12–1. For most applications, the required sound intensity is less than 100 db. The acoustical power associated with a given sound intensity, however, is a more complex problem. Some typical conditions will now be examined.

A relationship between the acoustical power level and the sound intensity level is shown in Fig. 12–2. The solid lines represent indoor conditions where the room volume is given in cubic feet. The dashed lines represent outdoor condi-

TABLE 12–1 *Typical Sound Intensity Levels*

Source	Sound Intensity
Threshold of pain	130 db
Concert hall music	100 db (max.)
Human speech	50–80 db
Ordinary conversation	65 db (average)
Whisper	10–20 db
Threshold of sound	0 db

tions where the distance from the speaker is measured in feet along the axis of the speaker. Minimum values are given in Fig. 12–2 since the curves are based on ideal speaker conditions.* Often, a higher acoustical power level is required in practice to achieve a given sound intensity level. From Fig. 12–2, a room having a volume of 2000 cubic feet would require a minimum acoustical power of 0.31 W to produce a sound intensity of 100 db.

The output power required from the power amplifier to produce the desired acoustical power is determined by the efficiency of the speaker. A common type of speaker is the moving coil or electrodynamic type. This type usually consists of a paper cone that is flexibly supported at the edges. The cone is driven at its apex by a moving coil called the voice coil. The voice coil is placed in a strong magnetic field produced either by a permanent magnet or by an electromagnet. The audio frequency output signal of the power amplifier causes a current to flow in the voice coil, which in turn causes motion in the speaker cone. The power efficiency of an electrodynamic speaker ranges from 1 to 5 percent with 3 percent being a typical value. The cost of constructing an amplifier with a high power level is generally less than constructing an electrodynamic speaker with a high efficiency.** Assuming an electrodynamic speaker system with an efficiency of 3 percent, an acoustical power of 0.31 W requires approximately 10 W from the power amplifier.

The impedance of the voice coil, Z_{sp}, is also important since it represents the load impedance on the power amplifier. The impedance of a typical electrodynamic speaker is shown in Fig. 12–3. Note the resonant condition at approximately 50 Hz.*** The speaker is generally used for frequencies above this resonant peak. Nominal values of Z_{sp} in the mid-frequency range (500 Hz) are 3.2, 4, 8, and 16 ohms.

The low value of Z_{sp} coupled with the requirement for large values of output power places special demands on the power amplifier. For example, a power amplifier delivering 10 W rms power to an 8-ohm speaker must provide 8.9 V rms and 1.12 A rms. These values are equivalent to 12.6 V peak (or 25.2 V peak-

* For the indoor case, the ideal speaker conditions are associated with optimum reverberation times. For the outdoor case, the ideal speaker conditions are associated with an infinite baffle.

** Other speaker systems, such as the horn speaker, have efficiencies ranging from 30 to 50 percent. However, these speakers require a relatively large space and are generally limited to public address systems, theater speaker systems, and other large area applications.

*** This peak can be removed with a properly designed baffle for the speaker.

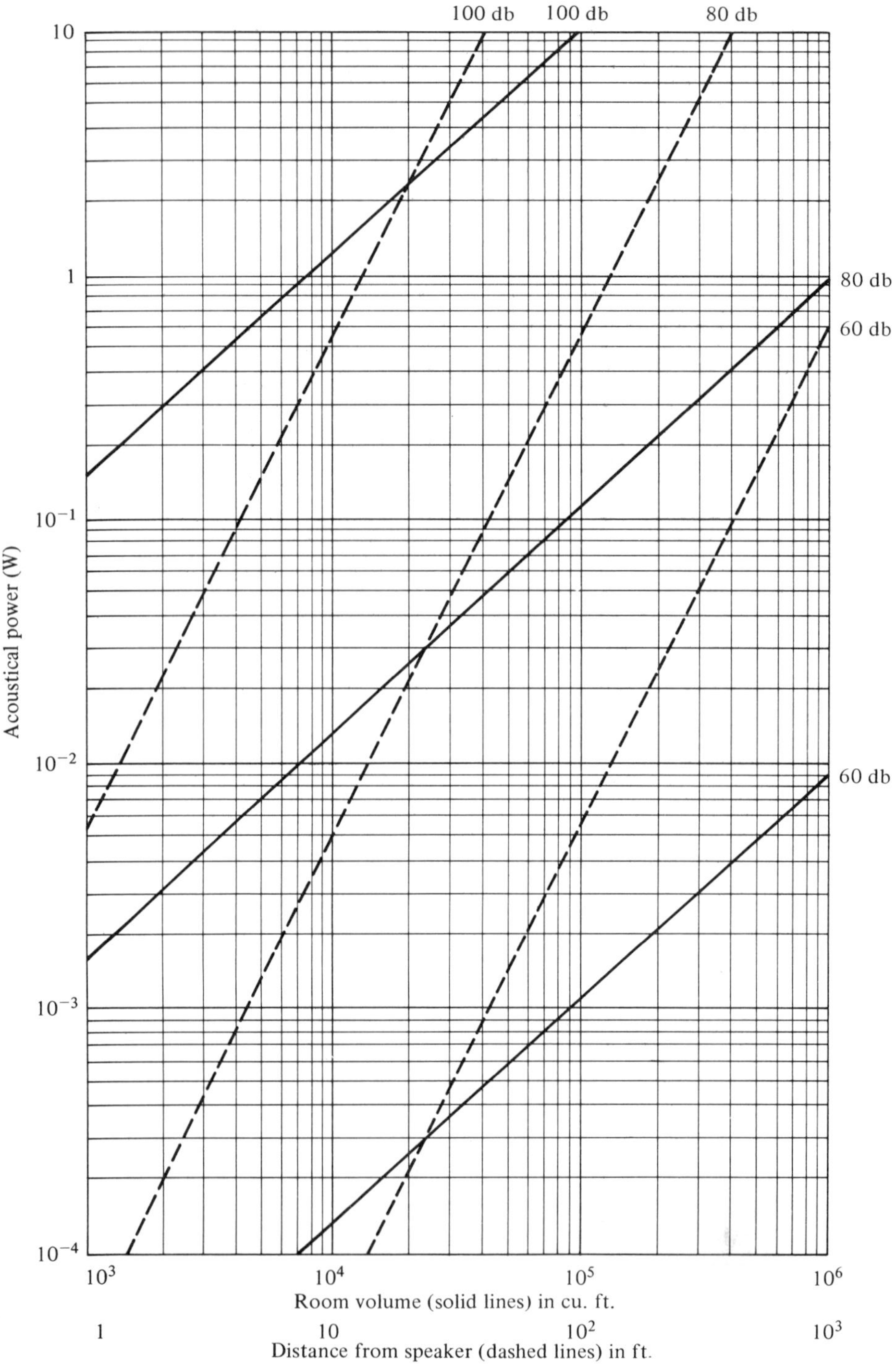

FIGURE 12–2 *Relationship between acoustical power and sound intensity for various conditions. The solid lines represent indoor conditions where the room volume is given in cubic feet. The dashed lines represent outdoor conditions with the distance from the speaker (along the axis of the speaker) given in feet. The sound intensity level is given in db. The power levels are the minimum required since the curves are based on the assumption of ideal speaker conditions.*

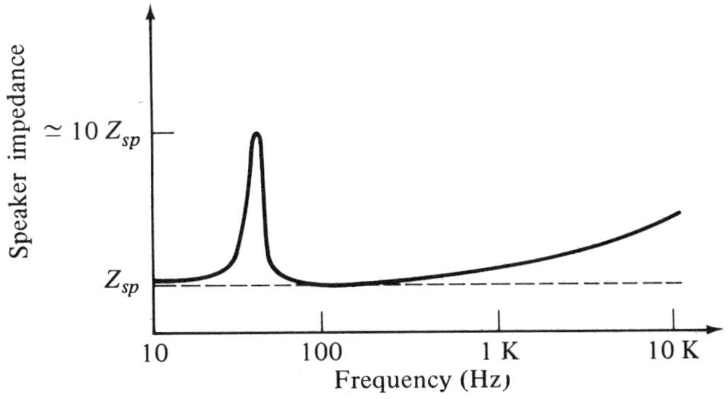

FIGURE 12–3 *Impedance of a typical electrodynamic speaker as a function of frequency.*

to-peak) and 1.6 A peak (or 3.2 A peak-to-peak). The principal problem encountered in the design of power amplifiers is that of obtaining large values of output voltage and output current while maintaining linearity and fidelity between the input and output signals.

The input impedance of a preamplifier depends primarily on the characteristics of the device used in the first section. Generally, input devices can be divided in two groups: (1) low-impedance input devices which are generally inductive, and (2) high-impedance input devices which are generally capacitive. Dynamic or moving coil microphones, magnetic tape heads, and magnetic phonograph cartridges fall into the first category while capacitive microphones and crystal phonograph cartridges fall into the second category. In some circumstances, the input impedance of the preamplifier should equal the impedance of the input device, especially for the low impedance case. However, high input impedance preamplifiers are widely used. Feedback amplifiers using voltage summing at the input are particularly useful.

A very high value of input impedance is required for the capacitive devices, particularly for capacitive microphones. The low-frequency response of this system depends on the capacitance of the input device and the input resistance of the preamplifier. Assuming that the microphone has a capacitance of 100 pF, then the input impedance of the preamplifier must be 50 M if the low-frequency response is to extend down to approximately 30 Hz. A crystal phonograph cartridge generally has a capacitance in the 1000 pF range. Consequently, an input resistance of only 5 M is needed to achieve a low corner frequency of 30 Hz.

Both bipolar and FET devices are used in preamplifier circuits. Aside from the input impedance requirement, the choice between a bipolar and an FET as the first stage of the preamplifier depends on noise considerations. A general treatment of noise is beyond the scope of this text. However, as a general guide, FET amplifiers produce less noise when the input device has a high impedance (greater than approximately 50 K), while bipolar transistor amplifiers produce less noise for lower values.

Our next task is to determine the characteristics of amplifiers and control circuits that are required for a given input device. Ideally, all frequencies in the audio range should be amplified equally by the sound reproduction system. However, in the initial recording, different frequency ranges are assigned different weights. For example, in one method of producing phonograph records, a lateral groove having an amplitude that is inversely proportional to the frequency is cut in the record material. Thus, the low-frequency or bass tones have relatively wide lateral swings while the high-frequency or treble tones have very small lateral swings. Modern long playing records ($33\frac{1}{3}$ rpm) have groove widths as small as 0.7 mil (or 0.0007 in.). Large lateral swings of the stylis may cause the needle to cut into an adjacent groove. Consequently, a more practical approach is to reduce the lateral movement of the cutting needle in the low-frequency range relative to that in the high-frequency range. This effect can be achieved by attenuating the low-frequency signals relative to the high-frequency signals when recording. On playback, the original relative amplitudes are produced by amplifying the bass tones more than the treble tones. This tone modification or tone control is generally referred to as *equalization*. An equalizing circuit is an important part of a sound reproduction system. Several common characteristics of equalizing circuits will now be considered.

Several equalization characteristics are shown in Fig. 12–4. These characteristics are generally based on the recommendations of professional societies such as the Record Industry Association of America (RIAA) and the National Association of Broadcasters (NAB). The mid-band reference frequency is generally 1 kHz. Frequencies below 1 kHz are called *bass tones* and those above 1 kHz are called *treble tones*. Equalizing networks generally are designed to boost (amplify) or cut (attenuate) either the bass tones or the treble tones. These networks can be classified as follows: (1) bass boost, (2) bass cut, (3) treble boost, and (4) treble cut. Equalization characteristics can be achieved by two methods: (1) a frequency selective network placed between stages, and (2) a frequency selective network in the feedback path of a feedback amplifier. Both cases are considered below.

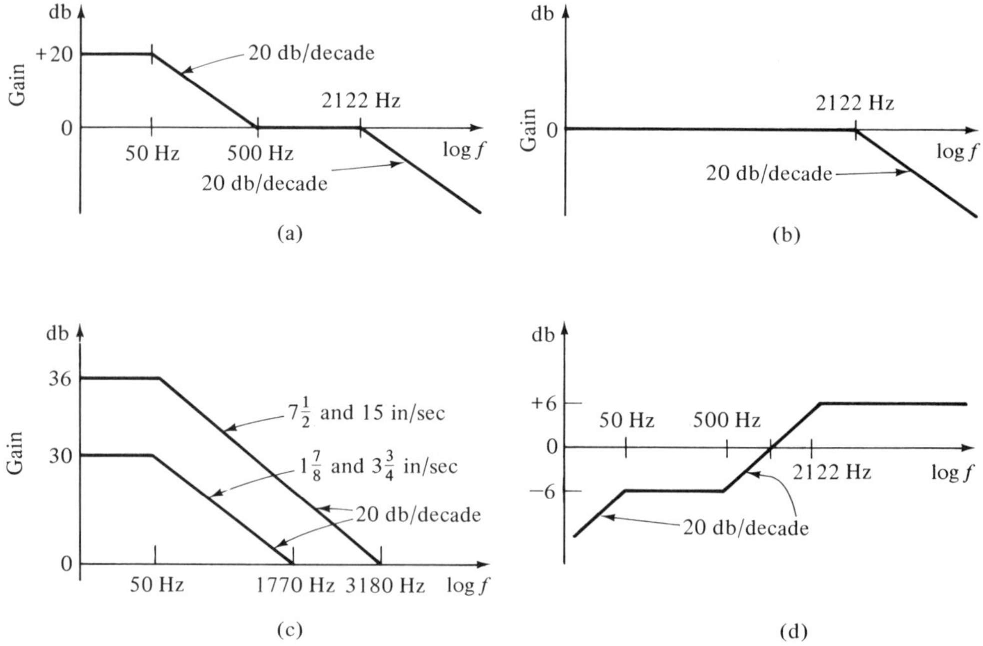

FIGURE 12–4 *Frequency response of some standard equalizing networks: (a) RIAA playback equalization for magnetic phonograph cartridge, (b) standard FM equalization, (c) NAB tape playback equalization characteristics, and (d) RIAA playback equalization for crystal cartridges.*

Three equalizing networks are shown in Fig. 12–5. The circuit shown in Fig. 12–5(a) operates as a bass boost circuit. In this circuit, the CE stage, Q_1, has an equivalent load impedance of

$$Z_{Leq1} = [R_L \,||\, (R_1 + R_2)] \frac{1 + s(R_1 \,||\, R_2)C_2}{1 + s[R_2 \,||\, (R_1 + R_2)]C_2} \tag{12–1}$$

where R_1, R_2, and C_2 are the components of the equalizing networks, and R_L is the equivalent load impedance for Q_1 with the equalizing networks disconnected. A comparison of the time constants in Eq. (12–1) indicates that the pole occurs at a lower frequency than the zero. By proper choice of components, the tape playback curves shown in Fig. 12–4(c) can be achieved by using the network shown in Fig. 12–5(a). Note that bass boost is achieved by attenuating the higher frequencies.

The tone control circuit shown in Fig. 12–5(b) operates as a treble cut circuit.* The equivalent load impedance for Q_1 is

$$Z_{Leq1} = \frac{R_L(1 + sR_1C_1)}{1 + s(R_L + R_1)C_1} \tag{12–2}$$

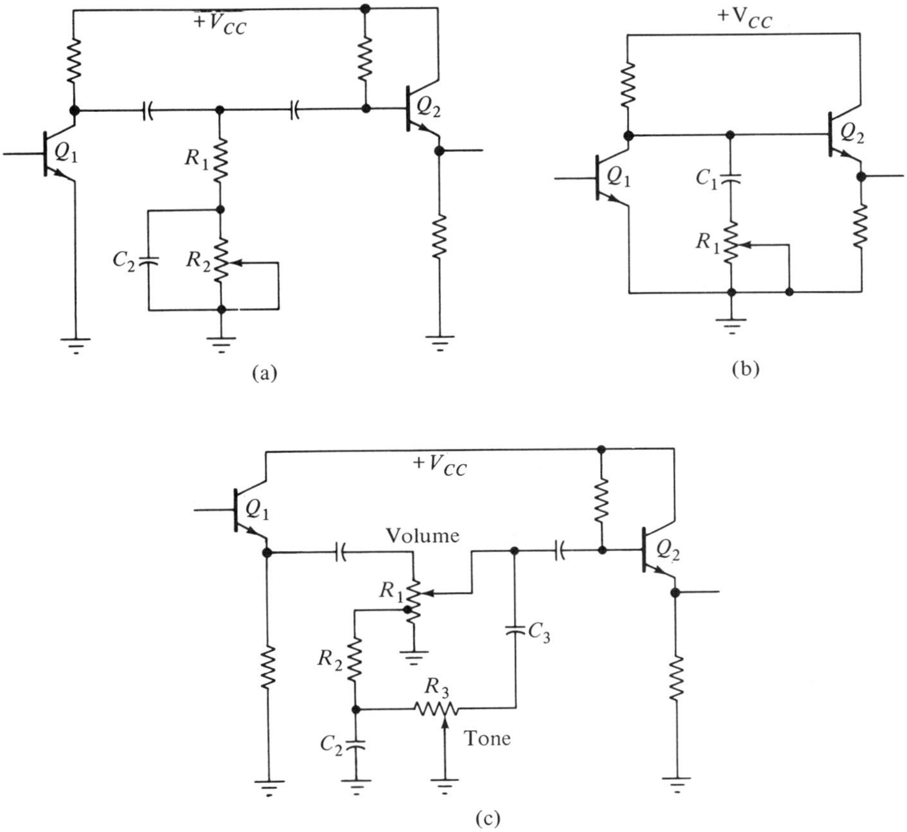

FIGURE 12–5 *Three examples of equalizing networks.*

* This circuit is identical to the *RC* lag compensation network in Sec. 11.7. Because of the low values of frequency generally encountered in this type of audio equalization, the capacitance to ground and hence the original pole at this node, can generally be neglected.

where R_1 and C_1 are the components of the equalizing network, and R_L is the equivalent load impedance for Q_1 with the equalizing network connected. For the case in which R_L is much greater than R_1, the pole in Eq. (12–2) occurs at a much lower frequency than the zero. By proper choice of the values of R_1 and C_1, the standard FM curve shown in Fig. 12–4(b) can be obtained using the circuit shown in Fig. 12–5(b). The value of R_1 is chosen to make the zero in Eq. (12–2) occur at a much higher frequency than the pole.

A combination volume control and tone control circuit is shown in Fig. 12–5(c). Potentiometer R_1 is a volume control. The bass boost section is formed by R_2, C_2, and the left side of R_3. The treble cut section is formed by C_3 and the right side of R_3. Because of the interconnections in the circuit, the overall transfer function does not have a simple form. By a proper choice of values for the controls in this circuit, the RIAA playback characteristic for magnetic phono cartridges shown in Fig. 12–4(a) can be obtained.

Several equalizing or tone control networks used in the feedback path of a feedback amplifier are shown in Fig. 12–6. The circuit in Fig. 12–6(a) is a bass boost circuit since, at low frequencies, C_2 is an open circuit, and the ratio V_o/V_i depends on the open-loop gain of the amplifier.

The circuit in Fig. 12–6(b) is a treble cut circuit. Assuming that the feedback network controls the closed-loop gain, then

$$\frac{V_o}{V_i} \simeq \left[\frac{R_1 + R_2 + R_3}{R_1}\right]\left[\frac{1 + s[R_3 \,||\, (R_1 + R_2)]C_3}{1 + sR_3C_3}\right] \tag{12–3}$$

For frequencies much below the pole in this equation,

$$V_o/V_i \simeq (R_1 + R_2 + R_3)/R_1 \tag{12–4}$$

and for frequencies much above the zero,

$$V_o/V_i \simeq (R_1 + R_2)/R_1 \tag{12–5}$$

The circuit shown in Fig. 12–6(c) is a combination bass and treble control circuit. In the very low-frequency region, C_3 effectively open circuits the feedback path, and the gain, V_o/V_i, depends on the open-loop characteristics of the amplifier. A similar condition occurs in the high-frequency region where C_2 essentially short circuits the feedback path. Since the open-loop gain is larger than the closed-loop gain in a feedback amplifier, the amplifier gain increases both in the low-frequency and in the high-frequency regions.

The circuit shown in Fig. 12–6(d) provides both bass and treble control. Assuming that the feedback network controls the closed-loop gain, then

$$\frac{V_o}{V_i} \simeq \left(\frac{R_1 + R_2 + R_3}{R_3}\right)\left(\frac{1 + s(R_1 \,||\, R_2)(C_1 + C_2)}{1 + sR_2C_2}\right)\left(\frac{1 + sR_3\left(\dfrac{C_1C_2}{C_1 + C_2}\right)}{1 + sR_1C_1}\right) \tag{12–6}$$

The circuit shown in Fig. 12–6(e) provides both bass boost and cut, and treble boost and cut. The upper feedback path is useful in the low-frequency region. Bass boost or cut is determined by the adjustment of R_2. The lower feedback path is effective in the high-frequency region. Treble boost or cut is determined by the setting of R_6.

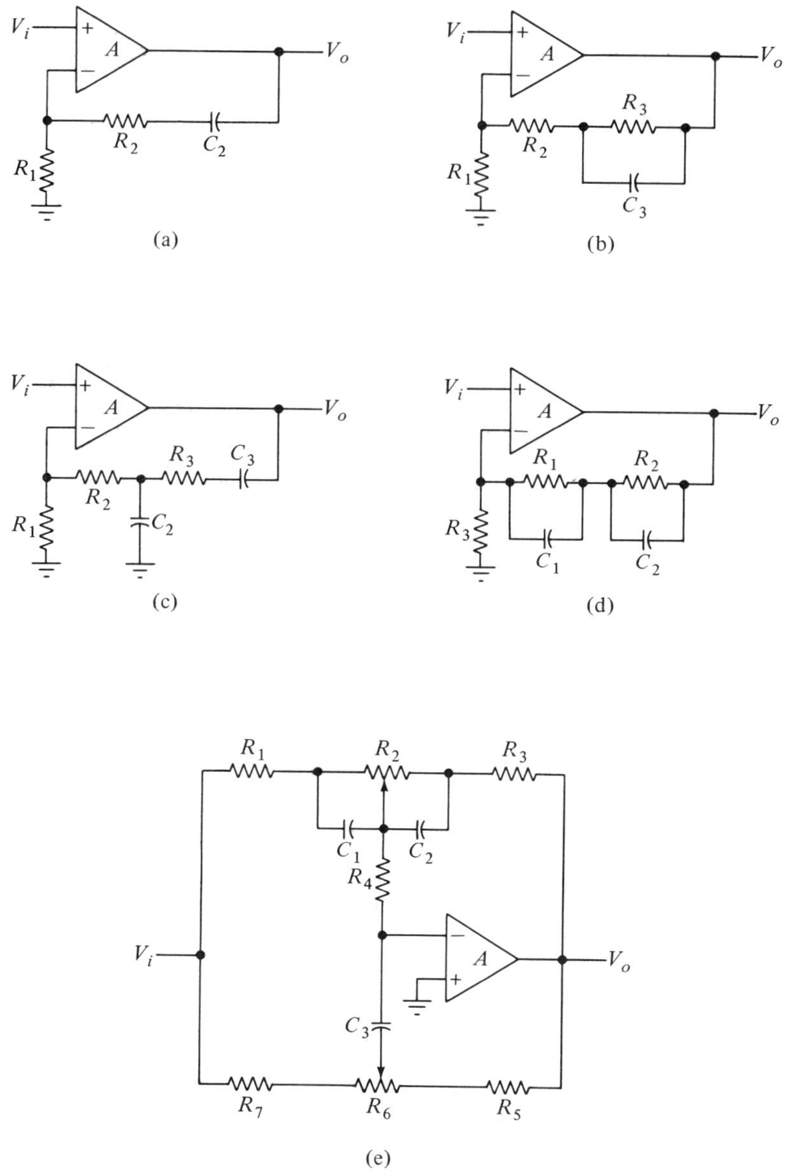

FIGURE 12–6 *Examples of equalizing circuits used in feedback networks.*

In this section, various aspects of a sound reproduction system have been considered. A complete treatment is not practical because of the wide variety of circuits encountered in practice. A wide variety results not only from a designer's choice and innovation, but also from efforts to avoid patent infringements. This short treatment was intended to develop a perspective of an audio system so that the characteristics of power amplifiers could be appreciated. In the remainder of this chapter, problems associated with audio power amplification will be treated.

12.2 Power Transistor Characteristics and Limitations

An output power of 100 W delivered to an 8-ohm speaker requires a peak voltage of 40 V and a peak current of 5 A. These large values are in sharp contrast to the millivolt signals encountered in many of the circuits discussed in previous chapters. Transistors designed to handle the large currents and voltages are necessarily different from those used in small-signal applications. The objectives of this section are to briefly study the characteristics, principal limitations, and the thermal problems of power transistors.

In Sec. 5.3, we noted that β is a function of collector current. The value of I_C for the maximum value of β is dependent on the effective emitter length. Since a reasonably large value of β is desirable for values of I_C in the 1 to 10 A range, the size of the emitter region in a power transistor must be much larger than the emitter region in a small transistor. However, a large junction area is associated with large values of C_{ob} and I_{CO}. Values of C_{ob} in the 100 to 2000 pF range are not uncommon in large power transistors. Values of I_{CO} in a power transistor can be as large as several milliamperes.* Silicon power transistors are generally preferable to germanium power transistors because of their inherently low values of I_{CO}.

In addition to larger values of C_{ob} and I_{CO}, most power transistors have smaller values of f_T and β when compared to small-signal units. Values of f_T may range as low as 0.5 MHz with 1 to 50 MHz being typical values. Note that this range is approximately an order of magnitude less than typical values of f_T for small-signal units. Also, at the higher current levels, values of β range from 20 to 70.

Another interesting result occurs when transistors are operated at high current levels. A transistor is said to be operating in the high-current region when the concentration of the emitter majority carriers injected into the base region (where they become minority carriers) reaches a value comparable with the base region majority carrier concentration.** One result of high-current operation is that the dynamic resistance of the base-emitter junction changes from r_e to approximately $2r_e$.

Since transistors used in power amplifiers are operated at high values of voltage and current, we need to study the various limitations and ratings of transistors. Maximum transistor ratings are generally divided into four groups: voltage, current, power, and thermal. Each type of rating is discussed below.

Three types of maximum or breakdown voltage ratings which are often specified for bipolar transistors are shown in Fig. 12–7. Generally, the largest breakdown voltage rating is BV_{CBO}. The value of BV_{CBO} corresponds to the avalanche breakdown voltage of the base-collector junction measured with the emitter lead open. The value of BV_{CEO} is generally less than BV_{CBO} and is measured with the base terminal open. The third voltage rating, $BV_{CE(sus)}$ or the sustaining voltage, represents the maximum collector-emitter voltage of a

---

* I_{CO}, which corresponds to I_S in a diode, is directly proportional to the junction area as shown in Chap. 2. Leakage currents are also important in power transistors due to the large junction area.

** A detailed discussion of high current operation is given in Chap. 7 of the text by Lindmayer and Wrigley listed in the references.

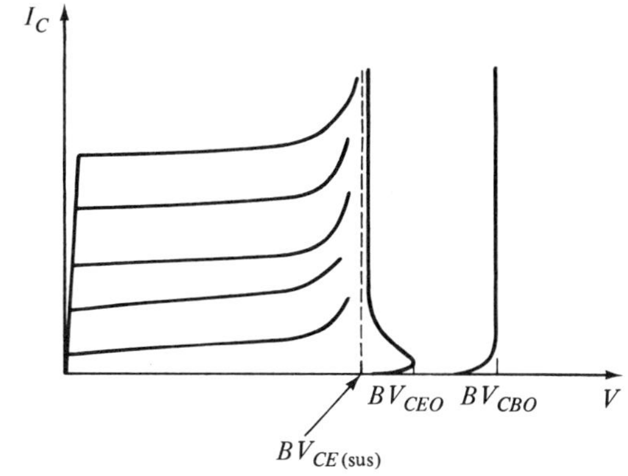

FIGURE 12–7 *Various transistor voltage ratings.*

common-emitter stage with the base current held constant. This type of breakdown is complicated by the presence of additional base majority carriers resulting from the generation of electron-hole pairs in the base-collector depletion region. For safe operation, transistors connected in the CE configuration should be operated at voltages less than $BV_{CE(sus)}$. Numerous other breakdown voltage ratings are sometimes specified. The design engineer should check the definition of these various ratings to obtain an understanding of the particular test conditions.

Current ratings are also important. Generally, a maximum value of collector current is specified by the device manufacturer. This value represents the upper limit of I_C for which the device is useful and is called $I_{C(max)}$. The value of $I_{C(max)}$ may be limited by a minimum value of β, by the current carrying capability of the internal leads, etc. Transistor performance generally falls off rapidly for values of I_C greater than $I_{C(max)}$.

The value of I_C must be less than $I_{C(max)}$ and the value of V_{CE} must be less than the breakdown voltage. In addition the value of the product of the average collector current and the average collector voltage is limited. Most manufacturers specify a value of $P_{D(max)}$ which represents the maximum power the transistor can dissipate. The effect of the $P_{D(max)}$ rating on the choice of the operating point is shown in Fig. 12–8 and is illustrated in the following example.

EXAMPLE 12–1 A silicon power transistor has the following maximum ratings: $I_{C(max)}$ equal to 5 A, BV_{CEO} equal to 50 V, and $P_{D(max)}$ equal to 30 W. Find the maximum permissible value of the average collector voltage when the transistor is operated at an average collector current of 3 A.

Solution: The value of BV_{CEO} limits the value of V_{CE} to less than 50 V. However, the value of $P_{D(max)}$ limits the average value of V_{CE} to 30 W/3 A or 10 V.

The problem of power dissipation is more complex than indicated by the previous example. The maximum average power that a transistor can dissipate is limited by the temperature that the base-collector junction can withstand, T_{jmax}. Typical values for T_{jmax} are 100°C for a germanium device and 200°C for the silicon device. Good engineering practice requires a derating of 10°C or more from this value. Also, the value of $P_{D(max)}$ specified for a transistor is given

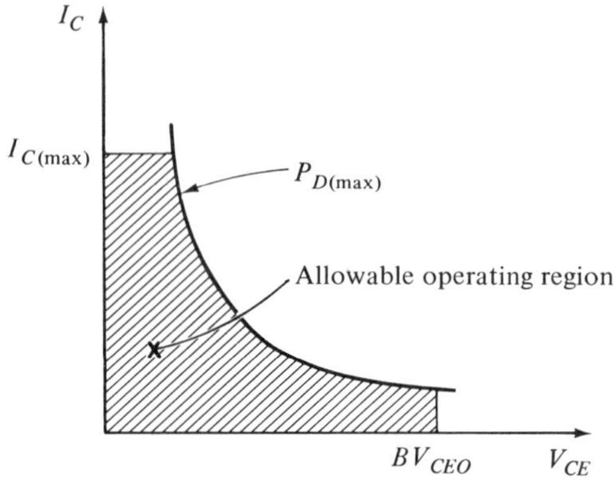

FIGURE 12–8 *Limitations imposed by the* $P_{D(max)}$ *rating of transistor on the choice of the* Q *point.*

at a particular value of transistor case temperature, usually 25°C. As the power dissipated by the transistor increases, the case temperature increases, and the permissible value of $P_{D(\max)}$ is reduced. Rather than use the value of $P_{D(\max)}$, most thermal calculations are based on the value of $T_{j\max}$ and the total thermal resistance from the junction to air, θ_{ja}, where the unit of thermal resistance is given in degrees centigrade per watt.

A thermal equivalent circuit that can be used in calculating the junction temperature of a transistor is shown in Fig. 12–9. The steady state* junction temperature is

$$T_j = P_D(\theta_{jc} + \theta_{cs} + \theta_{sa}) + T_A \tag{12–7}$$

where T_j is the junction temperature in °C, P_D is the power dissipated in the device, θ_{jc} is the thermal resistance from junction to case, θ_{cs} is the thermal resistance from the case to the heat sink, θ_{sa} is the thermal resistance from the heat sink to the ambient, and T_A is the ambient temperature. When a heat sink is not used, Eq. (12–7) reduces to

$$T_j = P_D(\theta_{jc} + \theta_{ca}) + T_A \tag{12–8}$$

where θ_{ca} is the thermal resistance from the transistor case to air. Typical values for θ_{ca} are 30°C/W for a TO-3 case, 150°C/W for a TO-5 case, 300°C/W for a TO-18 case, and 60°C/W for a TO-66 case. Semiconductor manufacturers usually specify values of $T_{j\max}$ and θ_{jc} for power transistors.

Often the collector of a transistor is connected directly to the case to reduce the value of θ_{jc}. This arrangement may require that the transistor be electrically insulated from the heat sink. Typical insulating materials are mica, anodized aluminum, and beryllium which have thermal resistance values of 0.2 to 0.4°C/W Improved thermal contact between the case, insulator, and heat sink is obtained with compounds such as silicon grease or zinc oxide filler. A typical value of θ_{cs} is 0.5°C/W. When the transistor is operated without a heat sink, the important thermal resistance, θ_{ca}, depends primarily on the transistor case configuration. The following example illustrates the importance of a heat sink.

* The transient thermal response of a transistor is a complex exponential function that can be represented as a series combination of several parallel *RC* networks, where *R* is the thermal resistance, and *C* is the thermal capacity of the system. Transient thermal response is discussed in the "RCA Silicon Power Circuits Manual" in the references listed at the end of the chapter.

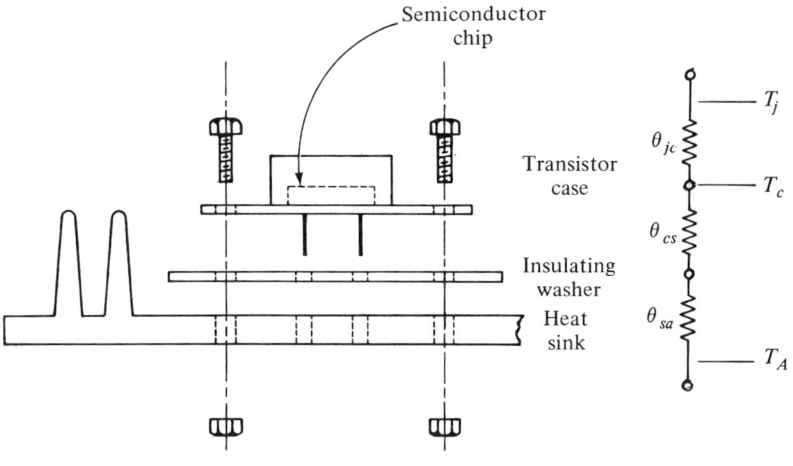

FIGURE 12–9 *A thermal equivalent circuit of a transistor with insulating washer and heat sink.*

EXAMPLE 12–2 A silicon power transistor in a TO-3 case with θ_{jc} equal to 1.5°C/W is operated at a power dissipation of 15 watts. Calculate the transistor case temperature and junction temperature when the transistor is operated (a) in free air, and (b) with a heat sink having a θ_{sa} of 2.2°C/W. The transistor is insulated from the heat sink with a mica washer. The ambient temperature is 25°C.

Solution: (a) Without a heat sink: Using Eq. (12–8),

$$T_j = P_D(\theta_{jc} + \theta_{ca}) + T_A$$
$$= (15\text{ W})[(1.5 + 30)°\text{C/W}] + 25°\text{C} \simeq 500°\text{C}$$
$$T_C = T_j - P_D\theta_{jc} \simeq 500°\text{C} - (15\text{ W})(1.5°\text{C/W}) \simeq 477.5°\text{C}$$

(b) With heat sink: Using Eq. (12–7),

$$T_j = P_D(\theta_{jc} + \theta_{cs} + \theta_{sa}) + T_A$$
$$\simeq (15\text{ W})[1.5 + 0.5 + 2.2)°\text{C/W}] + 25°\text{C} \simeq 88°\text{C}$$
$$T_C = T_j - P_D\theta_{jc} \simeq 88°\text{C} - (15\text{ W})(1.5°\text{C/W}) \simeq 65.5°\text{C}$$

(c) Experimental results: No attempt was made to measure the case temperature without a heat sink since T_j is greater than $T_{j\max}$ for a silicon transistor. With the heat sink connected, the case temperature was approximately 59°C using a simple thermometer and silicon grease. This value is less than the calculated value of T_C due to the thermal resistance between the thermometer and the case as shown in Fig. 12–10(c). The measurement indicates a value of 0.4°C/W for θ_{ct}.

Before we design a heat sink, we will briefly examine two other problems encountered in power transistors: (1) thermal runaway, (2) second breakdown. The first effect results from the temperature dependence of I_{CO}. The value of I_{CO} doubles approximately for each 10°C rise in junction temperature. As indicated in Sec. 4.3, an increase in I_{CO} of ΔI_{CO} causes an increase in I_C of $S_I \Delta I_{CO}$ where S_I can be found from Eq. (4–32). Consequently, an increase in I_{CO} increases the power dissipated at the collector of the transistor, thereby causing a further increase in the junction temperature. Unless the bias circuit stabilizes I_C, thermal runaway can result. The problem of thermal runaway is particularly severe with

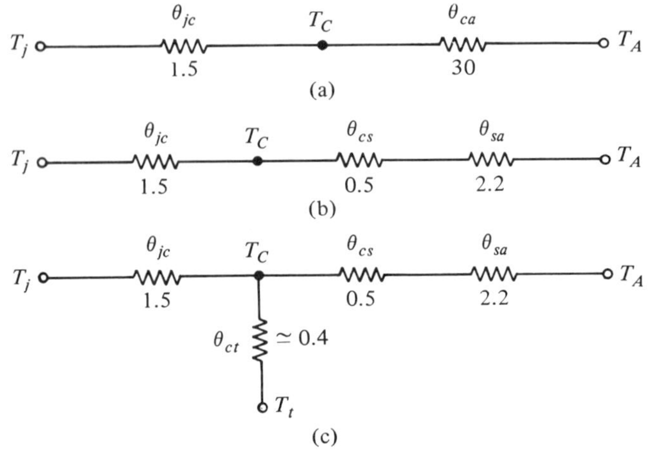

FIGURE 12–10 *Thermal equivalent circuit for Ex. 12–2: (a) without heat sink, (b) with heat sink, and (c) measurement scheme. In (c), θ_{ct} represents thermal resistance from the transistor case to the thermometer, and T_t represents the thermometer reading.*

germanium power transistors due to their large value of I_{CO} at room temperature. The use of silicon power transistors along with adequate heat sinks greatly reduces the chances of thermal runaway.*

Although a transistor may be operated within the maximum voltage, current, temperature, and power dissipation ratings, a transistor can still be destroyed by a mechanism called the *second breakdown* effect. This phenomenon is energy dependent and thus a function of time as well as voltage and current. Basically, second breakdown results from "hot spots" formed within the device due to nonuniform current densities. Most manufacturers provide curves which show the limitations imposed by second breakdown. A typical curve is shown in Fig. 12–11 and is discussed below.

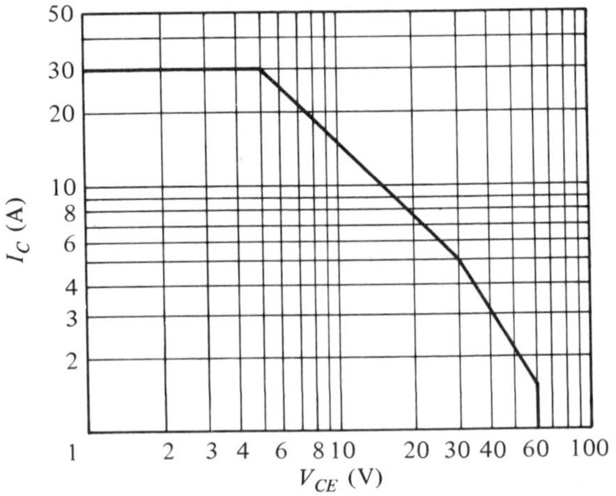

FIGURE 12–11 *An example of a safe operating area curve for a power transistor with* T_c *= 25°. The log-log plot is typical of a safe-area curve.*

* The use of small unbypassed emitter resistors also helps to maintain thermal stability in power amplifiers. This approach is illustrated in later sections dealing with specific power amplifier configurations.

The maximum ratings for the transistor type whose curve is shown in Fig. 12–11 are

$$P_{D(\max)} = 150 \text{ W (25°C case temperature)}$$
$$V_{CEO} = 60 \text{ V} \qquad I_{C(\max)} = 30 \text{ A} \tag{12–9}$$

For values of V_{CE} less than 5 V, the value of I_C is limited by $I_{C(\max)}$. For values of V_{CE} in the range of 5 V to 30 V, the value of I_C is limited by the 150 W value of $P_{D(\max)}$. For values of V_{CE} in the range 30 V to 60 V, the value of I_C is limited by second breakdown. For example, at $V_{CE} = 50$ V, the maximum value of I_C is limited by the 150-W dissipation at 3 A. However, the safe operating area curve shows that I_C must be less than 2 A which illustrates the limitation of second breakdown. Values of V_{CE} are restricted to less than 60 V by avalanche breakdown.

The safe-area curve shown in Fig. 12–11 is based on a case temperature of 25°C. As the case temperature increases, the effects of thermal resistance and the limitations of $T_{j\max}$ must be considered as illustrated in Ex. 12–2. Stated differently, a power transistor must be operated within the limitations imposed by $I_{C(\max)}$, $P_{D(\max)}$, BV_{CE}, second breakdown, and $T_{j\max}$.

The principal characteristic of a heat sink is its thermal resistance, θ_{sa}. Other characteristics such as size, volume, and weight may also be important. The relationship between total surface area and θ_{sa} for various types of free air heat sinks is shown in Fig. 12–12. Values of θ_{sa} less than about 1°C/W generally require forced air or water cooling. The following example illustrates the choice of a heat sink for a particular application.

EXAMPLE 12–3 A silicon power transistor whose safe operating area curve is shown in Fig. 12–11 is to be operated at an average collector current of 3 A, an average collector voltage V_{CE} of 20 V, and an ambient temperature of 25°C. The thermal characteristics of the transistor are $T_{j\max}$ equal to 200°C and θ_{jc} equal to 1°C/W. What type of heat sink should be used?

Solution: (a) Safe operating area: From Fig. 12–11, the operating point of 3 A and 20 V is well within the safe operating area. The power that the transistor must dissipate is

$$P_D = I_C V_{CE} = (3 \text{ A})(20 \text{ V}) = 60 \text{ W}$$

(b) Heat sink characteristics: Using Eq. (12–7), derating the junction temperature 10°C, and assuming that θ_{cs} is 0.5°C/W,

$$190°\text{C} = (60 \text{ W})[(1 + 0.5 + \theta_{sa})°\text{C/W}] + 25°\text{C}$$
$$\theta_{sa} = \frac{190 - 25 - (60)(1.5)}{60} \simeq 1.25°\text{C/W}$$

From Fig. 12–12, a flat, vertical-fin, black aluminum heat sink with a surface area of approximately 200 in.^{2} could be used.

In this section, we have examined some of the important characteristics of power transistors. Their small-signal parameters (such as C_{ob}, f_T, β, I_{CO}, etc.) are generally inferior to those of small-signal devices because of the large junction area of power transistors. The transistor must be operated within its maximum power dissipation, voltage, current, and temperature ratings. Generally, heat sinks must be used with power transistors to keep the operating temperature below $T_{j\max}$.

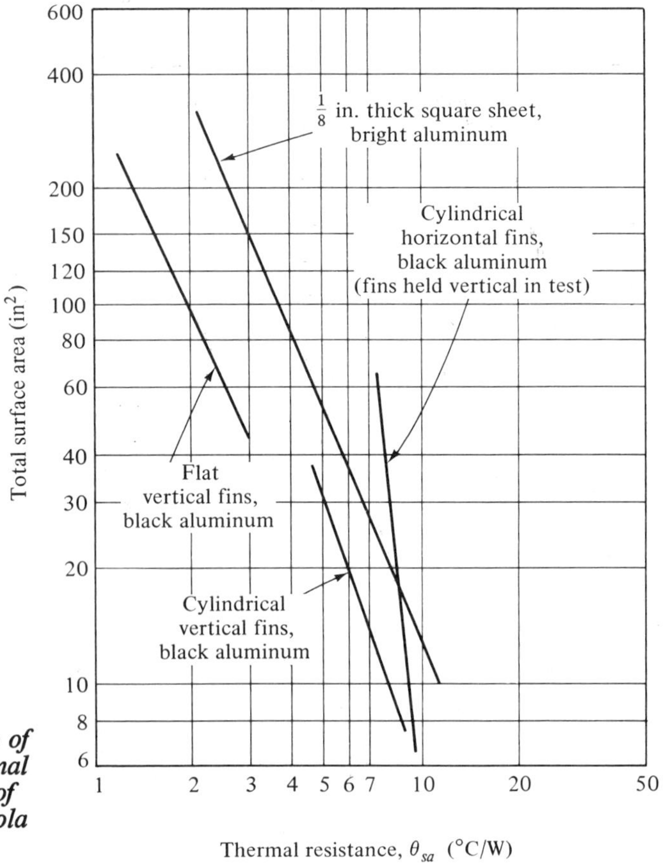

FIGURE 12–12 *Relationship of the total surface area and thermal resistance θ_{sa} for various types of heat sinks. (Courtesy of Motorola Semiconductor Products Inc.)*

12.3 Characteristics and Classifications of Ideal Power Amplifiers

In Sec. 12.1 we studied the overall characteristics of an audio system and determined, among other things, some typical values of output voltages and currents encountered in power amplifiers. In the last section, we examined some of the characteristics and limitations of power transistors as well as the problem of heat sinks. The objective of this section is to discuss some of the characteristics of power amplifier stages and the terms used to describe their performance.

A power transistor* can be operated in the CE, CB, or CC configuration. However, power amplifier stages can also be described in terms of their bias conditions relative to the amplitude of the input signal as Class A, Class AB, Class B, or Class C amplifiers. The identification of the various classes of operation can be made with the aid of Fig. 12–13.

A Class A amplifier is one in which current flows for all values of the input signal. Most of the amplifiers considered in the preceding chapters were Class A

---

* Most of the discussion in this section is in terms of bipolar transistors. A parallel treatment can be developed for both FETs and vacuum tubes. Most new circuit designs of power amplifiers use bipolar transistors, although many older circuits use vacuum tubes. Since few FET power transistors are commercially available, the FET in power applications will not be treated.

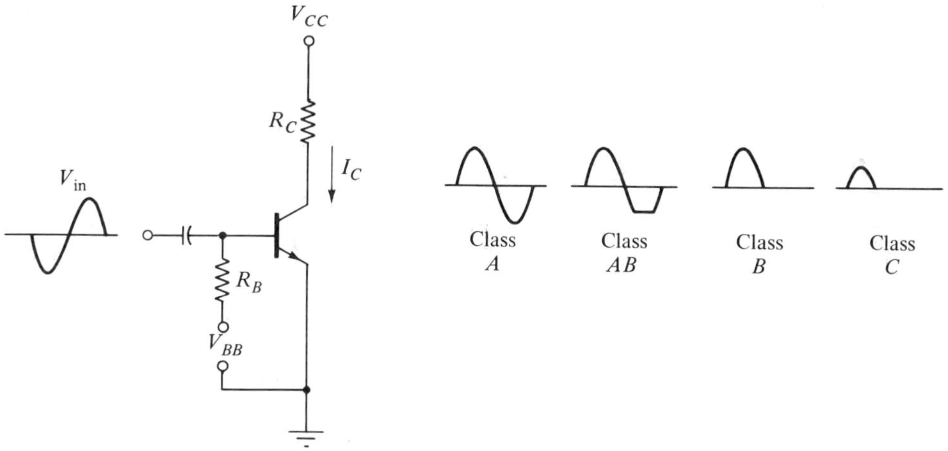

FIGURE 12–13 *Collector current waveform characteristics of the various classes of amplifier operation.*

amplifiers. In Class B operation, the collector current flows for only half of each cycle of the input signal. Class AB operation* is in between Class A and Class B since the collector current flows for more than one-half cycle but less than a full cycle of the input signal. In Class C operation, the collector current flows for less than one-half cycle of the input signal.

The waveforms in Fig. 12–13 provide some insight into the characteristics of each class of operation. The output signal is least distorted in Class A and most distorted in Class C operation. The opposite is true for the conversion efficiency η, of the various classes where

$$\eta \equiv \frac{\text{Output ac power}}{\text{Input dc power}} \times 100\% \tag{12–10}$$

Our next task is to develop an expression for the conversion efficiency in each class of operation. Distortion present in the output signal is treated in the next section.

A Class A CE amplifier with a resistive load is shown in Fig. 12–14(a). To obtain the maximum output signal, the Q point is chosen so that V_{CE} is equal to one-half of the supply voltage. Also, the load line is chosen tangent to the $P_{D(\max)}$ curve to obtain the maximum output power without overheating the transistor. Neglecting the small saturation voltages, the Q point values are

$$V_Q = V_{CC}/2 \qquad I_Q = V_{CC}/2R_L \tag{12–11}$$

For sinusoidal signals, the ac output power is

$$\begin{aligned} P_{\text{out(ac)}} &= V_{\text{rms}} I_{\text{rms}} = \left(\frac{V_{\max} - V_{\min}}{2\sqrt{2}}\right)\left(\frac{I_{\max} - I_{\min}}{2\sqrt{2}}\right) \\ &= \frac{(V_{\max} - V_{\min})(I_{\max} - I_{\min})}{8} \end{aligned} \tag{12–12}$$

* In vacuum tube circuits, a subscript "1" is used to indicate that no grid current flows, and the subscript "2" indicates that grid current does flow during the positive peak of the input signal. Thus, Class AB operation for a vacuum tube may be designated as AB_1 or AB_2.

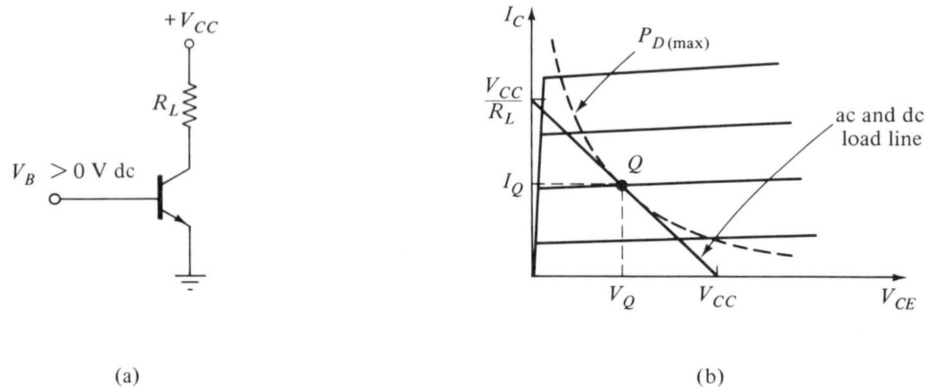

FIGURE 12–14 *A Class A amplifier with a resistive load: (a) simplified circuit diagram, and (b) output characteristic curves showing the operating point and load lines.*

For the circuit in Fig. 12–14, Eq. (12–12) becomes

$$P_{\text{out(ac)}} = \frac{(V_{CC} - 0)(V_{CC}/R_L - 0)}{8} = \frac{V_{CC}^2}{8R_L} \tag{12–13}$$

The dc power to the circuit from the power supply is

$$P_{\text{in(dc)}} = V_{CC}I_{dc} = V_{CC}I_Q = \frac{V_{CC}^2}{2R_L} \tag{12–14}$$

Combining Eqs. (12–10), (12–13) and (12–14), the conversion efficiency for this amplifier is

$$\eta = \frac{P_{\text{out(ac)}}}{P_{\text{in(dc)}}} \times 100\% = \frac{2}{8} \times 100\% = 25\% \tag{12–15}$$

While the ideal conversion efficiency of a Class A amplifier with a resistive load is 25 percent, practical values are generally less than 15 to 20 percent.

The theoretical efficiency of a Class A amplifier can be doubled if the dc current does not flow in R_L. This condition is accomplished by using a unity turns ratio transformer as shown in Fig. 12–15(a).* The transformer primary is assumed to have no dc resistance which yields a vertical dc load line. The maximum output power is obtained when the ac load line is tangent to the $P_{D(\max)}$ curve as shown in Fig. 12–15(b). Ideally, the output voltage can have a peak-to-peak value of $2V_{CC}$.** The ideal conversion efficiency of a transformer coupled Class A amplifier is derived as follows:

$$P_{\text{out(ac)}} = \frac{(2V_{CC} - 0)(2V_{CC}/R_L - 0)}{8} = \frac{V_{CC}^2}{2R_L} \tag{12–16}$$

$$P_{\text{in(dc)}} = (V_{CC})(V_{CC}/R_L) = V_{CC}^2/R_L \tag{12–17}$$

$$\eta = \frac{P_{\text{out(ac)}}}{P_{\text{in(dc)}}} \times 100\% = \frac{1}{2} \times 100\% = 50\% \tag{12–18}$$

Values in the range of 35 to 40 percent can be achieved in practice without excessive distortion.

---

* Unity turns ratio is assumed for convenience and is not the general case. The turns ratio is usually chosen to convert the actual load into a more desirable collector load. Then the ac load becomes $(n_2/n_1)^2R_L$.

** The ability to provide an instantaneous output voltage larger than the supply voltage is due to the energy stored in the inductance of the transformer.

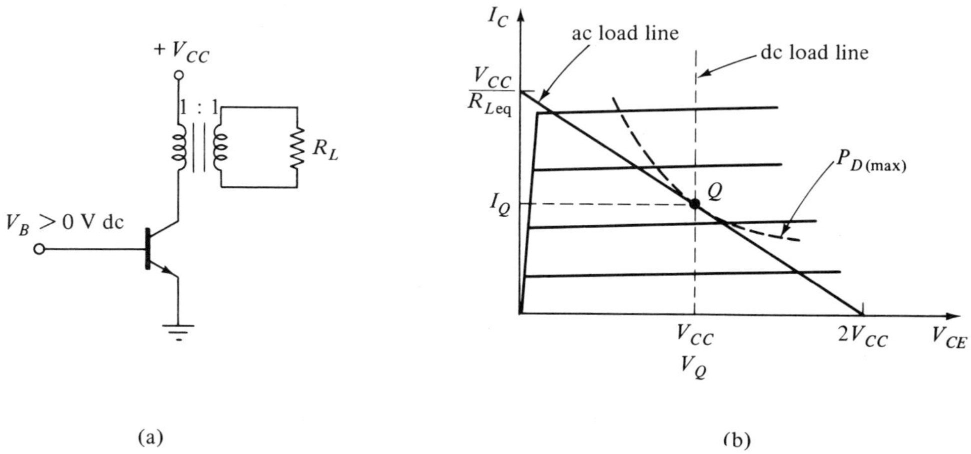

FIGURE 12–15 *A Class A amplifier with transformer coupling to the load resistor: (a) simplified circuit diagram, (b) output characteristic curves showing Q point and load lines.*

A Class B CE amplifier with a resistive load is shown in Fig. 12–16(a). The value of V_{BE} is 0 V or slightly positive but less than the cut-in voltage (0.5 V for silicon). Collector current flows for only half of each cycle of the input signal. Consequently, the ac output power is

$$P_{out(ac)} = \frac{1}{2} V_{rms} I_{rms} = \frac{1}{2}\left(\frac{V_m}{\sqrt{2}}\right)\left(\frac{I_m}{\sqrt{2}}\right) = \frac{V_{CC}^2}{4R_L} \qquad \textbf{(12–19)}$$

The dc current flowing in the transistor is equal to the average value of a half rectified sine wave or

$$I_{dc} = I_m/\pi = V_{CC}/(\pi R_L) \qquad \textbf{(12–20)}$$

The ideal conversion efficiency with a Class B amplifier is

$$\eta = \frac{P_{out(ac)}}{P_{in(dc)}} \times 100\% = \frac{V_{CC}^2/(4R_L)}{V_{CC}^2/(\pi R_L)} \times 100\% = \frac{\pi}{4} \times 100\% = 78.5\% \qquad \textbf{(12–21)}$$

Practical values range from 55 to 65 percent.

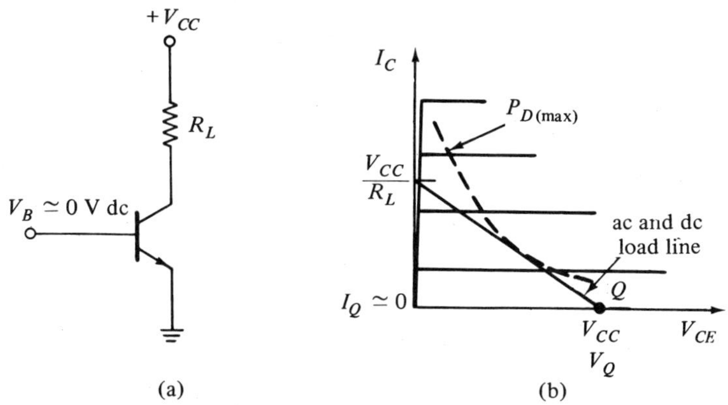

FIGURE 12–16 *A Class B amplifier with a resistive load: (a) simplified circuit diagram, and (b) output characteristic curves showing the Q point and load lines.*

A Class C amplifier with a tuned circuit load is shown in Fig. 12–17. The parallel *LC* circuit, called a "tank" circuit, is resonant at the frequency

$$f_r = 1/(2\pi\sqrt{LC}) \qquad \textbf{(12–22)}$$

Class C amplifiers are generally used as single-frequency amplifiers (in a transmitter, for example), and their conversion efficiency can theoretically approach 100 percent. Conversion efficiencies in excess of 85 percent can be achieved in practice. When a Class B amplifier is operated with a tuned circuit load, the amplifier is referred to as a *linear tuned amplifier.*

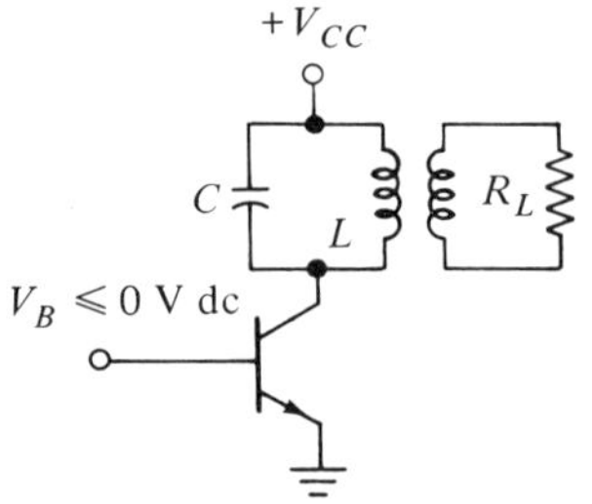

FIGURE 12–17 *A Class C amplifier with a tuned circuit load.*

The characteristics of Class A, AB, and B amplifiers will now be compared since they are the types used in audio power amplifiers. Class A operation has the least distortion and also the lowest conversion efficiency. This class is also used extensively in applications requiring less than a few watts of output power (for example, in the output stage of small radio receivers). Since a power transistor must dissipate an amount of power equal to the difference between $P_{\text{in(dc)}}$ and $P_{\text{out(ac)}}$, output power levels greater than a few watts place high dissipation requirements on the transistor used in Class A amplifiers due to the low conversion efficiency. This effect is illustrated in the following example.

EXAMPLE 12–4 A Class A amplifier is required to furnish an output power of 2 W. Estimate the power that the transistor must dissipate when the load is (a) directly connected to the transistor, and (b) transformer coupled to the transistor.

Solution: (a) Class A operation with resistive load: From Eq. (12–15), the ideal value of η is 25 percent with 15 percent being a reasonable estimate for a practical amplifier. Thus,

$$\begin{aligned} P_{\text{in(dc)}} &= P_{\text{out(ac)}}/\eta \simeq (2\text{ W})/(0.15) \simeq 13.3\text{ W} \\ P_D &= P_{\text{in(dc)}} - P_{\text{out(ac)}} \simeq 13.3 - 2 \simeq 11.3\text{ W} \end{aligned}$$

(b) Class A operation with transformer coupled load: From Eq. (12–18), the theoretical value of η is 50 percent with 35 percent being a reasonable estimate for a practical amplifier. Thus,

$$\begin{aligned} P_{\text{in(dc)}} &= P_{\text{out(ac)}}/\eta \simeq (2\text{ W})/(0.35) \simeq 5.7\text{ W} \\ P_D &= P_{\text{in(dc)}} - P_{\text{out(ac)}} \simeq 5.7 - 2 \simeq 3.7\text{ W} \end{aligned}$$

The principal disadvantage of Classes AB and B amplifiers is the distortion present because the collector current flows for less than a full cycle. However, two transistors can be combined so that each conducts for alternate half cycles, thus providing conduction for the full input cycle. The two basic arrangements are: (1) push-pull, and (2) complementary. Each circuit will now be discussed.

A Class B push-pull amplifier is shown in Fig. 12–18. The transistors are assumed to be ideal in that the cut-in voltage or V_{BE} bias is neglected. If the input base-to-ground voltage of one transistor is 180° out of phase with respect to the other, the output transformer delivers a good voltage sine wave to the load for a sine wave input. Each transistor acts effectively as a half-wave amplifier and carries the load current for one-half of each cycle.

As shown in Fig. 12–18, the input transformer supplies two voltages having equal amplitudes but 180° out of phase when referred to the center tap which is grounded. Thus, one transistor is cut off while the other one is conducting. If the top transistor is conducting, current flows through the output transformer in the direction of the collector of the upper transistor, through the transistor, and back through the battery in the center wire of the circuit. Little or no current flows in the lower half of the primary during this half-cycle period because the lower transistor, while in the off condition, acts like an open circuit. The changing primary current sets up a changing flux density in the core, which in turn induces a voltage in the secondary winding. This voltage produces a current in the load. During the next half-cycle period, the top transistor is off, and the lower unit causes a current to flow through the lower part of the transformer, as shown in Fig. 12–18. The flux induced in the core has a direction opposite that produced by the current in the top unit. Therefore, the polarity of the secondary voltage reverses to produce the other half of the output sine wave.

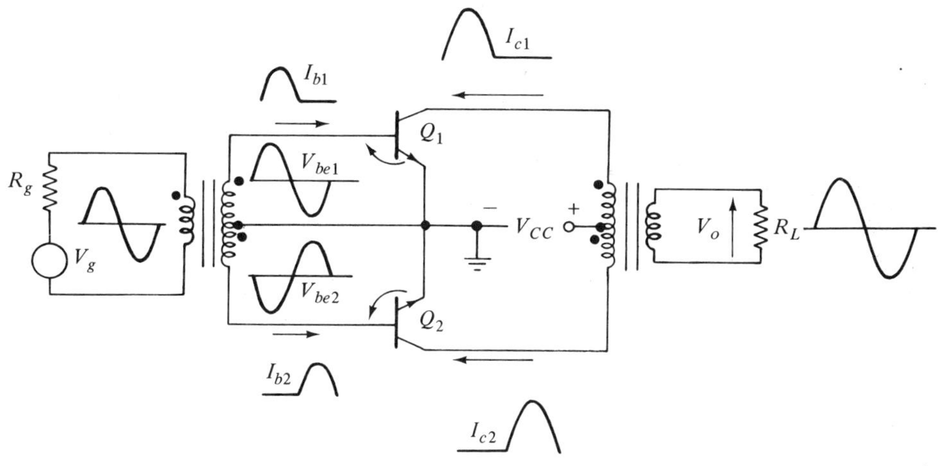

FIGURE 12–18 *A Class B push-pull amplifier with ideal transistors. The current directions are shown for NPN transistors.*

A Class B complementary amplifier is shown in Fig. 12–19. The operation of this circuit is discussed in detail in connection with Fig. 11–29. In this circuit, the NPN conducts on the positive portion of the input signal, and the PNP conducts on the negative portion of the input signal.

In this section, we have briefly examined some of the characteristics of the various classes of the amplifier operation. Class A operation is the least efficient but has the lowest distortion. Classes AB and B are more efficient but also have higher levels of distortion. Generally, Classes AB and B are used in either the

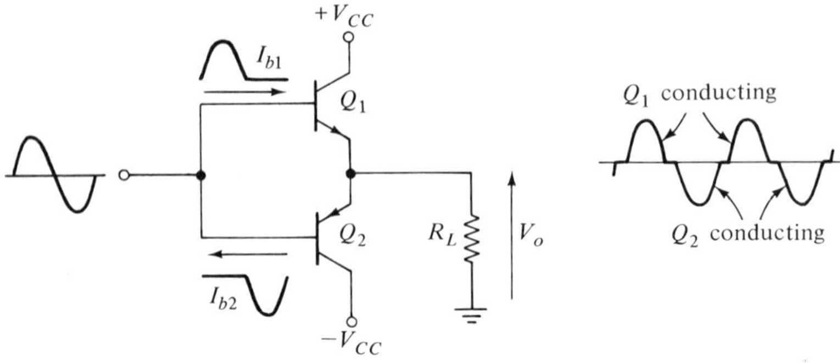

FIGURE 12–19 *A Class B complementary amplifier.*

push-pull arrangements or in complementary arrangements. Practical Class A amplifiers are treated in Sec. 12.5, push-pull amplifiers in Sec. 12.6, and complementary amplifiers in Sec. 12.7. Before considering these circuits, however, we will introduce briefly some of the aspects of distortion.

12.4 The Distortion Problem

The basic design goal of a linear amplifier is to provide an output signal that is directly proportional to the input signal regardless of the amplitude, frequency, or polarity of the input signal. Any deviation of the output from this ideal can be classified as distortion. Several examples of distortion were encountered in previous chapters. In the large-signal analysis of the CS FET amplifier in Sec. 3.4, the positive-going output signal had a different maximum amplitude than the negative-going output signal for a symmetrical input signal. This type of distortion was due to the nonlinear characteristic curves of an FET. A similar type of distortion may be produced by any active device. The frequency response characteristics of an amplifier can also be considered as a type of distortion since the amplifier may not amplify input signals in the low- and high-frequency regions the same amount as signals in the mid-band range. Also, the peaking, ringing, and overshoot encountered in feedback amplifiers are forms of distortion. In this section we will examine three types of distortion produced by power amplifiers: (1) harmonic distortion, (2) crossover distortion, and (3) intermodulation distortion.

When a single-frequency signal is applied to the input of an amplifier, the output waveform may contain frequency components that include the fundamental and several harmonics.* The generation of harmonic components is due to the nonlinear transfer characteristics of an active device as shown in Fig. 12–20. In this figure let the input variable lie on the x-axis and the output variable lie on the y-axis. In Fig. 12–20(a), the transfer admittance relating the input voltage and output current is linear. Therefore, a sinusoidal input voltage will produce a sinusoidal output current. An equation relating the two variables is

---

* The harmonics of a fundamental frequency are those frequencies that are multiples of the fundamental. For example, harmonics of a 1-kHz fundamental frequency are 2 kHz, 3 kHz, 4 kHz, etc.

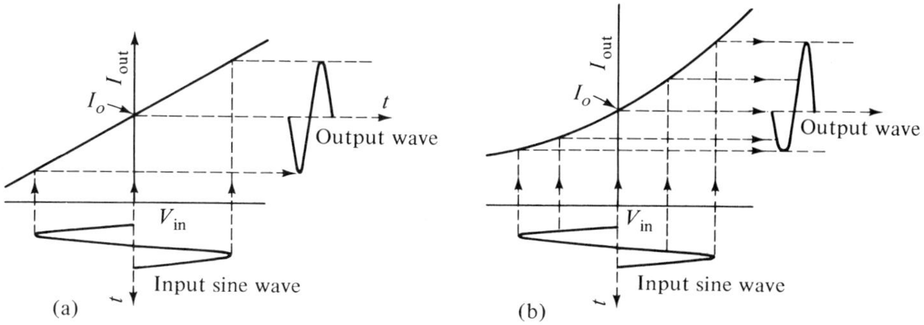

FIGURE 12–20 *Transfer functions and their effects on the output variable: (a) linear and (b) nonlinear.*

$$I = a_o + a_1 V_{in} \tag{12–23}$$

where a_o is the output current when V_{in} is zero, and a_1 is the slope of the transfer admittance curve.

In Fig. 12–20(b), the transfer admittance is a nonlinear function, and a sinusoidal input voltage does not produce a sinusoidal output current. Consequently, the output current wave is distorted compared with the input wave. Since this current is related to the output voltage through a straight line (the load line), the output voltage contains distortion. Our immediate problem is to evaluate this distortion.

The output wave contains a fundamental and several harmonic components. For a given waveform, the amplitudes and phases of the harmonics have fixed values relative to the fundamental. In this study, only the amplitudes are important. The number of significant harmonics required to synthesize a wave is determined by the curvature of the transfer function. In general, a curve similar to the one shown in Fig. 12–20(b) would produce only a few harmonics, whereas a function with a sharp bend would produce a large number.

Let us now develop a method for finding the harmonic content of the output wave. In Fig. 12–20(b) the two variables can be related with a power series which has the following form:

$$i = a_o + a_1 v_i + a_2 v_i^2 + \ldots + a_n v_i^n \tag{12–24}$$

If only the first three terms were used, the amplitudes of the first (fundamental) and second harmonics could be determined. With five terms, all components up to and including the fourth harmonic could be determined. The constants, a_o, a_1, etc. in Eq. (12–24) can be evaluated by substituting corresponding pairs of I_o and I_i into this equation. If four terms are needed, then the range of the input variable may be divided into four equal parts,* as shown in Fig. 12–20(b). Five corresponding pairs of voltage and current can be obtained, resulting in five simultaneous equations. The constants may be determined from this set of equations.

While these constants indicate the relative importance of the corresponding harmonic, they do not yield the amplitudes directly. Let us now consider the

* The increments are made equal for convenience, not by necessity, in dealing with cosine waves.

case of a Class A power amplifier in which the input voltage is

$$v_i = V_{max} \cos \omega t \tag{12–25}$$

Substituting $V_{max} \cos \omega t$ into Eq. (12–24) for v_i and expanding, we obtain the instantaneous output current for a given value of V_{max},

$$i = I_o + A_o + A_1 \cos \omega t + A_2 \cos 2\omega t + A_3 \cos 3\omega t + A_4 \cos 4\omega t + \ldots \tag{12–26}$$

where I_o is the dc output current with no signal present, A_o is the added dc or the rectified component when the signal is present, and the other coefficients are harmonic amplitudes. These coefficients may be evaluated in any particular situation by considering values of ωt that produce equal increments of change in the input voltage. For a cosine function these values are 0°, 60°, 90°, 120°, and 180°.

The transfer admittance for a typical power transistor is shown in Fig. 12–21. Using a cosine part of the input voltage wave, when $\omega t = 0$, we have $i = I_{max}$. Then Eq. (12–26) reduces to

$$I_{max} = I_o + A_o + A_1 + A_2 + A_3 + A_4 \tag{12–27}$$

For $\omega t = 60°$,

$$I_1 = I_o + A_o + \frac{A_1}{2} - \frac{A_2}{2} - A_3 - \frac{A_4}{2} \tag{12–28}$$

For $\omega t = 90°$,

$$I_o = I_o + A_o - A_2 + A_4 \tag{12–29}$$

For $\omega t = 120°$,

$$I_2 = I_o + A_o - \frac{A_1}{2} - \frac{A_2}{2} + A_3 - \frac{A_4}{2} \tag{12–30}$$

For $\omega t = 180°$,

$$I_{min} = I_o + A_o - A_1 + A_2 - A_3 + A_4 \tag{12–31}$$

The five amplitudes may now be expressed in terms of the output current values as follows:

$$A_o = \frac{(I_{max} + I_{min}) + 2(I_1 + I_2)}{6} - I_o \tag{12–32}$$

$$A_1 = \frac{(I_{max} - I_{min}) + (I_1 - I_2)}{3} \tag{12–33}$$

$$A_2 = \frac{I_{max} + I_{min} - 2I_o}{4} \tag{12–34}$$

$$A_3 = \frac{(I_{max} - I_{min}) - 2(I_1 - I_2)}{6} \tag{12–35}$$

$$A_4 = \frac{(I_{max} + I_{min}) - 4(I_1 + I_2) + 6I_o}{12} \tag{12–36}$$

Since the harmonics occur at different frequencies, their amplitudes cannot be added directly. The equivalent amplitude of the unwanted signal, A_D, is approximately equal to

$$A_D \simeq \sqrt{A_2^2 + A_3^2 + A_4^2} \tag{12–37}$$

Then the total distortion is

$$D_T = (A_D/A_1) \times 100\% \tag{12–38}$$

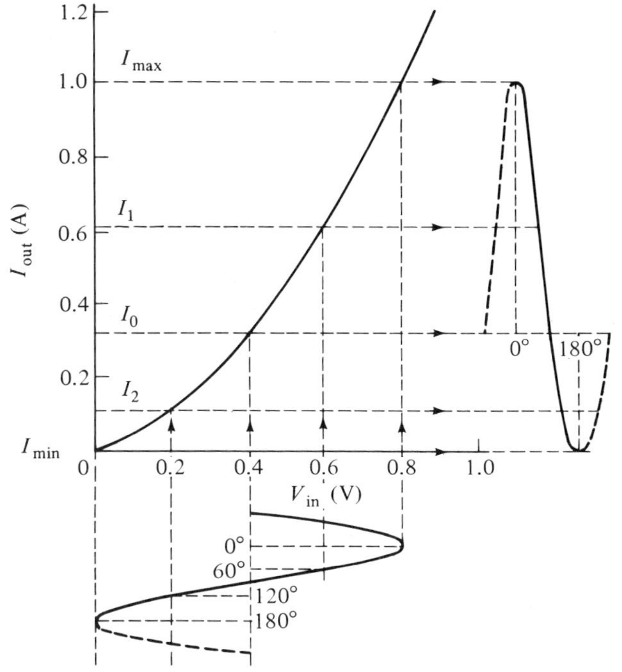

FIGURE 12–21 *The transfer characteristic curve for a typical power transistor.*

The distortion due to each harmonic may also be calculated separately. The following example illustrates the calculation of harmonic distortion.

EXAMPLE 12–5 The input voltage shown in Fig. 12–21 has a peak amplitude of 0.4 V and swings around a bias of 0.4 V dc. Find the total output dc current, the peak value of the fundamental current, and estimate the value of the total harmonic distortion.

Solution: (a) dc current: The dc current at the bias point is approximately 0.32 A. Since the area under the top of the output wave is larger than that of the lower part, a dc component is introduced by the nonlinear transfer curve. From Eq. (12–32), this component is

$$A_o \simeq \frac{(1.0 + 0) + 2(0.61 + 0.1)}{6} - 0.32 \simeq 0.08 \text{ A}$$

The dc current as indicated by a meter would increase from 0.32 A (or I_o) for no input signal to 0.4 A for an input signal of 0.4 V. An increase in the dc current level with an increase in signal drive indicates the presence of even harmonics.

(b) Fundamental current: The amplitude of the fundamental is given by Eq. (12–33) as

$$A_1 \simeq \frac{(1.0 - 0) + (0.61 - 0.1)}{3} \simeq 0.50 \text{ A}$$

(c) Harmonic components: From Eqs. (12–34), (12–35), and (12–36),

$$A_2 \simeq [1.0 + 0 - 2(0.32)]/4 \simeq 90 \text{ mA}$$

$$A_3 \simeq \frac{(1 - 0) - 2(0.61 - 0.1)}{6} \simeq 3.3 \text{ mA}$$

$$A_4 \simeq \frac{(1.0 + 0) - 4(0.61 + 0.1) + 6(0.32)}{12} \simeq 6.7 \text{ mA}$$

(d) Total harmonic distortion: From Eq. (12–37),

$$A_D \simeq (90^2 + 3.3^2 + 6.7^2)^{1/2} \simeq 90.4 \text{ mA}$$

From Eq. (12–38),

$$D_T = \frac{A_D}{A_1} \times 100\% \simeq \frac{90.4 \text{ mA}}{500 \text{ mA}} \times 100\% \simeq 18.1\%$$

Several difficulties are encountered in the application of distortion analysis to an actual circuit. First, an accurate curve of I_C as a function of V_{BE} must be obtained for the particular transistor under study. Due to the unit-to-unit variation between devices, even of the same type, accurate characteristic curves are difficult to obtain. Sometimes device manufacturers supply an average curve for a particular type of transistor that can be used for an approximate distortion analysis.

A second difficulty results from the temperature dependence of the I_C versus V_{BE} curve. As the junction temperature increases due to the power dissipated in the transistor, the transfer curve changes shape. Consequently, the various coefficients obtained from a distortion analysis are temperature dependent.

A third difficulty occurs in negative-feedback amplifiers. Generally, the input-to-output transfer function of a feedback amplifier is very linear, making the distortion very low. The principal exception occurs at the extremes of the input signal amplitude where bias values of the amplifier may limit the amplitude of the output signal. This type of distortion in feedback amplifiers is illustrated in Fig. 12–22.

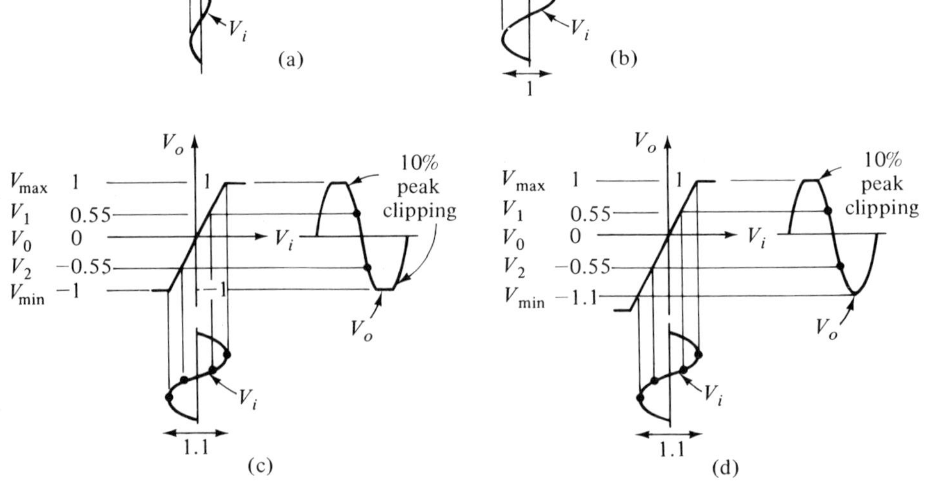

FIGURE 12–22 *Effect of the transfer characteristic in a feedback amplifier on amplitude distortion: (a) linear operation, (b) maximum signal amplitude for linear operation, (c) symmetrical distortion, and (d) nonsymmetrical distortion.*

The voltage transfer characteristic for a feedback amplifier is shown in Fig. 12–22(a). The linear portion of the curve has a slope of 2, indicating a gain V_o/V_i of 2. The maximum values of V_o are normalized to $+1$ and -1. Linear operation, or low distortion,* is obtained for output signal levels in the range of -1 to $+1$ corresponding to normalized input signal levels of -0.5 to $+0.5$.

The maximum undistorted output signal is ± 1 unit or 2 units peak-to-peak as shown in Fig. 12–22(b). The corresponding input signal is one unit peak-to-peak. Increasing the input signal 10 percent to 1.1 causes clipping on both positive and negative output peaks as shown in Fig. 12–22(c). The amount of harmonic distortion present can be estimated using Eqs. (12–32) through (12–38) where the values of V_{max}, V_1, V_o, V_2, and V_{min} are used instead of I_{max}, I_1, I_o, I_2, I_{min}, respectively. This calculation is illustrated in the following example.

EXAMPLE 12–6 A feedback amplifier with a voltage gain of 2 has a linear gain for all output signal values less than ± 10 V. The maximum possible output before clipping occurs is $+10$ V and -10 V. Estimate the percentage 2nd and 3rd harmonic distortion present in the output for an input signal of ± 5.5 V.

Solution: Figure 12–22(c) applies since the input is 10 percent larger than the maximum input signal for linear operation. Using the normalized output values in Fig. 12–22(c) and Eqs. (12–32) through (12–35),

$$A_o = \frac{(V_{max} + V_{min}) + 2(V_1 + V_2)}{6} - V_o$$

$$\simeq \frac{(1 - 1) + (0.55 - 0.55)}{6} - 0 \simeq 0$$

$$A_1 = \frac{(V_{max} - V_{min}) + (V_1 - V_2)}{3} \simeq \frac{3.1}{3} \simeq 1.03$$

$$A_2 = \frac{V_{max} + V_{min} - 2V_o}{4} \simeq \frac{1 - 1 - 2(0)}{4} \simeq 0$$

$$A_3 = \frac{(V_{max} - V_{min}) - 2(V_1 - V_2)}{6} \simeq -\frac{0.2}{6} \simeq 0.033$$

Since A_2 is zero, the second harmonic distortion in the output is zero. The percentage third harmonic distortion is

$$D_3 = (A_3/A_1) \times 100\% \simeq (0.033/1.03) \times 100\% \simeq 3.2\%$$

Two observations can be made relative to the preceding example. First, only odd harmonic components are present when the output is distorted symmetrically. Second, the amount of harmonic distortion produced by feedback amplifiers is low even when the output wave is obviously distorted.

The transfer function shown in Fig. 12–22(d) is different from the other transfer functions shown in Fig. 12–22. For linear operation, the maximum normalized positive signal is 1, and the maximum normalized negative value is -1.4. An input signal of 1.1 peak-to-peak will produce clipping only on the positive peak as shown in Fig. 12–22(d). Repeating Ex. 12–6 for this case yields 2.3 percent second harmonic distortion and 1.6 percent third harmonic distortion. When the output signal is clipped unsymmetrically, both even and odd harmonics may be present in the output.

---

* For large values of loop gain, the distortion in a feedback amplifier operating in its linear region can be as low as 0.05 percent. Measured distortion levels are given later in the chapter.

The distortion resulting from operating a single device in the Class AB or Class B mode can be reduced by using two devices in a push-pull or a complementary arrangement. However, a new form of distortion called *crossover distortion* can occur in these combination circuits. Crossover distortion results when both output devices approach cutoff as the output signal approaches zero. This type of distortion is shown in Fig. 12–23.

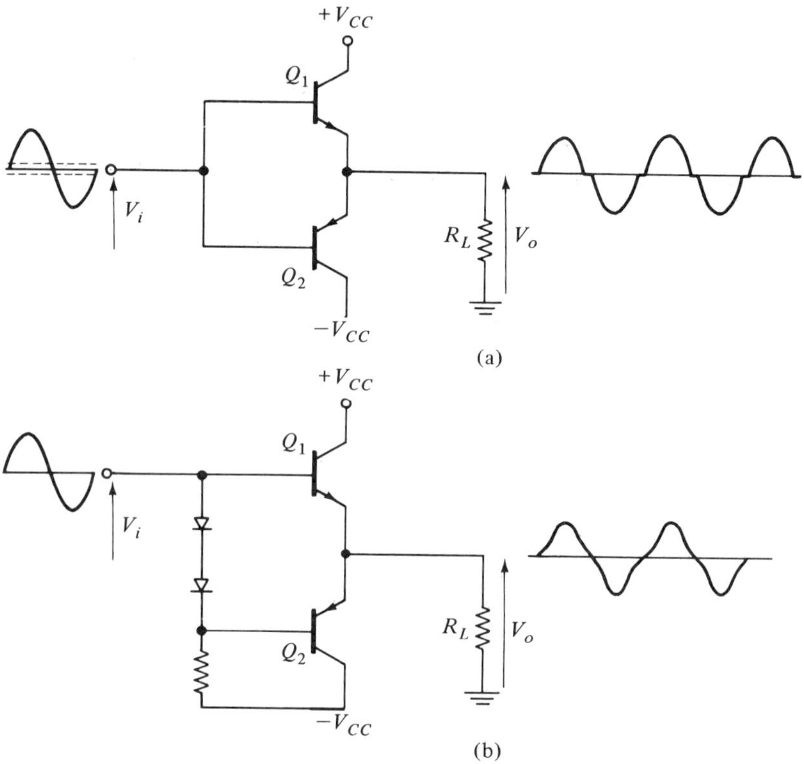

FIGURE 12–23 *Crossover distortion in a complementary amplifier:* (*a*) *Class B operation and* (*b*) *Class AB operation.*

Crossover distortion in Class B operation is shown in Fig. 12–23(a). In effect, a horizontal slice is removed from the input wave. The width of this slice is equal to the cut-in voltage on each side of zero. This dead zone results because no output current will flow until the input signal forward biases one of the base-emitter junctions. For silicon devices, the input signal required to produce an appreciable flow of emitter current is equal to 0.5 V or 1 V peak-to-peak. The resulting dead zone in the output signal can exceed 1 V peak-to-peak.

The severe crossover distortion present in a Class B amplifier can be reduced by operating each transistor in Class AB. The two diodes connected between the base terminals of the transistors, as shown in Fig. 12–23(b), are used for this purpose. Some crossover distortion can still result since the turn-off, and turn-on characteristics of a transistor do not exactly match. One common approach used in power amplifiers to reduce this distortion, is to combine complementary (or push-pull) output stages with negative feedback. Operating the complementary stage as Class AB results in a reasonably high value of conversion efficiency while negative feedback greatly reduces the distortion. Examples of this approach are given in later sections.

The final type of distortion to be considered is *intermodulation* (IM) distortion. This type of distortion results from a nonlinear transfer characteristic when two or more input frequencies are present.

Assuming that two input frequencies are present, then

$$v_i = V_1 \cos \omega_1 t + V_2 \cos \omega_2 t \tag{12–39}$$

Substituting Eq. (12–39) into (12–24) and simplifying gives

$$i = a_o + \frac{a_2}{2}(V_1^2 + V_2^2) + a_1 V_1 \cos \omega_1 t + a_1 V_2 \cos \omega_2 t + \frac{a_2 V_1^2}{2} \cos 2\omega_1 t + \frac{a_2 V_2^2}{2} \cos 2\omega_2 t + \frac{a_2 V_1 V_2}{2} \cos (\omega_1 + \omega_2)t + \frac{a_2 V_1 V_2}{2} \cos (\omega_1 - \omega_2)t \tag{12–40}$$

Six output frequencies are present in this equation: ω_1, ω_2, $2\omega_1$, $2\omega_2$, $(\omega_1 + \omega_2)$, and $(\omega_1 - \omega_2)$. Assuming that the two original input frequencies are 100 Hz and 1000 Hz, the frequencies in the output are 100, 1000, 200, 2000, 900, and 1100 Hz. The signals at 200 and 2000 Hz represent amplitude distortion while the signals at 900 and 1100 Hz represent IM distortion. The 900- and 1100-Hz signals are also termed "sum and difference" signals. Since many input signals contain multiple frequency components, IM distortion can be a significant limitation.

In this section we have examined three types of distortion that may occur in power amplifiers. Harmonic or amplitude distortion results from the nonlinear transfer characteristics of an amplifier and is particularly important in nonfeedback amplifiers. Harmonic distortion also occurs in feedback amplifiers when the input signal has sufficient amplitude to drive the output beyond its linear range. Generally, the signal is clipped under these conditions. Crossover distortion is present in complementary and push-pull amplifiers. This type of distortion may be reduced by operating each device in Class AB and by using negative feedback. Intermodulation distortion results from a nonlinear transfer function when two or more input frequencies are present. This type of distortion produces sum and difference signals. All forms of distortion represent a departure from ideal linear amplifier operation.

12.5 Class A Power Amplifiers

The basic configuration and characteristics of a Class A amplifier were introduced in Sec. 12.3. When the load is coupled to the transistor through a transformer, the conversion efficiency, η, has a maximum theoretical value of 50 percent. Practical circuits will have lower values of η fo reasonably low values of distortion.

Class A power amplifiers are widely used in circuits delivering less than approximately 2 W of output power. Many low cost AM and FM receivers, as well as TV sets, utilize this type of power amplifier as the output circuit of the audio section. The objective of this section is to develop a design approach for

Class A, transformer coupled, power amplifiers. The problem of thermal stability for variations in ambient temperature often encountered in consumer products is also introduced.

One simple form of a Class A, transformer coupled, power amplifier is shown in Fig. 12–24. Often the driving stage for this type of amplifier is a CC stage either directly coupled or capacitively coupled to the power amplifier. In this case the value of R_g would correspond to $R_{out(CC)}$, which can be very low. The design goals generally given for this type of amplifier include the required output power for a specified load impedance and ambient temperature.

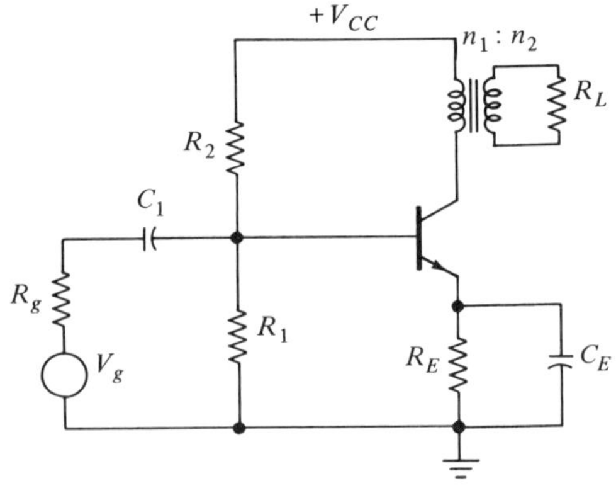

FIGURE 12–24 *One form of a Class A, transformer coupled, power amplifier.*

One of the first steps in the design is the choice of either the power transistor or the value of V_{CC}. Within some constraints, the circuit shown in Fig. 12–24 can be designed using any one of a large number of power transistor types. Since the output power depends on the product of voltage and current, a given value of output power can be obtained with a high-voltage low-current arrangement, or a low-voltage high-current arrangement. Often, however, the value of V_{CC} is determined by the available power supply voltage. For example, a battery powered unit may be limited to 6 V or 12 V dc while the usual dc voltage available for airborne equipment is 28 V. Many portable or small screen TV sets use 90 V in the video section, while some line-operated equipment* uses 100 to 125 V. Depending on the application, the design engineer may or may not be free to choose the dc power supply value. When the power supply is predetermined in a Class A transformer coupled amplifier, the value of V_{CEO} (or $V_{CE(sus)}$) of the power transistor chosen should be equal to or greater than 2 V_{CC} because the maximum collector voltage swing can be twice the power supply value as noted in Sec. 12.3. In this section, however, we will assume that a particular power transistor has been chosen rather than a value for V_{CC}. The specification sheet provides the following data on the power transistor:

TABLE 12–2 *Characteristics of the power transistor used in the design of the amplifier shown in Fig. 12–25.*

$BV_{CE(sus)} = 60$ V	$\beta = 50$ to 150
$I_{C(max)} = 0.7$ A	$I_{CO} = 0.25\ \mu$A at 25°C
$T_{j(max)} = 200$°C	$\theta_{jc} = 35$°C/W
$P_{D(max)} = 5$ W @ 25°C case temperature	

* Line operated equipment is generally operated directly from 110 V ac without a transformer in the power supply.

In the following design, we will assume the following requirements for the power amplifier:

$$P_o = 1\text{ W} \qquad R_L = 8\ \Omega \qquad T_A = 20°\text{C to } 60°\text{C} \tag{12–41}$$

The range of ambient temperature from 20°C to 60°C (or 68°F to 140°F) is typical of the temperature requirements for consumer products (AM and FM receivers and TV sets). Once the power transistor and amplifier specifications are known, the design can be completed by using the following step-by-step procedure.

Generally, V_{CC} is equal to or less than $BV_{CE(sus)}/2$ to ensure operation within the maximum voltage rating of the transistor. For this design, the initial choice for V_{CC} is 20 V.

The nonideal characteristics of a transformer which couples the load to the power transistor should also be considered. Transformers of the type used in low-power audio amplifiers have efficiencies ranging from 65 to 85 percent. In addition, the primary winding has a finite dc resistance and a significant leakage inductance. Assuming 75 percent efficiency, the output power that the transistor must deliver to the transformer, $P_{o(Q)}$, is

$$P_{o(Q)} = \frac{P_o}{\eta_{(\text{transformer})}} \simeq \frac{1\text{ W}}{0.75} \simeq 1.3\text{ W} \tag{12–42}$$

From Eq. (12–16), the required value of the load impedance on the transistor with a 20-V dc supply is

$$R_{L\text{eq}} = \frac{V_m^2}{2P_{o(Q)}} \simeq \frac{V_{CC}^2}{2P_{o(Q)}} \simeq \frac{20^2}{2(1.3)} \simeq 150\ \Omega \tag{12–43}$$

The turns ratio of the transformer must reflect the 8-ohm speaker on the transformer secondary as approximately 150 ohms on the primary. Solving for the turns ratio,

$$\frac{N_1}{N_2} = \left(\frac{R_{\text{pri}}}{R_{\text{sec}}}\right)^{1/2} = \left(\frac{150}{8}\right)^{1/2} \simeq 4.3 \tag{12–44}$$

Often audio transformers are characterized by their impedance rather than their turns ratio. Thus the 4.3/1 turns ratio may be indicated as a 150 ohm/8 ohm impedance ratio. The characteristics of a commercially available transformer that closely meet the design values are

$$Z_{\text{pri}}/Z_{\text{sec}} = 156/8 \qquad R_{\text{pri}} = 19\ \Omega$$
$$\eta_{\text{transformer}} = 75\% \qquad R_{\text{sec}} = 1.5\ \Omega \tag{12–45}$$

These values will be used for the remainder of the design.

The peak value of collector current required to produce 1.3 W is

$$I_m = \sqrt{2P_{o(Q)}/R_{L\text{eq}}} \simeq \sqrt{(2)(1.3)/156} \simeq 129\text{ mA} \tag{12–46}$$

To avoid excessive distortion, the dc collector current is chosen as 150 mA, or greater than I_m. For the worst case condition, the corresponding dc base current is calculated by using the minimum value of β resulting in I_B equal to 3 mA.

Before choosing values for the bias resistors, R_E, R_1, and R_2, we will review briefly the thermal stability characteristics of the circuit shown in Fig. 12–24. The general expression for the collector current as well as the thermal stability of this circuit are discussed in Sec. 4.3. Although β, V_{BE}, and I_{CO} all are temperature dependent, we will consider only the effects of the I_{CO} variations in this design. The stability factor associated with the I_{CO} temperature variations is (from Chap. 4),

$$S_I = \frac{\Delta I_C}{\Delta I_{CO}} \simeq 1 + \frac{\beta(R_B/R_E)}{1 + \beta + (R_B/R_E)} \qquad \textbf{(12–47)}$$

In the preceding equation the value of S_I indicates the change in collector current, ΔI_C, for a given ΔI_{CO}, and R_B is $R_1 \,||\, R_2$. A small value of S_I is desired for good thermal stability. Increasing the value of R_E will reduce the value of S_I. However, increasing R_E reduces V_{CE} due to the $I_C R_E$ voltage drop. A reduction of V_{CE} reduces the maximum output power. Thus the choice of R_E is a compromise. As trial values for this design, choose R_E as 10 ohms and S_I as 15.

Before determining the values for R_1 and R_2 we will examine the effects of R_E and the dc resistance of the transformer primary winding, R_{pri}, on the output power. The peak value of the ac output voltage for a symmetrical signal is limited to approximately V_{CE}. The value of V_{CE} is less than V_{CC} due to the dc drop across R_E and R_{pri}. For V_{CC} equal to 20 V,

$$V_{CE} = V_{CC} - I_C(R_E + R_{pri}) \simeq 20\text{ V} - (150\text{ mA})(10 + 29.8) \simeq 14\text{ V} \qquad \textbf{(12–48)}$$

The corresponding output power is

$$P_o = P_{o(Q)}\eta_{\text{transformer}} \simeq \left(\frac{V_{CE}^2}{2R_{Leq}}\right)(\eta) \simeq \frac{(14)^2 0.75}{2 \times 156} \simeq 0.47\text{ W} \qquad \textbf{(12–49)}$$

This value is approximately one-half the design value of P_o. Increasing V_{CC} to 26 V, V_{CE} becomes approximately 20 V, and P_o becomes approximately 1 W. The remaining calculations will be based on a V_{CC} value of 26 V.

The values of R_1 and R_2 can now be chosen. Solving Eq. (12–47) with an R_E of 10 ohms, S_I of 15, and β of 150,* R_B is approximately 155 ohm.** The required base voltage is

$$V_B = I_C R_E + V_{BE} \simeq (150\text{ mA})(10) + 0.7\text{ V} \simeq 2.2\text{ V} \qquad \textbf{(12–50)}$$

The two conditions used to determine R_1 and R_2 are

$$V_B \simeq \left(\frac{R_1}{R_1 + R_2}\right) V_{CC} \qquad R_B \simeq 155 = R_1 \,||\, R_2 \qquad \textbf{(12–51)}$$

Using these equations to solve for R_1 and R_2,

$$R_1 \simeq 169\,\Omega \qquad R_2 \simeq 1830\,\Omega \qquad \textbf{(12–52)}$$

The nearest standard values of carbon composition resistors are 180 ohms and 1.8 K.

The next step is to determine the heat sink requirements. The power dissipated by the transistor has a maximum value with no input signal applied and is***

$$P_D = V_{CE} I_C \simeq (20\text{ V})(150\text{ mA}) \simeq 3\text{ W} \qquad \textbf{(12–53)}$$

The most critical value of ambient temperature is 60°C. The specification sheet gives $T_{j(\max)}$ as 200°C. Derating $T_{j(\max)}$ 10°C, and assuming 0.5°C/W for the washer that insulates the transistor from the heat sink, θ_{sa} can be found as follows:

$$T_j = P_D(\theta_{jc} + \theta_{cs} + \theta_{sa}) + T_A$$
$$190°\text{C} = (3\text{ W})[(35 + 0.5 + \theta_{sa})°\text{C/W}] + 60°\text{C} \qquad \textbf{(12–54)}$$

---

* From Eq. (12–47), the largest value of S_I results when β has its maximum value.

** As a first approximation, $R_B \simeq S_I R_E$.

*** This value of P_D may prove inaccurate since the temperature dependence of I_C is not included.

$$\theta_{sa} \simeq \frac{190 - 60 - (3)(35.5)}{3} \simeq 8°\text{C/W} \qquad \textbf{(12–55)}$$

From Fig. 12–12, an aluminum heat sink having 25 square inches of surface area is adequate to obtain this value of θ_{sa}.

The conversion efficiency should include the effects of the bias resistors. The power dissipated in R_1 and R_2 is

$$P_{D1} = V_{CC}^2/(R_1 + R_2) \simeq (26)^2/(180 + 1800) \simeq 0.33 \text{ W} \qquad \textbf{(12–56)}$$

The power dissipated by the transistor, R_E and R_{pri}, plus the output power is

$$P_{D2} = V_{CC}I_C \simeq (26 \text{ V})(150 \text{ mA}) \simeq 3.9 \text{ W} \qquad \textbf{(12–57)}$$

Thus the total conversion efficiency is

$$\eta_{\text{total}} \simeq \frac{P_o}{P_{D1} + P_{D2}} \times 100\% \simeq \frac{1 \text{ W}}{4.2 \text{ W}} \times 100\% \simeq 24\% \qquad \textbf{(12–58)}$$

The harmonic distortion can be calculated if an accurate set of static characteristic curves is available. Since these curves are seldom available, no attempt will be made to calculate the harmonic distortion. Measured values of distortion are given later.

The next step is to calculate the input signal required to produce the maximum output power. Generally, this step requires a knowledge of the voltage gain, a value that depends on the transistor input impedance. An estimate* of R_{in} is

$$R_{\text{in(CE)}} \simeq r_{b'} + (\beta + 1)(2r_e) \qquad \textbf{(12–59)}$$

With a collector current of 150 mA, $2r_e$ is approximately 0.35 ohm. Consequently, the value of $r_{b'}$ will often determine the input resistance of the transistor. Unless $R_{\text{in(CE)}}$ is specified, generally as a value for h_{ie} at a given current level, a reasonable value for R_{in} cannot be calculated. Consequently, the input signal required to produce the rated output power in a Class A power amplifier should be determined experimentally.**

Often the frequency response of a Class A, transformer coupled, power amplifier depends on the frequency response of the transformer. In the low-frequency region, the response of the transformer is limited by the primary incremental inductance. The high-frequency response is limited by the leakage inductance. In addition, the response may contain a peak due to a resonance condition produced by the transformer inductance and the circuit capacitance. Generally, the frequency response should be determined and adjusted experimentally if necessary. Values for the input coupling capacitor and the emitter resistor bypass capacitor are chosen as large as practical.

The practical aspects of the design of a Class A, transformer coupled, power amplifier have been considered. However, the multivariable nature of the problem together with the lack of exact information indicates that the circuit should be analyzed to determine how closely the design specifications given in Eq. (12–41) are satisfied. Also, the amplifier performance should be evaluated experimentally. The circuit diagram of the amplifier is shown in Fig. 12–25, and the parameters of the power transistor, transformer, and heat sink are given in

---

* The use of $2r_e$ in place of r_e in Eq. (12–59) is due to high current operation and is discussed in Sec. 12.2.

** One exception occurs when the emitter resistor, R_E, is unbypassed. Then the gain is stabilized by ac feedback and is approximately equal to R_{Leq}/R_E. The input resistance for this case is approximately βR_E.

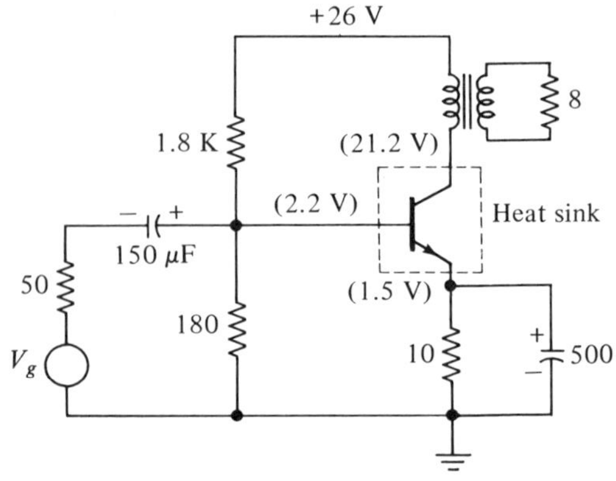

FIGURE 12–25 *The amplifier from Fig. 12–24 showing design values. (Component values are given in Table 12–3.)*

Table 12–3. The following example illustrates the analysis of the performance characteristics of this amplifier.

TABLE 12–3 *Component values for the amplifier shown in Fig. 12–25 and analyzed in Ex. 12–7. (Values obtained from the specification sheet are indicated with an asterisk.)*

Transistor:	$\beta \simeq 115$	$I_{CO} \simeq 90$ nA at $T_C = 25°C$
	$BV_{CE(sus)} \simeq 75$ V	$I_{C(max)} = 0.7$ A*
	$T_{jmax} = 200°C$*	$P_D = 5$ W at $T_C = 25°C$*
	$\theta_{jc} = 35°C/W$*	
Transformer:	$\eta = 75\%$	Impedance ratio: 156/8 Ω
	$R_{pri} = 29.8\ \Omega$	$R_{sec} = 0.8\ \Omega$
Heat sink:	$\theta_{sa} \simeq 8°C/W$	

EXAMPLE 12–7 Calculate the values of P_o, η, T_j, and T_C for the amplifier shown in Fig. 12–25 for an ambient temperature of 20°C. The component parameters are given in Table 12–3.

Solution: (a) Bias conditions: Neglecting I_B,

$$V_B \simeq V_{CC}\left(\frac{R_1}{R_1 + R_2}\right) \simeq (26\text{ V})\left(\frac{180}{180 + 1.8\text{ K}}\right) \simeq 2.3\text{ V}$$

$$V_E = V_B - V_{BE} \simeq 2.3\text{ V} - 0.7\text{ V} \simeq 1.6\text{ V}$$

$$I_C = V_E/R_E \simeq 1.6\text{ V}/10 \simeq 160\text{ mA}$$

$$V_C = V_{CC} - I_C R_{pri} \simeq 26\text{ V} - (160\text{ mA})(29.8) \simeq 21.2\text{ V}$$

$$V_{CE} = V_C - V_E \simeq 21.2\text{ V} - 1.6\text{ V} \simeq 19.6\text{ V}$$

(b) Amplifier performance: Power delivered to the primary of the transformer, $P_{o(Q)}$, is

$$P_{o(Q)} \simeq V_{CE}^2/2R_{Leq} \simeq (19.6)^2/(2)(156) \simeq 1.23\text{ W}$$

$$P_o = P_{o(Q)}\eta_{transformer} \simeq (1.23\text{ W})(0.75) \simeq 0.92\text{ W}$$

$$P_{in(dc)} = V_{CC}I_C + V_{CC}^2/(R_1 + R_2) \simeq 4.5\text{ W}$$

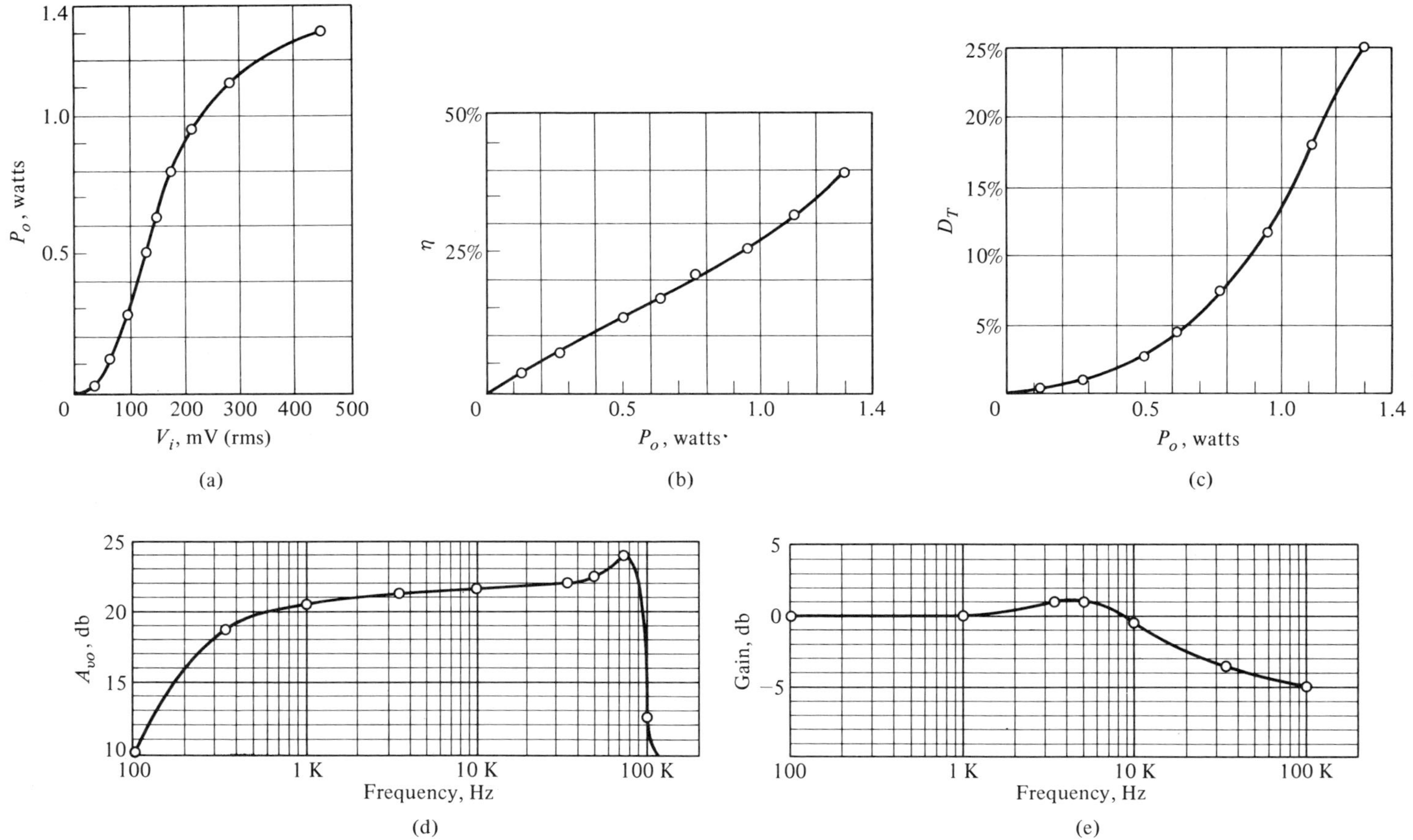

FIGURE 12–26 *Experimental results for the amplifier in Ex. 12–7: (a)* P_o *versus* V_i*, (b)* η *versus* P_o*, (c)* D_T *versus* P_o*, (d) the frequency response of the entire amplifier, and (e) the frequency response of the transformer alone.*

The conversion efficiency for the complete amplifier is

$$\eta = \frac{P_o}{P_{\text{in(dc)}}} \times 100\% \simeq \frac{0.92\ \text{W}}{4.5\ \text{W}} \times 100\% \simeq 20.5\%$$

The conversion efficiency for the transistor alone is

$$\eta_{(Q)} = \frac{P_{o(Q)}}{V_{CE}I_C} \times 100\% \simeq \frac{1.23\ \text{W}}{(19.6\ \text{V})(160\ \text{mA})} \times 100\% \simeq 39.3\%$$

(c) Thermal characteristics at $T_A = 20°\text{C}$: The power dissipated by the transistor is

$$P_{D(Q)} = V_{CE}I_C \simeq (19.6\ \text{V})(160\ \text{mA}) \simeq 3.15\ \text{W}$$

From Eq. (12–7),

$$T_j = P_D(\theta_{jc} + \theta_{cs} + \theta_{sa}) + T_A \simeq (3.15)(35 + 0.5 + 8) + 20° \simeq 157°\text{C}$$
$$T_C = T_j - P_D\theta_{jc} = 157°\text{C} - (3)(35) \simeq 52°\text{C}$$

The value of I_{CO} at 25°C is approximately 90 nA. Since I_{CO} doubles approximately for each 10°C rise in temperature, the value of I_{CO} at 157°C is approximately 0.8 mA. The value of R_B is

$$R_B = R_1 \,||\, R_2 \simeq 180 \,||\, 1.8\ \text{K} \simeq 164\ \Omega$$

From Eq. (12–47),

$$S_I \simeq 1 + \frac{(115)(164/10)}{1 + 115 + (164/10)} \simeq 14$$

The change in I_C due to the junction temperature to 150°C is

$$\Delta I_C \simeq S_I I_{CO} \simeq (14)(0.8\ \text{mA}) \simeq 11\ \text{mA}$$

Thus the collector current at $T_A = 20°\text{C}$ is approximately

$$I_C \simeq 160\ \text{mA} + 11\ \text{mA} \simeq 171\ \text{mA}$$

Since the value of I_C due to the change in I_{CO} is only 7 percent larger than the value calculated neglecting the I_{CO} effect, the values of P_o, η, T_j, and T_C calculated in (b) are reasonably accurate.

(d) Experimental results: The amplifier was tested at ambient temperature of 20°C. The bias values are shown in Fig. 12–25. Graphs of the various performance characteristics are shown in Fig. 12–26. Except for the frequency response plot, the input signal frequency was 1 kHz. From Fig. 12–26(a), an input signal of 230 mV is required to produce an output power of 1 W.

In this section, we have studied the design of a Class A, transformer coupled, power amplifier. We also examined some of the difficulties encountered in this type of design. In the next three sections, we will examine the performance of feedback power amplifiers whose performance is generally superior to those of nonfeedback power amplifiers.

12.6 Push-Pull Power Amplifiers

As indicated in Sec. 12.3, a push-pull amplifier uses two output transistors, each operated in Class AB or B. For a sine wave input, each transistor acts effectively as a half-wave amplifier and carries the total load for approximately one-half of each cycle. The half-wave amplifier signals are then combined to produce a complete sine wave output. In this section, we will examine various push-pull amplifier configurations with the objective of understanding the functions of the various components.

A block diagram of a push-pull power amplifier is shown in Fig. 12–27. The basic components include the driver, phase inverter, output stages, and load impedance. Most high quality push-pull power amplifiers also include a feedback network. The function of each component is discussed below.

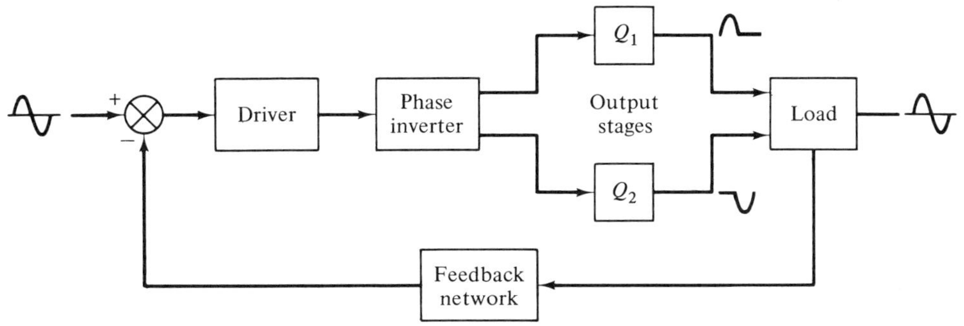

FIGURE 12–27 *A block diagram of a push-pull power amplifier.*

Two types of output coupling used in a push-pull power amplifier are shown in Fig. 12–28. Transformer coupling is shown in (a), and capacitor coupling in (b). In transformer coupling, each collector current flows during alternate half cycles and sets up a changing flux in the iron core of the transformer. The resulting flux density varies sinusoidally as a function of time and produces a sinusoidal voltage in the secondary or load circuit.

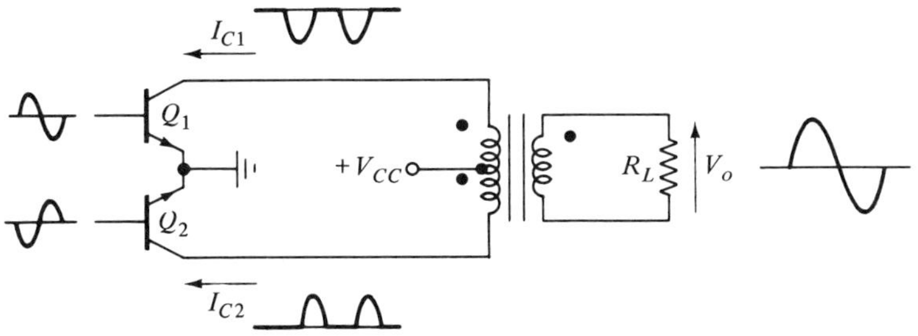

(a)

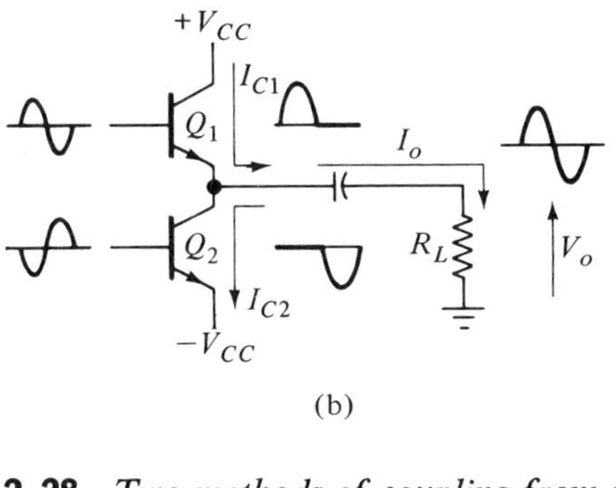

(b)

FIGURE 12–28 *Two methods of coupling from the output stages to the load impedance in a push-pull power amplifier: (a) transformer coupling, and (b) capacitive coupling. If the dc voltage at the emitter terminal of* Q_1 *in (b) is zero, the load can be directly coupled at the output stages.*

The circuit shown in Fig. 12–28(b) does not require an output transformer. Since the two ac base voltages are 180° out of phase, the two collector currents are also 180° out of phase and conduct for alternating half cycles. With capacitive coupling to the load, a single power supply voltage can be used by returning the emitter of Q_2 to ground. When two power supply voltages are used, the dc voltage of the emitter of Q_1 (collector of Q_2) can be adjusted to zero, so that direct coupling to the load is practical.

For either circuit in Fig. 12–28, the signals at the two base terminals must be equal in amplitude and 180° out of phase. The function of a phase inverter circuit is to provide two base voltages that meet these specifications.

Three types of phase inverter circuits are shown in Fig. 12–29. In Fig. 12–29(a), a transformer is used to obtain the two desired voltages with respect to the center tap. In Fig. 12–29(b), a single transistor stage with equal emitter and collector resistors is used. The amplitudes of both output signals are nearly equal to that of the input signal if R_1 is much greater than r_e. A differential amplifier can also be used as a phase inverter since, for a single input signal, the two collector signals are approximately equal in amplitude but are 180° out of phase. Although transistors are used in Figs. 12–29(b) and (c), vacuum tubes can also be used.* Either type of output circuit shown in Fig. 12–28 can be used with any phase inverter circuit shown in Fig. 12–29.

Negative feedback is generally used in high quality push-pull power amplifiers to reduce the distortion caused by the power transistors and the transformers. Voltage sampling at the output is almost universally used since a typical load impedance is only a few ohms.** However, both voltage summing and current summing at the input are practical.

Our next task is to examine two specific push-pull power amplifier circuits, one a transistor amplifier and the other a vacuum tube amplifier. The function of each component as well as the amplifier performance will be discussed.

The circuit diagram of a 15-W transistor push-pull power amplifier is shown in Fig. 12–30. The feedback configuration is a voltage sample-voltage sum type. The gain with feedback is

$$V_o/V_i \simeq (R_{f1} + R_{f2})/R_{f1} \simeq (100 + 1.8\text{ K})/100 \simeq 19 \qquad \textbf{(12–60)}$$

The value of R_i is approximately

$$R_i \simeq R_{11} \,||\, R_{21} \simeq 2.2\text{ K} \,||\, 75\text{ K} \simeq 2.1\text{ K} \qquad \textbf{(12–61)}$$

The value of R_o is very small and depends on the r_{ce} values for Q_3 and Q_4, the output transformer turns ratio, and the loop gain.

Our next task is to examine the function of the various components of the amplifier shown in Fig. 12–30. The driver section consists of a CE-CC amplifier (Q_1 and Q_2). A transformer type phase inverter (X_1) and also a transformer coupled output circuit (X_2) are used. Resistors R_{11} and R_{21} determine the base voltage of Q_1. Resistor R_{C1} is the collector resistor for Q_1. The emitter resistor of Q_2, R_{E2}, is ac bypassed by C_3. Neglecting the winding resistance of the secondary of X_1, the base voltage of Q_3 and Q_4 is established by R_{B2} and R_{B1}. The output stages, Q_3 and Q_4, can be operated Class AB by adjusting R_{B1}. Resistors R_{E3}

---

* Historically, push-pull operation was developed for vacuum tube amplifiers.

** The output impedance of a feedback amplifier is reduced by voltage sampling at the output.

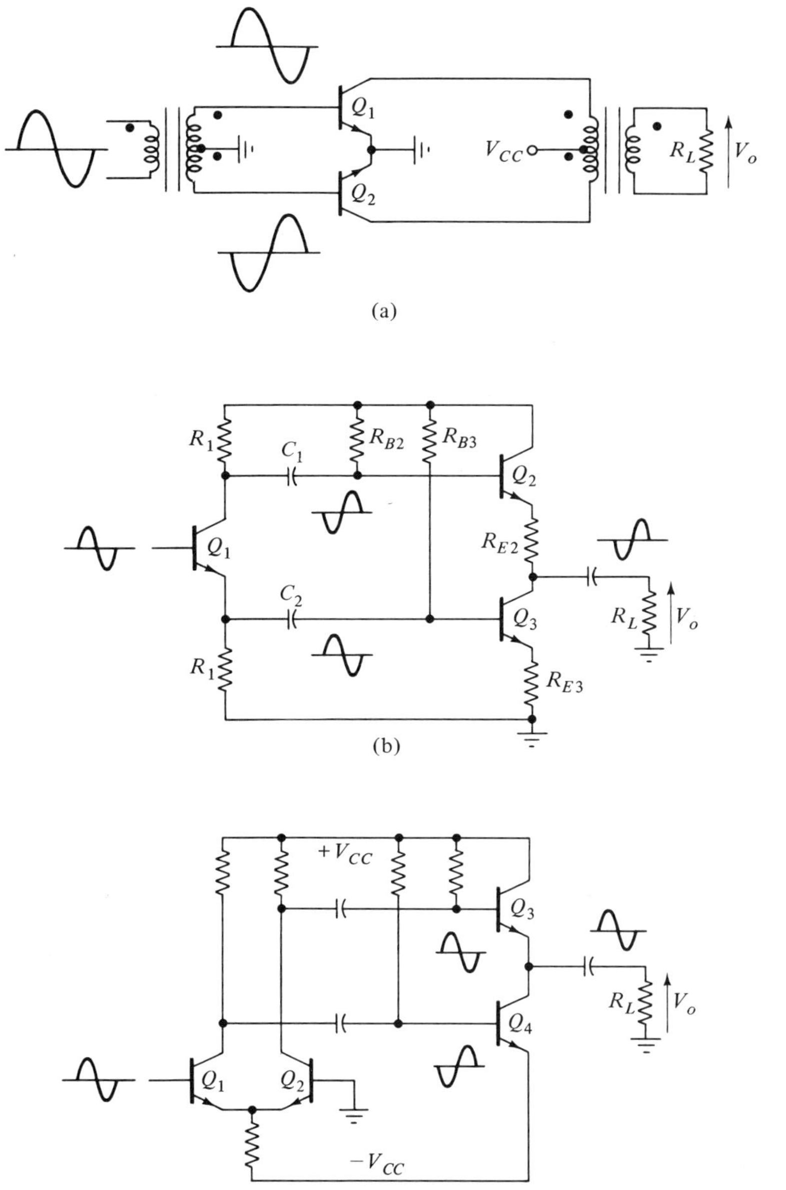

FIGURE 12–29 *Three types of phase inverter circuits: (a) transformer, (b) phase-splitter, and (c) differential amplifier.*

and R_{E4} are used to provide thermal stability. The power supply is a simple full-wave rectifier (X_3, D_1, and D_2) with a capacitive filter (C_5). The power supply for the input stages is isolated from the output stages by an L-section filter (L_1 and C_6). The high-frequency response of the amplifier is controlled by transformers X_1 and X_2 and capacitors C_2 and C_4. The low-frequency response is controlled by the input coupling capacitor, C_1, and the emitter bypass capacitor C_3. The negative-feedback circuit also influences the frequency response.

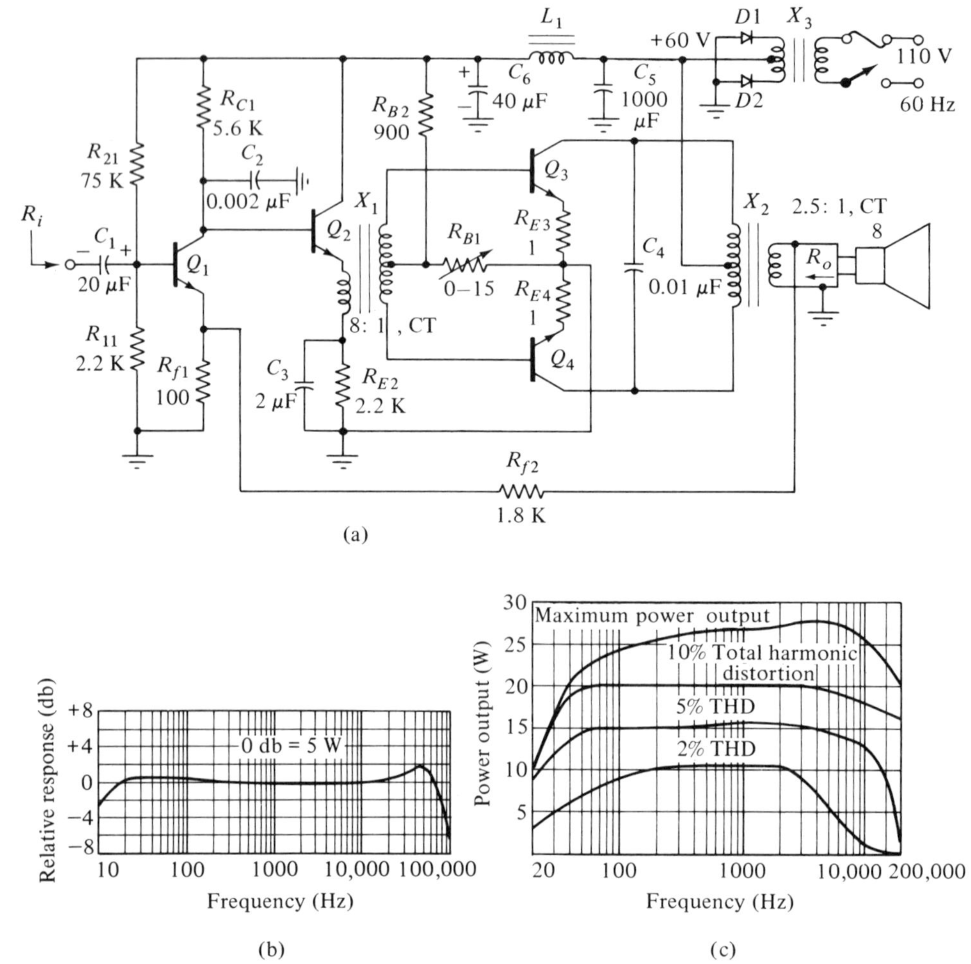

FIGURE 12–30 *A 15-W transistor push-pull power amplifier: (a) circuit diagram, (b) frequency response for a power output of 5 W and (c) total harmonic distortion and output power as a function of frequency. (From U. A. Stover, ed.,* Circuit Design for Audio, AM/FM, and TV, *McGraw-Hill Book Company, 1967.) By permission of Texas Instruments, Inc.*

The maximum power output can be estimated as follows. The 8-ohm speaker load appears in the collector circuit of Q_3 and Q_4, as

$$R_{Leq3} \simeq R_{Leq4} \simeq (n_2/n_1)^2 R_L \simeq (2.5)^2(8) \simeq 50\ \Omega \qquad \textbf{(12–62)}$$

Neglecting the primary winding resistance of X_2 and the resistors R_{E3} and R_{E4}, the total ac power delivered to the primary of X_2 by Q_3 and Q_4 is approximately

$$P_{oP} \simeq \frac{V_{CC}^2}{2R_{Leq3,4}} \simeq \frac{(60)^2}{(2)(50)} \simeq 36\text{ W} \qquad \textbf{(12–63)}$$

Assuming that X_2 has an efficiency of 75 percent, then the ac power delivered to the load is

$$P_{oS} \simeq (P_{oP})(0.75) \simeq (36\text{ W})(0.75) \simeq 27\text{ W} \qquad \textbf{(12–64)}$$

An estimate of the distortion present in the output signal is complicated by two factors. First, the distortion produced by the transformers X_1 and X_2 is difficult to predict. Second, negative feedback reduces the distortion. In most cases the distortion should be measured.

Experimental results for the amplifier shown in Fig. 12–30(a) are shown in Figs. 12–30(b) and (c). From Fig. 12–30(b), the 3-db frequencies for an output power of 5 W are approximately 10 Hz and 80 kHz. Note that the response is peaked in both the low-frequency region and the high-frequency region. These peaks are due to the transformers X_1 and X_2 as well as the negative feedback. Distortion data are given in Fig. 12–30(c). The maximum output power (with a high degree of distortion) was approximately 26 W. The maximum output power for 10 percent total harmonic distortion was approximately 20 W.

The circuit diagram of a 12-W vacuum tube push-pull power amplifier is shown in Fig. 12–31. The feedback network configuration is a voltage sample-voltage sum type which stabilizes the gain V_o/V_i. The driver section consists of a CK stage, T_1. A phase-splitter type phase inverter, T_2, is used to provide the drive signals for the output push-pull stages, T_3 and T_4. Note that to make the output signals of the phase splitter equal, the equivalent load resistance in the plate and cathode of T_2 must be equal, or

$$R_{P\text{eq}2} \simeq R_{K\text{eq}2} \simeq 100\text{ K} \parallel 330\text{ K} \simeq 77\text{ K} \tag{12–65}$$

Resistors R_2 and R_3, which are in series with the grid leads of the output tubes, limit the control grid dissipation if the grid is driven positively. The screen grids of T_3 and T_4 are connected to the plate supply voltage, while the suppressor grids are connected to the cathode internally. The cathode currents of both T_3 and T_4 flow through R_K which is ac bypassed by C_K. The push-pull output stages are transformer coupled to the load resistor by X_1. The power supply is a simple full-wave rectifier, T_5, with a shunt capacitor filter, C_5. Additional filtering is provided for the driver and phase inverter circuits by the RC filter composed of R_4 and C_7. The high-frequency response of the loop gain is controlled by the output coupling transformer, X_1, the RC lag network R_1C_2, the capacitors C_{P2} and C_{K2}, and the feedback network. The low-frequency response is controlled by the coupling capacitors C_1, C_3, and C_4; the bypass capacitor, C_{K2}; the output coupling transformer, X_1; and the negative-feedback circuit.

A plot of the total harmonic distortion as a function of output power is shown in Fig. 12–32(a). The rated output power was obtained with 10 percent distortion. Peak clipping on the output signal became evident at approximately 4.5-W output (6 V rms). The distortion increased rapidly for output power levels above 4.5 W. This performance is consistent with the previous discussion on the effects of feedback on distortion. The distortion at low power levels (less than 4 W) was approximately 0.6 percent. This distortion contained a 120-Hz signal indicating an imbalance* in the two output tube currents. The distortion data were measured for an input signal of 1 kHz. The frequency response with

* The dc currents flowing in the output tubes are not filtered very well. Consequently, several volts of ripple may appear across each half of the primary of the output transformer. These voltages are 180° out of phase and cancel in the secondary if the tube currents are equal. To reduce the 120-Hz signal in the output, the tubes must be well matched or be balanced experimentally.

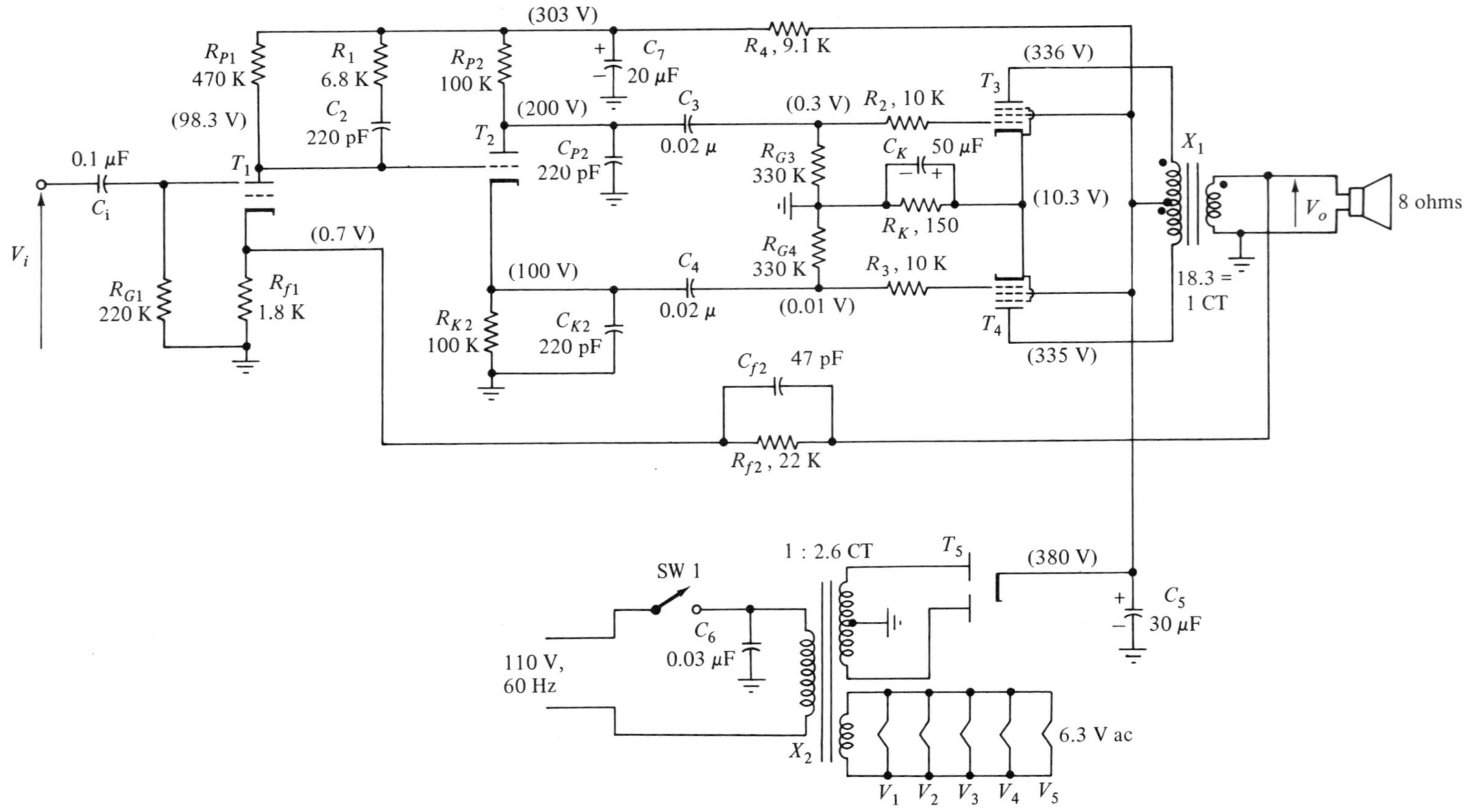

FIGURE 12–31 *The circuit diagram of a 12-W vacuum tube push-pull amplifier. Values in parentheses are measured dc bias values.*

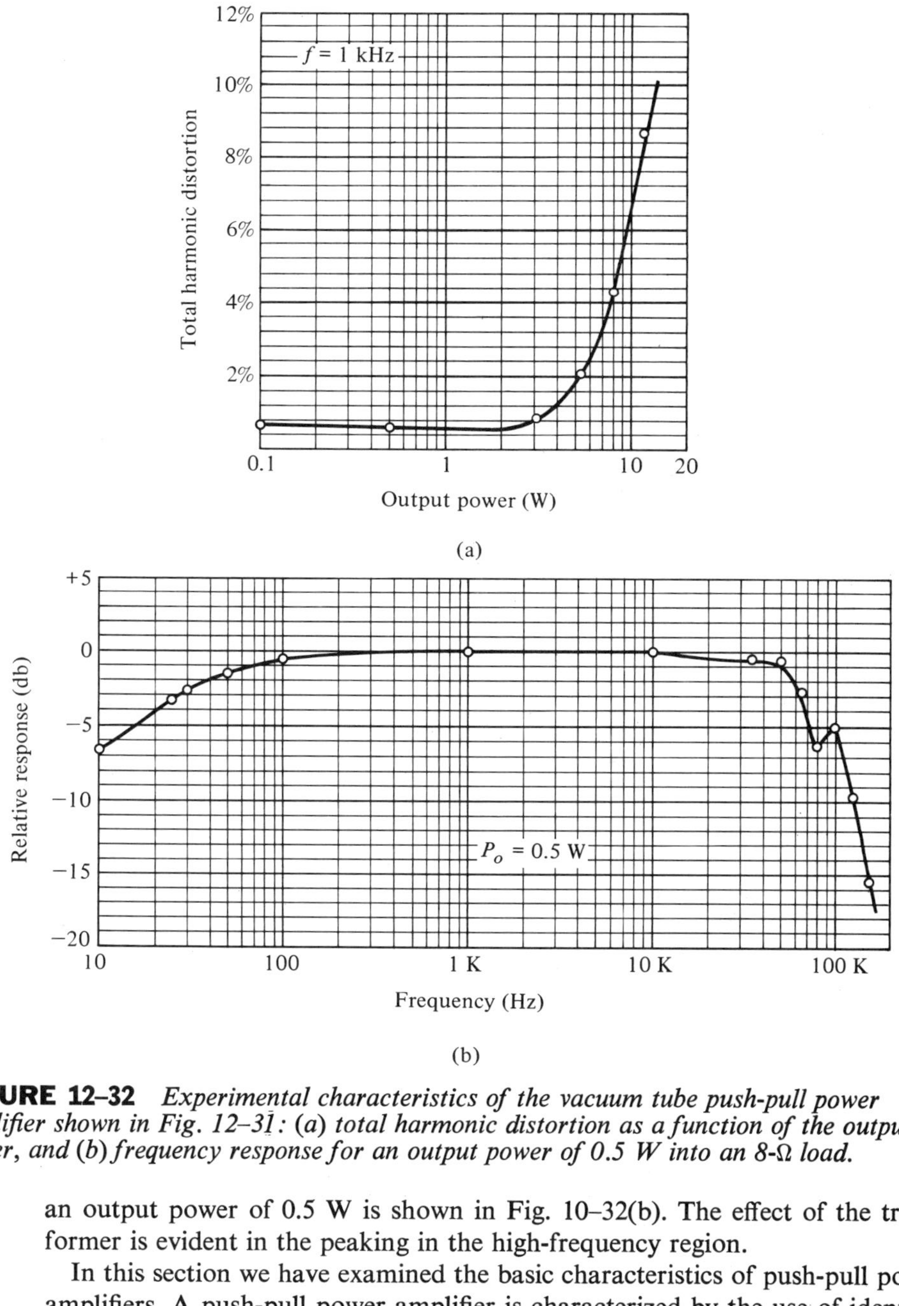

FIGURE 12–32 *Experimental characteristics of the vacuum tube push-pull power amplifier shown in Fig. 12–31: (a) total harmonic distortion as a function of the output power, and (b) frequency response for an output power of 0.5 W into an 8-Ω load.*

an output power of 0.5 W is shown in Fig. 10–32(b). The effect of the transformer is evident in the peaking in the high-frequency region.

In this section we have examined the basic characteristics of push-pull power amplifiers. A push-pull power amplifier is characterized by the use of identical output stages (usually CE or CK stages), a phase inverter, and an input circuit. Both transistor and vacuum tube push-pull amplifiers are widely used in practice. Most high quality push-pull amplifiers use negative feedback to stabilize the gain, reduce the distortion, and improve the frequency response. In the next section, we will examine the characteristics of power amplifiers which take advantage of the NPN-PNP complementary characteristics that are available in bipolar transistors.

12.7 Complementary Power Amplifiers

The basic characteristics of complementary power amplifiers were introduced in Sec. 12.3.* Like the push-pull power amplifiers in the last section, two output transistors are arranged so that they conduct on alternate half cycles. Unlike the push-pull arrangement, a complementary power amplifier does not require a separate phase inverter circuit. Phase inversion is effectively performed by the NPN-PNP arrangement because the NPN unit conducts on the positive part and the PNP on the negative part of the driving signal. Consequently, interstage and output coupling transformers are almost never used in complementary power amplifiers. Generally, CC output stages are used because of their low output impedance (better drive capability) and good linearity.

The objective of this section is to examine the characteristics and configurations of the output or power stages of a complementary power amplifier. The three specific topics to be studied are: (1) complementary output configurations, (2) problems encountered under peak signal conditions, and (3) the mechanism of thermal stability. The analysis and design of complete complementary amplifiers are discussed in the following sections.

A partial circuit diagram of one type of complementary amplifier is shown in Fig. 12–33. The output or power stages, Q_2 and Q_3, are connected in the complementary emitter follower configuration. The required base voltages for Class AB operation are established by diodes D_1 and D_2. Transistor Q_1 is the driver unit and must swing the entire peak-to-peak output voltage. Voltage sample-current sum feedback is provided by R_{f2}. The remainder of the circuit is not specified and is lumped into the block marked input circuit.

The circuit shown in Fig. 12–33 can function as a power amplifier provided the necessary output voltage and current can be supplied to R_L. For 10 W of output power delivered to an 8-Ω load, a peak current of 1.6 A and a peak voltage of 12.6 V are required. Assuming that the output transistors have βs of 20, the corresponding peak value of base current for Q_2 (and Q_3) is approximately 80

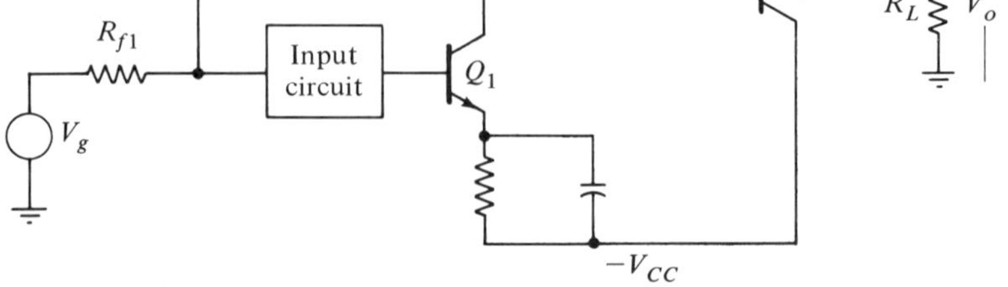

FIGURE 12–33 *Partial circuit diagram of Class AB complementary power amplifier.*

* The complementary emitter follower is the basic circuit used in complementary power amplifiers and was discussed in Chap. 11.

mA. Since the CE stage, Q_1, operates Class A, the bias current, I_{C1}, should be greater than 80 mA. This high value of I_{C1} can be reduced by modifying the output stages to provide higher values of current gain. Our next task is to consider alternate output configurations often encountered in practice.

Several variations of the complementary emitter follower are shown in Fig. 12–34. One of the two-stage circuits shown in Fig. 12–34 replaces a single CC stage (Q_2 or Q_3) in Fig. 12–33. The two configurations shown in Figs. 12–34(a) and (b) are called *Darlington circuits* or *Darlington pairs* and correspond to a compound transistor.* Many device manufacturers have Darlington pairs in a single package with three external leads. The overall current gain of a Darlington pair is approximately equal to the product of the current gains of the individual transistors.

Two modified Darlington circuits, using both NPN and PNP transistors, are shown in Figs. 12–34(c) and (d). These two circuits are called *complementary Darlington pairs*.** By comparing the ac currents in Figs. 12–34(a) and (c), we

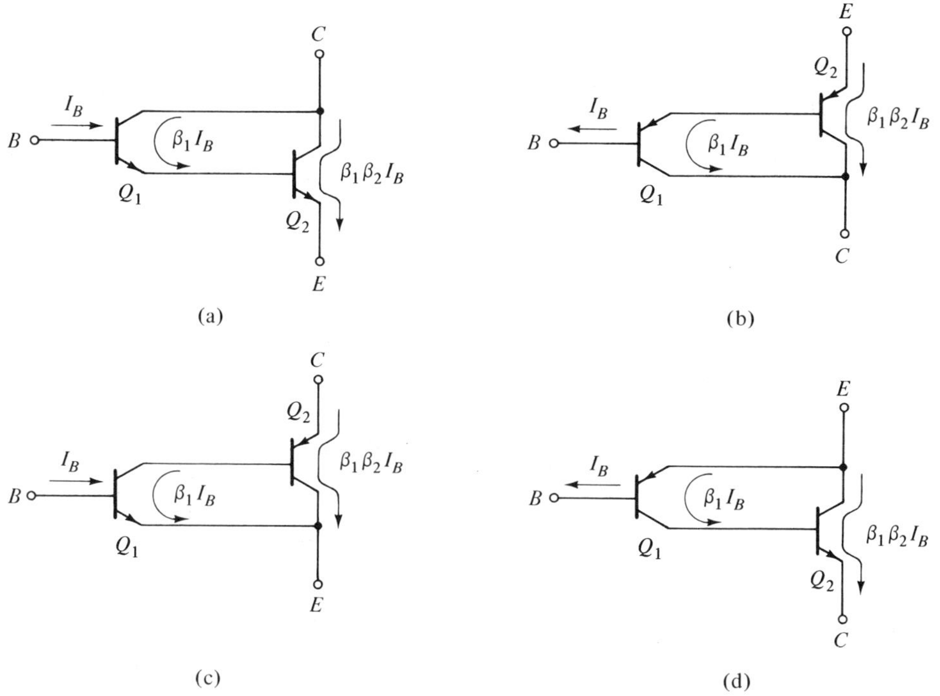

FIGURE 12–34 *Various output configurations encountered in complementary amplifiers: (a) NPN Darlington pair, (b) PNP Darlington pair, (c) NPN complementary Darlington pair, and (d) PNP complementary Darlington pair. (The equivalent base, emitter, and collector terminals are indicated by "B," "E," and "C," respectively.)*

* As shown in Fig. 12–34, the emitter current of the first transistor is equal to the base current of the second transistor. This is not the general case. An additional emitter bias resistor can be connected to the first transistor to allow its emitter current to be much larger than the base current of the second transistor. Such an arrangement is illustrated later in this section.

** The complementary Darlington pair is a special case of a negative-feedback amplifier of the voltage sample-voltage sum type with R_{f2} equal to zero and R_{f1} equal to infinity.

observe that each circuit acts as a compound NPN transistor. Stated differently, when an ac current flows into the base terminal of the first transistor in either Fig. 12–34(a) or (c), current flows into the equivalent collector terminal and out of the equivalent emitter terminal. Consequently, either the NPN Darlington pair or the NPN complementary Darlington pair can be used to replace the single NPN output transistor shown in Fig. 12–33.* Also, either form of the PNP Darlington pair can be used to replace the single PNP output transistor shown in Fig. 12–33. The use of compound transistors is illustrated in Fig. 12–35.

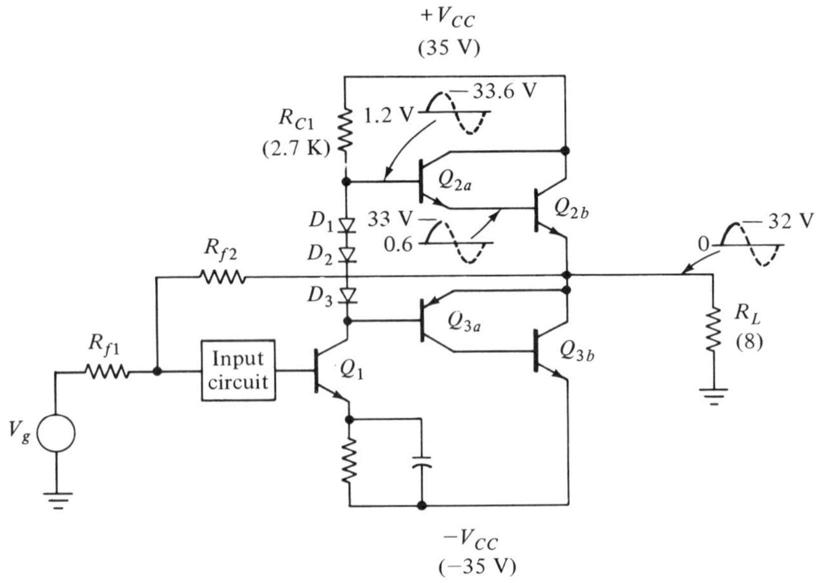

FIGURE 12–35 *Partial circuit diagrams of Class AB complementary power amplifier using Darlington pairs. This circuit is a modification of the circuit shown in Fig. 12–33. Component values are from Ex. 12–8.*

The partial circuit diagram of a Class AB complementary amplifier shown in Fig. 12–33 can be modified to use Darlington pairs for the output transistors as shown in Fig. 12–35. The output NPN transistor is formed by an NPN Darlington pair, Q_{2a} and Q_{2b}. The output PNP transistor is formed by a PNP complementary Darlington pair, Q_{3a} and Q_{3b}. Three bias diodes are used (D_1, D_2, and D_3) to forward bias the base-emitter junctions of Q_{2a}, Q_{2b}, and Q_{3a}. The remainder of the circuit is identical to that shown in Fig. 12–33. Our next task is to examine the circuit operation under peak signal conditions. First, however, we will illustrate in the following example the problem of obtaining the necessary peak current drive.

EXAMPLE 12–8 Can the amplifier shown in Fig. 12–35 deliver a 32-V peak signal to an 8-ohm load? Component values are given in Fig. 12–35. Assume that each transistor has a β of 50 and that the output bias voltage is initially set at 0 V dc.

---

* The complementary Darlington circuit is a transistor version of the "stacked-up cathode follower" reported by E. L. C. White. Both the White Cathode Follower and the White Emitter Follower are illustrated in the chapter problems associated with this section.

Solution: When V_o is equal to its peak positive value of 32 V, the voltage at the base of Q_{2a} is approximately 33.6 V. Then the voltage across R_{C1} is

$$V_{RC1} \simeq 35\text{ V} - 33.6\text{ V} \simeq 1.4\text{ V}$$

and the current through R_{C1} is limited to

$$I_{RC1} \simeq V_{RC1}/R_{C1} \simeq 1.4\text{ V}/2.7\text{ K} \simeq 0.52\text{ mA}$$

The required maximum value of the output current is

$$I_m = V_m/R_L = 32\text{ V}/8 = 4\text{ A}$$

The corresponding base current in Q_{2a} is

$$I_{B2b} = I_m/\beta^2 \simeq (4\text{ A})/50^2 \simeq 1.6\text{ mA}$$

The base current of Q_2 must flow through R_{C1}. However, the maximum value of the current in R_{C1} is only 0.52 mA when V_o equals +32 V. Thus the amplifier does not have adequate current drive to deliver a 32-V peak positive signal to an 8-ohm load. A 32-V peak negative signal can be delivered to an 8-ohm load provided the base-collector junction of Q_1 remains reversed-biased.

Operation of the circuit shown in Fig. 12–35 under peak positive signal conditions requires special consideration. The transistors that conduct on the positive portion of the output signal are the NPN Darlington pair, Q_{2a} and Q_{2b}. A partial circuit diagram showing these two transistors under peak positive signal conditions is shown in Fig. 12–36(a). The value of the peak output current, I_m, is established by the output power requirement and by the load resistor value. The values of β_{2b} and β_{2a} determine the value of the maximum base current drive, I_{b2m}, of Q_{2a}. This current must flow through R_{C1}. Therefore, the maximum output voltage is limited to

$$V_m = V_{CC} - I_{b2m}R_{C1} - V_{BE2a} - V_{BE2b} \qquad \textbf{(12–66)}$$

By making R_{C1} small, the allowable value of V_m can be increased. However, this change increases the standby or bias current since the voltage across R_{C1} is approximately V_{CC} when no input signal is present. Also, a small value of R_{C1} would greatly reduce the gain of the CE stage, Q_1. The problem of providing the required drive current for Q_{2a} under peak positive signal conditions can be solved by two methods: (1) bootstrapping R_{C1} and (2) using a constant-current source in place of R_{C1}. Both methods are discussed below.

The ac base current in Q_{2a} can be increased by using the *bootstrap circuit* shown in Fig. 12–36(b). Resistor R_{C1} is normally divided into two equal parts, R_{C11} and R_{C12}. Feedback is obtained by connecting a capacitor between the top of R_L and the junction of R_{C11} and R_{C12}, or point A in Fig. 12–36(b). As the signal voltage goes positive, the voltage at point A also goes positive. Since the voltage gain of Q_{2a} and Q_{2b} is only slightly less than unity (both are CC stages), the voltage drop across R_{C11} remains nearly constant. Thus, the peak ac voltage at point A can exceed the power supply voltage under peak positive signal conditions. The current needed to drive the base of Q_{2a} to obtain the peak output current,* I_m, is obtained from the feedback circuit. The value of the bootstrap

---

* In a manner of speaking, the circuit "lifts itself by its bootstrap" to produce the required drive current, hence the name *bootstrap circuit.* A bootstrap circuit is a positive-feedback circuit. Positive feedback is acceptable in a linear circuit since the loop gain, which is approximately equal to the product of the voltage gains of Q_{2a} and Q_{2b}, is less than unity. Too much positive feedback results in instability.

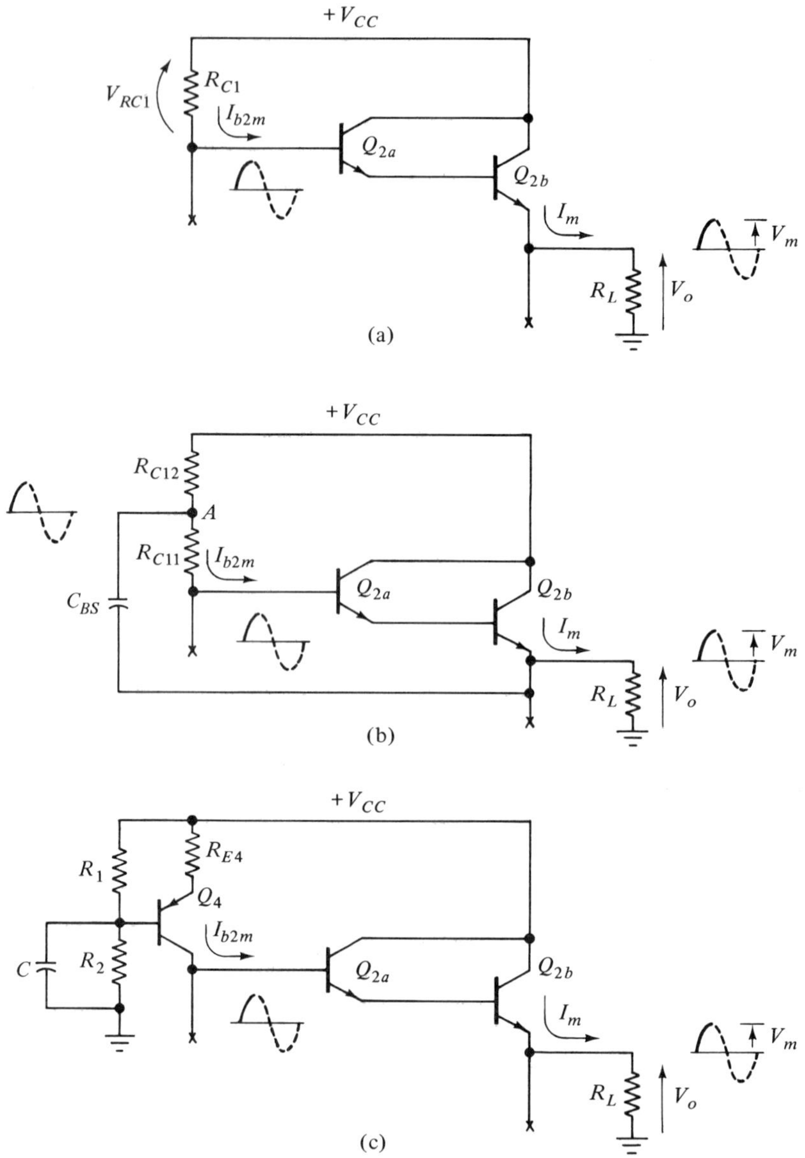

FIGURE 12–36 *Circuits used to illustrate the peak output signal problem: (a) basic circuit, (b) use of a "bootstrap" circuit, and (c) use of a constant-current source.*

capacitor, C_{BS}, is usually in the 10 to 100 μF range. Bootstrap techniques are used in a variety of practical circuits.

Replacing R_{C1} with a constant-current source, Q_4 in Fig. 12–36(c), is a second method used to provide the required drive current under peak signal conditions. The collector current in Q_4 is essentially constant as its collector voltage changes. The constant-current source remains in operation as long as its base-collector junction is reverse-biased, or until V_{CE} approaches zero. Thus, the maximum positive output voltage at the base of Q_{2a} is limited by the base bias voltage of

Q_4. In general, a constant-current bias arrangement is preferable to a bootstrap arrangement. However, both are illustrated in the next section. The following example illustrates the use of a constant-current source to bias a Darlington pair.

EXAMPLE 12–9 Determine the bias conditions for the constant-current source, Q_4, in Fig. 12–36(c) if 64 W of power is to be delivered to an 8-ohm load. Assume that V_{CC} is 35 V and that all transistors have βs of 50.

Solution: (a) Peak signal values: $V_m \simeq 32$ V and $I_m \simeq 4$ A. These values correspond to those given in Ex. 12–8. Thus, the peak voltage at the collector terminal of Q_4 is approximately 33.6 V, and the peak value of the base current for Q_2 is 1.6 mA.

(b) Bias values for the constant-current source: The base voltage of Q_4 should be chosen more positive than the peak voltage at the collector terminal of Q_4 or 33.6 V. Assuming that R_1 is 1 K and R_2 is 33 K, then

$$V_{B4} \simeq (V_{CC})\left(\frac{R_1}{R_1 + R_2}\right) \simeq (35\text{ V})\left(\frac{33\text{ K}}{1\text{ K} + 33\text{ K}}\right) \simeq 34\text{ V}$$

The voltage at the emitter of Q_4 is

$$V_{E4} = V_{B4} + V_{BE4} \simeq 34\text{ V} + 0.6\text{ V} \simeq 34.6\text{ V}$$

The collector current in Q_4 is then determined by the value of R_{E4} as

$$I_{C4} \simeq I_{E4} \simeq (V_{CC} - V_{E4})/R_{E4} \simeq (0.4\text{ V})/R_{E4}$$

The value of I_{C4} should be greater than the peak value of the base current of Q_2, or 1.6 mA. If R_{E4} is 150 ohms, then I_{C4} is approximately 2.7 mA.

The partial circuit diagram shown in Fig. 12–37 is a modification of the amplifier shown in Fig. 12–35. The principal modifications are the constant-current source, Q_4, and the resistors R_2, R_3, R_{E2}, and R_{E3}. The effect of the constant-current source was discussed above in connection with Fig. 12–36(c). The added resistors affect both the linearity and thermal response. These effects are discussed below.

In the classical Darlington circuits shown in Figs. 12–34(a) and (b), the base current of the second stage is equal to the emitter current of the first stage. When no ac signal is present, the output power transistor should be biased at a low value of collector current in order to obtain a high value of conversion efficiency. The corresponding base current is very small which means that the collector current of the input stage of the Darlington pair is also small. Often this latter condition is undesirable because β may be too small at these low current levels. Resistors R_2 and R_3 allow the input transistors of the Darlington pairs to be biased at higher current values (generally a few milliamps) to increase the βs.

As in all power amplifiers, thermal stability is one of the dominant problems. In addition to heat sinks, the complementary power amplifier in Fig. 12–37 has three features which help produce thermal stability: (1) overall dc feedback, (2) diode thermal compensation, and (3) unbypassed emitter resistors.

The feedback and feedforward circuits are dc coupled; i.e., blocking capacitors are not used. In a dc coupled feedback amplifier employing voltage sampling at the output terminal, the value of the output dc voltage is stabilized against the effects of parameter variations. Since an increase in junction temperature

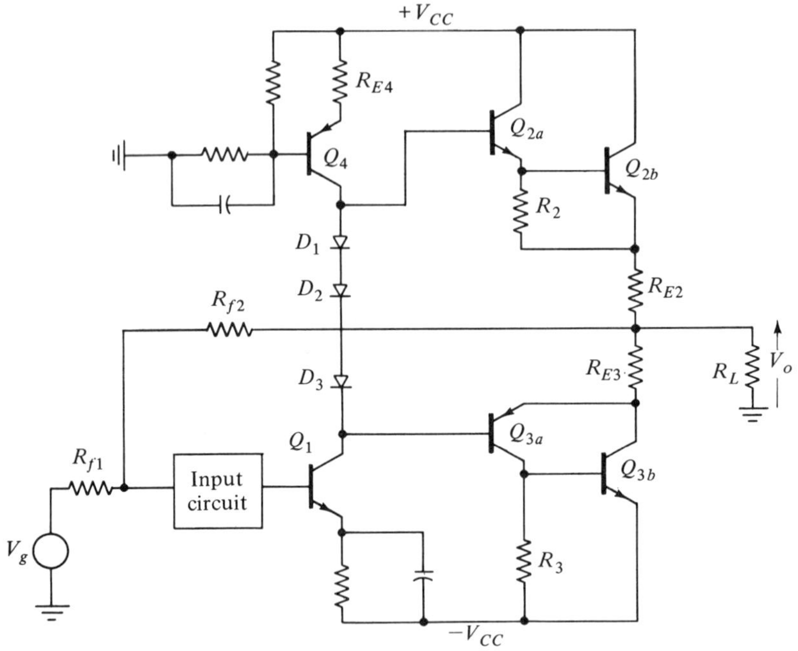

FIGURE 12–37 *Partial circuit diagram of a Class AB complementary power amplifier using a constant-current source load for the driver transistor. This circuit is a modification of the circuit shown in Fig. 12–35.*

increases the emitter currents, the dc output voltage could be a function of temperature. Negative feedback reduces the drift in the output voltage that is caused by thermal changes and also improves the thermal stability of the amplifier. A stable value for the output dc voltage is particularly important when the load is directly coupled to the amplifier.

Diodes D_1, D_2, and D_3 operate in conjunction with R_{E2} and R_{E3} to improve the thermal stability of the amplifier. An increase in temperature results in two effects: (1) the forward voltage of the diode is reduced approximately 2 mV/°C for a given current and (2) the emitter currents in the output transistors increase. This latter effect produces an increase in the voltage drop across R_{E2} and R_{E3}. Note that both effects tend to reduce the forward bias of the base-emitter junctions of the output transistors, thus improving the thermal stability of the amplifier. Often one or more of the diodes are mounted on the output transistor heat sinks to provide better thermal tracking. Since the possibility of overcompensation exists, the complete amplifier should be adjusted to obtain the desired degree of thermal stability. When the diodes are mounted on the heat sinks with the output transistors, smaller values of R_E can be used.

An analysis of the complementary Darlington pair shows that the value of R_{E3} should be chosen so that*

$$R_{E3} > (V_{\mathrm{CE}})(\theta_{ja})\left(|\Delta V_{\mathrm{BE}}/°\mathrm{C}|\right) \qquad \textbf{(12–67)}$$

where V_{CE} is the bias voltage across Q_{3b}; θ_{ja} is the total thermal resistance from the junction of Q_{3b} to ambient, and $\Delta V_{\mathrm{BE}}/°\mathrm{C}$ is approximately -2 mV/°C.

* See G. C. Haas in the references listed at the end of the chapter for the derivation of Eq. (12–67).

Thermal stability is generally ensured if R_{E3} is at least twice the minimum given by Eq. (12–67). Also, due to the basic similarity between the NPN Darlington pair and the PNP complementary Darlington pair, the value of R_{E2} should be equal to R_{E3}. Thus, a reasonable design choice is

$$R_{E2} \simeq R_{E3} \simeq (2)(V_{CE2,3})(\theta_{ja2,3})(2 \times 10^{-3}) \qquad \textbf{(12–68)}$$

The following example illustrates the values of R_{E2} and R_{E3} encountered in practice.

EXAMPLE 12–10 The amplifier shown in Fig. 12–37 is operated with $V_{CC} = \pm 35$ V. With R_L equal to 8 ohms, and the maximum value of P_o equal to 64 W, calculate the values and dissipation ratings of R_{E2} and R_{E3}. Assume that θ_{jc} for the output power transistors is 1.5°C/W and that θ_{sa} for the heat sink is 3°C/W. A mica washer is used for insulation.

Solution: (a) Values of R_{E2} and R_{E3}: From Fig. 12–37, the value of V_{CC} and V_{CE} are approximately the same or 35 V. The total thermal resistance from the junction to the ambient is

$$\theta_{ja} \simeq \theta_{jc} + \theta_{cs} + \theta_{sa} \simeq 1.5 + 0.5 + 3 \simeq 5°\text{C/W}$$

From Eq. (12–68),

$$R_{E2} \simeq R_{E3} \simeq (2)(35\text{ V})(5°\text{C/W})(2 \times 10^{-3}) \simeq 0.7\ \Omega$$

A standard value of 0.68 ohm can be used.

(b) Power Rating: When 64 W is delivered to an 8-ohm load, $V_m \simeq 32$ V and $I_m \simeq 4$ A. The power that R_{E2} and R_{E3} must dissipate is

$$P_D = I_{\text{rms}}^2 R_E \simeq \left(\frac{I_m}{2}\right)^2 (R_E) \simeq \left(\frac{4}{2}\right)^2 (0.68) \simeq 2.72\text{ W}$$

In a conservative design, a 5-W resistor would be used.

(c) Effects of R_{E2} and R_{E3} on the performance of the amplifier: The price paid for the thermal stability obtained by using R_{E2} and R_{E3} is a reduction in the maximum power output to R_L because the peak output voltage is reduced by the $I_E R_E$ drop. Assuming that the peak voltage that can be developed at the emitter terminal of Q_{2b} is limited to 32 V, then the peak value of the output voltage across R_L is

$$V_m \simeq 32\text{ V} - I_m R_E \simeq 32\text{ V} - (4\text{ A})(0.68) \simeq 29.3\text{ V}$$

The corresponding maximum output power is

$$P_{o(\max)} \simeq (V_m)^2/(2R_L) \simeq (29.3)^2/(2)(8) \simeq 54\text{ W}$$

In this section we have studied some of the basic characteristics of power amplifiers. The two basic configurations for the output stages are the Darlington pair and the complementary Darlington pair. Since either form can be used for the NPN or PNP portion of the complementary output, four basic output configurations are possible: (1) NPN Darlington-PNP Darlington, (2) NPN Darlington-PNP complementary Darlington, (3) NPN complementary Darlington-PNP Darlington, and (4) NPN complementary Darlington-PNP complementary Darlington. Numerous variations of these basic configurations are found in practice, particularly in the bias networks.

The problem of adequate current drive under peak signal conditions was also examined. Bootstrapping circuits as well as constant-current sources helped to solve this problem. Although both are encountered in practice, the constant-current source is the preferred approach.

Finally, the problem of thermal stability in complementary power amplifiers was studied. Four methods are generally used to produce thermal stability: (1) adequate heat sinks for the output stages, (2) overall dc feedback, (3) diode thermal compensation, and (4) unbypassed emitter resistors in the two output stages.

The material in this section will be used in the practical amplifiers found in the next two sections. Also, additional characteristics of complementary amplifiers will be studied.

12.8 Examples of Complementary Power Amplifiers

The basic characteristics and some possible output configurations of complementary power amplifiers were considered in the last section. In this section we will examine the operation of several typical complementary power amplifiers. This study will be divided into two major parts. First, several amplifier configurations will be introduced with the goal of understanding the function of each component, the limitations on the maximum output power, and the performance characteristics. Second, a complete analysis of a complementary power amplifier will be illustrated with a detailed example. Two additional topics will be considered briefly: (1) the use of buffer networks for loudspeaker operation and (2) a simple short-circuit protection circuit.

In the initial study of complementary power amplifiers, the overall amplifier performance parameters will be considered (the approximate values of V_o/V_g, R_i, R_o, and $P_{o(\max)}$). In addition, the bias arrangement, loop gain expression, and frequency response limitations will be discussed.

Consider the amplifier shown in Fig. 12–38. The output stages, Q_3 and Q_5, form an NPN Darlington pair, and Q_2 and Q_4 form a PNP Darlington pair. This output configuration is sometimes called a *full complementary* output. The driver and input sections are combined into a single CE stage, Q_1. The negative-feedback circuit is a voltage sample-current sum type. Using the approximate feedback analysis developed in Chap. 10 and assuming a large value for $|T|$, the mid-band performance characteristics are

$$V_o/V_g \simeq -R_{f2}/R_{f1} \simeq -47\text{ K}/4.7\text{ K} \simeq -10$$

$$R_{if} \simeq R_{f1} \simeq 4.7\text{ K} \qquad R_{of} \simeq 0 \tag{12–69}$$

Before the maximum positive or negative value of V_o can be estimated, we must examine the bias condition of the circuit. All bias values in the circuit can be controlled with the potentiometer R_2 which is part of the dc feedback network.* A reasonable first estimate is that the dc output voltage, $V_{o(\text{dc})}$, (at the circuit side of C_5) is adjusted to a value of approximately $V_{CC}/2$ or 12 V to allow nearly equal positive and negative output signal amplitudes. Then, with no input signal present, the bias values are

$$V_{B3} \simeq V_{o(\text{dc})} + I_{E5}R_{E5} + V_{BE5} + V_{BE3}$$
$$\simeq 12\text{ V} + 0 + 0.6\text{ V} + 0.6\text{ V} \simeq 13.2\text{ V} \tag{12–70}$$

$$I_{C1} \simeq (V_{CC} - V_{B3})/(R_{C11} + R_{C12})$$
$$\simeq (24\text{ V} - 13.2\text{ V})/(2.7\text{ K} + 2.7\text{ K}) \simeq 2\text{ mA} \tag{12–71}$$

$$V_{E1} \simeq I_{C1}R_{E1} \simeq (2\text{ mA})(100) \simeq 0.2\text{ V} \tag{12–72}$$

* Note in Fig. 12–38 that R_2 is ac bypassed by C_3 and is not part of the mid- and high-frequency feedback network.

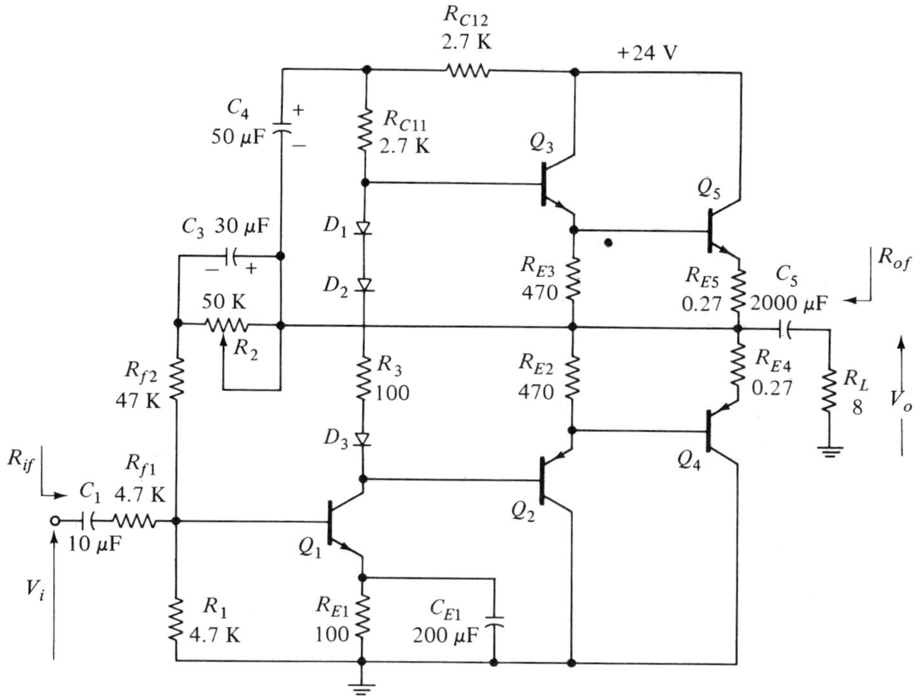

FIGURE 12–38 *One example of a complementary power amplifier.* (*Courtesy Southwest Technical Products Corp.*)

$$V_{B1} \simeq V_{E1} + V_{BE1} \simeq 0.2\text{ V} + 0.6\text{ V} \simeq 0.8\text{ V} \qquad \textbf{(12–73)}$$

$$V_{C1} \simeq V_{o(\text{dc})} - I_4 R_{E4} - V_{BE4} - V_{BE2}$$
$$\simeq 12\text{ V} - 0 - 0.6\text{ V} - 0.6\text{ V} \simeq 10.8\text{ V} \qquad \textbf{(12–74)}$$

As discussed in Sec. 12.4 in connection with Fig. 12–22, the distortion in a feedback amplifier with a large loop gain can remain very low as long as the feedback loop is active or connected.* Feedback is generally effective as long as all transistors operate in the linear region; i.e., the base-emitter junction is forward-biased, and the base-collector junction is reverse-biased. From Fig. 12–38, the most negative value of V_o is limited by the base voltage Q_1, and the most positive value is limited by the collector voltage of Q_3.** Neglecting the increase in V_{BE5} with an increase in I_{C5} and the drop across R_{E5} the maximum positive output signal is equal to the base-collector bias voltage of Q_3, or

$$V_{m(\text{pos})} \simeq V_{C3} - V_{B3} \simeq 24\text{ V} - 13.2\text{ V} \simeq 10.8\text{ V} \qquad \textbf{(12–75)}$$

Also, neglecting the increase in V_{BE4} with an increase in I_{C4} and the drop across R_{E4}, the maximum negative output signal is equal to the base-collector bias voltage of Q_1 or

$$V_{m(\text{neg})} \simeq V_{C1} - V_{B1} \simeq 10.8\text{ V} - 0.8\text{ V} \simeq 10\text{ V} \qquad \textbf{(12–76)}$$

* Two important conditions are: (1) that the value of $|T|$ is large, and (2) that adequate current drive is available.

** The collector voltage of Q_3, V_{CC}, limits the positive output voltage only if the bootstrap network operates properly by maintaining a constant current in R_{C11}. Exceeding either the positive or negative output voltage limitations results in a forward-biased, base-collector junction.

Consequently, from Eqs. (12–75) and (12–76), the output voltage that can be delivered to the load without clipping due to these bias values is approximately 10 V maximum, 20 V peak-to-peak, or 7 V rms.* The corresponding values of output power and peak output current are

$$P_{o(\max)} \simeq (V_m^2)/(2R_L) \simeq (10\text{ V})^2/(2)(8) \simeq 6.3\text{ W} \tag{12–77}$$

$$I_m \simeq V_m/R_L \simeq 10\text{ V}/8 \simeq 1.25\text{ A} \tag{12–78}$$

To summarize, the feedback complementary power amplifier shown in Fig. 12–38 with V_{CC} equal to 24 V should be capable of delivering 6.3 W of rms power to an 8-ohm load with a low value of harmonic distortion. This performance is based on two assumptions: (1) adequate current drive, and (2) adequate loop gain.** Each requirement is examined below.

When the output voltage goes negative, transistors Q_1, Q_2, and Q_4 turn on. Thus the peak output current of 1.25 A is not difficult to obtain. However, when the output voltage goes positive, transistors Q_3 and Q_5 turn on while Q_1 turns off. As discussed in the last section in connection with Fig. 12–36, the function of the bootstrap capacitor, C_4, in Fig. 12–38, is to maintain a nearly constant current through R_{C11}. This current is equal to the bias current I_{C1}, or 2 mA. Thus, with I_m equal to 1.25 A (the value calculated for the circuit in Fig. 12–38), the product of β_3 and β_5 at high current levels should exceed 625.

Our next task is to determine an expression for the loop gain for the amplifier shown in Fig. 12–38. Since Q_3 and Q_5 (or Q_2 and Q_4) operate as CC stages, the voltage gain from the collector of Q_1 to the output is approximately unity. Due to the bootstrap circuit, the effective value of R_{C11} is very large and can be neglected. Then the load impedance of Q_1 is r_{ce1} in parallel with $R_{\text{in}2}$ and $R_{\text{in}3}$. Under large signal conditions, we may consider only one pair of output transistors and neglect the other pair as turned off. Thus for a positive output signal,

$$R_{L\text{eq}1} \simeq r_{ce1} \,||\, R_{\text{in}3} \simeq r_{ce1} \,||\, (\beta_3\beta_5 R_L) \tag{12–79}$$

For a negative output signal,

$$R_{L\text{eq}1} \simeq r_{ce1} \,||\, R_{\text{in}2} \simeq r_{ce1} \,||\, (\beta_2\beta_4 R_L) \tag{12–80}$$

When calculating the voltage gain of Q_1, either the smallest or the average value of $R_{L\text{eq}1}$ can be used. Then from Fig. 12–38,

$$T \simeq (A_{vt1})(A_{vt3,5})\left(\frac{R_{f1} \,||\, R_1 \,||\, R_{\text{in}1}}{R_{f2} + R_{f1} \,||\, R_1 \,||\, R_{\text{in}1}}\right)$$

$$\simeq \left(-\frac{R_{L\text{eq}1}}{r_{e1}}\right)(1)\left(\frac{R_{f1} \,||\, R_1 \,||\, R_{\text{in}1}}{R_{f2} + R_{f1} \,||\, R_1 \,||\, R_{\text{in}1}}\right) \tag{12–81}$$

The frequency response of the amplifier shown in Fig. 12–38 depends on the transistor parameters. The analysis is also complicated for the bootstrap network which serves as a frequency dependent load for Q_1. A detailed analysis of this circuit can be made using node equations. In general, however, experimental results are used to determine the frequency response characteristics.

---

* These output voltage values could be increased slightly by adjusting R_2 to give equal amplitude clipping for both the positive- and the negative-going signals.

** Also assumed in the successful operation of any power amplifier is adequate heat sinking and frequency stability.

A second example of a complementary power amplifier is shown in Fig. 12–39. The output stages Q_6 and Q_7 form an NPN complementary Darlington pair, and Q_8 and Q_9 form a PNP complementary Darlington pair. This output configuration is sometimes called a *conjugate complementary* output. The driver stage, Q_4, is a CE configuration having a current source, Q_5, acting as a load. The input stages, Q_1 and Q_2, form a differential pair, with Q_3 acting as a constant-current source. The negative-feedback circuit is a voltage sample-voltage sum type. Using the approximate feedback analysis developed in Chap. 10 and assuming a large value of $|T|$, the mid-band performance characteristics are

$$A_{fv} \simeq \frac{V_o}{V_i} \simeq \frac{R_{f1} + R_{f2}}{R_{f1}} \simeq \frac{220 + 2.2\text{ K}}{220} \simeq 11 \qquad R_{if} \simeq 22\text{K}$$

$$A_{vg} \simeq \frac{20\text{ K}}{20\text{ K} + 2.2\text{ K}} \simeq \frac{10}{11} \qquad R_{of} \simeq 0 \tag{12–82}$$

$$A_{vo} = A_{vg}A_{fv} \simeq (10/11)(11) \simeq 10$$

The circuit bias values are controlled by overall dc feedback. Capacitor C_{f1} is an open circuit for the dc voltage, making the dc closed-loop gain unity. The input offset voltage has a very low value which is equal to $(I_{B1})(20\text{ K})$ or less, depending on the value of R_g. Thus, the output dc voltage is also approximately 0 V. Zener diode D_4 determines the current in both Q_3 and Q_5. By adjusting the value of R_4, the current in the four output transistors, Q_6 through Q_9, can be controlled. Resistors R_{E6} and R_{E8} improve the thermal stability. The two *RC* filter networks, $R_{E7}C_3$, and $R_{E9}C_4$, help isolate the output power transistors from the remainder of the circuit.

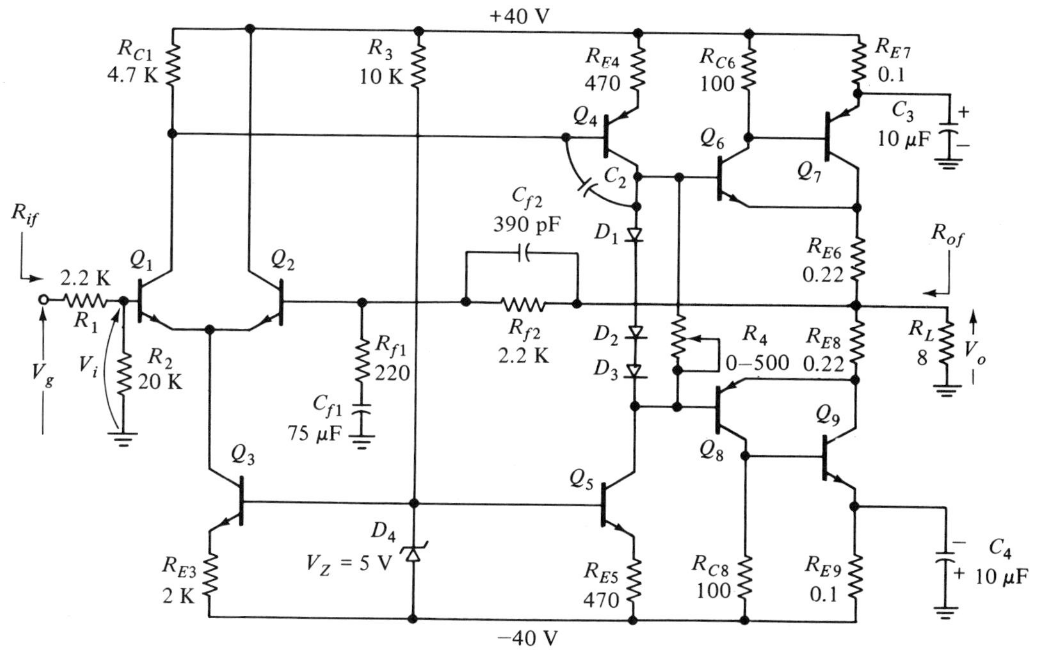

FIGURE 12–39 *A second example of the complementary power amplifier.*

The important bias voltages needed for maximum possible signal calculations are

$$V_{o(\mathrm{dc})} \simeq 0\ \mathrm{V} \qquad V_{B4} \simeq 35\ \mathrm{V}$$
$$V_{C4} \simeq 0.6\ \mathrm{V} \qquad V_{B5} \simeq -35\ \mathrm{V} \tag{12–83}$$
$$V_{C5} \simeq -0.6\ \mathrm{V}$$

The maximum positive output voltage is controlled by the base-to-collector voltage of Q_4. Allowing for a small drop across R_{E6} due to I_m,

$$V_{m(\mathrm{pos})} \simeq V_{B4} - V_{\mathrm{BE}6} - IR_{E6} \simeq 35\ \mathrm{V} - 0.8\ \mathrm{V} - 1\ \mathrm{V} \simeq 33.2\ \mathrm{V} \tag{12–84}$$

The maximum negative output voltage is controlled by the base-to-collector voltages of Q_5. Allowing for a small drop across R_{E8},

$$V_{m(\mathrm{neg})} \simeq V_{B5} + V_{\mathrm{BE}8} + IR_{E8} \simeq -35 + 0.8 + 1 \simeq -33.2\ \mathrm{V} \tag{12–85}$$

With these two limitations, the output voltage that can be delivered to the load without clipping due to the bias values is approximately 66 V peak-to-peak, or 23.3 V rms. The corresponding value of output power is

$$P_{o(\max)} \simeq V_m^2/2R_L \simeq (33)^2/(2)(8) \simeq 68\ \mathrm{W} \tag{12–86}$$

The peak output current is

$$I_m \simeq V_m/R_L \simeq (33\ \mathrm{V})/8 \simeq 4.1\ \mathrm{A} \tag{12–87}$$

To summarize, the feedback complementary power amplifier shown in Fig. 12–39 with V_{CC} equal to ± 40 V should be capable of delivering 68 W of rms power to an 8-ohm load with a low value of distortion. If some distortion is acceptable, higher values of output power are possible.

To deliver 68 W of output power, the peak output current must be 4.1 A. The required base current in Q_6 (assuming a positive output voltage) is approximately equal to $I_m/(\beta_6\beta_7)$. Since the bias value of I_{C4} is approximately 9.4 mA, the value of $\beta_6\beta_7$ should exceed

$$\beta_6\beta_7 > I_m/I_{C4} \simeq 4.1\ \mathrm{A}/9.4\ \mathrm{mA} \simeq 437 \tag{12–88}$$

A similar requirement exists for the minimum value of $\beta_8\beta_9$.

Our next task is to develop an expression for the loop gain of the amplifier shown in Fig. 12–39. The analysis is greatly simplified if we treat only one path from the collector terminal of Q_4 to the output; either the Q_6Q_7 pair or the Q_8Q_9 pair is on (and the other pair off) under large-signal conditions. An expression for the loop gain is

$$T \simeq (A_{vt4})(A_{vt6,7})\left(\frac{R_{f1} \parallel R_{\mathrm{in}2}}{R_{f1} \parallel R_{\mathrm{in}2} + R_{f2}}\right)(A_{vt2,1}) \tag{12–89}$$

The value of $A_{vt6,7}$ is approximately equal to unity. The voltage division term is approximately equal to (1/11), or the reciprocal of the gain. An expression for $A_{vt2,1}$ is

$$A_{vt2,1} \simeq \frac{R_{L\mathrm{eq}1}}{r_{e1} + r_{e2} + [2\ \mathrm{K}/(\beta_1 + 1)]} \tag{12–90}$$

The calculation of A_{vt4} requires the value of $R_{\mathrm{in}6}$. The input resistance of the complementary Darlington pair can be estimated with the aid of Fig. 12–40. Since the voltage gain of this circuit is very close to unity, the values of V_t and V_o are approximately equal. Thus, the value of R_i can be estimated as

$$R_i = \frac{V_t}{I_t} \simeq \frac{V_o}{I_t} \simeq \frac{\beta_1\beta_2 I_t R_L}{I_t} \simeq \beta_1\beta_2 R_L \tag{12–91}$$

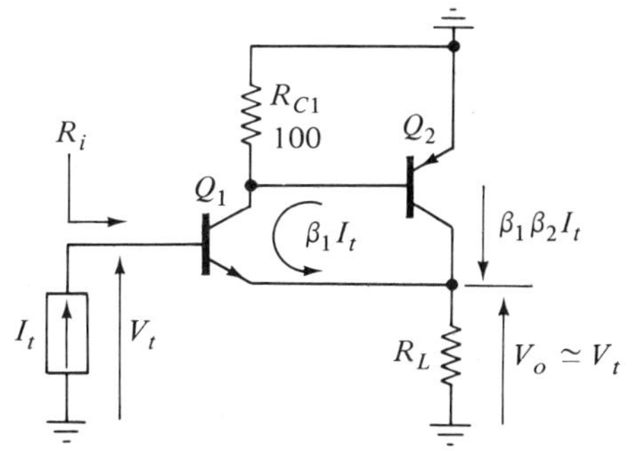

FIGURE 12–40 *Circuit diagram used to estimate the value of input resistance for a complementary Darlington pair.*

Eq. (12–89) can now be written as

$$T \simeq \left(-\frac{(\beta_6\beta_7 R_L) \parallel r_{ce4}}{r_{e4} + R_{E4}}\right)(1)\left(\frac{R_{f1}}{R_{f1} + R_{f2}}\right)\left(\frac{R_{C1} \parallel R_{\text{in}4}}{r_{e1} + r_{e2} + 2\text{ K}/(\beta_1 + 1)}\right) \quad \textbf{(12–92)}$$

Our next task is to determine the frequency response of the amplifier shown in Fig. 12–39. Except for C_{f1}, the amplifier is directly coupled and would respond to a dc signal. As noted earlier, however, the dc voltage gain is unity, and the mid-band voltage gain (V_o/V_i) is 11. In the low-frequency region, the expression for $1/B$ is

$$A_{f(\text{lo})} \simeq \frac{1}{B} \simeq \frac{1 + s(R_{f1} + R_{f2})C_{f1}}{1 + sR_{f1}C_{f1}} \quad \textbf{(12–93)}$$

Thus, the low-frequency pole occurs at

$$f_{\text{lo}} \simeq \frac{0.159}{R_{f1}C_{f1}} \simeq \frac{0.159}{(220)(75\ \mu\text{F})} \simeq 9.7\text{ Hz} \quad \textbf{(12–94)}$$

The low-frequency zero occurs at

$$f_z \simeq \frac{0.159}{(R_{f1} + R_{f2})C_{f1}} \simeq \frac{0.159}{(220 + 2.2\text{ K})(75\ \mu\text{F})} \simeq 0.9\text{ Hz} \quad \textbf{(12–95)}$$

The high-frequency response is controlled by the feedback network.* For this condition,

$$f_{\text{hi}(f)} \simeq \frac{0.159}{R_{f2}C_{f2}} \simeq \frac{0.159}{(2.2\text{ K})(390\text{ pF})} \simeq 185\text{ kHz} \quad \textbf{(12–96)}$$

Cascade high-frequency compensation is accomplished with C_2. The value of C_2 should be determined experimentally to eliminate high-frequency oscillation.

A third example of a complementary power amplifier is shown in Fig. 12–41. The output stages, Q_5 and Q_6, form an NPN Darlington pair while Q_7 and Q_8 form a PNP complementary Darlington pair. This output configuration is sometimes called a *quasi-complementary* output. The driver stage, Q_3, is a CE circuit with a constant-current load, Q_4. The input stages, Q_1 and Q_2, form a differential pair. The negative-feedback circuit is a voltage sample-voltage sum type. The analysis of this amplifier is illustrated in the following example.

EXAMPLE 12–11 For the power amplifier shown in Fig. 12–41, calculate the gain V_o/V_i, maximum output power with low distortion, junction and case temperatures of the output transistors, conversion efficiency, and frequency response. Assume that the output dc voltage is adjusted to 0 V dc by R_{EE} and that the total stand-by

---

* The effect of feedback compensation is treated in Chap. 11.

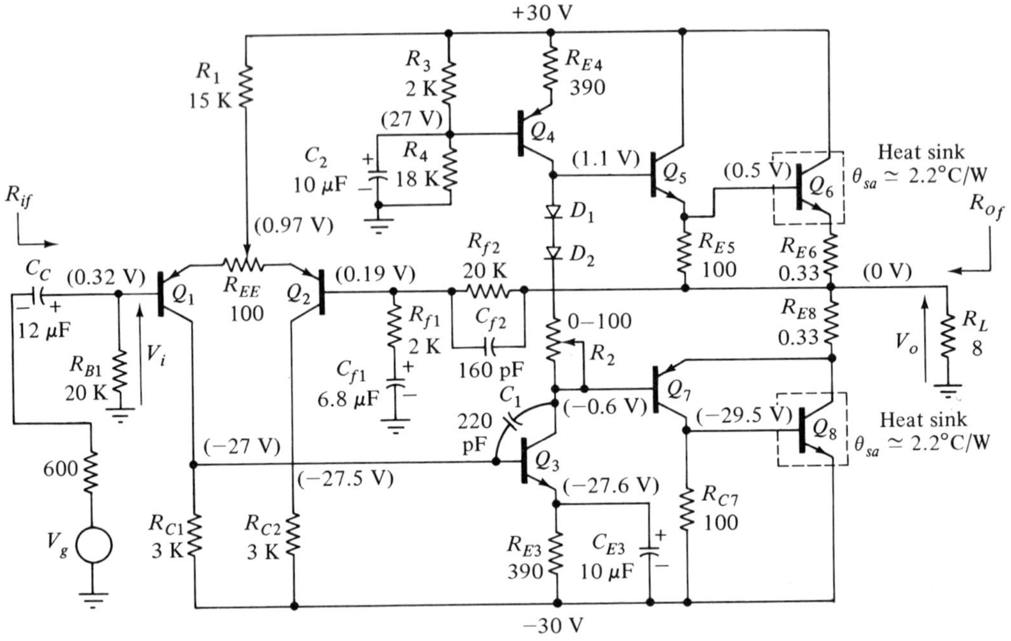

FIGURE 12–41 *The complementary power amplifier from Ex. 12–11. Values in parentheses are measured bias voltages.*

current is adjusted to less than 50 mA from each supply by R_2. Assume also that each output transistor has a θ_{jc} of 1.5°C/W and is operated with a heat sink having a θ_{sa} of 2.2°C/W. The ambient temperature is 25°C. The transistor characteristics are:

Transistor	β	f_T (MHz)	C_{ob} (pF)	$T_{j(\max)}$ (°C)
Q_1, Q_2	65	200	4.5	200
Q_3, Q_5	60	40	9	200
Q_4	150	200	8	125
Q_7	40	15	15	200°C
Q_6, Q_8	30	0.6	500	200°C

Solution: (a) Small-signal approximate analysis: Assuming $|T| \gg 1$,

$$A_{vf} = V_o/V_i \simeq (R_{f1} + R_{f2})/R_{f1} = (2\text{ K} + 20\text{ K})/2\text{ K} = 11$$

$$R_{if} \simeq 20\text{ K} \qquad R_{of} \simeq 0$$

$$A_{vg} = \frac{V_i}{V_g} = \frac{R_{if}}{R_{f1} + R_g} \simeq 0.97 \qquad A_{vo} = A_{vg}A_{vf} \simeq 10.7$$

(b) Bias values:

$$V_{B4} = V_{CC}\left(\frac{R_4}{R_3 + R_4}\right) \simeq 27\text{ V} \qquad V_{E4} = V_{B4} + V_{BE4} \simeq 27.6\text{ V}$$

$$I_{E4} = (V_{CC} - V_{E4})/R_{E4} \simeq 6.15\text{ mA} \qquad I_{E3} \simeq I_{E4} \simeq 6.15\text{ mA}$$

$$V_{C1} = V_{B3} = -V_{CC} + I_{E3}R_{E3} + V_{BE3} \simeq -27\text{ V}$$

$$I_{C1} = [V_{C1} - (-V_{CC})]/R_{C1} \simeq 1\text{ mA} \qquad I_{B1} \simeq I_{C1}/\beta_1 \simeq 15.3\ \mu\text{A}$$

$$V_{B1} \simeq I_{B1}R_{B1} \simeq 0.3 \text{ V} \qquad V_{E1} = V_{B1} + V_{\text{BE1}} \simeq 0.9 \text{ V}$$
$$I_{E1} + I_{E2} \simeq (V_{\text{CC}} - V_{E1})/R_1 \simeq (30 \text{ V} - 0.9 \text{ V})/15 \text{ K} \simeq 1.94 \text{ mA}$$
$$I_{E2} = 1.94 \text{ mA} - 1 \text{ mA} \simeq 0.94 \text{ mA} \qquad V_{C2} = -V_{\text{CC}} + I_{E2}R_{C2} \simeq -27.2 \text{ V}$$

(c) Maximum output signal without excessive distortion: The maximum positive output voltage is limited by the voltage V_{B4}. Allowing for a small drop across R_{E6} due to I_m,

$$V_{m(\text{pos})} \simeq V_{B4} - V_{\text{BE5}} - V_{\text{BE6}} - I_m R_{E6}$$
$$\simeq 27 \text{ V} - 0.8 \text{ V} - 1 \text{ V} - 1 \text{ V} \simeq 24.2 \text{ V}$$

The maximum negative output signal is limited by the voltage V_{B3}. Allowing for a small drop across R_{E8} due to I_m,

$$V_{m(\text{neg})} \simeq V_{B3} + V_{\text{BE7}} + I_m R_{E8} \simeq -25.2 \text{ V}$$

The corresponding maximum rms output voltage, without excessive distortion, is

$$V_{o(\max)} = \frac{V_{m(\text{pos})} - V_{m(\text{neg})}}{2\sqrt{2}} \simeq \frac{24.2 \text{ V} - (-25.2 \text{ V})}{2\sqrt{2}} \simeq 17.5 \text{ V rms}$$

(d) Maximum output power without excessive distortion:

$$P_{o(\max)} \simeq V_{o(\max)}^2/R_L = (17.5 \text{ V})^2/8 \simeq 38.2 \text{ W}$$

(e) Required current drive: The maximum or peak current required to deliver 38.2 W to an 8-ohm load is

$$I_m \simeq \sqrt{2P_o/R_L} \simeq \sqrt{(2)(38.2 \text{ W})/8} \simeq 3.1 \text{ A}$$

When the output voltage is at its maximum positive value, the required base current for Q_5 is

$$I_{b5(\max)} \simeq I_m/(\beta_5\beta_6) \simeq 3.1 \text{ A}/(60 \times 30) \simeq 1.72 \text{ mA}$$

When the output voltage is at its maximum negative value, the required base current for Q_7 is

$$I_{b7(\max)} \simeq I_m/(\beta_7\beta_8) \simeq 3.1 \text{ A}/(40 \times 30) \simeq 2.53 \text{ mA}$$

Since I_{C3} is approximately 6.15 mA, adequate current drive is available.

(f) Case and junction temperature of the output transistors at 38.2 W output power: The case and junction temperature depends on the power dissipated by each output transistor. The dc current flowing in the +30-V supply is approximately equal to the dc current of a half-rectified sine wave or I_m/π. The power taken from the +30-V supply is

$$P_{\text{in}(+30\text{V})} \simeq V_{\text{CC}}(I_m/\pi) \simeq (30 \text{ V})(3.1/\pi) \simeq 29.6 \text{ W}$$

The portion of this power delivered to the 8-ohm load is 1/2 the ac output power. The power dissipated by Q_6 is

$$P_{D6} \simeq P_{\text{in}(+30\text{V})} - (1/2)P_{\text{out(ac)}} \simeq 29.6 \text{ W} - (1/2)(38.2 \text{ W}) \simeq 10.5 \text{ W}$$

The case temperature of Q_6 is

$$T_{C6} \simeq P_{D6}(\theta_{cs} + \theta_{sa}) + T_A \simeq (10.5 \text{ W})[(0.5 + 2.2)°\text{C/W}] + 25°\text{C} \simeq 53.4°\text{C}$$

The junction temperature is

$$T_{j6} \simeq P_{D6}(\theta_{jc} + \theta_{cs} + \theta_{sa}) + T_A \simeq 60°\text{C}$$

These values should be slightly low since the stand-by current produces some dissipation in the output transistors. The case and junction temperatures of Q_8 should be the same as those of Q_6.

(g) Conversion efficiency: The total dc input power is the sum of the dc power furnished by the +30-V and −30-V supply or

$$P_{\text{in(dc)}} = P_{\text{in(+30V)}} + P_{\text{in(-30V)}} \simeq 2\,P_{\text{in(+30V)}} \simeq (2)(29.6\text{ W}) \simeq 59.2\text{ W}$$

Then the efficiency is

$$\eta = \frac{P_{\text{out(ac)}}}{P_{\text{in(dc)}}} \times 100\% \simeq \frac{38.2\text{ W}}{59.2\text{ W}} \times 100\% = 64.5\%$$

(h) Mid-band value of T: Beginning at the collector terminal of Q_3 and using the Q_7Q_8 path to the output,

$$T \simeq (A_{vt7,8})\left(\frac{R_{\text{in2}} \parallel R_{f1}}{R_{\text{in2}} \parallel R_{f1} + R_{f2}}\right)(A_{d1,2})(A_{vt3})$$

The factors in the above equation are:

$$\frac{R_{\text{in2}} \parallel R_{f1}}{R_{\text{in2}} \parallel R_{f1} + R_{f2}} \simeq \frac{2\text{ K}}{2\text{ K} + 20\text{ K}} \simeq \frac{1}{11} \qquad A_{vt7,8} \simeq 1$$

$$A_{d1,2} \simeq \frac{R_{L\text{eq}1}}{r_{e1} + r_{e2} + R_{\text{EE}}} \simeq \frac{R_{C1} \parallel R_{\text{in2}}}{r_{e1} + r_{e2} + R_{\text{EE}}} \simeq \frac{3\text{ K} \parallel 250}{26 + 26 + 200} \simeq 0.92$$

$$A_{vt3} \simeq -\frac{R_{L\text{eq}3}}{r_{e3}} \simeq -\frac{\beta_7\beta_8 R_L}{r_{e3}} \simeq -\frac{(40)(30)(8)}{4.2} \simeq -2290$$

Thus,

$$T_{\text{mid}} \simeq (1)\left(\frac{1}{11}\right)(0.92)(-2290) \simeq -189 \quad \text{or} \quad 45.5\text{ db}$$

(i) High-frequency response of T: The principal poles in the high-frequency response of T occur at the base of Q_3 and at the base of Q_2. Also, the pole at the base of Q_6 (or Q_8) is important due to the low value of f_{T6}. At the base of Q_3:

$$\begin{aligned}\tau_1 &= R_{L\text{eq}1}(C_{\text{out1}} + C_{\text{in3}}) \simeq R_{L\text{eq}1}[(C_{ob3} + C_1)(1 - A_{vt3})]\\ &\simeq (230)[9\text{ pF} + 220\text{ pF})(1 + 2290)] \simeq 120\ \mu\text{sec}\\ f_{\text{hi1}} &\simeq 1/(2\pi\tau_1) \simeq (0.159)/(120 \times 10^{-6}) \simeq 1.3\text{ kHz}\end{aligned}$$

At the base of Q_2,

$$\begin{aligned}\tau_2 &\simeq (R_{\text{in2}} \parallel R_{f1} \parallel R_{f2})(C_{\text{in2}} + C_{f2}) \simeq (1.8\text{ K})(170\text{ pF}) \simeq 0.3\ \mu\text{sec}\\ f_{\text{hi2}} &\simeq 1/(2\pi\tau_2) \simeq (0.159)/(0.3 \times 10^{-6}) \simeq 530\text{ kHz}\end{aligned}$$

The pole at the base of Q_6 is difficult to calculate. Note that R_{E5} is effectively bootstrapped since one side is connected to the output. Therefore, very little ac voltage is developed across R_{E5}, causing it to appear much larger than 100 ohms. The resistance looking into the emitter terminal of Q_5 is also large due to the large value of equivalent base resistance, $R_{o3} \parallel R_{o4}$. Thus the pole at the base of Q_6 is dominated by the response of Q_6.

$$\begin{aligned}\tau_3 &\simeq R_{\text{in6}}C_{\text{in6}} \simeq [(\beta_6 + 1)(r_{e6} + R_{E6} + R_L)][C_{ob6} + C_{b'g6}]\\ &\simeq r_{b'g6}\,C_{b'g6} \simeq 1/(2\pi f_{\beta 6})\\ f_{\text{hi3}} &= 1/(2\pi\tau_3) \simeq f_{\beta 6} = f_{T6}/\beta_6 \simeq \frac{600\text{ kHz}}{30} \simeq 20\text{ kHz}\end{aligned}$$

The feedback compensation network introduces a high-frequency zero in T at

$$\begin{aligned}\tau_{z1} &= R_{f2}C_{f2} = (20\text{ K})(160\text{ pF}) \simeq 3.2\ \mu\text{sec}\\ f_{z1} &= 1/(2\pi\tau_{z1}) \simeq (0.159)/(3.2 \times 10^{-6}) \simeq 50\text{ kHz}\end{aligned}$$

The high-frequency Bode plot of T is shown in Fig. 12–42(a). From this figure, $f_c \simeq$ 100 kHz and the P.M. $\simeq 60°$.

(j) Closed-loop, high-frequency response: Since the closed-loop pole associated with the $R_{f2}C_{f2}$ time constant, 50 kHz, is less than the value of f_c, 100 kHz, the closed-loop, high-frequency response is controlled by the feedback compensation.

$$f_{\text{hi}(f)} \simeq 1/(2\pi R_{f2}C_{f2}) \simeq 50 \text{ kHz}$$

(k) Closed-loop, low-frequency response: The low corner frequency associated with the input coupling capacitor is

$$\tau_4 \simeq (R_{if} + R_g)C_C \simeq (600 + 20 \text{ K})(12\ \mu\text{F}) \simeq 246 \text{ msec}$$

$$f_{\text{lo1}} \simeq 1/(2\pi\tau_4) = (0.159)/(246 \times 10^{-3}) \simeq 0.64 \text{ Hz}$$

The low corner frequency associated with the base of Q_2 is

$$\tau_5 = R_{f1}C_{f1} = (2 \text{ K})(6.8\ \mu\text{F}) \simeq 13.6 \text{ msec}$$

$$f_{\text{lo2}} \simeq 1/(2\pi\tau_5) \simeq (0.159)/(13.6 \times 10^{-3}) \simeq 11.7 \text{ Hz}$$

The overall low corner frequency of the amplifier is controlled by the feedback network.

(l) Experimental results: The measured bias values are shown in Fig. 12–41. A plot of the total harmonic distortion versus output power is shown in Fig. 12–42(b). The percent distortion increased in the low power range due to crossover distortion and in the high power range (over 40 W) due to peak clipping. Peak clipping started at 17.8 V rms output (39.4 W) which agrees favorably with the calculated value of 17.5 V rms output (38.2 W).

At 17.5 V rms output or 38.2 W output, the measured dc current from each power supply was 0.96 A. The corresponding conversion efficiency was approximately 66 percent as compared to the calculated value of 64.5 percent. The case temperature of both output transistors was approximately 52°C. The plot of conversion efficiency as a function of output power is shown in Fig. 12–42(c).

The measured closed-loop frequency response is shown in Fig. 12–42(d). The value of $f_{\text{hi}(f)}$ is approximately 43 kHz, and the value of $f_{\text{lo}(f)}$ is approximately 12 Hz.

Often a complementary power amplifier is used with a speaker system. In this application, a buffer network is usually placed between the output terminal of the amplifier and the speaker load. One common type of buffer network is shown in Fig. 12–43. The series R_1C network has two functions. First, this network is an RC lag circuit* and provides cascade compensation for the loop gain. Second, this network provides an ac load for the amplifier if the speaker is disconnected. The significance of this second effect can be appreciated by studying the results of removing the 10-ohm load from the amplifier shown in Fig. 12–41. Then the value of R_L would be approximately equal to $(R_{f1} + R_{f2})$ or 22 K. The load impedance for Q_3 is greatly increased causing a corresponding increase in the mid-band value of the loop gain. The net effect is just the opposite of gain compensation; i.e., a large increase in T_{mid} can result in closed-loop oscillations and possible device failure. Consequently, adding an RC lag network from the output to ground in a complementary power amplifier is good practice. Typical component values are 10 ohms and 0.1 μF.

The function of the parallel R_2L network is to isolate the speaker from the output of the amplifier. In Chap. 8 we noted that a CC stage with a capacitive load can become unstable due to a negative component of input resistance.

---

* A discussion of cascade compensation using an RC lag network may be found in Chap. 11.

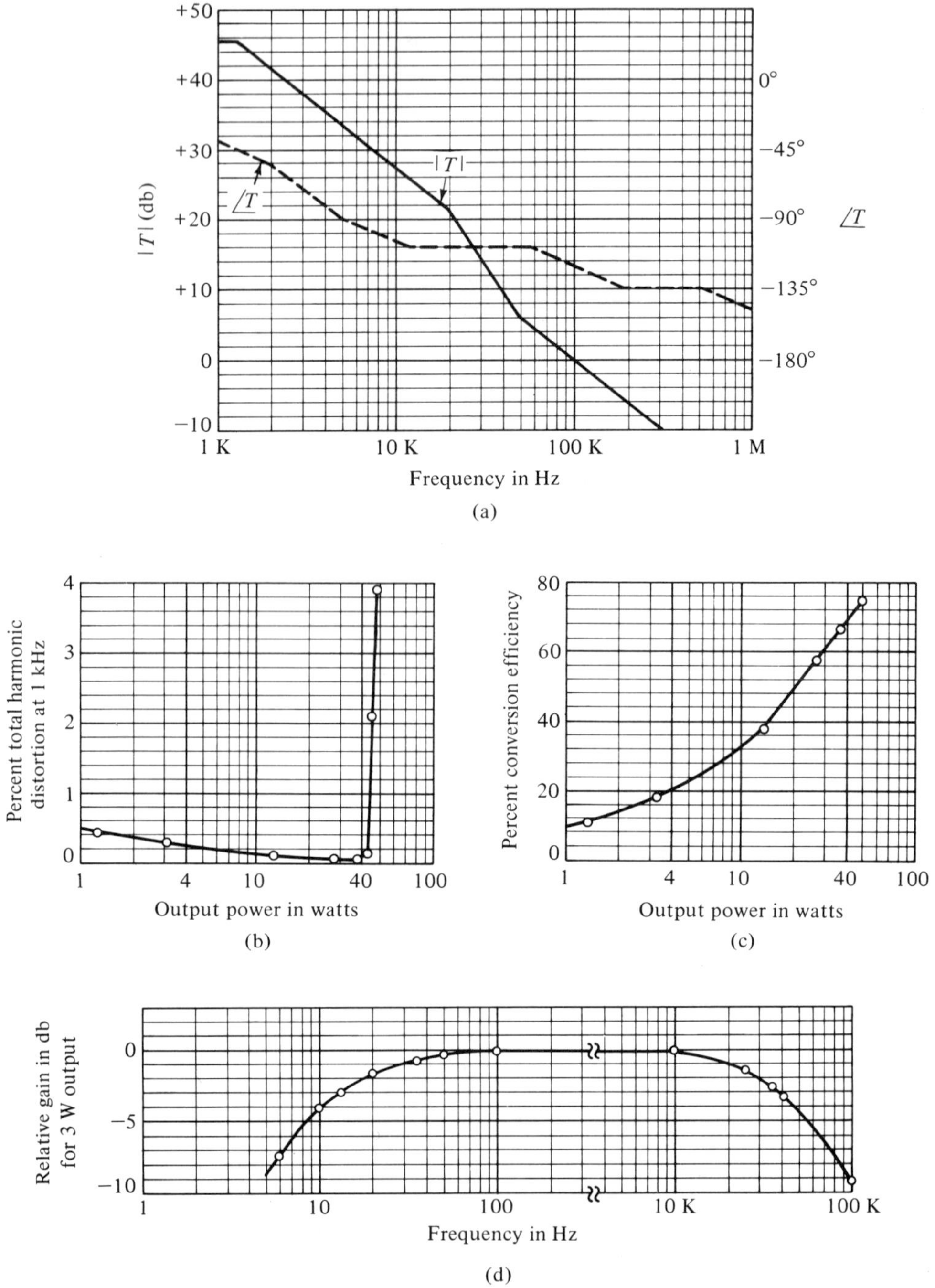

FIGURE 12–42 *Characteristics of the amplifier studied in Ex. 12–11: (a) Bode plot of* **T**, *(b) distortion as a function of output power, (c) conversion efficiency as a function of output power, and (d) the frequency response curve. (Curves (b), (c), and (d) were obtained from experimental data.)*

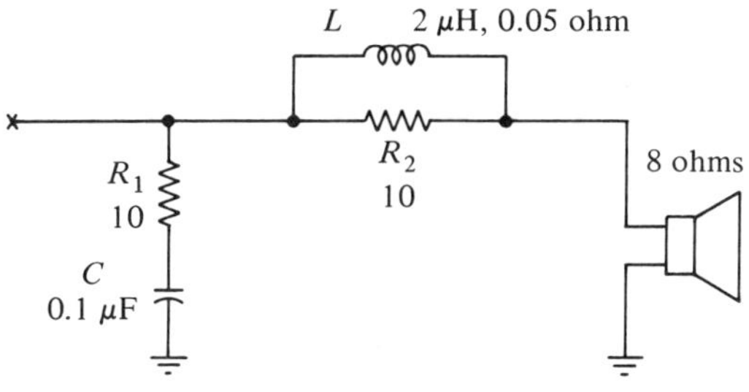

FIGURE 12–43 *A typical buffer network connected between the output of a complementary power amplifier and a loudspeaker load.*

Since the impedance of a speaker can be capacitive over a frequency range,* the coil L ensures that the load on the CC output stages is not capacitive. Often the coil is formed by winding one layer of thin enameled wire (#26) along the length of a 1-W carbon composition resistor. A typical value of R_2 is 10 ohms. A typical value of L is 2 μH.

The final topic to be discussed in this section is that of short-circuit protection. One simple network that provides short-circuit protection consisting of four diodes (D_3 through D_6) and three fuses, is shown in Fig. 12–44. Normally, the four diodes are reverse-biased and do not affect the circuit operation as shown in Fig. 12–44(a). However, if the output becomes short-circuited, the diodes limit the short-circuit current by limiting the voltage across R_{E3} (or R_{E5}). When the output voltage of the driver (Q_1) goes positive, and the output is shorted, diodes D_5 and D_6 become forward-biased and limit the positive-going drive signal. From Fig. 12–44(b) the short-circuit current is limited to

$$I_{\text{SC}} \simeq \frac{V_{RE3}}{R_{E3}} \simeq \frac{V_{D1} + V_{D2} + V_{D5} + V_{D6} - V_{\text{BE3}}}{R_{E3}} \simeq \frac{2\text{ V}}{R_{E3}} \qquad \textbf{(12–97)}$$

In a similar manner, diodes D_3 and D_4 limit the current when the output voltage of the driver goes negative, and the output is short-circuited.

The rating of each fuse must be sufficiently large so that the amplifier can operate under normal conditions but sufficiently low to provide protection in case the load is shorted. The current in the two output power transistors is essentially a half-rectified sine wave which has a dc value of I_m/π and an rms value of $I_m/2$. Since most fuses respond to rms current, the ratings of F_1 and F_2 should be slightly larger than $I_m/2$ where I_m is the peak value of the output current when maximum rated power is delivered to the load. The rating of F_3 should be slightly greater than $I_m/\sqrt{2}$, the rms value of the output current.

This section has been concerned with the characteristics of complementary power amplifiers. The principal output configurations consist of several arrangements of Darlington pairs and/or complementary Darlington pairs. Since the response of these amplifiers is often governed by negative feedback, their performance characteristics can often be approximated in a simple manner. The complete analysis of this type of amplifier was also illustrated. In addition, a typical buffer network that is placed between the output terminal of an amplifier and a speaker type load and also a simple short-circuit protection arrangement were considered.

* For the moving-coil cone loudspeaker, the coil impedance appears capacitive only over a range of frequencies, typically 70 to 200 Hz.

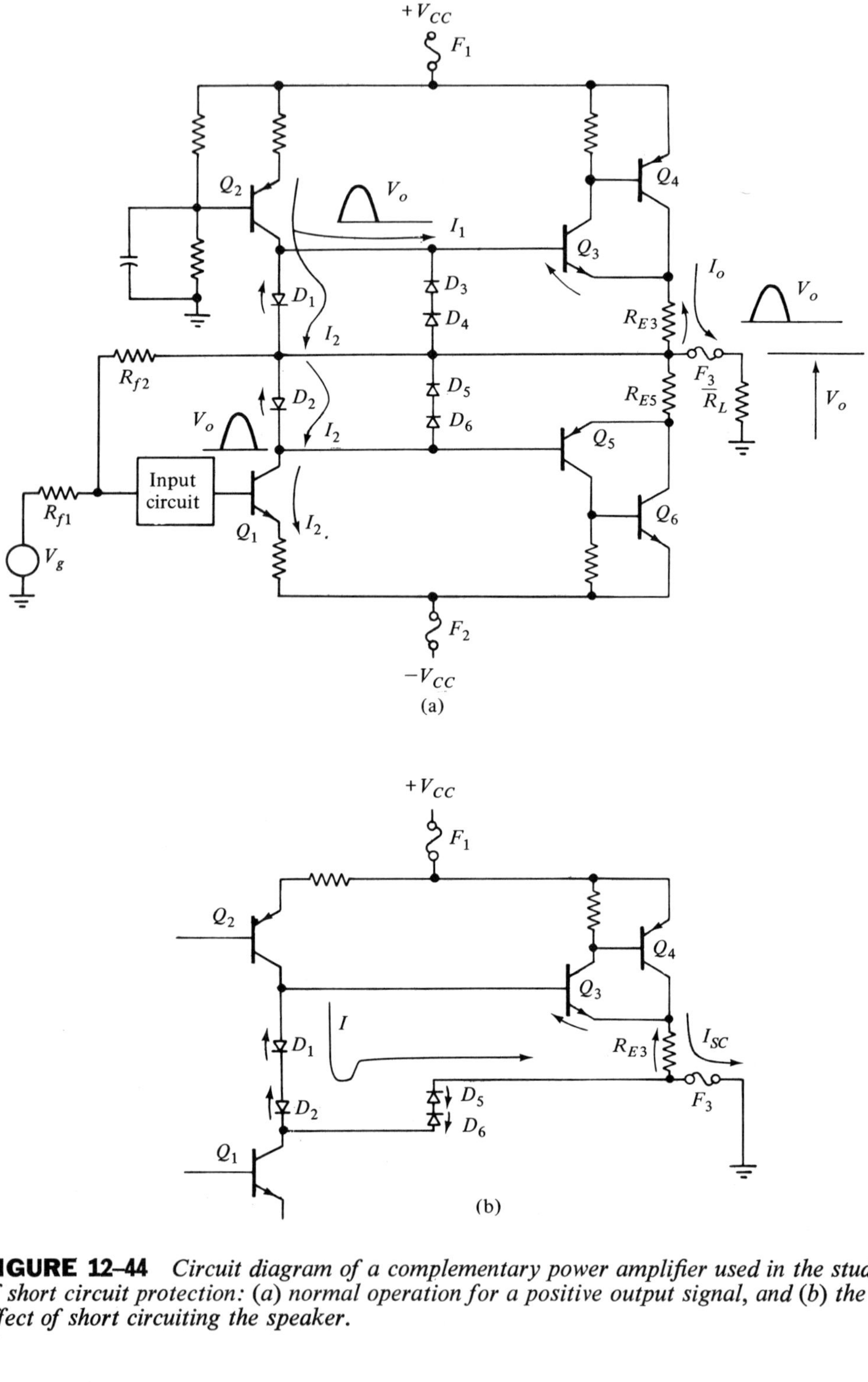

FIGURE 12–44 *Circuit diagram of a complementary power amplifier used in the study of short circuit protection: (a) normal operation for a positive output signal, and (b) the effect of short circuiting the speaker.*

12.9 IC Power Amplifiers

Although the principal uses of linear ICs in power circuits have been in low-power applications, some ICs are available that can be classified as power amplifiers. The objective of this section is to examine briefly the characteristics and limitations of IC power amplifiers. For the purpose of this study we will divide IC power amplifiers into three classifications: (1) true monolithic power amplifiers, (2) hybrid IC power amplifiers, and (3) low-power monolithic ICs operated with discrete output stages. Each class is discussed below.

Two difficulties encountered in the design of a true monolithic power amplifier are: (1) the removal of heat from the chip, and (2) the lack of NPN-PNP complementary transistors. Thermal problems generally limit the output power level to the 1- to 5-W range. To improve the thermal characteristics of monolithic circuits, metal tabs are often provided for direct connection to a heat sink.

The circuit diagram shown in Fig. 12–45 is typical of those used in many IC power amplifiers. The only PNP transistor used in the circuit is Q_3 which forms part of a complementary Darlington pair. Since the β of many monolithic PNP units is low, two NPN units are used to increase the gain in the PNP complementary Darlington pair. Note that the load resistor of Q_2 is divided into two parts, R_1 and R_2, with the center point available at an external terminal. This arrangement allows an external bootstrap capacitor to be connected between this point and the output terminal. With the development of PNP monolithic devices having high values of r_{ce} and β, resistors R_1 and R_2 could be replaced with a constant-current source.

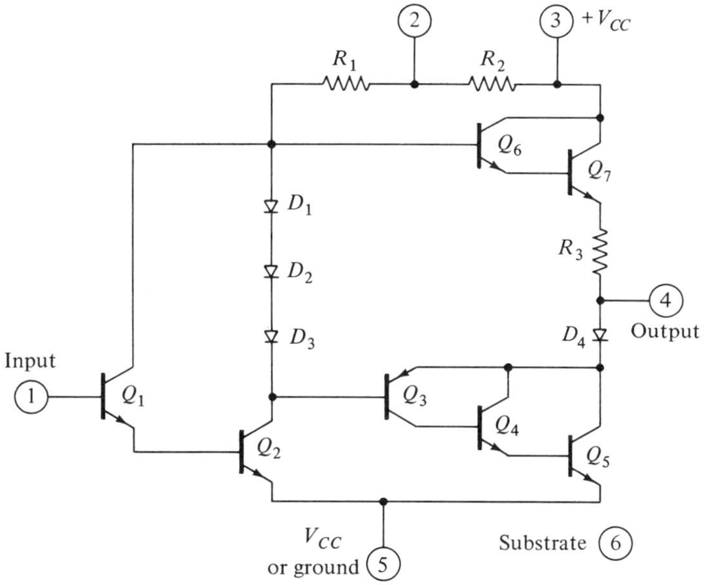

FIGURE 12–45 *Circuit diagram of a representative monolithic power amplifier.*

The performance of monolithic power amplifiers can be improved with negative feedback. Although the circuit shown in Fig 12–45 can be used in the voltage sample-current sum feedback configuration, other ICs are available which allow the voltage sample-voltage sum feedback configuration. The principal difference is that Q_1 and Q_2 are arranged as a differential pair rather than a Darlington pair. Typical performance characteristics of monolithic power amplifiers would be in the range of 2 W of output power into a 16-ohm load, 50 percent conversion efficiency, 0.1 to 1 percent total harmonic distortion at 25 percent rated output power, and 3 to 10 percent total harmonic distortion at rated output power.

A second type of IC power amplifier is the hybrid unit. As discussed in Sec. 5.9, a hybrid IC consists of several separate components mounted on the same substrate and connected internally by thick film, thin film, or wire bonding techniques. The choice of the substrate material and the design of the package are important in the removal of heat from the junction. Output power ratings in excess of 100 W rms have been achieved for units measuring less than 2.5″ × 2″ × 0.5″ (excluding leads and heat sinks). Since the output power transistors can be individually selected, complementary output circuits can be used. Thus, most of the material in Secs. 12.7 and 12.8 on complementary power amplifiers can be applied to hybrid IC power amplifiers. The principal limitation on this type of amplifier is the heat sink requirement. While the entire amplifier may be small in size, the heat sink must be sufficiently large to maintain junction temperatures below 150°C to 200°C.

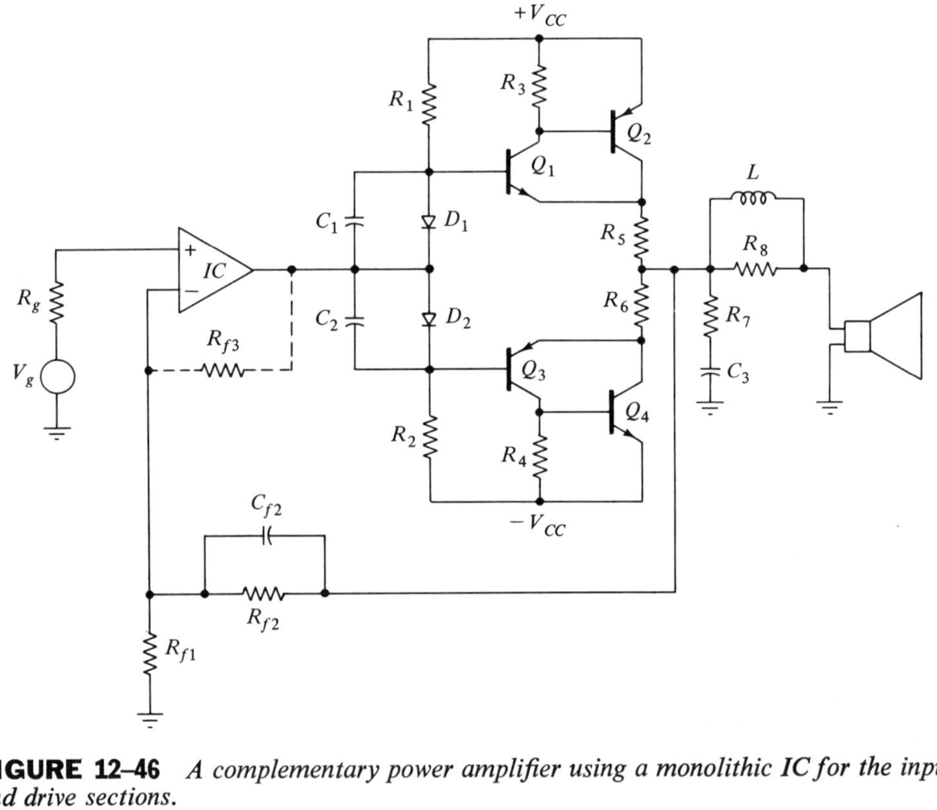

FIGURE 12–46 *A complementary power amplifier using a monolithic IC for the input and drive sections.*

A third use of ICs in power amplifiers is illustrated in Fig. 12–46 where the output or power stages are formed by discrete transistors. The IC can be a low-power operational amplifier type and can function as the input and drive sections of the amplifier. The IC must be able to swing the peak-to-peak output voltage since the voltage gain of the output section is approximately unity. However, the output current required from the IC can be small since its load impedance is large. Stated differently, the voltage gain of the circuit is provided by the IC, and the power gain is provided by the external discrete transistors.

One difficulty encountered with the circuit shown in Fig. 12–46 is high-frequency oscillation. Most low-power monolithic ICs are very high-frequency units. Although the frequencies of interest may be in the audio range, the IC can sustain oscillation in the megahertz range. When using a wide-band IC, even in low-frequency applications, the physical layout of the circuit is important. All leads should be kept as short as possible to reduce stray inductance, and power supplies should be well filtered. In addition, stability may be improved by limiting the value of the open-loop gain of the IC by an external feedback loop through R_{f3}.

In this section some characteristics and limitations of IC power amplifiers have been outlined. True monolithic power amplifiers can generally be used in any application suitable for Class A operation. Hybrid IC power amplifiers are limited by the heat-sink requirements and cost.

12.10 DC Regulated Power Supplies

The successful operation of many electronic circuits requires the availability of a dc power supply with a well regulated output voltage. The supply voltage should not change as the current drain changes from a few milliamps to several amperes. The simple diode power supply with a shunt-capacitor-type filter discussed in Chap. 2 is generally not adequate for use with a high-performance power amplifier. The objective of this section is to study the characteristics and configurations of some regulated dc power supplies suitable for use with high-performance power amplifiers.

A block diagram of a regulated power supply is shown in Fig. 12–47. Both the output voltage and the output current can be sampled and the data used to control the *pass* element. Generally, the pass element consists of a power transistor whose base current is varied to control the output voltage. The voltage sampling circuit is generally one form of a voltage sample-voltage sum feedback amplifier, and the current sampling circuit is generally a resistor, since the voltage

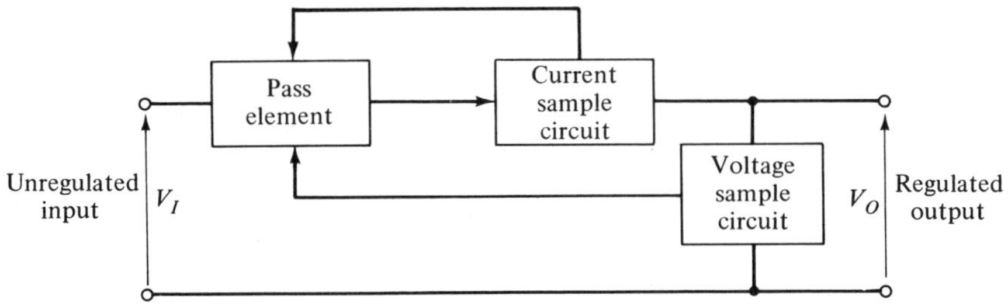

FIGURE 12–47 *Block diagram of a regulated power supply.*

drop across it is proportional to the output current. Our next task is to examine some typical configurations used to regulate the output voltage. Current regulating circuits are considered later in this section.

The regulator circuit whose block diagram is shown in Fig. 12–48(a) is called a *series regulator* since the pass element,* Q_1, appears in series with the load current. This diagram shows that the amplifier A and transistor Q_1 form a voltage sample-voltage sum feedback amplifier whose gain is determined by $(R_1 + R_2)/R_1$. For proper operation, the unregulated input dc level, V_I, must be greater than the regulated output dc level, V_O, so that the base-collector junction of Q_1 is reverse-biased. The equilibrium value of V_O is at a value so that the feedback signal, V_O', equals the reference signal, V_{REF}. Often V_{REF} is obtained across a Zener or reference diode.

The amplifier A can have numerous configurations. Some precision regulated power supplies use high performance differential amplifiers similar to the IC shown in Fig. 10–23. Also, the simple differential amplifier shown in Fig 12–48(b) is used. In this circuit, if V_O increases slightly, the value of V_{BE1} decreases, thus reducing I_L and V_O. An even simpler circuit is shown in Fig. 12–48(c) where the reference diode is placed in the emitter circuit of Q_2. Integrated circuits are available that combine both V_{REF} and A in a single package. This type of IC can then be connected to a pass transistor to form a precision voltage regulator.

One problem encountered with a regulated power supply is that of short-circuit protection. The ideal characteristic of a regulated power supply is to provide a constant output voltage at any current level. However, the power dissipation rating of the pass transistor places an upper limit on the load current. Thus, the load current must be limited to some finite value even under short-circuit conditions. Our next task is to examine circuits that can provide short-circuit protection for a regulated power supply.

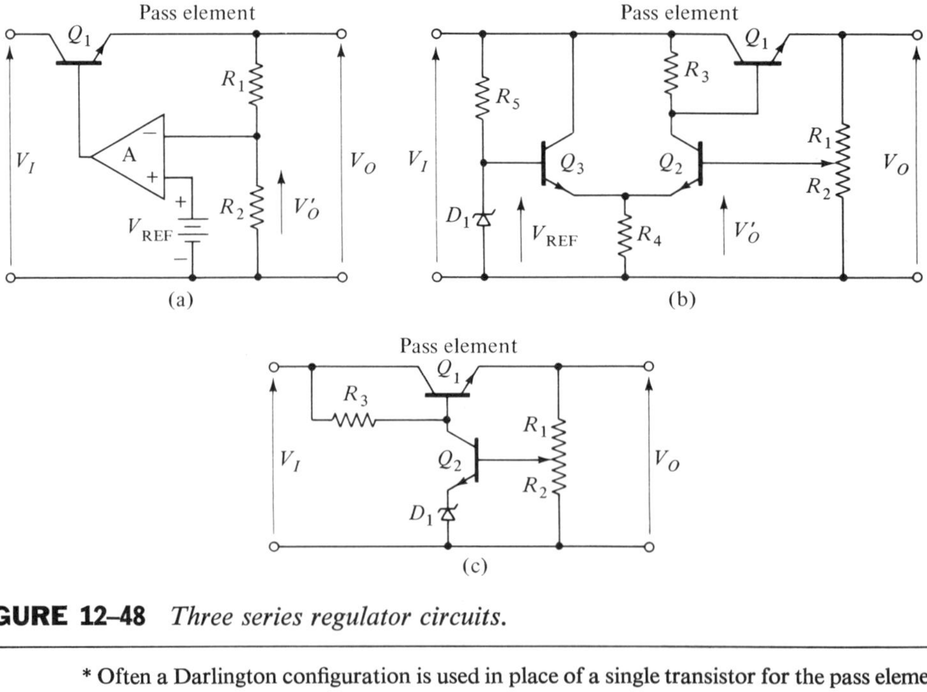

FIGURE 12–48 *Three series regulator circuits.*

* Often a Darlington configuration is used in place of a single transistor for the pass element.

Two circuits that can be used to limit the short-circuit current in a regulated power supply are shown in Fig. 12–49. In each case, a small resistor is connected in the circuit so that the load current flows through it. As the load current increases, the voltage drop across R_S increases. In Fig. 12–49(a), this voltage tends to forward bias D_2 and D_3 while in Fig. 12–49(b), the BE junction of Q_3 is forward-biased. In either case, the base current in Q_1 is reduced as the voltage drop across R_S increases. In Fig. 12–49(a), the short-circuit current is limited to $(V_{D2} + V_{D3} - V_{BE1})/R_S$. Varying the value of R_S provides a simple means of controlling the short-circuit current.

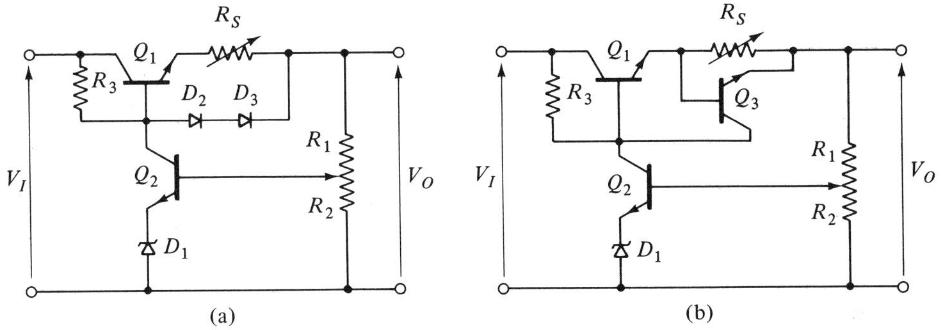

FIGURE 12–49 *Two circuits used to limit the output current of a regulated power supply.*

The circuit diagram of a complete dc regulated power supply is shown in Fig. 12–50. Note that the pass transistor is formed by the Darlington pair, Q_1 and Q_2. The regulated output voltage can be varied by means of the potentiometer R_{12}. The short-circuit current is determined by the value of R_S.

In this section we have examined the characteristics and some of the configurations found in regulated power supplies. In this study only transistor circuits were considered although similar vacuum tube circuits are sometimes found in practice, particularly in high-voltage applications. The improved performance of a regulated supply over an unregulated supply results from the use of negative feedback.

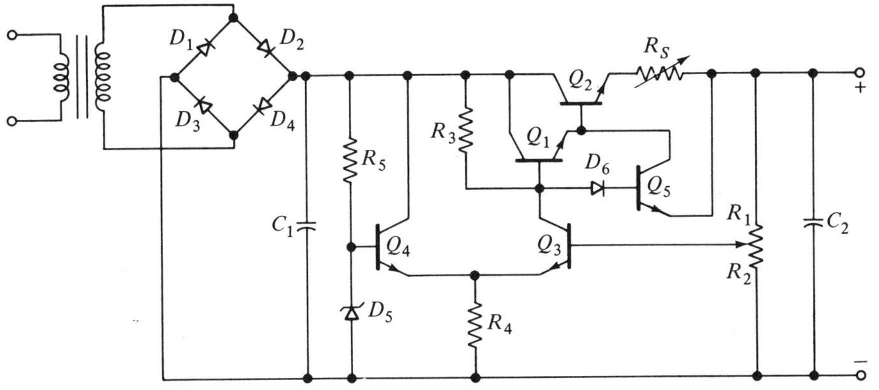

FIGURE 12–50 *A regulated power supply with short-circuit protection.*

12.11 Summary

The material in this chapter has been concerned with various aspects of the characteristics, application, analysis, and design of audio power amplifiers. Audio power amplifiers are an important part of sound reproduction systems along with the preamplifier and the speaker. Equalizing networks, such as bass boost (or cut) and treble cut (or boost) are also part of most sound reproduction systems. The characteristic that distinguishes an audio power amplifier from other amplifiers is its ability to deliver relatively large values of voltage and current to small values of load impedance (typically 8 ohms) in the audio frequency range.

Often an audio power amplifier can be formed by adding a power output stage to a small-signal amplifier. The transistors used in the output stage must be capable of handling relatively large values of voltage and current. The problem of heat sinks is also important since transistors are particularly susceptible to damage from heat.

Power amplifiers are generally divided into several classes of operation that depend on the bias conditions. The ideal characteristics of the Classes A, AB, B, and C were studied to determine the theoretical limits on their performance characteristics. Ideally, a power amplifier should have a high value of dc to ac conversion efficiency (η) and a low value of distortion. Of particular importance is the combined operation of two or more active devices in Class AB operation in either the push-pull or the complementary configurations.

Several types of distortion are important in a power amplifier including harmonic, crossover, and intermodulation. Each type of distortion results from a nonlinearity in the system. Negative feedback tends to reduce these nonlinearities.

The choice of the proper type or class of audio amplifiers depends on the application. Power output levels less than a few watts can be achieved with a Class A arrangement or with an IC. Larger values of output power generally require push-pull or complementary operation. Also, to obtain good performance, the larger power levels require that a well regulated dc power supply be used.

Problems

Section 12.1

12–1 A speech is to be given in a room measuring 50′ × 40′ × 10′. Estimate the electrical power that must be supplied to an 8-ohm moving-coil speaker from a power amplifier for the speech to be heard throughout the room.

12–2 A capacitive microphone has a capacitance of 75 pF. What is the minimum value of the preamplifier input resistance that will allow acceptable operation from 10 Hz to 10 kHz?

12–3 Choose a set of component values for the circuit shown in Fig. 12–5(a) that will result in the 15 in./sec NAB tape equalization curve. Assume that R_L is 10 K.

12–4 Choose a set of component values for the circuit shown in Fig. 12–6(d) that will result in the standard RIAA equalization response for magnetic phonograph cartridges.

12–5 Discuss the operation of the circuit shown in Fig. 12–6(e) in terms of the variation in the reactances of C_1, C_2, and C_3 as a function of frequency.

Section 12.2

12–6 A transistor is operated in the CE configuration with a V_{CC} of 24 V, a V_{CE} of 12 V, and an R_L of 8 ohms. (a) Estimate the minimum safe voltage, current, and power dissipation ratings of the transistor. (b) If the transistor is a silicon unit in a TO-3 case, can it operate without a heat sink? Use circuit shown in Fig. 12–15 (a).

12–7 What maximum power can the transistor studied in Ex. 12–2 dissipate with the heat sink connected?

12–8 Repeat Ex. 12–3 for the case in which the transistor is operated at an average collector current of 3 A and an average collector voltage of 10 V.

Section 12.3

12–9 Repeat Ex. 12–4 for 10 W output power.

12–10 A push-pull amplifier of the type shown in Fig. 12–18 is required to furnish an output power of 20 W. (a) Estimate the power that each transistor must dissipate. (b) Repeat for the complementary amplifier shown in Fig. 12–19.

12–11 The output waveform of the Class B complementary amplifier shown in Fig. 12–19 is not a perfect sine wave. How can the circuit be modified to produce a more nearly sine wave output signal for a sine wave input signal? How will the modification affect the conversion efficiency?

Section 12.4

12–12 Repeat Ex. 12–5 for the case of an input voltage having a peak amplitude of 0.5 V swinging around a bias of 0.5 V dc.

12–13 Repeat Ex. 12–6 using the transfer characteristic curve shown in Fig. 12–22(d).

12–14 If the transfer characteristic curve of a transistor is linear (I_C versus V_{BE}) for all values of V_{BE} greater than zero, which of the three types of distortion discussed in this section would be present?

Section 12.5

12–15 In Fig. 12–24, let $R_1 = 220$ ohms, $R_2 = 3.9$ K, $R_E = 10$ ohms, $C_E = 150\ \mu$F, $V_{CC} = 36$ V, $R_L = 8$ ohms, and $R_g = 100$ ohms. The transistor characteristics are: $\beta = 100$, $BV_{CE(sus)} = 80$ V, $T_{j(max)} = 200$°C, $\theta_{jc} = 20$°C/W, $I_{CO} = 0.25\ \mu$A at $T_C = 25$°C, $I_{C(max)} = 0.8$ A, and $P_D = 10$ W at $I_C = 25$°C. The transformer characteristics are: $\eta = 75$ percent, impedance ratio $= 200/8$, $R_{pri} = 10$ ohms, and $R_{sec} = 1$ ohm. The heat sink is characterized by $\theta_{sa} = 5$°C/W. Calculate the values of $P_{o(max)}$, η, T_j, and T_C for this amplifier for an ambient temperature range of 25°C to 50°C.

12–16 Design the circuit shown in Fig. 12–24 to satisfy the specifications given in Eq. (12–41) using a 90-V power supply.

12–17 What modifications would be necessary to convert the amplifier shown in Fig. 12–25 into one capable of delivering 2 W to an 8-ohm load?

Section 12.6

12–18 Redesign the amplifier shown in Fig. 12–30 so that transformers X_1 and X_2 can be eliminated.

12–19 What modifications are necessary to convert the amplifier shown in Fig. 12–30 into one capable of delivering 50 W to an 8-ohm load?

12–20 The vacuum tube power amplifier shown in Fig. 12–31 is to be converted into a solid state amplifier of the same configuration and having approximately the same performance characteristics. (a) Draw the modified amplifier making any necessary bias changes, (b) calculate the transistor power, current, and voltage ratings, and (c) calculate the required transformer turns ratios.

12–21 For the amplifier shown in Fig. 12–51, (a) discuss the configuration and function of each component, (b) estimate the maximum power that can be delivered to the load with low distortion, (c) estimate the values of input impedance, output impedance, and voltage gain, and (d) estimate the closed-loop, high- and low-frequency responses.

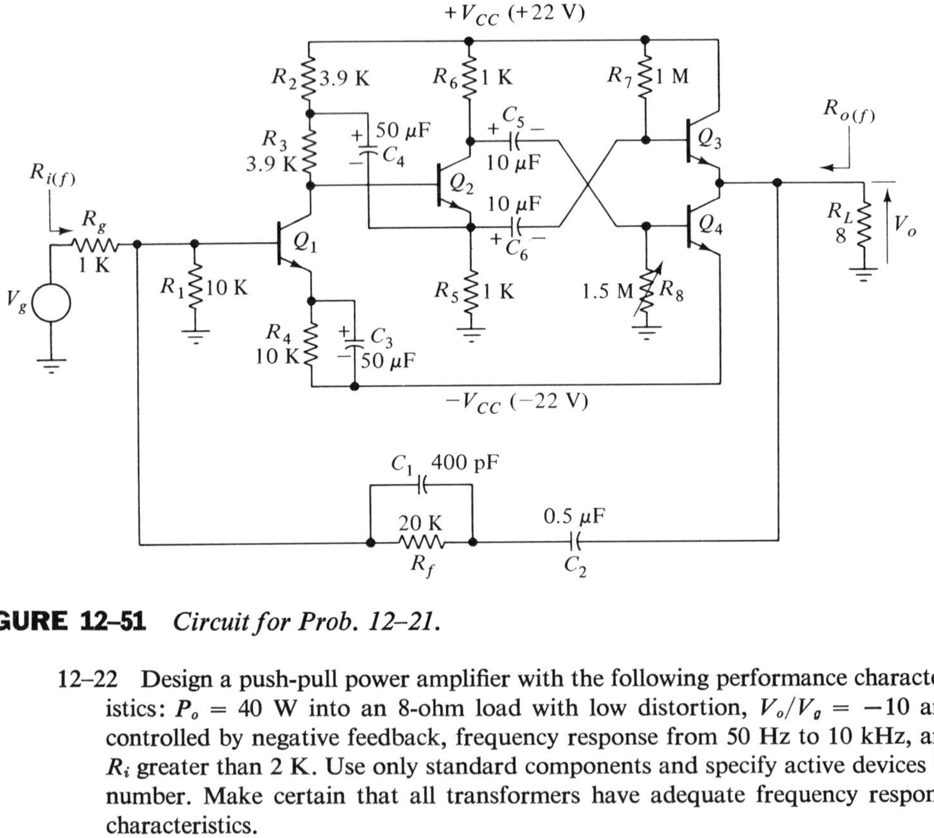

FIGURE 12–51 *Circuit for Prob. 12–21.*

12–22 Design a push-pull power amplifier with the following performance characteristics: $P_o = 40$ W into an 8-ohm load with low distortion, $V_o/V_g = -10$ and controlled by negative feedback, frequency response from 50 Hz to 10 kHz, and R_i greater than 2 K. Use only standard components and specify active devices by number. Make certain that all transformers have adequate frequency response characteristics.

Section 12.7

12–23 Discuss the operation and characteristics (gain, input impedance, output impedance, etc.) of the White Emitter Follower circuit shown in Fig. 12–52(a).

12–24 Repeat Prob. 12–23 for the White Cathode Follower circuit shown in Fig. 12–52(b).

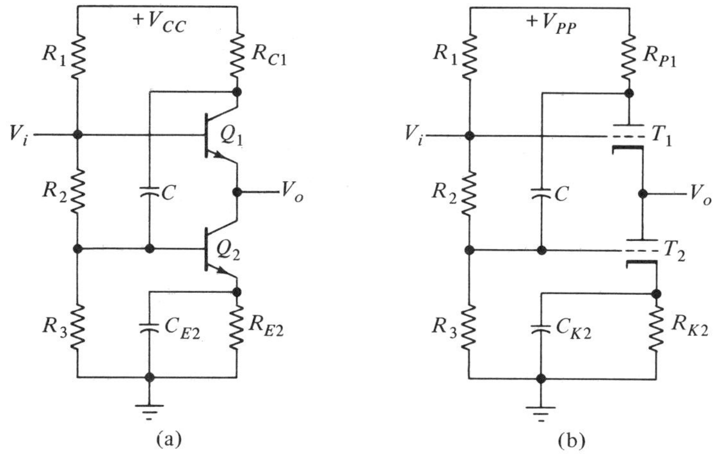

FIGURE 12–52 *Circuits for (a) Prob. 12–23 and (b) Prob. 12–24.*

12–25 Modify the circuit shown in Fig. 12–33 using a bootstrap load for Q_1 and complementary Darlington pairs for both the NPN and the PNP output transistors. Estimate the equivalent load impedance of Q_1.

12–26 Calculate the value of V_{CC} required by the amplifier shown in Fig. 12–37 to deliver 100 W to an 8-ohm load. Also calculate the values and dissipation ratings of R_{E2} and R_{E3}.

Section 12.8

12–27 Modify the amplifier shown in Fig. 12–38 by replacing the bootstrap network with a constant-current type load for Q_1.

12–28 In terms of the device physics, why will forward biasing the base-collector junction of a bipolar transistor render it useless as a linear amplifier?

12–29 Calculate the value of T_{mid} for the amplifier shown in Fig. 12–38 assuming all transistors have a β equal to 50.

12–30 The amplifier shown in Fig. 12–39 is operated over a temperature range of 25°C to 50°C. Calculate the heat sink requirements for the four output transistors if $\theta_{jc6,8}$ equals 15°C/W and $\theta_{jc7,9}$ equals 1°C/W. Let all βs equal 50.

12–31 Estimate the values of V_o/V_g, R_i, R_o, $P_{o(\max)}$, $f_{\text{hi}(f)}$, and $f_{\text{lo}(f)}$ for the amplifier shown in Fig. 12–53. Also discuss the function of each component in the amplifier.

12–32 For the amplifier shown in Fig. 12–53, calculate the values of T_{mid}, η, θ_{ca}, R_{E4}, R_{E5}, R_{E8}, and R_{E9} for proper operation. All active devices are silicon, values of β_1 through β_5 are 100 and β_6 through β_9 are 50, $\theta_{jc6,7}$ is 10°C/W, and $\theta_{jc8,9}$ is 0.9°C/W.

12–33 Design a power amplifier with the following specifications: $V_o/V_i = +50$ at midband, $V_o/V_i = +1$ at dc, $R_i \geq 20$ K, bandwidth extends from 10 Hz to 20 kHz and is determined by the feedback network, $|T| > 40$ db for the bandwidth from 10 Hz to 10 kHz, 80 W rms delivered to an 8-ohm load, 100 W rms delivered to a 4-ohm load, and $\eta \geq 50$ percent. Ensure that the amplifier is stable against self-oscillation and can operate in ambient temperatures ranging from 20°C to 40°C. Provide an output buffer network and short-circuit protection.

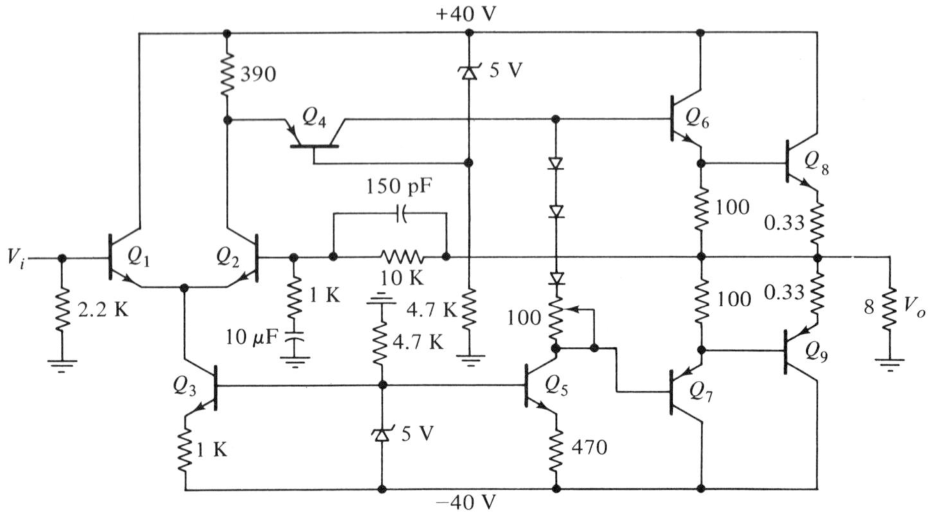

FIGURE 12–53 *Circuit for Probs. 12–31 and 12–32.*

Section 12.9

12–34 What difficulty would be encountered in fabricating the monolithic IC shown in Fig. 12–45 with a constant current source in place of R_1 and R_2?

12–35 Sketch the circuit diagram of a power amplifier which is suitable for monolithic fabrication and which uses voltage sample-voltage sum negative feedback.

Section 12.10

12–36 Discuss the function of each component in Fig. 12–50 and the general operation of the circuit.

12–37 A short circuit is suddenly applied to the output terminals of the power supply shown in Fig. 12–50. How is the output current limited?

References

Bongiorno, J., "Crossover Distortion in Transistor Power Amplifiers," *Audio*, 53, No. 5 (May, 1969), 26–30.

Greenburg, R., "Selecting Commerical Power Transistor Heat Sinks," Application Note AN-135. Phoenix, Arizona: Motorola Semiconductor Products, Inc., 1968.

Haas, G. C., "Design Factors and Considerations in Full Complementary Symmetry Audio Power Amplifiers," *Journal of the Audio Engineering Society*, 16, No. 3 (July, 1968), 321–326.

Langford-Smith, F., ed., *Radiation Designer's Handbook*. Sydney, Australia: Amalgamated Wireless Valve Co. Ptg., Ltd., 1953.

Lindmayer, J. and C. Y. Wrigley, *Fundamentals of Semiconductor Devices*. New York: D. Van Nostrand Company, Inc., 1965.

Littauer, R., *Pulse Electronics*. New York: McGraw-Hill Book Company, 1965.

Pierce, J. F., *Transistor Circuit Theory and Design*. Columbus, Ohio: Charles E. Merrill Publishing Company, 1963.

"RCA Silicon Power Circuits Manual." Harrison, New Jersey: Radio Corporation of America, 1967.

"Semiconductor Power Circuits Handbook." Phoenix, Arizona: Motorola Semiconductor Products, Inc., 1968.

Stover, U. A., ed., *Circuit Design for Audio, AM/FM, and TV*. New York: McGraw-Hill Book Company, 1967.

White, E. L. C., "Thermionic Valve Amplifier Circuit Arrangement." Great Britain Patent No. 564,250, Sept. 20, 1944.

APPENDIX A

Two-Port Networks

Most methods of analysis used for linear electronic circuit calculations are based on the theory of two-port networks. The objective of this appendix is to review briefly the most important characteristics of two-port circuit theory.

A block diagram of a two-port network is shown in Fig. A–1. The assumed positive direction of the input and output voltages are indicated. Assuming a linear system, the four terminal quantities, v_1, v_2, i_1, and i_2 can be used to describe the performance of the two-port network regardless of its configuration. Two of the four terminal quantities are chosen as the independent variables. The remaining two terminal quantities become the dependent variables and are related to the independent variables by various parameters.

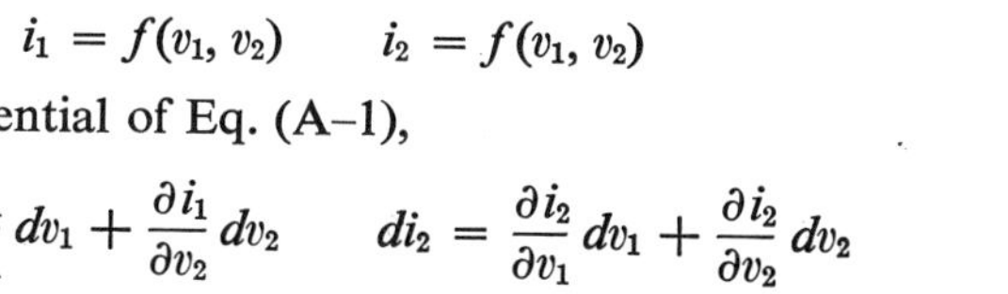

FIGURE A–1 *A two-port network.*

When v_1 and v_2 are chosen as the independent variables, i_1 and i_2 can be expressed as

$$i_1 = f(v_1, v_2) \qquad i_2 = f(v_1, v_2) \tag{A–1}$$

Forming the total differential of Eq. (A–1),

$$di_1 = \frac{\partial i_1}{\partial v_1} dv_1 + \frac{\partial i_1}{\partial v_2} dv_2 \qquad di_2 = \frac{\partial i_2}{\partial v_1} dv_1 + \frac{\partial i_2}{\partial v_2} dv_2 \tag{A–2}$$

If we consider the signals as ac quantities and assume operation around a Q-point in the linear region of an active device that makes up the two-port, then the differentials in Eq. (A–2) can be replaced with ac quantities and the partial derivatives with constants having the units of admittance. Then Eq. (A–2) becomes

$$I_1 = y_i V_1 + y_r V_2 \qquad I_2 = y_r V_1 + y_o V_2 \tag{A–3}$$

The two-port y-parameters y_i, y_r, y_f, and y_o are called the *short-circuit admittance parameters*. Their values can be determined experimentally by measuring the appropriate I/V ratio with the appropriate voltage, either V_1 or V_2, equal to zero. For example, from Eq. (A–3), y_i is I_1/V_1 when V_2 is zero (V_2 short-circuited) and corresponds to the input admittance of the two-port network. From this discussion, the admittance parameters can be summarized as follows:

$$\begin{aligned} y_i &= \left.\frac{I_1}{V_1}\right|_{V_2=0} = \text{short-circuit input admittance} \\ y_r &= \left.\frac{I_1}{V_2}\right|_{V_1=0} = \text{short-circuit reverse transfer admittance} \\ y_f &= \left.\frac{I_2}{V_1}\right|_{V_2=0} = \text{short-circuit forward transfer admittance} \\ y_o &= \left.\frac{I_2}{V_2}\right|_{V_1=0} = \text{short-circuit output admittance} \end{aligned} \tag{A–4}$$

Note from Eq. (A–4) that when I_2 corresponds to I_d, and V_1 corresponds to V_{gs}, y_f corresponds to g_m for an FET.

When I_1 and V_2 are chosen as the independent variables, a new set of parameters are derived to describe the characteristics of the two-port networks shown in Fig. A–1. The first step in the derivation is:

$$v_1 = f(i_1, v_2) \qquad i_2 = f(i_1, v_2) \tag{A–5}$$

Taking the total differential,

$$dv_1 = \frac{\partial v_1}{\partial i_1} di_1 + \frac{\partial v_1}{\partial v_2} dv_2 \qquad di_2 = \frac{\partial i_2}{\partial i_1} di_1 + \frac{\partial i_2}{\partial v_2} dv_2 \tag{A–6}$$

Assuming ac signals and linear operation,

$$V_1 = h_i I_1 + h_r V_2 \qquad I_2 = h_f I_1 + h_o V_2 \tag{A–7}$$

The two-port parameters h_i, h_r, h_f, and h_o are called the *hybrid parameters* because of the mixed units (ohms, mhos, and pure numbers). From this discussion, the hybrid parameters are:

$$\begin{aligned} h_i &= \left.\frac{V_1}{I_1}\right|_{V_2=0} = \text{short-circuit input impedance} \\ h_r &= \left.\frac{V_1}{V_2}\right|_{I_1=0} = \text{open-circuit reverse voltage ratio} \\ h_f &= \left.\frac{I_2}{I_1}\right|_{V_2=0} = \text{short-circuit forward current gain} \\ h_o &= \left.\frac{I_2}{V_2}\right|_{I_1=0} = \text{open-circuit output admittance} \end{aligned} \tag{A–8}$$

Note from Eq. (A–8) that for the CE stage at mid-band, h_f corresponds to β, h_o corresponds to $1/r_{ce}$, and h_i corresponds to $(r_{b'} + r_{b'e})$. The CE parameters are generally indicated by a second subscript e as h_{ie}, h_{re}, h_{fe} (or β), and h_{oe}. Similarly, the second subscript b is used for the CB configuration (h_{fb} is $-\alpha$), and c is used for the CC configuration (h_{fc} is $-[\beta + 1]$).

Four other sets of parameters can be developed to describe the small-signal performance of a two-port network. However, they are omitted from this summary.

In summary, active networks as well as passive networks can be represented by two-port networks. The only requirements are that the active network be linear and satisfy the equations for the two-port. For example, the *h*-parameter equivalent circuit shown in Fig. A–2(a) satisfies Eq. (A–7) while the *y*-parameter equivalent circuit shown in Fig. A–2(b) satisfies Eq. (A–3).

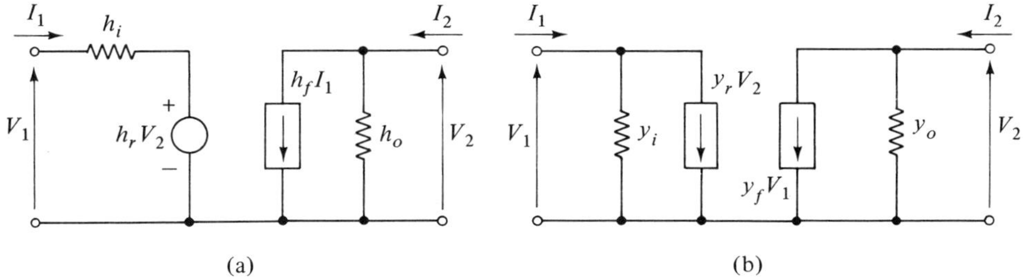

FIGURE A–2 *Two-port equivalent circuits for active networks: (a) using* h*-parameters, and (b) using* y*-parameters.*

APPENDIX B

Bode Plots

Transfer functions, gains, input impedances, and output impedances are often expressed as the ratio, $G(s)$, of two polynomials in the complex variable s as

$$G(s) = N(s)/D(s) \tag{B–1}$$

where $N(s)$ is the numerator polynomial, and $D(s)$ is the denominator polynomial. The roots of $N(s)$ are called *zeros* of $G(s)$, and the roots of $D(s)$ are called *poles* of $G(s)$. For real frequencies,

$$s = j\omega = j2\pi f \tag{B–2}$$

Often, $N(s)$ and $D(s)$ can be represented in factored form. A general polynomial may have four basic factors: (1) a constant, (2) a root at the origin, (3) a simple root, and (4) a complex conjugate pair of roots or a quadratic term.

One convenient method for graphically representing a function of the form of Eq. (B–1) is the Bode plot. A complete Bode plot consists of a magnitude plot and a phase plot as a function of frequency on a semilogarithmic scale. The magnitude is expressed in db where

$$G \text{ in db} = 20 \log |G| \tag{B–3}$$

and G is a voltage ratio, current ratio, or voltage-current mixed ratio. The coefficient in Eq. (B–3) changes to 10 when G is a power ratio. The phase of G is given in degrees. In this appendix we will examine the four basic terms listed above when they appear in either $N(s)$ or $D(s)$.

The Bode plot of a constant is shown in Fig. B–1. Both the magnitude plot and the phase plot are independent of frequency. For constants greater than unity, G is greater than 0 db. For constants less than unity, G has negative db values since the logarithm of a number can be expressed as minus the logarithm of the reciprocal of that number. The phase of a constant is either 0° or 180° depending on the sign of the constant. For example, the Bode plot of the constant 0.1 would consist of a magnitude term equal to −20 db and a phase equal to 0°. The Bode plot of the constant −10 would consist of a magnitude term equal to +20 db and a phase angle of 180°.

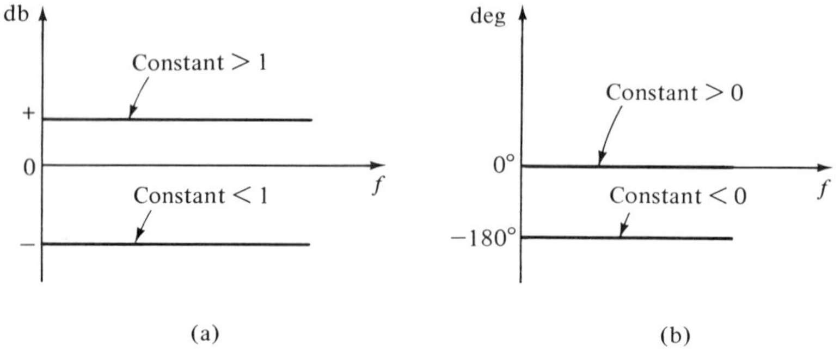

FIGURE B–1 *Bode plot of a constant: (a) magnitude and (b) phase.*

A zero at the origin can be expressed as (jf/f_1) and a pole as $1/(jf/f_1)$. Multiple roots at the origin can also occur and have the form $(jf/f_1)^n$ or $1/(jf/f_1)^n$. The magnitude plot of a zero or pole at the origin forms a straight line on a semilogarithmic scale as shown in Fig. B–2(a). The magnitude plot passes through 0 db at f equal to f_1 since $20 \log |1|$ is 0 db. The slope of the magnitude plot is $\pm(20n)$ db/decade* where n is the order of the root; the positive sign is associated with a zero, and the negative sign is associated with a pole. The phase plot is independent of frequency and has a value of $\pm n$ (90°) as shown in Fig. B–2(b).

A simple pole of $G(s)$ corresponds to a simple root of $D(s)$. A simple pole can be expressed in several different forms including: (1) $1/(a + s)$, (2) $1/(1 + \tau s)$, (3) $1/(1 + j\omega/\omega_1)$, and (4) $1/(1 + jf/f_1)$. The four forms are essentially equivalent since they are different only by a constant or the replacement of s by $j\omega$. The second and fourth forms of the pole find the widest use in this text.

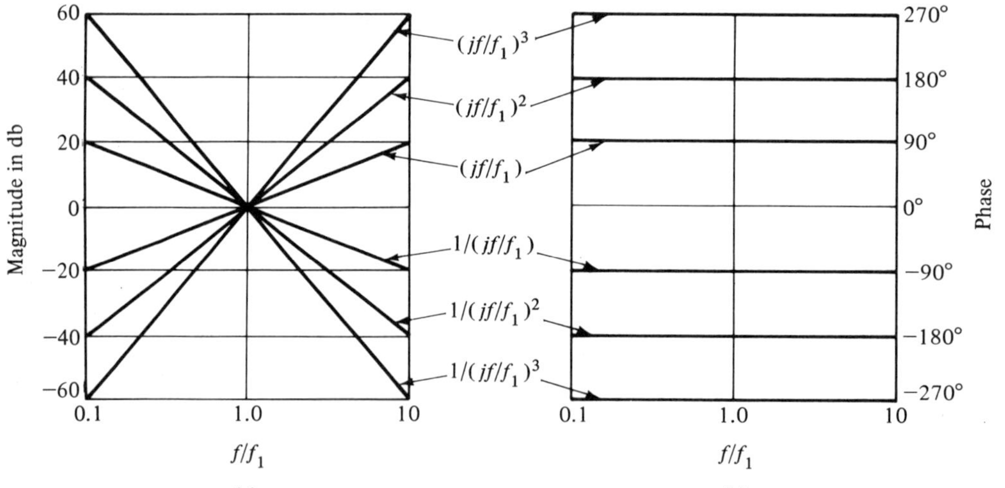

FIGURE B–2 *Bode plot of poles and zeros at the origin: (a) magnitude and (b) phase.*

* A decade in frequency corresponds to a factor of 10; 10 kHz is one decade above 1 kHz and two decades below 1 MHz. An octave in frequency corresponds to a factor of 2; 10 kHz is one octave above 5 kHz and two octaves below 40 kHz. A slope of 20 db/decade is approximately the same as a slope of 6 db/octave.

The Bode plot of a pole of the form

$$G = 1/(1 + jf/f_1) \qquad \textbf{(B–4)}$$

is shown in Fig B–3. For values of f much less than f_1, G is approximately equal to unity, and the magnitude plot becomes a straight line of 0 db value and 0 db/decade slope. This line is the low-frequency asymptote of the magnitude plot. The phase in the low-frequency region is 0°. For values of f much greater than f_1, G is approximately equal to $1/(jf/f_1)$, and the magnitude plot becomes a straight line with a −20 db/decade slope. This line is the high-frequency asymptote of the magnitude plot. The phase in the high-frequency region is −90°. For values of f near f_1, the magnitude and phase of G are found using Eq. (B–4). Some values of G in this intermediate frequency range are listed in Table B–1.

TABLE B–1 *A comparison between the exact and the approximate magnitude and phase responses of a simple pole given by Eq. (B–4).*

f/f_1	Actual Value of G (db)	Asymptote Value of G (db)	Magnitude Correction Factor (db)	Actual Value of $\angle G$	Approximate Value of $\angle G$	Phase Correction Factor
0.01	0	0	0	0°	0°	0°
0.1	−0.04	0	≃0	−5.7°	0°	−5.7°
0.5	−1	0	−1	−26.6°	−31.6°	+5°
1	−3	0	−3	−45.0°	−45.0°	0°
2	−7	−6	−1	−63.4°	−58.4°	−5°
10	−20.04	−20	≃0	−84.3°	−90°	+5.7°
100	−40	−40	0	−90°	−90°	0°

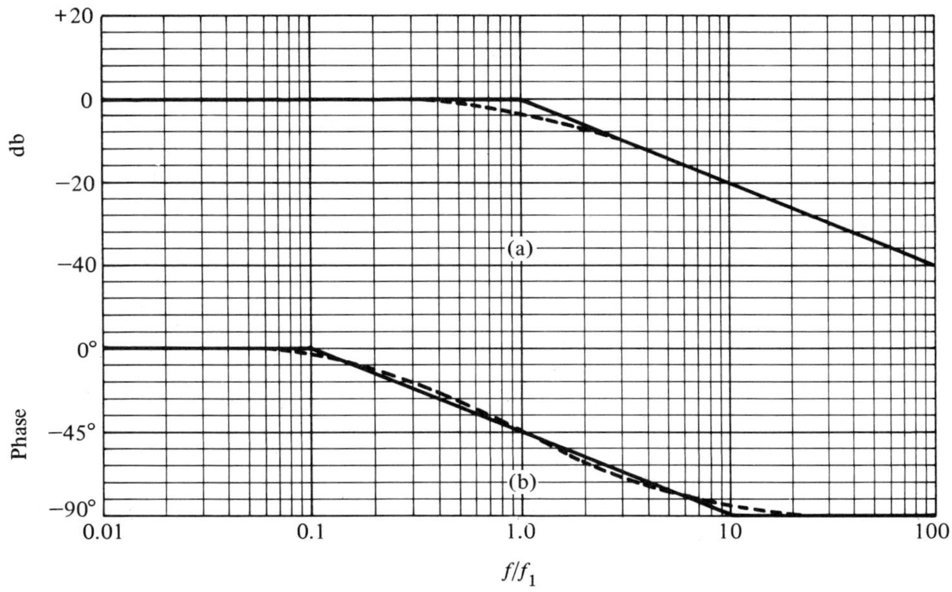

FIGURE B–3 *Bode plot of a simple pole: (a) magnitude and (b) phase.*

The magnitude response of a simple pole can be sketched using the following procedure: (1) locate the corner frequency f_1, (2) sketch the low-frequency asymptote with a slope of 0 db/decade for all frequencies less than f_1, (3) sketch the high-frequency asymptote with a slope of -20 db/decade passing through 0 db at the frequency f_1, and (4) use the magnitude correction factors from Table B–1 to obtain a more accurate magnitude plot.

The phase response of a simple pole can be sketched using the following procedure: (1) locate the corner frequency f_1, (2) the phase response is approximately 0° for frequencies less than 0.1 f_1, (3) the phase response is approximately $-90°$ for frequencies greater than 10 f_1, (4) connect the 0° and $-90°$ segments of the phase response with a line having a slope of $-45°$/decade passing through $-45°$ at the frequency f_1, and (5) use the phase correction factors from Table B–1 to obtain a more accurate phase plot.

A simple zero of $G(s)$ corresponds to a root of $N(s)$. Zeros have the form of the reciprocal of the expression for a pole. The Bode plot shown in Fig. B–4 is for a zero of the form

$$G = 1 + jf/f_1 \qquad \textbf{(B–5)}$$

Comparing Eq. (B–5) with (B–4) and Fig. B–4 with B–3, we can see that the Bode plot of a zero is a mirror image of a Bode plot of a pole. Thus, the procedure outlined above for sketching the magnitude and phase responses of a pole can also be used to sketch the magnitude and phase responses of a zero except for a change in sign.

Multiple poles or zeros can also occur. Their responses can be obtained by combining the response of n single poles (or zeros) where n is the order of the multiple pole. For example, the multiple pole $1/(1 + jf/f_1)^3$ has magnitude asymptotes of 0 db/decade slope for frequencies less than f_1 and -60 db/decade slope for frequencies greater than f_1. The total phase shift is $-270°$.

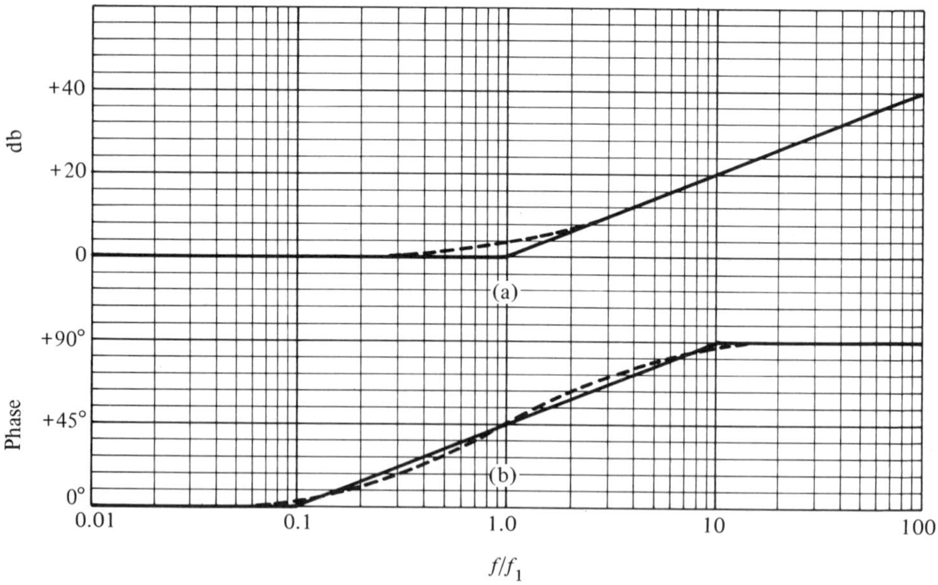

FIGURE B–4 *Bode plot of a simple zero: (a) magnitude and (b) phase.*

Roots of $N(s)$ and $D(s)$ can also occur in complex conjugate pairs rather than as simple poles or zeros. A quadratic pole factor can be expressed as

$$G = \frac{1}{(jf/f_n)^2 + 2\xi(jf/f_n) + 1} \quad \textbf{(B–6)}$$

where f_n is the undamped natural frequency, and ξ is the damping ratio. As shown in Fig. B–5, the Bode plot of a quadratic factor depends on both f_n and ξ. The magnitude asymptotes are 0 db/decade for frequencies much less than f_n and -40 db/decade for frequencies much greater than f_n. The phase changes

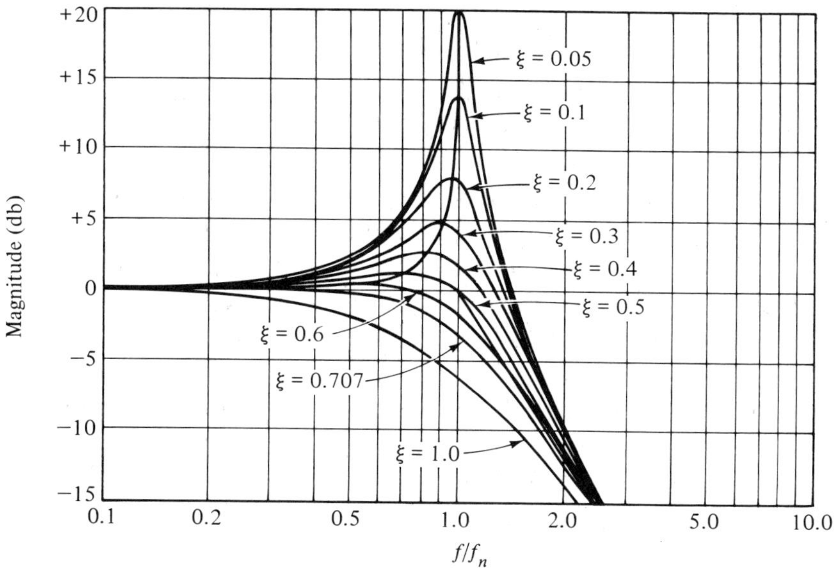

(a)

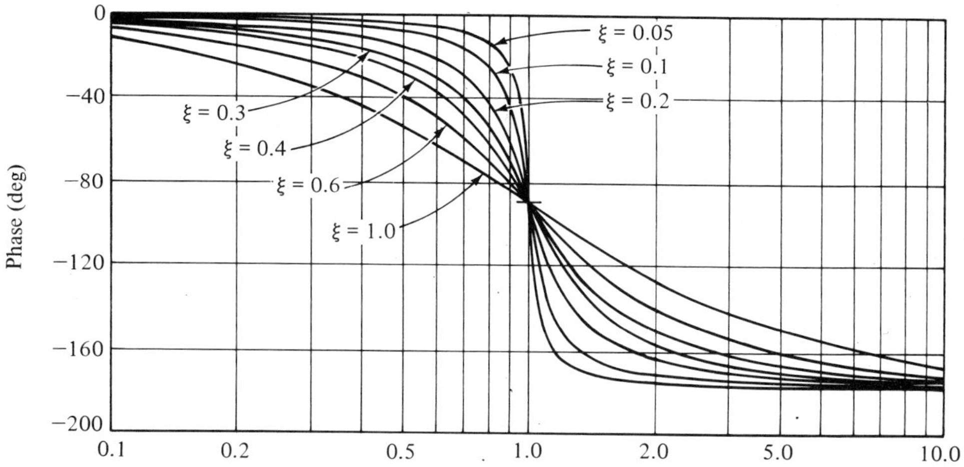

(b)

FIGURE B–5 *Bode plot of a quadratic pole: (a) magnitude and (b) phase.*

from 0° to −180°. Straight line approximations are generally not as useful for quadratic terms as they are for simple poles and zeros. Generally, the universal response curves shown in Fig. B–5 are used to construct the Bode plot of a quadratic pole. By simply reversing the signs on the magnitude and phase scales, Fig. B–5 can also be used to plot the response of a conjugate pair of zeros.

The overall response of a function $G(s)$ given by Eq. (B–1) is found by combining the responses of the individual terms of $G(s)$. Since all magnitude terms are expressed in db, the combined response at a given frequency is equal to the algebraic sum of the individual magnitude responses at that frequency. The combined phase response is obtained in a similar manner. Thus, the overall Bode plot of a function is constructed by first constructing the Bode plot of each individual factor of $G(s)$, both poles and zeros. Then the overall response is obtained by point-by-point adding the individual responses.

APPENDIX C

Second-Order Effects of Multiple Poles on the Performance Characteristics of Bipolar Amplifiers

The objective of this appendix is to outline the development of the equations used in Sec. 8.4 to describe some of the second-order effects of multiple poles on the performance characteristics of bipolar amplifiers. The CE, CC, and CB configurations are discussed in that order, primarily using nodal-analysis methods.

C.1 The Common-Emitter Case

Consider the CE amplifier shown in Fig. C–1(a). A high-frequency equivalent circuit is shown in Fig. C–1(b). A modified form of the high-frequency equivalent circuit is shown in Fig. C–1(c). This equivalent circuit is structured by neglecting $r_{b'}$ and by replacing the generator circuit, both R_g and V_g, by its Norton equivalent circuit. A modified form of Fig. C–1(c) is shown in Fig. C–1(d) where

$$R_b \equiv R_g \,||\, R_B \,||\, r_{b'e} = R_g \,||\, R_i \tag{C–1}$$

The immediate objective is to develop a voltage transfer function, V_o/V_g, for the CE amplifier shown in Fig. C–1(d). The nodal equations for the circuit in Fig. C–1(d) are:

$$\left.\begin{aligned} V_g/R_g &= [1/R_b + s(C_{b'e} + C_{b'c})]V_{b'e} - sC_{b'c}V_o \\ -g_m V_{b'e} &= -sC_{b'c}V_{b'e} + [1/R_{Leq} + s(C_{b'c} + C_L)]V_o \end{aligned}\right\} \tag{C–2}$$

These equations can be rewritten as

$$\left.\begin{aligned} V_g/R_g &= [1/R_b + s(C_{b'e} + C_{b'c})]V_{b'e} - sC_{b'c}V_o \\ 0 &= (g_m - sC_{b'c})V_{b'e} + [1/R_{Leq} + s(C_{b'c} + C_L)]V_o \end{aligned}\right\} \tag{C–3}$$

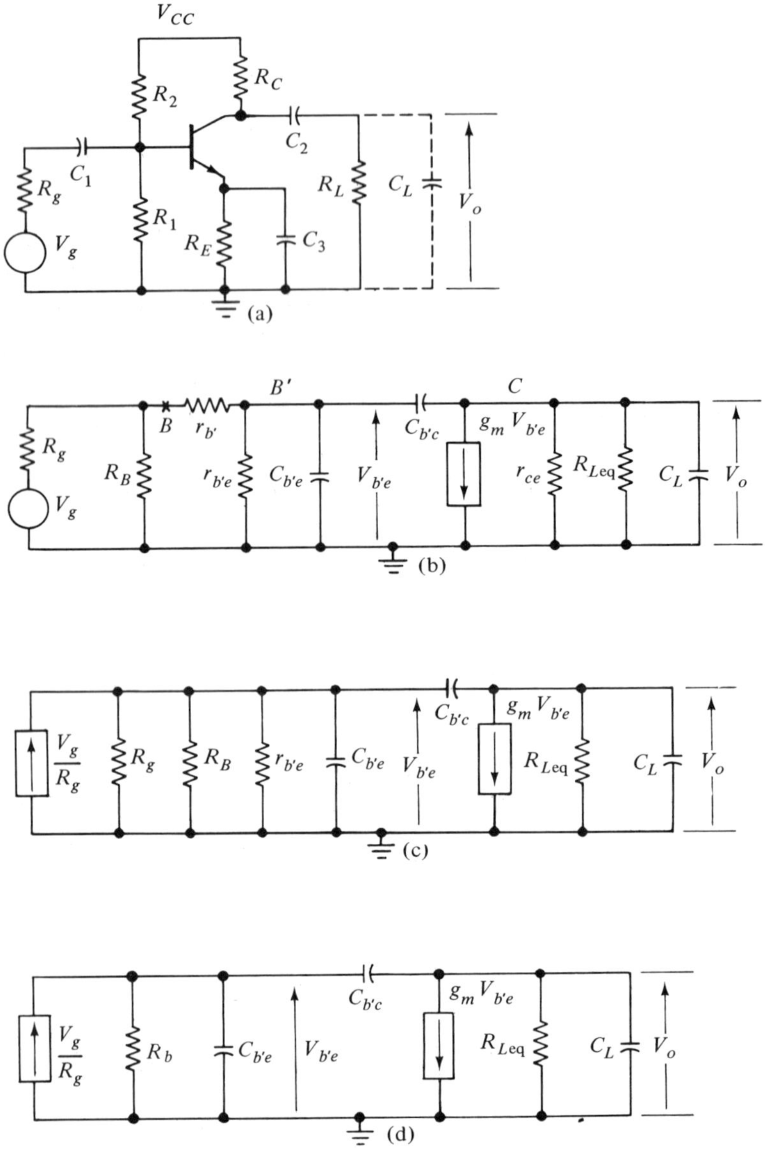

FIGURE C–1 *A CE amplifier: (a) a circuit diagram, (b) a high-frequency equivalent circuit, (c) a modified high-frequency equivalent circuit, and (d) the equivalent circuit used for nodal analysis.*

Solving for V_o/V_g,

$$\frac{V_o}{V_g} = -\frac{(g_m - sC_{b'c})/R_g}{a_1 s^2 + a_2 s + a_3} \tag{C–4}$$

where

$$\left.\begin{aligned} a_1 &= C_{b'e}C_{b'c} + C_{b'e}C_L + C_{b'c}C_L \\ a_2 &= (C_{b'e} + C_{b'c})/R_{Leq} + (C_{b'c} + C_L)/R_b + g_m C_{b'c} \\ a_3 &= 1/(R_b R_{Leq}) \end{aligned}\right\} \tag{C–5}$$

The mid-band voltage gain of a CE stage is

$$A_{\text{mid}} = -\left(\frac{R_i}{R_i + R_g}\right)(g_m R_{L\text{eq}}) \tag{C–6}$$

where

$$R_i = R_B \,||\, r_{b'e} \tag{C–7}$$

After some algebraic manipulation, Eq. (C–4) can be written in a manner to show that mid-band gain as

$$\frac{V_o}{V_g} = \frac{-[R_i/(R_i + R_g)][g_m R_{L\text{eq}}][1 - sC_{b'c}/g_m]}{b_1 s^2 + b_2 s + b_3} \tag{C–8}$$

where

$$\left.\begin{aligned} b_1 &= R_b R_{L\text{eq}}(C_{b'e}C_L + C_{b'c}C_L + C_{b'e}C_{b'c}) \\ b_2 &= R_b[C_{b'e} + C_{b'c}(1 - g_m R_{L\text{eq}})] + R_{L\text{eq}}(C_{b'c} + C_L) \\ b_3 &= 1 \end{aligned}\right\} \tag{C–9}$$

The overall transfer function for the mid- and high-frequencies regions can now be written as

$$\frac{V_o}{V_g} = (A_{\text{mid}})\left(\frac{1 - sC_{b'c}/g_m}{b_1 s^2 + b_2 s + b_3}\right) \tag{C–10}$$

where A_{mid} is given by Eq. (C–6) and the denominator coefficients by Eq. (C–9).

The approximate method of factoring a polynomial outlined in Sec. 8.4 can be applied to the denominator of the CE transfer function given in Eq. (C–10). Assuming that the roots of the denominator (poles of the transfer function) are widely separated,

$$\tau_1 \gg \tau_2 \tag{C–11}$$

$$\tau_1 \simeq b_2/b_3 \simeq R_b[C_{b'e} + C_{b'c}(1 + g_m R_{L\text{eq}})] + R_{L\text{eq}}(C_{b'c} + C_L) \tag{C–12}$$

$$\tau_2 \simeq b_1/b_2 \simeq \frac{R_b R_{L\text{eq}}(C_{b'e}C_L + C_{b'c}C_L + C_{b'e}C_{b'c})}{R_b[C_{b'e} + C_{b'c}(1 + g_m R_{L\text{eq}})] + R_{L\text{eq}}(C_{b'c} + C_L)} \tag{C–13}$$

where τ_1 is associated with the lower high corner frequency, and τ_2 is associated with the higher corner frequency. Eq. (C–10) can now be rewritten as

$$\frac{V_o}{V_g} \simeq \frac{(A_{\text{mid}})(1 - sC_{b'c}/g_m)}{(1 + s\tau_1)(1 + s\tau_2)} \tag{C–14}$$

which is the same as Eq. (8–11). Also, Eqs. (C–12) and (C–13) correspond to Eqs. (8–12) and (8–13) when R_b is replaced by $R_i \,||\, R_g$ [see Eq. (C–1)], and $g_m R_{L\text{eq}}$ is replaced by A_{vt}.

C.2 The Common-Collector Case

Consider the CC amplifier shown in Fig. C–2(a). A high-frequency equivalent circuit is shown in Fig. C–2(b). A modified form of the high-frequency equivalent circuit is shown in Fig. C–2(c). This equivalent circuit is structured by neglecting $r_{b'}$, replacing the generator circuit (R_g and V_g) by its Norton equivalent circuit, and connecting the collector to the common ground terminal. A modified form of Fig. C–2(c) is shown in Fig. C–2(d) where R_b is equal to $R_g \,||\, R_B$, and all of the capacitance from the base terminal to ground, including $C_{b'c}$, is represented by C_b.

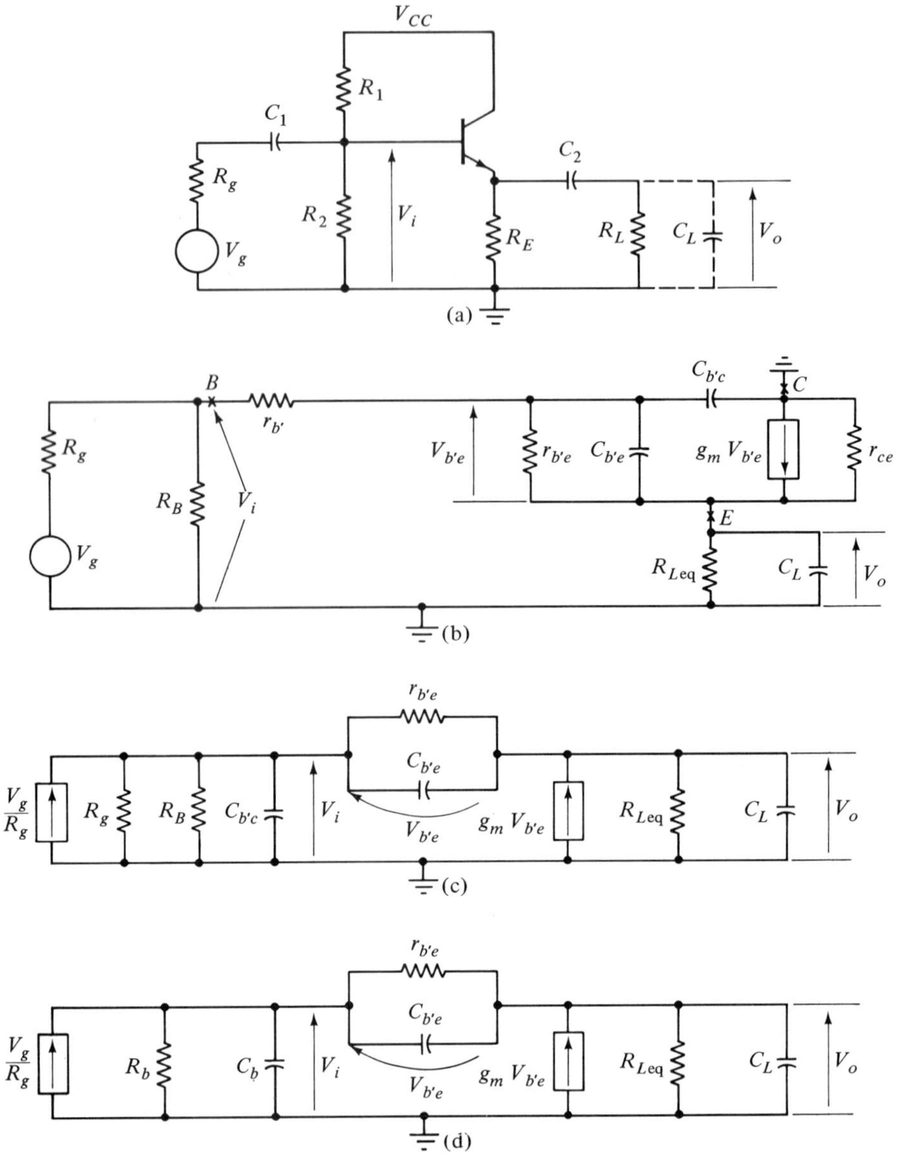

FIGURE C–2 *A CC amplifier: (a) a complete circuit, (b) a high-frequency equivalent circuit, (c) a modified high-frequency equivalent circuit, and (d) the equivalent circuit used for nodal analysis.*

Our immediate objective is to develop the transfer function, V_o/V_g, for the CC amplifier shown in Fig. C–2(d). The nodal equations for the circuit in Fig. C–2(d) are

$$\left.\begin{aligned} V_g/R_g &= [1/R_g + 1/r_{b'e} + s(C_b + C_{b'e})]V_i - [1/r_{b'e} + sC_{b'e}]V_o \\ g_m(V_i - V_o) &= -[1/r_{b'e} + sC_{b'e}] + [1/r_{b'e} + 1/R_{Leq} + s(C_L + C_{b'e})]V_o \end{aligned}\right\} \tag{C–15}$$

Solving Eq. (C–15) for V_o/V_g,

$$\frac{V_o}{V_g} = \frac{A_{\text{mid}}(1 + sr_eC_{b'e})}{a_1s^2 + a_2s + a_3} \tag{C–16}$$

where

$$A_{\text{mid}} = A_{vg}A_{vt} \simeq \left(\frac{R_i}{R_i + R_g}\right)\left(\frac{R_{L\text{eq}}}{r_e + R_{L\text{eq}}}\right) \tag{C–17}$$

$$R_i \simeq (\beta + 1)(r_e + R_{L\text{eq}}) \tag{C–18}$$

$$a_1 = (R_i \,||\, R_g)(r_e \,||\, R_{L\text{eq}})(C_bC_L + C_bC_{b'e} + C_{b'e}C_L) \tag{C–19}$$

$$a_2 = (R_i \,||\, R_g)(C_b + C_{b'g}) + (R_{\text{out(CC)}} \,||\, R_{L\text{eq}})(C_{\text{out(CC)}} + C_L) \tag{C–20}$$

$$a_3 = 1$$

$$C_{b'g} = \frac{1}{[2\pi(r_e + R_{L\text{eq}})f_T]} \tag{C–21}$$

$$C_{\text{out(CC)}} = \frac{1}{\{2\pi[r_e + R_g/(\beta + 1)]f_T\}} \tag{C–22}$$

Eq. (C–16) shows that the zero of the transfer function occurs near f_T.

The approximate method for factoring a polynomial that was discussed in Sec. 8.4 can be applied to the denominator of the CC transfer function given in Eq. (C–16). Assuming that the roots of the denominator (poles of the transfer function) are widely separated, Eqs. (C–16) through (C–22) yield

$$\tau_1 \gg \tau_2 \tag{C–23}$$

$$\tau_1 \simeq a_2/a_3 = (R_i \,||\, R_g)(C_{b'c} + C_{b'g}) + (R_{\text{out(CC)}} \,||\, R_{L\text{eq}})(C_{\text{out(CC)}} + C_L) \tag{C–24}$$

$$\tau_2 \simeq \frac{a_1}{a_2} = \frac{(R_i \,||\, R_g)(r_e \,||\, R_{L\text{eq}})(C_bC_L + C_bC_{b'e} + C_{b'e}C_L)}{(R_i \,||\, R_g)(C_b + C_{b'g}) + (R_{\text{out(CC)}} \,||\, R_{L\text{eq}})(C_{\text{out(CC)}} + C_L)} \tag{C–25}$$

where τ_1 is associated with the lower high corner frequency, and τ_2 and is associated with the higher high corner frequency. Eq. (C–16) can now be rewritten as

$$\frac{V_o}{V_g} \simeq \frac{(A_{\text{mid}})(1 + sr_eC_{b'e})}{(1 + s\tau_1)(1 + s\tau_2)} \tag{C–26}$$

which is the same as Eq. (8–18). Also, Eq. (C–24) results when Eqs. (8–20) and (8–21) are substituted into Eq. (8–19).

The second time constant will now be considered for three possible conditions: (1) τ_{in} much greater than τ_{out}, (2) τ_{out} much greater than τ_{in}, and (3) τ_{in} and τ_{out} approximately the same value.

When τ_{in} dominates the response, Eq. (C–24) requires that

$$(R_i \,||\, R_g)(C_b + C_{b'g}) \gg (R_{\text{out(CC)}} \,||\, R_{L\text{eq}})(C_{\text{out(CC)}} + C_L) \tag{C–27}$$

In practice, the condition given in Eq. (C–27) generally occurs when C_L is small, and when R_g is much greater than r_e. Assuming that C_L is small, then from Eqs. (C–25) and (C–27), the second time constant, τ_2, which is associated with the output node, can be expressed as

$$\tau_2 \simeq \frac{(r_e \,||\, R_{L\text{eq}})(C_bC_{b'e})}{C_b + C_{b'g}} \tag{C–28}$$

Regardless of the values of the various terms in Eq. (C–28), the value of τ_2 will be less than $r_eC_{b'e}$. Consequently, the high corner frequency associated with τ_2 will occur at some frequency greater than f_T.

When τ_{out} dominates the response, Eq. (C–24) requires that

$$(R_{out(CC)} \parallel R_{Leq})(C_{out(CC)} + C_L) \gg (R_i \parallel R_g)(C_b + C_{b'g}) \qquad \textbf{(C–29)}$$

In practice, the condition given in Eq. (C–29) generally occurs when C_L is large and when R_g is the same order of magnitude as r_e. From Eqs. (C–25) and (C–29), the smallest time constant, τ_2, which is now associated with the input node, can be expressed as

$$\tau_2 \simeq \frac{(R_i \parallel R_g)(r_e \parallel R_{Leq})(C_LC_b + C_LC_{b'e})}{(R_{out(CC)} \parallel R_{Leq})(C_{out(CC)} + C_L)} \qquad \textbf{(C–30)}$$

Assuming that C_L is much greater than $C_{out(CC)}$ and using the expression for R_i given in Eq. (C–18) and also the definition of $r_{b'e}$, Eq. (C–30) simplifies to

$$\tau_2 \simeq (r_{b'e} \parallel R_g)(C_b + C_{b'e}) \qquad \textbf{(C–31)}$$

Since R_g must have a value in the same order of magnitude as r_e for Eq. (C–29) to apply, the value of τ_2 approaches $r_eC_{b'e}$ when C_b is equal to $C_{b'c}$. Consequently, the high corner frequency associated with τ_2 will generally be in the neighborhood of f_T.

When the value of the approximate input and output high-frequency time constants are nearly the same, an analysis of the response is more complicated. However, the lower or dominant high corner frequency is still given by the sum of the two approximate time constants. The value of the time constant associated with the second or higher high corner frequency can be obtained from Eq. (C–25)(the ratio of adjacent coefficients) or from Eq. (C–16)(the general transfer function). However, the second or higher high corner frequency will generally occur in the neighborhood of f_T.

Consider the CC amplifier shown in Fig. C–3(a). The impedance looking into the base terminal, $Z_{in(CC)}$, can be found from the simplified equivalent circuit shown in Fig. C–3(b) where a test current generator, I_{t1}, has replaced the base bias network and the input generator. The equivalent circuit used for nodal analysis is shown in Fig. C–3(c) where the CE hybrid-π equivalent circuit has been used to represent the transistor. From Fig. C–3(c), the input impedance is given by

$$Z_{in(CC)} = V_{t1}/I_{t1} = r_{b'} + V_{b'g}/I_{t1} \qquad \textbf{(C–32)}$$

The nodal equations for the circuit in Fig. C–3(c) are

$$\begin{aligned} I_{t1} &= [1/r_{b'e} + s(C_{b'c} + C_{b'e})]V_{b'g} - (1/r_{b'e} + sC_{b'e})V_o \\ g_m(V_{b'g} - V_o) &= -[1/r_{b'e} + sC_{b'c}]V_{b'g} + [1/r_{b'e} + 1/R_E + s(C_{b'e} + C_E)]V_o \end{aligned} \qquad \textbf{(C–33)}$$

Solving Eq. (C–33) for $V_{b'g}/I_t$ using Cramer's rule, the expression for $Z_{in(CC)}$ becomes

$$Z_{in(CC)} = r_{b'} + \frac{(\beta + 1)(r_e + R_E)[1 + s(r_e \parallel R_E)(C_{b'e} + C_E)]}{b_1s^2 + b_2s + b_3} \qquad \textbf{(C–34)}$$

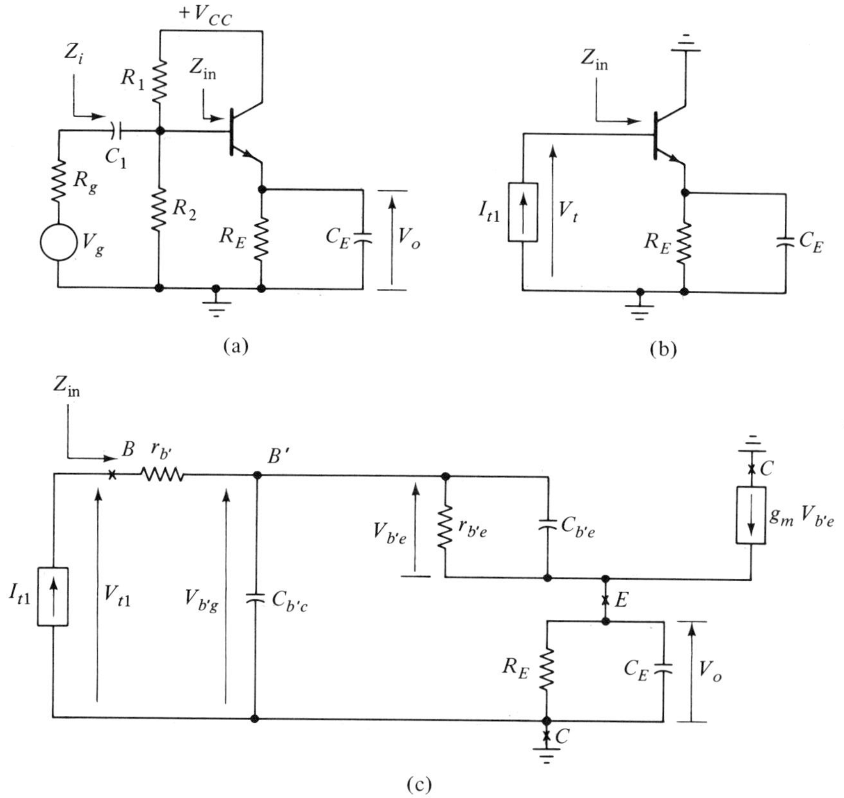

FIGURE C–3 *Circuit used in the development of* $Z_{in(CC)}$*: (a) a complete circuit, (b) a simplified equivalent circuit, and (c) the equivalent circuit used for nodal analysis.*

where

$$\left.\begin{aligned} b_1 &= r_{b'e}R_E(C_{b'c}C_{b'e} + C_{b'c}C_E + C_{b'e}C_E)] \\ b_2 &= (\beta + 1)(r_e + R_E)C_{b'c} + R_E C_E + r_{b'e}C_{b'e} \\ b_3 &= 1 \end{aligned}\right\} \tag{C–35}$$

Replacing s by $j\omega$, Eq. (C–34) becomes

$$Z_{\text{in(CC)}} = r_{b'} + \frac{a_1 + j\omega a_2}{(b_3 - \omega^2 b_1) + j\omega b_2} \tag{C–36}$$

where

$$\left.\begin{aligned} a_1 &= (\beta + 1)(r_e + R_E) \\ a_2 &= r_{b'e}R_E(C_{b'e} + C_E) \end{aligned}\right\} \tag{C–37}$$

To separate $Z_{\text{in(CC)}}$ into resistive and reactive components, the second term in Eq. (C–36) is multiplied by its complex conjugate yielding

$$\begin{aligned} Z_{\text{in(CC)}} &= r_{b'} + \frac{a_1 + j\omega a_2}{(b_3 - \omega^2 b_1) + j\omega b_2} \times \frac{(b_3 - \omega^2 b_1) - j\omega b_2}{(b_3 - \omega^2 b_1) - j\omega b_2} \\ &= R_{\text{in(CC)}} + jX_{\text{in(CC)}} \end{aligned} \tag{C–38}$$

where

$$R_{\text{in(CC)}} = r_{b'} + \frac{a_1 - \omega^2(a_1b_1 - a_2b_2)}{(1 - \omega^2 b_1)^2 + \omega^2 b_2^2} \tag{C–39}$$

$$X_{\text{in(CC)}} = -\frac{\omega(a_1b_1 - a_2) + \omega^3 a_2 b_1}{(1 - \omega^2 b_1)^2 + \omega^2 b_2^2} \tag{C–40}$$

From Eq. (C–39), when $\omega = 0$,

$$R_{\text{in(CC)}} = r_{b'} + a_1 = r_{b'} + (\beta + 1)(r_e + R_E) \tag{C–41}$$

and when $\omega = \infty$,

$$R_{\text{in(CC)}} = r_{b'} \tag{C–42}$$

Also, the second term in Eq. (C–39) will be negative when

$$a_1b_1 > a_2b_2 \tag{C–43}$$

making it possible for $R_{\text{in(CC)}}$ to be negative in some frequency range.

Next, we will outline the derivation of an approximate expression for the maximum value of negative resistance, NR_{max}, and the frequency where this maximum value occurs, $f_{m(NR)}$.

The value of $f_{m(NR)}$ can be found using Eq. (C–39) by setting the partial derivative with respect to ω equal to zero. Thus, when the

$$\partial R_{\text{in}}/\partial\omega = 0 \tag{C–44}$$

the frequency at which the negative resistance reaches a maximum is

$$f_{m(NR)} \simeq f_\beta[M + (M + 1)\sqrt{M}]^{1/2} \tag{C–45}$$

where

$$f_\beta = f_T/\beta \tag{C–46}$$

$$M = r_{b'e}C_{b'e}/R_EC_E \tag{C–47}$$

The following approximations were used in the derivation of Eq. (C–45): (1) $C_{b'c}$ is negligible, (2) β is much greater than unity, and (3) R_E is much greater than r_e. The value of the maximum negative resistance, NR_{max}, is found from Eq. (C–39) with ω equal to ω_m or

$$NR_{\text{max}} = r_{b'} + \frac{a_1 - \omega_m^2(a_1b_1 - a_2b_2)}{(1 - \omega_m^2 b_1)^2 + \omega_m^2 b_2^2} \tag{C–48}$$

where

$$\omega_m = 2\pi f_{m(NR)} \tag{C–49}$$

The results of this development are summarized in Eq. (8–23).

The final expression to be derived for the CC stage is its output impedance, $Z_{\text{out(CC)}}$. An expression for the $Z_{\text{out(CC)}}$ can be derived from the circuit shown in Fig. C–4. This circuit is obtained from Fig. C–2(b) as follows: (1) replace the components to the right of the emitter terminal by a test voltage generator, V_t, (2) replace V_g with a short circuit, (3) represent the parallel combinations of R_g and R_B by R_b, (4) represent the capacitance from the base terminal to ground including $C_{b'c}$ and any stray capacitance by C_b, and (5) neglect r_{ce}. A calculation of $Z_{\text{out(CC)}}$ is simplified by making the following definitions:

$$\left.\begin{aligned} Z_b &\equiv R_b \,||\, (1/sC_b) = R_b/(1 + sR_bC_b) \\ Z_{b'e} &\equiv r_{b'e} \,||\, (1/sC_{b'e}) = r_{b'e}/(1 + sr_{b'e}C_{b'e}) \end{aligned}\right\} \tag{C–50}$$

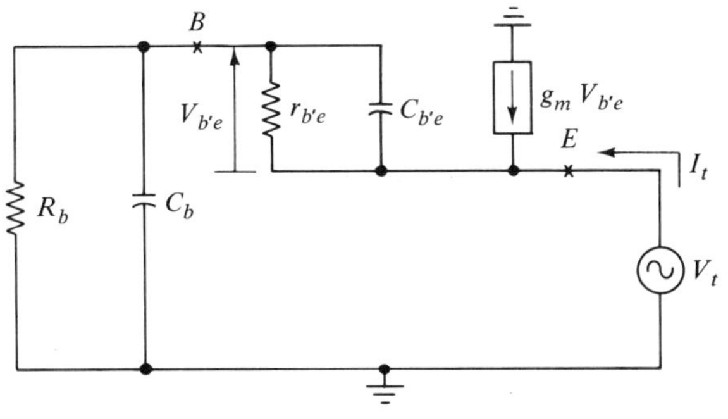

FIGURE C–4 *Circuit used to calculate the output impedance of a CC stage.*

The current flowing in V_t is given by

$$I_t = I_1 - g_m V_{b'e} = \frac{V_t}{Z_{b'e} + Z_b} - g_m V_{b'e} \tag{C–51}$$

From Fig. C–4, the value of $V_{b'e}$ is

$$V_{b'e} = -V_t \left(\frac{Z_{b'e}}{Z_{b'e} + Z_b}\right) \tag{C–52}$$

Consequently, Eq. (C–51) can be written as

$$I_t = \frac{V_t}{Z_{b'e} + Z_b} + \frac{g_m Z_{b'e} V_t}{Z_{b'e} + Z_b} = V_t \left(\frac{1 + g_m Z_{b'e}}{Z_{b'e} + Z_b}\right) \tag{C–53}$$

Solving for $Z_{\text{out(CC)}}$,

$$Z_{\text{out(CC)}} = \frac{V_t}{I_t} = \frac{Z_{b'e} + Z_b}{1 + g_m Z_{b'e}} \tag{C–54}$$

By substituting Eq. (C–50) into (C–54) and simplifying, $Z_{\text{out(CC)}}$ can be written as

$$\begin{aligned} Z_{\text{out(CC)}} &\simeq \left(r_e + \frac{R_b}{\beta + 1}\right)\left(\frac{1 + s(R_b r_{b'e})(C_b + C_{b'e})}{(1 + sR_bC_b)(1 + sr_eC_{b'e})}\right) \\ &\simeq \left(r_e + \frac{R_b}{\beta + 1}\right)\left(\frac{1 + jf/f_1}{(1 + jf/f_2)(1 + jf/f_T)}\right) \end{aligned} \tag{C–55}$$

where

$$\begin{aligned} f_1 &\equiv 1/[2\pi(R_b \,||\, r_{b'e})(C_b + C_{b'e})] \\ f_2 &\equiv 1/(2\pi R_b C_b) \qquad f_T \simeq 1/(2\pi r_e C_{b'e}) \end{aligned} \tag{C–56}$$

Eqs. (C–55) and (C–56) correspond to Eqs. (8–24) and (8–25).

C.3 The Common-Base Case

The circuit shown in Fig. C–5(a) will be used to develop the expression for the CB output resistance. The output resistance of the CB stage with both R_e and R_b present is given by the ratio V_t/I_t. From Fig. C–5(b),

$$\begin{aligned} I_t &= I_1 + I_2 - g_m V_{b'e} \\ &= \frac{V_t}{r_{ce} + R_e \,||\, R_{\text{ine}}} + \frac{V_t}{r_{b'c}} - \left(\frac{g_m R_e r_e}{R_{\text{ine}} + R_e}\right)\left(\frac{V_t}{r_{ce} + R_e \,||\, R_{\text{ine}}}\right) \end{aligned} \tag{C–57}$$

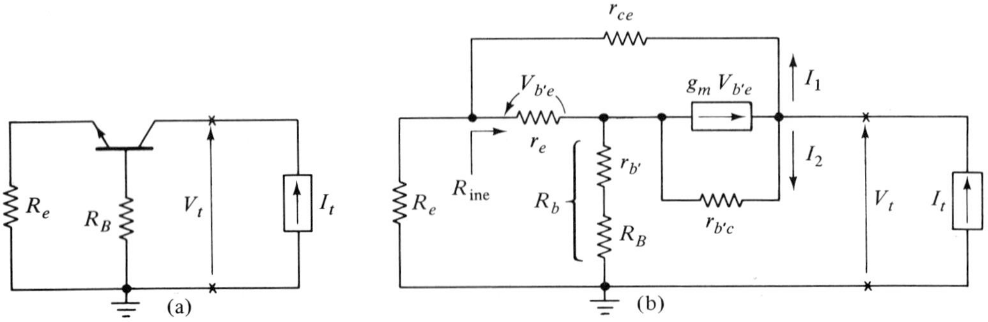

FIGURE C–5 *Circuit used to calculate* $R_{out(CC)}$: *(a) complete circuit, and (b) an equivalent circuit using the mid-band CB hybrid-π equivalent circuit to represent the transistor.*

where

$$R_b = r_{b'} + R_B$$

$$R_{\text{ine}} = r_e + [R_b/(\beta + 1)] \quad \textbf{(C–58)}$$

and the value of $r_{b'c}$ is assumed to be very large and controls the value of I_2. Solving Eq. (C–57) for $R_{\text{out(CB)}}$,

$$R_{\text{out(CB)}} = \frac{V_t}{I_t} = r_{b'c} \,||\, (r_{ce}) \left(\frac{1 + (R_e \,||\, R_{\text{ine}})/r_{ce}}{1 - [\beta/(\beta + 1)][R_e/(R_e + R_{\text{ine}})]} \right) \quad \textbf{(C–59)}$$

which corresponds to Eq. (8–28).

Index